ENC**Y**CLOPEDIA OF MATHEMATICS A**N**

C000049806

EDITED BY G.-C. R**C** ...

Editorial Board

R.S. Doran, T.-Y. Lam, E. Lutwak

Volume 42

Model theory

ENCYCLOPEDIA OF MATHEMATICS AND ITS APPLICATIONS

ENCYCLOPEDIA OF MATHEMATICS AND ITS APPLICATIONS

Model Theory

WILFRID HODGES

Professor of Mathematics
School of Mathematical Sciences
Queen Mary and Westfield College
University of London

CAMBRIDGE
UNIVERSITY PRESS

CAMBRIDGE UNIVERSITY PRESS
Cambridge, New York, Melbourne, Madrid, Cape Town, Singapore, São Paulo

Cambridge University Press
The Edinburgh Building, Cambridge CB2 8RU, UK

Published in the United States of America by Cambridge University Press, New York

www.cambridge.org
Information on this title: www.cambridge.org/9780521304429

First published 1993
This digitally printed version 2008

A catalogue record for this publication is available from the British Library

Library of Congress Cataloguing in Publication data
Hodges, Wilfrid.
Model theory / Wilfrid Hodges.
p.   cm. – (Encyclopedia of mathematics and its applications: v. 42)
Includes bibliographical references (p.   ) and index.
ISBN 0 521 30442 3 (hardback)
1. Model theory.   I. Title.   II. Series.
QA9.7.H64   1993
511′.8–dc20   91-25082 CIP

ISBN 978-0-521-30442-9 hardback
ISBN 978-0-521-06636-5 paperback

CONTENTS

INTRODUCTION

Should I begin by defining 'model theory'? This might be unsafe – do the readers get their money back if the definition doesn't match the contents? But here is an attempt at a definition: Model theory is the study of the construction and classification of structures within specified classes of structures.

A 'specified class of structures' is any class of structures that a mathematician might choose to name. For example it might be the class of abelian groups, or of Banach algebras, or sets with groups which act on them primitively. Thirty or forty years ago the founding fathers of model theory were particularly interested in classes specified by some set of axioms in first-order predicate logic – this would include the abelian groups but not the Banach algebras or the primitive groups. Today we have more catholic tastes, though many of our techniques work best on the first-order axiomatisable classes. One result of this broadening is that model theorists are usually much less interested than they used to be in the syntactical niceties of formal languages – if you want to know about formal languages today, you should go first to a computer scientist.

'Construction' means building structures, or sometimes families of structures, which have some feature that interests us. For example we might look for a graph which has very many automorphisms, or a group in which many systems of equations are solvable; or we might want a family of boolean algebras which can't be embedded in each other. 'Classifying' a class of structures means grouping the structures into subclasses in a useful way, and then proving that every structure in the collection does belong in just one of the subclasses. An archetypal example from algebra is the classification of vector spaces over a fixed field: we classify them by showing that each vector space has a unique dimension which determines it up to isomorphism. Model theory studies construction and classification in a broad setting, with methods that can be applied to many different classes of structure.

To give this book a shape, and to make it different from books which other people might write, I chose to concentrate on construction rather than classification. But that was twelve years ago, when (encouraged by Paul Cohn) I first sat down to write this book. Since then spells as editor, deputy head, dean etc. have destroyed any schedule that I ever had for writing the book, and of course the subject has moved on. The result was that far too much material accumulated. Some of it I diverted into three other books. Slightly over one megabyte has been shunted off to a file named Reject – I say this to appease readers who are annoyed to find no mention of their favourite topics or papers. The rejected material covers model theory of fields, atomic compact structures and infinitary languages, among several other things. These are valuable topics, but I simply ran out of space. (Other topics are missing out of brute ignorance – for example constructive model theory.)

Of the three other books, one has already appeared under the title *Building models by games*. It was originally a part of the present Chapter 8 (and it included the material that Paul Cohn had asked to see). A second book provisionally has the title *Structure and classification*, and will include much more stability theory. But several authors have already been kind enough to refer to the present book for some items in stability theory, and so I have taken the subject far enough to include those items. The third book will develop quasivarieties and Horn theories. Since this material has become more important for specification languages and logic programming than it ever was for model theory, that book will be aimed more at computer scientists.

After all these adjustments, the present book still has the emphases that I originally intended, though there is a lot more in it than I planned. Nearly every chapter is designed around some model-theoretic method of construction. In Chapter 1 it is diagrams, Chapter 3 includes Skolem hulls, Chapter 5 discusses interpretations as a method of construction, Chapter 6 tackles elementary amalgamation, Chapter 7 discusses omitting types and the Fraïssé construction, Chapter 8 is about existential closure, Chapter 9 deals with products, Chapter 10 builds saturated structures by unions of chains, Chapter 11 is about the Ehrenfeucht–Mostowski construction. That leaves Chapters 2, 4 and 12: Chapter 2 covers essential background material on languages, while Chapters 4 and 12 contain some recent developments of a geometric kind which I included because they are beautiful, important or both.

In the fourth century BC there was a bizarre philosophical debate about whether thought goes in straight lines or circles. Aristotle very sensibly supported the straight line theory, because (he maintained) proofs are linear.

Plato said circles, for astrological reasons which I wouldn't even wish to understand. But writing this book has convinced me that, just this once, Plato was right. There is no way that one can sensibly cover all the material in the book so that the later bits follow from the earlier ones. Time and again the more recent or sophisticated research throws up new information about the basic concepts. The later sections of several chapters, particularly Chapters 4 and 5, contain recent results which depend on things in later chapters. I trust this will cause no trouble; there are plenty of signposts in the text.

It would be hopeless to try to acknowledge all the people who have contributed to this book; they run into hundreds. But I warmly thank the people who read through sections – either of the final book or of parts now discarded – and gave me comments on them. They include Richard Archer, John Baldwin, Andreas Baudisch, Oleg Belegradek, Jeremy Clark, Paul Eklof, David Evans, Rami Grossberg, Deirdre Haskell, Lefty Kreouzis, Dugald Macpherson, Anand Pillay, Bruno Poizat, Philipp Rothmaler, Simon Thomas. I also thank David Tranah of Cambridge University Press for his encouragement and patience.

I owe a particular debt to Ian Hodkinson. There is no chance whatever that this book is free from errors; but thanks to his eagle eye and sound judgement, the number of mistakes is less than half what it would otherwise have been. His comments have led to improvements on practically every page. I say no more about his generous efforts for fear of getting him into trouble with his present employers, who must surely regard this as time misspent.

Finally a dedication. If this book is a success, I dedicate it to my students and colleagues, past and present, in the field of logic. Many of them appear in the pages which follow; but of those who don't, let me mention here two thoughtful and generous souls, Geoffrey Kneebone and Chris Fernau, both now retired, who ran the logic group of London University at Bedford College when I first came to London. If the book is not a success, I dedicate it to the burglars in Boulder, Colorado, who broke into our house and stole a television, two typewriters, my wife Helen's engagement ring and several pieces of cheese, somewhere about a third of the way through Chapter 8.

Acknowledgements. The passage of Hugh MacDiarmid, *On a raised beach* at the head of Chapter 2 is reprinted by permission of Martin Brian & O'Keefe Ltd, Blackheath. The lines from Eugène Ionesco, *La Cantatrice chauve* at the head of Chapter 3 are reprinted by permission of Editions Gallimard, Paris. The radiolarian skeleton at the head of Chapter 12 is reprinted by permission of John Wiley & Sons, Inc., New York.

NOTE ON NOTATION

Some exercises are marked with an asterisk *. This means only that I regard them as not the main exercises; maybe they assume specialist background, or they are very difficult, or they are off centre.

I assume Zermelo–Fraenkel set theory, ZFC. In particular I always assume the axiom of choice (except where the axiom itself is under discussion). I never assume the continuum hypothesis, existence of uncountable inaccessibles etc., without being honest about it.

The notation $x \subseteq y$ means that x is a subset of y; $x \subset y$ means that x is a proper subset of y. I write $\mathrm{dom}(f)$, $\mathrm{im}(f)$ for the domain and image of a function f. 'Greater than' means greater than, never 'greater than or equal to'. $\mathscr{P}(x)$ is the power set of x.

Ordinals are von Neumann ordinals, i.e. the predecessors of an ordinal α are exactly the elements of α. I use symbols α, β, γ, δ, i, j etc. for ordinals; δ is usually a limit ordinal. A cardinal κ is the smallest ordinal of cardinality κ, and the infinite cardinals are listed as ω_0, ω_1 etc. I use symbols κ, λ, μ, ν for cardinals; they are not assumed to be infinite unless the context clearly requires it (though I have probably slipped on this point once or twice). Natural numbers m, n etc. are the same thing as finite cardinals.

'Countable' means of cardinality ω. An infinite cardinal λ is a **regular** cardinal if it can't be written as the sum of fewer than λ cardinals which are all smaller than λ; otherwise it is **singular**. Every infinite successor cardinal κ^+ is regular. The smallest singular cardinal is $\omega_\omega = \sum_{n<\omega} \omega_n$. The **cofinality** $\mathrm{cf}(\alpha)$ of an ordinal α is the least ordinal β such that α has a cofinal subset of order-type β; it can be shown that this ordinal β is either finite or regular. If α and β are ordinals, $\alpha\beta$ is the ordinal product consisting of β copies of α laid end to end. If κ and λ are cardinals, $\kappa\lambda$ is the cardinal product. The context should always show which of these products is intended.

Some facts of cardinal arithmetic are assembled at the beginning of section 10.4.

Sequences are well-ordered (except for indiscernible sequences in Chapter 11, and it is explicit there what is happening). I use the notation $\bar{x}$, $\bar{a}$ etc. for sequences $(x_0, x_1, \ldots)$, $(a_0, a_1, \ldots)$ etc., but loosely: the nth term of a sequence $\bar{x}$ may be x_n or $x(n)$ or something else, depending on the context, and some sequences start at x_1. Sequences of finite length are called **tuples**. The terms of a sequence are sometimes called its **items**, to avoid the ambiguity in the term 'term'. A sequence is said to be **non-repeating** if no item occurs twice or more in it. If $\bar{a}$ is a sequence $(a_0, a_1, \ldots)$ and f is a map, then $f\bar{a}$ is $(fa_0, fa_1, \ldots)$. The length of a sequence σ is written $\mathrm{lh}(\sigma)$. If σ is a sequence of length m and $n \leqslant m$, then $\sigma|n$ is the initial segment consisting of the first n terms of σ. The set of sequences of length γ whose items all come from the set X is written $^{\gamma}X$. Thus $^{n}2$ is the set of ordered n-tuples of 0's and 1's; $^{<\gamma}X$ is $\bigcup_{\alpha<\gamma}{}^{\alpha}X$. I write η, ζ, θ etc. for linear orderings; η^* is the ordering η run backwards.

I don't distinguish systematically between tuples and strings. If $\bar{a}$ and $\bar{b}$ are strings, $\bar{a}\,\widehat{}\,\bar{b}$ is the concatenated string consisting of $\bar{a}$ followed by $\bar{b}$; but often for simplicity I write it $\bar{a}\bar{b}$. There is a clash between the usual notation of model theory and the usual notation of groups: in model theory xy is the string consisting of x followed by y, but in groups it is x times y. One has to live with this; but where there is any ambiguity I have used $x\,\widehat{}\,y$ for the concatenated string and $x \cdot y$ for the group product.

Model-theoretic notation is defined as and when we need it. The most basic items appear in Chapter 1 and the first five sections of Chapter 2.

'I' means I, 'we' means we.

1

Naming of parts

Every person had in the beginning one only proper name, except the savages of Mount Atlas in Barbary, which were reported to be both nameless and dreamless.

William Camden

In this first chapter we meet the main subject-matter of model theory: structures.

Every mathematician handles structures of some kind – be they modules, groups, rings, fields, lattices, partial orderings, Banach algebras or whatever. This chapter will define basic notions like 'element', 'homomorphism', 'substructure', and the definitions are not meant to contain any surprises. The notion of a (Robinson) 'diagram' of a structure may look a little strange at first, but really it is nothing more than a generalisation of the multiplication table of a group.

Nevertheless there is something that the reader may find unsettling. Model theorists are forever talking about symbols, names and labels. A group theorist will happily write the same abelian group multiplicatively or additively, whichever is more convenient for the matter in hand. Not so the model theorist: for him or her the group with ' · ' is one structure and the group with '+' is a different structure. Change the name and you change the structure.

This must look like pedantry. Model theory is an offshoot of mathematical logic, and I can't deny that some distinguished logicians have been pedantic about symbols. Nevertheless there are several good reasons why model theorists take the view that they do. For the moment let me mention two.

In the first place, we shall often want to compare two structures and study the homomorphisms from one to the other. What is a homomorphism? In the particular case of groups, a homomorphism from group G to group H is a map that carries multiplication in G to multiplication in H. There is an obvious way to generalise this notion to arbitrary structures: a homomorphism from structure A to structure B is a map which carries each operation of A to *the operation with the same name in B*.

Secondly, we shall often set out to build a structure with certain properties. One of the maxims of model theory is this: *name the elements of your*

1

structure first, then decide how they should behave. If the names are well chosen, they will serve both as a scaffolding for the construction, and as raw materials.

Aha – says the group theorist – I see you aren't really talking about *written* symbols at all. For the purposes you have described, you only need to have formal labels for some parts of your structures. It should be quite irrelevant what kinds of thing your labels are; you might even want to have uncountably many of them.

Quite right. In fact we shall follow the lead of A. I. Mal'tsev [1936] and put no restrictions at all on what can serve as a name. For example any ordinal can be a name, and any mathematical object can serve as a name of itself. The items called 'symbols' in this book need not be written down. They need not even be dreamed.

1.1 Structures

We begin with a definition of 'structure'. It would have been possible to set up the subject with a slicker definition – say by leaving out clauses (1.2) and (1.4) below. But a little extra generality at this stage will save us endless complications later on.

A **structure** A is an object with the following four ingredients.

(1.1)   A set called the **domain** of A, written $\text{dom}(A)$ or $\text{dom}\,A$ (some people call it the **universe** or **carrier** of A). The elements of $\text{dom}(A)$ are called the **elements** of the structure A. The **cardinality** of A, in symbols $|A|$, is defined to be the cardinality $|\text{dom}\,A|$ of $\text{dom}(A)$.

(1.2)   A set of elements of A called **constant elements**, each of which is named by one or more **constants**. If c is a constant, we write c^A for the constant element named by c.

(1.3)   For each positive integer n, a set of n-ary relations on $\text{dom}(A)$ (i.e. subsets of $(\text{dom}\,A)^n$), each of which is named by one or more n-ary **relation symbols**. If R is a relation symbol, we write R^A for the relation named by R.

(1.4)   For each positive integer n, a set of n-ary operations on $\text{dom}(A)$ (i.e maps from $(\text{dom}\,A)^n$ to $\text{dom}(A)$), each of which is named by one or more n-ary **function symbols**. If F is a function symbol, we write F^A for the function named by F.

Except where we say otherwise, any of the sets (1.1)–(1.4) may be empty. As mentioned in the chapter introduction, the constant, relation and function 'symbols' can be any mathematical objects, not necessarily written symbols;

but for peace of mind one normally assumes that, for instance, a 3-ary relation symbol doesn't also appear as a 3-ary function symbol or a 2-ary relation symbol. We shall use capital letters A, B, C, . . . for structures.

Sequences of elements of a structure are written $\bar{a}$, $\bar{b}$ etc. A **tuple in** A (or **from** A) is a finite sequence of elements of A; it is an n-**tuple** if it has length n. Usually we leave it to the context to determine the length of a sequence or tuple.

This concludes the definition of 'structure'.

Example 1: *Graphs.* A **graph** consists of a set V (the set of **vertices**) and a set E (the set of **edges**), where each edge is a set of two distinct vertices. An edge $\{v, w\}$ is said to **join** the two vertices v and w. We can picture a finite graph by putting dots for the vertices and joining two vertices v, w by a line when $\{v, w\}$ is an edge:

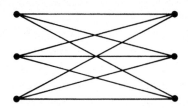

One natural way to make a graph G into a structure is as follows. The elements of G are the vertices. There is one binary relation R^G; the ordered pair (v, w) lies in R^G if and only if there is an edge joining v to w.

Example 2: *Linear orderings.* Suppose $\leqslant$ linearly orders a set X. Then we can make $(X, \leqslant)$ into a structure A as follows. The domain of A is the set X. There is one binary relation symbol R, and its interpretation R^A is the ordering $\leqslant$. (In practice we would usually write the relation symbol as $\leqslant$ rather than R.)

Example 3: *Groups.* We can think of a group as a structure G with one constant 1 naming the identity 1^G, one binary function symbol $\cdot$ naming the group product operation $\cdot^G$, and one unary function symbol $^{-1}$ naming the inverse operation $^{(-1)^G}$. Another group H will have the same symbols 1, $\cdot$, $^{-1}$; then 1^H is the identity element of H, $\cdot^H$ is the product operation of H, and so on.

Example 4: *Vector spaces.* There are several ways to make a vector space into a structure, but here is the most convenient. Suppose V is a vector space over a field of scalars K. Take the domain of V to be the set of vectors of V. There is one constant element 0^V, the origin of the vector space. There is one binary operation, $+^V$, which is addition of vectors. There is a 1-ary operation $-^V$ for additive inverse; and for every scalar k there is a 1-ary operation k^V to represent multiplying a vector by k. Thus each scalar serves as a 1-ary function symbol. (In fact the symbol '$-$' is redundant, because $-^V$ is the same operation as $(-1)^V$.)

When we speak of vector spaces below, we shall assume that they are structures of this form (unless anything is said to the contrary). The same goes for modules, replacing the field K by a ring.

Two questions spring to mind. First, aren't these examples a little arbitrary? For example, why did we give the group structure a symbol for the multiplicative inverse $^{-1}$, but not a symbol for the commutator $[\,,]$? Why did we put into the linear ordering structure a symbol for the ordering $\leq$, but not one for the corresponding strict ordering $<$?

The answer is yes; these choices were arbitrary. But some choices are more sensible than others. We shall come back to this in the next section.

And second, *exactly* what is a structure? Our definition said nothing about the way in which the ingredients (1.1)–(1.4) are packed into a single entity.

True again. But this was a deliberate oversight – the packing arrangements will never matter to us. Some writers define A to be an ordered pair $\langle \mathrm{dom}\,(A), f \rangle$ where f is a function taking each symbol S to the corresponding item S^A. The important thing is to know what the symbols and the ingredients are, and this can be indicated in any reasonable way.

For example a model theorist may refer to the structure

$$\langle \mathbb{R}, +, -, \cdot, 0, 1, \leq \rangle.$$

With some common sense the reader can guess that this means the structure whose domain is the set of real numbers, with constants 0 and 1 naming the numbers 0 and 1, a 2-ary relation symbol $\leq$ naming the relation $\leq$, 2-ary function symbols $+$ and $\cdot$ naming addition and multiplication respectively, and a 1-ary function symbol naming minus.

Signatures

The **signature** of a structure A is specified by giving

(1.5)   the set of constants of A, and for each separate $n > 0$, the set of n-ary relation symbols and the set of n-ary function symbols of A.

We shall assume that the signature of a structure can be read off uniquely from the structure.

The symbol L will be used to stand for signatures. Later it will also stand for languages – think of the signature of A as a kind of rudimentary language for talking about A. If A has signature L, we say A is an L-**structure**.

A signature L with no constants or function symbols is called a **relational signature**, and an L-structure is then said to be a **relational structure**. A signature with no relation symbols is sometimes called an **algebraic signature**.

Exercises for section 1.1

1. According to Thomas Aquinas, God is a structure G with three elements '*pater*', '*filius*' and '*spiritus sanctus*', in a signature consisting of one asymmetric binary relation ('*relatio opposita*') R, read as '*relatio originis*'. Aquinas asserts also that the three elements can be uniquely identified in terms of R^G. Deduce – as Aquinas did – that if the pairs (*pater, filius*) and (*pater, spiritus sanctus*) lie in R^G, then exactly one of the pairs (*filius, spiritus sanctus*) and (*spiritus sanctus, filius*) lies in R^G.

2. Let X be a set and L a signature; write $\kappa(X, L)$ for the number of distinct L-structures which have domain X. Show that if X is a finite set then $\kappa(X, L)$ is either finite or at least 2^ω.

1.2 Homomorphisms and substructures

The following definition is meant to take in, with one grand sweep of the arm, virtually all the things that are called 'homomorphism' in any branch of algebra.

Let L be a signature and let A and B be L-structures. By a **homomorphism** f from A to B, in symbols $f: A \rightarrow B$, we shall mean a function f from dom (A) to dom (B) with the following three properties.

(2.1)   For each constant c of L, $f(c^A) = c^B$.

(2.2)   For each $n > 0$ and each n-ary relation symbol R of L and n-tuple $\bar{a}$ from A, if $\bar{a} \in R^A$ then $f\bar{a} \in R^B$.

(2.3)   For each $n > 0$ and each n-ary function symbol F of L and n-tuple $\bar{a}$ from A, $f(F^A(\bar{a})) = F^B(f\bar{a})$.

(If $\bar{a}$ is $(a_0, \ldots, a_{n-1})$ then $f\bar{a}$ means $(fa_0, \ldots, fa_{n-1})$; cf. Note on notation.) By an **embedding** of A into B we mean a homomorphism $f: A \rightarrow B$ which is injective and satisfies the following stronger version of (2.2).

(2.4)   For each $n > 0$, each n-ary relation symbol R of L and each n-tuple $\bar{a}$ from A, $\bar{a} \in R^A \Leftrightarrow f\bar{a} \in R^B$.

An **isomorphism** is a surjective embedding. Homomorphisms $f: A \rightarrow A$ are called **endomorphisms** of A. Isomorphisms $f: A \rightarrow A$ are called **automorphisms** of A.

For example if G and H are groups and $f: G \to H$ is a homomorphism, then (2.1) says that $f(1^G) = 1^H$, and (2.3) says that for all elements a, b of G, $f(a \cdot {}^G b) = f(a) \cdot {}^H f(b)$ and $f(a^{(-1)^G}) = f(a)^{(-1)^H}$. This is exactly the usual definition of homomorphism between groups. Clause (2.2) adds nothing in this case since there are no relation symbols in the signature. For the same reason (2.4) is vacuous for groups. So a homomorphism between groups is an embedding if and only if it is an injective homomorphism.

We sometimes write 1_A for the identity map on $\operatorname{dom}(A)$. Clearly it is a homomorphism from A to A, in fact an automorphism of A. We say that A is **isomorphic** to B, in symbols $A \cong B$, if there is an isomorphism from A to B.

The following facts are nearly all immediate from the definitions.

Theorem 1.2.1. *Let L be a signature.*

(a) *If A, B, C are L-structures and $f: A \to B$ and $g: B \to C$ are homomorphisms, then the composed map gf is a homomorphism from A to C. If moreover f and g are both embeddings then so is gf.*

(b) *If A, B are L-structures and $f: A \to B$ is a homomorphism then $1_B f = f = f1_A$.*

(c) *Let A, B, C be L-structures. Then 1_A is an isomorphism. If $f: A \to B$ is an isomorphism then the inverse map $f^{-1}: \operatorname{dom}(B) \to \operatorname{dom}(A)$ exists and is an isomorphism from B to A. If $f: A \to B$ and $g: B \to C$ are isomorphisms then so is gf.*

(d) *The relation $\cong$ is an equivalence relation on the class of L-structures.*

(e) *If A, B are L-structures, $f: A \to B$ is a homomorphism and there exist homomorphisms $g: B \to A$ and $h: B \to A$ such that $gf = 1_A$ and $fh = 1_B$, then f is an isomorphism and $g = h = f^{-1}$.*

Proof. (e) $g = g1_B = gfh = 1_A h = h$. Since $gf = 1_A$, f is an embedding. Since $fh = 1_B$, f is surjective. $\square$

Substructures

If A and B are L-structures with $\operatorname{dom}(A) \subseteq \operatorname{dom}(B)$ and the inclusion map $i: \operatorname{dom}(A) \to \operatorname{dom}(B)$ is an embedding, then we say that B is an **extension** of A, or that A is a **substructure** of B, in symbols $A \subseteq B$. Note that if i is the inclusion map from $\operatorname{dom}(A)$ to $\operatorname{dom}(B)$, then condition (2.1) above says that $c^A = c^B$ for each constant c, condition (2.2) says that $R^A = R^B \cap (\operatorname{dom} A)^n$ for each n-ary relation symbol R, and finally condition (2.3) says that $F^A = F^B |(\operatorname{dom} A)^n$ (the restriction of F^B to $(\operatorname{dom} A)^n$) for each n-ary function symbol F.

When does a set of elements of a structure form the domain of a substructure? The next lemma gives us a criterion.

Lemma 1.2.2. *Let B be an L-structure and X a subset of $\operatorname{dom}(B)$. Then the following are equivalent.*
(a) *$X = \operatorname{dom}(A)$ for some $A \subseteq B$.*
(b) *For every constant c of L, $c^B \in X$; and for every $n > 0$, every n-ary function symbol F of L and every n-tuple $\bar{a}$ of elements of X, $F^B(\bar{a}) \in X$.*
If (a) and (b) hold, then A is uniquely determined.

Proof. Suppose (a) holds. Then for every constant c of L, $c^B = c^A$ by (2.1); but $c^A \in \operatorname{dom}(A) = X$, so $c^B \in X$. Similarly, for each n-ary function symbol F of L and each n-tuple $\bar{a}$ of elements of X, $\bar{a}$ is an n-tuple in A and so $F^B(\bar{a}) = F^A(\bar{a}) \in \operatorname{dom}(A) = X$. This proves (b).

Conversely, if (b) holds, then we can define A by putting $\operatorname{dom}(A) = X$, $c^A = c^B$ for each constant c of L, $F^A = F^B | X^n$ for each n-ary function symbol F of L, and $R^A = R^B \cap X^n$ for each n-ary relation symbol R of L. Then $A \subseteq B$; moreover this is the only possible definition of A, given that $A \subseteq B$ and $\operatorname{dom}(A) = X$. $\qquad\square$

Let B be an L-structure and Y a set of elements of B. Then it follows easily from Lemma 1.2.2 that there is a unique smallest substructure A of B whose domain includes Y; we call A the **substructure of B generated by Y**, or the **hull** of Y in B, in symbols $A = \langle Y \rangle_B$. We call Y a **set of generators** for A. A structure B is said to be **finitely generated** if B is of form $\langle Y \rangle_B$ for some finite set Y of elements.

When the context allows, we write just $\langle Y \rangle$ instead of $\langle Y \rangle_B$. Sometimes it's convenient to list the generators Y as a sequence $\bar{a}$; then we write $\langle \bar{a} \rangle_B$ for $\langle Y \rangle_B$.

For many purposes we shall need to know the cardinality of the structure $\langle Y \rangle_B$. This can be estimated as follows. We define the **cardinality** of L, in symbols $|L|$, to be the least infinite cardinal $\geqslant$ the number of symbols in L. (In fact we shall see in Exercise 2.1.7 below that $|L|$ is equal to the number of first-order formulas of L, up to choice of variables; this is one reason why $|L|$ is taken to be infinite even when L contains only finitely many symbols.)

Warning. Occasionally it's important to know that a signature L contains only finitely many symbols. In this case we say that L is **finite**, in spite of the definition just given for $|L|$. See Exercise 6.

Theorem 1.2.3. *Let B be an L-structure and Y a set of elements of B. Then $|\langle Y \rangle_B| \leqslant |Y| + |L|$.*

Proof. We shall construct $\langle Y \rangle_B$ explicitly, thus proving its existence and uniqueness at the same time. We define a set $Y_m \subseteq \mathrm{dom}(B)$ for each $m < \omega$, by induction on m:

$$Y_0 = Y \cup \{c^B : c \text{ a constant of } L\},$$
$$Y_{m+1} = Y_m \cup \{F^B(\bar{a}) : \text{for some } n > 0, \ F \text{ is an } n\text{-ary func-}$$
tion symbol of L and $\bar{a}$ is an n-tuple of elements of $Y_m\}.$

Finally we put $X = \cup_{m < \omega} Y_m$. Clearly X satisfies condition (b) of Lemma 1.2.2, so by that lemma there is a unique substructure A of B with $X = \mathrm{dom}(A)$. If A' is any substructure of B with $Y \subseteq \mathrm{dom}(A')$, then by induction on m we see that each Y_m is included in $\mathrm{dom}(A')$ (by the implication (a) $\Rightarrow$ (b) in Lemma 1.2.2), and hence $X \subseteq \mathrm{dom}(A')$. Therefore A is the unique smallest substructure of W whose domain includes Y, in short $A = \langle Y \rangle_B$.

Now we estimate the cardinality of A. Put $\kappa = |Y| + |L|$. Clearly $|Y_0| \leqslant \kappa$. For each fixed n, if Z is a subset of $\mathrm{dom}(B)$ of cardinality $\leqslant \kappa$, then the set

$$\{F^B(\bar{a}) : F \text{ is an } n\text{-ary function symbol of } L \text{ and } \bar{a} \in Z^n\}$$

has cardinality at most $\kappa \cdot \kappa^n = \kappa$, since κ is infinite. Hence if $|Y_m| \leqslant \kappa$, then $|Y_{m+1}| \leqslant \kappa + \kappa = \kappa$. Thus by induction on m, each $|Y_m| \leqslant \kappa$, and so $|X| \leqslant \omega \cdot \kappa = \kappa$. Since $|\langle Y \rangle_B| = |X|$ by definition, this proves the theorem. $\square$

Choice of signature

As we remarked earlier, one and the same mathematical object can be interpreted as a structure in several different ways. The same function or relation can be named by different symbols: for example the identity element in a group can be called e or 1. We also have some choice about which elements, functions or relations should be given names at all. A group should surely have a name $\cdot$ for the product operation; ought it also to have names $^{-1}$ and 1 for the inverse and identity? How about $[\,,]$ for the commutator, $[a, b] = a^{-1}b^{-1}ab$?

A healthy general principle is that, other things being equal, *signatures should be chosen so that the notions of homomorphism and substructure agree with the usual notions for the relevant branch of mathematics.*

For example in the case of groups, if the only named operation is the product $\cdot$, then the substructures of a group will be its subsemigroups, closed under $\cdot$ but not necessarily containing inverses or identity. If we name $\cdot$ and the identity 1, then the substructures will be the submonoids. To ensure that substructure equals subgroup, we also need to put in a symbol for $^{-1}$. Given that symbols for $\cdot$ and $^{-1}$ are included, we would get exactly the same substructures and homomorphisms if we also included a symbol for the commutator; most model theorists use Ockham's razor and leave it out.

Thus for some classes of objects there is a natural choice of signature. For groups the natural choice is to name $\cdot$, 1 (or e) and $^{-1}$; we call this the **signature of groups**. We shall always assume (unless otherwise stated) that rings have a 1. So the natural choice of signature for rings is to name $+$, $-$, $\cdot$, 0 and 1; we call this the **signature of rings**. The **signature of partial orderings** has just the symbol $\leqslant$. The **signature of lattices** has just $\wedge$ and $\vee$.

What if we want to add new symbols, for example to name a particular element? The next definitions take care of this situation.

Reduction and expansion

Suppose L^- and L^+ are signatures, and L^- is a subset of L^+. Then if A is an L^+-structure, we can turn A into an L^--structure by simply forgetting the symbols of L^+ which are not in L^-. (We don't remove any elements of A, though some constant elements in A may cease to be constant elements in the new structure.) The resulting L^--structure is called the L^--**reduct** of A or the **reduct** of A to L^-, in symbols $A|L^-$. If $f: A \to B$ is a homomorphism of L^+-structures, then the same map f is also a homomorphism $f: A|L^- \to B|L^-$ of L^--structures.

When A is an L^+-structure and C is its L^--reduct, we say that A is an **expansion** of C to L^+. In general C may have many different expansions to L^+. There is a useful but imprecise notation for expansions. Suppose the symbols which are in L^+ but not in L^- are constants c, d and a function symbol F; then c^A, d^A are respectively elements a, b of A, and F^A is some operation f on $\mathrm{dom}\,(A)$. We write

$$(2.5) \qquad\qquad A = (C, a, b, f)$$

to express that A is an expansion of C got by adding symbols to name a, b and f. The notation is imprecise because it doesn't say what symbols are used to name a, b and f respectively; but often the choice of symbols is unimportant or obvious from the context.

Exercises for section 1.2

1. Verify all parts of Theorem 1.2.1.

2. Let L be a signature and **K** a class of L-structures. Suppose that A and B are in **K**, and for every structure C in **K** there are unique homomorphisms $f: A \to C$ and $g: B \to C$. Show that there is a unique isomorphism from A to B.

3. Let A, B be L-structures, X a subset of $\mathrm{dom}\,(A)$ and $f: \langle X \rangle_A \to B$ and $g: \langle X \rangle_A \to B$ homomorphisms. Show that if $f|X = g|X$ then $f = g$.

4. Prove the following, in which all structures are assumed to be L-structures for some fixed signature L.

(a) Every homomorphism $f: A \to C$ can be factored as $f = hg$ for some surjective homomorphism $g: A \to B$ and extension $h: B \to C$:

(2.6)

The unique structure B is called the **image** of g, img *for short. More generally we say that an L-structure B is a* **homomorphic image** *of A if there exists a surjective homomorphism $g: A \to B$.*

(b) Every embedding $f: A \to C$ can be factored as $f = hg$ where g is an extension and h is an isomorphism. *This is a small piece of set theory which often gets swept under the carpet. It says that when we have embedded A into C, we can assume that A is a substructure of C. The embedding of the rationals in the reals is a familiar example.*

5. Let A and B be L-structures with A a substructure of B. A **retraction** from B to A is a homomorphism $f: B \to A$ such that $f(a) = a$ for every element a of A. Show (a) if $f: B \to A$ is a retraction, then $f^2 = f$, (b) if B is any L-structure and f an endomorphism of B such that $f^2 = f$, then f is a retraction from B to a substructure A of B.

6. Let L be a finite signature with no function symbols. (a) Show that every finitely generated L-structure is finite. (b) Show that for each $n < \omega$ there are up to isomorphism only finitely many L-structures of cardinality n.

7. Show that we can give fields a signature which makes substructure = subfield and homomorphism = field embedding, provided we are willing to let 0^{-1} be 0. *Most mathematicians seem to be unwilling, and so the normal custom is to give fields the same signature as rings.*

*8. Give an example of a field F (using the signature of rings, with symbols $+$, $-$, $\cdot$, 0, 1) which is not finitely generated, although there is an element b such that F is the smallest subfield of F containing b. (*So if fields are taken as a kind of ring, our notion of 'finitely generated' doesn't agree with the usual one in field theory.*)

*9. (Unnatural choices of signature) (a) Suppose we were to make partial orderings into structures by naming the relation $<$ but not the relation $\leqslant$. What would the homomorphisms between partial orderings be? (E g. would they all be injective?)
(b) Suppose we were to make commutative rings with 1 into structures by naming $+$, $-$, $\cdot$, 0, 1 and adding a 1-ary relation symbol for the Jacobson radical. What would the homomorphisms between these rings be?

10. Many model theorists require the domain of any structure to be non-empty. How would this affect Lemma 1.2.2 and the definition of $\langle Y \rangle_B$?

11. Give an example of a structure of cardinality ω_2 which has a substructure of cardinality ω but no substructure of cardinality ω_1.

1.3 Terms and atomic formulas

In Chapter 2 we shall introduce a number of formal languages for talking about L-structures. All these languages will be built up from the atomic formulas of L, which we must now define.

Each atomic formula will be a string of symbols including symbols of L. Since the symbols in L can be any kinds of object and not necessarily written expressions, the idea of a 'string of symbols' has to be taken with a pinch of set-theoretic coding. The same applies to the more elaborate 'formulas' introduced in Chapter 2 (cf. Exercise 2.1.16).

Terms

Every language has a stock of **variables**. These are symbols written v, x, y, z, t, x_0, x_1 etc., and one of their purposes is to serve as temporary labels for elements of a structure. Any symbol can be used as a variable, provided it is not already being used for something else. The choice of variables is never important, and for theoretical purposes many model theorists restrict them to be the expressions v_0, v_1, v_2, . . . with natural number or ordinal subscripts.

The **terms** of the signature L are strings of symbols defined as follows (where the symbols '(', ')' and ',' are assumed not to occur anywhere in L – henceforth points like this go without saying).

(3.1) Every variable is a term of L.

(3.2) Every constant of L is a term of L.

(3.3) If $n > 0$, F is an n-ary function symbol of L and $t_1, . . ., t_n$ are terms of L then the expression $F(t_1, . . ., t_n)$ is a term of L.

(3.4) Nothing else is a term of L.

A term is said to be **closed** (computer scientists say **ground**) if no variables occur in it. The **complexity** of a term is the number of symbols occurring in it, counting each occurrence separately. (The important point is that if t occurs as part of s then s has higher complexity than t.)

If we introduce a term t as $t(\bar{x})$, this will always mean that $\bar{x}$ is a sequence $(x_0, x_1, . . .)$, possibly infinite, of distinct variables, and every variable which occurs in t is among the variables in $\bar{x}$. Later in the same context we may write $t(\bar{s})$, where $\bar{s}$ is a sequence of terms $(s_0, s_1, \ldots)$; then $t(\bar{s})$ means the term got from t by putting s_0 in place of x_0, s_1 in place of x_1, etc., throughout t. (For example if $t(x, y)$ is the term $y + x$, then $t(0, 2y)$ is the term $2y + 0$ and $t(t(x, y), y)$ is the term $y + (y + x)$.)

To make variables and terms stand for elements of a structure, we use the following convention. Let $t(\bar{x})$ be a term of L, where $\bar{x} = (x_0, x_1, \ldots)$. Let A be an L-structure and $\bar{a} = (a_0, a_1, \ldots)$ a sequence of elements of A; we assume that $\bar{a}$ is at least as long as $\bar{x}$. Then $t^A(\bar{a})$ (or $t^A[\bar{a}]$ when we need a more distinctive notation) is defined to be the element of A which is named by t when x_0 is interpreted as a name of a_0, and x_1 as a name of a_1, and so on. More precisely, using induction on the complexity of t,

(3.5)   if t is the variable x_i then $t^A[\bar{a}]$ is a_i,

(3.6)   if t is a constant c then $t^A[\bar{a}]$ is the element c^A,

(3.7)   if t of the form $F(s_1, \ldots, s_n)$ where each s_i is a term $s_i(\bar{x})$, then $t^A[\bar{a}]$ is the element $F^A(s_1{}^A[\bar{a}], \ldots, s_n{}^A[\bar{a}])$.

(Cf. (1.2), (1.4) above.) If t is a closed term then $\bar{a}$ plays no role and we simply write t^A for $t^A[\bar{a}]$.

Atomic formulas

The **atomic formulas** of L are the strings of symbols given by (3.8) and (3.9) below.

(3.8)   If s and t are terms of L then the string $s = t$ is an atomic formula of L.

(3.9)   If $n > 0$, R is an n-ary relation symbol of L and $t_1, \ldots, t_n$ are terms of L then the expression $R(t_1, \ldots, t_n)$ is an atomic formula of L.

(Note that the symbol '=' is not assumed to be a relation symbol in the signature.) An **atomic sentence** of L is an atomic formula in which there are no variables.

Just as with terms, if we introduce an atomic formula ϕ as $\phi(\bar{x})$, then $\phi(\bar{s})$ means the atomic formula got from ϕ by putting terms from the sequence $\bar{s}$ in place of all occurrences of the corresponding variables from $\bar{x}$.

If the variables $\bar{x}$ in an atomic formula $\phi(\bar{x})$ are interpreted as names of elements $\bar{a}$ in a structure A, then ϕ makes a statement about A. The statement may be true or false. If it's true, we say ϕ is **true of $\bar{a}$ in** A, or that $\bar{a}$ **satisfies** ϕ **in** A, in symbols

$$A \vDash \phi[\bar{a}] \qquad \text{or equivalently} \qquad A \vDash \phi(\bar{a}).$$

We can give a formal definition of this relation $\vDash$. Let $\phi(\bar{x})$ be an atomic formula of L with $\bar{x} = (x_0, x_1, \ldots)$. Let A be an L-structure and $\bar{a}$ a sequence $(a_0, a_1, \ldots)$ of elements of A; we assume that $\bar{a}$ is at least as long as $\bar{x}$. Then

(3.10) if ϕ is the formula $s = t$ where $s(\bar{x})$, $t(\bar{x})$ are terms, then
$A \vDash \phi[\bar{a}]$ iff $s^A[\bar{a}] = t^A[\bar{a}]$,

(3.11) if ϕ is the formula $R(s_1, \ldots, s_n)$ where $s_1(\bar{x}), \ldots, s_n(\bar{x})$ are terms, then $A \vDash \phi[\bar{a}]$ iff the ordered n-tuple
$(s_1^A[\bar{a}], \ldots, s_n^A[\bar{a}])$ is in R^A.

(Cf. (1.3) above.) When ϕ is an atomic sentence, we can omit the sequence $\bar{a}$ and write simply $A \vDash \phi$ in place of $A \vDash \phi[\bar{a}]$.

We say that A is a **model** of ϕ, or that ϕ is **true in** A, if $A \vDash \phi$. When T is a set of atomic sentences, we say that A is a **model** of T (in symbols, $A \vDash T$) if A is a model of every atomic sentence in T.

Theorem 1.3.1. *Let A and B be L-structures and f a map from* $\mathrm{dom}(A)$ *to* $\mathrm{dom}(B)$.

(a) *If f is a homomorphism then, for every term $t(\bar{x})$ of L and tuple $\bar{a}$ from A, $f(t^A[\bar{a}]) = t^B[f\bar{a}]$.*

(b) *f is a homomorphism if and only if, for every atomic formula $\phi(\bar{x})$ of L and tuple $\bar{a}$ from A,*

(3.12) $$A \vDash \phi[\bar{a}] \Rightarrow B \vDash \phi[f\bar{a}].$$

(c) *f is an embedding if and only if, for every atomic formula $\phi(\bar{x})$ of L and tuple $\bar{a}$ from A,*

(3.13) $$A \vDash \phi[\bar{a}] \Leftrightarrow B \vDash \phi[f\bar{a}].$$

Proof. (a) is easily proved by induction on the complexity of t, using (3.5)–(3.7).

(b) Suppose first that f is a homomorphism. As a typical example, suppose $\phi(\bar{x})$ is $R(s, t)$, where $s(\bar{x})$ and $t(\bar{x})$ are terms. Assume $A \vDash \phi[\bar{a}]$. Then by (3.11) we have

(3.14) $$(s^A[\bar{a}], t^A[\bar{a}]) \in R^A.$$

Then by part (a) and the fact that f is a homomorphism (see (2.2) above),

(3.15) $$(s^B[f\bar{a}], t^B[f\bar{a}]) = (f(s^A[\bar{a}]), f(t^A[\bar{a}])) \in R^B.$$

Hence $B \vDash \phi[f\bar{a}]$ by (3.11) again. Essentially the same proof works for every atomic formula ϕ.

For the converse, again we take a typical example. Assume that (3.12) holds for all atomic ϕ and sequences $\bar{a}$. Suppose $(a_0, a_1) \in R^A$. Then writing $\bar{a}$ for (a_0, a_1), we have $A \vDash R(x_0, x_1)[\bar{a}]$. Then (3.12) implies $B \vDash R(x_0, x_1)[f\bar{a}]$, which by (3.11) implies that $(fa_0, fa_1) \in R^B$ as required. Thus f is a homomorphism.

(c) is proved like (b), but using (2.4) in place of (2.2).   □

A variant of Theorem 1.3.1(c) is often useful. By a **negated atomic formula** of L we mean a string $\neg \phi$ where ϕ is an atomic formula of L. We read the symbol $\neg$, as 'not' and we define

(3.16)   '$A \vDash \neg \phi[\bar{a}]$' holds   iff   '$A \vDash \phi[\bar{a}]$' doesn't hold.

where A is any L-structure, ϕ is an atomic formula and $\bar{a}$ is a sequence from A. A **literal** is an atomic or negated atomic formula; it's a **closed literal** if it contains no variables.

Corollary 1.3.2. *Let A and B be L-structures and f a map from $\mathrm{dom}\,(A)$ to $\mathrm{dom}\,(B)$. Then f is an embedding if and only if, for every literal $\phi(\bar{x})$ of L and sequence $\bar{a}$ from A,*

(3.17) $A \vDash \phi[\bar{a}] \Rightarrow B \vDash \phi[f\bar{a}].$

Proof. Immediate from (c) of the theorem and (3.16).   □

The term algebra

The Delphic oracle said 'Know thyself'. Terms can do this – they can describe themselves. Let L be any signature and X a set of variables. We define the **term algebra of L with basis X** to be the following L-structure A. The domain of A is the set of all terms of L whose variables are in X. We put

(3.18) $c^A = c$ for each constant c of L,

(3.19) $F^A(\bar{t}) = F(\bar{t})$ for each n-ary function symbol F of L and n-tuple $\bar{t}$ of elements of $\mathrm{dom}\,(A)$,

(3.20) R^A is empty   for each relation symbol R of L.

The term algebra of L with basis X is also known as the **absolutely free L-structure with basis X**, for reasons that will appear in section 9.2 below.

Exercises for section 1.3

1. Let B be an L-structure and Y a set of elements of B. Show that $\langle Y \rangle_B$ consists of those elements of B which have the form $t^B[\bar{b}]$ for some term $t(\bar{x})$ of L and some tuple $\bar{b}$ of elements of Y. [Use the construction of $\langle Y \rangle_B$ in the proof of Lemma 1.2.2.]

2. (a) If $t(x, y, z)$ is $F(G(x, z), x)$, what are $t(z, y, x)$, $t(x, z, z)$, $t(F(x, x), G(x, x), x)$, $t(t(a, b, c), b, c)$?
(b) Let A be the structure $\langle \mathbb{N}, +, \cdot, 0, 1 \rangle$ where $\mathbb{N}$ is the set of natural numbers 0,

1,.... Let $\phi(x, y, z, u)$ be the atomic formula $x + z = y \cdot u$ and let $t(x, y)$ be the term $y \cdot y$. Which of the following are true? $A \vDash \phi[0, 1, 2, 3]$; $A \vDash \phi[1, 5, t^A[4, 2], 1]$; $A \vDash \phi[9, 1, 16, 25]$; $A \vDash \phi[56, t^A[9, t^A[0, 3]], t^A[5, 7], 1]$. [Answers: $F(G(z, x), z)$, $F(G(x, z), x)$, $F(G(F(x, x), x), F(x, x))$, $F(G(F(G(G(a, c), a), c), F(G(a, c), a))$; no, yes, yes, no.]

3. Let A be an L-structure, $\bar{a} = (a_0, a_1, \ldots)$ a sequence of elements of A and $\bar{s} = (s_0, s_1, \ldots)$ a sequence of closed terms of L such that, for each i, $s_i^A = a_i$. Show (a) for each term $t(\bar{x})$ of L, $t(\bar{s})^A = t^A[\bar{a}]$, (b) for each atomic formula $\phi(\bar{x})$ of L, $A \vDash \phi(\bar{s}) \Leftrightarrow A \vDash \phi[\bar{a}]$.

4. Let A and B be L-structures, $\bar{a}$ a sequence of elements which generate A and f a map from dom (A) to dom (B). Show that f is a homomorphism if and only if for every atomic formula $\phi(\bar{x})$ of L, $A \vDash \phi[\bar{a}]$ implies $B \vDash \phi[f\bar{a}]$.

5. (Unique parsing lemma). Let L be a signature and t a term of L. Show that t can be constructed in only one way. [In other words, (a) if t is a constant then t is not also of the form $F(\bar{s})$, (b) if t is $F(s_0, \ldots, s_{m-1})$ and $G(r_0, \ldots, r_{n-1})$ then $F = G$, $m = n$ and for each $i < n$, $s_i = r_i$. For each occurrence σ of a symbol in t, define $^\#(\sigma)$ to be the number of occurrences of '(' to the left of σ, minus the number of occurrences of ')' to the left of σ; use $^\#(\sigma)$ to identify the commas connected with F in t.]

6. Let A be the term algebra of L with basis X. Show that for each term $t(\bar{x})$ of L and each tuple $\bar{s}$ of elements of dom (A), $t^A[\bar{s}] = t(\bar{s})$.

7. Let A be the term algebra of signature L with basis X, and B any L-structure. Show that for each map $f: X \rightarrow$ dom (B) there is a unique homomorphism $g: A \rightarrow B$ which agrees with f on X. Show also that if $t(\bar{x})$ is any term in dom (A) then $t^B[f\bar{x}] = g(t)$.

*8. (Unification theorem). Let L be a signature and A the term algebra of L with basis X. Let $s(\bar{x})$, $t(\bar{y})$ be terms in dom (A) with no variables in common. An endomorphism f of A is said to be a **unifier of** s and t if $f(s) = f(t)$; s and t are **unifiable** if such an endomorphism exists. (a) Show that if s and t are unifiable, then they have a unifier f such that every unifier g of s and t can be written as $g = hf$ for some endomorphism h of A. Such a unifier f is called a **most general unifier** of s and t. (b) Formulate and prove a theorem about the uniqueness of most general unifiers.

1.4 Parameters and diagrams

The conventions for interpreting variables are one of the more irksome parts of model theory. We can avoid them, at a price. Instead of interpreting a variable as a name of the element b, we can add a new constant for b to the signature. The price we pay is that the language changes every time another element is named. When constants are added to a signature, the new constants and the elements they name are called **parameters**.

Suppose for example that A is an L-structure, $\bar{a}$ is a sequence of elements of A, and we want to name the elements in $\bar{a}$. Then we choose a sequence $\bar{c}$ of distinct new constant symbols, of the same length as $\bar{a}$, and we form the signature $L(\bar{c})$ by adding the constants $\bar{c}$ to L. In the notation of (2.5), $(A, \bar{a})$ is an $L(\bar{c})$-structure, and each element a_i is $c_i^{(A,\bar{a})}$.

Likewise if B is another L-structure and $\bar{b}$ a sequence of elements of B of the same length as $\bar{c}$, then there is an $L(\bar{c})$-structure $(B, \bar{b})$ in which these same constants c_i name the elements of $\bar{b}$. The next lemma is about this situation. It comes straight out of the definitions, and it is often used silently.

Lemma 1.4.1. *Let A, B be L-structures and suppose $(A, \bar{a})$, $(B, \bar{b})$ are $L(\bar{c})$-structures. Then a homomorphism $f: (A, \bar{a}) \to (B, \bar{b})$ is the same thing as a homomorphism $f: A \to B$ such that $f\bar{a} = \bar{b}$. Likewise an embedding $f: (A, \bar{a}) \to (B, \bar{b})$ is the same thing as an embedding $f: A \to B$ such that $f\bar{a} = \bar{b}$.* $\qquad\square$

In the situation above, if $t(\bar{x})$ is a term of L then $t^A[\bar{a}]$ and $t(\bar{c})^{(A,\bar{a})}$ are the same element; and if $\phi(\bar{x})$ is an atomic formula then $A \vDash \phi[\bar{a}]$ $\Leftrightarrow (A, \bar{a}) \vDash \phi(\bar{c})$. These two notations – the square-bracket and the parameter notation – serve the same purpose, and it's a burden on one's patience to keep them separate. The following compromise works well in practice. We use the elements a_i as *constants naming themselves*. The expanded signature is then $L(\bar{a})$, and we write $t^A(\bar{a})$ and $A \vDash \phi(\bar{a})$ with an easy conscience. One should take care when $\bar{a}$ contains repetitions, or when two separate $L(\bar{c})$-structures are under discussion. To play safe I retain the square-bracket notation for the next few sections.

Let $\bar{a}$ be a sequence of elements of A. We say that $\bar{a}$ **generates** A, in symbols $A = \langle \bar{a} \rangle_A$, if A is generated by the set of all elements in $\bar{a}$. Suppose that A is an L-structure, $(A, \bar{a})$ is an $L(\bar{c})$-structure and $\bar{a}$ generates A. Then (cf. Exercise 1 below) every element of A is of the form $t^{(A,\bar{a})}$ for some closed term t of $L(\bar{c})$; so every element of A has a name in $L(\bar{c})$. The set of all closed literals of $L(\bar{c})$ which are true in $(A, \bar{a})$ is called the (Robinson) **diagram** of A, in symbols diag(A). The set of all atomic sentences of $L(\bar{c})$ which are true in $(A, \bar{a})$ is called the **positive diagram** of A, in symbols diag$^+(A)$.

Note that diag(A) and diag$^+(A)$ are not uniquely determined, because in general there are many ways of choosing $\bar{a}$ and $\bar{c}$ so that $\bar{a}$ generates A. But this never matters. There is always at least one possible choice of $\bar{a}$ and $\bar{c}$: simply list all the elements of A without repetition.

Diagrams and positive diagrams should be thought of as a generalisation of the multiplication tables of groups. If we know either diag(A) or diag$^+(A)$, we know A up to isomorphism. The name 'diagram' is due to Abraham Robinson, who was the first model theorist to use diagrams systematically. It's a slightly unfortunate name, because it invites confusion with diagrams in the sense of pictures with arrows, as in category theory. Actually the next result, which we shall use many times, is about diagrams in both senses.

Lemma 1.4.2 (*Diagram lemma*). *Let A and B be L-structures, $\bar{c}$ a sequence of constants, and $(A, \bar{a})$ and $(B, \bar{b})$ $L(\bar{c})$-structures. Then* (a) *and* (b) *are equivalent.*
(a) *For every atomic sentence ϕ of $L(\bar{c})$, if $(A, \bar{a}) \vDash \phi$ then $(B, \bar{b}) \vDash \phi$.*
(b) *There is a homomorphism $f\colon \langle \bar{a} \rangle_A \to B$ such that $f\bar{a} = \bar{b}$.*
The homomorphism f in (b) *is unique if it exists; it is an embedding if and only if*
(c) *for every atomic sentence ϕ of $L(\bar{c})$, $(A, \bar{a}) \vDash \phi \Leftrightarrow (B, \bar{b}) \vDash \phi$.*

Proof. Assume (a). Since the inclusion map embeds $\langle \bar{a} \rangle_A$ in A. Theorem 1.3.1(c) says that in (a) we can replace A by $\langle \bar{a} \rangle_A$. So without loss we can assume that $A = \langle \bar{a} \rangle_A'$. By Lemma 1.4.1, to prove (b) it suffices to find a homomorphism $f\colon (A, \bar{a}) \to (B, \bar{b})$. We define f as follows. Since $\bar{a}$ generates A, each element of A is of the form $t^{(A, \bar{a})}$ for some closed term t of $L(\bar{c})$. Put

(4.1) $$f(t^{(A, \bar{a})}) = t^{(B, \bar{b})}.$$

The definition is sound, since $s^{(A, \bar{a})} = t^{(A, \bar{a})}$ implies $(A, \bar{a}) \vDash s = t$, so $(B, \bar{b}) \vDash s = t$ by (a) and hence $s^{(B, \bar{b})} = t^{(B, \bar{b})}$. Then f is a homomorphism by (a) and Theorem 1.3.1(b), which proves (b). Since any homomorphism f from $(A, \bar{a})$ to $(B, \bar{b})$ must satisfy (4.1), f is unique in (b). The converse (b) $\Rightarrow$ (a) follows at once from Theorem 1.3.1(b).
The argument for embeddings and (c) is similar, using Theorem 1.3.1(c). $\Box$

Lemma 1.4.2 doesn't mention Robinson diagrams outright, so let me make the connection explicit. Suppose $\bar{a}$ generates A. Then the implication (a) $\Rightarrow$ (b) in the lemma says that A can be mapped homomorphically to a reduct of B whenever $B \vDash \mathrm{diag}^+(A)$. Similarly the last part of the lemma says that if $B \vDash \mathrm{diag}(A)$ then A can be embedded in a reduct of B.

Exercises for section 1.4

1. Let B be an L-structure. If $\bar{b}$ is a sequence of elements of B and $\bar{c}$ are parameters such that $(B, \bar{b})$ is an $L(\bar{c})$-structure, show that $\langle \bar{b} \rangle_B$ consists of those elements of B which have the form $t^{(B, \bar{b})}$ for some closed term t of $L(\bar{c})$.

2. Let A, B be L-structures and $\bar{a}$ a sequence of elements of A. Let g be a map from the elements of $\bar{a}$ to dom(B) such that for every atomic sentence ϕ of $L(\bar{a})$, $(A, \bar{a}) \vDash \phi$ implies $(B, g\bar{a}) \vDash \phi$. Show that g has a unique extension to a homomorphism $g': \langle \bar{a} \rangle_A \to B$. (In practice one often identifies g and g'.)

3. Let $(A, \bar{a})$ and $(B, \bar{b})$ be $L(\bar{c})$-structures which satisfy exactly the same atomic sentences of $L(\bar{c})$. Suppose also that the L-structures A, B are generated by $\bar{a}$, $\bar{b}$ respectively. Show that there is an isomorphism $f: A \to B$ such that $f\bar{a} = \bar{b}$.

1.5 Canonical models

In the previous section we saw how one can translate any structure into a set of atomic sentences. It turns out that there is a good route back home again: we can convert any set of atomic sentences into a structure.

Let L be a signature, A an L-structure and T the set of all atomic sentences of L which are true in A. Then T has the following two properties.

(5.1)   For every closed term t of L, the atomic sentence $t = t$ is in T.

(5.2)   If $\phi(x)$ is an atomic formula of L and the equation $s = t$ is in T, then $\phi(s) \in T$ if and only if $\phi(t) \in T$.

Any set T of atomic sentences which satisfies (5.1) and (5.2) will be said to be =-**closed** (in L).

Lemma 1.5.1. *Let T be an =-closed set of atomic sentences of L. Then there is an L-structure A such that*
(a) *T is the set of all atomic sentences of L which are true in A,*
(b) *every element of A is of the form t^A for some closed term t of L.*

Proof. Let X be the set of all closed terms of L. We define a relation $\sim$ on X by

(5.3) $s \sim t$ iff $s = t \in T$.

We claim that $\sim$ is an equivalence relation. (i) By (5.1), $\sim$ is reflexive. (ii) Suppose $s \sim t$; then $s = t \in T$. But if $\phi(x)$ is the formula $x = s$, then $\phi(s)$ is $s = s$ which is in T by (5.1); so by (5.2), T also contains $\phi(t)$, which is $t = s$. Hence $t \sim s$. (iii) Suppose $s \sim t$ and $t \sim r$. Then let $\phi(x)$ be $s = x$. By assumption both $\phi(t)$ and $t = r$ are in T, so by (5.2) T also contains $\phi(r)$, which is $s = r$. Hence $s \sim r$. This proves the claim.

For each closed term t let $t^\sim$ be the equivalence class of t under $\sim$, and let Y be the set of all equivalence classes $t^\sim$ with $t \in X$. We shall define an L-structure A with dom$(A) = Y$.

First, for each constant c of L we put $c^A = c^\sim$. Next, if $0 < n < \omega$ and F is an n-ary function symbol of L, we define F^A by

(5.4) $F^A(s_0^\sim, \ldots, s_{n-1}^\sim) = (F(s_0, \ldots, s_{n-1}))^\sim.$

One must check that (5.4) is a sound definition. Suppose $s_i \sim t_i$ for each $i < n$. Then by n applications of (5.2) to the sentence $F(s_0, \ldots, s_{n-1}) = F(s_0, \ldots, s_{n-1})$, which is in T by (5.1), we find that the equation $F(s_0, \ldots, s_{n-1}) = F(t_0, \ldots, t_{n-1})$ is in T. Hence $F(s_0, \ldots, s_{n-1})^\sim = F(t_0, \ldots, t_{n-1})^\sim$. This shows that the definition (5.4) is sound.

Finally we define the relation R^A, where R is any n-ary relation symbol of L, by

(5.5) $(s_0^\sim, \ldots, s_{n-1}^\sim) \in R^A$ iff $R(s_0, \ldots, s_{n-1}) \in T.$

(5.5) is justified in the same way as (5.4), and it completes the description of the L-structure A.

Now it is easy to prove by induction on the complexity of t, using (3.6) and (3.7) from section 1.3, that for every closed term t of L,

(5.6) $t^A = t^\sim.$

From this we infer that if s and t are any closed terms of L then

(5.7) $A \vDash s = t \iff s^A = t^A \iff s^\sim = t^\sim \iff s = t \in T.$

Together with a similar argument for atomic sentences of the form $R(t_0, \ldots, t_{n-1})$, using (3.11) of section 1.3, this shows that T is the set of all atomic sentences of L which are true in A. Also by (5.6) every element of A is of the form t^A for some closed term t of L. $\qquad\square$

Now if T is any set of atomic sentences of L, there is a least set U of atomic sentences of L which contains T and is =-closed in L. We call U the **=-closure** of T in L. Any L-structure which is a model of U must also be a model of T since $T \subseteq U$.

Theorem 1.5.2. *For any signature L, if T is a set of atomic sentences of L then there is an L-structure A such that*
(a) $A \vDash T$,
(b) *every element of A is of the form t^A for some closed term t of L,*
(c) *if B is an L-structure and $B \vDash T$ then there is a unique homomorphism $f : A \to B$.*

Proof. Apply the lemma to the =-closure U of T to get the L-structure A. Then (a) and (b) are clear. Now $A = \langle \varnothing \rangle_A$ by (b). So (c) will follow from the diagram lemma (Lemma 1.4.2) if we can show that every atomic sentence true in A is true in all models B of T. By the choice of A, every atomic sentence true in A is in U. The set of all atomic sentences true in a model B of T is an =-closed set containing T, and so it must contain the =-closure of T, which is U. $\qquad\qquad\qquad\qquad\qquad\qquad\qquad\qquad\qquad\qquad\quad$ □

By clause (c), the model A of T in Theorem 1.5.2 is unique up to isomorphism. (See Exercise 1.2.2.) We call it the **canonical model** of T. Note that it will be the empty L-structure if and only if L has no constant symbols.

Sometimes – for example in logic programming – one doesn't include equations as atomic formulas. In this situation the canonical model is much easier to construct, because there is no need to factor out an equivalence relation. Thus we get what has become known as the **Herbrand universe** of a set of atomic sentences. (But see footnote 27 of Davis [1983], who doubts whether this name is appropriate.)

Here is an example at the opposite extreme, where all the atomic sentences are equations.

Example 1: *Adding roots of polynomials to a field.* Let F be a field and $p(X)$ an irreducible polynomial over F in the indeterminate X. We can regard $F[X]$ as a structure in the signature of rings with constants added for X and all the elements of F. Let T be the set of all equations which are true in $F[X]$. (For example if a is $b \cdot (c + d)$ in $F[X]$, then T contains the equation '$a = b \cdot (c + d)$'. Also rings satisfy the law $1 \cdot x = x$, so T contains the equation $1 \cdot t = t$ for every closed term t.) Then T is a set of atomic sentences, and the equation '$p(X) = 0$' is another atomic sentence. Let C be the canonical model of the set $T \cup \{p(X) = 0\}$. Then C is a homomorphic image of $F[X]$, because it is a model of T and every element is named by a closed term. In particular C is a ring; moreover '$p(X) = 0$' holds in C, and so every element of the ideal I in $F[X]$ generated by $p(X)$ goes to 0 in C. Let θ be any root of p; then the field extension $F[\theta]$ is a model of $T \cup \{p(X) = 0\}$ too, with X read as a name of θ. So by Theorem 1.5.2(c), $F[\theta]$ is a homomorphic image of C. Since $F[\theta] = F[X]/I$, it follows that C is isomorphic to $F[\theta]$.

Exercises for section 1.5

1. Show that the property of being =-closed is not altered if we change 'if and only if' in (5.2) to '$\Rightarrow$'.

2. Let T be a set of closed literals of signature L. Show that (a) is equivalent to (b). (a) Some L-structure is a model of T. (b) If $\neg \phi$ is a negated atomic sentence in T, then ϕ is not in the =-closure of the set of atomic sentences in T.

3. Let T be a finite set of closed literals of signature L. Show that (a) of Exercise 2 is equivalent to (c) some finite L-structure is a model of T. [Suppose $A \vDash T$. Let X be the set of all closed terms which occur as parts of sentences in T, including those inside other terms. Choose $a_0 \in \text{dom}(A)$. Define a structure A' with domain $\{t^A: t \in X\} \cup \{a_0\}$ so that $A' \vDash T$.]

4. Let A be an L-structure and $\bar{a}$ a sequence listing the elements of A without repetition. If $f: A \to B$ is a homomorphism of L-structures, we define the **kernel** of f to be the set of atomic sentences $\phi(\bar{a})$ of $L(\bar{a})$ such that $B \vDash \phi(f\bar{a})$. Show that for any L-structure A and set T of atomic sentences of $L(\bar{a})$ the following are equivalent. (a) $\text{diag}^+(A) \subseteq T$ and T is =-closed. (b) T is the kernel of some homomorphism $f: A \to B$. (c) T is the kernel of some surjective homomorphism $f: A \to B$. Show also that, if (c) holds, then B in (c) is unique in the sense that if the homomorphism $g: A \to C$ also has kernel T then there is a unique isomorphism $h: B \to C$ such that $hf = g$.

History and bibliography

Section 1.1. The prehistory of the notion of a structure involves Aquinas, Leibniz, de Morgan, C. S. Peirce and many others. Structures were finally dragged into history by Schröder [1895]. Labelling of parts became accepted only gradually. Frege mounted a holy war against it (e.g. Frege [1971], written 1899–1906), and his objection (basically that the symbols are ambiguous because they have different interpretations in different structures) is still repeated by some philosophers. The use of uncountably many symbols is implicit in Gödel [1931a] and explicit in Mal'tsev [1936].

Until about a dozen years ago, most model theorists named structures in horrible Fraktur lettering. Recent writers sometimes adopt a notation according to which all structures are called M, M', M^*, $\bar{M}$, M_0, M_i or occasionally N. I hope I cause no offence by using a more freewheeling notation.

Exercise 1: In the thirteenth century Thomas Aquinas gave this argument ([1921] part 1 q. xxxvi art. 2) in order to justify the action of the Council of Toledo in AD 589. They had added the word *filioque* to the Nicene Creed, thus begetting the split between the Roman and Byzantine churches.

Section 1.2. William Ockham was a fourteenth-century logician whose surname comes from a village south-west of London. His razor ('Entia non sunt multiplicanda praeter necessitatem', 'Don't invoke more entities than you have to') is a summary of his general attitude, but apparently not a quotation from his works.

Section 1.3.   Jacques Herbrand [1930] studied term algebras.

Exercise 5: Kleene [1952b] §17. Exercise 8: J. A. Robinson [1965]; paragraph
2.4 of Herbrand [1930] contains an algorithm which finds g, though this fact
is not stated. The unification theorem has generated a large amount of
research in computerised deduction; see Kirchner [1990] for a survey.

Section 1.4.   The idea of diagrams goes back to the 1921 *Tractatus
logico-philosophicus* of Wittgenstein [1961] (4.26): 'Die Angabe aller wahren
Elementarsätze beschreibt die Welt vollständig. Die Welt ist vollständig
beschrieben durch die Angaben aller Elementarsätze plus der Angabe, welche
von ihnen wahr und welche falsch sind.' After Wittgenstein several writers
(e.g. Carnap) used diagrams as a substitute for structures. The systematic use
of the diagram lemma begins with Abraham Robinson (e.g. [1950], [1956a],
[1956b], [1956c]). The distinction between variables with changing interpretations and constants with a fixed interpretation was quite vague until the mid
1950s, and there were several conventions for dealing with it (for example in
Mostowski [1952b], Robinson [1956b], Kemeny [1956]). In mathematical
practice one has to make the distinction, but there are degrees of variable-
ness, not just two levels.

Section 1.5.   In [1882] Leopold Kronecker boasted that he had succeeded in
proving the '*arithmetical* existence' (his italics) of algebraic numbers. He
didn't explain what he meant by this. But the proof he had just given, viz.
factoring out maximal ideals in polynomial rings, provides a clue. Instead of
finding the algebraic numbers inside some real or complex field which was
assumed to exist 'in nature', he had constructed an algebraic number
explicitly and concretely out of a description of the number. Example 1 shows
that canonical models are a straightforward generalisation of Kronecker's
construction.

Canonical models are a special case of free models in a variety or a
quasivariety; cf. section 9.2 below and the references given there.

2

Classifying structures

> I must get into this stone world now.
> Ratchel, striae, relationships of tesserae,
>   Innumerable shades of grey . . .
> I try them with the old Norn words – hraun
> Duss, rønis, queedaruns, kollyarum . . .
> > *Hugh MacDiarmid, On a raised beach.*

Now that we have structures in front of us, the most pressing need is to start classifying them and their features. Classifying is a kind of defining. Most mathematical classification is by axioms or defining equations – in short, by formulas. This chapter could have been entitled 'The elementary theory of mathematical classification by formulas'.

Notice three ways in which mathematicians use formulas. First, a mathematician writes the equation '$y = 4x^2$'. By writing this equation one names a set of points in the plane, i.e. a set of ordered pairs of real numbers. As a model theorist would put it, the equation *defines a 2-ary relation on the reals*. We study this kind of definition in section 2.1.

Or second, a mathematician writes down the laws

(*)   For all x, y and z, $x \leqslant y$ and $y \leqslant z$ imply $x \leqslant z$;
       for all x and y, exactly one of $x \leqslant y$, $y \leqslant x$, $x = y$ holds.

By doing this one *names a class of relations*, namely those relations $\leqslant$ for which (*) is true. Section 2.2 lists some more examples of this kind of naming. They cover most branches of algebra.

Third, a mathematician defines a **homomorphism** from a group G to a group H to be a map from G to H such that $x = y \cdot z$ implies $f(x) = f(y) \cdot f(z)$. Here the equation $x = y \cdot z$ *defines a class of maps*. See section 2.5 for more examples.

Many of the founding fathers of model theory wanted to understand how language works in mathematics, and how it ought to. (One might name Frege, Padoa, Russell, Gödel and Tarski among others.) But a successful branch of mathematics needs more than just a desire to correct and catalogue things. It needs a programme of problems which are interesting and not quite too hard to solve. The first systematic programme of model theory was known as 'elimination of quantifiers'. Its aim was to find, in any concrete mathematical situation, the simplest set of formulas that would give all the classifications that one needs. Thoralf Skolem set this programme moving in 1919. One recent offshoot of the programme is the study of unification in computer science.

2.1 Definable subsets

We begin with a single structure A. What are the interesting sets of elements of A? More generally, what are the interesting relations on elements of A?

Example 1: *Algebraic curves.* Regard the field $\mathbb{R}$ of reals as a structure. An algebraic curve in the real plane is a set of ordered pairs of elements of $\mathbb{R}$ given by an equation $p(x, y) = 0$, where p is a polynomial with coefficients from $\mathbb{R}$. The parabola $y = x^2$ is perhaps the most quoted example; this equation can be written without naming any elements of $\mathbb{R}$ as parameters.

Example 2: *Recursive sets of natural numbers.* For this we use the structure $\mathbb{N} = (\omega, 0, 1, +, \cdot, <)$ of natural numbers. Any recursive subset X of ω can be defined, for example by an algorithm for computing whether any given number is in X. But unlike Example 1, the definition will usually be much too complicated to be written as an atomic formula. There is no need to use parameters in this case, since every element of $\mathbb{N}$ is named by a closed term of the signature of $\mathbb{N}$.

Example 3: *Connected components of graphs.* Let G be a graph as in Example 1 of section 1.1, and g an element of G. The **connected component** of g in G is the smallest set of Y of vertices of G such that (1) $g \in Y$ and (2) if $a \in Y$ and a is joined to b by an edge, then $b \in Y$. This description defines Y, using g as a parameter. But again there is generally no hope of expressing the definition as an atomic formula. Also it will probably be hopeless to try to define Y without mentioning any element as a parameter.

We shall describe some simple relations on a structure, and then we shall describe how to generate more complicated relations from these. Each definable relation will be defined by a formula, and the formula will show how the relation is built up from simpler relations.

This approach pays dividends in several ways. First, the formulas give a way of describing the relations. Second, we can prove theorems about all definable sets by induction on the complexity of their defining formulas. Third, we can prove theorems about those relations which are defined by formulas of particular kinds (as for example in section 2.4 below).

For our starting point, we take the relations expressed by atomic formulas.

Some notation will be helpful here. Given an L-structure A and an atomic formula $\phi(x_0, \ldots, x_{n-1})$ of L, we write $\phi(A^n)$ for the set of n-tuples $\{\bar{a}: A \vDash \phi(\bar{a})\}$. For example if R is a relation symbol of the signature L, then the relation R^A is of the form $\phi(A^n)$: take $\phi(x_0, \ldots, x_{n-1})$ to be the formula $R(x_0, \ldots, x_{n-1})$.

We allow parameters too. Let $\psi(x_0, \ldots, x_{n-1}, \bar{y})$ be an atomic formula of L and $\bar{b}$ a tuple from A. Then $\psi(A^n, \bar{b})$ means the set $\{\bar{a}: A \vDash \psi(\bar{a}, \bar{b})\}$. For example if A consists of the real numbers and $\psi(x, y)$ is the formula $x > y$, then $\psi(A, 0)$ is the set of all positive reals.

To build up more complicated relations, we introduce a formal language $L_{\infty\omega}$ based on the signature L, as follows. (Read on for a couple of pages to see what the subscripts $_\infty$ and $_\omega$ mean.)

Building a language

Let L be a signature. The language $L_{\infty\omega}$ will be infinitary, which means that some of its formulas will be infinitely long. It's a language in a formal or set-theoretic sense. The symbols of $L_{\infty\omega}$ are those of L together with some logical symbols, variables and punctuation signs. The logical symbols are the following:

(1.1)   =   'equals',
         ¬   'not',
         $\bigwedge$ 'and',
         $\bigvee$ 'or',
         ∀   'for all elements . . .',
         ∃   'there is an element . . .'.

We define the **terms**, the **atomic formulas** and the **literals** of $L_{\infty\omega}$ to be the same as those of L. The class of **formulas** of $L_{\infty\omega}$ is defined to be the smallest class X such that

(1.2)   all atomic formulas of L are in X,

(1.3)   if ϕ is in X then the expression $\neg\phi$ is in X, and if $\Phi \subseteq X$ then the expressions $\bigwedge\Phi$ and $\bigvee\Phi$ are both in X,

(1.4)   if ϕ is in X and y is a variable then $\forall y\, \phi$ and $\exists y\, \phi$ are both in X.

The formulas which go into the making of a formula ϕ are called the **subformulas** of ϕ. The formula ϕ is counted as a subformula of itself; its **proper subformulas** are all its subformulas except itself.

The quantifiers $\forall y$ ('for all y') and $\exists y$ ('there is y') bind variables just as in elementary logic. Just as in elementary logic, we can distinguish between free and bound occurrences of variables. The **free variables** of a formula ϕ

are those which have free occurrences in ϕ. We sometimes introduce a formula ϕ as $\phi(\bar{x})$, for some sequence $\bar{x}$ of variables; this means that the variables in $\bar{x}$ are all distinct, and the free variables of ϕ all lie in $\bar{x}$. Then $\phi(\bar{s})$ means the formula that we get from ϕ by putting the terms s_i in place of the free occurrences of the corresponding variables x_i. This extends the notation that we introduced in section 1.3 for atomic formulas.

Likewise for any L-structure A and sequence $\bar{a}$ of elements of A, we extend the notation '$A \vDash \phi[\bar{a}]$' or '$A \vDash \phi(\bar{a})$' ('$\bar{a}$ satisfies ϕ in A') to all formulas $\phi(\bar{x})$ of $L_{\infty\omega}$ as follows, by induction on the construction of ϕ. These definitions are meant to fit the intuitive meanings of the symbols as given in (1.1) above.

(1.5)   If ϕ is atomic, then '$A \vDash \phi[\bar{a}]$' holds or fails just as in (3.10) and (3.11) of section 1.3.

(1.6) $A \vDash \neg\phi[\bar{a}]$ iff it is not true that $A \vDash \phi[\bar{a}]$.

(1.7) $A \vDash \bigwedge\Phi[\bar{a}]$ iff for every formula $\psi(\bar{x}) \in \Phi$, $A \vDash \psi[\bar{a}]$.

(1.8) $A \vDash \bigvee\Phi[\bar{a}]$ iff for at least one formula $\psi(\bar{x}) \in \Phi$, $A \vDash \psi[\bar{a}]$.

(1.9)   Suppose ϕ is $\forall y\, \psi$, where ψ is $\psi(y, \bar{x})$. Then $A \vDash \phi[\bar{a}]$ iff for all elements b of A, $A \vDash \psi[b, \bar{a}]$.

(1.10) Suppose ϕ is $\exists y\, \psi$, where ψ is $\psi(y, \bar{x})$. Then $A \vDash \phi[\bar{a}]$ iff for at least one element b of A, $A \vDash \psi[b, \bar{a}]$.

If $\bar{x}$ is an n-tuple of variables, $\phi(\bar{x}, \bar{y})$ is a formula of $L_{\infty\omega}$ and $\bar{b}$ is a sequence of elements of A whose length matches that of $\bar{y}$, we write $\phi(A^n, \bar{b})$ for the set $\{\bar{a}: A \vDash \phi(\bar{a}, \bar{b})\}$. Then $\phi(A^n, \bar{b})$ is the **relation defined in** A by the formula $\phi(\bar{x}, \bar{b})$.

Example 3 *continued.* x_0 is in the same component as g if either x_0 is g, or x_0 is joined by an edge to g (in symbols $R(x_0, g)$), or there is x_1 such that $R(x_0, x_1)$ and $R(x_1, g)$, or there are x_1 and x_2 such that $R(x_0, x_1)$, $R(x_1, x_2)$ and $R(x_2, g)$, or $\ldots$. In other words, the connected component of g is defined by the formula

(1.11) $$\bigvee(\{x_0 = g\} \cup \{\exists x_1 \ldots \exists x_n \bigwedge (\{R(x_i, x_{i+1}):$$
$$i < n\} \cup \{R(x_n, g)\}): n < \omega\})$$

with parameter g. The formula (1.11) is not easy to read, but it is quite precise.

Levels of language

We can define the **complexity** of a formula ϕ, $\text{comp}(\phi)$, so that it is greater than the complexity of any proper subformula of ϕ. Using ordinals, one possible definition is

(1.12) $\text{comp}(\phi) = \sup\{\text{comp}(\psi) + 1: \psi \text{ is a proper subformula of } \phi\}$.

But the exact notion of complexity never matters. What does matter is that we shall now be able to prove theorems about relations definable in $L_{\infty\omega}$, by using induction on the complexity of the formulas that define them.

In fact complexity is not always the most useful measure. The subscripts $_{\infty\omega}$ suggest other classifications. The second subscript, $_\omega$, means that we can put only finitely many quantifiers together in a row. By the same token, $L_{\infty 0}$ is the language consisting of those formulas of $L_{\infty\omega}$ in which no quantifiers occur; we call such formulas **quantifier-free**. Every atomic formula is quantifier-free. (Thus the curves in Example 1 above were defined by quantifier-free formulas.)

Occasionally we shall want to go beyond the confines of $L_{\infty\omega}$ by applying a quantifier $\forall$ or $\exists$ to infinitely many variables at once: $\forall(x_i: i \in I)$ or $\exists(x_i: i \in I)$. The language we get by adding these quantifiers to $L_{\infty\omega}$ is written $L_{\infty\infty}$.

The first subscript $_\infty$ means that we can join together arbitrarily many formulas by $\bigwedge$ or $\bigvee$. The **first-order language** of L, in symbols $L_{\omega\omega}$, consists of those formulas in which $\bigwedge$ and $\bigvee$ are only used to join together finitely many formulas at a time, so that the whole formula is finite. More generally if κ is any regular cardinal (such as the first uncountable cardinal ω_1), then $L_{\kappa\omega}$ is the same as $L_{\infty\omega}$ except that $\bigwedge$ and $\bigvee$ are only used to join together fewer than κ formulas at a time. For example the formula (1.11) lies in $L_{\omega_1\omega}$, since $\bigvee$ is taken over a countable set and $\bigwedge$ is taken over finite sets.

Thus we can pick out many smaller languages inside $L_{\infty\omega}$, by choosing subclasses of the class of formulas of $L_{\infty\omega}$. In fact $L_{\infty\omega}$ itself is much too large for everyday use; most of this book will be concerned with the first-order level $L_{\omega\omega}$, making occasional forays into $L_{\omega_1\omega}$. We say that a set X of formulas of $L_{\infty\omega}$ is **first-order-closed** if (1) X satisfies (1.2), (1.3) and (1.4) where Φ in (1.3) is required to be finite, and (2) every subformula of a formula in X is also in X. All the languages $L_{\kappa\omega}$ are first-order-closed. First-order-closed sublanguages of $L_{\infty\omega}$ are sometimes known as **fragments** of $L_{\infty\omega}$.

We shall use L as a symbol to stand for languages as well as signatures. Since a language determines its signature, there is no ambiguity if we talk about L-structures for a language L. Also if L is a first-order language, it is clear what is meant by $L_{\infty\omega}$, $L_{\kappa\omega}$ etc.; these are infinitary languages extending L. If a set X of parameters are added to L, forming a new language $L(X)$, we shall refer to the formulas of $L(X)$ as **formulas of L with parameters from** X.

Let L be a first-order language and A an L-structure. If $\phi(\bar{x})$ is a first-order formula, then a set or relation of the form $\phi(A^n)$ is said to be

first-order definable without parameters, or more briefly ∅-**definable** (pronounced 'zero-definable'.) A set or relation of the form $\psi(A^n, \bar{b})$, where
$\psi(\bar{x}, \bar{y})$ is a first-order formula and $\bar{b}$ is a tuple from some set X of elements
of A, is said to be X-**definable** and **first-order definable with parameters**.
(When people say simply 'first-order definable', you should check whether
they allow parameters. Some do, some don't.)

The following abbreviations are standard:

(1.13) $x \neq y$ for $\neg x = y$,

$(\phi_1 \wedge \ldots \wedge \phi_n)$ for $\bigwedge\{\phi_1, \ldots, \phi_n\}$ (finite conjunction),

$(\phi_1 \vee \ldots \vee \phi_n)$ for $\bigvee\{\phi_1, \ldots, \phi_n\}$ (finite disjunction),

$\displaystyle\bigwedge_{i \in I} \phi_i$ for $\bigwedge\{\phi_i : i \in I\}$,

$\displaystyle\bigvee_{i \in I} \phi_i$ for $\bigvee\{\phi_i : i \in I\}$,

$(\phi \rightarrow \psi)$ for $(\neg \phi) \vee \psi$ ('if ϕ then ψ'),

$(\phi \leftrightarrow \psi)$ for $(\phi \rightarrow \psi) \wedge (\psi \rightarrow \phi)$ ('ϕ iff ψ'),

$\forall x_1 \ldots x_n$ or $\forall \bar{x}$ for $\forall x_1 \ldots \forall x_n$,

$\exists x_1 \ldots x_n$ or $\exists \bar{x}$ for $\exists x_1 \ldots \exists x_n$,

$\bot$ for $\bigvee \varnothing$ (empty disjunction, false everywhere).

Normal conventions for dropping brackets are in force: brackets around
$(\phi \wedge \psi)$ or $(\phi \vee \psi)$ can be omitted when either $\rightarrow$ or $\leftrightarrow$ stands immediately
outside these brackets. Thus $\phi \wedge \psi \rightarrow \chi$ always means $(\phi \wedge \psi) \rightarrow \chi$, not
$\phi \wedge (\psi \rightarrow \chi)$.

With these conventions, formula (1.11) would normally be written

$$(1.14) \qquad x_0 = g \vee \bigvee_{n < \omega} \exists x_1 \ldots x_n\left(\left(\bigwedge_{i < n} R(x_i, x_{i+1})\right) \wedge R(x_n, g)\right),$$

which is slightly easier to read.

One can devise other languages by using other notions besides those in
(1.1). See section 2.8 below for some examples, such as the second-order
language L^{II} based on the signature L; see also the Svenonius sentences of
Example 2 in section 3.4. A family of languages which differ from each other
only in signature is called a **logic**. Thus **first-order logic** consists of the
languages $L_{\omega\omega}$ as L ranges over all signatures; **second-order logic** consists of
all the languages L^{II}, and so on. **Generalised model theory** is the study of
different logics in terms of what can be defined in them. (See for example the
huge anthology Barwise & Feferman [1985] on '*Model-theoretic logics*'.)

Remarks on variables

Mostly we shall be interested in formulas $\phi(x_0, \ldots, x_{n-1})$ with just finitely
many free variables; we call such formulas **variable-finite**. Every first-order

formula is automatically variable-finite, since it only has finite length. As we noted earlier, a variable-finite formula $\phi(\bar{x})$ of $L_{\infty\omega}$ defines a relation $\phi(A^n)$ in any L-structure A. One can stretch this notation a little and write $\Phi(A^n)$ for $(\bigwedge\Phi)(A^n)$, where $\Phi(\bar{x})$ is a set of variable-finite formulas $\phi(\bar{x})$. Also $\Phi(\bar{a})$ is the set of all formulas $\phi(\bar{a})$ with $\phi(\bar{x}) \in \Phi$.

We say that two variable-finite formulas $\phi(\bar{x})$ and $\psi(\bar{x})$ are **equivalent in** A if $\phi(A^n) = \psi(A^n)$, or equivalently, if $A \vDash \forall\bar{x}(\phi \leftrightarrow \psi)$. Thus two formulas are equivalent in A iff they define the same relation in A. Likewise sets of formulas $\Phi(\bar{x})$ and $\Psi(\bar{x})$ are **equivalent in** A if $\Phi(A^n) = \Psi(A^n)$.

The formula $\phi(x_0, \ldots, x_{n-1})$ and the formula $\phi(y_0, \ldots, y_{n-1})$ are equivalent in any structure. Likewise $\forall y\, R(x, y)$ is equivalent to $\forall z\, R(x, z)$. Thus we have too many formulas trying to do the same job. To cut down the clutter, we shall say that one formula is a **variant** of another formula if the two formulas differ only in the choice of variables, i.e. if each can be got from the other by a consistent replacement of variables.

Variance is an equivalence relation on the class of formulas. We shall always take the **cardinality** of a language L, $|L|$, to be the number of equivalence classes of variable-finite formulas of L under the relation of being variants. For first-order language this agrees with the definition of $|L|$ for a signature L in section 1.3; see Exercise 7 below.

These dreary matters of syntax do have one important consequence. In a given structure A, two different first-order formulas can define the same relation; so the family of first-order definable relations on A may be much less rich than the language used to define them. How can we tell exactly what the different first-order definable relations on A are? (The same question arises for $L_{\omega_1\omega}$ and other sublanguages of $L_{\infty\omega}$, of course.) There is no uniform machinery for answering this question; often it takes months of research and inspiration. I close this section with two very different kinds of example.

Few definable subsets: minimality

Lemma 2.1.1. *Let L be a signature, A an L-structure, X a set of elements of A and Y a relation on* $\text{dom}(A)$*. Suppose Y is definable by some formula of signature L with parameters from X. Then for every automorphism f of A, if f fixes X pointwise (i.e. $f(a) = a$ for all a in X), then f fixes Y setwise (i.e. for every element a of A, $a \in Y \Leftrightarrow fa \in Y$).*

Proof. For formulas of $L_{\infty\omega}$ this can be proved by induction on complexity; see Exercise 10. But I urge the reader to see the lemma as a fundamental

insight about mathematical structure, which must apply equally well to
formulas in other logics besides $L_{\infty\omega}$.   □
   For example, we have the following.

Theorem 2.1.2. *Let L be the empty signature and A an L-structure so that A is
simply a set. Let X be any subset of A, and let Y be a subset of* dom (A)
*which is definable in A by a formula of some logic of signature L, using
parameters from X. Then Y is either a subset of X, or the complement in*
dom (A) *of a subset of X.*

Proof. Immediate from the lemma.   □

   Note that in this theorem, all finite subsets of X and their complements in
A can be defined by first-order formulas with parameters in X. The set
$\{a_0, \ldots, a_{n-1}\}$ is defined by the formula $x = a_0 \lor \ldots \lor x = a_{n-1}$ (which is $\bot$
if the set is empty); negate this formula to get a definition of the
complement.
   The situation in Theorem 2.1.2 is commoner than one might expect. We
say that a structure A is **minimal** if A is infinite but the only subsets of
dom (A) which are first-order definable with parameters are either finite or
cofinite (i.e. complements of finite sets). More generally a set $X \subseteq$ dom (A)
which is first-order definable with parameters is said to be **minimal** if X is
infinite, and for every set Z which is first-order definable in A with
parameters, either $X \cap Z$ or $X \backslash Z$ is finite. The minimal sets are the bottom
level of a hierarchy, where we classify sets according to how many times they
can be split up into smaller definable pieces; cf. Exercises 10–12 in section
3.4 below. Related ideas are at the heart of stability theory.

Many definable subsets: arithmetic

The definable subsets of the natural numbers have been very closely
analysed, because of their importance for recursion theory.
   We take $\mathbb{N}$ as in Example 2; let L be its signature. We write $(\forall x < y)\phi$,
$(\exists x < y)\phi$ as shorthand for $\forall x(x < y \to \phi)$ and $\exists x(x < y \land \phi)$ respectively.
The quantifiers $(\forall x < y)$ and $(\exists x < y)$ are said to be **bounded**. (The notion is
from Hilbert [1926].) We define a hierarchy of first-order formulas of L, as
follows.

(1.15)   A first-order formula of L is said to be a Π_0^0 formula, or
        equivalently a Σ_0^0 formula, if all quantifiers in it are bounded.

(1.16)   A formula is said to be a Π_{k+1}^0 formula if it is of form $\forall \bar{x}\, \psi$
        for some Σ_k^0 formula ψ. (The tuple $\bar{x}$ may be empty.)

(1.17)   A formula is said to be a Σ_{k+1}^0 formula if it is of form $\exists \bar{x} \, \psi$
for some Π_k^0 formula ψ. (The tuple $\bar{x}$ may be empty.)

Thus for example a Σ_3^0 formula consists of three blocks of quantifiers, $\exists \bar{x} \forall \bar{y} \exists \bar{z}$, followed by a formula with only bounded quantifiers. Because the blocks are allowed to be empty, every Π_k^0 formula is also a Σ_{k+1}^0 formula and a Π_{k+1}^0 formula; the higher classes gather up the lower ones.

Let $\bar{x}$ be $(x_0, \ldots, x_{n-1})$. A set R of n-tuples of natural numbers is called a Π_k^0 relation (resp. a Σ_k^0 relation) if it is of the form $\phi(\mathbb{N}^n)$ for some Π_k^0 formula (resp. some Σ_k^0 formula) $\phi(\bar{x})$. We say that R is a Δ_k^0 relation if it is both a Π_k^0 relation and a Σ_k^0 relation. A relation is said to be **arithmetical** if it is Σ_k^0 for some k (so the arithmetical relations are exactly the first-order definable ones).

Intuitively the hierarchy measures how many times we have to run through the entire set of natural numbers if we want to check by (1.5)–(1.10) whether a particular tuple belongs to the relation R. An important theorem of Kleene [1943] says that the Δ_1^0 relations are exactly the recursive ones, and the Σ_1^0 relations are exactly the recursively enumerable ones. Another theorem of Kleene [1943] says that for each $k < \omega$ there is a relation R which is Σ_{k+1}^0 but neither Σ_k^0 nor Π_k^0. So the hierarchy keeps growing.

Exercises for section 2.1

1. By exhibiting suitable formulas, show that the set of even numbers is a Σ_0^0 set in $\mathbb{N}$. Show the same for the set of prime numbers.

2. Let A be the partial ordering (in a signature with $\leqslant$) whose elements are the positive integers, with $m \leqslant^A n$ iff m divides n. (a) Show that the set $\{1\}$ and the set of primes are both $\varnothing$-definable in A. (b) A number n is **square-free** if there is no prime p such that p^2 divides n. Show that the set of square-free numbers is $\varnothing$-definable in A.

3. Let A be the graph whose vertices are all the sets $\{m, n\}$ of exactly two natural numbers, with a joined to b iff $a \cap b \neq \varnothing \wedge a \neq b$. Show that A is not minimal, but it has infinitely many minimal subsets.

An L-structure is said to be **O-minimal** *(read 'Oh-minimal' – the O is for Ordering) if L contains a symbol $\leqslant$ which linearly orders* $\mathrm{dom}(A)$ *in such a way that every subset of* $\mathrm{dom}(A)$ *which is first-order definable with parameters is a union of finitely many intervals of the forms (a, b), $\{a\}$, $(-\infty, b)$, (a, ∞) where a, b are elements of A. A theory is* **O-minimal** *if all its models are.*

4. Let A be a linear ordering with the order-type of the rationals (see Example 2 of section 1.1 for the signature). Show that A is O-minimal. [Use Lemma 2.1.1; Example 3 in section 3.2 below may help.]

5. Let A be an infinite-dimensional vector space over a finite field. Show that A is minimal, and that the only $\varnothing$-definable sets in A are $\varnothing$, $\{0\}$ and dom(A). *In Exercise 2.7.9 we shall remove the condition that the field of scalars is finite.*

6 (Time-sharing with formulas). Let $\bar{x}$ be a k-tuple of variables and let $\phi_i(\bar{x}, \bar{y})$ $(i < n)$ be formulas of a first-order language L. Show that there is a formula $\psi(\bar{x}, \bar{y}, \bar{w})$ of L such that for every L-structure A with at least two elements, and every tuple $\bar{a}$ in A, the set of all relations of the form $\psi(A^k, \bar{a}, \bar{b})$ with $\bar{b}$ in A is exactly the set of all relations of the form $\phi_i(A^k, \bar{a})$ with $i < n$. [Let $\bar{w}$ be $(w_0, \ldots, w_n)$ and arrange that $\psi(A^k, \bar{a}, c_0, \ldots, c_n)$ is $\phi_i(A^k, \bar{a})$ when $c_n = c_i$ and $c_n \neq c_j$ $(i \neq j)$.]

7. Show that if L is a first-order language then $|L|$ is equal to $\omega +$ (the number of symbols in the signature of L).

8. Let ϕ be a formula of $L_{\infty\omega}$. Show by induction on the complexity of ϕ that, if ϕ is variable-finite, then so is every subformula of ϕ.

9. Let ϕ be a variable-finite formula of $L_{\lambda^+\omega}$, where L is a countable signature and λ is an infinite cardinal. (a) Show by induction on the complexity of ϕ that ϕ has at most λ subformulas. (b) Show that there is a language K which is first-order-closed, consists of variable-finite formulas and contains ϕ, such that $|K| = \lambda$.

10. Prove Lemma 2.1.1 for variable-finite formulas of $L_{\infty\omega}$ by induction on the complexity of the formulas.

*11. Let L be the signature of abelian groups and p a prime. Let A be the direct sum of infinitely many copies of $\mathbb{Z}(p^2)$, the cyclic group of order p^2. Show (a) the subgroup of elements of order $\leqslant p$ is $\varnothing$-definable and minimal, (b) the set of elements of order p^2 is $\varnothing$-definable but not minimal.

*12. Let the structure A consist of an infinite set with no relations, functions or constants. Let n be any integer $\geqslant 2$. Describe the $\varnothing$-definable n-ary relations in A; in particular show that there are only finitely many.

*13. Let K be a field of characteristic 0, n a positive integer and G the group $\mathrm{GL}_n(K)$ of invertible linear transformations on K^n. Show that the following subsets of G are $\varnothing$-definable: (a) the set of all scalar matrices; (b) the set of all matrices which are similar to a diagonal matrix with distinct scalars down the main diagonal; (c) the set of diagonalisable matrices.

*14. Let G be a group, and let us call a subgroup H **definable** if H is first-order definable in G with parameters. Suppose that G satisfies the descending chain condition on definable groups of finite index in G. Show that there is a unique smallest definable subgroup of finite index in G. Show that this subgroup is in fact

$\emptyset$-definable, and deduce that it is a characteristic subgroup of G. *This subgroup is known as G°, by analogy with the connected component G° in an algebraic group – which is in fact a special case.*

*15. We write $x | y$ for 'x divides y'. Show that the relation $x = yz$ is $\emptyset$-definable in the structure $\langle \omega, +, | \rangle$. [First define the relation $x = y(y + 1)$. Then consider the equation $(y + z)(y + z + 1) = y(y + 1) + z(z + 1) + x + x$.]

*16. Show that the statement 'A is a structure and ϕ is a sentence of the corresponding language $L_{\infty\omega}$ such that $A \vDash \phi$' can be written as a Σ_1 formula $\psi(A, \phi)$ of the language of set theory. (This involves writing down precise set-theoretic definitions of 'structure' and '$L_{\infty\omega}$'.)

2.2 Definable classes of structures

Until the mid 1920s, formal languages were used for the most part either in a purely syntactic way, or for talking about definable sets and relations in a single structure. The chief exception was geometry, where Hilbert and others had used formal axioms to classify geometric structures. Today it's a commonplace that we can classify structures by asking what formal axioms are true in them – and so we can speak of *definable (or axiomatisable) classes of structures*. Model theory studies such classes.

We begin with some definitions. A **sentence** is a formula with no free variables. A **theory** is a set of sentences. (Strictly one should say 'class', since a theory in $L_{\infty\omega}$ could be a proper class. But normally theories are sets.)

If ϕ is a sentence of $L_{\infty\omega}$ and A is an L-structure, then clauses (1.5)–(1.10) in the previous section define a relation '$A \vDash \phi[]$', i.e. 'the empty sequence satisfies ϕ in A', taking $\bar{a}$ to be the empty sequence. We omit [] and write simply '$A \vDash \phi$'. We say that A is a **model** of ϕ, or that ϕ is **true in** A, when '$A \vDash \phi$' holds. Given a theory T in $L_{\infty\omega}$, we say that A is a **model of** T, in symbols $A \vDash T$, if A is a model of every sentence in T.

Let T be a theory in $L_{\infty\omega}$ and **K** a class of L-structures. We say that T **axiomatises K**, or is a **set of axioms** for **K**, if **K** is the class of all L-structures which are models of T. Obviously this determines **K** uniquely, and so we can write **K** $= \text{Mod}(T)$ to mean that T axiomatises **K**. Note that T is also a theory in $L_{\infty\omega}^+$ where L^+ is any signature containing L, and that $\text{Mod}(T)$ in L^+ is a different class from $\text{Mod}(T)$ in L. So the notion of 'model of T' depends on the signature. But we can leave it to the context to find a signature; if no signature is mentioned, choose the smallest L such that T is in $L_{\infty\omega}$.

Likewise if T is a theory, we say that a theory U **axiomatises** T (or is **equivalent to** T) if $\text{Mod}(U) = \text{Mod}(T)$. In particular if A is an L-structure and T is a first-order theory, we say that T **axiomatises** A if the first-order sentences true in A are exactly those which are true in every model of T. (The next section will examine this notion more closely.)

Let L be a language and **K** a class of L-structures. We define the L-**theory** of **K**, $\mathrm{Th}_L(\mathbf{K})$, to be the set (or class) of all sentences ϕ of L such that $A \vDash \phi$ for every structure A in **K**. We omit the subscript $_L$ when L is first-order: the **theory** of **K**, $\mathrm{Th}(\mathbf{K})$, is the set of all *first-order* sentences which are true in every structure in **K**.

We say that **K** is L-**definable** if **K** is the class of all models of some sentence in L. We say that **K** is L-**axiomatisable**, or **generalised L-definable**, if **K** is the class of models of some theory in L. For example **K** is **first-order definable** if **K** is the class of models of some first-order sentence, or equivalently, of some finite set of first-order sentences. Note that **K** is generalised first-order definable if and only if **K** is the class of all L-structures which are models of $\mathrm{Th}(\mathbf{K})$.

First-order definable and first-order axiomatisable classes are also known as EC and EC_Δ classes respectively. The E is for Elementary and the delta for Intersection (cf. Exercise 2 below – the German for intersection is Durchschnitt).

When one writes theories, there is no harm in using standard mathematical abbreviations, so long as they can be seen as abbreviations of genuine terms or formulas. For example we write

$$(2.1)$$

$x + y + z$	for	$(x + y) + z,$
$x - y$	for	$x + (-y),$
n	for	$1 + \ldots + 1$ (n times), where n is a positive integer,
nx	for	$\begin{cases} x + \ldots + x \ (n \text{ times}), \text{ where } n \text{ is a} \\ \qquad\qquad\qquad\qquad \text{positive integer,} \\ 0 \qquad\qquad\qquad\quad \text{where } n \text{ is } 0, \\ -(-n)x \qquad\qquad \text{where } n \text{ is a} \\ \qquad\qquad\qquad\qquad \text{negative integer,} \end{cases}$
xy	for	$x \cdot y,$
x^n	for	$x \ldots x$ (n times), where n is a positive integer,
$x \leqslant y$	for	$x < y \vee x = y,$
$x \geqslant y$	for	$y \leqslant x.$

The following notation is useful too; it allows us to say 'There are exactly n elements x such that $\ldots$', for any $n < \omega$. Let $\phi(x, \bar{z})$ be a formula. Then we define $\exists_{\geqslant n} x\, \phi$ ('At least n elements x satisfy ϕ') as follows, by induction on n.

(2.2) $\exists_{\geqslant 0} x\ \phi$ is $\forall x\ x = x.$
        $\exists_{\geqslant 1} x\ \phi$ is $\exists x\ \phi.$
        $\exists_{\geqslant n+1} x\ \phi$ is $\exists x(\phi(x, \bar{z}) \wedge \exists_{\geqslant n} y(\phi(y, \bar{z}) \wedge y \neq x))$ (for
                                            $n \geqslant 1$).

Then we put $\exists_{\leqslant n} x\ \phi$ for $\neg \exists_{\geqslant n+1} x\ \phi$, and finally $\exists_{=n} x\ \phi$ is $\exists_{\geqslant n} x\ \phi \wedge \exists_{\leqslant n} x\ \phi$. So for example the first-order sentence $\exists_{=n} x\ x = x$ expresses that there are exactly n elements.

Axioms for particular structures

Even in geometry, axioms were first used for describing a particular structure, not for defining a class of structures. When a theory T is written down in order to describe a particular structure A, we say that A is the **intended model** of T. It often happens – as it did in geometry – that people decide to take an interest in the unintended models too.

Example 1: *The term algebra.* Let L be an algebraic signature, X a set of variables and A the term algebra of L with basis X (see section 1.3). Then we can describe A by the set of all sentences of the following forms.

(2.3) $c \neq d$ where c, d are distinct constants.

(2.4) $\forall \bar{x}\ F(\bar{x}) \neq c$ where F is a function symbol and c a constant.

(2.5) $\forall \bar{x} \bar{y}\ F(\bar{x}) \neq G(\bar{y})$ where F, G are distinct function symbols.

(2.6) $\forall x_0 \ldots x_{n-1} y_0 \ldots y_{n-1}\ (F(x_0, \ldots, x_{n-1})$

$$= F(y_0, \ldots, y_{n-1}) \to \bigwedge_{i<n} x_i = y_i).$$

(2.7) $\forall x_0 \ldots x_{n-1}\ t(x_0, \ldots, x_{n-1}) \neq x_i$ where $i < n$ and t is any term containing x_i but distinct from x_i.

(2.8)   [Use this axiom only when L is finite.] Write $\mathrm{Var}(x)$ for the formula $\bigwedge\{x \neq c\colon c$ a constant of $L\} \wedge \bigwedge\{\forall \bar{y}\ x \neq F(\bar{y})\colon F$ a function symbol of $L\}$. Then if X has finite cardinality n, we add the axiom $\exists_{=n} x\ \mathrm{Var}(x)$. If X is infinite, we add the infinitely many axioms $\exists_{\geqslant n} x\ \mathrm{Var}(x)\ (n < \omega)$.

These axioms are all first-order. Each of them says something which is obviously true of A.

One can show that these axioms (2.3)–(2.8) axiomatise A. (This is by no means obvious. See section 2.7 below). Unfortunately they don't suffice to pin down the structure A itself, even up to isomorphism. For example let L

consist of one 1-ary function symbol F and one constant c, and let X be empty. Then (2.3)–(2.8) reduce to the following.

(2.9) $\forall x\, F(x) \neq c$. $\forall xy\, (F(x) = F(y) \rightarrow x = y)$.
        $\forall x\, F(F(F(\ldots (F(x) \ldots) \neq x$ (for any positive number of Fs).
        $\forall x(x = c \vee \exists y\, x = F(y))$.

We can get a model B of (2.9) by taking the intended term model A and adding all the integers as new elements, putting $F^B(n) = n + 1$ for each integer n:

(2.10)

This model B is clearly not isomorphic to A. One can think of new elements of B as terms that can be analysed into smaller terms infinitely often.

Example 2: *The first-order Peano axioms*. This is another example of a first-order theory with an intended model. Gödel studied a close variant of it in his [1931b]. For him it was a set of sentences which are true of the natural number structure $\mathbb{N}$ (cf. Example 2 in section 2.1); his Incompleteness Theorem in [1931b] says that the theory fails to axiomatise $\mathbb{N}$.

(2.11) $\forall x\, x + 1 \neq 0$.

(2.12) $\forall xy(x + 1 = y + 1 \rightarrow x = y)$.

(2.13) $\forall \bar{z}(\phi(0, \bar{z}) \wedge \forall x(\phi(x, \bar{z}) \rightarrow \phi(x + 1, \bar{z})) \rightarrow \forall x\, \phi(x, \bar{z}))$
                                    for each first-order formula $\phi(x, \bar{z})$.

(2.14) $\forall x\, x + 0 = x$; $\forall xy\, x + (y + 1) = (x + y) + 1$.

(2.15) $\forall x\, x \cdot 0 = 0$; $\forall xy\, x \cdot (y + 1) = x \cdot y + x$.

(2.16) $\forall x\, \neg(x < 0)$; $\forall xy(x < (y + 1) \leftrightarrow x < y \vee x = y)$.

Clause (2.13) is an example of an **axiom schema**, i.e. a set of axioms consisting of all the sentences of a certain pattern. This schema expresses that if X is a set which is first-order definable with parameters, and (1) $0 \in X$ and (2) if $n \in X$ then $n + 1 \in X$, then every number is in X. This is the **first-order induction schema**. The axioms (2.14)–(2.16) are the **recursive definitions** of $+$, $\cdot$ and $<$; given the meanings of 0 and 1 and the function $x \mapsto x + 1$, there is only one way of defining $+$, $\cdot$ and $<$ in $\mathbb{N}$ so as to make these axioms true.

   The axioms (2.11)–(2.16) are known as **first-order Peano arithmetic**, or P for short. It's natural to ask whether P has any models besides the intended one. In Chapter 6 we shall see that the compactness theorem (Theorem 6.1.1) gives the answer Yes at once. But before anybody realised this, Skolem [1934] had found some unintended models of these axioms, by a very ingenious construction which was later recognised as a kind of ultrapower (see section 9.5 below). Models of P which are not isomorphic to the intended one are known as **nonstandard models**. Like randomly discovered poisonous plants, they turn out to have important and wholly benign applications that nobody dreamed of beforehand; see section 11.4 below.

A list of axiomatisable classes

There follows a list of some classes which are definable or axiomatisable. The list is for reference, not for light reading. Some less straightforward examples are discussed in section 2.8 below.

In most cases the sentences given in the list are just the standard definition of the class, thrown into formal symbols. We shall refer to these sentences as the **theory of** the class: thus the axioms (2.21) form the **theory of left R-modules**. Besides providing examples, the list shows what signatures are commonly used for various classes. For example rings normally have signature $+$, $\cdot$, $-$ (1-ary), 0 and 1. (A ring has 1 unless otherwise stated.)

(2.17) **Groups (multiplicative):**
   $\forall xyz \ (xy)z = x(yz), \ \forall x \ x \cdot 1 = x, \ \forall x \ x \cdot x^{-1} = 1$.

(2.18) **Groups of exponent n (n a fixed positive integer):**
   (2.17) together with $\forall x \ x^n = 1$.

(2.19) **Abelian groups (additive):**
   $\forall xyz \ (x + y) + z = x + (y + z), \ \forall x \ x + 0 = x, \ \forall x \ x - x = 0,$
   $\forall xy \ x + y = y + x$.

(2.20) **Torsion-free abelian groups:**
   (2.19) together with $\forall x \ (nx = 0 \rightarrow x = 0)$ for each positive integer n.

(2.21) **Left R-modules where R is a ring:**
   As in vector spaces (Example 4 of section 1.1) the module elements are the elements of the structures. Each ring element r is used as a 1-ary function symbol, so that $r(x)$ represents rx. The axioms are (2.19) together with

$\forall xy \ r(x + y) = r(x) + r(y)$	for all $r \in R$,
$\forall x \ (r + s)(x) = r(x) + s(x)$	for all $r,s \in R$,
$\forall x \ (rs)(x) = r(s(x))$	for all $r,s \in R$,
$\forall x \ 1(x) = x$.	

(2.22) **Rings**:
        (2.19) together with
        $\forall xyz\ (xy)z = x(yz)$,
        $\forall x\ x1 = x$,
        $\forall x\ 1x = x$,
        $\forall xyz\ x(y + z) = xy + xz$,
        $\forall xyz\ (x + y)z = xz + yz$.

(2.23) **Von Neumann regular rings**:
        (2.22) together with $\forall x \exists y\ xyx = x$.

(2.24) **Fields**:
        (2.22) together with $\forall xy\ xy = yx,\ 0 \neq 1$,
        $\forall x(x \neq 0 \rightarrow \exists y\ xy = 1)$.

(2.25) **Fields of characteristic p (p prime)**:
        (2.24) together with $p = 0$.

(2.26) **Algebraically closed fields**:
        (2.24) together with
        $\forall x_1 \ldots x_n \exists y\ y^n + x_1 y^{n-1} + \ldots + x_{n-1}y + x_n = 0$,
        for each positive integer n.

(2.27) **Real-closed fields**:
        (2.24) together with
        $\forall x_1 \ldots x_n\ x_1^2 + \ldots + x_n^2 \neq -1$ (for each positive integer n),
        $\forall x \exists y\ (x = y^2 \lor -x = y^2)$,
        $\forall x_1 \ldots x_n \exists y\ y^n + x_1 y^{n-1} + \ldots + x_{n-1}y + x_n = 0$
        $\hspace{8cm}$ (for all odd n).

(2.28) **Lattices**:
        $\forall x\ x \land x = x,\ \forall xy\ x \land y = y \land x$,
        $\quad \forall xy\ (x \land y) \lor y = y,\ \forall xyz\ (x \land y) \land z = x \land (y \land z)$,
        $\forall x\ x \lor x = x,\ \forall xy\ x \lor y = y \lor x$,
        $\quad \forall xy\ (x \lor y) \land y = y,\ \forall xyz\ (x \lor y) \lor z = x \lor (y \lor z)$.

        (In lattices we write $x \leqslant y$ as an abbreviation of $x \land y = x$.
        Note that in sentences about lattices, the symbols $\land$ and $\lor$
        have two meanings: the lattice meaning and the logical
        meaning. Brackets can help to keep them distinct.)

(2.29) **Boolean algebras**:
        (2.28) together with
        $\forall xyz\ x \land (y \lor z) = (x \land y) \lor (x \land z)$,
        $\forall xyz\ x \lor (y \land z) = (x \lor y) \land (x \lor z)$,
        $\forall x\ x \lor x^* = 1,\ \forall x\ x \land x^* = 0,\ 0 \neq 1$.

(2.30) **Atomless boolean algebras**:
        (2.29) together with $\forall x \exists y(x \neq 0 \rightarrow 0 < y \land y < x)$.
        $\quad$ ($y < x$ is shorthand for $y \leqslant x \land y \neq x$.)

(2.31) **Linear orderings**:
$\forall x\, x \not< x$, $\forall xy(x = y \lor x < y \lor y < x)$,
$\forall xyz(x < y \land y < z \to x < z)$.

(2.32) **Dense linear orderings without endpoints**:
(2.31) together with $\forall xy(x < y \to \exists z(x < z \land z < y))$,
$\forall x \exists z\, z < x$, $\forall x \exists z\, x < z$.

All the classes above are generalised first-order definable. Here is a class with an infinitary definition:

(2.33) **Locally finite groups**:
(2.17) together with
$\forall x_1 \ldots x_n$

$$\bigvee_{m < \omega} (\exists y_1 \ldots y_m \bigwedge_{t(\bar{x}) \text{ a term}} (t(\bar{x}) = y_1 \lor \ldots \lor t(\bar{x}) = y_m)).$$

Exercises for section 2.2

1. Let L be a first-order language and T a theory in L. Show: (a) if T and U are theories in L then $T \subseteq U$ implies $\mathrm{Mod}(U) \subseteq \mathrm{Mod}(T)$, (b) if **J** and **K** are classes of L-structures then $\mathbf{J} \subseteq \mathbf{K}$ implies $\mathrm{Mod}(\mathbf{K}) \subseteq \mathrm{Mod}(\mathbf{J})$, (c) $T \subseteq \mathrm{Th}(\mathrm{Mod}(T))$ and $\mathbf{K} \subseteq \mathrm{Mod}(\mathrm{Th}(\mathbf{K}))$, (d) $\mathrm{Th}(\mathrm{Mod}(T)) = T$ if and only if T is of the form $\mathrm{Th}(\mathbf{K})$, and likewise $\mathrm{Mod}(\mathrm{Th}(\mathbf{K})) = \mathbf{K}$ if and only if $\mathbf{K}$ is of the form $\mathrm{Mod}(T)$.

2. Let L be a first-order language and for each $i \in I$ let $\mathbf{K}_i$ be a class of L-structures. Show that $\mathrm{Th}(\bigcup_{i \in I} \mathbf{K}_i) = \bigcap_{i \in I} \mathrm{Th}(\mathbf{K}_i)$.

3. Let L be a first-order language and for each $i \in I$ let T_i be a theory in L. (a) Show that $\mathrm{Mod}(\bigcup_{i \in I} T_i) = \bigcap_{i \in I} \mathrm{Mod}(T_i)$. In particular if T is any theory in L, $\mathrm{Mod}(T) = \bigcap_{\phi \in T} \mathrm{Mod}(\phi)$. (b) Show that the statement $\mathrm{Mod}(\bigcap_{i \in I} T_i) = \bigcup_{i \in I} \mathrm{Mod}(T_i)$ holds when I is finite and each T_i is of the form $\mathrm{Th}(\mathbf{K}_i)$ for some class $\mathbf{K}_i$. *It can fail if either of these conditions is dropped.*

4. Let L be any signature containing a 1-ary relation symbol P and a k-ary relation symbol R. (a) Write down a sentence of $L_{\omega_1 \omega}$ expressing that at most finitely many elements x have the property $P(x)$. (b) When $n < \omega$, write down a sentence of $L_{\omega \omega}$ expressing that at least n k-tuples $\bar{x}$ of elements have the property $R(\bar{x})$. *One abbreviates this sentence to $\exists_{\geqslant n} \bar{x}\, R(\bar{x})$.*

5. Let L be a first-order language and A a finite L-structure. Show that every model of $\mathrm{Th}(A)$ is isomorphic to A. [Caution: L may have infinitely many symbols.]

6. For each of the following classes, show that it can be defined by a single first-order sentence. (a) Nilpotent groups of class k ($k \geqslant 1$). (b) Commutative rings with identity. (c) Integral domains. (d) Commutative local rings (i.e. commutative rings with a single maximal ideal). (e) Ordered fields. (f) Distributive lattices.

7. For each of the following classes, show that it can be defined by a set of first-order sentences. (a) Divisible abelian groups. (b) Fields of characteristic 0. (c) Formally real fields. (d) Separably closed fields.

8. Show that the class of simple groups is definable by a sentence of $L_{\omega_1 \omega}$.

9. For each of the following classes, show that it can be defined by a single sentence of $L_{\omega_1 \omega}$. (a) Unique factorisation domains. (b) Principal ideal domains. (c) Dedekind domains. (d) Semisimple rings. (e) Left coherent rings. (f) Left artinian rings. (g) Noetherian local commutative rings. (h) Groups G such that if H, K are any two isomorphic finitely generated subgroups of G then H is congruent to K in G.

10. Let L be a signature with a symbol $<$, and T the theory in L which expresses that $<$ is a linear ordering. (a) Define, by induction on the ordinal α, a formula $\theta_\alpha(x)$ of $L_{\infty\omega}$ which expresses (in any model of T) 'The order-type of the set of predecessors of x is α'. [The idea of (2.2) may help.] (b) Write down a set of axioms in $L_{\infty\omega}$ for the class of orderings of order-type α. Check that if α is infinite and of cardinality κ, your axioms can be written as a single sentence of $L_{\kappa^+\omega}$.

11. Let P be a prime and A an abelian p-group. The **height** of an element a is defined by induction: (1) every element has height $\geqslant 0$; (2) a has height $\geqslant \alpha + 1$ if there is an element b of height $\geqslant \alpha$ such that $pb = a$; (3) if δ is a limit ordinal, a has height $\geqslant \delta$ if a has height $\geqslant \alpha$ for every $\alpha < \delta$. By induction on α, define a formula $\mathrm{ht}_\alpha(x)$ of $L_{\infty\omega}$ which defines the set of elements of A of height at least α.

You can make a sensible attempt at the next exercise, but at the moment you have no way of checking whether your answers are correct and complete. In truth there is no algorithm for checking this anyway. Section 2.7 suggests one way of verifying your answers; but down that path lie many hours of labour.

*12. Write $\mathbb{Q}$ for the set of rational numbers. (a) Let $R(x, y, z)$ be the relation on the rational numbers defined by 'y lies strictly between x and z in the ordering of the rationals'. Axiomatise the structure $\langle \mathbb{Q}, R \rangle$. (b) Let $S(x, y, z)$ be the relation on the rational numbers defined by 'either $x < y < z$ or $y < z < x$ or $z < x < y$'. Axiomatise the structure $\langle \mathbb{Q}, S \rangle$.

*13. *Show that (2.11)–(2.16) imply $\forall xy(x < y \leftrightarrow \exists z \, y = (x + z) + 1)$. So the recursive definition (2.16) can be replaced by an explicit definition (on which see section 2.6 below).*

2.3 Some notions from logic

The work of the two previous sections allows us to define several important notions from logic. Any general text of logic (such as Ebbinghaus, Flum & Thomas [1984]) will give background information about these notions. We begin with some definitions that will be in force throughout the book, and we finish the section with an important lemma about constructing models. The definitions are worth reading but they are not worth memorising – remember that this book has an index.

Truth and consequences

Let L be a signature, T a theory in $L_{\infty\omega}$ and ϕ a sentence of $L_{\infty\omega}$. We say that ϕ is a **consequence** of T, or that T **entails** ϕ, in symbols $T \vdash \phi$, if every model of T is a model of ϕ. (In particular if T has no models then T entails ϕ.)

Warning: we don't require that if $T \vdash \phi$ then there is a proof of ϕ from T. In any case, with infinitary languages it's not always clear what would constitute a proof. Some writers use '$T \vdash \phi$' to mean that ϕ is deducible from T in some particular formal proof calculus, and they write '$T \vDash \phi$' for our notion of entailment (a notation which clashes with our '$A \vDash \phi$'). For first-order logic the two kinds of entailment coincide by the completeness theorem for the proof calculus in question; see for example Shoenfield [1967] section 4.2.

We say that ϕ is **valid**, or is a **logical theorem**, in symbols $\vdash \phi$, if ϕ is true in every L-structure. We say that ϕ is **consistent** if ϕ is true in some L-structure. Likewise we say that a theory T is **consistent** if it has a model.

We say that two theories S and T in $L_{\infty\omega}$ are **equivalent** if they have the same models, i.e. if $\mathrm{Mod}(S) = \mathrm{Mod}(T)$. We also have a relativised notion of equivalence: when T is a theory in $L_{\infty\omega}$ and $\phi(\bar{x})$, $\psi(\bar{x})$ are formulas of $L_{\infty\omega}$, we say that ϕ is **equivalent to** ψ **modulo** T if for every model A of T and every sequence $\bar{a}$ from A, $A \vDash \phi(\bar{a}) \Leftrightarrow A \vDash \psi(\bar{a})$. Thus $\phi(\bar{x})$ is equivalent to $\psi(\bar{x})$ modulo T if and only if $T \vdash \forall \bar{x}(\phi \leftrightarrow \psi)$. (This sentence is not in $L_{\infty\omega}$ if ϕ and ψ have infinitely many free variables, but the sense is clear.) There is a metatheorem saying that if ϕ is equivalent to ψ modulo T, and χ' comes from χ by putting ψ in place of ϕ somewhere inside χ, then χ' is equivalent to χ modulo T. Results like this are proved for first-order logic in elementary texts, and the proofs for other languages are no different. We can generalise relative equivalence and talk of two sets of formulas $\Phi(\bar{x})$ and $\Psi(\bar{x})$ being **equivalent modulo** T, meaning that $\bigwedge \Phi$ is equivalent to $\bigwedge \Psi$ modulo T.

A special case is where T is empty: $\phi(\bar{x})$ and $\psi(\bar{x})$ are said to be **logically equivalent** if they are equivalent modulo the empty theory. In the terminology of section 2.1, this is the same as saying that they are equivalent in every L-structure. The reader will know some examples: $\neg \forall x \, \phi$ is logically equivalent to $\exists x \, \neg \phi$, and $\exists x \bigvee_{i \in I} \psi_i$ is logically equivalent to $\bigvee_{i \in I} \exists x \, \psi_i$.

As another example, a formula ϕ is said to be a **boolean combination** of formulas in a set Φ if ϕ is in the smallest set X such that (1) $\Phi \subseteq X$ and

(2) X is closed under $\wedge$, $\vee$ and $\neg$. We say that ϕ is in **disjunctive normal form over** Φ if ϕ is a finite disjunction of finite conjunctions of formulas in Y, where Y is Φ together with the negations of all formulas in Φ. *Every boolean combination* $\phi(\bar{x})$ *of formulas in a set* Φ *is logically equivalent to a formula* $\psi(\bar{x})$ *in disjunctive normal form over* Φ. (The same is true if we replace $\wedge$ and $\vee$ by $\bigwedge$ and $\bigvee$ respectively, dropping the word 'finite'; in this case we speak of **infinite boolean combinations** and **infinitary disjunctive normal form**.)

A formula is **prenex** if it consists of a string of quantifiers (possibly empty) followed by a quantifier-free formula. *Every first-order formula is logically equivalent to a prenex first-order formula.* (The result fails if one allows signatures to contain 0-ary relation symbols; see Exercise 7 below. This is one of the two embarrassing consequences of the fact that we allow structures to have empty domains. The other embarrassment is that $\vdash \exists x\, x = x$ holds if and only if the signature contains a constant. In practice these points never matter.)

Lemma 2.3.1. *Let T be a theory in a first-order language L, and Φ a set of formulas of L. Suppose*
(a) *every atomic formula of L is in Φ,*
(b) *Φ is closed under boolean combinations, and*
(c) *for every formula $\psi(\bar{x}, y)$ in Φ, $\exists y\, \psi$ is equivalent modulo T to a formula $\phi(\bar{x})$ in Φ.*
Then every formula $\chi(\bar{x})$ of L is equivalent modulo T to a formula $\phi(\bar{x})$ in Φ. (If (c) is weakened by requiring that $\bar{x}$ is non-empty, then the same conclusion holds provided $\bar{x}$ in $\chi(\bar{x})$ is also non-empty.)

Proof. By induction on the complexity of χ, using the fact that $\forall y\, \phi$ is equivalent to $\neg \exists x \neg \phi$. $\qquad\qquad\qquad\qquad\qquad\qquad\qquad\qquad\qquad\qquad\square$

An *n*-type of a theory T is a set $\Phi(\bar{x})$ of formulas, with $\bar{x} = (x_0, \ldots, x_{n-1})$, such that for some model A of T and some n-tuple $\bar{a}$ of elements of A, $A \vDash \phi(\bar{a})$ for all ϕ in Φ. We say then that A **realises** the *n*-type Φ, and that $\bar{a}$ **realises** Φ in A. We say that A **omits** Φ if no tuple in A realises Φ. A set Φ is a **type** if it is an n-type for some $n < \omega$. (These notions are central to model theory. See section 6.3 below for discussion and examples.)

Often we work in a language L which is smaller than $L_{\infty\omega}$, for example a first-order language. Then all formulas in a type will automatically be assumed to come from L.

Let L be a language and A, B two L-structures. We say that A is **L-equivalent** to B, in symbols $A \equiv_L B$, if for every sentence ϕ of L,

$A \vDash \phi \Leftrightarrow B \vDash \phi$. This means that A and B are indistinguishable by means of L. Two structures A and B are said to be **elementarily equivalent**, $A \equiv B$, if they are first-order equivalent. We write $\equiv_{\infty\omega}$, $\equiv_{\kappa\omega}$ for equivalence in $L_{\infty\omega}$, $L_{\kappa\omega}$ respectively.

If L is a language and A is an L-structure, the L-**theory** of A, $\mathrm{Th}_L(A)$, is the class of all sentences of L which are true in A. Thus $A \equiv_L B$ if and only if $\mathrm{Th}_L(A) = \mathrm{Th}_L(B)$. The **complete theory** of A, $\mathrm{Th}(A)$ without a language L specified, always means the complete first-order theory of A.

There is another usage of the word 'complete'. Let L be a first-order language and T a theory in L. We say that T is **complete** if T has models and any two of its models are elementarily equivalent. This is equivalent to saying that for every sentence ϕ of L, exactly one of ϕ and $\neg \phi$ is a consequence of T. Of course if A is any L-structure then $\mathrm{Th}(A)$ is complete in this sense; and conversely any complete theory in L is equivalent to a theory of the form $\mathrm{Th}(A)$ for some L-structure A.

We say that a theory T is **categorical** if T is consistent and all models of T are isomorphic. It will appear in section 6.1 that the only categorical first-order theories are the complete theories of finite structures, so the notion is not too useful. Instead we make the following definitions. Let λ be a cardinal. We say that a class **K** of L-structures is λ-**categorical** if there is, up to isomorphism, exactly one structure in **K** which has cardinality λ. Likewise a theory T is λ-**categorical** if the class of all its models is λ-categorical.

In a loose but very convenient turn of phrase, we say that a single structure A is λ-**categorical** if $\mathrm{Th}(A)$ is λ-categorical. This is a fairly recent change in terminology, and it reflects a shift of interest from theories to individual structures.

In section 1.4 we remarked that if $(A, \bar{a})$ is an $L(\bar{c})$-structure with A an L-structure, then for every atomic formula $\phi(\bar{x})$ of L, $A \vDash \phi[\bar{a}]$ if and only if $(A, \bar{a}) \vDash \phi(\bar{c})$. This remains true for all formulas $\phi(\bar{x})$ of $L_{\infty\omega}$, and it justifies us in using the compromise notation $A \vDash \phi(\bar{a})$ to represent either. (Thus $\bar{a}$ are either elements of A satisfying $\phi(\bar{x})$, or added constants naming themselves in the true sentence $\phi(\bar{a})$.)

Lemma 2.3.2 (*Lemma on constants*). *Let L be a signature, T a theory in $L_{\infty\omega}$ and $\phi(\bar{x})$ a formula in $L_{\infty\omega}$. Let $\bar{c}$ be a sequence of distinct constants which are not in L. Then $T \vdash \phi(\bar{c})$ if and only if $T \vdash \forall \bar{x} \, \phi$.*

Proof. Exercise. ☐

Lastly, suppose the signature L has just finitely many symbols. Then we can identify each of these symbols with a natural number, and each term and first-order formula of L with a natural number. If this is done in a reasonable

way, then syntactic operations such as forming the conjunction of two
formulas, or substituting a term for a free variable in some formula, become
recursive functions. In this situation we say we have a **recursive language**.
Then it makes sense to talk of **recursive sets of terms** and **recursive sets of
formulas**; these notions don't depend on the choice of coding. A theory T in
L is said to be **decidable** if the set of its consequences is recursive. With a
little more care, these definitions also make sense when the signature of L is
an infinite recursive set.

Hintikka sets

Each structure A has a first-order theory $\text{Th}(A)$. Does each first-order theory
have a model? Clearly not. In fact a theorem of Church [1936] implies that
there is no algorithm to determine whether or not a given first-order sentence
has a model.

Nevertheless we found in section 1.5 that a set of atomic sentences always
has a model. That fact can be generalised, as we shall see now. The results
below are important for constructing models; we shall use them in Chapters 6
and 7 below.

Consider an L-structure A which is generated by its constant elements. Let
T be the class of all sentences of $L_{\infty\omega}$ which are true in A. Then T has the
following properties.

(3.1)   For every atomic sentence ϕ of L, if $\phi \in T$ then $\neg\phi \notin T$.

(3.2)   For every closed term t of L, the sentence $t = t$ is in T.

(3.3)   If $\phi(x)$ is an atomic formula of L, s and t are closed terms of
         L and $s = t \in T$, then $\phi(s) \in T$ if and only if $\phi(t) \in T$.

(3.4)   If $\neg\neg\phi \in T$ then $\phi \in T$.

(3.5)   If $\bigwedge\Phi \in T$ then $\Phi \subseteq T$; if $\neg\bigwedge\Phi \in T$ then there is $\psi \in \Phi$
         such that $\neg\psi \in T$.

(3.6)   If $\bigvee\Phi \in T$ then there is $\psi \in \Phi$ such that $\psi \in T$. (In particular
         $\bot \notin T$.) If $\neg\bigvee\Phi \in T$ then $\neg\psi \in T$ for all $\psi \in \Phi$.

(3.7)   Let ϕ be $\phi(x)$. If $\forall x\,\phi \in T$ then $\phi(t) \in T$ for every closed
         term t of L; if $\neg\forall x\,\phi \in T$ then $\neg\phi(t) \in T$ for some closed
         term t of L.

(3.8)   Let ϕ be $\phi(x)$. If $\exists x\,\phi \in T$ then $\phi(t) \in T$ for some closed
         term t of L; if $\neg\exists x\,\phi \in T$ then for every closed term t of L,
         $\neg\phi(t) \in T$.

A theory T with the properties (3.1)–(3.8) is called a **Hintikka set** for L.

Theorem 2.3.3. *Let L be a signature and T a Hintikka set for L. Then T has a* *model in which every element is of the form t^A for some closed term t of L. In fact the canonical model of the set of atomic sentences in T is a model of T.*

Proof. Write U for the set of atomic sentences in T, and let A be the canonical model of U. We assert that for every sentence ϕ of $L_{\infty\omega}$,

(3.9)   if $\phi \in T$ then $A \vDash \phi$, and if $\neg\phi \in T$ then $A \vDash \neg\phi$.

(3.9) is proved as follows, by induction on the construction of ϕ, using the definition of $\vDash$ in clauses (1.5)–(1.10) of section 2.1.

By (3.2) and (3.3), U is =-closed in L (see section 1.5). Hence if ϕ is atomic, (3.9) is immediate by (3.1) and the definition of A.

If ϕ is of the form $\neg\psi$ for some sentence ψ, then by induction hypothesis (3.9) holds for ψ. This immediately gives the first half of (3.9) for ϕ. For the second half, suppose $\neg\phi \in T$; then $\psi \in T$ by (3.4) and hence $A \vDash \psi$ by (3.9) for ψ. But then $A \vDash \neg\phi$.

Suppose next that ϕ is $\forall x\, \psi$. If $\phi \in T$ then by (3.7), $\psi(t) \in T$ for every closed term t of L, so $A \vDash \psi(t)$ by induction hypothesis. Since every element of the canonical model is named by a closed term, this implies $A \vDash \forall x\, \psi$. If $\neg\phi \in T$ then by (3.7) again, $\neg\psi(t) \in T$ for some closed term t, and hence $A \vDash \neg\psi(t)$. Therefore $A \vDash \neg\forall x\, \psi$.

It should now be clear how the remaining cases go. From (3.9) it follows that A is a model of T. □

Theorem 2.3.3 reduces the problem of finding a model to the problem of finding a particular kind of theory. The next results shows where we might look for theories of the right kind.

Theorem 2.3.4. *Let L be a first-order language (or more generally a first-order-closed language). Let T be a theory in L such that*
(a) *every finite subset of T has a model,*
(b) *for every sentence ϕ of L, either ϕ or $\neg\phi$ is in T,*
(c) *for every sentence $\exists x\, \psi(x)$ in T there is a closed term t of L such that $\psi(t)$ is in T,*
(d) *for every sentence $\bigvee\Phi$ in T with Φ infinite, there is $\psi \in \Phi$ such that $\psi \in T$, and, for every sentence $\neg\bigwedge\Phi$ in T with Φ infinite, there is $\psi \in \Phi$ such that $\neg\psi \in T$.*
Then T is a Hintikka set for L. (Of course clause (d) has no effect if L is first-order.)

Proof. First we claim that

(3.10) if U is a finite subset of T and ϕ is a sentence of L such that
$U \vdash \phi$, then $\phi \in T$.

For let U and ϕ be a counterexample. Then $\phi \in T$, so by (b), $\neg \phi \in T$. It follows by (a) that there is a model of $U \cup \{\neg \phi\}$, contradicting the assumption that $U \vdash \phi$. This proves the claim, using just (a) and (b).

Now from (a) we deduce (3.1), and from the claim (3.10) we deduce (3.2), (3.3), (3.4), the first halves of (3.5) and (3.7) and the second halves of (3.6) and (3.8).

When Φ is infinite, we have the first half of (3.6) by (d). On the other hand suppose Φ is a finite set $\{\psi_0, \ldots, \psi_{n-1}\}$, and $\bigvee \Phi \in T$ but $\psi_i \notin T$ for all $i < n$. Then by (b), $\neg \psi_i \in T$ for all $i < n$, and so by (a) the set $\{\bigvee \Phi, \neg \psi_0, \ldots, \neg \psi_{n-1}\}$ has a model; which is absurd. So (3.6) holds. Analogous arguments prove the second half of (3.5).

Finally (c) implies the first half of (3.8), and by (3.10) this implies the second half of (3.7).   □

Exercises for section 2.3

1. Show that a theory T in a first-order language L is closed under taking consequences if and only if $T = \text{Th}(\text{Mod}(T))$.

2. Let T be the theory of vector spaces over a field K. (Take T to be (2.21) from the previous section, with $R = K$.) Show that T is λ-categorical whenever λ is an infinite cardinal $> |K|$.

3. Let $\mathbb{N}$ be the natural number structure $(\omega, 0, 1, +, \cdot, <)$; let its signature be L. Write a sentence of $L_{\omega_1\omega}$ whose models are precisely the structures isomorphic to $\mathbb{N}$. (*So there are categorical sentences of $L_{\omega_1\omega}$ with infinite models.*)

4. Prove the lemma on constants (Lemma 2.3.2).

5. Show that if L is a first-order language with finitely many relation, function and constant symbols, then there is an algorithm to determine, for any finite set T of quantifier-free sentences of L, whether or not T has a model. [Use Exercises 1.5.2 and 1.5.3.]

6. For each $n < \omega$ let L_n be a signature and Φ_n a Hintikka set for L_n. Suppose that for all $m < n < \omega$, $L_m \subseteq L_n$ and $\Phi_m \subseteq \Phi_n$. Show that $\cup_{n<\omega}\Phi_n$ is a Hintikka set for the signature $\cup_{n<\omega}L_n$.

7. Let L be a first-order language. (a) Show that if there is an empty L-structure A and ϕ is a prenex sentence which is true in A, then ϕ begins with a universal quantifier. (b) Show (without assuming that every structure is non-empty) that every formula $\phi(\bar{x})$ of L is logically equivalent to a prenex formula $\psi(\bar{x})$ of L. [You only need a new argument when $\bar{x}$ is empty.] (c) Sometimes it is convenient to allow L to contain 0-ary relation symbols (i.e. sentence letters) p; we interpret them so that for

each L-structure A, p^A is either truth or falsehood, and in the definition of $\vDash$ we put $A \vDash p \Leftrightarrow p^A =$ truth. Show that in such a language L there can be a sentence which is not logically equivalent to a prenex sentence.

8. Let L be a first-order language. An L-structure A is said to be **locally finite** if every finitely generated substructure of A is finite. (a) Show that there is a set Ω of quantifier-free types (i.e. types consisting of quantifier-free formulas) such that for every L-structure A, A is locally finite if and only if A omits every type in Ω. (b) Show that if L has finite signature, then we can choose the set Ω in (a) to consist of a single type.

2.4 Maps and the formulas they preserve

Let $f: A \to B$ be a homomorphism of L-structures and $\phi(\bar{x})$ a formula of $L_{\infty\omega}$. We say that f **preserves** ϕ if for every sequence $\bar{a}$ of elements of A,

(4.1) $A \vDash \phi(\bar{a}) \Rightarrow B \vDash \phi(f\bar{a})$.

In this terminology, Theorem 1.3.1 and its corollary said that homomorphisms preserve atomic formulas, and that a homomorphism is an embedding if and only if it preserves literals. (Set theorists speak of a formula ϕ being **absolute** under f if (4.1) holds with $\Rightarrow$ replaced by $\Leftrightarrow$; thus atomic formulas are absolute under embeddings.)

The notion of preservation can be used two ways round. In this section we classify formulas in terms of the maps which preserve them. The next section will classify maps in terms of the formulas which they preserve.

Classifying formulas by maps

Our main results will say that certain types of map preserve all formulas with certain syntactic features. Later (see sections 6.5 and 10.3) we shall be able to show that in a broad sense these results are best possible for first-order formulas.

A formula ϕ is said to be an $\forall_1$ formula (pronounced 'A1 formula'), or **universal**, if it is built up from quantifier-free formulas by means of $\bigwedge$, $\bigvee$ and universal quantification (at most). It is said to be an $\exists_1$ formula (pronounced 'E1 formula'), or **existential**, if it is built up from quantifier-free formulas by means of $\bigwedge$, $\bigvee$ and existential quantification (at most).

This definition is the bottom end of a hierarchy. We shall barely need the higher reaches of the hierarchy, but here it is for the record.

(4.2)   Formulas are said to be $\forall_0$, and $\exists_0$, if they are quantifier-free.

(4.3)   A formula is an $\forall_{n+1}$ formula if it is in the smallest class of
         formulas which contains the $\exists_n$ formulas and is closed under
         $\bigwedge$, $\bigvee$ and adding universal quantifiers at the front.

(4.4)   A formula is an $\exists_{n+1}$ formula if it is in the smallest class of
         formulas which contains the $\forall_n$ formulas and is closed under
         $\bigwedge$, $\bigvee$ and adding existential quantifiers at the front.

$\forall_2$ formulas are sometimes known as $\forall\exists$ formulas. Note that just like the
arithmetical hierarchy in section 2.1, the classes of formulas increase as we go
up: every quantifier-free formula is $\forall_1$ and $\exists_1$, and all $\forall_1$ or $\exists_1$ formulas are
$\forall_2$. (Some writers use this classification only for prenex formulas.)

   If a formula is formed from other formulas by means of just $\wedge$ and $\vee$, we
say it is a **positive boolean combination** of these other formulas. If just $\bigwedge$ and
$\bigvee$ are used, we talk of a **positive infinite boolean combination**. Note that the
class of $\forall_n$ formulas and the class of $\exists_n$ formulas of $L_{\infty\omega}$, for any $n < \omega$, are
both closed under positive infinite boolean combinations.

   In the next theorem and those which follow, the formulas are assumed to
be in $L_{\infty\omega}$. But if the reader has in mind some other kind of language for
talking about L-structures, he or she will very likely find that the theorems
adapt to this other language too.

Theorem 2.4.1. *Let $\phi(\bar{x})$ be an $\exists_1$ formula of signature L and $f: A \rightarrow B$ an
embedding of L-structures. Then f preserves ϕ.*

Proof. We first show that if $\phi(\bar{x})$ is a quantifier-free formula of L and $\bar{a}$ is a
sequence of elements of A, then

(4.5) $A \vDash \phi(\bar{a}) \Leftrightarrow B \vDash \phi(f\bar{a})$.

This is proved by induction on the complexity of ϕ. If ϕ is atomic, we have it
by Theorem 1.3.1(c). If ϕ is $\neg\psi$, $\bigwedge\Phi$ or $\bigvee\Phi$, then the result follows by
induction hypothesis and (1.6)–(1.8) of section 2.1.

   We prove the theorem by showing that for every $\exists_1$ formula $\phi(\bar{x})$ and
every sequence $\bar{a}$ of elements of A,

(4.6) $A \vDash \phi(\bar{a}) \Rightarrow B \vDash \phi(f\bar{a})$.

For quantifier-free ϕ this follows from (4.5), and $\bigwedge$ and $\bigvee$ raise no new
questions. There remains the case where $\phi(\bar{x})$ is $\exists y\, \psi(y, \bar{x})$; cf. (1.10) of
section 2.1. If $A \vDash \phi(\bar{a})$ then for some element c of A, $A \vDash \psi(c, \bar{a})$. So by
induction hypothesis, $B \vDash \psi(fc, f\bar{a})$ and hence $B \vDash \phi(f\bar{a})$ as required. $\square$

   We say that a formula $\phi(\bar{x})$ is **preserved in substructures** if whenever A
and B are L-structures, A is a substructure of B and $\bar{a}$ is a sequence of
elements of A such that $B \vDash \phi(\bar{a})$, then $A \vDash \phi(\bar{a})$ too. We say that a theory T
is an $\forall_1$ theory if all the sentences in T are $\forall_1$ formulas.

Corollary 2.4.2. (a) $\forall_1$ *formulas are preserved in substructures.*

(b) *If T is an $\forall_1$ theory then the class of models of T is closed under taking substructures.*

Proof. (a) Every $\forall_1$ formula is logically equivalent to the negation of an $\exists_1$ formula. (b) follows at once. ☐

Part (b) of the corollary can be checked against the theories listed in section 2.2. It depends on the choice of language, of course. In the signature with just the symbol $\cdot$, a substructure of a group need not be a group; so there is no $\forall_1$ axiomatisation of the class of groups in this signature, and one has to make do with $\forall_2$ axioms instead. (Cf. Exercise 7 for similar examples.)

A formula of $L_{\infty\omega}$ is said to be **positive** if $\neg$ never occurs in it (and so by implication $\rightarrow$ and $\leftrightarrow$ never occur in it either – but it can contain $\perp$). We call a formula $\exists_1^+$ or **positive existential** if it is both positive and existential. The proof of the next theorem uses no new ideas.

Theorem 2.4.3. *Let $\phi(\bar{x})$ be a formula of signature L and $f: A \rightarrow B$ a homomorphism of L-structures.*
(a) *If ϕ is an $\exists_1^+$ formula then f preserves ϕ.*
(b) *If ϕ is positive and f is surjective, then f preserves ϕ.*
(c) *If f is an isomorphism then f preserves ϕ.* ☐

There are innumerable other similar results for other types of homomorphism. See for example Exercises 8, 9, 12, 15.

Chains and direct limits

Let L be a signature and $(A_i: i < \gamma)$ a sequence of L-structures. We call $(A_i: i < \gamma)$ a **chain** if for all $i < j < \gamma$, $A_i \subseteq A_j$. If $(A_i: i < \gamma)$ is a chain, then we can define another L-structure B as follows. The domain of B is $\bigcup_{i<\gamma} \text{dom}(A_i)$. For each constant c, c^{A_i} is independent of the choice of i, so we can put $c^B = c^{A_i}$ for any $i < \gamma$. Likewise if F is an n-ary function symbol of L and $\bar{a}$ is an n-tuple of elements of B, then $\bar{a}$ is in $\text{dom}(A_i)$ for some $i < \gamma$, and without ambiguity we can define $F^B(\bar{a})$ to be $F^{A_i}(\bar{a})$. Finally if R is an n-ary relation symbol of L, we put $\bar{a} \in R^B$ if $\bar{a} \in R^{A_i}$ for some (or all) A_i containing $\bar{a}$. By construction, $A_i \subseteq B$ for every $i < \gamma$. We call B the **union** of the chain $(A_i: i < \gamma)$, in symbols $B = \bigcup_{i<\gamma} A_i$.

We say that a formula $\phi(\bar{x})$ of L is **preserved in unions of chains** if whenever $(A_i: i < \gamma)$ is a chain of L-structures, $\bar{a}$ is a sequence of elements of A_0 and $A_i \vDash \phi(\bar{a})$ for all $i < \gamma$, then $\bigcup_{i<\gamma} A_i \vDash \phi(\bar{a})$.

Theorem 2.4.4. *Let $\psi(\bar{y}, \bar{x})$ be an $\exists_1$ formula of signature L with $\bar{y}$ finite. Then $\forall \bar{y}\, \psi$ is preserved in unions of chains of L-structures.*

Proof. Let $(A_i : i < \gamma)$ be a chain of L-structures and $\bar{a}$ a sequence of elements of A_0 such that $A_i \vDash \forall \bar{y}\, \psi(\bar{y}, \bar{a})$ for all $i < \gamma$. Put $B = \bigcup_{i<\gamma} A_i$. To show that $B \vDash \forall \bar{y}\, \psi(\bar{y}, \bar{a})$, let $\bar{b}$ be any tuple of elements of B. Since $\bar{b}$ is finite, there is some $i < \gamma$ such that $\bar{b}$ lies in A_i. By assumption, $A_i \vDash \psi(\bar{b}, \bar{a})$. Since $A_i \subseteq B$, it follows from Theorem 2.4.1 that $B \vDash \psi(\bar{b}, \bar{a})$. $\square$

Any $\forall_2$ first-order formula can be brought to the form $\forall \bar{y}\, \psi$ with ψ existential. So Theorem 2.4.4 says in particular that all $\forall_2$ first-order formulas are preserved in unions of chains. For example the axioms (2.32) in section 2.2 (for dense linear orderings without endpoints) are $\forall_2$ first-order, and it follows at once that the union of a chain of dense linear orderings without endpoints is a dense linear ordering without endpoints.

Direct limits are a generalisation of unions of chains. An **upward directed poset** is a partially ordered set $(P, \leqslant)$ with the property that if $p, q \in P$ then there is $r \in P$ with $p \leqslant r$ and $q \leqslant r$. Let L be a signature; then a **directed diagram** in L consists of an upward directed poset $(P, \leqslant)$, an L-structure A_p for each $p \in P$, and for each pair of elements p, q in P with $p \leqslant q$, a homomorphism $h_{pq} : A_p \to A_q$, such that

(4.7)   for each $p \in P$, $h_{pp} : A_p \to A_p$ is the identity map,

(4.8)   if $p \leqslant q \leqslant r$ in P, then $h_{pr} = h_{qr} h_{pq}$.

We define the **direct limit** of this directed diagram to be the L-structure B constructed as follows.

Assume for simplicity that if $p \neq q$ then the domains of A_p and A_q are disjoint, and write X for $\bigcup_{p \in P} \mathrm{dom}(A_p)$. (Otherwise, let X be the disjoint union.) Define a relation $\sim$ on X as follows:

(4.9)   If $a \in \mathrm{dom}(A_p)$ and $b \in \mathrm{dom}(A_q)$, then $a \sim b$ iff there is r such that $p \leqslant r$, $q \leqslant r$ and $h_{pr}(a) = h_{qr}(b)$.

Using (4.7) and (4.8), $\sim$ is an equivalence relation. Write $a^{\sim}$ for the equivalence class of an element a of X. The domain of B shall be the set of equivalence classes $a^{\sim}$ with $a \in X$. For each $p \in P$ we define a map $h_p : \mathrm{dom}(A_p) \to \mathrm{dom}(B)$ by putting $h_p(a) = a^{\sim}$. Then

(4.10) $h_q h_{pq} = h_p$ whenever $p \leqslant q$ in P,

since if $a \in \mathrm{dom}(A_p)$ then $h_{pq}(a) \sim a$ by (4.9).

We want to make B an L-structure so that the maps h_p are homomorphisms. There is a natural way to do this, as follows. Suppose R is a relation symbol of L and $\bar{b}$ is a tuple in $\mathrm{dom}(B)$. We put

(4.11) $\bar{b} \in R^B \Leftrightarrow$ there are $p \in P$ and $\bar{a}$ in A_p such that $h_p(\bar{a}) = \bar{b}$
         and $A_p \vDash R(\bar{a})$.

Suppost next that F is a function symbol of L and $\bar{b}$ is a tuple in $\mathrm{dom}\,(B)$. Since $(P, \leqslant)$ is upward directed, there are some $p \in P$ and some $\bar{a}$ in A_p such that $\bar{b} = h_p\bar{a}$. Then put

(4.12) $F^B(\bar{b}) = h_p(F^A p(\bar{a}))$.

We have to check that this definition is sound. Suppose $q \in P$ and $\bar{c}$ is a tuple in A_q such that $h_q(\bar{c}) = \bar{b}$. Then there is $r \geqslant p, q$ such that $h_{pr}(\bar{a}) = h_{qr}(\bar{c})$. Using (4.10) we have

(4.13) $h_p(F^{A_p}(\bar{a})) = h_r h_{pr}(F^{A_p}(\bar{a})) = h_r F^{A_r}(h_{pr}(\bar{a}))$

                                $= h_r F^{A_r}(h_{qr}(\bar{c})) = h_q(F^{A_q}(\bar{c}))$.

This justifies (4.12). Likewise if c is a constant, we put $c^B = h_p(c^{A_p})$ for any $p \in P$.

Theorem 2.4.5. *In the situation above,*
(a) *the maps $h_p: A_p \to B$ are homomorphisms such that $h_p = h_q h_{pq}$ whenever $p \leqslant q$ in P,*
(b) *if C is any L-structure and homomorphisms $f_p: A_p \to C$ are given so that $f_p = f_q h_{pq}$ whenever $p \leqslant q$ in P, then there is a unique homomorphism $g: B \to C$ such that for each $p \in P, f_p = g \cdot h_p$,*
(c) *if the maps h_{pq} are embeddings, then the maps h_p are also embeddings,*
(d) *if the maps h_{pq} are surjective, then so are the maps h_p.*

Proof. Proofs of this kind of theorem are only of benefit to people who write them, not to people who read them. So I leave it as an exercise. $\qquad \square$

A structure B as in Theorem 2.4.5 is called the **direct limit**, or with more respect for category theory the **directed colimit**, of the directed diagram $((A_p: p \in P), (h_{pq}: p \leqslant q \text{ in } P))$. By (c) it is unique up to isomorphism. The maps h_p are called the **limit homomorphisms**.

We say that a formula $\phi(\bar{x})$ of L is **preserved in direct limits** if for every direct limit in L as above, every $p \in P$ and every tuple $\bar{a}$ in A such that $A_q \vDash \phi(h_{pq}(\bar{a}))$ for all $q \geqslant p$, we have $B \vDash \phi(h_p(\bar{a}))$.

Theorem 2.4.6. *Let L be a signature, and let Φ be the smallest class of formulas of signature L such that*
(a) *every positive $\exists_1$ formula and every negated atomic formula of L is in Φ,*
(b) *Φ is closed under finite disjunctions and arbitrary conjunctions,*
(c) *if $\phi(\bar{x}, \bar{y})$ is a formula in Φ and $\bar{x}$ is a tuple of variables, then $\forall \bar{x}\, \phi$ is in Φ.*
Then every formula in Φ is preserved in direct limits.

If we restrict to direct limits of directed diagrams in which the maps are all embeddings (resp. surjective homomorphisms), then in (a) 'positive $\exists_1$' can be replaced by '$\exists_1$' (resp. 'positive').

Proof. As always, by induction on the complexity of formulas.   □

The union of a chain is the direct limit of a directed diagram of embeddings, where the partial ordering $(P, \leqslant)$ is an ordinal. So Theorem 2.4.4 is a special case of the last sentence of Theorem 2.4.6.

Exercises for section 2.4

1. Let L be a first-order language. (a) Suppose ϕ is an $\forall_1$ sentence of L and A is an L-structure. Show that $A \vDash \phi$ if and only if $B \vDash \phi$ for every finitely generated substructure B of A. (b) Show that if A and B are L-structures, and every finitely generated substructure of A is embeddable in B, then every $\forall_1$ sentence of L which is true in B is true in A too.

2. Suppose the first-order language L has just finitely many relation symbols and constants, and no function symbols. Show that if A and B are L-structures such that every $\forall_1$ sentence of L which is true in B is true in A too, then every finitely generated substructure of A is embeddable in B.

3. Let L be a first-order language and T a theory in L, such that every $\forall_1$ formula $\phi(\bar{x})$ of L is equivalent modulo T to an $\exists_1$ formula $\psi(\bar{x})$. Show that every formula $\phi(\bar{x})$ of L is equivalent modulo T to an $\exists_1$ formula $\psi(\bar{x})$. [Put ϕ into prenex form and peel away the quantifier blocks, starting at the inside.]

4. Show that every L-structure is a direct limit of its finitely generated substructures, with inclusion maps as the limit homomorphisms.

5. Let ϕ be a sentence of $L_{\infty\infty}$ of the form $\forall \bar{x} \exists \bar{y} \, \psi$ where ψ is quantifier-free and $\bar{x}$ is a tuple of variables. Suppose A is an L-structure such that every finite subset of $\text{dom}(A)$ is contained in a substructure B of A with $B \vDash \phi$. Show that $A \vDash \phi$.

6. Show that when the maps in a directed diagram are inclusions, we can take the limit homomorphisms to be inclusions too.

7. In section 2.2 above there are axiomatisations of several important classes of structure. Show that, using the signatures given in section 2.2, it is not possible to write down sets of axioms of the following forms: (a) a set of $\exists_1$ axioms for the class of groups; (b) a set of $\forall_1$ axioms for the class of atomless boolean algebras; (c) a single $\exists_2$ first-order axiom for the class of dense linear orderings without endpoints.

*Let L be a signature containing the 2-ary relation symbol $<$. If A and B are L-structures, we say that B is an **end-extension** of A if $A \subseteq B$ and whenever a is an element of A and $B \vDash b < a$ then b is an element of A too. We say that an embedding $f: A \to B$ of L-structures is an **end-embedding** if B is an end-extension of the image of f. We define $(\forall x < y)$ and $(\exists x < y)$ as in section 2.1: $(\forall x < y)\phi$ is $\forall x(x < y \to \phi)$ and $(\exists x < y)\phi$ is $\exists x(x < y \wedge \phi)$, and the quantifiers $(\forall x < y)$ and $(\exists x < y)$ are said to be* **bounded**.

8. Let L be as above. A Π_0^0 formula is one in which all quantifiers are bounded. A Σ_1^0 formula is a formula in the smallest class of formulas which contains the Π_0^0 formulas and is closed under $\wedge$, $\vee$ and existential quantification. Show that end-embeddings preserve Σ_1^0 formulas.

9. Let L be a signature containing a 1-ary symbol P. By a P-**embedding** we mean an embedding $e: A \to B$, where A and B are L-structures, such that e maps P^A onto P^B. Let Φ be the smallest class of formulas of $L_{\infty\omega}$ such that (i) every quantifier-free formula is in Φ, (ii) Φ is closed under $\wedge$ and $\vee$, (iii) if ϕ is in Φ and x is a variable then $\exists x\, \phi$ and $\forall x(Px \to \phi)$ are in Φ. Show that every P-embedding preserves all the formulas in Φ.

*Let L be a signature. By a **descending chain of L-structures** we mean a sequence $(A_i: i < \gamma)$ of L-structures such that $A_j \subseteq A_i$ whenever $i < j < \gamma$.*
10. Show that if $(A_i: i < \gamma)$ is a descending chain of L-structures, then there is a unique L-structure B which is a substructure of each A_j and has domain $\bigcap_{i<\gamma} \mathrm{dom}\,(A_i)$. (We call this structure the **intersection** of the chain, in symbols $\bigcap_{i<\gamma} A_i$.)

*We say that a formula ϕ of $L_{\infty\omega}$ is **preserved in intersections of descending chains** if for every descending chain $(A_i: i < \gamma)$ of L-structures and every tuple $\bar{a}$ of elements of $\bigcap_{i<\gamma} A_i$, if $A_j \vDash \phi(\bar{a})$ for every $j < \gamma$ then $\bigcap_{i<\gamma} A_i \vDash \phi(\bar{a})$.*
11. (a) Show that, if ϕ is a formula of $L_{\infty\omega}$ of the form $\forall \bar{x} \exists_{=n} y\, \psi(\bar{x}, y, \bar{z})$, where ψ is quantifier-free, then ϕ is preserved in intersections of descending chains of L-structures. *(b) Write a set of first-order axioms of this form for the class of real-closed fields. (c) Can axioms of this form be found for the class of dense linear orderings without endpoints?

*Let **K** be the class of all L-structures that carry a partial ordering (named by the relation symbol $<$) which has no maximal element. When A and B are structures in **K**, we say that A is a **cofinal substructure** of B (and that B is a **cofinal extension** of A) if $A \subseteq B$ and for every element b of B there is an element a in A such that $B \vDash b < a$. A formula $\phi(\bar{x})$ of L is **preserved in cofinal substructures** if whenever A is a cofinal substructure of B and $\bar{a}$ is a tuple in A such that $B \vDash \phi(\bar{a})$, then $A \vDash \phi(\bar{a})$.*
12. Let L be a signature and Φ the smallest class of formulas of $L_{\infty\omega}$ such that (1) all literals of L are in Φ, (2) Φ is closed under $\wedge$ and $\vee$, and (3) if $\phi(x, \bar{y})$ is any formula in Φ, then so are the formulas $\forall x\, \phi$ and $\exists z \forall x(z < x \to \phi)$. Show that every formula in Φ is preserved in cofinal substructures.

13. Prove Theorem 2.4.5.

14. Let L and L^+ be signatures with $L \subseteq L^+$. Let $((A_p: p \in P), (h_{pq}: p \leqslant q \text{ in } P))$ be a directed diagram of L^+-structures with direct limit B. Show that $((A_p|L: p \in P), (h_{pq}: p \in P))$ is a directed diagram of L-structures, and its direct limit is $B|L$ (with the same limit homomorphisms).

*We say that a relation symbol R is **positive in** the formula ϕ of $L_{\infty\omega}$ if ϕ is in the smallest class X of formulas such that (1) every literal of L which doesn't contain R is in X, (2) every atomic formula of L is in X, and (3) X is closed under $\bigwedge$, $\bigvee$ and quantification. (For example R is positive in $\forall x(Qx \to Rx)$, but not in $\forall x(Rx \to Qx)$.)*
15. Let L be a signature and L^+ the signature got by adding to L a new n-ary relation symbol P. Let $\bar{x}$ be an n-tuple of variables and $\phi(\bar{x})$ a formula of L^+ in which P is positive. Let A be an L-structure; suppose X and Y are n-ary relations on $\text{dom}(A)$ with $X \subseteq Y$. It is clear that the identity map on A forms an embedding $e: (A, X) \to (A, Y)$ of L^+-structures. (a) Show that e preserves ϕ. (b) For any n-ary relation X on $\text{dom}(A)$ we define $\pi(X)$ to be the relation $\{\bar{a}: (A, X) \vDash \phi(\bar{a})\}$. Show that if $X \subseteq Y$ then $\pi(X) \subseteq \pi(Y)$.

16. Let Ω be a set and $\pi: \mathcal{P}(\Omega) \to \mathcal{P}(\Omega)$ a map which is **monotone**, i.e. such that $X \subseteq Y$ implies $\pi(X) \subseteq \pi(Y)$. We define the **least fixed point** of π, lfp (π), to be the intersection of all subsets X of Ω such that $\pi(X) \subseteq X$. Justify this name by showing (a) lfp $(\pi) = \pi(\text{lfp}(\pi))$, (b) if Y is any set such that $Y = \pi(Y)$ then lfp $(\pi) \subseteq Y$. [For (a), putting $F = \text{lfp}(\pi)$, show first that $\pi(F) \subseteq F$. Then show that $\pi(\pi F) \subseteq \pi F$, so that $F \subseteq \pi(F)$.]

17. We apply Exercise 16 to Exercise 15 by putting $\Omega = (\text{dom } A)^n$. Show that, if ϕ in Exercise 15 is a first-order $\exists_1$ formula, then lfp $(\pi) = \bigcup_{m<\omega}(\pi^m(0))$ where 0 is the empty n-ary relation on dom (A). What if ϕ is not $\exists_1$? *See section 2.8 on fixed point logics.*

2.5 Classifying maps by formulas

Let L be a signature, $f: A \to B$ a homomorphism of L-structures and Φ a class of formulas of $L_{\infty\omega}$. We call f a Φ-**map** if f preserves all the formulas in Φ.

By far the most important example is when Φ is the class of all first-order formulas. A homomorphism which preserves all first-order formulas must be an embedding (by Corollary 1.3.2); we call such a map an **elementary embedding**.

We say that B is an **elementary extension** of A, or that A is an **elementary substructure** of B, in symbols $A \preccurlyeq B$, if $A \subseteq B$ and the inclusion map is an elementary embedding. (This map is then described as an **elementary inclusion**.) We write $A \prec B$ when A is a proper elementary substructure of B.

Note that $A \preccurlyeq B$ implies $A \equiv B$. But there are examples to show that $A \subseteq B$ and $A \equiv B$ together don't imply $A \preccurlyeq B$; see Exercise 2.

Theorem 2.5.1 *(Tarski–Vaught criterion for elementary substructures).* *Let L be a first-order language and let A, B be L-structures with $A \subseteq B$. Then the following are equivalent.*
(a) *A is an elementary substructure of B.*
(b) *For every formula $\psi(\bar{x}, y)$ of L and all tuples $\bar{a}$ from A, if $B \vDash \exists y\ \psi(\bar{a}, y)$ then $B \vDash \psi(\bar{a}, d)$ for some element d of A.*

Proof. Let $f: A \rightarrow B$ be the inclusion map. (a) $\Rightarrow$ (b): If $B \vDash \exists y\ \psi(\bar{a}, y)$ then $A \vDash \exists y\ \psi(\bar{a}, y)$ since f is elementary; hence there is d in A such that $A \vDash \psi(\bar{a}, d)$, and we reach $B \vDash \psi(\bar{a}, d)$ by applying f again. (b) $\Rightarrow$ (a): take the proof of Theorem 2.4.1 in the previous section; our condition (b) is exactly what is needed to make that proof show that f is elementary. $\square$

Theorem 2.5.1 by itself isn't very useful for detecting elementary substructures in nature (though see Exercises 3, 4 below). Its main use is for constructing elementary substructures, as in Exercise 5.

If Φ is a set of formulas, we say that a chain $(A_i: i < \gamma)$ of L-structures is a **Φ-chain** when each inclusion map $A_i \subseteq A_j$ is a Φ-map. In particular an **elementary chain** is a chain in which the inclusions are elementary.

Theorem 2.5.2 *(Tarski–Vaught theorem on unions of elementary chains).* *Let $(A_i: i < \gamma)$ be an elementary chain of L-structures. Then $\bigcup_{i<\gamma} A_i$ is an elementary extension of each A_j $(j < \gamma)$.*

Proof. Put $A = \bigcup_{i<\gamma} A_i$. Let $\phi(\bar{x})$ be a first-order formula of signature L. We show by induction on the complexity of ϕ that for every $j < \gamma$ and every tuple $\bar{a}$ of elements of A_j,

(5.1) $\qquad\qquad A_j \vDash \phi(\bar{a}) \Leftrightarrow A \vDash \phi(\bar{a}).$

When ϕ is atomic, we have (5.1) by Theorem 1.3.1(c). The cases $\neg \psi, (\psi \wedge \chi)$ and $(\psi \vee \chi)$ are straightforward. Suppose then that ϕ is $\exists y\ \psi(\bar{x}, y)$. If $A \vDash \phi(\bar{a})$ then there is some b in A such that $A \vDash \psi(\bar{a}, b)$. Choose $k < \gamma$ so that b is in $\text{dom}(A_k)$ and $k \geqslant j$. Then $A_k \vDash \psi(\bar{a}, b)$ by induction hypothesis, hence $A_k \vDash \phi(\bar{a})$. Then $A_j \vDash \phi(\bar{a})$ as required, since the chain is elementary. This proves right to left in (5.1); the other direction is easier. The argument for $\forall y\ \psi(\bar{x}, y)$ is similar. $\square$

Lemma 2.5.3 *(Elementary diagram lemma).* *Suppose L is a first-order language, A and B are L-structures, $\bar{c}$ is a tuple of distinct constants not in L,*

$(A, \bar{a})$ *and* $(B, \bar{b})$ *are* $L(\bar{c})$*-structures, and* $\bar{a}$ *generates* A. *Then the following are equivalent*.
(a) *For every formula* $\phi(\bar{x})$ *of* L, *if* $(A, \bar{a}) \vDash \phi(\bar{c})$ *then* $(B, \bar{b}) \vDash \phi(\bar{c})$.
(b) *There is an elementary embedding* $f: A \rightarrow B$ *such that* $f\bar{a} = \bar{b}$.

Proof. (b) clearly implies (a). For the converse, define f as in the proof of Lemma 1.4.2. If $\bar{a}'$ is any tuple of elements of A and $\phi(\bar{z})$ is any formula of L, then by choosing a suitable sequence $\bar{x}$ of variables, we can write $\phi(\bar{z})$ as $\psi(\bar{x})$ so that $\phi(\bar{a}')$ is the same formula as $\psi(\bar{a})$. Then $A \vDash \phi(\bar{a}')$ implies $A \vDash \psi(\bar{a})$, which by (a) implies $B \vDash \psi(f\bar{a})$ and hence $B \vDash \phi(f\bar{a}')$. So f is an elementary embedding.   □

We define the **elementary diagram** of an L-structure A, in symbols eldiag(A), to be Th$(A, \bar{a})$ where $\bar{a}$ is any sequence which generates A. By (a) $\Rightarrow$ (b) in the lemma we have the following fact, which will be used constantly for constructing elementary extensions: *if D is a model of the elementary diagram of the L-structure A, then there is an elementary embedding of A into the reduct $D|L$*.
     The notion of a Φ-map has other applications.

Example 1: *Pure extensions*. This example is familiar to abelian group theorists and people who work with modules. Let A and B be left R-modules, and A a submodule of B. We say that A is **pure** in B, or that B is a **pure extension** of A, if the following holds:

(5.2)   for every finite set E of equations with parameters in A, if E has a solution in B then E already has a solution in A.

Now the statement that a certain finite set of equations with parameters $\bar{a}$ has a solution can be written $\exists \bar{x}(\psi_1(\bar{x}, \bar{a}) \wedge \ldots \wedge \psi_k(\bar{x}, \bar{a}))$ with $\psi_1, \ldots, \psi_k$ atomic; a first-order formula of this form is said to be **positive primitive**, or **p.p.** for short. So we can define a **pure embedding** to be one which preserves the negations of all positive primitive formulas. (Lemma A.5.12 will characterise when one field is pure in another.)

Exercises for section 2.5

Our first exercise is a slight refinement of Theorem 2.5.1.
1. Let L be a first-order language and B an L-structure. Suppose X is a set of elements of B such that, for every formula $\psi(\bar{x}, y)$ of L and all tuples $\bar{a}$ of elements of X, if $B \vDash \exists y\, \psi(\bar{a}, y)$ then $B \vDash \psi(\bar{a}, d)$ for some element d in X. Show that X is the domain of an elementary substructure of B.

2. Give an example of a structure A with a substructure B such that $A \cong B$ but B is not an elementary substructure of A. [Take A to be $(\omega, <)$.]

3. Let B be an L-structure and A a substructure with the following property: if $\bar{a}$ is any tuple of elements of A and b is an element of B, then there is an automorphism f of B such that $f\bar{a} = \bar{a}$ and $fb \in \mathrm{dom}\,(A)$. Show that if $\phi(\bar{x})$ is any formula of $L_{\infty\omega}$ and $\bar{a}$ a tuple in A, then $A \vDash \phi(\bar{a}) \Leftrightarrow B \vDash \phi(\bar{a})$. (In particular $A \preccurlyeq B$.)

4. Suppose B is a vector space and A is a subspace of infinite dimension. Show that $A \preccurlyeq B$.

The next exercise will be refined in section 3.1 below.

5. Let L be a countable first-order language and B an L-structure of infinite cardinality μ. Show that for every infinite cardinal $\lambda < \mu$, B has an elementary substructure of cardinality λ. [Choose a set X of λ elements of B, and close off X so that Exercise 1 applies.]

6. Let $(A_i : i < \gamma)$ be a chain of structures such that for all $i < j < \gamma$, A_i is a pure substructure of A_j. Show that each structure A_j $(j < \gamma)$ is a pure substructure of the union $\bigcup_{i<\gamma} A_i$.

7. Show that, in a direct limit of a directed diagram of structures and elementary embeddings, the limit homomorphisms are elementary embeddings.

8. Let us say that Φ is **closed under moving** $\neg$ **inwards** if, whenever Φ contains a formula $\neg \bigwedge_{i\in I}\phi_i$, $\neg \bigvee_{i\in I}\phi_i$, $\neg\forall \bar{x}\,\phi$ or $\neg\exists \bar{x}\,\phi$, it also contains the corresponding formula $\bigvee_{i\in I}\neg\phi_i$, $\bigwedge_{i\in I}\neg\phi_i$, $\exists \bar{x}\neg\phi$ or $\forall \bar{x}\neg\phi$ respectively. (For example each language $L_{\kappa\omega}$, including the first-order language $L_{\omega\omega}$, is closed under moving $\neg$ inwards. So also is the class of $\forall_n$ formulas, for any $n < \omega$.) Let L be a signature, Φ a class of formulas of L which is closed under subformulas and under moving $\neg$ inwards, and $\psi(\bar{y}, \bar{x})$ a formula in Φ with $\bar{y}$ finite. Show that $\forall \bar{y}\,\psi(\bar{y}, \bar{x})$ is preserved in unions of Φ-chains.

The next two exercises are not particularly hard, but they use things from later sections.

*9. Show that if n is a positive integer, then there are a first-order language L and L-structures A and B such that $A \subseteq B$ and for every n-tuple $\bar{a}$ in A and every formula $\phi(x_0, \ldots, x_{n-1})$ of L, $A \vDash \phi(\bar{a}) \Leftrightarrow B \vDash \phi(\bar{a})$, but B is not an elementary extension of A. [Use Fraïssé limits (Chapter 7) to construct a countable structure C^+ in which a $(2n + 2)$-ary relation symbol E defines a random equivalence relation on the sets of $n + 1$ elements, and a $(2n + 2)$-ary relation symbol R randomly arranges these equivalence classes in a linear ordering whose first element is defined by the $(n + 1)$-ary relation symbol P. Form C by dropping the symbol P from C^+. Find an embedding $e: C \to C$ whose image misses the first element of the ordering.]

*10. Suppose R is a ring and A, B are left R-modules with $A \subseteq B$, and, for every element a and A and every p.p. formula $\phi(x)$ without parameters, $A \vDash \phi(a) \Leftrightarrow B \vDash \phi(a)$. Show that A is pure in B. [Show by induction on n that if $\phi(x_0, \ldots, x_n)$ is a pure formula and $\bar{a}$ an $(n + 1)$-tuple in A such that $B \vDash \phi(\bar{a})$ then $A \vDash \phi(\bar{a})$. If $B \vDash \phi(\bar{a})$ then $B \vDash \exists x_n\,\phi(\bar{a}|n, x_n)$, so by induction hypothesis there is c in A such that

$A \vDash \phi(\bar{a}|n, c)$, so by subtraction (cf. proof of Corollary A.1.2) $B \vDash \phi(0, \ldots, 0, a_n - c)$; hence $A \vDash \phi(0, \ldots, 0, a_n - c)$, and addition yields $A \vDash \phi(\bar{a})$.]

2.6 Translations

In model theory as elsewhere, there can be several ways of saying the same thing. Say it in the wrong way and you may not get the results you intended.

This section is an introduction to some kinds of paraphrase that model theorists find useful. *These paraphrases never alter the class of definable relations on a structure – they only affect formulas which can be used to define them.*

1: Unnested formulas

Let L be a signature. By an **unnested atomic formula** of signature L we mean an atomic formula of one of the following forms:

(6.1) $x = y$;

(6.2) $c = y$ for some constant c of L;

(6.3) $F(\bar{x}) = y$ for some function symbol F of L;

(6.4) $R\bar{x}$ for some relation symbol R of L.

We call a formula **unnested** if all of its atomic subformulas are unnested.

Unnested formulas are handy when we want to make definitions or proofs by induction on the complexity of formulas. The atomic case becomes particularly simple: we never need to consider any terms except variables, constants and terms $F(\bar{x})$ where F is a function symbol. There will be examples in sections 3.3 (back-and-forth games), 5.3 (interpretations), 9.6 (Feferman–Vaught theorem) and 10.4 (Keisler games).

Theorem 2.6.1 *Let L be a signature. Then every atomic formula $\phi(\bar{x})$ of L is logically equivalent to unnested first-order formulas $\phi^{\forall}(\bar{x})$ and $\phi^{\exists}(\bar{x})$ of signature L such that $\phi^{\forall}$ is an $\forall_1$ formula and $\phi^{\exists}$ is an $\exists_1$ formula.*

Proof by example. The formula $F(G(x), z) = c$ is logically equivalent to

(6.5) $\forall uw(G(x) = u \wedge F(u, z) = w \rightarrow c = w)$

and to

(6.6) $\exists uw(G(x) = u \wedge F(u, z) = w \wedge c = w)$. $\square$

One can be a bit more precise. The formula $\phi^{\exists}$ in the theorem is positive primitive (as defined in the previous section). Also when we define Horn formulas in section 9.1 below, it will be clear that $\phi^{\forall}$ is strict universal Horn.

Corollary 2.6.2. *Let L be a first-order language. Then every formula $\phi(\bar{x})$ of L is logically equivalent to an unnested formula $\psi(\bar{x})$ of L. More generally every formula of $L_{\infty\omega}$ is logically equivalent to an unnested formula of $L_{\infty\omega}$.*

Proof. Use the theorem to replace all atomic subformulas by unnested first-order formulas. □

If ϕ in the corollary is an $\exists_1$ formula, then by choosing wisely between $\theta^\forall$ and $\theta^\exists$ for each atomic subformula θ of ϕ, we can arrange that ψ in the corollary is an $\exists_1$ formula too. In fact we can always choose ψ to lie in the same place in the $\forall_n$, $\exists_n$ hierarchy (see (4.2)–(4.4) in section 2.4 above) as ϕ, unless ϕ is quantifier-free.

2: Definitional expansions and extensions

Let L and L^+ be signatures with $L \subseteq L^+$, and let R be a relation symbol of L^+. Then an **explicit definition of R in terms of L** is a sentence of the form

(6.7) $$\forall \bar{x} \, (R\bar{x} \leftrightarrow \phi(\bar{x}))$$

where ϕ is a formula of L. Likewise if c is a constant and F is a function symbol of L^+, **explicit definitions of c, F in terms of L** are sentences of the form

(6.8) $\quad \forall y \, (c = y \leftrightarrow \phi(y))$,

$\qquad \forall \bar{x} y \, (F(\bar{x}) = y \leftrightarrow \psi(\bar{x}, y))$

where ϕ, ψ are formulas of L. Note that the sentences in (6.8) have consequences in the language L. They imply respectively

(6.9) $\quad \exists_{=1} y \, \phi(y)$,

$\qquad \forall \bar{x} \exists_{=1} y \, \psi(\bar{x}, y)$.

We call the sentences (6.9) the **admissibility conditions** of the two explicit definitions in (6.8).

Explicit definitions have two main properties, as follows.

Theorem 2.6.3 (*Uniqueness of definitional expansions*). *Let L and L^+ be signatures with $L \subseteq L^+$. Let A and B be L^+-structures, R a relation symbol of L^+ and θ an explicit definition of R in terms of L. If A and B are both models of θ, and $A|L = B|L$, then $R^A = R^B$. Similarly for constants and function symbols.*

Proof. Immediate. □

Theorem 2.6.4 *(Existence of definitional expansions)*. *Let L and L^+ be signatures with $L \subseteq L^+$. Suppose that for each symbol S of $L^+\backslash L$, θ_S is an explicit definition of S in terms of L; let U be the set of these definitions.*

(a) *If C is any L-structure which satisfies the admissibility conditions (if any) of the definitions θ_S, then we can expand C to form an L^+-structure C^+ which is a model of U.*

(b) *Every formula $\chi(\bar{x})$ of signature L^+ is equivalent modulo U to a formula $\chi^*(\bar{x})$ of signature L.*

(c) *If χ and all the sentences θ_S are first-order, then so is χ^*.*

Proof. The definitions tell us exactly how to interpret the symbols S in C^+, so we have (a). For (b), use Theorem 2.6.1 to replace every atomic formula in χ by an unnested formula, and observe that the explicit definitions translate each unnested atomic formula directly into a formula of signature L. Then (c) is clear too.   □

A structure C^+ as in Theorem 2.6.4(a) is called a **definitional expansion** of C. If L and L^+ are signatures with $L \subseteq L^+$, and T is a theory of signature L, then a **definitional extension** of T to L^+ is a theory equivalent to $T \cup \{\theta_S: S$ a symbol in $L^+\backslash L\}$ where for each symbol S in $L^+\backslash L$,

(6.10) θ_S is an explicit definition of S in terms of L, and

(6.11)   if S is a constant or function symbol and χ is the admissibility condition for θ_S then $T \vdash \chi$.

Theorems 2.6.3 and 2.6.4 tell us that if T^+ is a definitional extension of T to L^+, then every model C of T has a unique expansion C^+ which is a model of T^+, and C^+ is a definitional expansion of C.

Let T^+ be a theory in the language L^+, and L a language $\subseteq L^+$. We say that a symbol S of L^+ is **explicitly definable in T^+ in terms of** L if T^+ entails some explicit definition of S in terms of L. So, up to equivalence of theories, T^+ is a definitional extension of a theory T in L iff (1) T and T^+ have the same consequences in L and (2) every symbol of L^+ is explicitly definable in T^+ in terms of L.

Definitional extensions are useful for replacing complicated formulas by simple ones. For example in set theory they allow us to write $\phi(x \cup y)$ instead of the less readable formula $\exists z(\phi(z) \wedge \forall t(t \in z \leftrightarrow t \in x \vee t \in y))$.

Warning – particularly for software developers using first-order logic. It's important that the symbols being defined in (6.7) and (6.8) don't occur in the formulas ϕ, ψ. If the symbols were allowed to occur in ϕ or ψ, Theorems 2.6.3 and 2.6.4 would both fail.

There is a deathtrap here. One sometimes meets things called 'definitions', which look like explicit definitions except that the symbol being defined occurs on both sides of the formula. These may be implicit definitions, which are harmless – at least in first-order logic; see Beth's theorem, Theorem 6.6.4 below. On the other hand if they have purple fins underneath and pink spots on top, they are almost certainly *recursive definitions*. The recursive definitions of plus and times in arithmetic, (2.14)–(2.15) in section 2.2, are a typical example; one can rewrite them to look dangerously like explicit definitions. *In general, recursive definitions define symbols on a particular structure, not on all models of a theory. There is no guarantee that they can be translated into explicit definitions in a first-order theory.* The notion of a recursive definition lies outside model theory and I shall say no more about it here.

Before we leave definitional extensions, here is an extended example which will be useful in the next section.

Suppose L_1 and L_2 are signatures; to simplify the next definition let us assume they are disjoint. Let T_1 and T_2 be first-order theories of signature L_1, L_2 respectively. Then we say that T_1 and T_2 are **definitionally equivalent** if there is a first-order theory T in signature $L_1 \cup L_2$ which is a definitional extension both of T_1 and of T_2.

When theories T_1 and T_2 are definitionally equivalent as above, we can turn a model A_1 of T_1 into a model A_2 of T_2 by first expanding A_1 to a model of T and then restricting to the language L_2; we can get back to A_1 from A_2 by doing the same in the opposite direction. In this situation we say that the structures A_1 and A_2 are **definitionally equivalent**.

Example 1: *Term algebras in another language.* In Example 1 of section 2.2 we wrote down some sentences (2.3)–(2.7) which are true in every term algebra of a fixed algebraic signature L. Call these sentences T_1 and let L_1 be their first-order language. Let L_2 be the first-order language whose signature consists of the following symbols:

(6.12) 1-ary relation symbols Is_c (for each constant c of L_1) and Is_F (for each function symbol F of L_1);

(6.13) a 1-ary function F_i for each function symbol F of L_1 and each $i < \text{arity}(F)$.

We claim that T_1 is definitionally equivalent to the following theory T_2 in L_2.

(6.14) $\exists_{=1} y \, \text{Is}_c(y)$ for each constant symbol c of L.

(6.15) $\forall x_0 \ldots x_{n-1} \exists_{=1} y \, (\text{Is}_F(y) \wedge \bigwedge_{i<n} F_i(y) = x_i)$

for each function symbol F of L.

(6.16) $\forall x \, \neg(\text{Is}_c(x) \wedge \text{Is}_d(x))$ where c, d are distinct constant or function symbols.

(6.17) $\forall x(\neg\text{Is}_F(x) \rightarrow F_i(x) = x)$ for each function symbol F_i.

(6.18) $\forall x(t(F_i(x)) = x \rightarrow \neg\text{Is}_F(x))$ for each function symbol F_i and term $t(y)$ of L_2.

To show this, we must write down explicit definitions U_1 of the symbols of L_2 in terms of L_1, and explicit definitions U_2 of the symbols of L_1 in terms of L_2, in such a way that T_i implies the admissibility conditions for U_i $(i = 1, 2)$, and $T_1 \cup U_1$ is equivalent to $T_2 \cup U_2$. Here are the definitions of L_2 in terms of L_1, where c is any constant of L_1, F is any function symbol of L_1 with arity n and $i < n$.

(6.19) $\forall y \, (\text{Is}_c(y) \leftrightarrow y = c)$.

$\forall y \, (\text{Is}_F(y) \leftrightarrow \exists \bar{x} \, F\bar{x} = y)$.

$\forall xy \, (F_i x = y \leftrightarrow (\exists y_0 \ldots y_{i-1} y_{i+1} \ldots y_{n-1}$

$F(y_0, \ldots, y_{i-1}, y, y_{i+1}, \ldots, y_{n-1}) = x)$

$\vee \, (x = y \wedge \neg\exists \bar{y} \, F\bar{y} = x))$.

And here are the explicit definitions of L_1 in terms of L_2, where c is any constant of L_1 and F any function symbol of L_1:

(6.20) $\forall y(y = c \leftrightarrow \text{Is}_c(y))$;

$\forall x_0 \ldots x_{n-1} y$

$$(F(x_0, \ldots, x_{n-1}) = y \leftrightarrow (\text{Is}_F(y) \wedge \bigwedge_{i<n} F_i(y) = x_i)).$$

These two theories T_1 and T_2 give opposite ways of looking at the term algebra: T_1 generates the terms from their components, while T_2 recovers the components from the terms. One curious feature is that T_2 uses only 1-ary function and relation symbols, while there is no bound on the arities of the symbols in T_1. The theory T_2 will be useful for analysing T_1 in the next section.

3: Atomisation

Here we have a theory T in a language L, and a set Φ of formulas of L which are not sentences. The object is to extend T to a theory T^+ in a larger language L^+ in such a way that every formula in Φ is equivalent modulo T^+ to an atomic formula. In fact the set of new sentences $T^+\backslash T$ will turn out to depend only on L and not on T.

This device is sometimes called Morleyisation. But it has been well known since 1920 when Skolem introduced it, and it has nothing particularly to do with Morley. So I thought it best to use a more descriptive name. 'Skolemisation' already means something else; see section 3.1 below.

Theorem 2.6.5 (*Atomisation theorem*). *Let L be a first-order language. Then there are a first-order language $L^{\Theta} \supseteq L$ and a theory Θ in L^{Θ} such that*
(a) *every L-structure A can be expanded in just one way to an L^{Θ}-structure A^{Θ} which is a model of Θ,*
(b) *every formula $\phi(\bar{x})$ of L^{Θ} is equivalent modulo Θ to a formula $\psi(\bar{x})$ of L, and also (when $\bar{x}$ is not empty) to an atomic formula $\chi(\bar{x})$ of L^{Θ},*
(c) *every homomorphism between non-empty models of Θ is an elementary embedding,*
(d) *$|L^{\Theta}| = |L|$.*

Proof. For each formula $\phi(x_0, \ldots, x_{n-1})$ of L with $n > 0$, introduce a new n-ary relation symbol R_{ϕ}. Take L^{Θ} to be the first-order language got from L by adding all the symbols R_{ϕ}, and take Θ to be the set of all sentences of the form

(6.21) $$\forall \bar{x}(R_{\phi}\bar{x} \leftrightarrow \phi(\bar{x})).$$

Then (d) is immediate. The theory Θ is a definitional extension of the empty theory in L, so we have (a) and the first part of (b). The second part of (b) then follows by (6.21).

Now by (b), every formula of L which is not a sentence is equivalent modulo Θ to an atomic formula. If ϕ is a sentence of L then $\phi \wedge x = x$ is equivalent modulo Θ to an atomic formula $\chi(x)$; any homomorphism between non-empty models of Θ which preserves χ must also preserve ϕ. So (c) follows by Theorem 1.3.1(b). $\qquad\square$

One can apply the same technique to a particular set Φ of formulas of L, if one wants to study homomorphisms which preserve the formulas in Φ. If A and B are models of Θ, then every embedding (in fact every homomorphism) from A to B must preserve the formulas in Φ, by Theorem 1.3.1(b).

Theorem 2.6.6. *Let Θ be the theory constructed in the proof of Theorem 2.6.5. Then for every theory T in L^{Θ}, $T \cup \Theta$ is equivalent to an $\forall_2$ theory.*

Proof. By (b) of the theorem, every formula of L^{Θ} with at least one free variable is equivalent modulo Θ to an atomic formula of L^{Θ}. So $T \cup \Theta$ is equivalent to a theory $T' \cup \Theta$ where every sentence of T' is $\forall_1$ at worst. It suffices now to show that Θ itself is equivalent to an $\forall_2$ theory.

Let Θ' be the set of all sentences of the following forms:

(6.22) $\forall \bar{x}(\phi(\bar{x}) \leftrightarrow R_{\phi}(\bar{x}))$ where ϕ is an atomic formula of L;

(6.23) $\forall \bar{x}(R_\phi(\bar{x}) \wedge R_\psi(\bar{x}) \leftrightarrow R_{\phi \wedge \psi}(\bar{x}))$;

and likewise with $\vee$ for $\wedge$;

(6.24) $\forall \bar{x}(\neg R_\phi(\bar{x}) \leftrightarrow R_{\neg \phi}(\bar{x}))$;

(6.25) $\forall \bar{x}(\forall y \, R_{\phi(\bar{x},y)}(\bar{x}, y) \leftrightarrow R_{\forall y \, \phi(\bar{x},y)}(\bar{x}))$;

and likewise with $\exists$ for $\forall$.

After a slight rearrangement of the sentences (6.25), Θ' is an $\forall_2$ theory. We show that Θ is equivalent to Θ'. Clearly Θ implies all the sentences in Θ'. Conversely assume that Θ' holds. Then (6.21) follows by induction on the complexity of ϕ. $\square$

A first-order theory is said to be **model-complete** if every embedding between its models is elementary. Atomisation shows that we can turn any first-order theory into a model-complete theory in a harmless way. But the real interest of the notion of model-completeness is that a number of theories in algebra have this property without any prior tinkering. We shall think about this in Chapter 8.3.

Theorems 2.6.5 and 2.6.6 adapt straightforwardly to the situation where Φ is a set of variable-finite formulas of $L_{\infty\omega}$ which is closed under subformulas, and relation symbols have been introduced for the formulas in Φ. This gives us the following theorem of Chang, which makes it possible to apply the results of first-order model theory to infinitary theories.

Theorem 2.6.7 (Chang's reduction). *Let L be a signature and ϕ a sentence of $L_{\infty\omega}$. Then there are a signature $L^\phi \supseteq L$ and a theory T^ϕ in L^ϕ with the following properties:*

(a) *every model of ϕ can be expanded in just one way to a model of T^ϕ;*

(b) *every model of T^ϕ is a model of ϕ;*

(c) *every subformula $\psi(\bar{x})$ of ϕ (with $\bar{x}$ not empty) is equivalent modulo T^ϕ to an atomic formula $\chi(\bar{x})$ of L^ϕ;*

(d) *every sentence in T^ϕ is either a first-order $\forall_2$ sentence or an $\forall_1$ sentence of the form $\forall \bar{x} \bigvee_{i<\kappa} \psi_i(\bar{x})$ where each ψ_i is atomic.*

Proof. We proceed as in the proofs of Theorems 2.6.5 and 2.6.6, adding relation symbols R_θ only for subformulas θ of ϕ with at least one free variable. The resulting sentences (6.22), (6.24) and (6.25) are all $\forall_2$ first-order sentences. To deal with the analogue of (6.23), we split the sentence

(6.26) $$\forall \bar{x}\left(\bigwedge_{i<\kappa} R_{\theta_i}(\bar{x}) \leftrightarrow R_{\bigwedge_i \theta_i}(\bar{x})\right)$$

into pieces as follows: $\forall \bar{x}(R_{\bigwedge_i \theta_i}(\bar{x}) \to R_{\theta_j}(\bar{x}))$ (for each $j < \kappa$) and $\forall \bar{x}(\bigvee_{i<\kappa} R_{\neg \theta_i}(\bar{x}) \vee R_{\bigwedge_i \theta_i}(\bar{x}))$. $\square$

See Corollary 3.1.6 and sections 11.5 and 11.6 for applications of Chang's reduction.

Exercises for section 2.6

We can eliminate function symbols in favour of relation symbols:

1. Let L be a signature. Form a signature L^r from L as follows: for each positive n and each n-ary function symbol F of L, introduce an $(n+1)$-ary relation symbol R_F. If A is an L-structure, let A^r be the L^r-structure got from A by interpreting each R_F as the relation $\{(\bar{a},b): A \vDash (F\bar{a}=b)\}$ (the **graph** of the function F^A). (a) Define a translation $\phi \mapsto \phi^r$ from formulas of L to formulas of L^r, which is independent of A. Formulate and prove a theorem about this translation and the structures A, A^r. (b) Extend (a) so as to translate every formula of L into a formula which contains no function symbols and no constants.

In fact we can eliminate relation symbols in favour of function symbols too, provided that the structures have at least two elements. Unfortunately this generally alters the class of definable relations, so I exclude it from this section. See this device in use at Lemma 10.4.8 below.

A formula ϕ is said to be **negation normal** *if in ϕ the symbol $\neg$ never occurs except immediately in front of an atomic formula. (Recall that $\psi \to \chi$ is an abbreviation for $\neg \psi \lor \chi$.)*

2. Show that if L is a first-order language, then every formula $\phi(\bar{x})$ of L is logically equivalent to a negation normal formula $\phi^*(\bar{x})$ of L. (Your proof should adapt at once to show the same for $L_{\infty\omega}$ in place of L.) Show that if ϕ was unnested then ϕ^* can be chosen unnested.

3. Let L be a first-order language, R a relation symbol of L and ϕ a formula of L. Show that the following are equivalent. (a) R is positive in some formula of L which is logically equivalent to ϕ. (b) ϕ is logically equivalent to a formula of L in negation normal form in which R never has $\neg$ immediately before it.

Here is a more perverse rewriting, which depends on the properties of a particular theory.

4. Let T be the theory of linear orderings. For each positive integer n, write a first-order sentence which expresses (modulo T) 'There are at least n elements', and which uses only two variables, x and y.

5. Let T_0, T_1 and T_2 be first-order theories. Show that if T_2 is a definitional extension of T_1 and T_1 is a definitional extension of T_0, then T_2 is a definitional extension of T_0.

The method of the following exercise is known as **Padoa's method**. *It is not limited to first-order languages. Compare it with Lemma 2.1.1 above. See also the discussion after Theorem 6.6.4.*

6. Let L and L^+ be signatures with $L \subseteq L^+$; let T be a theory of signature L^+ and S a symbol of signature L^+. Suppose that there are two models A, B of T^+ such that $A|L = B|L$ but $S^A \neq S^B$. Deduce that S is not explicitly definable in T in terms of L (and hence T is not a definitional extension of any theory of signature L).

7. Let L^+ be the first-order language of arithmetic with symbols $0, 1, +, \cdot$, and let T be the complete theory of the natural numbers in this language. Let L be the language L^+ with the symbol $+$ removed. Show that $+$ is not explicitly definable in T in terms of L.

*Let L and L^+ be first-order languages with $L \subseteq L^+$, and let T, T^+ be theories in L, L^+ respectively. We say that T^+ is a **conservative extension** of T if for every sentence ϕ of L, $T \vdash \phi \Leftrightarrow T^+ \vdash \phi$.*

*8. (a) Show that if every L-structure which is a model of T can be expanded to form a model of T^+, then T^+ is a conservative extension of T. In particular every definitional extension is conservative. (b) Prove that the converse of (a) fails. [Let T say that $<$ is a linear ordering with first element 0, every element has an immediate successor and every element except 0 has an immediate predecessor. Let T^+ be Peano arithmetic. Show that T is complete in its language, so that T^+ is a conservative extension of T. Show that every countable model of T^+ has order-type either ω or $\omega + (\omega^* + \omega) \cdot \eta$ where ω^* is the reverse of ω and η is the order-type of the rationals; so not every countable model of T expands to a model of T^+.]

Even the definition of addition (a recursive definition, not an explicit one) can lead to new first-order consequences.

*9. Let L be the language with constant symbol 0 and 1-ary function symbol S; let L^+ be L with a 2-ary function symbol $+$ added. Let T^+ be the theory $\forall x \, x + 0 = x$, $\forall xy \, x + Sy = S(x + y)$. Show that T^+ is not a conservative extension of the empty theory in L.

*10. Show that the theory of boolean algebras is definitionally equivalent to the theory of commutative rings with $\forall x \, x + x = 0$.

*11. Let L be the signature with a 3-ary relation symbol *between* and a 4-ary relation symbol *congruent*. We make the real numbers into an L-structure $\mathbb{R}$ by reading *between* (x, y, z) as 'y lies strictly between x and z', and *congruent* (x, y, z, t) as 'the distance $|x - y|$ is equal to the distance $|z - t|$'. Show that there is an automorphism of $\mathbb{R}$ which changes the interpretation of *between* but keeps the interpretation of *congruent* fixed (so that betweenness is not definable in terms of congruence).

2.7 Quantifier elimination

The first systematic programme for model theory appeared in the decade after the first world war. This programme is known as **elimination of quantifiers**. Let me summarise it.

Take a first-order language L and a class $\mathbf{K}$ of L-structures. The class $\mathbf{K}$ might be, for example, the class of all dense linear orderings, or it might be the singleton $\{\mathbb{R}\}$ where $\mathbb{R}$ is the field of real numbers. We say that a set Φ of formulas of L is an **elimination set** for $\mathbf{K}$ if

(7.1) for every formula $\phi(\bar{x})$ of L there is a formula $\phi^*(\bar{x})$ which is a boolean combination of formulas in Φ, and ϕ is equivalent to ϕ^* in every structure in $\mathbf{K}$.

The programme can be stated briefly: given $\mathbf{K}$, *find an elimination set for* $\mathbf{K}$. There are analogous programmes for other languages, but the first-order case is the most interesting.

Of course there always is at least one elimination set Φ for any class $\mathbf{K}$ of L-structures: take Φ to be the set of all formulas of L. But with care and attention we can often find a much more revealing elimination set than this.

For example, here are two results which we owe to Tarski's Warsaw seminar in the late 1920s. (A linear ordering is **dense** if for all elements $x < y$ there is z such that $x < z < y$; cf. (2.31) and (2.32) in section 2.2.)

Theorem 2.7.1. *Let L be the first-order language whose signature consists of the 2-ary relation symbol $<$, and let $\mathbf{K}$ be the class of all dense linear orderings. Let Φ consist of formulas of L which express each of the following.*

(7.2) *There is a first element.*
       There is a last element.
       x is the first element.
       x is the last element.
       $x < y$.

Then Φ is an elimination set for $\mathbf{K}$.

Proof. This will be Exercise 1 below. ☐

Theorem 2.7.2. *Let L be the first-order language of rings, whose symbols are $+, -, \cdot, 0, 1$. Let $\mathbf{K}$ be the class of real-closed fields. Let Φ consist of the formulas*

(7.3) $$\exists y \, y^2 = t(x)$$

where t ranges over all terms of L not containing the variable y. Then Φ is an elimination set for $\mathbf{K}$. (Note that (7.3) expresses $t(x) \geqslant 0$.)

Proof. We shall see an algebraic proof of this in Theorem 8.4.4 below. ☐

The name 'quantifier elimination' refers either to the process of reducing a formula to a boolean combination of formulas in Φ, or to the process of discovering the appropriate set Φ in the first place. One should distinguish the method of quantifier elimination from the *property of quantifier elimination*, which is a property that some theories have. A theory T **has quantifier elimination** if the set of quantifier-free formulas forms an elimination set for the class of all models of T. (Cf. section 8.4 below, and note that some of the formulas in (7.2) and (7.3) are not quantifier-free.)

The point of quantifier elimination

Suppose we have an elimination set Φ for the class **K**. What does it tell us?

(a) *Classification up to elementary equivalence.* Suppose A and B are structures in the class **K**, and A is not elementarily equivalent to B. Then there is some boolean combination of sentences in Φ which is true in A but false in B. It follows at once that some sentence ϕ in Φ is true in one of A and B but false in the other. The conclusion is that we can classify the structures in **K**, up to elementary equivalence, by looking to see which *sentences* in Φ are true in these structures.

If the sentences in Φ each express some 'algebraic' property of structures (a vague notion this, but sharp enough to be helpful), then we have reduced elementary equivalence in **K** to a purely 'algebraic' notion. For example Tarski showed that two algebraically closed fields are elementarily equivalent if and only if they have the same characteristic.

(b) *Completeness proofs.* As a special case of (a), suppose that **K** is the class $\mathrm{Mod}(T)$ of all models of some first-order theory T. Suppose that all the sentences in Φ are either deducible from T or inconsistent with T. Then it follows that all models of T are elementarily equivalent, and so T is a complete theory.

Theorem 2.7.2 is a case in point. The sentences in Φ can all be written as $s = t$ or $s \leqslant t$ where s, t are closed terms of L. But every real-closed field has characteristic 0. In fields of characteristic 0, each closed term t has an integer value independent of the choice of the field, and so we can prove or refute the sentences $s = t, s \leqslant t$ from the axioms for real-closed fields. Thus Theorem 2.7.2 shows that the theory of real-closed fields is complete.

(c) *Decidability proofs.* This is a special case of (b) in turn. Suppose L is a recursive language (see section 2.3 above). The theory T in L is decidable if and only if there is an algorithm to determine whether any given sentence of L is a consequence of T. The **decision problem** for a theory T in L is the problem of finding such an algorithm (or showing that there isn't one).

Now suppose that **K** is $\mathrm{Mod}(T)$ and the map $\phi \mapsto \phi^*$ in (7.1) is recursive. Suppose also that we have an algorithm which tells us, for any sentence ψ in the elimination set Φ, either that ψ is provable from T or that it is refutable from T. Then by putting everything together, we derive an algorithm for determining which sentences are consequences of T; this is a positive solution of the decision problem for T. Again real-closed fields are a case in point.

(d) *Description of definable relations.* Suppose Φ is an elimination set for **K** and A is a structure in **K**. Let D be the set of all relations on A which have the form $\psi(A^n)$ for some formula $\psi(x_0, \ldots, x_{n-1})$ in Φ. Then the $\varnothing$-definable relations on A are precisely the boolean combinations of relations in D.

(e) *Description of elementary embeddings.* If Φ is an elimination set for **K**, then the elementary maps between structures in **K** are precisely those homomorphisms which preserve ψ and $\neg\psi$ for every formula ψ in Φ. For example, by Theorem 2.7.1, every embedding between dense linear orderings with no endpoints is elementary.

Points (a), (d) and (e) were vital for the future of model theory. What they said was that in certain important classes of structure, the natural model-theoretic classifications could be paraphrased into straightforward algebraic notions. This allowed logicians and algebraists to talk to each other and merge their methods.

The main snag of quantifier elimination is that the method proceeds entirely on the level of deducibility from a set of axioms. This makes it heavily syntactic, and it can prevent us using good algebraic information about the class **K**. In particular the method doesn't allow us to exploit anything we know about maps between structures in **K**. For example, to prove the following result of Tarski by the method of quantifier elimination we would need to undertake some rather heavy study of equations; but the more structural argument of Example 2 in section 8.4 below makes it almost a triviality.

Theorem 2.7.3. *The theory of algebraically closed fields has quantifier elimination.* $\qquad\qquad\qquad\qquad\qquad\qquad\qquad\qquad\qquad\qquad\qquad\qquad\qquad\qquad\square$

For this reason the example done in detail below is not one of the well-known algebraic results from Tarski's school. Most of those can be handled by smoother methods today. Instead I choose an example where the structures are syntactic objects themselves, so that the method meshes well with the problem.

But first a brief word on strategy. We have a first-order language L and a class **K** of L-structures. We also have a theory T which is a candidate for an axiomatisation of **K**, and a set of formulas Φ which is a candidate for an elimination set. If **K** is defined as $\mathrm{Mod}(T)$, then of course T does axiomatise **K**. But if **K** was given and T is a guess at an axiomisation, we may find during the course of the quantifier elimination that we have to adjust T.

The following straightforward lemma very much eases the burden of showing that Φ is an elimination set. We write Φ^- for the set $\{\neg\phi: \phi \in \Phi\}$.

Lemma 2.7.4. *Suppose that*

(7.4) *every atomic formula of L is in Φ, and*

(7.5) *for every formula $\theta(\bar{x})$ of L which is of form $\exists y \bigwedge_{i<n} \psi_i(\bar{x}, y)$*
       with each ψ_i in $\Phi \cup \Phi^-$, there is a formula $\theta^*(\bar{x})$ of L which
       (i)   is a boolean combination of formulas in Φ, and
       (ii) is equivalent to θ in every structure in K.

Then Φ is an elimination set for K.

Proof. See Lemma 2.3.1. $\square$

So to find an elimination set, we must discover a way of getting rid of the quantifier $\exists y$ in (7.5). Hence the name 'quantifier elimination'. As Lemma 2.7.4 suggests, we start with an arbitrary finite subset $\Theta(y, \bar{x})$ of $\Phi \cup \Phi^-$, and we aim to find a boolean combination $\psi(\bar{x})$ of formulas in Φ so that $\exists y \bigwedge \Theta$ is equivalent to ψ modulo T. Typically the move from Θ to ψ takes several steps, depending on what kinds of formula appear in Θ. If we run into a dead end, we can add sentences to T and formulas to Φ until the process moves again.

Example: term algebras

We consider the class **K** of term algebras of an algebraic signature L_1: see Example 1 in section 2.2 and Example 1 of section 2.6. The theory T_1, which consists of the sentences (2.3)–(2.7) of section 2.2, is true in every algebra in **K**. Our elimination will be easier to carry out if we switch to the language L_2 and the theory T_2 of section 2.6. Since T_2 is definitionally equivalent to T_1, everything can be translated back into the language of T_1 if needed.

If L_1 (and hence L_2) has finite signature, we can write for each positive integer k a sentence α_k of L_2 which says 'There are at least k elements satisfying all $\neg \mathrm{Is}_c$ and $\neg \mathrm{Is}_F$'. Let β be the sentence $\exists x\, x = x$.

Theorem 2.7.5. *Let K be the class of term algebras in the signature L_2 described above. Let Φ be the set of atomic formulas of L_2, together with the sentences α_k if L_1 has finite signature, and the sentence β if L_1 has no constants. Then Φ is an elimination set for the class of all models of T_2, and hence for K.*

Proof. Our task is the following. We have a finite set $\Theta_0(\bar{x}, y)$ which consists of literals of L_2 (and possibly some sentences β, α_k or their negations), and we must eliminate the quantifier $\exists y$ from the formula $\exists y \bigwedge \Theta_0$. We can assume without loss that

(7.6)   No sentence α_k or β or its negation is in Θ_0.

Reason: The variable y is not free in α_k, and so $\exists y (\alpha_k \wedge \psi)$ is logically equivalent to $\alpha_k \wedge \exists y \, \psi$. Likewise with β.

Also we can assume without loss the following:

(7.7)   there is no formula ψ such that both ψ and $\neg \psi$ are in Θ_0; moreover $y \neq y$ is not in Θ_0.

Reason: Otherwise $\exists y \bigwedge \Theta_0$ reduces at once to $\bot$.

We can replace Θ_0 by a set Θ_1 which satisfies (7.6), (7.7) and

(7.8)   if t is a term of L_2 in which y doesn't occur, then the formulas $y = t$ and $t = y$ are not in Θ_1.

Reason: $\exists y (y = t \wedge \psi(y, \bar{x}))$ is logically equivalent to $\exists y (y = t \wedge \psi(t, \bar{x}))$, hence to $\psi(t, \bar{x}) \wedge \exists y \, y = t$, or equivalently $\psi(t, \bar{x})$. $\exists y \, y = t$ is equivalent to $t = t$.

We can replace Θ_1 by one or more sets Θ_2 which satisfy (7.6)–(7.8) and

(7.9)   the variable y never occurs inside another term.

Reason: Suppose a term $s(F_i(y))$ appears somewhere in Θ_1. Now $\exists y \, \psi$ is equivalent to $\exists y (\mathrm{Is}_F(y) \wedge \psi) \vee \exists y (\neg \mathrm{Is}_F(y) \wedge \psi)$, so we can suppose that exactly one of $\mathrm{Is}_F(y)$ and $\neg \mathrm{Is}_F(y)$ appears in Θ_1. If $\neg \mathrm{Is}_F(y)$ appears, we can replace $F_i(y)$ by y according to axiom (6.17). If $\mathrm{Is}_F(y)$ appears and F is n-ary, we make the following changes. First if G is any function symbol of L_1 distinct from F, we replace any expression $G_j(y)$ in Θ_1 by y (again quoting (6.17)). We introduce n new variables $y_0, \ldots, y_{n-1}$, we replace each $F_j(y)$ by y_j and we add the formulas $F_j(y) = y_j$. Then $\exists y \bigwedge \Theta_1$ becomes equivalent to an expression $\exists y_0 \ldots y_{n-1} \exists y (\mathrm{Is}_F(y) \wedge \bigwedge_{j < n} F_j(y) = y_j \wedge \bigwedge \Theta_2)$ where Θ_2 satisfies (7.9). By (6.15) this reduces to $\exists y_0 \ldots y_{n-1} \bigwedge \Theta_2$. Here Θ_2 has more variables y_j to dispose of, but these all occur in shorter terms than those involving y in Θ_1. Hence we can deal with the variables $y_{n-1}, \ldots, y_0$ in turn, using an induction on the lengths of terms.

At this point Θ_2 consists of formulas of the forms $y \neq t$ or $t \neq y$ (where y doesn't occur in t), $y = y$, $\mathrm{Is}_c(y)$, $\neg \mathrm{Is}_c(y)$, $\mathrm{Is}_F(y)$ or $\neg \mathrm{Is}_F(y)$; as in (7.6), we can eliminate any literals in which y doesn't appear. We can replace Θ_2 by a set Θ_3 satisfying (7.6)–(7.9) and

(7.10)   there is no constant c such that $\mathrm{Is}_c(y)$ is in Θ_3, and no function symbol F such that $\mathrm{Is}_F(y)$ is in Θ_3.

Reason: $\exists y (\mathrm{Is}_c(y) \wedge y \neq t \wedge \neg \mathrm{Is}_F(y))$, say, is equivalent to $\neg \mathrm{Is}_c(t)$ by

(6.14) and (6.16); $\exists y \, \mathrm{Is}_c(y)$ is equivalent to $\neg\bot$ by (6.14). Function symbols need a more complicated argument. Let F be an n-ary function symbol.

We claim T_2 implies that for any k elements ($k > 0$) there is an element distinct from all of them which satisfies $\mathrm{Is}_F(y)$. By (6.15), T_2 implies that if $\mathrm{Is}_F(x_0)$ then there is a unique x_1 such that $\mathrm{Is}_F(x_1)$ and $F_i(x_1) = x_0$ for all $i < n$; by (6.18), $x_0 \neq x_1$. Likewise by (6.15) there is x_2 such that $\mathrm{Is}_F(x_2)$ and $F_i(x_2) = x_1$ for all $i < n$, and then by (6.18) again, $x_2 \neq x_1$ and $x_2 \neq x_0$. Etc. etc.; this proves the claim. With (6.16), the claim allows the reduction to (7.10), unless the problem is to eliminate the quantifier from $\exists y \, \mathrm{Is}_F(y)$. When L_1 has at least one constant c, the formula $\exists y \, \mathrm{Is}_F(y)$ reduces to $\neg\bot$ by (6.15); but in general it is equivalent to β.

We are almost home. When L_1 has infinite signature, $\exists y \bigwedge \Theta_3$ reduces to β by (6.16). There remains only the case where L_1 has finitely many symbols. As at the start of the reason for (7.9), we can suppose that for each symbol S of L_1, Θ_3 contains either $\mathrm{Is}_S(y)$ or $\neg \mathrm{Is}_S(y)$, and we have seen how to deal with the first of these formulas. So assume henceforth that Θ_3 contains $\neg \mathrm{Is}_S(y)$ for each symbol S of L_1. By the same reasoning we can suppose that for each term t appearing in Θ_3 (even inside other terms), one of the formulas $\mathrm{Is}_S(t)$ and $\neg \mathrm{Is}_S(t)$ is in Θ_3. Also we can suppose that for each pair of terms s, t appearing in Θ_3, either $s = t$ or $s \neq t$ also appears in Θ_3. Now $\bigwedge \Theta_3$ asserts (among other things) that there are at least k distinct items, including y, which satisfy $\neg \mathrm{Is}_S(x)$ for all S. Such an element y can be found if and only if α_k holds; so $\exists y \bigwedge \Theta_3$ reduces to a conjunction of α_k and the formulas in Θ_3 which don't mention y.   □

Notice that if we had overlooked the formulas α_k or β, or one of the axioms that should have been in T_2, then this procedure would have shown up our mistake and suggested how to correct it.

Corollary 2.7.6. *Let* **K** *be (as above) the class of all term algebras of L_1, regarded as L_2-structures. If L_1 has at least one constant symbol, or no symbols at all, then* $\mathrm{Th}(\mathbf{K})$ *is equivalent to T_2. If L_1 has function symbols but no constant symbols then* $\mathrm{Th}(\mathbf{K})$ *is equivalent to $T_2 \cup \{\beta \to \alpha_1\}$.*

Proof. Certainly every sentence of T_2 is in $\mathrm{Th}(\mathbf{K})$. In the other direction, the harder case is where L_1 is finite. Let ϕ be any sentence in $\mathrm{Th}(\mathbf{K})$. By the theorem, ϕ is equivalent modulo T_2 to a boolean combination of sentences β, α_k. (The signature L_2 has no closed terms.) By logic, each α_{k+1} entails α_k and α_1 entails β. If L_1 has at least one constant symbol then β is provable from T_2 but there are no other implications between β and the α_k; so in this case ϕ must be provable from T_2.

If L_1 has function symbols but no constant symbol then the term algebra is empty unless α_1 holds; so $\beta \to \alpha_1$ lies in Th($\mathbf{K}$). It is not a consequence of T_2, since we can construct a model of T_2 in which α_1 fails, by taking an infinitely decomposable 'term' as in (2.10) of section 2.2. Examples show that no other implications hold between β and the α_k. I leave to the reader the case where L_1 is empty. ▢

Corollary 2.7.7. *The theory of term algebras of a given finite algebraic signature, in either the language L_1 or the language L_2 above, is decidable.*

Proof. Any sentence of L_1 can be translated effectively into a sentence ϕ of L_2 by the explicit definitions (6.20) of section 2.6. Then we can compute a sentence ϕ^* of L_2 which is equivalent to ϕ modulo T_2 and is a boolean combination of sentences in Φ, where Φ is as in Theorem 2.7.5. The argument for the previous corollary shows that we can effectively check whether ϕ^* is a consequence of T_2 or $T_2 \cup \{\beta \to \alpha_1\}$ as appropriate. ▢

Exercises for section 2.7

For these exercises, be advised that the method of quantifier elimination is not intrinsically hard, but it does use up hours and paper.

1. Prove Theorem 2.7.1.

2. Let the signature L consist of finitely many 1-ary relation symbols $R_0, \ldots, R_{n-1}$. For each map $s: n \to 2$ let $\phi^s(x)$ be the conjunction $R_0^{s(0)}(x) \wedge \ldots \wedge R_{n-1}{}^{s(n-1)}(x)$, where R_i^j is R_i if $j = 1$ and $\neg R_i$ if $j = 0$. If $\mathbf{K}$ is the class of all L-structures, show that an elimination set for $\mathbf{K}$ is given by the formulas $\phi^s(x)$ and the sentences $\exists_{=k} x \, \phi^s(x)$ where $s: n \to 2$ and $k < \omega$.

3. Let L be the first-order language of boolean algebras (see (2.29) in section 2.2). Let Ω be a set, and let A be the power-set algebra of Ω, regarded as an L-structure with $\wedge$ for $\cap$, $\vee$ for $\cup$, $*$ for complement in Ω, 0 for $\varnothing$ and 1 for Ω. Let $\mathbf{K}$ be $\{A\}$. For each positive integer k, write $\alpha_k(y)$ for the formula 'y has at least k elements'. (This can be written as 'There are at least k atoms $\leq y$', where an **atom** of a boolean algebra is an element $b > 0$ such that there is no element c with $b > c > 0$.) Let Φ be the set of all atomic formulas of L and all formulas of the form $\alpha_k(t)$ where t is a term of L. Show that Φ is an elimination set for $\mathbf{K}$.

4. A boolean algebra B is said to be **atomic** if the supremum of the set of atoms in B is the top element of B. Let T be the theory of the class of atomic boolean algebras. Show (a) a boolean algebra B is atomic if and only if for every element $b > 0$ there is an atom $a \leq b$, (b) if B is an atomic boolean algebra then every formula $\phi(\bar{x})$ of the first-order language of boolean algebras is equivalent in B to a boolean combination of formulas which say exactly how many atoms are below elements $y_0 \wedge \ldots \wedge y_{n-1}$,

where each y_i is either x_i or x_i^* (complement), (c) T is the theory of the class of finite boolean algebras, (d) T is decidable.

5. Let **K** be the class of boolean algebras B such that the set of atoms of B has a supremum in B. (a) Show that **K** is a first-order axiomatisable class. (b) Use the method of quantifier elimination to show that up to elementary equivalence there are exactly ω boolean algebras in **K**; describe these algebras.

6. Let L be the signature consisting of one 2-ary relation symbol $<$. Let **K** be the class of L-structures $(X, <')$ where X is a non-empty set and $<'$ is a linear ordering of X in which every element has an immediate predecessor and an immediate successor. (a) Use the method of quantifier elimination to show that any two structures in **K** are elementarily equivalent. (b) Give necessary and sufficient conditions for a map from A to B to be an elementary embedding, where A and B are in **K**. (c) Show that for every infinite cardinal λ there are 2^λ non-isomorphic structures in **K** with cardinality λ. [Construct models $\Sigma_{i<\lambda}((\omega^* + \omega) \cdot \rho_i)$, where each ρ_i is either ω or $(\omega^* + \omega)$; show that each ρ_i is recoverable from the model.]

We shall handle the next result differently in section 3.3.
7. Let the signature L consist of constants 0, 1 and a 2-ary function symbol $+$. Let **K** consist of one structure, namely the natural numbers $\mathbb{N}$ regarded as an L-structure in the obvious way. Using the method of elimination of quantifiers, find a set of axioms for Th $(\mathbb{N})$ and show that Th $(\mathbb{N})$ is a decidable theory. [One elimination set consists of equations and formulas which express '$t(\bar{x})$ is divisible by n' where n is a positive integer.]

8. Let the signature L consist of one 1-ary function symbol F and one constant symbol 0. Let **K** be the class of L-structures which obey the second-order induction axiom, namely, for every set X of elements, $((0 \in X \wedge \forall y(y \in X \to F(y) \in X)) \to \exists y(y \in X))$. Use the method of quantifier elimination to find (a) a set of axioms for the first-order theory Th $(\mathbf{K})$ of **K**, and (b) a classification of the models of **K**, up to elementary equivalence.

9. Let K be a field; let **J** be the class of (left) vector spaces over K, in the language of left K-modules (i.e. with symbols $+, -, 0$ and for each scalar r a 1-ary function symbol $r(x)$ to represent multiplication of a vector by r; see Example 4 in section 1.1 above). (a) Show that the set of linear equations $r_0(x_0) + \ldots + r_{n-1}(x_{n-1}) = 0$, with $n < \omega$ and $r_0, \ldots, r_{n-1}$ scalars, together with the set of sentences $\exists_{=k} x\, x = x$ $(k < \omega)$, is an elimination set for **J**. (b) Deduce that every infinite vector space in **J** is a minimal structure (in the sense of section 2.1).

*An abelian group is **divisible** if for every non-zero element b and every positive integer n there is an element c such that $nc = b$. It is **ordered** if it carries a linear ordering relation $<$ such that $a < b$ implies $a + c < b + c$ for all a, b, c.*
10. Use the method of quantifier elimination to axiomatise the class of non-trivial

divisible ordered abelian groups. Show (a) all groups in this class are elementarily equivalent, (b) if A and B are non-trivial divisible ordered abelian groups and A is a subgroup of B then A is an elementary substructure of B.

11. Let T_1 be the theory of term algebras of a given algebraic signature L. (a) Show that if A is any finitely generated model of T_1, then A is isomorphic to a term algebra of L. (b) Show that if B is an L-structure then B is a model of T_1 if and only if every finitely generated substructure of B is isomorphic to a term algebra of L. (*Hence T_1 is known as the* **theory of locally free L-structures**.)

2.8 Further examples

There is nothing sacrosanct about the logic $L_{\infty\omega}$. It just happens to contain most of the formulas that we need in this book. Read on to see some other logics which have found uses.

In section 2.2 we saw that the standard definitions of many classes of structure can be written down straight away as axioms. At the end of this section I briefly discuss a few axiomatisable classes where one has to do some work in order to see that the axioms are correct.

Second-order formulas

In the days before infinitary languages came into common use, 'first-order' meant 'not higher-order'. A **higher-order** language is one that has quantifiers which range over sets of elements, or sets of sets of elements, or sets of sets of . . . , etc. A typical example of a higher-order language is the **second-order language** L^{II} of a given signature L. This language is like the first-order language $L_{\omega\omega}$, except that (a) for each positive integer i it has i-**ary relation variables** $X_0^i, X_1^i, \ldots$, which can sit in the same places in a formula as i-ary relation symbols, and (b) if ϕ is a formula then so are $\forall X \phi$ and $\exists X \phi$, where X is any relation variable.

The quantifiers $\forall X, \exists X$ are known as **second-order quantifiers** to distinguish them from the **first-order quantifiers** $\forall x, \exists x$ ranging over elements. The second-order quantifiers are interpreted thus.

(8.1) $A \vDash \forall X_j^i \phi(\bar{a})$ iff   for every i-ary relation R on $\mathrm{dom}(A)$, $(A, R) \vDash \phi(\bar{a})$ where X_j^i is read as a name of R.

(8.2) $A \vDash \exists X_j^i \phi(\bar{a})$ iff   there is an i-ary relation R on $\mathrm{dom}(A)$ such that $(A, R) \vDash \phi(\bar{a})$ where X_j^i is read as a name of R.

To save our eyesight and the printer's patience, we usually leave out the superscripts ^{i}, since they can always be read off from the formula. Thus second-order variables are X, Y, Z, X_0, X_1, etc.

For example the sentence $\mathrm{Geq}(Q, P)$, viz.

(8.3) $\exists X(\forall x(Px \to \exists z(Xxz \land Qz)) \land \forall xyz(Xxz \land Xyz \to x = y))$,

expresses that the number of elements satisfying Qx is at least as great as the number satisfying Px. Hence the sentence

(8.4) $\exists Y(\forall x(Yx \to Px) \land \exists x(Px \land \neg Yx) \land \mathrm{Geq}(Y, P))$

says that infinitely many elements satisfy Px.

The language L^{II} is very expressive, so that none of the usual algebraic constructions preserve all formulas of L^{II}. This cuts out much of the scope for a model theory of L^{II}. But still there are things to be done. In the first place, we can study the relations definable by formulas of L^{II} on particular structures that interest us. And secondly, we can pull in our horns and look at subsets of the full logic L^{II}.

Example 1: *Monadic second-order logic*, L^{mon}, is the same as L^{II} except that the second-order variables range only over sets and not over relations of higher arity. In monadic second-order logic we can express that a linear ordering is well-ordered:

(8.5) $\forall Z(\exists x\, Zx \to \exists x(Zx \land \forall y(Zy \to x \leqslant y)))$.

We can also quantify over the ideals or the subrings of a ring.

Much of the interest of L^{mon} revolves around a striking theorem of Rabin [1969]. Let L be the signature with one constant 0 and two 1-ary function symbols F and G. The term algebra of L with empty basis is called the **tree of two successor functions**.

Theorem 2.8.1. *The L^{mon}-theory of the tree of two successor functions is decidable.*

The proof uses automata. Several other important decidability results can be reduced to this theorem. See Gurevich [1985] for a survey of monadic second-order theories.

Example 2: *Weak second-order logic*, $L_{\mathrm{w}}^{\mathrm{II}}$. This logic has quantifiers ranging over finite sequences of elements, with symbols to express concatenation of sequences and membership of sequences. A simpler variant is **weak monadic second-order logic**, $L_{\mathrm{w}}^{\mathrm{mon}}$. This is the same as L^{mon}, except that the quantifiers are restricted to range over finite subsets of the domain. Using (8.4) above, everything that can be said in $L_{\mathrm{w}}^{\mathrm{mon}}$ can also be said in L^{II}. But $L_{\mathrm{w}}^{\mathrm{mon}}$ is much weaker; for example we shall see in section 11.5 that it can't define well-orderings. In $L_{\mathrm{w}}^{\mathrm{mon}}$ one can express that an element x of a group has finite order:

(8.6) $\qquad \exists W(Wx \wedge \forall yz(Wy \wedge Wz \rightarrow W(yz)))$.

The compactness theorem (Theorem 6.1.1) will show at once that this is not expressible by a first-order formula. (Cf. Exercise 6.3.12.)

All these logics fall into a common framework, which can be described loosely as follows. Let Γ be some operation which can be applied to all L-structures in a uniform way, so that $\Gamma(A)$ is A with a set of new elements and relations added. (For example let $\Gamma^{\mathrm{mon}}(A)$ be a structure consisting of A together with all subsets of $\mathrm{dom}\,(A)$ and the membership relation $\in$ between elements and subsets of A. Let $\Gamma_{\mathrm{w}}^{\mathrm{mon}}(A)$ be the same but with just the finite sets added.) Then we get a new logic which says about A exactly what can be said by *first-order formulas about* $\Gamma(A)$. Thus the L^{mon}-theory of A is essentially the same thing as the first-order theory of $\Gamma^{\mathrm{mon}}(A)$.

Fixed point logics

We get a very different extension of first-order logic if we ask to express transitive closures of relations. The impetus for this extension came from computer science.

Let L be a first-order language, $n < \omega$ and P an n-ary relation symbol not in L. Write L^+ for L with P (and possibly other symbols) added. Then every L^+-structure B can be written as $(B|L, P^B, \ldots)$. Let $\phi(x_0, \ldots, x_{n-1})$ be a formula of L^+ in which P is positive (see section 2.4 above). For any L-structure A we define a map $\pi_{P,\phi} = \pi$ from $\mathscr{P}((\mathrm{dom}\,A)^n)$ to $\mathscr{P}((\mathrm{dom}\,A)^n)$ as follows:

(8.7) $\qquad$ if $X \subseteq (\mathrm{dom}\,A)^n$ then $\pi(X) = \{\bar{a} : (A, X, \ldots) \vDash \phi(\bar{a})\}$.

By Exercise 2.4.15, π is a **monotone** operator, i.e. $X \subseteq Y$ implies $\pi(X) \subseteq \pi(Y)$. So by Exercise 2.4.16, π has a least fixed point $X = \pi(X)$; we write this least fixed point $\mathrm{lfp}\,(\pi)$.

If α is an ordinal, we define π^α as follows: $\pi^0(X) = X$, $\pi^{\alpha+1}(X) = \pi(\pi^\alpha(X))$, and when α is a limit ordinal, $\pi^\alpha(X) = \bigcup_{\beta<\alpha} \alpha^\beta(X)$. One can check that the least fixed point of π is $\pi^\gamma(\varnothing)$ where γ is the least ordinal such that $\pi^\gamma(\varnothing) = \pi^{\gamma+1}(\varnothing)$. We saw in Exercise 2.4.17 that when ϕ is an $\exists_1$ first-order formula, this ordinal is at most ω, so that $\mathrm{lfp}\,(\pi) = \pi^\omega$. In general the ordinal can be much bigger than ω (see Exercise 11).

We define a language L^{FP} as follows. For each positive integer n we choose an n-ary relation symbol P_n which is not in L; let Π be the set of these new symbols. The formulas of L^{FP} are the same as those of L, except that (a) the symbols in Π can occur as (unquantified) relation variables, and (b) if P is any n-ary relation symbol in Π and $\phi(x_0, \ldots, x_{n-1}, \bar{y})$ a formula of L^P in which P is positive, then the expression

$$(8.8) \qquad\qquad \text{LFP}_{P,\bar{x}}\,\phi$$

is a formula $\psi(\bar{x}, \bar{y})$ of L^P. This formula ψ is interpreted as follows. Let A be an L-structure and $\bar{a}$, $\bar{c}$ tuples of elements of A. Then

$$(8.9) \quad A \vDash \text{LFP}_{P,\bar{x}}\,\phi(\bar{a}, \bar{c}) \qquad \text{iff} \qquad \bar{a} \in \text{lfp}\,(\pi) \text{ where } \pi = \pi_{P,\phi(\bar{x},\bar{c})}.$$

The expression $\text{LFP}_{P,\bar{x}}$ binds all occurrences of P in $\text{LFP}_{P,\bar{x}}\phi$.

For example let L be the signature of graphs, and consider the formula $\phi(x, y)$ of L^P where P is a new 1-ary relation symbol,

$$(8.10) \qquad\qquad x = y \vee \exists z(Pz \wedge Rzx).$$

Then if G is a graph and g an element of G, we have

$$(8.11) \quad \pi^m_{P,\phi(x,g)}(\varnothing) = \{a\colon a \text{ can be joined to } g \text{ by a path with } m - 1 \text{ edges}\}.$$

So for any infinite α, $\pi^\alpha_{P,\phi(x,g)}(\varnothing)$ is the connected component of g in G, and hence

$$(8.12) \qquad G \vDash (\text{LFP}_{P,x}\,\phi)(a, g) \quad \text{iff} \quad a \text{ is in the connected component of } G.$$

(This illustrates Exercise 2.4.17: ϕ is an $\exists_1$ formula and so the least fixed point is $\pi^\omega(\varnothing)$.) We saw in section 2.7 that no first-order formula will serve for (8.12).

The model theory of L^{FP} is not yet well developed. But given the interest of computer scientists, this is surely only a matter of time.

Examples on axiomatisability

We return to ordinary first-order logic. Some classes of structure are defined by conditions that seem to call for a strong higher-order logic. For example in homological algebra one often meets conditions of the form 'Given such-and-such a diagram of structures involving A, there exists a homomorphism such that . . .'. Sometimes these conditions can be translated into a first-order theory, given a little work.

For example the usual definition of 'injective module' involves maps between modules and is far from being first-order. Nevertheless an abelian group is injective as $\mathbb{Z}$-module if and only if it is divisible, and the class of divisible abelian groups is straightforwardly first-order axiomatisable. (The implication from divisible to injective needs the axiom of choice; without choice there are models of set theory in which no non-trivial abelian groups are injective.)

To analyse this example, let R be a ring. Left R-modules will be regarded as L-structures for the signature L of (2.21) in section 2.2 above.

Suppose I is a left ideal of R and A is a left R-module. We say that A is *I*-**injective** if every homomorphism $f: I \to A$ extends to a homomorphism $g: R \to A$. (Here and below, f and g are module homomorphisms.) The module A is said to be **injective** if it is I-injective for every left ideal I of R. This is not exactly the usual definition of injective modules, but a well-known theorem of R. Baer says that it is equivalent to the usual definition. (See Cohn [1977] p. 107 or Jacobson [1980] Prop. 3.15.)

Suppose now that I is generated by a tuple $\bar{r} = (r_0, \dots, r_{n-1})$ of elements. Write $\Delta_I(x_0, \dots, x_{n-1})$ for the set of all equations '$t(\bar{x}) = 0$' such that $I \vDash t(\bar{r}) = 0$. The set Δ_I has two aspects. First, from the model-theoretic side, every atomic formula of L is equivalent (modulo the theory of left R-modules) to an equation of the form $t(\bar{y}) = 0$. So apart from having variables in place of parameters, Δ_I can be identified with the positive diagram diag$^+ I$, which we defined in section 1.4.

Second, thinking along algebraic lines, let $\bigoplus_{i<n} Re_i$ be the free left R-module on generators $e_0, \dots, e_{n-1}$, and let $h: \bigoplus_i Re_i \to I$ be the homomorphism which takes each e_i to r_i. The kernel of h is precisely the set of elements $\sum_i s_i e_i$ such that the equation '$\sum_i s_i x_i = 0$' is in Δ_I. We say that I is **finitely presented** if it is finitely generated, say by $\bar{r}$, and, for h as above, $\ker h$ is finitely generated. (The latter is independent of the choice of $\bar{r}$, see Bourbaki [1972] I.8 Lemma 9.) If $\ker h$ is finitely generated, let Γ_I be the finite set of equations in Δ_I which correspond to the generators of $\ker h$. Then Γ_I is equivalent to Δ_I modulo the theory of left R-modules.

Lemma 2.8.2. *If I is a left ideal of R generated by a tuple of elements $\bar{r}$ and A is a left R-module, then A is I-injective if and only if*

(8.13) $A \vDash \forall x_0 \dots x_{n-1}(\bigwedge \Delta_I(\bar{x}) \to \exists y (r_0 y = x_0 \wedge \dots \wedge r_{n-1} y = x_{n-1}))$,

where Δ_I is as above.

Proof. First suppose that A is I-injective. Let $\bar{a}$ be a tuple in A such that $A \vDash \bigwedge \Delta_I(\bar{a})$. Then by the diagram lemma (Lemma 1.4.2) there is a homomorphism $f: I \to A$ such that $f\bar{r} = \bar{a}$. Since A is I-injective, f extends to some $g: R \to A$. Let b be $g(1)$. Then each $r_i b$ is $r_i g(1) = g(r_i) = f(r_i) = a_i$. This proves (8.13). Conversely suppose (8.13) holds, and let $f: I \to A$ be a homomorphism. Put $\bar{a} = f\bar{r}$. Since homomorphisms preserve atomic formulas, $A \vDash \bigwedge \Delta_I(\bar{a})$. Hence according to (8.13) there is an element b of A such that $r_i b = a_i$ for each $i < n$. Then we get a homomorphism $g: R \to A$ extending f if we put $g(1) = b$.   □

The ring R is said to be **left coherent** if every finitely generated left ideal of R is finitely presented. If R is a left noetherian ring then for each $n < \omega$,

$\bigoplus_{i<n} Re_i$ is a noetherian R-module, so that R is left coherent. From Lemma 2.8.2 and the discussion above we quickly deduce the following.

Theorem 2.8.3. *Let R be a ring.*
(a) *If I is a finitely presented left ideal of R then the class of I-injective left R-modules is first-order definable.*
(b) *If R is left coherent then the class of left R-modules which are I-injective for all finitely generated left ideals I is first-order axiomatisable.*
(c) *If R is left noetherian then the class of injective left R-modules is first-order axiomatisable.* □

In section 9.5 we shall see that converses of all parts of Theorem 2.8.3 are true too.

Exercises for section 2.8

1. Let A be the tree of two successor functions F and G. We define an ordering $\leqslant$ of elements of A by $a \leqslant b$ iff $b = t(a)$ for some term $t(x)$ built up from F and G. We define an ordering $\preccurlyeq$ by $a \preccurlyeq b$ iff either $a \leqslant b$ or there is c with $F(c) \leqslant a$ and $G(c) \leqslant b$. (a) Show that both $\leqslant$ and $\preccurlyeq$ are definable in L^{mon}. (b) Show that $\preccurlyeq$ contains the order-type of the rationals. Deduce from Theorem 2.8.1 that the monadic second-order theory of the class of linear orderings of cardinality $\leqslant \omega$ is decidable. *By the downward Löwenheim–Skolem theorem, Corollary 3.1.5, it follows that the first-order theory of the class of linear orderings is decidable.* (c) Repeat (b) with 'linear orderings' replaced by 'well-orderings'. The same remark holds.

In one of many meanings of the word, a **tree** *is a partially ordered set $(P, \leqslant)$ in which for every element p the set $p^{\smallfrown} = \{q \in P : q < p\}$ is well-ordered by $<$. The* **height** *of p is the order-type of $p^{\smallfrown}$. The* **height** *of the tree is the least ordinal greater than the heights of all elements of the tree.*
2. (a) Show that the class of trees is definable by a single sentence of monadic second-order logic of the form $\forall Z\, \phi$ with ϕ first-order. (b) Show that, for each ordinal $\alpha \geqslant \omega$, the class of trees of height α is definable by a sentence of $L_{|\alpha|^+\omega}$.

The L^{II} properties of a structure are in general not set-theoretically absolute.
3. Let $\mathbb{N}$ be the structure $\langle \omega, 0, 1, +, \cdot, < \rangle$ and L the signature of $\mathbb{N}$. The theory of $\mathbb{N}$ in L^{II} is called **complete second-order arithmetic**. Show that there is a sentence of L^{II} which expresses 'Every real number is constructible in the sense of Gödel'. [See p. 18 of Mansfield & Weitkamp [1985] for the formula.] Deduce that complete second-order arithmetic varies from one universe of sets to another.

4. In **logic with permutation quantifiers**, $L^{1\text{-}1}$, the second-order variables are 2-ary relation variables, and they are interpreted to range over *permutations* of the domain. Thus for example $\exists X\, \phi(X)$ in $L^{1\text{-}1}$ can be rewritten in L^{II} as

$$\exists X(\phi(X) \wedge \forall x \exists_{=1} y \, Xxy \wedge \forall y \exists_{=1} x \, Xxy).$$

Show that if P is a 1-ary relation symbol, one can write down a sentence of L^{1-1} which expresses 'The set $\{x: Px\}$ has cardinality ω'. [Use Example 2 of section 5.4 below.]

5. Show (a) if A is an L-structure, then every relation definable in A by a formula of L_w^{II} (with or without parameters) is also definable by a formula of $L_{\omega_1 \omega}$ (correspondingly with or without parameters) and by a formula of L^{II} (ditto), (b) if $\mathbb{N}$ is $\langle \omega, 0, 1, +, \cdot, < \rangle$ and L is its signature, then every subset of $\mathbb{N}$ is definable by a formula of $L_{\omega_1 \omega}$ without parameters, whereas any subset of $\mathbb{N}$ which is definable by a formula of L_w^{II}, even with parameters, is already $\varnothing$-definable.

6. Show that if L is the first-order language of abelian groups, then there is a formula $\phi(x)$ of L_w^{mon} such that for every abelian group A, $\phi(A)$ is the torsion subgroup of A.

*7. Let F be an infinite field and X an indeterminate; write $L, L[X]$ for the first-order languages of the rings $F, F[X]$ respectively. Show that independently of the choice of F, every statement about F in L_w^{II} can be translated into a statement about $F[X]$ in $L[X]$, and vice versa; and likewise with L_w^{mon} for L_w^{II}.

If L is a first-order language and α is an ordinal, we form the language $L(Q_\alpha)$ as follows. We introduce a new quantifier symbol Q_α. If ϕ is a formula and x a variable, then $Q_\alpha x \, \phi$ is a formula (so that $Q_\alpha x$ is a quantifier binding x). We interpret $Q_\alpha x \, \phi(x)$ to mean 'There are at least ω_α elements b such that $\phi(b)$'.

8. (a) Show that for every formula $\phi(\bar{x})$ of $L(Q_0)$ there is a formula $\phi^*(\bar{x})$ of L_w^{mon} which expresses the same as ϕ. (b) Show that the converse fails, by proving that in an algebraically closed field of infinite transcendence degree, every set definable in $L(Q_0)$ with parameters is already first-order definable, but the set of elements of infinite order in the multiplicative group is definable in L_w^{mon}. [By the compactness theorem this set is not first-order definable.]

The next exercise is a surprising example of an infinitary property that turns out to be first-order expressible.

9. A lattice is said to be **relatively completely distributive** if it satisfies the equation $\bigwedge_{i \in I} \bigvee_{j \in J} a_{i,j} = \bigvee_{f: I \to J} \bigwedge_{i \in I} a_{i, f(i)}$ whenever $a_{i,j}$ are elements such that all the indicated infs and sups exist. Show that a lattice A is relatively completely distributive if and only if it is a model of the following first-order sentences ϕ_1, ϕ_2 and ϕ_3 (where we use bounded quantifiers for shorthand). ϕ_1 is $(\forall u < v < w)(\exists z > u)(\forall x \geq u)$ $((x \vee v) \geq w \to x \geq z)$. ϕ_2 is as ϕ_1 but with $<$, $\leq$ etc. for $>$, $\geq$ etc. ϕ_3 is

$$(\forall u < v) \exists wxyz((u \leq w < x \leq v)$$
$$\wedge \, (u \leq y < z \leq v) \wedge \forall t(u \leq t \leq v \to ((t \vee w) \geq x) \vee ((t \wedge z) \leq y))).$$

*10. Given linear orderings A and B, the product $A \times B$ is the partially ordered structure with domain $\text{dom}(A) \times \text{dom}(B)$, and with $(a_1, b_1) \leq (a_2, b_2)$ iff $a_1 \leq a_2$ and $b_1 \leq b_2$. Show that there is a sentence ϕ of the monadic second-order theory of partial orderings, such that if A and B are linear orderings, $A \times B \vDash \phi$ if and only if $A \cong B$.

It follows that in general the monadic second-order theory of $A \times B$ is not determined by those of A and B.

11. In the language of linear orderings, let $\phi(x)$ be the formula $\forall y(y < x \rightarrow P(y))$. Show that for every ordinal α there is a structure A_α such that if π is the monotone operator $\pi_{P,\phi}$ on A_α, then $\pi^\alpha(\varnothing) \neq \pi^{\alpha+1}(\varnothing)$.

History and bibliography

Section 2.1. First-order definable sets and relations in a structure were first studied by E. Schröder [1895], without any formal language attached. The notion of an inductively defined formal language is from Whitehead & Russell [1910]; Post [1921] and Wittgenstein [1961] (also from 1921) both emphasised how formulas of propositional logic could be interpreted by induction on the construction of the formulas. Tarski's famous paper [1935] on the concept of truth (from 1930/1) set out for the first time clauses like (1.5)–(1.10), though in this paper Tarski avoided the notion of interpreting the formulas in arbitrary structures of a given signature. (See Hodges [1986] for an analysis.) The first fully satisfactory statement of the model-theoretic 'truth definition' seems to be that in Tarski & Vaught [1957], though the definition was in general use for some years before that.

Probably the first completely explicit constructions of infinitary languages were by Scott & Tarski [1958] and Karp [1964]. The notion of infinitary truth-functions was around already in the 1920s. For example Frank Ramsey wrote in 1926: '... the logical sum of a set of propositions is the proposition that one of the set at least is true, and it doesn't appear to matter whether the set is finite or infinite.' (p. 225 of Ramsey [1978]).

The redundancy involved in having to choose one variable rather than another can be a pain. But proposals to eliminate this feature of logic (such as those of Schönfinkel [1924] or Quine [1960]) may eliminate too much of the freedom of logical notation as well.

Strongly minimal structures and sets (i.e. structures and sets which are minimal and remain minimal after elementary extension) were introduced by Marsh [1966] and developed by Baldwin & Lachlan [1971]; they form the bottom level of the hierarchy of Morley rank (Morley [1965a]; see Lascar [1987a] section 4.3 for a modern account).

Exercise 4: O-minimal structures were defined by Pillay & Steinhorn [1984, 1986], in the light of ideas in van den Dries [1982a], in order to generalise some model-theoretic properties of the field of real numbers. Exercise 13:

Mal'tsev [1961a]. Exercise 14: Zil'ber [1977a]. Exercise 15: R. Robinson [1951]. Exercise 16: Barwise [1975] section III.1.

Section 2.2. The notion of a set of statements which defines a class of structures goes back to late nineteenth century writers – notably in the foundations of geometry; see Hilbert [1899]. Russell [1903] Chapter 49 was perhaps the first writer to give a broadly correct account of how the statements define the class, though for a polished explanation one has to wait until Bernays [1942]. One of the first mathematicians to apply the idea outside geometry was Huntington [1904a], who axiomatised boolean algebras. Tarski undertook a systematic study of axiomatically defined classes in his Warsaw seminar from 1926 onwards; see the references to section 2.7. The axioms for term algebras in Example 1 are from Mal'tsev [1961b], [1962]; Clark [1978] came on them independently in connection with the 'negation as failure' interpretation of Prolog.

Exercise 9: (f) Baldwin [1978], (h) Kopperman & Mathias [1968]. Exercise 12: Huntington [1924].

Section 2.3. The notion of validity is a modern descendent of earlier ideas of logical or *a priori* truth; Bolzano [1837] was a milestone. It was Tarski [1930a] who made the notion of consequence fundamental in place of the notion of validity. Our definition of 'T entails ϕ' for particular formal languages, not using any proof calculi, is also due to Tarski. But it's noticeable how Tarski, even when he used the notion, avoided calling it entailment or consequence (the one exception seems to be Tarski, Mostowski & Robinson [1953] p. 8). Possibly this was because he was not prepared to claim that it really was an instance of the general 'concept of logical consequence' which he had analysed in [1936] – and indeed there is room for philosophical debate on this point (see Etchemendy [1988]).

The notations $\vdash$ and $\vDash$ have a complex history. Frege [1879] wrote $\vdash$ at the left of a sentence to express that he was asserting the sentence and not merely 'producing an idea in the reader'. Later logicians found the device unconvincing (as if someone should say 'I'm asserting that . . .' in front of every sentence). Rosser [1935] suggested using '$\vdash\phi$' for 'ϕ is provable' (relative to some proof calculus), and Kleene [1934] generalised this to '$T \vdash \phi$' for 'ϕ is provable from T'. Addison, Henkin & Tarski [1965] (Foreword) proposed '$T \vDash \phi$' for the model-theoretic notion of consequence. But they spoiled this excellent suggestion by also proposing '$\vDash_A \phi$' for 'ϕ is true in A'; inevitably the A moved upstairs to form '$A \vDash \phi$', which clashes with '$T \vDash \phi$'. I think my use of $\vdash$ and $\vDash$ follows the standard usage of model theorists today.

Hintikka sets were introduced by Hintikka [1955], and equivalent ideas appear in Beth [1955] and Schütte [1956].

Exercise 5: Ackermann [1954]. Exercise 7: reduction to prenex form can be traced back to Peirce [1885].

Section 2.4. The notion of a formula being preserved by an embedding is from Abraham Robinson's PhD thesis [1949] – he talks of formulas being **persistent** under various operations on structures. Marczewski [1951] pointed out Theorem 2.4.3(b). Most of the results of this section are trivial. See sections 6.5 and 10.3 for some of their non-trivial converses in the first-order case.

Exercise 1: Tarski [1954]. Exercise 16: Knaster [1928], Tarski [1955].

Section 2.5. Tarski & Vaught [1957] defined the notion of an elementary embedding, after Robinson had displayed some examples in his early papers. This paper of Tarski and Vaught gave Theorems 2.5.1 and 2.5.2 and Exercises 1, 3 (essentially), 4 and 5.

Exercise 10: Rothmaler [1991].

Section 2.6. Padoa's method appears in Padoa [1901], who also discusses the admissibility conditions for explicit definitions. Section 7 of Hilbert & Bernays [1934] contains a careful discussion of explicit and recursive definitions. In ZFC definitions by recursion on a well-founded relation can be converted into explicit definitions (e.g. Montague [1955]); but this is very much a peculiarity of ZFC. On Example 1 (the term algebra) see the references to section 2.7 below. Abraham Robinson [1956b] introduced the notion of model-completeness; see section 8.3 below. Atomisation is due to Skolem [1920]. Theorem 2.6.7 is from Chang [1968b]; a similar device appeared earlier in Engeler [1961].

Exercise 7: This followed from the fact that $\mathrm{Th}(\mathbb{N})$ is undecidable (by Gödel [1931b]) while $\mathrm{Th}(\mathbb{N}|L)$ is decidable (Skolem [1930]; cf. Exercise 9.6.14 below); but an algebraic proof by automorphisms of $\mathbb{N}|L$ is much easier. Exercise 10: Stone [1935, 1936]. Exercise 11: Tarski & Lindenbaum [1927].

Section 2.7. Exercise 2 is due to Löwenheim [1915], in a rather different setting. Exercise 3 is Skolem's [1919] reworking of Löwenheim's result as an instance of application (d). Langford [1926] proved Exercise 6(a) as an instance of application (b). In his Warsaw seminar on 'deductive systems' in 1926–8, Tarski began to apply the method of quantifier elimination to a wide range of classes of structures. By 1930 he had the whole of Exercise 6 (see the appendix to [1935 + 6]) and a version of Theorem 2.7.2 [1931]; from his remarks on dense orderings in his [1935 + 6] it's clear he must also have had Theorem 2.7.1. Exercise 7 was the work of his student Presburger [1930]. Application (a) is from Tarski [1935 + 6]. Application (c) is mentioned by Skolem [1928] and Tarski [1931]. It was left to Abraham Robinson to emphasise application (e); see his [1951].

Tarski saw quantifier elimination as part of a systematic approach to 'general metamathematics'. See for example his series of abstracts [1949a–d] which sum up his concerns of the preceding twenty years. It seems that his main target during this period was to classify the complete extensions of certain important theories, within the framework which he set out in his [1930a]. Over the same period he reworked his results on the field of real numbers several times with different emphases. The full statement and proof of Theorem 2.7.2 had to wait until [1951]. Theorem 2.7.3 is stated in Tarski [1949d]; Macintyre [1984] wrote out a proof by quantifier elimination. A series of papers in the *Journal of Symbolic Logic* 1987/8 review Tarski's work; see particularly the survey by Vaught [1987].

Theorem 2.7.5 and its corollaries, with Exercise 11, are due to Mal'tsev [1961b], [1962]. In fact Mal'tsev carried out a quantifier elimination in the language L_1. The simpler approach through L_2 was noticed independently by Kunen [1987] and Belegradek [1988]. Mal'tsev's proof has the merit that it can be adapted, for example to the case where each function is constant under permutation of its variables; see Mal'tsev [1962].

Exercises 4,5: Tarski [1935 + 6]. Exercise 8: H. Gupta and L. Henkin (unpublished). Exercise 10: (a) essentially Tarski [1931], (b) A. Robinson [1956b].

Section 2.8. Some of these generalisations of first-order logic are much older than one might guess. Before about 1920 there was no general agreement on either notation or semantics in mathematical logic, and one has to be careful not to read modern notions into the old terminology. Nevertheless Peirce [1885], a very modern writer, has a system which is well on the way to being a many-sorted version of second-order logic L^{II}. Whitehead & Russell [1910] had a system with quantifiers ranging over definable relations; this idea has had little direct impact in model theory, though it was the inspiration for

Gödel's constructible universe [1944]. Hilbert & Ackermann [1928] studied a version of L^{II}. Weak second-order logic was proposed by Tarski [1958].

The model theory of second-order logic is mainly confined to studying the relations definable by second-order formulas in particular structures. The two main approaches have been that of Feferman & Vaught (see section 9.6 below) and the automata-theoretic approach of Büchi [1960], [1962] which led to Rabin's [1969] proof of Theorem 2.8.1. Rabin's argument was simplified by Gurevich & Harrington [1982] and independently Muchnik [1985].

Fixed point logic grew out of work on relational databases, notably Chandra & Harel [1982]. See for example Blass, Gurevich & Kozen [1985], and Immerman [1986].

Theorem 2.8.3 is from Eklof & Sabbagh [1971]. The book of Jensen & Lenzing [1989], despite its name, is mainly about what classes of algebraic objects are first-order axiomatisable; on this it has a wealth of information. Models of ZF with no non-trivial injective abelian groups were given by Hodges and Blass; see Blass [1979] and related material in Pope [1982].

Exercise 1: Rabin [1977]. The decidability of the first-order theory of linear orderings is due to Ehrenfeucht [1959], and that of ordinals to Mostowski and Tarski; see Doner, Mostowski & Tarski [1978]. Exercise 4: the expressive power of L^{1-1} is studied in Shelah [1973c], Baldwin & Shelah [1985], Baldwin [1985]. Exercise 7: Bauval [1985]. Exercise 8: the quantifiers Q_α were introduced by Mostowski [1957]; see Part B of Barwise & Feferman [1985] for the consequences. The quantifier 'For two-thirds of all elements x' appears already in Peirce [1885]. Exercise 9: Ball [1984]. Exercise 10: Shelah [1975d].

3

Structures that look alike

M Martin: J'ai une petite fille, ma petite fille, elle habite avec moi, chère Madame. Elle a deux ans, elle est blonde, elle a un oeil blanc et un oeil rouge, elle est très jolie, elle s'appelle Alice, chère Madame.

Mme Martin: Quelle bizarre coïncidence! moi aussi j'ai une petite fille, elle a deux ans, un oeil blanc et un oeil rouge, elle est très jolie et s'appelle Alice, cher Monsieur!

M Martin, *même voix traînante, monotone:* Comme c'est curieux et quelle coïncidence!

Eugène Ionesco, La cantatrice chauve, © *Editions GALLIMARD 1954.*

If we consider a first-order language L, the number of isomorphism types of L-structures is vastly greater than the number of theories in L (counted up to equivalence of theories). So there must be some huge family of non-isomorphic L-structures which it is impossible to tell apart by sentences of L.

In this chapter we prove a variety of theorems which have the general form: if some sentence is true here, then it must be true there too.

3.1 Theorems of Skolem

In earlier days there were people who disliked uncountable structures and wanted to show that they are unnecessary for mathematics. Thoralf Skolem was one such. He proved that for every infinite structure B of countable signature there is a countable substructure of B which is elementarily equivalent to B. He inferred from this that there are countable models of Zermelo–Fraenkel set theory, and hence countable models of the sentence 'There are uncountably many reals'. He hoped that this paradoxical result would scare people away from set-theoretical foundations. If anything it had the opposite effect.

The quickest way to prove Skolem's result is as follows. Let B be any infinite structure with countable signature. By Theorem 1.2.3 we can build a chain $(A_n: n < \omega)$ of countable substructures of B, in such a way that

(1.1)   for each first-order formula $\phi(y, \bar{x})$, each $n < \omega$ and each
tuple $\bar{a}$ of elements of A_n such that $B \vDash \exists y\, \phi(y, \bar{a})$, if there

is b in B such that $B \vDash \phi(b, \bar{a})$ then there is such an element
b in A_{n+1}.

Put $A = \bigcup_{n<\omega} A_n$. Clearly A is countable. Also A is an elementary substructure of B by the Tarski–Vaught criterion, Theorem 2.5.1, and so $A \equiv B$. (We
took a similar route in Exercise 2.5.5 above.)

Skolem proceeded differently. He added functions to B in such a way that
every substructure of B which is closed under these functions is automatically
an elementary substructure, and then he invoked Theorem 1.2.3. The added
functions are called *Skolem functions*. Skolem's proof by this route is a little
more complicated than the one sketched in the paragraph above, but it is also
more versatile. We shall exploit some of its extra power by proving a 'local
theorem', Corollary 3.6.3, at the end of the chapter.

Theories with Skolem functions

Suppose T is a theory in a first-order language L. Then a **skolemisation** of T
is a theory $T^+ \supseteq T$ in a first-order language $L^+ \supseteq L$, such that

(1.2)   every L-structure which is a model of T can be expanded to
          a model of T^+, and

(1.3)   for every formula $\phi(\bar{x}, y)$ of L^+ with $\bar{x}$ non-empty, there is a
          term t of L^+ such that T^+ entails the sentence
          $\forall \bar{x}(\exists y\, \phi(\bar{x}, y) \rightarrow \phi(\bar{x}, t(\bar{x})))$.

The terms t of (1.3) (and the functions which they define in models of T^+)
are called **Skolem functions** for T^+.

We say that T **has Skolem functions** (or that T is a **Skolem theory**) if T is
a skolemisation of itself; in other words, if (1.3) holds with $L = L^+$ and
$T = T^+$. Note that if T^+ is a skolemisation of T, then T^+ has Skolem
functions. Note also that these notions depend on the language: if $L \subseteq L'$
and T is a Skolem theory in L, T will generally not be a Skolem theory in
L'. On the other hand if T has Skolem functions and T' is a theory with
$T \subseteq T'$, both in the first-order language L, then it's immediate that T' has
Skolem functions too.

Suppose T is a theory which has Skolem functions, in a first-order
language. Let A be an L-structure and X a set of elements of A. The **Skolem
hull** of X is defined to be $\langle X \rangle_A$, the substructure of A generated by X.

Theorem 3.1.1. *Suppose T is a theory in a first-order language L, and T has
Skolem functions.*

(a) *Modulo T, each formula $\phi(\bar{x})$ of L (with $\bar{x}$ not empty) is equivalent to
a quantifier-free formula $\phi^*(\bar{x})$ of L.*

(b) *If A is an L-structure and a model of T, and X is a set of elements of A such that the Skolem hull $\langle X \rangle_A$ is non-empty, then $\langle X \rangle_A$ is an elementary substructure of A.*

Proof. In (1.3), the formula $\phi(\bar{x}, t(\bar{x}))$ logically implies $\exists y\, \phi(\bar{x}, y)$, so we could have written $\leftrightarrow$ in place of $\rightarrow$. Hence (a) follows at once from Lemma 2.3.1, taking Φ to be the set of quantifier-free formulas.

To prove (b), put $B = \langle X \rangle_A$. Let $\bar{b}$ be a tuple of elements of B and $\phi(\bar{x}, y)$ a formula of L such that $A \vDash \exists y\, \phi(\bar{b}, y)$. Then by (1.3) there is a term t such that $A \vDash \phi(\bar{b}, t(\bar{b}))$. But the element $t^A(\bar{b})$ is in B since B is closed under the functions of L. By the Tarski–Vaught criterion (Theorem 2.5.1) it follows that B is an elementary substructure of A. $\quad\square$

Adding Skolem functions

Sadly, in a state of nature there are very few Skolem theories. They have to be constructed by artifice.

Theorem 3.1.2 (Skolemisation theorem). *Let L be a first-order language. Then there are a first-order language $L^\Sigma \supseteq L$ and a set Σ of sentences of L^Σ such that*
(a) *every L-structure A can be expanded to a model A^Σ of Σ,*
(b) *Σ is a Skolem theory in L^Σ,*
(c) *$|L^\Sigma| = |L|$.*

Proof. For each formula $\chi(\bar{x}, y)$ of L (where $\bar{x}$ is not empty), introduce a new function symbol $F_{\chi, \bar{x}}$ of the same arity as $\bar{x}$. The language L' will consist of L with these new function symbols added. The set $\Sigma(L)$ will consist of all the sentences

(1.4) $\forall \bar{x}\, (\exists y\, \chi(\bar{x}, y) \rightarrow \phi(\bar{x}, F_{\chi, \bar{x}}(\bar{x})))$.

We claim

(1.5)   every L-structure A can be expanded to a model of $\Sigma(L)$.

If A is empty it is already a model of $\Sigma(L)$. If it is not empty, we expand it to an L'-structure A' as follows. Let $\chi(\bar{x}, y)$ be any formula of L with $\bar{x}$ non-empty, and let $\bar{a}$ be a tuple of elements of A. If there is an element b such that $A \vDash \chi(\bar{a}, b)$, choose one such element b and put $F^{A'}_{\chi, \bar{x}}(\bar{a}) = b$. (Here we generally need the axiom of choice.) If there is no such element, let $F^{A'}_{\chi, \bar{x}}(\bar{a})$ be, say, the first element in $\bar{a}$. Then certainly A' is a model of all the sentences in $\Sigma(L)$. Thus (1.5) is proved.

The theory Σ is built by iterating the construction of $\Sigma(L)$ ω times. We define a chain of languages $(L_n : n < \omega)$ and a chain of theories $(\Sigma_n : n < \omega)$

by induction on n. We put $L_0 = L$ and take Σ_0 to be the empty theory. Then we define L_{n+1} to be $(L_n)'$ as above, and we define $\Sigma_{n+1} = \Sigma_n \cup \Sigma(L_n)$. Finally we define $L^\Sigma = \bigcup_{n<\omega} L_n$ and $\Sigma = \bigcup_{n<\omega} \Sigma_n$. Now (a) is true by making repeated expansions as in (1.5). For (b), every formula χ of L^Σ lies in some Σ_n, so the required sentence (1.4) is in Σ_{n+1}. Finally (c) is clear. ☐

The proof of the theorem can be coaxed into giving a little more information. This is useful if we want to add Skolem functions without disturbing the forms of the formulas (Horn formulas, for example).

Corollary 3.1.3. *In the theorem we can add the following. Suppose $\phi(\bar{x})$ is a prenex formula of L^Σ, of the form $Q\psi(\bar{x}, \bar{y})$ where Q is a string of quantifiers and ψ is quantifier-free. Then ϕ is equivalent (in all non-empty models of Σ) to an $\forall_1$ formula $\theta(\bar{x})$ such that*
(a) *θ is of the form $Q'\psi(\bar{x}, \bar{t})$ where Q' consists of the universal quantifiers in Q, and $\bar{t}$ is a tuple of terms of L^Σ,*
(b) *$\vdash \forall \bar{x}(\theta \to \phi)$.*

Proof. Exercise. ☐

However, the main application of the skolemisation theorem is to give us elementary substructures, as follows.

Corollary 3.1.4. *Let T be a theory in a first-order language L. Then T has a skolemisation T^+ in a first-order language L^+ with $|L^+| = |L|$.*

Proof. Put $T^+ = T \cup \Sigma$. In particular Σ is a skolemisation of the empty theory in L. ☐

Corollary 3.1.5 (*Downward Löwenheim–Skolem theorem*). *Let L be a first-order language, A an L-structure, X a set of elements of A, and λ a cardinal such that $|L| + |X| \leqslant \lambda \leqslant |A|$. Then A has an elementary substructure B of cardinality λ with $X \subseteq$ dom (B).*

Proof. Expand A to a model A^Σ of Σ in L^Σ. Let Y be a set of λ elements of B, with $X \subseteq Y$. Let B' be the Skolem hull $\langle Y \rangle_A$, and let B be the reduct $B'|L$. By Theorem 1.2.3, $|B| \leqslant |Y| + |L^\Sigma| = \lambda + |L| = \lambda = |Y| \leqslant |B|$. Since Σ is a Skolem theory, $B' \leqslant A^\Sigma$ by Theorem 3.1.1(b). Hence $B \leqslant A$. ☐

Example 1: *Simple subgroups of simple groups.* Let G be an infinite simple group. We show that for every infinite cardinal $\lambda \leqslant |G|$, G has a subgroup of cardinality λ which is simple. The language of groups is countable, so that by the downward Löwenheim–Skolem theorem, G has an elementary sub-structure H of cardinality λ. Clearly H is a subgroup of G. To show that H is simple it suffices to prove that if a, b are two elements of H and $b \neq 1$, then a is in the normal subgroup of H generated by b. Since G is simple, this is certainly true with G in place of H. Suppose for example that

$$G \vDash \exists y\, \exists z (a = y^{-1}by \cdot z^{-1}b^{-1}z).$$

Since $H \preccurlyeq G$, the same sentence is true in H. Hence there are c, d in H such that $a = c^{-1}bc \cdot d^{-1}b^{-1}d$ as required.

Lachlan [1978] constructed a countable L-structure B with an elementary substructure A, where L has a 2-ary relation symbol R, such that no expansions $A^+ \preccurlyeq B^+$ have a Skolem function for the quantifier in the formula $\exists y\, Rxy$. It was surprisingly hard to find an example of this phenomenon.

Corollary 3.1.6 (Weak downward Löwenheim–Skolem theorem for $L_{\lambda^+\omega}$ – see Exercise 10). *Let L be a signature, λ an infinite cardinal and ϕ a sentence of $L_{\lambda^+\omega}$. Let A be a model of ϕ and X a set of at most λ elements of A. Then there is a substructure B of A such that $X \subseteq \mathrm{dom}(B)$, B has cardinality at most λ, and B is a model of ϕ.*

Proof. This follows at once from Corollary 3.1.5, Chang's reduction in Theorem 2.6.7, and the fact that every elementary substructure of A omits all the types which A omits. □

Definable Skolem functions

We say that a theory has **definable Skolem functions** if it has a definitional extension with Skolem functions (see section 2.6 above). We say that a structure A **has definable Skolem functions** if $\mathrm{Th}(A)$ has definable Skolem functions.

Put more simply, we say that a theory T in a first-order language L has **definable Skolem functions** if for every formula $\phi(\bar{x}, y)$ of L with $\bar{x}$ not empty, there is a formula $\psi(\bar{x}, y)$ of L such that

(1.6) $T \vdash \forall \bar{x}(\exists y\, \phi(\bar{x}, y) \rightarrow (\exists_{=1} y\, \psi(\bar{x}, y) \wedge \forall y(\psi(\bar{x}, y) \rightarrow \phi(\bar{x}, y))))$.

Then we can use ψ to define a function F by

(1.7) $\forall \bar{x} y (F(\bar{x}) = y \leftrightarrow (\exists z\, \phi(\bar{x}, z) \wedge \psi(\bar{x}, y)) \vee (\neg \exists z\, \phi(\bar{x}, z)$

$\wedge\; y = (\text{first variable in } \bar{x})))$.

These definitions make T into a theory T^+ with Skolem functions. (Caution: we need to check that there are Skolem functions for formulas including the new function symbols too. But these new function symbols can be eliminated since they are explicitly definable: see Theorem 2.6.4(b).)

Example 2: *First-order Peano arithmetic* (see Example 2 in section 2.2). By the induction axiom, if there is an element that satisfies $\phi(y)$ then there is a first such element. So we can take ψ in (1.6) to be the formula

(1.8) $\phi(\bar{x}, y) \wedge \forall z(z < y \rightarrow \neg \phi(\bar{x}, z))$.

Thus first-order Peano arithmetic has definable Skolem functions. For exactly the same reason, set theory with the axiom of constructibility ($V = L$) has definable Skolem functions: if there is an element satisfying $\phi(y)$, then take the first such element in the definable well-ordering of the universe.

Example 3: *Real-closed fields.* Let T be the theory of real-closed fields in the signature of rings (see (2.27) in section 2.2). We can add a symbol $\leqslant$ for the ordering, defining $x \leqslant y$ as $\exists z \, z^2 = y - x$. Tarski's quantifier elimination theorem for T tells us that every formula $\phi(\bar{x})$ is equivalent modulo T to a boolean combination of formulas of the form $t(\bar{x}) \geqslant 0$ where t is a term (see Theorem 2.7.2). In particular a formula $\phi(\bar{x}, y)$ can be written as a boolean combination of formulas of the form

(1.9) $p_n(\bar{x})y^n + p_{n-1}(\bar{x})y^{n-1} + \ldots + p_1(\bar{x})y + p_0(\bar{x}) \geqslant 0$,

where $p_0, \ldots, p_n$ are polynomials with integer coefficients. We can rewrite (1.9) as follows:

(1.10) y lies in the union of a certain finite set of intervals of form
         $[s_0(\bar{x}), s_1(\bar{x})], (-\infty, s_1(\bar{x})]$ or $[s_0(\bar{x}), \infty)$, where $s_0(\bar{x}), s_1(\bar{x})$
         are points definable from $\bar{x}$ (and the defining formulas
         depend only on $p_0, \ldots, p_n$).

The upshot is that every formula $\phi(\bar{x}, y)$ says that y lies in some finite union of intervals whose endpoints are definable from $\bar{x}$; the intervals may be closed or open at either end. (Thus T is O-minimal; cf. Exercise 2.1.4.)

So to show that T has definable Skolem functions, the crucial step is to prove that we can define an element of any interval which has definable endpoints. If the interval is bounded, for example $(a, b]$ with $a < b$, then we can define the point $(a + b)/2$ which lies inside the interval. Suppose the interval is (a, ∞); then $a + 1$ is a point inside the interval; and likewise with $(-\infty, b)$. Finally suppose the interval is $(-\infty, \infty)$; then 0 is a point inside the interval. So the theory of real-closed fields has definable Skolem functions.

Exercises for section 3.1

1. Let L be a first-order language and L' a language which comes from L by adding constants. Show that if T is a Skolem theory in L, then T is a Skolem theory in L' too (and hence so is any theory $T' \supseteq T$ in L').

2. Use the downward Löwenheim–Skolem theorem and the result of Example 3 in the next section, to show that if A and B are dense linear orderings without endpoints, then $A \equiv B$.

3. Show that, if T is a first-order theory which has Skolem functions, then T is model-complete. Give an example of a first-order theory which is model-complete but doesn't have Skolem functions.

4. Let L be a first-order language with at least one constant. Show that if T is a Skolem theory in L, then T has elimination of quantifiers.

5. (a) Prove Corollary 3.1.3. [The argument of Lemma 2.3.1 removed all the quantifiers; here remove just the existential ones.] (b) Use the corollary to show that there is an algorithm which finds, for any first-order sentence ϕ, an $\forall_1$ first-order sentence ϕ^* (possibly with larger signature) such that ϕ has a non-empty model if and only if ϕ^* has a non-empty model. Deduce that if we had an algorithm for telling which $\forall_1$ first-order sentences have models, we could use it to test which first-order sentences have models (which is impossible by Church [1936]).

The next exercise shows that not all the sentences (1.4) are needed for Theorem 3.1.2. (However, they are needed for Corollary 3.1.3.)

6. Suppose T is a theory in a first-order language L, and, for every quantifier-free formula $\phi(\bar{x}, y)$ of L with $\bar{x}$ non-empty, there is a term t of L such that T entails the sentence $\forall \bar{x} (\exists y\, \phi(\bar{x}, y) \to \phi(\bar{x}, t(\bar{x})))$. (a) Show that T has Skolem functions. [Use Lemma 2.3.1 to show that modulo T, every formula with at least one free variable is equivalent to a quantifier-free formula.] (b) Show that for any theory T' in L with $T \subseteq T'$, T' is equivalent to an $\forall_1$ theory.

*We say that a structure A **has Skolem functions** if $\mathrm{Th}(A)$ has Skolem functions.*

7. Let L be a first-order language and A an L-structure which has Skolem functions. Suppose X is a set of elements which generate A, and $<$ is a linear ordering of X (not necessarily expressible in L). Show that every element of A has the form $t^A(\bar{c})$ for some term $t(\bar{x})$ of L and some tuple $\bar{c}$ from X which is strictly increasing in the sense of $<$.

8. Let $\mathbf{K}$ be the class of boolean algebras which are isomorphic to power set algebras of sets. Show that $\mathbf{K}$ is not first-order axiomatisable. [Use Corollary 3.1.5.]

9. A linear ordering $(X, <)$ is said to be **complete** if every non-empty set of elements of X which has an upper bound has a least upper bound. Show that there is no sentence ϕ of $L_{\infty\omega}$ such that the complete linear orderings are the models of ϕ. [Use Corollary 3.1.6.]

10. Let L be a signature, λ an infinite cardinal and K a first-order-closed fragment of $L_{\lambda^+\omega}$ of cardinality λ consisting of variable-finite formulas. (Cf. Exercise 2.1.9.) State

and prove an analogue of Theorem 3.1.2 for K. Deduce that if A is an L-structure and X is a set of at most λ elements of A, then there is a substructure B of A such that $X \subseteq \mathrm{dom}\,(B)$, B has cardinality at most λ, and for every formula $\phi(\bar{x})$ of K and tuple $\bar{b}$ in B, $A \vDash \phi(\bar{b}) \Leftrightarrow B \vDash \phi(\bar{b})$. *This will be used in section 11.6.*

11. Show that the ring of integers has definable Skolem functions. [Every non-negative integer is a sum of four squares.]

*12. Consider the statements S1, 'If L is a countable first-order language, then every theory in L which has an infinite model has a countable model', and S2, 'If L is a countable signature and A an infinite L-structure, then A has a countable elementary substructure'. Show that S1 is provable from ZF alone (without the axiom of choice), but S2 is not.

13. Let L be a finite relational signature and A an infinite L-structure. Suppose there is a simple group which acts transitively on A. *(This means that the automorphism group of A contains a subgroup G which is simple, such that if a, b are any two elements of A then some automorphism in G takes a to b.)* Show that A has a countable elementary substructure on which some simple group acts transitively.

The next argument assumes things from a later chapter. I include it as a beautiful example of how purely algebraical results can be proved neatly with a little bit of model theory.

*14. Call G a *-**group** if G is an infinite locally finite simple group, and for some prime p, every p-subgroup of G is soluble. It is known that a *-group is linear if and only if it is countable. (Left to right is by Winter [1968], right to left is by Theorem 4.8 in Kegel & Wehrfritz [1973] and the classification of finite simple groups.) Prove that in fact every *-group is countable. [Use the downward Löwenheim–Skolem theorem and the fact (cf. the proof of Corollary 6.6.8 below) that for every $n < \omega$, the class of linear groups of degree n is first-order axiomatisable.]

3.2 Back-and-forth equivalence

Compare the two relations $\cong$ (isomorphism) and $\equiv$ (elementary equivalence) between structures. In one sense isomorphism is a more intrinsic property of structures, because it's defined directly in terms of structural properties, whereas $\equiv$ involves a language. But in another sense elementary equivalence is more intrinsic, because the existence of an isomorphism can depend on some subtle questions about the surrounding universe of sets.

We can sharpen this second point with the help of some set theory. If M is a transitive model of set theory containing vector spaces A and B of dimensions ω and ω_1 over the same countable field, then A and B are not isomorphic in M, but they are isomorphic in an extension of M got by 'collapsing the cardinal ω_1 down to ω'. By contrast the question whether two structures A and B are elementarily equivalent depends only on A and B, and not on the sets around them (cf. Exercise 2.1.16).

In the early 1950s, Roland Fraïssé discovered a family of equivalence relations which hover somewhere between $\cong$ and $\equiv$. His equivalence relations are purely structural – there are no languages involved. On the other hand they are independent of the surrounding universe of sets. The trick is to look at isomorphisms, but only between a finite number of elements at a time. In the next section we shall find that Fraïssé's equivalence relations do often give us a way of proving that two structures are elementarily equivalent. Sometimes they give us proofs of isomorphism too, as we shall see in a few pages.

Back-and-forth games

Let L be a signature and let A and B be L-structures. We imagine two people, called $\forall$ and $\exists$ (male and female respectively, say $\forall$belard and $\exists$loise), who are comparing these structures. To add a note of conflict we imagine that $\forall$ wants to prove that A is different from B, while $\exists$ tries to show that A is the same as B. So their conversation has the form of a game. Player $\forall$ wins if he manages to find a difference between A and B before the game finishes; otherwise player $\exists$ wins.

The game is played as follows. An ordinal γ is given, which is the length of the game. Usually but not always, γ is ω or a finite number. The game is played in γ steps. At the ith step of a play, player $\forall$ takes one of the structures A, B and chooses an element of this structure; then player $\exists$ chooses an element of the other structure. So between them they choose an element a_i of A and an element b_i of B. Apart from the fact that player $\exists$ must choose from the other structure from player $\forall$ at each step, both players have complete freedom to choose as they please; in particular either player can choose an element which was chosen at an earlier step. Player $\exists$ is allowed to know which element player $\forall$ has chosen, and more generally each player is allowed to see and remember all previous moves in the play. (As the game theorists would put it, this is a **game of perfect information**.) At the end of the play, sequences $\bar{a} = (a_i : i < \gamma)$ and $\bar{b} = (b_i : i < \gamma)$ have been chosen. The pair $(\bar{a}, \bar{b})$ is known as the **play**.

We count the play $(\bar{a}, \bar{b})$ as a **win for player** $\exists$, and we say that player $\exists$ **wins the play**, if there is an isomorphism $f : \langle \bar{a} \rangle_A \to \langle \bar{b} \rangle_B$ such that $f\bar{a} = \bar{b}$. A play which is not a win for player $\exists$ counts as a **win for player** $\forall$.

Example 1: Rationals versus integers. Suppose $\gamma \geq 2$. Let A be the additive group $\mathbb{Q}$ of rational numbers and let B be the additive group $\mathbb{Z}$ of integers. Then player $\forall$ can win by playing as follows. He chooses a_0 to be any non-zero element of $\mathbb{Q}$. Then player $\exists$ must choose b_0 to be a non-zero integer; otherwise she loses the game at once. Now there is some integer n which doesn't divide b_0 in $\mathbb{Z}$. Let player $\forall$ choose a_1 in $\mathbb{Q}$ so that $na_1 = a_0$. There is no way that player $\exists$ can choose an element b_1 of $\mathbb{Z}$ so that $nb_1 = b_0$. It follows that, if $\gamma \geqslant 2$, player $\forall$ can always arrange to win the game on $\mathbb{Q}$ and $\mathbb{Z}$.

Let us write $A \equiv_0 B$ to mean that for every atomic sentence ϕ of L, $A \vDash \phi \Leftrightarrow B \vDash \phi$. (Clearly it makes no difference if we replace 'atomic' by 'quantifier-free'.) Then

(2.1)   player $\exists$ wins the play $(\bar{a}, \bar{b})$ if and only if $(A, \bar{a}) \equiv_0 (B, \bar{b})$.

This is equivalent to our definition of a win for $\exists$, by Theorem 1.3.1(c).

The game we have just described is called the **Ehrenfeucht–Fraïssé game of length γ on A and B**, in symbols $\mathrm{EF}_\gamma(A, B)$.

The more A is like B, the better chance player $\exists$ has of winning these games. In fact if player $\exists$ knows an isomorphism $i : A \to B$ then she can be sure of winning every time. All she has to do is to follow the rule

(2.2)   choose $i(a)$ whenever player $\forall$ has just chosen an element a
          of A, and $i^{-1}(b)$ whenever player $\forall$ has just chosen b from
          B.

We can express this point more precisely by using a notion from game theory, viz. the notion of a **winning strategy**.

A **strategy** for a player in a game is a set of rules which tell the player exactly how to move, depending on what has happened earlier in the play. We say that the player **uses** the strategy σ in a play if each of his or her moves in the play obeys the rules of σ. We say that the strategy σ is a **winning strategy** if the player wins every play in which he or she uses σ. For example the rule in (2.2) is a winning strategy for player $\exists$.

We write $A \sim_\gamma B$ to mean that player $\exists$ has a winning strategy in the game $\mathrm{EF}_\gamma(A, B)$. Thus for example $\mathbb{Q} \nsim_2 \mathbb{Z}$ by Example 1 above.

Before we forget empty structures altogether, we should stipulate that for any positive ordinal γ, if at least one of A, B is empty, then $A \sim_\gamma B$ if and only if both are empty.

Lemma 3.2.1. *Let L be a signature and let A, B be L-structures.*
    (a) *If $A \cong B$ then $A \sim_\gamma B$ for all ordinals γ.*
    (b) *If $\beta < \gamma$ and $A \sim_\gamma B$ then $A \sim_\beta B$.*
    (c) *If $A \sim_\gamma B$ and $B \sim_\gamma C$ then $A \sim_\gamma C$; in fact $\sim_\gamma$ is an equivalence relation on the class of L-structures.*

Proof. We have already proved (a) in the discussion above.

I leave (b) as an exercise and move on to (c). It's clear from the definition that $\sim_\gamma$ is reflexive and symmetric on the class of L-structures. (True, the definition of the game $EF_\gamma(A, B)$ was phrased as if A and B must be different structures. But the reader can handle this.) We prove transitivity. Suppose $A \sim_\gamma B$ and $B \sim_\gamma C$, so that player $\exists$ has winning strategies σ and τ for $EF_\gamma(A, B)$ and $EF_\gamma(B, C)$ respectively. Let the two players sit down to a match of $EF_\gamma(A, C)$. We have to find a winning strategy for player $\exists$.

Here we use a trick which is common in game theory. We make one of the players play a private game on the side, at the same time as the main game. In fact player $\exists$ will play two private games, one of $EF_\gamma(A, B)$ and one of $EF_\gamma(B, C)$. In the public game of $EF_\gamma(A, C)$ she will proceed as follows. Every time player $\forall$ chooses an element a_i of A, she first imagines to herself that player $\forall$ has made this move in $EF_\gamma(A, B)$; then σ tells her to pick an element b_i of B. Next she imagines that b_i was a choice of player $\forall$ in the private game of $EF_\gamma(B, C)$, and she uses her strategy τ to choose a corresponding c_i in C. This element c_i will be her answer to a_i in the public game. If player $\forall$ chose an element c_i of C, then she would respond in the same way but moving in the other direction, from C through B to A.

At the end of the contest, the players have constructed sequences $\bar{a}$ from A, $\bar{b}$ from B and $\bar{c}$ from C. The play of the public game $EF_\gamma(A, C)$ is $(\bar{a}, \bar{c})$. Now, in the private game $EF_\gamma(A, B)$, player $\exists$ used her winning strategy σ, and so the play $(\bar{a}, \bar{b})$ is a win for $\exists$. Similarly $(\bar{b}, \bar{c})$ is a win for $\exists$ in $EF_\gamma(B, C)$. So

(2.3) $(A, \bar{a}) \equiv_0 (B, \bar{b}) \equiv_0 (C, \bar{c})$

and hence $(A, \bar{a}) \equiv_0 (C, \bar{c})$. This shows that $(\bar{a}, \bar{c})$ is a win for $\exists$ in $EF_\gamma(A, C)$, and hence the strategy we described for her is winning. Thus $A \sim_\gamma B$.   □

Back-and-forth systems

Two L-structures A and B are said to be **back-and-forth equivalent** if $A \sim_\omega B$, i.e. if player $\exists$ has a winning strategy for the game $EF_\omega(A, B)$.

There is a useful criterion for two structures to be back-and-forth equivalent. A **back-and-forth system** from A to B is a set I of pairs $(\bar{a}, \bar{b})$ of tuples, with $\bar{a}$ from A and $\bar{b}$ from B, such that

(2.4)   if $(\bar{a}, \bar{b})$ is in I then $\bar{a}$ and $\bar{b}$ have the same length and $(A, \bar{a})$
         $\equiv_0 (B, \bar{b})$,

(2.5) I is not empty,

(2.6)   for every pair $(\bar{a}, \bar{b})$ in I and every element c of A there is
         an element d of B such that the pair $(\bar{a}c, \bar{b}d)$ is in I, and

(2.7)   for every pair $(\bar{a}, \bar{b})$ in I and every element d of B there is
         an element c of A such that the pair $(\bar{a}c, \bar{b}d)$ is in I.

Note that by (2.4) and Theorem 1.3.1(c), if $(\bar{a}, \bar{b})$ is in I then there is an
isomorphism $f: \langle\bar{a}\rangle_A \to \langle\bar{b}\rangle_B$ such that $f\bar{a} = \bar{b}$; f is unique since $\bar{a}$ generates
$\langle\bar{a}\rangle_A$. We write I^* for the set of all such functions f corresponding to pairs
of tuples in I. The conditions (2.4)–(2.7) imply some similar conditions on
the set $J = I^*$:

(2.4′)   each $f \in J$ is an isomorphism from a finitely generated
          substructure of A to a finitely generated substructure of B,

(2.5′) J is not empty,

(2.6′)   for every $f \in J$ and c in A there is $g \supseteq f$ such that $g \in J$ and
          $c \in \operatorname{dom} g$,

(2.7′)   for every $f \in J$ and d in B there is $g \supseteq f$ such that $g \in J$ and
          $d \in \operatorname{im} g$.

And conversely, if J is any set obeying the conditions (2.4′)–(2.7′), then
there is a back-and-forth system I such that $J = I^*$. Namely, take I to be the
set of all pairs of tuples $(\bar{a}, \bar{b})$ such that $\bar{a}$ is from A, $\bar{b}$ is from B and J
contains a map $f: \langle\bar{a}\rangle_A \to \langle\bar{b}\rangle_B$ such that $f\bar{a} = \bar{b}$.

Some writers refer to a set J satisfying (2.4′)–(2.7′) as a 'back-and-forth
system from A to B'. The clash between their terminology and ours is quite
harmless; the two notions are near enough the same.

Lemma 3.2.2. *Let L be a signature and let A, B be L-structures. Then A and
B are back-and-forth equivalent if and only if there is a back-and-forth system
from A to B.*

Proof. Suppose first that A is back-and-forth equivalent to B, so that player
$\exists$ has a winning strategy σ for the game $\mathrm{EF}_\omega(A, B)$. Then we define I to
consist of the pairs of tuples which are of the form $(\bar{c}|n, \bar{d}|n)$ for some $n < \omega$
and some play $(\bar{c}, \bar{d})$ in which player $\exists$ uses σ.

   The set I is a back-and-forth system from A to B. First, putting $n = 0$ in
the definition of I, we see that I contains the pair of 0-tuples $(\langle\rangle, \langle\rangle)$. This
establishes (2.5). Next, (2.6) and (2.7) express that σ tells player $\exists$ what to do
at each step of the game. And finally (2.4) holds because the strategy σ is
winning.

   In the other direction, suppose that there exists a back-and-forth system I
from A to B. Define the set I^* of maps as above, and choose an arbitrary
well-ordering of I^*. Consider the following strategy σ for player $\exists$ in the
game $\mathrm{EF}_\omega(A, B)$:

(2.8)   at each step, if the play so far is $(\bar{a}, \bar{b})$ and player $\forall$ has just
chosen an element c from A, find the first map f in I^* such
that $\bar{a}$ and c are in the domain of f and $f\bar{a} = \bar{b}$, and then
choose d to be fc; likewise in the other direction if player $\forall$
has just chosen an element d from B.

By (2.5′)–(2.7′) there always will be a map f in I^* as required, so that (2.8)
really does define a strategy. Suppose the resulting play is $(\bar{a}, \bar{b})$. Then by
(2.4′) and Theorem 1.3.1(c) we have $(A, \bar{a}) \equiv_0 (B, \bar{b})$, and so player $\exists$ wins.

$\square$

Example 2: *Algebraically closed fields.* Let A and B be algebraically closed
fields of the same characteristic and infinite transcendence degree. We shall
show that A is back-and-forth equivalent to B. Let J be the set of all
isomorphisms $e: A' \to B'$ where A', B' are finitely generated subfields of A,
B respectively. (A **finitely generated subfield** of A is the smallest subfield of
A containing some given finite set of elements of A. It need not be finitely
generated as a ring; see Exercise 1.2.8.) Clearly J satisfies (2.4′). J is not
empty since the prime subfields of A, B are isomorphic. Thus (2.5′) is
satisfied. Suppose $f: A' \to B'$ is in J and c is an element of A. We want to
find a matching element d in B. There are two cases. First suppose c is
algebraic over A'. Then c is determined up to isomorphism over A' by its
minimal polynomial $p(x)$ over A'. Now f carries $p(x)$ to a polynomial $fp(x)$
over B', and B contains a root d of $fp(x)$ since it is algebraically closed.
Thus f extends to an isomorphism $g: A'(c) \to B'(d)$. Second, suppose c is
transcendental over A'. Since B' is finitely generated and B has infinite
transcendence degree, there is an element d of B which is transcendental
over B'. Thus again f extends to an isomorphism $g: A'(c) \to B'(d)$. Either
way, condition (2.6′) is satisfied. By symmetry, so is (2.7′). So J defines a
back-and-forth system from A to B, and hence A is back-and-forth equivalent to B by Lemma 3.2.2.

If $A \subseteq B$ in the example above, then we can say a little more. For every
finitely generated subfield C of A, there is a system J as above, such that
every map in J pointwise fixes C. In terms of back-and-forth systems, this
says that if $\bar{e}$ is a tuple of elements which generate C, then there is a
back-and-forth system I from A to B in which every pair has the form
$(\bar{e}\bar{a}, \bar{e}\bar{b})$.

The next result says that for countable structures, back-and-forth equivalence is the same thing as isomorphism.

Theorem 3.2.3. *Let L be any signature (not necessarily countable) and let A
and B be L-structures.*

(a) *If $A \cong B$ then A is back-and-forth equivalent to B.*

(b) *Suppose A, B are at most countable. If A is back-and-forth equivalent to B then $A \cong B$. In fact, if $\bar{c}$, $\bar{d}$ are tuples from A, B respectively, such that $(\bar{c}, \bar{d})$ is a winning position for player $\exists$ in $\mathrm{EF}_\omega(A, B)$, then there is an isomorphism from A to B which takes $\bar{c}$ to $\bar{d}$.*

Proof. (a) is a special case of Lemma 3.2.1(a).

(b) Since the game $\mathrm{EF}_\omega(A, B)$ has infinite length, if A and B are at most countable then player $\forall$ can list all the elements of A and of B among his choices. Let player $\exists$ play to win, and let $(\bar{a}, \bar{b})$ be the resulting play. Since player $\exists$ wins, the diagram lemma (Lemma 1.4.2) gives an isomorphism $f: A = \langle \bar{a} \rangle_A \to \langle \bar{b} \rangle_B = B$. The last sentence is proved the same way, but starting the play at $(\bar{c}, \bar{d})$.   ☐

Example 3: *Dense linear orderings without endpoints.* An old theorem of Cantor states that if A and B are countable dense linear orderings without endpoints (see (2.32) in section 2.2) then $A \cong B$. This follows at once from Theorem 3.2.3(b), when we show that A is back-and-forth equivalent to B. The required back-and-forth system consists of all pairs of tuples $(\bar{a}, \bar{b})$ such that, for some $n < \omega$, $\bar{a} = (a_0, \ldots, a_{n-1})$ is a tuple of elements of A, $\bar{b} = (b_0, \ldots, b_{n-1})$ is a tuple of elements of B, and, for all $i < j < n$, $a_i \geqslant a_j \Leftrightarrow b_i \geqslant b_j$.

Example 4: *Atomless boolean algebras.* Let A and B be countable atomless boolean algebras (see (2.30) in section 2.2). Then $A \cong B$. Again we show this by Theorem 3.2.3(b). Let J be the set of all isomorphisms from finite subalgebras of A to finite subalgebras of B. Then (2.4′) and (2.5′) clearly hold. For (2.6′), suppose $f \in J$ and let $a_0, \ldots, a_{k-1}$ be the atoms of the boolean algebra A' which is the domain of f. Then the isomorphism type of any element c of A over A' is determined once we are told, for each $i < k$, whether $c \wedge a_i$ is 0, a_i or neither. Since B is atomless, there is an element d of B such that for each $i < k$,

(2.9) $d \wedge f(a_i)$ is 0 (resp. $f(a_i)$) $\Leftrightarrow c \wedge a_i$ is 0 (resp. a_i).

So f can be extended to an isomorphism whose domain includes c. Thus J satisfies (2.6′), and by symmetry it satisfies (2.7′) too. We have proved that $A \cong B$.

Incidentally the results of Examples 3 and 4 are as false as they possible could be when we replace 'countable' by an uncountable cardinal κ. By results in section 11.3 below (for example Theorem 11.3.10) there are 2^κ non-isomorphic dense linear orderings and 2^κ non-isomorphic atomless boolean algebras, all of cardinality κ.

Exercises for section 3.2

1. Prove Lemma 3.2.1(b).

2. (a) Show that, in the game $EF_\gamma(A, B)$, a strategy σ for a player can be written as a family $(\sigma_i: i < \gamma)$ where for each $i < \gamma$, σ_i is a function which picks the player's ith choice $\sigma_i(\bar{x})$ as a function of the sequence $\bar{x}$ of previous choices of the two players. (b) Show that σ can also be written as a family $(\sigma_i': i < \gamma)$ where for each $i < \gamma$, σ_i' is a function which picks the player's ith choice $\sigma_i'(\bar{y})$ as a function of the sequence $\bar{y}$ of previous choices of the *other* player. (c) How can the functions σ_i be found from the functions σ_i', and vice versa? [To simplify the statements, assume that the domains of A and B are disjoint.]

3. (a) The game $P_\omega(A, B)$ is defined exactly like $EF_\omega(A, B)$ except that player $\forall$ must always choose from structure A and player $\exists$ from structure B. Show that, if A is at most countable, then player $\exists$ has a winning strategy for $P_\omega(A, B)$ if and only if A is embeddable in B. (b) What if player $\forall$ must choose from structure A in even-numbered steps and from structure B in odd-numbered steps (and player $\exists$ vice versa)?

4. The game $H_\omega(A, B)$ is defined exactly like $EF_\omega(A, B)$ except that player $\exists$ wins the play $(\bar{a}, \bar{b})$ iff for every atomic formula ϕ, $A \vDash \phi(\bar{a}) \Rightarrow B \vDash \phi(\bar{b})$. Show that, if A and B are at most countable, then player $\exists$ has a winning strategy for $H_\omega(A, B)$ if and only if B is a homomorphic image of A.

5. Suppose A and B are two countable dense linearly ordered sets without endpoints, both partitioned into classes $P_0, \ldots, P_{n-1}$ so that each class P_i occurs densely in both orderings. Show that A is isomorphic to B.

6. (a) If ζ is a linear ordering, let ζ^+ be the ordering which we get by replacing each point of ζ by a pair of points a, b with $a < b$. Show that if ζ and ξ are dense linear orderings without endpoints then ζ^+ and ξ^+ are back-and-forth equivalent. (b) Show that, if ζ and ξ are dense linear orderings which have first points but no last points, then ζ is back-and-forth equivalent to ξ.

7. Let A, B and C be fields; suppose $A \subseteq B$, $A \subseteq C$, and both B and C are algebraically closed and of infinite transcendence degree over A. Let (B, A) be the structure consisting of B with a 1-ary relation symbol P added so as to pick out A; and likewise with (C, A) and C. Show that (B, A) is back-and-forth equivalent to (C, A).

Suppose L, L' are signatures with no function symbols, and with no symbols in common. We form the signature $L + L'$ as follows: the symbols of $L + L'$ are those of L, those of L' and two new 1-ary relation symbols P and Q. The **disjoint sum** *of an L-structure A and an L'-structure B is the $(L + L')$-structure $A + B$ whose domain is the disjoint union of $\mathrm{dom}(A)$ and $\mathrm{dom}(B)$; the symbols of L are interpreted on* dom(A) *exactly as in A, and those of L' are interpreted on* dom(B) *as in B; P and Q are interpreted as names of* dom(A) *and* dom(B) *respectively.*

8. Show that if A, B are respectively back-and-forth equivalent to A', B', then their disjoint sum $A + B$ is back-and-forth equivalent to $A' + B'$.

9. Let L be the first-order language whose signature consists of one 2-ary relation symbol R. For each set X containing at least two elements, we form an L-structure $\mathrm{Grass}_2(X)$ as follows. The domain of $\mathrm{Grass}_2(X)$ is the set of all sets of the form $\{a, b\}$ where a, b are distinct elements of X. We put $R^{\mathrm{Grass}_2(X)}(\{a, b\}, \{c, d\})$ iff $\{a, b\}$ is disjoint from $\{c, d\}$. Show that if X and Y are infinite sets then $\mathrm{Grass}_2(X)$ is back-and-forth equivalent to $\mathrm{Grass}_2(Y)$. *('Grass' is short for Grassmannian – the name in this context is due to Cherlin.)*

10. Show that if A and B are back-and-forth equivalent rings, then A is left noetherian if and only if B is left noetherian.

3.3 Games for elementary equivalence

In section 3.2 we noticed that there are Ehrenfeucht–Fraïssé games $\mathrm{EF}_k(A, B)$ of finite length k. But we did nothing with them. In this section we shall see that after a small piece of cosmetic surgery, these finite games are a powerful tool of first-order model theory. They have hundreds of applications – I shall discuss just two.

First comes the surgery. In place of $\mathrm{EF}_k(A, B)$ we devise a game $\mathrm{EF}_k[A, B]$, which is played exactly like $\mathrm{EF}_k(A, B)$ but with a different criterion for winning. The players between them make k pairs of choices, and at the end of the play, when tuples $\bar{c}$ from A and $\bar{d}$ from B have been chosen, player $\exists$ wins the game $\mathrm{EF}_k(A, B)$ iff

(3.1) for every *unnested* atomic formula ϕ of L, $A \vDash \phi(\bar{c}) \Leftrightarrow B \vDash \phi(\bar{d})$.

If the signature L contains no function symbols or constants, then every formula of L is unnested anyway, so that $\mathrm{EF}_k(A, B)$ and $\mathrm{EF}_k[A, B]$ are exactly the same. So readers who are only interested in linear orderings can read square brackets as round brackets throughout. The games $\mathrm{EF}_k[A, B]$ are called **unnested Ehrenfeucht–Fraïssé games**.

We write $A \approx_k B$ to mean that player $\exists$ has a winning strategy for the game $\mathrm{EF}_k[A, B]$. Then $\approx_k$ is an equivalence relation on the class of L-structures – this is immediate from the argument of Lemma 3.2.1(c).

It will be useful to allow the structures to carry some parameters with them. Thus if $n < \omega$ and $\bar{a}$, $\bar{b}$ are n-tuples of elements of A, B respectively, we write $(A, \bar{a}) \approx_k (B, \bar{b})$ to mean that player $\exists$ has a winning strategy for the game $\mathrm{EF}_k[(A, \bar{a}), (B, \bar{b})]$. The condition for player $\exists$ to win this game, when the play has chosen k-tuples $\bar{c}$, $\bar{d}$ from A, B respectively, is just that

(3.2)   for every unnested atomic formula ϕ of L,
$$A \vDash \phi(\bar{a}, \bar{c}) \Leftrightarrow B \vDash \phi(\bar{b}, \bar{d}).$$

This is a restatement of the condition (3.1) with $(A, \bar{a})$ and $(B, \bar{b})$ in place of A and B.

Lemma 3.3.1. *Let A and B be structures of the same signature. Suppose $n, k < \omega$; suppose $\bar{a}$, $\bar{b}$ are n-tuples of elements of A, B respectively. Then the following are equivalent.*

(a) $(A, \bar{a}) \approx_{k+1} (B, \bar{b})$.

(b) *For every element c of A there is an element d of B such that $(A, \bar{a}, c) \approx_k (B, \bar{b}, d)$; and for every element d of B there is an element c of A such that $(A, \bar{a}, c) \approx_k (B, \bar{b}, d)$.*

Proof. First suppose (a) holds. Let c be an element of A. Then player $\exists$ can regard c as player $\forall$'s first choice in a play of $\mathrm{EF}_{k+1}[(A, \bar{a}), (B, \bar{b})]$. Let her use her winning strategy σ to choose d as her reply to c. Now if the two players decide to play the game $\mathrm{EF}_k[(A, \bar{a}, c), (B, \bar{b}, d)]$, player $\exists$ can win by regarding this second game as the last k steps in the play of $\mathrm{EF}_{k+1}[(A, \bar{a}), (B, \bar{b})]$, using σ to choose her moves. This proves the first half of (b), and the second half follows by symmetry.

Conversely suppose (b) holds. Then player $\exists$ can win the game $\mathrm{EF}_{k+1}[(A, \bar{a}), (B, \bar{b})]$ as follows. If player $\forall$ opens by choosing some element c of A, then player $\exists$ chooses d as in (b), and for the rest of the game she follows her winning strategy for $\mathrm{EF}_k[(A, \bar{a}, c), (B, \bar{b}, d)]$. Likewise if player $\forall$ started with an element d of B. $\qquad\square$

We turn to the fundamental theorem about the equivalence relations $\approx_k$ between structures of the form $(A, \bar{a})$ with A an L-structure. It will say among other things that for each k there are just finitely many equivalence classes of $\approx_k$, and that each equivalence class is definable by a formula of L. It will also put a bound on the complexity of these defining formulas, in terms of the following notion.

For any formula ϕ of the first-order language L, we define the **quantifier rank** of ϕ, $\mathrm{qr}(\phi)$, by induction on the construction of ϕ, as follows.

(3.3)   If ϕ is atomic then $\mathrm{qr}(\phi) = 0$.

(3.4) $\mathrm{qr}(\neg\psi) = \mathrm{qr}(\psi)$.

(3.5) $\mathrm{qr}(\bigwedge\Phi) = \mathrm{qr}(\bigvee\Phi) = \max\{\mathrm{qr}(\psi): \psi \in \Phi\}$.

(3.6) $\mathrm{qr}(\forall x\, \psi) = \mathrm{qr}(\exists x\, \psi) = \mathrm{qr}(\psi) + 1$.

Thus $\mathrm{qr}(\phi)$ measures the nesting of quantifiers in ϕ.

Theorem 3.3.2 *(Fraïssé–Hintikka theorem). Let L be a first-order language with finite signature. Then we can effectively find for each k, n < ω a finite set $\Theta_{n,k}$ of unnested formulas $\theta(x_0, \ldots, x_{n-1})$ of quantifier rank at most k, such that*

(a) *for every L-structure A, all k, n < ω and each n-tuple $\bar{a} = (a_0, \ldots, a_{n-1})$ of elements of A, there is exactly one formula θ in $\Theta_{n,k}$ such that $A \vDash \theta(\bar{a})$.*

(b) *for all k, n < ω and every pair of L-structures A, B, if $\bar{a}$ and $\bar{b}$ are respectively n-tuples of elements of A and B, then $(A, \bar{a}) \approx_k (B, \bar{b})$ if and only if there is θ in $\Theta_{n,k}$ such that $A \vDash \theta(\bar{a})$ and $B \vDash \theta(\bar{b})$.*

(c) *for every k < ω and every unnested formula $\phi(\bar{x})$ of L with n free variables $\bar{x}$ and quantifier rank at most k, we can effectively find a disjunction $\theta_0 \vee \ldots \vee \theta_{m-1}$ of formulas $\theta_i(\bar{x})$ in $\Theta_{n,r}$ which is logically equivalent to ϕ.*

Proof. Here I describe the sets $\Theta_{n,k}$. When we know what these sets are, an induction on k quickly gives (a). Property (b) is not hard to prove directly (the reader should try it). But we shall prove it as Theorem 3.5.2 below, as a special case of an important fact about closed games. Property (c) is an exercise.

We write ϕ^1 for ϕ and ϕ^0 for $\neg\phi$. We write $^m2$ for the set of maps $s: m \to 2$, taking m as $\{0, \ldots, m-1\}$ and 2 as $\{0,1\}$.

We begin with $k = 0$ and a fixed $n < \omega$. There are just finitely many unnested atomic formulas $\phi(x_0, \ldots, x_{n-1})$ of L; list them as $\phi_0, \ldots, \phi_{m-1}$. Take $\Theta_{n,0}$ to be the set of all formulas of the form $\phi_0^{s(0)} \wedge \ldots \wedge \phi_{m-1}^{s(m-1)}$ as s ranges over $^m2$. Thus $\Theta_{n,0}$ lists all the possible unnested quantifier-free types of n-tuples of elements of an L-structure. (In general it includes some impossible types too, bearing in mind formulas with a conjunct $x_0 \neq x_0$. But no matter.)

When $\Theta_{n+1,k}$ has been defined, we list the formulas in it as $\chi_0(x_0, \ldots, x_n)$, $\ldots, \chi_{j-1}(x_0, \ldots, x_n)$. Then we define $\Theta_{n,k+1}$ to be the set of all formulas

$$(3.7) \cdot \qquad \bigwedge_{i \in X} \exists x_n \chi_i(x_0, \ldots, x_n) \wedge \forall x_n \bigvee_{i \in X} \chi_i(x_0, \ldots, x_n)$$

as X ranges over the subsets of j. (Thus each formula in $\Theta_{n,k+1}$ lists the ways in which the n-tuple can be extended to an $(n+1)$-tuple, in terms of the formulas of quantifier rank k satisfied by the $(n+1)$-tuple.) $\qquad\square$

I shall refer to the formulas in the sets $\Theta_{n,k}$ as **formulas in game-normal form**, or more briefly **game-normal formulas**. By (c) of the theorem, every first-order formula ϕ is logically equivalent to a disjunction of formulas in game-normal form with at most the same free variables as ϕ. If ϕ was unnested, the game-normal formulas can be chosen to be of the same quantifier rank as ϕ; but note that the process of reducing a formula to unnested form (Theorem 2.6.1) will generally raise the quantifier rank.

Corollary 3.3.3. *Let L be a first-order language of finite signature. Then for any two L-structures A and B the following are equivalent.*
(a) $A \equiv B$.
(b) *For every $k < \omega$, $A \approx_k B$.*

Proof. By the theorem, (b) says that A and B agree on all unnested sentences of finite quantifier rank. So (a) certainly implies (b). By Corollary 2.6.2, every first-order sentence is logically equivalent to an unnested sentence of finite quantifier rank; so (b) implies (a) too. $\square$

Application 1: elimination sets

Ehrenfeucht–Fraïssé games form a useful tool for quantifier elimination. They give us a way of finding elimination sets by thinking about the structures themselves, rather than about the theories of the structures.

Suppose $\mathbf{K}$ is a class of L-structures. For each structure A in $\mathbf{K}$, write tup(A) for the set of all pairs $(A, \bar{a})$ where $\bar{a}$ is a tuple of elements of A; write tup($\mathbf{K}$) for the union of the sets tup(A) with A in $\mathbf{K}$. By an (**unnested**) **graded back-and-forth system** for $\mathbf{K}$ we mean a family of equivalence relations $(E_k : k < \omega)$ on tup($\mathbf{K}$) with the following properties:

(3.8) if $\bar{a}$, $\bar{b}$ are in tup(A), tup(B) respectively and $\bar{a} E_0 \bar{b}$, then for every unnested atomic formula $\phi(\bar{x})$ of L, $A \vDash \phi(\bar{a})$ iff $B \vDash \phi(\bar{b})$;

(3.9) if $\bar{a}$, $\bar{b}$ are in tup(A), tup(B) respectively, $\bar{a} E_{k+1} \bar{b}$ and c is any element of A, then there is an element d of B such that $\bar{a}c E_k \bar{b}d$.

Lemma 3.3.4. *Suppose $(E_k : k < \omega)$ is a graded back-and-forth system for $\mathbf{K}$. Then $(A, \bar{a}) E_k (B, \bar{b})$ implies $(A, \bar{a}) \approx_k (B, \bar{b})$.*

Proof. By (3.9), player $\exists$ can choose so that after the 0-th step in the game $\mathrm{EF}_k[(A, \bar{a}), (B, \bar{b})]$ we have $\bar{a}c_0 E_{k-1} \bar{b}d_0$, after the 1-th step we have $\bar{a}c_0c_1 E_{k-2} \bar{b}d_0d_1$, and so on until $\bar{a}\bar{c} E_0 \bar{b}\bar{d}$ after k steps. But then player $\exists$ wins by (3.8). $\square$

Lemma 3.3.5. *Suppose $(E_k : k < \omega)$ is a graded back-and-forth system for $\mathbf{K}$. Suppose that, for each n and k, E_k has just finitely many equivalence classes on n-tuples, and each of these classes is definable by a formula $\chi_{k,n}(\bar{x})$. Then the set of all formulas $\chi_{k,n}$ (k, $n < \omega$) forms an elimination set for $\mathbf{K}$.*

Proof. We have to show that each formula $\phi(\bar{x})$ of the language L is logically equivalent to a boolean combination of formulas $\chi_{k,n}(\bar{x})$ $(k, n < \omega)$. By Corollary 2.6.2 we can suppose that ϕ is unnested, and so by Theorem 3.3.2(c), ϕ is logically equivalent to a boolean combination of game-normal formulas $\theta(\bar{x})$. Now by Lemma 3.3.4, each equivalence class under $\approx_k$ is a union of equivalence classes of E_k. It follows by Theorem 3.3.2(b) that each game-normal formula $\phi(\bar{x})$ is equivalent to a disjunction of formulas $\chi_{k,n}(\bar{x})$.

$\blacksquare$

This is all the general theory that we need. Now we turn to an example.

Consider the ordered group of integers. The appropriate language for this group is a language L whose symbols are $+$, $-$, 0, 1 and $<$. The ordered group of integers forms an L-structure which we shall write as $\mathbb{Z}$.

Our aim is to find an elimination set for $\text{Th}(\mathbb{Z})$. Where to look? We cast around for possible equivalence relations E_k. Roughly speaking, we shall formalise the notion '$\bar{a}$ can't be distinguished from $\bar{b}$ without mentioning more than m elements'.

More precisely, suppose $\bar{x}$ is $(x_0, \ldots, x_{n-1})$ and m is a positive integer. Then by an m-**term** $t(\bar{x})$ we shall mean a term $\Sigma_{i<m} s_i$ where each s_i is either 0 or 1 or x_j or $-x_j$ for some $j < n$. Let $\bar{a} = (a_0, \ldots, a_{n-1})$ and $\bar{b} = (b_0, \ldots, b_{n-1})$ be two n-tuples of elements of $\mathbb{Z}$. We say that $\bar{a}$ is m-**equivalent** to $\bar{b}$ if for every m-term $t(\bar{x})$ the following hold in $\mathbb{Z}$:

(3.10) $$t(\bar{a}) > 0 \Leftrightarrow t(\bar{b}) > 0;$$

(3.11) $\qquad t(\bar{a})$ is congruent to $t(\bar{b})$ $(\text{mod}\, q)$ (for each integer q, $1 \leqslant q \leqslant m$).

Note that if $\bar{a}$ is m-equivalent to $\bar{b}$, then $\bar{a}$ is m'-equivalent to $\bar{b}$ for all $m' < m$.

Lemma 3.3.6. *Suppose $\bar{a}$ and $\bar{b}$ are n-tuples of elements of $\mathbb{Z}$ which are 3-equivalent. Then for every unnested atomic formula $\phi(\bar{x})$ of L, $\mathbb{Z} \vDash \phi(\bar{a})$ iff $\mathbb{Z} \vDash \phi(\bar{b})$.*

Proof. For example, $\mathbb{Z} \vDash a_0 + a_1 = a_2 \quad \Leftrightarrow \quad \mathbb{Z} \nvDash (a_0 + a_1 - a_2 > 0 \vee - a_0 - a_1 + a_2 > 0) \quad \Leftrightarrow \quad \mathbb{Z} \nvDash (b_0 + b_1 - b_2 > 0 \vee -b_0 - b_1 + b_2 > 0) \quad \Leftrightarrow \quad \mathbb{Z} \vDash b_0 + b_1 = b_2.$ $\blacksquare$

Lemma 3.3.7. *Suppose m is a positive integer, and $\bar{a}$ and $\bar{b}$ are n-tuples of elements of $\mathbb{Z}$ which are m^{2m}-equivalent. Then for every element c of $\mathbb{Z}$ there is an element d of $\mathbb{Z}$ such that the tuples $\bar{a}c$, $\bar{b}d$ are m-equivalent.*

Proof. Take an element c, and consider all the true sentences of the form

(3.12) $$t(\bar{a}) + ic \equiv j \, (\text{mod}\, q)$$

$\qquad\qquad$ where $t(\bar{x})$ is an $(m - 1)$-term, $0 < i < m$ and $j < q \leqslant m$.

Since $\bar{a}$ and $\bar{b}$ are m^{2m}-equivalent, $t(\bar{a})$ and $t(\bar{b})$ are certainly congruent modulo $m!$. Let α be the remainder when c is divided by $m!$. Then if d is any element of $\mathbb{Z}$ which is congruent to α modulo $m!$, we have

(3.13) $t(\bar{b}) + id \equiv j \pmod{q}$ whenever $t(\bar{a}) + ic \equiv j \pmod{q}$.

This tells us how to find a d to take care of the conditions (3.11).

Turning to the conditions (3.10), consider the set of all true statements of the forms

(3.14) $t(\bar{a}) + ic > 0, \; t(\bar{a}) + ic \leqslant 0$
                    where $t(\bar{x})$ is an $(m-1)$-term and $0 < i < m$.

After multiplying by suitable integers, we can bring these inequalities to the forms

(3.15) $t(\bar{a}) + m!c > 0, \; t(\bar{a}) + m!c \leqslant 0$
                    where $t(\bar{x})$ is an $m!.(m-1)$-term.

Taking greatest and least values in the obvious way, we can reduce (3.15) to a condition of the form

(3.16) $-t_1(\bar{a}) < m!c \leqslant -t_2(\bar{a})$,

together with a set of inequalities $\Phi(\bar{a})$ which don't mention c. (Possibly we reach a single inequality in (3.16), if $m!c$ is bounded only on one side.) So by (3.16), there is a number x in $\mathbb{Z}$ such that

(3.17) $-t_1(\bar{a}) < x \leqslant -t_2(\bar{a})$, and x is congruent to $m!\alpha \pmod{(m!)^2}$.

Now $-t_1(\bar{a})$ is at most an $m!.(m-1)$-term, and so by assumption it is congruent modulo $(m!)^2$ to $-t_1(\bar{b})$; similarly with $-t_2(\bar{a})$. Hence there is also a number y in $\mathbb{Z}$ such that

(3.18) $-t_1(\bar{b}) < y \leqslant -t_2(\bar{b})$, and y is congruent to $m!\alpha \pmod{(m!)^2}$.

Put $d = y/m!$. Then d is congruent to α modulo $m!$. We have

(3.19) $-t_1(\bar{d}) < m!d \leqslant -t_2(\bar{d})$

(cf. (3.16)), and the inequalities $\Phi(\bar{b})$ also hold since they use at worst $m! \cdot 2(m-1)$-terms. So tracing backwards along the path from (3.14) to (3.16), we have all the corresponding

(3.20) $t(\bar{a}) + ic > 0 \Leftrightarrow t(\bar{b}) + id > 0$
                    where $t(\bar{x})$ is an $(m-1)$-term and $0 < i < m$.

Thus d serves for the lemma. $\square$

We define $m_0, m_1, \ldots$ inductively by

(3.21) $m_0 = 3, \; m_{i+1} = m_i^{2m_i}$.

We define the equivalence relations E_k by $(\mathbb{Z}, \bar{a}) \, E_k \, (\mathbb{Z}, \bar{b})$ if $\bar{a}$ is m_k-equivalent to $\bar{b}$. By Lemmas 3.3.6 and 3.3.7, $(E_k: k < \omega)$ is a graded back-and-forth system for $\{\mathbb{Z}\}$. So by Lemma 3.3.5 we have found an elimination set for $\mathrm{Th}(\mathbb{Z})$. The formulas in the elimination set are all fairly simple: each of them is either an inequality or a congruence to some fixed modulus.

Theorem 3.3.8. Th($\mathbb{Z}$) *is decidable*.

Proof. First we show that for any tuple $\bar{a}$ in $\mathbb{Z}$ and any $k < \omega$ we can compute a bound $\delta(\bar{a}, k)$ such that

(3.22)   for every c there is d with $|d| < \delta(\bar{a}, k)$ such that $(\mathbb{Z}, \bar{a}c) E_k (\mathbb{Z}, \bar{a}d)$.

The calculations in the proof of Lemma 3.3.7 show that $\delta(\bar{a}, k)$ can be chosen to be $m^{2m} \cdot \mu$ where m is m_k and μ is $\max\{|a_i|: a_i \text{ occurs in } \bar{a}\}$.

It follows by induction on k that if $\phi(\bar{x})$ is a formula of L of quantifier rank k, and $\bar{a}$ is a tuple of elements of $\mathbb{Z}$, then we can compute in a bounded number of steps whether or not $\mathbb{Z} \vdash \phi(\bar{a})$. Suppose for example that ϕ is $\exists y\, \psi(\bar{x}, y)$, where ψ has quantifier rank $k - 1$. If there is an element c such that $\mathbb{Z} \vDash \psi(\bar{a}, c)$, then there is such an element $c < \delta(\bar{a}, k - 1)$. So we only need check the truth of $\mathbb{Z} \vDash \psi(\bar{a}, c)$ for these finitely many c; by induction hypothesis this takes only a finite number of steps.   □

The proof of Theorem 3.3.8 gives a primitive recursive bound $f(n)$ on the number of steps needed to check the truth of a sentence of length n. The bound $f(n)$ rises very fast with n. Exercising a little more thrift in the proof of Lemma 3.3.7, one can bring the bound down to something of the order of $2^{2^{2^{\kappa n}}}$ for some constant κ. But Fischer & Rabin [1974] show that any decision procedure for Th($\mathbb{Z}$) needs at least 2^{2^n} steps to settle the truth of sentences with n symbols (in the limit as n tends to infinity).

Corollary 3.3.9. *The linear orderings* $\omega^* + \omega$ *and* ω *both have decidable theories.*

Proof. For $\omega^* + \omega$ we have already proved it: a sentence ϕ of the first-order language of linear orderings is true in $\omega^* + \omega$ if and only if it is true in $\mathbb{Z}$. For ω we note that statements about ω can be translated into statements about the non-negative elements of $\mathbb{Z}$.   □

Application 2: replacements preserving ≡

The following theorem is a special case of more far-reaching results on cartesian products of structures; cf. section 9.6 below.

Theorem 3.3.10. *Let* G_1, G_2 *and* H *be groups. Assume* $G_1 \equiv G_2$. *Then* $G_1 \times H \equiv G_2 \times H$.

Proof. By Corollary 3.3.3 it suffices to show that if $k < \omega$ and $G_1 \approx_k G_2$ then $G_1 \times H \approx_k G_2 \times H$. Assume henceforth that $G_1 \approx_k G_2$. Then player $\exists$ has a winning strategy σ for the game $\text{EF}_k[G_1, G_2]$.

Let the two players meet to play the game $EF_k[G_1 \times H, G_2 \times H]$. This will be one of those many occasions when player $\forall$ will guide her choices by playing another game on the side. The side game will in fact be $EF_k[G_1, G_2]$. Whenever player $\forall$ offers an element, say the element $a \in G_1 \times H$, player $\exists$ will first split it into a product $a = g \cdot h$ with $g \in G_1$ and $h \in H$. Then she will pretend that player $\forall$ has just chosen g in the side game, and she will use her strategy σ to choose a reply $g' \in G_2$ in the side game. Her public reply to the element a will then be the element $b = g' \cdot h \in G_2 \times H$. Likewise the other way round if player $\forall$ chose from $G_2 \times H$.

At the end of the game let the play be $(g_0 \cdot h_0, \ldots, g_{k-1} \cdot h_{k-1}; g_0' \cdot h_0, \ldots, g_{k-1}' \cdot h_{k-1})$. Player $\exists$ has won the side game. Now the unnested atomic formulas of the language L of groups are the formulas of the form $x = y$, $1 = y$, $x_0 \cdot x_1 = y$ and $x^{-1} = y$. So for all $i, j, l < k$ we have

(3.23)
$$g_i = g_j \text{ iff } g_i' = g_j',$$
$$1 = g_i \text{ iff } 1 = g_i',$$
$$g_i \cdot g_j = g_l \text{ iff } g_i' \cdot g_j' = g_l',$$
$$g_i^{-1} = g_j \text{ iff } g_i'^{-1} = g_j'.$$

By the definition of cartesian products of groups, this implies that for all $i, j, l < k$ we also have

(3.24)
$$g_i \cdot h_i = g_j \cdot h_j \text{ iff } g_i' \cdot h_i = g_j' \cdot h_j,$$
$$1 = g_i \cdot h_i \text{ iff } 1 = g_i' \cdot h_i,$$
$$g_i \cdot h_i \cdot g_j \cdot h_j = g_l \cdot h_l \text{ iff } g_i' \cdot h_i \cdot g_j' \cdot h_j = g_l' \cdot h_l,$$
$$(g_i \cdot h_i)^{-1} = g_j \cdot h_j \text{ iff } (g_i' \cdot h_i)^{-1} = g_j' \cdot h_j.$$

(Thus for example $g_i \cdot h_i = g_j \cdot h_j \Leftrightarrow (g_i = g_j$ and $h_i = h_j) \Leftrightarrow (g_i' = g_j'$ and $h_i = h_j) \Leftrightarrow g_i' \cdot h_i = g_j' \cdot h_j$.) So player $\exists$ wins the public game too, which proves the theorem. $\qquad\qquad\square$

The proof of Theorem 3.3.10 uses very few facts about groups. It would work equally well in any case where a part of a structure can be isolated and replaced: for example if an interval in a linear ordering is replaced by an elementarily equivalent linear ordering, the whole resulting ordering is elementarily equivalent to the original one. Also we could take an infinite product of groups and make replacements at all factors simultaneously.

Exercises for section 3.3

1. Prove Theorem 3.3.2(c).

2. Prove the following variant of Theorem 3.3.2. Let L be a first-order language with finite signature, and suppose $k, n < \omega$. Let $\bar{x}$ be a tuple of variables. Then there is a

finite set $\Psi = \{\psi_i(\bar{x}: i < m\}$ of $\exists_n$ formulas of L such that (a) each ψ_i is an unnested formula of quantifier rank $\leq k$, (b) $\vdash \forall x$(at least one of $\psi_0, \ldots, \psi_{m-1}$ is true), (c) for every unnested $\exists_n$ formula $\phi(\bar{x})$ with quantifier rank $\leq k$ we can effectively find a subset Φ of Ψ such that ϕ is logically equivalent to $\bigvee\Phi$. [When $k \geq 1$, use the formulas $\Theta_{n+1,k-1}$ but add only existential quantifiers.]

3. Show that in the statement of Theorem 3.3.2 the formulas in $\Theta_{n,k}$ can all be taken to be $\exists_{k+1}$ formulas. Show also that they can all be taken to be $\forall_{k+1}$ formulas. [Prove both by simultaneous induction.]

4. Show that every unnested first-order formula of quantifier rank k is logically equivalent to an unnested first-order formula of quantifier rank k which is in negation normal form.

5. Find a simple set of axioms for Th$(\mathbb{Z}, +, <)$. [They should say that $\mathbb{Z}$ is an ordered group with a least positive element 1, and for every positive integer n there should be an axiom expressing that for all x, exactly one of $x, x+1, \ldots, x+n-1$ is divisible by n. Rework the proof of Theorem 3.3.8, using the class of all models of your axioms in place of $\mathbb{Z}$.]

6. Let L be the first-order language of linear orderings. (a) Show that if $h < 2^k$ then there is a formula $\phi(x, y)$ of L of quantifier rank $\leq k$ which expresses (in any linear ordering) '$x < y$ and there are at least h elements strictly between x and y'. (b) Let A be the ordering of the integers, and write $s(a, b)$ for the number of integers strictly between a and b. Show that if $a_0 < \ldots < a_{n-1}$ and $b_0 < \ldots < b_{n-1}$ in A, then $(A, a_0, \ldots, a_{n-1}) \approx_k (A, b_0, \ldots, b_{n-1})$ iff   for all $m < n-1$ and   all $i < 2^k$, $s(a_m, a_{m+1}) = i \Leftrightarrow s(b_m, b_{m+1}) = i$.

7. Show that if G, G', H and H' are groups with $G \leq G'$ and $H \leq H'$, then $G \times H \leq G' \times H'$.

The quantifier $Q_\alpha x$ was defined before Exercise 2.8.8.
8. Let A and B be L-structures and $\kappa = \omega_\alpha$ an infinite cardinal. We define games $G_k[A, B]$ just as $\mathrm{EF}_k[A, B]$, except that, at each step, player $\forall$ chooses either (i) an element c or (ii) a set X of at least κ elements of one structure. In case (i), player $\exists$ must then reply with an element d of the other structure. In case (ii), player $\exists$ must choose a set Y of at least κ elements of the other structure, player $\forall$ must choose an element d of Y and finally player $\exists$ must choose an element d of X; the pair (c, d) are then recorded in the play. The condition for player $\exists$ to win is the same as for $\mathrm{EF}_k[A, B]$. Show that A is $L(Q_\alpha)$-equivalent to B if and only if for every $k < \omega$, player $\exists$ has a winning strategy for the game $G_k[A, B]$.

9. Let α be an ordinal and L a first-order language containing a 2-ary relation symbol E. Show that there is no sentence of $L(Q_\alpha)$ which expresses 'E is an equivalence relation with at least ω_α equivalence classes'.

10. Show that there is no formula of first-order logic which expresses '$\langle a, b \rangle$ is in the transitive closure of R', even on finite structures. (For infinite structures it is easy to show there is no such formula.)

11. Let A and B be structures of the same signature. Immerman's **pebble game** on A, B of length k with p pebbles is played as follows. Pebbles $\pi_0, \ldots, \pi_{p-1}, \rho_0, \ldots, \rho_{p-1}$ are given. The game is played like $\mathrm{EF}_k[A, B]$, except that at each step, player $\forall$ must place one of the pebbles on his choice (one of the π_i if he chose from A, one of the ρ_i if he chose from B), then player $\exists$ must put the corresponding ρ_i (π_i) on her choice. (At the beginning the pebbles are not on any elements; later in the game the players may have to move pebbles from one element to another.) The condition for player $\exists$ to win is that after every step, if $\bar{a} = (a_0, \ldots, a_{p-1})$ is the sequence of elements of A with pebbles $\pi_0, \ldots, \pi_{p-1}$ resting on them (where we ignore any pebbles not resting on an element), and likewise $\bar{b} = (b_0, \ldots, b_{p-1})$ the elements of B labelled by $\rho_0, \ldots, \rho_{p-1}$, then for every unnested atomic $\phi(x_0, \ldots, x_{p-1})$, $A \vDash \phi(\bar{a}) \Leftrightarrow B \vDash \phi(\bar{b})$. Show that player $\exists$ has a winning strategy for this game if and only if A and B agree on all first-order sentences which have quantifier rank $\leqslant k$ and use at most p distinct variables.

*12. Let L be a first-order language with no function symbols. Show that if ϕ is a sentence of L in which just two variables x, y occur (though they can occur any number of times), and ϕ has a model, then ϕ has a finite model.

13. Let A and B be structures of the same signature. Show that A is back-and-forth equivalent to B if and only if player $\exists$ has a winning strategy for the game $\mathrm{EF}_\omega[A, B]$.

3.4 Closed games

An extraordinary number of basic ideas in model theory can be expressed in terms of games. There is a close connection between games and ordinal ranks. In this section we shall explore that connection; it will have applications throughout the book.

Imagine a game G of length ω between two players $\forall$ and $\exists$, of the following form. The equipment for the game consists of

(4.1) a partitioning of ω into two sets $M_\forall$ and $M_\exists$,

(4.2) a set X,

(4.3) for each $i < \omega$, a set W_i of ordered i-tuples of elements of X.

At the ith step ($i < \omega$), one of the players must choose an element of X; we write x_i for the chosen element. The player who makes the choice is player $\forall$ if $i \in M_\forall$, and player $\exists$ otherwise. After ω moves, a sequence $\bar{x} = (x_0, x_1, \ldots)$ of elements of X has been chosen. We call this sequence a **play** of the game G. The play $\bar{x}$ counts as a **win for player** $\exists$ (and so we say that player $\exists$ **wins the play**) if

(4.4)   for every $i < \omega$ the i-tuple $\bar{x}|i$ is in W_i;

otherwise $\bar{x}$ is a win for player $\forall$. Each player is allowed to know all the previous moves of either player. There may be some further restrictions on which elements of X a player can choose.

Games that fit this general description are known as **closed games**.

Example 1: Well-founded partial orders. Put $M_\exists = \omega$, let X be a partially ordered set and let each W_i be the set of all i-tuples $(x_0, \ldots, x_{i-1})$ such that $x_0 > \ldots > x_{i-1}$ in X. This defines a closed game G. Player $\exists$ plays G by choosing an infinite sequence $x_0, x_1, \ldots$ of elements of X, and she wins if the sequence is strictly decreasing. Meanwhile player $\forall$ twiddles his thumbs. Twiddling his thumbs is a winning strategy for him if and only if the partial ordering of X is well-founded (i.e. contains no infinite descending sequence).

Example 2: Svenonius games. Let L be a first-order language, and for each $n < \omega$ let $\phi_n(x_0, \ldots, x_{n-1})$ be a formula of L. Write ϕ for the expression

(4.5) $$\forall x_0 \exists x_1 \forall x_2 \exists x_3 \ldots \bigwedge_{n < \omega} \phi_n.$$

For every L-structure A, ϕ defines a closed game $Sv(\phi, A)$ as follows. Player $\forall$ chooses an element a_0 of A, then player $\exists$ chooses an element a_1, then player $\forall$ chooses a_2, and so on. After ω moves, player $\exists$ wins if

(4.6) $A \vDash \phi_n(a_0, \ldots, a_{n-1})$ for every $n < \omega$.

(Thus W_n is the set of n-tuples $\bar{a}$ such that $A \vDash \phi_n(\bar{a})$.) We write $A \vDash \phi$ to mean that player $\exists$ has a winning strategy for the game $Sv(\phi, A)$. Sentences ϕ of this form are known as **Svenonius sentences**, and the games $Sv(\phi, A)$ are called **Svenonius games**.

Svenonius games call for a remark. Let L and L^+ be countable first-order languages with $L \subseteq L^+$, and let T be a theory in L^+. In Lemma 10.4.8 below we shall construct a Svenonius sentence of L which is true in a countable L-structure A if and only if A has an expansion which is a model of T. Conversely let σ be a Svenonius sentence of L. Then a structure A is a model of σ if and only if player $\exists$ has a winning strategy for the game $Sv(\sigma, A)$. We can write out a strategy for player $\exists$ as a countable set of functions to be added to A, and the condition for player $\exists$ to win is then a theory in the language L with these functions added. So for countable structures, *Svenonius sentences have the same expressive strength as statements of the form 'I can be expanded to a model of theory T' where T is a countable theory*. In short, the game quantifier is a kind of second-order existential quantifier. One could pursue this further and find that the game quantifier is really a generalisation of the Suslin sieve, which is used for defining analytic sets of reals. See the survey by Kolaitis [1985].

In section 10.3 we shall learn to play games with longer well-ordered strings of quantifiers. There's a natural way of giving a sense to formulas with quantifier strings which are linearly ordered but not well-ordered. However, Shelah [1970] showed that anything expressible in this way can already be expressed using well-ordered quantifier strings.

Example 3: *Ehrenfeucht–Fraïssé games.* The   Ehrenfeucht–Fraïssé   games $EF_\omega(A, B)$ of section 3.2 were closed games. We could bring them to exactly the format of (4.1)–(4.3) above; but it's more convenient to count a pair of successive choices (first player $\forall$'s and then player $\exists$'s) as a single step in the play. So each W_n will be a set of $2n$-tuples. To be precise, W_n is the set of all $2n$-tuples $\bar{x} = (x_0, \ldots, x_{2n-1})$ such that for each $m < n$, if x_{2m} comes from A then x_{2m+1} comes from B and vice versa, and if we list the choices from A (resp. B) in order as $\bar{a}$ (resp. $\bar{b}$), then $(A, \bar{a}) \equiv_0 (B, \bar{b})$. For W_n it's irrelevant which player chose from A and which chose from B. So there is no loss if we describe the position $(x_0, \ldots, x_{2n-1})$ as a pair of n-tuples, $(\bar{a}, \bar{b})$, where $\bar{a}$ lists in turn the items x_m which came from A and $\bar{b}$ lists those which came from B.

When k is finite, we can look at the games $EF_k(A, B)$, $EF_k[A, B]$ of sections 3.2, 3.3 in the same way. The n-th position in a play of $EF_k(A, B)$ is a pair $(\bar{a}, \bar{b})$ of n-tuples from A and B respectively.

Example 4: *Cut-and-choose games on structures.* Let L be a first-order language and A an L-structure. The **Cantor–Bendixson game** on A measures how easy it is to chop up definable relations on A into smaller pieces. Let n be a positive integer and Σ the set of all formulas $\phi(x_0, \ldots, x_{n-1})$ of L with parameters from A. Let ψ be a formula in Σ. The Cantor–Bendixson game $CB(\psi, A)$ is played as follows. We write ψ as $\psi_{\langle\rangle}$. We count one move of player $\exists$ and the following move of player $\forall$ as a single step, much as in Example 3.

Player $\exists$ opens by choosing a countable family of formulas ψ_i ($i < \omega$) with the properties

(4.7) $$A \vDash \forall \bar{x} \, (\psi_i(\bar{x}) \to \psi_{\langle\rangle}(\bar{x})) \quad (i < \omega),$$

(4.8) $$A \vDash \forall \bar{x} \, \neg(\psi_i(\bar{x}) \wedge \psi_j(\bar{x})) \quad (i < j < \omega).$$

In effect, player $\exists$ chooses a countable family of disjoint definable subsets of $\psi_{\langle\rangle}(A^n)$. Then player $\forall$ chooses some $k_0 < \omega$. At the next step, player $\exists$ chooses a countable family of formulas $\psi_{k_0,i}$ ($i < \omega$) so that

(4.9) $$A \vDash \forall \bar{x} \, (\psi_{k_0,i}(\bar{x}) \to \psi_{k_0}(\bar{x})) \quad (i < \omega),$$

(4.10) $$A \vDash \forall \bar{x} \, \neg \, (\psi_{k_0,i}(\bar{x}) \wedge \psi_{k_0,j}\bar{x})) \quad (i < j < \omega);$$

then player $\forall$ chooses some $k_1 < \omega$. At the third step, player $\exists$ chooses formulas that refine ψ_{k_0, k_1}; and so on for ω steps. Player $\forall$ wins if he chooses some k_n so that

$$(4.11) \qquad A \vDash \forall \bar{x} \neg \psi_{k_0, \ldots, k_n}(\bar{x}).$$

In other words, player $\exists$ loses if at some point she chooses the empty relation (or strictly, a formula defining the empty relation).

For example, suppose $n = 1$ and A is the ordering of the rationals. Then player $\exists$ can win $CB(x = x, A)$; at each step she chooses a countable family of disjoint non-empty open intervals $\{c : A \vDash a < c < b\}$ lying inside the relevant interval.

Ranks of positions

By a **position** of a closed game G we mean tuple of possible moves in the game, starting at the beginning of the game. For example if $\bar{x}$ is a play of G, then for every $n < \omega$, $\bar{x}|n$ is a position of G. We say that a position $\bar{x}$ is **winning** for a player if the player has a strategy which tells him or her how to proceed, once the play has reached $\bar{x}$, in such a way that he or she is guaranteed to win. So for example the initial position $\langle \rangle$ of length 0 is winning for player $\exists$ if and only if player $\exists$ has a winning strategy for the game.

We can rank the positions of G according to how near player $\exists$ is to winning. Each position will have a rank which is either -1, an ordinal or ∞; ∞ counts as greater than every ordinal. The ranks are defined as follows. Let $\bar{x}$ be a position of length n.

$(4.12) \qquad \operatorname{rank}(\bar{x}) \geqslant 0 \qquad$ iff $\bar{x}|m \in W_m$ for all $m \leqslant n$.

$(4.13) \qquad \operatorname{rank}(\bar{x}) \geqslant \alpha + 1 \qquad$ iff however player $\forall$ moves, player $\exists$ can be sure that after the next choice y, $\operatorname{rank}(\bar{x}y) \geqslant \alpha$.

$(4.14) \qquad \operatorname{rank}(\bar{x}) \geqslant \delta \text{ (limit)} \qquad$ iff for all $\alpha < \delta$, $\operatorname{rank}(\bar{x}) \geqslant \alpha$.

If the choice of y belongs to player $\forall$, the condition in (4.13) says that $\operatorname{rank}(\bar{x}y) \geqslant \alpha$ for all y; if to player $\exists$, then it says that $\operatorname{rank}(\bar{x}y) \geqslant \alpha$ for some y.

Example 1 *continued.* In the game G on a partially ordered set X, the position $(x_0, \ldots, x_i)$ has rank $\geqslant 0$ iff $x_0 > \ldots > x_i$. Henceforth let us consider only positions which have rank $\geqslant 0$. The position $(x_0, \ldots, x_i)$ has rank $\geqslant 1$ iff there is some element $y < x_i$; so it has rank 0 iff x_i is a minimal element of X. Likewise it has rank 1 iff x_i is not minimal but has only minimal elements below it. In general $(x_0, \ldots, x_i)$ has rank $\geqslant \alpha + 1$ iff for every element y

below x_i, $(x_0, \ldots, x_i, y)$ has rank $\geq \alpha$. Note that in this game, the rank of a position $(x_0, \ldots, x_i)$ of rank ≥ 0 is always equal to $\mathrm{rank}(x_i)$, so we lose nothing if we think of the rank function as being defined on X.

In Example 4 too, the rank of a position depends only on the last formula $\psi_{k_0, \ldots, k_i}$ to be chosen; so we can think of the ranks as being ranks of formulas. Thus the **Cantor–Bendixson rank** of a formula $\psi(\bar{x})$ in Σ, $\mathrm{RCB}(\psi)$, is defined as follows.

(4.15) $\mathrm{RCB}(\psi) \geq 0$ iff $A \vDash \exists \bar{x}\, \psi$.

(4.16) $\mathrm{RCB}(\psi) \geq \alpha + 1$ iff there are $\psi_i\, (i < \omega)$ in Σ such that $\mathrm{RCB}(\psi_i) \geq \alpha$ for each $i < \omega$, and $A \vDash \forall \bar{x}(\psi_i \to \psi) \bigwedge \forall \bar{x} \neg (\psi_i \bigwedge \psi_j)$ for all $i < j < \omega$.

(4.17) $\mathrm{RCB}(\psi) \geq \delta$ (limit)   iff $\mathrm{RCB}(\psi) \geq \alpha$ for all $\alpha < \delta$.

For example ψ has Cantor–Bendixson rank 0 iff $\psi(A^n)$ is finite and not empty.

Thus in both Examples 1 and 4 the rank of a position is essentially given by the rank of its last item. This is an unusual feature of these two games; though whenever an ordinal rank of elements of some set is defined, it's usually a safe bet that there is a closed game with this feature lurking in the background.

Lemma 3.4.1. *If α is an ordinal such that every position of rank $\geq \alpha$ is also of rank $\geq \alpha + 1$, then every position of rank $\geq \alpha$ is winning for player $\exists$.*

Proof. Suppose $\bar{x} = (x_0, \ldots, x_{n-1})$ is a position of rank $\geq \alpha$. Then $\bar{x}$ also has rank $\geq \alpha + 1$ by assumption, and so player $\exists$ can ensure that $(x_0, \ldots, x_n)$ has rank $\geq \alpha$ too. For the same reason repeated ω times, player $\exists$ can ensure that $(x_0, \ldots, x_{i-1})$ has rank $\geq \alpha$ for each $i\, (n \leq i < \omega)$. Therefore if $\bar{z}$ is the resulting play, $\mathrm{rank}(\bar{z}|i) \geq 0$ for each $i\, (n \leq i < \omega)$. By (4.12) it follows that $\bar{z}|i \in W_i$ for each i, and so player $\exists$ wins.

Theorem 3.4.2. *A position $\bar{x}$ has rank ∞ if and only if it is winning for player $\exists$.*

Proof. First we prove left to right. Put $\alpha = \sup\{\mathrm{rank}(\bar{y}) + 1 : \bar{y}$ is a position of rank $< \infty\}$; there must be such an ordinal since there is only a set of possible positions. By the lemma, every position of rank $\geq \alpha$ is winning for player $\exists$.

Conversely we show by induction on β that if $\bar{x} = (x_0, \ldots, x_{n-1})$ is a winning position for player $\exists$ and β is an ordinal, then $\mathrm{rank}(\bar{x}) \geq \beta$. The interesting case is where β is a successor ordinal, say $\beta = \gamma + 1$. Let σ be a strategy for player $\exists$ which is winning from $\bar{x}$ onwards, and imagine the game continues according to σ. Then the position $(\bar{x}, x_n)$ will still be winning for player $\exists$, regardless of how player $\forall$ chooses. So by induction hypothesis, player $\exists$ can ensure that $\text{rank}(\bar{x}, x_n) \geqslant \gamma$. Then $\text{rank}(\bar{x}) \geqslant \beta$ by (4.13). $\square$

A game is said to be **determined** if one of the players has a winning strategy.

Corollary 3.4.3 *(Gale–Stewart theorem)*. *Every closed game is determined.*

Proof. If the initial position $\langle \rangle$ has rank ∞, then by the theorem, player $\exists$ has a winning strategy for the game. On the other hand suppose $\text{rank}(\langle \rangle) < \infty$. As the play proceeds, player $\exists$ can never make the rank go up (this is clear from (4.13)), and (by (4.13) again) player $\forall$ can drive the rank downwards so long as it is > -1. Since the ordinals are well-ordered, player $\forall$ can eventually push the rank down to -1 after a finite number of steps, and so he wins. $\square$

The next result will be useful in Chapter 11.

Corollary 3.4.4. *A position $\bar{z}$ is winning for player $\exists$ if and only if there are a linearly ordered set I with least element -1, and a map ρ from the set of positions to I such that for every $n < \omega$ and every position $\bar{x}$ of length n,*

(4.18) *if $\rho(\bar{x}) > -1$ then $\bar{x}|m \in W_m$ for all $m \leqslant n$,*

(4.19) *if $i \in I$ and $\rho(\bar{x}) > i$, then however player $\forall$ moves, player $\exists$ can be sure that, after the next choice y, $\rho(\bar{x}y) \geqslant i$,*

(4.20) *I contains some infinite descending sequence of elements $< \rho(\bar{z})$.*

Proof. Suppose first that $\bar{z}$ is winning for player $\exists$. In the scale of ranks, replace the top element ∞ by an infinite descending sequence $\infty_0 > \infty_1 > \ldots$, all greater than every ordinal; this defines the ordered set I. Put $\rho(\bar{x}) = \text{rank}(\bar{x})$ if $\text{rank}(\bar{x})$ is an ordinal or -1, and $\rho(\bar{x}) = \infty_{(\text{length of } \bar{x})}$ otherwise. Then (4.18) holds by (4.12), (4.19) holds by (4.13) and Theorem 3.4.2, and (4.20) holds by Theorem 3.4.2.

Conversely suppose I and ρ are given as stated; let $i_0 > i_1 > \ldots$ be an infinite descending sequence of elements of I below $\rho(\bar{z})$. Then by (4.19), starting from $\bar{z}$, player $\exists$ can play so as to ensure that if $\bar{x}$ is the resulting play, then for every $m \geqslant n$, $\rho(\bar{x}|m) \geqslant i_k$ for some $k < \omega$. By (4.18) it follows that $\bar{x}|m \in W_m$ for all $m < \omega$, and so player $\exists$ wins. $\square$

Formulas that define ranks

Often we find that there are formulas which express '$\bar{a}$ is a position of ranks at least α' in some closed game. We shall need this fact mainly for Ehrenfeucht–Fraïssé games in the next section. But I begin with Svenonius games, because they are a little simpler and the principle emerges more clearly.

Theorem 3.4.5. *Let ϕ be a Svenonius sentence of signature L, as in (4.5). Then for every $n < \omega$ and every ordinal α there is a formula $\theta^{n,\alpha}(x_0, \ldots, x_{n-1})$ of $L_{\infty\omega}$ such that for every L-structure A and n-tuple $\bar{a}$ from A,*

$$(4.21) \qquad A \vDash \theta^{n,\alpha}(\bar{a}) \quad \Leftrightarrow \quad \mathrm{rank}(\bar{a}) \geqslant \alpha \text{ in } \mathrm{Sv}(\phi, A).$$

Proof. We define $\theta^{n,\alpha}$ by induction on α, following (4.12)–(4.14):

$$(4.22) \qquad \theta^{n,0} \text{ is } \bigwedge_{i \leqslant n} \phi_i(x_0, \ldots, x_{i-1});$$

(4.23) $\theta^{n,\alpha+1}$ is either $\forall x_n \, \theta^{n+1,\alpha}$ or $\exists x_n \, \theta^{n+1,\alpha}$, according as the next move belongs to player $\forall$ or player $\exists$;

$$(4.24) \qquad \theta^{n,\delta} \ (\delta \text{ limit}) \text{ is } \bigwedge_{\alpha < \delta} \theta^{n,\alpha}. \qquad \qquad \square$$

We write ϕ^α for the sentence $\theta^{0,\alpha}$, and we call it the α-**th approximation to** ϕ. Thus ϕ^α is true in A if and only if the initial position in $\mathrm{Sv}(\phi, A)$ has rank at least α. From this and Theorem 3.4.2, we can read off the next result without further proof.

Corollary 3.4.6. *Let ϕ be a Svenonius sentence.*
  (a) *ϕ is equivalent to the conjunction $\bigwedge_{\alpha \text{ ordinal}} \phi^\alpha$ (a conjunction of a proper class of formulas!); in particular $\phi \vdash \phi^\alpha$ for every ordinal α.*
  (b) *For any ordinals $\alpha \leqslant \beta$, $\phi^\beta \vdash \phi^\alpha$.* $\qquad \square$

Exercises for section 3.4

1. Let G be a closed game, $\alpha < \beta$ ordinals and $\bar{x}$ a position of rank β. Show that there is a position $\bar{x}\bar{y}$ which extends $\bar{x}$ and has rank α.

2. Show that in any closed game G the following are equivalent, for any position $\bar{x}$ and any finite number n. (a) $\bar{x}$ has rank $\leqslant n$. (b) Player $\forall$ can guarantee to win the game in at most $n + 1$ further steps.

The next exercise gives a more direct proof of the Gale–Stewart theorem, Corollary 3.4.3.

3. Show that if G is any game of length $\leq \omega$ between two players $\forall$ and $\exists$, then either player $\forall$ has a winning strategy, or player $\exists$ has a strategy which will guarantee that no position during the game is winning for player $\forall$. Observe that in a closed game, if player $\forall$ never reaches a winning position, then player $\exists$ wins.

Not every game is determined.

4. Players $\forall$ and $\exists$ play the game $G(X)$ as follows, where $X \subseteq {}^\omega 2$. Player $\forall$ chooses x_0, then player $\exists$ chooses x_1, then player $\forall$ chooses x_2, etc.; after ω steps, player $\exists$ wins iff $(x_0, x_1, \ldots) \in X$. Show that X can be chosen so that neither player has a winning strategy. [List all possible strategies for either player as σ_i $(i < 2^\omega)$. List the elements of ${}^\omega 2$ as $(y_i : i < 2^\omega)$; by induction on i, put y_i into X or leave it out, in such a way as to defeat strategy σ_i.]

5. Let X be a partially ordered set. We define a rank function rk: $X \to$ ordinals $\cup \{\infty\}$ as follows: $\text{rk}(x) \geq \alpha + 1$ iff there is an element $y < x$ with $\text{rk}(y) \geq \alpha$; $\text{rk}(x) \geq \delta$ (limit) iff $\text{rk}(x) \geq \alpha$ for all $\alpha < \delta$. Show (by induction on the ordinal α) that a descending sequence $(x_0, \ldots, x_{n-1})$ has rank α in the game of Example 1 iff $\text{rk}(x_{n-1}) = \alpha$.

6. If σ is the Svenonius sentence (4.5), let $\tau(x_0, x_1)$ be the Svenonius formula

$$\forall x_2 \exists x_3 \ldots \bigwedge_{n < \omega} \phi_n.$$

Show that for every L-structure A, $A \vDash \sigma \leftrightarrow \forall x_0 \exists x_1 \tau$.

7. Consider the Svenonius sentence (4.5); call it σ, and let σ^* be the sentence

$$\exists x_0 \forall x_1 \exists x_2 \forall x_3 \ldots \bigvee_{n < \omega} \neg \phi_n.$$

(a) Define a game whch gives a meaning to the statement '$A \vDash \sigma^*$'. [In general this game will not be closed.] (b) Show that for every L-structure A, $A \vDash \sigma \leftrightarrow \neg \sigma^*$.

8. Consider structures $[\alpha, \beta]$ consisting of two ordinals, in the signature with ordering $\leq_1$ for the first ordinal and $\leq_2$ for the second. Show that there is a Svenonius formula $\phi(x, y)$ which defines the relation '$\alpha \leq \beta$' on these structures. [It describes a game in which player $\forall$ chooses one by one the elements of a descending sequence in α, and player $\exists$ must match them with a descending sequence in β.]

The next exercise anticipates Lemma 5.6.1 below.

9. Let L be a first-order language, A an L-structure, n a positive integer and Σ the set of all formulas $\phi(x_0, \ldots, x_{n-1})$ of L with parameters in A. Show that for every formula ϕ in Σ the following are equivalent. (a) ϕ has Cantor–Bendixson rank ∞. (b) For every sequence $\bar{s} \in {}^{<\omega}\omega$ there is a formula $\phi_{\bar{s}}$ in Σ such that (i) $\phi_{\langle\rangle}$ is ϕ, (ii) for every sequence $\bar{s}$ and $i < \omega$, $A \vDash \forall \bar{x}(\phi_{\bar{s}^\frown i} \to \phi_{\bar{s}})$, (iii) for every sequence $\bar{s}$ and all $i < j < \omega$, $\forall \bar{x} \neg(\phi_{\bar{s}^\frown i} \wedge \phi_{\bar{s}^\frown j})$, and (iv) for every sequence $\bar{s}$, $A \vDash \exists \bar{x} \, \phi_{\bar{s}}$. (c) The same as (b), but with ${}^{<\omega}2$ in place of ${}^{<\omega}\omega$ (and $i, j < 2$ in place of $i, j < \omega$). [The tree of formulas in (b) is just a winning strategy for player $\exists$. To get (b) from (c), rearrange the tree given in (c).]

10. Let A be the additive group of rationals and the ϕ the formula $x = x$. Show that ϕ has Cantor–Bendixson rank ∞. [Use (c) of the previous exercise; at the kth step divide into those elements which are and those which aren't divisible by the kth prime number.]

11. In Example 4, suppose $n = 1$ and $\psi(A)$ is not empty. (a) Show that $\psi(A)$ has Cantor–Bendixson rank 0 iff $\psi(A)$ is finite. (b) Show that if $\psi(A)$ is a minimal set, or a union of finitely many minimal sets, then it has Cantor–Bendixson rank 1 (see section 2.1).

12. Construct an example of a set $\psi(A)$ of Cantor–Bendixson rank 2. [Try the set of elements of order 4 in a direct sum of infinitely many cyclic groups of order 4.]

*A linear ordering η is said to be **scattered** if the ordering of the rational numbers is not embeddable in η.*
*13. The **Hausdorff game** $\mathrm{Haus}(\eta)$ on a linear ordering η is played as follows. First player $\exists$ chooses an element a_0 of η, so that $\eta = \zeta_1 + a_0 + \zeta_2$. Player $\forall$ chooses one of ζ_1 and ζ_2, call it ζ; then player $\exists$ chooses an element a_1 of ζ, so that $\zeta = \xi_1 + a_1 + \xi_2$. Then player $\forall$ chooses one of ξ_1 and ξ_2, etc. If player $\forall$ chooses an empty interval, then he wins and the game grinds to a halt; otherwise player $\exists$ wins after ω steps. (a) Show that player $\exists$ has a winning strategy for $\mathrm{Haus}(\eta)$ if and only if η is not scattered. (b) Show that the class $\mathbf{S}$ of scattered orderings is the smallest class of linear orderings such that (i) the empty ordering and the one-element ordering are in $\mathbf{S}$, and (ii) if ζ_i $(i < \alpha)$ are in $\mathbf{S}$ for some ordinal α, then the sums $\zeta_0 + \zeta_1 + \ldots$ and $\ldots + \zeta_1 + \zeta_0$ are in $\mathbf{S}$. [The game defines ranks of scattered orderings; prove the result by induction on rank.]

3.5 Games and infinitary languages

We shall look at the games $\mathrm{EF}_\omega(A, B)$ in detail, using the ideas of section 3.4. Our results will include some important links between these games and the language $L_{\infty\omega}$.

Recall from Example 3 of section 3.4 that we count a pair of moves in the game $\mathrm{EF}_\omega(A, B)$ as a single step. Accordingly a **position of length** n in the game $\mathrm{EF}_\omega(A, B)$ is a pair $(\bar{a}, \bar{b})$ consisting of an n-tuple $\bar{a}$ from A and an n-tuple $\bar{b}$ from B. (We ignore the question of which picked each a_i and which picked b_i; the winner doesn't depend on who chose the elements.) We can rank positions just as in (4.12)–(4.14) of section 3.4.

The theorem corresponding to Theorem 3.4.5 is a little more complicated to state, because there are two structures involved. We hold B fixed and talk about definability in A. The result is as follows.

Theorem 3.5.1. *Let B be an L-structure. Then for every ordinal α, every $n < \omega$ and every n-tuple $\bar{b}$ from B there is a formula $\theta^{n,\bar{b},\alpha}(x_0, \ldots, x_{n-1})$ (or,* *more pedantically, $\theta^{n,(B,\bar{b}),\alpha}$) of L such that if A is an L-structure and $\bar{a}$ is an n-tuple from A then*

(5.1) $A \vDash \theta^{n,\bar{b},\alpha}(\bar{a}) \Leftrightarrow$ *the position $(\bar{a}, \bar{b})$ has rank $\geq \alpha$ in* $\mathrm{EF}_\omega(A, B)$.

Proof. Again we write the formulas to match the definition, by induction on α.

(5.2) $\theta^{n,\bar{b},0}$ is $\bigwedge \{\psi(x_0, \ldots, x_{n-1}): \psi$ is a literal and $B \vDash \psi(\bar{b})\}$.

(5.3) $\theta^{n,\bar{b},\alpha+1}$ is $\bigwedge\limits_{b \text{ in } B} \exists x_n \, \theta^{n+1,\bar{b}b,\alpha} \wedge \forall x_n \bigvee\limits_{b \text{ in } B} \theta^{n+1,\bar{b}b,\alpha}$.

(5.4)   When δ is a limit, $\theta^{n,\bar{b},\delta}$ is $\bigwedge\limits_{\alpha<\delta} \theta^{n,\bar{b},\alpha}$.

In (5.3) the first conjunct says that, however player $\forall$ chooses from B, player $\exists$ can make a good reply from A; the second conjunct says the same with A and B the other way round.   □

The notation in Theorem 3.5.1 is rather top-heavy. We shall omit the superscript ^{n} in $\theta^{n,(B,\bar{b}),\alpha}$ when the context allows. Note that the theorem holds equally well for unnested Ehrenfeucht–Fraïssé games: we simply require ψ in (5.2) to be unnested. (Yes, perceptive reader, in the unnested case the formulas $\theta^{n,\bar{b},k}$ are remarkably like the game-normal formulas in $\Theta_{n,k}$ from section 3.3. There is a slight difference, though we never need to know this: the game-normal formulas included some redundant parts, such as unsatisfiable disjuncts, in order to get effectivity in Theorem 3.3.2.)

Theorem 3.5.1 will unlock doors for us. It sets up the crucial connection between the games $\mathrm{EF}_\omega(A, B)$ and the language $L_{\infty\omega}$. To understand the force of it, we need to define a ranking of formulas of $L_{\infty\omega}$, which will correspond to the ranking of positions in the games.

Let ϕ be a formula of $L_{\infty\omega}$. Then we define the **quantifier rank** of ϕ, $\mathrm{qr}(\phi)$, by induction on the construction of ϕ. The definition is word for word the same as for first-order formulas, (3.3)–(3.6) in section 3.3. For example *each formula $\theta^{(B,\bar{b}),\alpha}$ has quantifier rank exactly α.*

The class of formulas of $L_{\infty\omega}$ of quantifier rank $\leq \alpha$ is written $L^\alpha_{\infty\omega}$. $L^\alpha_{K\omega}$ is $L_{K\omega} \cap L^\alpha_{\infty\omega}$. We write $A \equiv^\alpha_{\infty\omega} B$ to mean that A and B are $L^\alpha_{\infty\omega}$-equivalent, i.e. that for every sentence ϕ of $L^\alpha_{\infty\omega}$, $A \vDash \phi \Leftrightarrow B \vDash \phi$.

We say that two L-structures A and B are α-**equivalent**, in symbols $A \sim^\alpha B$, if the initial position in the game $\mathrm{EF}_\omega(A, B)$ has rank $\geq \alpha$. The next theorem will show that $\sim^\alpha$ is an equivalence relation, as the name implies. Recall from section 3.2 that A is back-and-forth equivalent to B (in symbols $A \sim_\omega B$) if and only if player $\exists$ has a winning strategy for $\mathrm{EF}_\omega(A, B)$. So, by Theorem 3.4.2, $A \sim_\omega B$ if and only if for every ordinal α, $A \sim^\alpha B$. Thus the relations $\sim^\alpha$ are a hierarchy of approximations to back-and-forth equivalence.

Warning. In general the relation $\sim^\alpha$ just defined is not the same as the relation $\sim_\alpha$ defined in section 3.2 for $\alpha \leqslant \omega$. The relation $\sim^\omega$ is usually much weaker than $\sim_\omega$. See Exercise 4.

Theorem 3.5.2. *Let A and B be L-structures, $\bar{a}$ and $\bar{b}$ n-tuples from A and B respectively, and α an ordinal. Then the following are equivalent.*
(a) $A \vDash \theta^{(B,\bar{b}),\alpha}(\bar{a})$.
(b) $(A, \bar{a}) \equiv^\alpha_{\infty\omega} (B, \bar{b})$.
(c) $(A, \bar{a}) \sim^\alpha (B, \bar{b})$.

Proof. By Theorem 3.5.1, (a) says that $(\bar{a}, \bar{b})$ is a position of rank $\geqslant \alpha$ in $\mathrm{EF}_\omega(A, B)$. Now the game $\mathrm{EF}_\omega((A, \bar{a}), (B, \bar{b}))$ is exactly the same as the tail end of the game $\mathrm{EF}_\omega(A, B)$, starting from a point where the players have just reached position $(\bar{a}, \bar{b})$. So (a) is equivalent to (c). (A more formal proof would go by induction on α.)

(b) $\Rightarrow$ (a). Consider the position $(\bar{b}, \bar{b})$ in the game $\mathrm{EF}_\omega(B, B)$. Obviously it's a winning position for player $\exists$ (let her copy player $\forall$ for the rest of the game), and so it has rank $\geqslant \alpha$ by Theorem 3.4.2. Hence $B \vDash \theta^{(B,\bar{b}),\alpha}(\bar{b})$. But $\theta^{(B,\bar{b}),\alpha}$ has quantifier rank at most α. So (b) implies (a).

(a) $\Rightarrow$ (b). We prove that if $\phi(\bar{x})$ is any formula of $L^\alpha_{\infty\omega}$ and $(\bar{a}, \bar{b})$ is any position of rank $\geqslant \alpha$ in the game $\mathrm{EF}_\omega(A, B)$, then $A \vDash \phi(\bar{a}) \Leftrightarrow B \vDash \phi(\bar{b})$. The proof is for all α simultaneously, by induction on the construction of ϕ.

If ϕ is atomic, the result follows from the definition of winning: $(\bar{a}, \bar{b})$ is in W_n if and only if the same atomic sentences are true in $(A, \bar{a})$ as in $(B, \bar{b})$ (see Example 3 and (4.12) in section 3.4, and (5.2) above).

If ϕ is of the form $\neg\psi$, $\bigwedge\Phi$ or $\bigvee\Phi$, then we have the result straightforwardly from the induction hypothesis.

Next let ϕ be $\exists y\, \psi(\bar{x}, y)$. Write γ for $\mathrm{qr}(\psi)$. Then, by the definition of quantifier rank, $\gamma < \alpha$. Suppose $A \vDash \phi(\bar{a})$; then there is an element c in A such that $A \vDash \psi(\bar{a}, c)$. By (4.13) of section 3.4, since $(\bar{a}, \bar{b})$ has rank $\geqslant \gamma + 1$, there is an element d in B such that $(\bar{a}c, \bar{b}d)$ has rank $\geqslant \gamma$, and hence by induction hypothesis $A \vDash \psi(\bar{a}, c) \Leftrightarrow B \vDash \psi(\bar{b}, d)$. So $B \vDash \exists y\, \psi(\bar{b}, y)$ as required. Likewise in the other direction from B to A.

Finally suppose ϕ is $\forall y\, \psi$. We reduce this to the previous cases by writing $\neg\exists y\neg$ for $\forall y$. $\qquad\qquad\square$

Theorems of Karp and Scott

Two of the central results of infinitary model theory follow almost at once from Theorem 3.5.2.

Corollary 3.5.3 *(Karp's theorem). Let A and B be L-structures. Then A and B are back-and-forth equivalent if and only if they are $L_{\infty\omega}$-equivalent.*

Proof. Immediate from (b) $\Leftrightarrow$ (c) of the theorem with $\bar{a}$, $\bar{b}$ empty, together with Theorem 3.4.2. $\qquad\square$

For example, by Karp's theorem and Example 2 in section 3.2, any two algebraically closed fields of the same characteristic and infinite transcendence degree are $L_{\infty\omega}$-equivalent. We can say more. By the remark at the end of that example, if A and B are algebraically closed fields of infinite transcendence degree, $A \subseteq B$ and $\bar{e}$ is a tuple in A, then $(A, \bar{e})$ is $L_{\infty\omega}$-equivalent to $(B, \bar{e})$. This implies at once that $A \preccurlyeq B$.

In section 8.4 we shall use a completely different method to show that if A and B are any two algebraically closed fields and $A \subseteq B$ then $A \preccurlyeq B$, regardless of the transcendence degree.

Corollary 3.5.4 *(Scott's isomorphism theorem). Let L be a countable signature and B a countable L-structure. Then there is a sentence σ_B of $L_{\omega_1\omega}$ such that the models of σ_B are exactly the structures A which are back-and-forth equivalent to B. In particular B is (up to isomorphism) the only countable model of σ_B.*

Proof. For each $n < \omega$, there are only countably many pairs $(\bar{b}, \bar{b}')$ of n-tuples of elements of B; so there are only countably many ranks of such pairs as positions in $\mathrm{EF}_\omega(B, B)$. It follows that there is a least ordinal α_n such that for each such pair $(\bar{b}, \bar{b}')$, if $\mathrm{rank}(\bar{b}, \bar{b}') \geq \alpha_n$ then $\mathrm{rank}(\bar{b}, \bar{b}') = \infty$. The ordinal α_n is countable since every ordinal $< \alpha_n$ is the rank of some pair (cf. Exercise 3.4.1). Let α be the supremum of the α_n ($n < \omega$).

Let σ_B be the sentence

(5.5) $\theta^{(B),\alpha}$

$$\wedge \bigwedge_{n<\omega} \forall x_0, \ldots, x_{n-1} \bigwedge \{\theta^{(B,\bar{b}),\alpha}(\bar{x}) \rightarrow \theta^{(B,\bar{b}),\alpha+1}(\bar{x}):$$

$$\bar{b} \text{ an } n\text{-tuple of elements of } B\}.$$

Here $\theta^{(B),\alpha}$ is the sentence which is true in A if and only if $A \sim^\alpha B$. So certainly $B \vDash \theta^{(B),\alpha}$. The second conjunct of σ_B is true in B by the choice of α. Hence $B \vDash \sigma_B$. It follows by the previous corollary that if A is back-and-forth equivalent to B then $A \vDash \sigma_B$ too.

Suppose now that A is any model of σ_B, and that $(\bar{a}, \bar{b})$ is any position of rank $\geq \alpha$ in the game $\mathrm{EF}_\omega(A, B)$. Then $A \vDash \theta^{(B,\bar{b}),\alpha}(\bar{a})$, so $A \vDash \theta^{(B,\bar{b}),\alpha+1}(\bar{a})$ by the second conjunct of σ_B, and thus $(\bar{a}, \bar{b})$ has rank $\geq \alpha + 1$. It follows by Lemma 3.4.1 that every position of rank $\geq \alpha$ in $\mathrm{EF}_\omega(A, B)$ is winning for player $\exists$. But, by the first conjunct in σ_B, the initial position in $\mathrm{EF}_\omega(A, B)$ has rank $\geqslant \alpha$. Thus A is back-and-forth equivalent to B. The last sentence of the corollary follows by Theorem 3.2.3(b).   □

The sentence σ_B of the corollary is the **(canonical) Scott sentence** of B. The ordinal α in its proof is the **Scott height** (or **Scott rank**) of B. It measures the complexity of the family of definable relations in B.

Karp's theorem (Corollary 3.5.3) implies that if two structures of cardinality ω are $L_{\infty\omega}$-equivalent, then they are isomorphic. If κ is a regular uncountable cardinal and T is an unsuperstable complete countable first-order theory (for example the complete theory of an infinite linear ordering), then Shelah [1987a] shows that T has a family of 2^κ non-isomorphic models of cardinality κ which are $L_{\infty\kappa}$-equivalent to each other. See Hyttinen & Tuuri [1991] for more on this theme.

Exercises for section 3.5

1. Let $\phi(\bar{x})$ be a formula of $L_{\infty\omega}$ of quantifier rank α, with $\bar{x}$ finite. Show that ϕ is logically equivalent to a formula $\bigvee\Theta$, where every formula in Θ is of the form $\theta^{(B,\bar{b}),\alpha}$ for some B and $\bar{b}$.

2. Let $\mathrm{EF}_\omega^+(A, B)$ be the same as $\mathrm{EF}_\omega(A, B)$ except that player $\exists$ wins iff for every atomic formula $\phi(\bar{x})$, $A \vDash \phi(\bar{a}) \Rightarrow B \vDash \phi(\bar{b})$. Show that player $\exists$ has a winning strategy for $\mathrm{EF}_\omega^+(A, B)$ if and only if for every positive sentence ψ of $L_{\infty\omega}$, $A \vDash \psi \Rightarrow B \vDash \psi$.

3. Let $\mathrm{EF}_\omega^\exists(A, B)$ be the same as $\mathrm{EF}_\omega(A, B)$ except that player $\forall$ must always choose from A. Show that player $\exists$ has a winning strategy for $\mathrm{EF}_\omega^\exists(A, B)$ if and only if for every $\exists_1$ sentence ψ of $L_{\infty\omega}$, $A \vDash \psi \Rightarrow B \vDash \psi$.

4. (a) Let A and B be structures of the same signature. Show that if $k < \omega$ then $A \sim_k B$ iff $A \sim^k B$. Deduce that if the signature is finite and relational, then $A \sim^\omega B$ iff $A \equiv B$. (b) Hence find an example of A, B so that $A \sim^\omega B$ but not $A \sim_\omega B$.

5. Write out a full proof of the equivalence of (a) and (c) in Theorem 3.5.2, by induction on α.

6. Show that if A is back-and-forth equivalent to B and A is at most countable then A is elementarily embeddable in B.

*A ring R is said to be **right primitive** if R has a proper right ideal M such that, for every nonzero two-sided ideal I of R, $I + M = R$; or equivalently (see Jacobson [1980] section 4.1) if R has a faithful irreducible representation.*
7. Show that the class of right primitive rings is not definable by a sentence of $L_{\infty\omega}$. [Define $R(\kappa, \lambda)$ to be the ring $F[X_i][Y_j]$ ($i < \kappa$, $j < \lambda$) where F is a countable field, the

X_i are central indeterminates and the Y_j are non-commuting indeterminates. Show that $R(\kappa, \lambda)$ is right primitive if and only if $\kappa \leq \lambda$; but for all infinite κ, $R(\kappa^+, \kappa)$ is $L_{\infty\kappa}$-equivalent to $R(\kappa, \kappa)$.]

The following ridiculously loose-limbed argument proves left to right in Karp's theorem on the basis of a tiny amount of information about $L_{\infty\omega}$. It's a typical example of a style known as **soft model theory**.

*8. Let A and B be back-and-forth equivalent structures in a universe V of set theory. (a) Form a boolean extension V^* in which the cardinalities of A and B are collapsed to ω. Verify that A and B are still back-and-forth equivalent in V^*. (b) Deduce that $A \equiv_{\infty\omega} B$ in V^*, and hence also in V by the absoluteness of $\vDash$ (cf. Exercise 2.1.16).

*9. Let F be a function defined by a Σ_1 formula of set theory, such that in ZFC it is provable that for every structure A in the domain of F, if $A \cong B$ then $F(B)$ is defined and isomorphic to $F(A)$. Show that if $F(A)$ is defined and $A \equiv_{\infty\omega} B$ then $F(A) \equiv_{\infty\omega} F(B)$. *For example, if A and B are algebraically closed fields of the same characteristic and infinite transcendence degree, then the polynomial rings $A[X]$ and $B[X]$ are $L_{\infty\omega}$-equivalent.*

10. Let L be the signature consisting of one 1-ary function symbol F. An L-structure $A = (X, f)$ is **loopless** if there is no element b such that $f^n(b) = b$ for some $n \geq 1$. We catalogue the loopless L-structures as follows. Say that an element a is **connected to** an element b if one can reach b from a in a finite number of steps by passing from an element c to $f(c)$ or vice versa. This is an equivalence relation; its equivalence classes are the **connected components** of A. Reduce to the case where A has just one connected component. If b is any element of A, define for each ordinal α the α-**type** of b as follows. All elements have the same 0-type. For each α, the $(\alpha + 1)$-type of b specifies (1) the α-type of b, (2) the α-type of $f(b)$ and (3) how many elements c of each α-type have the property that $f(c) = b$. At limit ordinals δ, the δ-type of b specifies its α-type for each $\alpha < \delta$. Show that if $\kappa = |X|$ then there is some ordinal $\sigma < \kappa^+$ such that any two elements of X with the same σ-type have the same $(\sigma + 1)$-type. Show that A is defined up to isomorphism by (i) the ordinal σ and (ii) a listing of the σ-types which occur in A.

*11. Give an example of a sentence of $L_{\omega_1\omega}$ which has countable models of arbitrarily high Scott height $< \omega_1$, but no uncountable models.

3.6 Clubs

We finish this chapter with a refinement of the downward Löwenheim–Skolem theorem. It says that a structure is a model of a sentence ϕ if and only if 'enough' small substructures are models of ϕ. Theorems of this type are known as **local theorems**. See also section 6.6 for some local theorems based on the compactness theorem for first-order logic.

Let L be a countable signature and B an uncountable L-structure. Write $\mathcal{P}_\omega(B)$ for the set of all countable substructures of $\mathrm{dom}(B)$. A subset $\mathbf{C}$ of $\mathcal{P}_\omega(B)$ is called a **club** (= **c**losed **u**n**b**ounded set) if it has the following properties.

(6.1)   (Closure) If $(A_i : i < \omega)$ is a countable increasing chain of structures $A_i \in \mathbf{C}$, then $\bigcup_{i < \gamma} A_i \in \mathbf{C}$.

(6.2)   (Unboundedness) If A is any countable substructure of B, then some extension of A is in $\mathbf{C}$.

(Some prefer the more reassuring name **cub**.) We say that a subset of $\mathcal{P}_\omega(B)$ is **fat** if it contains a club. The set of all fat subsets of $\mathcal{P}_\omega(B)$ is called the **closed unbounded filter on** $\mathcal{P}_\omega(B)$ (or **over** B). (Using Theorem 3.6.2 below it is clear that the closed unbounded filter is a filter; see section 6.2 for the definition of filters.)

There is another way of describing the closed unbounded filter. Suppose $\mathbf{D}$ is a subset of $\mathcal{P}_\omega(B)$, and two players $\forall$ and $\exists$ play the following game on B: player $\forall$ chooses a finite set X_0 of elements of B, then player $\exists$ chooses a finite set X_1 of elements of B which contains X_0, then $\forall$ chooses a finite $X_2 \supseteq X_1$, and so on for ω steps. Player $\exists$ wins if $\bigcup_{i<\omega} X_i$ is the domain of a structure in $\mathbf{D}$; otherwise player $\forall$ wins.

Lemma 3.6.1. *Let $\mathbf{D}$ be a subset of $\mathcal{P}_\omega(B)$. Then the following are equivalent.*
(a) *$\mathbf{D}$ is fat.*
(b) *Player $\exists$ has a winning strategy for the game above.*
(c) *There is an expansion B^+ of B, still of countable signature, such that $\mathbf{D}$ contains all the L-reducts of countable substructures of B^+.*

Proof. (a) $\Rightarrow$ (b). Assume $\mathbf{D}$ is fat, so that $\mathbf{D}$ contains a club $\mathbf{C}$. Player $\exists$ wins by playing as follows. When player $\forall$ has just chosen X_{2n}, choose a structure C_{2n} which extends C_{2n-2}, contains X_{2n} and is in $\mathbf{C}$; this is possible by (6.2). (C_{-2} can be arbitrary.) Then make sure that each element of C_{2n} appears in some X_{2m+1} with $m \geqslant n$. Infinitely many C_{2n} appear during the play, each of them infinite, so this needs some jiggling. But player $\exists$ has countably many moves to take care of countably many elements, and this gives her enough breathing-room. At the end of the play, $\bigcup_{i<\omega} X_i = \bigcup_{n<\omega} \mathrm{dom}(C_{2n})$, which is in $\mathbf{C}$ by (6.1).

(b) $\Rightarrow$ (c). A winning strategy σ for player $\exists$ tells her what elements to play when a given finite sequence of finite sets has been played. We can turn this strategy into a set of functions on B, so that if the sequence $X_0, \ldots, X_{2n}$ calls forth the response Y from σ, then for every element c of Y, some function applied to a tuple listing X_{2n} will take the value c. If a countable substructure A of B is closed under all these functions, and player $\forall$ tries simply to list all the elements of A as his moves, then player $\exists$ using σ will never take the game outside dom(A). So the set of elements chosen in the play will be precisely dom(A); since the strategy σ is winning for $\exists$, the structure A must lie in **D**.

(c) $\Rightarrow$ (a). Exercise.   □

Theorem 3.6.2. *The closed unbounded filter is closed under countable inter-sections.*

Proof. Suppose $\mathbf{C}_m$ $(m < \omega)$ are clubs and $\mathbf{C} = \bigcap_{m<\omega}\mathbf{C}_m$. Then **C** is clearly closed under unions of countable chains, since each $\mathbf{C}_m$ is. If A is a countable substructure of **C**, then we can choose a chain of structures $A \subseteq A_0 \subseteq A_1 \subseteq \ldots$ in such a way that for each $m < \omega$, infinitely many of the A_i lie in $\mathbf{C}_m$. Then $\bigcup_{i<\omega}A_i$ lies in each $\mathbf{C}_m$ and hence in **C**.   □

Corollary 3.6.3. *For each first-order sentence ϕ of L, $B \vDash \phi$ if and only if the set $\{A: A$ is a countable substructure of B and $A \vDash \phi\}$ is fat.*

Proof. Suppose first that $B \vDash \phi$. Add Skolem functions to B, forming B^+. The set **C** of all L-reducts of countable substructures of B^+ is fat (in fact a club), and by Theorem 3.1.1(b) every structure in **C** is an elementary substructure of B and hence a model of ϕ.

Conversely if $B \vDash \neg\phi$, then, by the same argument, there is a fat set **D** consisting of models of $\neg\phi$. But now the set $\{A: A$ is a countable substructure of B and $A \vDash \phi\}$ can't possibly be fat, since then its intersection with **D** would be fat by Theorem 3.6.2, which is absurd.   □

This is only a very small taste of what one can do with clubs. The exercises say a little more – for example Corollary 3.6.3 extends to sentences of $L_{\omega_1\omega}$. The definitions of $\mathscr{P}_\omega(B)$ and fat sets are usually given for the case where the structure B is just a set, so that substructures are subsets of B. A subset of $\mathscr{P}_\omega(B)$ is called **thin** if it's the complement of a fat set, and **stationary** if it's not thin. One should think of 'fat', 'thin' and 'stationary' as the analogues of 'measure 1', 'measure 0' and 'positive measure' respectively.

There is a well-known notion of stationary subsets of an ordinal (see Facts 11.3.2 and 11.3.3). By generalising from $\mathscr{P}_\omega(B)$ to $\mathscr{P}_\lambda(B)$ where λ is any infinite cardinal, one can make that a special case of our notion (cf. Exercises 3, 4 below).

Exercises for section 3.6

1. Let B be an uncountable set. (a) Let $\mathbf{C}_b$ $(b \in B)$ be clubs in $\mathscr{P}_\omega(B)$. We define the **diagonal intersection** $\triangle_{b\in B}\mathbf{C}_b$ to be the set $\{D \in \mathscr{P}_\omega(B): D \in \mathbf{C}_b$ for all $b \in D\}$. Show

that $\Delta_{b \in B} C_b$ is a club. (b) *A form of Födor's lemma*: Let **S** be a stationary set in $\mathcal{P}_\omega(B)$, and suppose that $f: \mathbf{S} \to B$ is such that for every A in **S**, $f(A) \in A$. Show that f is constant on some stationary set. [Use (a).] (c) Let **S** be a stationary set in $\mathcal{P}_\omega(B)$, and let g be a map defined on **S**, such that for each $A \in \mathbf{S}$, $g(A)$ is a finite subset of B. Show that there is a stationary set $\mathbf{R} \subseteq \mathbf{S}$ such that g is constant on **R**.

*2. Show that the club filter on $\mathcal{P}_\omega(\omega_1)$ is not an ultrafilter. $[|\mathcal{P}_\omega(\omega_1)|$ is not measurable!]

In the next five exercises, λ is any finite cardinal.
3. Show that if we replace ω and 'countable' everywhere by λ and 'of cardinality λ', the following still hold. (a) The implication (c) $\Rightarrow$ (a) in Lemma 3.6.1. (b) Theorem 3.6.2. (c) Exercise 1.

4. Show that the ordinals $< \lambda^+$ form a club in $\mathcal{P}_\lambda(\lambda^+)$.

5. Let **S** be a stationary set in $\mathcal{P}_\lambda(\lambda^+)$. (a) Let f be a map from **S** to λ^+. Show that there is a stationary set $\mathbf{R} \subseteq \mathbf{S}$ such that f is either constant or injective on **R**. (b) Let f, g be maps from **S** to λ^+ such that $f(X) \neq g(X)$ for all $X \in \mathbf{S}$. Show that there is a stationary set $\mathbf{R} \subseteq \mathbf{S}$ such that $f(\mathbf{R})$ is disjoint from $g(\mathbf{R})$.

6. Suppose a group G of cardinality $< \lambda$ is a direct sum of copies of a group H of cardinality $\leqslant \lambda$. Show that in $\mathcal{P}_\lambda(G)$ there is a fat set of subgroups which are direct sums of copies of H.

7. Let $|L| \leqslant \lambda$ and let B be an L-structure of cardinality $> \lambda$. Let α be an automorphism of B. Show that in $\mathcal{P}_\lambda(B)$ there is a club of substructures of B which are closed under α.

*8. A group G is said to be **characteristically simple** if the only subgroups H of G such that $\alpha(H) = H$ for every automorphism α of G are F and $\{1\}$. Show that if G is an uncountable characteristically simple group, then there is a club of countable characteristically simple subgroups of G.

*9. Let G be a group of uncountable cardinality, and let **K** be the class of groups which are direct sums of non-abelian simple groups. Show that if there is a fat subset **F** of $\mathcal{P}_\omega(G)$ with $\mathbf{F} \subseteq \mathbf{K}$, then G is in **K**.

*10. A group G is said to be **residually finite** if for every element $a \neq 1$ in G there is a homomorphism $h: G \to H$ with H finite, such that $f(a) \neq 1$. (a) Show that the class of residually finite groups is defined by an $\forall_1$ sentence of $L_{\omega_1\omega}$, where L is the language of groups. [Define the class of groups G which are not residually finite: for some element $g \neq 1$, if $n < \omega$ then by the compactness theorem (Theorem 6.1.1) there is some p.p. formula $\phi(g)$ true in G, such that any group satisfying $\phi \wedge g \neq 1$ must have at least n elements.] (b) Deduce that if G is a group and every countable subgroup of G is contained in a residually finite subgroup of G, then G is residually finite.

History and bibliography

Section 3.1. Skolem introduced Skolem functions in [1920], and used them to prove that if T is a theory in a countable first-order language, and T has a model, then T has a model which is at most countable. (In fact he noted that the same argument gives the same result for some infinitary theories.) Löwenheim [1915] had proved this when T consists of a single first-order sentence. See Skolem [1922] for a proof (not using the axiom of choice) that, if a countable first-order theory has a model, then it has a countable model; the same paper also contains Skolem's amazing views about the implications of his result for set theory. The downward Löwenheim–Skolem theorem in the form of Corollary 3.1.5 was first stated in Tarski & Vaught [1957]. It is sometimes known as the Downward Löwenheim–Skolem–Tarski Theorem; there are tales of a logician who turned up at Cornell and asked to meet Professor Downward. Example 3 appears in van den Dries [1984].

Exercise 9: Cole & Dickmann [1972]. Exercise 10: Skolem [1922]. Exercise 14: Thomas [1983b].

Section 3.2. Cantor's theorem on countable dense linear orderings without endpoints is from Cantor [1895], but the back-and-forth proof is due to Huntington [1904b] and Hausdorff [1914] (see Plotkin [1990] for the history). The back-and-forth idea was soon used in various other special cases, notably by Ulm [1933] in connection with abelian groups. The notion reached model theory through the results of section 3.3. Back-and-forth methods are often ascribed to Cantor, Bertrand Russell and C. H. Langford [1926], but there is no evidence to support any of these attributions.

Exercise 5: Skolem [1920]. Exercise 10: Sabbagh & Eklof [1971].

Section 3.3. The finite Ehrenfeucht–Fraïssé games were studied in Fraïssé [1955], [1956b], Ehrenfeucht [1961] and independently Taïmanov [1962]; the formulation in terms of games is due to Ehrenfeucht. Theorem 3.3.2(a, b) is from Fraïssé [1956b] and Ehrenfeucht [1961]. Hintikka [1953] described the game-normal formulas (he named them **distributive normal forms**) and supplied Theorem 3.3.2(c). Theorem 3.3.8 is from the Master's thesis of Presburger [1930] (more precisely he gave a quantifier elimination for the theory of $(\omega, +, <)$, after his supervisor Tarski had suggested the solution to Exercise 5). Langford [1926] proved Corollary 3.3.9. For sharper upper bounds on the complexity of an algorithm for $\mathrm{Th}(\mathbb{Z})$, see Ferrante & Rackoff [1979], who use a variant of the method of this section. Mekler [1984] has an interesting variant of the model-theoretic method behind Theorem 3.3.8; he calls it **neighbourhood systems**. Theorem 3.3.10 is from Mostowski [1952a].

Exercise 2: on indirect evidence I believe some form of this must be in Weinstein [1965]. Exercise 8: Vinner [1972]. Exercise 9: Keisler [1970]. Exercise 11: Immerman [1982]. Exercise 12: Mortimer [1975].

Section 3.4. The whole of this section is thinly disguised descriptive set theory. The technology of ranks was first developed by Sierpiński and Lusin in a series of papers on analytic sets in the 1920s (see section 1.VI of Kuratowski [1958]). The notion of closed games and Corollary 3.4.3 are from Gale & Stewart [1953]. Svenonius sentences are from Svenonius [1965]; they are discussed fully in Moschovakis [1974]. Cut-and-choose games like that of Example 4 lie at the heart of stability theory; they were introduced by Morley [1965a], who also gave a version of Exercise 9.

Exercise 3: here one should mention the deep and difficult theorem of Martin [1975]: if we put the discrete topology on ω and give $^\omega\omega$ the product topology, then the game $G(X)$ is determined whenever X is a Borel subset of $^\omega\omega$. Exercise 7: Shelah [1971d]. Exercise 13: Hausdorff [1908].

Section 3.5. Karp's theorem, Corollary 3.5.3, was proved by Karp [1965] without the apparatus of ranks. Theorems 3.5.1 and 3.5.2 are the immediate analogues for arbitrary α of theorems of Fraïssé [1956a] for finite α. Corollary 3.5.4 was proved by Scott [1965], and reportedly also by Engeler (unpublished); the formulation (5.5) is from Chang [1968b]. Scott heights were studied in connection with **admissible model theory**, where one works with the structures inside some small model of Kripke–Platek set theory; see for example Barwise [1975], Makkai [1977] and Nadel [1985]. Kueker [1981] gives a characterisation of $L_{\infty\omega_1}$-equivalence of structures of cardinality ω_1, in terms of how the structures are built up from smaller structures.

Exercise 7: Lawrence [1981b] and Mekler [1982]. Exercise 8: Barwise [1972]. Exercise 9: Nadel [1972]. Exercise 10: Shelah (unpublished). Exercise 11: Makkai [1981].

Section 3.6. The club filter is an offshoot of work on large cardinals in set theory round about 1970 (e.g. in Magidor [1971]), but it has venerable ancestors going back at least to Mahlo [1911]. Corollary 3.6.3 is from Mycielski [1965] and Lemma 3.6.1 from Kueker [1977]. Local theorems are part of the stock in trade of group theory; for the use of clubs in this connection see Eklof [1980], Eklof & Mekler [1990], Hickin & Phillips [1973], Hodges [1981a].

Exercise 1: (a, b) are in Jech [1973a] and (c) in Hickin & Phillips [1973]. Exercise 2: Ulam [1930]. Exercise 5: (a) and (b) are attributed to Węglorz and Shelah respectively, in Mekler, Pelletier & Taylor [1981]. Exercise 8: Kopperman & Mathias [1968] and Phillips [1971]. Exercise 9: Hickin [1973]. Exercise 10: Mekler [1980].

4

Automorphisms

Pargeting on a wall in Saffron Walden

The twenty-three year old Felix Klein in his famous Erlanger Programm [1872] proposed to classify geometries by their automorphisms. He hit on something fundamental here: in a sense, *structure is whatever is preserved by automorphisms*. One consequence – if slogans can have consequences – is that a model-theoretic structure implicitly carries with it all the features which are set-theoretically definable in terms of it, since these features are preserved under all automorphisms of the structure.

Model theorists have recently become very interested in these implicit features. Partly it's a matter of technique: how does one show that this or that item is or is not definable in terms of a given structure? But there is no doubt that the original stimulus was Boris Zil'ber's [1980b] realisation that the automorphism groups of models of an uncountably categorical first-order theory can (in part) be found already sitting inside the models. So it became important to know what groups are interpretable in these models. Zil'ber (loc. cit. with a correction in [1981]), Hrushovski [1990] and Laskowski [1988]

131

all used ideas along these lines to solve problems of general model theory which don't mention groups.

With his usual foresight, Shelah had already provided some of the machinery needed for handling this. Stability theory has been invaluable; so have the notions of imaginary elements and canonical bases (which make sense even outside stable theories). At this stage in the book we are not yet ready to define what a stable theory is. But some of the more startling results in stability theory can be proved here in a rudimentary form, using strongly minimal sets. This treatment may even be helpful, where it brings out the essentially geometric ideas behind the arguments. In this spirit I finish the chapter with the beautiful theorem of Zil'ber and Hrushovski on groups interpretable in modular geometries.

During the Logic Colloquium at Granada in 1987 I ran into John Baldwin in one of the rooms of the Alhambra. I remarked that the claim in Coxeter & Moser [1957] 4.1 that all seventeen two-dimensional crystallographic groups occur in the decorations of the Alhambra seems to be an exaggeration (I couldn't find more than nine or ten, though it's certainly more than one meets in the decorations of Saffron Walden). Never mind, said John, here's a theorem: every crystallographic group is superstable with finite Lascar rank (Baldwin [1989]).

4.1 Automorphisms

Let A be an L-structure. Every automorphism of A is a permutation of $\mathrm{dom}(A)$. By Theorem 1.2.1(c), the collection of all automorphisms of A is a group under composition. We write $\mathrm{Aut}(A)$ for this group, regarded as a permutation group on $\mathrm{dom}(A)$, and we call it the **automorphism group** of A.

Automorphism groups have traditionally been studied by group theorists and geometers, in settings remote from model theory. We need some translations between model theory and group theory. This section will make a start.

Aut(A) as a permutation group

For any set Ω, the group of all permutations of Ω is called the **symmetric group** on Ω, in symbols $\mathrm{Sym}(\Omega)$. Several important properties of a structure A are really properties of its automorphism group as a subgroup of $\mathrm{Sym}(\mathrm{dom}\,A)$. In the next few definitions, suppose G is a subgroup of $\mathrm{Sym}(\Omega)$.

First, if X is a subset of Ω, then the **pointwise stabiliser** of X in G is the set $\{g \in G: g(a) = a$ for all $a \in X\}$. This set forms a subgroup of G, and we write it $G_{(X)}$ (or $G_{(\bar{a})}$ where $\bar{a}$ is a sequence listing the elements of X). The **setwise stabiliser** of X in G, $G_{\{X\}}$, is the set $\{g \in G: g(X) = X\}$, which is also a subgroup of G. In fact we have $G_{(X)} \subseteq G_{\{X\}} \subseteq G$.

If a is an element of Ω, the **orbit** of a under G is the set $\{g(a): g \in G\}$. The orbits of all elements of Ω under G form a partition of Ω. If the orbit of every element (or the orbit of one element – it comes to the same thing) is the whole of Ω, we say that G is **transitive on** Ω.

We say that a structure A is **transitive** if $\mathrm{Aut}(A)$ is transitive on $\mathrm{dom}(A)$. The opposite case is where A has no automorphisms except the identity 1_A; in this case we say that A is **rigid**. Here are two examples.

Example 1: *Ordinals*. Let the structure A be an ordinal $(\alpha, <)$, so that $<$ well-orders the elements of A. Then A is rigid. For suppose f is an automorphism of A which is not the identity. Then there is some element a such that $f(a) \neq a$; replacing f by f^{-1} if necessary, we can suppose that $f(a) < a$. Since f is a homomorphism, $f^2(a) = f(f(a)) < f(a)$, and so by induction $f^{n+1}(a) < f^n(a)$ for each $n < \omega$. Then $a > f(a) > f^2(a) > \ldots$, contradicting that $<$ is a well-ordering.

Example 2: *Shelah's all-purpose counterexample*. Let D be the direct sum of countably many cyclic groups of order 2. (Or equivalently, let D be a countable-dimensional vector space over the two-element field $\mathbb{F}_2$.) On D we define a relation

(1.1) $$R(x, y, z, w) \Leftrightarrow x + y = z + w.$$

The structure A consists of the set D with the relation R. If d is any element of D, there is a permutation e_d of the set D, defined by $e_d(a) = a + d$. This permutation is clearly an automorphism of A taking 0 to d – which shows that A is a transitive structure. Also if we fix d, we can make D into an abelian group with d as the identity, by defining an addition operation $+_d$ in terms of R:

(1.2) $$x +_d y = z \Leftrightarrow R(x, y, z, d).$$

So A is what remains of the group D when we forget which element is 0. It is known to geometers as the **countable-dimensional affine space over** $\mathbb{F}_2$. (See Example 3 in section 4.5 for affine spaces over any field.)

Returning to our group G of permutations of Ω, we write Ω^n for the set of all ordered n-tuples of elements of Ω (where n is a positive integer). Then G automatically acts as a set of permutations of Ω^n too, putting $g(a_0, \ldots, a_{n-1}) = (ga_0, \ldots, ga_{n-1})$. So we can talk about the **orbits** of G on Ω^n. When n is greater than 1 and Ω has more than one element, then G is certainly not transitive on Ω^n. But model theory has a special regard for the following possibility, which is not far off transitivity.

We say that G is **oligomorphic** (on Ω) if for every positive integer n, the number of orbits of G on Ω^n is finite. We say that a structure A is **oligomorphic** if $\mathrm{Aut}(A)$ is oligomorphic on $\mathrm{dom}(A)$. (Cameron [1990] credits the name to Roger Green.) In section 7.3 below we shall see that for countable structures, oligomorphic is the same thing as ω-categorical; this will give us dozens of examples. But for the moment, consider the ordered set $A = (\mathbb{Q}, <)$ of rational numbers. If $\bar{a}$ and $\bar{b}$ are any two n-tuples whose elements are in the same relative order in $\mathbb{Q}$, then there is an automorphism of A which takes $\bar{a}$ to $\bar{b}$. The number of possible $<$-orders of the elements of an n-tuple $(a_0, \ldots, a_{n-1})$ is at most, say, $(2n - 1)!$ (thus a_1 is either $< a_0$ or $= a_0$ or $> a_0$; then there are at most five cases for a_2; etc). So A is oligomorphic.

We turn next to a pair of notions from group theory which don't always match their model-theoretic counterparts. (But sometimes they do match, and this will be important.) These are the notions of definitional and algebraic closure. To avoid confusions I shall use capital letters e.g. 'DCL', for the group-theoretic notions, and lower case 'dcl' for the corresponding notions from model theory. In practice one normally uses lower case for both.

Suppose we are given a group G of permutations of a set Ω. Let X be a set of elements of Ω and a an element of Ω. We say that a is **DEFINABLE over** X if every element of G which pointwise fixes X also fixes a; in other words, if the orbit of a under $G_{(X)}$ is a singleton. We say that a is **ALGEBRAIC over** X if the orbit of a under $G_{(X)}$ is finite. The **DEFINITIONAL CLOSURE** of X, $\mathrm{DCL}(X)$ (resp. the **ALGEBRAIC CLOSURE** of X, $\mathrm{ACL}(X)$) is the set of all elements of Ω which are DEFINABLE (resp. ALGEBRAIC) over X.

For examples of these notions, open any text of Galois theory. In fields, an element is ALGEBRAIC over a set X if and only if it is algebraic (in the Galois-theory sense) over the field generated by X. In a field of characteristic p, an element a is DEFINABLE over any subfield containing a^p.

The corresponding model-theoretic notions are as follows. Let L be a language and A an L-structure with domain Ω. Suppose $X \subseteq \Omega$ and a is an element of Ω. We say that a is **L-definable over** X if there is a formula $\phi(x)$ of L with parameters from X, such that $A \vDash \phi(a) \wedge \exists_{=1} x\, \phi(x)$; we say that a is **L-algebraic over** X if there is a formula $\phi(x)$ of L with parameters from X, such that $A \vDash \phi(a) \wedge \exists_{\leqslant n} x\, \phi(x)$ for some $n < \omega$. The set of all elements which are L-definable over X (resp. L-algebraic over X) is written $\mathrm{dcl}_L(X)$ (resp. $\mathrm{acl}_L(X)$). When L is a first-order language, we omit L and write simply $\mathrm{dcl}(X)$, $\mathrm{acl}(X)$; these sets are respectively the **definitional closure** and the **algebraic closure** of the set X in A. We also write $\mathrm{acl}(X)$ etc. as $\mathrm{acl}_A(X)$

etc., when we want to indicate the structure A. We say that a set X is
algebraically closed if $\mathrm{acl}(X) = X$; and similarly with the other notions DCL
etc. Sometimes I shall write shorthand, for example $\mathrm{acl}(a, b)$ for $\mathrm{acl}(\{a, b\})$.

Lemma 4.1.1. *Let L be a language and A an L-structure with domain Ω, and
let G be $\mathrm{Aut}(A)$ as a permutation group on Ω.*
   (a) $\mathrm{dcl}_L(X) \subseteq \mathrm{DCL}(X)$ *and* $\mathrm{acl}_L(X) \subseteq \mathrm{ACL}(X)$.
   (b) *If $X \subseteq \Omega$ then $X \subseteq \mathrm{DCL}(X) \subseteq \mathrm{ACL}(X)$.*
   (c) *If $X \subseteq Y \subseteq \Omega$ then $\mathrm{DCL}(X) \subseteq \mathrm{DCL}(Y)$ and $\mathrm{ACL}(X) \subseteq \mathrm{ACL}(Y)$.*
   (d) *If $X \subseteq \Omega$ then $X \subseteq \mathrm{dcl}_L(X) \subseteq \mathrm{acl}_L(X)$.*
   (e) *If $X \subseteq Y \subseteq \Omega$ then $\mathrm{dcl}_L(X) \subseteq \mathrm{dcl}_L(Y)$ and $\mathrm{acl}_L(X) \subseteq \mathrm{acl}_L(Y)$.*

Proof. All trivial.   □

Lemma 4.1.2. *Take L to be first-order in Lemma 4.1.1.*
   (a) *If $X \subseteq \mathrm{dcl}(Y)$ then $\mathrm{dcl}(X) \subseteq \mathrm{dcl}(Y)$.*
   (b) $\mathrm{dcl}(X) = \mathrm{dcl}(\mathrm{dcl}(X))$.
   (c) *If $a \in \mathrm{dcl}(X)$ then $a \in \mathrm{dcl}(Y)$ for some finite $Y \subseteq X$.*
   (a')–(c') *The same as* (a)–(c) *but with* acl *for* dcl.

Proof. (c) holds because a first-order formula has only finitely many parameters. Now consider (a). If $a \in \mathrm{dcl}(X)$ then there are a formula
$\phi(x, y_0, \ldots, y_{n-1})$ of L and parameters $b_0, \ldots, b_{n-1}$ in X such that

(1.3) $A \vDash \phi(a, b_0, \ldots, b_{n-1}) \wedge \exists_{=1} x\, \phi(x, b_0, \ldots, b_{n-1})$.

Since $X \subseteq \mathrm{dcl}(Y)$, there is for each b_i a formula $\psi_i(x)$ with parameters in Y
such that $A \vDash \psi_i(b_i) \wedge \exists_{=1} x\, \psi_i(x)$. So writing $\theta(x)$ for

$$\exists y_0 \ldots y_{n-1}(\textstyle\bigwedge_{i<n} \psi_i(y_i) \wedge \phi(x, y_0, \ldots, y_{n-1})),$$

we have $A \vDash \theta(a) \wedge \exists_{=1} x\, \theta(x)$. This shows that $a \in \mathrm{dcl}(Y)$. Thus (a) holds.
Finally to prove (b), we have $\mathrm{dcl}(\mathrm{dcl}(X)) \subseteq \mathrm{dcl}(X)$ by (a), while
$\mathrm{dcl}(X) \subseteq \mathrm{dcl}(\mathrm{dcl}(X))$ by Lemma 4.1.1(d).
   Similar arguments establish (a')–(c').   □

Nothing so far is at all deep. The next item is more substantial; it sets up a
translation between permutation groups and $L_{\omega_1 \omega}$ for countable structures.

Lemma 4.1.3. *Let L be $L_{\omega_1 \omega}$ with a countable signature. Let A be a countable
L-structure, and consider $\mathrm{Aut}(A)$ as a permutation group on $\mathrm{dom}(A)$. Then
for all sets X of elements of A we have*
(a) $\mathrm{DCL}(X) = \mathrm{dcl}_L(X)$,
(b) $\mathrm{ACL}(X) = \mathrm{acl}_L(X)$.

Proof. Suppose a is DEFINABLE over X; we must show that a is L-definable over X. List the elements of X as $\bar{b}$, and let $\sigma(a, \bar{b})$ be the Scott sentence of the structure $(A, a, \bar{b})$ (see Corollary 3.5.4). Then, for any element c, $A \vDash \sigma(c, \bar{b})$ if and only if there is an automorphism of A which fixes X pointwise and takes a to c. So $\sigma(x, \bar{b})$ puts a into $\mathrm{dcl}_L(X)$. The same formula works for (b). $\square$

Aut(A) as a topological group

Suppose G is a group of permutations of a set Ω. If we know that G is Aut(A) for some structure A with domain Ω, what does this tell us about G? The answer needs some topology.

Let H be a subgroup of G. We say that H is **closed** in G if the following holds:

(1.4)   suppose $g \in G$ and for every tuple $\bar{a}$ of elements of Ω there is
            h in H such that $g\bar{a} = h\bar{a}$; then $g \in H$.

We say that the group G is **closed** if it is closed in the symmetric group Sym(Ω). Note that if G is closed and H is closed in G, then H is closed; this is immediate from the definitions.

Theorem 4.1.4. *Let Ω be a set; let G be a subgroup of* Sym(Ω) *and H a subgroup of G. Then the following are equivalent.*
(a) *H is closed in G.*
(b) *There is a structure A with* dom(A) = Ω *such that $H = G \cap$ Aut(A).*
In particular a subgroup H of Sym(Ω) *is of form* Aut(B) *for some structure B with domain Ω if and only if H is closed*

Proof. (a) $\Rightarrow$ (b). For each $n < \omega$ and each orbit Δ of H on Ω^n, choose an n-ary relation symbol R_Δ. Take L to be the signature consisting of all these relation symbols, and make Ω into an L-structure A by putting $R_\Delta^A = \Delta$. Every permutation in H takes R_Δ to R_Δ, so that $H \subseteq G \cap$ Aut(A). For the converse, suppose g is an automorphism of A and $g \in G$. For each n-tuple $\bar{a}$, if $\bar{a}$ lies in $\Delta = R_\Delta^A$ then so does $g\bar{a}$, and so $g\bar{a} = h\bar{a}$ for some h in H. Since H is closed in G, it follows that g is in H.

(b) $\Rightarrow$ (a). Assuming (b), we show that H is closed in G. Let g be an element of G such that for each finite subset W of Ω there is $h \in H$ with $g|W = h|W$. Let $\phi(\bar{x})$ be an atomic formula of the signature of A, and $\bar{a}$ a tuple of elements of A. Then choose W above so that it contains $\bar{a}$. We have $A \vDash \phi(\bar{a}) \Leftrightarrow A \vDash \phi(h\bar{a})$ (since $h \in$ Aut(A)) $\Leftrightarrow A \vDash \phi(g\bar{a})$. Thus g is an automorphism of A. $\square$

When H is closed, the structure A constructed in the proof of (a) $\Rightarrow$ (b) above is called the **canonical structure** for H. By the proof, A can be chosen to be an L-structure with $|L| \leq |\Omega| + \omega$.

The word 'closed' suggests a topology, and here it is. We say that a subset S of $\text{Sym}(\Omega)$ is **basic open** if there are tuples $\bar{a}$ and $\bar{b}$ in Ω such that $S = \{g \in \text{Sym}(\Omega): g\bar{a} = \bar{b}\}$ (write this set as $S(\bar{a}, \bar{b})$). In particular $\text{Sym}(\Omega)_{(\bar{a})}$ is a basic open set. An **open subset** of $\text{Sym}(\Omega)$ is a union of basic open subsets. If $\Omega = \text{dom}(A)$, we define a **(basic) open subset** of $\text{Aut}(A)$ to be the intersection of $\text{Aut}(A)$ with some (basic) open subset of $\text{Sym}(\Omega)$.

Lemma 4.1.5. *Let A be a structure and write G for $\text{Aut}(A)$.*

(a) *The definitions above define a topology on G; it is the topology induced by that on $\text{Sym}(\Omega)$. Under this topology, G is a topological group, i.e. multiplication and inverse in G are continuous operations.*

(b) *A subgroup of G is open if and only if it contains the pointwise stabiliser of some finite set of elements of A.*

(c) *A subset F of G is closed under this topology if and only if it is closed in the sense of (1.4) above (with F for H).*

(d) *A subgroup H of G is dense in G if and only if H and G have the same orbits on $(\text{dom } A)^n$ for each positive integer n.*

Proof. (a) A permutation g takes $\bar{a}_1$ to $\bar{b}_1$ and $\bar{a}_2$ to $\bar{b}_2$ if and only if it takes $\bar{a}_1\bar{a}_2$ to $\bar{b}_1\bar{b}_2$; so the intersection of two basic open sets is again basic open. The first sentence of (a) follows at once by general topology. For the second sentence, $g \in S(\bar{a}, \bar{b})$ if and only if $g^{-1} \in S(\bar{b}, \bar{a})$, which proves the continuity of inverse. Finally if $gh \in S(\bar{a}, \bar{b})$, write $\bar{c}$ for $h\bar{a}$; then $g \in S(\bar{c}, \bar{b})$, $h \in S(\bar{a}, \bar{c})$ and $S(\bar{c}, \bar{b}) \cdot S(\bar{a}, \bar{c}) \subseteq S(\bar{a}, \bar{b})$; so multiplication is continuous.

(b) For each tuple $\bar{a}$ the pointwise stabiliser $G_{(\bar{a})}$ is $G \cap S(\bar{a}, \bar{a})$, which is open. Every subgroup containing $G_{(\bar{a})}$ is a union of cosets of $G_{(\bar{a})}$, hence it is open too. In the other direction, suppose H is an open subgroup containing a non-empty basic open set $G \cap S(\bar{a}, \bar{b})$. Then H contains $G_{(\bar{a})}$, since every element of $G_{(\bar{a})}$ can be written as gh with $g \in G \cap S(\bar{b}, \bar{a}) \subseteq H$ and $h \in G \cap S(\bar{a}, \bar{b}) \subseteq H$.

(c) A set $F \subseteq G$ is closed in the topology if and only if for every g in $\text{Aut}(A)$, if each basic neighbourhood of g meets F then g lies in F. This is exactly (1.4).

(d) A subgroup H of G is dense if and only if for every g in $\text{Aut}(A)$, each basic neighbourhood of g meets H. $\qquad\qquad\qquad\square$

As we pass from A to $\text{Aut}(A)$ as permutation group, then to $\text{Aut}(A)$ as topological group and finally to $\text{Aut}(A)$ as abstract group, we keep throwing away information. How much of this information can be recovered? In some cases, precious little of it – consider Example 1 above. On the whole, the larger the automorphism group of a structure, the better the chances of reconstructing the structure from the automorphism group. In a series of papers, Mati Rubin has shown that an impressive number of structures are essentially determined by their abstract automorphism groups. For example in [1979] he showed the following.

Theorem 4.1.6. *Call a boolean algebra B* **homogeneous** *if for every non-zero element b of B there is an isomorphism from B to the algebra of elements $\leq b$ (taking 1^B to b and preserving the lattice operations). If A and B are homogeneous boolean algebras and* $\mathrm{Aut}(A) \cong \mathrm{Aut}(B)$ *as abstract groups, then* $A \cong B$. ☐

In [1989a] he extends this to several other classes of boolean algebras. The case of boolean algebras is particularly worth studying, because one step towards reconstructing a structure A is to reconstruct the boolean algebra of its definable subsets (in some appropriate sense of definable). See also Rubin [1989b] on topological spaces, [199?a] on ω-categorical structures and [199?b] on trees. Example 2 in section 5.4 below will give a small hint of how one can set about proving this kind of theorem.

Exercises for section 4.1

1. Show that for every abstract group G there is a structure with domain G whose automorphism group is isomorphic to G. [Let X be a set of generators of G, and for each $x \in X$ introduce a function $f_x\colon g \to g \cdot x$ on the set G. Consider the structure consisting of the set G and the functions f_x. *This structure is essentially the* **Cayley graph** *of the group G.*]

2. Show that if the structure B is an expansion of A, then there is a continuous embedding of $\mathrm{Aut}(B)$ into $\mathrm{Aut}(A)$. (b) Show that if B is a definitional expansion, then this embedding is an isomorphism.

3. Show that if G is a group of permutations of a set Ω, $\bar{a}$ is a tuple of elements of Ω and h is a permutation of Ω, then $G_{(h\bar{a})} = h(G_{(\bar{a})})h^{-1}$.

4. Give examples to show that in Lemma 4.1.1 we can't replace any of the inclusions by equalities.

5. (a) Show that if A is a structure and X a set of elements of A, then $\mathrm{acl}(X)$ is the domain of a substructure of A. (b) Show that if $A \leq B$ and X is a set of elements of A, then $\mathrm{acl}(X)$ in the sense of A is the same as $\mathrm{acl}(X)$ in the sense of B. In particular every elementary substructure of B is algebraically closed in B.

6. Show that if G is an oligomorphic group of permutations of a set Ω, and X is a finite subset of Ω, then $G_{(X)}$ is also oligomorphic.

7. Show that if $\mathrm{Aut}(A)$ is oligomorphic, then for every finite set X of elements of A, $\mathrm{ACL}(X)$ is finite. *In section 7.3 below we shall see that if $\mathrm{Aut}(A)$ is oligomorphic then* ACL *and* acl *coincide on finite subsets of* A.

8. Suppose G is a subgroup of $\mathrm{Sym}(\Omega)$. (a) Show that the topology on G is Hausdorff. (b) Show that the basic open sets are exactly the right cosets of basic open subgroups; show that they are also exactly the left cosets of basic open subgroups.

9. Show that every open subgroup of $\mathrm{Aut}(A)$ is closed. Give an example to show that the converse fails.

10. Show that a subgroup of $\mathrm{Aut}(A)$ is open if and only if it has non-empty interior.

11. Show that if A is an infinite structure then $\mathrm{Aut}(A)$ has a dense subgroup of cardinality at most $\mathrm{card}(A)$.

12. Suppose K, H and G are subgroups of $\mathrm{Sym}(\Omega)$ with K a dense subgroup of H and H a dense subgroup of G. Show that K is a dense subgroup of G.

*13. Find a complete first-order theory T such that for every group G there is a model of T whose automorphism group is isomorphic to G. *(It is known that T can be the theory of a certain distributive lattice.)*

*14. Let G be a closed subgroup of $\mathrm{Sym}(\Omega)$. Show that G is a compact space if and only if every orbit of G on Ω is finite.

*15. (a) If A is a countable structure, show that $\mathrm{Aut}(A)$ forms a complete metric space under the following metric μ. List the elements of A as a_n $(n < \omega)$. For any $f, g \in \mathrm{Aut}(A)$ put $d(f, g) = 1/2^n$ where $f(a_0) = g(a_0), \ldots, f(a_{n-1}) = g(a_{n-1})$ and $f(a_n) \neq g(a_n)$; $d(f, f) = 0$. Put $\mu(f, g) = \max\{d(f, g), d(f^{-1}, g^{-1})\}$. (b) Show that d also defines a metric on $\mathrm{Aut}(A)$, but in general this metric space is not complete.

A set X in a topological space is said to have the **Baire property** *if for some open set Y, the symmetric difference between X and Y is meagre (i.e. of first category in the sense of Baire).*
*16. Show that for any structure A and any subgroup H of $\mathrm{Aut}(A)$, if H has the Baire property then H is either meagre or open.

*17. Show that if G is a topological group, A is a countable structure and there is an isomorphism $\alpha: G \to \mathrm{Aut}(A)$ of abstract groups which is also continuous, then α is a homeomorphism. [Use the two previous exercises.]

*Let L be a first-order language and A an L-structure. By a **definable automorphism** of A we mean an automorphism of A which is definable in A by a formula of L with parameters. We write* Defaut(A) *for the set of definable automorphisms of A.*

*18. (a) Show that Defaut(A) forms a group under composition. (b) Show that for every group G there is an L-structure A (for some appropriate L) such that $G \cong$ Aut(G) = Defaut(G). [See Exercise 1 above.] (c) Show that if A and B are elementarily equivalent L-structures and ϕ is an $\forall_1$ first-order sentence of the language of groups, then Defaut(A) $\models \phi$ iff Defaut(B) $\models \phi$. (d) Give an example of elementarily equivalent L-structures A, B such that some $\forall_2$ first-order sentence of L is true in Defaut(A) but not in Defaut(B).

4.2 Subgroups of small index

We write $(G:H)$ for the index of the subgroup H in the group G. This section will prove several theorems on subgroups of small index in a permutation group.

Covering by cosets

Our first result is about groups in general, but we shall see in a moment that it has a useful link with permutation groups. Model-theoretic applications will appear at Exercises 6.4.6, 7.1.8 and Theorems 8.6.8 and A.1.1.

Lemma 4.2.1 (*B. H. Neumann's lemma*). *Let G be a group, and let $H_0, \ldots, H_{n-1}$ be subgroups of G and $a_0, \ldots, a_{n-1}$ elements of G. Suppose that G is the union of the set of cosets $X = \{H_0 a_0, \ldots, H_{n-1} a_{n-1}\}$, but not the union of any proper subset of X. Then $(G:\bigcap_{i<n} H_i) \leq n!$; in particular each of the groups H_i has finite index in G.*

Proof. Put $H = \bigcap_{i<n} H_i$. We shall show that

(2.1) $(H_j:H) \leq (n-1)!$ for each $j < n$.

Given (2.1), each $H_i a_i$ is the union of at most $(n-1)!$ cosets of H, so that G is the union of at most $n!$ cosets, proving the lemma. (2.1) follows at once from the case $m = 1$ of the following claim:

(2.2)   if $1 \leq m \leq n$ and $K_0, \ldots, K_{m-1}$ are $H_{i(0)}, \ldots, H_{i(m-1)}$ for
         some $i(0) < \ldots < i(m-1)$, then $(K_0 \cap \ldots \cap K_{m-1}:H)$
         $\leq (n-m)!$.

We shall prove (2.2) by downward induction on m.

When $m = n$ there is nothing to prove. Suppose then that (2.2) has been proved for $m + 1$; let $K_0, \ldots, K_{m-1}$ be groups as in the hypothesis of (2.2), and put $K = \bigcap_{j<m} K_j$. By the minimality of X, there is some element g which is not in $\bigcup_{j<m} H_{i(j)} a_{i(j)}$. Then for each $j < m$, $H_{i(j)} a_{i(j)}$ is disjoint from

$H_{i(j)}g$ and hence from Kg, so that $H_{i(j)}a_{i(j)}g^{-1}$ is disjoint from K. Now consider any $i < n$. The set $K \cap (H_i a_i g^{-1})$ is either empty or a coset $(K \cap H_i)y_i$; we have seen that it is empty whenever $i = i(j)$ for some j. Hence K is the union of at most $n - m$ cosets $(K \cap H_i)y_i$, each of which is (by induction hypothesis) the union of at most $(n - m - 1)!$ cosets of H. The claim is proved, and with it the lemma. $\square$

Corollary 4.2.2 (Π. *M. Neumann's lemma*). *Let A be an L-structure, and suppose $\bar{a}$ is a non-empty tuple of elements of A such that no element in $\bar{a}$ lies in a finite orbit of* $\mathrm{Aut}(A)$. *Then the orbit of $\bar{a}$ under* $\mathrm{Aut}(A)$ *contains an infinite set of pairwise disjoint tuples.*

Proof. Let G be $\mathrm{Aut}(A)$ and let $X = \{b_0, \ldots, b_{k-1}\}$ be any finite set of elements of A. We shall show that there is some g in G such that $g\bar{a}$ is disjoint from X.

Assume not. Let $\bar{a}$ be the n-tuple $(a_0, \ldots, a_{n-1})$. For each $i < k$, write H_i for the stabiliser G_{a_i} of a_i. For each $i < n$ and each $j < k$, choose (if possible) an element g_{ij} of G such that $g_{ij}a_i = b_j$. Our assumption is that every element of G lies in some coset $g_{ij}H_i$. By B. H. Neumann's lemma it follows that at least one of the subgroups H_i has finite index in G, so that the orbit of a_i is finite; contradiction. $\square$

Automorphisms of countable structures

If A is a countable structure, we can build up automorphisms of A from finite approximations. From this observation it's a short step to the next theorem.

Theorem 4.2.3. *Let G be a closed group of permutations of ω and H a closed subgroup of G. Then the following are equivalent.*
(a) *H is open in G.*
(b) *$(G : H) \leq \omega$.*
(c) *$(G : H) < 2^\omega$.*

Proof. (a) $\Rightarrow$ (b). Suppose (a) holds. Then there is some tuple $\bar{a}$ of elements of ω such that the stabiliser $G_{(\bar{a})}$ of $\bar{a}$ lies in H. Suppose now that g, j are two elements of G such that $g\bar{a} = j\bar{a}$; then $j^{-1}g \in G_{(\bar{a})} \subseteq H$ and so the cosets gH, jH are equal. Since there are only countably many possibilities for $g\bar{a}$, the index $(G{:}H)$ must be at most countable.

(b) $\Rightarrow$ (c) is trivial.

(c) $\Rightarrow$ (a). We suppose that H is not open in G, and we construct continuum many left cosets of H in G.

We define by induction sequences $(\bar{a}_i: i < \omega)$, $(\bar{b}_i: i < \omega)$ of tuples of elements of ω and a sequence $(g_i: i < \omega)$ of elements of G such that the following hold for all i.

(2.3) $\bar{b}_0 = \langle \rangle$; $\bar{b}_{i+1}$ is a concatenation of all the sequences

$$(k_0 \ldots k_i)(\bar{a}_0\,\hat{}\, \ldots\, \hat{}\, \bar{a}_i)$$

where each k_j is in $\{1, g_0, \ldots, g_i\}$;

(2.4) $g_i \bar{b}_i = \bar{b}_i$;

(2.5)   there is no $h \in H$ such that $h\bar{a}_i = g_i \bar{a}_i$;

(2.6) i is an item in $\bar{a}_i$.

When $\bar{b}_i$ has been chosen, we have by assumption that $G_{(\bar{b}_i)} \not\subseteq H$, and so there is some automorphism $g_i \in G$ which fixes $\bar{b}_i$ (giving (2.4)) and is not in H. Since H is closed in G, it follows that there is a tuple $\bar{a}_i$ such that $h\bar{a}_i \neq g_i \bar{a}_i$ for all h in H. This ensures (2.5); adding i to $\bar{a}_i$ if necessary, we have (2.6) too.

Now for any subset S of $\omega \backslash \{0\}$, define $g_i^S = g_i$ if $i \in S$, and $= 1$ if $i \notin S$. Put $f_i^S = g_i^S \ldots g_0^S$. For each $j > i$ we have $f_j^S \bar{a}_i = f_i^S \bar{a}_i$, by (2.3) and (2.4). So by (2.6) we can define a map $g_S: \omega \to \omega$ by:

(2.7)   for each $i < \omega$, $g_S(i) = f_j^S(i)$ for all $j \geq i$.

Since the maps f_i^S are automorphisms, g_S is injective. But also g_S is surjective; for consider any $i \in \omega$ and put $j = (f_i^S)^{-1}(i)$. If $j \leq i$ then $g_S(j) = f_i^S(f_i^S)^{-1}(i) = i$. If $(f_i^S)^{-1}(i) = j > i$, then $g_S(j) = f_j^S(f_i^S)^{-1}(i) = g_j^S \ldots g_{i+1}^S(i) = i$ by (2.3), (2.4) and (2.6). So g_S is a permutation of ω. Since the f_i^S are in the closed group G and, for each tuple $\bar{a}$ in ω, g_S agrees on $\bar{a}$ with some f_i^S, it follows that g_S is in G.

There are 2^ω distinct subsets S of $\omega \backslash \{0\}$. It remains only to show that the corresponding permutations g_S lie in different right cosets of H.

Suppose $S \neq T$. Then there is some least $i > 0$ which is, say, in S but not in T. By (2.5) there is no element of H which agrees with g_i on $\bar{a}_i$. Put $f = f_{i-1}^S = f_i^T$. Now consider $f^{-1}\bar{a}_i$, and choose some $j \geq i$ such that all the items in $f^{-1}\bar{a}_i$ are $\leq j$. We have, for all h in H,

(2.8) $g_S(f^{-1}\bar{a}_i) = f_j^S f^{-1}(\bar{a}_i) = g_j^S \ldots g_{i+1}^S g_i(\bar{a}_i) = g_i(\bar{a}_i) \neq h(\bar{a}_i)$

$= hg_j^T \ldots g_{i+1}^T(\bar{a}_i) = hf_j^T f^{-1}(\bar{a}_i) = (hg_T)(f^{-1}\bar{a}_i)$.

So $g_S \notin Hg_T$, which finishes the proof.   □

At first sight this theorem is not very model-theoretic. We can translate it into model theory as follows.

Let A be a countable L^+-structure, suppose $L^- \subseteq L^+$ and let B be the L^--reduct $A|L^-$ of A. Then $H = \mathrm{Aut}(A)$ is a subgroup of $G = \mathrm{Aut}(B)$. Let

g be any element of G, and consider the structure gA; gA is exactly like A except that for each symbol S of L^+, $S^{gA} = g(S^A)$. In particular the domain of gA is $\mathrm{dom}(A)$, and $g(S^A) = S^A$ for each symbol S in L^-, so that the reduct $(gA)|L^-$ is exactly B again.

Suppose now that k is another element of G. When is gA equal to kA? The answer is: when $g(S^A) = k(S^A)$ for each symbol S, or in other words, when $k^{-1}g$ is an automorphism of A – or, in other words again, when the cosets gH and kH in G are equal. This shows that *the index of* $\mathrm{Aut}(A)$ *in* $\mathrm{Aut}(B)$ *is equal to the number of different ways in which the symbols of* $L^+ \backslash L^-$ *can be interpreted in* B *so as to give a structure isomorphic to* A.

Theorem 4.2.4 *(Kueker–Reyes theorem).* *Let* L^- *and* L^+ *be signatures with* $L^- \subseteq L^+$. *Let* A *be a countable* L^+-*structure and let* B *be the reduct* $A|L^-$. *Put* $G = \mathrm{Aut}(B)$. *Then the following are equivalent.*
(a) *There is a tuple* $\bar{a}$ *of elements of* A *such that* $G_{(\bar{a})} \subseteq \mathrm{Aut}(A)$.
(b) *There are at most countably many distinct expansions of* B *which are isomorphic to* A.
(c) *The number of distinct expansions of* B *which are isomorphic to* A *is less than* 2^ω.
(d) *There is a tuple* $\bar{a}$ *of elements of* A *such that for each atomic formula* $\phi(x_0, \ldots, x_{n-1})$ *of* L^+ *there is a formula* $\psi(x_0, \ldots, x_{n-1}, \bar{y})$ *of* $L_{\omega_1\omega}^-$ *such that* $A \vDash \forall \bar{x}(\phi(\bar{x}) \leftrightarrow \psi(\bar{x}, \bar{a}))$.

Proof. Our translation of Theorem 4.2.3 gives the equivalence of (a), (b) and (c) at once. It remains to show that (a) is equivalent to (d).

From (d) to (a) is clear. For the converse, suppose $G_{(\bar{a})} \subseteq \mathrm{Aut}(A)$, and let $\phi(x_0, \ldots, x_{n-1})$ be an atomic formula of L^+. Without loss we can suppose that ϕ is unnested, and for simplicity let us assume too that ϕ is $R(x_0, \ldots, x_{n-1})$ where R is some n-ary relation symbol. For each n-tuple $\bar{c}$ in $\phi(A^n)$ let $\sigma_{\bar{c}}(\bar{a}, \bar{c})$ be the Scott sentence of the structure $(B, \bar{a}, \bar{c})$ (see section 3.5). Now, if $\bar{c}$ is in $\phi(A^n)$ and $\bar{d}$ is an n-tuple such that $A \vDash \sigma_{\bar{c}}(\bar{a}, \bar{d})$, then $(B, \bar{a}, \bar{c}) \cong (B, \bar{a}, \bar{d})$, so $(B, \bar{a}, \bar{c}, R^A) \cong (B, \bar{a}, \bar{d}, R^A)$ by (a) and hence $A \vDash \phi(\bar{d})$. It follows that $A \vDash \forall \bar{x}(\phi(\bar{x}) \leftrightarrow \bigvee_{\bar{c} \in \phi(A^n)} \sigma_{\bar{c}}(\bar{a}, \bar{x}))$. □

Corollary 4.2.5. *Let* A *be a countable structure. Then the following are equivalent.*
(a) $|\mathrm{Aut}(A)| \leqslant \omega$.
(b) $|\mathrm{Aut}(A)| < 2^\omega$.
(c) *There is a tuple* $\bar{a}$ *in* A *such that* $(A, \bar{a})$ *is rigid.*

Proof. The implications (c) $\Rightarrow$ (a) $\Rightarrow$ (b) are immediate. The implication (b) $\Rightarrow$ (c) follows from the theorem by adding a constant for each element of

A. Alternatively, use (c) $\Rightarrow$ (a) from Theorem 4.2.3 with $G = \text{Aut}(A)$, $H = \{1\}$. $\quad\square$

The small index property

Let A be a countable L-structure. We say that A has the **small index property** if every subgroup of $\text{Aut}(A)$ with index at most ω is open. (Some people say '$< 2^{\omega}$' in place of 'at most ω'; in practice it makes little difference.)

What is the interest of this notion? Every basic open subgroup of $\text{Aut}(A)$ has at most countable index, since there are only countably many tuples of elements in A. So when A has the small index property, the open subgroups of $\text{Aut}(A)$ can be identified from $\text{Aut}(A)$ as an abstract group, and hence the whole topology on $\text{Aut}(A)$ is recoverable from the abstract group $\text{Aut}(A)$.

Lemma 4.2.6. *Suppose A and B are countable structures and $\alpha: \text{Aut}(A) \to \text{Aut}(B)$ is an embedding of abstract groups. If A has the small index property then α is continuous.*

Proof. The left cosets of stabilisers of tuples in B form a basis for the open sets in $\text{Aut}(B)$. Let X be a left coset of the stabiliser H of some tuple in B. If X is disjoint from the image of α, then $\alpha^{-1}(X)$ is $\varnothing$, which is certainly open. On the other hand, suppose some element g lies in $X \cap \text{im}(\alpha)$. Then X is the coset gH, and so $\alpha^{-1}(X) = \alpha^{-1}(g) \cdot \alpha^{-1}(H)$. Thus it suffices to show that $\alpha^{-1}(H)$ is open. Now since H is the stabiliser of some tuple $\bar{b}$ in B, the index of H in $\text{Aut}(B)$ is the cardinality of the orbit of $\bar{b}$, which is at most countable since B is countable. Therefore $H \cap \text{im}(\alpha)$ has at most countable index in $\text{im}(\alpha)$, and so $\alpha^{-1}(H)$ has at most countable index in $\text{Aut}(A)$ since α is an embedding. But then $\alpha^{-1}(H)$ is open by the small index property. $\quad\square$

A number of countable structures are known to have the small index property, but in the present state of the art we only have proofs for one structure at a time. As a sample I give what is probably the most important case, where A is a countable set without further structure. We begin with a lemma.

Lemma 4.2.7. *Let Ω be a countable set, $G = \text{Sym}(\Omega)$, and let X, Y be subsets of Ω such that $\Omega \backslash (X \cup Y)$ is infinite. Then $G_{(X \cap Y)} = \langle G_{(X)} \cup G_{(Y)} \rangle$.*

Proof. The group $\langle G_{(X)} \cup G_{(Y)} \rangle$ generated by $G_{(X)}$ and $G_{(Y)}$ is trivially a subgroup of $G_{(X \cap Y)}$, so we need only prove $\subseteq$. Take any $g \in G_{(X \cap Y)}$. We can assume that g also pointwise fixes some infinite set W which is disjoint from $X \cup Y$. For otherwise we can argue as follows. The set $\Omega \backslash (X \cup Y)$, being infinite, has infinite overlap either with $g^{-1}X$ or with $\Omega \backslash g^{-1}X$. In the former case it has infinite overlap with $\Omega \backslash g^{-1}Y$, since g fixes $X \cap Y$ pointwise. So by symmetry we can suppose that

$$Z = (\Omega \backslash (X \cup Y)) \cap (\Omega \backslash g^{-1}X) = \Omega \backslash (X \cup Y \cup g^{-1}X)$$

is infinite. Choose $W \subseteq Z$ so that both W and $Z \backslash W$ are infinite. Since W and gW are both disjoint from X, there is some $h \in G_{(X)}$ such that $h|W = g|W$. Thus $h^{-1}g$ fixes W pointwise, and so we can replace g by $h^{-1}g$.

Now let $k \in G_{(X)}$ be a permutation of order 2 which swaps $Y \backslash X$ with some subset of W and pointwise fixes the rest of Ω. Then $k^{-1}gk$ lies in $G_{(X \cap Y)}$ and in $G_{(kW)}$, so $k^{-1}gk \in G_{(Y)}$. The lemma follows.   □

Theorem 4.2.8. *Let Ω be a set of cardinality ω and let G be $\mathrm{Sym}(\Omega)$. Suppose H is a subgroup of G. Then the following are equivalent.*
(a) *$(G:H)$ is at most countable.*
(b) *$(G:H) < 2^\omega$.*
(c) *There is a finite subset X of Ω such that $G_{(X)} \subseteq H$.*

Proof. (a) trivially implies (b). Also (c) implies (a) since there are only countably many choices for X. It remains to show that (b) implies (c).

We begin by partitioning Ω into countably many disjoint infinite sets Δ_i $(i < \omega)$. Let G^* be the subgroup of G consisting of all permutations g such that $g\Delta_i = \Delta_i$ for every i. For each i let H_i be $\{h|\Delta_i : h \in G^* \cap H\}$. Then

$$(2.9) \qquad \prod_{i < \omega} (\mathrm{Sym}(\Delta_i):H_i) \leqslant (G^*:G^* \cap H) \leqslant (G:H) < 2^\omega.$$

It follows that there is at least one i such that $H_i = \mathrm{Sym}(\Delta_i)$. Fixing this i, let S_i be $G_{(\Omega \backslash \Delta_i)}$ $(\cong \mathrm{Sym}(\Delta_i))$, and consider $S_i \cap H$. If $g \in S_i$ and $h \in S_i \cap H$ then there is g' in H which agrees with g on Δ_i, and so $g^{-1}hg = g'^{-1}hg' \in S_i \cap H$; thus $S_i \cap H$ is normal in S_i. Since $(S_i:S_i \cap H) < 2^\omega$, it follows that $S_i \subseteq H$ (cf. for example Scott [1964] p. 306 for the normal subgroups of symmetric groups).

Let Δ' be a subset Δ_i such that Δ' and $\Delta_i \backslash \Delta'$ are both infinite. There is (see Exercise 4) a family $(D_j : j < 2^\omega)$ of infinite subsets of Δ' such that if $j \neq k$ then $D_j \cap D_k$ is finite. For each $j < 2^\omega$, let π_j be a permutation of Ω of order 2 which maps D_j to $\Omega \backslash \Delta_i$ and vice versa, and pointwise fixes $\Delta_i \backslash D_j$. At least two of the cosets $\pi_j H$ $(j < 2^\omega)$ are equal, since $(G:H) < 2^\omega$. Suppose then that $j < k$ but $\pi_j H = \pi_k H$, i.e. $\pi_k^{-1}\pi_j \in H$. Write Z for the set $\pi_k^{-1}\pi_j \Delta_i$. Since $G_{(\Omega \backslash \Delta_i)} = S_i \subseteq H$, it follows that H also contains $G_{(\Omega \backslash Z)}$; so H contains $\langle G_{(\Omega \backslash \Delta_i)} \cup G_{(\Omega \backslash Z)} \rangle$. Computing, we see that $\Delta_i \cup Z$ is $\Omega \backslash X$ where X is the finite set $\pi_k^{-1}(D_j \cap D_k)$, while $\Delta_i \cap Z$ is an infinite subset of Δ_i (because it contains $\Delta_i \backslash \Delta'$). Since $\Delta_i \cap Z$ is infinite, the lemma applies and tells us that H contains $G_{(X)}$. □

We can strengthen Theorem 4.2.8 a little. Consider the following condition (which is a weak form of the conclusion of Lemma 4.2.7). We suppose that A is a countable structure and G is Aut(A).

(2.10)   If X and Y are finite ALGEBRAICALLY closed sets of elements of A, then $G_{(X \cap Y)} = \langle G_{(X)} \cup G_{(Y)} \rangle$.

The inclusion from right to left is trivial; the substance of (2.10) is that every automorphism fixing $X \cap Y$ pointwise can be built up from finitely many automorphisms which fix either X or Y pointwise. When (2.10) holds, we say that G satisfies the **intersection condition**.

Theorem 4.2.9. *Let A be a structure and $G = $ Aut(A). Suppose G satisfies the intersection condition, H is a subgroup of G, and there is some finite ALGEBRAICALLY closed set W of elements of A such that $G_{(W)} \subseteq H$. Then there is a unique smallest ALGEBRAICALLY closed finite set V of elements of A such that $G_{(V)} \subseteq H$. Moreover for this set V we have $H \subseteq G_{\{V\}}$.*

Proof. If W and W' are ALGEBRAICALLY closed finite sets with $G_{(W)} \subseteq H$ and $G_{(W')} \subseteq H$, then $G_{(W \cap W')} = \langle G_{(W)} \cup G_{(W')} \rangle \subseteq H$; so for V we can take an ALGEBRAICALLY closed set of smallest cardinality with $G_{(V)} \subseteq H$. Now suppose $h \in H$. Then $G_{(hV)} = h G_{(V)} h^{-1} \subseteq h H h^{-1} = H$, so that $V \subseteq hV$ and hence $V = hV$, so that $h \in G_{\{V\}}$. □

This gives an immediate improvement of Theorem 4.2.8.

Corollary 4.2.10. *Let Ω be a countable set, $G = $ Sym(Ω) and H a subgroup of at most countable index in G. Then there is a finite set X such that $G_{(X)} \subseteq H \subseteq G_{\{X\}}$.*

Proof. By Lemma 4.2.7, G satisfies the intersection condition. □

We say that a structure A has the **strong small index property** if Corollary 4.2.10 holds with Aut(A) in place of G.

I close with some other interesting examples.

Fact 4.2.11. *The following structures have the small index property.*
(a) *The rationals as an ordered set.*
(b) *The countable atomless boolean algebra.*
(c) *Any vector space of countable dimension over a finite or countable field.*

(d) *Any countable-dimensional orthogonal or symplectic space over a finite or countable field.*

(e) *Any countable abelian group of bounded exponent.*

Examples (a) and (b) have the strong small index property too, as do (c) and (d) when the field of scalars is finite. There are countable ω-categorical structures with the small index property but not the strong small index property. One example is an equivalence relation with two classes, both of cardinality ω; Droste, Holland & Macpherson [1989] give another. Further information appears at Exercise 7.3.16, Exercise 7.4.15 and Theorem 12.2.17.

Exercises for section 4.2

1. Show that if A is a countable structure which is $L_{\omega_1\omega}$-equivalent to some uncountable structure, then A has 2^ω automorphisms.

2. Show that if A is a countable structure and every orbit of $\text{Aut}(A)$ on elements of A is finite, then $|\text{Aut}(A)|$ is either finite or 2^ω.

3. Let G be a closed subgroup of $\text{Sym}(\omega)$ and H any subgroup of G. (a) Show that the closure of H in G is a subgroup of G. (b) Show that if $(G:H) < 2^\omega$ then there is some tuple $\bar{a}$ such that $H_{(\bar{a})}$ is a dense subgroup of $G_{(\bar{a})}$.

4. Show that there is a family of 2^ω infinite subsets of ω, $(X_i: i < 2^\omega)$, such that if $i < j$ then $X_i \cap X_j$ is finite. [Take for each real number a monotone increasing sequence of rationals which converges to it.]

5. Suppose A is a countable L^+-structure, $L \subseteq L^+$ and $\text{Aut}(A)$ has at most countable index in $\text{Aut}(A|L)$. Show that A has the small index property if and only if $A|L$ has the small index property.

Affine spaces in general are defined in Example 3 of section 4.5 below; see also Exercise 4.5.12. Affine geometries are defined in Example 3 of section 4.6; see also Exercise 4.6.10.

6. (a) Let A be an affine space of countable dimension over a finite or countable field. Show that A has the small index property. [Use Fact 4.2.11(c) and the preceding exercise.] (b) Show the same for an affine geometry of countable dimension over a finite field. [One needs a finite set of parameters to fix the field.]

7. Let A and B be countable structures. (a) Show that if A and B have the small index property, then the disjoint sum $A + B$ also has the small index property. [Reduce to the case where H projects onto $\text{Aut}(A)$ and $\text{Aut}(B)$. Then consider the subgroup $\{g_1 \in \text{Aut}(A): \text{for all } g_2 \in \text{Aut}(B), g_1 g_2 \in H\}$ of $\text{Aut}(A).$] (b) Show that if $\text{Aut}(A)$ and $\text{Aut}(B)$ satisfy the intersection condition, then so does $\text{Aut}(A + B)$.

8. Let G be the symmetric group on a countable set. Show that every element of $G_{(X \cap Y)}$ lies either in $G_{(X)}G_{(Y)}G_{(X)}$ or in $G_{(Y)}G_{(X)}G_{(Y)}$. Show that in general neither $G_{(X)}G_{(Y)}$ nor $G_{(Y)}G_{(X)}$ will suffice.

9. Let A be a countable-dimensional vector space over a finite field. Show that $\mathrm{Aut}(A)$ satisfies the intersection condition.

10. Let G be the group of order-automorphisms of the rationals. Show that G satisfies the intersection condition.

4.3 Imaginary elements

Every structure carries on its back some other structures which are 'definable in terms of it'. Thus a vector space has a dual space, a boolean algebra has a Stone space, an integral domain has a field of fractions, any structure has an automorphism group, and so on. The elements of these other structures are in some sense 'ideal' elements of the structure we began with. Often in model theory we find ourselves looking at a particular case and trying to decide which kinds of ideal element can be reached by which kinds of definition. For example, which automorphisms of a structure can be specified by first-order formulas with parameters from the structure?

The Galois approach

Suppose a structure A is given. Broadly speaking, if B is any structure which can be defined in terms of A, then B can be defined *set-theoretically* in terms of A.

Think of the elements of A as atoms without any set-theoretic structure inside themselves. Then one can build up a universe of sets over the domain of A, thus.

(3.1) $V_0(A) = \mathrm{dom}(A)$.

(3.2)   For each ordinal α, $V_{\alpha+1}(A) = \mathscr{P}(V_\alpha(A)) \cup V_\alpha(A)$.

(3.3)   For each limit ordinal δ, $V_\delta(A) = \bigcup_{\alpha < \delta} V_\alpha(A)$.

The union of all the $V_\alpha(A)$ as α ranges through the ordinals is a class $V(A)$ called the **universe of sets over** A.

Let G be the automorphism group of A. Then G acts on $V(A)$ thus, using induction on the rank of sets.

(3.4)   If $a \in \mathrm{dom}(A)$, $g(a)$ is already given
          If X is a set $\in V(A)$, then $g(X) = \{g(y): y \in X\}$.

For every element X of $V(A)$, we define $G_X = \{g \in G: g(X) = X\}$ (the **setwise stabiliser** of X in G), and $G_{(X)} = \{g \in G: g(y) = y$ for every $y \in X\}$

(the **pointwise stabiliser** of X in G). These agree with our notation for pointwise and setwise stabilisers in section 4.1 above (since $G_{\{X\}}$ is clearly the same group as G_X). We say that X is **fixed under** G if $g(X) = X$ for all $g \in G$.

For example, suppose L is a first-order language and A is an L-structure. *We always assume that the symbols of L are sets not involving A, so that they stay fixed under G* (cf. Exercise 1). If R is a relation symbol of L, then its interpretation R^A also stays fixed under G. The same holds if R is a relation in some definitional expansion of A (see section 2.6).

Definition (3.4) allows us to talk freely about the action of Aut(A) on objects that lie in $V(A)$, without further explanation.

Example 1: *Automorphisms acting on types.* Suppose the L-structure B is an extension of A (not necessarily inside $V(A)$), $\bar{b}$ is some tuple in B and X is a set of elements of A. As we shall see in section 6.3, the **type** of $\bar{b}$ over X with respect to B is the set of all formulas $\phi(\bar{x}, \bar{a})$ with ϕ in L and $\bar{a}$ in X, such that $B \vDash \phi(\bar{b}, \bar{a})$. Let $p(\bar{x})$ be the type of $\bar{b}$ over X. Then Aut(A) acts on p as follows: $g(p) = \{\phi(\bar{x}, g\bar{a}): \phi(\bar{x}, \bar{a}) \in p\}$.

In a very broad sense, every element of $V(A)$ is 'definable from A'. Model theorists are mostly interested in those elements of $V(A)$ which are definable from just finitely many elements of A. To make this precise, let us say that a set $S \subseteq \mathrm{dom}(A)$ is a **support** of the element X of $V(A)$ if $G_{(S)} \subseteq G_X$. (Thus for example dom(A) is a support of every element of $V(A)$.)

We say that X has **finite support** (over A) if some finite set is a support of it. We say that X is **hereditarily of finite support** (over A) if X, the elements of X, the elements of elements of X, etc. etc., all have finite support.

Lemma 4.3.1. *Let A be an L-structure. Then A itself is hereditarily of finite support, and so is every relation which is first-order definable with parameters in A.*

Proof. Strictly this depends on the set-theoretic definitions of L-structure, n-tuple etc. But I only once in my life saw definitions which make the lemma false. (I wrote to the author and he withdrew them.) $\qquad\square$

If an object X in $V(A)$ is not of finite support, then X is not first-order definable in A. But we can say more: X is not even set-theoretically definable, unless we cheat and use the axiom of choice or information about the internal structure of the elements of A.

Lemma 4.3.2. *Let L be a countable first-order language and A an L-structure. Then there is a transitive model M of ZF (Zermelo–Fraenkel set theory*

without the axiom of choice) *containing an L-structure B such that*
(a) $A \cong B$,
(b) *if X is* (*in M*) *a countable sequence of relations* (*of any finite arity*) *on*
     $\mathrm{dom}(A)$, *then X is of finite support*.

Proof. This is most easily proved from the Jech–Sochor embedding theorem
(Jech [1973b]). There are the usual set-theoretic caveats, of which I mention
two. First, the proof assumes that ZFC has a model – which is not itself a
theorem of ZFC unless ZFC is inconsistent. (There are ways around this.)
Second, a property which B has in M will not necessarily be a property which
A has in the real world, unless the property is *set-theoretically absolute*.
Properties of the form 'Such-and-such a first-order sentence is true in A' are
always set-theoretically absolute (see Exercise 2.1.16). So for example, if A is
a field, then B is a field in the sense of M. For further details on both these
caveats, see any textbook of set theory.   □

This lemma tells us that some of our model-theoretic proofs of undefin-
ability are in fact proofs that the axiom of choice is needed to prove the
existence of something.

Example 2: *Amorphous sets.* An amorphous set is an infinite set whose subsets
are all either finite or cofinite. The proof of Theorem 2.1.2 in fact proved
that, without the axiom of choice, there can be amorphous sets. I must add
that I see little future for model theory without the axiom of choice.

Example 3: *Unique prime factorisation.* Let K be the union of an infinite
tower of unramified abelian extensions starting with some algebraic number
field (by Shafarevich – see Roquette [1967]), and let A be the ring of integers
of K. One can show that A is an integral domain in which there are no prime
elements, and that the only ideals of A which have finite support are
principal. Hence by Lemma 4.3.2 it is consistent with ZF that there is a
principal ideal domain in which not every non-zero element is a product of
primes.

Warning. Sometimes we specify a structure B which is based on A, by saying
what the isomorphism type of B is over A. For example the algebraic closure
B of a field A exists; we know exactly what it is up to isomorphism over A.
But the elements of B don't have to be related to the elements of A in any
particular way – for example they are not tuples or sets of elements of A.
This is a very common situation in model theory, and it needs a quite
different treatment from that above. I return to this question at the end of
section 12.3 below. Meanwhile note that the approach of Lemmas 4.3.1 and

4.3.2 does apply if the existence of B implies the existence of some particular kind of subset of A. For example if A is a formally real field and B is a real-closed field extending A, then from B we can define an ordering of A. So, in a universe of sets where A is a formally real field with no ordering of finite support, A has no real closure of finite support either.

This is enough set theory for the moment. It's time to return to those elements of $V(A)$ which are first-order definable in A.

Imaginary elements and A^{eq}

Let L be a first-order language and A an L-structure. An **equivalence formula** of A is a formula $\phi(\bar{x}, \bar{y})$ of L, without parameters, such that the relation $\{(\bar{a}, \bar{b}): A \vDash \phi(\bar{a}, \bar{b})\}$ is a non-empty equivalence relation E_ϕ. We write ∂_ϕ for the set $\{\bar{a}: A \vDash \phi(\bar{a}, \bar{a})\}$. We write $\bar{a}/\phi$ for the E_ϕ-equivalence class of the tuple $\bar{a} \in \partial_\phi$. Items of the form $\bar{a}/\phi$, where ϕ is an equivalence formula and $\bar{a}$ a tuple, are known as **imaginary elements** of A.

Note that in every L-structure A, there is a natural correspondence between the elements a of A and the imaginary elements $a/(x = y)$. (Every real element is imaginary!) Likewise every tuple $(a_0, \ldots, a_{n-1})$ of elements of A can be identified with a single element $\bar{a}/\theta$ where $\theta(\bar{x}, \bar{y})$ is the formula $\bigwedge_{i<n} x_i = y_i$. Note also that, if $\phi(x_0, \ldots, x_{n-1})$ is any formula of L, there are imaginary elements corresponding to the set $\phi(A^n)$ and its complement, thanks to the equivalence formula $\phi(\bar{x}) \leftrightarrow \phi(\bar{y})$.

Saharon Shelah suggested a way of regarding imaginary elements of a structure as genuine elements. Let A be an L-structure. Shelah defined another structure A^{eq} (pronounced 'A E Q') which contains A as a definable part. The definition of A^{eq} is as follows.

Let L be a signature and A an L-structure. Let $\theta(\bar{x}, \bar{y})$ be an equivalence formula of A. We write I_θ for the set of all classes $\bar{a}/\theta$ with $\bar{a} \in \partial_\theta$, and $f_\theta: \partial_\theta \to I_\theta$ for the map $\bar{a} \mapsto \bar{a}/\theta$. The elements of A^{eq} are the equivalence classes $\bar{a}/\theta$ with θ an equivalence formula of A and $\bar{a} \in \partial_\theta$. We identify each element a of A with the equivalence class $a/(x = y)$. For each equivalence formula θ there are a 1-ary relation symbol P_θ and a relation symbol R_θ, interpreted in A^{eq} so that P_θ names I_θ and R_θ names (the graph of) the function f_θ. Also the symbols of L are interpreted in A^{eq} just as in A, on the elements in $I_{(x=y)}$. (Since $I_{(x=y)}$ is not the whole domain of A^{eq}, the function symbols of L will have to be replaced by relation symbols; see Exercise 2.6.1. In talking about A^{eq}, we can keep the function symbols as a shorthand.)

This defines the structure A^{eq}. We write L^{eq} for its signature. The sets I_θ are called the **sorts** of A^{eq}.

Theorem 4.3.3. *Let A be a structure of signature L.*

(a) $\text{dom}(A)$ *is $\varnothing$-definable in A^{eq}, and so are all the relations R^A, functions F^A and constants c^A of A.*

(b) *For every element a of A^{eq} there are a formula $\phi(\bar{x}, y)$ of L^{eq} and a tuple $\bar{b}$ in A such that a is the unique element of A^{eq} satisfying $\phi(\bar{b}, y)$. (In the language of section 4.1, every element of A^{eq} is definable over A.)*

(c) *For every first-order formula $\phi(\bar{x})$ of L^{eq} there is a first-order formula $\phi^{\downarrow}(\bar{x})$ of L such that, for every tuple $\bar{a}$ of elements of A, $A^{\text{eq}} \vDash \phi(\bar{a})$ $\Leftrightarrow A \vDash \phi^{\downarrow}(\bar{a})$.*

(d) *If $e \colon A \to B$ is an elementary embedding, then e extends to an elementary embedding $e^{\text{eq}} \colon A^{\text{eq}} \to B^{\text{eq}}$.*

(e) *If X is a set of elements of A, then $\text{acl}_A(X) = \text{acl}_{A^{\text{eq}}}(X) \cap \text{dom}(A)$.*

Proof. Left to the reader, who should probably read section 5.1 below before attempting it. $\qquad\qquad\qquad\qquad\qquad\qquad\qquad\qquad\qquad\qquad\qquad\qquad\qquad\qquad\square$

Example 4: *Projective planes*. Suppose A is a three-dimensional vector space over a finite field, and let L be the first-order language of A. Then we can write a formula $\theta(x, y)$ of L which expresses 'vectors x and y are non-zero and are linearly dependent on each other'. The formula θ is an equivalence formula of A, and the sort I_θ is the set of points of the projective plane P associated with A. Likewise we can write a formula $\eta(x_1, x_2, y_1, y_2)$ which expresses 'x_1 and x_2 are linearly independent; so are y_1 and y_2; and x_1 and x_2 together generate the same plane as y_1 and y_2'. Again η is an equivalence formula of A, and the sort I_η is the set of lines of P. Using the relations R_θ and R_η we easily define the incidence relation of P. So statements about the projective plane P can be translated rather directly into statements about A^{eq}.

Perhaps it was overkill to define the whole of A^{eq} at once. In most applications we are only concerned with a finite number of the sorts I_θ. A **finite slice** of A^{eq} is a structure defined exactly the same way as A^{eq}, but using only finitely many of the equivalence formulas θ (always including the equivalence formula $x = y$).

Bases of types

André Weil proved a suggestive theorem about fields.

Fact 4.3.4. *Let A be an algebraically closed field and B a subfield of A. Suppose $\bar{a}$ is a tuple of elements of A. Then there is a subfield C of B which is finitely generated (as a field), such that for every automorphism g of B,*

(3.5)   *g fixes the type of $\bar{a}$ over B with respect to A if and only if g
        pointwise fixes C.*

(See Example 1 above.)

Proof. Strictly this is a model-theoretic consequence of Weil's result. See
Weil [1946] p. 19 or Lang [1958] p. 62 for Weil's proof, and the proof of
Theorem 4.4.6 below for a more algebraic statement.   □

In some sense the tuple $\bar{a}$ casts a shadow over the field B – namely its type
over B – and the subfield C allows us to describe that shadow *inside B and
without mentioning the tuple $\bar{a}$.*

Let us say that a field C which satisfies (3.5) is a **base** of the type of $\bar{a}$ over
B (with respect to A). Analogues of Weil's result come to light fairly often,
but in general the base has to be taken in A^{eq} rather than in A.

Thus, suppose A is an L-structure, X is a set of elements of A and $\bar{a}$ is a
tuple of elements of A. We can define the **type** of $\bar{a}$ over X (with repect to
A) to be the set of all formulas $\phi(\bar{x}, \bar{b})$ (with ϕ in L and $\bar{b}$ in X) such that
$A \vDash \phi(\bar{a}, \bar{b})$. Write $p(\bar{x})$ for this type. A **base** for $p(\bar{x})$ is a set W of elements
$\bar{b}/\psi$ of A^{eq}, with $\bar{b}$ in X, such that for every automorphism g of A, $g(p) = p$
if and only if g pointwise fixes W.

Our definition of base is rather sensitive to the automorphism group of A.
Normally one only uses this notion when A has a large automorphism group
(for example when it is big in the sense of section 10.1 below); otherwise
statements about 'all automorphisms' are not very significant. But in the
interesting cases, we can find a base regardless of how many automorphisms
A has.

For example, the type $p(\bar{x})$ over X is said to be **definable** if for every
formula $\phi(\bar{x}, \bar{y})$ of L there are a formula $d\phi(\bar{y}, \bar{z})$ of L and a tuple $\bar{b}$ in X
such that

(3.6)   For every tuple $\bar{a}$ in X, $\phi(\bar{x}, \bar{a}) \in p \Leftrightarrow A \vDash d\phi(\bar{a}, \bar{b})$.

Definable types do occur in nature. Thus in models of a stable theory, all
types are definable; see Corollary 6.7.11.

Theorem 4.3.5. *Let L be a first-order language, and suppose the type $p(\bar{x})$ of
some tuple $\bar{a}$ over X with respect to the L-structure A is definable. Then p has
a base W in A^{eq}. We can choose W so that $W \subseteq \mathrm{dcl}(X)$ in A^{eq}.*

Proof. For each formula $\phi(\bar{x}, \bar{y})$ of L we let $\phi^*(\bar{z}, \bar{z}')$ be the following
equivalence formula:

(3.7) $\forall \bar{y} \, (d\phi(\bar{y}, \bar{z}) \leftrightarrow d\phi(\bar{y}, \bar{z}'))$

where $d\phi$ is as in (3.6). If we know the ϕ^*-equivalence class of the tuple $\bar{b}$ in (3.6), we can immediately recover the set of formulas $\phi(\bar{x}, \bar{b}')$ which lie in p. Let W be the set of all elements e of A^{eq} such that for some formula $\phi(\bar{x}, \bar{y})$ of L and some $d\phi$ and $\bar{b}$ from X as in (3.6), e is $\bar{b}/\phi^*$. Clearly e is definable over $\bar{b}$ and hence over X.

We claim that W is a base of p. For suppose g is an automorphism of A. If g fixes W pointwise, then for each formula $\phi(\bar{x}, \bar{y})$ of L there are $d\phi$ in L and $\bar{b}$ in X such that (3.6) holds, and by assumption g fixes $\bar{b}/\phi^*$. So $A \vDash \phi^*(\bar{b}, g\bar{b})$, and hence for every tuple $\bar{d}$ in X we have, applying g to (3.6),

$$(3.8) \quad A \vDash \phi(g\bar{a}, \bar{d}) \Leftrightarrow A \vDash d\phi(\bar{d}, g\bar{b}) \Leftrightarrow A \vDash d\phi(\bar{d}, \bar{b}) \Leftrightarrow A \vDash \phi(\bar{a}, \bar{d}).$$

Thus $g\bar{a}$ and $\bar{a}$ have the same type over X. The same argument in reverse shows that, if $g\bar{a}$ and $\bar{a}$ have the same type over X, then g fixes each element of W. $\qquad\square$

The set W defined in the proof of Theorem 4.3.5 is called the **canonical base** of the type p, in symbols $\mathrm{cb}(p)$. (Harmlessly one could define $\mathrm{cb}(p)$ to be $\mathrm{dcl}(W)$ in A^{eq}, as Shelah does.)

One source of definable types is algebraicity. Let X be a set of elements of an L-structure A. As in section 4.1, a tuple $\bar{a}$ is **algebraic** over X if and only if there are some formula $\phi(\bar{x}, \bar{b})$ of L with parameters $\bar{b}$ from X, and some positive integer n such that $A \vDash \phi(\bar{a}, \bar{b}) \wedge \exists_{=n}\bar{x}\,\phi(\bar{x}, \bar{b})$. Likewise we say that the type p is **algebraic** if p contains a formula $\phi(\bar{x}, \bar{b})$ with $\bar{b}$ in X, such that $A \vDash \exists_{=n}\bar{x}\,\phi(\bar{x}, \bar{b})$ for some $n < \omega$. Thus $\bar{a}$ is algebraic over X if and only if its type $p(\bar{x})$ over X is algebraic.

Theorem 4.3.6. *In an L-structure A, suppose $\bar{a}$ is algebraic over X. Then the type p of $\bar{a}$ over X is definable, and the canonical base of p lies in $\mathrm{dcl}_{A^{\mathrm{eq}}}(\bar{a})$.*

Proof. Choose a formula $\phi(\bar{x}, \bar{b})$ in p such that $A \vDash \exists_{=n}\bar{x}\,\phi(\bar{x}, \bar{b})$, and choose it to make n as small as possible. Then for every formula $\psi(\bar{x}, \bar{z})$ of L and every tuple $\bar{d}$ from X, we have

$$(3.9) \qquad \psi(\bar{x}, \bar{d}) \in p \Leftrightarrow A \vDash \forall\bar{x}(\phi(\bar{x}, \bar{b}) \rightarrow \psi(\bar{x}, \bar{d})).$$

The canonical base is definable over the equivalence class $\bar{b}/\chi$ where $\chi(\bar{y}, \bar{z})$ is the formula $\forall\bar{x}(\phi(\bar{x}, \bar{y}) \leftrightarrow \phi(\bar{x}, \bar{z}))$. This equivalence class is clearly definable from $\bar{a}$. $\qquad\square$

For example let the domain of A be the set of points of the real plane, and give A just two relations, Rvw, 'v and w lie in the same line parallel to the y-axis', and Pv, 'v lies in the x-axis'. Let X be the set of points with positive y-coordinate. Then every point a in the x-axis is algebraic over X, since it's the unique element satisfying the formula $Pv \wedge Rvb$ where b is any element due north of a. But the element b is not a base for the type of a over X, because we can move b northwards keeping the type fixed. So we pass to A^{eq}, which contains the vertical line l passing through a. This line l is determined by the type of a over X, and we can reconstruct the type from l.

Finite cover property

The following notion plays an important role in the study of uncountably categorical structures. I mention it here because one of its more memorable formulations is in terms of A^{eq}.

Let T be a complete theory in a first-order language L. We say that T **doesn't have the finite cover property** (for short, doesn't have the f.c.p.) if for every formula $\theta(\bar{x}, \bar{y}, \bar{z})$ of L with $\bar{x}$, $\bar{y}$ of the same length n, there is a formula $\phi(\bar{z})$ of L with the following property:

(3.10)   in any model A of T, if $\bar{a}$ is a tuple of elements and $\theta(\bar{x}, \bar{y}, \bar{a})$ defines an equivalence relation E on the set of n-tuples of elements of A, then E has infinitely many classes if and only if $A \vDash \phi(\bar{a})$.

Warning. This is the definition commonly used in practice, but it is not quite equivalent to the original definition by Keisler [1967], which is given in Exercise 10 below. Our formulation is from Shelah [1978a] Chapter II.4. who shows that it's equivalent to Keisler's when T is stable. According to Keisler's definition an unstable theory always has the f.c.p (cf. Exercise 6.7.12).

There is another way of phrasing the f.c.p. If L is a first-order language and A is an L-structure, we say that **infinity is definable in** A if for every formula $\phi(\bar{x}, \bar{y})$ of L there is a formula $\sigma(\bar{y})$ of L such that for each tuple $\bar{a}$ in A,

(3.11) $A \vDash \sigma(\bar{a}) \Leftrightarrow \phi(A^n, \bar{a})$ is infinite.

Then infinity is definable in A^{eq} if and only if A fails to have the finite cover property. For more on the f.c.p., see Corollary 9.5.6 below.

Exercises for section 4.3

1. We define $V_0 = \varnothing$, $V_{\alpha+1} = \mathcal{P}(V_\alpha) \cup V_\alpha$, V_δ (δ a limit ordinal) $= \bigcup_{\alpha < \delta} V_\alpha$, $V = \bigcup_{\alpha \text{ an ordinal}} V_\alpha$. V is the **universe of pure sets**, and V_α is the **set of pure sets hereditarily of rank** $< \alpha$. Show that if A is a structure, then $V \subseteq V(A)$ and every automorphism of A fixes all elements of V. [Use induction on α.]

2. Let A be an infinite-dimensional vector space. Show (a) the subspaces of A which have finite support over A are exactly the finite-dimensional ones, (b) the automorphisms of A which have finite support are exactly the dilations (i.e. transformations which multiply all vectors by some fixed scalar), (c) the statement 'Every vector space has a basis' is not provable from ZF alone.

3. Let A be the countable atomless boolean algebra (see Example 4 in section 3.2). Show (a) if $(X_i: i < \omega)$ is a descending or ascending chain of ideals of A, which has finite support over A, then $(X_i: i < \omega)$ is eventually constant, (b) the only automorphism of A with finite support is the identity, (c) the statement 'In a commutative ring, if every ascending chain of ideals is eventually constant then every non-empty set of ideals has a maximal element' is not provable from ZF alone.

4. Take your favourite application of the axiom of choice in algebra, and use Lemma 4.3.2 to show that the theorem in question is not provable just from ZF.

5. Show that if E is an equivalence relation on tuples in a set I_θ in A^{eq}, and E is definable in A^{eq} without parameters, then E is equivalent in a natural way to an equivalence relation on tuples of elements of A which is definable in A without parameters. (So there is nothing to be gained by passing to $(A^{\text{eq}})^{\text{eq}}$.)

6. Let A be an L-structure and B a finite slice of A^{eq}. (a) Show that the restriction map $g \mapsto g|\text{dom}(A)$ defines an isomorphism from $\text{Aut}(B)$ to $\text{Aut}(A)$. (b) Show that if $\text{Aut}(A)$ is oligomorphic, then so is $\text{Aut}(B)$.

7. Let A be a vector space, X a subspace of A and $\bar{a}$ a tuple from A. Show that $\text{dcl}(\bar{a}) \cap X$ is a base of the type of $\bar{a}$ over X. [By quantifier elimination the type is determined by what formulas of the form $\sum_{i<n} r_i x_i + d$ it contains, with $r_0, \ldots, r_{n-1}$ scalars and $d \in X$.]

*8. Let A be a universal 1-ary function as in Example 1 of section 8.3. Let X be a set of elements of A and $\bar{a}$ a tuple from A. (a) Show that the type of $\bar{a}$ over X is definable, and its canonical base lies within $\text{dcl}(\bar{a})$ in the sense of A^{eq}. (b) Let B be the same as A but with '$F(x) = y$' replaced by a symmetric binary relation '$Rxy \wedge Ryx$'. (This amounts to removing the arrows on the edges.) Show that although the type of $\bar{a}$ over X with respect to B is still definable, it is not always true that the canonical base of this type lies in $\text{acl}(\bar{a})$ in the sense of A^{eq}. *This provides an example of a 1-based structure with a reduct which is not 1-based; for lack of background I leave this comment unexplained.*

9. Let L be a first-order language and A an L-structure such that for every formula $\theta(x, \bar{y})$ of L there is a formula $\phi(\bar{y})$ of L such that if $\bar{b}$ is any tuple in A, then $A \vDash \phi(\bar{b})$ if and only if $\theta(A, \bar{b})$ is infinite. (a) Show that infinity is definable in A. (b) Deduce that infinity is definable in algebraically closed fields.

*10. Let L be a first-order language and T a complete theory in L. Show that, if T has the finite cover property (in Shelah's sense), then there is a formula $\phi(x, \bar{y})$ of L such that, for arbitrarily large finite n, T implies that there are $\bar{a}_0, \ldots, \bar{a}_{n-1}$ for which $\neg \exists x \bigwedge_{i \in n} \phi(x, \bar{a}_i)$ holds, but $\exists x \bigwedge_{i \in W} \phi(x, \bar{a}_i)$ holds for each proper subset W of n. This conclusion is Keisler's form of the finite cover property. [The easy part is to deduce it with $\bar{x}$ in place of x. Then use induction on the length of $\bar{x}$.]

Let L be a first-order language and m a positive integer. The language $L(Q_0^m)$ is formed by adding to L the **Ramsey** quantifier $Q_0^m x_0 \ldots x_{m-1}$ where $A \vDash Q_0^m \bar{x} \phi(\bar{x}) \Leftrightarrow$ there is an infinite set X of elements of A such that, for all m-tuples $\bar{a}$ in X, $A \vDash \phi(\bar{a})$.
*11. Let T be a complete theory in the first-order language L. We say that Q_0^m is **eliminable in** T if for every formula $\phi(\bar{y})$ of $L(Q_0^m)$ there is a formula $\phi^*(\bar{y})$ of L which is equivalent to ϕ modulo T. Show that the following are equivalent. (a) For all positive integers m, Q_0^m is eliminable in T. (b) T doesn't have the finite cover property (in Shelah's sense).

4.4 Eliminating imaginaries

Throughout this section, L is a first-order language.

Compact topological spaces don't need to be compactified – they already contain the needed limit points. In rather the same way, there are structures which don't need to have imaginary elements added to them, because they already contain representatives of their imaginary elements.

Following Bruno Poizat, we say that an L-structure A has **elimination of imaginaries** if for every equivalence formula $\theta(\bar{x}, \bar{y})$ of A and each tuple $\bar{a}$ in A there is a formula $\phi(\bar{x}, \bar{z})$ of L such that the equivalence class $\bar{a}/\theta$ of $\bar{a}$ can be written as $\phi(A^n, \bar{b})$ for some *unique* tuple $\bar{b}$ from A. We allow $\bar{z}$ and $\bar{b}$ to be empty here, in which case $\bar{a}/\theta$ is $\varnothing$-definable.

More usefully, we say that A has **uniform elimination of imaginaries** if the same holds, except that ϕ depends only on θ and not on $\bar{a}$. Another way of saying this is that for every equivalence formula $\theta(\bar{x}, \bar{y})$ of A there is a function F which is definable without parameters, taking tuples as values, such that for all $\bar{a}_1$ and $\bar{a}_2$ in A, $\bar{a}_1$ is θ-equivalent to $\bar{a}_2$ if and only if $F(\bar{a}_1) = F(\bar{a}_2)$.

The uniformity is helpful, because it creates bijections between sets of tuples and sets of definable classes in A, thus.

Theorem 4.4.1. *Let A be an L-structure. Then the following are equivalent.*
(a) *A has uniform elimination of imaginaries.*
(b) *For every formula $\psi(\bar{x}, \bar{y})$ of L, with $\bar{x}$ of length n, there is a formula $\chi(\bar{x}, \bar{z})$ of L such that for each tuple $\bar{a}$ in A there is a unique tuple $\bar{b}$ in A such that $\psi(A^n, \bar{a}) = \chi(A^n, \bar{b})$.*

Proof. (a) $\Rightarrow$ (b). Write $\theta(\bar{y}, \bar{y}')$ for the formula

(4.1) $\forall \bar{x}(\psi(\bar{x}, \bar{y}) \leftrightarrow \psi(\bar{x}, \bar{y}'))$.

Then θ is an equivalence formula, so by (a) there is a formula $\phi(\bar{y}, \bar{z})$ of L such that for each $\bar{b}$ in A, $\theta(A^m, \bar{b}) = \phi(A^m, \bar{c})$ for some unique tuple $\bar{c}$. Let $\chi(\bar{x}, \bar{z})$ be the formula $\exists \bar{y} \, (\psi(\bar{x}, \bar{y}) \wedge \phi(\bar{y}, \bar{z}))$.

   (b) $\Rightarrow$ (a) is immediate.   □

We say that a theory T in L **has elimination of imaginaries** if all the models of T have elimination of imaginaries. Likewise T **has uniform elimination of imaginaries** if every model of T has uniform elimination of imaginaries; but in fact we can say something tidier in this case.

Theorem 4.4.2. *Let T be a theory in L. The following are equivalent.*
(a) *T has uniform elimination of imaginaries.*
(b) *If $\theta(\bar{x}, \bar{y})$ is any formula of L such that $T \vdash {}'\theta$ defines an equivalence relation', then there is a formula $\phi(\bar{x}, \bar{z})$ of L such that $T \vdash \forall \bar{y} \exists_{=1} \bar{z} \forall \bar{x} (\theta(\bar{x}, \bar{y}) \leftrightarrow \phi(\bar{x}, \bar{z}))$.*
(c) *For every formula $\theta(\bar{x}, \bar{y})$ of L there is a formula $\phi(\bar{x}, \bar{z})$ of L such that $T \vdash ({}'\theta$ defines an equivalence relation' $\rightarrow \quad \forall \bar{y} \exists_{=1} \bar{z} \forall \bar{x} (\theta(\bar{x}, \bar{y}) \leftrightarrow \phi(\bar{x}, \bar{z})))$.*

Proof. (c) $\Rightarrow$ (a) $\Rightarrow$ (b) are immediate. Assuming (b) we get (c) as follows: when $\bar{x}$ and $\bar{y}$ both have the same length n, write $\theta^*(\bar{x}, \bar{y})$ for the formula

(4.2)   (${}'\theta$ defines an equivalence relation' $\rightarrow \theta$).

Then in any L-structure A, θ^* defines an equivalence relation E; but if θ is not an equivalence formula of A then E is the trivial equivalence relation on $(\text{dom } A)^n$ with just one class. Choose ϕ by putting θ^* for θ in (b). Then ϕ works for (c).   □

Examples of uniform elimination of imaginaries

Let us say that an L-structure A is **uniformly 1-eliminable** if clause (b) of Theorem 4.4.1 holds for all formulas $\psi(x, \bar{z})$ of L with just one variable x free. It's sometimes easy to check that a structure is uniformly 1-eliminable. For example any real-closed field is uniformly 1-eliminable by Example 3 in section 3.1. This is not quite immediate. We have to check that whatever $\bar{a}$ is, the set $\theta(A, \bar{a})$ is a union of at most k intervals for some number k depending on θ; then the required formula ϕ depends on k. I leave details to the reader.

   In fact Knight, Pillay & Steinhorn [1986] show that every structure elementarily equivalent to an O-minimal structure is also O-minimal. From this and the compactness theorem one can deduce that every O-minimal structure with at least two constant elements is uniformly 1-eliminable.

Lemma 4.4.3. *Suppose the L-structure A is uniformly 1-eliminable and has definable Skolem functions. Then A has uniform elimination of imaginaries.*

Proof. Let $\theta(\bar{x}, \bar{y})$ be an equivalence formula on n-tuples, and $\bar{a} = (a_0, \ldots, a_{n-1})$ an n-tuple from A. By uniform 1-eliminability there is a formula $\psi(x, \bar{z})$ of L such that for some unique $\bar{b}$,

$$\psi(A, \bar{b}) = \{c : A \vDash \exists y_1 \ldots y_{n-1}\, \theta(\bar{a}, c, y_1, \ldots, y_{n-1})\}.$$

Using $\bar{b}$, the definable Skolem functions find us a unique element c_0 of this class. Now continue the same way, finding a unique c_1 such that $A \vDash \exists y_2 \ldots \exists y_{n-1}\, \theta(\bar{a}, c_0, c_1, y_2, \ldots, y_{n-1})$; etc. The resulting formula depends only on θ and not on $\bar{a}$. □

Thus for example, still using Example 3 in section 3.1, we deduce the following.

Theorem 4.4.4. *Every real-closed field has uniform elimination of imaginaries. More generally, so does every O-minimal structure with definable Skolem functions and at least two constant elements.* □

Before we look at any more examples, let me comment on the role of the two constant elements. We say that a theory T has **semi-uniform elimination of imaginaries** if for every formula $\psi(\bar{x}, \bar{y})$ of L there is a finite set $\{\phi_i(\bar{x}, \bar{z}_i) : i < m\}$ of formulas of L such that

(4.3) $T \vdash \forall \bar{y}$ (For a unique $i < m$, $\exists_{=1}\bar{z}\forall\bar{x}(\psi(\bar{x}, \bar{y}) \leftrightarrow \phi_i(\bar{x}, \bar{z})))$.

We say that an L-structure A has **semi-uniform elimination of imaginaries** if $\mathrm{Th}(A)$ has semi-uniform elimination of imaginaries.

Lemma 4.4.5. *Suppose T is a theory in L, there are closed terms s, t of L such that $T \vdash s \neq t$, and T has semi-uniform elimination of imaginaries. Then T has uniform elimination of imaginaries.*

Proof. An exercise, using Exercise 2.1.6 (time-sharing with formulas). □

For example if A is a dense linearly ordered set without endpoints, then A is O-minimal but no elements of A are named by closed terms. Let $\theta(x, y_0, y_1)$ be the formula $y_0 = y_1$. Then $\theta(A, \bar{a})$ is always either empty or $\mathrm{dom}(A)$, so that it is definable by one of the two formulas $x = x$, $x \neq x$. But there is no way of using a unique tuple to pick out one of these formulas.

Our next sufficient condition for elimination of imaginaries is stolen from Chapter 10 below.

Corollary 10.5.5. *Let T be a theory in a first-order language L. Suppose that for every model A of T, every equivalence formula θ of A and every tuple $\bar{a}$ in A, there is a tuple $\bar{b}$ in A such that*

(4.4) *for every automorphism g of A, g fixes the set $\theta(A^n, \bar{a})$ (setwise) if and only if $g\bar{b} = \bar{b}$.*

Then T has semi-uniform elimination of imaginaries. □

The proof of the next theorem also needs some results from elsewhere, though they are standard items of algebraic geometry.

Theorem 4.4.6. *The theory of algebraically closed fields has uniform elimination of imaginaries.*

Proof. Let A be a field and n a positive integer. By a **closed set** in A we mean a set of the form

(4.5) $\{\bar{a}$ an n-tuple in A: $p(\bar{a}) = 0$ for all $p(\bar{x})$ in $P\}$

where P is a finite set of polynomials with coefficients in A. This defines the **Zariski topology** on A^n. We say that the set (4.5) is **algebraically definable over** a set W if all the polynomials in P can be chosen to have their coefficients in W. A **constructible set** is a boolean combination of closed sets.

Henceforth we assume A is algebraically closed.

> *First Fact. Every n-ary relation on A which is first-order definable with parameters is a constructible set.*

This is elimination of quantifiers for algebraically closed fields; cf. Theorem 2.7.3.

> *Second Fact. If X and Y are constructible sets with the same closure (in the Zariski topology), then X meets Y.*

This is an exercise (see Exercise 7 below) from the fact that the Zariski topology is **noetherian**, i.e. the family of closed sets satisfies the descending chain condition.

> *Third Fact. For every closed set Z there is a unique smallest subfield B of A (the **field of definition** of Z) such that Z is algebraically definable over B. Moreover if g is any automorphism of A, then $g(Z) = Z$ if and only if g fixes B pointwise.*

This is the theorem of Weil which we met in Fact 4.3.4 above. See Weil [1946] p. 19, Lang [1958] p. 62.

We prove the theorem. Let T be the theory of algebraically closed fields, A a model of T and $\theta(\bar{x}, \bar{y})$ an equivalence formula of A. Consider an equivalence class $X = \bar{a}/\theta$, and let Z be its closure in the Zariski topology. By the First Fact above, X is a constructible set. If g is any automorphism of A taking Z to Z, then g takes X to a constructible set Y with closure Z. By the Second Fact, X meets Y, and so X and Y must be the same equivalence class of θ. Thus for any automorphism g of A, $g(Z) = Z$ iff $g(X) = X$. By the Third Fact there is a unique smallest subfield B of A such that Z is algebraically definable over B; moreover if g is an automorphism of A, then $g(Z) = Z$ iff g fixes B pointwise. Let $\bar{b}$ be a tuple in B such that Z is algebraically definable over $\bar{b}$. Then Z is definable from $\bar{b}$ in the usual model-theoretic sense. So the hypotheses of Corollary 10.5.5 above are satisfied, and hence T has semi-uniform elimination of imaginaries. The rest is by Lemma 4.4.5, since $0 \neq 1$.   □

Weak elimination

Life is unkind and forces us to consider one more variation of elimination of imaginaries. Some of the arguments below need methods from later in this book.

We say that an L-structure A has **weak elimination of imaginaries** if for every equivalence formula $\theta(\bar{x}, \bar{y})$ of A and each tuple $\bar{a}$ in A there are a formula $\phi(\bar{x}, \bar{z})$ of L and a *finite set X of tuples* such that the equivalence class $\bar{a}/\theta$ of $\bar{a}$ can be written as $\phi(A^n, \bar{b})$ if and only if $\bar{b}$ lies in X. Weak elimination of imaginaries can apply to theories as well as structures, and it can be uniform or semi-uniform, just as before. The analogue of Theorem 4.4.2 holds, except that now the compactness theorem (Theorem 6.1.1 below) is needed to get a uniform finite bound.

One example is the empty theory T in the language L with empty signature. Suppose A is a model of T – i.e. a set with no further structure – and X is an n-ary relation which is definable in A with parameters. Then we can write X as $\phi(A^n, \bar{a})$ for some formula ϕ of L and some tuple $\bar{a}$ from A. If we choose ϕ and $\bar{a}$ to make $\bar{a}$ as short as possible, then the set of items in $\bar{a}$ is determined by X, but unfortunately their order in $\bar{a}$ need not be. Take for example the set $\{a, b\}$; we can define it equally well using either the tuple ab or the tuple ba. In fact the theory T has weak semi-uniform elimination of imaginaries. Exercise 7.3.16 will prove this.

Theorem 4.4.7. *Let T be the theory of vector spaces over a fixed field. Then T has weak semi-uniform elimination of imaginaries.*

Proof. I give a quick sketch. Most of the proof of Theorem 4.4.6 goes over.

If V is a vector space and n a positive integer, we can topologise the set V^n by taking as basic closed sets the sets $\{\bar{a}: V \vDash \phi(\bar{a})\}$ where ϕ is the conjunction of a finite set of equations with parameters in V. By quantifier elimination, every relation $\subseteq V^n$ which is first-order definable with parameters from V is constructible. The topology has the descending chain condition on closed sets, and so any two constructible sets with the same closure meet each other. Exactly as in the proof of Theorem 4.4.6, we reduce to the case of eliminating imaginaries from the definition of a closed set. However, we have no analogue of Weil's theorem.

Borrowing the terminology of the geometers (as in Exercise 7), we can express a closed set X in just one way as an irredundant union of finitely many irreducible closed sets. The irreducible closed sets are basic closed sets as defined above, and for these it's easy to find unique defining parameters. (For example if the closed set is defined by the equation $x_0 + 5x_1 = 3a - 2b$, take $3a - 2b$ as parameter.) The only reason that this gives us a set of tuples for defining X, and not a single tuple, is that the order of the irreducible closed sets in the union is not fixed. $\qquad\qquad\square$

Exercises for section 4.4

1. Show that if the L-structure A has uniform elimination of imaginaries, and $\bar{a}$ is a sequence of elements of A, then $(A, \bar{a})$ also has uniform elimination of imaginaries.

2. (a) Show that ZFC has elimination of imaginaries. [For each equivalence class, choose the set of all elements of least rank in the class.] (b) Show that first-order Peano arithmetic has uniform elimination of imaginaries.

3. Give an example of a structure with definable Skolem functions but without uniform elimination of imaginaries. [The elements of A are the pairs $\langle m, n \rangle$ of integers. A has functions $F(\langle m, n \rangle) = \langle m + 1, n \rangle$, $G(\langle m, n \rangle) = \langle m, n + 1 \rangle$, and a relation $E(\langle m, n \rangle, \langle m', n' \rangle) \equiv m = m'$.]

4. Show that if V is a vector space of dimension at least 2 over a finite field of at least three elements, then V doesn't have elimination of imaginaries.

5. Prove Lemma 4.4.5.

6. Let $\theta(\bar{x}, \bar{y})$ be the equivalence formula $x_0 = x_1 \leftrightarrow y_0 = y_1$, and let A^* be the slice of A^{eq} got by adding to A the equivalence classes of θ. Show that if A has semi-uniform elimination of imaginaries, then A^* has uniform elimination of imaginaries.

7. Let Ω be a space with a noetherian topology (i.e. there are no infinite descending chains of closed sets). A **constructible** set is a boolean combination of closed sets.

(a) Show that every constructible set is a finite union $\bigcup_{i<k}(F_i \cap G_i)$ where each F_i is closed and each G_i is open. (b) A closed set is **irreducible** if it isn't the union of two smaller closed sets. A union $\bigcup_{i\in I}X_i$ is **irredundant** if $\bigcup_{i\in I}X_i \neq \bigcup_{i\in J}X_i$ whenever J is a proper subset of I. Show that every closed set X can be written in just one way as an irredundant union of irreducible closed sets. (These are the **irreducible components** of X.) [For existence, apply the d.c.c. to the set of closed sets which can't be written as such a finite union.] (c) Show that in an irreducible closed set, any two open subsets have non-empty intersection. (d) Show that if X is a constructible set which is dense in a closed set Z, then X is dense in every irreducible component of Z. [If G is open, consider the intersection of G with the complements of the other irreducible components of Z.] (e) Show that if X is a constructible set which is dense in an irreducible closed set Z, then writing X as in (a), at least one of the sets F_i contains Z. (f) Show that if X and Y are constructible sets which have the same closure, then X meets Y.

The next exercise uses the compactness theorem, Theorem 6.1.1 below.
8. Suppose T is a first-order theory, and every model of T has elimination of imaginaries. Show that T has semi-uniform elimination of imaginaries.

9. Show that the linear ordering $(\eta + \omega^* + \omega) \cdot (\omega^* + \omega)$ (where η is the ordering of the rationals) doesn't have semi-uniform elimination of imaginaries. [θ says that two elements are in the same discrete component.]

4.5 Minimal sets

The remainder of this chapter is about the classification of 'very simple' structures. It explores a part of stability theory, but a rudimentary part. Many of the complexities of stability theory were invented for generalising facts about minimal sets to more general settings. (Thus 'regular types' generalise the dimension analysis of this section, and 'forking' generalises the notion of that name in section 4.7.) In order to reach interesting results, I shall occasionally borrow theorems from later chapters. None of those theorems need the results of this chapter for their proofs!

From now till the end of the chapter, L is a first-order language and A an L-structure.

As in section 2.1 above, we say that a definable set X of elements of A is **minimal** if X is infinite but for every formula $\phi(x)$ of L, maybe with parameters from A, one of the sets $X \cap \phi(A)$ and $X \backslash \phi(A)$ is finite. We say that the structure A itself is a **minimal structure** if $\text{dom}(A)$ is a minimal set. (**Warning.** This has nothing to do with the notion of 'minimal models' in section 7.2 below.)

In section 2.1 we lacked the equipment to find any interesting examples. But the quantifier elimination of section 2.7 puts several in our hands.

Example 1: *Algebraically closed fields.* In an algebraically closed field A, every formula $\phi(x)$ of L is equivalent to a boolean combination of polynomial equations $p(x) = 0$, where p has coefficients from A. (See Theorem 2.7.3.) The solution set of $p(x)$ is either the whole of A (when p is the zero polynomial) or a finite set (otherwise). It follows at once that A is minimal.

Example 2: *Vector spaces.* Let V be an infinite vector space over a field K. Every formula $\phi(x)$ of L is equivalent to a boolean combination of linear equations $rx = a$ with r in K and a in V. (See Exercise 2.7.9 above or Example 1 in section 8.4 below.) The set of vectors satisfying the equation $rx = a$ is either the whole of V or a singleton or the empty set. Thus again A is minimal.

Example 3: *Affine spaces.* Once again let V be an infinite vector space over a finite field K. This time we put a different structure on V, to create a structure A. The elements of A are the vectors in V. For each field element α we introduce a 2-ary function symbol F_α with the following interpretation:

$$(5.1) \qquad\qquad F_\alpha(a, b) = \alpha a + (1 - \alpha)b.$$

We also introduce a 3-ary function symbol G with the interpretation

$$(5.2) \qquad\qquad G(a, b, c) = a - b + c.$$

The signature of A consists of these symbols F_α ($\alpha \in K$) and G. In particular there is no symbol for 0. A structure A formed in this way is called an **affine space**. Its substructures are the cosets (under the additive group of V) of the subspaces of V; these substructures are known as the **affine flats**. The structure A is certainly minimal, since each of its definable relations is already a definable relation in the vector space V. It may be worth remarking that in this example and Example 2, nothing changes if the field is a skew field.

We say that a formula $\psi(x)$ of L, maybe with parameters in the structure A, is **minimal** (for A) if $\psi(A)$ is minimal. We say that ψ and $\psi(A)$ are **strongly minimal** (for A) if for every elementary extension B of A, $\psi(B)$ is minimal in B. In particular the structure A itself is **strongly minimal** if every elementary extension of A is a minimal structure. The classes of structures mentioned in Examples 1–3 above are all closed under taking elementary extensions, and so these are all strongly minimal.

Fact 4.5.1. *If $\psi(x)$ is a formula of L, then the following are equivalent.*
(a) $\psi(A)$ *is strongly minimal.*
(b) *For every structure B which is elementarily equivalent to A, $\psi(B)$ is minimal in B.*

Proof. (b) implies (a) by the definitions. In the other direction, suppose B is a counterexample to (b). The elementary amalgamation theorem (Theorem 6.4.1 below) will show that there is an elementary embedding of B into some elementary extension C of A, and it follows that $\psi(C)$ is not minimal in C either.   □

This fact has two consequences. The first is terminological: let us say that a formula ψ of L is **strongly minimal for** a theory T in L if ψ defines a strongly minimal set in every model of T. Then $\psi(A)$ is strongly minimal in A if and only if ψ is strongly minimal for $\mathrm{Th}(A)$. The second is a matter of style: it can be helpful to hide the parameters used to define a strongly minimal set, by adding them to the language as new constants. I shall often do this without comment.

Algebraic dependence and minimal sets

In section 4.1 we met the notion of algebraic dependence. This notion is particularly well behaved in minimal sets. For example, we have the following result.

Lemma 4.5.2. *(Exchange lemma) Let X be a set of elements of A and Ω an X-definable minimal set in A. Suppose a is an element of A and b is an element of Ω. If $a \in \mathrm{acl}(X \cup \{b\})\backslash\mathrm{acl}(X)$, then $b \in \mathrm{acl}(X \cup \{a\})$.*

Proof. We deny this and aim for a contradiction. Adding the elements of X as parameters to L, we can replace X by $\varnothing$ in the lemma. Since $a \in \mathrm{acl}(b)$, there are a formula $\phi(x, y)$ of L and a positive integer n such that

$$(5.3) \qquad\qquad A \vDash \phi(a, b) \wedge \exists_{=n} x\, \phi(x, b).$$

Hence, since Ω is minimal and $b \in \Omega\backslash\mathrm{acl}(a)$, there is a finite subset Y of Ω such that

$$(5.4) \qquad \text{for all } b' \in \Omega\backslash Y,\ A \vDash \phi(a, b') \wedge \exists_{=n} x\, \phi(x, b').$$

Since $a \notin \mathrm{acl}(\varnothing)$, there is an infinite set Z of elements a' in A such that

$$(5.5) \quad \text{for all but } |Y| \text{ elements } b' \text{ of } \Omega,\ A \vDash \phi(a', b') \wedge \exists_{=n} x\, \phi(x, b').$$

If $a_0, \ldots, a_n$ are distinct elements of Z, then by (5.5) there is some element b' of Ω such that $A \vDash \bigwedge_{i \leqslant n} \phi(a_i, b') \wedge \exists_{=n} x\, \phi(x, b')$; this is a contradiction. □

We put this result together with the facts that we proved about algebraic closure in section 4.1.

Theorem 4.5.3. *Let Ω be a minimal set in A, and let U be a set of elements of A such that Ω is U-definable. Then for all subsets X and Y of Ω we have*

(5.6) $X \subseteq \mathrm{acl}(X)$

(5.7) $X \subseteq \mathrm{acl}(Y) \Rightarrow \mathrm{acl}(X) \subseteq \mathrm{acl}(Y)$,

(5.8) $a \in \mathrm{acl}(X) \Rightarrow a \in \mathrm{acl}(Z)$ *for some finite* $Z \subseteq X$,

(5.9)   (Exchange law) *For any set* $W \supseteq U$ *of elements of A, and any two elements a, b of* Ω, *if a is algebraic over* $W \cup \{b\}$ *but not over* W, *then b is algebraic over* $W \cup \{a\}$.

Proof. (5.6)–(5.8) are from Lemmas 4.1.1 and 4.1.2, and (5.9) is from Lemma 4.5.2.   □

Together these four laws show that algebraic dependence in a minimal set Ω behaves very much like linear dependence in a vector space. In fact if we read '$a \in \mathrm{acl}(X)$' as 'a is dependent on X', then (5.6), (5.7) and (5.9) together are equivalent to the 'axiomatisation of the concept of linear dependence' which van der Waerden ([1949] p. 100) proposed. As it happens, infinite vector spaces are a special case of Theorem 4.5.3, since in an infinite vector space the algebraic closure of a set X of vectors is exactly the subspace spanned by X.

This allows us to commit piracy on the language of linear algebra, and steal terms like 'independent' and 'basis' for use in any minimal set Ω. We call a subset Y of Ω **closed** if it contains every element of Ω which is dependent on it; this is the analogue of a subspace. The set Ω itself is closed. Every subset Z of Ω lies in a smallest closed set, $\Omega \cap \mathrm{acl}(Z)$; we call this set the **closure** of Z, and we say that Z **spans** this set. If Y is a closed set, then a **basis** of Y is a maximal independent set of elements of Y. The usual vector space arguments go over without any alteration at all, to give the following facts.

Theorem 4.5.4. *Let* Ω *be a* $\varnothing$-*definable minimal set in A.*

(a) *Every set* $X \subseteq \Omega$ *contains a maximal independent set W of elements, and any such set W is a basis of the closure of X.*

(b) *A subset W of a closed set X is a basis of X if and only if it is minimal with the property that* $X \subseteq \mathrm{acl}(W)$.

(c) *Any two bases of a closed set have the same cardinality.*

(d) *If X and Y are closed sets with* $X \subseteq Y$, *then any basis of X can be extended to a basis of Y.*

Proof. It's strangely hard to find a full proof of (c) in the elementary algebra texts. For example P. M. Cohn reports the infinite case as a theorem of universal algebra [1981] (Proposition II.5.5) but not as a fact about vector spaces [1974] (section 4.4).   □

The **dimension** of a closed set X in Ω is defined to be the cardinality of any basis of X. This is well-defined by (c).

Example 1 *continued.* In an algebraically closed field, 'dependence' means algebraic dependence. A basis in our sense is the same as a transcendence basis, and the dimension is the transcendence degree over the prime field.

Example 3 *continued.* In an infinite affine space the closed sets are the affine flats. **Warning.** Geometers count the dimension of an affine flat differently from us. Their dimension, the **geometric dimension**, is one less than ours. For example an affine line has dimension 2 for us, because it needs two independent points to specify it. But the geometers give it geometric dimension one, because lines in vector spaces have dimension 1.

Automorphisms of minimal sets

Suppose that A and B are L-structures – and recall that L is first-order throughout this section. By an **elementary map** between A and B we mean a map $f: X \to \mathrm{dom}(B)$ where $X \subseteq \mathrm{dom}(A)$, such that for every tuple $\bar{a}$ of elements of X and every formula $\phi(\bar{x})$ of L, $A \vDash \phi(\bar{a})$ iff $B \vDash \phi(f\bar{a})$. (This includes the empty tuple: an empty map from A to B is elementary if and only if $A \equiv B$.) An **elementary bijection** $f: X \to Y$ between A and B is an elementary map f between A and B which is a bijection from X to Y.

The next two lemmas show how one can build up elementary maps between minimal sets.

Lemma 4.5.5. *Let* $f: X \to Y$ *be an elementary bijection between the L-structures A, B. Then f can be extended to an elementary bijection* $g: \mathrm{acl}_A(X) \to \mathrm{acl}_B(Y)$.

Proof. By Zorn's lemma there is a maximal elementary bijection $f': X' \to Y'$ with $X \subseteq X' \subseteq \mathrm{acl}_A(X)$ and extending f. Since every element of X' is algebraic over X, it's easily checked that $Y' \subseteq \mathrm{acl}_B(Y)$ too. We show first that $X' = \mathrm{acl}_A(X)$.

Suppose $X' \neq \mathrm{acl}_A(X)$, and let a be an element of $\mathrm{acl}_A(X) \backslash X'$. Since $a \in \mathrm{acl}_A(X')$, we can choose a formula $\phi(x)$ with parameters in X' so that $A \vDash \phi(a) \wedge \exists_{=n} x\, \phi(x)$ for some n; we choose ϕ to make n as small as possible. Since f' is elementary, $B \vDash \exists_{=n} x\, \phi(x)$, so we can find an element b in $\mathrm{acl}(Y')$ such that $B \vDash \phi(b)$. If $\bar{c}$ is any tuple of elements on X' and $\psi(x, \bar{z})$ any formula of L, we claim that

(5.10) $$A \vDash \psi(a, \bar{c}) \Rightarrow B \vDash \psi(b, f'\bar{c}).$$

For suppose $A \vDash \psi(a, \bar{d})$. Then by choice of n as minimal, $A \vDash \forall x(\phi(x) \rightarrow \psi(x, \bar{c}))$, so $B \vDash \forall x(\phi(x) \rightarrow \psi(x, f'\bar{c}))$ by applying f', and hence $B \vDash \psi(b, f'\bar{c})$ as required. The claim shows that we can extend f' by taking a to b, contradicting the maximality of X'.

Thus $X' = \mathrm{acl}_A(X)$, and so $\mathrm{acl}_A(X)$ is the domain of f'. We must still show that the image of f' contains $\mathrm{acl}_B(Y)$. But every element of $\mathrm{acl}_B(Y)$ is one of a finite set of elements satisfying some formula over Y, and the domain of f' must contain the same number of elements satisfying the corresponding formula over X. $\qquad \square$

Now let us suppose that A contains a minimal set Ω. After adding parameters we can assume that Ω is $\varnothing$-definable, and so without loss we can suppose that $\Omega = P^A$ for some 1-ary relation symbol $P(x)$ of L. An **independent sequence in** Ω **over** a set of elements X is a sequence $(a_i : i < \gamma)$ of distinct elements of Ω, such that for each $i < \gamma$, $a_i \notin \mathrm{acl}(X \cup \{a_j : j \neq i\})$. The sequence is simply **independent** if it is independent over the empty set.

Lemma 4.5.6. *Suppose A and B are elementarily equivalent L-structures, and P^A, P^B are minimal sets in A, B respectively. Let X, Y be sets of elements of A, B respectively, and $f : X \rightarrow Y$ an elementary bijection. Suppose $(a_i : i < \gamma)$ is an independent sequence in P^A over X, and $(b_i : i < \gamma)$ is an independent sequence in P^B over Y. Then f can be extended to an elementary map g which takes each a_i to b_i.*

Proof. Write g_i for the extension of f which takes a_j to b_j whenever $j < i$. We show by induction on i that each map g_i is elementary. First, g_0 is f, which is elementary by assumption. Next, if δ is a limit ordinal and g_i is elementary for each $i < \delta$, then g_δ is elementary too, since first-order formulas mention only finitely many elements.

Finally suppose $i = k + 1$ and g_k is elementary. Let $\phi(\bar{c}, \bar{d}, x)$ be a formula of L with $\bar{c}$ in X and $\bar{d}$ in $\{a_j : j < k\}$. Then one of $P^A \cap \phi(\bar{c}, \bar{d}, A)$ and $P^A \setminus \phi(\bar{c}, \bar{d}, A)$ is finite. But a_k is independent of X and $\bar{d}$, and so a_k satisfies whichever one of $\phi(\bar{c}, \bar{d}, x)$, $\neg \phi(\bar{c}, \bar{d}, x)$ picks out an infinite subset of P^A. If it be $\phi(\bar{c}, \bar{d}, x)$, then since g_k is elementary, $\phi(g_k\bar{c}, g_k\bar{d}, B)$ is infinite too, and so $B \vDash \phi(g_k\bar{c}, g_k\bar{d}, g_i a_k)$. This shows that g_i is elementary. $\qquad \square$

These two lemmas give us a way of building up automorphisms of a minimal set: we choose two maximal independent sets, use Lemma 4.5.6 to find an elementary bijection from one to the other, and then cover the whole minimal set by means of Lemma 4.5.5.

Thus the following grand old theorem of Steinitz [1910] is really a result in the model theory of minimal structures.

Corollary 4.5.7. *Let A and B be algebraically closed fields of the same characteristic and the same uncountable cardinality λ. Then A is isomorphic to B.*

Proof. Since they have the same characteristic, A and B satisfy the same quantifier-free sentences. Hence quantifier elimination (Theorem 2.7.3) tells us that A and B are elementarily equivalent. By Example 1, A and B are minimal structures. Let X and Y be bases of A and B respectively. Since the language is countable, the cardinality of $\mathrm{acl}_A(X)$ is at most $|X| + \omega$; but A is uncountable and equal to $\mathrm{acl}_A(X)$, so that $|X| = \lambda$. Similarly $|Y| = \lambda$. The empty map between A and B is elementary since $A \equiv B$; so by Lemma 4.5.6 there is an elementary bijection from X to Y. By Lemma 4.5.5 this map extends to an elementary map from A onto B, this is clearly an isomorphism. $\qquad\square$

Exercises for section 4.5

1. Show that a minimal set remains minimal if we add parameters; likewise with 'strongly minimal' for 'minimal'.

2. Give an example to show that the notion 'algebraic over' need not obey the Steinitz exchange law. More precisely, find a structure A with elements a, b such that a is algebraic over $\{b\}$ but not over the empty set, while b is not algebraic over $\{a\}$.

3. Deduce the following properties of algebraic closure from the axioms (5.6)–(5.9). (a) $\mathrm{acl}(X) = \mathrm{acl}(\mathrm{acl}(X))$. (b) $X \subseteq Y \Rightarrow \mathrm{acl}(X) \subseteq \mathrm{acl}(Y)$.

The next exercise can be handled tediously by the method of quantifier elimination, or swiftly by Lindström's test (Theorem 8.3.4) and Theorem 8.4.1.
4. Let G be a group. We define a G-**set** to be an L-structure A as follows. The signature is a family of 1-ary functions $(F_g: g \in G)$; the laws $\forall x\, F_g F_h(x) = F_{gh}(x)$, $\forall x\, F_1(x) = x$ hold in A. We say A is a **faithful** G-set if for all $g \neq h$ in G and all elements a, $F_g^A(a) \neq F_h^A(a)$. (a) Show that the class of faithful G-sets is first-order axiomatisable, in fact by $\forall_1$ sentences. (b) Show that if A is an infinite faithful G-set, then A decomposes in a natural way into a set of connected components, and each component is isomorphic to a Cayley graph of the group G (see Exercise 4.1.1 above). (c) Deduce that every infinite faithful G-set is strongly minimal, and its dimension is the number of components.

5. Let A be an L-structure, Ω a $\varnothing$-definable minimal set in A, $(a_i: i < \gamma)$ a sequence of elements of Ω, X a set of elements of A and $\bar{b}$ a sequence listing the elements of X. Show that the following are equivalent: (a) $(a_i: i < \gamma)$ is independent over X; (b) for each $i < \gamma$, $a_i \notin \mathrm{acl}(X \cup \{a_j: j < i\})$; (c) $a_i \neq a_j$ whenever $i < j$, and $\{a_i: i < \gamma\}$ is an independent set in $(A, \bar{b})$.

6. Suppose $\phi(x)$ is a formula of L and $\phi(A)$ is a minimal set in A with dimension κ. If $|L| \leqslant \lambda \leqslant \kappa$, show that A has an elementary substructure B in which $\phi(B)$ has dimension λ.

The conclusion (c) below is a trivial consequence of the upward Löwenheim–Skolem theorem (Corollary 6.1.4), but we can already get it without using that theorem.
7. (a) Show that if A is a minimal structure of infinite dimension, then A has a proper elementary extension B which is isomorphic to A. (b) Show that the union of an elementary chain of minimal structures is again minimal. (c) Deduce, without using the compactness theorem, that every minimal structure of infinite dimension has arbitrarily large elementary extensions.

8. Show that if $\psi(A)$ is a minimal set in A, then $\psi(A)$ is strongly minimal if and only if A has an elementary extension B in which $\psi(B)$ is minimal and of infinite dimension. In particular, every minimal structure of infinite dimension is strongly minimal.

9. Show that if A is an L-structure and P^A is a minimal set of uncountable cardinality, then A is strongly minimal. [If L is countable, use the previous exercise. If L is uncountable, apply that exercise to each countable sublanguage.]

10. Show that the structure $(\omega, <)$ is minimal but not strongly minimal.

11. Let A be a structure (Ω, E) where E is an equivalence relation whose equivalence classes are all finite, and E has just one class of cardinality n for each positive integer n. Show that A is minimal but not strongly minimal.

12. Let A be the affine space derived from a vector space V. A **translation** of A is a map $\tau_b: a \mapsto a + b$, where b is a fixed element of A. (a) Show that the translations of A form a normal subgroup Trans(A) of Aut(A) which is isomorphic to $(V, +)$. (b) Show that Aut(A)/Trans(A) is isomorphic to Aut(V).

*13. Let A be a strongly minimal structure of dimension d with $0 < d < \omega$, and for each i with $1 \leqslant i \leqslant d$ let A_i be an elementary substructure of A with dimension i. Suppose $|\text{Aut}(A)|$ is infinite. Show that $|\text{Aut}(A_i)| = 2^\omega$ for each i.

4.6 Geometries

A **combinatorial geometry**, or for short a **geometry**, consists of a set Ω and an operation cl taking subsets of Ω to subsets of Ω, which obeys the following laws.

(6.1)   For every set $X \subseteq \Omega$, $X \subseteq \text{cl}(X)$.

(6.2)   For all sets $X, Y \subseteq \Omega$, if $X \subseteq \text{cl}(Y)$ then $\text{cl}(X) \subseteq \text{cl}(Y)$.

(6.3)   For all sets X and elements a, b of Ω, if $a \in$ cl$(X \cup \{b\})$\cl$(X)$ then $b \in$ cl$(X \cup \{a\})$.

(6.4)   cl$(\varnothing) = \varnothing$, and for every element a of Ω, cl$(\{a\}) = \{a\}$.

The elements of Ω are known as **points**; cl(X) is read as 'closure of X', and sets of the form cl(X) are said to be **closed**.

This definition (Crapo & Rota [1970]) was intended for dealing mainly with finite sets, and indeed I have left out a clause which says that every closed set is the closure of a finite set. Instead we add another condition:

(6.5)   for all sets X and points a, if $a \in$ cl(X) then $a \in$ cl(Y) for some finite $Y \subseteq X$.

A **pregeometry** (some say **matroid**) is defined the same way as a geometry, but without clause (6.4).

Strictly, geometries aren't structures in the sense of section 1.1 above, because the operation cl acts on sets of elements and not on elements. However, we can turn them into structures by adding infinitely many relation symbols R_n (n a positive integer), so that $R_n(a, b_0, \ldots, b_{n-1})$ holds if and only if $a \in$ cl$\{b_0, \ldots, b_{n-1}\}$. Condition (6.5) says that the whole closure operation cl can be reconstructed from these relations.

If we put acl for cl, the axioms of a pregeometry agree exactly with the laws (5.6)–(5.9) for algebraic closure in a minimal set, in section 4.5 above. It follows that for a pregeometry, just as for a minimal set, we can take over the terminology of linear algebra and speak of spanning sets, independent sets and bases. Any two maximal independent subsets of the same set X have the same cardinality, and this is known as the **dimension** of X, dim(X). The whole set Ω has a dimension, which is called the **dimension of the pregeometry**. (Cf. Theorem 4.5.3 and Exercise 4.5.3.)

As we noted just now, every minimal set forms a pregeometry. Vector spaces are not geometries, because acl$(\varnothing)$ contains the origin. If G is a group with more than one element, faithful G-sets (see Exercise 4.5.4) don't form a geometry, because $|$acl$(a)| = |G|$ for every element a. Affine spaces are geometries. Here are three other important classes of geometries.

Example 1: *Sets.* Let Ω be a set, and define cl$(X) = X$ for every subset X of Ω. Then (Ω, cl) is a geometry.

Example 2: *Projective geometries.* Suppose V is a vector space over a field k. We form the associated projective geometry $P = (\Omega, \text{cl})$ of V as follows. The set Ω of **points** of P is the set of straight lines of V through the origin; we can think of such a line as consisting of its non-zero points. A **flat** of P is the set of all points of P which lie in some given subspace of V. The **closure**

cl(X) of a set X of points is the smallest flat which contains X. The projective geometry P is a geometry.

First warning. As with affine spaces in Example 3 of section 4.5, geometers usually count the dimension differently from us. A projective flat has **geometric dimension** n if it consists of the lines which lie in an $(n + 1)$-dimensional subspace of V – passing from V to P drops the dimension by 1. On our definition this flat has dimension $n + 1$, because it needs $n + 1$ independent generators. For example a projective line has geometric dimension 1, but model theorists give it dimension 2. Incidentally the projective geometries of interest to us will all have geometric dimension at least 3, so they are desarguesian.

Second warning. When a desarguesian projective geometry is coordinatised, it gains some extra features, such as harmonic ratios on a line. If the coordinatising field is rigid, then these extra features don't change the automorphism group. But in general the field has automorphisms which induce automorphisms of the geometry; if one fixes the field pointwise, one cuts out these automorphisms. We speak of a **projective geometry** if we have in mind just the points and the closure operation. If the geometry is coordinatised and automorphisms must keep the field elements fixed, we speak of a **projective space**. The automorphism group of a projective geometry Q is known as the **projective general semilinear group** and written PΓL(Q); the projective space automorphisms form a normal subgroup PGL(Q), the **projective general linear group**.

Example 3: *Affine geometries.* These are the geometries corresponding to the affine spaces of Example 3 in section 4.5. The closure of a set of points is the smallest affine flat containing them. Just as with projective geometries, the automorphism group AΓL(B) (the **affine general semilinear group**) of an affine geometry B has as a normal subgroup the group AGL(B) of affine space automorphisms (the **affine general linear group**). There is more information on these and their projective counterparts in Gruenberg & Weir [1977] and Tsuzuku [1982].

We noted that a vector space is a pregeometry if we take cl(X) to be the subspace spanned by X. The method used in Example 2 for turning this pregeometry into a geometry works for all pregeometries. Given a pregeometry (Ω, cl), we construct a geometry (Ω', cl') as follows. The points of Ω' are the sets cl($\{a\}$) where $a \notin \text{cl}(\varnothing)$. For any subset X of Ω' we put cl'(X) = {cl($\{b\}$): $b \in \text{cl}(\bigcup X)$}. We call (Ω', cl') the **associated geometry** of (Ω, cl).

Example 4: *The associated geometry of a minimal set.* Suppose Ω is a strongly minimal set in a structure A. Then Ω forms a pregeometry with acl for cl. So Ω has its associated geometry. When the relation '$a \in \text{acl}(b)$' is first-order definable, the points of the associated geometry form a definable set in A^{eq}.

We say that a geometry (Ω, cl) is **disintegrated** if it has the property that for every set X, $\text{cl}(X) = X$ (cf. Example 1). We say that (Ω, cl) is **projective** (over the field k) if there is a projective space P over k, such that Ω is the set of points of P, and the closed sets of (Ω, cl) are precisely the flats of P (see Example 2). We call it **affine** (over k) if the same holds but with an affine space in place of P (see Example 3). We call a pregeometry **disintegrated** (or **projective**, or **affine**) if its associated geometry is disintegrated (or projective, or affine).

Homogeneous locally finite geometries

Continuing our definitions, we say that a pregeometry is **locally finite** if the closure of each finite set of points is finite. For example an infinite set is locally finite. If k is a finite field, then an infinite vector space over k, an infinite projective geometry over k and an infinite affine geometry over k are all locally finite.

All the geometries mentioned so far have one further property:

(6.6)   if X is the closure of a finite set and a, b are any two
        elements not in X, then there is an automorphism of (Ω, cl)
        which fixes X pointwise and takes a to b.

Geometries with this property are said to be **homogeneous**. In fact the associated geometry of a minimal set is always homogeneous, using Lemmas 4.5.5 and 4.5.6.

Combinatorial geometries suddenly became important for model theory when Boris Zil'ber conjectured Theorem 4.6.1 (and then proved it, along with other people).

Theorem 4.6.1 (*Cherlin–Mills–Zil'ber theorem*). *Let (Ω, cl) be a homogeneous locally finite combinatorial geometry of infinite dimension. Then one of the following three cases holds.*
(a) *(Ω, cl) is disintegrated.*
(b) *(Ω, cl) is isomorphic to a projective geometry over a finite field.*
(c) *(Ω, cl) is isomorphic to an affine geometry over a finite field.*

Proof. By any route the proof is not a casual affair. See the references for some background information. $\square$

All the geometries described in (a)–(c) of the theorem have another property in common. Note first of all that in any pregeometry the closed sets form a lattice, with $X \wedge Y = X \cap Y$ and $X \vee Y = \text{cl}(X \cup Y)$. We say that a combinatorial geometry (Ω, cl) is **modular** if for all closed sets X, Y of finite dimension in Ω,

$$(6.7) \qquad \dim(X) + \dim(Y) = \dim(X \vee Y) + \dim(X \wedge Y).$$

We say that (Ω, cl) is **locally modular** if (6.7) holds whenever $X \cap Y$ is not empty. (With $\geq$ for $=$, (6.7) is true for any pregeometry; see Exercise 1 below.)

It will be easier to discuss these notions if we introduce the notion of a localisation. Let (Ω, cl) be a geometry and W a set of points. Then we define the **localisation of** (Ω, cl) **at** W, in symbols $(\Omega, \text{cl})_{(W)}$, to be the following geometry (Ω', cl'). The set Ω' is the set of all sets of form $\text{cl}(W \cup \{b\})$ with $b \notin \text{cl}(W)$. If X is a subset of Ω', its closure $\text{cl}'(X)$ is the set of elements of Ω' which are $\subseteq \text{cl}(\bigcup X)$. I leave it as an exercise to verify that (Ω', cl') is a geometry. There is a natural correspondence between the closed subsets X of Ω which contain W, and closed subsets $X^* = \{p \in \Omega' : p \subseteq X\}$ of Ω'; $\dim(X) = \dim(X^*) + \dim(W)$.

A geometry (Ω, cl) is locally modular if and only if every localisation of (Ω, cl) at a singleton is modular. Again this is an exercise.

Theorem 4.6.2. *Disintegrated and projective geometries are modular. Affine geometries are locally modular but not modular.*

Proof. In a disintegrated geometry, the dimension of a set X is its cardinality, and for any two closed sets X and Y, $X \vee Y = X \cup Y$. From this one easily checks (6.7). I leave the projective case as an exercise; essentially it follows from Grassmann's law:

(6.8) If U, V are subspaces of a vector space,
       then $\dim(U) + \dim(V) = \dim(U + V) + \dim(U \cap V)$.

Finally let (Ω, cl) be an affine geometry. Let p be a point in Ω, and localise at $\{p\}$. The resulting geometry is a projective geometry, and so it is modular. This implies (6.7) back in the affine geometry, whenever X and Y are sets containing p. We get a counterexample to (6.7) with $X \cap Y$ empty by taking X and Y to be parallel lines. Thus the affine geometry is locally modular. $\square$

Not every homogeneous geometry is locally modular. Algebraically closed fields give a counterexample.

Theorem 4.6.3. *If* (Ω, cl) *is the associated geometry of an algebraically closed field of infinite transcendence degree over the prime field, then no localisation of* (Ω, cl) *at a finite set is modular.*

Proof. Let D be an algebraically closed field of infinite transcendence degree. I work in the field itself rather than in its associated geometry. We shall localise at a finite set W of dimension n; list the points of W as $\bar{e}$. Since D has infinite transcendence degree, we can find elements a, b, c, d which are algebraically independent over $\bar{e}$. Put $f = (a - c)/(b - d)$. Inside D, let A be the algebraic closure of $\{a, b, \bar{e}\}$ and B the algebraic closure of $\{\bar{e}, f, a - bf\}$. Then $A \vee B$ is the algebraic closure of $\{a, b, \bar{e}, f\}$, which has dimension $n + 3$. Each of A and B has dimension $n + 2$. It suffices to show that $A \wedge B$ has dimension $\neq (n + 2) + (n + 2) - (n + 3) = n + 1$.

Since $\bar{e} \in A \wedge B$, $A \wedge B$ has dimension at least n. Let g be any element of $A \wedge B$; we shall show that g is algebraically dependent on $\bar{e}$, so that $A \wedge B$ has dimension exactly n. Take an automorphism α of D which carries $(a, b, c, d, \bar{e})$ to $(c, d, a, b, \bar{e})$. Then $\alpha(f) = f$ and $\alpha(a - bf) = c - df = a - bf$; hence α can be chosen so that it fixes the field B pointwise. Thus we have $\alpha(g) = g$ since $g \in B$. But also $\alpha(g)$ is in the algebraic closure E of $\{c, d, \bar{e}\}$, since $g \in A$. So g lies in $A \cap E$, which is the algebraic closure of $\bar{e}$.

□

For some time it seemed that this might be a complete catalogue: the geometry of an infinite-dimensional minimal set either is locally modular or comes from an algebraically closed field (possibly with some extra structure added). One form of 'Zil'ber's conjecture' said just this. But in 1988 Ehud Hrushovski found a counterexample; it is too complicated to describe here.

Hrushovski recently announced some further results in this area. He constructs a strongly minimal set which has as reducts two algebraically closed fields of different characteristics. He also shows that if a strongly minimal structure A is an expansion of an algebraically closed field K, $\bar{a}$ is a sequence of algebraic elements of K, and algebraic closure is the same in $(A, \bar{a})$ as in $(K, \bar{a})$, then $(A, \bar{a})$ and $(K, \bar{a})$ have exactly the same $\varnothing$-definable relations.

Zil'ber's configuration

One of the pieces of evidence that gave rise to Zil'ber's conjecture was a striking result of Zil'ber which we shall prove as Theorem 4.8.1: if the geometry of a strongly minimal set is modular but not disintegrated, then we can reconstruct an infinite group from it. (For example in the affine case, we have the group of translations of the underlying affine space.) This suggested that if a geometry is even further from being disintegrated, we might be able to find more complicated algebraic structures hidden in it. Hrushovski's counterexample was designed to be as barren as possible of any interesting algebraic ingredients – though this is a general impression and not a theorem!

An important ingredient of Zil'ber's argument is the picture known as **Zil'ber's configuration**:

(6.9)

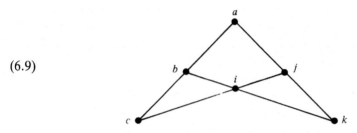

It represents six points of a pregeometry, a, b, c, i, j, k. A line joining three points means that each of these points is in the closure of the other two. No other dependencies between the six points are allowed. In particular none of the six points is in $cl(\varnothing)$.

One easily finds situations where Zil'ber's configuration occurs.

Theorem 4.6.4. *Let A be a strongly minimal structure of dimension $\geqslant 3$, and suppose that A is group-like (i.e. some reduct of A is a group). Then Zil'ber's configuration occurs in the pregeometry of A. (The group is abelian by Reineke's theorem, Theorem A.4.10.)*

Proof. There is an independent set $\{a, b, j\}$. Put $c = b \cdot a$, $k = a \cdot j$, $i = b \cdot a \cdot j = b \cdot k = c \cdot j$.   □

Theorem 4.6.5. *If a homogeneous geometry is infinite-dimensional, locally modular and not disintegrated, then Zil'ber's configuration occurs in it.*

Proof. Localising at one point, we can suppose that the geometry is modular. Since the geometry is not disintegrated, there is a finite set V such that $cl(V)$ is not $\bigcup\{cl\{a\}: a \in V\}$. Choose V as small as possible with this property. Then V has dimension at least 2; let W be a set consisting of all but two elements of a basis of V, and localise at W. The result is a modular geometry in which there are two independent points a, b whose closure includes some point c not in $cl\{a\} \cup cl\{b\}$. Choose a point $j \notin cl\{a, b\}$. Then by homogeneity (consider an automorphism fixing a and taking b to j) there is $k \in cl\{a, j\}\backslash(cl\{a\} \cup cl\{j\})$.

Consider now $X = cl\{b, k\}$, $Y = cl\{c, j\}$. We have $\dim(X) = 2$; for otherwise $b \in cl\{k\} \subseteq cl\{a, j\}$, and hence $j \in cl\{a, b\}$ by the exchange property, contradicting the choice of j. Also $\dim(Y) = 2$, since otherwise

$j \in \text{cl}\{c\} \subseteq \text{cl}\{a, b\}$. We claim that $X \cap Y \neq \emptyset$. For otherwise $4 = \dim(X) + \dim(Y) = \dim(X \vee Y)$ by modularity; which is impossible since a, b and j form a basis for $X \vee Y$. This proves the claim.

By the claim, let i be a point in $X \cap Y$. Then clearly we have the lines $\{b, i, k\}$ and $\{c, i, j\}$. Now for each line in the configuration, we can check that if one of the points shown as being off the line is in fact dependent on the line, then all six points are dependent on the line and the total dimension collapses to 2, which is impossible. $\qquad\square$

In section 4.7 below we shall a prove a converse of Theorem 4.6.4. Unfortunately even the simplest case of this converse, starting with a strongly minimal structure A, leads us quickly into A^{eq}. But A^{eq} is not strongly minimal. Fortunately every element of A^{eq} is algebraic over A, and this will give us the leverage we need. The next section will set out the main tools.

Exercises for section 4.6

1. Let X and Y be closed sets in a pregeometry. Show that $\dim(X) + \dim(Y) \geq \dim(X \cap Y) + \dim(X \cup Y)$. [Take a basis $\bar{z}$ of $X \cap Y$, and extend it to bases $\bar{z}\bar{x}$ of X and $\bar{z}\bar{y}$ of Y. Then $\bar{z}\bar{x}\bar{y}$ spans $X \cup Y$, though in general it need not be independent.]

2. (a) Show that the closed sets of a pregeometry form a complete lattice (i.e. a lattice with sups and infs of arbitrary sets), with $\inf(X_i : i \in I) = \bigcap_{i \in I} X_i$ and $\sup(X_i : i \in I) = \text{cl}(\bigcup_{i \in I} X_i)$ for any family $(X_i : i \in I)$ of closed sets. (b) Show that the dimension of a closed set X is equal to the cardinality of any maximal chain $(Y_i : i < \alpha)$ of elements $< X$ in the lattice. (c) Show also that the lattice satisfies the **semi-modular law**: if X, Y and Z are closed sets, X and Y each cover Z and $X \neq Y$ then $X \vee Y$ covers both X and Y. (We say that X **covers** Z if $X > Z$ and no element lies properly between them.)

3. Prove that the geometry of a projective space is modular. [See Grassmann's law, (6.8) above, noting that the geometry of the projective space of a vector space V is exactly the associated geometry of V.]

4. Show that for any group G, the geometry of a faithful G-set is disintegrated.

5. Show that for any geometry (Ω, cl) the following are equivalent. (a) (Ω, cl) is modular. (b) If X and Y are finite-dimensional closed sets with $X \leq Y$, then for any finite-dimensional closed set Z, $(X \vee Z) \wedge Y \leq X \vee (Z \wedge Y)$. (c) The same as (b), but without the restriction to finite-dimensional sets.

6. Let (Ω, cl) be a geometry, W a subset of Ω and (Ω', cl') the localisation $(\Omega, \text{cl})_{(W)}$ of (Ω, cl) at W. (a) Check that (Ω', cl') is a geometry. (b) Let Λ, Λ' be respectively the lattice of closed sets of (Ω, cl) and of (Ω', cl'). Show that there is a lattice

isomorphism between Λ' and the sublattice of Λ consisting of the closed sets containing W. (c) Show that if (Ω, cl) is modular, then so is (Ω', cl'). (d) Show that (Ω, cl) is locally modular if and only if for every point $p \in \Omega$, $(\Omega, \text{cl})_{\{p\}}$ is modular.

7. Let Ω be a strongly minimal set in an L-structure A, and $\bar{a}$ a tuple of elements of Ω. Let G be the associated geometry of Ω. Show that the associated geometry of Ω as a strongly minimal set in $(A, \bar{a})$ is the localisation of G at the set of closures of the points which are in $\text{cl}(\bar{a})$ but not in $\text{cl}(\varnothing)$.

*8. Let Λ be a complete lattice. A **point** of Λ is a minimal element > 0 in Λ. We say Λ is **relatively complemented** if every element of Λ is the join of a set of points. (a) Show that if Λ is the lattice of closed subsets of a geometry, then Λ is relatively complemented. (b) Show that if Λ is modular and relatively complemented, and every join of two points has at least three points below it, then Λ is isomorphic to the lattice of flats of some projective space.

9. Let (Ω, cl) be a geometry and X a subset of Ω. A sequence $(a_0, \ldots, a_{n-1})$ of points in Ω is **independent over** X if for every $i < n$, $a_i \notin \text{acl}(X \cup \{a_0, \ldots, a_{i-1}, a_{i+1}, \ldots, a_{n-1}\})$. Show that if (Ω, cl) is homogeneous, X is the closure of a finite set and the two sequences $(a_0, \ldots, a_{n-1})$ and $(b_0, \ldots, b_{n-1})$ are independent over X, there is an automorphism of (Ω, cl) which fixes X pointwise and takes each a_i to b_i.

10. Let Q be a projective geometry and B an affine geometry, both derived in the usual way from a vector space over a field k. Show that $\text{P}\Gamma\text{L}(Q)/\text{PGL}(Q) \cong \text{A}\Gamma\text{L}(Q)/\text{AGL}(Q) \cong \text{Aut}(k)$.

11. In Theorem 4.6.4, show that we can add to the configuration a seventh point which lies on a line through a and i and on a line through c and k. [Use the fact that the group is abelian.]

4.7 Almost strongly minimal theories

Our aim in this section is to study the structure of finite slices of A^{eq} where A is a strongly minimal structure. If B is a finite slice of A^{eq}, there is a first-order sentence which is true in B and expresses that every element of B is in the algebraic closure of A. So the following setting will take care of our needs.

Let T be a theory in L and $\phi(x)$ a formula of T. We say that T is **almost strongly minimal (over ϕ)** if T is complete, ϕ is strongly minimal for T, and in every model A of T, $\text{dom}(A) = \text{acl}\,\phi(A)$. We say that an L-structure A is **almost strongly minimal (over Ω)** if there is a formula $\phi(x)$ of L such that $\Omega = \phi(A)$ and $\text{Th}(A)$ is almost strongly minimal over ϕ.

Warning with apology. The definition just given is not correct, but the correction is tiresome and best forgotten in most arguments. We say that a theory T^+ is a **principal extension** of T if there are a tuple $\bar{c}$ of new constants and a formula $\theta(\bar{z})$ such that (1) T^+ is equivalent to $T \cup \{\theta(\bar{c})\}$, and (2) T^+ is consistent and complete. (In particular $T \vdash \exists \bar{z}\, \theta(\bar{z})$, so every model of T can be expanded to a model of T^+.) The correct definition is that T is **almost strongly minimal** if some principal extension of T is almost strongly minimal in the sense of the previous paragraph, over a formula $\phi(x, \bar{c})$; we say then that T is **almost strongly minimal over** $\phi(x, \bar{z})$. For the first few results below, I shall use the correct definition; but thereafter I shall slip back into the simplified version.

Warning without apology. Even without the refinement just mentioned, the definition of almost strongly minimal theories is rather sensitive. For example if the structure A has a $\varnothing$-definable strongly minimal set Ω and $\text{dom}(A)$ $= \text{acl}(\Omega)$, it need not follow that $\text{Th}(A)$ is almost strongly minimal; see Exercise 1. Also it can happen that an almost strongly minimal theory T over a formula $\phi(x)$ has another strongly minimal formula $\psi(x)$ but is not almost strongly minimal over ψ. This is quite an interesting possibility, and there are some remarks on it at the end of the section.

We turn to some examples.

Theorem 4.7.1. *Every strongly minimal structure is almost strongly minimal.*

Proof. Immediate.   □

Observe that none of the structures in Examples 1 to 3 of section 4.5 required us to pass to a principal extension.

Example 1: *A structure which is not almost strongly minimal.* Let κ be an infinite cardinal and p a prime, and let A be the abelian group $\bigoplus_{i<\kappa}\mathbb{Z}(p^2)b_i$, i.e. the direct sum of κ cyclic groups $\mathbb{Z}(p^2)b_i$ of order p^2. Then A carries a strongly minimal set Ω, namely the socle (the set of elements a such that $pa = 0$). But A is not algebraic over Ω; in the terminology of section 4.1, it is not even ALGEBRAIC over Ω. To see this, take any element c of the socle, and let α_c be the automorphism of A which carries b_0 to $b_0 + c$ and b_i $(i \neq 0)$ to b_i. Then α_c fixes the socle pointwise, but it can carry b_0 to infinitely many different elements according to the choice of c. The only other $\varnothing$-definable strongly minimal set in A is the socle less 0.

I mention the next example because it has been very influential, not because we have the means to handle it here.

Example 2: *Simple algebraic groups.* Let G be an infinite simple algebraic group over an algebraically closed field. Then G is almost strongly minimal. Zil'ber [1977a] proved this as a corollary of his indecomposability theorem (Theorem 5.7.12).

Basic properties

If T is almost strongly minimal over θ and A is a model of T, the **dimension** of A means the dimension of $\theta(A)$ in T. For all we know so far, this may depend on the choice of θ. But Theorem 4.7.16 below will show that another choice of θ can only change the dimension by a finite amount.

Lemma 4.7.2. *If T is almost strongly minimal, A is a model of T and $\bar{a}$ is a tuple in A, then $\mathrm{Th}(A, \bar{a})$ is almost strongly minimal. Moreover there is some k depending only on $\mathrm{Th}(A, \bar{a})$, such that $\dim(A) = \dim(A, \bar{a}) + k$ (with the same strongly minimal set in both A and $(A, \bar{a})$).*

Proof. Suppose the principal extension $T^+ = T \cup \{\theta(\bar{c})\}$ of T is almost strongly minimal over the formula $\phi(x, \bar{c})$ of L^+. Choose a tuple $\bar{d}$ in A such that $A \vDash \theta(\bar{d})$; then $\phi(A, \bar{d})$ is strongly minimal in A and $\mathrm{dom}(A) = \mathrm{acl}(\phi(A, \bar{d}))$. So there is some tuple $\bar{i}$ in $\phi(A, \bar{d})$ such that $\bar{a}$ is algebraic over $\bar{i}$. Choose $\bar{d}$ and $\bar{i}$ so as to make $\bar{i}$ as short as possible (so that it forms an independent sequence over $\bar{d}$); let k be the length of $\bar{i}$. Find a formula $\psi(\bar{y}, \bar{x}, \bar{z})$ such that $A \vDash \psi(\bar{a}, \bar{i}, \bar{d})$ and for some finite n there are exactly n tuples $\bar{a}'$ such that $A \vDash \psi(\bar{a}', \bar{i}, \bar{d})$; choose ψ to make n as small as possible.

We shall show that $\mathrm{Th}(A, \bar{a}, \bar{i}, \bar{d})$ is a principal extension of $\mathrm{Th}(A, \bar{a})$. Since $\phi(A, \bar{d})$ remains strongly minimal in $(A, \bar{a}, \bar{i}, \bar{d})$, this gives everything except the dimension count.

Let $\theta^*(\bar{y}, \bar{x}, \bar{z})$ be a formula which says $\theta(\bar{z}) \wedge {}'\phi(x, \bar{z})$ for each x in $\bar{x}' \wedge \psi(\bar{y}, \bar{x}, \bar{z}) \wedge$ 'there are exactly n tuples $\bar{y}'$ such that $\psi(\bar{y}', \bar{x}, \bar{z})'$. I claim that if B is any L-structure with tuples $\bar{b}$, $\bar{j}$, $\bar{e}$ such that $(B, \bar{b}) \equiv (A, \bar{a})$ and $B \vDash \theta^*(\bar{b}, \bar{j}, \bar{e})$, then $(B, \bar{b}, \bar{j}, \bar{e}) \equiv (A, \bar{a}, \bar{i}, \bar{d})$. Certainly $(B, \bar{e}) \equiv (A, \bar{d})$ and $\phi(B, \bar{e})$ is strongly minimal in B, by choice of θ; by θ^* we have that $\bar{j}$ lies in $\phi(B, \bar{e})$. Now $\bar{j}$ must be an independent sequence over $\bar{e}$; for if not, some formula $\chi(\bar{j}, \bar{e})$ would express a dependence in $\bar{j}$, giving

(7.1) $B \vDash \exists \bar{x} \bar{z} (\theta(\bar{z}) \wedge \chi(\bar{x}, \bar{z}) \wedge \psi(\bar{b}, \bar{x}, \bar{z}) \wedge$

           'there are exactly n tuples $\bar{y}$ such that $\psi(\bar{y}, \bar{x}, \bar{z})'$).

Since $(B, \bar{b}) \equiv (A, \bar{a})$, the same holds of $\bar{a}$ in A, which contradicts the minimality of $\bar{i}$. By Lemma 4.5.6 applied to $(B, \bar{e})$ and $(A, \bar{d})$, we deduce that $(B, \bar{j}, \bar{e}) \equiv (A, \bar{i}, \bar{d})$. Now since ψ was chosen to make n minimal, it follows that $(B, \bar{b}, \bar{j}, \bar{e}) \equiv (A, \bar{a}, \bar{i}, \bar{d})$ as well (cf. the proof of Lemma 4.5.5).

The strongly minimal set in $(A, \bar{a}, \bar{i}, \bar{d})$ is still $\phi(A, \bar{d})$, but its dimension has dropped by exactly k since we have turned the elements $\bar{i}$ into constants in the language. $\square$

We can go at once to the most significant fact about almost strongly minimal theories. (By Example 1 the converse fails.)

Theorem 4.7.3 *Let L be a first-order language, and let T be an almost strongly minimal theory in L. Then T is λ-categorical for every cardinal $\lambda > |L|$.*

Proof. We begin with the uniqueness. Let A and B be models of T, both of cardinality $\lambda > |L|$. Suppose the principal extension $T^+ = T \cup \{\theta(\bar{c})\}$ of T is almost strongly minimal over the formula $\phi(x, \bar{c})$ of L^+. Then A and B contain tuples $\bar{a}$ and $\bar{b}$ such that $(A, \bar{a})$, $(B, \bar{b})$ are models of T^+ (and hence elementarily equivalent), and $\phi(A, \bar{a})$, $\phi(B, \bar{b})$ are strongly minimal sets in $(A, \bar{a})$, $(B, \bar{b})$ respectively. Let X, Y be bases of these two sets. Since $\text{dom}(A) = \text{acl}(X)$ and $\text{dom}(B) = \text{acl}(Y)$, a cardinality count shows that $|X| = |Y|$. Then by Lemma 4.5.6 there is an elementary bijection from X to Y, and by Lemma 4.5.5 it extends to an elementary bijection f from $(A, \bar{a})$ onto $(B, \bar{b})$. This map f is an isomorphism from A to B.

For the existence, we need to know that T has a model A of cardinality λ. This follows at once from the upward Löwenheim–Skolem theorem (Corollary 6.1.4) below; but if we know that T has at least one model which has infinite dimension, we can also use the argument of Exercise 4.5.7. $\square$

Corollary 4.7.4. *Let T be an almost strongly minimal theory, A a model of T and $\bar{a}$, $\bar{b}$ tuples in A such that $(A, \bar{a}) \equiv (A, \bar{b})$ and both have infinite dimension. Then $(A, \bar{a}) \cong (A, \bar{b})$.*

Proof. By Lemma 4.7.2, $(A, \bar{a})$ and $(A, \bar{b})$ are both almost strongly minimal and have the same dimension. Then we build up an isomorphism from $(A, \bar{a})$ to $(A, \bar{b})$ just as in the proof of Theorem 4.7.3. $\square$

Theorem 4.7.3 has some other useful corollaries, whose proofs need more advanced methods.

Corollary 4.7.5. *Let the language L be countable. Let the L-structure A be strongly minimal of infinite dimension, and let X be a finite set of elements in A. Then $\text{ACL}(X) = \text{acl}(X)$ and $\text{DCL}(X) = \text{dcl}(X)$ (see section 4.1).*

Proof. In view of Exercise 10.1.5(a) and Corollary 10.4.12 below, it suffices to show that A is saturated. But this follows from the theorem and Theorem 12.2.12(a) below.   □

Corollary 4.7.6. *An almost strongly minimal theory doesn't have the finite cover property.*

Proof. This follows at once by Corollary 9.5.6 below.   □

Corollary 4.7.7. *Suppose A is almost strongly minimal, with strongly minimal set $\Omega = P^A$. Let $\phi(\bar{x}, \bar{y})$ be a formula of L. Then there is a formula $\theta(\bar{x})$ such that for all tuples $\bar{a}$ the following are equivalent.*
(a) $A \vDash \theta(\bar{a})$.
(b) *For every tuple $\bar{c}$ in Ω which is independent over $\mathrm{acl}(\bar{a})$, $A \vDash \phi(\bar{a}, \bar{c})$.*

Proof. If $\bar{y}$ is a single variable, then (b) is equivalent to

(7.2)   For infinitely many elements c of Ω, $A \vDash \phi(\bar{a}, c)$.

Since infinity is definable in A (see the end of section 4.3 above), a formula $\theta(\bar{x})$ exists for this case. For the general case, use induction on the length of $\bar{c}$.   □

Lemma 4.7.8. *Suppose the L-structure A is almost strongly minimal with $\emptyset$-definable strongly minimal set Ω, and X is a set of elements of A such that $X \cap \Omega$ is infinite. Then the following are equivalent.*
(a) *X is algebraically closed in A.*
(b) *X is the domain of some elementary substructure of A.*
(c) *X is the algebraic closure of some set of elements of Ω.*

Proof. (a) $\Rightarrow$ (b). Assume X is algebraically closed. We use the Tarski–Vaught criterion of Exercise 2.5.1. Let $\phi(\bar{x}, y)$ be a formula of L and $\bar{a}$ a tuple of elements of X such that $A \vDash \phi(\bar{a}, b')$ for some element b' of A. Suppose first that b' lies in Ω. Either $\Omega \cap \phi(\bar{a}, A)$ or $\Omega \backslash \phi(\bar{a}, A)$ is finite. If the former, then the non-empty set $\Omega \cap \phi(\bar{a}, A)$ lies in $\mathrm{acl}(X)$ by definition of algebraic closure, and so it has an element in X. If the latter, then $\Omega \cap \phi(\bar{a}, A)$ meets X since $X \cap \Omega$ is infinite. Either way there is an element b in X such that $A \vDash \phi(\bar{a}, b)$.

Suppose on the other hand that b' is not in Ω. Then there are elements $\bar{c} = (c_0, \ldots, c_{m-1})$ in Ω such that $b' \in \mathrm{acl}(\bar{c})$; say $A \vDash \chi(b', \bar{c}) \wedge \exists_{=n} x \, \chi(x, \bar{c})$. Writing $P(x)$ for the formula defining Ω, apply the previous argument to the formula

(7.3) $\exists y_0 \ldots \exists y_{m-1} \exists x (P(y_0) \wedge \ldots \wedge P(y_{m-1})$

$\wedge \chi(x, y_0, \ldots, y_{m-1}) \wedge \phi(\bar{a}, x))$, peeling off the quantifiers $\exists y_i$ one by one. This finds elements $\bar{d}$ in X such that some element b in $\mathrm{acl}(\bar{d})$ satisfies $\phi(\bar{a}, x)$; but then $b \in X$ since X is algebraically closed.

Thus in both cases there is an element b in X with $A \vDash \phi(\bar{a}, b)$, and so the Tarski–Vaught criterion is satisfied.

(b) $\Rightarrow$ (c). Assuming (b), let Y be $\Omega \cap X$. Then $\mathrm{acl}(Y) \subseteq X$ since X is the domain of an elementary substructure of A. (This was Exercise 4.1.5(b).) On the other hand, since A is almost strongly minimal, every element of X is algebraic over some elements of Ω, and hence over some elements of Y too, by (b) again; so $X \subseteq \mathrm{acl}(Y)$.

(c) trivially implies (a). $\square$

Weight

Weights are a device for spreading the geometry of the strongly minimal set across the whole structure. Roughly speaking, the 'weight' of a tuple $\bar{a}$ is the number of degrees of freedom in $\bar{a}$, measured by the dimension of the closure of $\bar{a}$ in the strongly minimal set.

Until further notice the setting will be as follows.

(7.4) L is a first-order language; A is an infinite-dimensional almost strongly minimal L-structure over the strongly minimal set $\Omega = P^A$.

(7.5) A has an elementary substructure B which contains infinitely many elements of Ω; L^+ is the language got from L by adding constants to name all the elements of B, thus making A into an L^+-structure A^+; A^+ (which is also almost strongly minimal; see Exercise 4.5.1) has infinite dimension.

Assuming (7.4) holds, we can take a basis of A, split it into two infinite pieces X and Y, and take B to be $\mathrm{acl}(X)$ (by Lemma 4.7.8). Thus if (7.4) is true, we can make (7.5) true too.

We shall compute algebraic closures in A^+ unless otherwise stated. For any finite set X in A^+, the **support** of X, $\mathrm{supp}(X)$, is the set $\mathrm{acl}(X) \cap \Omega$. (This is a different notion of 'support' from that in section 4.3 above.) The next lemma shows at once why assumption (7.5) is helpful.

Lemma 4.7.9. *If X is any finite set of elements of A^+, then $X \subseteq \mathrm{acl}\,\mathrm{supp}(X)$, and so $\mathrm{acl}(X) = \mathrm{acl}\,\mathrm{supp}(X)$.*

Proof. Let a be an element of X. Since A^+ is almost strongly minimal, there are a tuple $\bar{d}$ in Ω, a formula $\psi(\bar{x}, y)$ of L^+ and a positive integer n such that $A^+ \vDash \psi(\bar{d}, a) \wedge \exists_{=n} y\, \psi(\bar{d}, y)$. Since $\mathrm{acl}(a)$ includes all the elements of B

by (7.6), it has infinite overlap with Ω, and so by Lemma 4.7.8 it is the domain of an elementary substructure C of A. So there are elements $\bar{c}$ in P^C such that $A^+ \vDash \psi(\bar{c}, a) \wedge \exists_{=n} y\, \psi(\bar{c}, y)$. But $P^C = \mathrm{acl}(a) \cap \Omega \subseteq \mathrm{supp}(X)$, which implies that $a \in \mathrm{acl\,supp}(X)$. $\qquad\square$

Let us say that a set X is **of finite weight** if its support has finite dimension. When X has finite weight, we define the **weight** of X, weight(X), to be $\dim\mathrm{supp}(X)$. We also write weight$(\bar{a})$ for weight(X), where $\bar{a}$ is a tuple listing the elements of X. In some sense, weight(X) measures the number of degrees of freedom in X. If X happens to be a subset of Ω, then clearly weight$(X) = \dim(X)$.

The **relative weight** of X over Y, weight(X/Y), is defined to be weight$(X \cup Y) - $ weight(Y). In particular weight$(\bar{a}/Y) = 0$ iff $\bar{a}$ is algebraic over Y.

Lemma 4.7.10. *Let $\bar{a}$, $\bar{b}$ be tuples and Y a set of elements of A^+ which has finite weight. Then (writing $\bar{a}Y$ or $Y\bar{a}$ as shorthand for the set consisting of Y and the elements of $\bar{a}$)*

(7.6)   weight$(\bar{a}\bar{b}/Y) = $ weight$(\bar{a}/Y) + $ weight$(\bar{b}/\bar{a}Y)$.

Proof. Immediate from the definition. $\qquad\square$

We say that $\bar{a}$ is **free from** $\bar{b}$ **over** Y (or that $\bar{a}$ **doesn't fork against** $\bar{b}$ **over** Y), in symbols $\bar{a} \underset{Y}{\cup} \bar{b}$, if

(7.7)   weight$(\bar{a}\bar{b}/Y) = $ weight$(\bar{a}/Y) + $ weight$(\bar{b}/Y)$.

We say simply that $\bar{a}$ is **free from** $\bar{b}$ if $\bar{a} \underset{\varnothing}{\cup} \bar{b}$; we write this as $\bar{a} \cup \bar{b}$.

To get some intuition what (7.7) means, first add the elements of Y to L^+ as parameters, so that weight over Y becomes weight over $\varnothing$. Then if $\bar{a}$ and $\bar{b}$ are tuples of elements of Ω, (7.7) says that the only algebraic dependencies between elements of the tuple $\bar{a}\bar{b}$ are those within $\bar{a}$ and those within $\bar{b}$.

The next lemma is simple arithmetic.

Lemma 4.7.11 (a) $\bar{a} \underset{Y}{\cup} \bar{b}$ *iff* weight$(\bar{a}\bar{b}/\bar{b}Y) = $ weight$(\bar{a}/Y)$.
  (b) ('Forking symmetry') *If $\bar{a} \underset{X}{\cup} \bar{b}$ then $\bar{b} \underset{X}{\cup} \bar{a}$.*
  (c) *If $\bar{a} \underset{X}{\cup} \bar{b}$ and $X \subseteq Y$ then $\bar{a} \underset{Y}{\cup} \bar{b}$.*
  (d) $(\bar{a} \underset{X}{\cup} \bar{b}$ *and* $\bar{a} \underset{X\bar{b}}{\cup} \bar{c})$ *iff* $\bar{a} \underset{X}{\cup} \bar{b}\bar{c}$.
  (e) *If $\bar{a}$ has weight 1, then $\bar{a} \cup \bar{b}$ if and only if $\bar{a}$ is not in* $\mathrm{acl}(\bar{b})$.
  (f) *If $\bar{a} \underset{X}{\cup} \bar{b}$ and $\bar{c}$ is in* $\mathrm{acl}(X\bar{b})$, *then $\bar{a} \underset{X}{\cup} \bar{c}$.*

Proof. Exercise. $\qquad\square$

Note that by (d), if $\bar{a} \underset{X}{\cup} \bar{b}\bar{c}$ then $\bar{a} \underset{X}{\cup} \bar{b}$. Note also that $\bar{a} \cup \bar{b}$ and $\bar{a} \cup \bar{c}$

do not imply $\bar{a} \perp \bar{b}\bar{c}$. (Counterexample in a vector space: b and c are independent vectors and $a = b + c$.)

Let $\bar{a}$ be a tuple in A^+. We write $\Delta(\bar{a})$ for the orbit of $\bar{a}$ under $\text{Aut}(A^+)$. Given two sets U, V of tuples in A^+, we write $U \oplus V$ for the set of all pairs $(\bar{a}, \bar{b})$ with $\bar{a} \in U$, $\bar{b} \in V$ and $\bar{a} \perp \bar{b}$.

Lemma 4.7.12. *Suppose $\bar{a} \perp_X \bar{b}$ and X is a closed set of finite weight.*

(a) If $\bar{b}'$ is in the $\text{Aut}(A^+)_{(X)}$-orbit of $\bar{b}$ and $\bar{a} \perp \bar{b}'$, then there is an automorphism $g \in \text{Aut}(A^+)$ which fixes $X \cup \text{acl}(\bar{a})$ pointwise and takes $\bar{b}$ to $\bar{b}'$.

(b) For all $\bar{a}'$, $\bar{b}'$ respectively in the $\text{Aut}(A^+)_{(X)}$-orbits of $\bar{a}$ and $\bar{b}$ with $\bar{a}' \perp \bar{b}'$, there is $g \in \text{Aut}(A^+)_{(X)}$ which takes $\bar{a}$ to $\bar{a}'$ and $\bar{b}$ to $\bar{b}'$ respectively. In particular, for all tuples $\bar{a}$, $\bar{b}$ in A^+, $\text{Aut}(A^+)$ acts transitively on $\Delta(\bar{a}) \oplus \Delta(\bar{b})$.

Proof. (a) Since X is closed and contains B, it forms an elementary substructure of A, so the proof when $X = \text{acl}(\emptyset)$ gives the general case. Thus we can ignore X.

Let $\bar{c}$, $\bar{d}$ respectively be bases of $\text{supp}(\bar{a})$ and $\text{supp}(\bar{b})$. We must find an automorphism g of A^+ which fixes $\text{acl}(\bar{a})$ and takes $\bar{b}$ to $\bar{b}'$. By assumption there is an automorphism h of A^+ which takes $\bar{b}$ to $\bar{b}'$. Let $\bar{d}'$ be $h\bar{d}$. Since $\bar{a} \perp \bar{b}$ and $\bar{a} \perp \bar{b}'$, we have $\bar{c} \perp \bar{d}$ and $\bar{c} \perp \bar{d}'$. By Lemma 4.5.6 there is an automorphism g_1 of A^+ which fixes $\text{acl}(\bar{c}) = \text{acl}(\bar{a})$ and takes $\bar{d}$ to $\bar{d}'$. So all we need now is an automorphism g_2 of A^+ which fixes $\text{acl}(\bar{a})$ and $\bar{d}'$, and takes $g_1\bar{b}$ to $\bar{b}'$. For this it suffices by Corollary 4.7.4 to show that

(7.8) $(A^+, \bar{a}*, \bar{d}', g_1\bar{b}) \equiv (A^+, \bar{a}*, \bar{d}', \bar{b}')$,

where $\bar{a}*$ is a sequence listing the elements of $\text{acl}(\bar{a})$. If (7.8) fails, there is a formula χ of L such that $A \models \chi(g_1\bar{b}, \bar{d}', \bar{a}*) \wedge \neg\chi(\bar{b}', \bar{d}', \bar{a}*)$; note that we have changed from A^+ to A, and that only a finite part of $\bar{a}*$ is mentioned in this formula, so that we can replace $\bar{a}*$ by a finite subsequence $\bar{a}**$. Now in A, both $\bar{b}'$ and $g_1\bar{b}$ are in the algebraic closure of $Y \cup \{\bar{d}'\}$ for some finite subset Y of B. Since $\bar{c}$ is independent of $Y \cup \{\bar{d}'\}$, there is by Lemma 4.5.6 an automorphism f of A which fixes $\text{acl}_A(Y \cup \{\bar{d}'\})$ pointwise and takes $\bar{a}**$ into B. Then $A \models \chi(g_1\bar{b}, \bar{d}', f\bar{a}*) \wedge \neg\chi(\bar{b}', \bar{d}', f\bar{a}*)$. Applying g_1^{-1} and h^{-1}, $A \models \chi(\bar{b}, \bar{d}, f\bar{a}*) \wedge \neg\chi(\bar{b}, \bar{d}, f\bar{a}*)$, contradiction.

(b) follows at once.   □

Lemma 4.7.13. *If $\bar{a}$, $\bar{b}$ are any tuples in A^+, then there is $\bar{b}' \in \Delta(\bar{b})$ such that $\bar{a} \perp \bar{b}'$.*

Proof. Let $\bar{c}$, $\bar{d}$ be the supports of $\bar{a}$, $\bar{b}$. Move $\bar{d}$ to be independent of $\bar{c}$.   □

Lemma 4.7.14. *If $\bar{a} \in \mathrm{acl}(X\bar{b})$ and $\bar{a} \downarrow_X \bar{b}$, then $\bar{a} \in \mathrm{acl}(X)$.*

Proof. Using Lemma 4.7.10 and the definition of $\downarrow$, $0 = \mathrm{weight}(\bar{a}/X\bar{b}) = \mathrm{weight}(\bar{a}\bar{b}/X) - \mathrm{weight}(\bar{b}/X) = \mathrm{weight}(\bar{a}/X)$. $\qquad\qquad\square$

Lemma 4.7.15. *Let $\bar{a}$ and $\bar{b}$ be tuples and $\phi(\bar{x}, \bar{y})$ a formula. Then there is a formula $\theta(\bar{x})$ such that for all $\bar{a}' \in \Delta(\bar{a})$,*

(7.9) $\quad A \vDash \theta(\bar{a}') \Leftrightarrow$ *for every $\bar{b}' \in \Delta(\bar{b})$, if $\bar{a}' \downarrow \bar{b}'$ then $A \vDash \phi(\bar{a}', \bar{b}')$.*

Proof. Let $\bar{c}$, $\bar{d}$ be bases of the supports of $\bar{a}$, $\bar{b}$ respectively. The type of $\bar{a}$ over the empty set is determined by some formula $\psi(\bar{x}, \bar{c})$, and likewise the type of $\bar{b}$ over the empty set is determined by a formula $\chi(\bar{w}, \bar{d})$. What we want is a formula $\theta(\bar{x})$ which expresses

(7.10) $\qquad \forall \bar{y}(\psi(\bar{x}, \bar{y}) \rightarrow$ For every tuple $\bar{z}$ in Ω which is independent over $\bar{y}$,

$$\forall \bar{w}(\chi(\bar{w}, \bar{z}) \rightarrow \phi(\bar{x}, \bar{w}))).$$

The necessary formula is guaranteed by Corollary 4.7.7. $\qquad\qquad\square$

Two strongly minimal sets

We close the section with some remarks about comparing two strongly minimal sets in one almost strongly minimal structure. The assumption (7.5) is no longer in force.

Theorem 4.7.16. *Let A be an almost strongly minimal structure and Ω, Ξ two $\varnothing$-definable strongly minimal sets in A. Then there is a tuple $\bar{a}$ of elements of A such that there is an inclusion-preserving bijection between the family $\mathrm{Cl}(\Omega)$ of closed subsets of Ω in $(A, \bar{a})$ and the family $\mathrm{Cl}(\Xi)$ of closed subsets of Ξ in $(A, \bar{a})$.*

Proof. Let Ω and Ξ be respectively P^A and Q^A. Without loss we can suppose that A is almost strongly minimal over at least one of these two sets, say Ω. If Ω has finite dimension and Ξ has dimension 0, then we prove the theorem by taking $\bar{a}$ to be a basis of Ω. So we can assume either that Ω has infinite dimension or that Ξ has dimension > 0. In the former case, by an argument using the compactness theorem (cf. the upward Löwenheim–Skolem theorem, Corollary 6.1.4 below), one can show that A has an elementary extension B in which Q^B contains at least one element b which is not algebraic over $\varnothing$. Then there are a finite number of elements $a_0, \ldots, a_n$ of P^B such that $b \in \mathrm{acl}(a_0, \ldots, a_n)$. Since Ω has infinite dimension, we can apply an automorphism that moves $(a_0, \ldots, a_n)$ into Ω and b into Ξ, so that Ξ has dimension > 0 anyway.

Thus we can assume henceforth that Ξ contains an element which is not algebraic over $\varnothing$, but is algebraic over elements $a_0, \ldots, a_n$ of Ω. Choose n as small as possible, so that $a_0, \ldots, a_n$ are independent in Ω, and add $a_0, \ldots, a_{n-1}$ as a tuple $\bar{a}$ of parameters to the language, to form an expanded language L^+. Henceforth we regard A as an L^+-structure. Thus b is algebraic over a_n but not over $\varnothing$. So by the exchange lemma (Lemma 4.5.2), a_n is algebraic over b. There are a formula $\phi(x, y)$ of L^+ and integers k, m such that $A \vDash \phi(a_n, b)$, there are exactly k elements a' of Ω such that $A \vDash \phi(a', b)$, and there are exactly m elements b' of Ξ such that $A \vDash \phi(a_n, b')$.

We claim that for all but finitely many elements a' of Ω, $\phi(a', A) \cap \Xi$ has cardinality m. For otherwise there are only finitely many a' in Ω such that $\phi(a', A) \cap \Xi$ has cardinality m, and this implies that $a_n \in \mathrm{acl}(\varnothing)$, which is false.

Also we claim that for all but finitely many elements b' of Ξ, $\phi(A, b') \cap \Omega$ has cardinality k. For otherwise there are only finitely many such b', and this would imply that $b \in \mathrm{acl}(\varnothing)$.

Now for each closed subset X of Ω, write X^* for the closed subset $\Xi \cap \mathrm{acl}(X)$ of Ξ. This maps $\mathrm{Cl}(\Omega)$ to $\mathrm{Cl}(\Xi)$, and the map clearly preserves inclusion. We have to show that it is a bijection.

First, suppose X_1 and X_2 are distinct closed sets in Ω with $X_1^* = X_2^*$. Then some element a' lies, say, in $X_1 \backslash X_2$. Since $a' \notin X_2$ and X_2 is closed, there are exactly m elements b' of Ξ such that $A \vDash \phi(a', b')$; these elements lie in X_1^*. Since a' is algebraic over any of these elements, $a' \in \mathrm{acl}(X_1^*) = \mathrm{acl}(X_2^*) \subseteq \mathrm{acl}(X_2)$; contradiction. Thus the map $X \mapsto X^*$ is injective.

Next, let Y be any closed set in Ξ, and let X be $\Omega \cap \mathrm{acl}(Y)$. We claim that $Y = X^*$. It's immediate that $X^* \subseteq Y$. For the converse, if b' is any element of $Y \backslash \mathrm{acl}(\varnothing)$, then $A \vDash \phi(a', b')$ for some $a' \in X$, and then $b' \in \Xi \cap \mathrm{acl}(a') \subseteq X^*$; so $Y \subseteq X^*$. $\qquad\square$

Exercises for section 4.7

1. Give an example of a structure A which contains a $\varnothing$-definable strongly minimal set Ω, such that $\mathrm{dom}(A) = \mathrm{acl}(\Omega)$ but $\mathrm{Th}(A)$ is not almost strongly minimal. [Take a structure consisting of an infinite set with no relations on it, and a copy of the ordered set ω.]

2. Prove Lemma 4.7.12.

3. Show that in an almost strongly minimal structure, if $\bar{a}$ lies in $\mathrm{acl}(X)$ then $\bar{a} \downvDash_X \bar{b}$ for all $\bar{b}$.

4. If L is countable and A is a countable almost strongly minimal structure with strongly minimal set Ω, show that $\mathrm{Th}(A)$ is ω-categorical if and only if every finite subset of Ω has finite algebraic closure.

The next exercise shows that some assumption on X is needed in Lemma 4.7.8.
5. Let A be an algebraically closed field. Form a structure B by adding to A as new elements all the straight lines in the plane A^2, together with the relation $R(a, b, l)$ expressing that the point (a, b) of A^2 lies on the line l. (a) Show that B is almost strongly minimal, with A as strongly minimal set. (b) Explain how B can be regarded as part of A^{eq}. (c) Show that every line has weight at most 2 over $\varnothing$; give examples of lines of weight 0, 1 and 2 over $\varnothing$. (d) Show that if l is a line of weight 2 over $\varnothing$, then no tuple of elements of A is mutually algebraic with l over $\varnothing$.

6. A countable L-structure A is said to be **smoothly approximated** if $\mathrm{Aut}(A)$ is oligomorphic (or equivalently by Theorem 7.3.1, A is ω-categorical) and A can be written as the union of a chain of finite structures $(A_i : i < \omega)$ such that for each i and all tuples $\bar{a}$, $\bar{b}$ in A_i, if $\bar{a}$ and $\bar{b}$ lie in the same orbit under $\mathrm{Aut}(A)$ then they lie in the same orbit under $\mathrm{Aut}(A)_{\{\mathrm{dom}A_i\}}$. (a) Prove that if A is smoothly approximated, then for every sentence ϕ of L which is true in A, there is $n < \omega$ such that $A_i \vDash \phi$ for all $i \geqslant n$. [Use induction on complexity of negation-normal formulas.] (b) Deduce that $\mathrm{Th}(A)$ is not finitely axiomatisable.

7. Show that if A is a countable ω-categorical almost strongly minimal structure, then A is smoothly approximated. [Every tuple $\bar{a}$ is algebraic over an independent tuple $\bar{b}\bar{c}$ in Ω such that $\bar{b}$ lies in $\mathrm{acl}(\bar{a})$ and $\bar{c}$ outside $\mathrm{acl}(\bar{a})$; then $\bar{c}$ can be moved freely.]

4.8 Zil'ber's configuration

Our target in this section is an influential and very pretty result of Boris Zil'ber: if the geometry of an infinite-dimensional strongly minimal set Ω contains the Zil'ber configuration, then there is an infinite group which we can define in terms of Ω. From the construction it will be clear that the group is very closely related to the set Ω. But in this book we lack the notions from stability theory which are needed to say just what the relationship is.

The proof given below is due to Ehud Hrushovski, and it provides examples of some of the devices of section 4.3 at work.

Throughout the section we assume the following:

(8.1) L is a first-order language; A is an infinite-dimensional
          almost strongly minimal L-structure with a $\varnothing$-definable
          strongly minimal set Ω.

Recall from section 4.6 that Zil'ber's configuration is this picture:

(8.2)

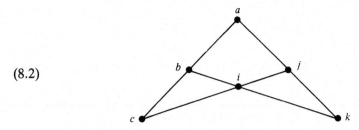

This configuration **occurs in** a pregeometry (Ω, cl) if a, b, c, i, j, k are points of $\Omega \backslash \mathrm{cl}(\varnothing)$, none is in the closure of any other, and a point in the set $\{a, b, c, i, j, k\}$ is in the closure of two others in the set if and only if some line in the configuration contains all three points.

Theorem 4.8.1. *Suppose that Zil'ber's configuration occurs in the pregeometry of A. Then in A^{eq} we can define an infinite group; the definition uses perhaps infinitely many formulas.*

We shall add two further assumptions. As at (7.5) in the previous section, we shall assume the following:

(8.3) A has an elementary substructure B which contains infinitely
many elements of Ω; L^{+} is the language got from L by
adding constants to name all the elements of B, thus making
A into an L^{+}-structure A^{+}; A^{+} has infinite dimension.

This is justified in the same way as in the previous section. If Zil'ber's configuration occurs in A, we can choose B so that the configuration remains in place in A^{+}. We shall also assume

(8.4)   at the finite number of places in the argument where we
introduce imaginary elements, the required sorts from A^{eq}
already exist inside A.

This was the purpose of working in an almost strongly minimal structure. Every element of A^{eq} is definable over A, by Theorem 4.3.3(b).

In our calculations with weight and forking, we always work in L^{+} and A^{+}. This is equivalent to working in L but reckoning over B. For example $\bar{a} \underset{}{\downarrow} \bar{b}$ (in A^{+}) means the same as $\bar{a} \underset{\mathrm{dom}(B)}{\downarrow} \bar{b}$ (in A). I shall use letters $\bar{a}$, $\bar{b}$ etc. to stand for elements of $(A^{+})^{\mathrm{eq}}$; they may or may not be tuples of elements of A^{+}.

The proof of Theorem 4.8.1 will use the following condition.

(8.5)   **Unique definability condition.** There are elements $\bar{a}$, $\bar{b}$, $\bar{c}$, $\bar{i}$,
$\bar{j}$, $\bar{k}$ of A^{+} such that $\mathrm{acl}(\bar{a}) = \mathrm{acl}(a), \ldots, \mathrm{acl}(\bar{k}) = \mathrm{acl}(k)$, and
moreover $\bar{j} \in \mathrm{dcl}(\bar{a}, \bar{k})$, $\bar{k} \in \mathrm{dcl}(\bar{a}, \bar{j})$, $\bar{i} \in \mathrm{dcl}(\bar{b}, \bar{k})$ and
$\bar{k} \in \mathrm{dcl}(\bar{b}, \bar{i})$.

Note that since $a, \ldots, k$ are elements of Ω, each of the elements $\bar{a}, \ldots, \bar{k}$ has weight 1. Thus for example $\bar{a} \mathbin{\underset{\smile}{\cup}} \bar{j}$ since $\bar{a} \notin \mathrm{acl}(\bar{j})$ (cf. Lemma 4.7.11(e)).

Example 1: *Definability in affine space.* Suppose A is an affine space and l is a line in A. Then as soon as two distinct points x, y of l have been given, every other point z of l is definable as $\alpha x + (1 - \alpha)y$ for some unique scalar α. So in this case the unique definability condition holds at once.

Example 2: *Failures of unique definability.* Here are two examples where the unique definability condition fails. First suppose A is not an affine space but an affine geometry (see Example 3 in section 4.6), and suppose there is some automorphism g of the field of scalars which takes some scalar α to $g\alpha \neq \alpha$. Then the point $\alpha x + (1 - \alpha)y$ is not uniquely determined in A by x and y. This is not a serious failure, because we have already repaired it by passing from A to A^+. Take a line l which lies in B, and let l' be a line through x which is parallel to l. We can define a bijection from l to l' using parallel lines, and then a bijection from l' to the line xy. Since every point of l is named by a constant, this defines each point of the line xy in terms of x and y.

For a more serious failure of (8.5), suppose A is a projective geometry where every line contains at least four points. If l is any line not meeting B, then the automorphism group of A^+ takes any three distinct points on l to any other three distinct points; so if we fix two points x and y on l, this determines l but doesn't allow us to tell apart any other two points on l.

The argument for Theorem 4.8.1 splits into two parts. In the first part we show that the group can be defined if the unique definability condition holds. In the second part we show that this condition can always be met by adding parts of A^{eq} to A. The second part is also the reason why we need the parameters from B.

First part: we prove the theorem assuming the unique definability condition

By the unique definability condition there are formulas $\phi(\bar{x}, \bar{y}, \bar{z})$ and $\psi(\bar{u}, \bar{v}, \bar{z})$ of L such that

(8.6) $\bar{k}$ is the unique tuple in A for which $A \vDash \phi(\bar{a}, \bar{j}, \bar{k})$,
       $\bar{j}$ is the unique tuple in A for which $A \vDash \phi(\bar{a}, \bar{j}, \bar{k})$,
       $\bar{k}$ is the unique tuple in A for which $A \vDash \psi(\bar{b}, \bar{i}, \bar{k})$,
       $\bar{i}$ is the unique tuple in A for which $A \vDash \psi(\bar{b}, \bar{i}, \bar{k})$.

Write $\Delta(\bar{d})$ for the orbit of a tuple $\bar{d}$ under $\mathrm{Aut}(A^+)$. We begin by using ϕ to define a family of 'translations' which are defined on a subset of the set $\Delta(\bar{j})$.

Each 'translation' will be indexed by a pair of elements $\bar{a}_1$, $\bar{a}_2$ of $\Delta(\bar{a})$, and written $\mu(\bar{a}_1, \bar{a}_2)$. Thus, let $\bar{a}_1$ and $\bar{a}_2$ be any two elements of $\Delta(\bar{a})$ such that $\bar{a}_1 \Downarrow \bar{a}_2$. We define $\mu(\bar{a}_1, \bar{a}_2)$ to be the following map on $\Delta(\bar{j})\backslash\mathrm{acl}(\bar{a}_1, \bar{a}_2)$.

(8.7)   Given $\bar{e} \in \Delta(\bar{j})\backslash\mathrm{acl}(\bar{a}_1, \bar{a}_2)$, find the unique $\bar{f} \in \Delta(\bar{k})$ such that $A \vDash \phi(\bar{a}_1, \bar{e}, \bar{f})$. Then find the unique $\bar{e}' \in \Delta(\bar{j})$ such that $A \vDash \phi(\bar{a}_2, \bar{e}', \bar{f})$. Put $\mu(\bar{a}_1, \bar{a}_2)(\bar{e}) = (\bar{e}')$.

Before we check that this definition makes sense, consider what it says in the case of an affine space. We can take all the sets $\Delta(\bar{a})$, $\Delta(\bar{b})$ etc. to be $\mathrm{dom}(A)$. Suppose $\phi(x, y, z)$ is the formula $z = \alpha x + (1 - \alpha)y$. Then if a_1 and a_2 are any two points of the affine plane, the map $\mu(a_1, a_2)$ acts on e as in the diagram:

(8.8)

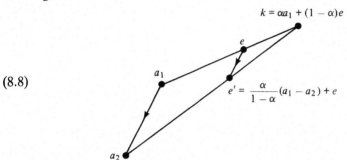

$$k = \alpha a_1 + (1 - \alpha)e$$

$$e' = \frac{\alpha}{1 - \alpha}(a_1 - a_2) + e$$

Thus $\mu(a_1, a_2)$ defines a translation parallel to that taking a_1 to a_2, and of length $(\alpha/(\alpha - 1)) \times$ (distance from a_1 to a_2). By our definition (8.7), $\mu(a_1, a_2)$ is defined only on points that are off the line $a_1 a_2$.

Turning to the justification of (8.7), we first note that $\bar{a} \Downarrow \bar{j}$ by the configuration (see Lemma 4.7.11(e)), and $\bar{a}_1 \Downarrow \bar{e}$ by assumption, so by Lemma 4.7.12, $(A^+, \bar{a}, \bar{j}) \equiv (A^+, \bar{a}_1, \bar{e})$. This guarantees that $\bar{f}$ is well-defined in (8.7). But then to find $\bar{e}'$ by the same argument, we need to know that $\bar{a}_2 \Downarrow \bar{f}$. The next lemma provides this.

Lemma 4.8.2. *Suppose $\bar{a}_1$ and $\bar{a}_2$ are elements of $\Delta(\bar{a})$ such that $\bar{a}_1 \Downarrow \bar{a}_2$, and $\bar{e}$ is an element of $\Delta(\bar{j})\backslash\mathrm{acl}(\bar{a}_1, \bar{a}_2)$; let $\bar{f}$ be the unique element of $\Delta(\bar{k})$ such that $A \vDash \phi(\bar{a}_1, \bar{e}, \bar{f})$. Then $\bar{a}_2 \Downarrow \bar{f}$.*

Proof. Note first that since weight($\bar{e}$) = 1, the condition that $\bar{e} \notin \mathrm{acl}(\bar{a}_1, \bar{a}_2)$ is equivalent to $\bar{e} \Downarrow \bar{a}_1 \hat{\ } \bar{a}_2$ (cf. Lemma 4.7.11(e) again).

Now to show that $\bar{a}_2 \Downarrow \bar{f}$, we apply Lemma 4.7.11(d) to get $\bar{e} \Downarrow \bar{a}_2$, then Lemma 4.7.11(b, c) to get $\bar{a}_2 \Downarrow_{\bar{a}_1} \bar{e}$, which with Lemma 4.7.11(d) and $\bar{a}_1 \Downarrow \bar{a}_2$ gives $\bar{a}_2 \Downarrow \bar{a}_1 \hat{\ } \bar{e}$. Finally by Lemma 4.7.11(f), since $\bar{f} \in \mathrm{acl}(\bar{a}_1, \bar{e})$, we have $\bar{a}_2 \Downarrow \bar{f}$.   □

The map $\mu(\bar{a}_1, \bar{a}_2)$ is not defined everywhere on $\Delta(\bar{j})$; it has a singularity at

$\Delta(\bar{j}) \cap \mathrm{acl}(\bar{a}_1, \bar{a}_2)$. We overcome this by an analogue of analytic continuation. We define a relation $\sim$ on $\Delta(\bar{a}) \oplus \Delta(\bar{a})$ by

(8.9) $(\bar{a}_1, \bar{a}_2) \sim (\bar{a}_3, \bar{a}_4)$ iff for every $\bar{e} \in \Delta(\bar{j})\backslash\mathrm{acl}(\bar{a}_1, \bar{a}_2, \bar{a}_3, \bar{a}_4)$,
        $\mu(\bar{a}_1, \bar{a}_2)(\bar{e}) = \mu(\bar{a}_3, \bar{a}_4)(\bar{e})$.

The aim is that $\sim$ should be expressed by an equivalence formula in A^+, so that we can take imaginary elements. The next lemma justifies this.

Lemma 4.8.3. (a) *There is a formula $\sigma(x, y, z, t)$ of L^+ such that for all $\bar{a}_1$, $\bar{a}_2$ in $\Delta(\bar{a})$ with $\bar{a}_1 \Downarrow \bar{a}_2$, and all $\bar{e}$, $\bar{e}'$ in $\Delta(\bar{j})$ with $\bar{e} \Downarrow \bar{a}_1 {}^\frown \bar{a}_2$,
$A^+ \vDash \sigma(\bar{a}_1, \bar{a}_2, \bar{e}, \bar{e}') \Leftrightarrow \mu(\bar{a}_1, \bar{a}_2)(\bar{e}) = (\bar{e}')$.*
(b) $\sim$ *is an equivalence relation on $\Delta(\bar{a}) \oplus \Delta(\bar{a})$.*
(c) *There is an equivalence formula $\theta(x_1, x_2, y_1, y_2)$ of L^+ such that for all $\bar{a}_1$, $\bar{a}_2$, $\bar{a}_3$, $\bar{a}_4$ in $\Delta(\bar{a})$, if $\bar{a}_1 \Downarrow \bar{a}_2$ and $\bar{a}_3 \Downarrow \bar{a}_4$ then $A^+ \vDash$
$\theta(\bar{a}_1, \bar{a}_2, \bar{a}_3, \bar{a}_4) \Leftrightarrow (\bar{a}_1, \bar{a}_2) \sim (\bar{a}_3, \bar{a}_4)$.*

Proof. (a) The formula is $\exists u \, (\phi(x, z, u) \wedge \phi(y, t, u))$.

(b) We claim that 'every' in (8.9) can be replaced by 'some'. Suppose $\bar{e}$, $\bar{e}' \in \Delta(\bar{j})\backslash\mathrm{acl}(\bar{a}_1, \bar{a}_2, \bar{a}_3, \bar{a}_4)$ and $\mu(\bar{a}_1, \bar{a}_2)(\bar{e}) = \mu(\bar{a}_3, \bar{a}_4)(\bar{e})$. Then applying Lemma 4.7.12 to $\Delta(\bar{a}_1 {}^\frown \bar{a}_2 {}^\frown \bar{a}_3 {}^\frown \bar{a}_4) \oplus \Delta(\bar{j})$, we find an automorphism of A^+ which fixes $\bar{a}_1 {}^\frown \bar{a}_2 {}^\frown \bar{a}_3 {}^\frown \bar{a}_4$ and takes $\bar{e}$ to $\bar{e}'$. So $\mu(\bar{a}_1, \bar{a}_2)(\bar{e}') = \mu(\bar{a}_3, \bar{a}_4)(\bar{e}')$ too. This proves the claim.

Now suppose $(\bar{a}_1, \bar{a}_2) \sim (\bar{a}_3, \bar{a}_4) \sim (\bar{a}_5, \bar{a}_6)$. Then if we take any $\bar{e} \in \Delta(\bar{j})\backslash\mathrm{acl}(\bar{a}_1, \bar{a}_2, \bar{a}_3, \bar{a}_4, \bar{a}_5, \bar{a}_6)$, we have $\mu(\bar{a}_1, \bar{a}_2)(\bar{e}) = \mu(\bar{a}_3, \bar{a}_4)(\bar{e}) = \mu(\bar{a}_5, \bar{a}_6)(\bar{e})$. By the claim, this proves that $\sim$ is transitive. Since $\sim$ is clearly reflexive and symmetric, it is an equivalence relation.

(c) θ should say

(8.10) For all $z \in \Delta(\bar{j})$, if $x_1 {}^\frown x_2 {}^\frown y_1 {}^\frown y_2 \Downarrow z$ then
        $(\forall w(\sigma(x_1, x_2, z, w) \leftrightarrow \sigma(y_1, y_2, z, w)))$.

This is expressible by a formula θ of L^+, by Lemma 4.7.15. From the form of (8.10), θ must be an equivalence formula. □

We write $[\bar{a}_1, \bar{a}_2]$ for the equivalence class $\bar{a}_1 {}^\frown \bar{a}_2/\sim$, and H for the set of equivalence classes $[\bar{a}_1, \bar{a}_2]$ such that $\bar{a}_1 \Downarrow \bar{a}_2$ in $\Delta(\bar{a})$. Each class $[\bar{a}_1, \bar{a}_2]$ acts like $\mu(\bar{a}_1, \bar{a}_2)$ on $\Delta(\bar{j})\backslash\mathrm{acl}(\bar{a}_1, \bar{a}_2)$. We call $[\bar{a}_1, \bar{a}_2]$ the **germ** of $\mu(\bar{a}_1, \bar{a}_2)$.

Lemma 4.8.4. *There is a set Ξ of formulas $\xi(x)$ of L^+ such that $H = (\bigwedge \Xi)(A^+)$.*

Proof. The formulas Ξ together should say that x is of the form $x_1 {}^\frown x_2/\sim$ where x_1 and x_2 each realise the type of $\bar{a}$ over B, and there are no formulas creating an algebraic dependency between x_1 and x_2. □

Lemma 4.8.5. *Put* $\alpha = [\bar{a}_1, \bar{a}_2]$. *Then* $\mu(\bar{a}_1, \bar{a}_2)(\bar{j}) \in \operatorname{dcl}(\alpha, \bar{j})$ *whenever* $\bar{a}_1{}^{\frown} \bar{a}_2 \Downarrow \bar{j}$.

Proof. Using Lemma 4.8.3, the formula defining $y = \mu(\bar{a}_1, \bar{a}_2)(\bar{j})$ in terms of α and $\bar{j}$ says

> For all x_1 such that $x_1 \Downarrow \bar{j}$, and for all x_2 such that $x_2 \Downarrow x_1 \bar{j}$,
> if $\alpha = x_1{}^{\frown} x_2/\!\!\sim$ then $\sigma(x_1, x_2, \bar{j}, y)$.

Such a formula exists by Lemma 4.7.15. $\qquad\square$

The following technical lemma is crucial.

Lemma 4.8.6. *Suppose* $\bar{a}_1$, $\bar{a}_2 \in \Delta(\bar{a})$ *and* $\bar{a}_1 \Downarrow \bar{a}_2$. *Then* $[\bar{a}_1, \bar{a}_2] \Downarrow \bar{a}_1$, $[\bar{a}_1, \bar{a}_2] \Downarrow \bar{a}_2$ *and* $[\bar{a}_1, \bar{a}_2]$ *has weight* 1.

Proof. Write α for $[\bar{a}_1, \bar{a}_2]$. By Lemma 4.7.13 there is $\bar{a}' \in \Delta(\bar{a})$ such that $\bar{a}' \Downarrow \bar{a}^{\frown} \bar{c}^{\frown} \bar{j}$. This implies $\bar{a}' \Downarrow_{\bar{a}^{\frown} \bar{c}} \bar{j}$ by Lemma 4.7.11(d), and so $\bar{j} \Downarrow \bar{a}^{\frown} \bar{a}'^{\frown} \bar{c}$ since $\bar{j} \Downarrow \bar{a}^{\frown} \bar{c}$ by the configuration. Since $\operatorname{Aut}(A^+)$ is transitive on $\Delta(\bar{a}) \oplus \Delta(\bar{a})$, we can suppose that $\bar{a}_1 = \bar{a}$ and $\bar{a}_2 = \bar{a}'$. Then $\bar{a}_1 \Downarrow \bar{b}^{\frown} \bar{k}$ from the configuration and $\bar{a}_2 \Downarrow \bar{b}^{\frown} \bar{k}$ by Lemma 4.7.11(f), and so by Lemma 4.7.12(a) there is an automorphism g of A^+ which pointwise fixes $\operatorname{acl}(\bar{b}, \bar{k})$ and takes $\bar{a}_1$ to $\bar{a}_2$; put $\bar{c}_1 = \bar{c}$ and $\bar{c}_2 = g\bar{c}$. Observe that by the definition of α, $\alpha(\bar{j}) = g(\bar{j})$.

(8.11)

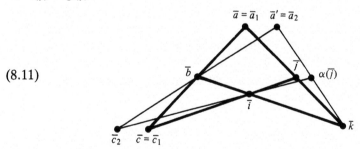

From this picture we draw out two facts.

(8.12) $$\alpha(\bar{j}) \in \operatorname{acl}(\bar{c}_1, \bar{c}_2, \bar{j}).$$

This is immediate from the diagram.

(8.13) $$\bar{j} \Downarrow \bar{a}_1{}^{\frown} \bar{a}_2{}^{\frown} \bar{b}^{\frown} \bar{c}_1{}^{\frown} \bar{c}_2.$$

This holds because $\bar{j} \Downarrow \bar{a}^{\frown} \bar{a}'^{\frown} \bar{c}$ and $\operatorname{acl}(\bar{a}_1, \bar{a}_2, \bar{c}_1, \bar{c}_2) = \operatorname{acl}(\bar{a}, \bar{a}', \bar{c})$; see Lemma 4.7.11(f).

We claim that $\alpha \in \operatorname{acl}(\bar{c}_1, \bar{c}_2)$. For this it suffices to show that $\alpha \in \operatorname{acl}(\bar{c}_1, \bar{c}_2, \bar{j})$. (By (8.13) and Lemma 4.7.11(d), $\bar{a}_1{}^{\frown} \bar{a}_2 \Downarrow_{\bar{c}_1{}^{\frown} \bar{c}_2} \bar{j}$, so that $\alpha \Downarrow_{\bar{c}_1{}^{\frown} \bar{c}_2} \bar{j}$ since $\alpha \in \operatorname{acl}(\bar{a}_1, \bar{a}_2)$. Now apply Lemma 4.7.14.) In fact we shall show that $\alpha \in \operatorname{ACL}(\bar{c}_1, \bar{c}_2, \bar{j})$; by Corollary 4.7.5 this is enough. Let s be any automorphism of A which pointwise fixes $\mathrm{acl}(\bar{c}_1, \bar{c}_2, \bar{j})$. We shall prove that $\alpha = s\alpha$.

Using Lemma 4.7.13, let $\bar{j}'$ be any element of $\Delta(\bar{j})$ such that $\bar{j}' \Downarrow \bar{a}_1 {}^\frown \bar{a}_2 {}^\frown \bar{c}_1 {}^\frown \bar{c}_2 {}^\frown \alpha {}^\frown s\alpha$. By Lemma 4.7.12(a) there is an automorphism t which fixes $\mathrm{acl}(\bar{c}_1, \bar{c}_2, \bar{j}')$ pointwise and takes α to $s\alpha$. Now $\bar{j} \Downarrow \bar{a}_1 {}^\frown \bar{a}_2 {}^\frown \bar{c}_1 {}^\frown \bar{c}_2$ by (8.13) and $\bar{j}' \Downarrow \bar{a}_1 {}^\frown \bar{a}_2 {}^\frown \bar{c}_1 {}^\frown \bar{c}_2$ by choice of $\bar{j}'$, so by Lemma 4.7.12(a) and (8.12) we have $\alpha(\bar{j}') \in \mathrm{acl}(\bar{c}_1, \bar{c}_2, \bar{j}')$. Applying t,

$$\alpha(\bar{j}') = t(\alpha(\bar{j}')) = (t\alpha)(\bar{j}') = (s\alpha)(\bar{j}').$$

Thus α and $s\alpha$ agree on some, and hence every, element of $\Delta(\bar{j})$ which doesn't fork against them. We infer that $\alpha = s\alpha$ as required.

To show that α has weight $\leqslant 1$, suppose otherwise. By the claim just proved, we have $\alpha \in \mathrm{acl}(\bar{a}_1, \bar{a}_2) \cap \mathrm{acl}(\bar{c}_1, \bar{c}_2)$. Since $\bar{a}_1 {}^\frown \bar{a}_2$ and $\bar{c}_1 {}^\frown \bar{c}_2$ both have weight 2, it must follow that $\mathrm{acl}(\bar{a}_1, \bar{a}_2) = \mathrm{acl}(\bar{c}_1, \bar{c}_2)$, and so $\bar{a}_1 {}^\frown \bar{a}_2 {}^\frown \bar{c}_1 {}^\frown \bar{c}_2 {}^\frown \bar{j}$ has weight 3; but then $\bar{a}_2$ is algebraic over $\bar{a}_1 {}^\frown \bar{c}_1 {}^\frown \bar{j}$, contrary to the choice of $\bar{a}_2$.

Finally to show that α has weight $\geqslant 1$, suppose not. Then α is in $\mathrm{acl}(\varnothing)$, and so $\alpha(\bar{j})$ is algebraic over $\bar{j}$. The diagram shows at once that this is false. $\qquad\square$

Lemma 4.8.7. If $\bar{a}_1$, $\bar{a}_2$, $\bar{a}_3$ are such that $\mathrm{weight}(\bar{a}_1 {}^\frown \bar{a}_2 {}^\frown \bar{a}_3) = 3$, then $[\bar{a}_1, \bar{a}_2] \Downarrow [\bar{a}_2, \bar{a}_3]$.

Proof. By the weight $\bar{a}_1 \Downarrow \bar{a}_2 {}^\frown \bar{a}_3$, and hence $\bar{a}_1 \Downarrow \bar{a}_2 {}^\frown [\bar{a}_2, \bar{a}_3]$ since $[\bar{a}_2, \bar{a}_3]$ lies in $\mathrm{acl}(\bar{a}_2, \bar{a}_3)$. Now $[\bar{a}_1, \bar{a}_2]$ and $[\bar{a}_2, \bar{a}_3]$ each have weight 1 by Lemma 4.8.6, and so if they fork against each other they have the same algebraic closure. Hence $\bar{a}_1 \Downarrow \bar{a}_2 {}^\frown [\bar{a}_1, \bar{a}_2]$. But $\bar{a}_2 {}^\frown [\bar{a}_1, \bar{a}_2]$ has weight 2 by Lemma 4.8.6 again, and it follows that $\bar{a}_1 {}^\frown \bar{a}_2 {}^\frown [\bar{a}_1, \bar{a}_2]$ has weight 3, which is impossible since it has the same algebraic closure as $\bar{a}_1 {}^\frown \bar{a}_2$. $\qquad\square$

Lemma 4.8.8. Given any two elements α, β of H such that $\alpha \Downarrow \beta$, there are $\bar{a}_1$, $\bar{a}_2$, $\bar{a}_3$ such that $\alpha = [\bar{a}_1, \bar{a}_2]$, $\beta = [\bar{a}_2, \bar{a}_3]$ and $\mathrm{weight}(\bar{a}_1, \bar{a}_2, \bar{a}_3) = 3$.

Proof. By the previous lemma, since $\mathrm{Aut}(A^+)$ acts transitively on $H \oplus H$. $\square$

The elements of H act on $\Delta(\bar{j})$, and so it makes sense to compose their actions. We write $\alpha \cdot \beta$ for an element γ of H such that for all $\bar{j}' \in \Delta(\bar{j})$, $\bar{j}' \Downarrow \alpha {}^\frown \beta {}^\frown \gamma$ implies $\gamma(\bar{j}') = \alpha(\beta(\bar{j}'))$. There need not be any such element γ. The next lemma says that there is when $\alpha \Downarrow \beta$.

Lemma 4.8.9. (a) If α, β are elements of H and $\alpha \Downarrow \beta$, then $\alpha \cdot \beta$ is a well-defined element of H; moreover $\alpha \cdot \beta \Downarrow \alpha$ and $\alpha \cdot \beta \Downarrow \beta$.

(b) *There is a formula* $\rho(x, y, z)$ *such that if* α, β *are elements of* H *with* $\alpha \mathbin{\bigcup} \beta$, *then* $\alpha \cdot \beta$ *is the unique element* γ *of* H *such that* $A^+ \vDash \rho(\alpha, \beta, \gamma)$.

Proof. (a) By Lemma 4.8.8 there are $\bar{a}_1$, $\bar{a}_2$, $\bar{a}_3$ such that $\alpha = [\bar{a}_1, \bar{a}_2]$ and $\beta = [\bar{a}_2, \bar{a}_3]$. Take any $\bar{j}' \in \Delta(\bar{j})$ such that $\bar{j}' \mathbin{\bigcup} \bar{a}_1 {}^\frown \bar{a}_2 {}^\frown \bar{a}_3$. Then $\mu(\bar{a}_1, \bar{a}_2)(\bar{j}') \mathbin{\bigcup} \bar{a}_2$ by the definition of μ. Also $\mu(\bar{a}_1, \bar{a}_2)(\bar{j}') \mathbin{\bigcup} \bar{a}_3$ since $\bar{a}_3 \mathbin{\bigcup} \bar{j}' {}^\frown \bar{a}_1 {}^\frown \bar{a}_2$. Then the definition of μ immediately gives that $\mu(\bar{a}_1, \bar{a}_3)(\bar{j}')$ is defined and equal to $\mu(\bar{a}_2, \bar{a}_3)\mu(\bar{a}_1, \bar{a}_2)(\bar{j}')$.

Let γ be the product $\alpha \cdot \beta$, and suppose for contradiction that $\gamma \in \mathrm{acl}(\alpha)$. Then the element $\mu(\bar{a}_1, \bar{a}_3)(\bar{j}')$ above is algebraic over $\bar{a}_1$, $\bar{a}_2$ and $\bar{j}'$. Varying $\bar{a}_3$, this is absurd. A similar argument shows that $\gamma \notin \mathrm{acl}(\beta)$.

The proof of (b) is yet another application of Lemma 4.7.15.   □

In order to get closure under composition, we take germs once more. We define another equivalence relation $\approx$ on $H \oplus H$ by

(8.14) $(\gamma, \delta) \approx (\gamma', \delta')$ iff for every tuple $\bar{e} \in \Delta(\bar{j})$ such that $\gamma \cdot \delta(\bar{e})$
and $\gamma' \cdot \delta'(\bar{e})$ are both defined, $\gamma \cdot \delta(\bar{e}) = \gamma' \cdot \delta'(\bar{e})$.

We define G to be the set $(H \oplus H)^{\approx}$ of equivalence classes of $\approx$.

Lemma 4.8.10. *G is a group; there is a set Ξ' of formulas $\xi'(x)$ of L^+ such that $G = (\bigwedge \Xi')(A^+)$, and there is a formula which expresses the product in G.*

Proof. H is closed under inverse, since the definitions imply $[\bar{a}_2, \bar{a}_1] = [\bar{a}_1, \bar{a}_2]^{-1}$. Hence G is closed under inverse. To show that G is closed under product, it suffices to show that any product of three elements of H can be written as a product of two. Consider α, β, γ in H such that $\alpha \mathbin{\bigcup} \beta$ and $\beta \mathbin{\bigcup} \gamma$. There is δ in H such that $\delta \mathbin{\bigcup} \alpha {}^\frown \beta {}^\frown \gamma$. Since $\mathrm{Aut}(A)$ is transitive on $H \oplus H$, Lemma 4.8.9 gives us an element δ' of H such that $\beta = \delta \cdot \delta'$ and β is independent of δ. Then $\alpha \cdot \beta \cdot \gamma = \alpha \cdot \delta \cdot \delta' \cdot \gamma$. Since $\delta \mathbin{\bigcup} \alpha$, $\alpha \cdot \delta$ is in H. Also $\delta' \mathbin{\bigcup} \gamma$; for otherwise β is algebraic over $\{\gamma, \delta\}$, so weight $(\gamma {}^\frown \delta) = \mathrm{weight}(\beta {}^\frown \gamma {}^\frown \delta)$, contradicting the choice of δ.

The argument to find Ξ' is much the same as Lemma 4.8.4. For the product formula one uses Lemma 4.8.9(b).   □

Second part: assuming the hypothesis of the theorem, we show that the unique definability condition can be met by making small adjustments to the configuration

Given a Zil'ber configuration $Z = (a, b, c, i, j, k)$, we say that a 6-tuple $Z' = (a', b', c', i', j', k')$ of elements of A^+ is a **configuration equivalent to** Z if $\mathrm{acl}(a) = \mathrm{acl}(a')$, $\mathrm{acl}(b) = \mathrm{acl}(b')$, ..., $\mathrm{acl}(k) = \mathrm{acl}(k')$.

Lemma 4.8.11. *Suppose a configuration Z occurs in the strongly minimal set Ω of A. Then we can find an equivalent configuration Z' such that $j' \in \mathrm{dcl}(a'^{\wedge}k')$, $k \in \mathrm{dcl}(k')$, and b, c, i are equal to b', c', i' respectively.*

Proof. First a comment on the strategy. In the configuration Z we know that $j \in \mathrm{acl}(a^{\wedge}k)$; we would like to have $j \in \mathrm{dcl}(a^{\wedge}k)$. The natural way to achieve this would be to replace j by the finite set of all elements realising the same type as j over a and k. Alternatively we might replace j by the canonical base of the type of j over the set $\{a, k\}$ (which exists by Theorem 4.3.6, and is in the definable closure of $\{a, k\}$ by Theorem 4.3.5) – this is more sophisticated but stands a better chance of generalising to other situations. The difficulty is that on either choice, the new element need not be algebraic over the original j, and so the new configuration may not be equivalent to the old one.

We use a cunning way out. The trick is to form a' and k' by adding to a and k some elements of their algebraic closures over B, in such a way that the new element j' will sit on the intersection of *two* lines, which is a point. Then every element realising the type of j' over a' and k' must be algebraic over j', simply by counting dimensions.

To work the trick, use Lemma 4.5.6 to find an automorphism g of A (not of A^{+}) which fixes the line $\mathrm{acl}(a, k)$ pointwise and takes b to an element gb of B. Put $a' = a^{\wedge}gc$, $k' = k^{\wedge}gi$. Clearly $\mathrm{acl}(a') = \mathrm{acl}(a)$, since gc is algebraic over a and gb, and hence over a (since gb lies in B and hence is named by a constant). Similarly $\mathrm{acl}(k') = \mathrm{acl}(k)$, and it's immediate that $k \in \mathrm{dcl}(k')$. Let j' be the canonical base of the type p of j over $a'^{\wedge}k'$. Then $j' \in \mathrm{dcl}(a'^{\wedge}k')$. We have to show that $\mathrm{acl}(j) = \mathrm{acl}(j')$.

Let e be any element realising the type p. Clearly $\mathrm{acl}(a, k) \supseteq \mathrm{acl}(j, e) \subseteq \mathrm{acl}(gc, gi)$. We claim that $e \in \mathrm{acl}(j)$, i.e. that $j^{\wedge}e$ has weight 1. For if not, then $\mathrm{acl}(a, k) = \mathrm{acl}(j, e) = \mathrm{acl}(gc, gi)$, contradicting the configuration. Thus p is an algebraic type, and so by Theorem 4.3.6 its canonical base is algebraic over j. This shows $\mathrm{acl}(j) \supseteq \mathrm{acl}(j')$. But from j' we can recover the type p, and in particular we can tell that j is algebraic over a' and k' by some particular formula. So j' pins down j to one of a finite number of elements, and hence $j \in \mathrm{acl}(j')$. This shows $\mathrm{acl}(j) \subseteq \mathrm{acl}(j')$. $\qquad\square$

Lemma 4.8.12. *Given a configuration Z, we can find an equivalent configuration Z'' in which $j'' \in \mathrm{dcl}(a'', k'')$ and $i'' \in \mathrm{dcl}(b'', k'')$.*

Proof. Use the previous lemma twice. Note that $j'' = j' \in \mathrm{dcl}(a', k') = \mathrm{dcl}(a'', k') \subseteq \mathrm{dcl}(a'', k'')$ since $k' \in \mathrm{dcl}(k'')$. $\qquad\square$

It remains to get k to be definable over $a^{\wedge}j$ and over $b^{\wedge}i$.

Lemma 4.8.13. *Suppose the configuration Z has the property which Z'' has in the previous lemma. Then we can find an equivalent configuration Z' which also has this property, but with $k' \in \mathrm{dcl}(a'^\frown j')$ and $k' \in \mathrm{dcl}(b'^\frown i')$.*

Proof. Let k' be the canonical base of the type q of k over $a^\frown b^\frown i^\frown j$. Then as in the proof of Lemma 4.8.11, q is an algebraic type because it marks the intersection of the lines aj and bi; and by the same argument as before, k' and k have the same algebraic closure. We have to check that if we replace k by k', j will still be definable over $a^\frown k'$ and i will still be definable over $b^\frown k'$. By symmetry we need only prove one of these.

We prove that $j \in \mathrm{dcl}(a^\frown k')$. As canonical base, k' determines the type q. In particular it tells us that for some formula $\chi(x, y, z)$ of L, q contains the formula $\chi(x, a, j) \wedge \exists_{=1} z\, \chi(x, a, z)$. But j is the unique element with this property. This finds j uniquely in terms of a and k', as required. $\square$

With the help of the theory of stable groups, one can refine Theorem 4.8.1. Starting from G, one finds an infinite abelian group whose domain and product are definable by single formulas.

History and bibliography

Section 4.1. For some twenty years Shelah has been using Example 2, often heavily disguised, to refute all kinds of false conjecture; two recent specimens are Shelah & Hart [1990] and Theorem 14 in Shelah, Tuuri and Väänänen [199?]. It was only after the Cherlin–Mills–Zil'ber theorem (Theorem 4.6.1) that model theorists in general woke up to affine spaces; in another context Coxeter ([1961], p. 191) remarked '... most readers will discover that they have often been working in the affine plane without realizing that it could be so designated'.

Cameron recently published an elegant monograph [1990] on oligomorphic structures; see also Chapter 7 below. The topology on $\mathrm{Aut}(A)$ was used by Ahlbrandt & Ziegler [1986] and more recently by Hodges, Hodkinson & Macpherson [1990], Evans & Hewitt [1990] and Lascar [1991]. Model theorists have used several different notions of algebraicity in different contexts. See Bacsich [1973] for a discussion. Theorem 4.1.4 is in Reyes [1970]; related ideas in Krasner [1938] have inspired recent research.

Exercise 13: Baranskiĭ [1983]. Exercise 16: Banach in Kuratowski [1958] p. 81f. Exercise 17: Lascar [1991]. Exercise 18: (b, c) Denes [1973], (d) Ziegler [1975].

Section 4.2. Neumann père proved his Lemma 4.2.1 in [1954]. He showed that each H_i has finite index in G; the proof given here (which established the bound $n!$) is from Tomkinson [1987]. Then Neumann fils used the lemma to deduce Corollary 4.2.2 in his [1976]; see also the direct proof in Birch *et al.* [1976], and Theorem 6.4.5 below for an equivalent model-theoretic statement which goes back to Park [1964a]. Theorem 4.2.4 (when $L^+ \backslash L^-$ is finite) is from Kueker [1968] and Reyes [1970]; the full statement of Theorem 4.2.4 can be deduced from this special case by a coding argument. The group-theoretic formulation in Theorem 4.2.3 is from Evans [1987]. Corollary 4.2.5 is from Kueker [1968]. Lemma 4.2.6 is due to Hodkinson in Hodges, Hodkinson & Macpherson [1990]. The strong small index property of $\text{Sym}(\omega)$ is due to Rabinovich [1977]; it was rediscovered by Semmes [1981] and Пeter Neumann, and the proof here follows Dixon, Neumann & Thomas [1986]. In Fact 4.2.11, (a) and (b) are due to Truss [1989], (c) to Evans [1986b], (d) to Evans [1991] and (e) to Evans in Evans, Hodges & Hodkinson [1991]. The intersection condition is familiar to set theorists from Mostowski's [1939] work on independence proofs. This section owes much to conversations with David Evans and Ian Hodkinson.

Exercise 1: Kueker [1968]. Exercise 2: One of my first students, Edward Hodge, noted this in a letter to me a couple of weeks before his sad death. Exercise 4: Sierpiński [1928]. Exercise 10: Mostowski [1939].

Section 4.3. The extension of automorphisms of a structure to the universe above that structure is essentially due to Fraenkel [1922] and Mostowski [1939]. The formulation of Lemma 4.3.2 is from Hodges [1976b], and Example 3 from Hodges [1976c].

The construction of A^{eq} is from Shelah [1978a] section III.6, where he also introduces canonical bases for definable types; see Harnik & Harrington [1984] for a discussion of A^{eq}. Zil'ber [1977b] uses finite slices of A^{eq} instead. Definable types were introduced by Gaifman in 1967, in connection with models of arithmetic; see Gaifman [1976]. See section 6.7 below for the notion of definable types in stable theories. The finite cover property is due to Keisler [1967]; Shelah [1978a] section II.4 proves that Shelah's definition is equivalent to Keisler's for stable theories. See Poizat [1983b], [1984] for discussions of Shelah's version in stable theories.

Exercise 2: Läuchli [1962]. Exercise 3: Hodges [1974]. Exercise 8: the example is discussed in Evans, Pillay & Poizat [1990]; an example was given also by Belegradek. Exercise 10: Shelah [1978a] Theorem II.4.4. Exercise 11: Baldwin & Kueker [1980].

Section 4.4. Elimination of imaginaries is from Poizat [1983c], [1985], together with Theorem 4.4.6. The argument of Lemma 4.4.3 is due to van den Dries [1984]. I am not sure who first noticed Theorem 4.4.7 – possibly Pillay.

Exercise 3: Macpherson.

Section 4.5. Strongly minimal sets were introduced by Marsh [1966]. They were taken over and publicised by Baldwin & Lachlan [1971] as a tool for proving Morley's theorem (Theorem 12.2.1 below, for countable languages).

Exercise 9: Zil'ber & Smurov [1988]. Exercise 10: Marker [1987] proved the remarkable fact that this structure has an expansion which is strongly minimal. Exercise 13: Nurtazin [1990].

Section 4.6. Zil'ber essentially conjectured Theorem 4.6.1 in [1980a]. Zil'ber himself proved the theorem in Zil'ber [1984a], [1984b], by an argument which in part rests on analogies with intersection theory in algebraic geometry. His best form of this proof (Zil'ber [1988]) now gives the same result for all geometries of dimension at least 8. (The idea of transporting arguments from algebraic geometry into model theory by way of Morley rank goes back to Zil'ber's research in the early 1970s at Novosibirsk under M. A. Taĭtslin.) Independently of Zil'ber's proof, Cherlin (in Cherlin, Harrington & Lachlan [1985]) and Mills (unpublished) noticed that Theorem 4.6.1 follows quite easily from the classification of finite simple groups, applied to Jordan sets; see also Π. M. Neumann [1985]. Further proofs have been given by Evans [1986a] (a proof which uses coherent configurations and covers dimensions from 23 upwards) and Hrushovski [199?b] (a quite short model-theoretic argument for the infinite-dimensional case).

Zil'ber's conjecture [1984c] was that 'Every uncountably categorical pseudo-plane is definable in an algebraically closed field, and the field is definable in the pseudoplane'. This implies in particular that every strongly minimal set whose geometry is not locally modular has an algebraically closed field interpretable in it (in the sense of section 5.3 below). Hrushovski's counter-example is in [1988]. Hrushovski [199?a] contains the strongly minimal set carrying two fields of different characteristic. On Zil'ber's configuration, see the references to section 4.8 below. Theorem 4.6.3 is by Shelah in Ash & Rosenthal [1980].

Examples 2, 3 and 4 are all examples of uncountably categorical free structures in a variety; structures of this form were catalogued by Givant [1979] and Palyutin [1975].

Exercise 2: Mac Lane [1938]. Exercise 8: (a) Birkhoff [1935a], Whitney [1935], (b) Prenowitz [1943], Frink [1946].

Section 4.7.   Almost strongly minimal sets were studied by Baldwin [1972].
Weights appear in section V.3 of Shelah [1978a] – the general definition given
there is more complicated than mine, because I consider only the special case
of almost strongly minimal structures.

Exercise 6: The notion of smoothly approximated structures appears in
Cherlin, Harrington & Lachlan [1985]; see Kantor, Liebeck & Macpherson
[1989] for further discussion and a classification theorem. Exercise 7: Makowsky [1974] proved that if A is ω-categorical and almost strongly minimal then
Th(A) is not finitely axiomatisable. The conclusion was generalised to totally
categorical theories by Zil'ber [1980b] (with a correction in [1981]), and (in
part independently) to ω-categorical ω-stable theories by Cherlin, Harrington
& Lachlan [1985].

Section 4.8.   Zil'ber's configuration appears in Lemma 3.3 of Zil'ber [1984a],
and a version of Theorem 4.8.1 in Zil'ber [1984b]. My proof follows
Hrushovski, who generalised Zil'ber's argument greatly; see Hrushovski
[1987], Evans & Hrushovski [1991] and the exposition by Bouscaren [1989c].
The configuration is sometimes known as 'Hrushovski's configuration' because Zil'ber described it in Russian instead of drawing a picture (to save
trouble with the printers, he told me).

5

Interpretations

She turnd hersell into an eel,
To swim into yon burn,
And he became a speckled trout,
To gie the eel a turn.

Then she became a silken plaid,
And stretched upon a bed,
And he became a green covering,
And gaind her maidenhead.

Scots ballad from F. J. Child, The English and Scottish Popular Ballads.

Since 1797 we have known (thanks to Caspar Wessel) that the complex numbers can be thought of as points in the real plane. Since 1637 we have known (from Descartes) that the points in the real plane can be identified by their cartesian coordinates. So in two steps we see that a complex number is really an ordered pair of real numbers. This insight is called an *interpretation* of the complex numbers in the reals. It takes two reals for each complex number, so the interpretation is said to be *two-dimensional*.

An interpretation of an L-structure B in a K-structure A involves four things: the structures A and B, and the languages K and L. There is a map that takes A to B; in fact we shall see in section 5.3 that this map is a functor taking any structure A' enough like A to a structure B' rather like B. (For example the construction of complex numbers from real numbers works equally well if one starts from any real-closed field.) There is also a map from L to K: it translates statements about B into statements about A.

Interpretations are about different ways of looking at one and the same thing. So it should cause no surprise that there are several different ways of looking at interpretations. Sometimes the important thing is the syntactic map from L to K, and the map taking A to B is just a guide towards forming mental pictures. One example of this is the use of interpretations to prove the undecidability of theories; see section 5.5. There are signs that interpretations are becoming helpful in software development: the syntactic map is one step in building up the written program, and the map on structures expresses the

intended use of the program. But this application is in its infancy and I have
not pursued it here.

More often in model theory, the real interest lies in the structures
themselves, and the syntax is merely a tool for handling them. For example
Cherlin's conjecture is that every totally transcendental simple group is an
algebraic group over an algebraically closed field (see section A.4 below).
One step towards proving Cherlin's conjecture would be to show that if G is
an infinite simple group of finite Morley rank, then some algebraically closed
field is interpretable in G. The question lies in stability theory, but I hope
that the treatment of totally transcendental groups in sections 5.6 and 5.7 will
be useful as an introduction to it.

Interpretations appear in philosophical analysis too, and cause warfare
there. Are mental phenomena interpretable in physical ones? Can statements
of morals be translated into statements of fact? Are gods really a form of
social control? Fortunately the mathematical theory behind these controver-
sies is entirely peaceful and harmonious.

5.1 Relativisation

Often in algebra one considers a pair of structures, for example a field and its
algebraic closure, or a group that acts on a set. Model theory is not very good
at handling pairs of structures. Instead a model theorist will usually try to
represent the two structures as parts of some larger structure.

This may be a complicated matter. Let us start with the simplest case,
where one structure can be picked out inside another one by a single 1-ary
relation symbol.

Consider two signatures L and L' with $L \subseteq L'$. Let C be an L'-structure
and B a substructure of the reduct $C|L$. Then we can make the pair of
structures C, B into a single structure as follows. Take a new 1-ary relation
symbol P, and write L^+ for L' with P added. Expand C to an L^+-structure
A by putting $P^A = \mathrm{dom}(B)$. Then we can recover C and B from A by

(1.1) $C = A|L'$,

$B = $ the substructure of $A|L$ whose domain is P^A.

Thus C is a reduct of A. We call B a **relativised reduct** of A, meaning that to
get B from A we have to 'relativise' the domain to a definable subset of
$\mathrm{dom}(A)$ as well as removing some symbols.

From now on, we forget about C. The setting is that L and L^+ are
signatures with $L \subseteq L^+$, and P is a 1-ary relation symbol in $L^+ \backslash L$.

Let A be an L^+-structure. Lemma 1.2.2 gave necessary and sufficient
conditions for P^A to be the domain of a substructure of $A|L$. When these
conditions are satisfied, the substructure is uniquely determined. We write it

A_P, and we call it the P-**part** of A. Otherwise A_P is not defined. (From Lemma 1.2.2 one can write these necessary and sufficient conditions as a set of first-order sentences that A must satisfy. We call them the **admissibility conditions** for relativisation to P.) Of course A_P depends on the language L as well as A and P.

The next theorem says that facts about A_P can be translated systematically into facts about A.

Theorem 5.1.1 (*Relativisation theorem*). *Let L and L^+ be signatures such that $L \subseteq L^+$, and P a 1-ary relation symbol in $L^+ \backslash L$. Then for every formula $\phi(\bar{x})$ of $L_{\infty\omega}$ there is a formula $\phi^P(\bar{x})$ of $L^+_{\infty\omega}$ such that the following holds:*

if A is an L^+-structure such that A_P is defined, and $\bar{a}$ is a sequence of elements from A_P, then

(1.2) $\qquad\qquad A_P \vDash \phi(\bar{a})$ *if and only if* $A \vDash \phi^P(\bar{a})$.

Proof. We define ϕ^P by induction on the complexity of ϕ:

(1.3) $\quad \phi^P$ is ϕ $\qquad$ when ϕ is atomic;

(1.4) $\quad \left(\bigwedge_{i \in I} \psi_i \right)^P$ is $\bigwedge_{i \in I} (\psi_i^P)$, and likewise with $\bigvee$ for $\bigwedge$;

(1.5) $\quad (\neg \phi)^P$ is $\neg(\psi^P)$;

(1.6) $\quad (\forall y \, \psi(\bar{x}, y))^P$ is $\forall y (Py \to \psi^P(\bar{x}, y))$, and
$\qquad (\exists y \, \psi(\bar{x}, y))^P$ is $\exists y (Py \wedge \psi^P(\bar{x}, y))$.

Then (1.2) follows at once by induction on the complexity of ϕ. $\qquad\qquad\square$

The formula ϕ^P in this theorem is called the **relativisation** of ϕ to P. Note that if ϕ is first-order then so is ϕ^P. In fact the passage from ϕ to ϕ^P preserves the form of ϕ rather faithfully.

Corollary 5.1.2. *Let L and L^+ be signatures with $L \subseteq L^+$ and P a 1-ary relation symbol in $L^+ \backslash L$. If A and B are L^+-structures such that $A \preccurlyeq B$ and A_P is defined, then B_P is defined and $A_P \preccurlyeq B_P$.*

Proof. Exercise. $\qquad\qquad\square$

Example 1: *Linear groups.* Suppose G is a group of $n \times n$ matrices over a field F. Then we can make G and F into a single structure A as follows. The signature of A has 1-ary relation symbols *group* and *field*, 3-ary relation symbols *add* and *mult*, and n^2 2-ary relation symbols *coeff*$_{ij}$ ($1 \leqslant i, j \leqslant n$). The sets *group*^{A} and *field*^{A} consist of the elements of G and F respectively. The relations *add*^{A} and *mult*^{A} express addition and multiplication in F. For each matrix $g \in G$, the ijth entry in g is the unique element f such that $coeff_{ij}(g, f)$ holds. There's no need to put in a symbol for multiplication in G, because it can be defined in terms of the field operations, using the symbols $coeff_{ij}$ (see Exercise 4). Note that in this example there are no function or constant symbols, and so B_P and B_Q are automatically defined for any structure B of the same signature as A.

Here follows a typical application of relativisation. Let us say that a group G is **completely reducible** of degree n over the field F if G is isomorphic to a multiplicative group G_1 of linear transformations of an n-dimensional vector space V over F, and every subspace of V which is closed under G_1 has a complement in V which is also closed under G_1.

Proposition 5.1.3. *Suppose G is an infinite completely reducible group of degree n (over some field F). Then some countable elementary subgroup of G is also completely reducible of degree n over some field.*

Proof. We can suppose that G itself is a group of linear transformations of the vector space $^n F$ of $n \times 1$ column matrices over F. Make G and F into a two-part structure A as in Example 1 above. It's not hard to see that there is a first-order sentence ϕ which is true in A and expresses 'F is a field, G is a group of $n \times n$ matrices over F, and every subspace of $^n F$ which is closed under G has a complement in $^n F$ which is also closed under G'. Now use the downward Löwenheim–Skolem theorem (Corollary 3.1.5) to find a countable elementary substructure A' of A. By Corollary 5.1.2, A'_{group} is an elementary substructure of $A_{group} = G$; and clearly A'_{group} is countable, since it is infinite and it lies inside A'. Since $A' \vDash \phi$, we have that A'_{group} is a completely reducible group of degree n over the field A'_{field}.   □

Sometimes a structure B is picked out inside a structure A, not by a 1-ary relation symbol P but by a formula $\theta(x)$. When θ is in the first-order language of A, then again we call B a **relativised reduct** of A. The case considered above, where $\theta(x)$ is Px, becomes a special case. If θ also contains parameters from A, we call B a **relativised reduct with parameters**. One can adapt the relativisation theorem straightforwardly by putting θ in place of P everywhere.

Example 2: ω as a relativised reduct. Logicians are often pleased when they find the natural numbers as a part of some other structure. It gives them access to results of several kinds – see for example sections 5.5 (undecidability) and 11.4 (nonstandard methods) below. One simple example is where A is a transitive model of Zermelo–Fraenkel set theory and $\theta(x)$ is the formula '$x \in \omega$'. Then the ordering $<$ on ω coincides with $\in$, and we can write set-theoretic formulas that define $+$ and $\cdot$. Note that in this example ω satisfies a rather strong form of the Peano axioms, as follows:

(1.7)   0 is not of the form $x + 1$; if $x, y \in \omega$ and $x + 1 = y + 1$ then
         $x = y$;

(1.8)   for every formula $\phi(x)$ of the first-order language of A,
         possibly with parameters from A, if $\phi(0)$ and
         $\forall x (x \in \omega \wedge \phi(x) \rightarrow \phi(x + 1))$ both hold in A then $\forall x (x \in \omega$
         $\rightarrow \phi(x))$ holds in A.

(1.8) is the induction axiom schema for subsets of ω which are first-order definable (with parameters) in A. Of course this includes the subsets of ω which are first-order definable in the structure $\langle \omega, < \rangle$ itself, by the relativisation theorem. But it may contain a great many more subsets besides – perhaps all the subsets of ω. That depends on our choice of A.

Example 3: _Relativised reducts of the rationals as ordered set._ Let A be the following structure: the domain of A is the set $\mathbb{Q}$ of rational numbers, and the relations of A are all those which are $\varnothing$-definable from the usual ordering $<$ of the rationals. What are the relativised reducts of A (without parameters)? First note that $\mathrm{Aut}(A)$ is exactly $\mathrm{Aut}(\mathbb{Q}, <)$ since A is a definitional expansion of $(\mathbb{Q}, <)$. Next, $\mathrm{Aut}(A)$ is transitive on $\mathbb{Q}$, and it follows that any subset of $\mathbb{Q}$ which is definable without parameters is either empty or the whole of $\mathbb{Q}$. So we can forget the relativisation. Thirdly, if B is any reduct of A then $\mathrm{Aut}(A) \subseteq \mathrm{Aut}(B) \subseteq \mathrm{Sym}(\mathbb{Q})$, and $\mathrm{Aut}(B)$ is closed in $\mathrm{Sym}(\mathbb{Q})$ by Theorem 4.1.4. And finally, $\mathrm{Aut}(\mathbb{Q}, <)$ is oligomorphic and its orbits on n-tuples are all $\varnothing$-definable; so every orbit of $\mathrm{Aut}(B)$ on n-tuples is a union of finitely many orbits of $\mathrm{Aut}(\mathbb{Q}, <)$ and hence is defined by some relation of A. It follows that, up to definitional equivalence, the relativised reducts of A correspond exactly to the closed groups lying between $\mathrm{Aut}(A)$ and $\mathrm{Sym}(\mathbb{Q})$.

It can be shown that apart from $\mathrm{Aut}(A)$ and $\mathrm{Sym}(\mathbb{Q})$, there are just three such groups. The first is the group of all permutations of A which either preserve the order or reverse it. The second is the group of all permutations which preserve the cyclic relation '$x < y < z$ or $y < z < x$ or $z < x < y$'; this corresponds to taking an initial segment of $\mathbb{Q}$ and moving it to the end. The third is the group generated by these other two: it consists of those permutations which preserve the relation 'exactly one of x, y lies between z and w'.

One often finds some further devices attached to relativisation. For example there are the **relativised quantifiers** $(\forall x \in P)$ and $(\exists x \in P)$. These are defined thus:

(1.9) $(\forall x \in P)\phi$ means $\forall x (Px \rightarrow \phi)$; $(\exists x \in P)\phi$ means $\exists x (Px \wedge \phi)$.

Some logicians introduce variables x_P which range over those elements which satisfy Px; for example in set theory one has variables α, β etc. that range over the class of ordinals. Some writers use **sortal signatures**, whose structures carry partial functions; the functions are defined only when the arguments come from some given sets, known as **sorts**. I shall not pursue this further.

Exercises for section 5.1

1. Show that if A is a relativised reduct of B and B is a relativised reduct of C, then A is a relativised reduct of C.

2. Describe the admissibility formulas for relativisation to P.

3. Prove Corollary 5.1.2.

4. (a) In Example 1, write out a formula $\psi(x, y, z)$ which expresses that the matrix z is the product of the matrices x and y. (b) Write out the sentence ϕ of Proposition 5.1.3.

5. Show that the structure $\langle \omega, + \rangle$ is a relativised reduct of the ring of integers. [Sums of squares.]

6. Show that if B is a relativised reduct of A, then there is an induced continuous homomorphism $h: \mathrm{Aut}(A) \to \mathrm{Aut}(B)$.

*7. Let A be an algebraically closed field of characteristic $p \neq 0$ and $A(t)$ the field of rational functions over A. Show that the field A is a relativised reduct of the function field $A(t)$. [If $n > 2$ and n is relatively prime to p, then the curve $x^n + y^n = 1$ has no points in $A(t)^2$ but outside A^2; see Shafarevich [1977] pp. 7f.]

*8. Suppose B is a relativised reduct of A and every finite partial automorphism of B extends to an automorphism of A. (a) Show that if $\mathrm{dom}(A)\backslash\mathrm{dom}(B)$ is finite, every automorphism of B extends to an automorphism of A. (b) Show that if every element of $\mathrm{dom}(A)\backslash\mathrm{dom}(B)$ has finite orbit under $\mathrm{Aut}(A)$, then every automorphism of B extends to an automorphism of A. [This follows from (a) by the compactness theorem, Theorem 6.1.1.] (c) Give an example where A is countable and not every automorphism of B extends to an automorphism of A. [B the nodes of a binary tree and $A\backslash B$ a dense set of branches.]

5.2 Pseudo-elementary classes

Many classes of mathematical structures are defined in terms of some feature that can be added to them.

Example 1: *Orderable groups.* An **ordered group** is a group G which carries a linear ordering $<$ such that if g, h and k are any elements of G, then

(2.1) $\qquad\qquad g < h$ implies $k \cdot g < k \cdot h$ and $g \cdot k < h \cdot k$.

A group is **orderable** if a linear ordering can be added so as to make it into an ordered group. Clearly an orderable group can't have elements $\neq 1$ of finite order. But this is not a sufficient condition for orderability (unless the group happens to be abelian; see Exercise 6.2.13).

For the whole of this section, let L be a first-order language. A **pseudo-elementary class** (for short, a **PC class**) of L-structures is a class of structures of the form $\{A|L: A \vDash \phi\}$ for some sentence ϕ in a first-order language $L^+ \supseteq L$. A **PC$_\Delta$ class** of L-structures is a class of the form $\{A|L: A \vDash U\}$ for some theory U in a first-order language $L^+ \supseteq L$.

For example the class of orderable groups is a PC class. But we get a more intriguing example by turning Example 1 on its head:

Example 2: *Ordered abelian groups.* Let L be the first-order language of linear orderings (with symbol $<$) and let U be the theory of ordered abelian groups. Then the class $\mathbf{K} = \{A|L: A \vDash U\}$ is the class of all linear orderings which are orderings of abelian groups. This is a PC class, since U can be written as a finite theory and hence as a single sentence. By a result of Morel [1968], the countable order-types which are in $\mathbf{K}$ are precisely those of the form ζ^α or $\zeta^\alpha \cdot \eta$ where ζ is the order-type of the integers, η the order-type of the rationals, α is an ordinal $< \omega_1$, and ζ^α is the order-type defined as follows. Write $^{(\alpha)}\mathbb{Z}$ for the set of all sequences $(n_i: i < \alpha)$ where n_i is an integer for each $i < \alpha$, and $n_i \neq 0$ for only finitely many i. If $m = (m_i: i < \alpha)$ and $n = (n_i: i < \alpha)$ are two distinct elements of $^{(\alpha)}\mathbb{Z}$, write $m < n$ iff $m_i < n_i$ for the greatest i at which $m_i \neq n_i$. Then ζ^α is the order-type of $(^{(\alpha)}\mathbb{Z}, <)$. It follows that $\mathbf{K}$ is not first-order axiomatisable. For example the linear ordering ζ is elementarily equivalent to $\zeta(1 + \eta)$ (e.g. by Exercise 2.7.6(a) or Lindström's test, Theorem 8.3.4), but the first is in $\mathbf{K}$ and the second is not.

It also follows that $\mathbf{K}$ contains, up to isomorphism, just ω_1 linear orderings of cardinality ω. This is interesting, because an old conjecture of Vaught states that the number of countable models of a countable first-order theory (up to isomorphism) is always either 2^ω or $\leq \omega$. Morley [1970a] showed that if $\mathbf{J}$ is a class of the form $\{A|L: A \vDash U\}$ for some countable first-order theory U, then the number of countable structures in $\mathbf{J}$, up to isomorphism, is either $\leq \omega_1$ or $= 2^\omega$. Since there are plenty of examples for cardinalities $\leq \omega$ and $= 2^\omega$, the example $\mathbf{K}$ above shows that Morley's result is best possible. See section 7.2 below for more on Vaught's conjecture.

One can generalise these notions, using relativised reducts A_P as in section 5.1. We define a **PC$'_\Delta$ class** of L-structures to be a class of the form $\{A_P : A \vDash U$ and A_P is defined$\}$ for some theory U in a language $L^+ \supseteq L \cup \{P\}$. By the admissibility conditions, every PC$'_\Delta$ class can be written as $\{A_P : A \vDash U'\}$ for some theory U' in L^+.

A natural example of a PC$'_\Delta$ class is the class of multiplicative groups of fields. Here U is the theory of fields together with a symbol P which picks out the non-zero elements, and L has only the symbol for multiplication. Theorem A.2.9 below will show that this class is not first-order axiomatisable.

It turns out that PC$'_\Delta$ is not really a generalisation of PC$_\Delta$; it's exactly the same thing.

Theorem 5.2.1. *The PC$'_\Delta$ classes are exactly the PC$_\Delta$ classes. More precisely, let **K** be a class of L-structures.*

(a) *If **K** is a PC$'_\Delta$ class $\{A_P : A \vDash U$ and A_P is defined$\}$ for some theory U in a first-order language L^+, then **K** is also a PC$_\Delta$ class $\{A|L : A \vDash U^*\}$ for some theory U^* in a first-order language L^* with $|L^*| \leqslant |L^+|$.*

(b) *If **K** is a PC$'$ class and all structures in **K** are infinite, then **K** is a PC class.*

Proof. To make a PC$_\Delta$ class into a PC$'_\Delta$ class, add the symbol P with the axiom $\forall x\, Px$.

The full proof of the other direction in (a) is surprisingly subtle, and so I postpone it for a moment. But if **K** is a PC$'_\Delta$ class in which every structure has cardinality $\geqslant |L^+|$, then we can show that **K** is a PC$_\Delta$ class by the following handy argument. The same argument proves (b).

Let **K** be $\{A_P : A \vDash U$ and A_P is defined$\}$, and suppose that each structure in **K** has cardinality $\geqslant |L^+|$. Then by the downward Löwenheim–Skolem theorem, each structure in **K** is of the form $B = A_P$ for some model A of U with $|A| = |B|$. For each symbol S in the signature of L^+, introduce a copy S^*. Add a 1-ary function symbol F. If A and B are as above, take an arbitrary bijection $f : \mathrm{dom}(A) \to \mathrm{dom}(B)$, and interpret each symbol S^* on $\mathrm{dom}(B)$ as the image of S^A under f. Interpret F as the restriction of f to $\mathrm{dom}(B)$. Using the symbols of L together with the symbols S^* and F, we can write down a theory U^* which expresses that the interpretations of the symbols S^* make $\mathrm{dom}(B)$ into a model D of U, and F maps B isomorphically onto the P^*-part. Then $\mathbf{K} = \{D|L : D \vDash U^*\}$, and so **K** is a PC$_\Delta$ class.

We turn to the full proof of part (a). For each atomic formula $\phi(x_0, \ldots, x_{n-1})$ of L^+, whose free variables are an initial segment of the variables $x_0, x_1, \ldots$, we introduce a new n-ary relation symbol R_ϕ. Let L^0

be the first-order language got by adding these new symbols R_ϕ to L, and L^- the language whose signature consists of just the symbols R_ϕ. Suppose A is an L-structure and X is a set of generators of A. We expand A to an L^0-structure A^0 as follows: for each atomic formula ϕ of L^+ and each tuple $\bar{b}$ in A^0 we put

$$(2.2) \qquad A^0 \vDash R_\phi(\bar{b}) \text{ iff } \bar{b} \text{ lies in } X \text{ and } A \vDash \phi(\bar{b}).$$

We write $A \downarrow X$ for the substructure of $A^0|L^-$ with domain X. By (2.2) the structure $A \downarrow X$ contains a complete description of A, and so we can recover A up to isomorphism from $A \downarrow X$.

In fact we can sharpen this. Write Σ for the set of those $\forall_1$ sentences of L^- which are true in every structure of form $A \downarrow X$.

Lemma 5.2.2. *Every model of Σ is of form $A \downarrow X$ for some L^+-structure A and set of generators X.*

Proof. Suppose the L-structure B is a model of Σ. We shall construct an L^+-structure A with $A \downarrow X$ isomorphic to B.

Let Y be the set of expressions $t(\bar{b})$ where $t(\bar{x})$ is a term of L^+ (possibly a constant) and $\bar{b}$ is a tuple of elements of B. We define an equivalence relation $\sim$ on Y as follows: if $s(\bar{a})$ and $t(\bar{b})$ are in Y, let ϕ be the formula $s(\bar{x}) = t(\bar{y})$, and put $s(\bar{a}) \sim t(\bar{b})$ iff $B \vDash R_\phi(s(\bar{a}), t(\bar{b}))$. It is easily checked that $\sim$ is an equivalence relation, using the fact that B is a model of Σ. We shall define an L^0-structure A' whose domain is the set of equivalence classes $t^\sim$ of expressions t in Y.

If a and b are distinct elements of A', then $B \vDash \neg R_{x=y}(a, b)$ since Σ contains the sentence $\forall uv(R_{x=y}(u, v) \to u = v)$; so $a^\sim \neq b^\sim$. This allows us to put $A' \vDash R_\phi(\bar{a}^\sim)$ iff $B \vDash R_\phi(\bar{a})$; elsewhere R_ϕ is false. When ψ is an atomic formula of L^+, we define $A' \vDash \psi(s(\bar{a})^\sim, t(\bar{b})^\sim)$ iff $B \vDash R_\phi(\bar{a}, \bar{b})$, where ϕ is $\psi(s(\bar{x}), t(\bar{y}))$; again Σ guarantees that this definition is sound. This defines A'. We take A to be the L^+-structure $A'|L^+$.

Let X be the set of elements $b^\sim$ with b an element of B. Suppose $t(\bar{x})$ is a term of L^+ and $\bar{b}$ is a tuple of elements of B. We claim that $t^{A'}(\bar{b}^\sim) = t(\bar{b})^\sim$. For let $\psi(\bar{x}, y)$ be the formula $t(\bar{x}) = y$ and let ϕ be the formula $t(\bar{x}) = t(\bar{y})$ (which is $\psi(\bar{x}, t(\bar{y}))$). Then

$$A' \vDash t^{A'}(\bar{b}^\sim) = t(\bar{b})^\sim \Leftrightarrow A' \vDash \psi(\bar{b}^\sim, t(\bar{b})^\sim) \Leftrightarrow B \vDash R_\phi(\bar{b}, \bar{b});$$

but Σ must contain the sentence $\forall \bar{x}\, R_\phi(\bar{x}, \bar{x})$. This proves the claim. It follows that X generates A' as an L^+-structure.

If $\bar{b}$ are elements of B and ϕ is an atomic formula of L^+, then $A' \vDash R_\phi(\bar{b}^\sim) \Leftrightarrow B \vDash R_\phi(\bar{b}) \Leftrightarrow A \vDash \phi(\bar{b}) \Leftrightarrow A^0 \vDash R_\phi(\bar{b}^\sim)$. This shows that A' is in fact A^0. The map $b \mapsto b^\sim$ is an isomorphism from B to $A \downarrow X$.

$\square$ Lemma 5.2.2

If B is a model of Σ, we write ΓB for the L^+-structure A of Lemma 5.2.2. Thus $(\Gamma B) \downarrow (\text{dom } B) = B$.

Lemma 5.2.3. *In the situation above, let U be an $\forall_1$ theory in L^+. Then there is a theory Φ in L^- such that for every model B of Σ, $\Gamma B \vDash U$ iff $B \vDash \Phi$.*

Proof. We shall prove an analogue of the relativisation theorem, Theorem 5.1.1. More precisely, if $\phi(y_0, \ldots, y_{k-1})$ is any $\forall_1$ formula of L^+ and $t_0, \ldots, t_{k-1}$ are terms of L^+, we shall find a formula $\phi_\Gamma^{t_0 \cdots t_{k-1}}(\bar{x}_0, \ldots, \bar{x}_{k-1})$ of $L_{\infty\omega}^-$ (in fact a conjunction of formulas of L^-) such that for all tuples $\bar{b}_0, \ldots, \bar{b}_{k-1}$ of elements of B,

$$(2.3) \quad \Gamma B \vDash \phi(t_0(\bar{b}_0)^\sim, \ldots, t_{k-1}(\bar{b}_{k-1})^\sim) \Leftrightarrow B \vDash \phi_\Gamma^{t_0 \cdots t_{k-1}}(\bar{b}_0, \ldots, \bar{b}_{k-1}).$$

The lemma follows by taking Φ to be the set of all sentences θ_i such that for some $\phi \in U$, $\phi_\Gamma = \bigwedge_{i \in I} \theta_i$.

We go by induction on the complexity of ϕ. Suppose $t_0, \ldots, t_{k-1}$ are terms and ϕ is atomic. Then for all $\bar{b}_0, \ldots, \bar{b}_{k-1}$ from B we have

$$(2.4) \quad \Gamma B \vDash \phi(t_0(\bar{b}_0)^\sim, \ldots, t_{k-1}(\bar{b}_{k-1})^\sim) \Leftrightarrow B \vDash R_{\phi'}(\bar{b}_0, \ldots, \bar{b}_{k-1})$$

for some suitable $R_{\phi'}$ as in the proof of the previous lemma.

There is no problem with boolean combinations of atomic formulas.

Suppose finally that ϕ is of the form

$$\forall z_0 \ldots z_{m-1}\, \psi(y_0, \ldots, y_{k-1}, z_0, \ldots, z_{m-1})$$

where ψ is quantifier-free. Then we take $\phi_\Gamma^{\bar{t}}(\bar{x}_0, \ldots, \bar{x}_{k-1})$ to be the conjunction of all the formulas

$$\forall \bar{z}_0 \ldots \bar{z}_{m-1}(\psi_\Gamma^{\bar{t}, s_0, \ldots, s_{m-1}}(\bar{x}_0, \ldots, \bar{x}_{k-1}, \bar{z}_0, \ldots, \bar{z}_{m-1}))$$

as $s_0(\bar{z}_0), \ldots, s_{m-1}(\bar{z}_{m-1})$ range through all terms of L^+.   □ Lemma 5.2.3

Now we finish the proof.

By skolemising (Theorem 3.1.2) we can assume without loss that U is an $\forall_1$ theory. Expand L^+ to L^0 as above. In the notation of the two lemmas, let L^* be the first-order language with signature $L^- \cup L$, and let U^* be the L^*-theory consisting of Σ, Φ and all the sentences

$$(2.5) \quad \forall \bar{x}(\phi(\bar{x}) \leftrightarrow R_\phi(\bar{x})) \quad \text{where } \phi \text{ is an atomic formula of } L,$$

$$(2.6) \quad \forall \bar{u}(R_{P(t(\bar{x}))}(\bar{u}) \to \exists v\, R_{t(\bar{x})=y}(\bar{u}, v)) \quad \text{where } t(\bar{x}) \text{ is any term of } L.$$

We shall show that **K** is the class of those L-structures which can be expanded to models of U^*.

If C is in **K** then C is A_P for some model A of U. Form A^0, and expand C to an L^*-structure C^* by adding the relations R_ϕ as in $A \downarrow P^A$. This latter structure is a model of $\Sigma \cup \Phi$ in any case, by the two lemmas. Also for each atomic formula $\phi(\bar{x})$ of L we have $C \vDash \phi(\bar{b}) \Leftrightarrow A^0 \vDash \phi(\bar{b}) \Leftrightarrow C^* \vDash R_\phi(\bar{b})$, so that C^* is a model of (2.5). Finally C^* is a model of (2.6) since every element of P^A is in C^*.

Conversely suppose C is $C^*|L$ for some model C^* of U^*. Since C^* is a model of $\Sigma \cup \Phi$, we can form $A^0 = \Gamma(C^*|L^-)$, and A^0 is a model of U; we put $A = A^0|L^+$. Then A is also a model of U, and it follows that A_P is in **K**. Every element of C^* satisfies $R_{P(x)}$, and so it lies in P^A. In the other direction, suppose a is an element of P^A. Then a is of the form $t^A(\bar{b})$ for some elements $\bar{b}$ of C. It follows by (2.5) and (2.6) that $t^A(\bar{b})$ is equal to some element of C; so a lies in C. Thus A_P is exactly C, and so C is in **K**.

$\square$ Theorem 5.2.1

Section 6.6 below has more to say about pseudo-elementary classes.

Exercises for section 5.2

1. Show that the downward Löwenheim–Skolem theorem holds for PC_Δ classes in the following sense: if $L \subseteq L^+$, U is a theory in L^+ and **K** is the class of all L-reducts of models of U, then for every structure A in **K** and every set X of elements of A, there is an elementary substructure of A of cardinality $\leqslant |X| + |L^+|$ which contains all the elements of X.

2. In Example 2, show that each ordering ζ^α is isomorphic to the reverse ordering $(\zeta^\alpha)^*$.

3. Show that the class of multiplicative groups of real-closed fields is first-order axiomatisable.

4. Show that in Theorem 5.2.1(b) the condition that all structures in **K** are infinite can't be dropped. [For a PC class **K** the set $\{|A|: A \in \mathbf{K}, |A| < \omega\}$ is primitive recursive.]

*5. A ring R is said to be **(right) primitive** if there exists a faithful irreducible right R-module. Show that the class of right primitive rings is a PC class. [To make it a PC' class is straightforward; see Theorem 2.1.1 in Herstein [1968] for a way of reaching PC without going through Theorem 5.2.1.]

*6. Construct a group which has no elements of finite order but is not orderable. [A useful step is Hölder's theorem: if G is an ordered group with no non-trivial proper subgroup which forms a convex subset of the ordering, then G is isomorphic to a subgroup of the additive group of real numbers, and hence is abelian.]

5.3 Interpreting one structure in another

We begin by interpreting one structure in another. Later in this chapter we shall interpret one theory in another.

Let K and L be signatures, A a K-structure and B an L-structure, and n a positive integer. An (n-dimensional) **interpretation** Γ **of B in** A is defined to consist of three items,

(3.1)   a formula $\partial_\Gamma(x_0, \ldots, x_{n-1})$ of signature K,

(3.2)   for each unnested atomic formula $\phi(y_0, \ldots, y_{m-1})$ of L, a formula $\phi_\Gamma(\bar{x}_0, \ldots, \bar{x}_{m-1})$ of signature K in which the $\bar{x}_i$ are disjoint n-tuples of distinct variables,

(3.3)   a surjective map $f_\Gamma\colon \partial_\Gamma(A^n) \to \mathrm{dom}\,(B)$,

such that for all unnested atomic formulas ϕ of L and all $\bar{a}_i \in \partial_\Gamma(A^n)$,

(3.4) $B \vDash \phi(f_\Gamma\bar{a}_0, \ldots, f_\Gamma\bar{a}_{m-1}) \Leftrightarrow A \vDash \phi_\Gamma(\bar{a}_0, \ldots, \bar{a}_{m-1})$.

The formula ∂_Γ is the **domain formula** of Γ; the formulas ∂_Γ and ϕ_Γ (for all unnested atomic ϕ) are the **defining formulas** of Γ. The map f_Γ is the **coordinate map** of Γ. It assigns to each element $f_\Gamma\bar{a}$ of B the 'coordinates' $\bar{a}$ in A; in general one element may have several different tuples of coordinates.

Unless anything is said to the contrary, we assume that the defining formulas of Γ are all first-order. For example we say that B is **interpretable** in A if there is an interpretation of B in A with all its defining formulas first-order. We say that B is **interpretable in** A **with parameters** if there is a sequence $\bar{a}$ of elements of A such that B is interpretable in $(A, \bar{a})$.

We shall write $=_\Gamma$ for ϕ_Γ when ϕ is the formula $y_0 = y_1$. Wherever possible we shall abbreviate $(\bar{a}_0, \ldots, \bar{a}_{m-1})$ and $(f\bar{a}_0, \ldots, f\bar{a}_{m-1})$ to $\bar{a}$ and $f\bar{a}$ respectively.

Example 1: *Relativised reductions.* Using the notation of section 5.1, suppose B is the relativised reduct A_P. Then there is a one-dimensional interpretation Γ of B in A as follows.

$\partial_\Gamma(x) := Px$.

$\phi_\Gamma := \phi(\bar{x})$, for each unnested atomic formula $\phi(\bar{y})$.

The coordinate map $f_\Gamma\colon P^A \to \mathrm{dom}\,(A)$ is simply the inclusion map. We call the interpretation Γ a **relativised reduction**.

There is a close relationship between interpretations and A^{eq} (see section 4.3). In fact some writers invoke the following theorem as an excuse for working with A^{eq} instead of interpretations; it's purely a matter of taste.

Theorem 5.3.1. *Let A be a K-structure with more than one element, and B an L-structure. Then B is interpretable in A if and only if B is isomorphic to a relativised reduct of a definitional expansion of a finite slice of A^{eq}.*

Proof. Suppose first that there is an n-dimensional interpretation Γ of B in A. Let θ be the formula $\partial_\Gamma(\bar{x}_1) \wedge \partial_\Gamma(\bar{x}_2) \wedge =_\Gamma(\bar{x}_1, \bar{x}_2)$. Then θ is an equivalence formula, so that I_θ is a sort of A^{eq}. The imaginary elements $\bar{a}/\theta$ are in one-one correspondence with the elements $f_\Gamma(\bar{a})$ ($\bar{a} \in \partial_\Gamma(A^n)$) of B. I leave it to the reader to check that the relations of B correspond to $\varnothing$-definable relations on A^{eq}, using the symbol R_θ.

In the other direction, suppose B is a relativised reduct A^*_χ where A^* is a definitional extension of a finite slice of A^{eq}. By time-sharing (Exercise 2.1.6, using two or more elements of A) we can suppose that the domain of B lies inside some one set I_θ where $\theta(\bar{x}_1, \bar{x}_2)$ is an n-ary equivalence formula. Then the required interpretation Γ has

$$\partial_\Gamma(\bar{x}) := P^*(\bar{x}),$$

$$=_\Gamma := \theta^*$$

where $P^*(\bar{x})$ and θ^* are formulas of K^{eq} which are equivalent to $P(\bar{x})$ and θ; and likewise with the remaining symbols of Γ.   □

Example 2: *Rationals and integers*. The familiar interpretation of the rationals in the integers is a two-dimensional interpretation Γ, as follows.

$$\partial_\Gamma(x_0, x_1) := x_1 \neq 0,$$

$$=_\Gamma(x_{00}, x_{01}; x_{10}, x_{11}) := x_{00} \cdot x_{11} = x_{01} \cdot x_{10},$$

$$plus_\Gamma(x_{00}, x_{01}; x_{10}, x_{11}; x_{20}, x_{21})$$
$$:= x_{21} \cdot (x_{00} \cdot x_{11} + x_{01} \cdot x_{10}) = x_{01} \cdot x_{11} \cdot x_{20}.$$

$$times_\Gamma(x_{00}, x_{01}; x_{10}, x_{11}; x_{20}, x_{21})$$
$$:= x_{00} \cdot x_{10} \cdot x_{21} = x_{01} \cdot x_{11} \cdot x_{20}.$$

The coordinate map takes each pair (m, n) with $n \neq 0$ to the rational number m/n. The formulas ψ_Γ for the remaining unnested atomic formulas ψ express addition and multiplication of rationals in terms of addition and multiplication of integers, just as in the algebra texts.

Example 3: *Algebraic extensions*. Let A be a field, $p(X)$ an irreducible polynomial of degree n over A and ξ a root of $p(X)$ in some field extending A. Then there is an n-dimensional interpretation Γ of $A[\xi]$ in A. If $\bar{a} = (a_0, \ldots, a_{n-1})$ is an n-tuple of elements of A, write $q_{\bar{a}}(X)$ for the polynomial $X^n + a_{n-1}X^{n-1} + \ldots + a_1 X + a_0$. Then $\partial_\Gamma(A^n)$ is the whole of A^n and $f_\Gamma(\bar{a})$ is $q_{\bar{a}}(\xi)$. The formula $=_\Gamma(\bar{a}, \bar{b})$ will say that $p(X)$ divides $(q_{\bar{a}}(X) - q_{\bar{b}}(X))$; I leave it to the reader to verify that $=_\Gamma$ can be written as a p.p. formula. The formulas $(y_0 + y_1 = y_2)_\Gamma$ and $(y_0 \cdot y_1 = y_2)_\Gamma$ follow the usual definitions of addition and multiplication of polynomials, and again these give us p.p. formulas. In fact ∂_Γ is quantifier-free and ϕ_Γ is p.p. for every unnested atomic ϕ. Just as in Example 2, we are looking at a familiar algebraic object from a slightly peculiar angle.

If Γ is an interpretation of an L-structure B in a K-structure A, then there are certain sentences of signature K which must be true in A just because Γ is an interpretation, regardless of what A and B are. These sentences say
(i) $=_\Gamma$ defines an equivalence relation on $\partial_\Gamma(A^n)$,
(ii) for each unnested atomic formula ϕ of L, if $A \vDash \phi_\Gamma(\bar{a}_0, \ldots, \bar{a}_{n-1})$ with $\bar{a}_0, \ldots, \bar{a}_{n-1}$ in $\partial_\Gamma(A^n)$, then also $A \vDash \phi_\Gamma(\bar{b}_0, \ldots, \bar{b}_{n-1})$ when each $\bar{b}_i$ is an element of $\partial_\Gamma(A^n)$ which is $=_\Gamma$-equivalent to $\bar{a}_i$,
(iii) if $\phi(y_0)$ is a formula of L of form $c = y_0$, then there is $\bar{a}$ in $\partial_\Gamma(A^n)$ such that for all $\bar{b}$ in $\partial_\Gamma(A^n)$, $A \vDash \phi_\Gamma(\bar{b})$ if and only if $\bar{b}$ is $=_\Gamma$-equivalent to $\bar{a}$,
(iv) a clause like (iii) for each function symbol.
These first-order sentences are called the **admissibility conditions** of Γ. They generalise the admissibility conditions for a relativised reduct. Note that they depend only on parts (3.1) and (3.2) of Γ, and not on the coordinate map.

Suppose Γ interprets B in A. Then in a sense, A knows everything there is to know about B, and so we can answer questions about B by reducing them to questions about A. The next theorem develops this idea. Bearing in mind Example 1, this theorem is best seen as a refinement of the relativisation theorem (Theorem 5.1.1).

Theorem 5.3.2 (Reduction theorem). *Let A be a K-structure, B an L-structure and Γ an n-dimensional interpretation of B in A. Then for every formula $\phi(\bar{y})$ of the language $L_{\infty\omega}$ there is a formula $\phi_\Gamma(\bar{x})$ of the language $K_{\infty\omega}$ such that for all $\bar{a}$ from $\partial_\Gamma(A^n)$,*

$$(3.5) \qquad\qquad B \vDash \phi(f_\Gamma\bar{a}) \Leftrightarrow A \vDash \phi_\Gamma(\bar{a}).$$

(Recall our notational conventions. Since Γ is n-dimensional, $\bar{a}$ will be a tuple of n-tuples.)

Proof. By Corollary 2.6.2, every formula of $L_{\infty\omega}$ is equivalent to a formula of $L_{\infty\omega}$ in which all atomic subformulas are unnested. So we can prove the theorem by induction on the complexity of formulas, and clause (3.4) in the definition of interpretations already takes care of the atomic formulas. For compound formulas we define

$$(3.6) \qquad (\neg\phi)_\Gamma = \neg(\phi_\Gamma),$$

$$(3.7) \qquad (\bigwedge_{i\in I} \phi_i)_\Gamma = \bigwedge_{i\in I} (\phi_i)_\Gamma, \quad \text{and likewise with } \bigvee \text{ for } \bigwedge,$$

$$(3.8) \qquad (\forall y\, \phi)_\Gamma = \forall x_0 \ldots x_{n-1}(\partial_\Gamma(x_0, \ldots, x_{n-1}) \to \phi_\Gamma),$$

$$(3.9) \qquad (\exists y\, \phi)_\Gamma = \exists x_0 \ldots x_{n-1}(\partial_\Gamma(x_0, \ldots, x_{n-1}) \wedge \phi_\Gamma). \qquad \square$$

This fundamental but trivial theorem calls for several remarks.

Remark 1. The map $\phi \mapsto \phi_\Gamma$ of the theorem depends only on parts (3.1) and (3.2) of Γ, and not at all on the coordinate map f_Γ. We shall exploit this fact through the rest of the chapter. Parts (3.1) and (3.2) of the definition of Γ form an **interpretation of L in K**, and the map $\phi \mapsto \phi_\Gamma$ of the reduction theorem is the **reduction map** of this interpretation.

Remark 2. We have been rather careless about variables. For example if ϕ in the theorem is $\phi(z)$, what are the variables of ϕ_Γ? I avoid answering this question. For purposes of the reduction theorem, all formulas of $L_{\infty\omega}$ are of form $\phi(y_0, y_1, \ldots)$ and all formulas of $K_{\infty\omega}$ are correspondingly of form $\psi(x_{00}, \ldots, x_{0,n-1}; x_{10}, \ldots, x_{1,n-1}; \ldots)$. Also I assume tacitly that if ϕ and θ are the same unnested atomic formula of $L_{\infty\omega}$ up to permutation of variables, then ϕ_Γ and θ_Γ are the same formula up to a corresponding permutation of their free variables.

Remark 3. If ∂_Γ and the formulas ϕ_Γ (for unnested atomic ϕ) are $\exists_1^+$ first-order formulas, then for every $\exists_1^+$ first-order formula ψ of L, ψ_Γ is also $\exists_1^+$ first-order. This is true even when ψ contains nested atomic formulas, since the removal of nesting introduces existential quantifiers at worst (see Theorem 2.6.1). More generally but more vaguely, if we have any reasonable measure of the complexity of formulas, then ϕ_Γ will be only a bounded amount more complex than ϕ.

Remark 4. If L is a recursive language (see section 2.3 above) and the map $\phi \mapsto \phi_\Gamma$ in (3.2) is recursive, then we call Γ a **recursive interpretation**. For a recursive interpretation the reduction map (restricted to first-order formulas) is recursive too. With inessential changes in the proof of Theorem 5.3.1 we can arrange that the reduction map is also 1–1 on first-order formulas. Recursion theorists will deduce that if Γ is a recursive interpretation of B in A, then Th(B) is 1–1 reducible to Th(A). (See Rogers [1967] p. 80 for 1–1 reducibility.)

Remark 5. Let A be a K-structure and B an L-structure. One often meets an interpretation Γ of L in K and a first-order formula $\psi(\bar{x})$ of K such that $A \vDash \exists \bar{x} \, \psi$, and for every tuple $\bar{a}$ with $A \vDash \psi(\bar{a})$, Γ interprets B in $(A, \bar{a})$. In this case we shall say that Γ is an **interpretation of B in A with definable parameters**. Now the reduction theorem tells us that if ϕ is any *sentence* of L and $\bar{a}$ is any tuple in A such that $A \vDash \psi(\bar{a})$, then $B \vDash \phi$ if and only if $A \vDash \phi_\Gamma(\bar{a})$. So $B \vDash \phi(\bar{a})$ if and only if $A \vDash \forall \bar{x}(\psi \rightarrow \phi_\Gamma)$. Thus if Γ is recursive, then Th(B) is again 1–1 reducible to Th(A). For examples of this phenomenon, see Example 1 in section 5.4 and Exercise 5.5.6.

The associated functor

Theorem 5.3.3. *Let Γ be an n-dimensional interpretation of a signature L in a signature K, and let* Admis(Γ) *be the set of admissibility conditions of Γ. Then for every K-structure A which is a model of* Admis(Γ), *there are an L-structure B and a map $f: \partial_\Gamma(A^n) \to \text{dom}(B)$ such that*
(a) *Γ with f forms an interpretation of B in A, and*
(b) *if g and C are such that Γ and g form an interpretation of C in A, then there is an isomorphism $i: B \to C$ such that $i(f\bar{a}) = g(\bar{a})$ for all $\bar{a} \in \partial_\Gamma(A^n)$.*

Proof. Let A be a model of Γ. Then we build an L-structure B as follows. Define a relation $\sim$ on $\partial_\Gamma(A^n)$ by

(3.10) $\bar{a} \sim \bar{a}'$ iff $A \vDash =_\Gamma(\bar{a}, \bar{a}')$.

By (i) of the admissibility conditions, $\sim$ is an equivalence relation. Write $\bar{a}^\sim$ for the equivalence class of $\bar{a}$. The domain of B will be the set of all equivalence classes $\bar{a}^\sim$ with $\bar{a}$ in $\partial_\Gamma(A^n)$. (Readers who don't allow empty structures should add $\exists \bar{x}\, \partial_\Gamma(\bar{x})$ to the admissibility conditions, here and henceforth.)

For every relation symbol R of L, we define the relation R^B by

(3.11) $(\bar{a}_0^\sim, \ldots, \bar{a}_{m-1}^\sim) \in R^B$ iff $A \vDash \phi_\Gamma(\bar{a}_0, \ldots, \bar{a}_{m-1})$
         where $\phi(y_0, \ldots, y_{m-1})$ is $Ry_0 \ldots y_{m-1}$.

By (ii) of the admissibility conditions, this is a sound definition. The definitions of c^B and F^B are similar, relying on (iii) and (iv) of the admissibility conditions. This defines the L-structure B. We define $f: \partial_\Gamma(A^n) \to \text{dom}(B)$ by $f\bar{a} = \bar{a}^\sim$. Then f is surjective, and B has been defined so as to make (3.4) true. Hence Γ and f are an interpretation of B in A. This proves (a).

To prove (b), suppose Γ and g are an interpretation of C in A. For each tuple $\bar{a} \in \partial_\Gamma(A^n)$, define $i(f\bar{a})$ to be $g\bar{a}$. We claim this is a sound definition of an isomorphism $i: B \to C$. If $f\bar{a} = f\bar{a}'$ then $A \vDash =_\Gamma(\bar{a}, \bar{a}')$ and hence $g\bar{a} = g\bar{a}'$ by (3.4); thus the definition of i is sound. A similar argument in the other direction shows that i is injective; moreover i is surjective since g is surjective by (3.3). Then (3.4) for f and g shows that i is an embedding. This proves the claim, and with it the theorem. $\square$

We write ΓA for the structure B of the theorem. The reduction theorem (Theorem 5.3.2) applies to ΓA as follows:

(3.12)   for all formulas $\phi(\bar{y})$ of L, all K-structures A satisfying the admissibility conditions of Γ, and all tuples $\bar{a} \in \partial_\Gamma(A^n)$,
         $\Gamma A \vDash \phi(\bar{a}^\sim) \Leftrightarrow A \vDash \phi_\Gamma(\bar{a})$.

Let Γ be as in the theorem, let A and A' be models of the admissibility conditions of Γ, and let $e: A \to A'$ be an elementary embedding. Then for every tuple $\bar{a} \in \partial_\Gamma(A^n)$, $e\bar{a}$ is in $\partial_\Gamma(A'^n)$. Moreover if $\bar{c}$ is another tuple in $\partial_\Gamma(A^n)$ and $A \models =_\Gamma(\bar{a}, \bar{c})$, then $A' \models =_\Gamma(e\bar{a}, e\bar{c})$. It follows that there is a well-defined map Γe from $\mathrm{dom}(\Gamma A)$ to $\mathrm{dom}(\Gamma A')$, defined by $(\Gamma e)(\bar{a}^\sim) = (e\bar{a})^\sim$. One easily verifies that $\Gamma(1_A)$ is the map $1_{\Gamma A}$. Also if $e_1: A \to A'$ and $e_2: A' \to A''$ are elementary embeddings, then $\Gamma(e_2 e_1) = (\Gamma e_2)(\Gamma e_1)$.

Furthermore, Γe is an elementary embedding of ΓA into $\Gamma A'$. For let $\bar{a}$ be a sequence of tuples from $\partial_\Gamma(A^n)$ and ϕ a formula of L. Then we have

(3.13) $\Gamma A \models \phi(\bar{a}^\sim) \Leftrightarrow A \models \phi_\Gamma(\bar{a}) \Rightarrow A' \models \phi_\Gamma(e\bar{a}) \Leftrightarrow \Gamma A' \models \phi((\Gamma e)\bar{a}^\sim).$

Here the left-hand equivalence is by the reduction theorem ((3.12) above), the central implication is because e is elementary, and the right-hand equivalence is by the reduction theorem and the definition of Γe.

In fact the definition of Γe makes sense whenever A, A' are models of the admissibility conditions of Γ and $e: A \to A'$ is any homomorphism which preserves the formulas ∂_Γ and $=_\Gamma$. If e also preserves all the formulas ϕ_Γ for unnested atomic formulas ϕ of L, then (3.13) is good for these formulas ϕ too, and so Γe is a homomorphism from ΓA to $\Gamma A'$. To sum up, we have the following.

Theorem 5.3.4. *Let Γ be an interpretation of a signature L in a signature K, with admissibility conditions $\mathrm{Admis}(\Gamma)$.*

(a) Γ induces a functor, written $\mathrm{Func}(\Gamma)$, from the category of models of $\mathrm{Admis}(\Gamma)$ and elementary embeddings, to the category of L-structures and elementary embeddings.

(b) If the formulas ∂_Γ and ϕ_Γ (for unnested atomic ϕ) are $\exists_1^+$ formulas, then we can extend the functor $\mathrm{Func}(\Gamma)$ in (a), replacing 'elementary embeddings' by 'homomorphisms'. $\qquad\square$

We call the functor $\mathrm{Func}(\Gamma)$ of Theorem 5.3.4, in either the (a) or the (b) version, the **associated functor** of the interpretation Γ. Usually we shall write it just Γ; there is little danger of confusing the interpretation with the functor.

Suppose Γ is the associated functor of an interpretation of L in K. Then whenever ΓA is defined, we have a group homomorphism $\alpha \mapsto \Gamma\alpha$ from $\mathrm{Aut}(A)$ to $\mathrm{Aut}(\Gamma A)$. What can be said about this homomorphism?

Theorem 5.3.5. *Let Γ be an interpretation of L in K, and let A be an L-structure such that ΓA is defined. Then the induced homomorphism h: $\mathrm{Aut}(A) \to \mathrm{Aut}(B)$ is continuous.*

Proof. It suffices to show that if F is a basic open subgroup of $\mathrm{Aut}(B)$, then there is an open subgroup E of $\mathrm{Aut}(A)$ such that $h(E) \subseteq F$. Let F be $\mathrm{Aut}(B)_{(\bar{b})}$ for some tuple $\bar{b}$ of elements of B. Let X be a finite set of elements of A such that each element in $\bar{b}$ is of form $f_\Gamma(\bar{a})$ for some tuple $\bar{a}$ of elements of X. Then by the definition of h, $h(\mathrm{Aut}(A)_{(X)}) \subseteq \mathrm{Aut}(B)_{(\bar{b})}$. $\square$

Suppose K, L and M are signatures, Γ is an interpretation of L in K and Δ is an interpretation of M in L. Then we can define an interpretation Ξ of M in K as follows:

(3.14) $\quad \partial_\Xi := (\partial_\Delta)_\Gamma,$

$\qquad \phi_\Xi := (\phi_\Delta)_\Gamma$ for each unnested atomic formula ϕ of M.

We call this interpretation the **composite** of Γ and Δ, in symbols $\Delta \circ \Gamma$. Writing $\mathrm{Func}(\Gamma)$ for the associated functor of Γ, one can check that $\mathrm{Func}(\Delta \circ \Gamma)$ is naturally isomorphic to the composite functor

$$\mathrm{Func}(\Delta) \circ \mathrm{Func}(\Gamma).$$

Finally note that if K is any signature, then there is an interpretation Γ of K in K which does nothing: ϕ_Γ is ϕ and ΓA is A. We call Γ the **identity interpretation** on K.

Exercises for section 5.3

1. Show that if Γ is an n-dimensional interpretation of L in K, then for every K-structure A for which ΓA is defined, $|\Gamma A| \leqslant |A|^n$.

2. Let A, B and C be structures. Show that if B is interpretable in A and C is interpretable in B, then C is interpretable in A.

3. Let Γ be an interpretation of L in K. Let $(A_i : i < \alpha)$ be an elementary chain of K-structures such that ΓA_i is defined for each $i < \alpha$. Show that $\Gamma(\bigcup_{i<\alpha} A_i)$ is defined and equal to the direct limit of the directed diagram of structures ΓA_i $(i < \alpha)$ and elementary embeddings $\Gamma e_{ij} : \Gamma A_i \to \Gamma A_j$ induced by the inclusions $e_{ij} : A_i \to A_j$.

4. Write down an interpretation Γ such that for every abelian group A, ΓA is the group $A/5A$. By applying Γ to the inclusion $\mathbb{Z} \to \mathbb{Q}$, show that Theorem 5.3.4(b) fails if we replace $\exists_1^+$ by $\exists_1$ and 'homomorphisms' by 'embeddings'. (But see Exercise 5.4.3.)

5. Let A be an L-structure with at least two elements. Show that the direct sum $A + A$ (see Exercise 3.2.8) is interpretable in A.

6. Let G be a group and A a normal abelian subgroup of G such that G/A is finite. Show that G is interpretable in A with parameters. [The cosets of A can be written as $h_1 A + \ldots + h_n A$. Interpret in $A + \ldots + A$ (n summands).]

*7. Let R be a ring, $\mathbf{K}$ the class of right R-modules and M a finitely presented left R-module. Show that there is an interpretation Γ such that for each module A in $\mathbf{K}$, ΓA is the $\mathbb{Z}$-module $A \otimes_{\mathbb{Z}} M$. [Suppose $0 \to N \to R^n \to M \to 0$ is an exact sequence of left R-modules, where N has generators $(r_{i1}, \ldots, r_{in})$, $i < m$. Then $A \otimes_{\mathbb{Z}} R^n$ is M^n with typical element $(a_1, \ldots, a_n) = a_1 \otimes 1 + \ldots + a_n \otimes 1$. Factor out the subgroup of M^n consisting of elements of form $\Sigma_{i < m}(a_i \otimes r_{i1} + \ldots + a_i \otimes r_{in})$.]

*8. Let R, S be rings and T a functor from the category Mod_R of right R-modules to the category Mod_S of right S-modules. Show that if T is a category equivalence (or even a retraction; see Cohn [1976] pp. 31ff) then T is the associated functor of an $\exists_1^+$ interpretation of Mod_R in Mod_S. [Use the previous exercise and the Morita theorems.]

*Suppose the languages K and L have no relation symbols. A **polynomial interpretation** of L in K is a map Δ which assigns to each constant c of L a closed term c_Δ of K, and to each m-ary function symbol F of L a term $F_\Delta(x_0, \ldots, x_{m-1})$ of K.*

9. Show how a polynomial interpretation Δ of L in K induces an interpretation Γ of L in K, in which for every equation ϕ, ϕ_Γ is also an equation. Show that for every K-structure A, ΓA exists and is a reduct of a definitional extension of A. (We write ΔA for ΓA.)

10. Write down a polynomial interpretation Δ such that for every ring A, ΔA is the Lie ring of A.

11. Let A be a field and n a positive integer. (a) Let R be the set of all n-tuples $\bar{a} = (a_0, \ldots, a_{n-1})$ of elements of A such that the polynomial $X^n + a_{n-1}X^{n-1} + \ldots + a_1 X + a_0$ is irreducible over A, and if α is a root of this polynomial then the field $A[\alpha]$ is Galois over A. Show that R is $\varnothing$-definable over A. [Split 'Galois' into 'normal' and 'separable'. Note that every element of $A[\alpha]$ can be regarded as an equivalence class of n-tuples of elements of A, by an equivalence relation which is definable in terms of $\bar{a}$.] (b) Let G be a finite group. Let R_G be the set of all n-tuples as above, such that the Galois group $A[\alpha]/A$ is isomorphic to G. Show that R_G is also $\varnothing$-definable over A.

*12. If E is an elliptic curve over an algebraically closed field A, then E carries the structure of an abelian group G. (See Olson [1973].) Show that this group is interpretable in A with parameters.

5.4 Shapes and sizes of interpretations

Alfred Tarski (Tarski, Mostowski & Robinson [1953]) used interpretations as a device for proving the undecidability of theories – we shall see a few examples in the next section. He took some steps towards cataloguing the different types of interpretation. Unfortunately his classification is no longer helpful: it's too closely based on the syntactic maps $\phi \mapsto \phi_\Gamma$, and it leaves out many of the interpretations which are in common use today.

In fact there is no systematic terminology for describing interpretations. Here I shall make some proposals towards one. At the same time I take the opportunity to mention some of the purposes that interpretations have been used for.

As a general point, *an interpretation of a signature L in a signature K maps K-structures to L-structures*. So it **preserves** a property π of structures if for every K-structure A with π, ΓA (when defined) also has π. It **preserves** a property σ of complete theories if whenever the theory $\text{Th}(A)$ of a K-structure A has σ, then $\text{Th}(\Gamma A)$ (when defined) also has σ. It is not clear what would be meant by saying that an interpretation preserves some property of incomplete theories.

(a) Syntactic classifications

Let us assume throughout that Γ is an interpretation of the signature L in the signature K. Some kinds of classification come for free. For example we can say that Γ is a **first-order** interpretation if its defining formulas are first-order, a **quantifier-free** interpretation if its defining formulas are quantifier-free, and so on. In the literature and in this book, interpretations are usually assumed to be first-order unless something is said to the contrary.

Because of his interest in undecidability proofs, Tarski (in Tarski, Mostowski & Robinson [1953]) required that the map $\phi \mapsto \phi_\Gamma$ on unnested atomic formulas is recursive. This requirement must be dropped in general; in the previous section we have already agreed to say that Γ is **recursive** when it does hold.

We have also classified interpretations by their dimension. This classification is unproblematic and in common use. Tarski's interpretations were all one-dimensional. Higher dimensions appear in Mal'tsev [1960] – though of course the interpretation of the complex numbers in the reals is two-dimensional and about two hundred years old.

In Tarski's interpretations, ∂_Γ^A was normally the whole domain of A; when it was a defined subset, Tarski spoke of 'relative interpretability'. Since the relativised case is the usual one, it needs no special name. When ∂_Γ^A is defined as $\bar{x} = \bar{x}$, I shall say that the interpretation is **unrelativised**.

In Tarski's interpretations the equivalence relation $=_\Gamma$ is identity; in other words, the coordinate map is always injective. The more general case appeared in Raphael Robinson [1951] and is common currency today. I shall say that Γ is **injective** if the coordinate map is always injective. (There is never any need to describe the functor Γ as injective, so the terminology is unambiguous.)

The next result is in constant use for studying structures interpretable in an algebraically closed field (bearing in mind that algebraically closed fields have uniform elimination of imaginaries by Theorem 4.4.6).

Theorem 5.4.1. *Suppose A is a structure with uniform elimination of imaginaries, and B is interpretable in A. Then B is interpretable in A by an injective interpretation.*

Proof. Suppose Γ is an interpretation of B in A. Take $\theta(\bar{x}, \bar{y})$ so that each equivalence class $\bar{a}/=_\Gamma$ of $=_\Gamma$ can be written as $\theta(A^n, \bar{b})$ for a unique tuple $\bar{b}$. Rewrite the defining formulas of Γ so that each class $\bar{a}/=_\Gamma$ is replaced by the corresponding $\bar{b}$.   □

(b) How the classes of structures are related

Suppose **V** is a class of K-structures and **W** a class of L-structures. I propose that an **interpretation of W in V** is simply an interpretation of L in K. Likewise if T and U are theories of signature K and L respectively, then an **interpretation of U in T** is simply an interpretation of L in K. Tarski added the requirement that $\Gamma A \vDash U$ for every model A of T. In many applications today this requirement fails, so we must drop it.

Nevertheless conditions like Tarski's are still useful, and we need some terminology for them. If Γ is an interpretation of L in K, the **domain** of Γ is the class of all K-structures which satisfy the admissibility conditions of Γ; in other words, the class of structures A such that ΓA is defined. I shall say that Γ is **defined on** a class **V** if **V** lies within the domain of Γ.

Suppose Γ is introduced as an interpretation of **W** in **V**. I shall say that Γ is **left total** when Tarski's condition is met, i.e.

(4.1)   for every $A \in \mathbf{V}$, $\Gamma A \in \mathbf{W}$.

This implies that Γ is defined on **V**. I shall say that Γ is **right total** when the dual and equally interesting condition holds:

(4.2)   for every structure B in **W** there is A in **V** with $\Gamma A \cong B$.

I shall say that Γ is **total** when it is both right and left total. Left and right are fixed by remembering that the functor Γ goes from **V** to **W**.

When **W** and **V** are respectively the classes of models of theories T and U, then I shall speak of (left, right) total interpretations of U in T. (In this case Ershov [1974] proposes 'exact' rather than total. But 'exact' already means something quite different for functors.) We can mix the terminology too: for example a total interpretation of L in T is a total interpretation of the class of all L-structures in the class of all models of T.

Right total interpretations often appear in proofs of undecidability. Right total interpretations are useful for model theory too, because they tell us that

all structures in some class can be coded up inside structures of another kind
– then we know that we can concentrate on structures of the second kind
without losing any information. Section 5.5 below discusses all this.

I shall say that the interpretation Γ of **W** in **V** is **separating** if

(4.3)   whenever A and B are structures in **V** such that ΓA and ΓB
          are defined and isomorphic, then A and B are isomorphic.

This notion has been used by universal algebraists (e.g. Burris [1975], and see
section 9.7 below) as a way of constructing many non-isomorphic structures in
some given class. Actually this technique often uses word-constructions rather
than interpretations (see section 9.3 below), but the same terminology
applies.

(c) Homotopies and bi-interpretations

We turn to a cluster of important notions which need one extra piece of
definability and one steady head.

Let **V** and **W** be respectively classes of K-structures and of L-structures,
and let Γ, Δ be left total interpretations of **W** in **V**. Following Ahlbrandt &
Ziegler [1986], I shall say that Γ and Δ are **homotopic** if there is a formula
$\chi(\bar{x}, \bar{y})$ of K with the following property:

(4.4)   for any structure A in **V**, the relation $\{(\bar{a}, \bar{b}): A \vDash \chi(\bar{a}, \bar{b})\}$
          defines a bijection between $\mathrm{dom}(\Gamma A)$ and $\mathrm{dom}(\Delta A)$, and this
          bijection induces an isomorphism from ΓA to ΔA.

The formula χ is called a **homotopy** from Γ to Δ. (When nothing is said,
assume χ is first-order.)

In particular a **self-interpretation** is a left total interpretation of the class **V**
in itself which is homotopic to the identity interpretation.

Again following Alhbrandt & Ziegler [1986], a **bi-interpretation** between **V**
and **W** consists of a left total interpretation Γ of **W** in **V** and a left total
interpretation Δ of **V** in **W**, such that $\Delta \circ \Gamma$ is a self-interpretation of **V** and
$\Gamma \circ \Delta$ is a self-interpretation of **W**. We say that **W** and **V** are **bi-interpretable**
if there is a bi-interpretation between **W** and **V**. In particular A and B are
bi-interpretable if $\{A\}$ and $\{B\}$ are bi-interpretable.

Example 1: *Desarguesian projective planes.* A **projective plane** is a structure
$\langle P \cup \Lambda, P, \Lambda, I \rangle$ where P is the set of points, Λ is the set of lines and I is
the incidence relation between points and lines, subject to the usual laws:
(a) any two distinct points are incident with exactly one line in common; (b)
any two distinct lines are incident with exactly one point in common; (c) there
are at least four points, no three of them collinear (i.e. incident with the same
line). Let L be the signature appropriate for projective planes, and K the

signature of rings. If the K-structure A is a skew field, we can construct the projective plane over A, ΠA, as follows. Regarding A^3 as a three-dimensional vector space over A, we take the points and lines of ΠA to be respectively the one-dimensional and two-dimensional subspaces of A^3; a point p is incident with a line l iff $p \subseteq l$. Then ΠA is a desarguesian projective plane (i.e. a projective plane satisfying Desargues' axiom). The reader can verify that Π can be written as an interpretation of ΠA in A (see Exercise 10).

There is an interpretation in the other direction, but with parameters. If B is any desarguesian projective plane, we can choose four points $\bar{b} = (b_0, \ldots, b_3)$ of B, no three of them collinear. Then by any of several methods of coordinatisation (e.g. Chapter V of Hughes & Piper [1973]) we can define a skew field $\Delta(B, \bar{b})$ in terms of $(B, \bar{b})$. Again I leave it to the reader to check that Δ can be written as an interpretation. Note that Δ is an interpretation with definable parameters, in the sense of Remark 5 of section 5.3.

We adjust the interpretation Π a little. Let Γ be an interpretation of L in K such that for every skew field A, ΓA is $(\Pi A, b_0, \ldots, b_3)$ where b_0, b_1 and b_2 are respectively the X, Y and Z axes in A^3, and b_3 is the straight line through the origin and the point $(1, 1, 1)$. Let $\mathbf{V}$ be the class of skew fields and $\mathbf{W}$ the class of structures $(B, \bar{b})$ where B is a desarguesian projective plane and $\bar{b}$ is a 4-tuple of points no three of which are collinear. Then Γ and Δ are a bi-interpretation from $\mathbf{V}$ to $\mathbf{W}$. The proof of this statement runs through the definitions of Γ and Δ; it is long but uneventful.

Since interpretations preserve ω-categoricity (Theorem 7.3.8), and there are no ω-categorical fields (Theorem A.5.17), it follows that there are no ω-categorical desarguesian projective planes. It's a tantalising open problem whether there are any ω-categorical projective planes at all.

The next example illustrates two themes. First it shows how in one simple case we can reconstruct a structure from its automorphism group.

Second, it is an example of a K-structure A, an L-structure B, an interpretation Γ of L in K and an interpretation Δ of K in L, such that (a) $\Gamma A = B$, (b) Δ is a relativised reduction and (c) $\Delta \circ \Gamma$ is a self-interpretation of A. In section 12.5 below we shall describe this situation by saying that B is **finitely coordinatisable** over A. (Rubin [1989a] says instead that B is **strongly interpretable** in A.) By (b) and (c), the structure B contains an isomorphic copy A' of A, so that the graph of the coordinate map $f_\Gamma: \partial_\Gamma(A^n) \to \operatorname{dom}(B)$ can be regarded as a relation on $\operatorname{dom}(B)$. If this relation is $\varnothing$-definable in B, then (Γ, Δ) is a bi-interpretation between A and B (and in the terminology of section 12.5, B is **finitely coordinatised** over A). See Exercises 12.5.13 and 12.5.14.

Example 2: *Interpreting a set in its symmetric group.* Given an infinite set Ω and its symmetric group $\mathrm{Sym}(\Omega)$ as an abstract group, we can recover from $\mathrm{Sym}(\Omega)$ both Ω (up to a natural bijection) and the action of $\mathrm{Sym}(\Omega)$ on Ω, as follows. We suppose $\mathrm{Sym}(\Omega)$ is a K-structure where K is the first-order language with a 2-ary function symbol $\cdot$ for group multiplication. The language L for the pair $(\Omega, \mathrm{Sym}(\Omega))$ will have relation symbols *Point*, *Permutation*, *Product* and *Action* to represent respectively the set of points in Ω, the set of permutations in $\mathrm{Sym}(\Omega)$, multiplication in $\mathrm{Sym}(\Omega)$ and the action of $\mathrm{Sym}(\Omega)$ on Ω.

Writing $C(x)$ for the centraliser $\{y : x \cdot y = y \cdot x\}$ of x, let *Pair*(x) say that $x^2 = 1, x \neq 1$ and for all y, if $y^2 = 1$ and $C(x) \subseteq C(y)$ then either $x = y$ or $y = 1$. Then *Pair*(x) expresses that x is a transposition (a, b). Let *Overlap*(x, y) say that *Pair*(x) and *Pair*(y) and $x \cdot y \neq y \cdot x$; this expresses that for some distinct elements a, b and c of Ω, $x = (a, b)$ and $y = (b, c)$. Write *Equiv*(x, y, z, w) for the statement that *Overlap*(x, y) and *Overlap*(z, w), and every element v (apart from at most one) which satisifes *Overlap*(v, x) and *Overlap*(v, y) also satisfies *Overlap*(v, z) and *Overlap*(v, w). This expresses that the one point of Ω moved by both x and y is also the one point moved by both z and w. Now define the interpretation Γ by

(4.5) $\partial_\Gamma(x, y) := x = x \wedge y = y,$

   $Point_\Gamma(x, y) := Overlap(x, y),$

   $Permutation_\Gamma(x, y) := \neg\, Overlap(x, y),$

   $=_\Gamma(x_1, x_2, y_1, y_2) := Equiv(x_1, x_2, y_1, y_2)\ \vee$

      $(\neg\, Overlap(x_1, x_2) \wedge \neg\, Overlap(y_1, y_2) \wedge x_1 = y_1),$

   $Product_\Gamma(x_1, x_2, y_1, y_2, z_1, z_2) := \neg\, Overlap(x_1, x_2)\ \wedge$

      $\neg\, Overlap(y_1, y_2) \wedge \neg\, Overlap(z_1, z_2) \wedge z_1 = x_1 \cdot y_1,$

   $Action_\Gamma(x_1, x_2, y_1, y_2, z_1, z_2) := \neg\, Overlap(x_1, x_2)\ \wedge$

      $Overlap(y_1, y_2) \wedge Overlap(z_1, z_2)$

      $\wedge\ Equiv(z_1, z_2, x_1 \cdot y_1 \cdot x_1^{-1}, x_1 \cdot y_2 \cdot x_1^{-1}).$

Then for any infinite set Ω, $\Gamma(\mathrm{Sym}(\Omega))$ is a structure $B = (B_1, B_2)$ consisting of a set B_1 and a permutation group B_2 which acts on B_1. The permutation group B_2 is isomorphic to $\mathrm{Sym}(\Omega)$. (I cheated here by using different symbols $\cdot$ and *Product* for the two multiplications; but this was purely to make the formulas easier to read.) Write Δ for the relativised reduction taking (B_1, B_2) to B_2.

The interpretation $\Delta \circ \Gamma$ is a self-interpretation; in other words, if $\Gamma(\mathrm{Sym}(\Omega)) = (B_1, B_2)$, then *we can define the isomorphism between B_2 and* $\mathrm{Sym}(\Omega)$ *already within* $\mathrm{Sym}(\Omega)$. The following formula says that (y_1, y_2) is a representative of the element of B which stands for x:

(4.6) $x = y_1 \wedge \neg\,Overlap(y_1, y_2).$

In fact (Γ, Δ) is a bi-interpretation, since one can show that the coordinate map of Γ is $\varnothing$-definable in $\Gamma(\mathrm{Sym}(\Omega))$.

While we have this example in front of us, note some things that we can say about the infinite set Ω within the structure $(\Omega, \mathrm{Sym}(\Omega))$. A set X of two or more elements of Ω can be represented by a permutation which moves all and only the elements of X. An element $g \neq 1$ of $\mathrm{Sym}(\Omega)$ is cyclic if and only if every element in $\mathrm{Sym}(\Omega)$ which commutes with g either moves every element moved by g or moves no element moved by g. And lastly, Ω is countable if and only if there is a cyclic permutation which moves every element of Ω. Hence there is a first-order sentence σ in the language of $(\Omega, \mathrm{Sym}(\Omega))$, which expresses that Ω is countable. Then σ_Γ says the same thing in the language K of $\mathrm{Sym}(\Omega)$. This proves that as abstract groups, the symmetric groups on two infinite sets need not be elementarily equivalent.

Exercises for section 5.4

1. Let A be a K-structure, L a signature, U a first-order theory in L and Γ an interpretation of L in K. Show that if $\Gamma A \vDash U$ then Γ is a left total interpretation of U in $\mathrm{Th}\,(A)$.

2. For each of the following properties, show that if Γ is an interpretation of $\mathbf{V}$ in $\mathbf{U}$ and Δ is an interpretation of $\mathbf{W}$ in $\mathbf{V}$, and Γ and Δ both have the property, then $\Delta \circ \Gamma$ has the property too. (a) Injective. (b) Right total. (c) Left total. (d) Separating. (e) Interpretation with definable parameters.

3. Let Γ be a left total interpretation of the theory U in the theory T. Suppose the formula ∂_Γ is $\exists_1$, and for each unnested atomic formula ϕ the formula ϕ_Γ is equivalent modulo T to both an $\exists_1$ formula and an $\forall_1$ formula. Show that for every embedding $e\colon A \to C$ of models of T, $\Gamma e\colon \Gamma A \to \Gamma C$ is an embedding.

4. Let K be the language of linear orderings and $\mathbf{V}$ the class of linear orderings. For some positive integer n let L be the language of linear orderings together with 1-ary relation symbols $P_0, \ldots, P_{n-1}$, and $\mathbf{W}$ the class of L-structures which are linear orderings each of whose elements satisfies exactly one of $P_0(x), \ldots, P_{n-1}(x)$. Show that there is a right total interpretation of $\mathbf{W}$ in $\mathbf{V}$. [Interpret an element satisfying $P_i(x)$ by an interval $i + 1 + \omega^* + \omega$.]

5. Let (Γ, Δ) be a bi-interpretation between $\mathbf{V}$ and $\mathbf{W}$. Show (a) Γ is total from $\mathbf{W}$ to $\mathbf{V}$ and Δ is total from $\mathbf{V}$ to $\mathbf{W}$, (b) for any infinite structure A in $\mathbf{V}$, $|A| = |\Gamma A|$, (c) Γ and Δ are both separating.

6. Suppose Γ and Δ are a bi-interpretation between $\mathbf{U}$ and $\mathbf{V}$, and Θ and Ξ are a bi-interpretation between $\mathbf{V}$ and $\mathbf{W}$. Show that $\Theta \circ \Gamma$ and $\Delta \circ \Xi$ are a bi-interpretation between $\mathbf{U}$ and $\mathbf{W}$.

7. If Γ is a left total interpretation of $\mathbf{V}$ in $\mathbf{W}$, show that for each structure A in $\mathbf{V}$, Γ induces a continuous homomorphism $\eta_A^\Gamma \colon \operatorname{Aut}(A) \to \operatorname{Aut}(\Gamma A)$, with the property that if $f \colon A \to B$ is a homomorphism between structures in $\mathbf{V}$ for which Γf is defined, then $\Gamma f \circ \eta_A^\Gamma = \eta_B^\Gamma \circ f$. *Thus in category terms, η^Γ is a natural transformation from the identity functor on $\mathbf{V}$ to the associated functor of Γ. I suggest we call it the* **unit** *of Γ.*

8. (a) Show that if Γ and Δ are homotopic interpretations defined on A, and $i \colon \Gamma A \to \Delta A$ is the isomorphism defined by the homotopy, inducing the homomorphism $i' \colon \operatorname{Aut}(\Gamma A) \to \operatorname{Aut}(\Delta A)$, then $\eta_A^\Delta = i' \circ \eta_A^\Gamma$. (b) Show that if A and B are bi-interpretable by a bi-interpretation (Γ, Δ), then Γ and Δ induce mutually inverse topological isomorphisms between $\operatorname{Aut}(A)$ and $\operatorname{Aut}(B)$.

9. Let K and L be signatures which each contain at least one function symbol or relation symbol. Show that there are a K-structure A and an L-structure B such that A is interpretable in B, B is interpretable in A, but A and B are not bi-interpretable. [Use (b) of the previous exercise.]

10. In the setting of Example 1, show the following. (a) Given a skew field A, the projective plane over A is interpretable in A. [Use 6-tuples of elements of A. A non-zero tuple $(a, b, c, 0, 0, 0)$ represents the line through the origin and the point (a, b, c); suitable tuples (a, b, c, a', b', c') represent the subspace of A^3 spanned by (a, b, c) and (a', b', c').] (b) Given a desarguesian projective plane B and a 4-tuple $\bar{b}$ of points no three of which are collinear, then a skew field which coordinatises B is interpretable in $(B, \bar{b})$. (c) Γ and Δ as in Example 1 form a bi-interpretation. [In the Hughes–Piper [1973] setting, say we choose the X, Y and Z axes as the points X, Y, 0 respectively, and the line through $(1, 1, 1)$ as the point I. The coordinates are chosen as the lines in the YX plane, excluding the Y axis. Identify each element λ of A with the line through $(0, \lambda, 1)$. In the other direction, the interpretation of A in B provides each point and line with a set of coordinates which enable us to reconstruct it from A.]

11. Show that in Example 2, the interpretation Γ of $(\Omega, \operatorname{Sym}(\Omega))$ in $\operatorname{Sym}(\Omega)$ works provided Ω has at least five elements.

12. In Example 2 with Ω of cardinality $\geqslant 5$, we can represent a set of two or more points in Ω by a permutation which moves exactly those points; hence we can represent any subset of Ω by a pair of permutations. Show that the following are expressible by first-order formulas in $(\Omega, \operatorname{Sym}(\Omega))$ and hence also in $\operatorname{Sym}(\Omega)$. (a) X and Y are disjoint subsets of Ω. (b) X and Y are disjoint subsets of Ω, and the permutation f of order 2 defines an injection from X into Y. (c) X is an infinite subset of Ω. (d) X and Y form a partition of Ω into two sets of the same cardinality. (e) (d) holds, and Z is a subset of Y which has cardinality ω_1. (f) (d) holds, and Z is a subset of Y which has cardinality ω_n (for any fixed $n < \omega$). (g) (For any fixed $n < \omega$) Ω has cardinality ω_n.

13. Let Ω be an infinite set. If p and q are partitions of Ω, we put $p \leqslant q$ iff every partition set of p lies inside some partition set of q. This defines the lattice of

partitions of Ω, Part(Ω). (a) Show that a partition p satisfies the law $\forall q, r(q \leqslant r \Rightarrow (q \vee p) \wedge r \leqslant q \vee (p \wedge r))$ if and only if p contains at most one set which is not a singleton. (b) Show that for every positive integer n there is a formula $Set_n(x)$ which defines the class of partitions in which one partition set has n elements and the rest are singletons. (c) Let Ω_1 be the structure consisting of $\Omega, \mathcal{P}(\Omega)$ and Part(Ω), together with the membership relations between Ω and $\mathcal{P}(\Omega)$, and between $\mathcal{P}(\Omega)$ and Part(Ω). Show that Ω_1 is interpretable in the lattice Part(Ω). (d) Show that for each $n < \omega$ there is a first-order sentence χ_n such that Part$(\Omega) \vDash \chi_n$ if and only if Ω has cardinality ω_n. (e) Show that some structure Ω_2 of the following form is interpretable in Part(Ω) with definable parameters: $\Omega_2 = (\Omega_1, \Omega^2, \pi, f)$ where f is a bijection from Ω to Ω_2 and π is a relation such that $\pi(a, b, c)$ holds iff $a, b \in \Omega$ and $c = \langle a, b \rangle$. (f) Deduce that Th(Part(Ω)) determines the second-order theory of Ω, and vice versa.

14. Show that if the structures A and B are bi-interpretable by polynomial interpretations, then A and B are definitionally equivalent.

*15. Let K be an algebraic signature and A a K-structure. We say that a tuple $\bar{a}$ of elements of A is **independent** if the following holds: If there are two distinct terms s_1, s_2 of K such that $s_1^A(\bar{a}) = s_2^A(\bar{a})$, then for all terms t_1 and t_2 of K, $t_1^A(\bar{a}) = t_2^A(\bar{a})$. Assume (i) if a is any element of A which is not a constant element, then (a) is an independent tuple, (ii) if $\bar{a}$ is a non-empty tuple which is independent, and b an element such that $\bar{a}\,\hat{}\,b$ is not independent, then $b \in \langle \bar{a} \rangle_A$, (iii) there is a constant in L, and there is a 3-ary term $t(x, y, z)$ which is not equivalent in A to any term $s(x)$ or $s(y)$ or $s(z)$. Show that there are a skew field F, a vector space V over F and a subspace W of V, such that A is polynomially equivalent to V with constants added for the elements of W.

*16. Let F be a skew field of characteristic $\neq 2$ and α an ordinal number. Show that the structure $\langle \alpha + 1, < \rangle$ is interpretable in the general linear group GL(ω_α, F) (the automorphism group of an ω_α-dimensional vector space over F).

17. Let Γ be an interpretation of the signature L in the signature K. Let T be a theory in $K_{\infty\omega}$ such that Γ is defined on the class of models of T. Show (a) if ϕ_i ($i \in I$) and ψ are sentences of $L_{\infty\omega}$ such that $\vdash \bigwedge_{i \in I} \phi_i \to \psi$, then $T \vdash \bigwedge_{i \in I} (\phi_i)_\Gamma \to \psi_\Gamma$, (b) the class $\{\phi$ in $L_{\infty\omega} : T \vdash \phi_\Gamma\}$ is closed under logical consequence.

5.5 Theories that interpret anything

One can classify theories according to how strong they are for interpretations. But there are different classifications for different purposes. In one sense complete first-order arithmetic, T_{arith}, is a strong theory, because every countable consistent first-order theory has a model which is interpretable in a model of T_{arith}. (Gödel's proof of the completeness theorem was really a proof of this.) But from another point of view, the empty theory T_{br} in the language of one binary relation is a strong theory, because every structure in a countable language can be interpreted in some model of T_{br}.

Classes which interpret any structure

Our first result says that two's company: provided that the signatures involved are finite, every structure can be wrapped up into a single binary relation.

Theorem 5.5.1. *Let K be the first-order language whose signature consists of the 2-ary relation symbol R, and let L be a first-order language with finite signature. Then there is a sentence χ of K such that*
(a) *every model of χ is a graph (in the sense of Example 1, section 1.1),*
(b) *the class of models of χ is bi-interpretable with the class of all L-structures which have more than one element.*
Moreover both the interpretations in (b) preserve embeddings.

Proof. First we show how to convert each L-structure B into a graph $A = \Delta(B)$. We draw

(5.1)

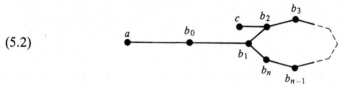

to mean that a and b are distinct elements of A and $A \vDash Rab \wedge Rba$. For each $n \geq 3$, we say that an element a of A is n-**tagged** if A contains elements

(5.2)

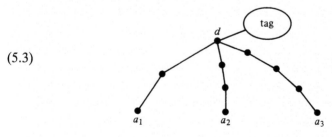

(where the picture means that the elements shown are distinct and R holds between them only as shown). Since L has finite signature, we can list the symbols in its signature as $S_0, \ldots, S_{l-1}$.

We build up A as follows. First, every element of B is a 5-tagged element of A; thus we add four new tagging elements for each element of B. Next we consider a symbol S_i. Suppose for example that S_i is a 3-ary relation symbol. Then we take the set of triples $(S_i)^B$, and for each triple (a_1, a_2, a_3) in this set we add elements as follows:

(5.3)

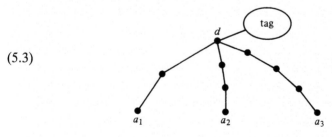

where the tag at the top makes the element d $(i + 6)$-tagged. Note that the numbers of beads on the strings down from d show the positions of a_1, a_2, a_3 in the tuple. (The diagram (5.3) has to be adjusted in the obvious way if a_1, a_2, a_3 are not all distinct.) If S_i had been an m-ary function symbol we would have proceeded the same way but regarding S_i as an $(m + 1)$-ary relation symbol. Likewise a constant can be treated as a 1-ary relation symbol. The collection of all the nodes representing elements of B and tuples in relations of B, together with the edges and tags described above, forms the graph A.

We can classify the elements of A into **species**, and describe each species by an $\exists_1$ formula. Thus let $\rho^+(x, y)$ be the formula $R(x, y) \wedge R(y, x)$, and $\rho^-(x, y)$ the formula $x \neq y \wedge \neg R(x, y) \wedge \neg R(y, x)$. Let $Tag_n(x, y_0, \ldots, y_n, z)$ be the formula

$$(5.4) \quad \rho^+(x, y_0) \wedge \rho^+(y_0, y_1) \wedge \rho^+(y_1, y_2) \wedge \ldots \wedge \rho^+(y_{n-1}, y_n)$$

$$\wedge \rho^+(y_n, y_1) \wedge \rho^+(y_2, z)$$

$$\wedge \bigwedge_{i \leq n} \rho^-(x, y_i) \wedge \rho^-(x, z) \wedge \bigwedge_{i+1 < j < n} \rho^-(y_i, y_j)$$

$$\wedge \bigwedge_{\substack{i \neq 1 \\ i+1 < n}} \rho^-(y_i, y_n) \wedge \bigwedge_{\substack{i \neq 2 \\ i \leq n}} \rho^-(y_i, z).$$

Then there are eight species $E_0, \ldots, E_7$ defined by formulas $E_i(w)$ $(i < 8)$, where $E_i(w)$ is

$$(5.5) \qquad \exists w_0 \ldots w_{i-1} w_{i+1} \ldots w_7 \, Tag_5(w_0, \ldots, w_{i-1}, w, w_{i+1}, \ldots, w_7).$$

Thus an element of A is of species E_0 if and only if it is an element of B; species $E_1, \ldots, E_7$ all correspond to other parts of the tag (5.1). Likewise there are species for elements occurring as parts of a configuration (5.3), excluding a_1, a_2 and a_3 which are of species E_0. The formula χ says

(5.6) R is a symmetric irreflexive relation;
      every element has one of the species $E_0, E_1, \ldots$ (etc. through all the possibilities);
      for each $i < l$, if S_i is an n-ary relation symbol and $\bar{a}$ is an n-tuple of elements of species E_0, then there is at most one configuration (5.3) with $\bar{a}$ at the bottom;
      the corresponding condition when S_i is either a function symbol or a constant;
      there are at least two elements of species E_0.

One can check that if B is any L-structure with at least two elements, then the structure $A = \Delta(B)$ described above is a model of χ; also that every model of χ is isomorphic to some structure $\Delta(B)$ where B has at least two elements.

We must show that Δ can be written as an interpretation. Suppose the total number of possible species of element is p, and the maximum arity of relation symbols in L (counting n-ary function symbols and constants as $(n+1)$-ary relation symbols and 1-ary relation symbols respectively) is q; these two numbers depend only on L. Then each element e of A can be represented by a set of $(q + p + 1)$-tuples $(\bar{a}, \bar{b}, c)$ of elements of B, as follows: $\bar{a}$ lists the elements of B which give rise to e, and $\bar{b}^\wedge c$ indicates what the species of e is, by the rule

(5.7) e is the ith species iff c is equal to the ith item in $\bar{b}$.

This is where we need the assumption that B has at least two elements. We have described an interpretation Δ of L in K. Clearly Δ preserves embeddings.

Suppose $A = \Delta(B)$. We can recover the elements of B from A by putting

(5.8) $\partial_\Gamma(x_0) := x_0$ is of species E_0.

The elements of B stay single elements in A, and so we put

(5.9) $=_\Gamma(x_0, x_1) := (x_0 = x_1)$.

If S_i is a 3-ary relation and $\phi(y_0, y_1, y_2)$ is $S_i(y_0, y_1, y_2)$, then we put

(5.10) $\phi_\Gamma(a_0, a_1, a_2) := $ 'There are elements as in (5.3), so that d is $(i + 6)$-tagged'.

Likewise for the other symbols of L. This defines an interpretation Γ, and it is easily verified that ΓA is isomorphic to B, and that Γ preserves embeddings.

Finally we need to check that $\Delta \circ \Gamma$ and $\Gamma \circ \Delta$ are self-interpretations. I leave this to the reader, with one remark. For the case of $\Gamma \circ \Delta$, we have to be able to recover uniquely each element of a tag or a configuration (5.3) attached to elements of B. This is the purpose of the element c in (5.2), which makes the tag rigid. $\square$

Let us say that a class **V** of structures of signature K is **universal for bi-interpretability** if for every finite signature L there is a bi-interpretation between **V** and the class of L-structures with at least two elements. We have just shown that a certain class of graphs is universal for bi-interpretability. Many other classes have this property. For example, define the **height** of a lattice Λ to be the supremum of the natural numbers n such that Λ contains a finite chain $a_0 < \ldots < a_{n-1}$. In a lattice of finite height, the **height** of an element a is the least n such that there are $a_0 < \ldots < a_n = a$.

Theorem 5.5.2. *There is a first-order definable class of lattices of height $\leqslant 4$ which is universal for bi-interpretability.*

Proof. Let θ be the sentence of the language K of lattices (see (2.28) in section 2.2) which says

(5.11)   I am a lattice of height ≤ 4, and I have unique and distinct
         top and bottom elements.

We claim that the class of models of θ is bi-interpretable with the class of
graphs.

Let L be the first-order language of graphs. Suppose B is a graph
(X, R^B), so that R^B is an irreflexive symmetric binary relation on X. We
define a lattice $\Delta(B)$. Take two new objects *top* and *bottom*. For each pair
(a, b) in R^B, introduce a new object $edge_{ab}$. The domain of $\Delta(B)$ consists of
top, *bottom*, the elements of X and the objects $edge_{ab}$; *top* and *bottom* are
respectively top and bottom elements of the lattice, and the only other
relationships are that each object $edge_{ab}$ is $a \vee b$:

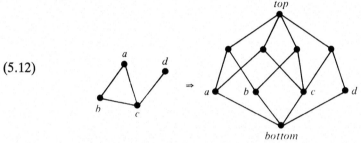

(5.12)

Then $\Delta(B)$ is clearly a model of θ. The reader can check that Δ can be
written as an interpretation, and there is an interpretation Γ such that (Γ, Δ)
is a bi-interpretation between the class of models of θ and the class of graphs.

Now let χ be the sentence of Theorem 5.5.1. By that theorem and Exercise
5.4.6, if L' is any finite signature then there is a bi-interpretation between the
class of models of $\theta \wedge \chi_\Gamma$ and the class of L-structures.   □

When a first-order axiomatisable class **V** is universal for bi-interpretability,
this tells us that many model-theoretic questions can be systematically
reduced to the case of structures in **V**. For example Theorem 5.5.1 and
Theorem 7.3.8 together tell us that we can classify all countable ω-categorical
structures of a finite signature, provided we can classify those which are
graphs. (This is bad news for graphs.) Likewise Theorem 5.5.2 tells us that
Vaught's conjecture for theories of the form $\mathrm{Th}(A)$, where A is a lattice of
height ≤ 4, would imply Vaught's conjecture for all first-order theories of
finite signature. (See section 7.2 for Vaught's conjecture.)

What when the language L has countable signature, not finite? In this case
we can convert any L-structure B into a graph $A = \Delta(B)$ more or less as
before. Unfortunately the operation $A \mapsto \Gamma(A)$ for recovering B is not an
interpretation but a word-construction (see section 9.3). Word-constructions
preserve isomorphism but not ω-categoricity. Hence Vaught's conjecture for
countable first-order theories reduces to the special case of lattices of height
4; but we have no reduction of the classification of countable ω-categorical
structures to the special case of graphs.

Alan Mekler [1981] gave an analogue of Theorems 5.5.1 and 5.5.2 using nilpotent groups in place of graphs or lattices. In this case the operation corresponding to Δ is a word-construction, but Mekler was able to show that the groups are rather faithful copies of the structures which they interpret. See section A.3 below.

Structures which interpret any class

We turn to a different kind of universality.

Theorem 5.5.3. *Let U be a theory in a recursive first-order language L, and suppose that the set of consequences of U in L is a Δ_n^0 set for some positive integer n. If U has a model then U has a model B whose elements are natural numbers, and whose elementary diagram is a Δ_n^0 set. In particular if $\phi(\bar{x})$ is any formula of L, then the set $\{\bar{b}: B \vDash \phi(\bar{b})\}$ is a Δ_n^0 set. (The relation $=^B$ may be an equivalence relation rather than identity.)*

Proof. We assume U has a non-empty model; otherwise the result is trivial. Let L^+ be L with the natural numbers added as new constants. List the sentences of L^+ recursively as $(\phi_n : n < \omega)$. Then define U^* to be the following theory: for each $n < \omega$, $\phi_n \in U^*$ iff

(5.13)   THERE IS a sequence $(s_0, \ldots, s_n)$ such that for each $i < n$,
      (1) s_i is either a sentence θ_i or a pair of sentences (θ_i, χ_i), and θ_i is either ϕ_i or $\neg \phi_i$,
      (2) if θ_i is ϕ_i then $U \vdash \bigwedge \Phi \to \phi_i$, and if θ_i is $\neg \phi_i$ then $U \nvdash \bigwedge \Phi \to \phi_i$, where Φ is the set of sentences θ_j and χ_j with $j < i$,
      (3) if θ_i is a sentence $\exists x\, \psi(x)$, then χ_i is $\psi(m)$ for the first natural number m not occurring as a constant in θ_i or any s_j $(j < i)$,
      (4) θ_n is ϕ_n.

By the proof of the completeness theorem, U^* is a complete and deductively closed theory containing U. The definition of U^* has the form $\exists x\, \sigma$, where everything in σ is recursive except for the clauses in (2) calling for the sets of consequences and non-consequences of U. These sets are Δ_n^0. (The sets are taken in the language L^+, but we can effectively reduce to consequences in L by the lemma on constants, Lemma 2.3.2.) Hence U^* is a Σ_n^0 set. But since U^* is complete and deductively closed, U^* is also the set of negations of sentences not in U^*; so U^* is a Π_n^0 set too, and thus U^* is Δ_n^0.   □

In particular if T is an r.e. theory, then T has a model whose elementary diagram is a Δ_2^0 set. Thus every r.e. theory has a Δ_2^0 completion.

By Theorem 5.5.3, if U is a consistent theory in a recursive language, and

T is a theory such that $\mathbb{N}$ is interpretable in some model of T, then some model of U is interpretable in some model of T. We shall use this in Theorem 5.5.7 below in order to prove undecidability results. But there are other reasons why it's interesting to find interpretations of $\mathbb{N}$ in structures. For example it implies that the structures are unstable (section 6.7), and it can give a purchase to nonstandard arguments (section 11.4).

A classic result in this area is that $\mathbb{N}$ is a relativised reduct of the field of rationals. Any proof of this result uses a substantial amount of number theory. One approach is by way of the following fact.

Fact 5.5.4. *In the language of rings, there is a finite set of formulas $\phi_i(x, \bar{y})$ ($i < n$) such that*
(a) *if A is a global field and V a valuation ring of A, then there are $i < n$ and parameters $\bar{a}$ in A such that $\phi_i(A, \bar{a}) = V$,*
(b) *for any $i < n$ and any parameters $\bar{b}$ in any global field B, $\phi_i(B, \bar{b})$ is either a valuation ring of B or the whole of B.*

Corollary 5.5.5. $\mathbb{N}$ *is a relativised reduct of the field $\mathbb{Q}$ of rational numbers.*

Proof. The ring $\mathbb{Z}$ of integers in $\mathbb{Q}$ is the intersection of all the valuation rings of $\mathbb{Q}$. So using the Fact above, $\mathbb{Z}$ is defined by the formula $\bigwedge_{i<n} \forall \bar{y} \, \phi_i(x, \bar{y})$. Then $\mathbb{N}$ is the substructure of $\mathbb{Z}$ consisting of the elements which are the sum of four squares. $\qquad\qquad\square$

Applications to undecidability

Undecidability is not a topic in model theory. But until above twelve years ago, the main excuse that mathematical logicians had for studying interpretations was to prove undecidability results. I mention two of the main methods, and leave some others to the exercises.

The first method uses a theory known as **(Raphael) Robinson's** Q. This is a finite theory in the language of arithmetic with symbols for zero (0), successor (S), $+$ and $\cdot$, and it runs as follows.

(5.14) $\forall x \forall y (Sx = Sy \to x = y)$.

      $\forall x \, 0 \neq Sx$.

      $\forall x (x \neq 0 \to \exists y \, x = Sy)$.

      $\forall x \, x + 0 = x$.

      $\forall x \forall y \, x + Sy = S(x + y)$.

      $\forall x \, x \cdot 0 = 0$.

      $\forall x \forall y \, x \cdot Sy = (x \cdot y) + x$.

A theory T in a recursive first-order language L is said to be **essentially undecidable** is T is consistent, and for every consistent theory T' in L with $T \subseteq T'$, T' is undecidable.

Fact 5.5.6. *Robinson's Q is essentially undecidable.*

Proof. Tarski, Mostowski & Robinson [1953] p. 60 Theorem 9.   □

A theory T in a first-order language K is said to be **hereditarily undecidable** if every subset of the deductive closure of T is an undecidable theory in K.

Theorem 5.5.7. *Let U be a finite and essentially undecidable theory in a first-order language L of finite signature. Let K be a recursive first-order language and T a theory in K. Let Γ be an interpretation of L in K which interprets some model of U in some model of T. Then T is hereditarily undecidable.*

Proof. We can suppose that U is a single sentence χ. Let Γ be an interpretation of a model B of χ in a model A of T, and let σ be the conjunction of the admissibility conditions for Γ. Note that by the reduction theorem (Theorem 5.3.2), $A \vDash \chi_\Gamma$. Let Φ be the set of sentences $\{\phi$ in $L: T \cup \{\sigma\} \vdash \chi_\Gamma \to \phi_\Gamma\}$. Then $\chi \in \Phi$. By Exercise 5.4.17, Φ is closed under logical consequences. If $\phi \in \Phi$ then $A \vDash \chi_\Gamma \to \phi_\Gamma$, whence $A \vDash \phi_\Gamma$ and so $B \vDash \phi$ by the reduction theorem again; so Φ is consistent. Thus Φ is undecidable. But the map $\phi \mapsto (\sigma \wedge \chi_\Gamma \to \phi_\Gamma)$ is recursive. So if T was decidable, we could compute what sentences are in Φ, and hence what sentences are deducible from Φ. This shows that T is undecidable. But the same argument works for any subset of the deductive closure of T, and so T is hereditarily undecidable.   □

Together, Fact 5.5.6 and Theorem 5.5.7 tell us that if some model of Robinson's Q is interpretable in a structure A, then every subset of $\mathrm{Th}(A)$ is undecidable. For example any subset of the complete theory of the field of rational numbers is undecidable (in the language of rings). Exercise 7 is a typical undecidability result proved by this method.

The second method rests on the following result of Trakhtenbrot.

Fact 5.5.8. *There is a first-order language L of finite signature such that the two sets of sentences of L, $\{\psi: \vdash \psi\}$, $\{\psi: \psi$ has no finite models$\}$ are recursively inseparable. (In other words, there is no recursive set which contains the first set and is disjoint from the second.)*

Proof. See Börger [1989] section FI.1 for a proof by encoding register machines.   □

Let us say that a theory U in a recursive first-order language L is **strongly undecidable** if the sets $\{\phi \text{ in } L: U \vdash \phi\}$ and $\{\phi \text{ in } L: U \cup \{\neg\phi\} \text{ has a finite model}\}$ are recursively inseparable. Trakhtenbrot's result above tells us that L can be chosen with finite signature, so that the empty theory in L is strongly undecidable. If T is a strongly undecidable theory, then T is undecidable, and the theory of all finite models of T is also undecidable.

Theorem 5.5.9. *Let K be a first-order language and T a theory in K. Let L be a first-order language of finite signature and U a finitely axiomatisable strongly undecidable theory in L. Let Γ be an interpretation of L in K such that*

(5.15) *if B is any finite model of U then there is a finite model A of T with $\Gamma A \cong B$.*

Then T is strongly undecidable.

Proof. Let σ be the conjunction of the admissibility conditions for Γ, and let χ be a sentence which axiomatises U. Suppose for contradiction that there is a recursive set X containing $\{\psi: T \vdash \psi\}$ but disjoint from $\{\psi: T \cup \{\neg\psi\} \text{ has a finite model}\}$. Define Y to be the set $\{\phi \text{ in } L: \sigma \wedge \chi_\Gamma \to \phi_\Gamma \in X\}$ of sentences of L. Since X is recursive, Y is recursive too. If ϕ is a sentence of L such that $U \vdash \phi$, then for every model A of $T \cup \{\sigma, \chi_\Gamma\}$, ΓA is defined and a model of χ. So $\Gamma A \vDash \phi$ and hence $A \vDash \phi_\Gamma$ by the reduction theorem, so that $T \vdash \sigma \wedge \chi_\Gamma \to \phi_\Gamma$ and hence $\phi \in Y$. Thus $\{\phi: U \vdash \phi\} \subseteq Y$. If $\neg\phi$ is a sentence of L which is true in some finite model B of U, then by (5.15) there is a finite model A of T with $B \cong \Gamma A$, and we have $A \vDash \sigma \wedge \chi_\Gamma \wedge \neg\phi_\Gamma$, which shows that $T \cup \{\neg(\sigma \wedge \chi_\Gamma \to \phi_\Gamma)\}$ has a finite model and so $\phi \notin Y$. Hence Y contradicts the choice of U. $\qquad\square$

In Theorems 5.5.1 and 5.5.2, the only reason we restricted to L-structures with more than one element was to make Δ an interpretation. Drop that restriction, and the interpretations Γ in those two theorems meet the conditions of Theorem 5.5.9. Thus for example if K is the first-order language of graphs, there is no algorithm which distinguishes the sentences of K which are true in every graph from those which are false in some finite graph. Similarly for lattices of height $\leqslant 4$. Exercise 5 is a typical undecidability result proved by this method.

Exercises for section 5.5

1. Let K be the first-order language whose signature consists of one 2-ary relation symbol. Show that for every first-order language L, the problem of determining what

sentences of L are logical theorems can be effectively reduced to the same question for K.

2. Let K be the signature consisting of two 1-ary function symbols F and G. Show that there is a first-order definable class of K-structures which is universal for bi-interpretability.

3. Let K and L be first-order languages and T, U theories in K, L respectively. (a) Show that if U is undecidable and Γ is a total interpretation of U in T, then T is undecidable. (b) Show that if U is finitely axiomatisable and undecidable, and Γ is a right total interpretation of U in T, then T is undecidable.

4. Let K and L be first-order languages and T, U theories in K, L respectively. Suppose L has finite signature and L' is L with finitely many constants added. Show that if U is hereditarily undecidable, and Γ is a right total interpretation of U in T (as theory in L'), then T is hereditarily undecidable (as theory in L).

*5. (a) Let p be an odd prime. Show that there is a right total interpretation Γ of the class of graphs in the class of rings of characteristic p, such that if B is a finite graph then $B \cong \Gamma A$ for some finite ring A. [Let I be the set of nodes of a graph B in a language with binary relation symbol R. Introduce indeterminates ξ_i, η_i $(i \in I)$. Write Θ_p for the prime ring of characteristic p and A' for the ring $\Theta_p[\xi_i, \eta_i]_{i \in I}$. Let J be the ideal generated by all elements of the following forms: $\eta_i \eta_j$ where $B \vDash Rij$; $\xi_i \xi_j \eta_k$; $\xi_i \eta_j \eta_k$; ξ_i^2; η_i^3; $\xi_i \eta_i$. Write $\Delta(B)$ for A'/J. In $\Delta(B)$ let J_1 be the ideal generated by the elements $\xi_i + J, \eta_i + J$. Show that $J_1 = \{a: a^3 = 0\}$. Show that for all a in $J_1, a^2 = 0$ iff $a \equiv n_i \xi_i + \gamma \pmod{J}$ for some $i \in I, n_i \in \Theta_p$ and $\gamma \in J_1^2$. Recover I as the set defined by $x^2 = 0$, modulo the equivalence relation 'For some $q \neq 0$ in $\Theta_p, x - qy$ is of the form z^2 for some z satisfying $z^3 = 0$'. Recover R by the formula 'There are u, v such that $ux = vy = uv = 0$ and $u^2 \neq 0 \neq v^2$'.] (b) Deduce that the theory of commutative rings is strongly undecidable.

*6. (a) Let R be a ring. The set $UT_3(R)$ of upper unitriangular 3×3 matrices over R consists of the 3×3 matrices over R which are of the form

$$(5.16) \qquad M(a, b, c) = \begin{pmatrix} 1 & a & b \\ 0 & 1 & c \\ 0 & 0 & 1 \end{pmatrix}.$$

Show that if R is of prime characteristic $p > 2$ then $UT_3(R)$ with matrix multiplication forms a nilpotent group of class 2 and exponent p. (b) Let R be a ring of prime characteristic p, and in $UT_3(R)$ let P, Q be the two matrices $M(1, 0, 0), M(0, 0, 1)$. Show that we can recover R from $UT_3(R)$ by the following interpretation Γ.

$\partial_\Gamma(x) :=$ the centraliser of P.

$=_\Gamma(x, y) := x^{-1} \cdot y$ is central.

$+_\Gamma(x, y, z) := x \cdot y = z$ modulo the centre.

$\cdot_\Gamma(x, y, z) :=$ there is w such that $[w, Q] = 1, [y, Q] = [P, w]$
and $[x, w] = [z, Q]$.

(c) Deduce that the theory of nilpotent groups of class 2 and exponent p is undecidable. (d) Show that if R is commutative Γ can be written as an interpretation with definable parameters (see Remark 5 in section 5.3). [The defining formula should say that P and Q are elements allowing this interpretation of a ring.]

*7. Show that if F is a field and $\bar{X}$ is a tuple of at least three indeterminates, then $\mathbb{N}$ is interpretable in the formal power series ring $F[[\bar{X}]]$. Deduce that $\mathrm{Th}\,(F[[\bar{X}]])$ is hereditarily undecidable.

The next exercise needs the compactness theorem, Theorem 6.1.1 below.
8. Let T and U be theories in the first-order languages K and L respectively. Suppose that U is finite, and for every model A of T, some model of U is one-dimensionally interpretable in A. Show that there is a left total one-dimensional interpretation of U in T.

The next exercise is for application in section A.3 below.
9. Show that the graphs which are models of χ in Theorem 5.5.1 have the following properties. (a) If a and b are any two distinct vertexes, then there is a vertex which is joined by an edge to a but not to b. (b) If $a_0, \ldots, a_{n-1}$ are distinct vertexes such that each a_i $(i < n - 1)$ is joined by an edge to a_{i+1} and a_{n-1} is joined by an edge to a_0, then $n \neq 3, 4$.

5.6 Totally transcendental structures

In the final two sections of this chapter we show that certain structures are *not* interpretable in certain other structures. Results of this form are nearly always difficult to prove. The best results use techniques which have only become available in the last ten years. So these two sections will make free use of some facts which I shall prove only in later chapters, and some which are not proved at all in this book. First-time readers proceed at their own risk.

To prove that B is not interpretable in A, we have to find some property which A has and B has not, and which is preserved by interpretations. Thus if A is strongly minimal, we need some property which any structure B interpretable in a strongly minimal structure must have. Provided that B is infinite, one such property is that of *having finite Morley rank* – read on for the definition. I believe it's still an open question whether the converse holds, namely that every structure of finite Morley rank is interpretable in some strongly minimal structure.

Throughout this section, L is a fixed first-order language. 'Definable' means 'definable in L with parameters'.

Cantor-Bendixson rank revisited

We shall use Cantor-Bendixson rank, which we have already met in Example 4 of section 3.4. For convenience I repeat the definition; (6.1)–(6.3) below are clearly equivalent to (4.15)–(4.17) of section 3.4.

Let A be an L-structure. For any formula $\psi(\bar{x})$ of L with parameters from A, the **Cantor–Bendixson rank** $\mathrm{RCB}_A(\psi)$ of the formula ψ (with respect to A) is either -1 or an ordinal or ∞, and is defined as follows (where $\bar{x}$ is an n-tuple of variables).

(6.1) $\mathrm{RCB}_A(\psi) \geqslant 0$ iff $\psi(A^n)$ is not empty.

(6.2) $\mathrm{RCB}_A(\psi) \geqslant \alpha + 1$ iff   there are formulas $\psi_i(\bar{x})$ $(i < \omega)$
                                            of L with parameters from A,
                                            such that the sets
                                            $\psi(A^n) \cap \psi_i(A^n)$ $(i < \omega)$ are
                                            pairwise disjoint and
                                            $\mathrm{RCB}_A(\psi \wedge \psi_i) \geqslant \alpha$ for each
                                            $i < \omega$.

(6.3) $\mathrm{RCB}_A(\psi) \geqslant \delta$ (limit)   iff   for all $\alpha < \delta$, $\mathrm{RCB}_A(\psi) \geqslant \alpha$.

Since $\mathrm{RCB}_A(\phi(\bar{x}, \bar{a}))$ depends on the relation $\phi(A^n)$ and not on the formula chosen to define it, we can also write $\mathrm{RCB}_A(\phi(A^n))$ for $\mathrm{RCB}_A(\phi)$. This defines the Cantor–Bendixson rank of every relation definable in A with parameters.

Lemma 5.6.1. *Let A be an L-structure and $\phi(\bar{x})$ a formula of L with parameters from A. Then the following are equivalent.*
(a) $\mathrm{RCB}_A(\phi) = \infty$.
(b) *For every finite sequence $\bar{s}$ of 0's and 1's, there is a formula $\psi_{\bar{s}}(\bar{x})$ of L with parameters from A, such that*
    (i) *for each $\bar{s}$, $A \vDash \exists \bar{x}\, \psi_{\bar{s}}(\bar{x})$,*
    (ii) *for each $\bar{s}$, $A \vDash \forall \bar{x}(\psi_{\bar{s}^\frown 0} \to \psi_{\bar{s}}) \wedge \forall \bar{x}(\psi_{\bar{s}^\frown 1} \to \psi_{\bar{s}})$,*
    (iii) *for each $\bar{s}$, $A \vDash \neg \exists \bar{x}(\psi_{\bar{s}^\frown 0} \wedge \psi_{\bar{s}^\frown 1})$,*
    (iv) *$\psi_{\langle \rangle}$ is ϕ.*
This can be illustrated thus:

(6.4)

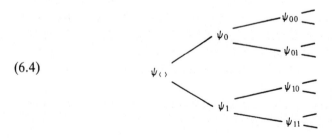

Proof. (a) $\Rightarrow$ (b). There is an ordinal α such that every formula of L with parameters in A has rank either $< \alpha$ or $= \infty$. So if $\psi_{\bar{s}}$ has been chosen and has rank ∞ by (6.2), there is an infinite set of formulas $\psi_{\bar{s}^{\frown}i}(\bar{x})$ ($i < \omega$) of L with parameters in A, such that the sets $\psi_{\bar{s}^{\frown}i}(A^n)$ are pairwise disjoint subsets of $\psi_{\bar{s}}(A^n)$ and all have rank $\geqslant \alpha$, i.e. rank ∞. We get the tree in (b) by discarding the formulas $\psi_{\bar{s}^{\frown}i}$ with $i \geqslant 2$.

(b) $\Rightarrow$ (a). Suppose there is a tree of formulas as in (b). We show by induction on the ordinal β that every formula $\psi_{\bar{s}}$ in the tree has Cantor–Bendixson rank $\geqslant \beta$. For $\beta = 0$ this holds by (i), and at limit ordinals there is nothing to prove. Finally suppose $\beta = \gamma + 1$. Write ϕ_i for the formula $\psi_{\bar{s}^{\frown}0\ldots01}$ with i 0's after $\bar{s}$. By (ii) and (iii) the sets $\phi_i(A^n)$ are pairwise disjoint subsets of $\psi_{\bar{s}}(A^n)$, and by induction hypothesis each ϕ_i has rank $\geqslant \gamma$: so $\psi_{\bar{s}}$ has rank $\geqslant \beta$. $\qquad\square$

The next lemma gathers together the fundamental properties of Cantor–Bendixson rank.

Lemma 5.6.2. *Let L be a first-order language and A an L-structure. Suppose $\phi(\bar{x})$, $\psi(\bar{x})$ are formulas of L with parameters in A, and $\chi(\bar{x}, \bar{y})$ is a formula of L without parameters.*
(a) *If $A \vDash \forall \bar{x}(\phi \to \psi)$ then $\mathrm{RCB}_A(\phi) \leqslant \mathrm{RCB}_A(\psi)$. More generally if $\psi'(\bar{x}')$ is a formula of L with parameters in A, and there is a formula of L with parameters in A which defines an injective map from $\phi(A^n)$ into $\psi'(A^m)$ for some $m < \omega$, then $\mathrm{RCB}_A(\phi) \leqslant \mathrm{RCB}_A(\psi')$.*
(b) *Suppose $\bar{a}$ is a tuple from A. If $A \preccurlyeq B$ then $\mathrm{RCB}_A(\chi(\bar{x}, \bar{a})) \leqslant \mathrm{RCB}_B(\chi(\bar{x}, \bar{a}))$. More generally if $f: A \to B$ is an elementary embedding then $\mathrm{RCB}_A(\chi(\bar{x}, \bar{a})) \leqslant \mathrm{RCB}_B(\chi(\bar{x}, f\bar{a}))$.*
(c) *If $A \preccurlyeq B$ and X, Y are definable relations in A, B respectively with $X \subseteq Y$, then $\mathrm{RCB}_A(X) \leqslant \mathrm{RCB}_B(Y)$.*
(d) *$\mathrm{RCB}_A(\phi \vee \psi) = \max(\mathrm{RCB}_A(\phi), \mathrm{RCB}_A(\psi))$.*
(e) *If A in (b) is ω-saturated (to be defined in section 10.1 below), then $\mathrm{RCB}_A(\phi) = \mathrm{RCB}_B(\phi)$.*
(f) *If A is ω-saturated and $(A, \bar{a}) \equiv (A, \bar{b})$, then $\mathrm{RCB}_A(\chi(\bar{x}, \bar{a})) = \mathrm{RCB}_A(\chi(\bar{x}, \bar{b}))$.*

Proof. (a) By induction on α, prove that if $\mathrm{RCB}_A(\phi) \geqslant \alpha$ then $\mathrm{RCB}_B(\psi) \geqslant \alpha$.

(b) Again by induction on α, prove that if $\mathrm{RCB}_A(\chi(\bar{x}, \bar{a})) \geqslant \alpha$ then $\mathrm{RCB}_B(\chi(\bar{x}, \bar{a})) \geqslant \alpha$.

(c) We show by induction on α that if $\mathrm{RCB}_A(X) \geqslant \alpha$ then $\mathrm{RCB}_B(Y) \geqslant \alpha$. When $\alpha = 0$ this is clear, and there is nothing to prove at limit ordinals. Suppose then that $\alpha = \beta + 1$. Since $\mathrm{RCB}_A(X) \geqslant \alpha$, by (6.2) there are disjoint definable subsets $\psi_i(A^n)$ $(i < \omega)$ of X, all of Cantor–Bendixson rank $\geq \beta$. Then $\psi_i(B^n)$ $(i < \omega)$ are also pairwise disjoint, and they have Cantor–Bendixson rank $\geq \beta$ by (b). By induction hypothesis the sets $Y \cap \psi_i(B)$ have Cantor–Bendixson rank $\geq \beta$, and so by (6.2) again, $\mathrm{RCB}_B(Y) \geq \alpha$ too.

(d) We have $\geq$ by (a). For the converse we use induction on $\max(\mathrm{RCB}_A(\phi), \mathrm{RCB}_A(\psi))$. Suppose $\mathrm{RCB}_A(\phi \vee \psi) = \alpha$ but $\mathrm{RCB}_A(\phi)$, $\mathrm{RCB}_A(\psi)$ are both $\leq \beta < \alpha$. Then (putting $n = $ length of $\bar{x}$) there are infinitely many disjoint definable subsets X_i $(i < \omega)$ of $(\phi \vee \psi)(A^n)$, each of Cantor–Bendixson rank $\geq \beta$. Since $\mathrm{RCB}_A(\phi) \leq \beta$, there is $k_\phi < \omega$ such that $\mathrm{RCB}_A(X_i \cap \phi(A^n)) < \beta$ whenever $i \geq k_\phi$, and likewise there is $k_\psi < \omega$ such that $\mathrm{RCB}_A(X_i \cap \psi(A^n)) < \beta$ whenever $i \geq k_\psi$. Putting $k = \max(k_\phi, k_\psi)$, both $\mathrm{RCB}_A(X_k \cap \phi(A^n))$ and $\mathrm{RCB}_A(X_k \cap \psi(A^n))$ are $< \beta$ though X_k has Cantor–Bendixson rank $\geq \beta$. This contradicts the induction hypothesis.

(e) is equally straightforward, but since it uses a notion from a later chapter, I postpone its proof to Exercise 10.1.22. (f) is immediate from (e) and the fact (Corollary 10.2.6(b)) that every structure has a strongly ω-homogeneous elementary extension B, so that B has an automorphism carrying $\bar{a}$ to $\bar{b}$. $\qquad\square$

Morley rank

Let $\phi(\bar{x})$ be a formula of L with parameters from the L-structure A. We define the **Morley rank** of ϕ with respect to A, in symbols $\mathrm{RM}_A(\phi)$, to be the sup of all the values $\mathrm{RCB}_B(\phi)$ as B ranges over all elementary extensions of A. Just as with Cantor–Bendixson rank, we extend this notion to definable relations by writing $\mathrm{RM}_A(\phi(A^n))$ for $\mathrm{RM}_A(\phi)$.

Fortunately we need only look at one suitably chosen elementary extension of A to calculate $\mathrm{RM}_A(\phi)$.

Lemma 5.6.3. *Let A be an L-structure and $\phi(\bar{x})$ a formula of L with parameters from A. Let B be an ω-saturated elementary extension of A. (By Corollary 10.2.2, A has such an elementary extension.) Then $\mathrm{RM}_A(\phi) = \mathrm{RCB}_B(\phi)$. In particular Morley rank and Cantor–Bendixson rank coincide in any ω-saturated structure.*

Proof. Let α be any ordinal. Suppose A has an elementary extension C such that $\mathrm{RCB}_C(\phi) \geq \alpha$; we have to show that $\mathrm{RCB}_B(\phi) \geq \alpha$. The elementary amalgamation theorem, Theorem 6.4.1, will tell us that there are an elementary extension D of C and an elementary embedding $g: B \to D$ which is the identity on A. By Lemma 5.6.2(b), $\mathrm{RCB}_D(\phi) \geq \alpha$; by Lemma 5.6.2(e), $\mathrm{RCB}_D(\phi) = \mathrm{RCB}_B(\phi)$. $\qquad\square$

Hence all our work on Cantor–Bendixson rank translates at once to Morley rank.

Lemma 5.6.4. *Lemma 5.6.2 holds with* RM *in place of* RCB *everywhere. For the sake of cross-references, Lemma 5.6.4(a)–(f) shall be Lemma 5.6.2(a)–(f) but with* RM *for* RCB.

Proof. By Lemmas 5.6.2 and 5.6.3. For part (b) we need to know that f can be extended to an elementary embedding between suitable ω-saturated extensions of A and B; for this, take an ω-saturated elementary extension of the combined structure (A, B, f) and quote Theorem 10.1.9. For part (c) we need to know that A and B have ω-saturated elementary extensions A^*, B^* with $A^* \preccurlyeq B^*$, such that if $X = \phi(A^n)$ and $Y = \psi(B^n)$, then $\phi(A^{*n}) \subseteq \psi(B^{*n})$. I leave this as Exercise 4. □

Morley rank is about splitting up into infinitely many pieces. How about splitting up into finitely many? There are two things to be said here. First, splitting up into finitely many pieces *infinitely often* is just as bad as splitting up into infinitely many.

Lemma 5.6.5. *Let A be an L-structure and $\phi(\bar{x})$ a formula of L with parameters in A. Then* $\mathrm{RM}_A(\phi) = \infty$ *if and only if there are an elementary extension B of A and formulas $\psi_{\bar{s}}(\bar{x})$ (for all tuples $\bar{s}$ of 0's and 1's) with parameters in B, such that (i)–(iv) of (b) in Lemma 5.6.1 hold with B for A.*

Proof. If there is such a B, then by taking a further elementary extension if necessary, we can suppose that B is ω-saturated. □

And second, there is a finite upper bound to the number of times we can split a formula with Morley rank $\alpha < \infty$ into formulas of the same Morley rank α. More precisely, we have the following.

Lemma 5.6.6. *Let A be ω-saturated. Let $\phi(\bar{x})$ be a formula with parameters in A, of Morley rank $\alpha < \infty$. Then there is some greatest positive integer d such that $\phi(A^n)$ can be written as the union of disjoint sets $\phi_0(A^n), \ldots, \phi_{d-1}(A^n)$ all of Morley rank α. This integer d is not altered by passing to any elementary extension of A.*

Proof. Suppose there is no such d. Then for every positive integer d, $\phi(A^n)$ can be written as a union of disjoint sets $\phi_{d,0}(A^n), \ldots, \phi_{d,d-1}(A^n)$ of Morley rank α, forming a partition π_d of $\phi(A^n)$ into d sets. We shall adjust these formulas a little, by induction on d, starting at $d = 1$. For each $i \leqslant d$, the set

$\phi_{d+1,i}(A^n)$ is partitioned into d sets $(\phi_{d+1,i} \wedge \phi_{d,j})(A^n)$ $(j < d)$, and by Lemma 5.6.4(d), at least one of these partition sets has Morley rank α. Using this and Lemma 5.6.4(a), we can rewrite the formulas $\phi_{d+1,i}$, still of Morley rank α, so that the partition π_{d+1} refines the partition π_d.

Suppose this adjustment has been made for all d. By induction we can choose integers $d_0 < d_1 < \ldots$ and integers $i_0, i_1, \ldots$ so that, writing ψ_m for ϕ_{d_m,i_m}, each set $\psi_{m+1}(A^n)$ is a proper subset of $\psi_m(A^n)$. (See Exercise 5 if this gives trouble.) Then the sets $(\psi_m \wedge \neg \psi_{m+1})(A^n)$ $(m < \omega)$ are disjoint subsets of $\phi(A^n)$, all of Morley rank α, so that $\phi(A^n)$ has Morley rank $\geq \alpha + 1$; contradiction. This proves the existence of d.

Suppose B is an elementary extension of A, and let $\bar{a}$ be the parameters in ϕ. Then $\phi(B^n)$ is partitioned into d subsets $\phi_i(B^n)$ of Morley rank α. If it can also be partitioned into $d+1$ such sets $\psi_i(B_n, \bar{b}_i)$ $(i < d+1)$, then since A is ω-saturated, there are $\bar{a}_0, \ldots, \bar{a}_d$ in A such that $(A, \bar{a}, \bar{a}_0, \ldots, \bar{a}_d) \equiv (B, \bar{a}, \bar{b}_0, \ldots, \bar{b}_d)$, and hence by Lemma 5.6.4(f), the sets $\psi_i(A^n, \bar{a}_i)$ form a partition of $\phi(A^n)$ into $d+1$ sets of Morley rank α, contradicting the choice of d. □

The positive integer d as in Lemma 5.6.6 is called the **Morley degree** of ϕ (or of $\phi(A^n)$).

The **Morley rank** of the structure A is defined to be $\mathrm{RM}_A(x = x)$; this is equal to $\mathrm{RM}_A(\mathrm{dom}\, A)$. We say that A is **totally transcendental** if A has Morley rank $< \infty$. And of course we say that A **has finite Morley rank** if the Morley rank of A is finite. By Lemma 5.6.4(d), the Morley rank of A is the same as the Morley rank of any structure elementarily equivalent to A; so we can define the **Morley rank** of a complete theory T in L to be the Morley rank of any model of T. Likewise we say that T is **totally transcendental** if some (or every) model of T is totally transcendental.

If L is a countable language, then a complete theory T in L is totally transcendental if and only if it is ω-stable (to be defined in section 6.7 below). This is a theorem and not a definition (see Theorem 6.7.5), and it fails when L is uncountable.

Lemma 5.6.7. *Let A be an L-structure. Then the following are equivalent, for every ordinal α.*
(a) *A has Morley rank $\leq \alpha$.*
(b) *For some ω-saturated elementary extension B of A, $\mathrm{RCB}_B(x = x) \leq \alpha$.*
(c) *Every formula $\phi(x)$ of L with parameters in A, with just one variable x, has Morley rank $\leq \alpha$.*

Proof. (a) ⇔ (b) is by Lemma 5.6.3. (a) ⇔ (c) follows from Lemma 5.6.4(a).

□

Lemma 5.6.7(c) tells us that if A is totally transcendental, then every formula $\phi(x)$ with one free variable has Morley rank $< \infty$. What about formulas with several free variables? The next lemma answers this question.

Lemma 5.6.8 (*Erimbetov's inequality*). *Let A be an ω-saturated L-structure and $\phi(\bar{x}, \bar{y})$ a formula of L with parameters in A. Let α and β be ordinals such that*
(a) $RM_A(\exists \bar{y} \, \phi(\bar{x}, \bar{y})) = \alpha$,
(b) *for every tuple $\bar{a}$ in A, $RM_A(\phi(\bar{a}, \bar{y})) \leqslant \beta$.*
Then $RM_A(\phi(\bar{x}, \bar{y}))$ and $RM_A(\exists \bar{x} \, \phi(\bar{x}, \bar{y}))$ are both $\leqslant \beta \cdot (\alpha + 1)$ when $\beta \neq 0$, and $= \alpha$ when $\beta = 0$.

Proof. I omit the proof for $\exists \bar{x} \, \phi(\bar{x}, \bar{y})$; it can be found in Lachlan [1980]. Assuming it, we prove the result for $\phi(\bar{x}, \bar{y})$ as follows. Let $\chi(\bar{x}, \bar{y}, \bar{z})$ be the formula $\phi(\bar{x}, \bar{y}) \wedge \bar{x} = \bar{z}$. The inequality tells us that if $RM_A(\exists \bar{x} \bar{y} \, \chi(\bar{x}, \bar{y}, \bar{z})) = \alpha$ and for every tuple $\bar{a}$ in A, $RM_A(\chi(\bar{x}, \bar{y}, \bar{a})) \leqslant \beta$, then

$$RM_A(\exists \bar{z} \, \chi(\bar{x}, \bar{y}, \bar{z})) \leqslant \beta \cdot (\alpha + 1) \text{ (or } = \alpha \text{ when } \beta = 0).$$

But $\exists \bar{x} \bar{y} \, \chi(\bar{x}, \bar{y}, \bar{z})$ and $\exists \bar{z} \, \chi(\bar{x}, \bar{y}, \bar{z})$ are logically equivalent to $\exists \bar{y} \, \phi(\bar{z}, \bar{y})$ and $\phi(\bar{x}, \bar{y})$ respectively, while $\chi(\bar{x}, \bar{y}, \bar{a})$ and $\phi(\bar{a}, \bar{y})$ have the same Morley rank by Lemma 5.6.4(a). $\square$

We can apply this lemma to any formula ϕ with more than one free variable, by splitting up the variables into two groups $\bar{x}$ and $\bar{y}$. The formulas $\exists \bar{y} \, \phi$ and $\phi(\bar{a}, \bar{y})$ both have fewer free variables than ϕ. So by induction and simple arithmetic we have two useful consequences.

Lemma 5.6.9. *Let A be an L-structure and $\phi(\bar{x})$ any formula of L with parameters in A.*
   (a) *If A has Morley rank α and $\bar{x}$ has length n, then ϕ has Morley rank at most $(\alpha + 1)^n$.*
   (b) *If A is totally transcendental then ϕ has Morley rank $< \infty$.*
   (c) *If A has finite Morley rank, then so does ϕ.* $\square$

Warning. It makes no sense to speak of the Morley rank of a formula ϕ unless we specify the variables of ϕ. By writing $x = x$ as $\phi(x, y)$, we add an extra degree of freedom and so raise the Morley rank.

Structures of finite Morley rank

In section 12.2 we shall see that every ω_1-categorical theory in a countable first-order language has finite Morley rank. For the moment we prove an important special case of this.

Theorem 5.6.10. *Let A be an L-structure and X a non-empty definable set in A.*

(a) *X is finite if and only if X has Morley degree 0.*

(b) *X is strongly minimal if and only if its Morley rank and degree are both equal to 1. In particular a strongly minimal structure has Morley rank 1.*

Proof. Let $\phi(\bar{x})$ be a formula of L with parameters from A, such that $X = \phi(A)$.

(a) To say that ϕ has Morley rank 0 is to say that there is no elementary extension B of A in which $\phi(B)$ contains infinitely many disjoint non-empty definable sets. Since singletons are definable, this is to say that $\phi(B)$ is finite for every elementary extension B of A; equivalently, $\phi(A)$ is finite.

(b) To say that ϕ has Morley rank and degree 1 is to say that $\phi(A)$ is infinite (by (a)), but there is no elementary extension B of A with a definable subset Y such that Y and $\phi(B)\backslash Y$ are both infinite. This is the definition of strong minimality. $\square$

Interpretations between infinite structures preserve the property of having finite Morley rank.

Theorem 5.6.11. *Let K and L be first-order languages, A a K-structure and B an L-structure. Suppose B is interpretable in A. If A has Morley rank α, then the Morley rank of B is at most $(\alpha + 1)^n$ for some $n < \omega$. In particular, the following hold.*

(a) *If A is totally transcendental then so is B.*

(b) *If A has finite Morley rank then so has B.*

Proof. Suppose Γ is an n-dimensional interpretation of B in A with coordinate map $\sim$. The main work of the proof is to show that

(6.5)   for every formula $\phi(\bar{y}, \bar{z})$ of L and every tuple $\bar{a}$ in A,
$$\mathrm{RM}_B(\phi(\bar{y}, \bar{a}^\sim)) \leqslant \mathrm{RM}_A(\phi_\Gamma(\bar{x}, \bar{a})).$$

Let A' be an ω-saturated elementary extension of A. Then by Theorem 10.1.9, $\Gamma A'$ is ω-saturated, and by Theorem 5.3.4 there is an elementary embedding of $B = \Gamma A$ into $\Gamma A'$. So for computing Morley ranks, we can suppose that A and B are ω-saturated. Write $\phi^\Gamma(\bar{x}_0, \ldots, \bar{x}_{m-1}, \bar{a})$ for the formula $\partial_\Gamma(\bar{x}_0) \wedge \ldots \wedge \partial_\Gamma(\bar{x}_{m-1}) \wedge \phi_\Gamma(\bar{x}_0, \ldots, \bar{x}_{m-1}, \bar{a})$. We claim that for every ordinal α, if $\mathrm{RM}_B(\phi(\bar{x}, \bar{a}^\sim)) \geqslant \alpha$ then $\mathrm{RM}_A(\phi^\Gamma(\bar{x}, \bar{a})) \geqslant \alpha$.

If $\alpha = 0$, then $\mathrm{RM}_B(\phi(\bar{x}, \bar{a}^\sim)) \geqslant \alpha$ means $B \vDash \exists \bar{y}\, \phi(\bar{y}, \bar{a}^\sim)$, so $A \vDash \exists \bar{x}\, \phi^\Gamma(\bar{x}, \bar{a})$ by the reduction theorem (Theorem 5.3.2), (3.9) in section 5.3 and the definition of ϕ^Γ; whence $\mathrm{RM}_A(\phi^\Gamma) \geqslant 0$. If δ is a limit ordinal there is nothing to show. Suppose finally that $\alpha = \beta + 1$. Then using Lemma 5.6.3, since $\text{RCB}_B(\phi(\bar{y}, \bar{a}^\sim)) = \text{RM}_B(\phi(\bar{y}, \bar{a}^\sim)) \geqslant \alpha$, there are formulas $\psi_i(\bar{y}, \bar{b}_i)$ $(i < \omega)$ of L with parameters $\bar{b}_i$ in B, such that the relations defined by the formulas $\phi(\bar{y}, \bar{a}^\sim) \wedge \psi_i(\bar{y}, \bar{b}_i)$ are pairwise disjoint and of Morley rank $\geqslant \beta$. For each $\bar{b}_i$, choose $\bar{a}_i$ in A so that $\bar{b}_i = \bar{a}_i^\sim$. By induction hypothesis and (3.7) in section 5.3 it follows that the formulas $\phi^\Gamma(\bar{x}, \bar{a}) \wedge (\psi_i)^\Gamma(\bar{x}, \bar{a}_i)$ define relations of rank $\geqslant \beta$ in A. Moreover for all $i < j < \omega$,

(6.6) $\qquad B \vDash \forall \bar{y}(\phi(\bar{y}, \bar{a}^\sim) \rightarrow \neg(\psi_i(\bar{y}, \bar{b}_i) \wedge \psi_j(\bar{y}, \bar{b}_j)))$,

so by the reduction theorem it follows that

(6.7) $\qquad A \vDash \forall \bar{x}(\phi^\Gamma(\bar{x}, \bar{a}) \rightarrow \neg(\psi_i^\Gamma(\bar{x}, \bar{a}_i) \wedge \psi_j^\Gamma(\bar{x}, \bar{a}_j)))$

and thus the formulas $\phi^\Gamma(\bar{x}, \bar{a}) \wedge (\psi_i)^\Gamma(\bar{x}, \bar{a}_i)$ $(i < \omega)$ define pairwise disjoint relations on A. This proves that $\text{RM}_A(\phi^\Gamma(\bar{x}, \bar{a})) = \text{RCB}_A(\phi^\Gamma(\bar{x}, \bar{a})) \geqslant \alpha$. The claim is proved.

It follows by Lemma 5.6.4(a) that (6.5) holds. Now suppose that A has Morley rank α. By Lemma 5.6.9(a) the formula $\bar{x} =_\Gamma \bar{x}$ has Morley rank $\leqslant (\alpha + 1)^n$ in A, and so by (6.5) the formula $x = x$ has Morley rank $\leqslant (\alpha + 1)^n$ in B. $\qquad \square$

So by Theorem 5.6.10, *every structure interpretable in a strongly minimal structure has finite Morley rank.*

Complete types

Let A be an L-structure and X a set of elements of A. The notion of a **complete n-type over X with respect to** A is defined in section 6.3 below. If $p(\bar{x})$ is a complete n-type over X with respect to A, then every formula $\phi(\bar{x})$ in p is a formula of L with parameters in A, and so it has a Morley rank (possibly ∞, but definitely not -1). We define the **Morley rank** of p, $\text{RM}(p)$, to be the least value of $\text{RM}_A(\phi)$ as ϕ ranges over the formulas in p.

Lemma 5.6.12. *Let $\Phi(\bar{x})$ be a set of formulas of L with parameters all lying in a set X of elements of the L-structures A.*

(a) Let α be an ordinal such that for every finite subset Ψ of Φ, $\text{RM}_A(\bigwedge \Psi) \geqslant \alpha$. Then there is a complete type $p(\bar{x})$ over X with respect to A, which has Morley rank $\geqslant \alpha$ and contains Φ.

(b) If Φ contains a formula of Morley rank α and Morley degree d, then there are at most d complete types $p(\bar{x})$ over X which contain Φ and have Morley rank α.

Proof. (a) Let $\Phi'(\bar{x})$ be the set of all those formulas $\phi(\bar{x})$ of L with parameters from X, with the property that for some finite subset Ψ of Φ,

$RM_A(\bigwedge\Psi \wedge \neg\phi) < \alpha$. Certainly every formula of Φ is in Φ', since contradictions have Morley rank -1.

We claim that for every finite subset Ψ' of Φ', $A \vDash \exists\bar{x} \bigwedge\Psi'$. If ϕ is in Ψ', then there is a finite set Ψ_ϕ in Φ such that $RM_A(\bigwedge\Psi_\phi \wedge \neg\phi) < \alpha$. Taking Ψ to be the union of the sets Ψ_ϕ with ϕ in Ψ', we have $RM_A(\bigwedge\Psi \wedge \neg\phi) < \alpha$ for each ψ in Ψ', by Lemma 5.6.4(a), and hence $RM_A(\bigwedge\Psi \wedge \neg\bigwedge\Psi') < \alpha$ by Lemma 5.6.4(d). But $RM_A(\bigwedge\Psi) \geq \alpha$, so $RM_A(\bigwedge\Psi \wedge \bigwedge\Psi') \geq \alpha$ by Lemma 5.6.4(d) again, and hence $RM_A(\bigwedge\Psi') \geq \alpha \geq 0$. The claim is proved.

So by Theorem 6.3.1 below, Ψ' can be extended to a complete type $p(\bar{x})$ over X with respect to A. This type p contains Ψ. If ϕ is a formula of Morley rank $< \alpha$ with parameters from X, then $\neg\phi$ lies in Φ' and hence in Ψ, so that ϕ is not in p (since types are consistent!). Thus p has Morley rank $\geq \alpha$.

(b) Suppose Φ contains the formula $\phi(\bar{x})$ of Morley rank α and Morley degree d. If $p_0, \ldots, p_d$ are distinct complete types over X which contain Φ and have Morley rank α, then there are formulas $\psi_0(\bar{x}), \ldots, \psi_d(\bar{x})$ in $p_0, \ldots, p_d$ respectively such that for all $i < j \leq d$, $\psi_i(A^n)$ is disjoint from $\psi_j(A^n)$. Then the formulas $\phi \wedge \psi_i$ define pairwise disjoint sets of Morley rank α, so that ϕ has Morley degree at least $d + 1$; contradiction. $\qquad\square$

As a special case of Lemma 5.6.12, suppose A is an L-structure and X, Y are sets of elements of A with $X \subseteq Y$. Then every complete type $p(\bar{x})$ over X can be extended to a complete type $q(\bar{x})$ over Y with the same Morley rank. We can ensure $RM(q) \geq RM(p)$ by the lemma, and $RM(q) \leq RM(p)$ since $p \subseteq q$.

Example 1: *The regular types of strongly minimal sets.* Let A be an L-structure and $\phi(x)$ a strongly minimal formula of L. By Theorem 5.6.10 and Lemma 5.6.12 there is a complete type $p(x)$ over $\varnothing$ which contains the formula $\phi(x)$ and has Morley rank 1. In fact p is uniquely determined: if $\psi(x)$ is any formula of L, then exactly one of $(\phi \wedge \psi)(A), (\phi \wedge \neg\psi)(A)$ is infinite. Say it is $\phi \wedge \psi$; then $\phi \wedge \neg\psi$ can't be in p, so ψ must be in p. For reasons that lie in stability theory, p is called the **regular type** of the strongly minimal set $\phi(A)$.

For example if A is an algebraically closed field, then its unique regular type is the type over $\varnothing$ of a transcendental element.

Suppose $p(\bar{x})$ is a complete type over a set X with respect to an L-structure A. If p has Morley rank α, its **Morley degree** is the least integer d such that p contains a formula $\phi(\bar{x})$ of Morley rank α and Morley degree d. The type p is the unique complete type over X which has rank α and contains ϕ. For suppose $q(\bar{x})$ is another such type, and $\psi(\bar{x})$ is a formula which is in p but not in q. Then both $\phi \wedge \psi$ and $\phi \wedge \neg\psi$ have Morley rank

α, and so at least one of them must have lower Morley degree than ϕ, contradicting the choice of d. The Morley degree of a complete type may be any positive integer; see Exercise 10.

Clearly the regular type of a strongly minimal set has Morley degree 1.

Exercises for section 5.6

1. Let A be an L-structure and $\phi(\bar{x})$ a formula of L with parameters in A. (a) Show that if $\bar{b}$ is any sequence of elements of A, and B is $(A, \bar{b})$, then $\mathrm{RM}_A(\phi) = \mathrm{RM}_B(\phi)$. (b) Give an example to show that if A is a reduct of C, it need not be true that $\mathrm{RM}_A(\phi) = \mathrm{RM}_C(\phi)$.

2. Write out a full proof of Lemma 5.6.2(a, b).

3. Let A be an L-structure and $\phi(\bar{x})$ a formula of L with parameters $\bar{a}$ in A. Show that there is a set T of sentences of L with parameters $\bar{a}$, such that the following are equivalent. (a) There is a set of formulas $\psi_s(\bar{x})$ with parameters from some elementary extension B of A, such that the properties (i)–(iv) of Lemma 5.6.1(b) hold with B for A. (b) A is a model of T. *This allows a more direct proof that if A is totally transcendental, then so is any structure elementarily equivalent to A.*

4. Complete the proof of Lemma 5.6.4(c). [Expand B to B^+ by adding a 1-ary relation symbol P whose interpretation is $\mathrm{dom}(A)$. Take an ω-saturated elementary extension of B^+.]

The following combinatorial argument is known as **König's tree lemma**.
5. Let $(P, \leq)$ be an infinite partially ordered set with the following properties. (a) There is a unique bottom element. (b) Every element has at most finitely many elements below it, and they are linearly ordered. (c) Every element has at most finitely many immediate successors. Show that P contains an infinite linearly ordered subset. [The **rank** of an element is the number of elements below it. By induction on rank, choose an increasing chain of elements $p_0 < p_1 < \ldots$ so that each p_i has infinitely many elements above it.]

6. Let A be a K-structure, B an L-structure and Γ a one-dimensional interpretation of B in A. Suppose A is a minimal structure and B is infinite. Show that B is a minimal structure.

7. Give an example of a non-strongly-minimal structure which is interpretable in a strongly minimal structure. (By the previous exercise, the interpretation must have dimension at least 2.)

8. Let A be an L-structure, X a set of elements of A and $p(\bar{x})$ a complete type over X with respect to A. Write a direct definition of the Morley rank of p, along the

following lines: p has Morley rank $\geq \alpha + 1$ if and only if there are complete types q_i $(i < \omega)$ over some set Y in some elementary extension B of A such that . . . .

9. Let A be an L-structure, X a set of elements of A and $\bar{a}, \bar{b}$ tuples of elements of A. We use the notation of section 6.3 below for types. Show that if $\mathrm{tp}_A(\bar{a}/X)$ has Morley rank α and $\bar{b}$ is algebraic over X and $\bar{a}$, then $\mathrm{tp}_A(\bar{b}/X)$ has Morley rank $\leq \alpha$.

10. Give examples to show that for every positive integer d there are a structure A and a complete type $p(x)$ over $\varnothing$ with respect to A, such that p has Morley rank 1 and Morley rank d. [Equivalence relation with d classes, all infinite.]

11. Let $p(x)$ be a complete type over a set X with respect to the L-structure A. Suppose $\phi(x)$ and $\psi(x)$ are formulas of L with parameters in X, and suppose $\phi \in p$. Let $\theta(x, y)$ be a formula with parameters in X, which defines in A a bijection from $\phi(A)$ to $\psi(A)$. Show that there is a complete type $q(y)$ over X, such that for every formula $\sigma(y)$ of L with parameters in X, $\sigma \in q$ if and only if $\exists y(\theta(x, y) \wedge \sigma(y))$ is in p. Show that q has the same Morley rank as p.

12. Let A be an ω-saturated totally transcendental L-structure, and X an infinite subset of $\mathrm{dom}\,(A)$ which is first-order definable in A with parameters. Show that there is a strongly minimal set $Y \subseteq X$. [Compare Exercise 3.4.1.]

13. Let L be a first-order language, A a totally transcendental L-structure and X a set of elements of A. We write $L(X)$ for the first-order language got from L by adding parameters for the elements of X. Show that if $\phi(x)$ is a formula of $L(X)$ and $A \vDash \exists x\,\phi$, then there is a formula $\psi(x)$ of $L(X)$ such that $A \vDash \forall x(\psi \rightarrow \phi)$ and for every formula $\theta(x)$ of $L(X)$, exactly one of $A \vDash \exists x(\psi \wedge \theta)$, $A \vDash \exists x(\psi \wedge \neg\theta)$ holds. [If not, we can contradict Lemma 5.6.5.]

5.7 Interpreting groups and fields

There are many situations where one wants to know what groups and fields are interpretable in a structure. Questions of this kind are often remarkably hard to answer. Several of the more interesting results were found by techniques that go well beyond this book. In this section I shall prove one result that is easy to state and not particularly interesting in its own right.

Theorem 5.7.1. *No infinite field can be interpreted (with parameters) in a vector space.*

Its value lies in the methods that we need in order to prove it – notably Macintyre's characterisation of totally transcendental fields, Theorem 5.7.7 below. With this much equipment to hand, it seemed a pity not to finish the section with Zil'ber's indecomposability theorem and an application about interpreting algebraically closed fields. As I threatened in the previous section, I shall use some facts from later chapters and some other facts that I don't propose to prove at all.

We fix some terminology for this section. A **group-like structure** is a structure which has a group as a relativised reduct (without parameters, so that the group is $\varnothing$-definable). When we speak of a structure A as group-like, we always have a particular relativised reduct G in mind, and we call it the **group** of A; we write '$x \in G$' for the formula which defines G. For example every field F is group-like in at least two different ways – the group of F can be either the additive group or the multiplicative group of F.

'Definable' will mean 'first-order definable with parameters'. A **definable subgroup** of a group-like structure A is a set which is definable with parameters from A, and which forms a subgroup of the group of A; the parameters need not be in the group. If A is a group-like structure with group G, and B is an elementary extension of A, then the formula '$x \in G$' defines a group G^B in B, and $G \preccurlyeq G^B$ by Corollary 5.1.2.

Algebraic geometers might like to note why we introduce group-like structures. Real-life groups often carry extra structure besides the group operations. Thus a linear group is not just an abstract group; it arrives with a description of how it lies inside a group of matrices over some field. Algebraic groups can be thought of in several different ways involving various amounts of sheaf theory, but usually they come armed with a topology which they inherit from the field in which they are defined. Several of the ideas in this section were borrowed from the theory of algebraic groups, and in model theory the best we can do to represent the extra structure is to put our groups inside group-like structures.

Totally transcendental groups

In a group-like structure, total transcendence is a surprisingly strong condition.

Theorem 5.7.2. *Let A be a totally transcendental group-like structure.*

(a) *There are no infinite strictly decreasing chains of definable subgroups of A.*

(b) *Every intersection of a family of definable subgroups of A is equal to the intersection of a finite subfamily.*

Proof. (a) Let $(G_n : n < \omega)$ be such a chain. Then for every $n < \omega$, each coset of G_n splits into two or more cosets of G_{n+1}, contradicting Lemma 5.6.5.

(b) follows at once. $\square$

Corollary 5.7.3. *In the group G of a totally transcendental group-like structure there is a unique smallest definable subgroup G° of finite index; this subgroup is $\varnothing$-definable. We have $(G^\circ)^\circ = G^\circ$.*

Proof. Let $(H_i : i \in I)$ be the family of all definable subgroups of finite index in G. Write $G^\circ = \bigcap_{i \in I} H_i$. By the theorem, G° is equal to $\bigcap_{i \in J} H_i$ for some finite subset J of I. So G° is definable and of finite index in G. Suppose its index is n and its defining formula is $\phi(x, \bar{a})$. Then G° is also defined by the formula

(7.1) $\exists \bar{y}(\phi(x, \bar{y}) \wedge \text{`}\phi(x, \bar{y})$ defines a subgroup of index $n\text{'})$

which has no parameters. The final sentence of the lemma is straightforward.

$\square$

Following the analogy of algebraic groups, the subgroup G° is called the **connected component** of G. We say G is **connected** if $G = G^\circ$.

The next concept has unlocked doors. Let A be a totally transcendental group-like structure with group G of Morley rank α, and X a set of elements of A. We shall say that a complete 1-type $p(x)$ over X with respect to A is G-**generic over** X if p contains the formula '$x \in G$' and has Morley rank α. We say simply 'generic' when G is clear from the context.

Example 1: *Generic types of algebraically closed fields.* An algebraically closed field A has Morley rank 1, by Theorem 5.6.10(b). It is group-like for at least two different groups, the additive group A^+ of all elements and the multiplicative group $A^\times$ of non-zero elements. There is just one complete type of Morley rank 1 over the empty set, namely the type of a transcendental element. This type is both A^+-generic and $A^\times$-generic. If B is an algebraically closed subfield of A, then there is still just one type of Morley rank 1 over B, namely the type of an element not in B; this type is also both A^+-generic and $A^\times$-generic.

We shall define an action of group elements on types. Suppose A is a group-like structure with group G, and $\Phi(x)$ is a set of formulas (maybe with parameters) which include the formula '$x \in G$'. Then we define $g\Phi$ to be the set of formulas $\phi(g^{-1} \cdot x)$ such that $\phi(x) \in \Phi$. If Y is the set of elements of A which satisfy Φ, we write gY for the set of elements which satisfy $g\Phi$. Thus

(7.2) $g \cdot b \in g(\bigwedge \Phi(A)) \Leftrightarrow g \cdot b \in \bigwedge g\Phi(A) \Leftrightarrow b \in \bigwedge \Phi(A)$.

Two cases will interest us. First, if Y is a definable set then $gY = \{g \cdot h : h \in Y\}$.

The second case is where $p(x)$ is the complete type of some element b over some set of parameters with respect to A. Then $g(p)$ is the complete type of the element $g \cdot b$ over the same parameters. By Lemma 5.6.4(a), since left multiplication by g is a permutation of the set of elements satisfying '$x \in G$', the type $g(p)$ has the same Morley rank as p (cf. Exercise 5.6.11).

Note that if $p(x)$ is a complete type over a set of parameters containing g, and p contains the formula '$x \in G$', then it certainly contains the formula '$g^{-1} \cdot x \in G$' too, so that $g(p)$ contains '$x \in G$'. Clearly we do have a group action; i.e. $g(h(p)) = (gh)(p)$ for all g,h in G, and $1(p) = p$.

Warning. This action of G on types is not the action on types which we defined in Example 1 of section 4.3 above.

We define the **stabiliser** of the type p, $\mathrm{Stab}_G(p)$, to be the subgroup of G consisting of the elements g such that $g(p) = p$.

Lemma 5.7.4. *Let A be a totally transcendental group-like L-structure with group G. Let X be a set of elements of A containing G, and let p be a G-generic type over X. Then $\mathrm{Stab}_G(p) = G^\circ$.*

Proof. Let α be the Morley rank of G, and B an elementary extension of A in which some element b realises p. Since p is generic, $\mathrm{RM}(p) = \alpha$. Let $\phi(x)$ be a formula in p which has Morley rank α and Morley degree as small as possible. Then (see the end of section 5.6) p is the only complete type over X which has Morley rank α and contains ϕ. We shall see in Theorem 6.7.5 that A is stable, and hence by Theorem 6.7.8 that there is some formula $\chi(z)$ whose parameters are in X, such that

(7.3)   For all g in X, $B \vDash \phi(g \cdot b) \Leftrightarrow B \vDash \chi(g)$.

If g is any element of G, then we know that $g(p)$ has Morley rank α too. So $g(p) = p$ if and only if $g(p)$ contains $\phi(x)$. But $\phi(x) \in g(p) \Leftrightarrow \phi(g \cdot x) \in p \Leftrightarrow B \vDash \chi(g)$, by (7.3). Hence $\mathrm{Stab}_G(p)$ is defined by the formula $\chi(x)$.

Now by Lemma 5.6.12(b), the number of generic types over X with respect to A is at most the Morley degree of the formula '$x \in G$'; so the index of $\mathrm{Stab}_G(p)$ in G is finite. By the definition of G° it follows that $G^\circ \subseteq \mathrm{Stab}_G(p)$.

Let $G^\circ g_0, \ldots, G^\circ g_{k-1}$ be the right cosets of G° in G. Since the type p is complete, it contains exactly one of the formulas '$x \cdot g_i^{-1} \in G^\circ$' $(i < k)$ (expressing that $x \in G^\circ g_i$); let it be '$x \cdot g_0^{-1} \in G^\circ$'. Now if h is some element of G which is not in G°, then $h(p)$ contains the formula '$h^{-1} \cdot x \cdot g_0^{-1} \in G^\circ$', and it follows that $h(p) \neq p$. This proves that $\mathrm{Stab}_G(p) = G^\circ$. $\square$

Lemma 5.7.5. *Let A be a totally transcendental group-like L-structure with group G, and suppose G is connected. Then the formula '$x \in G$' has Morley degree 1.*

Proof. Suppose G has Morley rank α. By Lemma 5.6.6 there is no harm if we pass to an elementary extension of A, and so we can assume A is ω-saturated. Then we can partition G into d sets $\phi_0(A), \ldots, \phi_{d-1}(A)$ of Morley rank α and Morley degree 1, where the ϕ_i are formulas with parameters in A. Suppose for contradiction that $d > 1$. For each $i < d$ there is a G-generic type p_i over $\mathrm{dom}(A)$ containing ϕ_i, by Lemma 5.6.12(a). By the previous lemma, $\mathrm{Stab}_G(p_i) = G° = G$ for each $i < d$. Hence if g is any element of G, then $g\phi_i(x)$ is in p_i. Thus if $g_0, \ldots, g_{k-1}$ are any elements of G,

(7.4) $\qquad g_0\phi_i(A) \cap \ldots \cap g_{k-1}\phi_i(A)$ has Morley rank α.

By symmetry we can define an action of g on the right, and (7.4) will apply here too. Thus if $g \in G$ then $\phi_i(A)g \cap \phi_i(A)$ also has Morley rank α.

Now by (7.4) and the compactness theorem, there is an elementary extension B of A containing an element b which lies in $\bigcap\{g\phi_0(B) : g \in G\}$. Thus $\phi_1(A) \subseteq G \subseteq \phi_0(B)b^{-1}$. So $\phi_1(A) \subseteq \phi_0(B)b^{-1} \cap \phi_1(B)$. Since $\phi_1(A)$ has Morley rank α in A, it follows by Lemma 5.6.4(c) that $\phi_0(B)b^{-1} \cap \phi_1(B)$ has Morley rank α in B. But we know that $\phi_0(B)b^{-1}$ has Morley rank α and Morley degree 1, and just as in (7.4), its intersection with $\phi_0(B)$ also has Morley rank α. Since $\phi_0(B)$ and $\phi_1(B)$ are disjoint, this is a contradiction. Thus $d = 1$. $\qquad\qquad\square$

Lemma 5.7.6. *Let A be a totally transcendental group-like L-structure with group G, and let X be a set of elements of A containing G. Write Γ for the set of G-generic types of X. Then the Morley degree of the formula '$x \in G$' is $(G : G°)$, which is equal to $|\Gamma|$, and the action of G is transitive on Γ.*

Proof. Let α be the Morley rank of G. Consider the cosets $G°g_i$ $(i < m)$ of $G°$. The map $h \mapsto h \cdot g_i$ is a bijection from $G°$ to $G°g_i$, and so by Lemma 5.6.4(a) these cosets all have the same Morley rank. By Lemma 5.6.4(d) this rank must be α. Now Lemma 5.6.12(a) gives us a $G°$-generic type p over X. By Lemma 5.7.5 and Lemma 5.6.12(b), p is the unique $G°$-generic type over X.

Applying the action of G on Γ, we deduce that for each right coset $G°g_i$ of $G°$ in G, there is exactly one G-generic type which contains the formula '$x \in G°g_i$'. On the other hand every complete type over X containing '$x \in G$' must contain one of the formulas '$x \in G°g_i$' $(i < m)$. So there is one G-generic type per coset, and G acts transitively on the set of G-generic types.

The number of cosets of $G°$ in G is first-order expressible, and hence it doesn't change if we pass from A to an ω-saturated elementary extension B of A. So the Morley degree of G^B is $(G^B:(G^B)°) = (G:G°)$; by Lemma 5.6.6 this is the Morley degree of G too. $\qquad\square$

The next theorem was the first significant result about the connection between algebraic properties and stability properties.

Theorem 5.7.7 (*Macintyre's theorem on totally transcendental fields*). *Every totally transcendental field is algebraically closed.*

Proof. Let A be a totally transcendental field. By Corollary 10.2.2 we can assume without loss that A is ω-saturated. The connected component $(A^+)°$ of the additive group A^+ of A is a $\varnothing$-definable set. Multiplication by a non-zero element b of A is an automorphism of A^+, so $b \cdot (A^+)°$ is also a subgroup of finite index in A^+, whence $(A^+)° \subseteq b \cdot (A^+)°$, or in other words $b^{-1} \cdot (A^+)° \subseteq (A^+)°$. This shows that $(A^+)°$ is a subgroup of A^+ which is closed under multiplication by elements of A, and so it is an ideal of A. It must therefore be either A or $\{0\}$. Since A is infinite and $(A^+)°$ has finite index in A^+, we deduce that $(A^+)° = A^+$, and hence A^+ is a connected group. Therefore by Lemma 5.7.6, A has a unique generic type.

Now we switch to the multiplicative group $A^\times$, recalling that A^+ and $A^\times$ have the same generic types. By Lemma 5.7.6 again, $A^\times$ must be connected too.

Suppose n is a positive integer. Then the subgroup $(A^\times)^n$ of nth powers has finite index in $A^\times$, and so it is equal to $A^\times$. Thus every element of A is an nth power. Likewise suppose A has prime characteristic p. Then the set of all elements of the form $(a^p - a)$ is an additive subgroup of finite index in A^+, and hence it is the whole of A^+.

We claim that A contains all roots of unity. For if not, then there is a least n such that A fails to contain some primitive nth root θ of unity. A primitive square root of unity is -1, so n must be at least 3. Consider the extension $A(\theta)$ of A, which has degree $< n$ over A. By Galois theory there is a field B intermediate between A and $A(\theta)$ such that the Galois group of $A(\theta)$ over B is cyclic, say of order m; then $m < n$, and so B contains a primitive mth root of unity. But then the extension $A(\theta)/B$ is a Kummer extension (see Jacobson [1964] Chapter III or Cohn [1977] section 6.9 for the classification of abelian extensions). Now by Example 3 in section 5.3, B is interpretable in A, so by Theorem 5.6.11, B is also totally transcendental. But we have just seen that a totally transcendental field has no Kummer extensions; contradiction. The claim is proved.

Now we prove that A is algebraically closed. If not, let $A(\eta)$ be a proper

algebraic extension of finite degree over A. As above, there is an intermediate field C such that $A(\eta)/C$ has Galois group of prime order q. Now we
quote the theory of extensions of prime order. There are two cases. First, if q
is not the characteristic of A, then $A(\eta)$ is again a Kummer extension of C,
and we have a contradiction as before. Second, if q is the characteristic of A,
then $A(\eta)$ is formed by adjoining a root of an equation $x^q - x = a$ with a in
A, and again we have a contradiction. $\square$

Theorem 5.7.1: fields in vector spaces

Suppose A is a vector space over a field F, and B is an infinite field which is
interpretable in A by an interpretation Γ with parameters $\bar{a}$. By Exercise
5.3.1, A must be infinite. So by the upward Löwenheim–Skolem theorem
(Corollary 6.1.4), A has an elementary extension A' of uncountable dimen-
sion; say $A' = \bigoplus_{i<\lambda} Fv_i$ where $(v_i : i < \lambda)$ is a basis and λ is an uncountable
cardinal greater than the cardinality of the field F. Then the field $B = \Gamma A$ is
elementarily embeddable in $B' = \Gamma A'$ (by Theorem 5.3.4), and so there is no
loss in assuming that A is A'. By Example 2 in section 4.5 (and the
discussion after it), A is strongly minimal, so by Theorem 5.6.10(b), A is
totally transcendental. Hence by Theorem 5.6.11, B is totally transcendental,
and so by Macintyre's theorem (Theorem 5.7.7), B is an algebraically closed
field. Now by Example 3 in section 10.1 below, A is saturated; so by
Corollary 10.1.10, B is either finite or saturated of cardinality λ. Since B is
algebraically closed, it must be of cardinality λ, and hence it has infinite
transcendence degree. Therefore the multiplicative group G of B is an
uncountable abelian group whose torsion subgroup is countably infinite.

Now I quote the main result of Evans, Pillay & Poizat [1990]. Its proof
uses generic types for stable structures, and lies beyond what we can handle
in this book.

Fact 5.7.8. *Let A be a module and G a group which is interpretable in A,
possibly with parameters. Then there are a positive integer n and definable
subgroups $H_1 \subseteq H_2$ of A^n such that H_2/H_1 is isomorphic to a subgroup of
finite index in G.* $\square$

With the help of Fact 5.7.8, we shall show that no uncountable abelian
group interpretable in A has countably infinite torsion subgroup. For this we
need a description of the definable subgroups of V^n when V is a vector
space. This is something of a brain-teaser, since V^n has dimensions of two
different kinds: the dimension n over V, and the dimension of V itself as a
vector space.

Let V be a vector space over a field F, and let n be a positive integer.

Then the general linear group $GL_n(F)$ of invertible $n \times n$ matrices over F acts on the set of subsets X of $(\text{dom } V)^n$ by

(7.5) $$M(X) = \{M\bar{a} : \bar{a} \in X\}$$

where M is any $n \times n$ invertible matrix over F, and we read the n-tuple $\bar{a}$ as a column vector.

Lemma 5.7.9. *The set of definable subgroups of V^n is closed under the action of $GL_n(F)$; so is the set of subspaces of V^n regarded as a vector space over F.*

Proof. Let G be a definable subgroup of V^n. For any formula $\phi(\bar{x})$ and matrix M we easily write down a formula $\psi(\bar{x})$ such that for every n-tuple $\bar{a}$ in V, $V \vDash \psi(\bar{a}) \Leftrightarrow \phi(M\bar{a})$. It follows that $M(G)$ is definable. I leave to the reader the proof that $M(G)$ is a subgroup.   □ Lemma

Lemma 5.7.10. *If F has characteristic 0, then the only two definable subgroups of V are $\{0\}$ and V.*

Proof. Every definable subgroup G of V is defined by a boolean combination $\phi(x)$ of formulas $x = a$ (a in V), by Exercise 2.7.9. Thus $\phi(V)$ is either finite or cofinite. If it is finite then it must be $\{0\}$, since F has characteristic 0. If it is cofinite, then so are its cosets, so it must be the whole of V.   □ Lemma

Write W_k for the subgroup of V^n consisting of all elements of the form $(a_0, \ldots, a_{k-1}, 0, \ldots, 0)$. Note that since the additive group of a vector space is divisible, W_k is a divisible group.

Lemma 5.7.11. *If G is a definable subgroup of V^n, then $G = M(W_k)$ for some $k \leq n$ and some invertible $n \times n$ matrix M over F.*

Proof. Choose $k < \omega$ as large as possible such that for some invertible matrix N, $W_k \subseteq N(G)$. We claim that $W_k = N(G)$.

Suppose not; let $\bar{b} = (b_0, \ldots, b_{n-1})$ be an element of $N(X) \setminus W_k$. Subtracting some suitable element of W_k from $\bar{b}$ if necessary, we can suppose that $b_0 = \ldots = b_{k-1} = 0$. Since $b_k, \ldots, b_{n-1}$ are not all 0, there is some invertible matrix P which pointwise fixes W_k and takes $\bar{b}$ to some element $\bar{c} = (0, \ldots, 0, c, 0, \ldots, 0)$ with c in the kth place. Thus $W_k \subseteq PN(G)$ and $\bar{c} \in PN(G)$. Let H be the subgroup of V^n consisting of all elements $(d_0, \ldots, d_{n-1})$ with $d_0 = \ldots = d_{k-1} = d_{k+1} = \ldots = d_{n-1} = 0$. Then $\bar{0} \neq \bar{c} \in PN(G) \cap H$, and so by Lemma 5.7.10, $H \subseteq PN(G)$. Thus $W_k \oplus H \subseteq PN(G)$, contradicting the choice of k.

The claim is proved. For the lemma, take M to be N^{-1}.   □ Lemma

We return to the proof of Theorem 5.7.1. By Fact 5.7.8, our group G has a subgroup of finite index which is isomorphic to H_2/H_1 for two definable subgroups $H_1 \subseteq H_2$ of A^n, for some positive integer n. Now G has elements of infinite order; it follows at once that the field F has characteristic 0, since otherwise H_2 would have prime exponent. Thus Lemma 5.7.11 applies, and hence H_1 and H_2 are divisible torsion-free groups. But then H_2/H_1 has no torsion elements except 0, contradicting the fact that G has an infinite torsion subgroup. $\square$ Theorem 5.7.1

It should be mentioned that there is a more powerful proof of Theorem 5.7.1 using deeper geometric ideas. A vector space is 1-based, but an expansion of an algebraically closed field is never 1-based. This argument will show for example that no infinite field is interpretable in a module of finite Morley rank. In fact the notion of 1-based groups (otherwise known as weakly normal groups) makes an appearance in the proof of Fact 5.7.8, where it quotes Hrushovski & Pillay [1987].

Zil'ber's indecomposability theorem

We turn to a powerful and pretty result of Zil'ber about how to generate subgroups of a group of finite Morley rank.

For the rest of this section, A is a group-like structure of finite Morley rank, with group G. If H is a subgroup of G and X is a set of elements of G, we say that X is H-**indecomposable** if X meets either one or infinitely many of the left cosets of H in G. We say that X is **indecomposable** if X is H-indecomposable for every definable subgroup H of G. If X and Y are sets of elements of G, we write $X \cdot Y$ for the set of elements of the form $x \cdot y$ with $x \in X$ and $y \in Y$, and X^{-1} for the set of elements x^{-1} with $x \in X$.

Theorem 5.7.12 (Zil'ber's indecomposability theorem). *Let A be a group-like structure of finite Morley rank, and G the group of A. Suppose that for each $i \in I$, X_i is a definable indecomposable set of elements of G which contains 1. Let K be the subgroup generated by the union of the sets X_i ($i \in I$). Then there is a finite sequence $(i_0, \ldots, i_{n-1})$ of elements of I such that $K = X'_{i_0} \ldots X'_{i_{n-1}}$ where each X'_i is either X_i or X_i^{-1}.*

Proof. Among the sets of the form $X_{i_0} \ldots X_{i_{n-1}}$ ($n < \omega$ and $i_0, \ldots, i_{n-1} \in I$), choose one of maximal Morley rank. This is possible since G has finite Morley rank. Write this set as $X = X_{i_0} \ldots X_{i_{m-1}}$. There is a formula $\phi(x)$ which defines X. By Lemma 5.6.12 there is a G-generic type p over $\operatorname{dom}(G)$

which contains ϕ. Write H for $\mathrm{Stab}_G(p)$. We shall show that $K = X \cdot X^{-1}$ by proving a chain of inclusions: $H \subseteq X \cdot X^{-1} \subseteq K \subseteq H$.

First, $H \subseteq X \cdot X^{-1}$. For this let h be any element of H. The type p is equal to $h(p)$, which contains the formula $\phi(h^{-1} \cdot x)$ expressing that $x \in hX$. But p also contains the formula '$x \in X$'. Since p is a type, there must be some element g lying in both X and hX; thus $g = hh'$ with h' in X, and so $h = g \cdot h'^{-1} \in X \cdot X^{-1}$.

The inclusion $X \cdot X^{-1} \subseteq K$ is trivial. Finally to show that $K \subseteq H$, noting that H is a group, it suffices to prove that each set X_i ($i \in I$) is a subset of H. By Lemma 5.7.4, H has finite index in G, and so the left cosets of H cut X_i into just finitely many pieces. Since X_i is H-indecomposable and contains 1, it follows that X_i lies in H. $\qquad\square$

One typical consequence of Zil'ber's theorem is the following result on group actions.

Theorem 5.7.13. *Suppose A is a structure of finite Morley rank. Suppose that A has relativised reducts B, G (defined without parameters) such that*
(a) *B and G are infinite abelian groups,*
(b) *in A a faithful action of G on B is $\varnothing$-definable,*
(c) *every element of G acts on B as an endomorphism, and*
(d) *there is no proper infinite definable subgroup of B which is setwise fixed under the action of G.*
Then an algebraically closed field is interpretable in A without parameters.

Proof. We write B additively and G multiplicatively. There is a formula $\phi(x, y)$ of L such that for every element b of B, $\phi(A, b)$ is the set $G_{(b)}$ of elements of G which fix b. By Theorem 5.7.2 above, $\bigcap_{b \in B} G_{(b)}$ can be written as $\phi(A, b_0) \cap \ldots \cap \phi(A, b_{k-1})$ for some $b_0, \ldots, b_{k-1}$; write $\bar{b} = (b_0, \ldots, b_{k-1})$. So if g is any element of G, the action of g is determined by what it does to $\bar{b}$. Since G is infinite, there must be some b_i ($i < k$) in B whose orbit Gb_i under G is infinite; let it be b_0. Let X be $Gb_0 \cup \{0\}$. Then X is setwise fixed under the action of G.

By the theorem hypothesis, if H is a definable subgroup of B which is setwise fixed by the action of G, then H is either finite or the whole of B. It follows that X is H-indecomposable. We claim that X is indecomposable.

For if not, there is some definable subgroup K of B such that the distinct left cosets of K which meet X are $k_0 + K, \ldots, k_{m-1} + K$ with $2 \leqslant m < \omega$. By Theorem 5.7.2 again, the intersection $H = \bigcap_{g \in G} g(K)$ can be written as $g_0(K) \cap \ldots \cap g_{n-1}(K)$ for some $g_0, \ldots, g_{n-1}$ in G. Each $g_i(K)$ meets $g_i(X) = X$ in exactly m cosets, and so H meets X in some number k of cosets with $2 \leqslant k \leqslant m^n$. But H is a definable subgroup of B which is setwise fixed under the action of G, so this is a contradiction to the previous paragraph. The claim is proved.

Applying Zil'ber's indecomposability theorem to X, there is some positive integer l such that every element of the group $\langle X \rangle$ generated by X can be written as a sum $\varepsilon_0 g_0 b_0 + \ldots + \varepsilon_{l-1} g_{l-1} b_0$ with $g_0, \ldots, g_{l-1} \in G \cup \{0\}$ and each $\varepsilon_i \in \{1, -1\}$. Thus the group $\langle X \rangle$ is an infinite definable subgroup of B which is setwise fixed by G, and hence it is the whole of B.

Now we consider the ring R of endomorphisms of the group B. Each element g in G acts as an element $\rho(g)$ of R. Put $\rho(0) = 0$. Let R_0 be the subring of R generated by the endomorphisms $\rho(g)$ with g in G. Thus composition of endomorphisms corresponds to multiplication in G; hence R_0 is commutative since G is abelian. Also R_0 is infinite since G is infinite and acts faithfully on B.

Let e be an endomorphism $\in R_0$, and consider the element $e(b_0)$ of B. We have seen that $e(b) = \varepsilon_0 g_0 b_0 + \ldots + \varepsilon_{l-1} g_{l-1} b_0$ for some elements $g_0, \ldots, g_{l-1}$ of $G \cup \{0\}$ and some $\varepsilon_0, \ldots, \varepsilon_{l-1}$ in $\{1, -1\}$. In other words, e agrees with the endomorphism $\sum_{i<l} \varepsilon_i \rho(g_i)$ on b_0. Since G is commutative and B is generated by Gb_0, it follows that $e = \sum_{i<l} \varepsilon_i \rho(g_i)$. In other words, every endomorphism in R_0 can be written as the sum of $\leqslant l$ endomorphisms of the form $\pm\rho(g)$ with $g \in G$. From this we easily write down an interpretation of R_0 in A, so that R_0 has finite Morley rank by Theorem 5.6.11.

The kernel and the image of an endomorphism in R_0 are both of them definable subgroups of B which are setwise fixed by G; so they are either finite or the whole of B. Thus if the product of two endomorphisms is 0, one of them must already by 0; hence R_0 is an integral domain. Since R_0 is stable (Theorem 6.7.5), this implies that R_0 is a field, by Exercise 6.7.11. So by Macintyre's theorem (Theorem 5.7.7), R_0 is an algebraically closed field. $\square$

Theorem 5.7.13 appeared in an unpublished paper of Zil'ber, as part of the proof of the following very suggestive result, which whisks an algebraically closed field into existence from nowhere.

Fact 5.7.14. *If G is a connected and soluble but not nilpotent group, and G is interpretable in a model of an uncountably categorical theory, then some algebraically closed field is interpretable in G.* $\square$

This result was perhaps the first indication that model theorists and group theorists have serious business to do together. It gave rise to a flood of results and problems on interpretations between groups and fields; some of the chief currents are reported in the Appendix, section A.4 below.

Exercises for section 5.7

1. Show that if G is a definable group in the totally transcendental structure A and B is an elementary extension of A, then $(G°)^B = (G^B)°$.

2. Let A be a group-like structure of finite Morley rank, with group G. Show that there is a finite k such that G has no chain of definable subgroups $G_0 \subseteq \ldots \subseteq G_k$ in which each G_{i+1}/G_i is infinite. [The Morley rank bounds the length of the chain.]

3. Show that if A is a field and A has finite Morley rank, then in A there is no proper infinite subfield which is definable (with parameters). [Apply Macintyre's theorem to the subfield and then use the previous exercise.]

4. Suppose that A is group-like and totally transcendental, and the group G of A is connected. (a) Suppose also that in A there is a definable action of G on a finite set X. Show that G fixes X pointwise. [The index of the stabiliser of a point of X is finite.] (b) Show that if g is an element of G whose conjugacy class is finite, then g lies in the centre $Z(G)$ of G. (c) Show that if $Z(G)$ is finite then $G/Z(G)$ is centreless. (d) Show that there is some term Z_i of the upper central series of G, such that G/Z_i is centreless. [Use Exercise 2.]

5. Let A be a group-like L-structure of finite Morley rank with group G. We say that a formula $\phi(x)$ of L with parameters from A is **generic** if $\phi(A) \subseteq G$ and ϕ has the same Morley rank as G. (a) Let p be a complete type over a set of parameters in A, and suppose '$x \in G$' is in p. Show that p is generic if and only if all the formulas in it are generic. (b) Show that a formula ϕ is generic if and only if $G \subseteq g_0 \cdot \phi(A) \cup \ldots \cup g_{m-1} \cdot \phi(A)$ for some finite set of elements $g_0, \ldots, g_{m-1}$ of G. *This characterisation of 'generic formula' turns out to work as a definition of 'generic' for stable groups in general.*

6. Let A be an infinite abelian group of finite exponent. Describe the generic types over A.

7. Let A be a group-like structure with group G, and let X be a set of elements of G. Show that X is indecomposable if and only if X is H-indecomposable for every definable normal subgroup H of G. [See the claim in the proof of Theorem 5.7.13.]

8. Let A be a group-like structure of finite Morley rank with group G. Let H be a connected definable subgroup of G and X any set of elements of G. Show that $[X, H]$ (the subgroup generated by the set of commutators $[x, h]$ with $x \in X$ and $h \in H$) is a connected definable subgroup of G. [For each x in X, apply the previous exercise to show that the set $\{h^{-1} \cdot x \cdot h : h \in H\}$ is indecomposable, and apply the indecomposability theorem.]

*9. Let G be a connected nilpotent group with finite Morley rank. Show that if the centre of G has Morley rank $\leqslant 1$, then G is uncountably categorical.

*10. Show that if A is a torsion-free abelian group, then the only $\varnothing$-definable subgroups of A are the subgroups of the form mA for some natural number m.

History and bibliography

Section 5.1. In Langford [1926], written under the influence of *Principia Mathematica*, every structure is regarded as a relativised reduct of the set-theoretic universe. Thus relativisation was a fact of life already in the early days of model theory. A systematic treatment of relativisation appears in Tarski, Mostowski & Robinson [1953] §I.5. Example 3 follows from the 'Thèoréme de réduction' in Frasnay [1965]; it was found independently by Hodges (see Hodges, Lachlan & Shelah [1977] for a proof by Shelah) and Cameron [1976]. For further refinements see Macpherson [1986b], Frasnay [1984]. Thomas [1991] proves a corresponding result for the random graph.

Exercise 7: Videla [1987].

Section 5.2. The notions of PC and PC_Δ classes are due to Mal'tsev [1941], though not by those names; see section 6.6. The notation PC, PC_Δ, PC' and PC'_Δ was introduced by Tarski in [1954]. Theorem 5.2.1 is from Makkai [1964b].

Exercise 3: Fuchs [1960] Theorem 77.2. Exercise 4: Makkai [1964b]. Exercise 5: Sabbagh in Hirschelmann [1972]. Exercise 6: e.g. Botto Mura & Rhemtulla [1977].

Section 5.3. Interpretations are an old device in geometry, where they appeared in two guises during the mid nineteenth century. First there are consistency proofs: in [1868] Beltrami interpreted a region of a Lobachevskian plane in Euclidean space. (The title of Beltrami's paper may be the origin of the word *interpretation* in this sense.) This idea was developed by Klein, Poincaré and above all Hilbert [1899] as a way of proving the consistency of sets of geometrical axioms. And second there are coordinatisation theorems, where one interprets a field (or a passable imitation of a field) in a suitable lattice. The first step along these lines was von Staudt's [1847] 'calculus of throws', which was a version of the coordinatisation theorem of projective geometry. The interpretation of rings or groups in modular lattices is still one of the active research tools of lattice theory; see the survey in Day [1983].

Within model theory, interpretations were developed by Tarski from the late 1930s as a way of proving undecidability; see the next two sections and their references. The associated functor is from Hodges [1976a].

Exercise 6: Baur, Cherlin & Macintyre [1979]. Exercise 8: Point & Prest [1988]. Exercise 11: E.g. Schuppar [1980].

Section 5.4. The work of Tarski, Mostowski & Robinson [1953], R. Robinson [1951], Mal'tsev [1960] and Ershov [1974] was all aimed at showing that certain first-order theories were undecidable; see section 5.5. Shelah [1973b] defined the interpretation in Example 2, after Isbell in McKenzie [1971a] had made the observation that there is a first-order sentence which is true in Sym(Ω) and expresses that Ω is countable. Exercise 12 is a small taste of what Shelah expressed in Sym(Ω). In Shelah [1976a] he developed the same idea further; for example he interpreted the whole universe of set theory in the category of endomorphisms of abelian groups.

Exercise 4: Gurevich [1964]. Exercise 8: Ahlbrandt & Ziegler [1986]. Exercise 13: (a)–(d) Ježek [1985]; (e), (f) Pinus [1988]. Exercise 15: Urbanik [1963]. Exercise 16: Tolstykh [1990].

Section 5.5. The earliest version of Theorem 5.5.1 was Exercise 1, which appears in Löwenheim [1915]. Lavrov [1963] gives a version essentially the same as in the text. Theorem 5.5.2 is due to Taĭtslin [1962].

As background to Theorem 5.5.3, note that any structure can be regarded as ΓV where V is the universe of sets and Γ is a suitable interpretation with parameters. This was the setting of Langford [1926], and Mostowski [1952b] still thinks of models in this way. Theorem 5.5.3 is due to Hasenjaeger [1953] and Kleene [1952b] §72; Gödel [1930] proved the completeness theorem by constructing an interpretation in the natural numbers. Constructive models have largely been studied in the Soviet Union; see Chapter 6 of Ershov [1980a]. For developments in a different direction, see Knight [1986a], who develops a method of Harrington for building models whose elementary diagram has much higher degree of unsolvability than their diagram. Fact 5.5.4 is due to Rumely [1980] and Corollary 5.5.5 to Julia Robinson [1949]; see also J. Robinson [1959]. Julia Robinson uses the Hasse–Minkowski theorem on quadratic forms, while Rumely calls on Hasse's norm theorem and Artin reciprocity.

In the late 1930s Mostowski and Tarski discovered a finitely axiomatisable essentially undecidable theory, with the aim of applying it as in Theorem 5.5.7. Theorem 5.5.7 appears as Theorem 8(i) on pp. 23f of Tarski, Mostowski & Robinson [1953]; it was used by J. Robinson [1949] to prove the undecidability of the theory of the field of rationals. Fact 5.5.8 is from Trakhtenbrot [1953]. The approach of Theorem 5.5.9 is due to Ershov &

Taĭtslin [1963]; see Ershov, Lavrov, *et al.* [1965]. Lavrov [1963] proved the strong undecidability of the theory of graphs; its undecidability had been proved by Hartley Rogers [1956].

Exercise 3: Rabin [1965a]. Exercise 4: Ershov [1965a]. Exercise 5: Taĭtslin [1963]. Exercise 6: Mal'tsev [1960]; (d) Belegradek [199?]. Exercise 7: Delon [1981]. Exercise 8: Montague [1965].

Section 5.6. One of the most original moments in model theory was the day that Morley [1965a] introduced Morley rank and the related notions of total transcendence and Morley degree. He must have had the Cantor–Bendixson argument from descriptive set theory in mind, and in fact his treatment was thoroughly topological. But the name 'totally transcendental' comes from Example 1 applied to algebraically closed fields: every complete type $p(x)$ is either algebraic or (at worst) transcendental with Morley rank 1. Morley's account of Morley rank is still one of the clearest, and in one form or another it contains most of this section. But be warned that he assumes the language is countable, and some of his arguments (using cardinalities of Stone spaces) fail to generalise. Also Morley defines the rank of $\mathrm{Th}(A)$ to be $\mathrm{RM}_A(x = x) + 1$, and some writers (mostly Russian) still use this convention.

The notation RCB, RM is in deference to the Paris school, who put the noun before its qualifier; but I believe Hebrew does the same. Lemma 5.6.8 is due to Erimbetov [1975] and independently Shelah [1978a] Theorem V.7.8. Theorem 5.6.11 is used in Zil'ber [1977a].

Exercise 5: D. König [1926]. Exercise 13. Morley [1965a].

Section 5.7. Groups of finite Morley rank were studied by Zil'ber [1977a], [1977b] and independently by Cherlin [1979]. Cherlin's paper conjectured that every simple totally transcendental group is an algebraic group over an algebraically closed field. Some partial results are known, all of them positive; see Appendix A.4. For example the answer is Yes if the group has Morley rank at most 2 (Cherlin [1979]). Recently Hrushovski [1989a] showed that this result of Cherlin is closely tied to a fundamental question of pure stability theory; this would have been an amazing discovery in the 1970s, but today we expect such connections.

The notion of a generic type also came out of the papers of Zil'ber and Cherlin via Cherlin & Shelah [1980] and Poizat [1983a]. Zil'ber's indecomposability theorem (Theorem 5.7.12) is from [1977a], and Theorem 5.7.13 and Fact 5.7.14 are in Chapter 2 of Zil'ber [1977b]. The indecomposability theorem is an adaptation of a classical result of Chevalley ([1951] §7 Prop. 2; see Humphreys [1975] p. 55) on algebraic groups. Theorem 5.7.7 is from

Macintyre [1971b], though the use of generic types is from Poizat [1987a]; see Theorem A.5.5 for an extension. Theorem 5.7.1 was also noted by Macintyre [1971a].

Poizat introduced the terminology 'group' for group-like structure and 'pure group' for group. Poizat is a master of exposition, but this choice of names is bound to confuse people; I have not followed it. As noted in the text, Fact 5.7.8 is from Evans, Pillay & Poizat [1990]; though in the special case of vector spaces it was proved in about 1986 by Colin Elliott, using more classical arguments. (The classical proof is long and elaborate, which caused Colin to give up mathematics and become a successful barrister in the West Country.)

Exercise 4: Baur, Cherlin & Macintyre [1979], Nesin [1989]. Exercise 5: Poizat [1983a]. Exercise 7: Zil'ber [1977a]. Exercise 8: Zil'ber [1977b]. Exercise 9: Tsuboi [1988]. Exercise 10: Villemaire [1990].

6

The first-order case: compactness

A given species of bird would show the same ability of grasping . . . numbers
. . . but the ability differs with the species. Thus with pigeons it may be five
or six according to experimental conditions, with jackdaws it is six and with
ravens and parrots, seven.
O. Koehler, The ability of birds to 'count', *Bull. Animal Behaviour* **9** (1950) 41–5.

Ravens, so we read, can only count up to seven. They can't tell the difference between two numbers greater than or equal to eight. First-order logic is much the same as ravens, except that the cutoff point is rather higher: it's ω instead of 8.

This chapter is wholly devoted to the model theory of first-order languages. First-order model theory has always been the heart of model theory. The main reason for this is that first-order logic, for all its expressive power, is too weak to distinguish between one large number and another. The result is that there are a number of constructions which give models of a first-order theory, or turn a given model into a new one. In this chapter we study two such constructions. The first is a combination of the compactness theorem with (Robinson) diagrams. The second is amalgamation; it can be seen as an application of the first.

Granted, some fragments of first-order logic are even weaker than the full logic, so that they allow even more constructions. An obvious example is Horn logic: the class of models of a Horn theory is closed under direct products, as we shall see in section 9.1. But this would hardly be a good reason for concentrating our efforts on Horn theories. In the first place, regardless of the success of Prolog and equational specification languages, mathematics is full of notions which can be written in a natural way as first-order formulas but not as Horn formulas. In the second place, the compactness theorem is our single most powerful tool, and it lives at the level of first-order logic.

In fact the compactness theorem rests on the boolean structure of propositional logic. The quantifiers have very little to do with it – we can eliminate them by skolemising. I use this as an excuse to introduce boolean algebras. We shall need them for ultrafilters in Chapter 9, and their dual spaces keep popping up in stability theory as spaces of complete types.

The idea of amalgamation is very powerful, and I have used it whenever I can. Subtly different kinds of amalgamation produce preservation theorems and definability theorems. We can even base stability theory on amalgamations. (The more familiar notion of extending types is really amalgamation in disguise.) A thumbnail definition of stable theories might be 'first-order theories in whose models amalgamation is particularly well behaved'.

In a logical development this chapter would lead straight on to Chapters 10 and 11, which also explore the compactness theorem and the inability of first-order logic to tell one infinite cardinal from another. I have postponed these chapters in order to avoid being drawn too soon into matters of cardinal arithmetic – and to express solidarity with the ravens.

6.1 Compactness for first-order logic

If any theorem is fundamental in first-order model theory, it must surely be the compactness theorem.

Theorem 6.1.1 (*Compactness theorem for first-order logic*). *Let T be a first-order theory. If every finite subset of T has a model then T has a model.*

Proof. Let L be a first-order language and T a theory in L. Assume first that every finite subset of T has a non-empty model. We shall show that T can be extended to a Hintikka set T^+ (see section 2.3 above) in a larger first-order language L^+. By Theorem 2.3.3 it follows that some L^+-structure A is a model of T^+, so the reduct $A^+|L$ will be a model of T.

Write κ for the cardinality of L. Let $c_i \, (i < \kappa)$ be distinct constants not in L; we call these constants **witnesses**. Let L^+ be the first-order language got by adding the witnesses c_i to the signature of L. Then L^+ has κ sentences, and we can list them as $\phi_i \, (i < \kappa)$. We shall define an increasing chain $(T_i : i \leqslant \kappa)$ of theories in L^+, so that the following hold. (All models are assumed to be L^+-structures.)

(1.1)   For each $i \leqslant \kappa$, every finite subset of T_i has a model.

(1.2)   For each $i < \kappa$, the number of witnesses c_k which are used in

$$T_i \text{ but not in } \bigcup_{j < i} T_j \text{ is finite.}$$

The definition is by induction on i. We put $T_0 = T$ and at limit ordinals δ we take $T_\delta = \bigcup_{i < \delta} T_i$. Clearly these definitions respect (1.1) and (1.2); (1.1) is true at T_0 because of our assumption that every finite subset of T has a non-empty model.

For successor ordinals $i + 1$ we first define

$$(1.3) \quad T'_{i+1} = \begin{cases} T_i \cup \{\phi_i\} & \text{if every finite subset of this set has a model,} \\ T_i & \text{otherwise.} \end{cases}$$

If $\phi_i \in T'_{i+1}$ and ϕ_i has the form $\exists x\, \psi$ for some formula $\psi(x)$, then we choose the earliest witness c_j which is not used in T'_{i+1} (by (1.2) there is such a witness), and we put

$$(1.4) \qquad\qquad\qquad T_{i+1} = T'_{i+1} \cup \{\psi(c_j)\}.$$

If $\phi_i \notin T'_{i+1}$ or ϕ_i is not of the form $\exists x\, \psi$, we put $T_{i+1} = T'_{i+1}$. These definitions clearly ensure (1.2), but we must show that (1.1) remains true when (1.4) holds. Let U be a finite subset of T'_{i+1} and let A be any L^+-structure which is a model of $U \cup \{\exists x\, \psi\}$. Then there is an element a of A such that $A \vDash \psi(a)$. Take such an element a, and let B be the L^+-structure which is exactly like A except that $c_j^B = a$. Since the witness c_j never occurs in U, B is still a model of U, and since c_j never occurs in $\psi(x)$, $B \vDash \psi(a)$ and so $B \vDash \psi(c_j)$. Condition (1.1) is secured.

We claim that T_κ is a Hintikka set for L^+. By Theorem 2.3.4 it suffices to prove three things.

(a) Every finite subset of T_κ has a model. This holds by (1.1).

(b) For every sentence ϕ of L^+, either ϕ or $\neg\phi$ is in T_κ. To prove this, suppose ϕ is ϕ_i and $\neg\phi$ is ϕ_j. If $\phi \notin T_\kappa$ then $\phi_i \notin T_{i+1}$, and by (1.3) this means that there is a finite subset U of T_i such that $U \cup \{\phi\}$ has no model. By the same argument, if $\neg\phi \notin T_\kappa$ then there is a finite subset U' of T_j such that $U' \cup \{\neg\phi\}$ has no model. Now $U \cup U'$ is a finite subset of T_κ, so it has a model A. Either $A \vDash \phi$ or $A \vDash \neg\phi$, and we have a contradiction either way. Thus at least one of ϕ, $\neg\phi$ is in T_κ.

(c) For every sentence $\exists x\, \psi(x)$ in T_κ there is a closed term t of L^+ such that $\psi(t) \in T_\kappa$. For this, suppose $\exists x\, \psi(x)$ is ϕ_i. Since $\phi_i \in T_\kappa$, (1.4) applies, and so T_{i+1} contains a sentence $\psi(c_j)$ where c_j is a witness. Then $\psi(c_j)$ is in T_κ.

Thus T_κ is a Hintikka set T^+ for L^+ and it contains T; so T has a model. This completes the argument, except in the freak case where some finite subset of T has only the empty model. In this case the empty L-structure must be a model of all T. $\qquad\qquad\square$

Corollaries of compactness

The first two corollaries are immediate. In fact Corollary 6.1.2 is just another way of stating the theorem. (See Exercise 1.)

Corollary 6.1.2. *If T is a first-order theory, ψ a first-order sentence and $T \vdash \psi$, then $U \vdash \psi$ for some finite subset U of T.*

Proof. Suppose to the contrary that $U \not\vdash \psi$ for every finite subset U of T. Then every finite subset of $T \cup \{\neg \psi\}$ has a model, so by the compactness theorem $T \cup \{\neg \psi\}$ has a model. Therefore $T \not\vdash \psi$.   □

The next corollary is one of our few encounters with formal inference rules. If we overlook the fact that computers have finite memories, a set is recursively enumerable (r.e. for short) if and only if it can be listed by a computer. (Recursive languages were defined in section 2.3.)

Corollary 6.1.3. *Suppose L is a recursive first-order language, and T is a recursively enumerable theory in L. Then the set of consequences of T in L is also recursively enumerable.*

Proof. Using one's favourite proof calculus, one can recursively enumerate all the consequences in L of a finite set of sentences. Since T is r.e., we can recursively enumerate its finite subsets; Corollary 6.1.2 says that every consequence of T is a consequence of one of these finite subsets.   □

Since first-order logic can't distinguish between one infinite cardinal and another, there should be no surprise that every infinite structure has arbitrarily large elementary extensions. The compactness theorem will give us this result very quickly.

I follow custom in calling Corollary 6.1.4 the upward Löwenheim–Skolem theorem. But in fact Skolem didn't even believe it, because he didn't believe in the existence of uncountable sets.

Corollary 6.1.4 *(Upward Löwenheim–Skolem theorem). Let L be a first-order language of cardinality $\leqslant \lambda$ and A an infinite L-structure of cardinality $\leqslant \lambda$. Then A has an elementary extension of cardinality λ.*

Proof. Supplying names for all the elements of A, consider the elementary diagram of A, eldiag(A) for short (see section 2.5 above). Let c_i $(i < \lambda)$ be λ new constants, and let T be the theory

(1.5) $\mathrm{eldiag}(A) \cup \{c_i \neq c_j : i < j < \lambda\}$.

We claim that every finite subset of T has a model. For suppose U is a finite subset of T. Then for some $n < \omega$, just n of the new constants c_i occur in U. We can find a model of T by taking n distinct elements of A and letting each of the new constants in U stand for one of these elements; this is possible because A is infinite.

It follows by the compactness that T has a model B. Since B is a model of eldiag(A), there is an elementary embedding $e: A \to B|L$ (by the elementary diagram lemma, Lemma 2.5.3). Replacing elements of the image of e by the corresponding elements of A, we make $B|L$ an elementary extension of A. Since $B \vDash T$, we have $c_i^B \neq c_j^B$ whenever $i < j < \lambda$, and hence $B|L$ has at least λ elements. To bring the cardinality of $B|L$ down to exactly λ, we invoke the downward Löwenheim–Skolem theorem (Corollary 3.1.5).   □

Compactness in infinitary languages?

The compactness theorem fails for languages of form $L_{\omega_1\omega}$. For example let c_i ($i < \omega$) be distinct constants, and consider the theory

(1.6) $c_0 \neq c_1,\ c_0 \neq c_2,\ c_0 \neq c_2,\ \ldots,$

$$\bigvee_{0<i<\omega} c_0 = c_i.$$

This theory has no model, but every proper subset of it has a model.

Naturally one looks for weaker forms of compactness which might hold in some infinitary languages. For example, let L be a language and λ an infinite cardinal. We say that L is **weakly λ-compact** if for every theory T in L with $|T| = \lambda$, if each set of fewer than λ sentences in T has a model, then T has a model. Thus every first-order language is weakly ω-compact.

The next result shows that even with this weaker notion of compactness, very few languages $L_{\lambda\omega}$ are willing to be helpful.

Theorem 6.1.5. *Let λ be an infinite cardinal such that for every signature L the language $L_{\lambda\omega}$ is weakly λ-compact. Then λ is a regular limit cardinal.*

Proof. Suppose first that λ is not a limit cardinal. Then $\lambda = \kappa^+$ for some cardinal κ. Let L be the signature consisting of a 2-ary relation symbol $<$, a 2-ary function symbol F, and the ordinals $\alpha \leq \lambda$ as constants. Let T be the following theory in $L_{\lambda\omega}$.

(1.7)   All true sentences of the form '$\alpha < \beta$' with $\alpha, \beta \leq \lambda$.

(1.8)   '$<$ linearly orders all elements.'

(1.9) $\forall x (x < \kappa \to \bigvee_{\alpha<\kappa} x = \alpha)$.

(1.10) $\forall x$ ($F(x, -)$ is an injective map from $\{y : y < x\}$ to $\{y : y < \kappa\}$).

Then $|T| = \lambda$ and any set of at most κ sentences of T has a model, but T has no model. Hence $L_{\lambda\omega}$ is not weakly λ-compact.

A similar argument shows that if λ is singular then some language $L_{\lambda\omega}$ is not weakly λ-compact. I leave this as an exercise.   □

Now there are models of Zermelo–Fraenkel set theory in which the only regular limit cardinal is ω. So one can be a perfectly sane person and still believe that weak λ-compactness never occurs except when $\lambda = \omega$. If we want useful generalisations of the compactness theorem, maybe we are looking in the wrong place. Sections 11.5 and 11.6 will suggest another direction.

Exercises for section 6.1

1. Show that each of the following is equivalent to the compactness theorem for first-order logic. (a) For every theory T and sentence ϕ of a first-order language, if $T \vdash \phi$ then for some finite $U \subseteq T$, $U \vdash \phi$. (b) For every theory T and sentence ϕ of a first-order language, if T is equivalent to the theory $\{\phi\}$ then T is equivalent to some finite subset of T. (c) For every first-order theory T, every tuple $\bar{x}$ of distinct variables and all sets $\Phi(\bar{x})$, $\Psi(\bar{x})$ of first-order formulas, if $T \vdash \forall \bar{x}(\bigwedge \Phi \leftrightarrow \bigvee \Psi)$ then there are finite sets $\Phi' \subseteq \Phi$ and $\Psi' \subseteq \Psi$ such that $T \vdash \forall \bar{x}(\bigwedge \Phi' \leftrightarrow \bigvee \Psi')$.

2. (a) Let L be a first-order language, T a theory in L and Φ a set of sentences of L. Suppose that for all models A, B of T, if $A \vDash \phi \Leftrightarrow B \vDash \phi$ for each $\phi \in \Phi$, then $A \equiv B$. Show that every sentence ψ of L is equivalent modulo T to a boolean combination ψ^* of sentences in Φ. (b) Show moreover that if L and T are recursive then ψ^* can be effectively computed from ψ.

3. (Craig's trick) In a recursive first-order language L let T be an r.e. theory. Show that T is equivalent to a recursive theory T^*. [Write ϕ^n for $\phi \wedge \ldots \wedge \phi$ (n times). Try $T^* = \{\phi^n : n$ is the Gödel number of a computation putting ϕ into $T\}$.]

4. Let L be a first-order language, δ a limit ordinal (for example ω) and $(T_i : i < \delta)$ an increasing chain of theories in L, such that for every $i < \delta$ there is a model of T_i which is not a model of T_{i+1}. Show that $\bigcup_{i < \delta} T_i$ is not equivalent to a sentence of L.

5. Show that none of the following classes is first-order definable (i.e. by a single sentence; see section 2.2). (a) The class of infinite sets. (b) The class of torsion-free abelian groups. (c) The class of algebraically closed fields.

6. Let L be a first-order language and T a theory in L. (a) Suppose T has models of arbitrarily high cardinalities; show that T has an infinite model. (b) Let $\phi(x)$ be a formula of L such that for every $n < \omega$, T has a model A with $|\phi(A)| \geq n$. Show that T has a model B for which $\phi(B)$ is infinite.

7. Let L be the first-order language of fields and ϕ a sentence in L. Show that if ϕ is true in every field of characteristic 0, then there is a positive integer m such that ϕ is true in every field of characteristic $\geq m$.

*Recall that we say a theory T is λ-**categorical** if T has, up to isomorphism, exactly one model of cardinality λ.*

8. (a) Let L be a first-order language, T a theory in L and λ a cardinal $\geqslant |L|$. Show that if T is λ-categorical then T is complete. (b) Use (a) to give quick proofs of the completeness of (i) the theory of (non-empty) dense linear orderings without endpoints and (ii) the theory of algebraically closed fields of a fixed characteristic. Find two other nice examples.

9. Suppose L is a first-order language, A is an L-structure and $\phi(x)$ is a formula of L such that $\phi(A)$ is infinite. Show that for every cardinal $\lambda \geqslant \max(|A|, |L|)$, there is an elementary extension B of A in which $\phi(B) = \lambda$.

10. Suppose L is a first-order language, A is an L-structure and λ is a cardinal $\geqslant \max(|A|, |L|)$. Show that A has an elementary extension B of cardinality λ such that for every formula $\phi(x, \bar{y})$ of L and every tuple $\bar{b}$ of elements of B, $\phi(B, \bar{b})$ has cardinality either $= \lambda$ or $< \omega$. [This is a baby version of the constructions of saturated elementary extensions in section 10.2 below. Iterate the construction of the previous exercise to form an elementary chain of length λ, using the Tarski–Vaught theorem on elementary chains, Theorem 2.5.2.]

11. Show that if L is a first-order language of cardinality λ and A is an L-structure of cardinality $\mu \geqslant \lambda$, then A has an elementary extension B of cardinality μ with $|\text{dom}(B)\backslash\text{dom}(A)| = \mu$. If $v < \mu$, give an example to show that A need not have an elementary extension B of cardinality μ with $|\text{dom}(B)\backslash\text{dom}(A)| = v$.

Section 11.5 will explore some analogues of the next exercise for infinitary logics.

12. Let L be a first-order language and $\phi(x, y)$ a formula of L. Show that if A is an L-structure in which $\phi(A^2)$ is a well-ordering of an infinite set, then there is an L-structure B elementarily equivalent to A, in which $\phi(B^2)$ is a linear ordering relation which is not well-ordered. [Adapt the proof of Corollary 6.1.4.]

The following result in algebraic geometry was first proved by methods of logic.

*13. Let L be the first-order language of fields. (a) Show that for every positive integer n there is an $\forall_2$ sentence ϕ_n of L which expresses (in any field K) 'for any $k, m \leqslant n$, if $\bar{X} = (X_0, \ldots, X_{m-1})$ is an m-tuple of indeterminates and $p_0, \ldots, p_{k-1}$, $q_0, \ldots, q_{m-1}$ are any polynomials $\in K[\bar{X}]$ of degree $\leqslant n$ such that $q_0, \ldots, q_{m-1}$ define an injective mapping from V to V, where $V = \{\bar{a} \in K^m : p_0(\bar{a}) = \ldots = p_{k-1}(\bar{a}) = 0\}$, then this mapping is surjective'. (b) Show that each sentence ϕ_n is true in every finite field, and hence in any union of a chain of finite fields. [See Theorem 2.4.4.] (c) Show that each sentence ϕ_n is true in every algebraically closed field of prime characteristic. (d) Show that each sentence ϕ_n is true in every algebraically closed field. (e) Deduce that if V is an algebraic variety and F is an injective morphism from V to V, then F is surjective.

*14. Show that there is a field K such that $\text{Th}(K)$ is undecidable but every sentence in $\text{Th}(K)$ has a finite model. [Let U be the set of all first-order sentences true in every finite field. For any two disjoint finite sets X, Y of primes there is a prime q such that the Legendre symbol (p/q) has the value 1 when $p \in X$ and -1 when $p \in Y$. Now take any non-recursive set Z of primes, and use compactness to find a model of $U \cup \{\exists x \, x^2 = p : p \in Z\} \cup \{\neg\exists x \, x^2 = p : p \text{ a prime} \notin Z\}$.]

15. Show that if λ is a singular cardinal then, for suitable signatures L, the language $L_{\lambda\omega}$ is not weakly λ-compact.

16. Let L be a recursive first-order language. Show that if T is a recursively enumerable theory in L, and there are only finitely many inequivalent complete theories $\supseteq T$ in L, then T is decidable.

17. Let L be a recursive first-order language and T a theory in L. Show that if T is decidable then there is a decidable complete theory $T' \supseteq T$ in L.

18. Let L be a recursive first-order language and T a consistent recursively enumerable theory in L. Show that T is decidable if and only if there are models A_i $(i < \omega)$ of T such that $T = \text{Th}\{A_i : i < \omega\}$, and an algorithm which determines, for any $i < \omega$ and any sentence ϕ of L, whether or not $A_i \vDash \phi$. [For left to right use Exercise 17.]

6.2 Boolean algebras and Stone spaces

Propositional logic and boolean algebras both came from the hand of George Boole; in fact he made no clear distinction between the two. Today we can distinguish them better. But we can also connect them in more ways than Boole could have dreamed of.

In this section I concentrate on one of these newer connections: the compactness theorem is equivalent to a choice principle known as the boolean prime ideal theorem. Experts on boolean algebras can move on quickly to the next section.

Let $\mathscr{B}$ be a boolean algebra (see (2.29) in section 2.2 for the axioms). We write 1, 0 for the top and bottom elements of $\mathscr{B}$, and $\wedge$, $\vee$, * for the boolean operations of meet, join and complement. We write $\leqslant$ for the associated partial ordering: $a \leqslant b$ iff $a \wedge b = a$. An **atom** of $\mathscr{B}$ is an element $b > 0$ such that there is no element c with $0 < c < b$. When we need to avoid confusion between the elements of $\mathscr{B}$ and those of some other structure, I use Greek letters α, β etc. for the elements of $\mathscr{B}$.

Filters on a boolean algebra

A **filter** on $\mathscr{B}$ is a non-empty subset $\mathscr{F}$ of $\mathscr{B}$ such that

(2.1)   if $a \in \mathscr{F}$ and $a \leqslant b$ then $b \in \mathscr{F}$,

(2.2)   if $a, b \in \mathscr{F}$ then $a \wedge b \in \mathscr{F}$, and

(2.3) $0 \notin \mathscr{F}$.

Since $\mathcal{F}$ is non-empty, (2.1) implies that $1 \in \mathcal{F}$. When (2.1) and (2.2) hold but (2.3) fails, $\mathcal{F}$ is the whole of $\mathcal{B}$, and we describe $\mathcal{F}$ as the **improper filter** on $\mathcal{B}$. An **ultrafilter** on $\mathcal{B}$ is a filter $\mathcal{F}$ of $\mathcal{B}$ such that

(2.4)   for each element a of $\mathcal{B}$, exactly one of a, a^* is in $\mathcal{F}$.

For filters I use script letters $\mathcal{F}$ etc.; $\mathcal{U}$ is an ultrafilter. Ultrafilters can also be written p, q etc., particularly when they are thought of as points of a space of ultrafilters.

A filter is called **principal** if it is of the form $\{a: a \geqslant b\}$ for some fixed element b. A set W of elements of a boolean algebra $\mathcal{B}$ is said to have the **finite intersection property** if for every finite set $b_0, \ldots, b_{m-1}$ of elements of W, $b_0 \wedge \ldots \wedge b_{m-1} \neq 0$. Note that every filter has the finite intersection property, by (2.2) and (2.3).

Example 1: *Power-set algebras.* The set $\mathcal{P}(X)$ of all subsets of a non-empty set X forms a boolean algebra with $\subseteq$ as the partial ordering relation; we call this the **power-set algebra** on X. Filters and ultrafilters of $\mathcal{P}(X)$ are called **filters over** X and **ultrafilters over** X. The principal ultrafilters over X are the sets of the form $\{a: x \in a\}$ for some element x of X, as one easily checks.

Example 2: *The finite–cofinite algebra on a set.* Let X be any infinite set. A **cofinite** subset of X is a set $Y \subseteq X$ such that $X \backslash Y$ is finite. The set of all finite or cofinite subsets of X forms a boolean algebra with $\subseteq$ as the partial ordering relation; we call it the **finite–cofinite** algebra on X. In this boolean algebra, the set of all cofinite sets forms a non-principal ultrafilter.

Theorem 6.2.1 below is often known as the **boolean prime ideal theorem**, or BPI for short. To explain the name, a **prime ideal** of $\mathcal{B}$ is a set of the form $\{a$ in $\mathcal{B}: a \notin \mathcal{U}\}$ for some ultrafilter $\mathcal{U}$ of $\mathcal{B}$. (One can show that this set is equal to $\{a^*: a \in \mathcal{U}\}$.) When the theorem was first proved, it was stated in terms of prime ideals, but for our purposes it's better to think in terms of ultrafilters.

Before I prove the theorem, let me make some remarks about its proof.

It's impossible to prove the BPI from ZF (Zermelo–Fraenkel set theory without the axiom of choice). But there is a fairly easy direct proof of it by means of Zorn's lemma, which is a form of the axiom of choice. So the BPI is a kind of set-theoretic choice principle; in fact one can use it to prove various consequences of the axiom of choice. Modulo ZF, the BPI is equivalent to the compactness theorem. (This should be no surprise – we assumed the axiom of choice when we well-ordered the set of sentences of L^+ in the proof of the compactness theorem. See Exercise 5 below for the other direction.) So one can use the compactness theorem as a kind of choice principle too, and this is what is happening in Theorem 6.2.1. Exercises 12–15 are other examples.

However, Halpern and Lévy [1971] showed that the full axiom of choice is not deducible from ZF with the BPI: the BPI doesn't have the full strength of the axiom of choice. In fact the BPI turns out to be equivalent (modulo ZF) to each of the statements 'Every ring has a prime ideal' and 'Every product of compact Hausdorff spaces is compact', while the full axiom of choice is equivalent to the stronger statements 'Every ring has a maximal ideal' and 'Every product of compact spaces is compact'. (See Rubin & Rubin [1985] on equivalents of the axiom of choice.)

Theorem 6.2.1. *Let $\mathscr{B}$ be a boolean algebra and W a set of elements of $\mathscr{B}$ which has the finite intersection property. Then there is an ultrafilter $\mathscr{U}$ on $\mathscr{B}$ such that $W \subseteq \mathscr{U}$.*

Proof. We first prove the theorem under the assumption that $\mathscr{B}$ is finitely generated. In this case every element of $\mathscr{B}$ can be written as a boolean combination of the generators. This boolean combination can be brought to disjunctive normal form, and it easily follows that $\mathscr{B}$ must be finite. So W is finite; say $W = \{d_0, \ldots, d_{m-1}\}$. Then by the finite intersection property, $d = d_0 \wedge \ldots \wedge d_{m-1}$ is a non-zero element of $\mathscr{B}$. Among the finitely many elements of $\mathscr{B}$ which are $\leqslant d$, there must be an atom, say b. Define $\mathscr{U}$ to be the set $\{a: a \geqslant b\}$. Then $\mathscr{U}$ is an ultrafilter containing all of W.

Now we turn to the general case. Let L be the language of boolean algebras with each element of $\mathscr{B}$ added as a new constant, and one extra 1-ary relation symbol P. Let T be the theory

(2.5) $\quad \{P(a) \to P(b): a \leqslant b \text{ in } \mathscr{B}\}$

$\qquad \cup \{P(a) \wedge P(b) \to P(c): a \wedge b = c \text{ in } \mathscr{B}\}$

$\qquad \cup \{P(a) \leftrightarrow \neg P(b): a^* = b \text{ in } \mathscr{B}\} \cup \{P(a): a \in W\}.$

We claim that T has a model. For suppose not. Then by the compactness theorem, some finite subset U of T has no model. Let $\mathscr{C}$ be the subalgebra of $\mathscr{B}$ generated by all the elements b of $\mathscr{B}$ which occur somewhere in the sentences in U. Since $\mathscr{B}$ is a boolean algebra and U is finite, $\mathscr{C}$ must be a finite boolean algebra. Let $c_0, \ldots, c_{k-1}$ be the elements c of $\mathscr{C}$ such that '$P(c)$' $\in U$. Then since W has the finite intersection property, $c_0 \wedge \ldots \wedge c_{k-1} \neq 0$ in $\mathscr{B}$ and hence also in $\mathscr{C}$. Since $\mathscr{C}$ is finite, the first part of our proof shows that there is an ultrafilter $\mathscr{F}$ of $\mathscr{C}$ which contains $c_0 \wedge \ldots \wedge c_{k-1}$. Then, as one can check from (2.5), we form a model A of U by expanding $\mathscr{C}$ so that each element names itself and P names $\mathscr{F}$. This proves the claim.

Now let B be a model of T. Define a subset $\mathscr{U}$ of dom($\mathscr{B}$) by $b \in \mathscr{U}$ iff $B \vDash P(b)$. Then we can read off from (2.5) that $\mathscr{U}$ is an ultrafilter containing W. $\qquad \square$

Corollary 6.2.2. *Every infinite boolean algebra has a non-principal ultrafilter.*

Proof. Let $\mathcal{B}$ be an infinite boolean algebra, and let W be the set $\{a^*: a$ is an atom of $\mathcal{B}\}$. Then W has the finite intersection property; for otherwise there are atoms $a_0, \ldots, a_{n-1}$ of $\mathcal{B}$ such that

$$a_0 \vee \ldots \vee a_{n-1} = (a_0^* \wedge \ldots \wedge a_{n-1}^*)^* = 1,$$

which implies that $\mathcal{B}$ is finite. Let $\mathcal{U}$ be an ultrafilter containing W. Then $\mathcal{U}$ is not principal; for otherwise it would contain some atom a, which is impossible since it contains a^*. $\square$

Now let $\mathcal{B}$ be any boolean algebra. We can define a topological space $S(\mathcal{B})$ from $\mathcal{B}$ as follows. The elements of $S(\mathcal{B})$ are the ultrafilters of $\mathcal{B}$. We make $S(\mathcal{B})$ into a topological space by taking as a basis of closed sets the family of all sets of the form $\{p \in S(\mathcal{B}): \beta \in p\}$ as β ranges over the elements of $\mathcal{B}$. I omit the details.

Theorem 6.2.3 lists the central facts about these spaces $S(\mathcal{B})$. A subset of a topological space is **clopen** if it is simultaneously closed and open. A **Stone space** is a non-empty topological space which is compact Hausdorff, and in which the clopen sets form a basis of closed sets.

Theorem 6.2.3. (a) *If $\mathcal{B}$ is a boolean algebra, then $S(\mathcal{B})$ is a Stone space.*

(b) *If $\mathcal{S}$ is a Stone space, then the clopen subsets of $\mathcal{S}$ form a boolean algebra $B(\mathcal{S})$ with $\subseteq$ as the algebra ordering.*

(c) *For every boolean algebra $\mathcal{B}$ the boolean algebra $B(S(\mathcal{B}))$ is isomorphic to $\mathcal{B}$ by the correspondence $\beta \mapsto \{p \in S(\mathcal{B}): \beta \in p\}$.*

(d) *For every Stone space $\mathcal{S}$ the Stone space $S(B(\mathcal{S}))$ is a homeomorphic image of $\mathcal{S}$ under the homeomorphism $p \mapsto \{\beta \in B(\mathcal{S}): p \in \beta\}$.*

Proof. For the whole story see Halmos [1963] or Sikorski [1964]. $\square$

Theorem 6.2.3 is a form of the **Stone duality theorem** for boolean algebras. The boolean algebra $B(S(\mathcal{B}))$ is known as the **second dual** of $\mathcal{B}$.

Exercises for section 6.2

1. Let X be a set and $\mathcal{P}(X)$ the power-set algebra of X. If $\mathcal{U}$ is an ultrafilter on $\mathcal{P}(X)$, show that $\mathcal{U}$ is a principal ultrafilter if and only if $\bigcap \mathcal{U}$ is non-empty.

2. Let $\mathcal{B}$ be a boolean algebra, and suppose a well-ordering of dom$(\mathcal{B})$ is given. Without using the axiom of choice, show that $\mathcal{B}$ has an ultrafilter. [Copy the proof of Theorem 6.1.1.]

3. Show that if $\mathcal{B}$ is a boolean algebra and W is a set of elements of $\mathcal{B}$ which has the finite intersection property and is maximal with this property, then W is an ultrafilter on $\mathcal{B}$. Deduce Theorem 6.2.1 by Zorn's lemma.

4. Let $\mathcal{B}$ be a boolean algebra and $\mathcal{F}$ a filter on $\mathcal{B}$. We define a relation $\sim$ on the set of elements of $\mathcal{B}$ by $a \sim b$ iff $a^* \vee b$ and $a \vee b^*$ are both in $\mathcal{F}$. (a) Show that $\sim$ is an equivalence relation. (b) Show that the set of equivalence classes $a^\sim$ can be given the structure of a boolean algebra in such a way that the map $e: a \mapsto a^\sim$ is a homomorphism and $\mathcal{F}$ is the set of all elements a of $\mathcal{B}$ such that $e(a) = 1$. (We write this algebra as $\mathcal{B}/\mathcal{F}$.) (c) Show (without using any form of the axiom of choice) that Theorem 6.2.1 follows from the statement 'Every boolean algebra has an ultrafilter'. [If W is a set of elements of $\mathcal{B}$ with the finite intersection property, let W generate a filter $\mathcal{F}$ on $\mathcal{B}$. Find an ultrafilter on $\mathcal{B}/\mathcal{F}$ and pull it back to $\mathcal{B}$.]

5. Show (using only ZF and not any form of the axiom of choice) that the BPI implies the compactness theorem. [To get rid of the quantifiers, skolemise and add constants.]

6. Assume it has been proved that every map on the plane, with finitely many countries, can be coloured with just four colours so that two adjacent countries never have the same colour. Use the compactness theorem to show that the same holds even when the map has infinitely many countries (but each country has finitely many neighbours, of course).

7. Let $\mathcal{B}$ be a boolean algebra. (a) Show that $\mathcal{B}$ can be regarded as a commutative ring by reading $x \cdot y$ as $x \wedge y$, $x + y$ as $(x^* \wedge y) \vee (x \wedge y^*)$, $-x$ as x, 1 as 1 and 0 as 0. (Rings formed in this way from boolean algebras are called **boolean rings**.) (b) If X is a set of elements of $\mathcal{B}$, show that X is a prime ideal in the boolean algebra sense if and only if X is a prime ideal in the ring sense. (c) An **ideal** of $\mathcal{B}$ is a non-empty set I of elements of $\mathcal{B}$ such that (i) if $b \in I$ and $a \leqslant b$ then $a \in I$, (ii) if $a, b \in I$ then $a \vee b \in I$, and (iii) $1 \notin I$. (This is the dual to the definition of filter.) Show that a subset I of $\text{dom}(\mathcal{B})$ is an ideal of $\mathcal{B}$ if and only if it is a proper ideal in the ring-theoretic sense.

8. Show that if $\mathcal{B}$ is a boolean algebra, $\mathcal{B}_0$ is a subalgebra of $\mathcal{B}$ and a is an element of $\mathcal{B}$ which is not in $\mathcal{B}_0$, then there are ultrafilters $\mathcal{U}, \mathcal{V}$ on $\mathcal{B}$ such that $a \in \mathcal{U}$, $a \notin \mathcal{V}$, and $\mathcal{U}, \mathcal{V}$ agree on $\mathcal{B}_0$. [Adapt the proof of Theorem 6.2.1: introduce a second new 1-ary relation symbol Q, and add to (2.5) the corresponding sentences for Q (when b is in $\mathcal{B}_0$), together with the new sentences $P(b) \leftrightarrow Q(b)$ (whenever b is in $\mathcal{B}_0$) and $P(a) \wedge \neg Q(a)$.]

9. Show that if $\mathcal{B}$ is an infinite boolean algebra then $|\mathcal{B}| \leqslant |S(\mathcal{B})| \leqslant 2^{|\mathcal{B}|}$. [For the lower bound, use the previous exercise.]

*The next two exercises show that the bounds in Exercise 9 are best possible. A family $(A_\alpha: \alpha \in I)$ of subsets of a set S is said to be **independent** if for all distinct $i_0, \ldots, i_{m-1}$, $j_0, \ldots, j_{n-1} \in I$, $A_{i_0} \cap \ldots \cap A_{i_{m-1}} \cap (S \backslash A_{j_0}) \cap \ldots \cap (S \backslash A_{j_{n-1}}) \neq \emptyset$.*

10. Let λ be an infinite cardinal. (a) Show that there is a family $(f_\alpha: \alpha < 2^\lambda)$ of functions $f: \lambda \to \lambda$, so that for every finite subset X of 2^λ and every family of functions $g_\alpha: \lambda \to \lambda \, (\alpha \in X)$ there is $i < \lambda$ such that $g_\alpha(i) = f_\alpha(i)$ for all $\alpha \in X$. [List as $(Y_i, h_i) \, (i < \lambda)$ all the pairs (Y, h) where Y is a finite subset of λ and $h: \mathcal{P}(Y) \to \lambda$. For each $Z \subseteq \lambda$, define $f_Z: \lambda \to \lambda$ by $f_Z(i) = h_i(Z \cap Y_i)$.] (b) Show that there is an independent family of 2^λ subsets of λ. (c) Deduce that there are 2^{2^λ} ultrafilters over λ.

Let β be an ordinal. The **interval algebra** *of β is the subalgebra of $\mathcal{P}(\beta)$ generated by the intervals $[i, j)$ of β $(i \leqslant j < \beta)$.*
11. Let $\mathcal{B}$ be the interval algebra of an infinite ordinal β. Show that the number of ultrafilters on $\mathcal{B}$ is $|\beta|$.

12. Assume it has been proved that every finite commutative ring with $0 \neq 1$ has a prime ideal. Use the compactness theorem for first-order logic to deduce that the same holds without the restriction to finite rings.

13. An **ordered group** is a group whose set of elements is linearly ordered in such a way that $a < b$ implies $c \cdot a < c \cdot b$ and $a \cdot c < b \cdot c$ for all elements a, b, c. A group is **orderable** if it can be made into an ordered group by adding a suitable ordering. (a) Show that an orderable group can't have elements of finite order, except the identity. (b) Show from the structure theorem for finitely generated abelian groups that every finitely generated torsion-free abelian group is orderable. (c) Using the compactness theorem, show that if G is a group and every finitely generated subgroup of G is orderable then G is orderable. (d) Deduce that an abelian group is orderable if and only if it is torsion-free.

14. Let A be an integral domain, not necessarily commutative. Use the compactness theorem to show that if every finitely generated subring of A can be embedded in a division ring, then A can be embedded in a division ring.

15. Let A be a field. (a) Assume it has been proved that for every finite set of non-zero polynomials $p_0(X), \ldots, p_{n-1}(X)$ over A, there is an algebraic extension B of A such that each of the polynomials p_i has a root in B. Use the compactness theorem to deduce that A has an algebraic closure. (b) Use the compactness theorem to show that if B and C are algebraic closures of A then there is an isomorphism from B to C which is the identity on A.

Our next exercise shows that a form of the upward Löwenheim–Skolem theorem is equivalent to the full axiom of choice. Hartogs' theorem, which is provable from ZF without the axiom of choice, states that for every set b there is a least ordinal $\beta(b)$ such that no injective map $f: \beta(b) \to b$ exists.
16. (a) Show that if b is an infinite set and A is an abelian group whose set of elements is the disjoint union $b \cup \beta(b)$, then it is possible to define a well-ordering of b in terms of the natural ordering of $\beta(b)$ and the group operation of A. (b) Deduce (without assuming the axiom of choice) that if for every infinite set x there is an

abelian group whose set of elements is x, then the axiom of choice holds. (c) Show
that modulo ZF, the following form of the upward Löwenheim–Skolem theorem is
equivalent to the axiom of choice: if a countable theory T has an infinite model, then
for every infinite set x, T has a model whose set of elements is x.

6.3 Types

The upward Löwenheim–Skolem theorem (Corollary 6.1.4) was a little too
crude for comfort. It told us that an infinite structure A can be built up into a
larger structure, but it said nothing at all about what the new parts of the
structure look like. We want to ask for example whether this piece of A can
be kept small while that piece is expanded, whether automorphisms of A
extend to automorphisms of the elementary extension of A, and so forth.
Questions like these are often hard to answer, and they will keep us occupied
right through to Chapter 12.

The main notion used for analysing such questions is the notion of a **type**.
We met types briefly in section 2.3, and some of the later sections of
Chapters 4 and 5 used them. Now it's time to examine them more systematically. One can think of types as a common generalisation of two well-known
mathematical notions: the notion of a minimal polynomial in field theory, and
the notion of an orbit in permutation group theory.

Let L be a first-order language and A an L-structure. Let X be a set of
elements of A and $\bar{b}$ a tuple of elements of A. Let $\bar{a}$ be a sequence listing the
elements of X. The **complete type of $\bar{b}$ over** X (with respect to A, in the
variables $\bar{x}$) is defined to be the set of all formulas $\psi(\bar{x}, \bar{a})$ such that $\psi(\bar{x}, \bar{y})$
is in L and $A \vDash \psi(\bar{b}, \bar{a})$. More loosely, the complete type of $\bar{b}$ over X is
everything we can say about $\bar{b}$ in terms of X. (Although $\bar{a}$ will in general be
infinite, each formula $\psi(\bar{x}, \bar{y})$ of L has only finitely many free variables, so
that only a finite part of X is mentioned in $\psi(\bar{x}, \bar{a})$.)

We write the complete type of $\bar{b}$ over X with respect to A as $\mathrm{tp}_A(\bar{b}/X)$, or
$\mathrm{tp}_A(\bar{b}/\bar{a})$ where $\bar{a}$ lists the elements of X. The elements of X are called the
parameters of the complete type. Complete types are written p, q, r etc.;
one writes $p(\bar{x})$ if one wants to show that the variables of the type are $\bar{x}$. We
write $\mathrm{tp}_A(\bar{b})$ for $\mathrm{tp}_A(\bar{b}/\varnothing)$, the type of $\bar{b}$ over the empty set of parameters.
Note that if B is an elementary extension of A, then $\mathrm{tp}_B(\bar{b}/X) = \mathrm{tp}_A(\bar{b}/X)$.

Let $p(\bar{x})$ be a set of formulas of L with parameters from X. We shall say
that $p(\bar{x})$ is a **complete type over** X (with respect to A, in the variables $\bar{x}$) if
it is the complete type of some tuple $\bar{b}$ over X with respect to some
elementary extension of A. Putting it loosely again, a complete type over X
is everything we can say in terms of X about some possible tuple $\bar{b}$ of
elements of A – maybe the tuple $\bar{b}$ is really there in A, or maybe it only
exists in an elementary extension of A.

A **type** over X (with respect to A, in the variables $\bar{x}$) is a subset of a complete type over X. We shall write Φ, Ψ, $\Phi(\bar{x})$ etc. for types. A type is called an n-**type** $(n < \omega)$ if it has just n free variables. (Some writers insist that these variables should be $x_0, \ldots, x_{n-1}$ from some stock of variables x_0, $x_1, \ldots$. I shall not be so strict.)

We say that a type $\Phi(\bar{x})$ over X is **realised** by a tuple $\bar{b}$ in A if $\Phi \subseteq \mathrm{tp}_A(\bar{b}/X)$. If Φ is not realised by any tuple in A, we say that A **omits** Φ.

The next result, a corollary of the compactness theorem, tells us where to look for types. We say that a set $\Phi(\bar{x})$ of formulas of L with parameters in A is **finitely realised** in A if for every finite subset Ψ of Φ, $A \vDash \exists \bar{x} \bigwedge \Psi$.

Theorem 6.3.1. *Let L be a first-order language, A an L-structure, X a set of elements of A and $\Phi(x_0, \ldots, x_{n-1})$ a set of formulas of L with parameters from X. Then, writing $\bar{x}$ for $(x_0, \ldots, x_{n-1})$,*
(a) *$\Phi(\bar{x})$ is a type over X with respect to A if and only if Φ is finitely realised in A,*
(b) *$\Phi(\bar{x})$ is a complete type over X with respect to A if and only if $\Phi(\bar{x})$ is a set of formulas of L with parameters from X, which is maximal with the property that it is finitely realised in A.*
In particular if Φ is finitely realised in A, then it can be extended to a complete type over X with respect to A.

Proof. (a) Suppose first that Φ is a type over X with respect to A. Then there are an elementary extension B of A and an n-tuple $\bar{b}$ in B such that $B \vDash \bigwedge \Phi(\bar{b})$. So if Ψ is a finite subset of Φ, then $B \vDash \bigwedge \Psi(\bar{b})$ and hence $B \vDash \exists \bar{x} \bigwedge \Psi(\bar{x})$. Since $A \preccurlyeq B$ and the sentence is first-order, this implies $A \vDash \exists \bar{x} \bigwedge \Psi(\bar{x})$.

Conversely suppose Φ is finitely realised in A. Form eldiag(A) and take an n-tuple of distinct new constants $\bar{c} = (c_0, \ldots, c_{n-1})$. Let T be the theory

(3.1)   eldiag$(A) \cup \Phi(\bar{c})$.

We claim that every finite subset of T has a model. For let U be a finite subset of T, and let Ψ be the set of formulas $\psi(\bar{x})$ of Φ such that $\psi(\bar{c}) \in U$. By assumption $A \vDash \exists \bar{x} \bigwedge \Psi$, and hence there are elements $\bar{a}$ in A such that $A \vDash \bigwedge \Psi(\bar{a})$. By interpreting the constants $\bar{c}$ as names of the elements $\bar{a}$, we make A into a model of U. This proves the claim.

Now by the compactness theorem (Theorem 6.1.1), T has a model C. Since $C \vDash$ eldiag(A), the elementary diagram lemma (Lemma 2.5.3) gives us an elementary embedding $e: A \to C|L$, and by making the usual replacements we can assume that in fact $A \preccurlyeq C|L$. Let $\bar{b}$ be the tuple $\bar{c}^C$. Then $C \vDash \bigwedge \Phi(\bar{b})$ since $C \vDash T$. It follows that $\bar{b}$ satisfies $\Phi(\bar{x})$ in some elementary extension of A, and so Φ is a type over X with respect to A.

(b) If Φ is a complete type over X then Φ contains either ϕ or $\neg\phi$, for each formula $\phi(\bar{x})$ of L with parameters from X. This implies that Φ is a maximal type over X with respect to A. On the other hand if Φ is a maximal type over X with respect to A, then for some tuple $\bar{b}$ in some elementary extension B of A, $B \vDash \bigwedge \Phi(\bar{b})$, so that Φ is included in the complete type of $\bar{b}$ over X. By maximality it must equal this complete type. $\square$

By this theorem, if X is the empty set of parameters, then the question whether Φ is a type over X with respect to A depends only on $\mathrm{Th}(A)$. Types over the empty set with respect to A are also known as the **types of** $\mathrm{Th}(A)$. More generally let T be any theory in a first-order language. Then a **type of** T is a set $\Phi(\bar{x})$ of formulas of L such that $T \cup \{\exists \bar{x} \bigwedge \Psi\}$ is consistent for every finite subset $\Psi(\bar{x})$ of Φ; a **complete type** of T is a maximal type of T. If T happens to be a complete theory, then we can replace '$T \cup \{\exists \bar{x} \bigwedge \Psi\}$ is consistent' by the equivalent statement '$T \vdash \exists \bar{x} \bigwedge \Psi$'. All this agrees with the definitions in section 2.3.

Types and automorphisms

The next result sets up the connection between types and orbits. I quote it without proof, since everything in it follows almost at once from the definitions. Granted, we haven't defined λ-big structures yet. They will be defined in section 10.1, and in section 10.2 we shall prove the following fact (a corollary of Theorem 10.2.1):

(3.2)   *every L-structure A has λ-big elementary extensions for arbitrarily large cardinals λ.*

Theorem 6.3.2. *Let L be a first-order language, A an L-structure, X a set of elements of A and $\bar{a}$, $\bar{b}$ two n-tuples of elements of A. Write $G_{(X)}$ for the group of all automorphisms of A which pointwise fix X.*

   (a) *If there is g in $G_{(X)}$ such that $g\bar{a} = \bar{b}$, then $\bar{a}$ and $\bar{b}$ have the same complete type over X.*

   (b) *When A is λ-big for some infinite $\lambda > |X|$, the converse of (a) holds too.*

Proof. (a) is because automorphisms preserve all formulas. (b) is by Exercise 10.1.4(a) below. $\square$

Thus we can identify complete types with orbits as follows:

(3.3)   type over a set X with respect to a structure A = orbit of $\mathrm{Aut}(B)_{(X)}$, where B is any λ-big elementary extension of A for large enough λ. (In fact one can weaken 'λ-big' to 'strongly λ-homogeneous' throughout this discussion; see Exercise 10.1.4.)

Stone spaces

Let L be a first-order language, and let T be a theory in L which has models. For each natural number n, we write $S_n(T)$ for the set of complete n-types of T. For reasons that will appear in a moment, $S_n(T)$ is called the nth **Stone space** of the theory T.

One mostly meets two special cases of this notion.

Example 1: *Complete extensions of theories.* Suppose $n = 0$. The elements of $S_0(T)$ are the sets of the form $\mathrm{Th}(A)$ where A is a model of T. In other words $S_0(T)$ is, up to equivalence of theories, the set of all complete theories in L which contain T.

Example 2: *Complete types over a set of parameters.* Suppose B is an L-structure and X is a set of elements of B. Then a complete n-type over X with respect to B is the same thing as a complete n-type of $\mathrm{Th}(B^+)$, where B^+ is B expanded by adding constants for all the elements of X. We write $S_n(\mathrm{Th}(B^+))$ as $S_n(X; B)$. (When B is fixed, we write simply $S_n(X)$. This is the standard notation used throughout model-theoretic stability theory, except that some people write $S^n(X)$ instead.)

The nth Stone space of a theory T is in fact the Stone space of a boolean algebra generally known as the nth **Lindenbaum algebra** of T. (See section 6.2 above for the Stone spaces of boolean algebras.) I shall construct the Stone topology directly and leave the Lindenbaum algebra to the exercises. The space is much more interesting for us than the algebra, because it brings with it the charms and devices of topology. It brings the snares too – many people have made mistakes by translating their problems into topological terms when they should have stayed in first-order predicate calculus.

For any formula $\phi(\bar{x})$ of L, write $\|\phi\|$ for the set $\{p \in S_n(T): \phi \in p\}$. Note the trivial laws

$$(3.4) \quad \|\neg\phi\| = S_n(T)\backslash\|\phi\|, \quad \|\phi \wedge \psi\| = \|\phi\| \cap \|\psi\|,$$
$$\|\phi \vee \psi\| = \|\phi\| \cup \|\psi\|, \quad \|\bot\| = \varnothing.$$

To make $S_n(T)$ into a topological space, we take as basic closed subsets the sets of the form $\|\phi\|$ as $\phi(\bar{x})$ ranges over the formulas of L. By (3.4) this family of sets is closed under finite intersections.

By (3.4), the sets $\|\phi\|$ with $\phi(\bar{x})$ in L form a boolean algebra, with the partial ordering given by

(3.5) $$\|\phi\| \leqslant \|\psi\| \Leftrightarrow \|\phi\| \subseteq \|\psi\|.$$

Exercise 9 will show that this algebra is the second dual of the nth Lindenbaum algebra of the theory T.

Theorem 6.3.3. *Let T be a theory in a first-order language L, and let n be a natural number.*

(a) *The set $S_n(T)$ forms a Stone space of cardinality $\leqslant 2^{|L|}$.*

(b) *The closed subsets of $S_n(T)$ are the sets of the form $\{p \in S_n(T): \Phi(\bar{x}) \subseteq p\}$ where $\Phi(\bar{x})$ is a set of formulas of L.*

Proof. (a) If p and q are two distinct types in $S_n(T)$, then there must be some formula ϕ such that $\phi \in p$ and $\neg \phi \in q$, and so $p \in \|\phi\|$ and $q \in \|\neg \phi\|$; this shows that the space is Hausdorff. Since each basic closed set $\|\phi\|$ is the complement of a set $\|\neg \phi\|$, the basis sets are clopen. For compactness it suffices to show that if J is a set of basic closed sets and the intersection of any finite subset of J is not empty, then $\bigcap J$ is not empty; but this follows at once from the compactness theorem. The bound holds because every element of $S_n(T)$ is a set of formulas of L.

(b) A closed subset of $S_n(T)$ is an intersection of basic closed sets. In other words, it is a set of the form $\bigcap_{i \in I} \|\phi_i\|$ where $\phi_i(\bar{x})$ $(i \in I)$ are formulas of L. But this set is simply the set of all complete n-types p which contain all the formulas ϕ_i $(i \in I)$. Thus (b) holds. $\qquad\square$

The name 'compactness theorem' originally came from the fact that $S_n(T)$ is a compact space.

Cardinalities of spaces of types

When L is a countable language, the spaces $S_n(T)$ clearly have cardinality at most 2^ω, and one can show that they are metrisable. In particular it follows that if W is a closed subset of $S_n(T)$ then the cardinality of W is either 2^ω or $\leqslant \omega$. But a non-topological proof is just as easy. Here is one; it proves a bit more than I said. But readers not familiar with Morley rank (section 5.6 above) can ignore this – they should omit part (b) in both the theorems below.

Theorem 6.3.4. *Let L be a first-order language, T a complete theory in L and n a positive integer. Suppose $|S_n(T)| > |L|$.*

(a) *$S_n(T)$ has cardinality $\geqslant 2^\omega$.*

(b) *T is not totally transcendental.*

Proof. Put $\lambda = |L|$, and let Ψ be the set of all those formulas $\psi(\bar{x})$ of L which lie in at least λ^+ of the types in $S_n(T)$. Let V be the set of all types in $S_n(T)$ which are subsets of Ψ. Then V contains all but at most λ of the types in $S_n(T)$. Hence every formula in Ψ lies in at least λ^+ types $\in V$. We claim

(3.6) if $\psi \in \Psi$ then there is χ such that $\psi \wedge \chi$, $\psi \wedge \neg\chi$ are both in Ψ.

For let p, q be two distinct types $\in V$ which contain ψ. Since they are distinct, there is some formula χ such that $\psi \wedge \chi \in p$ and $\psi \wedge \neg\chi \in q$. This proves the claim.

Now by the claim we can find, for each sequence $\bar{s} \in {}^{<\omega}2$, a formula $\phi_{\bar{s}}(\bar{x}) \in \Psi$ so that

(3.7)   for each $\bar{s}$, $T \vdash \exists\bar{x}\, \phi_{\bar{s}}(\bar{x})$,

(3.8)   for each $\bar{s}$, $\vdash \forall\bar{x}(\phi_{\bar{s}\smallfrown 0} \rightarrow \phi_{\bar{s}})$ and $\vdash \forall\bar{x}(\phi_{\bar{s}\smallfrown 1} \rightarrow \phi_{\bar{s}})$,

(3.9)   for each $\bar{s}$, $\vdash \neg\exists\bar{x}(\phi_{\bar{s}\smallfrown 0} \wedge \phi_{\bar{s}\smallfrown 1})$.

For each $\bar{\imath} \in {}^{\omega}2$, let $\Phi_{\bar{\imath}}$ be the set of formulas $\Phi \cup \{\phi_{\bar{\imath}|m} : m < \omega\}$. By (3.7) and (3.8) (and Theorem 6.3.1), each $\Phi_{\bar{\imath}}$ is a type; so each $\Phi_{\bar{\imath}}$ can be extended to a complete type $p_{\bar{\imath}}$ in $S_n(T)$. By (3.9), distinct $\bar{\imath}$ give distinct types $p_{\bar{\imath}}$. Hence $S_n(T)$ has cardinality $\geqslant 2^{\omega}$, proving (a).

For (b), note that by (3.7)–(3.9) above and Lemma 5.6.1, the formula $\bar{x} = \bar{x}$ has Cantor–Bendixson rank ∞ for any model of T, and hence also Morley rank ∞. But then T is not totally transcendental, by Lemma 5.6.9. $\square$

Still considering a theory T in a first-order language L, we say that a type $\Phi(\bar{x})$ of T is **principal** if there is a formula $\psi(\bar{x}) \in \Phi$ such that $T \vdash \forall\bar{x}(\psi \rightarrow \phi)$ for every formula $\phi(\bar{x})$ in Φ. We refer to Φ as the **principal type generated by** ψ.

Theorem 6.3.5. *Let L be a first-order language and T a complete theory in L; let n be a positive integer. Suppose either that*
(a) *$S_n(T)$ is countable, or that*
(b) *T is totally transcendental.*
Then for every formula $\phi(x_0, \ldots, x_{n-1})$ such that $T \vdash \exists x_0 \ldots x_{n-1}\phi$, there is a principal type $\in S_n(T)$ containing ϕ.

Proof. Suppose not; let $\phi(\bar{x})$ be a counterexample, writing $\bar{x}$ for $(x_0, \ldots, x_{n-1})$. Write Ψ for the set of all formulas $\psi(\bar{x})$ of L such that $T \vdash \exists\bar{x}\, \psi \wedge \forall\bar{x}(\psi \rightarrow \phi)$. If ψ is in Ψ, then by assumption the principal type generated by ψ is not complete, and so

(3.10) If $\psi \in \Psi$ then there is χ such that $\psi \wedge \chi$, $\psi \wedge \neg\chi$ are both in Ψ.

But (3.10) is exactly the same as (3.6). So we can borrow the rest of the proof of Theorem 6.3.4 to show that $S_n(T)$ has cardinality $\geqslant 2^\omega$ and T is not totally transcendental; which is a contradiction. $\qquad\square$

Exercises for section 6.3

1. Let A be an L-structure and B an extension of A. Show that B is an elementary extension of A if and only if for every tuple $\bar{a}$ of elements of A, $\text{tp}_A(\bar{a}) = \text{tp}_B(\bar{a})$.

2. Show that if A is a structure, X a set of elements of A and $\bar{a}$ a tuple of elements in X, then $\text{tp}_A(\bar{a}/X)$ is a principal type. Deduce that the number of principal complete 1-types over X is always at least $|X|$.

3. Let A be an L-structure, X a set of elements of A, $\bar{a}$ a tuple of elements of A and e an automorphism of A which fixes X pointwise. Show that $\bar{a}$ and $e\bar{a}$ have the same complete type over X with respect to A.

For keeping track of the complete types of a theory, quantifier elimination is invaluable. Here are four examples:

4. Let A be the structure $(\mathbb{Q}, <)$ where $\mathbb{Q}$ is the set of rational numbers and $<$ is the usual ordering. Describe the complete 1-types over $\text{dom}(A)$. Which of them are principal?

5. Let A be an algebraically closed field and C a subfield of A; to save notation I write C also for $\text{dom}(C)$. (a) Show that two elements a, b of A have the same complete type over C if and only if they have the same minimal polynomial over C. (b) Show that $\text{tp}_A(b/C)$ is principal if and only if b is algebraic over C. (c) Show that for all n, $|S_n(C; A)| = \omega + |C|$.

6. Let A be a vector space over a field k and X a set of elements of A. Describe $S_n(X; A)$ for each $n < \omega$, and show that $|S_n(X; A)| \leqslant |k| + |X| + \omega$.

7. Let A be a structure consisting of an infinite set on which there are countably many named equivalence relations E_0, E_1, ...; suppose that for each i, every equivalence class of E_i is the union of infinitely many equivalence classes of E_{i+1}. (a) Show that there is one complete 1-type over the empty set. (b) If X is a set of elements of A and p is a complete 1-type over X, show that p is principal if and only if p contains a formula $x = a$ for some element a of X. (c) Show that a countable set of elements X can be found so that $S_n(X)$ is countable. (d) Show that a countable set of elements X can be found so that $S_n(X)$ has the cardinality of the continuum.

8. Let $\mathbb{Z}$ be the ring of integers, and let $\phi(x)$ be a formula in the language of $\mathbb{Z}$ which expresses 'x is a prime number'. (a) Show that $\mathbb{Z}$ has an elementary extension in which some element not in $\mathbb{Z}$ satisfies ϕ. [There is a type containing all the formulas $x > 0$, $x > 1$, ... and ϕ.] *Elements not in $\mathbb{Z}$ which satisfy ϕ are known as* **nonstandard primes**, *as opposed to the* **standard primes** *(i.e. the usual ones)*. (b) Show that $\mathbb{Z}$ has an

elementary extension in which some non-zero element is divisible by every standard prime.

The next exercise describes how to construct Lindenbaum algebras.

9. Let L be a first-order language, $\bar{x}$ a tuple $(x_0, \ldots, x_{n-1})$ of distinct variables and T a theory in L. Show that for any formulas $\phi(\bar{x})$ and $\psi(\bar{x})$ of L we have

$$\|\phi\| \leqslant \|\psi\| \Leftrightarrow T \vdash \forall \bar{x} \, (\phi \rightarrow \psi),$$

where $\|\phi\|$ and $\|\psi\|$ are the subsets of $S_n(T)$ described in the text. Using this result, define an equivalence relation on the set of formulas $\phi(\bar{x})$ of L, and show how the set of equivalence classes forms a boolean algebra isomorphic to the algebras $B(S_n(T))$ (see Theorem 6.2.3). (This algebra of equivalence classes of formulas is the nth **Lindenbaum algebra** of T.)

10. Suppose B is an abelian group and X a subgroup of B. The group B is said to be an **essential extension** of X if for every non-trivial subgroup A of B, $A \cap X \neq \{0\}$. Show that B is an essential extension of X if and only if B omits the type $\{nx \neq a : n \in \mathbb{Z}, a \in X \setminus \{0\}\}$ (which is a 1-type over X with respect to B).

The next three exercises illustrate how we can use the realisation of types to show that certain things aren't first-order expressible.

11. Let L be the language of groups. Show that there is no formula $\phi(x, y)$ of L such that if G is a group and g, h are elements of G then g has the same order as h if and only if $G \vDash \phi(g, h)$. [If a group G contains elements g, h of arbitrarily high finite orders, such that $G \vDash \neg \phi(g, h)$, then compactness gives us an elementary extension of G with two elements g', h' of infinite order such that $G \vDash \neg \phi(g', h')$.]

*A group G is **of bounded exponent** if there is some $n < \omega$ such that $g^n = 1$ for every element g of G.*

12. Let G be a group. Show that the following are equivalent. (a) G is not of bounded exponent. (b) G has an elementary extension in which some element has infinite order.

13. Show that 'simple group' is not first-order expressible, by finding groups G, H such that $G \equiv H$ and G is simple but H is not simple. [Let G be the simple group consisting of all permutations of ω which move just finitely many elements and are even permutations. Using relativisation (see section 5.1), take A to be a structure which contains G, ω and the action of G on ω, and let B be an elementary extension of A which contains an extension H of G in which some element moves infinitely many points. The subgroup of elements moving just finitely many points is normal.]

14. Let L be a first-order language with just finitely many function and constant symbols and no relation symbols, and let V be a set of variables. Assume that the function symbols of L include two distinct 2-ary function symbols F and G. Let A be the term algebra of L with basis V, as in section 1.3. Show that A has an elementary extension in which there is (for example) an element of the form

$F(x, F(x, G(x, F(x, F(x, G(x, \ldots). \ldots).$ *For Prolog programmers: what is the relation between this fact and the absence of the 'occur check' in the unification algorithm?*

15. Show that if L is a countable first-order language and B is an L-structure with a set of elements X such that $|S_n(X; B)| > |X|$, then there is an at most countable subset Y of X such that $|S_n(Y; B)| = 2^\omega$. [The proof of Theorem 6.3.4 may suggest an argument.]

16. Let L be a first-order language, T a complete theory in L and $\Phi(\bar{x})$ a set of formulas of L in the n variables $\bar{x} = (x_0, \ldots, x_{n-1})$. Let W be the set of all complete n-types of T which contain Φ. Show that if $|W| > |L|$ then W has cardinality $\geq 2^\omega$.

17. Let $e: A \to B$ be an elementary embedding. Let X, Y be sets of elements of A, B respectively, such that $e(X) \subseteq Y$. Show that for each $n < \omega$, the map e induces a continuous surjection $e': S_n(Y; B) \to S_n(X; A)$.

18. Find a complete theory T in a countable first-order language, such that $S_1(T)$ is uncountable but the conclusion of Theorem 6.3.5 holds. [Exercise 7.]

6.4 Elementary amalgamation

An amalgamation theorem is a theorem of the following shape. We are given two models B, C of some theory T, and a structure A (not necessarily a model of T) which is embedded into both B and C. The theorem states that there is a third model D of T such that both B and C are embeddable into D by embeddings which agree on A:

(4.1)

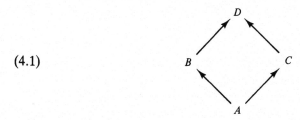

The embeddings may be required to preserve certain formulas.

It was Roland Fraïssé who first called attention to the diagram (4.1). Everything that has happened in model theory during the last thirty years has confirmed how important this diagram is. To see why, imagine some structure M and a small part A of M, and ask 'How does M sit around A?' For the answer, we need to know how A can be extended within M to structures B, C etc. But then we also need to know how any two of these extensions B, C of A are related to each other inside M: what formulas relate the elements of B to the elements of C? An amalgam (4.1) of B and C over A answers this last question.

There are two ways of using these ideas. One is to *build up* a structure M by taking smaller structures, extending them and then amalgamating the extensions. (Fraïssé did exactly this; see section 7.1 below, and section 10.2 on λ-saturated structures for a more general but slightly disguised version of the same idea.) The second way of using amalgamation is not to construct but to *classify*. We classify all the ways of extending the bottom structure A, and then we classify the ways of amalgamating these extensions. In favourable cases this leads to a structural classification of all the models of a theory. Stability theory follows this path; the reader will need to look at another book for details.

First-order model theory is peculiarly rich in amalgamation theorems. More than anything else this accounts for the flavour of the subject. By a theorem of Makowsky and Shelah (Makowsky [1985b] Theorem 3.3.1), if $\mathscr{L}$ is a logic whose sentences are finite, and $\mathscr{L}$ obeys Theorem 6.4.1 below, then $\mathscr{L}$ must also obey the compactness theorem. This is one of a crop of theorems, first harvested by Per Lindström, which say that any logic which shares certain broad features with first-order logic must essentially be first-order logic.

One amalgamation theorem is father to the rest. This is the elementary amalgamation theorem, Theorem 6.4.1 below. The other amalgamation results of first-order model theory differ from it in various ways: the maps are not necessarily elementary embeddings, they change the language, the amalgam is required to be strong, and so forth. This section will study the versions where all the maps are elementary. Sections 6.5 and 6.6 will tackle the remainder, so far as they are general results of model theory. Some particular classes of structures have their own particular amalgamation theorems, as we shall see in section 7.1.

Theorem 6.4.1 *(Elementary amalgamation theorem). Let L be a first-order language. Let B and C be L-structures and $\bar{a}$, $\bar{c}$ sequences of elements of B, C respectively, such that $(B, \bar{a}) \equiv (C, \bar{c})$. Then there exist an elementary extension D of B and an elementary embedding $g: C \to D$ such that $g\bar{c} = \bar{a}$. In a picture,*

(4.2)

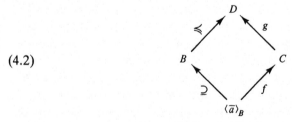

where $f: \langle \bar{a} \rangle \to C$ is the unique embedding which takes $\bar{a}$ to $\bar{c}$ (see the diagram lemma, Lemma 1.4.2).

Proof. Replacing C by an isomorphic copy if necessary, we can assume that $\bar{a} = \bar{c}$, and otherwise B and C have no elements in common. Consider the theory $T = \text{eldiag}(B) \cup \text{eldiag}(C)$, where each element names itself.

We claim that T has a model. By the compactness theorem (Theorem 6.1.1), it suffices to show that every finite subset of T has a model. Let T_0 be a finite subset of T. Then T_0 contains just finitely many sentences from eldiag(C). Let their conjunction be $\phi(\bar{a}, \bar{d})$ where $\phi(\bar{x}, \bar{y})$ is a formula of L and $\bar{d}$ consists of pairwise distinct elements which are in C but not in $\bar{a}$. (Of course only finitely many variables in $\bar{x}$ occur free in ϕ.) If T_0 has no model then eldiag(B) $\vdash \neg\phi(\bar{a}, \bar{d})$. By the lemma on constants (Lemma 2.3.2), since the elements $\bar{d}$ are distinct and they are not in B, eldiag(B) $\vdash \forall\bar{y}\neg\phi(\bar{a}, \bar{y})$. But then $(B, \bar{a}) \vDash \forall\bar{y}\neg\phi(\bar{a}, \bar{y})$, and so $(C, \bar{c}) \vDash \forall\bar{y}\neg\phi(\bar{c}, \bar{y})$ by the theorem assumption. This contradicts that $\phi(\bar{a}, \bar{d})$ is in eldiag(C). The claim is proved.

Let D^+ be a model of T and let D be the reduct $D^+|L$. By the elementary diagram lemma (Lemma 2.5.3), since $D^+ \vDash \text{eldiag}(B)$, we can assume that D is an elementary extension of B and $b^{D^+} = b$ for all elements b of B. Define $g(d) = d^{D^+}$ for each element d of C. Then by the elementary diagram lemma again, since $D^+ \vDash \text{eldiag}(C)$, g is an elementary embedding of C into D. Finally $g\bar{c} = \bar{a}^{D^+} = \bar{a}$. $\qquad\square$

Note that $\bar{a}$ can be empty in Theorem 6.4.1. In this case the theorem says that any two elementarily equivalent structures can be elementarily embedded together into some structure.

Note also that the theorem can be rephrased as follows. If $(B, \bar{a}) \equiv (C, \bar{c})$ and $\bar{d}$ is any sequence of elements of C, then there is an elementary extension B' of B containing elements $\bar{b}$ so that $(B', \bar{a}, \bar{b}) \equiv (C, \bar{c}, \bar{d})$. Chapter 10 below is an extended meditation on this fact.

One of the most important consequences of the elementary amalgamation theorem is that if A is any structure, we can simultaneously realise all the complete types with respect to A in one and the same elementary extension of A, as follows.

Corollary 6.4.2. *Let L be a first-order language and A an L-structure. Then there is an elementary extension B of A such that every type over $\text{dom}(A)$ with respect to A is realised in B.*

Proof. It suffices to realise all the maximal types over $\text{dom}(A)$ with respect to A. Let these be p_i ($i < \lambda$) with λ a cardinal. For each $i < \lambda$ let $\bar{a}_i$ be a tuple in an elementary extension A_i of A, such that p_i is $\text{tp}_{A_i}(\bar{a}_i/\text{dom} A)$. Define an elementary chain $(B_i : i \leqslant \lambda)$ by induction as follows. B_0 is A, and for each limit ordinal $\delta \leqslant \lambda$, $B_\delta = \bigcup_{i<\delta}B_i$ (which is an elementary extension of each

B_i by Theorem 2.5.2). When B_i has been defined and $i < \lambda$, choose B_{i+1} to be an elementary extension of B_i such that there is an elementary embedding $e_i : A_i \to B_{i+1}$ which is the identity on A; this is possible by Theorem 6.4.1. Put $B = B_\lambda$. Then for each $i < \lambda$, $e_i(\bar{a}_i)$ is a tuple in B_λ which realises p_i. $\quad\square$

In section 10.2 below we shall refine this corollary in several ways. The corollary fails in general when the language is not first-order. But it takes a little work to show this; see Exercise 8.6.10. Malitz & Reinhardt [1972] describe a counterexample to elementary amalgamation for $L(Q_1)$ and for $L_{\omega_1\omega}$.

Heir–coheir amalgams

Consider the case of Theorem 6.4.1 where $\bar{a}$ lists the elements of an elementary substructure A of B. In this case the theorem tells us that if A, B and C are L-structures and $A \preccurlyeq B$ and $A \preccurlyeq C$, then there are an elementary extension D of B and an elementary embedding $g : C \to D$ such that, putting $C' = gC$, the following diagram of elementary inclusions commutes:

(4.3)

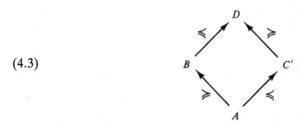

But actually we can say more in this case. Call the commutative diagram (4.3) of L-structures and elementary inclusions an **heir–coheir amalgam** if

(4.4)   for every first-order formula $\psi(\bar{x}, \bar{y})$ of L and all tuples $\bar{b}$, $\bar{c}$
from B, C' respectively, if $D \vDash \psi(\bar{b}, \bar{c})$ then there is $\bar{a}$ in A
such that $B \vDash \psi(\bar{b}, \bar{a})$.

We say also that (4.3) is an **heir–coheir amalgam of B and C over A**; in fact it is an **heir–coheir amalgam of B'' and C'' over A** whenever B'' and C'' are elementary extensions of A such that there are isomorphisms $i : B'' \to B$ and $j : C'' \to C'$ which are the identity on A.

Example 1: *Vector spaces.* Suppose A is an infinite vector space over a field K, and B and C are vector spaces with A as subspace. Put $B = B_1 \oplus A$ and $C = C_1 \oplus A$. We can amalgamate B and C over A by putting $D = B_1 \oplus C_1 \oplus A$. Suppose some equation $\sum_{i<m} \lambda_i b_i = \sum_{j<n} \mu_j c_j$ holds in D, where the b_i are in B and the c_j are in C. Let $\pi : D \to B_1 \oplus A$ be the projection along C_1. Then $\sum_{i<m} \lambda_i \pi(b_i) = \sum_{j<n} \mu_j \pi(c_j)$. But $\pi(b_i) = b_i$ and

$\pi(c_j)$ lies in A. Thus (4.4) holds when ψ is the formula $\sum_{i<m} \lambda_i x_i = \sum_{j<n} \mu_j y_j$. In fact since A is infinite, one can show that (4.4) holds whenever ψ is quantifier-free; but don't check this by linear algebra – it will fall out almost trivially from the next theorem. By quantifier elimination (see Exercise 2.7.9) it follows that D forms an heir–coheir amalgam of B and C over A. This example is typical: heir–coheir amalgams are 'as free as possible'.

The next theorem says that heir–coheir amalgams always exist when B and C are elementary extensions of A.

Theorem 6.4.3. *Let A, B and C be L-structures such that $A \preccurlyeq B$ and $A \preccurlyeq C$. Then there exist an elementary extension D of B and an elementary embedding $g: C \to D$ such that (4.3) (with $C' = gC$) is an heir–coheir amalgam.*

Proof. Much like the proof of Theorem 6.4.1. As before, we assume that $(\text{dom } B) \cap (\text{dom } C) = \text{dom}(A)$, so that constants behave properly in diagrams. Then we take T to be the theory

(4.5) eldiag$(B) \cup$ eldiag$(C) \cup$
  $\{\neg \psi(\bar{b}, \bar{c}): \psi$ is a first-order formula of L and $\bar{b}$ is a tuple in
  B such that $B \vDash \neg \psi(\bar{b}, \bar{a})$ for all $\bar{a}$ in $A\}$.

If T has no model, then by the compactness theorem there are a tuple $\bar{a}$ from A, a tuple $\bar{d}$ of distinct elements in C but not in A, a tuple $\bar{b}$ of elements of B, a sentence $\theta(\bar{a}, \bar{d})$ in eldiag(C) and sentences $\psi_i(\bar{b}, \bar{a}, \bar{d})$ $(i < k)$ such that $B \vDash \neg \psi_i(\bar{b}, \bar{a}', \bar{a}'')$ for all $\bar{a}'$, $\bar{a}''$ in A, and

(4.6) eldiag$(B) \vdash \theta(\bar{a}, \bar{d}) \to \psi_0(\bar{b}, \bar{a}, \bar{d}) \vee \ldots \vee \psi_{k-1}(\bar{b}, \bar{a}, \bar{d})$.

Quantifying out the constants $\bar{d}$ by the lemma on constants, we have

(4.7) $B \vDash \forall \bar{y}(\theta(\bar{a}, \bar{y}) \to \psi_0(\bar{b}, \bar{a}, \bar{y}) \vee \ldots \vee \psi_{k-1}(\bar{b}, \bar{a}, \bar{y}))$.

But also $C \vDash \exists \bar{y} \, \theta(\bar{a}, \bar{y})$, so $A \vDash \exists \bar{y} \, \theta(\bar{a}, \bar{y})$ and hence there is a tuple $\bar{a}''$ in A such that $A \vDash \theta(\bar{a}, \bar{a}'')$ and so $B \vDash \theta(\bar{a}, \bar{a}'')$. Applying (4.7), we infer that $B \vDash \psi_i(\bar{b}, \bar{a}, \bar{a}'')$ for some $i < k$; contradiction. The rest of the proof is as before. $\qquad\square$

If (4.3) is an heir–coheir amalgam, then the overlap of B and C in D is precisely A. For suppose $b = g(c)$ for some b in B and some c in C. Then by the heir–coheir property, $b = a$ for some a in A. Amalgams with this minimum-overlap property are said to be **strong**. In this terminology we have just shown that first-order logic has the **strong elementary amalgamation property**.

Example 1 *continued.* Here is a more abstract proof that D is an heir–coheir amalgam of B and C over A. Since A is infinite, $A \preccurlyeq B$ and $A \preccurlyeq C$ by quantifier elimination (Exercise 2.7.9). So by the theorem, some vector space D' forms an heir–coheir amalgam of B and C over A. Identifying B and C with their images in D', we can suppose that B and C generate D'; for if D'' is the subspace of D' generated by B and C, then $D'' \leqslant D'$ by quantifier elimination again. Now D' is a strong amalgam of B and C over A. But this means precisely that $D' = B_1 \oplus C_1 \oplus A$, so that D' is D. Thus D is an heir–coheir amalgam of B and C over A.

When do two extensions of a structure have a strong amalgam?

Example 2: *Algebraically closed fields.* Suppose that in (4.1), both B and C are the field of complex numbers, A is the field of reals and D is some algebraically closed field which amalgamates B and C over A. Let i, $-$i be the square roots of -1 regarded as elements of B, and j, $-$j the same elements thought of as in C. Then in D, i must be identified with either j or $-$j, and so the amalgam is not strong.

The moral of this example is that if $\langle a \rangle_B$ in Theorem 6.4.1 is not algebraically closed in B, then in general there is no hope of making the amalgam strong in that theorem. To turn this remark from a moral into a theorem, we need to formalise the notion of 'algebraically closed'.

Strong amalgams and algebraicity

Let B be an L-structure and X a set of elements of B. Recall a definition from section 4.1: we say that an element b of B is **algebraic over** X if there are a first-order formula $\phi(x, \bar{y})$ of L and a tuple $\bar{a}$ in X such that $B \vDash \phi(b, \bar{a}) \wedge \exists_{\leqslant n} x\, \phi(x, \bar{a})$ for some finite n. We write $\mathrm{acl}_B(X)$ for the set of all elements of B which are algebraic over X. If $\bar{a}$ lists the elements of X, we also write $\mathrm{acl}_B(\bar{a})$ for $\mathrm{acl}_B(X)$.

In section 4.1 we showed

(4.8) $X \subseteq \mathrm{acl}_B(X)$ (see Lemma 4.1.1(d)),

(4.9) $Y \subseteq \mathrm{acl}_B(X)$ implies $\mathrm{acl}_B(Y) \subseteq \mathrm{acl}_B(X)$ (see Lemma 4.1.2(a')),

(4.10)   if $B \leqslant C$ then $\mathrm{acl}_B(X) = \mathrm{acl}_C(X)$ (see Exercise 4.1.5(b)).

By (4.10) we can often write $\mathrm{acl}(X)$ for $\mathrm{acl}_B(X)$ without danger of confusion.

We say that a tuple $\bar{b}$ is **algebraic over** X if every element in $\bar{b}$ is algebraic over X. We say that a type $\Phi(\bar{x})$ over a set X with respect to B is **algebraic** if every tuple realising it is algebraic over X.

Theorem 6.4.5 below will tell us that we can make the amalgam strong in Theorem 6.4.1 whenever the bottom structure $\langle \bar{a} \rangle_B$ is algebraically closed in B (or in C, by symmetry). This theorem is deducible from Π. M. Neumann's lemma, Corollary 4.2.2. But we can reach the theorem just as quickly by a model-theoretic argument. Exercise 6 will beat a path between the theorem and Neumann's lemma.

Lemma 6.4.4. *Let B be an L-structure, X a set of elements of B listed as $\bar{a}$, and b an element of B. Suppose $b \notin \mathrm{acl}(X)$.*

(a) There is an elementary extension A of B with an element $c \notin \mathrm{dom}(B)$ such that $(B, \bar{a}, b) \equiv (A, \bar{a}, c)$.

(b) There is an elementary extension D of B with an elementary substructure C containing X such that $b \notin \mathrm{dom}(C)$.

Proof. (a) Let c be a new constant and let $p(x)$ be the complete type of b over X. It suffices to show that the theory

$$(4.11) \qquad \mathrm{eldiag}(B) \cup p(c) \cup \{c \neq d : d \in \mathrm{dom}(B)\}$$

has a model. But if it has none, then by the compactness theorem and the lemma on constants, there are finitely many elements $d_0, \ldots, d_{n-1}$ of B and a formula $\phi(x)$ of $p(x)$ (noting that $p(x)$ is closed under $\wedge$), such that

$$(4.12) \qquad \mathrm{eldiag}(B) \vdash \forall x (\phi(x) \rightarrow x = d_0 \vee \ldots \vee x = d_{n-1}).$$

Hence $B \vDash \phi(b) \wedge \exists_{\leqslant n} x\, \phi(x)$, so that $b \in \mathrm{acl}(X)$; contradiction.

(b) Take A and c as in part (a). Since $(A, \bar{a}, b) \equiv (A, \bar{a}, c)$, Theorem 6.4.1 gives us an elementary extension D of A and an elementary embedding $g: A \rightarrow D$ such that $g\bar{a} = \bar{a}$ and $gb = c$. Then D is an elementary extension of gB and $gb = c \notin \mathrm{dom}(B)$. So we get the lemma by taking gB, B for B, C respectively. $\qquad\qquad\square$

Theorem 6.4.5 (Strong elementary amalgamation over algebraically closed sets). *Let B and C be L-structures and $\bar{a}$ a sequence of elements in both B and C such that $(B, \bar{a}) \equiv (C, \bar{a})$. Then there exist an elementary extension D of B and an elementary embedding $g: C \rightarrow D$ such that $g\bar{a} = \bar{a}$ and $(\mathrm{dom}\, B) \cap g(\mathrm{dom}\, C) = \mathrm{acl}_B(\bar{a})$.*

Proof. Start by repeating the proof of Theorem 6.4.1, but adding to T all the sentences '$b \neq c$' where b is in B but not in $\mathrm{acl}_B(\bar{a})$ and c is in C but not in $\mathrm{acl}_C(\bar{a})$. Write T^+ for this enlarged theory. Suppose D and g are defined using T^+ in place of T. Then $g\bar{a} = \bar{a}$, and it easily follows that g maps $\mathrm{acl}_C(\bar{a})$ onto $\mathrm{acl}_B(\bar{a})$. Thus we have $\mathrm{acl}_B(\bar{a}) \subseteq (\mathrm{dom}\, B) \cap g(\mathrm{dom}\, C)$, and the sentences '$b \neq c$' guarantee the opposite inclusion. It remains only to show that T^+ has a model.

Assume for contradiction that T^+ has no model. Then by the compactness theorem there are finite subsets Y of $\mathrm{dom}(B) \backslash \mathrm{acl}_B(\bar{a})$ and Z of $\mathrm{dom}(C) \backslash \mathrm{acl}_C(\bar{a})$, such that for every elementary extension D of B and elementary embedding $g: C \to D$ with $g\bar{a} = \bar{a}$, $Y \cap g(Z) \neq \varnothing$. Choose D and g to make $Y \cap g(Z)$ as small as possible. To save notation we can assume that g is the identity, so that $C \preccurlyeq D$.

Since $Y \cap Z \neq \varnothing$, there is some $b \in Y \cap Z$. By the lemma, there is an elementary extension D' of D with an elementary substructure C' containing $\bar{a}$ such that $b \notin \text{dom}(C')$. Applying Theorem 6.4.3 to the elementary embedding $C' \preccurlyeq D'$ (the same embedding twice over), we find an elementary extension E of D' and an elementary embedding $e: D' \to E$ which is the identity on C', such that $(\text{dom } D') \cap e(\text{dom } D') = \text{dom}(C')$. Now $Y \cap e(Z) \subseteq Y \cap Z$; for if $d \in Y \cap e(Z)$ then d is in C' and hence $ed = d$. But b is in $(Y \cap Z) \setminus (Y \cap e(Z))$; for since b is in D' but not in C', $b \notin e(\text{dom } D')$ and hence $b \notin e(Z)$. Thus e contradicts the choice of $Y \cap g(Z)$ as minimal. $\square$

Exercises for section 6.4

1. Let T be the theory of dense linear orderings without endpoints. Describe the heir–coheir amalgams of models of T. [In D, any interval (b_1, b_2) with b_1, b_2 in B which contains elements of C must also contain elements of A. If b in B is $>$ all A, then b is $>$ all C too; and likewise with $<$ for $>$.]

2. Let T be the theory of the linear ordering of the integers. Describe the heir–coheir amalgams of models of T.

Let A, B and D be L-structures with $A \preccurlyeq B \preccurlyeq D$, and let $\bar{d}$ be a tuple in D. Put $p = \text{tp}_D(\bar{d}/A)$ and $p^+ = \text{tp}_D(\bar{d}/B)$. We say that p^+ is an heir of p if for every formula $\phi(\bar{x}, \bar{y})$ of L with parameters in A, and every tuple $\bar{b}$ in B such that $\phi(\bar{x}, \bar{b}) \in p^+$, there is $\bar{a}$ in A such that $\phi(\bar{x}, \bar{a}) \in p$. We say that p^+ is a coheir of p if for every formula $\phi(\bar{x}, \bar{y})$ of L and tuple $\bar{b}$ in B such that $\phi(\bar{x}, \bar{b}) \in p^+$, there is $\bar{a}$ in A such that $D \vDash \phi(\bar{a}, \bar{b})$.

3. Show that the following are equivalent, given the amalgam (4.3) above. (a) The amalgam is heir–coheir. (b) For every tuple $\bar{b}$ in B, $\text{tp}_D(\bar{b}/C)$ is an heir of $\text{tp}_D(\bar{b}/A)$. (c) For every tuple $\bar{c}$ in C', $\text{tp}_D(\bar{c}/B)$ is a coheir of $\text{tp}_D(\bar{c}, A)$. *The type notions introduce a false asymmetry between the two sides of the amalgam. But they do allow us to think about one tuple at a time, and this can be an advantage.*

4. Let $\Phi(\bar{x})$ be a type over a set X with respect to a structure A. Show that the following are equivalent. (a) Φ is algebraic. (b) Φ contains a formula ϕ such that $A \vDash \exists_{\leqslant n} \bar{x} \, \phi(\bar{x})$ for some finite n. (c) In every elementary extension of A, at most finitely many tuples realise Φ.

5. Let L be a first-order language and A an L-structure. Suppose X is a set of elements of A and $\bar{a}$ is a tuple of elements of A, none of which are algebraic over A. Show that some elementary extension B of A contains infinitely many pairwise disjoint tuples $\bar{a}_i$ ($i < \omega$) which all realise $\text{tp}_A(\bar{a}/X)$. [Iterate Theorem 6.4.5.]

6. Prove Theorem 6.4.5 from Π. M. Neumann's lemma (Corollary 4.2.2) as follows. By Exercise 5 we can suppose that every non-algebraic complete 1-type over $\langle \bar{a} \rangle_B$ is realised by infinitely many elements of B. Let G be the group of automorphisms of B that pointwise fix $\langle \bar{a} \rangle_B$. By Theorem 6.3.2 and (3.2) before it, we can suppose that any two elements of B realising the same complete type over $\langle \bar{a} \rangle_B$ lie in the same G-orbit. Now follow the proof of Theorem 6.4.5 as far as the choice of Y and Z. Form any elementary amalgam of B and C over $\langle \bar{a} \rangle_B$. Use Neumann's lemma to twist B around inside this amalgam so that the image of Y is disjoint from that of Z.

7. Let A be an L-structure and X a set of elements of A. Show that there is an elementary extension B of A with a descending sequence $(C_i : i < \omega)$ of elementary substructures such that $\mathrm{acl}_A(X) = \bigcap_{i < \omega} \mathrm{dom}(C_i)$. [Use Theorem 6.4.5 to build up a chain of strong amalgams

(4.13)

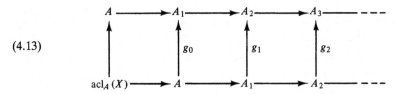

where the g_i are elementary embeddings and the horizontal maps are elementary inclusions. Put $B = \bigcup_{n < \omega} A_n$, $g = \bigcup_{n < \omega} g_n$ and $C_i = g^i B$. See Corollary 8.6.5.]

*8. Let L be a first-order language and T a theory in L. Show that the following are equivalent. (a) If A is a model of T, then the intersection of any two elementary substructures of A is again an elementary substructure of A. (b) If A is a model of T, then the intersection of any family of elementary substructures of A is again an elementary substructure of A. (c) If A is a model of T and $(B_i : i < \gamma)$ is a descending sequence of elementary substructures of A then the intersection of the B_i is again an elementary substructure of A. (d) If A is any model of T and X is a set of elements of A then $\mathrm{acl}_A(X)$ is an elementary substructure of A. (e) For any formula $\phi(x, \bar{y})$ of L there are a formula $\psi(x, \bar{y})$ of L and an integer n such that $T \vdash \forall x \bar{y}(\phi \rightarrow \exists x(\phi \wedge \psi) \wedge \exists_{\leqslant n} x\, \psi)$.

*9. In the context of the previous exercise, show that the following are equivalent. (a′) If A is a model of T and B, C are elementary substructures of A with non-empty intersection, then this intersection is an elementary substructure of A. (b′) As (e) in the previous exercise, but restricted to formulas $\phi(x, \bar{y})$ where $\bar{y}$ is not empty.

As far as I know, the syntactic characterisation of the class of theories described by the next exercise is an open problem.
*10. Let L be a first-order language and T a theory in L. Show that the following are equivalent. (a) If A is a model of T, then the intersection of any two elementary substructures of A is a model of T. (b) If A is a model of T, then the intersection of any family of elementary substructures of A is a model of T. (c) If A is a model of T and $(B_i : i < \gamma)$ is a descending sequence of elementary substructures of A then the

intersection of the B_i is a model of T. (d) If A is any model of T and X is a set of elements of A then $\mathrm{acl}_A(X)$ is a model of T.

*11. Let L be a first-order language, one of whose symbols is a relation symbol R, and write L^- for L with R removed. Let A be an L-structure, and let $\bar{b}$, $\bar{c}$ be sequences of elements of A such that (1) $(A|L^-, \bar{b}) \equiv (A|L^-, \bar{c})$, (2) R holds between any given elements in $\bar{b}$ if and only if it holds between the corresponding elements of $\bar{c}$, and (3) the sets of elements listed by $\bar{b}$, $\bar{c}$ respectively are algebraically closed in $A|L^-$. Show that there is an extension D of A such that $A|L^- \preccurlyeq D|L^-$ and $(D, \bar{b}) \equiv (D, \bar{c})$.

6.5 Amalgamation and preservation

We have a fair amount of work to do in this section and the next. But fortunately most of the proofs are routine variations, either of each other or of the arguments of section 6.4.

If A and B are L-structures, we write $A \Rrightarrow_1 B$ to mean that for every first-order existential sentence ϕ of L, if $A \vDash \phi$ then $B \vDash \phi$. Likewise we write $A \Rrightarrow_1^+ B$ to mean the same for every first-order $\exists_1^+$ sentence of L. Note that $\Rrightarrow_1$ implies $\Rrightarrow_1^+$. Note also that if $f: \langle \bar{a} \rangle_B \to C$ is a homomorphism, then the assertion $(C, f\bar{a}) \Rrightarrow_1^+ (B, \bar{a})$ implies that f is an embedding.

Theorem 6.5.1 (Existential amalgamation theorem). *Let B and C be L-structures, $\bar{a}$ a sequence of elements of B and $f: \langle \bar{a} \rangle \to C$ a homomorphism such that $(C, f\bar{a}) \Rrightarrow_1 (B, \bar{a})$. Then there exist an elementary extension D of B and an embedding $g: C \to D$ such that $gf\bar{a} = \bar{a}$. In a picture,*

(5.1)

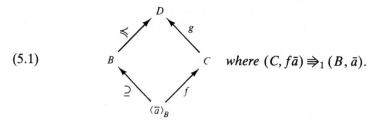

where $(C, f\bar{a}) \Rrightarrow_1 (B, \bar{a})$.

Proof. The assumptions imply that f is an embedding, so that we can replace C by an isomorphic copy and assume that f is the identity on $\langle \bar{a} \rangle_B$, and that $\langle \bar{a} \rangle_B$ is the overlap of $\mathrm{dom}(B)$ and $\mathrm{dom}(C)$. By the same argument as for Theorem 6.4.1, it suffices to show that the theory $T = \mathrm{eldiag}(B) \cup \mathrm{diag}(C)$ has a model. Again as in the proof of Theorem 6.4.1, if T has no model, then by the compactness theorem there is a conjunction $\phi(\bar{a}, \bar{d})$ of finitely many sentences in $\mathrm{diag}(C)$, such that $(B, \bar{a}) \vDash \neg \exists \bar{y} \, \phi(\bar{a}, \bar{y})$. Since $\phi(\bar{a}, \bar{y})$ is quantifier-free and $(C, \bar{a}) \Rrightarrow_1 (B, \bar{a})$, we infer that $(C, \bar{a}) \vDash \neg \exists \bar{y} \, \phi(\bar{a}, \bar{y})$. This contradicts that $\phi(\bar{a}, \bar{d})$ is true in C, and so the proof is complete. $\qquad\square$

Since we allow structures to be empty, the tuple $\bar{a}$ in Theorem 6.5.1 can be the empty tuple. This gives the following.

Corollary 6.5.2. *Let B and C be L-structures such that $C \Rrightarrow_1 B$. Then C is embeddable in some elementary extension of B.* □

Amalgamation theorems like Theorem 6.5.1 tend to spawn offspring of the following kinds: (i) criteria for a structure to be expandable or extendable in certain ways, (ii) syntactic criteria for a formula or set of formulas to be preserved under certain model-theoretic operations (results of this kind are called **preservation theorems**), (iii) interpolation theorems. Let me illustrate all these.

(i) We give a criterion for a structure to be extendable to a model of a given theory. If T is a theory in a first-order language L, then we write $T_\forall$ for the set of all $\forall_1$ sentences of L which are consequences of T.

Corollary 6.5.3. *If T is a theory in a first-order language L, then the models of $T_\forall$ are precisely the substructures of models of T.*

Proof. Any substructure of a model of T is certainly a model of $T_\forall$, by Corollary 2.4.2. Conversely let C be a model of $T_\forall$. To show that C is a substructure of a model of T, it suffices to find a model B of T such that $C \Rrightarrow_1 B$, and then quote Corollary 6.5.2.

We find B as follows. Let U be the set of all $\exists_1$ sentences ϕ of L such that $C \vDash \phi$. We claim that $T \cup U$ has a model. For if not, then by the compactness theorem there is some finite set $\{\phi_0, \ldots, \phi_{k-1}\}$ of sentences in U such that $T \vdash \neg \phi_0 \vee \ldots \vee \neg \phi_{k-1}$. Now $\neg \phi_0 \vee \ldots \vee \neg \phi_{k-1}$ is logically equivalent to an $\forall_1$ sentence θ, and $T \vdash \theta$, so that $\theta \in T_\forall$ and hence $C \vDash \theta$. But this is absurd, since $C \vDash \phi_i$ for each $i < k$. So $T \cup U$ has a model as claimed. Let B^+ be any model of $T \cup U$ and B the L-reduct of B^+. □

(ii) We characterise those formulas or sets of formulas which are preserved in substructures.

Theorem 6.5.4 (Łoś–Tarski theorem). *Let T be a theory in a first-order language L and $\Phi(\bar{x})$ a set of formulas of L. (The sequence of variables $\bar{x}$ need not be finite.) Then the following are equivalent.*

(a) *If A and B are models of T, $A \subseteq B$, $\bar{a}$ is a sequence of elements of A and $B \vDash \bigwedge \Phi(\bar{a})$, then $A \vDash \bigwedge \Phi(\bar{a})$. ('$\Phi$ is preserved in substructures for models of T.')*

(b) *Φ is equivalent modulo T to a set $\Psi(\bar{x})$ of $\forall_1$ formulas of L.*

Proof. (b) ⇒ (a) is by Corollary 2.4.2(a). For the converse, assume (a). We first prove (b) under the assumption that Φ is a set of sentences. Let Ψ be $(T \cup \Phi)_\forall$. By Corollary 6.5.3, confining ourselves to models of T, the models of Ψ are precisely the substructures of models of Φ. But by (a), every such substructure is itself a model of Φ. So Φ and Ψ are equivalent modulo T.

For the case where $\bar{x}$ is not empty, form the language $L(\bar{c})$ by adding new constants $\bar{c}$ to L. If $\Phi(\bar{x})$ is preserved in substructures for L-structures which are models of T, then it's not hard to see that $\Phi(\bar{c})$ must be preserved in substructures for $L(\bar{c})$-structures which are models of T. But $\Phi(\bar{c})$ is a set of sentences, so the previous argument shows that $\Phi(\bar{c})$ is equivalent modulo T to a set $\Psi(\bar{c})$ of $\forall_1$ sentences of $L(\bar{c})$. Then by the lemma on constants (Lemma 2.3.2), $T \vdash \forall\bar{x}(\bigwedge \Phi(\bar{x}) \leftrightarrow \bigwedge \Psi(\bar{x}))$, so that $\Phi(\bar{x})$ is equivalent to $\Psi(\bar{x})$ modulo T, in the language $L(\bar{c})$ and hence also in the language L. (If there are empty L-structures, they trivially satisfy $\forall\bar{x}(\bigwedge \Phi(\bar{x}) \leftrightarrow \bigwedge \Psi(\bar{x}))$.) ◻

If Φ in the Łoś–Tarski theorem is a single formula, then one more application of the compactness theorem boils Ψ down to a single $\forall_1$ formula. In short, modulo any first-order theory T, the formulas preserved in substructures are precisely the $\forall_1$ formulas. The Łoś–Tarski theorem is often quoted in this form.

Since $\exists_1$ formulas are up to logical equivalence just the negations of $\forall_1$ formulas, this version of the theorem immediately implies the following.

Corollary 6.5.5. *If T is a theory in a first-order language L and ϕ is a formula of L, then the following are equivalent:*
(a) *ϕ is preserved by embeddings between models of T;*
(b) *ϕ is equivalent modulo T to an $\exists_1$ formula of L.* ◻

Actually the full dual of Theorem 6.5.4 is true too, with sets of $\exists_1$ formulas rather than single $\exists_1$ formulas. This can be an exercise (Exercise 1).

(iii) The interpolation theorem associated with Theorem 6.5.1 is a fancy elaboration of the Łoś–Tarski theorem (which it obviously implies, in the case of single formulas).

Theorem 6.5.6. *Let T be a theory in a first-order language L and let $\phi(\bar{x})$, $\chi(\bar{x})$ be formulas of L. Then the following are equivalent.*
(a) *Whenever $A \subseteq B$, A and B are models of T, $\bar{a}$ is a tuple in A and $B \vDash \phi(\bar{a})$, then $A \vDash \chi(\bar{a})$.*
(b) *There is an $\forall_1$ formula $\psi(\bar{x})$ of L such that $T \vdash \forall\bar{x}(\phi \to \psi) \wedge \forall\bar{x}(\psi \to \chi)$. ($\psi$ is an 'interpolant' between ϕ and χ.)*

Proof. The obvious adaptation of the proof of Theorem 6.5.4 works. ◻

Variants of existential amalgamation

Theorem 6.5.1 has infinitely many variants for different classes of formulas, with only trivial changes in the proof. Each of these variants has its own preservation and interpolation theorems; some are described in the exercises. (See also section 8.4 below for some related results that involve quantifier-free formulas.) I quote two variants of Theorem 6.5.1 without proof.

Theorem 6.5.7. *Let L be a first-order language, and let B and C be L-structures, $\bar{a}$ a sequence of elements of C and $f: \langle \bar{a} \rangle_C \to B$ a homomorphism such that $(C, \bar{a}) \Rightarrow_1^+ (B, f\bar{a})$. Then there exist an elementary extension D of B and a homomorphism $g: C \to D$ which extends f.* ☐

Let L be a first-order language and let A, B be L-structures. We write $A \Rightarrow_2 B$ to mean that for every $\exists_2$ sentence ϕ of L, if $A \vDash \phi$ then $B \vDash \phi$; equivalently, for every $\forall_2$ sentence ϕ of L, if $B \vDash \phi$ then $A \vDash \phi$.

Theorem 6.5.8. *Let L be a first-order language, B and C L-structures, $\bar{a}$ a sequence of elements of B and $f: \langle \bar{a} \rangle_B \to C$ an embedding such that $(C, f\bar{a}) \Rightarrow_2 (B, \bar{a})$. Then there exist an elementary extension D of B and an embedding $g: C \to D$ such that g preserves all $\forall_1$ formulas of L.* ☐

Theorem 6.5.8 is used to characterise the formulas which are preserved in unions of chains.

Theorem 6.5.9 (**Chang–Łoś–Suszko theorem**). *Let T be a theory in a first-order language L, and $\Phi(\bar{x})$ a set of formulas of L. Then the following are equivalent.*
(a) *$\bigwedge \Phi$ is preserved in unions of chains $(A_i: i < \gamma)$ whenever $\bigcup_{i<\gamma} A_i$ and all the A_i $(i < \gamma)$ are models of T.*
(b) *Φ is equivalent modulo T to a set of $\forall_2$ formulas of L.*

Proof. (b) $\Rightarrow$ (a) is by Theorem 2.4.4. For the other direction, assume (a). Just as in the proof of Theorem 6.5.4, we can assume that Φ is a set of sentences. Let Ψ be the set of all $\forall_2$ sentences of L which are consequences of $T \cup \Phi$. We have to show that $T \cup \Psi \vdash \Phi$, and for this it will be enough to prove that every model of $T \cup \Psi$ is elementarily equivalent to a union of some chain of models of $T \cup \Phi$ which is itself a model of T.

Let A_0 be any model of $T \cup \Psi$. We shall construct an elementary chain $(A_i: i < \omega)$, extensions $B_i \supseteq A_i$ and embeddings $g_i: B_i \to A_{i+1}$, so that the following diagram commutes:

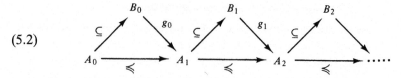

(5.2)

We shall require that for each $i < \omega$,

(5.3) $B_i \models T \cup \Phi$, and $(B_i, \bar{a}_i) \Rightarrow_1 (A_i, \bar{a}_i)$
         when $\bar{a}_i$ lists all the elements of A_i.

The diagram is constructed as follows. Suppose A_i has been chosen. Then since $A_0 \preccurlyeq A_i$, A_i is a model of all the $\forall_2$ consequences of $T \cup \Phi$. By exactly the argument of Corollary 6.5.3 (with $\forall_2$ in place of $\forall_1$ and Theorem 6.5.8 in place of Theorem 6.5.1), it follows that A_i can be extended to a structure B_i satisfying (5.3). Then by Theorem 6.5.1 and the second part of (5.3), there are an elementary extension A_{i+1} of A_i and an embedding $g_i: B_i \to A_{i+1}$ such that g is the identity on A_i.

Now in (5.2) we can replace each B_i by its image under g_i, and so assume that all the maps are inclusions. Then $\bigcup_{i<\omega} A_i$ and $\bigcup_{i<\omega} B_i$ are the same structure C. By the Tarski–Vaught elementary chain theorem (Theorem 2.5.2), $A_0 \preccurlyeq C$. So C is a model of T and the union of a chain of models B_i of $T \cup \Phi$, and A_0 is elementarily equivalent to C, as required. $\square$

Just as with the Łoś–Tarski theorem, compactness gives us a finite version: a formula ϕ of L is preserved in unions of chains (where all the structures are models of T) if and only if ϕ is equivalent modulo T to an $\forall_2$ formula of L.

Exercises for section 6.5

1. Let T be a theory in a first-order language L and $\Phi(\bar{x})$ a set of formulas of L. Show that the following are equivalent. (a) If A and B are models of T, $A \subseteq B$, $\bar{a}$ is a sequence of elements of A and $A \models \bigwedge \Phi(\bar{a})$, then $B \models \bigwedge \Phi(\bar{a})$. (b) Φ is equivalent modulo T to a set of $\exists_1$ formulas of L. [Assuming Φ is a set of sentences, let Ψ be the set of $\exists_1$ consequences of $T \cup \Phi$. Take B in Theorem 6.5.1 to be a model of $T \cup \Psi$, and C a model of $T \cup \Phi$ which we can embed into some elementary extension of B. Cf. Corollary 10.3.2 for a different proof using λ-saturated models.]

2. Let L be a first-order language and T a theory in L. Suppose A and B are models of T. Show that the following are equivalent. (a) There is a model C of T such that both A and B can be embedded in C. (b) ϕ and ψ are $\forall_1$ sentences of L such that $T \vdash \phi \lor \psi$, then either (i) A and B are both models of ϕ, or (ii) A and B are both models of ψ.

Let L be a first-order language, and let A, B be L-structures. For any $n < \omega$, we write $A \preccurlyeq_n B$ to mean that $A \subseteq B$ and for every $\exists_n$ formula $\phi(\bar{x})$ of L and every tuple $\bar{a}$ of *elements of A, $B \vDash \phi(\bar{a}) \Rightarrow A \vDash \phi(\bar{a})$. A $\preccurlyeq_n$-chain is a chain $(A_i : i < \gamma)$ in which* $i < j < \gamma$ *implies* $A_i \preccurlyeq_n A_j$.

3. Let L be a first-order language and let A_0, A_1 be L-structures with $A_0 \subseteq A_1$. Let n be a positive integer. Show that $A_0 \preccurlyeq_{2n-1} A_1$ if and only if there is a chain $A_0 \subseteq \ldots \subseteq A_{2n}$ in which $A_i \preccurlyeq A_{i+2}$ for each i:

(5.4)

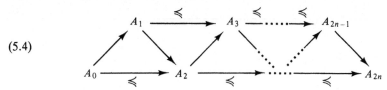

where all the arrows are inclusions.

4. Let L be a first-order language, T a theory in L, n an integer ≥ 2 and $\phi(\bar{x})$ a formula of L. Show that the following are equivalent. (a) ϕ is equivalent modulo T to an $\forall_n$ formula $\psi(\bar{x})$ of L. (b) If A and B are models of T such that $A \preccurlyeq_{n-1} B$, and $\bar{a}$ is a tuple of elements of A such that $B \vDash \phi(\bar{a})$, then $A \vDash \phi(\bar{a})$. (c) ϕ is preserved in unions of $\preccurlyeq_{n-2}$-chains of models of T. [Atomise all $\exists_{n-1}$ formulas.]

5. Let L be a first-order language and T a theory in L. Show that the following are equivalent. (a) T is equivalent to a set of $\forall_2$ sentences of L. (b) If A is an L-structure and for every tuple $\bar{a}$ of elements of A there is a substructure of A which contains $\bar{a}$ and is a model of T, then A is a model of T. (c) The class of models of T is closed under direct limits of directed diagrams of embeddings (see section 2.4).

6. Let L be a first-order language and T a theory in L. (a) Show that the following are equivalent. (i) T is equivalent to a set of $\forall_1$ sentences of L. (ii) The class of models of T is closed under taking finitely generated substructures and under unions of chains. [Use the previous exercise.] (b) Give an example to show that the condition on unions of chains can't be dropped in (ii).

7. Let L be a first-order language and T a theory in L. Show that the following are equivalent. (a) T is equivalent to a set of sentences of L of the form $\forall x \exists \bar{y} \, \phi(x, \bar{y})$ with ϕ quantifier-free. (b) If A is an L-structure and for every element a of A there is a substructure of A which contains a and is a model of T, then A is a model of T.

8. Let L be a first-order language and T a theory in L. Show that the following are equivalent. (a) Whenever A and B are models of T with $A \preccurlyeq B$, and $A \subseteq C \subseteq B$, then C is also a model of T. (b) Whenever A and B are models of T with $A \preccurlyeq_2 B$, and $A \subseteq C \subseteq B$, then C is also a model of T. (c) T is equivalent to a set of $\exists_2$ sentences.

9. Let T be a first-order theory, and suppose that whenever a model A of T has substructures B and C which are also models of T with non-empty intersection, then $B \cap C$ is also a model of T. Show that T is equivalent to an $\forall_2$ first-order theory. [Let $(D_i : i < \gamma)$ be a chain of models of T with union D. Show that D is a substructure of

a model B of T. Let U be the theory consisting of T, $\mathrm{diag}(B)$ and sentences which say 'The elements satisfying Rx are a model of T', 'Rd' for all d in D and '$\neg Rb$' for all b in B but not in D. Show that U has a model A, and take the intersection of B in A with the structure defined by R.]

6.6 Expanding the language

We continue where we left off at the end of section 6.5, except that the next amalgamation result is about expansions rather than extensions.

Theorem 6.6.1. *Let L_1 and L_2 be first-order languages, $L = L_1 \cap L_2$, B an L_1-structure, C an L_2-structure, and $\bar{a}$ a sequence of elements of B and of C such that $(B|L, \bar{a}) \equiv (C|L, \bar{a})$. Then there are an $(L_1 \cup L_2)$-structure D such that $B \preccurlyeq D|L_1$, and an elementary embedding $g: C \to D|L_2$, such that $g\bar{a} = \bar{a}$.*

Proof. We start by noting that an almost invisible alteration of the proof of Theorem 6.4.1 (elementary amalgamation) gives a weak version of the theorem we want, viz.: Under the hypotheses of Theorem 6.6.1, there are an elementary extension D of B and an elementary embedding $g: C|L \to D|L$, such that $g\bar{a} = \bar{a}$. (It suffices to show that $\mathrm{eldiag}(B) \cup \mathrm{eldiag}(C|L)$ has a model.)

Put $B_0 = B$, $C_0 = C$ and use the weak version of the theorem, alternately from this side and from that, to build up a commutative diagram

(6.1)

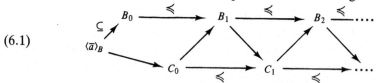

where the maps from B_i to C_i and from C_i to B_{i+1} are elementary embeddings of the L-reducts. The diagram induces an isomorphism $e: \bigcup_{i<\omega} B_i|L \to \bigcup_{i<\omega} C_i|L$. Now $\bigcup_{i<\omega} B_i|L$ and $\bigcup_{i<\omega} C_i|L$ are respectively an L_1-structure and an L_2-structure. Use e with $\bigcup_{i<\omega} C_i|L$ as a template to expand $\bigcup_{i<\omega} B_i|L$ to an $(L_1 \cup L_2)$-structure D. By the elementary chain theorem (Theorem 2.5.2), D is as required. $\qquad\square$

Interests change. When Theorem 6.6.1 first appeared in the 1950s, it wasn't even stated explicitly; but thoughtful readers could extract it from the proof of a purely syntactic result about first-order logic (Robinson's joint consistency lemma, Exercise 1 below).

True to form, Theorem 6.6.1 generates a brood of characterisation and interpolation theorems. If L and L^+ are first-order languages with $L \subseteq L^+$, and T is a theory in L^+, we write T_L for the set of all consequences of T in the language L.

Corollary 6.6.2. *Let L and L^+ be first-order languages with $L \subseteq L^+$ and T a theory in L^+. Let A be an L-structure. Then $A \vDash T_L$ if and only if for some model B of T, $A \preccurlyeq B|L$.*

Proof. As Corollary 6.5.3 followed from Theorem 6.5.1. $\square$

Next an interpolation theorem.

Theorem 6.6.3. *Let L_1, L_2 be first-order languages, $L = L_1 \cap L_2$ and T_1, T_2 theories in L_1, L_2 respectively, such that $T_1 \cup T_2$ has no model. Then there is some sentence ψ of L such that $T_1 \vdash \psi$ and $T_2 \vdash \neg \psi$.*

Proof. Take $\Psi = (T_1)_L$. By the compactness theorem it suffices to show that $\Psi \cup T_2$ has no model. For contradiction let C be a model of $\Psi \cup T_2$. By Corollary 6.6.2 there is an L_1-structure B such that $C|L \preccurlyeq B|L$ and $B \vDash T_1$. Then $B|L \equiv C|L$, and so by Theorem 6.6.1 there are an $(L_1 \cup L_2)$-structure D such that $B \preccurlyeq D|L_1$ and an elementary embedding $g \colon C \to D|L_2$. Now on the one hand $B \preccurlyeq D|L_1$, so that $D \vDash T_1$; but on the other hand g is elementary, so that $D \vDash T_2$. Thus $T_1 \cup T_2$ does have a model; contradiction. $\square$

In particular suppose ϕ and χ are sentences of L_1 and L_2 respectively, such that $\phi \vdash \chi$. Then there is a sentence ψ of $L_1 \cap L_2$ such that $\phi \vdash \psi$ and $\psi \vdash \chi$. This case of the theorem is known as **Craig's interpolation theorem**.

And of course there is the expected preservation theorem. It talks about formulas which are preserved under taking off symbols and putting them back on again.

Theorem 6.6.4. *Let L and L^+ be first-order languages with $L \subseteq L^+$, let T be a theory in L^+ and $\phi(\bar{x})$ a formula of L^+. Then the following are equivalent.*
(a) *If A and B are models of T and $A|L = B|L$, then for all tuples $\bar{a}$ in A, $A \vDash \phi(\bar{a})$ if and only if $B \vDash \phi(\bar{a})$.*
(b) *$\phi(\bar{x})$ is equivalent modulo T to a formula $\psi(\bar{x})$ of L.*

Proof. From Theorem 6.6.1 as Theorem 6.5.4 followed from Theorem 6.5.1. $\square$

Pause for a moment at the implication (a) $\Rightarrow$ (b) in the case where ϕ is an unnested atomic formula $R(x_0, \ldots, x_{n-1})$ or $F(x_0, \ldots, x_{n-1}) = x_n$. This particular case of Theorem 6.6.4 is known as **Beth's definability theorem**.

In more detail: we say that a relation symbol R is **implicitly defined** by T

in terms of L if whenever A and B are models of T with $A|L = B|L$, then $R^A = R^B$; and likewise with a function symbol F. As in section 2.6 above, we say that R is **explicitly defined** by T in terms of L if T has some consequence of the form $\forall \bar{x}(R\bar{x} \leftrightarrow \psi)$, where $\psi(\bar{x})$ is a formula of L; and similarly with F. It's immediate that if R (or F) is explicitly defined by T in terms of L, then it is implicitly defined by T in terms of L. Beth's theorem states the converse. In a slogan: *relative to a first-order theory, implicit definability equals explicit definability.*

In section 10.5 below we shall give Beth's theorem a fresh coat of paint, in terms of automorphisms.

The notion of implicit definability makes sense in a broader context. Let L and L^+ be languages (not necessarily first-order), T a theory in L^+ and R a relation symbol of L^+. Just as above, we say that R is **implicitly defined** by T in terms of L if whenever A and B are models of T with $A|L = B|L$, then $R^A = R^B$. A person who produces models A and B of T with $A|L = B|L$ but $R^A \neq R^B$ is said to be using **Padoa's method** for proving the undefinability of R by T in terms of L. There were some examples in section 2.6 above; see Exercises 2.6.6 and 2.6.7. If L and L^+ are not first-order, Beth's theorem may fail; see Makowsky & Shelah [1979].

We can draw a little more out of Theorem 6.6.3 by using Theorem 5.2.1. The next result is an interpolation theorem for interpretations.

Theorem 6.6.5. *Let K_1, K_2 and L be first-order languages. Suppose T_1, T_2 are theories in K_1, K_2 respectively, and Γ_1, Γ_2 are left total interpretations of L in T_1, T_2 respectively. Suppose there are no models A_1 of T_1 and A_2 of T_2 with $\Gamma A_1 \cong \Gamma A_2$. Then there is a sentence ψ of L such that $T_1 \vdash \psi_{\Gamma_1}$ and $T_2 \vdash \neg \psi_{\Gamma_2}$.*

Proof. Using Theorem 5.3.1, we can regard interpretations as relativised reductions. This trick allows us to write down, for each i $(i = 1, 2)$, a theory U_i in some first-order language M_i extending L, such that an L-structure B is of the form $\Gamma_i A$ for some model A of T_i if and only if B is of the form C_P for some model C of U_i. By Theorem 5.2.1 we can get rid of the relativisation, so that an L-structure B is of the form $\Gamma_i A$ for some model A of T_i if and only if B is of the form $C|L$ for some model C of U_i. Without loss of generality, M_1 and M_2 are disjoint except for L. In terms of U_1 and U_2, the hypothesis is that there are no models C_1 of U_1 and C_2 of U_2 with $C_1|L = C_2|L$; in other words $U_1 \cup U_2$ has no model. By Theorem 6.6.3 we infer that there is some sentence ψ of L such that $U_1 \vdash \psi$ and $U_2 \vdash \neg \psi$. Now if A is any model of T_1, then $\Gamma_1 A$ exists (since Γ_1 is left total) and is $C|L$ for some model C of U_1, and so $\Gamma_1 A \vDash \psi$, whence $A \vDash \psi_{\Gamma_1}$ by the reduction theorem (Theorem 5.3.2). Thus $T_1 \vdash \psi_{\Gamma_1}$, and similarly $T_2 \vdash \neg \psi_{\Gamma_2}$. $\qquad\square$

Corollary 6.6.6. *Let K and L be first-order languages, T a theory in K and Γ a left total interpretation of L in T. Suppose ϕ is a sentence of K such that for all models A, B of T, if $\Gamma A \cong \Gamma B$ then $A \vDash \phi \Leftrightarrow B \vDash \phi$. Then modulo T, ϕ is equivalent to ψ_Γ for some sentence ψ of L.*

Proof. Apply the theorem with $\Gamma_1 = \Gamma_2 = \Gamma$, $T_1 = T \cup \{\phi\}$ and $T_2 = T \cup \{\neg \phi\}$. $\qquad\square$

Local theorems

Our last theorem tackles the following question: If 'enough' finitely generated substructures of a structure A belong to a certain class **K**, must A belong to **K** too? Theorems which say that the answer is Yes are known as **local theorems**. (See section 3.6 for some other local theorems.)

In section 5.2 we defined PC_Δ and PC'_Δ classes. We are now in a position to apply the compactness theorem to them. Many of the local theorems of group theory can be proved in this way.

Theorem 6.6.7. *Let L be a first-order language and* **K** *a PC'_Δ class of L-structures. Suppose that* **K** *is closed under taking substructures. Then* **K** *is axiomatised by a set of $\forall_1$ sentences of L.*

Proof. The theorem refines the Łoś–Tarski theorem (Theorem 6.5.4), and its proof is a refinement of the earlier proof.

Let **K** be the PC'_Δ class $\{B_P: B \vDash U\}$. Write T^* for the set of all $\forall_1$ sentences ϕ of L such that $B_P \vDash \phi$ whenever $B \vDash U$. Every structure in **K** is a model of T^*. We prove that every model A of T^* is in **K**. For this, consider the theory

(6.2) $\text{diag}(A) \cup \{Pa: a \in \text{dom } A\} \cup U.$

We claim that (6.2) has a model. For if not, then by the compactness theorem there are a conjunction $\psi(\bar{x})$ of literals of L, and a tuple $\bar{a}$ of distinct elements $a_0, \ldots, a_{m-1}$ of A, such that $A \vDash \psi(\bar{a})$ and $U \vdash Pa_0 \wedge \ldots \wedge Pa_{m-1} \rightarrow \neg \psi(\bar{a})$. By the lemma on constants,

$$U \vdash \forall \bar{x}(Px_0 \wedge \ldots Px_{m-1} \rightarrow \neg \psi(\bar{x})).$$

Hence the sentence $\forall \bar{x} \neg \psi(\bar{x})$ is in T^*, so it must be true in A. This contradicts the fact that $A \vDash \psi(\bar{a})$. The claim is proved.

By the claim there is a model D of (6.2). By the diagram lemma, A is embeddable in D_P. But D_P is in **K** and **K** is closed under substructures. Since **K** is clearly closed under isomorphic copies, it follows that A is in **K**. $\qquad\square$

We apply Theorem 6.6.7 at once. The next result was perhaps the first purely algebraic theorem whose proof (by Mal'tsev in 1940) made essential use of model theory. Let n be a positive integer and G a group. We say that G has a **faithful n-dimensional linear representation** if G is embeddable in $GL_n(F)$, the group of invertible n-by-n matrices over some field F.

Corollary 6.6.8. *Let n be a positive integer and G a group. Suppose that every finitely generated subgroup of G has a faithful n-dimensional linear representation. Then G has a faithful n-dimensional linear representation.*

Proof. Let **K** be the class of groups with faithful n-dimensional linear representations. We note that **K** is closed under substructures. There is a theory U in a suitable first-order language, such that **K** is precisely the class $\{B_P : B \vDash U\}$. (If this is not obvious, consult Example 1 of section 5.1.) By the theorem, **K** is axiomatised by an $\forall_1$ theory T. If G is not in **K**, then there is some sentence $\forall \bar{x}\, \psi(\bar{x})$ in T, with ψ quantifier-free, such that $G \vDash \exists \bar{x} \neg \psi(\bar{x})$. Find a tuple $\bar{a}$ in G so that $G \vDash \neg \psi(\bar{a})$; then the subgroup $\langle \bar{a} \rangle_G$ is not in **K**. $\qquad\square$

Exercises for section 6.6

1. (Robinson's joint consistency lemma.) Let L_1 and L_2 be first-order languages and $L = L_1 \cap L_2$. Let T_1 and T_2 be consistent theories in L_1 and L_2 respectively, such that $T_1 \cap T_2$ is a complete theory in L. Show that $T_1 \cup T_2$ is consistent.

2. Let L and L^+ be first-order languages with $L \subseteq L^+$, and let $\phi(\bar{x})$ be a formula of L^+ and T a theory in L^+. Suppose that whenever A and B are models of T and $f: A|L \to B|L$ is a homomorphism, f preserves ϕ. Show that ϕ is equivalent modulo T to an $\exists_1^+$ formula of L.

3. Let L_1 and L_2 be first-order languages with $L = L_1 \cap L_2$. Suppose ϕ and ψ are sentences of L_1 and L_2 respectively, such that $\phi \vdash \psi$. Show that if every function or constant symbol of L_1 is in L_2, and ϕ is an $\forall_1$ sentence and ψ is an $\exists_1$ sentence, then there is a quantifier-free sentence θ of L such that $\phi \vdash \theta$ and $\theta \vdash \psi$. [If Θ is the set of quantifier-free sentences θ of L such that $\phi \vdash \theta$, consider any L_2-structure A which is a model of Θ; writing A_0 for the substructure of A consisting of elements named by closed terms, embed A_0 in $B|L$ for some model B of ϕ.]

4. Let L_1 and L_2 be first-order languages with $L = L_1 \cap L_2$ Suppose ϕ and ψ are sentences of L_1 and L_2 respectively, such that $\phi \vdash \psi$. Show that if ϕ and ψ are both $\forall_1$ sentences then there is an $\forall_1$ sentence θ of L such that $\phi \vdash \theta$ and $\theta \vdash \psi$.

5. Let L and L^+ be first-order languages with $L \subseteq L^+$, and suppose P is a 1-ary relation symbol of L^+. Let ϕ be a sentence of L^+, and T a theory in L^+ such that for every model A of T, P^A is the domain of a substructure A^* of $A|L$. Suppose that whenever A and B are models of T with $A \vDash \phi$ and $f: A^* \to B^*$ is a homomorphism, then $B \vDash \phi$. Show that ϕ is equivalent modulo T to a sentence of the form $\exists y_0 \ldots y_{k-1}(\bigwedge_{i<k} Py_i \wedge \psi(\bar{y}))$ where ψ is a positive quantifier-free formula of L.

6. Let L be the first-order language with relation symbols for 'x is a son of y', 'x is a daughter of y', 'x is a father of y', 'x is a mother of y', 'x is a grandparent of y'. Let T be a first-order theory which reports the basic biological facts about these relations (e.g. that everybody has a unique mother, nobody is both a son and a daughter, etc.). Show (a) in T, 'son of' is definable from 'father of', 'mother of' and 'daughter of', (b) in T, 'son of' is not definable from 'father of' and 'mother of', (c) in T, 'father of' is definable from 'son of' and 'daughter of', (d) in T, 'mother of' is not definable from 'father of' and 'grandparent of', (e) etc. ad lib.

7. Let L be a first-order language, L^+ the language got from L by adding one new relation symbol R, and ϕ a sentence of L^+. Suppose that every L-structure can be expanded in at most one way to a model of ϕ. Show that there is a sentence θ of L such that an L-structure A is a model of θ if and only if A can be expanded to a model of ϕ.

*8. Let L be a first-order language whose signature has finitely many symbols, including a constant 0. Let T be a theory in L with the following property: if A is any model of T and $f: A \to B$ and $g: A \to C$ are homomorphisms such that for all elements a, $f(a) = f(0) \Leftrightarrow g(a) = g(0)$, then for all elements b, c of A, $f(b) = f(c) \Leftrightarrow g(b) = g(c)$. A set N of elements of A is called **normal** if for some homomorphism $f: A \to B$, N is $\{a: f(a) = f(0)\}$. Let L^+ be L with an extra 1-ary relation symbol P. Show that there are a sentence ϕ of L^+ and a formula $\theta(x, y)$ of L^+ such that (a) for any model A of T, a set X of elements of A is normal if and only if $(A, X) \vDash \phi$, and (b) if $f: A \to B$ is any homomorphism and b, c are elements of A, then $f(b) = f(c)$ if and only if $(A, \{a: f(a) = f(0)\}) \vDash \theta(b, c)$.

9. Show that there exist two integral domains which are not elementarily equivalent but have the same additive groups. [Use Theorem 6.6.5, say with ordered rings whose non-negative elements are models of Peano arithmetic.]

*10. Let L and L^+ be first-order languages with $L \subseteq L^+$, and T a theory in L^+. Write T_L for the set of all consequences of T in L. (a) Show that if L^+ has no function or constant symbols and T is an $\forall_1$ theory then T_L is equivalent to an $\forall_1$ theory. (b) Show that if L^+ has no function or constant symbols and T is a complete $\exists_2$ theory then T_L is equivalent to an $\exists_2$ theory. [Cf. Exercise 6.5.8.]

11. Let T be a first-order theory. Show that the class of groups $\{G: G$ acts faithfully on some model of $T\}$ is axiomatised by an $\forall_1$ first-order theory in the language of groups.

12. Prove that if $Q\phi$ is a second-order sentence where ϕ is quantifier-free and Q is a string of quantifiers of the form $\forall R$, $\exists R$ (with R a relation variable) or $\forall x$, then $Q\phi$ is equivalent to a set of universal first-order sentences.

*For the next two exercises, a **partial order** of a group G is a partial order $\leqslant$ on the set of elements of G such that whenever $g \leqslant h$, we have $a \cdot g \cdot b \leqslant a \cdot h \cdot b$ for all a, b in G. An **ordered group** is a structure $(G, \leqslant)$ where G is a group and $\leqslant$ is a linear ordering which is a partial order of G.*

13. A group G is said to be **orderable** if there exists a linear ordering $\leqslant$ which is a partial order of G. Show that the class of orderable groups is axiomatised by a set of $\forall_1$ sentences in the first-order language of groups. Deduce that if every finitely generated subgroup of a group G is orderable, then G is orderable.

14. A group is said to be **freely orderable** if every partial order of G can be extended to a linear order of G. Show that if every finitely generated subgroup of a group G is freely orderable, then G is freely orderable. [Use Exercise 12.]

*15. A **commutator sequence** for a group G is a set S of normal subgroups of G, linearly ordered by an ordering $<_S$ with the properties (i) $H <_S K$ implies $H \supseteq K$, and (ii) for any two non-identity elements f, g of G there is H in S which contains the commutator $[f, g]$ but not both f and g. (a) Show that the class of groups which have a commutator sequence is $\forall_1$ axiomatisable. (b) Deduce that if every finitely generated subgroup of a group G is free, then G is not simple. [Finitely generated free groups have a commutator sequence.]

6.7 Stability

Saharon Shelah introduced the distinction between stable and unstable theories in [1969], as an aid to counting the number of non-isomorphic models of a theory in a given cardinality. A year or so later he set his sights on the *classification problem*. Roughly speaking, this problem is to classify all complete first-order theories into two kinds: those where there is a good structure theory for the class of all models, and those where there isn't. After writing a book and about a hundred papers on the subject, Shelah finally gave a solution of the classification problem in the case of countable languages in 1982. On any reckoning this is one of the major achievements of mathematical logic since Aristotle.

However, it is not our topic here. There are several accounts in print. (For example, in increasing order of length and detail: Shelah [1985], Hodges [1987], Lascar [1987a], Baldwin [1988], Shelah [1990].) Whatever Shelah's own interests, the results of stability theory (in Shelah's own words [1987b]) 'give, and are intended to give, instances for fine structure investigation with considerable tools to start with'. This section is an introduction to those considerable tools. There are several equivalent ways of distinguishing between stable and unstable. One (Theorem 6.7.2 below) is in terms of the cardinalities of spaces of types, and another (Theorem 11.3.1 with Corollary 6.7.12 below) is in terms of elementary amalgamations.

Let T be a complete theory in a first-order language L. Then T is **unstable** if there are a formula $\phi(\bar{x}, \bar{y})$ of L and a model A of T containing tuples of elements $\bar{a}_i$ $(i < \omega)$ such that

(7.1)   for all $i, j < \omega$, $A \vDash \phi(\bar{a}_i, \bar{a}_j) \Leftrightarrow i < j$.

T is **stable** if it is not unstable. We say that a structure A is **stable** or **unstable** according as $\text{Th}(A)$ is stable or unstable.

From the definition it's immediate that every finite structure is stable. Stability theory has very little to say about finite structures – though there are some interesting facts (for example in Lachlan [1987a]) about how infinite stable structures can be built up as the unions of chains of finite structures. *For the rest of this section we assume that T is a complete first-order theory in a first-order language L, and all the models of T are infinite.*

By the definition, any structure which contains an infinite set linearly ordered by some formula is unstable. Note that the set itself need not be definable, but there must be a first-order formula (possibly with parameters – see Exercise 1) which orders it.

Example 1: *Infinite boolean algebras.* Every infinite boolean algebra B contains an infinite set of elements which is linearly ordered by the algebra ordering 'x < y'. So $\text{Th}(B)$ is unstable. In general the linearly ordered set will not be first-order definable.

Theorem 6.7.1. *Suppose A is stable.*
  (a) *If $\bar{a}$ is a sequence of elements of A, then $(A, \bar{a})$ is stable.*
  (b) *Suppose the structure B is interpretable in A; then B is stable.*

Proof. (a) is trivial. For (b), let Γ be an interpretation of the L-structure B in the K-structure A with coordinate map f, and suppose B is unstable. Then there are a formula $\phi(\bar{x}, \bar{y})$ of L and tuples $\bar{b}_i$ $(i < \omega)$ in B such that for all $i, j < \omega$, $B \vDash \phi(\bar{b}_i, \bar{b}_j) \Leftrightarrow i < j$. For each i, choose $\bar{a}_i$ in A so that $f\bar{a}_i = \bar{b}_i$. Then by the reduction theorem (Theorem 5.3.2), $A \vDash \phi_\Gamma(\bar{a}_i, \bar{a}_j) \Leftrightarrow i < j$, and so A is unstable.   □

In particular every reduct of a stable structure is stable. It may be worth remarking here that stability is a robust property of theories – it is not affected by changing the language. Contrast this with the property of quantifier elimination, which comes and goes as we take definitional equivalents.

Sizes of Stone spaces

To find examples of stable theories, we turn to the cardinalities of spaces of types (see section 6.3 above). Let λ be an infinite cardinal. We say that T is λ-**stable** if for every model A of T and every set X of at most λ elements of A, $|S_1(X; A)| \leq \lambda$. We say that A is λ-**stable** if $\mathrm{Th}(A)$ is λ-stable. (**Warning**: this is the usual terminology, but Shelah [1978a] means something different when he says a structure is λ-stable.) A theory or structure which is not λ-stable is λ-**unstable**.

Theorem 6.7.2. *The following are equivalent:*
(a) *T is stable;*
(b) *For at least one infinite cardinal λ, T is λ-stable.*

The proof of (a) $\Rightarrow$ (b) needs some preparation, and I postpone it to Theorem 6.7.13 below. The proof of (b) $\Rightarrow$ (a) is not quite immediate either. It needs two lemmas. When P and Q are linear orderings and $P \subseteq Q$, we say that P is **dense in** Q if for any two elements b, c of Q with $b < c$ there is a in P such that $b < a < c$.

Lemma 6.7.3. *Let λ be any infinite cardinal. Then there are linear orderings P and Q with $P \subseteq Q$, such that $|P| \leq \lambda < |Q|$ and P is dense in Q.*

Proof. Let μ be the least cardinal such that $2^\mu > \lambda$; clearly $\mu \leq \lambda$. Consider the set $^\mu 2$ of all sequences of 0's and 1's of length μ; this set has cardinality $2^\mu > \lambda$. We can linearly order $^\mu 2$ as follows:

(7.2) $s < t \Leftrightarrow$ there is $i < \mu$ such that $s|i = t|i$ and $s(i) < t(i)$.

Let X be the set of all sequences in $^\mu 2$ which are 1 from some point onwards, and let Q be the linearly ordered set of all sequences in $^\mu 2$ which are not in X. Let P be the subset of Q consisting of those sequences which are 0 from some point onwards. By choice of μ, both X and P have cardinality $\Sigma_{\kappa < \mu} 2^\kappa \leq \lambda$, and so Q has cardinality $> \lambda$. Also one readily checks that P is dense in Q. $\square$ Lemma 6.7.3.

Lemma 6.7.4. *Let A be an L-structure, λ an infinite cardinal and X a set of at most λ elements of A. Suppose there is a positive integer n such that $|S_n(X; A)| > |X|$. Then A is λ-unstable.*

Proof. We go by induction on n. When $n = 1$ there is nothing to prove. Suppose then that $n > 1$, and for each $r \in S_n(X; A)$ choose an $(n-1)$-tuple $\bar{b}_r$ and an element c_r, both in some elementary extension B_r of A, so that the n-tuple $\bar{b}_r \hat{\ } c_r$ realises r in B_r. Since $\lambda \geq |L|$, we can assume that each B_r has cardinality λ. Write q_r for the type $\mathrm{tp}_{B_r}(\bar{b}_r/X)$, which is a complete $(n-1)$-type over X with respect to A. There are two cases.

First, suppose the number of distinct types q_r is at most λ. Then since λ^+ is a regular cardinal, there must be a set $S \subseteq S_n(X; A)$ of cardinality λ^+, such that all the types q_r with $r \in S$ are equal. Choose some fixed $s \in S$, and for each $r \in S$ let q'_r be $\{\phi(\bar{b}_s, x): \phi(x_0, \ldots, x_{n-1}) \in r\}$. Then it follows from Theorem 6.3.1 that the q'_r with $r \in S$ are λ^+ distinct 1-types over $X \cup \{\bar{b}_s\}$.

Second, suppose there are more than λ distinct types q_r. Then we have more than λ complete $(n-1)$-types over X, and the conclusion holds by induction hypothesis.   ☐ Lemma 6.7.4

Now we prove (b) $\Rightarrow$ (a) of the theorem. Assume T is unstable, so that there are a formula $\phi(\bar{x}, \bar{y})$ and a model of T as in (7.1). Replacing $\phi(\bar{x}, \bar{y})$ by $\phi(\bar{x}, \bar{y}) \wedge \neg \phi(\bar{y}, \bar{x})$ if necessary, we can suppose that ϕ defines an asymmetric relation on tuples in models of T. Let P and Q be as in Lemma 6.7.3. Take a new tuple of constants $\bar{c}_s$ for each $s \in P$, and consider the theory

$$(7.3) \qquad\qquad U = T \cup \{\phi(\bar{c}_s, \bar{c}_t): s < t \text{ in } P\}.$$

By the compactness theorem, U has a model B; we can choose B to be of cardinality $\leqslant \lambda$. Put $A = B|L$. For each element r of Q, consider the set of formulas $\Phi_r(\bar{x}) = \{\phi(\bar{c}_s, \bar{x}): s < r\} \cup \{\phi(\bar{x}, \bar{c}_t): r < t\}$. By Theorem 6.3.1, each set Φ_r is a type over $\mathrm{dom}(A)$ with respect to A, and we can extend it to a complete type $p_r(\bar{x})$. If $r < r'$ in Q, then by density there is s in P such that $r < s < r'$, and it follows that $p_r \neq p_{r'}$. So there are more than λ complete types p_r with $r \in Q$.

We have proved that there are more than λ complete n-types over $\mathrm{dom}(A)$, where n is the length of $\bar{x}$. By Lemma 6.7.4 this shows that T is λ-unstable. So (b) $\Rightarrow$ (a) of Theorem 6.7.2 is proved.   ☐ Theorem 6.7.2

Example 2: *Term algebras.* Let T be the complete theory of a term algebra of a language L. Then T is λ-stable whenever $\lambda^{|L|} = \lambda$, and hence T is stable. To prove this, we can suppose that T has infinite models. We use the language L_2 of Example 1 in section 2.6, and we assume the result of Theorem 2.7.5. By that theorem, if A is a model of T and $p(x)$ is a 1-type over $\mathrm{dom}(A)$ with respect to A, then p is completely determined once we know which formulas of the following forms it contains: $\mathrm{Is}_c(t(x))$, $\mathrm{Is}_F(t(x))$, $s(x) = t(x)$, $t(x) = a$, where c, F are constant and function symbols of L, a is an element of A and s, t are terms built up from function symbols F_i of L_2. Note that for each $t(x)$, p contains at most one formula $t(x) = a$ with a in A. By this and a little cardinal arithmetic, the number of distinct types p is at most $|A|^{|L|}$. Let λ be any cardinal such that $\lambda^{|L|} = \lambda$ (for example any cardinal of the form $\mu^{|L|}$), and let A be any model of T of cardinality λ. Then there are at most $|A|$ complete 1-types over $\mathrm{dom}(A)$. It follows that T is stable.

For totally transcendental theories T we can prove something stronger.

Theorem 6.7.5. *Let T be a complete first-order theory.*

(a) *If T is totally transcendental, then T is stable; in fact it is λ-stable for every $\lambda \geqslant |L|$.*

(b) *Conversely if T is ω-stable then T is totally transcendental.*

Proof. We proved (a) already in Theorem 6.3.4(b). For (b), suppose T is not totally transcendental. Then by Lemma 5.6.5 there are a model A of T and a tree of formulas as in (6.4) of section 5.6 above (with $x = x$ as ϕ). There are countably many formulas in the tree, so between them they use at most countably many parameters from A; let X be the set of these parameters. For each branch β of the tree, let p_β be some complete type over X which contains all the formulas in β; there is such a type by Theorem 6.3.1 and (i), (ii) of Lemma 5.6.1. By (iii) of that lemma, the types p_β are all distinct. But the tree has continuum many branches, so that T is not ω-stable. $\qquad\square$

A complete theory T is said to be **superstable** if there is a cardinal μ such that T is λ-stable for all $\lambda \geqslant \mu$. Totally transcendental implies superstable, and superstable implies stable. In particular every infinite vector space is superstable, and so is every algebraically closed field. See Exercise 6 for an example of a theory which is superstable but not totally transcendental, and Exercise 7 for an example of a theory which is stable but not superstable.

Stable formulas

For the rest of this section we are headed for the proof that a stable theory is λ-stable for some λ. The proof is roundabout, and there are some interesting sights on the way. The first step is a slight adjustment of (7.1).

Let T be a complete theory in a first-order language L. Let $\phi(\bar{x}, \bar{y})$ be a formula of L, with the free variables divided into two groups $\bar{x}, \bar{y}$. An *n-ladder* for ϕ is a sequence $(\bar{a}_0, \ldots, \bar{a}_{n-1}, \bar{b}_0, \ldots, \bar{b}_{n-1})$ of tuples in some model A of T, such that

(7.4) $\qquad$ for all $i, j < n$, $A \vDash \phi(\bar{a}_i, \bar{b}_j) \Leftrightarrow i \leqslant j$.

We say that ϕ is a **stable** formula (for T, or for A) if there is some $n < \omega$ such that no n-ladder for ϕ exists; otherwise it is **unstable**. The least such n is the **ladder index** of $\phi(\bar{x}, \bar{y})$ (of course it may depend on the way we split the variables).

Lemma 6.7.6. *The theory T is unstable if and only if there is an unstable formula in L for T.*

Proof. Suppose first that T has an unstable formula $\phi(\bar{x}, \bar{y})$. Thus ϕ has an n-ladder for each $n < \omega$. Taking new tuples of constants $\bar{a}_i$, $\bar{b}_i$ $(i < \omega)$, consider the theory

(7.5) $T \cup \{\phi(\bar{a}_i, \bar{b}_j): i \leqslant j < \omega\} \cup \{\neg \phi(\bar{a}_i, \bar{b}_j): j < i < \omega\}.$

Each finite subset of (7.5) has a model – take an n-ladder for some large enough n. So by the compactness theorem there is a model A of T in which there are tuples $\bar{a}_i$, $\bar{b}_i$ $(i < \omega)$ obeying the conditions in (7.4). Put $\bar{c}_i = \bar{a}_i \bar{b}_i$ for each $i < \omega$, and let $\psi(\bar{x}\bar{x}', \bar{y}\bar{y}')$ be the formula $\neg \phi(\bar{y}, \bar{x}')$. Then

(7.6) $A \vDash \psi(\bar{c}_i, \bar{c}_j) \Leftrightarrow A \vDash \neg \phi(\bar{a}_j, \bar{b}_i) \Leftrightarrow i < j,$

so that ψ linearly orders $(\bar{c}_i: i < \omega)$. Thus T is unstable. The converse is an exercise. $\square$

Theorem A.1.13 will quote this result to show that every module is stable. Using Lemma 6.7.6 in the other direction, stability implies a rather strong chain condition on groups. We say that a structure A is **group-like** if some relativised reduct of A is a group; the relativised reduct that we have in mind is called the **group of** A. By convention a **stable group** is a stable group-like structure. See section 5.7 above for some discussion of this notion.

Lemma 6.7.7 (Baldwin–Saxl lemma). *Let L be a first-order language and A an L-structure which is a stable group; let G be the group of A. Let $\phi(x, \bar{y})$ be a formula of L. Write $\mathbf{S}$ for the set of all subsets of $\text{dom}(A)$ which are subgroups of G of the form $\phi(A, \bar{b})$ with $\bar{b}$ in A; write $\bigcap \mathbf{S}$ for the set of all intersections of groups in $\mathbf{S}$.*
  (a) There is $n < \omega$ such that every group in $\bigcap \mathbf{S}$ can be written as the intersection of at most n groups in $\mathbf{S}$.
  (b) There is $m < \omega$ such that no chain in $\bigcap \mathbf{S}$ (under inclusion) has length $> m$.

Proof. Let n be the ladder index of ϕ. Suppose $n < k < \omega$ and there are subgroups $H_i = \phi(A, \bar{b}_i)$ $(i < k)$ such that no H_i contains all of $\bigcap_{j \neq i} H_j$. For each i choose $h_i \in \bigcap_{j \neq i} H_j \backslash H_i$, and write $a_0 = 1$, $a_{i+1} = h_0 \cdot \ldots \cdot h_i$. Then $A \vDash \phi(a_i, \bar{b}_j) \Leftrightarrow i \leqslant j$. This contradicts the choice of n. It follows that

(7.7)   every intersection of finitely many groups in $\mathbf{S}$ is already an intersection of at most n of them.

Now let $\bigcap_f \mathbf{S}$ be the set of all intersections of finitely many groups in $\mathbf{S}$. By (7.7) we can write each group in $\bigcap_f \mathbf{S}$ as a set of the form $\psi(A, \bar{c})$ where $\psi(x, \bar{z})$ is $\phi(x, \bar{y}_0) \wedge \ldots \wedge \phi(x, \bar{y}_{n-1})$. Since A is stable, ψ has a ladder index too; let it be m. Suppose there is a strictly ascending chain of subgroups of G,

$$(7.8) \qquad\qquad \psi(A, \bar{c}_0) \subset \ldots \subset \psi(A, \bar{c}_m).$$

Let a_0 be in $\psi(A, \bar{c}_0)$, and for each positive $i \leqslant m$ choose $a_i \in \psi(G, \bar{c}_i) \setminus \psi(G, \bar{c}_{i-1})$. Then $A \vDash \psi(a_i, \bar{c}_j) \Leftrightarrow i \leqslant j$, contradicting the choice of m. We have proved that every chain in $\bigcap_f S$ has length at most m.

But now it follows that $\bigcap_f S = \bigcap S$. For otherwise we could find subgroups $\phi(A, \bar{d}_i)$ $(i < \omega)$ such that the intersections $\bigcap_{j<i} \phi(A, \bar{d}_j)$ form a strictly decreasing sequence, contradicting the descending chain condition on $\bigcap_f S$. Both (a) and (b) follow at once. $\qquad\qquad\qquad\qquad\qquad\qquad\square$

The **centraliser** $C_G(X)$ of a set of elements X in G is the subgroup of all elements g of G which commute with everything in X. Then $C_G(X) = \bigcap_{g \in X} C_G(g)$, and $C_G(g) = \phi(A, g)$ where $\phi(x, y)$ is the formula of L which expresses that $x \cdot y = y \cdot x$. So the Baldwin–Saxl lemma implies at once that *a stable group satisfies the descending chain condition on centralisers*.

Definability of types

Suppose A is a stable L-structure, X is a set of elements of A, and $\phi(y_0, \ldots, y_{n-1})$ is a formula of L with parameters from A. We write $\phi(X^n)$ for the set of all n-tuples $\bar{a}$ of elements of X such that $A \vDash \phi(\bar{a})$. Maybe ϕ has parameters lying outside X. But according to the next result, stability will allow us to define $\phi(X^n)$ using only parameters that come from X. This is a very counterintuitive fact, and it shows how strong the assumption of stability is.

Theorem 6.7.8. *Let A be a stable L-structure and $\phi(\bar{x}, \bar{y})$ a stable formula for A, with $\bar{y} = (y_0, \ldots, y_{n-1})$. Let X be a set of elements of A and $\bar{b}$ a tuple in A. Then there is a formula $\chi(\bar{y})$ of L with parameters in X, such that $\phi(\bar{b}, X^n) = \chi(X^n)$.*

Let us interpret this result for a moment. Write $p(\bar{x})$ for the type $\mathrm{tp}_A(\bar{b}/X)$. Since $\phi(\bar{b}, X^n)$ depends only on ϕ and $p(\bar{x})$, and not on the particular choice of tuple realising p, we can write the formula χ as $\mathrm{d}_p \phi(\bar{y})$. Then we can rephrase the equation at the end of the theorem as follows:

$$(7.9) \qquad\qquad \text{for every } \bar{c} \text{ in } X, \ \phi(\bar{x}, \bar{c}) \in p \Leftrightarrow A \vDash \mathrm{d}_p \phi(\bar{c}).$$

A formula $\mathrm{d}_p \phi$ with the property of (7.9) is called a ϕ**-definition** of the type p.

The proof of the theorem depends on some combinatorial facts about ladders and trees.

We write $^n2$ for the set of sequences of length n whose terms are either 0 or 1, and $^{<n}2$ for $\bigcup_{j<n}{}^j2$. We write σ, τ for sequences; $\sigma|j$ is the initial segment of σ of length j. An n-**tree** for the formula $\phi(\bar{x}, \bar{y})$ is defined to consist of two families of tuples, $(\bar{b}_\sigma: \sigma \in {}^n2)$ and $(\bar{c}_\tau: \tau \in {}^{<n}2)$ such that for all $\sigma \in {}^n2$ and all $i < n$,

(7.10) $M \vDash \phi(\bar{b}_\sigma, \bar{c}_{\sigma|i}) \Leftrightarrow \sigma(i) = 0.$

The tuples $\bar{b}_\sigma$ are called the **branches** of the n-tree and the tuples $\bar{c}_\tau$ are called its **nodes**.

We say that a formula $\phi(\bar{x}, \bar{y})$ of L has **branching index** $\geqslant n$, in symbols $\mathrm{BI}(\phi) \geqslant n$, if there exists an n-tree for ϕ. This defines $\mathrm{BI}(\phi)$ uniquely as a natural number or ∞. The statement '$\mathrm{BI}(\phi) \geqslant n$' can be written as a sentence of L.

Lemma 6.7.9. *Let $\phi(\bar{x}, \bar{y})$ be a formula of L. If ϕ has branching index n, then ϕ has ladder index $< 2^{n+1}$. If ϕ has ladder index n, then ϕ has branching index $< 2^{n+2} - 2$.*

Proof. Typographical reasons force us to write $\bar{b}[i]$ for $\bar{b}_i$. For the first implication it is enough to note that if $\bar{b}[0], \ldots, \bar{b}[2^{n+1} - 1], \bar{c}[0], \ldots, \bar{c}[2^{n+1} - 1]$ form a 2^{n+1}-ladder for ϕ, then they can be turned into an $(n + 1)$-tree by taking the $\bar{b}[i]$'s as branches and the $\bar{c}[j]$'s as nodes, relabelling in an obvious way.

We prove the second implication by showing that if ϕ has branching index at least $2^{n+2} - 2$, then ϕ has ladder index at least $n + 1$; for ease of reading let us say rather that if ϕ has branching index at least $2^{n+1} - 2$, then ϕ has ladder index at least n.

If H is an $(n + 1)$-tree for ϕ and $i = 0$ or 1, we write $H_{(i)}$ for the n-tree whose nodes and branches are the nodes $\bar{c}[\tau]$ and branches $\bar{b}[\sigma]$ of H such that $\tau(0) = \sigma(0) = i$. We shall say that a map $f: {}^{<n}2 \to {}^{<m}2$ is a **tree-map** if for any two sequences σ, τ in $^{<n}2$, $f(\sigma)$ is an end-extension of $f(\tau)$ precisely when σ is an end-extension of τ. If H is an m-tree and N is a set of nodes of H, we say that N **contains an** n-**tree** if there is a tree-map $f: {}^{<n}2 \to {}^{<m}2$ such that for each $\tau \in {}^{<n}2$, $\bar{c}[f(\tau)]$ is in N. Clearly this implies that there is an n-tree J for ϕ whose nodes all lie in N and whose branches are branches of H; we shall say also that N **contains the** n-**tree** J.

We claim the following. Consider $n, k \geqslant 0$ and let H be an $(n + k)$-tree for ϕ. If the nodes of H are partitioned into two sets N, P, then either N contains an n-tree or P contains a k-tree.

The claim is proved by induction on $n + k$. The case $n = k = 0$ is trivial. Suppose $n + k > 0$, and let the tuples $\bar{c}[\tau]$ be the nodes of H. Suppose $\bar{c}[\langle\rangle] \in N$. (The argument when $\bar{c}[\langle\rangle] \in P$ is parallel.) For $i = 0, 1$ let Z_i be the set of all nodes of $H_{(i)}$. By induction hypothesis, if $i = 0$ or 1 then either $N \cap Z_i$ contains an $(n-1)$-tree or $P \cap Z_i$ contains a k-tree. If at least one of $P \cap Z_0$, $P \cap Z_1$ contains a k-tree, then so does P. If neither does, then both $N \cap Z_0$ and $N \cap Z_1$ contain $(n-1)$-trees, and so, since $\bar{c}[\langle\rangle] \in N$, N contains an n-tree. This proves the claim.

To complete the proof of the lemma, assume that ϕ has branching index at least $2^{n+1} - 2$. We shall show, by induction on $n - r$, that for $1 \leqslant r \leqslant n$ the following situation S_r holds: there are

(7.11) $\bar{b}'[0], \bar{c}'[0], \ldots, \bar{b}'[q-1], \bar{c}'[q-1], H, \bar{b}'[q], \bar{c}'[q], \ldots,$
          $\bar{b}'[n-r-1], \bar{c}'[n-r-1]$

such that

(7.12) H is a $(2^{r+1} - 2)$-tree for ϕ,

(7.13)   for all $i, j < n - r$, $A \vDash \phi(\bar{b}[i], \bar{c}[j]) \Leftrightarrow i \leqslant j$,

(7.14)   if $\bar{c}$ is a node of H then $A \vDash \phi(\bar{b}[i], \bar{c}) \Leftrightarrow i < q$,

(7.15)   if b is a branch of H then $A \vDash \phi(\bar{b}, \bar{c}[j]) \Leftrightarrow j \geqslant q$.

The initial case S_n states simply that there is a $(2^{n+1} - 2)$-tree for ϕ, which we have assumed. The final case S_1 implies that ϕ has ladder index at least n as follows. As H is a 2-tree, it has a node $\bar{c}$ and a branch $\bar{b}$ such that $M \vDash \phi(\bar{b}, \bar{c})$. Put $\bar{b}, \bar{c}$ in that order between $\bar{c}'[q-1]$ and $\bar{b}'[q]$ in (7.11); then conditions (7.13)–(7.15) show that the resulting list of tuples yields an n-ladder for ϕ.

It remains to show that if $r > 1$ and S_r holds, then so does S_{r-1}. Assume S_r and put $h = 2^r - 2$. By (7.12), H is a $(2h + 2)$-tree. For each branch $\bar{b}$ of H, write $H(\bar{b})$ for the set of those nodes $\bar{c}$ of H such that $A \vDash \phi(\bar{b}, \bar{c})$. There are two cases.

Case 1: there is a branch $\bar{b}$ of H such that $H(\bar{b})$ contains an $(h+1)$-tree. Then there are a node $\bar{c}$ in $H(\bar{b})$ and an h-tree H' for ϕ such that S_{r-1} holds when we replace H in (7.11) by $\bar{b}, \bar{c}, H'$ in that order.

Case 2: for every branch $\bar{b}$ of H, $H(\bar{b})$ contains no $(h+1)$-tree. Then let $\bar{c}$ be the bottom node $\bar{c}[\langle\rangle]$ of H, let $\bar{b}$ be any branch of $H_{(0)}$ and let N be the set of all nodes of $H_{(0)}$. The case assumption tells us that $H(\bar{b}) \cap N$ contains no $(h+1)$-tree. So by the claim applied to $H_{(0)}$, the set $N \backslash H(\bar{b})$ contains an h-tree H' for ϕ. Then S_{r-1} holds when we replace H in (7.11) by $H', \bar{b}, \bar{c}$ in that order.

So in either case S_{r-1} holds. This completes the induction.

$\square$ Lemma 6.7.9.

It follows at once that T is stable if and only if every formula $\phi(\bar{x}, \bar{y})$ of L has finite branching index.

Let $\phi(\bar{x}, \bar{y})$ be a formula of L and $\psi(\bar{x})$ a formula of L with parameters from some model A of T. We define the **relativised branching index** $\mathrm{BI}(\phi, \psi)$ to be $\geq n$ iff there is an n-tree for ϕ whose branches all satisfy ψ. Thus $\mathrm{BI}(\phi, \psi)$ is either a unique finite number or ∞. Clearly $\mathrm{BI}(\phi, \psi) \leq \mathrm{BI}(\phi)$, so that if ϕ is stable then $\mathrm{BI}(\phi, \psi)$ must be finite.

The statement '$\mathrm{BI}(\phi, \psi) \geq n$' can be written as a sentence of L with parameters from A. Since the variables $\bar{x}$ don't occur free in this sentence, it could be misleading to write it as '$\mathrm{BI}(\phi, \psi(\bar{x})) \geq n$'. So when it is necessary to mention the variables $\bar{x}$, I replace them with '$-$', thus: '$\mathrm{BI}(\phi, \psi(-)) \geq n$'.

The following lemma is crucial for proving Theorem 6.7.8.

Lemma 6.7.10. *If* $\mathrm{BI}(\phi, \psi)$ *is a finite number* n, *then for every tuple* $\bar{c}$ *of elements of* M, *either* $\mathrm{BI}(\phi, \psi \wedge \phi(-, \bar{c})) < n$ *or* $\mathrm{BI}(\phi, \psi \wedge \neg \phi(-, \bar{c})) < n$.

Proof. Suppose H_0 is an n-tree for ϕ whose branches satisfy $\psi(\bar{x}) \wedge \phi(\bar{x}, \bar{c})$, and H_1 is an n-tree for ϕ whose branches satisfy $\psi(\bar{x}) \wedge \neg \phi(\bar{x}, \bar{c})$. Then we can form an $(n+1)$-tree H for ϕ whose branches satisfy $\psi(\bar{x})$, by putting $H_{(0)} = H_0$, $H_{(1)} = H_1$ (in the notation of the proof of Lemma 6.7.9), and taking $\bar{c}$ as the bottom node. $\square$

Suppose now that X is a set of elements of a model A of T, $p(\bar{x})$ a complete type over X with respect to A, and $\phi(\bar{x}, \bar{y})$ a formula of L with finite branching index. Then the minimum value of $\mathrm{BI}(\phi, \psi)$, as $\psi(\bar{x})$ ranges over all formulas in p, is called the ϕ-**rank** of p.

Proof of Theorem 6.7.8. Let $p(\bar{x})$ be the type of $\bar{b}$ over X. Since ϕ is stable, the ϕ-rank of p must be some finite number n. Choose $\psi(\bar{x})$ in p so that $\mathrm{BI}(\phi, \psi) = n$, and let $\mathrm{d}_p \phi(\bar{y})$ be the formula '$\mathrm{BI}(\phi, \psi \wedge \phi(-, \bar{y})) \geq n$'.

We claim that (7.9) holds. From left to right, suppose $\phi(\bar{x}, \bar{c}) \in p$. Then $\psi(\bar{x}) \wedge \phi(\bar{x}, \bar{c}) \in p$ and hence $\mathrm{BI}(\phi, \psi \wedge \phi(-, \bar{c})) \geq n$, so that $A \vDash \mathrm{d}_p \phi(\bar{c})$. In the other direction, suppose $\phi(\bar{x}, \bar{c})$ is not in p. Then $\neg \phi(\bar{x}, \bar{c})$ is in p, and the same argument as before shows that $\mathrm{BI}(\phi, \psi \wedge \neg \phi(\bar{x}, \bar{c})) \geq n$. It follows by Lemma 6.7.10 that $\mathrm{BI}(\phi, \psi \wedge \phi(\bar{x}, \bar{c})) < n$, so that $A \vDash \neg \mathrm{d}_p \phi(\bar{c})$.
                                                                          $\square$ Theorem 6.7.8

Let $p(\bar{x})$ be a complete type over a set X of elements of a model A of T. By a **definition schema** of p we mean a map d which takes each formula $\phi(\bar{x}, \bar{y})$ of L to a ϕ-definition $\mathrm{d}\phi(\bar{y})$ of p, where any parameters in $\mathrm{d}\phi$ come from X. A type is said to be **definable** if it has a definition schema. The next result says that *for stable theories, all complete types are definable*.

Corollary 6.7.11 *Let the theory* T *be stable. Let* A *be a model of* T, *let* X *be a set of elements of* A *and* p *a type over* X. *Then* p *is definable.*

Proof. Immediate from Theorem 6.7.8.   □

The moral of the next corollary is that if one is forming an heir–coheir amalgam of models of a stable theory, there is just one way to do it. The formulas that hold between the elements of the structures being amalgamated are completely determined by the definition schemas of the types of these tuples. In a slogan, *for stable theories, heir–coheir amalgams are unique*. The converse holds too: if T is a complete theory and heir–coheir amalgams are unique for T, then T is stable. This will be Theorem 11.3.1 below.

Corollary 6.7.12. *Assuming T is stable, let A, B, C' and D be models of T which form an amalgam as in (4.3) of section 6.4 above. Let $\bar{b}$ be any tuple in B, let the type $p(\bar{x})$ be $\mathrm{tp}_B(\bar{b}/\mathrm{dom}\,A)$, and let d be a definition schema of p. Then d is also a definition schema of $\mathrm{tp}_D(\bar{b}/\mathrm{dom}\,C')$.*

Proof. We must show that

(7.16)   for every tuple $\bar{c}$ in C', $D \vDash \phi(\bar{b}, \bar{c})$ if and only if $C' \vDash \mathrm{d}\phi(\bar{c})$.

Let $\bar{c}$ be a counterexample. Then $D \vDash \phi(\bar{b}, \bar{c}) \leftrightarrow \neg\mathrm{d}\phi(\bar{c})$. By the definition of heir–coheir amalgams there is a tuple $\bar{a}$ in A such that $B \vDash \phi(\bar{b}, \bar{a}) \leftrightarrow \neg\mathrm{d}\phi(\bar{a})$. But this is impossible by the choice of d.   □

To complete the circle, we show (a) $\Rightarrow$ (b) in Theorem 6.7.1. In fact we show a little more.

Theorem 6.7.13. *Let T be a complete theory. The following are equivalent.*
(a) *T is stable.*
(b) *All types are definable for T.*
(c) *For every cardinal λ such that $\lambda = \lambda^{|L|}$, T is λ-stable.*

Proof. (a) $\Rightarrow$ (b) is by Corollary 6.7.11, and we proved (c) $\Rightarrow$ (a) at the beginning of the section. There remains (b) $\Rightarrow$ (c). Let A be a model of T and X a set of at most λ elements of A. By (b), each type p in $S_1(X; A)$ has a definition schema d_p, and if $p \neq q$ then $\mathrm{d}_p \neq \mathrm{d}_q$. So it suffices to count the number of possible definition schemas. Write $L(X)$ for L with the elements of X added as parameters. Then $|L(X)| \leq \lambda$, and each definition schema d_p is a map from a set of formulas of L to formulas of $L(X)$. So the number of schemas is at most $\lambda^{|L|} = \lambda$.   □

Exercises for section 6.7

1. (a) Show that in the definition of 'unstable theory' it makes no difference if we allow parameters in the formula ϕ of (7.1). (b) Show that if A is an L-structure and $\bar{a}$

is a sequence of elements of A, then A is superstable if and only if $(A, \bar{a})$ is superstable.

2. Show that if $\mathbb{Q}$ and $\mathbb{R}$ are respectively the field of rational numbers and the field of reals, then both $\mathrm{Th}(\mathbb{Q})$ and $\mathrm{Th}(\mathbb{R})$ are unstable. [For $\mathbb{Q}$, find an infinite set of numbers which are linearly ordered by the relation '$y - x$ is a non-zero sum of four squares'. The same argument works for any formally real field.]

A formula $\phi(\bar{x}, \bar{y})$ is said to have the **strict order property** *(for a complete theory T) if in every model of T, ϕ defines a partial ordering relation on the set of all n-tuples, which contains arbitrarily long finite chains. T has the* **strict order property** *if some formula has the strict order property for T.*
3. Show that every complete theory with the strict order property is unstable.

A formula $\phi(\bar{x}, \bar{y})$ is said to have the **independence property** *(for a complete theory T) if in every model A of T there is, for each $n < \omega$, a family of tuples $\bar{b}_0, \ldots, \bar{b}_{n-1}$ such that for every subset X of n there is a tuple $\bar{a}$ in A for which $A \vDash \phi(\bar{a}, \bar{b}_i) \Leftrightarrow i \in X$. T has the* **independence property** *if some formula has the independence property for T.*
4. Show that every complete theory with the independence property is unstable.

5. Give examples of (a) a complete theory which has the strict order property but not the independence property, (b) a complete theory which has the independence property but not the strict order property. *Shelah [1978a] Theorem II.4.7 shows that a complete theory is unstable if and only if it has either the strict order property or the independence property.*

6. Let the signature of L consist of countably many 1-ary relation symbols P_i $(i < \omega)$. Let A be the L-structure whose domain is the set of positive integers, with $A \vDash P_i(n) \Leftrightarrow n$ is divisible by the ith prime number. Show that $\mathrm{Th}(A)$ is superstable but not totally transcendental.

7. Let the signature of L consist of countably many 2-ary relation symbols E_i, and let the theory T say that each E_i is an equivalence relation, and each class of E_i is the union of infinitely many classes of E_{i+1}. Show that T is stable but not superstable.

8. Show that if T is unstable then it has an unstable formula.

9. Let L be a first-order language and T a complete theory in L. (a) Show that if $\phi(\bar{x}, \bar{y})$ is a stable formula for T, then so is $\neg \phi(\bar{x}, \bar{y})$. (b) Show that if $\phi(\bar{x}, \bar{y})$ and $\psi(\bar{x}, \bar{y})$ are stable formulas for T, then so is $(\phi \wedge \psi)(\bar{x}, \bar{y})$. [Use the finite Ramsey theorem, Corollary 11.1.4 below.]

10. Suppose $\phi(\bar{x}, \bar{y})$ is a stable formula for T. (a) Show that if $\bar{y}'$ is a tuple of variables which includes all those in $\bar{y}$, and $\psi(\bar{x}, \bar{y}')$ is equivalent to $\phi(\bar{x}, \bar{y})$ modulo

T, then ψ is also stable. (b) Show that if $\bar{x}'$ is $\bar{y}$ and $\bar{y}'$ is $\bar{x}$, and $\theta(\bar{x}', \bar{y}')$ is the formula $\phi(\bar{x}, \bar{y})$, then θ is stable.

11. Show that every stable integral domain is a field. [If t is any element $\neq 0$, t^n divides t^{n+1} for each positive n; so by stability there is some n such that t^{n+1} divides t^n too.]

The next exercise shows that unstable theories have the finite cover property in Keisler's sense; see Exercise 4.3.10.
12. Show that if T is an unstable complete theory in the first-order language L, then there is a formula $\phi(x, \bar{y})$ of L such that for arbitrarily large finite n, T implies that there are $\bar{a}_0, \ldots, \bar{a}_{n-1}$ for which $\neg \exists x \bigwedge_{i<n} \phi(x, \bar{a}_i)$ holds, but $\exists x \bigwedge_{i\in w} \phi(x, \bar{a}_i)$ holds for each proper subset w of n.

13. Let the L-structure A be a model of a stable theory T, let $p(\bar{x})$ be a complete type over X with respect to A, and let $\phi(\bar{x}, \bar{y})$ be a formula of L. Define the **strict ϕ-rank** of $p(\bar{x})$ to be the minimum value of $\mathrm{BI}(\phi, \psi)$ as $\psi(\bar{x})$ ranges over all formulas of p which are conjunctions of formulas of the form $\phi(\bar{x}, \bar{c})$ or $\neg \phi(\bar{x}, \bar{c})$. Prove Theorem 6.7.8 using strict ϕ-rank in place of ϕ-rank.

*14. Describe heir–coheir amalgams in term algebras.

15. Suppose A is an L-structure and B is interpretable in A. (a) Show that if λ is an infinite cardinal and A is λ-stable then B is λ-stable. [Use Theorem 10.1.9.] (b) Show that if A is superstable then B is superstable.

History and bibliography

Section 6.1.   The compactness theorem was proved for countable first-order theories by Gödel [1930], for arbitrary propositional theories by Gödel [1931a], and for arbitrary first-order theories by Mal'tsev [1936], [1941] and independently by Henkin [1949]. See Henkin & Mostowski [1959] on the shortcomings of Mal'tsev's statement and proof of the theorem (they can be patched up). The formulation in Corollary 6.1.2 (for countable languages) appears in the last paragraph of Gödel's doctoral dissertation [1929], but he removed it from the published version. It seems to have reached print only in the 1950s. The upward Löwenheim–Skolem theorem, in the form of Corollary 6.1.4, is from Tarski & Vaught [1957]. Already by 1928 Tarski had proved that every first-order theory with an infinite model has an uncountable model; see the Editors' Note at the end of Skolem [1934] and footnote 8 in Tarski & Vaught [1957]. Skolem [1955] rejected the result as meaningless; Tarski (see the end of Skolem [1958]) very reasonably responded that Skolem's formalist viewpoint ought to reckon the downward Löwenheim–

Skolem theorem meaningless just like the upward. Theorem 6.1.5 is an offshoot of results in Erdős & Tarski [1962].

Many of the logics described in Barwise & Feferman [1985] were found in the course of attempts to generalise the compactness theorem from first-order to other logics. See for example their Chapter IV on $L(Q)$, Chapter VIII on admissible logics and Chapter XVIII on compactness in general. See Hodges [1985] Chapters 7, 8 for an exposition of the compactness of $L(Q)$ and Magidor–Malitz logic.

Exercise 3: Craig [1953]. Exercise 4: Tarski [1930a] Theorem 24. Exercise 7: A. Robinson [1950]. Exercise 8: Łoś [1954a], Vaught [1954a]. Exercise 12: Henkin [1950]. Exercise 13: Ax [1971]. Exercise 14: J. Robinson [1965]. Exercise 16: Tarski in Huber-Dyson [1964]. Exercise 17: Tarski, Mostowski & Robinson [1953]; with the word 'decidable' removed, this is **Lindenbaum's lemma**, Theorem 56 in Tarski [1930a]. Exercise 18: Ershov [1963].

Section 6.2. An authoritative reference on boolean algebras is Monk [1989]. Boole [1847] showed that for any formal language closed under 'and', 'not' and exclusive 'or', the equivalence classes of formulas form a boolean ring, i.e. a commutative ring in which $x^2 = x$ for all x. The switch to boolean algebras with lattice operations $\wedge$ and $\vee$ is due to Jevons [1864] (who is better known for his contributions to mathematical economics), and Peirce [1870] introduced the algebra ordering $\leqslant$. These early works are all crude by modern standards; the first recognisable axiomatisation of boolean algebras appears in Huntington [1904a]. Boole's argument to show that classes obey the same laws as propositional logic was a precursor of Stone's duality theorem, Stone [1936]. The boolean prime ideal theorem (Theorem 6.2.1) is due to Tarski [1930b].

Exercise 5: Henkin [1954]. Exercise 6: Läuchli [1971]. Exercise 9: the lower bound is from Makinson [1969]. Exercise 10: (a) Engelking & Karłowicz [1965]; (b) Hausdorff [1936]; (c) Pospíšil [1937]. Exercise 12: Scott [1954]; cf. Rav [1977] for related results. Exercise 13: (c) Łoś [1954b], (d) Tarski [1929] (before the BPI and the compactness theorem were proved). Exercise 16: (b) is from Hajnal & Kertész [1972], and (c) follows at once from Vaught [1956].

Section 6.3. It was Tarski [1935 + 6] who first gave formal definitions of the boolean algebras associated with a *theory* as opposed to a logical language. Rasiowa & Sikorski [1963] named these algebras **Lindenbaum algebras** in memory of a close colleague of Tarski who died at the hands of the Nazis; Tarski ([1935 + 6] p. 370) credits Lindenbaum with raising the question when two theories have isomorphic 0-th Lindenbaum algebras. (See also Surma [1982].) Theorem 6.3.4(a) was found by Mostowski [1937], who was Tarski's student. Perhaps the first systematic treatment of types was by Vaught [1961], another student of Tarski. This paper contained Theorem 6.3.5(a). Theorems 6.3.4(b) and 6.3.5(b) are from Morley [1965a], together with Exercises 15 and 17.

Section 6.4.   Theorem 6.4.1 is due to Abraham Robinson [1956b] Theorem 4.2.2; see also his [1956a]. Theorem 6.4.3 is from Lascar & Poizat [1979]. (Curiously the 'heir' is from Lascar and the 'coheir' from Poizat; see Exercise 3 for a partial explanation.) Theorem 6.4.5 appears in Park's unpublished dissertation [1964a] (see Bacsich [1975]). Sadly Park died in 1990, after a fertile career as a computer scientist.

Some writers have contrasted 'Western' theorems about *elementary embeddings* with 'Eastern' theorems about *embeddings*. Allegedly the Western theorems, like those in this section, belong to Alfred Tarski's school at Berkeley, while the Eastern theorems like those in section 6.5 have more to do with Abraham Robinson at Yale. The references above and to section 6.5 show that this is not good history. Writing in London, it's terrible geography.

Exercises 1, 2: Poizat [1985] section 12f. Exercises 7–9: Park [1964a], [1964b]. Exercise 10: Cherlin & Volger [1984]. Exercise 11: Marcus [1975].

Section 6.5.   Marczewski [1951] asked what formulas are preserved by surjective homomorphisms. His question ushered in a period of some ten years' work devoted to preservation theorems. Theorem 6.5.4 is from Łoś [1955a]; Tarski [1954] had it for T empty. Theorem 6.5.9 was proved by Chang [1959] and Łoś & Suszko [1957] when T is empty; Abraham Robinson [1959a] gave the full theorem.

Exercise 2: Łoś & Suszko [1955]. Exercises 3, 4: Keisler [1960]. Exercise 9: A. Robinson [1963a] Theorem 3.5.1.

Section 6.6.   Theorem 6.6.1 is from A. Robinson [1956a]; Theorem 6.6.3 was first proved independently by Craig [1957] using a proof-theoretic argument. Theorem 6.6.5 and Corollary 6.6.6 are from Barwise [1973] (for countable languages, but his proof allows them to be infinitary). In 1983 Michael A. McRobbie of the Australian National University at Canberra compiled a substantial bibliography of interpolation theorems, but I never heard what became of it.

In a less formal context, the distinction between explicit and implicit definitions is due to Gergonne [1818], the same man who introduced the symbol ⊂ of set theory as an abbreviation for 'Contenu dans'. It has no serious connection with the muddled and self-contradictory notion of 'implicit definition' found in some philosophers of geometry, according to which the axioms of geometry completely define the 'relations' of a geometry without implying anything whatever about what things do or do not stand in these relations. (See Nagel [1939].) The term was rescued from this cruel and unusual punishment by Beth [1953], who proved Theorem 6.6.4.

Theorem 6.6.7 is implicit in Mal'tsev [1941] and explicit in Tarski [1954]. Corollary 6.6.8 is due to Mal'tsev [1940]; though Mal'tsev keeps the fact well hidden, this was the first application of the compactness theorem in algebra. For more on local theorems, see Mal'tsev [1959], McLain [1959] and Cleave [1969].

Exercise 1: A. Robinson [1956a]. Exercises 7, 8: Vaught [1962]. Exercise 10(b): Lachlan [1990]. Exercise 11: Rabin [1965b]. Exercise 12: Tarski [1954]. Exercise 13: Łoś [1954b]. Exercise 14: Mal'tsev [1959]. Exercise 15: Fuks-Rabinovich [1940]; this proof is from Mal'tsev [1941].

Section 6.7. See the references to sections 5.6 and 12.2 for totally transcendental theories and their role in the theory of categoricity. Theorem 6.7.5 is in Morley [1965a], though not in this terminology; Rowbottom [1964] introduced the word 'λ-stable'. Stable theories were introduced by Shelah [1969], as theories which are λ-stable for some λ; at the same time he defined superstable theories. The definition of stable theories in terms of linear orderings appeared in Shelah [1971c], together with the notion of stable formulas.

Lemma 6.7.7 is from Baldwin & Saxl [1976] (essentially; cf. Cherlin & Reineke [1976]) and Belegradek [1978a]. Theorem 6.7.8 is due to Lachlan [1972], and independently Shelah [1971c]; the defining formula used here is due to Shelah. Corollary 6.7.11 is immediate from Theorem 6.7.8; but in fact it was proved already in 1968 by Ressayre (unpublished) in the case of uncountably categorical countable theories and types over a model. Baldwin [1970] had an intermediate form. Lemma 6.7.9 and Exercise 9 are from Hodges [1981b]. Example 2 and Exercise 14 are developed by Belegradek [1988] and Bouscaren & Poizat [1988].

Shelah [1978a] is a compendium of stability theory in Shelah's own distinctive style; like the music of Charles Ives, it's impossible to get through but enormously worth the effort of trying. Other people reworked the foundations. In particular Daniel Lascar and Bruno Poizat in Paris in the late 1970s found an elegant way in via heirs, coheirs and definable types. Expositions in the French style are Pillay [1983a], Poizat [1985] and Lascar [1987a]. Baldwin [1988] is a general reference.

Exercise 2: Duret [1977]. Exercises 3–5: Shelah [1978a] section II.4. Exercise 12: Shelah [1978a] Theorem II.4.2.

7

The countable case

For eighthly he rubs himself against a post.
For ninthly he looks up for his instructions.
For tenthly he goes in quest of food.
Christopher Smart (1722–71), Jubilate Agno.

The cardinal number ω is the only infinite cardinal which is a limit of finite cardinals. This gives us two reasons why countable structures are good to build. First, a countable structure can be built as the union of a chain of finite pieces. And second, we have infinitely many chances to make sure that the right pieces go in. No other cardinal allows us this amount of control.

So it's not surprising that model theory has a rich array of methods for constructing countable structures. Pride of place goes to Roland Fraïssé's majestic construction, which bestrides this chapter. Next comes the omitting types construction (section 7.2); it was discovered in several forms by several people. At heart it is a Baire category argument, the same as in model-theoretic forcing (section 8.2 below). I once wrote a book (Hodges [1985]) about the dozen or so variants of forcing which appear in model theory.

There is another countable construction whose roots lie rather deeper in descriptive set theory – namely the construction which proves that the ordinal ω_1 is not definable by any sentence of $L_{\omega_1\omega}$. But the uncountable analogues of this construction are just as important as the countable case, and so I have postponed it to section 11.5 below, where we shall be in a better position to use it.

Besides constructions, this chapter contains one section on classification. In section 7.3 we shall see that the countable ω-categorical structures are exactly those whose automorphism groups are oligomorphic. These automorphism groups tend to be very rich, and they give plenty of scope for collaboration between model theorists and permutation group theorists.

7.1 Fraïssé's construction

In 1954 Roland Fraïssé published a paper which has become a classic of model theory. He pointed out that we can think of the class of finite linear orderings as a set of approximations to the ordering of the rationals, and he described a way of building the rationals out of these finite approximations. Fraïssé's construction is important because it works in many other cases too. Starting from a suitable set of finite structures, we can build their 'limit', and some of the structures built in this way have turned out to be remarkably interesting. Several will appear later in this book.

Ages

Let L be a signature and D an L-structure. The **age** of D is the class **K** of all finitely generated structures that can be embedded in D. Actually what interests us is not the structures in **K** but their isomorphism types. So we shall also call a class **J** the **age** of D if the structures in **J** are, *up to isomorphism*, exactly the finitely generated substructures of D. Thus it will make sense to say, for example, that D has 'countable age' – it will mean that D has just countably many isomorphism types of finitely generated substructure.

We call a class an **age** if it is the age of some structure. If **K** is an age, then clearly **K** is non-empty and has the following two properties.

(1.1)   **(Hereditary property**, HP for short) If $A \in \mathbf{K}$ and B is a finitely generated substructure of A then B is isomorphic to some structure in **K**.

(1.2)   **(Joint embedding property**, JEP for short) If A, B are in **K** then there is C in **K** such that both A and B are embeddable in C:

One of Fraïssé's theorems was a converse to this.

Theorem 7.1.1. *Suppose L is a signature and* **K** *is a non-empty finite or countable set of finitely generated L-structures which has the HP and the JEP. Then* **K** *is the age of some finite or countable structure.*

Proof. List the structures in **K**, possibly with repetitions, as $(A_i: i < \omega)$. Define a chain $(B_i: i < \omega)$ of structures isomorphic to structures in **K**, as follows. First put $B_0 = A_0$. When B_i has been chosen, use the joint embedding property to find a structure B' in **K** such that both B_i and A_{i+1} are embeddable in B'. Take B_{i+1} to be an isomorphic copy of B' which extends B_i (recalling Exercise 1.2.4(b)). Finally let C be the union $\bigcup_{i<\omega}B_i$. Since C is the union of countably many structures which are at most countable, C is at most countable. By construction every structure in $\mathbf{K}$ is embeddable in C. If A is any finitely generated substructure of C, then the finitely many generators of A lie in some B_i, so that A is isomorphic to a structure in $\mathbf{K}$ (by the hereditary property). So $\mathbf{K}$ is the age of C. $\qquad\square$

The theorem holds even if L has function symbols. But one way to guarantee that $\mathbf{K}$ is at most countable is to assume that L is a finite signature with no function·symbols – see Exercise 1.2.6 above. When there are no function symbols and only finitely many constant symbols, a finitely generated structure is the same thing as a finite structure.

All infinite linear orderings have exactly the same age, namely the finite linear orderings. (In Fraïssé's own charming terminology, any two infinite linear orderings are 'younger than each other'.) In what sense do the finite linear orderings 'tend to' the rationals rather than, say, the ordering of the integers?

In order to answer this, Fraïssé singled out one further property of the finite linear orderings, to set beside HP and JEP. This property, the amalgamation property, has been of crucial importance in model theory ever since; we met several variants in Chapter 6.

(1.3)   (**Amalgamation property**, AP for short) If A, B, C are in $\mathbf{K}$ and $e: A \to B, f: A \to C$ are embeddings, then there are D in $\mathbf{K}$ and embeddings $g: B \to D$ and $h: C \to D$ such that $ge = hf$:

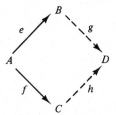

(**Warning**: In general JEP is not a special case of AP. Think of fields).

The class of all finite linear orderings has the amalgamation property. The simplest way to see this is to start with the case where A is a substructure of B and C, the maps e, f are inclusions and A is exactly the overlap of B and C. In this case we can form D as an extension of B, by adding the elements of C one by one in the appropriate places. (Formally, use an induction on the cardinality of C.) The general case then follows by diagram-chasing.

We call a structure D **ultrahomogeneous** if every isomorphism between finitely generated substructures of D extends to an automorphism of D.

(Usually one says **homogeneous**; in this book we have too many different notions that are known by this name.)

Fraïssé's main conclusion was as follows.

Theorem 7.1.2 (*Fraïssé's theorem*). *Let L be a countable signature and let* **K** *be a non-empty finite or countable set of finitely generated L-structures which has HP, JEP and AP. Then there is an L-structure D, unique up to isomorphism, such that*

(1.4)   *D has cardinality* ≤ ω,

(1.5)   **K** *is the age of D, and*

(1.6)   *D is ultrahomogenous.*

There seems to be no accepted name for the structure D of the theorem. Some people, with Lemma 7.1.3 below in mind, refer to it as the **universal homogeneous structure of age K**. I shall call it the **Fraïssé limit** of the class **K**. Of course it is only determined up to isomorphism.

The rest of this section is devoted to the proof of Theorem 7.1.2.

Uniqueness proof

We begin by saying that a structure D is **weakly homogeneous** if it has the property

(1.7)   if A, B are finitely generated substructures of D, $A \subseteq B$ and $f: A \to D$ is an embedding, then there is an embedding $g: B \to D$ which extends f:

If D is ultrahomogeneous then clearly D is weakly homogeneous.

Lemma 7.1.3. *Let C and D be L-structures which are both at most countable. Suppose the age of C is included in the age of D, and D is weakly homogeneous. Then C is embeddable in D; in fact any embedding from a finitely generated substructure of C into D can be extended to an embedding of C into D.*

Proof. Let $f_0: A_0 \to D$ be an embedding of a finitely generated substructure A_0 of C into D. We shall extend f_0 to an embedding $f_\omega: C \to D$ as follows.

Since C is at most countable, it can be written as a union $\bigcup_{n<\omega} A_n$ of a chain of finitely generated substructures, starting with A_0. By induction on n we define an increasing chain of embeddings $f_n: A_n \to D$. The first embedding f_0 is given. Suppose f_n has just been defined. Since the age of D includes that of C, there is an isomorphism $g: A_{n+1} \to B$ where B is a substructure of D. Then $f_n \cdot g^{-1}$ embeds $g(A_n)$ into D, and by weak homogeneity this embedding extends to an embedding $h: B \to D$. Let $f_{n+1}: A_{n+1} \to D$ be hg. Then $f_n \subseteq f_{n+1}$. This defines the chain of maps f_n. Finally take f_ω to be the union of the f_n $(n < \omega)$. □

This lemma justifies some terminology. We say that a countable structure D of age **K** is **universal** (for **K**) if every finite or countable structure of age $\subseteq$ **K** is embeddable in D. Lemma 7.1.3 tells us that countable weakly homogeneous structures are universal for their age.

When C and D are both weakly homogeneous and have the same age, we can throw the argument of this lemma to and fro between C and D to prove that C is isomorphic to D.

Lemma 7.1.4. (a) *Let C and D be L-structures with the same age. Suppose that C and D are both at most countable and are both weakly homogeneous. Then C is isomorphic to D. In fact if A is a finitely generated substructure of C and $f: A \to D$ is an embedding, then f extends to an isomorphism from C to D.*

(b) *A finite or countable structure is ultrahomogeneous (and hence is the Fraïssé limit of its age) if and only if it is weakly homogeneous.*

Proof. (a) Express C and D as the unions of chains $(C_n: n < \omega)$ and $(D_n: n < \omega)$ of finitely generated substructures. Define a chain $(f_n: n < \omega)$ of isomorphisms between finitely generated substructures of C and D, so that for each n, the domain of f_{2n} includes C_n and the image of f_{2n+1} includes D_n. The machinery for doing this is just as in the proof of the previous lemma. Then the union of the f_n is an isomorphism from C to D.

To get the last sentence of (a), take C_0 to be A and D_0 to be $f(A)$.

(b) We have already noted that ultrahomogeneous structures are weakly homogeneous. The converse follows at once from (a), taking $C = D$. □

What if C and D are not countable? Then part (a) of the lemma fails. For example let η be the order-type of the rationals, and consider the order-type $\eta \cdot \omega_1$ $(= \omega_1$ copies of η laid in a row) and its mirror image ξ. Both $\eta \cdot \omega_1$ and ξ are weakly homogeneous, and they have the same age, namely the set of all finite linear orderings. But clearly they are not isomorphic, since in $\eta \cdot \omega_1$ but not in ξ every element has uncountably many successors. The best we can say is the following.

Lemma 7.1.5. *Suppose* C *and* D *are weakly homogeneous* L-*structures with the same age. Then* C *is back-and-forth equivalent to* D, *so that* $C \equiv_{\infty\omega} D$. *If moreover* $C \subseteq D$ *then for every tuple* $\bar{c}$ *in* C, $(C, \bar{c}) \equiv_{\infty\omega} (D, \bar{c})$, *so that* $C \preccurlyeq D$.

The *proof* is by the proof of Lemma 7.1.4. Use Karp's theorem (Corollary 3.5.3) for the connection with $L_{\infty\omega}$. □

Existence proof

Lemma 7.1.4 takes care of the uniqueness of Fraïssé limits.

If the signature L is finite and has no function symbols, the statement 'The age of D is a subset of **K**' can be written as an $\forall_1$ first-order theory T. To see this, take the set of all those finite L-structures which *don't* occur in **K**, and for each such structure A write an $\forall_1$ sentence χ_A which says 'No substructure is isomorphic to A'. Then T is the set of all these sentences χ_A.

If L is not finite, or has function symbols, there is no guarantee that the statement 'The age is a subset of **K**' can be written as a first-order theory, even when the structures in **K** are all finite. (Consider for instance the class **K** of all finite groups; see Example 1 below.) This will cause fewer difficulties than one might have feared. But it does remind us that we are working with a class of structures where the upward Löwenheim–Skolem theorem need not apply; Theorem 7.1.2 will give us a structure of cardinality $\leqslant \omega$, but it may be much harder a find a similar structure of uncountable cardinality.

We first note an easy fact about ages.

Lemma 7.1.6. *Let* **J** *be a set of finitely generated* L-*structures, and* $(D_i : i < \alpha)$ *a chain of* L-*structures. If for each* $i < \alpha$ *the age of* D_i *is included in* **J**, *then the age of the union* $\bigcup_{i<\alpha} D_i$ *is also included in* **J**. *If each* D_i *has age* **J**, *then* $\bigcup_{i<\alpha} D_i$ *has age* **J**. □

Henceforth we assume that **K** is non-empty, has HP, JEP and AP, and contains at most countably many isomorphism types of structure. We can suppose without loss that **K** is closed under taking isomorphic copies.

We shall construct a chain $(D_i : i < \omega)$ of structures in **K**, such that the following holds:

(1.8)   if A and B are structures in **K** with $A \subseteq B$, and there is an
        embedding $f : A \to D_i$ for some $i < \omega$, then there are $j > i$
        and an embedding $g : B \to D_j$ which extends f.

We take D to be the union $\bigcup_{i<\omega} D_i$. Then the age of D is included in **K** by Lemma 7.1.6. In fact the age of D is exactly **K**. For suppose A is in **K**; then by JEP there is B in **K** such that $A \subseteq B$ and D_0 is embeddable in B. By

(1.8) the identity map on D_0 extends to an embedding of B in some D_j, so that B and A lie in the age of D. Thus (1.8) tells us that D is weakly homogeneous, and so by Lemma 7.1.4(b) it is ultrahomogeneous of age **K** as required.

It remains to construct the chain. Let **P** be a countable set of pairs of structures (A, B) such that $A, B \in \mathbf{K}$ and $A \subseteq B$; we can choose **P** so that it includes a representative of each isomorphism type of such pairs. Take a bijection $\pi: \omega \times \omega \to \omega$ such that $\pi(i, j) \geqslant i$ for all i and j. Let D_0 be any structure in **K**. The rest is by induction, as follows. When D_k has been chosen, list as $((f_{kj}, A_{kj}, B_{kj}): j < \omega)$ the triples (f, A, B) where $(A, B) \in \mathbf{P}$ and $f: A \to D_k$. Construct D_{k+1} by the amalgamation property, so that if $k = \pi(i, j)$ then f_{ij} extends to an embedding of B_{ij} into D_{k+1}.

$\square$ Theorem 7.1.2

All the conditions on **K** in Theorem 7.1.2 were necessary, according to the next theorem.

Theorem 7.1.7. *Let L be a countable signature and D a finite or countable structure which is ultrahomogeneous. Let* **K** *be the age of D. Then* **K** *is non-empty,* **K** *has at most countably many isomorphism types of structure, and* **K** *satisfies HP, JEP and AP.*

Proof. We already know everything except that **K** satisfies the amalgamation property. For this we can assume without loss that **K** contains all finitely generated substructures of D. Suppose that A, B, C are in **K** and $e: A \to B$, $f: A \to C$ are embeddings. Then there are isomorphisms $i_A: A \to A'$, $i_B: B \to B'$ and $i_C: C \to C'$ where A', B', C' are substructures of D. So $i_A \cdot e^{-1}$ embeds $e(A)$ into D, and by weak homogeneity there is an embedding $j_B: B \to D$ which extends $i_A \cdot e^{-1}$, so that the bottom left quadrilateral in (1.9) commutes; likewise with the bottom right quadrilateral:

(1.9)

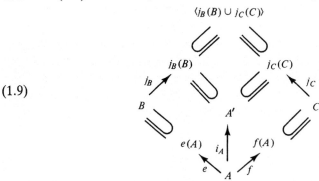

The top square commutes, and hence the outer maps in (1.9) give the needed amalgam.

$\square$

Strong amalgamation

Taken together, Theorem 7.1.2 and Theorem 7.1.7 say that the problem of constructing countable ultrahomogeneous structures reduces to a problem of finding well-behaved sets of finite structures. If we know more about the finite structures, we can generally say more about the ultrahomogeneous structure too.

For example section 7.4 will show how we can ensure that the ultra-homogeneous structure is ω-categorical. Meanwhile here is an easier example.

We say that a structure M has **no algebraicity** if for every set x of elements of M, the ALGEBRAIC CLOSURE of x (in the sense of section 4.1 above) is X itself. Obviously one only needs to check this for finite sets X. We say that a set **K** of structures has the **strong amalgamation property** (SAP) if (1.3) holds with the extra property that the intersection $(\text{im } g) \cap (\text{im } h)$ is exactly im ge (which is necessarily im hf too).

Theorem 7.1.8. *Suppose* **K** *is the age of a countable structure* M. *Then the following are equivalent.*
(a) **K** *has the strong amalgamation property.*
(b) M *has no algebraicity.*

Proof. Exercise 8 below.   ☐

Example 1: *An ultrahomogeneous universal locally finite group.* Let **K** be the class of finite groups. Then **K** has the strong amalgamation property. This can be proved using the permutation products of B. H. Neumann (see section 3 of his [1959]). Also it's clear that **K** contains countably many isomorphism types and has HP and JEP. The Fraïssé limit of **K** is known as **Philip Hall's universal locally finite group** (see Philip Hall [1959]). The group is simple, and any two isomorphic finite subgroups are conjugate.

We shall return to Fraïssé's construction with some more examples in section 7.4 below, after we have discussed ω-categoricity.

A good deal is known about the classification of countable ultrahomogeneous structures. Two high spots are the classification by Lachlan & Woodrow [1980] of countable ultrahomogeneous graphs, and the structure theory of Cherlin & Lachlan [1986] for the countable stable ultrahomogeneous structures of a finite relational signature, using the O'Nan–Scott theorem.

Exercises for section 7.1

1. Suppose **K** is a class of L-structures, and **K** contains a structure which is embeddable in every structure in **K**. Show that if **K** has the AP then **K** has the JEP too.

The next four exercises give algebraic objects which are Fraïssé limits.

2. Let p be a prime and let $\mathbf{K}$ be the class of all finite fields of characteristic p. Show that $\mathbf{K}$ has HP, JEP and AP, and that the Fraïssé limit of $\mathbf{K}$ is the algebraic closure of the prime field of characteristic p.

3. Let $\mathbf{K}$ be the class of finitely generated torsion-free abelian groups. Show that $\mathbf{K}$ has HP, JEP and AP, and that the Fraïssé limit of $\mathbf{K}$ is the direct sum of countably many copies of the additive group of rationals.

4. Let $\mathbf{K}$ be the class of finite boolean algebras. Show that $\mathbf{K}$ has HP, JEP and SAP, and that the Fraïssé limit of $\mathbf{K}$ is the countable atomless boolean algebra.

See section 9.2 below for the notion of finitely presented algebras. By a theorem of Graham Higman [1961] the finitely generated recursively presented groups are precisely the finitely generated subgroups of finitely presented groups.

*5. Let $\mathbf{K}$ be the class of finitely generated recursively presented groups. Show that $\mathbf{K}$ has HP, JEP and SAP.

6. Show, using the compactness theorem, that the first-order theory of Philip Hall's universal locally finite group (defined in Example 1) is not ω-categorical.

7. Show that the abelian group $Z(4) \oplus \bigoplus_{i<\omega} Z(2)$ is not elementarily equivalent to any ultrahomogeneous structure.

8. Prove Theorem 7.1.8. [For (b) $\Rightarrow$ (a), use П. M. Neumann's lemma (Corollary 4.2.2 above).]

If T is a first-order theory and we want to use Fraïssé's theorem to construct a countable model of T from finitely generated models of T, we should require that the class of models of T is closed under unions of chains. By Exercise 6.5.6, T is equivalent to an $\forall_1$ first-order theory if and only if the class of models of T is closed under taking unions of chains and finitely generated substructures. The next two exercises perform a similar service for the properties JEP and AP.

9. Show that if T is a theory in a first-order language L, the following are equivalent. (a) T has the joint embedding property. (b) For every pair of $\forall_1$ sentences $\phi(\bar{x})$, $\psi(\bar{x})$ of L, if $T \vdash \phi \vee \psi$ then either $T \vdash \phi$ or $T \vdash \psi$. (c) The class of all finitely generated substructures of models of T has JEP. [For (b) $\Rightarrow$ (a), consider two models A, B of T and show that $T \cup \mathrm{diag}(A) \cup \mathrm{diag}(B)$ has a model.]

10. Let T be a theory in a first-order language L. Show that T has the amalgamation property if and only if for all $\forall_1$ formulas $\phi_1(\bar{x})$ and $\phi_2(\bar{x})$ of L, if $T \vdash \forall \bar{x}(\phi_1 \vee \phi_2)$ then there are $\exists_1$ formulas $\psi_1(\bar{x})$ and $\psi_2(\bar{x})$ of L such that $T \vdash \forall \bar{x}(\psi_i \to \phi_i)$ $(i = 1, 2)$ and $T \vdash \forall \bar{x}(\psi_1 \vee \psi_2)$.

The next exercise suggests one way of applying Fraïssé's construction to classes which don't have the amalgamation property. The idea is to pick out a suitable subclass which does have it; see Exercise 12 for an example. Fraïssé's idea can be made to work with even less amalgamation than in Exercise 11; but it is probably better to handle such cases by the methods of Chapter 8 below.

11. Let $\mathbf{K}$ be a countable class of finitely generated L-structures which has JEP and AP, and $\mathbf{K}^{\downarrow}$ the class of all finitely generated substructures of structures in $\mathbf{K}$. (a) Show that there is a countable structure D, unique up to isomorphism, such that (i) age $(D) = \mathbf{K}^{\downarrow}$, (ii) D is the union of a chain of isomorphic copies of structures in $\mathbf{K}$, and (iii) every isomorphism between substructures of D which are (up to isomorphism) in $\mathbf{K}$ extends to an automorphism of D. (We can call such a structure D a **limit** of $\mathbf{K}$.) (b) Show that if $\mathbf{J}$ is another class of finitely generated L-structures which has JEP and AP, and $\mathbf{J}^{\downarrow} = \mathbf{K}^{\downarrow}$, then $\mathbf{J}$ and $\mathbf{K}$ have isomorphic limits.

12. In the signature L consisting of one 2-ary relation symbol R, let $\mathbf{K}$ be the class of all finite models of the theory: $\forall x \exists_{=1} y\, Rxy$, $\forall x \exists_{=1} y\, Ryx$, $\forall xyz\, \neg (Rxy \wedge Ryz)$. (a) Show that $\mathbf{K}$ has JEP and AP but not HP. Describe its limit. (b) Show that $\mathbf{K}^{\downarrow}$ has HP and JEP but not AP.

13. Let $\mathbf{K}$ be a class of finitely generated L-structures with HP, JEP and AP. Suppose $A \in \mathbf{K}$, B and C are countable structures of age $\mathbf{K}$, and A is a substructure of both B and C. Show that there are a structure D of age $\mathbf{K}$ such that $C \subseteq D$, and an embedding $f: B \to D$ which is the identity on A. [Decompose B and C into countable chains of structures in $\mathbf{K}$, and build up a chain of amalgams.] *In the proof of Theorem 8.6.9 below we shall see that this amalgamation result can fail if we replace A by a structure of age included in $\mathbf{K}$.*

14. Let $\mathbf{K}$ be a class of finitely generated L-structures with HP, JEP, AP, and suppose $\mathbf{K}$ contains countably many isomorphism types of structure. (a) Show that if $\mathbf{K}$ has SAP then every countable structure of age $\mathbf{K}$ has a proper extension which is also of age $\mathbf{K}$. (b) Show that if every ultrahomogeneous countable structure of age $\mathbf{K}$ has a proper extension of age $\mathbf{K}$, then there is a weakly homogeneous structure of age $\mathbf{K}$ and cardinality ω_1.

15. Let L be a finite signature with no function symbols, and $\mathbf{K}$ a class of finite L-structures which has HP, JEP and AP. Show that there is a first-order theory T in L such that (a) the countable models of T are exactly the countable ultrahomogeneous structures of age $\mathbf{K}$, and (b) every sentence in T is either $\forall_1$ or of the form $\forall \bar{x} \exists y\, \phi(\bar{x}, y)$ where ϕ is quantifier-free.

16. Give an example of a countable set $\mathbf{K}$ of finitely generated L-structures which has HP, JEP, AP, such that there are infinite structures with age $\mathbf{K}$ but no uncountable structures with age $\mathbf{K}$. [By the previous exercise and the upward Löwenheim–Skolem theorem, L must either be infinite or contain function symbols. Examples can be found both ways.]

I have no solution for the next exercise.
*17. Give an example to show that if $\mathbf{K}$ is an uncountable class of L-structures which has HP and JEP, there need not be a structure of age $\mathbf{K}$.

7.2 Omitting types

Let L be a first-order language and T a theory in L. Let $\Phi(\bar{x})$ be a set of formulas of L, with $\bar{x} = (x_0, \ldots, x_{n-1})$. As in section 2.3, we say that Φ is **realised in** an L-structure A if there is a tuple $\bar{a}$ of elements of A such that $A \vDash \bigwedge \Phi(\bar{a})$; we say A **omits** Φ if Φ isn't realised in A.

When does T have a model that omits Φ?

Suppose for example that

(2.1)   there is a formula $\theta(\bar{x})$ of L such that $T \cup \{\exists \bar{x} \, \theta\}$ has a model, and for every formula $\phi(\bar{x})$ in Φ, $T \vdash \forall \bar{x}(\theta \to \phi)$.

If T is a complete theory, then (2.1) implies that $T \vdash \exists \bar{x} \, \theta$, and so T certainly has no model that omits Φ. Our next theorem will imply that when the language L is countable, the converse holds too, even if T is not a complete theory: if every model of T realises Φ then (2.1) is true.

Example 1: _A type omitted._ Let L be a first-order language whose signature consists of 1-ary relation symbols P_i $(i < \omega)$. Let T be the theory in L which consists of all the sentences $\exists x \, P_0(x)$, $\exists x \neg P_0(x)$, $\exists x(P_0(x) \wedge P_1(x))$, $\exists x(P_0(x) \wedge \neg P_1(x))$, $\exists x(\neg P_0(x) \wedge P_1(x))$ etc. (through all the possible combinations). Then T is complete. (One can show this by quantifier elimination; Exercise 16 below suggests another proof.) If s is any subset of ω, let $\Phi_s(x)$ be the set $\{P_i(x): i \in s\} \cup \{\neg P_i(x): i \notin s\}$. Now T has a countable model A, which must omit at least one of the continuum many sets Φ_s $(s \subseteq \omega)$. So by symmetry, if $s \subseteq \omega$ there must be a countable model of T which omits Φ_s. However, a model of T can't omit all the sets Φ_s, or it would be empty.

Note that if Φ is Φ_s for some $s \subseteq \omega$, then there is obviously no formula θ as in (2.1) – it takes infinitely many first-order formulas to specify Φ_s.

When (2.1) holds, we say that θ is a **support** of Φ over T. When (2.1) holds and θ is a formula in Φ, we say that θ **generates** Φ over T. We say that a set of first-order formulas $\Phi(\bar{x})$ is a **supported type over** T if Φ has a support over T; we say that Φ is a **principal type over** T if Φ has a generator over T. The set Φ is said to be **unsupported** (resp. **non-principal**) over T if it is not a supported (resp. principal) type over T.

Note that if $p(\bar{x})$ is a complete type over the empty set, then a formula $\phi(\bar{x})$ of L is a support of p if and only if it generates p; so a complete type p is principal if and only if it is supported. We say that a formula $\phi(\bar{x})$ is **complete** (for T) if it generates a complete type of T.

Theorem 7.2.1 (*Countable omitting types theorem*). *Let L be a countable first-order language, T a theory in L which has a model, and for each $m < \omega$ let*

Φ_m *be an unsupported set over T in L. Then T has a model which omits all*
the sets Φ_m.

Proof. The theorem is trivial when T has an empty model, so that we can
assume that T has a non-empty model. Let L^+ be the first-order language
which comes from L by adding countably many new constants c_i $(i < \omega)$, to
be known as **witnesses**. We shall define an increasing chain $(T_i : i < \omega)$ of
finite sets of sentences of L^+, such that for every i, $T \cup T_i$ has a model. To
prime the pump we take T_{-1} to be the empty theory. Then (since T has a
non-empty model) $T \cup T_{-1}$ has a model which is an L^+-structure.

The intention is that the union of the chain, call it T^+, will be a Hintikka
set for L^+ (see section 2.3), and the canonical model of the atomic sentences
in T^+ will be a model of T which omits all the types Φ_m. To ensure that T^+
will have these properties, we carry out various tasks as we build the chain.
These tasks are as follows.

(2.2)   Ensure that for every sentence ϕ of L^+, either ϕ or $\neg \phi$ is
               in T^+.

(2.3)$_{\psi(x)}$ (For each formula $\psi(x)$ of L^+:) Ensure that if $\exists x\, \psi(x)$ is
               in T^+ then there are infinitely many witnesses c such that
               $\psi(c)$ is in T^+.

(2.4)$_m$ (For each $m < \omega$:) Ensure that for every tuple $\bar{c}$ of distinct
               witnesses (of appropriate length) there is a formula $\phi(\bar{x})$ in
               Φ_m such that the formula $\neg \phi(\bar{c})$ is in T^+.

If these tasks are all carried out, then Theorem 2.3.4 tells us that T^+ will be
a Hintikka set. Write A^+ for the canonical model of the atomic sentences in
T^+ (see Theorem 1.5.2). Then A^+ is a model of T^+ in which every element
is named by a closed term. By the tasks (2.3) where $\psi(x)$ are the formulas
$x = t$ (t a closed term), every element of A^+ is named by infinitely many
witnesses, and so every tuple of elements is named by a tuple of distinct
witnesses. Given this, the tasks (2.4) make sure that A^+ omits all the types
Φ_m. The required model of T will be the reduct $A^+|L$.

There are countably many tasks in the list: (2.2) is one, and we have a task
(2.3)$_{\psi(x)}$ for each formula $\psi(x)$ and a task (2.4)$_m$ for each $m < \omega$. Metaphor-
ically speaking, we shall hire countably many experts and give them one task
each. We partition ω into infinitely many infinite sets, and we assign one of
these sets to each expert. When T_{i-1} has been chosen, if i is in the set
assigned to some expert E, then E will choose T_i. It remains to tell the
experts how they should go about their business. (Experts in this kind of task
are female. This comes from a useful convention in model-theoretic games;
see the remarks after this proof.)

First consider the expert who handles task (2.2), and let X be her subset
of ω. Let her list as $(\phi_i : i \in X)$ all the sentences of L^+. When T_{i-1} has been

chosen with i in X, she should consider whether $T \cup T_{i-1} \cup \{\phi_i\}$ has a model. If it has, she should put $T_i = T_{i-1} \cup \{\phi_i\}$. If not, then every model of $T \cup T_i$ is a model of $\neg \phi_i$, and she can take T_i to be $T_{i-1} \cup \{\neg \phi_i\}$. In this way she can be sure of carrying out task (2.2) by the time the chain is complete.

Next consider the expert who deals with task $(2.3)_\psi$. She waits until she is given a set T_{i-1} which contains $\exists x\, \psi(x)$. Every time this happens, she looks for a witness c which is not used anywhere in T_{i-1}; there is such a witness, because T_{i-1} is finite. Then a model of $T \cup T_{i-1}$ can be made into a model of $\psi(c)$ by choosing a suitable interpretation for c. Let her take T_i to be $T_{i-1} \cup \{\psi(c)\}$. Otherwise she should do nothing. This strategy works, because her subset of ω contains arbitrarily large numbers.

Finally consider the expert who handles task $(2.4)_m$, where Φ_m is a type in n variables. Let Y be her assigned subset of ω. She begins by listing as $\{\bar{c}_i : i \in Y\}$ all the n-tuples $\bar{c}$ of distinct witnesses. When T_{i-1} has been given, with i in Y, she writes $\bigwedge T_{i-1}$ as a sentence $\chi(\bar{c}_i, \bar{d})$ where $\chi(\bar{x}, \bar{y})$ is in L and $\bar{d}$ lists the distinct witnesses which occur in T_{i-1} but not in $\bar{c}_i$. By assumption the theory $T \cup \{\exists \bar{x} \exists \bar{y}\, \chi(\bar{x}, \bar{y})\}$ has a model. But Φ_m is unsupported, and it follows that there is some formula $\phi(\bar{x})$ in Φ_m such that $T \not\vdash \forall \bar{x}(\exists \bar{y}\, \chi(\bar{x}, \bar{y}) \to \phi(\bar{x}))$. By the lemma on constants (Lemma 2.3.2) it follows that $T \not\vdash \chi(\bar{c}_i, \bar{d}) \to \phi(\bar{c}_i)$. Hence she can put $T_i = T_{i-1} \cup \{\neg \phi(\bar{c}_i)\}$. Thus she fulfils her task. $\qquad\square$

In the proof of Theorem 7.2.1, each expert has to make sure that the theory T^+ has some particular property π. The proof shows that the expert can make T^+ have π, provided that she is allowed to choose T_i for infinitely many i. We can express this in terms of a game, call it $G(\pi, X)$. There are two players, $\forall$ (male, sometimes called $\forall$belard) and $\exists$ (female, $\exists$loise), and X is an infinite subset of ω with $\omega \backslash X$ infinite and $0 \notin X$. The players have to pick the sets T_i in turn; player $\exists$ makes the choice of T_i if and only if $i \in X$. Player $\exists$ wins if T^+ has property π; otherwise $\forall$ wins. We say that π is **enforceable** if player $\exists$ has a winning strategy for this game. (One can show that the question whether π is enforceable is independent of the choice of X, provided that both X and $\omega \backslash X$ are infinite.) Some properties of T^+ are really properties of the canonical model A^+ – for example, that every element of A^+ is named by infinitely many witnesses. So we can talk of 'enforceable properties' of A^+ too.

The proof of Theorem 7.2.1 amounted to showing that the properties described in (2.2), $(2.3)_\psi$ and $(2.4)_m$ are enforceable. The main advantage of this viewpoint is that it breaks down the overall task into infinitely many smaller tasks, and these can be carried out independently without interfering with each other.

The sets Φ_m in the omitting types theorem are not necessarily complete types. When they are, we can improve the theorem by omitting not just countably many types at once, but λ where λ is any number less than 2^ω. See Corollary 7.2.5 below.

We turn to two strikingly different ways of applying the omitting types theorem. In Theorem 7.2.2 we apply it to the situation where there are only a few complete types, in order to get a model which is particularly well nailed down. In Corollary 7.2.5 we use omitting types to get a very large number of different models.

Atomic and prime models

We say that a structure A is **atomic** if for every tuple $\bar{a}$ of elements of A, the complete type of $\bar{a}$ in A is principal. A model A of a theory T is said to be **prime** if A be elementarily embedded in every model of T. (**Warning**: Don't confuse this with Abraham Robinson's notion of 'prime model' in section 8.3 below.)

Recall from section 6.3 that $S_n(T)$ is the set of complete first-order types $p(x_0, \ldots, x_{n-1})$ over the empty set with respect to models of T.

Theorem 7.2.2. *Let L be a countable first-order language and T a complete theory in L which has infinite models.*

(a) *If for every $n < \omega$, $S_n(T)$ is at most countable, then T has a countable atomic model.*

(b) *If A is a countable atomic L-structure which is a model of T, then A is a prime model of T.*

Proof. (a) There are only countably many non-principal complete types, so by Theorem 7.2.1 we can omit the lot of them in some model A of T. Since T is complete and has infinite models, A can be found with cardinality ω.

(b) Let B be any model of T. We shall show

(2.5)   if $\bar{a}, \bar{b}$ are n-tuples realising the same complete type in A, B respectively, and d is any element of B, then there is an element c of A such that $\bar{a}c$, $\bar{b}d$ realise the same complete $(n + 1)$-type in A, B respectively.

Since the complete type of $\bar{b}d$ is principal by assumption, it has a generator $\psi(\bar{x}, y)$. Since $\bar{a}$ and $\bar{b}$ realise the same complete type, and $B \vDash \exists y\, \psi(\bar{b}, y)$, we infer that $A \vDash \exists y\, \psi(\bar{a}, y)$, and hence there is an element c in A such that $A \vDash \psi(\bar{a}, c)$. Then $\bar{a}c$ realises the same complete type as $\bar{b}d$. This proves (2.5).

Now let $b_0, b_1, \ldots$ list all the elements of B. By induction on n, use (2.5) to find elements $a_0, a_1, \ldots$ of A so that for each n, $(A, a_0, \ldots, a_{n-1}) \equiv (B, b_0, \ldots, b_{n-1})$. Then the map $b_i \mapsto a_i$ is an elementary embedding of B into A, by the elementary diagram lemma (Lemma 2.5.3). $\qquad\square$

In short, if T is complete and all the sets $S_n(T)$ are countable then there is a 'smallest' countable model of T. The exercises discuss in what sense the prime model is 'smallest'. It happens that under the same assumptions there is also a 'biggest' countable model of T, into which all other countable models of T can be elementarily embedded; see Exercise 11.

While the proof of Theorem 7.2.2 is fresh at hand, let me adapt it to prove another useful result which has nothing to do with countable structures.

Theorem 7.2.3. *Let L be a countable first-order language. Let A and B be two elementarily equivalent L-structures, both of which are atomic. Then A and B are back-and-forth equivalent.*

Proof. We show that if $\bar{a}$ and $\bar{b}$ are tuples in A and B respectively, such that $(A, \bar{a}) \equiv (B, \bar{b})$, then for every element c of A there is an element d of B such that $(A, \bar{a}, c) \equiv (B, \bar{b}, d)$, and conversely for every element d of B there is an element c of A such that $(A, \bar{a}, c) \equiv (B, \bar{b}, d)$. The argument is exactly the same as for (2.5) above. $\qquad\square$

Making several different models

In our next application we build two models of the theory T at once. As we go, we make sure that as far as possible, no tuple in the first model looks like any tuple in the second.

Theorem 7.2.4. *Let L be a countable first-order language and T a complete theory in L which has an infinite model. Then there are two countable models A, B of T with the property that if $\Phi(\bar{x})$ is a complete type which is realised both by a tuple in A and by a tuple in B, then Φ is principal.*

Proof. The proof is a variation on the proof of Theorem 7.2.1. We set our experts to construct two models A, B of T at once, both using the same language L, the same countable set of witnesses and the same theory T. So two chains of finite theories are built simultaneously, say $(S_i : i < \omega)$ for A and $(T_i : i < \omega)$ for B. The two constructions proceed side by side with the same rules as before, but with one exception. We assign one of the infinite subsets of ω, say X, to a new expert whose task is to make sure that

(2.6)   for all tuples $\bar{c}, \bar{d}$ of distinct witnesses, if $\bar{a}, \bar{b}$ are the tuples of elements named by $\bar{c}, \bar{d}$ in A^+, B^+ respectively, and

$\Phi(\bar{x})$, $\Psi(\bar{x})$ are the complete types of $\bar{a}, \bar{b}$ in A, B respectively, then either $\Phi \neq \Psi$ or Φ is principal.

We have to show that (2.6) is enforceable.

Suppose the set X contains i, and S_{i-1} and T_{i-1} have just been chosen. We can write $\bigwedge S_{i-1}$ as $\psi(\bar{c}, \bar{e})$ and $\bigwedge T_{i-1}$ as $\chi(\bar{d}, \bar{f})$, where ψ and χ are in L, the tuple $\bar{e}$ lists the distinct witnesses which are in S_{i-1} but not in $\bar{c}$, and $\bar{f}$ lists those in T_{i-1} but not in $\bar{d}$.

There are two cases. Suppose first that $\exists \bar{y}\, \psi(\bar{x}, \bar{y})$ is a support of a complete type Φ over T. Then Φ is principal, so our expert can put $S_i = S_{i-1}$ and $T_i = T_{i-1}$.

Second, suppose $\exists \bar{y}\, \psi(\bar{x}, \bar{y})$ is not the support of a complete type over T. Then at least two complete types $\Delta(\bar{x}), \Xi(\bar{x})$ contain the formula $\exists \bar{y}\, \psi(\bar{x}, \bar{y})$. Let $\Psi(\bar{x})$ be any complete type which contains the formula $\exists \bar{z}\, \chi(\bar{x}, \bar{y})$. We can choose one of Δ and Ξ which is different from Ψ; call it Φ. There is a formula $\theta(\bar{x})$ which is in Φ but not in Ψ. Let our expert take S_i to be $S_{i-1} \cup \{\theta(\bar{c})\}$ and T_i to be $T_{i-1} \cup \{\neg \theta(\bar{d})\}$. □

There is no difficulty in stretching the idea of Theorem 7.2.4 to make the experts build three, four or any finite number of models which overlap only at principal types. With a pinch of diagonalisation we can extend the number to ω. To push it up still further to 2^ω needs a new idea.

Corollary 7.2.5. *Under the hypothesis of Theorem 7.2.4, there is a family* $(A_i : i < 2^\omega)$ *of countable models of T, with the property that if $\Phi(\bar{x})$ is a complete type which is realised in at least two of the models A_i $(i < 2^\omega)$, then Φ is principal.*

Proof. Here I wave my hands a little. We adopt the argument of Theorem 7.2.4, but as the construction proceeds, we keep changing our minds about the number of models that we are constructing. Or rather, we allow the experts at any stage to introduce a new chain of theories which is a duplicate copy of one of the chains being constructed. Thus the construction will take the form of a tree:

(2.7)

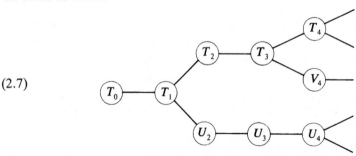

where each branch β separately will give rise to a chain $T_0 \subseteq T_1 \subseteq \ldots$ of conditions; write T^β for the union of this chain. From each set T^β we can read off a compiled structure A_β^+; we write A_β for the reduct $A_\beta^+ | L$.

By splitting the tree often enough, the experts can ensure that there are continuum many branches and hence continuum many models at the end. There is one expert whose job is to make sure that (2.6) holds for each pair of branches. She can do this, because at each step where she has to act, the number of branches is still finite, and thus she can arrange that (2.6) holds for each pair of branches at that step.   □

The results of this section were a powerful incentive to research in the 1960s, and a great deal of what has been done in model theory since then took its start from here. For example **Vaught's conjecture** states that the number of non-isomorphic countable models of a complete countable first-order theory is either $\leq \omega$ or $= 2^\omega$. Morley [1970a] succeeded in showing that this number is always either 2^ω or ω_1 or $\leq \omega$. (For PC_Δ classes this is best possible; see Example 2 in section 5.2.) But beyond this we have no idea whether Vaught's conjecture is true. It has been proved in a few special cases, with great difficulty and by special arguments. For example Rubin [1974] showed that if A is an infinite linear ordering then $\text{Th}(A)$ has either one or continuum many countable models (cf. Rubin [1977] for an extension to $L_{\omega_1\omega}$). For more on Vaught's conjecture see also Shelah, Harrington & Makkai [1984], Lascar [1985], Mayer [1988] and Steel [1978]. Recent unpublished work of Buechler [1990], [199?] and Newelski takes the question a stage deeper. Shelah [1990] p. xxi comments 'Some people think this is the most important question in model theory as its solution will give us an understanding of countable models which is the most important kind of models. We disagree with all those three statements.'

Exercises for section 7.2

The first two questions refer to the games $G(\pi, X)$ described after the proof of Theorem 7.2.1.

1. Let X be an infinite subset of $\omega \setminus \{0\}$ such that $\omega \setminus X$ is infinite, and let Y be the set of odd positive integers. Show that player $\exists$ has a winning strategy for the game $G(\pi, X)$ if and only if she has one for $G(\pi, Y)$.

2. Show that if π_i is an enforceable property for each $i < \omega$, then the property $\bigwedge_{i<\omega} \pi_i$ (which T^+ has if it has all the properties π_i) is also enforceable.

3. Let T be a complete theory in a countable first-order language L, and suppose that T has infinite models. Show that T has a countable model A such that for every tuple

$\bar{a}$ of elements of A there is a formula $\phi(\bar{x})$ with $A \vDash \phi(\bar{a})$, such that either (a) ϕ supports a complete type over T, or (b) no principal complete type over T contains ϕ.

4. (a) Let A be a finite or countable structure of countable signature. Show that A is atomic if and only if A is a prime model of $\mathrm{Th}(A)$. (b) Deduce that any two prime models of a countable complete theory are isomorphic.

5. Let L be a countable first-order language, and T a complete theory in L which has an infinite model. We say that **principal types are dense** for T if for every formula $\phi(\bar{x})$ of L, if $T \vdash \exists \bar{x}\, \phi$ then T has a principal complete type $p(\bar{x})$ (over the empty set) which contains ϕ. (The Stone topology of sections 6.2 and 6.3 justifies this terminology.) Show that T has a prime model if and only if principal types are dense for T.

We say that a model A of a theory T is **minimal** *if no proper elementary substructure of A is a model of T. (*Warning*: There is no connection with the 'minimal structures' of section 4.5.)*
6. Give examples of complete countable first-order theories T_1, T_2 such that (a) T_1 has a prime model but no minimal model, and (b) T_2 has at least two minimal models but no prime model. [For (b), try term algebras.] *In fact by Shelah [1978b], the number $\mu(T)$ of minimal models of a countable complete theory T can be chosen to be 2^ω or any number $\leq \omega$. As far as I know, the possibility of $\mu(T) = \omega_1$ is still open. Using Morley [1970a], there are no other possibilities.*

7. Show that if a first-order theory T has a prime model and a minimal model, then the prime and minimal models of T coincide.

*8. Show that the theory of the additive group of integers has a minimal model but no prime model.

9. (a) Show that every elementary substructure of an atomic structure is atomic, and the union of an elementary chain of atomic structures is atomic. (b) Show that if T has a countable atomic model which is not minimal then T has an uncountable atomic model.

*Anticipating Chapter 10, an L-structure A is said to be ω-***saturated*** if for every L-structure B and all tuples $\bar{a}, \bar{b}$ of elements of A, B respectively, if $(A, \bar{a}) \equiv (B, \bar{b})$ and d is an element of B, then there is an element c of A such that $(A, \bar{a}, c) \equiv (B, \bar{b}, d)$.*
10. (a) Show that if A and B are elementarily equivalent ω-saturated structures then A is back-and-forth equivalent to B. (b) Show that if A is ω-saturated and B is a countable structure elementarily equivalent to A, then B is elementarily embeddable in A.

11. Show that the following are equivalent, for any countable complete first-order theory T with infinite models. (a) T has a countable ω-saturated model. (b) T has a countable model A such that every countable model of T is elementarily embeddable in A. (c) For every $n < \omega$, $S_n(T)$ is at most countable.

12. Give an example of a countable first-order theory which has a countable prime model but no countable ω-saturated model. [Take infinitely many equivalence relations E_i so that each E_{i+1} refines E_i.]

13. Let L be a countable first-order language, T a theory in L, and $\Phi(\bar{x})$ and $\Psi(\bar{y})$ sets of formulas of L. Show that (a) implies (b): (a) for every formula $\sigma(\bar{x}, \bar{y})$ of L there is a formula $\psi(\bar{y})$ in Ψ such that for all $\phi_1(\bar{x}), \ldots, \phi_n(\bar{x})$ in Φ, if $T \cup \{\exists \bar{x} \bar{y}(\sigma \wedge \phi_1 \wedge \ldots \wedge \phi_n)\}$ has a model then $T \cup \{\exists \bar{x} \bar{y}(\sigma \wedge \phi_1 \wedge \ldots \wedge \phi_n \wedge \neg \psi)\}$ has a model; (b) T has a model which realises Φ and omits Ψ.

14. Show that if T is a consistent countable first-order theory and W is a set of fewer than 2^ω unsupported complete types over the empty set for W, then there is a model of T which omits all the types in W.

*15. Show by a suitable example that Zermelo–Fraenkel set theory (ZFC) doesn't imply the result of Exercise 14 if we remove the word 'complete'.

16. Show that the theory T of Example 1 is complete as follows. If A is a model of T and s is a subset of ω, $|\Phi_s(A)|$ is the number of elements of A which realise Φ_s. (a) Show that if A is a model of T, then A is determined up to isomorphism by the cardinals $|\Phi_s(A)|$ $(s \subseteq \omega)$. (b) Show that if $s \subseteq \omega$ and A is a model of T of cardinality $\leqslant 2^\omega$, then A has an elementary extension B of cardinality 2^ω with $|\Phi_s(B)| = 2^\omega$. (c) By iterating (b), show that every model of T of cardinality $\leqslant 2^\omega$ has an elementary extension C of cardinality 2^ω with $|\Phi_s(C)| = 2^\omega$ for each $s \subseteq \omega$.

7.3 Countable categoricity

A complete theory which has exactly one countable model up to isomorphism is said to be ω-**categorical**. We also say that a structure A is ω-**categorical** if Th(A) is ω-categorical. People with Hebrew on their word processors sometimes say $\aleph_0$-**categorical**.

Our first result is a handful of characterisations of ω-categoricity. It rests on the countable omitting types theorem (Theorem 7.2.1 in the previous section). One of the most useful characterisations is (b), which allows us to use the methods of Chapter 4 above.

Theorem 7.3.1 (*Theorem of Engeler, Ryll-Nardzewski and Svenonius*). *Let L be a countable first-order language and T a complete theory in L which has infinite models. Then the following are equivalent.*
(a) *Any two countable models of T are isomorphic.*
(b) *If A is any countable model of T, then* Aut(A) *is oligomorphic (i.e. for every $n < \omega$,* Aut(A) *has only finitely many orbits in its action on n-tuples of elements of A).*
(c) *T has a countable model A such that* Aut(A) *is oligomorphic.*

(d) *Some countable model of T realises only finitely many complete n-types for each $n < \omega$.*

(e) *For each $n < \omega$, $S_n(T)$ is finite.*

(f) *For each $\bar{x} = (x_0, \ldots, x_{n-1})$, there are only finitely many pairwise non-equivalent formulas $\phi(\bar{x})$ of L modulo T.*

(g) *For each $n < \omega$, every type in $S_n(T)$ is principal.*

Proof. (b) $\Rightarrow$ (c) is immediate, since T has a countable model (by the downward Löwenheim–Skolem theorem, Corollary 3.1.5). (c) $\Rightarrow$ (d) is also direct: automorphisms preserve all formulas.

(d) $\Rightarrow$ (e). Let A be a countable model of T realising only finitely many complete n-types for each $n < \omega$. For a fixed n, let $p_0, \ldots, p_{k-1}$ be the distinct types in $S_n(T)$ which are realised in A. For each of these types p_i there is a formula $\phi_i(\bar{x})$ of L which is in p_i but not in p_j when $j \neq i$. Since $A \vDash \forall \bar{x} \bigvee_{i<k} \phi_i(\bar{x})$, this sentence $\forall \bar{x} \bigvee_{i<k} \phi_i(\bar{x})$ must be a consequence of the complete theory T. Likewise if $\psi(\bar{x})$ is any formula of L, then A is a model of the sentence $\forall \bar{x} \bar{y} (\phi_i(\bar{x}) \wedge \phi_i(\bar{y}) \rightarrow (\psi(\bar{x}) \leftrightarrow \psi(\bar{y})))$, for every $i < k$, and these sentences must also be consequences of T. It follows that $p_0, \ldots, p_{k-1}$ are the only types in $S_n(T)$.

(e) $\Rightarrow$ (f). If two formulas $\phi(\bar{x})$ and $\psi(\bar{x})$ of L lie in exactly the same types $\in S_n(T)$, then ϕ and ψ must be equivalent modulo T. So if $S_n(T)$ has finite cardinality k, then there are at most 2^k non-equivalent formulas $\phi(\bar{x})$ of L modulo T.

(f) $\Rightarrow$ (g). For any $n < \omega$ and $\bar{x} = (x_0, \ldots, x_{n-1})$, take a maximal family of pairwise non-equivalent formulas $\phi(\bar{x})$ of L modulo T. Assuming (f), this family is finite. If p is any type $\in S_n(T)$, let θ be the conjunction of all formulas of the family which also lie in p. Then θ is a support of p.

(a) $\Rightarrow$ (g). Suppose (g) fails. Then for some $n < \omega$, there is a non-principal type q in $S_n(T)$. By the omitting types theorem (Theorem 7.2.1), T has a model A which omits q. By the definition of types, T also has a model B which realises q. Since T is complete and has infinite models, both A and B are infinite. By the downward Löwenheim–Skolem theorem we can suppose that both A and B are countable. Hence T has two countable models which are not isomorphic, and thus (a) fails.

(g) $\Rightarrow$ (a). By (g) all models of T are atomic, and hence back-and-forth equivalent by Theorem 7.2.3. It follows by Theorem 3.2.3(b) that all countable models of T are isomorphic.

(g) $\Rightarrow$ (b). Again we deduce from (g) that all models of T are atomic. But rather than quote Theorem 7.2.3, we extract part of its proof:

(3.1)   if A, B are countable models of T and $\bar{a}, \bar{b}$ are n-tuples in A, B respectively, such that $(A, \bar{a}) \equiv (B, \bar{b})$, then there is an isomorphism from A to B which takes $\bar{a}$ to $\bar{b}$.

Now let A be a countable model of T and let $\bar{a}, \bar{b}$ be n-tuples which realise the same complete type in A. Then by (3.1), $\bar{a}$ and $\bar{b}$ lie in the same orbit of Aut (A). So to deduce (b), we need only show that (g) implies (e).

By Corollary 6.2.2, every infinite boolean algebra has non-principal ultrafilters, and a non-principal ultrafilter in the boolean algebra of equivalence classes of formulas $\phi(x_0, \ldots, x_{n-1})$ modulo T is the same thing as a non-principal complete type over T. So by (g) there are only finitely many complete n-types, proving (e) as required. $\square$

Where do ω-categorical structures occur in nature?

As soon as Theorem 7.3.1 was proved, model theorists went searching through the mathematical archives for examples of ω-categorical structures. It was natural to ask, for example, what countable rings are ω-categorical. There was also a good hope of interesting mathematicians outside model theory, because one can explain what it means for Aut (A) to be oligomorphic without mentioning any notions from logic. The following corollary was a handy starting point.

Corollary 7.3.2. *If A is an ω-categorical structure, then A is locally finite. In fact there is a (unique) function $f: \omega \to \omega$, depending only on Th (A), with the property that for each $n < \omega$, $f(n)$ is the least number m such that every n-generator substructure of A has at most m elements.*

Proof. Let $\bar{a}$ be an n-tuple of elements of A. If c and d are two distinct elements of the substructure $\langle \bar{a} \rangle_A$ generated by $\bar{a}$, then the complete types of $\bar{a}c$ and $\bar{a}d$ over the empty set must be different, because they say how c and d are generated. So by (e) of the theorem for $n + 1$, $\langle \bar{a} \rangle_A$ is finite. This proves the first sentence.

Now let B be the unique countable structure elementarily equivalent to A. By (b) of the theorem, for each $n < \omega$ there are finitely many orbits of n-tuples in B; let $\bar{b}_0, \ldots, \bar{b}_{k-1}$ be representatives of these orbits, and write m_i for the number of elements of the substructure $\langle \bar{b}_i \rangle_B$ generated by $\bar{b}_i$. Then we can put $f(n) = \max(m_i : i < k)$. This choice of f works for A as well as B, since A and B realise exactly the same types in $S_n(T)$, namely all of them. $\square$

Example 1: ω-categorical groups. By the corollary, every countable ω-categorical group is locally finite and has finite exponent. For abelian groups, this is the end of the story: any abelian group A of finite exponent is a direct sum of finite cyclic groups, and we can write down a first-order theory which says how often each cyclic group occurs in the sum (where the number of times is either $0, 1, 2, \ldots$ or infinity – for further details, see the Szmielew invariants in section A.2 below). So an infinite abelian group is ω-categorical if and only if it has finite exponent.

For groups in general the situation is much more complicated. We don't yet have even a characterisation of those nilpotent groups of class 2 which are ω-categorical, though there are some interesting constructions which give ω-categorical groups in this class (see Apps [1983b], Felgner [1976], Saracino & Wood [1979]). In section A.4 there is some information about stable ω-categorical groups.

Another open question is whether there exists an ω-categorical countable projective plane (as a structure with points and lines for elements, and a relation for incidence).

The exercises and the references give some information about ω-categorical structures in other interesting classes. The results are often pretty, but they tend not to contribute much to a broad understanding of ω-categoricity. For that we need some more general method of constructing ω-categorical structures. The best method is certainly Fraïssé's construction, as we shall see in the next section. Meanwhile let us deduce a little more from Theorem 7.3.1.

Interpretations and ω-categorical structures

In the course of proving Theorem 7.3.1, we found the property which I numbered (3.1). This property is worth extracting.

Corollary 7.3.3. *Let L be a countable first-order language. Let A be an L-structure which is either finite, or countable and ω-categorical. Then for any positive integer n, a pair $\bar{a}, \bar{b}$ of n-tuples from A are in the same orbit under* Aut (A) *if and only if they satisfy the same formulas of L.*

Proof. As in the proof of Theorem 7.3.1, it suffices to show that A is atomic. We know this by (g) of the theorem when A is countable and ω-categorical. When A is finite, we deduce it by the argument of (d) $\Rightarrow$ (e) in the proof of the theorem. $\square$

This immediately implies the following result, which I promised in section 4.1.

Corollary 7.3.4. *Let A be a countable ω-categorical structure. Then for every finite set X of elements of A,* ACL $(X) = $ acl (X) *and* DCL $(X) = $ dcl (X). $\square$

Since there are finitely many types of $S_n(T)$, and all of them are principal, Corollary 7.3.3 can be rephrased as follows (recalling that a formula is **principal** if it generates a complete type).

Corollary 7.3.5. *Let L be a countable first-order language. Let A be an L-structure which is either finite, or countable and ω-categorical. Then for each n there are finitely many complete formulas $\phi_i(x_0, \ldots, x_{n-1})$ $(i < k_n)$ of L for Th(A), and the orbits of Aut(A) on $(\mathrm{dom}\, A)^n$ are exactly the sets $\phi_i(A^n)$ $(i < k_n)$.* $\square$

This tells us in particular that we can almost recover A from the permutation group Aut(A).

Theorem 7.3.6. *Let A be a countable ω-categorical L-structure with domain Ω, and let B be the canonical structure for Aut(A) on Ω (see section 4.1 above). Then the relations on Ω which are first-order definable in A without parameters are exactly the same as those definable in B without parameters; in other words, A and B are definitionally equivalent (see section 2.6).*

Proof. By definition of the canonical structure B, it has the same automorphism group as A; write G for this group. Write L' for the language of B; we assume it is disjoint from L. If R is an n-ary relation symbol of L, then R^A is a union of finitely many orbits of G on Ω, and so R can be defined by a disjunction of formulas of L' which define these orbits. The same argument works in the other direction too. $\square$

So for countable ω-categorical structures, identity of automorphism groups coincides with a strong form of two-way interpretability. Readers of Chapters 4 and 5 will wonder whether weaker relations between the automorphism groups correspond to weaker forms of interpretation. Yes, they do. The next theorem is a partial converse to Theorem 5.3.5 above.

Theorem 7.3.7. *Let K and L be countable first-order languages; let A and B be respectively a K-structure and an L-structure, and let $h\colon \mathrm{Aut}(A) \to \mathrm{Aut}(B)$ be a continuous homomorphism between the topological automorphism groups. Assume that*
(a) A is either finite with more than one element, or countable and ω-categorical, and
(b) the image of h has finitely many orbits in its action on B.
Then there is an interpretation Γ of L in K with $B \cong \Gamma(A)$, which induces h.

Proof. For simplicity suppose first that the image of h is transitive on dom (B). Let b be any element of B. Since h is continuous, $h^{-1}(\text{Aut}(B)_{(b)})$ is an open subgroup of $\text{Aut}(B)$, and hence it contains $\text{Aut}(A)_{(\bar{a})}$ for some n-tuple $\bar{a}$ in A. If α and α' are elements of $\text{Aut}(A)$ with $\alpha\bar{a} = \alpha'\bar{a}$, then $h(\alpha)b = h(\alpha')b$; so for each tuple $\alpha\bar{a}$ in the orbit of $\bar{a}$ under $\text{Aut}(A)$, we can unambiguously define $f(\alpha\bar{a}) = h(\alpha)b$; f will be the coordinate map of Γ. Now the set $\{\alpha\bar{a}: \alpha \in \text{Aut}(A)\}$ is an orbit of $\text{Aut}(A)$, and so by (a) and Corollary 7.3.5 it is $\theta(A^n)$ for some formula $\theta(\bar{x})$ of K. Likewise if $\phi(x_0, \ldots, x_{m-1})$ is an unnested atomic formula of L, let X be the set

(3.2) $\{(\alpha_0\bar{a})^\frown \ldots {}^\frown(\alpha_{m-1}\bar{a}): B \vDash \phi(h(\alpha_0)b, \ldots, h(\alpha_{m-1})b)\}.$

Then using the fact that h is a homomorphism, we have $\alpha X = X$ for every automorphism α of A. Hence X is a union of orbits, and so by ω-categoricity again we can write X as $\phi^*(A^{mn})$ for some formula ψ of K. The domain formula θ and the map $\phi \mapsto \phi^*$ define the interpretation Γ. Finally we handle the case where the image of h has finitely many orbits on B, by defining f separately for each orbit. Since A has at least two elements, we can arrange that the tuples $\bar{a}$ in A corresponding to different orbits on B are all distinct but of the same length; thus the definitions of f for the separate orbits fit together. $\square$

Finally we note that interpretations always preserve ω-categoricity (so that B in Theorem 7.3.7 must be ω-categorical):

Theorem 7.3.8. *Let K and L be countable first-order languages, Γ an interpretation of L in K, and A an ω-categorical K-structure. Then ΓA is ω-categorical.*

Proof. Let A' be a countable structure which is elementarily equivalent to A. Then $\Gamma A' \equiv \Gamma A$ by the reduction theorem (Theorem 5.3.2), and so it suffices to show that $\Gamma A'$ is ω-categorical. By the construction in Theorem 5.3.3, every element of $\Gamma A'$ is an equivalence class of the relation $=_\Gamma$ on dom (A'); write $\bar{a}^=$ for the equivalence class containing the tuple $\bar{a}$. Each automorphism α of A' induces an automorphism $\Gamma(\alpha)$ of $\Gamma A'$, by the rule $\Gamma(\alpha)(\bar{a}^=) = (\alpha\bar{a})^=$. Since $\text{Aut}(A')$ is oligomorphic, it follows at once that $\text{Aut}(\Gamma A')$ is oligomorphic too. $\square$

In particular, relativised reducts of ω-categorical structures are ω-categorical.

Exercises for section 7.3

1. Show that if A is an infinite structure and $\bar{a}$ is a tuple of elements of A, then $\text{Th}(A, \bar{a})$ is ω-categorical if and only if $\text{Th}(A)$ is ω-categorical.

2. Show that if A is a countable structure, then A is ω-categorical if and only if for every tuple $\bar{a}$ of elements of A, the number of orbits of $\mathrm{Aut}\,(A, \bar{a})$ on single elements of A is finite.

3. Show that if T is a first-order theory with countable models, then T is ω-categorical if and only if all the models of T are pairwise back-and-forth equivalent.

4. Let T be a complete and countable first-order theory. Show that T is ω-categorical if and only if every countable model of T is atomic.

5. Let L be a countable signature and let A and B be ω-categorical L-structures. Show that the product $A \times B$ is ω-categorical.

6. Give an example of an ω-categorical first-order theory T such no skolemisation of T is ω-categorical. [Rationals!]

7. Let B be a countable boolean algebra. Show that B is ω-categorical if and only if B has finitely many atoms.

*8. The **direct sum** of groups G_i $(i < \omega)$ is the subgroup of $\Pi_{i<\omega}G_i$ consisting of those elements g such that $g(i) = 1$ for all but finitely many $i < \omega$. Let G be the direct sum of groups $G_i\,(i < \omega)$, and suppose there is a finite k such that every group G_i has order $< k$. Show that G is ω-categorical if and only if every group which occurs infinitely often as a G_i is abelian.

*9. Let G be a countable ω-categorical group. Show (a) G has only finitely many characteristic subgroups, (b) if H is a characteristic subgroup of G such that H and G/H are both infinite, then both H and G/H are ω-categorical, (c) G has a finite characteristic series $\{1\} = G_0 < \ldots < G_m = G$ in which each factor is characteristic-ally simple and either finite or ω-categorical.

*10. Give an example of an ω-categorical group G with a normal subgroup H such that G/H is infinite and not ω-categorical.

*11. Let R be a locally nilpotent ω-categorical ring. Show that R is nilpotent.

*12. Let G be a countable, soluble, locally nilpotent and ω-categorical group. Show that G is nilpotent. [If G is not metabelian then each of $G', G/G''$ is a locally nilpotent ω-categorical group of smaller derived length than G; if G' and G/G'' are nilpotent then so is G. So we can assume G is metabelian. Taking any $a \in G'$, we show as follows that a lies somewhere in the upper central series of G. Regard $\langle a^G \rangle$ as a $\mathbb{Z}G$-module with G acting by conjugation, and let the ring R be the image of $\mathbb{Z}G$ in the endomorphism ring of $\langle a^G \rangle$; R is commutative. By ω-categoricity there is $d < \omega$ such that each element of R is a sum of at most d images of elements of G. Let E be the subring of R generated by the elements $r - 1$ where $r \in R$ are the images of elements of G; then E is locally nilpotent. Let Θ be the group of automorphisms of G

fixing a. Then Θ induces a group of automorphisms of E; using the bound d, this group of automorphisms is oligomorphic, and so E is ω-categorical. Now Exercise 11 implies a lies in a term of the upper central series of G'.]

*13. Let L be a countable language, A a countable ω-categorical L-structure and B a countable L-structure such that every finite substructure of B is embeddable in A. Show that B is embeddable in A. [Use König's tree lemma, Exercise 5.6.5 above.]

*14. Let A and B be countable ω-categorical structures. Show that $\{A\}$ is bi-interpretable with $\{B\}$ if and only if $\mathrm{Aut}(A)$ and $\mathrm{Aut}(B)$ are isomorphic as topological groups.

15. Let A be a countable ω-categorical structure. Show that if H is an open subgroup of $\mathrm{Aut}(A)$ then the number of groups K such that $H \subseteq K \subseteq \mathrm{Aut}(A)$ is finite; deduce that $\mathrm{Aut}(A)$ has at most countably many open subgroups. [If H is a basic open subgroup $\mathrm{Aut}(A)_{(\bar{a})}$ for some n-tuple $\bar{a}$, and K is any group between H and $\mathrm{Aut}(A)$, then $\{k\bar{a}: k \in K\}$ is a set of imprimitivity for $\mathrm{Aut}(A)$ on n-tuples. Hence each group K is characterised by the $2n$-types realised by pairs $\bar{a}^\frown k\bar{a}$ with k in K.]

16. Suppose A is a countable ω-categorical structure which has the small index property. (a) Show that A has the strong small index property if and only if $\mathrm{Aut}(A)$ satisfies the intersection condition (see section 4.2). [For one direction, given $K = \langle G_{(X)} \cup G_{(Y)} \rangle$, find an algebraically closed finite set Z such that $G_{(Z)} \subseteq K \subseteq G_{\{Z\}}$, and show $Z = X \cap Y$. For the other direction use Theorem 4.2.9.] (b) Show that if A has the strong small index property then A has weak semi-uniform elimination of imaginaries.

7.4 ω-categorical structures by Fraïssé's method

Fraïssé's construction from section 7.1 above has proved to be a very versatile way of building ω-categorical structures.

The trick is to make sure that if $\mathbf{K}$ is the class whose Fraïssé limit we are taking, the sizes of the structures in $\mathbf{K}$ are kept under control by the number of generators. For this, we say that a structure A is **uniformly locally finite** if there is a function $f: \omega \to \omega$ such that

(4.1)   for every substructure B of A, if B has a generator set of cardinality $\leq n$ then B itself has cardinality $\leq f(n)$.

We say that a class $\mathbf{K}$ of structures is **uniformly locally finite** if there is a function $f: \omega \to \omega$ such that (4.1) holds for every structure A in $\mathbf{K}$.

Note that if the signature of $\mathbf{K}$ is finite and has no function symbols, then $\mathbf{K}$ is uniformly locally finite. (See Exercise 1.2.6.)

Theorem 7.4.1. *Suppose that the signature L is finite and $\mathbf{K}$ is a countable uniformly locally finite set of finitely generated L-structures with HP, JEP and*

AP. Let M be the Fraïssé limit of K, and let T be the first-order theory
Th (M) *of M.*
   (a) *T is ω-categorical,*
   (b) *T has quantifier elimination.*

Proof. First we show that there is an $\forall_2$ theory U in L whose models are
precisely the weakly homogeneous structures (see (1.7) in section 7.1) of age
K. There are two crucial points here. The first is that by our assumption on
L, if A is any finite L-structure with n generators $\bar{a}$, then there is a
quantifier-free formula $\psi = \psi_{A,\bar{a}}(x_0, \ldots, x_{n-1})$ such that for any L-structure
B and n-tuple $\bar{b}$ of elements of B,

(4.3) $B \vDash \psi(\bar{b})$ if and only if
          there is an isomorphism from A to $\langle \bar{b} \rangle_B$ which takes $\bar{a}$ to $\bar{b}$.

In fact $\psi_{A,\bar{a}}$ is a conjunction of literals satisfied by $\bar{a}$ in A. The second is that
by the uniform local finiteness, for each $n < \omega$ there are only finitely many
isomorphism types of structures in **K** with n generators.

   These two facts can both be checked by inspection. Given them, we take
U_0 to be the set of all sentences of the form

(4.3) $\forall \bar{x}(\psi_{A,\bar{a}}(\bar{x}) \rightarrow \exists y \, \psi_{B,\bar{a}b}(\bar{x}, y))$

where B is a structure in **K** generated by a tuple $\bar{a}b$ of distinct elements, and
A is the substructure generated by $\bar{a}$. This includes the case where $\bar{a}$ is
empty, so that the sentence (4.3) reduces to $\exists y \, \psi_{B,b}(y)$. We take U_1 to be
the set of all sentences of the form

(4.4) $\forall \bar{x} \bigvee_{A,\bar{a}} \psi_{A,\bar{a}}(\bar{x})$

where the disjunction ranges over all pairs $A, \bar{a}$ such that A is in **K** and $\bar{a}$ is a
tuple of the same length as $\bar{x}$ which generates A. Uniform local finiteness
implies that this is a finite disjunction (up to logical equivalence). We write U
for the union $U_0 \cup U_1$. Clearly M is a model of U.

   Suppose D is any countable model of U. When $\bar{a}$ is empty, the sentences
(4.3) say that every one-generator structure in **K** is embeddable in D. In
general the sentences (4.3) say that

(4.5)   if A, B are finitely generated substructures of $D, A \subseteq B, B$
          comes from A by adding one more generator, and $f: A \rightarrow D$
          is an embedding, then there is an embedding $g: B \rightarrow D$ which
          extends f.

It's not hard to see, using induction on the number of generators, that these
two facts imply that every structure in **K** is embeddable in D; so together
with the sentences (4.4), they tell us that the age of D is exactly **K**. Using the
sentences (4.3) again, an induction on the size of $\text{dom}(B) \backslash \text{dom}(A)$ tells us

that D is weakly homogeneous. So by Lemma 7.1.4, D is isomorphic to M. Hence U is ω-categorical, and U is a set of axioms for T.

Suppose now that $\phi(\bar{x})$ is a formula of L with $\bar{x}$ non-empty, and let X be the set of all tuples $\bar{a}$ in M such that $M \vDash \phi(\bar{a})$. If $\bar{a}$ is in X, and $\bar{b}$ is a tuple of elements such that there is an isomorphism $e: \langle \bar{a} \rangle_M \to \langle \bar{b} \rangle_M$ taking $\bar{a}$ to $\bar{b}$, then e extends to an automorphism of M, so that $\bar{b}$ is in X too. It follows that ϕ is equivalent modulo T to the disjunction of all the formulas $\psi_{\langle \bar{a} \rangle, \bar{a}}(\bar{x})$ with $\bar{a}$ in X. This is a finite disjunction of quantifier-free formulas. Finally if ϕ is a sentence of L, then since T is complete, ϕ is equivalent modulo T to either $\neg \bot$ or $\bot$. So T has quantifier elimination. $\square$

Corollary 7.4.2. *Let L be a finite signature and M a countable L-structure. Then the following are equivalent.*
(a) *M is ultrahomogeneous and uniformly locally finite.*
(b) *$\mathrm{Th}\,(M)$ is ω-categorical and has quantifier elimination.*

Proof. (a) $\Rightarrow$ (b) is by Theorem 7.4.1. (b) $\Rightarrow$ (a) is by Corollaries 7.3.2 and 7.3.3 above. $\square$

There are so many attractive applications of Theorem 7.4.1 that it's hard to know which to mention first. I describe three examples in detail and leave others to the exercises.

First application: the random graph

A **graph** is a structure consisting of a set X with an irreflexive symmetric binary relation R defined on X (see Example 1 in section 1.1). The elements of X are called the **vertices**; an **edge** is a pair of vertices $\{a, b\}$ such that aRb. We say that two vertices a, b are **adjacent** if $\{a, b\}$ is an edge. A **path of length** n is a sequence of edges $\{a_0, a_1\}, \{a_1, a_2\}, \ldots, \{a_{n-2}, a_{n-1}\}$, $\{a_{n-1}, a_n\}$; the path is a **cycle** if $a_n = a_0$. A **subgraph** of a graph G is simply a substructure of G. We write L for the first-order language appropriate for graphs; its signature consists of just one binary relation symbol R.

Let **K** be the class of all finite graphs. The following facts hardly need proof.

Lemma 7.4.3. *The class **K** has HP, JEP and AP. Also **K** contains arbitrarily large finite structures and is uniformly locally finite. The signature of **K** contains only finitely many symbols.* $\square$

So by Theorems 7.1.2 and 7.4.1, **K** has a Fraïssé limit A; $\mathrm{Th}(A)$ is ω-categorical and has quantifier elimination. The structure A is a countable graph known as the **random graph**. We shall denote it by Γ.

Some Fraïssé limits are hard to describe in detail. Not so this one.

Theorem 7.4.4. *Let A be a countable graph. The following are equivalent.*
(a) *A is the random graph Γ.*
(b) *Let X and Y be disjoint finite sets of vertices of A; then there is an element $\notin X \cup Y$ which is adjacent to all vertices in X and no vertices in Y.*

Proof. (a) $\Rightarrow$ (b). Let A be the random graph Γ and let X, Y be disjoint finite sets of vertices of Γ. There is a finite graph G as follows: the vertices of G are the vertices in $X \cup Y$ together with one new vertex v, and vertices in $X \cup Y$ are adjacent in G if they are adjacent in Γ, while v is adjacent to all the vertices in X and none of the vertices in Y. Since Γ is the Fraïssé limit of **K**, there is an embedding $f: G \to \Gamma$. The restriction of f to $X \cup Y$ is an isomorphism between finite substructures of Γ, and so it extends to an automorphism g of Γ. Then $g^{-1}f(v)$ is the element described in (b).

(b) $\Rightarrow$ (a). Assume (b). We make the following claim.

(4.6)   Suppose $G \subseteq H$ are finite graphs and $f: G \to A$ is an embedding. Then f extends to an embedding $g: H \to A$.

This is proved by induction on the number n of vertices which are in H but not in G. Clearly we only need worry about the case $n = 1$. Let w be the vertex which is in H but not in G. Let X be the set of vertices $f(x)$ such that x is in G and adjacent to w in H, and let Y be the set of vertices $f(y)$ such that y is in G but not adjacent to w. By (b) there is a vertex v of A which is adjacent to all of X and none of Y. We extend f to g by putting $g(w) = v$. This proves the claim.

Taking G to be the empty structure, it follows that every finite graph is embeddable in A, so that the age of A is **K**. Taking G to be a substructure of A, it follows that A is weakly homogeneous. So by Lemma 7.1.4, A is the Fraïssé limit of **K**, proving (a). $\square$

Why 'random'? The name refers to another construction which also yields the same graph. Imagine some very indecisive man constructing a graph with vertices c_i ($i < \omega$). He runs through the vertices in order, starting at c_0. Whenever he comes to a vertex c_i, he has to decide, for each $j < i$, whether or not c_i is to be adjacent to c_j. So he tosses a coin for it. What sort of graph is he likely to finish up with?

This is a probabilistic question, and so we need to describe the appropriate measure. Writing 2 for the set $\{0, 1\}$ with the discrete topology, there is a natural σ-additive measure μ on the product space 2^ω, as follows. If X is an n-element subset of ω and $f: X \to 2$, then $\mu\{s \in 2^\omega: s(k) = f(k)$ for each

$k \in X\} = 1/2^n$; in particular the whole space has measure 1. There is a bijection ϕ between ω and the set of pairs (j, i) with $j < i < \omega$. So we can interpret each element s of 2^ω as a run of coin-tosses: if $\phi k = (j, i)$ then $s(k)$ will decide whether or not c_j is adjacent to c_i. In this way ϕ induces a bijection ψ between 2^ω and the set of all graphs with vertices $\{c_i : i < \omega\}$. We can use ψ to talk about the measure of a set of graphs.

Theorem 7.4.5. *With measure* 1, *a graph on vertices* $\{c_i : i < \omega\}$ *is random.*

Proof. A short exercise in probability; see Cameron [1990] sections 4.3 and 4.4 for a discussion. It would be intriguing if this theorem could be upgraded to a general method of model construction. This would give a measure-theoretic counterpart of forcing (see section 8.2 below); so Theorem 7.4.5 would say that Γ is 'enforceable in measure'. Cameron [1990] section 4.10 points out some snags; see also Bankston & Ruitenburg [1990]. $\qquad\qquad\square$

Recently it was proved that the random graph has the small index property; see Hodges, Hodkinson, Lascar & Shelah [199?].

Second application: the random structure

The main interest of our next example is that it leads directly to a theorem about finite models. Until recently the model theory of finite structures was fallow ground. A team of Russians from Gorki planted the first seed by proving Theorem 7.4.7 in 1969.

Let L be a non-empty finite signature, and let **K** be the class of all finite L-structures. Then clearly **K** has HP, JEP and AP, and (by Exercise 1.2.6) there are just countably many isomorphism types of structures in **K**. So **K** has a countable Fraïssé limit, which is known as the **random structure** of signature L. Let us call it Ran(L).

Let T be the set of all sentences of L of the form

(4.7) $\forall \bar{x}(\psi(\bar{x}) \rightarrow \exists y\, \chi(\bar{x}, y))$

such that for some finite L-structure B and some listing of the elements of B without repetition as $\bar{b}, c$, the formula $\psi(\bar{x})$ (resp. $\chi(\bar{x}, y)$) lists the literals satisfied by $\bar{b}$ (resp. $\bar{b}, c$) in B. Inspection shows that T consists of exactly the sentences (4.3) of the proof of Theorem 7.4.1. The sentences (4.4) of that proof are trivially true in this case, so they add nothing. It follows that T is a set of axioms for the theory of Ran(L). In particular T is complete.

Consider any $n < \omega$, any formula $\phi(x_0, \ldots, x_{n-1})$ of L and any tuple $\bar{a}$ of objects a_i with $i < n$. We write $\kappa_n(\phi)$ for the number of non-isomorphic L-structures B whose distinct elements are $a_0, \ldots, a_{n-1}$, such that $B \models \phi(\bar{a})$.

We write $\mu_n(\phi(\bar{a}))$ for the ratio $\kappa_n(\phi(\bar{a}))/\kappa_n(\forall x \, x = x)$, i.e. the proportion of those L-structures with elements $a_0, \ldots, a_{n-1}$ for which $\phi(\bar{a})$ is true.

Lemma 7.4.6. *Let ϕ be any sentence in T. Then $\lim_{n\to\infty} \mu_n(\phi) = 1$.*

Proof. Let ϕ be the sentence $\forall \bar{x}(\psi(\bar{x}) \to \exists y \, \chi(\bar{x}, y))$. We shall show that $\lim_{n\to\infty} \mu_n(\neg\phi) = 0$. Since $\mu_n(\neg\phi) = 1 - \mu_n(\phi)$, this will prove the lemma.

Suppose $\bar{x}$ is $(x_0, \ldots, x_{m-1})$, and $n > m$. Consider those structures B whose distinct elements are $a_0, \ldots, a_{n-1}$, such that $B \vDash \psi(a_0, \ldots, a_{m-1})$. What is the probability p that $B \vDash \forall y \, \neg\chi(a_0, \ldots, a_{m-1}, y)$? It must be the $(n - m)$-th power of the probability that $B \vDash \neg\chi(a_0, \ldots, a_{m-1}, a_m)$, since the $n - m$ elements $a_m, \ldots, a_{n-1}$ have equal and independent chances of serving for y. Since the signature of L is not empty, there is some positive real $k < 1$ such that $B \vDash \neg\chi(a_0, \ldots, a_{m-1}, a_m)$ with probability k, and so $p = k^{n-m}$.

Next, consider those L-structures C whose distinct elements are $a_0, \ldots, a_{n-1}$. We estimate the probability $\mu_n(\neg\phi)$ that $C \vDash \neg\phi$. This probability q is at most the probability that $C \vDash \psi(\bar{c}) \wedge \forall y \, \neg\chi(\bar{c}, y)$ for a tuple $\bar{c}$ of distinct elements of C, times the number of ways of choosing $\bar{c}$ in C. So $\mu_n(\neg\phi) \leqslant n^m \cdot k^{n-m} = \gamma \cdot n^m \cdot k^n$ where $\gamma = k^{-m}$.

Now since $0 < k < 1$, we have that $n^m \cdot k^n \to 0$ as $n \to \infty$ (e.g. Rudin [1964] Theorem 3.20). It follows that $\lim_{n\to\infty} \mu_n(\neg\phi) = 0$ as claimed. $\square$

Theorem 7.4.7 (Zero–one law). *Let ϕ be any first-order sentence of a finite relational signature. Then $\lim_{n\to\infty} \mu_n(\phi)$ is either 0 or 1.*

Proof. We have already seen that T is a complete theory. If ϕ is a consequence of T, then it follows from the lemma that $\lim_{n\to\infty} \mu_n(\phi)$ is 1. If ϕ is not a consequence of T, then T implies $\neg\phi$, and so $\lim_{n\to\infty} \mu_n(\neg\phi)$ is 1, whence $\lim_{n\to\infty} \mu_n(\phi)$ is 0. $\square$

Third application: induced structures

Let L be a first-order language whose symbols contain a 1-ary relation symbol P. If B is an L-structure, we say that a structure A is the **structure induced** by B on P if $\mathrm{dom}(A) = P^B$ and the $\varnothing$-definable relations of A are exactly the $\varnothing$-definable relations of B which lie in P^B. This determines A up to definitional equivalence. For shorthand we write $A = \mathrm{Ind}(B, P)$.

If B is ω-categorical and $A = \mathrm{Ind}(B, P)$, then A is ω-categorical too. But there is a catch: even when A and B are countable and ω-categorical, $\mathrm{Aut}(A)$ is not necessarily the restriction of $\mathrm{Aut}(B)$ to $\mathrm{dom}(A)$. The induced continuous homomorphism $h: \mathrm{Aut}(B) \to \mathrm{Aut}(A)$ has dense image in $\mathrm{Aut}(A)$, but it is not necessarily surjective. In other words, automorphisms of A may fail to lift to B.

Example 1: *Dense codense subsets of the rationals.* Let B be the ordered set of rationals, with P^B chosen so that both P^B and its complement are dense in the ordering. Then $\text{Aut}(A)$ acts transively on the set of 'irrational' Dedekind cuts of A. But some of these cuts contain points of B and some don't. (I thank Dugald Macpherson for this neat argument.)

Theorem 7.4.8. *There is a finite relational language L, containing a 1-ary relation symbol P, such that every countable ω-categorical structure (in any countable first-order language) is of the form $\text{Ind}(B, P)$ for some countable ω-categorical L-structure B.*

Proof. We take L to have the relation symbols P, λ and ρ (all 1-ary), H (2-ary) and S (4-ary).

Let A be a countable ω-categorical L'-structure; up to definitional equivalence we can assume that L' consists of countably many relation symbols R_n ($1 < n < \omega$), each R_n is $l(n)$-ary for some $l(n) < n$, and the sets R_n^A are precisely the orbits of $\text{Aut}(A)$ on tuples. Thus A has quantifier elimination. Let $\mathbf{J}$ be the age of A. By Corollary 7.4.2 and Theorem 7.1.7, $\mathbf{J}$ has HP, JEP and AP.

Let L^+ be the disjoint union of L and L'. Let T be the theory in L^+ which says 'If $\lambda(x)$ or $\rho(x)$ then not $P(x)$; if $H(x, y)$ then not $P(x)$ and not $P(y)$; if $S(x, y, z, w)$ then $P(z)$ and $P(w)$ but not $P(x)$ or $P(y)$; if $R_n(\bar{x})$ for some n, then each element of $\bar{x}$ satisfies P'. Suppose M is any model of T. By an *n-pair* in M we mean a set $\{a_0, \ldots, a_{l(n)-1}, c_0, \ldots, c_{n-1}\}$ of elements of M such that

(4.8) $P(a_i)$ holds for all $i < l(n)$, and $\neg P(c_i)$ holds for all $i < n$,

(4.9)   the elements c_i are distinct, and $H(c_i, c_j)$ holds iff $j \equiv i + 1$ (mod n),

(4.10) $\lambda(c_i)$ holds iff $i = 0$, $\rho(c_i)$ holds iff $i = l(n) - 1$,

(4.11) $S(c_h, c_i, a_k, a_m)$ holds iff $a_h = a_k$.

Note that if this set is given, we can uniquely recover the sequence of distinct elements $(c_0, \ldots, c_{l(n)-1})$ by (4.9) and (4.10). Then by (4.11) we can uniquely recover the sequence $(a_0, \ldots, a_{l(n)-1})$, which may contain repetitions. We say that the *n*-pair **labels** this sequence $(a_0, \ldots, a_{l(n)-1})$.

Let $\mathbf{K}$ be the class of all finite models B' of T such that for some finite substructure A' of A, the restriction of B' to P is isomorphic to A', and for every $n < \omega$, if B' contains an n-pair which labels the sequence $\bar{a}$, then $B' \vDash R_n(\bar{a})$.

It is easily checked that $\mathbf{K}$ has HP and JEP. For AP we can consider the case where D_1 and D_2 are in $\mathbf{K}$, their intersection is C, and each of D_1 and D_2 has just one element (d_1, d_2 respectively) which is not in C. If both d_1

and d_2 satisfy $P(x)$, then since $\mathbf{J}$ has AP, we can form within $\mathbf{J}$ an amalgam of the restrictions of D_1 and D_2 to P. Then we can expand this into an amalgam of D_1 and D_2 within $\mathbf{K}$, adding no new elements; by (4.11) we can make sure that the amalgam has no new n-pairs, by arranging that $S(x, y, d_1, d_2)$ never holds. (This is why we included c_i and a_m in (4.11).) Easier arguments cover the other cases. Thus $\mathbf{K}$ has HP, JEP and AP. Since $\mathbf{J}$ is countable up to isomorphism, so is $\mathbf{K}$.

Let M be the Fraïssé limit of $\mathbf{K}$, and let B be the reduct $M|L$. I leave it to the reader to check that the restriction of M to P is ultrahomogeneous of age $\mathbf{J}$, and hence is isomorphic to A; let us identify it with A. Whenever $A \vDash R_n(\bar{a})$ holds, the weak homogeneity of M ensures that there is an n-pair in A which labels $\bar{a}$; and conversely by the choice of $\mathbf{K}$, if some n-pair in M labels $\bar{a}$ then $A \vDash R_n(\bar{a})$. So the relations R_n^A are all definable in B. Finally each R_n^A is an orbit of $\text{Aut}(A)$ and hence also of $\text{Aut}(M)$, since A and M are ultrahomogeneous and share the same language on P^M. $\square$

Exercises for section 7.4

1. Show that if M is a countable ω-categorical structure, then there is a definitional expansion of M which is ultrahomogeneous. [Atomise.]

2. Let L be the first-order language whose signature consists of one 2-ary relation symbol R. For each integer $n \geqslant 2$, let A_{n-2} be the L-structure with domain n, such that $A_n \vDash \neg R(i, j)$ iff $i + 1 \equiv j \pmod{n}$. If S is any infinite subset of ω, we write $\mathbf{J}_S$ for the class $\{A_n : n \in S\}$ and $\mathbf{K}_S$ for the class of all finite L-structures C such that no structure in $\mathbf{J}_S$ is embeddable in C. Show (a) for each infinite $S \subseteq \omega$, the class $\mathbf{K}_S$ has HP, JEP and AP, and is uniformly locally finite, (b) if S and S' are distinct infinite subsets of ω then the Fraïssé limits of $\mathbf{K}_S$ and $\mathbf{K}_{S'}$ are non-isomorphic countable ω-categorical L-structures, (c) if L is the signature with one binary relation symbol, there are continuum many non-isomorphic ultrahomogeneous ω-categorical structures of signature L.

3. Let A be a countable ω-categorical L-structure. Show that there is a linear ordering $<$ of $\text{dom}(A)$ in the order-type of the rationals, such that $(A, <)$ is ω-categorical. [We can assume $\text{Th}(A)$ is model-complete. Let B be the substructure of A whose domain is the algebraic closure of the empty set; define a linear ordering $<^*$ on $\text{dom}(B)$. Let $\mathbf{K}$ be the class of all structures $(C, <)$ where C is a finite substructure of A with algebraically closed domain, and $<$ is a linear ordering of $\text{dom}(C)$ which agrees with $<^*$ on B. Show that $\mathbf{K}$ is uniformly locally finite and has JEP and AP. Then use Exercise 7.1.11.]

4. We define a graph A on the set of vertices $\{c_i : i < \omega\}$ as follows. When $i < j$, first write j as a sum of distinct powers of 2, and then make c_i adjacent to c_j iff 2^i occurs in this sum. Show that A is the random graph.

5. Show that if Γ is the random graph, then the vertices of Γ can be listed as $\{v_i: i < \omega\}$ in such a way that for each i, v_i is joined to v_{i+1}.

6. Show that if the set of vertices of the random graph Γ is partitioned into finitely many sets X_i $(i < n)$, then there is some $i < n$ such that the restriction of Γ to X_i is isomorphic to Γ.

The **complete graph on** n **vertices,** K_n, *is the graph with n vertices, such that vertex v is joined to vertex w iff $v \neq w$.*

7. Let n be an integer ≥ 3 and let $\mathbf{K}_n$ be the class of finite graphs which do not have K_n as a subgraph. (a) Show that $\mathbf{K}_n$ has HP, JEP and AP. (b) If Γ_n is the Fraïssé limit of $\mathbf{K}_n$, show that Γ_n is the unique countable graph with the following two properties: (i) every finite subgraph of Γ_n is in $\mathbf{K}_n$, and (ii) if X and Y are disjoint finite sets of vertices of Γ_n, and K_{n-1} is not embeddable in the restriction of Γ_n to X, then there is a vertex in Γ_n which is joined to every vertex in X and to no vertex in Y.

8. Show that if G is either the random graph Γ or the graph Γ_n for some $n \geq 3$, then G contains a set X of vertices such that every permutation of X extends to a unique automorphism of G.

*9. A **total cycle** on a countable structure A is an infinite cyclic permutation which moves all the elements of A. Show that the random graph Γ has 2^ω conjugacy classes of total cycles. [For every subset s of $\omega \backslash \{0\}$, there is a graph A_s with vertex set $\{v_i: i \in \mathbb{Z}\}$ such that v_i is joined to v_j iff $|i - j| \in s$. Show that the set $\{s: A_s \cong \Gamma\}$ is a Borel set of measure 1 in 2^ω. For each such s, choose an isomorphism from A_s to Γ, inducing a total cycle g_s on Γ; show that if $s \neq t$ then g_s is not conjugate to g_t.]

10. Show that the zero–one law (Theorem 7.4.7) still holds if the signature L is empty.

11. Let T be $\mathrm{Th}(\Gamma)$ where Γ is the random graph. Show that T has the independence property and hence is unstable. *See section 6.7 for these notions.*

*12. Let L be a first-order language whose signature is a finite non-empty set of relation symbols. For each $n < \omega$, let $\mathbf{K}_n$ be the set of isomorphism types of n-element L-structures. For any sentence ϕ of L, let $\nu_n(\phi)$ be the number of structures in $\mathbf{K}_n$ which are models of ϕ, divided by $|\mathbf{K}_n|$. Show that for every sentence ϕ of L, $\lim_{n \to \infty} \nu_n(\phi) = \lim_{n \to \infty} \mu_n(\phi)$.

Theorem 7.4.7 fails if the language has function symbols.

13. Let L be the first-order language whose signature consists of one 1-ary function symbol F. Show that $\lim_{n \to \infty} \mu_n(\forall x \, F(x) \neq x) = 1/e$. $[\mu_n(\forall x \, F(x) \neq x) = (n-1)^n/n^n.]$

*14. (a) Show that there is a group G of permutations of ω with the following properties. (i) For every finite subset X of ω there are distinct elements a, b, c of $\omega \backslash X$, and elements g, h of $G_{(X)}$, such that $g(a) = b$, $g(b) = c$, $h(b) = c$, $h(c) = a$. (ii) For every finite subset X of ω, $G_{(X)} = G_{\{X\}}$. (iii) For any two finite subsets X, Y of

ω, $\langle G_{(X)}, G_{(Y)} \rangle = G_{(X \cap Y)}$. [Let L be the language consisting of n-ary function symbols F_n for each positive integer n. Let $\mathbf{K}$ be the class of all finite L-structures A such that for each n, $F_n^A(a_0, \ldots, a_{n-1}) = F_n^A(a_{\pi(0)}, \ldots, a_{\pi(n-1)}) \in \{a_0, \ldots, a_{n-1}\}$ whenever $a_0, \ldots, a_{n-1}$ are distinct elements of A (and, say, $F_n^A(a_0, \ldots, a_{n-1}) = a_0$ otherwise). Show that $\mathbf{K}$ has HP, JEP, AP, and take G to be the automorphism group of the Fraïssé limit.] (b) With G as in (a), show that there are no linear ordering $<$ of ω and finite set $X \subseteq \omega$ such that every permutation in $G_{(X)}$ preserves $<$.

15. Let L be the first-order language whose symbols are, for each positive integer n, a $2n$-ary relation symbol E_n. Let $\mathbf{K}$ be the class of finite L-structures A such that for each n, E_n^A is an equivalence relation on the class of all n-element subsets of $\mathrm{dom}(A)$, with at most two equivalence classes. (a) Show that $\mathbf{K}$ has HP, JEP, AP and is uniformly locally finite. (b) Writing A for the Fraïssé limit of $\mathbf{K}$, show that the product of countably many cyclic groups of order 2 is a homomorphic image of $\mathrm{Aut}(A)$. (c) Show that A is a countable ω-categorical structure which doesn't have the small index property. [Exercise 7.3.15 may help.]

*16. (a) Write $\mathrm{Sym}(n)$ for the symmetric group on n letters, and let G be the product $\Pi_{0 < n < \omega} \mathrm{Sym}(n)$ as abstract group. Show that there is a countable ω-categorical structure A with a reduct A' such that $\mathrm{Aut}(A)$ is normal in $\mathrm{Aut}(A')$ and $\mathrm{Aut}(A')/\mathrm{Aut}(A) \cong G$. [Let L be the language with a $2n$-ary relation symbol E_n for each positive integer n, and let $\mathbf{K}$ be the class of finite L-structures B in which each E_n defines an equivalence relation on the set of n-element subsets of $\mathrm{dom}(B)$. Take A' to be the Fraïssé limit of $\mathbf{K}$, and A to be A' with extra symbols to fix the equivalence classes of each E_n.] (b) Show that if H is a closed subgroup of G, then with the same A as before, there is a reduct A'' of A such that $\mathrm{Aut}(A'')/\mathrm{Aut}(A) \cong H$.

*17. Let F be a finite field and V a vector space of countable dimension over F. Show that there is a linear ordering $<$ of V such that the automorphism group of $(V, <)$ is transitive on the set of n-dimensional subspaces, for each $n < \omega$. [Choose an arbitrary linear ordering of F, and let $\mathbf{K}$ be the set of structures $(W, <)$ where W is a finite-dimensional space over F and $<$ is the lexicographic ordering of W with respect to some basis of W. Show that $\mathbf{K}$ has HP, JEP and AP, and consider the Fraïssé limit.]

History and bibliography

Section 7.1. Fraïssé proved versions of Theorems 7.1.1, 7.1.2 and 7.1.7 in his [1954]. He restricted himself to the case where the language has just finitely many relation symbols and no constant or function symbols, but his arguments generalise straightforwardly. The name 'ultrahomogeneous' is from Woodrow [1979] with a slightly different definition. The rest is probably folklore. Droste & Göbel [199?] extract the category-theoretic content of Fraïssé's construction and give some applications in group theory and the theory of Scott domains. Kueker & Laskowski [199?] consider another generalisation of Fraïssé's construction.

Exercise 8: Schmerl [1980], Cameron [1990] p. 37. Exercise 9: A. Robinson [1971a]. Exercise 10: Bryars [1973].

Section 7.2. The results of this section are really all applications of the Baire category theorem for compact Hausdorff spaces, though the individual proofs never need to mention this fact. One can make the translation from category to games by way of the Banach–Mazur theorem; see Oxtoby [1971] Chapter 6. Rasiowa & Sikorski [1950] used Baire's theorem to prove the completeness theorem for first-order logic in a countable language. Theorem 7.2.1 is due to Ehrenfeucht (in Vaught [1961], and for complete types of a complete theory); Vaught [1961] used it to deduce Theorem 7.2.2. Vaught also ascribes the name 'atomic model' and a form of Theorem 7.2.3 to Svenonius (unpublished). Theorem 7.2.4 is from Grilliot [1972] and Corollary 7.2.5 from Shelah [1978a] Theorem IV.5.16.

Exercises 3–5, 10 and 11 are all from the seminal paper of Vaught [1961]. Exercise 6: for more on minimal and prime models, see Marcus [1972] and Pillay [1978]. Exercise 8: Baldwin, Blass, Glass & Kueker [1973]. Exercise 13: Chang & Keisler [1973] Ex. 2.2.4. Exercise 15: After giving a solution on p. 141 of my [1985], I found Shelah [1978a] Exercise IV.5.8 already knew the same solution; my apologies to him. See Newelski [1987] for a fuller discussion.

Section 7.3. Theorem 7.3.1 was proved by Engeler [1959], Ryll-Nardzewski [1959] and Svenonius [1959a]; the conditions involving automorphisms are due to Svenonius. Theorem 7.3.7 is from Ahlbrandt & Ziegler [1986]. Hodkinson & Macpherson [1988] characterise an extreme form of ω-categoricity: countable structures determined up to isomorphism by their age.

Besides the papers mentioned elsewhere in this section, the following papers have material on identifying the ω-categorical structures in various classes of structure. (1) Groups: Apps [1982], [1983a], [1983c], Baur, Cherlin & Macintyre [1979], Cherlin & Rosenstein [1978], Felgner [1977], [1978], Macpherson [1988]. (2) Rings: Macintyre & Rosenstein [1976], Baldwin & Rose [1977]. (3) Modules: Baur [1975]. (4) Linear orderings: Rosenstein [1969]. (5) Partial orderings: Schmerl [1977], [1984]. (6) Geometry: Wiegand [1978]; cf. Cameron [1990] p. 137.

Exercise 5: Grzegorczyk [1968]. Exercise 7: Waszkiewicz & Węglorz [1969]. Exercise 8: Rosenstein [1973]. Exercise 9: Wilson [1982]. Exercise 10: Rosenstein [1976]. Exercise 11: Cherlin [1980a, 1980b]. Exercise 12: Wilson

[1982] deduces this from the previous exercise. Exercise 13: Macpherson [1983]. Exercise 14: Ahlbrandt & Ziegler [1986]. Exercise 15: D. Evans in Hodges [1989]. Exercise 16: (a) D. Evans (unpublished); (b) Hodges, Hodkinson & Macpherson [1990].

Section 7.4. The random graph Γ is perhaps the simplest example of Fraïssé's construction. But the first explicit descriptions of Γ seem to be in Erdős & Rényi [1963], who constructed it as in Theorem 7.4.5, and Rado [1964] who found it as in Exercise 5. Henson [1971] showed that Rado's graph is the Fraïssé limit of the class of finite graphs, and gave Exercises 5–8. Truss [1985] studied the automorphism group of Γ. Bouscaren [1989a] used Γ in her characterisation of superstable theories with DOP. See Cameron [1990] for further information and references on Aut (Γ).

The zero–one law, Theorem 7.4.7, is from Glebskiĭ, Kogan, Liogon'kiĭ & Talanov [1969], and independently Fagin [1976]. See also Grandjean [1983] on complexity aspects, Blass, Gurevich & Kozen [1985] for an extension of the law to fixed-point logic, Kaufmann & Shelah [1985] for the failure of the law for monadic second-order logic, and Shelah & Spencer [1988] on zero–one laws for graphs constructed probabilistically.

Example 1 was given by Cameron [1990] p. 126 to show that Cantor's proof of the isomorphism of countable dense orderings without endpoints (which was forth only, not back-and-forth as in Example 3 of section 3.2) won't work for all ω-categorical theories. Theorem 7.4.8 is due to Hrushovski (unpublished). Hrushovski has recently used Fraïssé-like constructions to solve several major problems of stability theory – which unfortunately lie beyond the scope of this book.

Exercise 2: Henson [1972], Peretyat'kin [1973] (see also Ash [1971], Ehrenfeucht [1972], Glassmire [1971]). Exercise 3: Schmerl [1980]. Exercises 5–8: Henson [1971]. Exercise 9: Cameron [1984]. Exercises 12, 13: Fagin [1976]. Exercise 14: Läuchli [1964], as part of a proof that the axiom of choice for finite sets doesn't imply that every set is linearly orderable; this proof of (a) is from Pincus [1976]. Exercise 15: The structure is described in an example of Cherlin in Lascar [1982]; Hrushovski pointed out that a similar example lacks the small index property. Exercise 16: Evans and Hewitt [1990] use this in their construction of two countable ω-categorical structures whose automorphism groups are isomorphic as abstract groups but not as topological groups. Exercise 17: Thomas [1986].

8

The existential case

J'ay trouvé quelques élémens d'une nouvelle caracteristique, tout à fait différent de l'Algebre, et qui aura des grands avantages pour representer à l'esprit exactement et au naturel, quoyque sans figures, tout ce qui depend de l'imagination. ... L'utilité principale consiste dans les consequences et raisonnements, qui se peuvent fair par les operations des caracteres, qui ne sçauroient exprimer par des figures (et encor mois par des modelles). ... Cette caracteristique servira beaucoúp à trouver de belles constructions, parceque le calcul et la construction s'y trouvent tout à la fois.

Leibniz, Letters to Huygens (1679).

Abraham Robinson referred to these letters of Leibniz in his address to the International Congress of Mathematicians in 1950. It seemed to Robinson that Leibniz hinted at Robinson's own ambition: to make logic useful for 'actual mathematics, more particularly for the development of algebra and ... algebraic geometry'.

Robinson introduced many of the fundamental notions of model theory. Besides diagrams (section 1.4), preservation of formulas (section 2.5) and nonstandard methods (section 11.4), he gave us model-completeness, model companions and model-theoretic forcing (all in this chapter). These three notions have a flavour in common. Each of them allows us to develop a tidy piece of pure model theory for its own sake, and at the same time each of them fits neatly onto some piece of 'actual mathematics'. Often they allow us to give tidier and more conceptual proofs of known mathematical theorems.

Of course not everything in this chapter is the work of Abraham Robinson. The property of quantifier elimination harks back to the work of Skolem and Tarski in the years after the first world war. Existentially closed structures are one of those happy ideas which occur to several different mathematicians at different times and places, beginning with those Renaissance scholars who invented imaginary numbers.

The ideas in this chapter are perhaps the main methods of applied model theory. I have given plenty of examples along the way; but to understand these things properly one should apply the methods vigorously to some concrete branch of mathematics. In the appendix sections A.4 and A.5 point to some applications in group theory and field theory. As Leibniz says, 'L'utilité principale consiste dans les consequences'.

Why was Leibniz so happy to avoid 'modelles'? I imagine he meant wooden or paper maquettes. There's not much scope for such things in our kind of mathematics, but Peretyat'kin has some splendid working models of finitely axiomatisable uncountably categorical theories.

8.1 Existentially closed structures

Take a first-order language L without relation symbols and a class **K** of L-structures. For example L might be the language of rings and **K** the class of fields; or L might be the language of groups and **K** the class of groups. We say that a structure A in **K** is **existentially closed in K** (or more briefly, **e.c. in K**) if the following holds:

(1.1)   if E is a finite set of equations and inequations with para-
          meters from A, and E has a simultaneous solution in some
          extension B of A with B in **K**, then E has a solution already
          in A.

We can rewrite this definition in a model-theoretic form. The model-theoretic version has the advantage that it covers languages which have relation symbols too.

A formula is **primitive** if it has the form $\exists \bar{y} \bigwedge_{i<n} \psi_i(\bar{x}, \bar{y})$, where n is a positive integer and each formula ψ_i is a literal. (In a language without relation symbols, each literal is either an equation or an inequation; so a primitive formula expresses that a certain finite set of equations and inequations has a solution.) We say that a structure A in the class **K** of L-structures is **e.c. in K** if

(1.2)   for every primitive formula $\phi(\bar{x})$ of L and every tuple $\bar{a}$ in
          A, if there is a structure B in **K** such that $A \subseteq B$ and
          $B \vDash \phi(\bar{a})$, then already $A \vDash \phi(\bar{a})$.

This definition agrees with (1.1).

When **K** is the class of fields, an e.c. structure in **K** is known as an **e.c. field**; likewise **e.c. lattice** when **K** is the class of lattices, and so on. When **K** is the class of all models of a theory T, we refer to e.c. structures in **K** as **e.c. models of** T. (**Warning**: in the literature, 'e.c.' sometimes means 'equationally compact', which is quite a different notion. Cf. section 10.7 below.)

The next lemma explains the name 'existentially closed'. If L is a first-order language and A, B are L-structures, we write $A \leqslant_1 B$ to mean that for every existential formula $\phi(\bar{x})$ of L and every tuple $\bar{a}$ in A, $B \vDash \phi(\bar{a})$ implies $A \vDash \phi(\bar{a})$.

Lemma 8.1.1. *Let* **K** *be a class of L-structures and A a structure in* **K**. *Then A is e.c. in* **K** *if and only if* (1.2) *holds with 'primitive' replaced by 'existential'. In particular if A and B are structures in* **K**, $A \subseteq B$ *and A is e.c., then $A \leqslant_1 B$.*

Proof. By disjunctive normal form, every $\exists_1$ formula is logically equivalent to a disjunction of primitive formulas. $\qquad\qquad\qquad\qquad\qquad\qquad\qquad\qquad\qquad\quad\square$

Example 1: *Algebraically closed fields.* What are the e.c. fields? Certainly an e.c. field A must be algebraically closed. For suppose $F(y)$ is a polynomial of positive degree with coefficients from A. Using the language of rings, we can rewrite $F(y)$ as $p(\bar{a}, y)$ where $p(\bar{x}, y)$ is a term and $\bar{a}$ is a tuple of elements of A. Replacing F by an irreducible factor if necessary, we have a field $B = A[y]/(F)$ which extends A and contains a root of F. So $B \vDash \exists y \, p(\bar{a}, y) = 0$. Since A is an e.c. field, $A \vDash \exists y \, p(\bar{a}, y) = 0$ too, so that F has a root already in A. Thus every e.c. field is algebraically closed.

The converse holds as well:

(1.3)   if A is an algebraically closed field, then every finite system
         of equations and inequations over A which is solvable in
         some field extending A already has a solution in A.

This is one form of Hilbert's Nullstellensatz; cf. Exercise 11 for another form. If $E(\bar{x})$ is a finite system of equations and inequations over A, then we can write the statement 'E has a solution' as a primitive formula $\phi(\bar{a})$ where $\bar{a}$ are the coefficients of E in A. To prove (1.3), suppose E has a solution in some field B which extends A. Extend B to an algebraically closed field C (for example by Jacobson [1980] Theorem 8.1). Then $B \vDash \phi(\bar{a})$ by assumption, so $C \vDash \phi(\bar{a})$ since $\phi(\bar{a})$ is an $\exists_1$ formula (see Theorem 2.4.1), and hence $A \vDash \phi(\bar{a})$ since the theory of algebraically closed fields is model-complete (see Example 2 in section 8.4, or Theorem A.5.1 in the Appendix). Thus E already has a solution in A. In short, *the e.c. fields are precisely the algebraically closed fields.*

A remark of Rabinowitsch is worth recalling here. Note first that every equation with parameters $\bar{a}$ from a field A can be written as $p(\bar{a}, \bar{y}) = 0$, where $p(\bar{x}, \bar{y})$ is a polynomial whose indeterminates are variables from $\bar{x}$, $\bar{y}$, with integer coefficients. So we can assume that any primitive formula has the form

(1.4) $\exists \bar{y}(p_0(\bar{a}, \bar{y}) = 0 \wedge \ldots \wedge p_{k-1}(\bar{a}, \bar{y}) = 0$
         $\wedge \; q_0(\bar{a}, \bar{y}) \neq 0 \wedge \ldots \wedge q_{m-1}(\bar{a}, \bar{y}) \neq 0).$

In a field, $x \neq 0$ says the same as $\exists z \, x \cdot z = 1$. Hence (and this is Rabinowitsch's observation) we can eliminate the inequations in (1.4), bringing the new existential quantifiers $\exists z$ forward to join $\exists \bar{y}$. This reduces (1.4) to the form

(1.5) $$\exists \bar{y}(p_0(\bar{a}, \bar{y}) = 0 \wedge \ldots \wedge p_{k-1}(\bar{a}, \bar{y}) = 0).$$

In short: to show that a field A is existentially closed, we only need to consider *equations* in the definition (1.1), not both equations and inequations. But this is just a happy fact about fields; in general one has to live with the inequations in (1.1).

Are there interesting e.c. models of other theories besides that of fields? Yes, certainly there are. In the next section we shall see two very general methods for proving the existence of e.c. models. But before that, let me note that we have already found a bunch of e.c. models in Chapter 7.

Example 2: *Fraïssé limits*. Let L be a countable signature and $\mathbf{J}$ a countable set of finitely generated L-structures which has HP, JEP and AP (see section 7.1). Let $\mathbf{K}$ be the class of all L-structures with age $\subseteq \mathbf{J}$, and let A be the Fraïssé limit of $\mathbf{K}$. Then A is existentially closed in $\mathbf{K}$. For suppose B is in $\mathbf{K}$, $A \subseteq B$, $\bar{a}$ is a tuple of elements of A and $\psi(\bar{x}, \bar{y})$ is a conjunction of literals of L such that $B \vDash \exists \bar{y} \, \psi(\bar{a}, \bar{y})$. Take a tuple $\bar{b}$ in B such that $B \vDash \psi(\bar{a}, \bar{b})$; let C be the substructure of A generated by $\bar{a}$, and D the substructure of B generated by $\bar{a}\bar{b}$. Then $C \subseteq D$, and both C and D are in the age $\mathbf{J}$ of A. Since A is weakly homogeneous (see (1.7) in section 7.1), the inclusion map $f: C \to A$ extends to an embedding $g: D \to A$. So $A \vDash \psi(\bar{a}, g\bar{b})$, and hence $A \vDash \exists \bar{y} \, \psi(\bar{a}, \bar{y})$ as required. It's not hard to show that if $\mathbf{J}$ above is a set of finite structures, then the Fraïssé limit of $\mathbf{J}$ is the unique countable e.c. structure in $\mathbf{K}$ (see Exercise 1).

The influence of JEP and AP

Example 2 above is interesting because it's untypical. JEP and AP are quite strong properties, and we shall see that there are plenty of examples of e.c. structures in classes where the finitely generated structures fail to have one or other of these properties. Both JEP and AP exert a strong influence on the behaviour of e.c. structures. This is a convenient moment to prove · two theorems which illustrate the point.

If T is a first-order theory, we say that T has the **joint embedding property** (JEP) if, given any two models A, B of T, there is a model of T in which both A and B are embeddable.

Theorem 8.1.2. *Let L be a first-order language and T a theory in L. Suppose that T has JEP, and let A, B be e.c. models of T. Then every $\forall_2$ sentence of L which is true in A is true in B too.*

Proof. By assumption there is a model C of T in which both A and B are embeddable; we lose nothing by assuming that A and B are substructures of C. Since A is e.c. in the class of models of T, we have $A \leqslant_1 C$. It follows by the existential amalgamation theorem (Theorem 6.5.1) that there is an elementary extension D of A with $C \subseteq D$. Suppose $A \vDash \forall \bar{x} \exists \bar{y} \, \phi(\bar{x}, \bar{y})$ where ϕ is a quantifier-free formula of L, and let $\bar{b}$ be a tuple of elements of B. We must show that $B \vDash \exists \bar{y} \, \phi(\bar{b}, \bar{y})$. But $D \vDash \exists \bar{y} \, \phi(\bar{b}, \bar{y})$ since $A \leqslant D$. Now B is an e.c. model, so that $B \leqslant_1 D$, whence $B \vDash \exists \bar{y} \, \phi(\bar{b}, \bar{y})$ as required. $\qquad \square$

Theorem 8.1.3. *Let L be a first-order language. Let $\mathbf{K}$ be a class of L-structures which is closed under isomorphic copies, and suppose that the class of all substructures of structures in $\mathbf{K}$ has AP. Then for every $\exists_1$ formula $\phi(\bar{x})$ of L there is a quantifier-free formula $\chi(\bar{x})$ (possibly infinitary) which is equivalent to ϕ in all e.c. structures in $\mathbf{K}$. In particular if there is a first-order theory T such that the e.c. structures in $\mathbf{K}$ are exactly the models of T, then ϕ is equivalent modulo T to a quantifier-free formula of L.*

Proof. Let us say that a pair $(A, \bar{a})$ is **good** if A is an e.c. structure in $\mathbf{K}$, $\bar{a}$ is a tuple of elements of A, and $A \vDash \phi(\bar{a})$. For each good pair $(A, \bar{a})$, let $\theta_{(A, \bar{a})}(\bar{x})$ be the conjunction $\bigwedge \{ \psi(\bar{x}) : \psi$ is a literal of L and $A \vDash \psi(\bar{a}) \}$. Let $\chi(\bar{x})$ be the disjunction of all the formulas $\theta_{(A, \bar{a})}$ where $(A, \bar{a})$ ranges over good pairs. Clearly if B is any e.c. structure in $\mathbf{K}$ and $\bar{b}$ a tuple in B such that $B \vDash \phi(\bar{b})$, then $B \vDash \chi(\bar{b})$ since $(B, \bar{b})$ is a good pair. Conversely if B is an e.c. structure in $\mathbf{K}$ and $B \vDash \chi(\bar{b})$, then there is a good pair $(A, \bar{a})$ with an isomorphism $e : \langle \bar{a} \rangle_A \to B$ taking $\bar{a}$ to $\bar{b}$. By the amalgamation property for the class of substructures of structures in $\mathbf{K}$, there is a structure C in $\mathbf{K}$ with embeddings $g : A \to C$ and $h : B \to C$ such that $g\bar{a} = h\bar{b}$. Since $\mathbf{K}$ is closed under isomorphic copies, we can suppose that h is an inclusion map. Now $A \vDash \phi(\bar{a})$ by assumption, so $C \vDash \phi(g\bar{a})$ since ϕ is an $\exists_1$ formula. It follows that $B \vDash \phi(\bar{b})$ since B is e.c. in $\mathbf{K}$.

Finally suppose that the models of T are exactly the e.c. structures in $\mathbf{K}$. Then $T \vdash \forall \bar{x} (\phi \leftrightarrow \chi)$. Two applications of the compactness theorem reduce χ to a first-order quantifier-free formula. $\qquad \square$

Warning. Theorem 8.1.3 is false if we exclude empty structures. Consider the theory $\forall xy (P(x) \leftrightarrow P(y))$, and let ϕ be the formula $\exists x \, P(x)$. Readers who do exclude empty structures can rescue the theorem by requiring ϕ to have at least one free variable. See Exercise 7 for more details.

Theorem 8.1.3 gives useful information. Let $\mathbf{K}$ be the class of fields. Thanks to Example 1, we already know that the class of e.c. fields is first-order axiomatisable. It's a well-known fact of algebra that $\mathbf{K}$ has the amalgamation property (for example by taking composites as in Cohn [1977] section 6.10, or by the uniqueness of algebraic closures as in Jacobson [1980] section 8.1). From this we easily deduce that the class of integral domains has the amalgamation property too (by taking fields of fractions). But in the signature of rings, a substructure of a field is the same thing as an integral domain. Hence we have the following.

Corollary 8.1.4. *Let T be the theory of algebraically closed fields. Then T has quantifier elimination.*

Proof. By Theorem 8.1.3, the argument above shows that every $\exists_1$ formula is equivalent to a quantifier-free formula modulo T. This suffices, by Lemma 2.3.1. $\qquad\square$

We shall come back to applications of Theorem 8.1.3 in section 8.4 below.

Exercises for section 8.1

1. Let L be a signature with just finitely many symbols, and $\mathbf{J}$ an at most countable set of finite (NB: not just finitely generated) L-structures which has HP, JEP and AP. Let $\mathbf{K}$ be the class of all L-structures whose age is $\subseteq \mathbf{J}$. If C is a finite or countable structure in $\mathbf{K}$, show that C is existentially closed in $\mathbf{K}$ if and only if C is the Fraïssé limit of $\mathbf{J}$.

2. Show that there is a unique countable e.c. linear ordering, namely the ordering of the rationals.

3. Show (a) an e.c. integral domain is the same thing as an e.c. field, (b) an e.c. field is never an e.c. commutative ring.

4. Show that if A is an e.c. abelian group, then A is divisible and has infinitely many p-ary direct summands for each prime p. [Don't try to prove there are infinitely many torsion-free direct summands – it's false!]

5. Give an example of a first-order theory with e.c. models, where there are e.c. models which satisfy different quantifier-free sentences. [Try the theory of fields.]

6. Let T be the theory $\forall x \neg(\exists y\, Rxy \wedge \exists y\, Ryx)$. Describe the e.c. models of T. Give an example of an $\exists_1$ first-order formula which is not equivalent to a quantifier-free formula in e.c. models of T.

We say that a class $\mathbf{K}$ has **amalgamation over non-empty structures** *if the amalgamation property holds whenever the structures involved are all non-empty.*

7. In Theorem 8.1.3, suppose that we weaken the assumption of AP to amalgamation

over non-empty structures. Show that the conclusion still holds, provided that we require ϕ to have at least one free variable.

The following example is a freak, but I include it to prevent careless conjectures.
8. Let T be complete first-order arithmetic, i.e. the first-order theory of the natural numbers with symbols 0, 1, $+$, $\cdot$. Show that the natural numbers form an e.c. model of T.

*9. Show that every e.c. group is ultrahomogeneous; in fact every isomorphism between finitely generated subgroups extends to an inner automorphism. [This follows directly from HNN extensions; see Lyndon & Schupp [1977].]

10. Show that if A and B are e.c. commutative rings, then so is $A \times B$.

11. Using purely algebraic arguments, show the equivalence between (1.3) above and the following form of Hilbert's Nullstellensatz. Suppose A is an algebraically closed field, I is an ideal in the polynomial ring $A[x_0, \ldots, x_{n-1}]$ and $p(x_0, \ldots, x_{n-1})$ is a polynomial $\in A[x_0, \ldots, x_{n-1}]$ such that for all $\bar{a}$ in A, if $q(\bar{a}) = 0$ for all $q \in I$ then $p(\bar{a}) = 0$. Then for some positive integer k, $p^k \in I$. [$\Rightarrow$: if not, there is a prime ideal P of $A[x_0, \ldots, x_{n-1}]$ which contains I and not p; consider the field of fractions of the integral domain $A[x_0, \ldots, x_{n-1}]/P$. $\Leftarrow$: use Rabinowitsch to eliminate the inequations, and turn the equations into generators of an ideal I of $A[x_0, \ldots, x_{n-1}]$; if the equations have no solution in A then there is no $\bar{a}$ in A such that $q(\bar{a}) = 0$ for all $q \in I$.]

8.2 Two methods of construction

In this section we describe two ways of finding e.c. structures. The first applies in a very general setting, and it doesn't tell us much about the structure that we have built. The second applies only to $\forall_2$ first-order theories, but it allows much better control over the properties of the structure.

First construction: e.c. structures in inductive classes

Our first existence proof is based on the case of fields, and more precisely on Steinitz's proof that every field has an algebraically closed extension.

Let L be a signature and **K** a class of L-structures. We say that **K** is **inductive** if (1) K is closed under taking unions of chains, and (2) (for tidiness) every structure isomorphic to a structure in **K** is also in **K**.

For example if T is an $\forall_2$ first-order theory in L and **K** is the class of all models of T, then Theorem 2.4.4 says that **K** is inductive. Thus the class of all groups is inductive, as is the class of all fields.

Local properties give us some more examples. Let π be a structural property which an L-structure might have. We say that an L-structure A **has** π **locally** if all the finitely generated substructures of A have property π. Then the class of all L-structures which have π locally is an inductive class. Thus for example a structure is **locally finite** if all its finitely generated substructures are finite. The class of all locally finite groups is inductive. Likewise the class of all groups without elements of infinite order is inductive. Neither of these two classes is first-order axiomatisable.

Theorem 8.2.1. *Let* **K** *be an inductive class of L-structures and A a structure in* **K**. *Then there is an e.c. structure B in* **K** *such that $A \subseteq B$.*

Proof. We begin by showing a weaker result: there is a structure A^* in **K** such that $A \subseteq A^*$, and if $\phi(\bar{x})$ is an $\exists_1$ formula of L, $\bar{a}$ a tuple in A and there is a structure C in **K** such that $A^* \subseteq C$ and $C \vDash \phi(\bar{a})$, then $A^* \vDash \phi(\bar{a})$.

List as $(\phi_i, a_i)_{i<\lambda}$ all pairs $(\phi, \bar{a})$ where ϕ is an $\exists_1$ formula of L and $\bar{a}$ is a tuple in A. By induction on i, define a chain of structures $(A_i : i \leq \lambda)$ in **K** by

(2.1) $A_0 = A$,

(2.2)   when δ is a limit ordinal $\leq \lambda$, $A_\delta = \bigcup_{i<\delta} A_i$,

(2.3) $A_{i+1} = \begin{cases} \text{some structure } C \text{ in } \mathbf{K} \text{ such that } A_i \subseteq C \\ \quad \text{and } C \vDash \phi_i(\bar{a}_i), \text{ if there is such a structure } C, \\ A_i \text{ otherwise.} \end{cases}$

Put $A^* = A_\delta$. To show that A^* is as required, take any $\exists_1$ formula $\phi(\bar{x})$ of L and any tuple $\bar{a}$ of elements of A. Then $(\phi, \bar{a})$ is $(\phi_i, \bar{a}_i)$ for some $i < \lambda$. Suppose C is a structure in **K** such that $A^* \subseteq C$ and $C \vDash \phi(\bar{a})$. Then $A_i \subseteq C$, and so $A_{i+1} \vDash \phi(\bar{a})$ by (2.3). Since ϕ is an $\exists_1$ formula and $A_{i+1} \subseteq A^*$, we infer by Theorem 2.4.1 that $A^* \vDash \phi(\bar{a})$.

Now define a chain of structures $A^{(n)}$ $(n < \omega)$ in **K** as follows, by induction on n:

(2.4) $A^{(0)} = A$,

(2.5) $A^{(n+1)} = A^{(n)*}$.

Put $B = \bigcup_{n<\omega} A^{(n)}$. Then B is in **K** since **K** is inductive, and certainly $A \subseteq B$. Suppose $\phi(\bar{x})$ is an $\exists_1$ formula of L and $\bar{a}$ is a tuple from B, such that $C \vDash \phi(\bar{a})$ for some C which is in **K** and extends B. Since $\bar{a}$ is finite, it lies within $A^{(n)}$ for some $n < \omega$. Hence $A^{(n+1)} \vDash \phi(\bar{a})$ since $A^{(n+1)}$ is $A^{(n)*}$. But ϕ is an $\exists_1$ formula and $A^{(n+1)} \subseteq B$, so that again it follows that $B \vDash \phi(\bar{a})$. $\square$

Often we can say more about the size of the e.c. structure.

Corollary 8.2.2. *Let* **K** *be an inductive class, A a structure in* **K** *and λ an infinite cardinal $\geq |A|$. Suppose also that for every structure C in* **K** *and every*

set X of $\leqslant \lambda$ elements of C, there is a structure B in $\mathbf{K}$ such that $X \subseteq \operatorname{dom}(B)$, $B \leqslant C$ and $|B| \leqslant \lambda$. (For example, suppose $\mathbf{K}$ is the class of all models of some $\forall_2$ theory in a first-order language of cardinality $\leqslant \lambda$.) Then there is an e.c. structure B in $\mathbf{K}$ such that $A \subseteq B$ and $|B| \leqslant \lambda$.

Proof. Counting the number of pairs $(\phi, \bar{a})$, we find that we can use λ as the cardinal λ in the proof of the theorem. In (2.3) of that proof, choose each structure A_{i+1} so that it has cardinality $\leqslant \lambda$. For example if there is a structure C in $\mathbf{K}$ such that $A_i \subseteq C$ and $C \vDash \phi_i(\bar{a}_i)$, choose A_{i+1} in $\mathbf{K}$ so that $A_{i+1} \leqslant C$, $\operatorname{dom}(A_i) \subseteq \operatorname{dom}(A_{i+1})$ and $|A_{i+1}| \leqslant \lambda$. Then $A_i \subseteq A_{i+1}$ and $A_{i+1} \vDash \phi_i(\bar{a}_i)$ as required. With these choices, A^* also has cardinality $\leqslant \lambda$, and hence so does B in the theorem. $\qquad\qquad\square$

E.c. models of first-order theories

Theorem 8.2.1 and its corollary more or less exhaust what one can say about e.c. structures in arbitrary inductive classes. So for the rest of this section we turn to the most interesting case, namely the e.c. models of an $\forall_2$ first-order theory T.

In section 6.5, $T_\forall$ was defined to be the set of all $\forall_1$ sentences ϕ of L such that $T \vdash \phi$. By Corollary 6.5.3, every model of $T_\forall$ can be extended to a model of T.

Corollary 8.2.3. *Let L be a first-order language, T an $\forall_2$ theory in L and A an infinite model of $T_\forall$. Then there is an e.c. model B of T such that $A \subseteq B$ and $|B| = \max(|A|, |L|)$.*

Proof. By Corollary 6.5.3 there is a model C of T such that $A \subseteq C$. By the downward Löwenheim–Skolem theorem we can take C to be of cardinality $\max(|A|, |L|)$. The rest follows from Corollary 8.2.2. $\qquad\qquad\square$

There is a wealth of information about e.c. models of $\forall_2$ first-order theories. Our next result takes us straight to the inner personality of these models. It says that in such a model, no $\exists_1$ formula is ever false unless some true $\exists_1$ formula compels it to be. We shall need to know this for our second construction of e.c. models.

Theorem 8.2.4. *Let L be a first-order language, T an $\forall_2$ theory in L and A a model of T. Then the following are equivalent.*
(a) A is an e.c. model of T.
(b) A is a model of $T_\forall$, and for every $\exists_1$ formula $\phi(\bar{x})$ of L and every tuple $\bar{a}$

in A, if $A \vDash \neg \phi(\bar{a})$ then there is an $\exists_1$ formula $\chi(\bar{x})$ of L such that $A \vDash \chi(\bar{a})$ and $T \vdash \forall \bar{x}(\chi \to \neg \phi)$.

(c) A is an e.c. model of $T_\forall$.

Proof. (a) $\Rightarrow$ (b). Assume (a). Then certainly A is a model of $T_\forall$. Suppose $\phi(\bar{x})$ is an $\exists_1$ formula of L and $\bar{a}$ is a tuple in A such that $A \vDash \neg \phi(\bar{a})$. Then since A is an e.c. model of T, it follows that there is no model C of T such that $A \subseteq C$ and $C \vDash \phi(\bar{a})$. Add distinct new constants $\bar{c}$ to the language as names for the elements of $\bar{a}$. Since there is no model C as described, the diagram lemma (Lemma 1.4.2) tells us that the theory

(2.6) $\mathrm{diag}(A) \cup T \cup \{\phi(\bar{c})\}$

has no model. So by the compactness theorem (Theorem 6.1.1) there are a tuple $\bar{d}$ of distinct elements of A and a quantifier-free formula $\theta(\bar{x}, \bar{y})$ of L such that $A \vDash \theta(\bar{c}, \bar{d})$ and

(2.7) $T \vdash \theta(\bar{c}, \bar{d}) \to \neg \phi(\bar{c})$.

Now we use the lemma on constants (Lemma 2.3.2), noting that even if the tuple $\bar{a}$ contains repetitions, we had the foresight to introduce a tuple $\bar{c}$ of distinct constants. The result is that $T \vdash \forall \bar{x}(\exists \bar{y}\, \theta(\bar{x}, \bar{y}) \to \neg \phi(\bar{x}))$. To infer (b), let χ be $\exists \bar{y}\, \theta$.

(b) $\Rightarrow$ (c). Assume (b). Then A is a model of $T_\forall$. Suppose $\phi(\bar{x})$ is an $\exists_1$ formula of L and $\bar{a}$ is a tuple in A such that for some model C of $T_\forall$, $A \subseteq C$ and $C \vDash \phi(\bar{a})$. We must show that $A \vDash \phi(\bar{a})$. If not, then by (b) there is an $\exists_1$ formula $\chi(\bar{x})$ of L such that $A \vDash \chi(\bar{a})$ and $T \vdash \forall \bar{x}(\chi \to \neg \phi)$. Since χ is $\exists_1$ and $A \subseteq C$, we have $C \vDash \chi(\bar{a})$. The sentence $\forall \bar{x}(\chi \to \neg \phi)$ is $\forall_1$ (after some trivial rearrangement), and so it lies in $T_\forall$. Since C is a model of $T_\forall$, we infer that $C \vDash \neg \phi(\bar{a})$; contradiction.

(c) $\Rightarrow$ (a). Assume (c). First we must show that A is a model of T. Since T is an $\forall_2$ theory, a typical sentence in T can be written $\forall \bar{x} \exists \bar{y}\, \psi(\bar{x}, \bar{y})$ with ψ quantifier-free. Let $\bar{a}$ be any tuple in A; we must show that $A \vDash \exists \bar{y}\, \psi(\bar{a}, \bar{y})$. By Corollary 6.5.3, since $A \vDash T_\forall$ there is a model C of T such that $A \subseteq C$. Then $C \vDash \exists \bar{y}\, \psi(\bar{a}, \bar{y})$, and so $A \vDash \exists \bar{y}\, \psi(\bar{a}, \bar{y})$ since A is an e.c. model of $T_\forall$. Thus A is a model of T. It follows easily that A is an e.c. model of T, since every model of T extending A is in fact a model of $T_\forall$ too. $\square$

Second construction: $\forall_2$ theories

Our second construction is commonly known as **forcing**. It yields e.c. models of countable $\forall_2$ first-order theories. It's an adaptation of the proof of the countable omitting theorem (Theorem 7.2.1), which the reader should review.

Henceforth, let L be a countable first-order language and T an $\forall_2$ theory in L. We shall **force with** T. Since we shall be expanding the language with constants, we assume that T has a non-empty model.

Let L^+ be the first-order language which comes from L by adding countably many new constants c_i $(i < \omega)$, known as **witnesses**. We shall define an increasing chain $(T_i : i < \omega)$ of finite sets of atomic or negated atomic sentences of L^+, such that for every i, $T \cup T_i$ has a model. Finite sets S of atomic or negated atomic sentences such that $T \cup S$ has a model are known as **conditions**. The union $\bigcup_{i<\omega} T_i$ is called T^+. Note that every finite subset of T^+ is a condition.

As in the proof of the omitting types theorem, we call on a team of experts, and we give each expert E an infinite subset X_E of ω. When T_{i-1} has been chosen, if $i \in X_E$ then expert E will choose the condition T_i, which must include T_{i-1}. Formally we put $T_{-1} = \varnothing$. Each expert has a task, and she must choose her conditions so that the task is completed by the end of the construction. The tasks include the following.

$(2.8)_{\Psi(\bar{x})}$ For each finite set of Ψ of literals $\psi(\bar{x})$ of L^+, ensure that either $T \cup T^+ \vdash \forall \bar{x} \neg \bigwedge \Psi$, or there are infinitely many pairwise disjoint tuples $\bar{c}$ of distinct witnesses such that $\Psi(\bar{c}) \subseteq T^+$.

The argument to show that each task $(2.8)_\Psi$ can be carried out is very much the same as for $(2.3)_\psi$ in section 7.2. For the moment we impose no other tasks.

Let U be the set of atomic sentences in T^+, and let A^+ be the canonical model of U. We write A for the reduct $A^+|L$. We call A (or A^+) the **compiled structure**.

Theorem 8.2.5. *In forcing with an $\forall_2$ theory T, the compiled structure A is an existentially closed model of T.*

Proof. The proof is a little more fussy than for the omitting types theorem; one has to be more careful about the complexity of formulas.

We note at once that every element of A is named by infinitely many witnesses. This follows from the tasks $(2.8)_{\{\psi\}}$ where ψ is the formula $x = t$ for a closed term t, bearing in mind that every element of A is named by a closed term of L^+.

Next we note that U is an $=$-closed set. For example if T^+ contains the sentences $\phi(c)$ and $c = d$, then $T \cup T^+ \nvdash \neg\phi(d)$ (otherwise it would be inconsistent), and hence $\phi(d) \in T^+$ by task $(2.8)_{\{\phi(d)\}}$. It follows that

(2.9)   for every atomic sentence θ of L^+, $A^+ \vDash \theta \Leftrightarrow \theta \in T^+$,

by the definition of canonical models (see section 1.5).

To prove that A is an e.c. model of T, we use Theorem 8.2.4(b). The first step is to show that $A \vDash T_\forall$. For this, suppose $T \vdash \forall \bar{x} \chi$ where χ is quantifier-free; after putting χ into conjunctive normal form, we can suppose without loss that χ is of the form $\bigvee \Theta$ for some set Θ of literals of L. If $A \vDash \neg \forall \bar{x} \, \chi$, then there is a tuple $\bar{c}$ of distinct witnesses such that $A \vDash \bigwedge \{\neg \theta(\bar{c}) \colon \theta \in \Theta\}$. It follows that $\{\neg \theta(\bar{c}) \colon \theta \in \Theta\} \subseteq T^{+}$ by (2.9). Hence $\{\neg \theta(\bar{c}) \colon \theta \in \Theta\}$ is a condition, contradicting that $T \vdash \forall \bar{x} \, \chi$.

Finally let $\exists \bar{y} \, \phi(\bar{x}, \bar{y})$ be an $\exists_1$ formula of L (with ϕ quantifier-free), and $\bar{c}$ a tuple of distinct witnesses. We must show that if $A^{+} \vDash \neg \exists \bar{y} \, \phi(\bar{c}, \bar{y})$ then there is an $\exists_1$ formula $\chi(\bar{x})$ of L such that $A \vDash \chi(\bar{a})$ and $T \vdash \forall \bar{x}(\chi \to \neg \exists \bar{y} \, \phi)$. By disjunctive normal form (or Exercise 3 below) we can assume that ϕ is of the form $\bigwedge \Psi$ where Ψ is a conjunction of literals. Assuming that $A^{+} \vDash \neg \exists \bar{y} \, \phi(\bar{c}, \bar{y})$, it follows that $\Psi(\bar{c}, \bar{e}) \nsubseteq T^{+}$ for all tuples $\bar{e}$ of witnesses, and so by task $(2.8)_{\Psi(\bar{c}, \bar{y})}$ we know that

$$(2.10) \qquad\qquad T \cup T^{+} \vdash \forall \bar{y} \, \neg \bigwedge \Psi(\bar{c}, \bar{y}).$$

By the compactness theorem it follows from (2.10) that there is a conjunction $\sigma(\bar{c}, \bar{d})$ of sentences in T^{+}, with $\bar{d}$ listing the distinct witnesses not in $\bar{c}$, such that

$$(2.11) \qquad\qquad T \vdash \sigma(\bar{c}, \bar{d}) \to \neg \exists \bar{y} \, \phi(\bar{c}, \bar{y}).$$

Our required formula χ is $\exists \bar{z} \, \sigma(\bar{x}, \bar{z})$, using (2.11) and the lemma on constants. $\qquad\qquad\qquad\qquad\qquad\qquad\qquad\qquad\qquad\qquad\qquad\square$

When we gave the experts their portfolios, we said only that their tasks should *include* the tasks (2.8). This leaves room for some experts with other tasks. Let π be a property which the compiled structure A^{+} might have. We say that π is **enforceable** if an expert E, given an infinite subset X_E of ω for her workspace, can guarantee that A^{+} will have property π, regardless of what happens at other steps of the construction. (As in section 7.2, one only needs to check whether the expert can carry out this task when X_E is the set of odd positive integers.) For example, we say that a sentence ϕ of $L_{\omega_1 \omega}^{+}$ is **enforceable** if the expert can guarantee that $A^{+} \vDash \phi$.

Just as in the setting of section 7.2, the conjunction of countably many enforceable properties is enforceable.

One could paraphrase Theorem 8.2.5 as 'The property "A is an e.c. model of T" is enforceable'. Here is another important example.

Theorem 8.2.6 (*Omitting types theorem for $\forall$-types*). *Let L be a countable first-order language and T an $\forall_2$ theory in L (with non-empty models). Let $\Phi(\bar{x})$ be a set of $\forall_1$ formulas of L, such that there is no $\exists_1$ formula $\psi(\bar{x})$ of L satisfying the two conditions*

$$(2.12) \qquad\qquad T \cup \{\exists \bar{x} \, \psi\} \text{ has a model,}$$

$$(2.13) \qquad\qquad T \vdash \forall \bar{x}(\psi \to \phi) \text{ for every formula } \phi(\bar{x}) \text{ in } \Phi.$$

Then it is enforceable that there is no tuple of elements of A satisfying $\bigwedge \Phi$.

Proof. The proof uses exactly the same technology as the proofs of Theorem 7.2.1 and Theorem 8.2.5, and so I leave it as Exercise 9. □

We say that a model B of T is **enforceable** if the property 'The compiled model A is isomorphic to B' is enforceable. In this case the forcing construction can only yield one model, up to isomorphism. An extreme case is where there is only one countable e.c. model up to isomorphism – as for example with the theory of linear orderings. It can also happen that a theory has an enforceable model together with several other countable e.c. models; see Exercise 11 below. Many theories have no enforceable model; see Exercise 12 below.

Exercises for section 8.2

1. If **K** is an inductive class of L-structures, and **J** is the class of e.c. structures in **K**, show that **J** is closed under unions of chains.

2. A **near-linear space** is a structure with two kinds of elements, 'points' and 'lines', and a symmetric binary relation of 'incidence' relating points and lines in such a way that (i) any line is incident with at least two points, and (ii) any two distinct points are incident with at most one line. The near-linear space is a **projective plane** if (ii') any two distinct points are incident with exactly one line, and moreover (iii) any two distinct lines are incident with at least one point, and (iv) there is a set of four points, no three of which are all incident with a line. (a) Show that the class of near-linear spaces is inductive. (b) Show that every e.c. near-linear space is a projective plane. (c) Deduce that every near-linear space can be embedded in a projective plane.

3. Show that in Theorem 8.2.4(b) we can replace '$\exists_1$' by 'primitive' (both times).

4. Let L be a first-order language and T an $\forall_2$ theory in L. (a) Show that if B is a model of T and $A \leqslant_1 B$, then A is a model of T. (b) Show that if B is an e.c. model of T and $A \leqslant_1 B$, then A is an e.c. model of T.

5. Let L be a first-order language, T an $\forall_2$ theory in L and A an L-structure. Show that the following are equivalent. (a) A is an e.c. model of T. (b) A is a model of $T_\forall$, and for every model C of T such that $A \subseteq C$, we have $A \leqslant_1 C$. (c) For some e.c. model B of T, $A \leqslant_1 B$.

6. Let L be a countable first-order language and T an $\forall_2$ theory in L. Show that there is a sentence ϕ of $L_{\omega_1\omega}$ such that the models of ϕ are exactly the e.c. models of T.

7. Let L be a first-order language and T an $\forall_2$ theory in L. (a) Show that if A is an e.c. model of T and B is a model of T with $A \subseteq B$, then for every $\forall_2$ formula $\phi(\bar{x})$ of

L and every tuple $\bar{a}$ in A, if $B \models \phi(\bar{a})$ then $A \models \phi(\bar{a})$. (b) Show that if B in (a) is also an e.c. model of T then the same holds with $\forall_3$ in place of $\forall_2$.

8. Let L be a first-order language and U an $\forall_2$ theory in L. Show that among the $\forall_2$ theories T in L such that $T_\forall = U_\forall$, there is a unique maximal one under the ordering $\subseteq$. Writing U_0 for this maximal T, show that U_0 is the set of those $\forall_2$ sentences of L which are true in every e.c. model of U. (U_0 is known as the **Kaiser hull** of U.)

9. Prove Theorem 8.2.6.

10. Let L be a finite relational signature and $\mathbf{K}$ a non-empty class of L-structures which has HP, JEP and AP. Let T be the theory of structures whose age is $\subseteq \mathbf{K}$. Show that the Fraïssé limit of $\mathbf{K}$ is an enforceable model of T.

The next exercise illustrates a case where there is an enforceable model, but not every countable e.c. model is enforceable.
11. Let T be the theory of fields of a fixed characteristic. Show that T has an enforceable model, namely the algebraic closure of the prime field. [Consider the expert responsible for this task. If she wants to make the element c algebraic while the condition Φ is facing her, she should start by showing that Φ is satisfiable in the algebraic closure of the prime field. The Nullstellensatz guarantees this.]

And here is a theory with no enforceable model.
12. Let L be the signature consisting of infinitely many 1-ary relation symbols P_i ($i < \omega$), and let T be the empty theory. Show that T has no enforceable model. [For each set $X \subseteq \omega$, some rival expert can ensure that there is no element a such that $A \models P_i(a) \Leftrightarrow i \in X$. This is a special case of Theorem 8.2.6.]

*A model A of a theory T is said to be **simple** if for every homomorphism $h: A \to B$ with B a model of T, either h is an embedding or h is a constant map.*
13. Let L be a first-order language (not necessarily countable) and T an $\forall_2$ theory in L. (a) Show that if every model of T is embeddable in a simple model, then every e.c. model of T is simple. [Given models A, B of T with A e.c., and a homomorphism $h: A \to B$ and elements a_1, a_2 of A such that $h(a_1) \neq h(a_2)$, suppose $\phi(\bar{y})$ is atomic and $A \models \neg \phi(\bar{c})$; required to show $B \models \neg \phi(h\bar{c})$. Let $C \supseteq A$ be a simple model of T. Then $T \cup \mathrm{diag}^+(C) \cup \{a_1 \neq a_2, \phi(\bar{c})\}$ has no model. Now use compactness and the fact that A is e.c.] (b) Deduce that every e.c. group is simple. [See W. Scott [1964] 11.5.4 for embedding any group into a simple group.]

*For the remaining exercises, L is a countable first-order language and T is an $\forall_2$ theory in L which has a non-empty model. Forcing is always for T. We use letters p, q etc. for conditions. We say that a condition p **forces** a sentence ϕ of L^+ if an expert who chooses T_i for all odd i can guarantee that A^+ will be a model of ϕ provided that $p \subseteq T_0$. Thus a sentence is enforceable iff it is forced by the empty condition.*
*14. (a) Show that if p forces ϕ and $T \vdash \phi \to \psi$ then p forces ψ. (b) Show that if ϕ is

an $\forall_1$ sentence of L^+, then p forces ϕ if and only if $T \cup p \vdash \phi$. (c) Show that for every formula $\phi(\bar{x})$ of $L_{\omega_1\omega}$ and every tuple $\bar{c}$ of distinct witnesses not occurring in p, p forces $\forall \bar{x} \phi$ if and only if p forces $\phi(\bar{c})$. (d) Show that p forces $\phi \wedge \psi$ if and only if p forces ϕ and p forces ψ. (e) Show that for every sentence ϕ of L, p forces $\neg \phi$ if and only if there is no condition $q \supseteq p$ which forces ϕ.

The set of enforceable sentences of L is known as the **finite forcing companion** *of T, in symbols T^f.*

*15. Show that if T has JEP then T^f is a complete theory.

*16. Suppose T has JEP, and let ϕ be an $\forall_3$ sentence of L which is not enforceable. Show that ϕ is false in all e.c. models of T.

*17. We force with the theory of groups. Show that if H is a finitely generated group with unsolvable word problem, then it is enforceable that H is not embeddable in the compiled group. [Use Theorem 8.2.6. Let $\Phi(\bar{x})$ be the set of all equations and inequations true of the generators of H. If $\phi(\bar{x})$ is an $\exists_1$ formula satisfying (2.12) and (2.13), then we can solve the word problem of H by listing all equations and inequations $E(\bar{x})$ deducible from $T \cup \{\phi\}$.]

8.3 Model-completeness

In the early 1950s Abraham Robinson noticed that certain maps studied by algebraists are in fact elementary embeddings. If you choose a map at random, the chances of its being an elementary embedding are negligible. So Robinson reckoned that there must be a systematic reason for the appearance of these elementary embeddings, and he set out to find what the reason was. In the course of his quest he introduced the notions of model-complete theories, companionable theories and model companions. These notions have become essential tools for the model theory of algebra. In this section we shall examine them.

Model-completeness

In section 2.6 we defined a theory T in a first-order language L to be **model-complete** if every embedding between L-structures which are models of T is elementary.

Theorem 8.3.1. *Let T be a theory in a first-order language L. The following are equivalent.*
(a) *T is model-complete.*
(b) *Every model of T is an e.c. model of T.*
(c) *If L-structures A, B are models of T and $e: A \to B$ is an embedding then there are an elementary extension D of A and an embedding $g: B \to D$ such that ge is the identity on A.*

(d) *If $\phi(\bar{x}, \bar{y})$ is a formula of L which is a conjunction of literals, then $\exists \bar{y}\, \phi$ is equivalent modulo T to an $\forall_1$ formula $\psi(\bar{x})$ of L.*

(e) *Every formula $\phi(\bar{x})$ of L is equivalent modulo T to an $\forall_1$ formula $\psi(\bar{x})$ of L.*

Proof. (a) $\Rightarrow$ (b) is immediate from the definition of e.c. model.

(b) $\Rightarrow$ (c). Assume (b). Let $e\colon A \to B$ be an embedding between models of T, and let $\bar{a}$ be a sequence listing all the elements of A. Then $(B, e\bar{a}) \Rrightarrow_1 (A, \bar{a})$ since A is an e.c. model of T. (See section 6.5 for the notion $\Rrightarrow_1$.) The conclusion of (c) follows by the existential amalgamation theorem, Theorem 6.5.1.

(c) $\Rightarrow$ (d). We first claim that if (c) holds, then every embedding between models of T preserves $\forall_1$ formulas of L. For if $e\colon A \to B$ is such an embedding, $\bar{a}$ is in A and $\phi(\bar{x})$ is an $\forall_1$ formula of L such that $A \vDash \phi(\bar{a})$, then taking D and g as in (c) we have $D \vDash \phi(ge\bar{a})$ and so $B \vDash \phi(e\bar{a})$ since ϕ is an $\forall_1$ formula. This proves the claim. It follows by Corollary 6.5.5, taking negations, that every $\exists_1$ formula $\phi(\bar{x})$ of L is equivalent modulo T to an $\forall_1$ formula $\psi(\bar{x})$ of L; hence (d) holds.

(d) $\Rightarrow$ (e). Assume (d), and let $\phi(\bar{x})$ be any formula of L; we can assume that ϕ is in prenex form, say as $\exists \bar{x}_0 \forall \bar{x}_1 \ldots \forall \bar{x}_{n-2} \exists \bar{x}_{n-1} \theta_n(\bar{x}_0, \bar{x}_1, \ldots, \bar{x}_{n-1}, \bar{x})$ where θ_n is quantifier-free. By (d), $\exists \bar{x}_{n-1} \theta_n(\bar{x}_0, \bar{x}_1, \ldots, \bar{x}_{n-1}, \bar{x})$ is equivalent modulo T to a formula $\forall \bar{z}_{n-1} \theta_{n-1}(\bar{x}_0, \bar{x}_1, \ldots, \bar{x}_{n-2}, \bar{z}_{n-1}, \bar{x})$ with θ_{n-1} quantifier-free, and so ϕ is equivalent to $\exists \bar{x}_0 \forall \bar{x}_1 \ldots \forall \bar{x}_{n-2} \bar{z}_{n-1} \theta_{n-1}(\bar{x}_0, \bar{x}_1, \ldots, \bar{x}_{n-2}, \bar{z}_{n-1}, \bar{x})$. By (d) again, taking negations, the formula $\forall \bar{x}_{n-2} \bar{z}_{n-1} \theta_{n-1}(\bar{x}_0, \bar{x}_1, \ldots, \bar{x}_{n-2}, \bar{z}_{n-1}, \bar{x})$ is equivalent modulo T to a formula $\exists \bar{z}_{n-2} \theta_{n-2}(\bar{x}_0, \bar{x}_1, \ldots, \bar{z}_{n-2}, \bar{x})$ with θ_{n-2} quantifier-free, so that ϕ is equivalent modulo T to $\exists \bar{x}_0 \forall \bar{x}_1 \ldots \exists \bar{x}_{n-3} \bar{z}_{n-2} \theta_{n-2}(\bar{x}_0, \bar{x}_1, \ldots, \bar{z}_{n-2}, \bar{x})$. After n steps in this style, all the quantifiers will have been gathered up into a universal quantifier $\forall \bar{z}_0$.

(e) $\Rightarrow$ (a) follows from the fact (Corollary 2.4.2(a)) that $\forall_1$ formulas are preserved in substructures. $\qquad\square$

Corollary 8.3.2 (Robinson's test). *For the first-order theory T to be model-complete, it is necessary and sufficient that if A and B are any two models of T with $A \subseteq B$ then $A \preccurlyeq_1 B$.*

Proof. This is a restatement of (b) in the theorem. $\qquad\square$

The next fact is elementary, but it should be mentioned.

Theorem 8.3.3. *Let T be a model-complete theory in a first-order language L. Then T is equivalent to an $\forall_2$ theory in L.*

Proof. Every chain of models of T is elementary, and so its union is a model of T by Theorem 2.5.2. Hence T is equivalent to an $\forall_2$ theory by the Chang–Łoś–Suszko theorem, Theorem 6.5.9.   □

Ways of proving model-completeness

How does one show in practice that a theory is model-complete? It's not a property that can be checked by inspection; serious mathematical work may be involved.

One method is already to hand. Let L be a first-order language and T a theory in L. We say that T **has quantifier elimination** if for every formula $\phi(\bar{x})$ of L there is a quantifier-free formula $\phi^*(\bar{x})$ of L which is equivalent to ϕ modulo T. Every quantifier-free formula is an $\forall_1$ formula. So by condition (e) in Theorem 8.3.1, *every theory with quantifier elimination is model-complete*.

This observation gives us plenty of examples, thanks to the quantifier elimination technique of section 2.7 above. Thus the theory of infinite vector spaces over a fixed field is model-complete (Exercise 2.7.9); so is the theory of real-closed fields in the language of ordered fields (Theorem 2.7.2); and so is the theory of dense linear orderings without endpoints (Theorem 2.7.1).

This method is to hand, but it is also rather heavy. However, there are other methods. One of the simplest – when it applies – is the approach of Per Lindström. Recall that a theory is λ-**categorical** if it has models of cardinality λ and all its models of cardinality λ are isomorphic.

Theorem 8.3.4 *(Lindström's test). Let L be a first-order language and T an $\forall_2$ theory in L which has no finite models. If T is λ-categorical for some cardinal $\lambda \geqslant |L|$ then T is model-complete.*

Proof. We use Robinson's test. Suppose $\lambda \geqslant |L|$ and T is λ-categorical, and for contradiction assume that T has models A and B such that $A \subseteq B$, but there are an $\exists_1$ formula $\phi(\bar{x})$ of L and a tuple $\bar{a}$ in A such that $A \vDash \neg\phi(\bar{a})$ and $B \vDash \phi(\bar{a})$. Extend L to a language L^+ by adding a new 1-ary relation symbol P, and expand B to an L^+-structure B^+ by interpreting P as $\mathrm{dom}(A)$. Let T^+ be $\mathrm{Th}(B^+)$. Note that T^+ contains the sentence

(3.1) $$\exists\bar{x}(\phi(\bar{x}) \wedge \neg\phi^P(\bar{x}))$$

by the relativisation theorem (Theorem 5.1.1).

Since $|L^+| \leqslant \lambda$ and T has no finite models, an easy argument with the compactness theorem (see e.g. Exercise 6.1.10) shows that T^+ has a model D^+ of cardinality λ such that $|P^{D^+}| = \lambda$. Let D be $D^+|L$. Since $D^+ \equiv B^+$, there is an L-structure C with domain P^{D^+} such that $C \subseteq D$. Using the relativisation theorem, C is a model of T of cardinality λ. But by Corollary

8.2.3 there is an e.c. model of T of cardinality λ, and so C is an e.c. model of T since T is λ-categorical. It follows that for every tuple $\bar{c}$ in C, if $D \vDash \phi(\bar{c})$ then $C \vDash \phi(\bar{c})$. This contradicts the fact that D^+ is a model of (3.1). $\qquad\square$

For example the theory of dense linear orderings without endpoints ((2.32) in section 2.2) is $\forall_2$ by inspection and ω-categorical by Example 3 in section 3.2. So by Lindström's test it is model-complete.

Some wise soul said that one should learn the proofs of theorems, not their statements. Maybe he had in mind Lindström's test. The following example shows that the test needs rather less than λ-categoricity to make it work.

Example 1: *Universal 1-ary functions.* Let L be the first-order language whose signature consists of one 1-ary function symbol F. The theory T in L shall say the following.

(3.2)   There is at least one element. There are no loops (i.e. for all positive integers n and all elements x, $F^n(x) \neq x$). Every element x is of the form $F(y)$ for infinitely many elements y.

This can be illustrated thus:

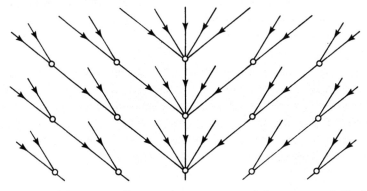

In any model of T, we call two elements a and b **connected** if there are positive integers m and n such that $F^m(a) = F^n(b)$. This is an equivalence relation, and its equivalence classes are the **connected components** of the model. A countable connected component is determined up to isomorphism. But there are countable models of T in which there are any number $\leq \omega$ of connected components, and so T is not ω-categorical. One easily sees that it isn't λ-categorical for any uncountable cardinal λ either.

Now T is an $\forall_2$ theory (for example because it is closed under union of chains; see Theorem 6.5.9). So the unique countable model with infinitely many connected components must be e.c., since it is not embeddable in any other countable model. Now read through the proof of Theorem 8.3.4, making one change: the two models C, D are to be countable, and we use compactness to ensure that C has infinitely many connected components.

Then C must be e.c. and so $C \leqslant_1 D$ as before. It follows that T is model-complete.

Completeness

A proof of model-completeness can sometimes be the first step towards a proof of completeness.

Let L be a first-order language and T a theory in L. Following Baldwin & Kueker [1981], let us say that a model A of T is an **algebraically prime** model if A is embeddable in every model of T. (Abraham Robinson said simply 'prime', but this terminology clashes with the notion of a prime model in stability theory.)

Theorem 8.3.5. *Let L be a first-order language and T a theory in L. If T is model-complete and has an algebraically prime model then T is complete.*

Proof. Let A and B be any two models of T, and C an algebraically prime model of T. Then C is embeddable in A and in B. Since A and C are both models of T and T is model-complete, it follows that $C \leqslant A$. Similarly $C \leqslant B$. It follows that $A \equiv C \equiv B$. ☐

The results above say nothing about whether the theory T is decidable. But there is an easy general argument to show that if T is complete and has a recursively enumerable set of axioms, then T is decidable – we simply list the consequences of T and wait for the relevant sentence or its negation to turn up (see Exercise 6.1.16). One should avoid trying to impress number theorists with this kind of algorithm: it is hopelessly inefficient.

Model companions

Let T be a theory in a first-order language L. We say that a theory U in L is a **model companion** of T if

(3.3) U is model-complete,

(3.4)   every model of T has an extension which is a model of U, and

(3.5)   every model of U has an extension which is a model of T.

(By Corollary 6.5.3, (3.4) and (3.5) together are equivalent to the equation $T_\forall = U_\forall$.) Maybe T has no model companion; if it does have one, we say that T is **companionable**.

Our next theorem shows that model companions are intimately connected with e.c. models.

Theorem 8.3.6. *Let T be an* $\forall_2$ *theory in a first-order language L.*

(a) *T is companionable if and only if the class of e.c. models of T is axiomatisable by a theory in L.*

(b) *If T is companionable, then up to equivalence of theories, its model companion is unique and is the theory of the class of e.c. models of T.*

Proof. Suppose first that T is companionable, with a model companion T'. We show that the e.c. models of T are precisely the models of T'. First assume A is a model of T'. Then by Theorem 8.2.1 some extension B of A is an e.c. model of T. Since T' is a model companion of T, some extension C of B is a model of T', and $A \leqslant C$. If $\phi(\bar{x})$ is an $\exists_1$ formula of L and $\bar{a}$ is a tuple of elements of A such that $B \vDash \phi(\bar{a})$, then $C \vDash \phi(\bar{a})$ since $B \subseteq C$, and hence $A \vDash \phi(\bar{a})$ since $A \leqslant C$. So $A \leqslant_1 B$, and it follows (see Exercise 8.2.4(b)) that A is an e.c. model of T. Conversely if A is an e.c. model of T, then some extension B of A is a model of T'. Since T' is equivalent to an $\forall_2$ theory (by Theorem 8.3.3), it easily follows that A is a model of T' (see Exercise 8.2.7(a)).

This proves (b) and left to right in (a). For the other direction in (a), suppose that the class of e.c. models of T is the class of all models of a theory U in L. Then (3.4) and (3.5) hold, so that $T_\forall = U_\forall$. Since the class of e.c. models of T is closed under unions of chains, we can assume by the Chang–Łoś–Suszko theorem (Theorem 6.5.9) that U is an $\forall_2$ theory. Every model A of U is an e.c. model of T, and hence of $T_\forall$ and of U by two applications of Theorem 8.2.4. By Theorem 8.3.1(a $\Leftrightarrow$ b) it follows that U is model-complete. So U is a model companion of T. □

In practice, one often finds that the easiest way to show that a theory T is not companionable is to show that some e.c. model has an elementary extension which is not e.c. See Example 1 and Exercise 2 in section 8.5 below.

The classic example of a model companion is the theory of algebraically closed fields, which is the model companion of the theory of fields. (In fact it is the model completion, a stronger notion which we shall meet in the next section.) Abraham Robinson hoped that the notion would be useful for identifying classes of structure which play the role of algebraically closed fields in other branches of algebra. Its best achievements seem to be in the model theory of fields; see section A.5 below.

From this point onwards, the study of model-completeness rapidly slips into specialised problems in this or that area of algebra. Macintyre [1977] is a very readable account of work in this style up to the mid 1970s.

Exercises for section 8.3

1. Show that if L is a first-order language, T and U are theories in L, $T \subseteq U$ and T is model-complete, then U is also model-complete.

2. Show that if a first-order theory T is model-complete and has the joint embedding property, then T is complete.

3. In the first-order language whose signature consists of one 1-ary function symbol F, let T be the theory which consists of the sentences $\forall x\, F^n(x) \neq x$ (for all positive integers n) and $\forall x \exists_{=1} y\, F(y) = x$. Apply Lindström's test to show that T is model-complete.

4. Let T be the theory of (non-empty) linear orderings in which each element has an immediate predecessor and an immediate successor, in a language with relation symbols $<$ for the ordering and $S(x, y)$ for the relation 'y is the immediate successor of x'. Show that T can be written as an $\forall_2$ theory. Show by Lindström's test that T is model-complete. Deduce that T is complete.

5. Give an example of a theory T in a first-order language L, such that T is not model-complete but every complete theory in L containing T is model-complete.

6. Suppose T is a theory in a first-order language, and every completion of T is equivalent to a theory of the form $T \cup U$ for some set U of $\exists_1$ sentences. Suppose also that every completion of T is model-complete. Show that T is model-complete.

7. Give an example of a theory T in a countable first-order language, such that T has 2^ω pairwise non-isomorphic algebraically prime models. [Let Ω be a signature consisting of countably many 1-ary function symbols. Write a theory T which says that for each $i < \omega$, the set of elements satisfying $P_i(x)$ is a term algebra of Ω. In a model A of T, consider the number of components of each P_i^A.]

*8. Show that the theory of divisible abelian groups with infinitely many elements of each finite order is the model companion of the theory of abelian groups. [Model-completeness can be proved by the generalised version of Lindström's test.]

There are many examples of non-companionable theories, but most of them depend on substantial mathematics. Here is an elementary example.

9. Let T be the theory which says the following. All elements satisfy exactly one of $P(x)$ and $Q(x)$; for every element a satisfying $P(x)$ there are unique elements b and c such that $R(b, a)$ and $R(a, c)$; $R(x, y)$ implies $P(x)$ and $P(y)$; there are no finite R-cycles; $S(x, y)$ implies $P(x)$ and $Q(y)$; if $R(x, y)$ and $Q(z)$ then $S(x, z)$ iff $S(y, z)$. Show that in an e.c. model of T, elements a, b satisfying $P(x)$ are connected by R if and only if there is no element c such that $S(a, c) \leftrightarrow \neg S(b, c)$. Deduce that T has e.c. models with elementary extensions which are not e.c., and hence that T is not companionable.

10. Give an example of an ω_1-categorical theory T in a countable first-order language L, such that no definitional expansion of T by adding finitely many symbols is model-complete. [$T = \text{Th}(A)$, where $\text{dom}(A) = \omega$ and A carries equivalence relations E_i $(i < \omega)$ as follows: the classes of E_i are $\{0, \ldots, 2^{i+1} - 2\}$ and $\{(2^{i+1} - 1) + 2^{i+1}k, \ldots, (2^{i+1} - 1) + 2^{i+1}(k + 1) - 1\}$ $(k < \omega)$.]

11. Give an example of an ω_1-categorical countable first-order theory which is not companionable. [First consider the following structure A. The elements are the pairs (m, n) of natural numbers with $m \geq -1$ and $n \geq 0$. There is a 1-ary relation symbol *Diagonal* picking out the elements (m, m). There are two symmetric 2-ary relation symbols *Horizontal* and *Vertical*. The relation *Horizontal* is an equivalence relation; its classes are the sets $\{(m, b): m \geq -1\}$ with b fixed. The relation *Vertical* holds between (m, n) and (m, m) whenever $m, n \geq 0$. Note that A has elementary extensions B and C such that $B \subseteq C$ but C is not an elementary extension of B. The required theory is $\text{Th}(A, \bar{a})$ where $\bar{a}$ is a sequence listing all the elements of A. Suppose T is a model companion. Then T says that infinitely many named elements are not paired with other elements by *Vertical*, so every model of T has an elementary extension with new such elements.]

8.4 Quantifier elimination revisited

A theory T in a first-order language L is said to have **quantifier elimination** if every formula $\phi(\bar{x})$ of L is equivalent modulo T to a quantifier-free formula $\psi(\bar{x})$ of L. In section 2.7 we described a procedure which can be used to show that certain theories T have quantifier elimination. The main idea of the procedure is to find a formula which has a certain form and is equivalent modulo T to some given formula.

Abraham Robinson urged a different approach. For many interesting theories T, we have a mass of good structural information about the models of T: for example decomposition theorems, facts about algebraic or other closures of sets of elements, or results about embedding one model in another. It's hard to use these facts in an argument which concentrates on deducibility from T in the first-order predicate calculus.

So Robinson's message was: to prove quantifier elimination, use model theory rather than syntax, when you can fit the model theory onto the known algebraic structure theory. Of course we can only follow suit when we know some model-theoretic criteria for a theory to have quantifier elimination. The next theorem states some.

Criteria for quantifier elimination

If A and B are L-structures, we write $A \equiv_0 B$ to mean that exactly the same quantifier-free sentences of L are true in A as in B; as before, we write

$A \Rrightarrow_1 B$ to mean that for every $\exists_1$ sentence ϕ of L, if $A \vDash \phi$ then $B \vDash \phi$. Recall that $T_\forall$ is the set of $\forall_1$ first-order consequences of T.

We say that a first-order theory T has the **amalgamation property** (AP) if the class **K** of all models of T has the AP; in other words, if the following holds:

(4.1) if A, B, C are models of T and $e: A \to B$, $f: A \to C$ are embeddings, then there are D in **K** and embeddings $g: B \to D$ and $h: C \to D$ such that $ge = hf$.

(Cf. (1.3) in section 7.1.)

Theorem 8.4.1. *Let L be a first-order language and T a theory in L. The following are equivalent.*

(a) *T has quantifier elimination.*

(b) *If A and B are models of T, and $\bar{a}$, $\bar{b}$ are tuples from A, B respectively such that $(A, \bar{a}) \equiv_0 (B, \bar{b})$, then $(A, \bar{a}) \Rrightarrow_1 (B, \bar{b})$.*

(c) *If A and B are models of T, $\bar{a}$ a sequence from A and $e: \langle \bar{a} \rangle_A \to B$ is an embedding, then there are an elementary extension D of B and an embedding $f: A \to D$ which extends e:*

(4.2)

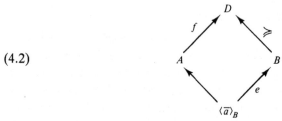

(d) *T is model-complete and $T_\forall$ has the amalgamation property.*

(e) *For every quantifier-free formula $\phi(\bar{x}, y)$ of L, $\exists y\, \phi$ is equivalent modulo T to a quantifier-free formula $\psi(\bar{x})$.*

Proof. (a) $\Rightarrow$ (b) is immediate.

(b) $\Rightarrow$ (c). Assume (b). For every tuple $\bar{a}'$ inside $\bar{a}$, the hypothesis of (c) implies that $(A, \bar{a}') \equiv_0 (B, e\bar{a}')$ and hence $(A, \bar{a}') \Rrightarrow_1 (B, e\bar{a}')$. Since every sentence of $L(\bar{a})$ (L with parameters $\bar{a}$ added) mentions just finitely many elements, it follows that $(A, \bar{a}) \Rrightarrow_1 (B, e\bar{a})$. Hence the conclusion of (c) holds by the existential amalgamation theorem (Theorem 6.5.1).

(c) $\Rightarrow$ (d). Assume (c). If $e: B \to A$ is any embedding between models of T, put $\langle \bar{a} \rangle_A = B$ in (4.2) and deduce that T is model-complete by Theorem 8.3.1(c) (with A, B transposed). To prove that $T_\forall$ has the amalgamation property, let $g: C \to A'$ and $e: C \to B'$ be embeddings between models of $T_\forall$. By Corollary 6.5.3, A' and B' can be extended to models A, B of T respectively. Now apply (c) with the embedding $C \to A$ in place of the inclusion $\langle \bar{a} \rangle_A \subseteq A$.

(d) $\Rightarrow$ (e) is by Theorem 8.1.3, taking **K** to be the class of all models of T, and noting that all models of T are e.c. by Theorem 8.3.1. (e) $\Rightarrow$ (a) is by Lemma 2.3.1. $\qquad\qquad\Box$

Warning. The theorem fails if L has no constants and we require that every L-structure is non-empty. (See the remarks after Theorem 8.1.3.)

An L-structure A is said to be **quantifier-eliminable (q.e.** for short) or to **have quantifier elimination** if A is a model of a theory in L which has quantifier elimination; this is equivalent to saying that Th(A) has quantifier elimination. If the structure A is finite, the corollary below gives a purely algebraic criterion for A to be q.e. As in section 7.1, we say that a structure A is **ultrahomogeneous** if every isomorphism between finitely generated substructures of A extends to an automorphism of A.

Corollary 8.4.2. *Let A be a finite L-structure. Then A is q.e. if and only if A is ultrahomogeneous.*

Proof. Since A is finite, Th(A) says how many elements A has. So in (4.2) with $A = B$, D must be A and f must be an isomorphism. $\qquad\qquad\Box$

Let us put these techniques to work by proving that some theories have quantifier elimination. In our first three illustrations, Lindström's test (Theorem 8.3.4) gives a quick proof that the theory is model-complete, and we use (d) of Theorem 8.4.1 to climb from model-completeness to quantifier elimination. The fourth example has to work harder to prove model-completeness, but it also finishes the argument by quoting (d) of Theorem 8.4.1.

Example 1: *Vector spaces.* Let R be a field and T the theory of infinite (left) vector spaces over R. The axioms of T consist of (2.21) in section 2.2 together with the sentences $\exists_{\geqslant n} x\, x = x$ for each positive integer n; inspection shows that these axioms are $\forall_2$ first-order sentences. Certainly T is λ-categorical for any infinite cardinal $\lambda > |R|$, and by definition T has no finite models. So by Lindström's test, T is model-complete. Elementary algebra shows that $T_\forall$, which is the theory of vector spaces over R, has the amalgamation property. So by (d) of the theorem, T has quantifier elimination.

Now by Corollary 8.4.2, every finite vector space is q.e.; so we have shown that *every vector space is q.e.*

Example 2: *Algebraically closed fields and finite fields.* By a theorem of Steinitz [1910] which we have already invoked as Corollary 4.5.7, any two algebraically closed fields of the same characteristic and the same uncountable cardinality are isomorphic. (**Warning**: our proof of Corollary 4.5.7, unlike Steinitz', used the fact that algebraically closed fields are strongly minimal, which in turn follows from the fact that the theory of algebraically closed fields has quantifier elimination! With algebraically closed fields, everything implies everything else.) Also algebraically closed fields are infinite. So by Lindström's test, the theory of algebraically closed fields of a fixed characteristic is model-complete, whence (see Exercise 8.3.6) the theory of algebraically closed fields is model-complete. We saw already in Corollary 8.1.4 how this implies that every algebraically closed field is q.e.; it's the same proof as by Theorem 8.4.1 (d $\Rightarrow$ e).

Finite fields are q.e. too – in other words, by Corollary 8.4.2, they are ultrahomogeneous. Let A be a finite field; let B and C be subfields of A with $f: B \to C$ an isomorphism. We must show that f extends to an automorphism of A. Since B and C have the same cardinality, their multiplicative groups are the same subgroup of the multiplicative group of A, which is a finite cyclic group; it follows that $B = C$. Thus we need only show that every automorphism of B extends to one of A. We do this by a counting argument. Write D for the prime field of the same characteristic p as B. Then D is rigid, so the Galois group $\mathrm{Gal}(B/D)$ is $\mathrm{Aut}(B)$; likewise $\mathrm{Gal}(A/D)$ is $\mathrm{Aut}(A)$. Write H for the group of all those automorphisms of B which extend an automorphism of A. Then using Galois theory,

$$|\mathrm{Aut}(A)_{(B)}| \cdot H = |\mathrm{Aut}(A)| = (A : D) = (A : B) \cdot (B : D)$$
$$= |\mathrm{Aut}(A)_{(B)}| \cdot |\mathrm{Aut}(B)|.$$

So $|\mathrm{Aut}(B)| = |H|$. Since $H \subseteq \mathrm{Aut}(B)$, this proves that $H = \mathrm{Aut}(B)$, i.e. every automorphism of B extends to one of A.

Example 3: *Universal 1-ary functions*. We introduced the theory T of such functions in section 8.3, and we showed that it's model-complete. A model of $T_\forall$ is simply a 1-ary function without loops. So one easily sees that $T_\forall$ has the amalgamation property, and hence T has quantifier elimination by (d) of the theorem above. Thus every formula of the language L of T is equivalent modulo T to a boolean combination of equations $F^m(x) = F^n(y)$.

Our fourth example needs a more substantial argument and a new heading.

Real-closed fields

The necessary and sufficient conditions of Theorem 8.4.1 are not always the best route to a proof of quantifier elimination. Sometimes a sufficient condition ties in better with the known algebra. For example, we have the following.

Corollary 8.4.3. *Let L be a first-order language and T a theory in L. Suppose that T satisfies the following conditions.*
(a) *For any two models A and B of T, if $A \subseteq B$, $\phi(\bar{x}, y)$ is a quantifier-free formula of L and $\bar{a}$ is a tuple of elements of A such that $B \vDash \exists y\, \phi(\bar{a}, y)$, then $A \vDash \exists y\, \phi(\bar{a}, y)$. ('T is 1-model-complete.')*
(b) *For every model A of T and every substructure C of A there is a model A' of T such that (i) $C \subseteq A' \subseteq A$, and (ii) if B is another model of T with $C \subseteq B$ then there is an embedding of A' into B over C.*
Then T has quantifier elimination.

Proof. Assuming (a) and (b), we prove Theorem 8.4.1(b). Suppose A and B are models of T, $\bar{a}$ and $\bar{b}$ are tuples of elements of A and B respectively, and $(A, \bar{a}) \equiv_0 (B, \bar{b})$. Let $\phi(\bar{x}, \bar{y})$ be a quantifier-free formula of L such that $A \vDash \exists \bar{y}\, \phi(\bar{a}, \bar{y})$; we must show that $B \vDash \exists \bar{y}\, \phi(\bar{b}, \bar{y})$. Without loss we can suppose that $\bar{b}$ is $\bar{a}$.

For this, suppose $A \vDash \phi(\bar{a}, \bar{c})$ where $\bar{c}$ is $(c_0, \ldots, c_{k-1})$. We claim that there is an element d_0 in some elementary extension B_0 of B, such that $(A, \bar{a}, c_0) \equiv_0 (B_0, \bar{a}, d_0)$.

Write $\Psi_0(\bar{x}, y)$ for the set of all quantifier-free formulas $\psi(\bar{x}, y)$ such that $A \vDash \psi(\bar{a}, c_0)$. Since $(A, \bar{a}) \equiv_0 (B, \bar{a})$, we can write $C = \langle \bar{a} \rangle_A = \langle \bar{a} \rangle_B$. By (b) there is a model A' of T such that $C \subseteq A' \subseteq A$ and there is an embedding of A' into B over C; without loss we can suppose that A' is a substructure of B. Since $\bar{a}$ is in A' and each formula $\psi(\bar{a}, y)$ in $\Psi_0(\bar{a}, y)$ has just one free variable, we infer by (a) that $A' \vDash \exists y\, \psi(a, y)$, since $A \vDash \exists y\, \psi(\bar{a}, y)$. But since $A' \subseteq B$, it follows that $B \vDash \exists y\, \psi(\bar{a}, y)$ too. Hence every finite subset of $\Psi_0(\bar{a}, y)$ is satisfied by an element of B. By compactness it follows that there is an elementary extension B_0 of B with an element d_0 which realises the type $\Psi_0(\bar{a}, y)$. Thus $(A, \bar{a}, c_0) \equiv_0 (B_0, \bar{a}, d_0)$, proving the claim.

Now we repeat to find an elementary extension B_1 of B_0 with an element d_1 such that $(A, \bar{a}, c_0, c_1) \equiv_0 (B_1, \bar{a}, d_0, d_1)$, and so on. Eventually we reach an elementary extension B_{n-1} of B and elements $\bar{d}$ such that $(A, \bar{a}, \bar{c}) \equiv_0 (B_{n-1}, \bar{a}, \bar{d})$. In particular we have $B_{n-1} \vDash \phi(\bar{a}, \bar{d})$, and so $B_{n-1} \vDash \exists \bar{y}\, \phi(\bar{a}, \bar{y})$. Thus $B \vDash \exists \bar{y}\, \phi(\bar{a}, \bar{y})$ as required. $\square$

To illustrate this, we give a Robinson-style proof of a result of Tarski which was quoted without proof as Theorem 2.7.2.

Theorem 8.4.4. *The theory T of real-closed fields in the first-order language L of ordered fields has quantifier elimination.*

Proof. The proof borrows two facts from the algebraists.

Fact 8.4.5. *The intermediate value theorem holds in real-closed fields for all functions defined by polynomials $p(x)$, possibly with parameters. (I.e. if $p(a) \cdot p(b) < 0$ then $p(c) = 0$ for some c strictly between a and b.)*

Proof. This is not deep. See Cohn [1977] p. 267, Jacobson [1974] p. 294.

☐ Fact

Fact 8.4.6. *If A is a real-closed field and C an ordered subfield of A, then there is a smallest real-closed field B such that $C \subseteq B \subseteq A$. Moreover if A' is any real-closed field $\supseteq C$ then B is embeddable in A' over A. (A' is called the* **real closure of C in A**.*)*

Proof. This is deep. The proof by Artin & Schreier [1927] used Sturm's lemma. Knebusch [1972] gave a proof using the theory of quadratic forms in place of Sturm.

☐ Fact

We use Corollary 8.4.3. First we prove (a). Let A and B be real-closed fields with $A \subseteq B$, and let $\phi(x)$ be a quantifier-free formula of L with parameters from A, such that $B \vDash \exists x\, \phi$. We must show that $A \vDash \exists x\, \phi$. After bringing ϕ to disjunctive normal form and distributing the quantifier through it, we can assume that ϕ is a conjunction of literals. Now $y \neq z$ is equivalent to $y < z \vee z < y$, and $\neg y < z$ is equivalent to $y = z \vee z < y$. So we can suppose that ϕ has the form

(4.3) $p_0(x) = 0 \wedge \ldots \wedge p_{k-1}(x) = 0 \wedge q_0(x) > 0 \wedge \ldots \wedge q_{m-1}(x) > 0$

where $p_0, \ldots, p_{k-1}, q_0, \ldots, q_{m-1}$ are polynomials with coefficients in A.

If ϕ contains a non-trivial equation $p_i(x) = 0$, then any element of B satisfying ϕ is algebraic over A and hence is already in A. Suppose on the other hand that $k = 0$. There are finitely many points $c_0 < \ldots < c_{n-1}$ in A which are zeros of one or more of the polynomials q_j $(j < m)$. By the intermediate value property (Fact 8.4.5), none of the q_j changes sign except at the points c_i $(i < n)$. So it suffices to take a point b of B such that $B \vDash \phi(b)$, and choose a point a in A which lies in the same interval of the c_is as b. This proves (a) of Corollary 8.4.3.

Next we prove (b). Let A and B be real-closed fields and C a common substructure of A and B. Then C is an ordered integral domain. It is not hard to show that the quotient field of C in A is isomorphic over C to the quotient field of C in B; so we can identify these quotient fields and suppose that C is itself an ordered field. By Fact 8.4.6 we can take A' to be the real closure of C in A, and (b) is proved.

☐ Theorem

Variants of quantifier elimination

There are some variants of Theorem 8.4.1 that consider only formulas of a particular kind, or only one formula at a time. These are mostly easy imitations of Theorem 8.4.1 itself. Here are a few specimens.

Let us say that the theory T in the first-order language L **has quantifier elimination for non-sentences** if every formula $\phi(\bar{x})$ of L with at least one free variable is equivalent modulo T to a quantifier-free formula $\psi(\bar{x})$ (i.e. with at most the same free variables $\bar{x}$). Unfortunately we shall see that there are theories which have quantifier elimination for non-sentences but not in general. (Even more unfortunately, some writers blur the point by saying that T has quantifier elimination 'if every formula is equivalent modulo T to a quantifier-free formula', without indicating the variables. This is bad practice – it leads to confusions.) The good news is that unlike Theorem 8.4.1, Theorem 8.4.7 remains true if we require every L-structure to have at least one element.

Theorem 8.4.7. *Let L be a first-order language and T a theory in L. The following are equivalent.*
(a) *T has quantifier elimination for non-sentences.*
(b) *If A and B are models of T, and $\bar{a}$, $\bar{b}$ are non-empty tuples from A, B respectively such that $(A, \bar{a}) \equiv_0 (B, \bar{b})$, then $(A, \bar{a}) \Rightarrow_1 (B, \bar{b})$.*
(c) *If A and B are models of T, $\bar{a}$ a non-empty sequence from A and $e: \langle \bar{a} \rangle_A \to B$ is an embedding, then there are an elementary extension D of B and an embedding $f: A \to D$ which extends e. ((4.2) will serve again as a picture.)*
(d) *The theory $T \cup \{\exists x\, x = x\}$ is model-complete and the theory $T_\forall \cup \{\exists x\, x = x\}$ has the amalgamation property.*

Proof. An easy variant of the proof of Theorem 8.4.1. $\qquad\qquad\square$

Lemma 8.4.8. *Let L be a first-order language and T a theory in L. For any formula $\phi(\bar{x})$ of L, the following are equivalent.*
(a) *ϕ is equivalent modulo T to some positive quantifier-free formula $\psi(\bar{x})$ of L.*
(b) *Whenever A and B are models of T, $\bar{a}$ is a sequence of elements of A such that $A \vDash \phi(\bar{a})$, and $f: \langle \bar{a} \rangle_A \to B$ is a homomorphism, then $B \vDash \phi(f\bar{a})$.*

Proof. (a) $\Rightarrow$ (b) is clear from Theorem 1.3.1. We prove the converse as follows. Adding constants, we can assume that ϕ is sentence. Let Φ be the set of all positive quantifier-free sentences ψ of L such that $T \vdash \phi \to \psi$. It suffices to prove that $T \cup \Phi \vdash \phi$. Let B be a model of $T \cup \Phi$. Write U for the theory

(4.4) $T \cup \{\phi\} \cup \{\neg\chi: \chi$ is an atomic sentence of L which is false in $B\}$.

If U had no model, there would be a disjunction χ of atomic sentences of L such that $T \vdash \phi \rightarrow \chi$ but χ is false in B, contradicting that B is a model of Φ. So U has a model A, so that $A \vDash \phi$. Let C be the substructure of A consisting of the elements named by closed terms. Then by the last part of U, there is a homomorphism $f: C \rightarrow B$. By (b) we infer that $B \vDash \phi$ as required.

$\square$

Theorem 8.4.9. *Let L be a first-order language and T a theory in L. The following are equivalent.*
(a) *Every $\exists_1^+$ formula $\phi(\bar{x})$ of L is equivalent modulo T to a positive quantifier-free formula $\psi(\bar{x})$ of L.*
(b) *If A and B are models of T, $\bar{a}$ is a sequence of elements of A and $f: \langle \bar{a} \rangle_A \rightarrow B$ is a homomorphism, then there are a pure extension D of B and a homomorphism $g: A \rightarrow D$ which extends f.*

Proof. To derive (b) from (a) we use Theorem 6.5.7 (in fact this gives us D as an elementary extension of B). In the other direction, (b) implies that whenever A and B are models of T, $\bar{a}$ is a sequence of elements of A and $f: \langle \bar{a} \rangle_A \rightarrow B$ is a homomorphism, then for every $\exists_1^+$ formula ϕ, $A \vDash \phi(\bar{a})$ implies $B \vDash \phi(f\bar{a})$. The lemma does the rest. $\square$

All these results have applications. Theorem 8.4.7 does useful work in the model theory of Horn theories (see Exercise 9.1.16), and Theorem 8.4.9 likewise in the model theory of modules (see Theorem A.1.4). Here is Lemma 8.4.8 in action.

Example 4: *Closed sets in real algebraic geometry.* Let A be a real-closed field and n a positive integer. We define a topology on $(\text{dom } A)^n$ by taking the product of the interval topology on $\text{dom}(A)$. Brumfiel [1979] p. 164 asked for a proof that every closed subset of $(\text{dom } A)^n$ which is first-order definable with parameters can be written as a positive boolean combination of sets of the form $\{\bar{a}: p(\bar{a}) \geq 0\}$ where $p(\bar{x})$ are polynomials with coefficients in A. To supply the proof that Brumfiel asked for, use a first-order language L which consists of the language of rings together with $\leq$. Let $\phi(\bar{x})$ be a formula of L with parameters in A, which defines a closed subset of $(\text{dom } A)^n$. Writing T for the theory of real-closed fields in L, and U for $T \cup \text{diag}(A)$, it suffices (by Lemma 8.4.8) to show that if B, C are any models of T, $\bar{a}$ is a sequence of elements of B such that $B \vDash \phi(\bar{a})$, and $f: \langle \bar{a} \rangle_B \rightarrow C$ is a homomorphism, then $C \vDash \phi(\bar{a})$. Geometrical methods yield this: see van den Dries [1982b] for details.

Exercises for section 8.4

1. In Theorem 8.4.1, show that T has quantifier elimination if and only if condition (c) holds whenever $\bar{a}$ is a tuple of elements of A.

2. Show that in Corollary 8.4.3, condition (a) can be replaced by (a'): if B is a model of T and A is a proper substructure of B, then there are an element b of B which is not in A, and a set $\Phi(x)$ of quantifier-free formulas of L with parameters in A, such that $B \vDash \bigwedge \Phi(b)$, Φ determines the quantifier-free type of b over A, and for every finite subset Φ_0 of Φ, $A \vDash \exists x \bigwedge \Phi_0$.

In section 3.3 above we saw a way to use Ehrenfeucht–Fraïssé games in order to find elimination sets. The following exercise translates that discussion into a criterion for quantifier elimination.

3. Let L be a first-order language with finite signature, and T a theory in L. Show that the following are equivalent. (a) T has quantifier elimination. (b) If A and B are any models of T, then for each $n < \omega$, any pair of tuples $(\bar{a}, \bar{b})$ from A, B respectively, such that $(A, \bar{a}) \equiv_0 (B, \bar{b})$, is a winning position for player $\exists$ in the game $G_n[A, B]$.

This exercise shows the link between the back-and-forth test of Exercise 3 and the amalgamation criterion of Theorem 8.4.1(c). It uses the notion of λ-saturation, which is defined in section 10.1 below.

4. Let L be a first-order language and T a theory in L. Show that the following are equivalent. (a) T has quantifier elimination. (b) If A and B are any ω-saturated models of T and $\bar{a}$ a tuple of elements of T such that $(A, \bar{a}) \equiv_0 (B, \bar{a})$, then $(A, \bar{a})$ and $(B, \bar{a})$ are back-and-forth equivalent. (c) If A is a model of T, B is a λ-saturated model of T for some infinite cardinal $\lambda > |A|$, and $\bar{a}$ is a tuple of elements of T such that $(A, \bar{a}) \equiv_0 (B, \bar{a})$, then there is an elementary embedding $f: A \to B$ such that $f\bar{a} = \bar{b}$.

5. Let L be a first-order language, T a theory in L and $\phi(\bar{x})$ a formula of L. Show that the following are equivalent. (a) ϕ is equivalent modulo T to a quantifier-free formula $\psi(\bar{x})$. (b) If A and B are any two models of T and $\bar{a}$, $\bar{b}$ are tuples of elements of A, B respectively such that $(A, \bar{a}) \equiv_0 (B, \bar{b})$, then $A \vDash \phi(\bar{a})$ implies $B \vDash \phi(\bar{b})$. (c) If A and B are any two models of T, $\bar{a}$ is a tuple of elements of A such that $A \vDash \phi(\bar{a})$, and $f: \langle \bar{a} \rangle_A \to B$ is an embedding, then $B \vDash \phi(f\bar{a})$. [(b) and (c) are different ways of saying exactly the same thing.]

The next exercise assumes the existence of strongly ω-homogeneous elementary extensions; see section 10.2 below.

6. Let L be a first-order language and T a complete theory in L. Show that T has quantifier elimination if and only if every model of T has an ultrahomogeneous elementary extension.

7. Let A be an integral domain. Show that if A is q.e. then A is a field. [If a, $b \neq 0$ then multiplication by b maps $\langle a \rangle_A$ isomorphically to $\langle a \cdot b \rangle_A$. Use Exercise 6.] *In fact every q.e. field is either finite or algebraically closed, by Theorem A.5.3 below.*

8. Let T be a first-order theory with a model companion U. Show that U has quantifier elimination if and only if $T_\forall$ has the amalgamation property.

9. Let T be a theory in a first-order language L, and U a model companion of T. Show that the following are equivalent. (a) T has the amalgamation property. (b) For every model A of T, $U \cup \text{diag}(A)$ is a complete theory in $L(A)$. (A theory U satisfying (a) or (b) is said to be a **model-completion** of T.)

10. Let L be the first-order language whose signature consists of one 1-ary function symbol. Show that the empty theory in L has a model completion.

11. An **ordered abelian group** is an abelian group with a 2-ary relation $\leq$ satisfying the laws '$\leq$ is a linear ordering' and $\forall xyz(x \leq y \rightarrow x + z \leq y + z)$. Let T_{oa} be the first-order theory of ordered abelian groups, and T_{doa} the first-order theory of ordered abelian groups which are divisible as abelian groups. (a) Show that T_{doa} is the model companion of T_{oa}. (b) Show that T_{doa} has quantifier elimination and is complete. (c) Show that $T_{doa} = \text{Th}(\mathbb{Q}, \leq)$ where $\mathbb{Q}$ is the additive group of rationals and $\leq$ is the usual ordering.

12. In the notation of the previous exercise, show that all ordered abelian groups satisfy the same $\exists_1$ first-order sentences. [Show this for $(\mathbb{Q}, \leq)$ and $(\mathbb{Z}, \leq)$, and trap all other groups between these.]

13. Let L be a first-order language and T an $\forall_1$ theory in L. Show that if $\phi(\bar{x})$ is a formula of L such that both ϕ and $\neg\phi$ are preserved by all embeddings between models of T, then ϕ is equivalent modulo T to a quantifier-free formula $\psi(\bar{x})$.

The next result is known to logic programmers (and others) as Herbrand's theorem. In fact it is the trivial case of a much deeper theorem of Herbrand about how one should understand proofs of first-order sentences (for which see Herbrand [1930]).

14. Let L be a first-order language, T an $\forall_1$ theory in L and $\phi(\bar{x})$ a quantifier-free formula of L. Show that if $T \vdash \exists \bar{x}\, \phi$, then for some $m < \omega$ there are tuples of terms $\bar{t}_0(\bar{y}), \ldots, \bar{t}_{m-1}(\bar{y})$ such that $T \vdash \forall \bar{y} \bigwedge_{i<m} \phi(\bar{t}_i(\bar{y}))$. *In particular the terms can be chosen to be ground terms when L has at least one constant; in this case the result is a kind of interpolation theorem, finding a quantifier-free interpolant between an $\forall_1$ premise and an $\exists_1$ conclusion.*

*15. Let L^+, L^- be respectively the first-order languages of groups with symbols $\{\cdot, {}^{-1}, 1\}$ and $\{\cdot, 1\}$. We say that a group G (as L^+-structure) has **elimination of inverses** if every quantifier-free formula of L^+ is equivalent in G to a quantifier-free formula of L^-. (a) A **monoidal identity** is an equation $s(x, y) = t(x, y)$ where s, t are terms of L^- and the equation is false in at least one group. Show that a group G has elimination of inverses if and only if G satisfies some sentence of the form $\forall xy(\phi_0 \vee \ldots \vee \phi_{n-1})$ where the ϕ_i are monoidal identities. (b) Show that every nilpotent group has elimination of inverses.

16. Let L be the first-order language of orderings (with $<$), and T the theory of dense linear orderings, i.e. linear orderings satisfying $\forall x \forall y \exists z (x < y \to x < z \land z < y)$. Show that T has quantifier elimination for non-sentences, but not quantifier elimination.

17. Let A be an atomless boolean algebra and $\bar{a}$ a tuple of elements of A. Show that $\mathrm{Th}(A, \bar{a})$ has quantifier elimination.

8.5 More on e.c. models

In this section we shall discuss two further characterisations of the e.c. models of a countable $\forall_2$ theory. The first characterisation – in terms of resultants – is important for understanding the definable relations of an e.c. model. The second is in terms of maximal $\exists$-types, and we need it for classifying and counting e.c. models.

Resultants

What systems of equations and inequations over A can be solved in some extension of A? Our next lemma answers this question, at least in general terms. Of course, for particular kinds of structure (fields, groups etc.) one has to work harder to get a specific answer.

Let T be an $\forall_2$ theory in a first-order language L, and let $\phi(\bar{x})$ be an $\exists_1$ formula of L. We write $\mathrm{Res}_\phi(\bar{x})$ for the set of all $\forall_1$ formulas $\psi(\bar{x})$ of L such that $T \vdash \forall \bar{x}(\phi \to \psi)$. The set Res_ϕ is called the **resultant** of ϕ.

Lemma 8.5.1. *Let L be a first-order language, T an $\forall_2$ theory in L and A an L-structure. Suppose $\phi(\bar{x})$ is an $\exists_1$ formula of L and $\bar{a}$ is a tuple from A. Then the following are equivalent.*
(a) *There is a model B of T such that $A \subseteq B$ and $B \vDash \phi(\bar{a})$.*
(b) $A \vDash \bigwedge \mathrm{Res}_\phi(\bar{a})$.

Proof. (a) $\Rightarrow$ (b). Suppose (a) holds, and let $\psi(\bar{x})$ be a formula in Res_ϕ. Then $B \vDash \psi(\bar{a})$ since B is a model of T. But ψ is an $\forall_1$ formula and $A \subseteq B$, so that $A \vDash \psi(\bar{a})$.

(b) $\Rightarrow$ (a). Assuming that (a) fails, we shall contradict (b). Introduce a tuple $\bar{c}$ of distinct new constants to name the elements $\bar{a}$. The diagram lemma (Lemma 1.4.2) tells us that if (a) fails, then the following theory has no model:

$$(5.1) \qquad\qquad T \cup \mathrm{diag}(A) \cup \{\phi(\bar{c})\}.$$

So by the compactness theorem there are a quantifier-free formula $\theta(\bar{x}, \bar{y})$ of L and distinct elements $\bar{d}$ of A such that $A \vDash \theta(\bar{a}, \bar{d})$ and

$$(5.2) \qquad\qquad T \vdash \phi(\bar{c}) \to \neg \theta(\bar{c}, \bar{d}).$$

Applying the lemma on constants (Lemma 2.3.2), we find

(5.3) $T \vdash \forall \bar{x}(\phi \rightarrow \forall \bar{y} \, \neg \theta)$

so that $\forall \bar{y} \, \neg \theta$ is a formula in Res_ϕ. But this contradicts (b), since $A \vDash \exists \bar{y} \, \theta(\bar{a}, \bar{y})$.   □

The formula $\bigwedge \mathrm{Res}_\phi(\bar{x})$ in (b) of the lemma is generally an infinitary formula. The following example is typical.

Example 1: *Nilpotent elements in commutative rings.* Let A be a commutative ring and a an element of A. When is there a commutative group $B \supseteq A$ containing a non-zero idempotent b (i.e. $b^2 = b$) which is divisible by a? In other words, when can A be extended to a commutative ring B in which the formula $\exists z \, (az \neq 0 \wedge (az)^2 = az)$ is true? The answer is that there is such a ring B if and only if a is not nilpotent (i.e. if and only if there is no $n < \omega$ such that $a^n = 0$). In one direction, if $ab \neq 0$ and $(ab)^2 = ab$, then for every positive integer n, $0 \neq (ab)^n = a^n b^n$ and so $a^n \neq 0$. In the other direction, if a is not nilpotent, consider the ring $A[x]/I$ where I is the ideal generated by $a^2 x^2 - ax$. To show that $A[x]/I$ will serve for B with x/I as b, we need to check that I doesn't contain either ax or any non-zero element of A. Suppose for example that

(5.4) $ax = \left(\sum_{i < n} c_i x^i \right)(a^2 x^2 - ax)$.

Multiplying out, $(-c_0 a - 1)x + \sum_{2 \leq i \leq n+1} (c_{i-2} a^2 - c_{i-1} a)x^i + c_n a^2 x^{n+2} = 0$. From this we infer

(5.5) $c_0 a = -1, \; c_{i-2} a^2 = c_{i-1} a \; (2 \leq i \leq n + 1), \; c_n a^2 = 0$.

Then $0 = c_n a^2 = c_{n-1} a^3 = \ldots = c_0 a^{n+2} = -a^{n+1}$, contradiction. The argument to show $A \cap I = \{0\}$ is similar but easier.

Thus in commutative rings, the resultant of the formula $\exists z(xz \neq 0 \wedge (xz)^2 = xz)$ is a set of $\forall_1$ formulas which is equivalent (modulo the theory of commutative rings) to the set $\{x^n \neq 0 : n > \omega\}$. There will be no harm in identifying the resultant with this set, or with the infinitary formula $\bigwedge_{n < \omega} x^n \neq 0$.

Theorem 8.5.2. *Let L be a first-order language, T an $\forall_2$ theory in L and A a model of T. The following are equivalent.*
(a) A is an e.c. model of T.
(b) For every $\exists_1$ formula $\phi(\bar{x})$ of L, $A \vDash \forall \bar{x}(\phi(\bar{x}) \leftrightarrow \bigwedge \mathrm{Res}_\phi(\bar{x}))$.

Proof. By definition of Res_ϕ, every model of T satisfies the implication $\forall \bar{x}(\phi(\bar{x}) \rightarrow \bigwedge \mathrm{Res}_\phi(\bar{x}))$. The implication in the other direction is just a rewrite of clause (b) in Theorem 8.2.4.   □

Example 1: *continued.* By the theorem, if A is an e.c. commutative ring, then an element of A is nilpotent if and only if it doesn't divide any nonzero idempotent. It follows that the condition 'x is nilpotent' can be expressed in A by a *first-order* formula, and moreover the same first-order formula works for any other e.c. commutative ring. This condition certainly isn't first-order for commutative rings in general (see Exercise 1). Moreover every e.c. commutative ring has an elementary extension in which there are non-nilpotent elements which don't divide any non-zero idempotent (by the same exercise). It follows by Theorem 8.3.6 that the *theory of commutative rings is not companionable.*

The forms of resultants are closely related to other properties of the theory T, as the next two theorems bear witness. Theorem 8.5.5 uses notions which will be introduced later in sections 9.1 and 9.2 for universal Horn formulas. I omit its proof (but see Exercise 9.1.18).

Theorem 8.5.3. *Let L be a first-order language and T an $\forall_2$ theory in L with a model companion T^*. Then for every $\exists_1$ formula $\phi(\bar{x})$ of L, $\mathrm{Res}_\phi(\bar{x})$ is equivalent modulo T^* to a single $\forall_1$ formula $\psi(\bar{x})$ of L.*

Proof. This follows from Theorem 8.5.2 and Theorem 8.3.6. Note that in general $\mathrm{Res}_\phi(\bar{x})$ won't be equivalent to a single $\forall_1$ formula of L modulo T; Exercise 4 gives a counterexample. $\square$

Theorem 8.5.4. *Let T be an $\forall_2$ theory in a first-order language L. Then the following are equivalent.*
(a) *$T_\forall$ has the amalgamation property.*
(b) *For every $\exists_1$ formula $\phi(\bar{x})$ of L, Res_ϕ is equivalent modulo T to a set $\Phi(\bar{x})$ of quantifier-free formulas of L.* $\square$

Proof. The direction (a) $\Rightarrow$ (b) follows from Exercise 5 and Theorem 8.1.3. In the other direction, assume (b) and suppose A, B and C are models of $T_\forall$ and A is a substructure of both B and C. We can extend B and C to e.c. models of T, call them B' and C' respectively. Suppose $\phi(\bar{x})$ is an $\exists_1$ formula of L and $\bar{a}$ is a tuple in A such that $C' \vDash \phi(\bar{a})$. Then, $A \vDash \bigwedge \mathrm{Res}_\phi(\bar{a})$, and so $B' \vDash \bigwedge \mathrm{Res}_\phi(\bar{a})$ since Res_ϕ is quantifier-free. Then $B' \vDash \phi(\bar{a})$, and the existential amalgamation theorem (Theorem 6.5.1) does the rest. $\square$

Theorem 8.5.5. *Let L be a first-order language and T a universal Horn theory in L; suppose T has the amalgamation property.*
(a) *For every p.p. formula $\phi(\bar{x})$ of L (with at least one free variable), $\mathrm{Res}_\phi(\bar{x})$ is equivalent modulo T to a set $\Phi(\bar{x})$ of quantifier-free Horn formulas of L.*

(b) *If moreover L is a recursive language and T is a recursive theory, then Φ can be chosen to be a recursively enumerable set of quantifier-free Horn formulas.*

(The restriction that ϕ has a free variable applies only if we exclude empty structures and L has no constants.)   □

Here are two examples of very different kinds.

Example 2: *Commutative rings without nilradical.* Let T be the theory of all commutative rings without nonzero nilpotent elements. Then by a theorem of Carson [1973] and Lipshitz & Saracino [1973], T is companionable. The model companion T^* is the theory of commutative von Neumann regular rings in which every monic polynomial has a root, and whose boolean algebra of idempotents is atomless. Since T is a universal Horn formula, we know by general principles that for each p.p. formula $\phi(\bar{x})$ of L the resultant Res_ϕ is equivalent modulo T^* to a universal Horn formula $\psi(\bar{x})$. Unfortunately T lacks amalgamation (Exercise 8), so that in general ψ can't be reduced to a quantifier-free formula. In fact T^* is also the model companion of the theory T' of commutative von Neumann regular rings. (Every model of T can be expanded to a model of T', and conversely every model of T' is already one of T.) The theory T' does have the amalgamation property (Cohn [1959]); but this is no use for getting ψ to be quantifier-free, since T' is not an $\forall_1$ theory.

Example 3: *Groups.* Let T be the theory of groups. By a theorem of Ziegler [1980], if $\Phi(\bar{x})$ is any recursively enumerable set of quantifier-free Horn formulas, there is an $\exists_1^+$ formula $\phi(\bar{x})$ such that Φ is equivalent to Res_ϕ modulo T. This is a very strong failure of companionability, and it casts a vivid shadow across the class of e.c. groups. See section A.4 in the appendix.

Maximal ∃-types

Let T be an $\forall_2$ theory in a countable first-order language L. The **∃-type** of a tuple $\bar{a}$ in a structure A is the set of $\exists_1$ formulas $\phi(\bar{x})$ of L such that $A \vDash \phi(\bar{a})$. By a **maximal ∃-type** (of T) we mean a set $\Phi(\bar{x})$ of $\exists_1$ formulas of L such that (1) $\Phi(\bar{x})$ is an ∃-type for T, and (2) if $\Psi(\bar{x})$ is any ∃-type of T and $\Phi \subseteq \Psi$, then $\Phi = \Psi$. We say that a tuple $\bar{a}$ **realises** the maximal ∃-type $\Phi(\bar{x})$ in the model A of T if $A \vDash \bigwedge \Phi(\bar{a})$.

Maximal ∃-types give us another characterisation of e.c. models.

Theorem 8.5.6. *Let T be an $\forall_2$ theory in a first-order language L, and let A be an L-structure. Then the following are equivalent.*

(a) A is an e.c. model of T.

(b) A is a model of $T_\forall$, and for every tuple $\bar{a}$ of elements of A, the set of $\exists_1$ formulas $\phi(\bar{x})$ such that $A \vDash \phi(\bar{a})$ is a maximal $\exists$-type of T.

Proof. (a) $\Rightarrow$ (b). Let A be an e.c. model of T containing a tuple $\bar{a}$ of elements, and assume for contradiction that there is in some model B of T a tuple $\bar{b}$ which satisfies all the $\exists_1$ formulas satisfied by $\bar{a}$ in A, and some more besides. Then by the existential amalgamation theorem (Theorem 6.5.1) there are an elementary extension D of B and an embedding $g: A \to D$ such that $g\bar{a} = \bar{b}$. If $\phi(\bar{x})$ is an $\exists_1$ formula of L such that $B \vDash \phi(\bar{b})$, then $D \vDash \phi(g\bar{a})$ and so $A \vDash \phi(\bar{a})$ since A is an e.c. model. This contradicts the choice of B and $\bar{b}$.

(b) $\Rightarrow$ (a). Assume (b), and let B be a model of T extending A, $\psi(\bar{x})$ an $\exists_1$ formula and $\bar{a}$ a tuple of elements of A such that $B \vDash \psi(\bar{a})$. Since $\bar{a}$ satisfies in B all the $\exists_1$ formulas which it satisfies in A, (b) implies that $A \vDash \psi(\bar{a})$ too. Thus A is an e.c. model of T. $\qquad\square$

If T is a countable $\forall_2$ theory with the joint embedding property, we can compare two e.c. models of T by seeing which maximal $\exists$-types are realised in them. It turns out that there are two extreme possibilities and nothing between them. The first possibility is that T has an enforceable model. The second is that for every enforceable property π, there are continuum many countable e.c. models of T with property π, such that no two of them realise the same maximal $\exists$-types.

The key to the dichotomy is an analogue of the notion of a supported type. We say that an $\exists_1$ formula $\phi(\bar{x})$ **isolates** the maximal $\exists$-type $\Phi(\bar{x})$ if Φ is the unique maximal $\exists$-type which contains ϕ. We say that a maximal $\exists$-type is **isolated** if it is isolated by some $\exists_1$ formula. The following example will give some idea of how the bad possibility works.

Example 4: *Making two models as different as possible.* Suppose the signature of L consists of just countably many 1-ary relation symbols R_i ($i < \omega$), and let T be the empty theory. Let A be any L-structure. Then A is a model of T. If b is an element of A, define the **colour** of b to be the set $\{i < \omega: A \vDash R_i(b)\}$. Colours correspond to maximal $\exists$-types, and since there are countably many relation symbols R_i, it's clear that none of these maximal $\exists$-types is isolated. Imagine now that a team of experts, say $\forall$ and $\exists$, settle down to construct L-structures A and B simultaneously by forcing. At any stage during the construction, just finitely many elements of each structure A and B have been mentioned, and only a finite amount of the colour of each element has been specified. So player $\exists$ can use her moves to make sure that no element of A has the same colour as any element of B. Note that although player $\exists$ can enforce that the structures A and B are completely different from each other, there is no way that she can make any particular colour appear in either structure.

In the proof below, I follow the notation of the second construction in section 8.2. As in section 7.2, it's sometimes helpful to think of the construction as a game between players $\exists$ and $\forall$. (**Warning.** In section 7.2 I said that the question whether a property π is enforceable is independent of the choice X of steps where player $\exists$ moves, provided both X and $\pi \backslash X$ are infinite. This was true for the construction of section 7.2, but for the forcing construction of this chapter it can fail: one of the players can steal an advantage by having first move. See Exercise 11 for an example. Fortunately if T has the joint embedding property, the statement from section 7.2 is still true. In fact it was true in section 7.2 because complete theories have the joint embedding property for elementary embeddings.)

Theorem 8.5.7 (*Forcing dichotomy theorem*). *Let L be a countable first-order language and T an $\forall_2$ theory in L which has JEP. The following are equivalent.*
(a) *There is an $\exists_1$ formula $\phi(\bar{x})$ of L such that $T \cup \{\exists \bar{x} \, \phi\}$ has a model but ϕ lies in no isolated maximal $\exists$-type of T.*
(b) *For every enforceable property π there are 2^ω non-isomorphic e.c. models A_β ($\beta < 2^\omega$) of T which have π and are finite or countable.*
(c) *T has no enforceable model.*
Moreover in case (b) *the family can be chosen so that every maximal $\exists$-type which appears in more than one of the structures A_β is isolated.*

Proof. We prove (a) $\Rightarrow$ (b) $\Rightarrow$ (c) $\Rightarrow$ (a). (b) $\Rightarrow$ (c) is clear: if there is an enforceable model C, then let π be the property of being isomorphic to C, and we get a contradiction to (b) at once.

(c) $\Rightarrow$ (a). Assume (a) fails, and suppose a team of experts is building a model of T by forcing. We set one of the experts, say E, to carry out the following tasks:
(5.6)   for each tuple $\bar{c}$ of distinct witnesses, make sure that the constructed theory T^+ includes $\Phi(\bar{c})$ for some isolated maximal $\exists$-type $\Phi(\bar{x})$.

There is one task for each tuple $\bar{c}$. Suppose E tackles the tuple $\bar{c}$ at the point where the condition T_i has just been chosen. We can write the conjunction of T_i as a quantifier-free sentence $\theta(\bar{c}, \bar{d})$ where $\bar{d}$ lists without repetition the witnesses appearing in T_i but not in $\bar{c}$. By assumption the formula $\exists \bar{y} \, \theta(\bar{x}, \bar{y})$ must lie in some isolated maximal $\exists$-type $\Phi(\bar{x})$; let $\phi(\bar{x})$ isolate this type, and let E choose T_{i+1} to be a condition implying $T_i \cup \{\phi(\bar{c})\}$. By Theorem 8.2.5 the compiled structure A is an e.c. model of T, and so by Theorem 8.5.6, the elements $\bar{c}^A$ realise a maximal $\exists$-type. But $A \vDash \phi(\bar{c})$, and so this type must be Φ. Thus the expert E ensures that

(5.7) every tuple of elements in A realises an isolated maximal $\exists$-type.

It remains to show that if A and B are any two at most countable e.c. models of T satisfying (5.7), then A is isomorphic to B. We do this by a back-and-forth argument, inductively building up sequences $\bar{a}$, $\bar{b}$ which list dom(A) and dom(B) respectively, such that $(A, \bar{a}) \equiv_1 (B, \bar{b})$. The starting point is that $A \equiv_1 B$ since T has the joint embedding property (Theorem 8.1.2). At the successor stages we use the argument of Theorem 7.2.2(b) above, but with $\exists_1$ formulas throughout.

There remains the proof of (a) $\Rightarrow$ (b). This is the hardest part of the argument. We suppose that (a) holds and π is an enforceable property. Write 'odds' for the set of all odd positive integers; so the game $G(\pi, \text{odds})$ is the game for constructing a model of T where player $\forall$ moves at steps $0, 2, \ldots$ and player $\exists$ at steps $1, 3, \ldots$, and player $\exists$ wins if the compiled structure has property π. Since π is enforceable, player $\exists$ has a winning strategy σ for $G(\pi, \text{odds})$.

After his first move, player $\forall$ will split the game into two copies, and player $\exists$ must answer in both simultaneously – though she can give different answers in the two cases. After his next move, player $\forall$ will split each of the two copies of the game into two, so that now player $\exists$ must make four replies. The same happens throughout the play, so that we have a branching tree of games. The tree has 2^ω branches, and player $\exists$ will use her winning strategy σ on each branch. Thus if β is a branch, it gives rise to an e.c. model A_β of T which has the property π. We shall see that if $\beta \neq \gamma$ then player $\forall$ can use his moves to ensure that no non-isolated maximal $\exists$-type is realised in both A_β and A_γ.

List as $(\bar{c}_i, \bar{d}_i)$ $(i < \omega)$ all the pairs of tuples of distinct witnesses, where $\bar{c}_i$ and $\bar{d}_i$ have the same length; make the list so that each pair appears infinitely often. At the $2i$th step, player $\forall$ is faced with some finite number of games, $G_0, \ldots, G_{k-1}$ (in fact $k = 2^i$, but we don't need to know this). Let T_{i-1}^j be the condition which has just been chosen by player $\exists$ in the game G_j. (Wait a moment for the definition of T_{-1}^0.) Player $\forall$ will do several things before he delivers his set of choices $T_i^0, \ldots, T_i^{k-1}$ for these games. His aim will be to choose these conditions so that

(5.8) if $(\bar{c}_i, \bar{d}_i)$ is the pair $(\bar{c}, \bar{d})$, then for any $m < n < k$ there are formulas in T_i^m and T_i^n which commit $\bar{c}$ in the first and $\bar{d}$ in the second either to the same isolated maximal $\exists$-type, or to distinct maximal $\exists$-types.

There are $k(k - 1)$ such pairs (m, n), so player $\forall$ will in fact make $k(k - 1)$ moves in turn, one to deal with each pair. All but the last of these moves will be in secret; the last is the one which he returns in the public game.

Suppose player $\forall$ has got as far as choosing conditions $S_i^0, \ldots, S_i^{k-1}$ for these k games, and he is about to choose $S'^0_i, \ldots, S'^{k-1}_i$, tackling the pair (m, n) with $m < n < k$. He writes $\bigwedge S_i^m$ as a formula $\theta(\bar{c}, \bar{e})$ where $\bar{e}$ lists without repetition the witnesses which are in S_i^m but not in $\bar{c}$; likewise he writes $\bigwedge S_i^n$ as $\xi(\bar{d}, \bar{f})$ where $\bar{f}$ lists without repetition the witnesses in S_i^n but not in $\bar{d}$. If both the formulas $\exists \bar{y}\, \theta(\bar{x}, \bar{y})$ and $\exists \bar{z}\, \xi(\bar{x}, \bar{z})$ isolate maximal $\exists$-types, then either they isolate the same one or they isolate different ones; either way player $\forall$ need do nothing, so let him choose each S'^j_i to be S_i^j. On the other hand if at least one of $\exists \bar{y}\, \theta(\bar{x}, \bar{y})$ and $\exists \bar{z}\, \xi(\bar{x}, \bar{z})$ doesn't isolate a maximal $\exists$-type, then he can choose distinct maximal $\exists$-types Θ, Ξ which contain $\exists \bar{y}\, \theta(\bar{x}, \bar{y})$, $\exists \bar{z}\, \xi(\bar{x}, \bar{z})$ respectively. Let $\theta'(\bar{x})$ be a formula in $\Theta \backslash \Xi$, and $\xi'(\bar{x})$ a formula in $\Xi \backslash \Theta$. Let player $\forall$ choose S'^m_i to imply $S_i^m \cup \{\theta'(\bar{c})\}$ and S'^n_i to imply $S_i^n \cup \{\xi'(\bar{d})\}$; for the remaining j let him put $S'^j_i = S_i^j$.

What was T^0_{-1}? At the beginning of the play, player $\forall$ started with a clean board in front of him. So he could take T^0_{-1} to be empty; but in fact we suppose he did something different. Let $\phi(\bar{x})$ be the formula in (a) of the theorem. We shall suppose that player $\forall$ chose a tuple $\bar{a}$ of distinct witnesses, and put T^0_{-1} a condition which implies $\phi(\bar{a})$.

This completes the construction. As we saw, player $\exists$ arranged that each branch β yields an e.c. model A_β of T with the property π. Now if β and γ are different branches, then they must differ from some step i_0 onwards. Now any two n-tuples of elements from A_β, A_γ can be written as $\bar{c}_i^{A_\beta}$, $\bar{d}_i^{A_\gamma}$ respectively, for some pair $(\bar{c}_i, \bar{d}_i)$ with $i > i_0$. So at the $2i$th step, player $\forall$ made sure by (5.8) that the two tuples realise either the same isolated maximal $\exists$-type, or distinct maximal $\exists$-types. Thus every maximal $\exists$-type which appears in two or more of the structures A_β must be isolated.

It remains to show that if $\beta \neq \gamma$ then A_β and A_γ are not isomorphic. But the choice of T^0_{-1} ensured that $\bar{a}^{A_\beta}$ doesn't realise any isolated maximal $\exists$-type; so the maximal $\exists$-type of $\bar{a}^{A_\beta}$ is not realised anywhere in A_γ. $\quad\square$

Exercises for section 8.5

1. In commutative rings, write (*) for the condition 'For each positive integer n, there is an element a such that $a^n = 0$ but $a^i \neq 0$ for all $i < n$'. (a) Show that if A is a commutative ring in which (*) holds, and $\phi(x)$ is any formula (maybe with parameters from A) such that $\phi(A)$ is the set of nilpotent elements in A, then A has an elementary extension in which some non-nilpotent element satisfies ϕ. (b) By iterating (a) to form an elementary chain, show that there is a commutative ring in which the set of nilpotent elements is not first-order definable with parameters. (c) Show that every e.c. commutative ring satisfies (*).

2. Let G be a group and a, b two elements of G. (a) Show that the following are equivalent. (i) There is a group $H \supseteq G$ with an element h such that $h^{-1}ah = b$. (ii)

The elements a and b have the same order. (b) Explain this as an instance of Lemma 8.5.1. (c) Prove that the theory of groups is not companionable.

*3. Show that every e.c. group contains a finitely generated subgroup with unsolvable word problem. [Let X, Y be recursively inseparable r.e. subsets of ω. Find a recursive family of equations $E_i(x, y)$ $(i < \omega)$ which can hold or fail to hold independently for the two generators x, y of a group (for example put $E_i(x, y) = [[x, y^{2i+1}], x])$). Use Ziegler's result from Example 3 to write an $\exists_1$ formula $\phi(x, y)$ which expresses that all the equations E_i with $i \in X$ hold and none of those with $i \in Y$ hold.]

4. Give an example of a companionable $\forall_1$ theory T in a first-order language L, and an $\exists_1$ formula $\phi(\bar{x})$ in L such that Res_ϕ is not equivalent modulo T to any finite set of $\forall_1$ formulas. [Consider the set of sentences $\forall xy \neg (P_0 x \wedge P_i y)$ $(0 < i < \omega)$, and let ϕ be the sentence $\exists x P_0 x$.]

The next exercise is a partial converse to Theorem 8.5.2.
5. Let T be an $\forall_2$ theory in a first-order language L. Suppose $\phi(\bar{x})$ is an $\exists_1$ formula of L, and $\chi(\bar{x})$ is a quantifier-free formula of $L_{\infty\omega}$ such that for every e.c. model A of T, $A \vDash \forall \bar{x}(\phi \leftrightarrow \chi)$. Show that χ is equivalent to Res_ϕ modulo T, and hence that Res_ϕ is equivalent modulo T to a set of quantifier-free formulas of L.

6. Let A be a commutative ring, a an element of A and n a positive integer. Show that the following are equivalent. (a) For every element b of A, if $ba^{n+1} = 0$ then $ba^n = 0$. (b) There is a commutative ring B which extends A and contains an element b such that $(a^2 b - a)^n = 0$.

*7. Show that there is a family of continuum many e.c. commutative rings which are pairwise not elementarily equivalent. [Let p be a prime. By Example 1 and Exercise 6 there is a first-order sentence χ_p which expresses, in any e.c. commutative ring, 'There is an element a such that $pa = 0$, and for each positive integer n there is an element b such that $ba^{n+1} = 0 \neq ba^n$'. Construct a ring $A(p)$ of the form $\mathbb{F}_p[x, y_i \ (1 \leqslant i < \omega)]/I$, where $\mathbb{F}_p$ is the p-element field and I is the ideal generated by the elements $y_i x^n$ $(i \leqslant n)$ and $y_i y_j$ $(1 \leqslant i, j < \omega)$. Now for each set Π of primes, introducing parameters for the elements x, y_i in each $A(p)$, we can write down an $\forall_1$ theory $T(\Pi)$ which says 'I am a commutative ring and I contain a copy of $A(p)$ for each $p \in \Pi$'. If $A \vDash T(\Pi)$ then certainly $A \vDash \chi_p$ for each $p \in \Pi$. If $p \notin \Pi$, show that $\neg \chi_p$ is enforceable.]

8. Show that the amalgamation property fails for the class of commutative rings without nonzero nilpotents. [Let A be $\mathbb{Q}[x]$; let B be a field containing A, and let C be $A[y]/(xy)$. A zero-divisor can never be amalgamated with an invertible element.]

9. Let T be a first-order theory. Show that the maximal $\exists$-types of T are exactly the maximal $\exists$-types of $T_\forall$.

10. Let T be a theory in a countable first-order language L. (a) Show that if T is an $\forall_2$ theory and for each $n < \omega$, T has only finitely many maximal $\exists$-types, then T is

companionable. (b) Show that if T is ω-categorical, then T is companionable and its model companion is also ω-categorical. [$T_{\forall_2}$ has only finitely many maximal $\exists$-types $\Phi(\bar{x})$ for each $\bar{x}$, and satisfies the JEP. So it has a unique countable e.c. model.]

11. Let T be the theory of fields, and suppose experts are constructing a model of T by forcing (see section 8.2). (a) Show that for each prime p, the expert who makes the first move can determine that the characteristic of the compiled field shall be p. (b) Show that it is enforceable that the compiled field is the algebraic closure of the prime field, and that the characteristic is not 0. [This needs the observation that any $\exists_1$ sentence true in some field is also true in some field of finite characteristic. One can deduce it from Exercise 6.1.7.]

12. Let L be a countable first-order language and T an $\forall_2$ theory in L, not necessarily with the joint embedding property. Show that *either* there is a finite or countable set **C** of models of T such that the property 'A is isomorphic to some structure in **C**' is enforceable (in which case the first expert or player can determine which structure in **C** it shall be), *or* (b) of Theorem 8.5.7 holds.

*13. Let L be a countable first-order language and T an $\forall_2$ theory in L with the joint embedding property. Show that (a)–(c) in Theorem 8.5.7 are equivalent to (d): There is a property π such that the game $G(\pi, \text{odds})$ is not determined.

8.6 Amalgamation revisited

Let **K** be a class of L-structures, for example the class of all models of a theory T. We say that a structure A is an **amalgamation base** for **K** (or T) if for all embeddings $e: A \to B$ and $f: A \to C$ of A into structures B and C which are in **K**, there are a structure D in **K** and embeddings $g: B \to D$ and $h: C \to D$ such that $ge = hf$:

(6.1)

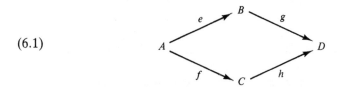

Likewise we say that A is a **strong amalgamation base** for **K** (or T) if the same holds with the further restriction that if $g(b) = h(c)$ then b is in the image of e. Note that the structure A need not be a member of **K** itself.

The amalgamation property (see section 7.1 above) is definable in terms of amalgamation bases, as follows: **K** has the amalgamation property if and only if every structure in **K** is an amalgamation base for **K**. The same goes for strong amalgamation.

E.c. models are strong amalgamation bases

The theory of amalgamation bases is really the existential analogue of the amalgamation theorems of section 6.4 above. We begin with the existential twin sister of Theorem 6.4.3 on heir–coheir amalgams. ($A \leqslant_1 B$ means that $A \subseteq B$ and for every tuple $\bar{a}$ in A and every $\exists_1$ first-order formula $\phi(\bar{x})$, $B \vDash \phi(\bar{a})$ implies $A \vDash \phi(\bar{a})$.)

Theorem 8.6.1. *Let L be a first-order language and A, B and C L-structures such that $A \subseteq B$ and $A \leqslant_1 C$. Then there exist an elementary extension D of B and an embedding $g: C \to D$ such that for every formula $\psi(\bar{x}, \bar{y})$ of L and all tuples $\bar{b}$ in B and $\bar{c}$ in C, if $D \vDash \psi(\bar{b}, g\bar{c})$ then there is a tuple $\bar{a}$ in A such that $B \vDash \psi(\bar{b}, \bar{a})$.*

Proof. A slight variant of the proof of Theorem 6.4.3. As there, we assume that $(\mathrm{dom}\ B) \cap (\mathrm{dom}\ C) = \mathrm{dom}\ A$. We consider the theory

(6.2)   eldiag(B) $\cup$ diag(C) $\cup \{\neg \psi(\bar{b}, \bar{c}): \psi(\bar{x}, \bar{y})$ is a formula of L
          and $\bar{b}$ a tuple in B such that $B \vDash \neg \psi(\bar{b}, \bar{a})$ for all tuples $\bar{a}$ in
          $A\}$.

It is enough to show that T has a model. In hopes of a contradiction, suppose T has no model. Then as in the proof of Theorem 6.4.3, compactness and the lemma on constants give us a tuple $\bar{a}$ from A, a tuple $\bar{b}$ from B, a quantifier-free formula $\theta(\bar{v}, \bar{y})$ of L such that $C \vDash \exists \bar{y}\ \theta(\bar{a}, \bar{y})$, and formulas $\psi_i(\bar{x}, \bar{v}, \bar{y})$ $(i < k)$ of L such that for all $\bar{a}'$, $\bar{a}''$ in A, $B \vDash \neg \psi_i(\bar{b}, \bar{a}', \bar{a}'')$, and

(6.3) $B \vDash \forall \bar{y}(\theta(\bar{a}, \bar{y}) \to \psi_0(\bar{b}, \bar{a}, \bar{y}) \vee \ldots \vee \psi_{k-1}(\bar{b}, \bar{a}, \bar{y}))$.

Since $A \leqslant_1 C$, we have $A \vDash \exists \bar{y}\ \theta(\bar{a}, \bar{y})$ and hence there is a tuple $\bar{a}''$ in A such that $A \vDash \theta(\bar{a}, \bar{a}'')$. So $B \vDash \theta(\bar{a}, \bar{a}'')$, and by (6.3) it follows that $B \vDash \psi_i(\bar{b}, \bar{a}, \bar{a}'')$ for some $i < k$. This is a contradiction, proving the theorem. $\square$

We can read off two immediate corollaries.

Corollary 8.6.2. *Let L be a first-order language and T an $\forall_2$ theory in L. Then every e.c. model of T is a strong amalgamation base for T.*

Proof. Let A be an e.c. model of T. Then $A \leqslant_1 C$ whenever C is a model of T extending A. Now apply the theorem with the formula $x = y$ as ψ. $\square$

Corollary 8.6.3. *Let T be a model-complete first-order theory. Then T has the strong amalgamation property.*

Proof. Every model of T is e.c., by Theorem 8.3.1. $\square$

Looking at the first corollary, one might be tempted to think that for an $\forall_2$ first-order theory the e.c. models are exactly the strong amalgamation bases. Not so. For example it's easy to see that every e.c. group must be infinite. But all groups are strong amalgamation bases for the theory of groups; this follows at once from the properties of free products with amalgamated subgroup (see Lyndon & Schupp [1977] p. 174).

So it becomes natural to ask which substructures of models of a first-order theory T are strong amalgamation bases for T. Theorem 8.6.8 below will answer this question completely.

Before we turn to the proof of Theorem 8.6.8, there are two other matters to settle. The first is to show that the restriction to $\forall_2$ theories in that theorem means no loss of generality at all.

Lemma 8.6.4. *Let T be a theory in a first-order language L. Then the strong amalgamation bases for T are exactly the strong amalgamation bases for $T_\forall$, and likewise the strong amalgamation bases for $T_{\forall_2}$.*

Proof. I give the argument for $T_\forall$; it works for $T_{\forall_2}$ as well. Suppose first that A is a strong amalgamation base for T, and let B, C in (6.1) be models of $T_\forall$ which extend A. (There is obviously no loss of generality in assuming that e and f in (6.1) are inclusion maps.) Then by Corollary 6.5.3, B and C can be extended to models B' and C' of T, and so we can find a strong amalgam D of B' and C' over A which is a model of T and hence of $T_\forall$. Clearly D serves as a strong amalgam of B and C too.

Conversely, suppose A is a strong amalgamation base for $T_\forall$, and B, C are models of T which extend A. Then there is some model D of $T_\forall$ which is a strong amalgam of B and C over A. Using Corollary 6.5.3, we can extend D to a model D' of T; all the added elements lie outside B and C, so that D' is still a strong amalgam. ☐

Our second preliminary is a red herring. The next result has nothing to do with amalgamation bases, but it is rather pretty and it follows quickly from Theorem 8.6.1.

Intersections of descending chains

We say that a first-order theory T has the **descending intersection property** if for every descending chain $(A_i : i < \gamma)$ of models of T, the intersection $\bigcap_{i<\gamma} A_i$ is either empty or a model of T.

Corollary 8.6.5. *Suppose L is a first-order language and T is a theory in L*

with the descending intersection property. Then T is equivalent to an $\forall_2$ theory in T.

Proof. Let B be a model of T and suppose $A \preccurlyeq_1 B$. We have to show that A is also a model of T. (See Exercise 6.5.4.) If A is empty then so is B; we can assume henceforth that A is not empty. We build up a commutative diagram as follows:

(6.4)

where the horizontal maps e_i $(i < \omega)$ are elementary extensions, and each square is a strong amalgam. The first square comes from Theorem 8.6.1 by taking B and C in the theorem to be respectively the north-western copy of B and the south-eastern. Then $B \preccurlyeq B_1$, and so we can use the same argument to get the second square, and so on for ω steps.

Now write B_0 for B, and put $C = \bigcup_{i<\omega} B_i$ and $g = \bigcup_{i<\omega} g_i$. Then g is an embedding from C into C. It is not necessarily elementary; but for any positive n, the image of g^n is a copy of C and hence a model of T. Define $g^0 C$ to be C. We have $gC \subseteq C$ and hence $g^{n+1}C \subseteq g^n C$ for all $n < \omega$. It follows that $(g^n C: n < \omega)$ is a descending chain of models of T. For the corollary we need only prove that $A = \bigcap_{n<\omega} g^n C$.

Since each of the maps g_i is the identity on A, g fixes A pointwise, and so $A \subseteq \bigcap_{n<\omega} g^n C$. For the converse, note first that the strong amalgams give us

(6.5) $B_0 \cap gC = A$, and for every $n < \omega$, $B_{n+1} \cap gC = gB_n$.

We claim that for every $n < \omega$, $B_n \cap g^{n+1}C = A$. For $n = 0$ this is the first equation of (6.5). Assuming it for n, we prove it for $n + 1$ as follows, using the second equation of (6.5):

$$B_{n+1} \cap g^{n+2}C = (B_{n+1} \cap gC) \cap g^{n+2}C = gB_n \cap g^{n+2}C$$
$$= g(B_n \cap g^{n+1}C) = gA = A$$

since g is the identity on A. This proves the claim. It follows that if c is in B_n but not in A, then $c \notin g^{n+1}C$, which proves the theorem. $\square$

Strong amalgamation bases characterised

There is an easy characterisation of amalgamation bases for an $\forall_2$ theory. But I believe it is not much use in practice, and so I leave its rather uninteresting proof as an exercise. See section 8.5 for the definition of $\exists$-types.

Theorem 8.6.6. *Let T be an $\forall_2$ first-order theory and A a substructure of a model of T. Then the following are equivalent.*

(a) *A is an amalgamation base for T.*

(b) *For every tuple $\bar{a}$ in A, there is a unique maximal $\exists$-type of T containing $\exists$-$\mathrm{tp}_A(\bar{a})$.* $\square$

The analogous result for strong amalgams is the existential analogue of Theorem 6.4.5.

Let L be a first-order language, B an L-structure and X a set of elements of B. We say that an element b of B is $\exists$-**algebraic over** X **in** B if there are an $\exists_1$ formula $\phi(x, \bar{y})$ of L and a tuple $\bar{a}$ of elements of X such that

$$B \vDash \phi(b, \bar{a}) \wedge \exists_{\leqslant n} x\, \phi(x, \bar{a})$$

for some finite n. Just as with (4.8)–(4.10) in section 6.4, one can easily check that

(6.6) b is $\exists$-algebraic over any set X containing b,

(6.7)   if Y is a set of elements of B, b is $\exists$-algebraic over Y and every element of Y is $\exists$-algebraic over X, then b is $\exists$-algebraic over X,

(6.8)   if $B \leqslant_1 C$ then for every element c of C, c is $\exists$-algebraic over X in C if and only if c is in B and is $\exists$-algebraic over X in B.

The $\exists$-**algebraic closure** of X in B, $\exists$-$\mathrm{acl}_B(X)$ or $\exists$-$\mathrm{acl}(X)$, is the set of all elements of B which are $\exists$-algebraic over X. By (6.6), $X \subseteq \exists$-$\mathrm{acl}(X)$. By (6.7), $\exists$-$\mathrm{acl}(\exists$-$\mathrm{acl}(X)) = \exists$-$\mathrm{acl}(X)$. By (6.8), $\exists$-$\mathrm{acl}_B(X) = \exists$-$\mathrm{acl}_C(X)$ whenever $B \leqslant_1 C$. If $\bar{a}$ lists the elements of X, we write $\mathrm{acl}(\bar{a}) = \mathrm{acl}(X)$.

In the next lemma, $A \equiv_1 B$ means that every $\exists_1$ first-order sentence which is true in A is true also in B, and vice versa.

Lemma 8.6.7. *Let L be a first-order language and T an $\forall_2$ theory in L. Suppose A is a model of T, and λ is an infinite cardinal. Then there is an e.c. model D of T with $A \subseteq D$, such that*

(a) *any two tuples in D which realise the same maximal $\exists$-type are in the same orbit of $\mathrm{Aut}(D)$, and*

(b) *if $\bar{a}$, $\bar{b}$ are sequences of fewer than λ elements of B such that $(D, \bar{a}) \equiv_1 (D, \bar{b})$, and c is an element of some e.c. model C of T with $D \subseteq C$, then there is d in D such that $(D, \bar{a}, c) \equiv_1 (C, \bar{a}, d)$.*

Proof. This is a theorem about big models. See Exercise 10.2.11. $\square$

Theorem 8.6.8. *Let L be a first-order language and T an $\forall_2$ theory in L. The following are equivalent.*

(a) *A is a strong amalgamation base for T.*

(b) *A is an amalgamation base for T, and for every e.c. model B of T extending A, dom(A) is ∃-algebraically closed in B.*

Proof. (a) ⇒ (b). Assume (a). Then certainly A is an amalgamation base. Let B be an e.c. model of T extending A. By strong amalgamation there is a model D of T extending B, together with an embedding $g: B \to D$ such that g is the identity on A and $(\text{dom } B) \cap g(\text{dom } B) = \text{dom}(A)$. By Theorem 8.2.1 we can take D to be an e.c. model. By (6.8) the ∃-algebraic closure of dom(A) in B is the same as its ∃-algebraic closure in D; this in turn lies in both B and gB, and hence it is exactly dom(A).

(b) ⇒ (a). Assume (b), and let B, C be models of T extending A. Let $\bar{a}$ be a sequence listing the elements of A. Taking an isomorphic copy of C if necessary, we can suppose that $\text{dom}(B) \cap \text{dom}(C)$ is exactly $\{\bar{a}\}$. After adding constants for $\bar{a}$ to the language L, let U be the theory

(6.9) $T \cup \text{diag}(B) \cup \text{diag}(C) \cup$

$\{b \neq c: b \in \text{dom } B \backslash \{\bar{a}\}$ and $c \in \text{dom } C \backslash \{\bar{a}\}\}$.

It suffices to show that U has a model. For contradiction, suppose it has none.

Then by the compactness theorem there are finite subsets X of $\text{dom}(B) \backslash \{\bar{a}\}$ and Y of $\text{dom}(C) \backslash \{\bar{a}\}$, and a finite subset $\Phi(\bar{c})$ of diag(C) (where $\bar{c}$ is a tuple of elements of Y), such that

(6.10) $T \cup \text{diag}(B) \vdash \forall \bar{x}(\bigwedge \Phi(\bar{x}) \to$ some element in $\bar{x}$ lies in $X)$.

Now since A is an amalgamation base, there are a model D of T with $B \subseteq D$ and an embedding $h: C \to D$ which is the identity on A. We can choose D so that it satisfies the conclusion of Lemma 8.6.7 with $\lambda \geqslant |A|^+$.

Since Φ is quantifier-free, we have $D \vDash \bigwedge \Phi(h\bar{c})$. By (6.10) it follows that $h\bar{c}$ has only finitely many images under $\text{Aut}(D)_{(\bar{a})}$. This brings Π. M. Neumann's lemma (Corollary 4.2.2) into play, and we deduce that some element d in $h\bar{c}$ has finite orbit under $\text{Aut}(D)_{(\bar{a})}$. By the choice of D, this implies that for every e.c. model D' extending D, there are only finitely many elements d' such that $(D, \bar{a}, d) \equiv_1 (D', \bar{a}, d')$. By compactness it follows that d is ∃-algebraic over $\bar{a}$, and hence d is in A since dom(A) is ∃-algebraically closed in D by assumption (b). But this is impossible, since d is in $h\bar{c}$ and $h\bar{c}$ lies entirely outside A.   □

An example of non-amalgamation

To close this section, here is an example of a class **K** in which the e.c. structures are not amalgamation bases. The class of e.c. structures in **K** is not first-order axiomatisable, though it has other pleasant properties like ω-categoricity and model-completeness.

Theorem 8.6.9. *Let L be the first-order language whose signature consists of one 2-ary function symbol F. Let* **J** *be the class of all finite L-structures, and let* **K** *be the class of L-structures with age* $\subseteq$ **J***.*

(a) **J** *is countable (up to isomorphism) and has HP, JEP and AP.*

(b) *A countable L-structure A is e.c. in* **K** *if and only if A is the Fraïssé limit of* **J***.*

(c) *If A and B are e.c. structures in* **K***,* $A \subseteq B$ *and* $\bar{a}$ *is a tuple of elements of A, then* $(A, \bar{a}) \equiv_{\infty\omega} (B, \bar{a})$.

(d) *The countable e.c. structure in* **K** *is not an amalgamation base for e.c. structures in* **K***.*

Proof. (a) is immediate, and (b) is by Example 2 in section 8.1 above. Then (c) follows by Lemma 7.1.5.

For (d) we show first that the amalgamation property fails in **K**. Let A be an L-structure whose elements are the countably many elements a_i ($i < \omega$), with $F^A(a_i, a_j) = a_i$ for all i. Define an L-structure $B \supseteq A$ as follows: B has one new element b, and we put

(6.11) $F^B(b, b) = b,$

(6.12) $F^B(b, a_i) = F^B(a_i, b) = a_{i+1}$ when i is even,

(6.13) $F^B(b, a_i) = F^B(a_i, b) = a_{i-1}$ when i is odd.

Likewise define an L-structure $C \supseteq A$ with one extra element c, which satisfies the analogues of equations (6.11)–(6.13) except that $F^C(c, a_0) = F^C(a_0, c) = a_0$, and for $i > 0$, 'even' and 'odd' are transposed in (6.12) and (6.13). Then A, B and C are in **K**, but any structure which amalgamates B and C over A has a finitely generated infinite substructure, and hence is not in **K**.

The structure A is clearly not existentially closed. We shall extend A, B and C to countable e.c. structures A', B', C' in such a way that the failure of amalgamation is maintained. In fact it's enough to extend A, B, C to structures A', B', C' respectively, all countable and in **K**, such that $A' \subseteq B'$ and $A' \subseteq C'$ and A' is e.c. For by Corollary 8.2.2 there are countable e.c. structures B^*, C^* in **K** extending B', C' respectively; any amalgamation of B^* and C^* over A' would restrict to an amalgamation of B and C over A, which we have seen is impossible.

First, suppose M, N_0 and N_1 are structures in **J** such that $M \subseteq N_i$ ($i < 2$) and $\mathrm{dom}(M) = \mathrm{dom}(N_0) \cap \mathrm{dom}(N_1)$. We form a structure P in **J** as follows. The domain of P is $\mathrm{dom}(N_0) \cup \mathrm{dom}(N_1)$. If a is in N_0 and b in N_1, but neither a nor b is in M, then we put $F^P(a, b) = F^P(b, a) = a$. Elsewhere we define F^P to agree with F^{N_0} or F^{N_1}. Then P is a strong amalgam of N_0 and N_1 over M, with the further property that every element of P lies in either N_0 or N_1. We call P the **thrifty amalgam** of N_0 and N_1 over M.

Since A is in **K**, we can write A as the union of a chain $(A_i : i < \omega)$ of finite structures. We shall define a chain $(A_i' : i < \omega)$ of finite structures such that $A_i \subseteq A_i'$ for each i, by induction on i as follows. A_0' is A_0. When A_i' has been chosen, we consider some finite structure $E_i \supseteq A_i'$, and we take A_{i+1}' to be the thrifty amalgam of E_i and A_{i+1} over A_i. We put $A' = \bigcup_{i<\omega} A_i'$. The reader can check that the structures E_i can be chosen in such a way that A' is an e.c. structure in **K**. Also it's clear that A' can be defined in such a way that $\mathrm{dom}(A') \cap \mathrm{dom}(B) = \mathrm{dom}(A') \cap \mathrm{dom}(C) = \mathrm{dom}(A)$.

It remains to define B' and C'. The definitions are exactly parallel, and I give the argument for B'. First we form a chain $(B_i' : i < \omega)$ of structures in **J** such that for each $i < \omega$ we have a diagram of inclusions:

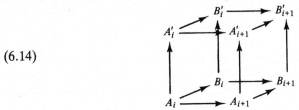

(6.14)

We proceed by induction on i. In the cube (6.14) the bottom face and the front face are given, and the lefthand face has just been constructed. The problem of finding B_{i+1}' is a three-dimensional amalgamation problem. These higher-dimensional amalgamations are not always possible. But here we can make the amalgamation possible by adding an inductive assumption:

(6.15)   Each B_i' is the thrifty amalgam of A_i' and B_i over A_i.

This works as follows.

We define the domain of B_{i+1}' to be $\mathrm{dom}(A_{i+1}') \cup \mathrm{dom}(B_{i+1})$; by our choice of A', $\mathrm{dom}(A_{i+1}') \cap \mathrm{dom}(B_{i+1}) = \mathrm{dom}(A_{i+1})$. What remains to be checked is that if a and b are two elements of $\mathrm{dom}(B_{i+1}')$, then either the value of $F(a, b)$ is already determined *in a unique way* by the structures already constructed in (6.14), or we have complete freedom to define it as we want; in the first case we have to check that the value agrees with the thrifty amalgam of A_{i+1}' and B_{i+1} over A_{i+1}. The uniqueness is guaranteed by the fact that the front and left faces of (6.14) are strong amalgams. I leave the details to the reader.

Finally when the chain $(B_i' : i < \omega)$ is formed, we can take B' to be $\bigcup_{i<\omega} B_i'$. $\square$

Exercises for section 8.6

1. Let T be a first-order theory. (a) Show that T has the amalgamation property if and only if $T_{\forall_2}$ has the amalgamation property. (b) Show that if $T_\forall$ has the amalgamation property then so has T, but not necessarily vice versa. [For (a) see the characterisation of AP in Exercise 7.1.10.]

2. Show that $\exists\text{-acl}_B(X)$ is always a substructure of B.

3. Given formulas $\phi(\bar{x}, \bar{z})$, $\psi(\bar{y}, \bar{z})$, we write $\text{Meet}(\phi, \psi)$ for the sentence $\forall \bar{x}\bar{y}\bar{z}(\phi \wedge \psi \to$ 'some element in $\bar{x}$ is also in $\bar{y}$'). Let L be a first-order language and T a theory in L. Show that the following are equivalent. (a) T has the strong amalgamation property. (b) If $\phi(\bar{x}, \bar{z})$ and $\psi(\bar{y}, \bar{z})$ are conjunctions of literals, and $T \vdash \text{Meet}(\phi, \psi)$, then there are quantifier-free formulas $\phi'(\bar{u}, \bar{z})$, $\psi'(\bar{v}, \bar{z})$ of L such that $T \vdash \forall \bar{z}(\exists \bar{u}\, \phi' \vee \exists \bar{v}\, \psi') \wedge \text{Meet}(\phi, \phi') \wedge \text{Meet}(\psi, \psi')$.

4. Let L be a first-order language and T an $\forall_2$ theory in L. Suppose B is a model of T and $A \subseteq B$. Show that the following are equivalent. (a) For every model C of T with $A \subseteq C$, there are a model D of T extending B and an embedding $g: C \to D$ which is the identity on A. (b) For every tuple $\bar{a}$ of elements of A, $\exists\text{-tp}_A(\bar{a})$ and $\exists\text{-tp}_B(\bar{a})$ lie in exactly the same maximal $\exists$-types of T.

5. Let T be an $\forall_2$ theory in a first-order language L. Show that the following are equivalent. (a) $T_\forall$ has the amalgamation property. (b) For every $\exists_1$ formula $\phi(\bar{x})$ of L, the resultant Res_ϕ is equivalent modulo T to a set $\Phi(\bar{x})$ of quantifier-free formulas of L.

6. Prove Theorem 8.6.6.

7. Let T be an $\forall_1$ first-order theory and A, B models of T. (a) Show that the class of strong amalgamation bases for T is closed under unions of chains. (b) Show that if A is a strong amalgamation base for T and $B \leqslant_1 A$ then B is a strong amalgamation base for T. [Given extensions C, D of B, first apply Theorem 8.6.1 to get a strong amalgam of A and C over B, and likewise with D.]

8. Show that in Corollary 8.6.5, we can assume that the class is PC_Δ rather than first-order axiomatisable. [Use Theorem 6.6.1. Note that a PC_Δ class which is closed under elementary substructures is in fact first-order axiomatisable, by Corollary 6.6.2.]

*9. A graph is said to be **N-free** if it has no subgraph isomorphic to the graph

(6.16)

Let T be the theory of N-free graphs. (a) Show that T doesn't have the amalgamation property. (b) Show that there is a finite set of $\exists_1$ formulas of the language of graphs, ϕ_i ($i < n$), such that if we introduce new relation symbols R_i ($i < n$) by the definitions $\forall \bar{x}(R_i\bar{x} \leftrightarrow \phi_i)$, then T together with these definitions does have the amalgamation property.

10. Let $\mathbf{K}$ be as in Theorem 8.6.9. (a) Show that the class of e.c. structures in $\mathbf{K}$ is defined by a single sentence of $L_{\omega_1\omega}$. (b) Let A' be a countable e.c. structure in $\mathbf{K}$. Describe two (first-order) types $p(x)$ and $q(x)$ over $\text{dom}(A')$ with respect to A', such that no e.c. structure in $\mathbf{K}$ is an elementary extension of A' in which both p and q are realised.

History and bibliography

Section 8.1. One could reasonably trace the notion of existential closure back to Cardan (1545) and his Renaissance colleagues who introduced the complex numbers. Modern developments were (i) a notion equivalent to existential closure as we defined it, in Hilbert's Nullstellensatz (Hilbert [1893]), (ii) study of algebraic or similar closures in classes other than the class of fields (Hensel [1907]; Artin & Schreier [1927]; W. Scott [1951]), and (iii) a proof of the existence of existentially closed extensions (see section 8.2). E.c. models were introduced into model theory by Rabin [1962] and Lindström [1964]. Abraham Robinson's group called them **existentially complete** (e.g. Robinson [1971b]). Rabinowitsch's remark is in [1930]. Robinson often wrote of the model-theoretic significance of Hilbert's Nullstellensatz, for example in his [1957].

Exercise 9: Macintyre [1972b]. Exercise 10: Hodkinson & Shelah [199?].

Section 8.2. Theorem 8.2.1 belongs by rights to Steinitz [1910]. The theory of e.c. models of $\forall_2$ first-order theories was opened up by Rabin [1962], who proved the essentials of Theorem 8.2.4. The construction in Theorem 8.2.5 is due to Abraham Robinson (Barwise & Robinson [1970]); my account is influenced by Ziegler [1980]. Theorem 8.2.6 was implicit in Macintyre's [1972a] beautiful proof of Exercise 17, and Keisler [1973] and Simmons [1973] made it explicit. See Hodges [1985] Chapters 3, 4 for a fuller treatment of this material.

The notion of enforceable models was studied by Hodges [1985], and independently by Cameron in lectures in the mid 1980s under the name of 'ubiquitous in category' (see Cameron [1990] p. 114). It is one of several possible answers to a long-standing but vague problem: what are the most representative countable models of a non-ω-categorical theory? See Pouzet [1972a–c], Simmons [1976].

Exercise 2: Marshall Hall [1943]. Exercise 8: Kaiser [1969]. Exercise 13: Bacsich [1972] and Sabbagh [1976] for the general result and B. H. Neumann [1952] for the case of groups. Exercise 14: This is A. Robinson's [1971a] 'weak forcing'. Exercise 15: Barwise & Robinson [1970]. Exercise 16: Henrard [1973], Hodges [1985] Theorem 4.3.4. Exercise 17: Macintyre [1972a].

Section 8.3 Abraham Robinson introduced the notion of model-completeness, with many examples, in his book [1956b]. That book also contains Robinson's test (Corollary 8.3.2) and Theorem 8.3.5. Lindström's test,

Theorem 8.3.4, is from Lindström [1964]. Model companions were studied by Eklof & Sabbagh [1971], who established the connection with e.c. models.

Exercise 7: Baldwin & Kueker [1981]. Exercise 10: Mortimer [1974]. Exercise 11: This example is from Saracino [1975]; Belegradek and Zil'ber also gave an example.

Section 8.4.   See the references to section 2.7 for the method of quantifier elimination. A. Robinson [1958] showed that the model-completion of an $\forall_1$ theory has quantifier elimination; this follows from Exercises 8 and 9. Theorem 8.4.1(d) is a variant which appears in Eklof & Sabbagh [1971]. Shoenfield [1971] gave a model-theoretic necessary and sufficient condition for quantifier elimination (see Exercise 4) which doesn't explicitly mention model-completeness; it uses saturation but is essentially equivalent to Theorem 8.4.1(c). Shoenfield [1977] applies this criterion to prove quantifier elimination for algebraically closed and real-closed fields. Corollary 8.4.3 is based on a criterion (Exercise 2) proposed by van den Dries [1988a], who comments 'A careful analysis of many cases shows that [this criterion] is the most useful'.

For a full proof of quantifier elimination for real-closed fields, using a version of Knebusch's argument, see Prestel [1984]. It seems Lemma 8.4.8 went neglected until about 1980, when van den Dries [1982b] applied it to get Example 4 among other good things.

A great deal of work has been done proving quantifier elimination theorems for various theories. I regret I lack the time and expertise to survey it.

Exercise 4(c) is essentially Shoenfield's [1971] criterion for quantifier elimination. Exercise 5: Feferman [1968a]. Exercise 9: Eklof & Sabbagh [1971] attribute this to Eli Bers, who is in fact an acronym for Barwise, Eklof, (Abraham) Robinson and Sabbagh, the leading people in Robinson's group at Yale in 1970. Exercise 10: Ivanov [1986] proves it when L has a finite number of function symbols. Exercise 11: A. Robinson & Zakon [1960]. Exercise 12: Gurevich [1964]. Exercise 15: Boffa [1986].

Section 8.5.   If T is the theory of algebraically closed fields, then every resultant is equivalent to a single formula by Theorem 8.5.3. These formulas were known in the nineteenth century, for example to Kronecker [1882]. B. H. Neumann [1943] discovered resultants for the theory of groups; his argument works for any variety. Abraham Robinson's group at Yale round about 1970 made the extension to arbitrary $\forall_2$ first-order theories. Example 1 and Exercise 6 are due to Cherlin [1973]. Theorem 8.5.7 and Exercise 12 are in Hodges [1985] Theorem 4.2.6, adapting Corollary 7.2.5 above.

Exercise 2: (a) Higman, Neumann & Neumann [1949]. (b) Eklof & Sabbagh [1971]. Exercise 3: Macintyre [1972b]. Exercise 7: Taĭtslin [1974]. Exercise 10: Pouzet [1972b] and Saracino [1973]. Exercise 13: Hodges [199?].

Section 8.6. As my memory serves me, Corollaries 8.6.2 and 8.6.3 were circulated in a preprint of Bacsich and Fisher in about 1971; Theorem 8.6.1 is an improvement of Lemma 3.2.8 of Hodges [1985]. Corollary 8.6.5 is from Park [1964b] and Vaught [1966]. Theorem 8.6.8 is due to Bacsich [1975]; see also Eklof [1974a]. Theorem 8.6.9 is a simple analogue of the result of Grossberg & Shelah [1983] that AP fails for the class of universal locally finite groups.

Exercise 3: Bryars [1973]; cf. Bacsich & Rowlands Hughes [1974]. Exercise 5: Wheeler [1979a]. Exercise 7: Bacsich [1975]. Exercise 8: Vaught [1966]. Exercise 9: Covington [1989]; see also Albert & Burris [1988] for a study of 'obstructions to amalgamation'.

9

The Horn case: products

Fair maid, white and red,
Comb me smooth, and stroke my head;
And every hair a sheave shall be,
And every sheave a golden tree.
George Peele, The Old Wife's Tale (1595)

Direct products are everywhere in algebra. They are the first general form of construction that the student meets – thanks to groups and vector spaces. From rings to automata they are part of the everyday scenery.

Why do so many types of mathematical object allow this construction? G. Birkhoff gave part of the answer when he showed that every class defined by identities is closed under direct products. J. C. C. McKinsey showed the same for classes defined by universal Horn sentences. A. I. Mal′tsev pointed out the intimate link between direct products and another common construction with a quite different history: presenting a structure by generators and relations.

Once these basic facts about direct products are established, we can set out in at least three directions. First, there is plenty more to say about generators and relations. They can be used to describe not just single structures, but constructions taking one class of structures to another. Many well-known constructions (such as tensor products and polynomial rings) have this form, and we can study the logical features which they share. This topic has recently had a shot of adrenalin through the development of software specification languages based on 'initial semantics'.

The second useful direction is the study of other constructions based on direct products. For example in boolean powers one takes a substructure of a direct product; in reduced products one takes a homomorphic image. A class may be closed under some of these adjusted products even when it is not closed under direct products themselves. For example every first-order axiomatisable class is closed under ultraproducts, and in the 1960s and 1970s a great deal of model theory was built around this fact. (I include a section on ultraproducts. But readers with more than an average interest in ultra-products should probably go straight to the very full account in Chang & Keisler [1973].)

And thirdly one can decompose. Direct products are a way of building up. But they are also a way of breaking down: one takes a structure and shows that it can be analysed into a direct product of factors which have some simple form. Universal algebraists and ring theorists have hugely extended the range of this kind of analysis by generalising from direct products to sheaves. Direct products are the special case of a sheaf over a discrete space.

Model theory doesn't have much to say about this kind of decomposition in general. But there is a useful technique for computing the first-order (or even monadic second-order) theory of a sheaf in terms of the theories of its stalks. The technique carries the names of Solomon Feferman and Robert Vaught, who wrote a pioneering paper on this theme.

9.1 Direct products

We begin by defining direct products. The definition below is the obvious generalisation of the notion of 'direct product' of groups, and it agrees with the usual notion of product in most other classes of algebraic structures.

Let L be a signature and I a non-empty set. Suppose that for each $i \in I$ an L-structure A_i is given. We define the **direct product** $\prod_{i \in I} A_i$ (or $\prod_I A_i$ for short) to be the L-structure B defined in the next paragraph. Direct products are also known as **Cartesian products**; for brevity we usually call them just **products**.

Write X for the set of all maps $a: I \to \bigcup_{i \in I} \mathrm{dom}\,(A_i)$ such that for each $i \in I$, $a(i) \in \mathrm{dom}\,(A_i)$. We put $\mathrm{dom}\,(B) = X$. For each constant c of L we take c^B to be the element a of X such that $a(i) = c^{A_i}$ for each $i \in I$. For each n-ary function symbol F of L and n-tuple $\bar{a} = (a_0, \ldots, a_{n-1})$ from X, we define $F^B(\bar{a})$ to be the element b of X such that for each $i \in I$, $b(i) = F^{A_i}(a_0(i), \ldots, a_{n-1}(i))$. For each n-ary relation symbol R of L and n-tuple $\bar{a}$ from X, we put $\bar{a}$ in R^B iff for every $i \in I$, $(a_0(i), \ldots, a_{n-1}(i)) \in R^{A_i}$. Then B is an L-structure, and we define $\prod_I A_i$ to be B. The structure A_i is called the ith **factor** of the product. If $I = \{0, \ldots, n-1\}$ we write $A_0 \times \ldots \times A_{n-1}$ for $\prod_I A_i$.

If $\bar{a}$ is a sequence of elements $(a_0, a_1, \ldots)$ of the product $\prod_I A_i$, then $\bar{a}(i)$ will always mean $(a_0(i), a_1(i), \ldots)$, and never a_i.

The product $\prod_I A_i$ has some homomorphisms naturally associated with it. For each $j \in I$, define a map $p_j: \mathrm{dom} \prod_I A_i \to \mathrm{dom}\,(A_j)$ by $p_j(a) = a(j)$. The map p_j is called the **(canonical) projection onto the jth factor**.

Theorem 9.1.1. *Let L be a signature, I a non-empty set and $(A_i)_{i \in I}$ a family of L-structures.*

(a) Each projection p_j is a homomorphism from $\prod_I A_i$ to A_j. If no A_i has empty domain then p_j is surjective.

(b) *If C is any L-structure and for each $j \in I, f_j: C \to A_j$ is a homomorphism, then there is a unique homomorphism $g: C \to \prod_I A_i$ such that for each $j \in I, f_j = p_j g$.*

(c) *If each f_j in (b) is an embedding, then g is an embedding.*

Proof. The proof is straightforward from the definitions, and I merely sketch some highlights. Write B for $\prod_I A_i$.

(a) $p_j(F^B(\bar{a})) = (F^B(\bar{a}))(j) = F^{A_j}(a_0(j), \ldots, a_{n-1}(j)) = F^{A_j}(p_j \bar{a})$. Suppose each A_i has an element b_i, and c is any element of A_j; then $c = p_j(a)$ where $a(i) = b_i$ $(i \neq j)$ and $a(j) = c$.

(b) First we show uniqueness. If $f_i = p_i g$ for each $i \in I$, then for each element a of C,

$$(1.1) \qquad\qquad (g(a))(i) = f_i(a) \qquad \text{for each } i \in I.$$

But (1.1) determines $g(a)$ uniquely. To show existence, use (1.1) to define $g: \mathrm{dom}\,(C) \to \mathrm{dom}\,(B)$ and verify that g is a homomorphism.

(c) If $g(a) \neq g(b)$ then there is some $j \in I$ such that $g(a)(j) \neq g(b)(j)$, and so $f_j(a) \neq f_j(b)$. This shows that if each f_j is injective then so is g. The rest of the proof of (c) is similar. $\qquad\qquad\qquad\qquad\qquad\qquad\qquad\qquad\qquad\qquad$ $\square$

A **subdirect product** of structures A_i $(i \in I)$ is a substructure C of $\prod_I A_i$ with the property that for every j in I, the image of C under the projection p_j is the whole of A_j; we call the structures A_j the **subdirect factors** of C. In view of Theorem 9.1.1(a), we only speak of the subdirect product of structures A_i when the structures A_i are all non-empty.

If all the factors of the product $\prod_I A_i$ are equal to some fixed L-structure A, then we say that the product is a **power** (or **direct power**) of A, and we write A^I. We write A^2 for $A \times A$, A^3 for $A \times A \times A$, and so on. By (b) and (c) of the theorem (with $C = A$ and each f_j the identity map), there is an embedding $d: A \to A^I$ such that for each element a of A, $d(a)$ is the constant map with value a. We call d the **diagonal embedding**. (Bear in mind the cartesian plane $\mathbb{R}^2$; d maps $\mathbb{R}$ onto the diagonal line $x = y$.)

It's sometimes convenient to extend the notion of product and allow a product of the empty family of L-structures. By definition this so-called **trivial product** is the L-structure 1 or 1_L defined as follows. The domain of 1_L has just one element a (chosen arbitrarily). For each constant c of $L, c^{1_L} = a$; for every function symbol F of $L, F^{1_L}(a, \ldots, a) = a$. For every relation symbol R of $L, (a, \ldots, a) \in R^{1_L}$. *In this book, products will always be non-trivial.* When I mean to include the non-trivial product, I shall say 'products and 1'.

If L is the signature of groups, then 1_L is the one-element group.

One can iterate the construction of direct products, by taking a product of products. But this never gives anything new. A product of products is already a product.

Theorem 9.1.2. *Suppose that $(A_i: i \in I)$ is a non-empty family of L-structures, and each A_i is of form $\prod\{B_j: j \in J_i\}$ for some disjoint non-empty sets J_i $(i \in I)$. Then putting $J = \bigcup_{i \in I} J_i$, $\prod_I A_i$ is naturally isomorphic to $\prod_J B_j$.*

Proof. Define a map $g: \operatorname{dom}\prod_I A_i \to \operatorname{dom}\prod_J B_j$ by $g(a)(j) = a(i)(j)$ for each $i \in I$, $j \in J_i$ and $a \in \operatorname{dom}\prod_I A_i$. It's easy to check that g is an isomorphism. $\square$

The next result may look like an item in a catalogue, but in fact it's an important feature of direct products and related constructions. Suppose for example that L and L' are disjoint signatures, and for each $i \in I$, A_i is an L-structure and A_i' is an L'-structure such that A_i and A_i' have the same domain. Then Theorem 9.1.3 tells us that $\prod_I A_i$ and $\prod_I A_i'$ have the same domain, so that we can construct an $(L \cup L')$-structure whose L-reduct is $\prod_I A_i$ and whose L'-reduct is $\prod_I A_i'$. This is a kind of amalgamation theorem.

Theorem 9.1.3. *Let L and L^- be signatures with $L^- \subseteq L$. Let $(A_i: i \in I)$ be a non-empty family of L-structures. Then $(\prod_I A_i)|L^- = \prod_I (A_i|L^-)$.*

Proof. Immediate from the definitions. $\square$

Horn formulas

We say that a theory or sentence T is **preserved in products** if every direct product of models of T is a model of T. Likewise we say that a class **V** of L-structures is **closed under products** if every direct product of structures in **V** is itself in **V**. We say that a theory or sentence is **preserved in products and 1** if it is preserved in products and true in 1.

A number of classes well-known to algebraists are closed under products and 1: think of groups, abelian groups, rings, etc. We shall see that this follows at once from the syntactic form of the axioms defining these classes.

A **basic Horn formula** is a formula of the form

(1.2) $\bigwedge \Phi \to \psi$

where Φ is a set of atomic formulas and ψ is either an atomic formula or $\bot$. We allow Φ to be empty; in this case (1.2) is just ψ. We call the formula (1.2) **strict** if ψ is an atomic formula and not $\bot$.

The formula $\phi_0 \wedge \ldots \wedge \phi_{n-1} \to \bot$ is logically equivalent to $\neg \phi_0 \vee \ldots \vee \neg \phi_{n-1}$. So a disjunction of negated atomic formulas is logically equivalent to a basic Horn formula (though not a strict one).

A **Horn formula** is a formula consisting of a finite (possibly empty) string of quantifiers, followed by a conjunction of basic Horn formulas. The set Φ in

(1.2) may be infinite, and so it will be convenient if we allow the quantifiers to be of form $\forall \bar{x}$ or $\exists \bar{x}$ for a possibly infinite sequence $\bar{x}$ of variables. A Horn formula is **strict** if its basic Horn subformulas are strict. Likewise a **Horn theory** is a set of Horn sentences.

A **universal Horn** (or $\forall_1$ Horn) theory is a theory consisting of sentences which are both $\forall_1$ and Horn. However, since $\forall \bar{x}(\phi \wedge \psi)$ is logically equivalent to $\forall \bar{x} \phi \wedge \forall \bar{x} \psi$, it is harmless and slightly more convenient to take universal Horn sentences to be of form $\forall \bar{x} \phi$ where ϕ is a basic Horn formula. A universal Horn theory is then a set of sentences of this form.

Refer back for a moment to section 2.2. All the axioms of theories (2.17)–(2.23) listed there are Horn. Theories (2.24)–(2.27) are not; neither are (2.30)–(2.32). The axioms (2.29) for boolean algebras are Horn, if we read $0 \neq 1$ as $0 = 1 \rightarrow \bot$. In section 2.8, the sentences (8.13) defining I-injective left R-modules for a left ideal I of R are Horn if we bring the quantifier $\exists y$ out into the quantifier prefix; in general these sentences are not first-order.

In Theorem 9.1.5 below we shall show that Horn theories are preserved in products. The next lemma is a step towards that result.

Lemma 9.1.4. *Let L be a signature, and $\psi(\bar{x})$ a formula of the form $\exists \bar{y} \bigwedge \Phi(\bar{x}, \bar{y})$ where Φ is a set of atomic formulas of L. Let $(A_i : i \in I)$ be a non-empty family of L-structures and $\bar{a}$ a sequence of elements of $\prod_I A_i$. Then*

$$(1.3) \qquad \prod_I A_i \vDash \psi(\bar{a}) \quad \text{if and only if} \quad \text{for every } i \in I, A_i \vDash \psi(\bar{a}(i)).$$

(The sequences $\bar{x}, \bar{y}, \bar{a}$ may be infinite.)

Proof. We can assume that the atomic formulas in Φ are unnested (see Theorem 2.6.1). Removal of nesting introduces new existential quantifiers, but we can absorb these into the quantifier prefix $\exists \bar{y}$.

We prove (1.3) by induction on the complexity of ψ. If ψ is a single atomic formula, then ψ is unnested and (1.3) reduces to the definition of $\prod_I A_i$. One quickly deduces that (1.3) holds when ψ is a conjunction of atomic formulas.

In the general case, suppose first that $\prod_I A_i \vDash \psi(\bar{a})$. Then there is a sequence $\bar{b}$ such that $\prod_I A_i \vDash \bigwedge \Phi(\bar{a}, \bar{b})$. So by induction hypothesis, $A_i \vDash \bigwedge \Phi(\bar{a}(i), \bar{b}(i))$ for each $i \in I$, whence also $A_i \vDash \psi(\bar{a}(i))$. The other direction uses the axiom of choice. Suppose that for each $i \in I$ there is a sequence $\bar{b}_i$ of elements of A_i such that $A_i \vDash \bigwedge \Phi(\bar{a}(i), \bar{b}_i)$. Choosing such a $\bar{b}_i$, we can form a sequence $\bar{b}$ of elements of $\prod_I A_i$ by putting $\bar{b}(i) = \bar{b}_i$ for each i. By induction hypothesis, $\prod_I A_i \vDash \bigwedge \Phi(\bar{a}, \bar{b})$ and so $\prod_I A_i \vDash \psi(\bar{a})$. $\square$

In particular a positive primitive sentence is true in a product of structures if and only if it is true in every factor.

Theorem 9.1.5. *Let L be a signature, and let $\chi(\bar{x})$ be a Horn formula of signature L. Let $(A_i : i \in I)$ be a non-empty family of L-structures and $\bar{a}$ a sequence of elements of $\prod_I A_i$. Then*

(1.4) $\text{If } A_i \vDash \chi(\bar{a}(i)) \text{ for every } i \in I, \text{ then } \prod_I A_i \vDash \chi(\bar{a}).$

Proof. First we prove the theorem for a basic Horn formula $\bigwedge \Phi(\bar{x}) \to \psi(\bar{x})$, where Φ is a set of atomic formulas and ψ is either atomic or $\bot$. Suppose χ is of this form, and $A_i \vDash \chi(\bar{a}(i))$ for every $i \in I$. We must show that $\prod_I A_i \vDash (\bigwedge \Phi \to \psi)(\bar{a})$. Suppose then that $\prod_I A_i \vDash \bigwedge \Phi(\bar{a})$. By the lemma it follows that $A_i \vDash \bigwedge \Phi(\bar{a}(i))$ for each $i \in I$. Hence $A_i \vDash \psi(\bar{a}(i))$ for each $i \in I$. If ψ is atomic, then it follows that $\prod_I A_i \vDash \psi(\bar{a})$ by right to left in the lemma, and so $\prod_I A_i \vDash \chi(\bar{a})$ as required. If ψ is $\bot$, then we have reached a contradiction; again it follows that $\prod_I A_i \vDash \chi(\bar{a})$.

Clearly if (1.4) holds when χ is basic Horn, then it holds also when χ is a conjunction of basic Horn formulas.

In the general case we prove (1.4) by induction on the number of alternating blocks $\exists \bar{y}, \forall \bar{z}$ in the quantifier prefix of χ. The proof of Lemma 9.1.4 showed how to deal with an existential quantifier $\exists \bar{y}$, and the argument for a universal quantifier $\forall \bar{z}$ is trivial.   □

Corollary 9.1.6. (a) *Every Horn theory is preserved in products.*

(b) *Every strict Horn theory is preserved in products and* 1.

Proof. (a) is immediate from the theorem. For (b) we only have to check that every strict Horn sentence is true in the trivial product 1_L. But it is easy to see that if ψ is any atomic formula of L, then $1_L \vDash \psi(a, \ldots, a)$; the result follows.   □

As a simple example, the axioms defining von Neumann regular rings ((2.23) in section 2.2 above) are strict Horn, and so the class of von Neumann regular rings is closed under products. A direct proof of this would be tedious though straightforward.

Lemma 9.1.4 and Theorem 9.1.5 together have another corollary, namely McKinsey's lemma below. McKinsey's lemma is fundamental in any branch of logic where Horn formulas appear. For example it implies that if T is a first-order universal Horn theory, then the positive quantifier-free consequences of T are exactly the positive quantifier-free consequences of the set of atomic consequences of T. The fixed point semantics for logic programming (van Emden & Kowalski [1976]) rests on this fact.

Corollary 9.1.7 (McKinsey's lemma). *Let $\mathbf{V}$ be a class of L-structures which is closed under products. Let $\Phi(\bar{x})$ be a set of Horn formulas of signature L and*

$\Psi(\bar{x})$ *a non-empty set of formulas of L which are atomic or* $\bot$. *If every structure in* **V** *is a model of* $\forall \bar{x}(\bigwedge \Phi \rightarrow \bigvee \Psi)$, *then there is* $\psi \in \Psi$ *such that every structure in* **V** *is a model of* $\forall \bar{x}(\bigwedge \Phi \rightarrow \psi)$.

Proof. Suppose to the contrary that for every ψ in Ψ there is a structure A_ψ in **V** with elements $\bar{a}_\psi$ such that $A_\psi \vDash (\bigwedge \Phi \wedge \neg \psi)(\bar{a}_\psi)$. Let B be the product $\prod_{\psi \in \Psi} A_\psi$ (which exists since Ψ is not empty). Let $\bar{b}$ be the sequence of elements of B such that $\bar{b}(\psi) = \bar{a}_\psi$ for each $\psi \in \Psi$. By Theorem 9.1.5, $B \vDash \bigwedge \Phi(\bar{b})$. Since **V** is closed under products, this implies that $B \vDash \theta(\bar{b})$ for some $\theta \in \Psi$. But then $A_\theta \vDash \theta(\bar{a}_\theta)$ by Lemma 9.1.4; contradiction. $\qquad\square$

Exercises for section 9.1

1. Show that Theorem 9.1.1(b) characterises $\prod_I A_i$ up to isomorphism in the following sense. Let $(A_i : i \in I)$ be a non-empty family of L-structures, B an L-structure and for each $j \in I$ let $q_j : B \rightarrow A_j$ be a homomorphism such that Theorem 9.1.1(b) holds with B, q_j in place of $\prod_I A_i, p_j$. Then $B \cong \prod_I A_i$.

2. Prove that for every L-structure A there is a unique homomorphism $f : A \rightarrow 1_L$. Show that this property characterises 1_L up to isomorphism.

3. If A, B and C are L-structures, prove the following isomorphisms: (a) $(A \times B) \times C \cong A \times B \times C$; (b) $A \times B \cong B \times A$; (c) $1_L \times A \cong A$.

4. Let $(A_i : i \in I)$ be a non-empty family of L-structures. Show that $|\prod_I A_i|$ is the product of the cardinals $|A_i|$ $(i \in I)$.

5. Show that for any L-structure A and chain $(B_i : i < \gamma)$ of L-structures, $A \times \bigcup_{i < \gamma} B_i = \bigcup_{i < \gamma} (A \times B_i)$.

6. Let $(A_i : i \in I)$ be a family of L-structures such that every A_i is finite. Show that $\prod_I A_i$ is locally finite.

7. Show that if T is a first-order Horn theory, then T has a skolemisation which is a universal Horn theory. [See Corollary 3.1.3.]

8. Let T be a set of universal first-order Horn sentences such that T has no model but every proper subset of T has a model. Show that T contains exactly one sentence which is not strict. [Use McKinsey's lemma.]

9. Let L be a signature with at least two constant symbols c, d and let A be an L-structure in which $c^A \neq d^A$. Show that every Horn formula $\phi(\bar{x})$ of L is equivalent in A to a strict Horn formula $\psi(\bar{x})$ of L.

10. Let L be a first-order language and V a class of L-structures which is closed under products. Suppose that V is not empty, and that no structure in V is empty. Show that there is a structure A in V such that for every positive sentence ϕ of L, if $A \vDash \phi$ then $B \vDash \phi$ for every $B \in V$.

*11. Let L be the first-order language with empty signature. Write a Horn sentence of L which says 'There are not exactly seven things'.

12. Show that if $A_i, B_i \, (i \in I)$ are L-structures such that $A_i \equiv_{\infty\omega} B_i$, then $\prod_I A_i \equiv_{\infty\omega} \prod_I B_i$.

13. Let L be a first-order language. We say that a formula of L is **special Horn** if it has the form $\forall \bar{x}(\phi \rightarrow \psi)$ where ϕ is positive and ψ is either atomic or $\bot$. We say that a theory T is **preserved in subdirect products** if every subdirect product of non-empty models of T is a model of T. Show that every theory consisting of special Horn sentences is preserved in subdirect products.

14. Let $\mathbf{K}$ be a PC'_Δ class and $\mathbf{J}$ the class of structures isomorphic to subdirect products of structures in $\mathbf{K}$. Show that $\mathbf{J}$ is a PC'_Δ class.

15. (a) Let A be an L-structure, and suppose that there are a positive primitive formula $\phi(\bar{x}, \bar{y})$ of L and tuples $\bar{a}_0, \bar{a}_1, \bar{b}_0, \bar{b}_1$ in A such that $A \vDash \neg \phi(\bar{a}_i, \bar{b}_j)$ if and only if $i = 1$ and $j = 0$. Show that for every infinite set I the power A^I is unstable; in fact there are a sequence $(\bar{c}_i : i < |I|)$ of tuples of elements of A^I and a p.p. formula ψ such that for all $i, j < |I|$, $A^I \vDash \psi(\bar{c}_i, \bar{c}_j) \Leftrightarrow i \leqslant j$. [See (7.7) in the proof of Theorem 9.7.9.] (b) Show that every non-abelian group meets the condition of (a). [Let $\phi(x_1, x_2, y_1, y_2)$ be the formula $x_1 \cdot y_2 = y_2 \cdot x_1$, and consider the pairs $(1, b)$, (a, a) where a, b are non-commuting elements.]

16. Let L be a first-order language and T a theory in L which is preserved in products. We write T^∞ for the theory of infinite models of T. Show that if T^∞ is model-complete then T^∞ has quantifier elimination for non-sentences. [See Theorem 8.4.7(c). If A, B are infinite models of T and $\langle \bar{a} \rangle_A = \langle \bar{a} \rangle_B$ is not empty, note first that $A \times \langle \bar{a} \rangle_A$ is a model of T^∞. Use model-completeness to amalgamate A^2 and $A \times B$ over $A \times \langle \bar{a} \rangle_A$, and $A \times B$ and B^2 and over $\langle \bar{a} \rangle_A \times B$, and hence A^2 and B^2 over $\langle \bar{a} \rangle_A$. Now use the diagonal embeddings of A into A^2 and B into B^2 in order to amalgamate A and B over $\langle \bar{a} \rangle_A$.]

17. Let L be a first-order language and T a model-complete theory in L which is preserved in powers. Show that if U is a complete theory in L which extends T, then U is model-complete and preserved in products.

18. Let L be a first-order language and T a theory in L which is preserved in products. Show that $T_\forall$ is equivalent to a set of universal Horn sentences. [Use McKinsey's lemma.]

19. Let L be a first-order language and T an $\forall_2$ theory in L which is preserved in products. (a) Show that for every positive formula $\phi(\bar{x})$ of L, $\mathrm{Res}_\phi(\bar{x})$ is equivalent modulo T to a set of universal Horn formulas. (b) Show that $T_\forall$ has the amalgamation property if and only if Res_ϕ in (a) can always be taken to be a set of quantifier-free Horn formulas. [For (a), use McKinsey's lemma.]

20. Let L be a first-order language and T a theory in L which is preserved in products. (a) Show that T has the amalgamation property if and only if, for every model A of T, every tuple $\bar{a}$ from A and every p.p. formula $\phi(\bar{x})$, if there is a model $C \supseteq A$ of T in which $\phi(\bar{a})$ holds, then for every model $B \supseteq A$ of T, there is a model D of T such that $B \subseteq D$ and $D \vDash \phi(\bar{a})$. (b) Deduce that if the theory of infinite models of T is model-complete, then T has the amalgamation property.

An L-structure A is said to be **residually finite** *if for every tuple $\bar{a}$ in A and every quantifier-free formula $\phi(\bar{x})$ of L (possibly infinitary) such that $A \vDash \phi(\bar{a})$, there is a homomorphism $f: A \to B$ with B finite, such that $B \vDash \phi(f\bar{a})$.*
21. (a) Show that in the definition of 'residually finite' we can assume without loss that $\phi(\bar{x})$ is negated atomic. (b) Show that an L-structure A is residually finite if and only if A is embeddable in a product of finite homomorphic images of A.

Polynomial interpretations were defined at Exercise 5.3.9.
22. Let Γ be a polynomial interpretation of the signature L in the signature K. Show that if $(A_i : i \in I)$ is a non-empty family of K-structures, then $\Gamma(\prod_{i \in I} A_i) = \prod_{i \in I} \Gamma(A_i)$.

9.2 Presentations

In this section we study the classes that can be axiomatised by universal Horn theories. Some notation from universal algebra will be useful. Let L be a signature and **K** a class of L-structures.

(2.1)   **P(K)** is the class of all direct products of structures in **K**, together with the trivial product **1**.

S(K) is the class of all substructures of structures in **K**.

H(K) is the class of all homomorphic images of structures in **K**.

I(K) is the class of all isomorphic copies of structures in **K**.

Thus for example **K = I(K)** iff **K** is closed under isomorphic copies. As in section 2.2, we write Th (**K**) for the set of all first-order sentences of L which are true in every structure in **K**.

Presentations defined

One can describe the cyclic group of order n as the group generated by a single element g subject to the condition $g^n = 1$. This is a simple example of a definition by presentation. What is it about groups that makes such a definition possible? Theorem 9.2.2 will answer this question.

Let L be a signature and $\mathbf{K}$ a class of L-structures. We define a **presentation**, or more strictly a **K-presentation**, to be a pair $\langle \bar{c}; \Phi \rangle$ (or more strictly $\langle \bar{c}; \Phi \rangle_{\mathbf{K}}$), where $\bar{c}$ is a sequence of distinct constants which are not in L and Φ is a set of atomic sentences of the signature $L(\bar{c})$ which is got by adding $\bar{c}$ to L. We call the constants $\bar{c}$ the **generators** of this presentation. By a **model** of $\langle \bar{c}; \Phi \rangle_{\mathbf{K}}$ we mean an $L(\bar{c})$-structure $(A, \bar{a})$ such that

(2.2) A is in $\mathbf{K}$ and $(A, \bar{a}) \vDash \bigwedge \Phi$.

We say that $\langle \bar{c}; \Phi \rangle_{\mathbf{K}}$ **presents** the $L(\bar{c})$-structure $(A, \bar{a})$, or more loosely that $\langle \bar{c}; \Phi \rangle_{\mathbf{K}}$ **presents** A, if

(2.3) $(A, \bar{a})$ is a model of $\langle \bar{c}; \Phi \rangle_{\mathbf{K}}$ generated by $\bar{a}$, and

(2.4)   for every model $(B, \bar{b})$ of $\langle \bar{c}; \Phi \rangle_{\mathbf{K}}$ there is a homomorphism
           $f: A \rightarrow B$ such that $f\bar{a} = \bar{b}$.

Since $\bar{a}$ generates A, the homomorphism f in (2.4) is necessarily unique.

A remark on generators. The sequence $\bar{c}$ in the presentation $\langle \bar{c}; \Phi \rangle$ is well-ordered to ensure that each parameter c_i names the corresponding element a_i in $(A, \bar{a})$. But otherwise the ordering of the elements in $\bar{c}$ plays no role. We could just as well define a presentation to be a pair $\langle X; \Phi \rangle$ where X is a *set* of new constants and Φ is a set of atomic sentences of $L(X)$; but in that case we would need to find a different way of tying the parameters in X to the elements of a structure. When the correlation of parameter to element is not important, we may as well use the notation $\langle X; \Phi \rangle$. (There is a similar harmless confusion in algebra textbooks, about whether a basis of a vector space is a set or a sequence.)

Lemma 9.2.1. *Let $\Phi(\bar{x})$ be a set of atomic formulas of L, $\langle \bar{c}; \Phi(\bar{c}) \rangle_{\mathbf{K}}$ a* **K**-*presentation and $(A, \bar{a})$ an $L(\bar{c})$-structure with A in* **K**. *Then the following are equivalent.*
(a) *$\langle \bar{c}; \Phi(\bar{c}) \rangle_{\mathbf{K}}$ presents $(A, \bar{a})$.*
(b) *$\bar{a}$ generates A; and for every atomic formula $\psi(\bar{x})$ of L, $A \vDash \psi(\bar{a})$ if and only if every structure in* **K** *is a model of $\forall \bar{x}(\bigwedge \Phi \rightarrow \psi)$.*

Proof. (a) $\Rightarrow$ (b). Assume (a). Then $\bar{a}$ generates A by (2.3). Let $\psi(\bar{x})$ be an atomic formula of L. If $A \vDash \psi(\bar{a})$ and B is a structure in $\mathbf{K}$ such that $B \vDash \bigwedge \Phi(\bar{b})$, then by (2.4) there is a homomorphism $f: A \rightarrow B$ such that $f\bar{a} = \bar{b}$, and so $B \vDash \psi(\bar{b})$ by Theorem 1.3.1(b); hence every structure in $\mathbf{K}$ is a model of $\forall \bar{x}(\bigwedge \Phi \rightarrow \psi)$. Conversely if every structure in $\mathbf{K}$ is a model of this sentence then so is A, and hence $A \vDash \psi(\bar{a})$ by (2.3).

(b) $\Rightarrow$ (a). Assume (b). If $\psi \in \Phi$ then $A \vDash \psi(\bar{a})$ by (b); thus we have (2.3).

If $(B, \bar{b})$ is a model of $\langle \bar{c}; \Phi(\bar{c}) \rangle_{\mathbf{K}}$ then B is in $\mathbf{K}$ and $B \vDash \bigwedge \Phi(\bar{b})$, and by (b) it follows that $(B, \bar{b}) \vDash \mathrm{diag}^+ A$. The diagram lemma, Lemma 1.4.2, gives us the required homomorphism $f \colon A \to B$ to prove (2.4). $\qquad\qquad \square$

For example let $\mathbf{K}$ be the class of groups, and consider the presentation $\langle g; \{g^n = 1\} \rangle$ (for some positive integer n). This presentation presents the n-element cyclic group C_n. In other words, C_n is generated by an element a such that $a^n = 1$, and if B is any group with an element b of order n, then there is a homomorphism from C_n to B taking a to b. Lemma 9.2.1 tells us that the element a in C_n satisfies exactly those equations $\psi(x)$ which follow from $x^n = 1$ and the axioms for groups.

If $\mathbf{K}$ is any class of L-structures and A is any structure in $\mathbf{K}$, then there is a $\mathbf{K}$-presentation which presents A: let $\bar{a}$ list the elements of A without repetition, and consider the presentation $\langle \bar{a}; \mathrm{diag}^+ A \rangle_{\mathbf{K}}$. Sometimes we can do better than this and find a presentation of a particular form.

For example we say that A is **finitely presented** if there are a tuple $\bar{c}$ and a finite set Φ such that $\langle \bar{c}; \Phi \rangle_{\mathbf{K}}$ is a presentation A. When $\mathbf{K}$ is the class of groups, this agrees with the usual notion of 'finitely presented group'. Likewise one says that a finitely generated group G is **recursively presented** if G is presented by some presentation $\langle \bar{c}; \Phi \rangle$ where $\bar{c}$ is finite and Φ is a recursive set of sentences of $L(\bar{c})$ (where L is the signature of groups). Model theory has not too much to say about these notions. I mention them to show how our definitions fit those in the group theory texts.

Let $\mathbf{K}$ be a class of L-structures. We say that $\mathbf{K}$ **admits presentations** if every $\mathbf{K}$-presentation presents a structure in $\mathbf{K}$; in other words, if definition by presentation always works in $\mathbf{K}$.

Theorem 9.2.2. *Let L be a signature and $\mathbf{K}$ a class of L-structures which is closed under isomorphic copies. Then the following are equivalent.*
(a) $\mathbf{K}$ *is closed under products, 1 and substructures.*
(b) $\mathbf{K}$ *admits presentations.*
(c) $\mathbf{K}$ *is the class of all models of some class of (possibly infinitary) strict universal Horn sentences of L.*

Proof. (a) $\Rightarrow$ (b). Suppose $\mathbf{K}$ is closed under products, 1 and substructures, and $\langle \bar{c}; \Phi(\bar{c}) \rangle$ is a $\mathbf{K}$-presentation. Let Ψ be the set consisting of $\bot$ and those atomic formulas $\psi(\bar{x})$ of L such that not every structure in $\mathbf{K}$ is a model of $\forall \bar{x}(\bigwedge \Phi(\bar{x}) \to \psi)$. Note that since $\mathbf{K}$ contains the trivial product, not every structure in $\mathbf{K}$ is a model of $\forall \bar{x}(\bigwedge \Phi \to \bot)$. By McKinsey's lemma (Corollary 9.1.7), there is a structure A in $\mathbf{K}$ with elements $\bar{a}$ such that $A \vDash (\bigwedge \Phi \wedge \neg \bigvee \Psi)(\bar{a})$. By the choice of Ψ, $(A, \bar{a})$ satisfies the second part of (b) in Lemma 9.2.1. Hence so does $(\langle \bar{a} \rangle_A, \bar{a})$ where $\langle \bar{a} \rangle_A$ is the substructure of A generated by $\bar{a}$. But $\langle \bar{a} \rangle_A$ is in **K** since **K** is closed under substructures. So by the lemma, $\langle \bar{c}; \Phi(\bar{c}) \rangle_{\mathbf{K}}$ presents $\langle \bar{a} \rangle_A$ as required.

(b) $\Rightarrow$ (c). Assume **K** admits presentations. Let T be the class of all strict universal Horn sentences χ of L such that every structure in **K** is a model of χ, and let B be a model of T. We must show that B is in **K**. Let $\bar{b}$ list the elements of B without repetition, and consider the presentation $\langle \bar{b}; \text{diag}^+ B \rangle_{\mathbf{K}}$. By assumption this presentation presents some $(A, \bar{a})$ with A in **K**. The diagram lemma gives a homomorphism $f: B \to A$ such that $f\bar{b} = \bar{a}$. Since $\bar{a}$ generates A, f is surjective. We claim that f is an embedding.

For this, first write $\text{diag}^+ B$ as $\Phi(\bar{b})$ where $\Phi(\bar{x})$ is a (possibly infinite) set of atomic formulas of L. Suppose $\psi(\bar{x})$ is an atomic formula of L such that $A \vDash \psi(\bar{a})$. By Lemma 9.2.1, T contains the sentence $\forall \bar{x}(\bigwedge \Phi \to \psi)$. Since B is a model of T, it follows that $B \vDash \psi(\bar{b})$. Hence the claim holds. We infer that f is an isomorphism, and so B is in **K** since **K** is closed under isomorphic copies.

(c) $\Rightarrow$ (a) follows from Corollary 2.4.2(b) and Corollary 9.1.6. $\qquad\qquad\square$

In (c) of the theorem, can we take the class of Horn sentences axiomatising **K** to be a set? Curiously the answer depends on large cardinal assumptions about the set-theoretic universe; see Exercise 19.

Let L be a first-order language. By a **quasivariety** (**in** L) we mean the class of models of a set of strict Horn *first-order* sentences of L.

The class of all L-structures is a quasivariety in L. By inspection of the axioms in section 2.2 above, the following classes are quasivarieties: (2.17) groups, (2.18) groups of exponent n, (2.19) abelian groups, (2.20) torsion-free abelian groups, (2.21) left R-modules, (2.22) rings, (2.28) lattices. Thus each of these classes admits presentations. The class of boolean algebras forms a quasivariety if we allow the one-element boolean algebra.

Theorem 9.2.3. *Let L be a signature and **K** a class of L-structures.*

(a) **ISP(K)** *is the smallest class of L-structures which contains **K** and is closed under products, 1, substructures and isomorphic copies; it is also the class of models of all the strict universal Horn sentences true in every structure in **K**.*

(b) *If either **K** or **ISP(K)** is first-order axiomatisable, then **ISP(K)** is a quasivariety.*

Proof. Let T be the class of all strict universal Horn sentences which are true in every structure in **K**.

(a) Let **J** be the smallest class of L-structures which contains **K** and is closed under products, 1, substructures and isomorphic copies. Then certainly **ISP(K)** $\subseteq$ **J**, and by Corollary 2.4.2(b) and Corollary 9.1.6 every structure in **J** is a model of T. So it suffices to show that every model of T lies in **ISP(K)**.

Let B be a model of T, and let $\bar{b}$ list the elements of B without repetition. Write $\Phi(\bar{x})$ for the set of all atomic formulas $\phi(\bar{x})$ of L such that $B \vDash \phi(\bar{b})$, and $\Psi'(\bar{x})$ for the set of all other atomic formulas of L; put $\Psi = \Psi' \cup \{\bot\}$, thus ensuring that Ψ is not empty. Since B is a model of T, for every formula $\psi(\bar{x})$ in Ψ there is a structure $(A_\psi, \bar{a}_\psi)$ with A_ψ in $\mathbf{K}$, such that $A_\psi \vDash (\bigwedge \Phi \wedge \neg \psi)(\bar{a}_\psi)$. Form the product $(C, \bar{c}) = \prod_\Psi (A_\psi, \bar{a}_\psi)$. Then $\mathrm{diag}(\langle \bar{c} \rangle_C, \bar{c}) = \mathrm{diag}(B, \bar{b})$, and so by the diagram lemma there is an embedding $e: B \to C$ taking $\bar{b}$ to $\bar{c}$. By Theorem 9.1.3 we have embedded B into a product of structures in $\mathbf{K}$, and so $B \in \mathbf{ISP(K)}$.

(b) Suppose U is a first-order axiomatisation of either $\mathbf{K}$ or $\mathbf{ISP(K)}$, and consider a sentence $\forall \bar{x}(\bigwedge \Phi \to \psi)$ in T. Let $\bar{c}$ be a sequence of new constants. Then $U \vdash \forall \bar{x}(\bigwedge \Phi \to \psi)$ and so $U \cup \Phi(\bar{c}) \vdash \psi(\bar{c})$. By the compactness theorem there is some finite subset Φ_0 of Φ such that $U \cup \Phi_0(\bar{c}) \vdash \psi(\bar{c})$. Then $\forall \bar{x}(\bigwedge \Phi_0 \to \psi)$ is a first-order sentence which is true everywhere in $\mathbf{K}$ and implies $\forall \bar{x}(\bigwedge \Phi \to \psi)$. Hence T is equivalent to a set of strict universal first-order Horn sentences. $\qquad \square$

We call $\mathbf{ISP(K)}$ the **quasivariety generated by K**.

Corollary 9.2.4. *Let L be a signature and $\mathbf{K}$ a class of L-structures. Then the following are equivalent.*
(a) *$\mathbf{K}$ is a quasivariety.*
(b) *$\mathbf{K}$ is closed under products, 1 and substructures, and is first-order axiomatisable.*

Proof. (a) $\Rightarrow$ (b) is by Theorem 9.2.2, and the converse is by Theorem 9.2.3 with $\mathbf{K} = \mathbf{ISP(K)}$. $\qquad \square$

The next theorem is a slight variant of Theorem 9.2.2. Clause (b) can be read as saying that if we are given consistent but incomplete information about a structure in $\mathbf{K}$, then there is a canonical way of choosing a structure in $\mathbf{K}$ that fits the information.

Theorem 9.2.5. *Let L be a signature and $\mathbf{K}$ a class of L-structures which is closed under isomorphic copies. Then the following are equivalent.*
(a) *$\mathbf{K}$ is closed under substructures and products.*
(b) *Every $\mathbf{K}$-presentation which has a model $(A, \bar{a})$ with A in $\mathbf{K}$ presents a structure in $\mathbf{K}$.*
(c) *$\mathbf{K}$ is the class of all models of some class of (possibly infinitary) universal Horn sentences of L.*

Proof. Left to the reader. $\qquad \square$

Free structures and bases

Let **K** be a class of L-structures, A a structure in **K** and $\bar{a}$ a sequence of distinct elements of A. We say that $(A, \bar{a})$ is **free** (in **K**) if $(A, \bar{a})$ is presented by the **K**-presentation $\langle \bar{a}; \varnothing \rangle$. If X is the set of all elements in the sequence $\bar{a}$, this definition is equivalent to the following two conditions together:

(2.5) A is in **K** and X generates A, and

(2.6)   for every structure B in **K** and every map $f: X \to \mathrm{dom}\,(B)$
         there is a (necessarily unique) homomorphism $g: A \to B$
         which extends f.

When these conditions hold, we call $\bar{a}$ or X a **basis** of A; we say that A is **free** in **K** if A has a basis, or in other words if some $(A, \bar{a})$ is free in **K**. These definitions agree with the usual use of 'free' as in 'free abelian group', and 'basis' as in 'basis of vector space'.

Lemma 9.2.6. *Let* **K** *be a class of L-structures, A a structure in* **K** *and X a set of elements of A. Then the following are equivalent.*
(a) *A is free with basis X.*
(b) *X generates A; and if $\bar{a}$ is any tuple of distinct elements of X and $\psi(\bar{x})$ an atomic formula of L such that $A \vDash \psi(\bar{a})$, then every structure in* **K** *is a model of $\forall \bar{x}\, \psi$.*

Proof. This is just a special case of Lemma 9.2.1. $\square$

Let A and B be free in **K** with bases X and Y respectively. If $f: X \to Y$ is a bijection then f extends to a homomorphism $g: A \to B$ and f^{-1} extends to a homomorphism $h: B \to A$; the usual argument with Theorem 1.2.1(e) then shows that g is an isomorphism. This has two important consequences. First, *every permutation of the basis X extends to an automorphism of A* – in particular the ordering of the basis in the sequence $\bar{a}$ in the definition of 'free' above is irrelevant.

Second, *a free structure A in* **K** *is determined up to isomorphism by the cardinality of a basis of A*. We write $\mathrm{Fr}_\lambda (\mathbf{K})$ for the structure in **K** which is free in **K** with a basis of cardinality λ (if there is such a structure). Note that by Exercise 9, it can happen that a free structure has two bases of different cardinalities. The structure $\mathrm{Fr}_0 (\mathbf{K})$ (if it exists) has a unique homomorphism to each structure in **K**; it is known as the **initial structure** in **K**.

Theorem 9.2.7. *Let L be a signature and* **K** *a class of L-structures which is closed under products, substructures and isomorphic copies.*
  (a) *If B is a structure in* **K** *which is generated by a set of X of cardinality λ, then $\mathrm{Fr}_\lambda (\mathbf{K})$ exists and B is a homomorphic image of $\mathrm{Fr}_\lambda (\mathbf{K})$.*

(b) *If* **K** *contains a structure with at least two elements then for every cardinal* λ *the free structure* $\mathrm{Fr}_\lambda(\mathbf{K})$ *exists.*

Proof. (a) Take a sequence $\bar{c}$ of λ distinct new constants, and let $(A, \bar{a})$ be the $L(\bar{c})$-structure presented by $\langle \bar{c}; \varnothing \rangle$; there is such a structure by Theorem 9.2.5. Then $(A, \bar{a})$ is free with basis $\bar{a}$ of cardinality λ, so long as there are no repetitions in $\bar{a}$. Suppose to the contrary that there are $i \neq j$ such that $a_i = a_j$. Then by Lemma 9.2.6 we infer that every structure in **K** is a model of $\forall \bar{x}\, x_i = x_j$, so that no structure in **K** has more than one element. But this is absurd; X has at least two elements since $i \neq j$. So $\bar{a}$ has no repetitions. Now we can take a bijection from the set of elements in $\bar{a}$ to X, and extend it to a homomorphism $g: A \to B$. Since X generates B, g is surjective.

(b) follows from (a): if **K** is closed under products and contains a structure with at least two elements, then **K** contains arbitrarily large structures. $\qquad\square$

This quickly leads to a famous result of G. Birkhoff.

Corollary 9.2.8. *Let L be a signature and* **K** *a class of L-structures. Then the following are equivalent.*
(a) **K** *is closed under products, substructures and homomorphic images.*
(b) $\mathbf{K} = \mathrm{HSP}(\mathbf{K})$.
(c) **K** *is axiomatised by a set of sentences of form* $\forall \bar{x}\, \psi$, ψ *atomic.*

Proof. (a) is trivially equivalent to (b). (c) implies (a) by Corollary 2.4.2(b), Theorem 2.4.3(b) and Corollary 9.1.6(b). It remains to prove (a) $\Rightarrow$ (c). Assume (a). Let T be the set of all sentences in $\mathrm{Th}(\mathbf{K})$ of form $\forall \bar{x}\, \psi$ with ψ atomic, and write **J** for the class of models of T. Clearly $\mathbf{K} \subseteq \mathbf{J}$; we must show that $\mathbf{J} \subseteq \mathbf{K}$.

Suppose now that **K** contains a structure with at least two elements. Then for any cardinal λ, $\mathrm{Fr}_\lambda(\mathbf{K})$ exists by Theorem 9.2.7(b). Since $\mathbf{K} \subseteq \mathbf{J}$, $\mathrm{Fr}_\lambda(\mathbf{K})$ is in **J**. But by the implication (c) $\Rightarrow$ (a), **J** is closed under products, substructures and isomorphic copies; so Lemma 9.2.6 shows that $\mathrm{Fr}_\lambda(\mathbf{K})$ is in fact $\mathrm{Fr}_\lambda(\mathbf{J})$. Hence **K** contains all the free structures of **J**. If A is any structure in **J**, then by Theorem 9.2.7(a), A is a homomorphic image of a free structure in **J**, and hence it is in **K** since **K** is closed under homomorphic images.

If **K** contains no structure with more than one element, then the same is true of **J** since T contains the sentence $\forall xy\, x = y$. Then every structure in **K** is a homomorphic image of $\mathrm{Fr}_0(\mathbf{K})$ or $\mathrm{Fr}_1(\mathbf{K})$. $\qquad\square$

When L is a signature containing no relation symbols, a class of L-structures which is axiomatised by a set of sentences of form $\forall \bar{x}\, \psi$ with ψ atomic is called a **variety (in L)**. We call a variety **non-trivial** if it has

structures with more than one element. The class **HSP(K)** is known as the
variety generated by the class of structures **K**.

A glance at the axioms in section 2.2 shows that the following classes are
varieties: (2.17) groups, (2.18) groups of exponent n, (2.19) abelian groups,
(2.21) left R-modules, (2.22) rings, (2.28) lattices. The class of torsion-free
abelian groups is a quasivariety but not a variety; it is not closed under
homomorphic images. The class of boolean algebras is a variety if we allow
the one-element boolean algebra.

There is a reason for debarring relation symbols in varieties. If L is an
algebraic signature and A is an L-structure, then a surjective homomorphism
$f: A \to B$ is determined up to isomorphism once we know the set $\{(a, a')$:
$fa = fa'\}$. This set is a substructure of the power A^2, and it is known as the
kernel of f. Universal algebraists classify the homomorphic images of A by
classifying the substructures of A^2. In the presence of relation symbols this
approach doesn't work, because homomorphic images need not be determined by their kernels.

Some writers don't allow relation symbols in quasivarieties either. I have
dropped this restriction because it seems pointless when homomorphic images
are no longer in the spotlight.

Exercises for section 9.2

1. Let L be a first-order language; let **K** be the class of models of some theory in L
(or more generally, a PC_Δ class of L-structures). Show that **K** is the class of models of
a universal Horn theory in L if and only if **K** is closed under products and
substructures.

2. Let L be a signature and **K** a class of L-structures; let T be the set of all sentences
in Th(**K**) of form $\forall \bar{x} \, \psi$ with ψ atomic. Show that **HSP(K)** is the class of all models of
T.

3. Show that if **K** is a class of L-structures such that **K** = **ISP(K)**, then **K** and **HSP(K)**
have the same free structures (with the same bases).

*The following exercise justifies the name 'absolutely free algebra' for term algebras; see
section 1.3.*
4. Let L be a signature, **V** the quasivariety of all L-structures, X a set of variables
and A the term algebra of L with basis X. Show that A is $\mathrm{Fr}_{|X|}(\mathbf{V})$ with basis X.
[Use Exercise 1.3.7.]

5. Let L be a signature and **V** a class of L-structures which contains a structure with
more than one element and is closed under products, substructures and homomorphic
images. (a) Show that **V** is uniquely determined by $\mathrm{Fr}_\omega(\mathbf{V})$, and also by the set
$\{\mathrm{Fr}_n(\mathbf{V}): n < \omega\}$. (b) Show by examples that if $n < \omega$ then **V** need not be determined
by $\mathrm{Fr}_n(\mathbf{V})$.

6. Let L be a signature and $\mathbf{W}$ a quasivariety in L. Show that $\mathbf{W}$ need not be uniquely determined by $\mathrm{Fr}_\omega(\mathbf{W})$.

The next exercise has computer science implications; see Bergstra & Meyer [1983].
7. Give an example of a finitely axiomatised quasivariety $\mathbf{K}$ such that $\mathbf{HSP(K)}$ is not finitely axiomatisable. [Consider $P(0)$, $\forall x(P(x) \to P(S(S(x))))$.]

8. Let $\mathbf{K}$ be a class of L-structures, and let A and B be free structures in $\mathbf{K}$ with infinite bases. Show that A is back-and-forth equivalent to B.

9. Let $\mathbf{V}$ be the class of all models of the following identities: $\forall x \, F(G(x), H(x)) = x$, $\forall xy \, G(F(x, y)) = x$, $\forall xy \, H(F(x, y)) = y$. (a) Show that $\mathbf{V}$ contains infinite structures. (b) Show that if m and n are any positive integers then $\mathrm{Fr}_m(\mathbf{V}) \cong \mathrm{Fr}_n(\mathbf{V})$.

10. Let $\mathbf{K}$ be a class of L-structures such that some structure in $\mathbf{K}$ is free with a basis of cardinality κ and also with a basis of cardinality λ. Show that if $\kappa \neq \lambda$ then both κ and λ are finite. [The usual proof for vector spaces works in general.]

G-sets were defined at Exercise 4.5.4.
11. Let G be any group. Show that the class of G-sets is a variety, and the free G-sets are exactly the faithful G-sets.

12. Let $\mathbf{V}$ be the class of all groups G such that the intersection $\bigcap_{n<\omega} G^{(n)}$ is $\{1\}$, where $G^{(0)} = G$ and $G^{(n+1)} = [G^{(n)}, G^{(n)}]$. Show that $\mathbf{V}$ is closed under products, 1 and substructures, but is not a quasivariety.

13. Let L be a first-order language with finite signature, and $\mathbf{K}$ a class of L-structures. Show that if A is a finitely presented structure in $\mathbf{K}$, and $\bar{a}$ is a tuple of elements which generates A, then A has a finite presentation of form $\langle \bar{a}, \Phi \rangle$.

14. Let L be a recursive first-order language and $\mathbf{K}$ a recursively axiomatised quasivariety of L-structures, and let A be a finitely generated structure in $\mathbf{K}$. Show that A is recursively presented if and only if A has a presentation of form $\langle \bar{a}; \Phi \rangle$ where $\bar{a}$ is a tuple and Φ is recursively enumerable. (See Craig's trick, Exercise 6.1.3.)

15. Let L be a recursive first-order language and $\mathbf{K}$ a variety axiomatised by a recursive theory T in L. Show (a) if A is a residually finite structure in $\mathbf{K}$ with finite presentation $\langle \bar{a}; \Phi \rangle$, then the set of atomic formulas $\psi(\bar{x})$ of L such that $A \vDash \psi(\bar{a})$ is recursive (i.e. A has **soluble word problem**), (b) the conclusion of (a) is independent of the choice of finite presentation for A, (c) if every finitely presented structure in $\mathbf{K}$ is residually finite, then the set of $\forall_1$ sentences in $\mathrm{Th}\,(\mathbf{K})$ is recursive (i.e. the **universal sentence problem** of $\mathbf{K}$ is soluble).

16. Let $\mathbf{K}$ be a class of L-structures and $m < n$ two positive integers. Show that the following are equivalent. (a) $\mathbf{K}$ has a free structure which has a basis of cardinality m and also a basis of cardinality n. (b) There are terms $t_0(\bar{x})$, $\ldots$, $t_{m-1}(\bar{x})$, $s_0(\bar{y})$, $\ldots$,

$s_{n-1}(\bar{y})$ such that $\mathrm{Th}(\mathbf{K})$ contains all the sentences $\forall \bar{y}(y_i = t_i(s_0(\bar{y}), \ldots, s_{n-1}(\bar{y})))$ $(i < m)$ and $\forall \bar{x}(x_j = s_j(t_0(\bar{x}), \ldots, t_{m-1}(\bar{x})))$ $(j < n)$.

17. Suppose (a) or (b) holds in the previous exercise. Show (a) $\mathbf{K}$ contains no finite structure with more than one element [use (b) to show $|A|^m = |A|^n$], (b) $\mathbf{K}$ contains a structure with unstable theory.

18. Let L be an algebraic signature containing at least a 2-ary function symbol F and constants c, d. Let $\mathbf{K}$ be the class of all L-structures which are models of the theory $\forall xy((x = y \to F(x, y) = c) \wedge (x \neq y \to F(x, y) = d) \wedge c \neq d)$. If A is a structure in $\mathbf{K}$, show that there is a strict universal Horn sentence ϕ_A such that for every structure B in $\mathbf{K}$, $B \vDash \phi_A$ iff A is not embeddable in B.

Vopěnka's Principle *states 'If L is a signature and $\mathbf{K}$ a proper class of L-structures, then there are A, B in $\mathbf{K}$ such that $A \neq B$ and A is embeddable in B'. (Vopěnka's Principle is a large cardinal assumption; for example it implies the existence of a large number of supercompact cardinals. See Jech [1978] p. 414.)*

19. Show that Vopěnka's Principle is equivalent to the statement 'If L is a signature and $\mathbf{K}$ is a class of L-structures which is closed under substructures, products, 1 and isomorphic copies, then $\mathbf{K}$ is the class of models of some *set* of strict universal Horn sentences of L'. [For $\Rightarrow$, consider a chain $\{\Phi_i: i$ an ordinal$\}$ of sets of $\forall_1$ strict Horn theories, such that if $i < j$ then Φ_j implies Φ_i but not vice versa; consider L-structures A_i which are models of Φ_i but not of Φ_{i+1}. For $\Leftarrow$ use the preceding exercise, noting that by set-theoretical reflection, if Φ is a class of $\forall_1$ sentences and ψ is an $\forall_1$ sentence which is equivalent to Φ, then Φ is equivalent to some subset of itself.]

In a quasivariety $\mathbf{W}$, the **coproduct** *$A*B$ of two structures A, B is the structure in $\mathbf{W}$ presented by $\langle \bar{a}\bar{b}; \mathrm{diag}^+ A \cup \mathrm{diag}^+ B \rangle$ where $\bar{a}$ and $\bar{b}$ (assumed disjoint) generate A and B respectively. By the diagram lemma there are natural homomorphisms from A and B to $A*B$.*

20. Let $\mathbf{W}$ be a quasivariety in a first-order language L. Show that the following are equivalent. (a) $\mathbf{W}$ has the joint embedding property. (b) The natural homomorphisms of coproducts in $\mathbf{W}$ are embeddings. (c) If ϕ and ψ are universal Horn sentences of L and $\mathrm{Th}(\mathbf{W}) \vdash \phi \vee \psi$, then either $\mathrm{Th}(\mathbf{W}) \vdash \phi$ or $\mathrm{Th}(\mathbf{W}) \vdash \psi$.

*21. (a) Show that in the variety of R-modules (R a ring), $A*B$ is $A \oplus B$. (b) Show that in the variety of boolean algebras, $A*B$ is the boolean algebra of $\mathrm{St}(A) \times \mathrm{St}(B)$, where $\mathrm{St}(C)$ means the Stone space of C, and that the natural homomorphisms are embeddings. (c) Show that in the variety of commutative rings (with 1, as always), the homomorphisms of coproducts are not necessarily embeddings. [Consider the characteristic.]

22. Let $\mathbf{K}$ be a class of L-structures. Prove the following. (a) If A is free in $\mathbf{K}$ with basis $X \cup Y$ where $X \cap Y = \varnothing$, and Z is a basis of $\langle X \rangle_A$ in $\mathbf{K}$, then $Z \cup Y$ is a basis of A. (b) Suppose $(A_i: i < \gamma)$ is a chain of L-structures with union $A \in \mathbf{K}$, and each A_i is free in $\mathbf{K}$ with basis X_i, such that if $i < j < \gamma$ then $X_i = X_j \cap \mathrm{dom}(A_i)$; then $\bigcup_{i<\gamma} X_i$ is a basis of A in $\mathbf{K}$.

*23. Write N_2^p for the variety of all groups which obey the laws $x^p = 1$ and $[x, [y, z]] = 1$. (These are the groups of exponent p which are nilpotent of class ≤ 2.) (a) Show that the free group F_κ on κ generators in N_2^p is constructed as follows. Introduce formal elements x_i $(i < \kappa)$ and $[x_i, x_j]$ $(i < j < \kappa)$. Let X be the set of elements of the elementary p-group $\bigoplus_{i<\kappa} Z_p x_i$ with basis the set of elements x_i $(i < \kappa)$, and let Y be the set of elements of the elementary p-group $\bigoplus_{i<j<\kappa} Z_p[x_i, x_j]$ with basis the set of elements $[x_i, x_j]$ $(i < j < \kappa)$. On the direct product $X \times Y$, define a group product $*$ as follows:

$$\left(\sum_{i<\kappa} \alpha_i x_i, c\right) * \left(\sum_{i<\kappa} \beta_i x_i, d\right) = \left(\sum_{i<\kappa} (\alpha_i + \beta_i)x_i, c + d - \sum_{i<j<\kappa} (\beta_i \cdot \alpha_j)[x_i, x_j]\right).$$

Then $F_\kappa = \langle X \times Y, * \rangle$ with basis the set of elements $(x_i, 0)$. (b) Show that the commutator in F_κ of any two basis elements $(x_i, 0)$, $(x_j, 0)$ $(i < j)$ is the element $(0, [x_i, x_j])$. Deduce that Y is the commutator subgroup of F_κ.

9.3 Word-constructions

In both algebra and datatype theory there are a large number of constructions which just miss being interpretations. To take a typical example from algebra, suppose we have a field F. Then there is a construction Γ which turns any set A into the vector space $\Gamma(A)$ over F with basis A. Clearly $\Gamma(A)$ is in some sense reducible to A; but each element of $\Gamma(A)$ depends on a finite number of elements of A, and there is no finite upper bound to this number. A typical example from datatype theory is the construction Γ which takes any structure A and forms a structure $\Gamma(A)$ consisting of A together with all finite strings of elements of A, with an operation of concatenation.

Word-constructions are a general framework for handling constructions of this kind. I shall report three properties of word-constructions. First (Corollary 9.3.2), they preserve equivalence in most infinitary languages. Usually they don't preserve elementary equivalence – this is the price they pay for not being interpretations. Second (Theorem 9.3.3), they are always set-theoretically definable, in an explicit and elementary way. (To make the contrast, section 12.3 will discuss some other common algebraic constructions which are in general not set-theoretically definable except 'up to isomorphism'.) And third (Theorem 11.3.6), they are handy for proving non-structure theorems. Word-constructions have some sporadic uses too: for example the proof of Theorem 5.2.1(a) (that PC'_Δ classes are PC_Δ classes) used a word-construction. But the general theory is not much help for these specific cases.

The full definition of word-constructions is unpleasantly syntactic, though it seems to be needed for our proofs. Fortunately there is a more algebraic sufficient condition for a construction to be a word-construction, and it covers most cases of interest (including all the datatype applications, so it seems). The condition is in terms of preservation of filtered colimits, and it appears as Theorem 9.3.6.

Replicas

Let K and K^+ be signatures with $K \subseteq K^+$, and P a 1-ary relation symbol which is in K^+ but not in K. Let $\mathbf{V}$ be a class of K-structures and $\mathbf{W}$ a quasivariety in K^+. We define a map Γ from $\mathbf{V}$ to $\mathbf{W}$ as follows.

Let A be any structure in $\mathbf{V}$. Then $\Gamma(A)$ is defined to be the structure in $\mathbf{W}$ presented by $\langle \mathrm{dom}(A); \mathrm{diag}^+ A \cup \{P(a): a \in \mathrm{dom}(A)\} \rangle_{\mathbf{W}}$. In some sense $\Gamma(A)$ is the 'best approximation' to A in $\mathbf{W}$.

Suppose $f: A \to B$ is a homomorphism between structures in $\mathbf{V}$. Then the map f, applied to the parameters in $\mathrm{dom}(A)$, sends the set $\mathrm{diag}^+ A \cup \{P(a): a \in \mathrm{dom}(A)\}$ into the set $\mathrm{diag}^+ B \cup \{P(b): b \in \mathrm{dom}(B)\}$. Thus

(3.1) $\qquad \Gamma(B) \vDash f(\mathrm{diag}^+ A \cup \{P(a): a \in \mathrm{dom}(A)\})$.

So by the definition of 'presents', there is a unique homomorphism $\Gamma(f): \Gamma(A) \to \Gamma(B)$ which agrees with f on $\mathrm{dom}(A)$.

Theorem 9.3.1. *As defined above, the map Γ is a functor from $\mathbf{V}$ to $\mathbf{W}$.*

Proof. Chase through the definitions, using the uniqueness of Γ on homomorphisms. $\qquad\qquad\qquad\qquad\qquad\qquad\qquad\qquad\qquad\qquad\qquad\qquad$ $\square$

There is no agreed name for functors Γ as defined above. To category theorists they are the left adjoints of forgetful functors between quasivarieties; but this is too heavy. With slight adjustments, some computer theorists know them as parametrised data types – but this is too specialised. Seeing that Mal'tsev [1958a] first described them, he should name them. He calls $\Gamma(A)$ the *replica* of A in $\mathbf{W}$, and accordingly I shall refer to these functors Γ as **replica functors**. The replica functor Γ is completely specified as soon as we know the classes $\mathbf{V}$ and $\mathbf{W}$ in the languages K and K^+.

Example 1: *Polynomial rings.* Let $\mathbf{V}$ be the class of commutative rings and $\mathbf{W}$ the class of commutative rings with distinguished elements X, Y. Then for any commutative ring $R, \Gamma(R)$ is the polynomial ring $R[X, Y]$ with R itself picked out by the relation symbol P. One can check directly that with this interpretation of P, the polynomial ring $R[X, Y]$ is presented in $\mathbf{W}$ by the presentation $\langle \mathrm{dom}(R); \mathrm{diag}^+ R \cup \{P(r): r \in \mathrm{dom}(R)\} \rangle$.

Example 2: *Abelianisations.* Let $\mathbf{V}$ be the class of groups and $\mathbf{W}$ the class of abelian groups. Then for any group $G, \Gamma(G)$ is the abelianisation $G^{\mathrm{ab}} = G/[G, G]$ (with all elements satisfying Px), and Γ is the usual abelianisation functor.

Example 3: *Finite sets.* Let $\mathbf{V}$ be any class of L-structures. Let K^+ have the symbols of K, together with P, a constant $\{\}_0$, for each positive integer n an

n-ary function symbol $\{x_0, \ldots, x_{n-1}\}_n$, and a 2-ary relation symbol *member*.
Let **W** be the quasivariety in K^+ defined by the axioms

(3.2) $\forall x_0 \ldots x_{n-1} \{x_0, \ldots, x_{n-1}\}_n = \{y_0, \ldots, y_{m-1}\}_m$
         (for all positive integers $n \leq m$ and each tuple of vari-
         ables $(y_0, \ldots, y_{m-1})$ which contains exactly the variables
         $x_0, \ldots, x_{n-1}$),

(3.3) $\forall x_0 \ldots x_{n-1}$ *member* $(x_i, \{x_0, \ldots, x_{n-1}\}_n)$
         (for each positive integer n and each $i < n$).

If A is any structure in **V**, let S be the smallest collection of sets such that the
empty set is in S, each singleton $\{a\}$ with $a \in \mathrm{dom}(A)$ is in S, and if σ is a
finite subset of S then σ is in S. Let B be the K^+-structure whose domain is
the disjoint union $\mathrm{dom}(A) \cup S$, with P interpreted as $\mathrm{dom}(A)$,
$\{x_0, \ldots, x_{n-1}\}_n$ interpreted as the finite set $\{x_0, \ldots, x_{n-1}\}$ and member
interpreted as set membership. Then again one can check directly that
$B = \Gamma(A)$.

It's not hard to generalise the notion of an interpretation so as to include
replica functors. The result is as follows.

Word-constructions

We broaden the setting. Let K and L be signatures, and let **V** be a class of
K-structures and **W** a quasivariety of signature L. We shall describe a way of
building up a structure $\Gamma(A)$ in **W** by forming 'words' in the elements of A.

By a **word-construction** Γ from **V** to **W** we mean a triple $(J_\Gamma, \Gamma^{\mathrm{gen}}, \Gamma^{\mathrm{rel}})$
such that (writing J for J_Γ) the following hold.

(3.4) J is a signature consisting of function symbols and constant
         symbols, none of which are in K or L. (J generalises the
         coordinate map of interpretations.)

(3.5) Γ^{gen} is a map which assigns to each term $t(x_0, \ldots, x_{n-1})$ of J
         a formula $\Gamma_t^{\mathrm{gen}}(x_0, \ldots, x_{n-1})$ of $K_{\infty\omega}$.

(3.6) Γ^{rel} is a map whose domain is the set of formulas
         $\phi(t_0(\bar{x}_0), \ldots, t_{k-1}(\bar{x}_{k-1}))$ where $\phi(y_0, \ldots, y_{k-1})$ is an atomic
         formula of L and $t_0, \ldots, t_{k-1}$ are terms of J; to each such
         formula $\psi(\bar{x}_0, \ldots, \bar{x}_{k-1})$, Γ^{rel} assigns a formula $\Gamma_\psi^{\mathrm{rel}}(\bar{x}_0,$
         $\ldots, \bar{x}_{k-1})$ of $K_{\infty\omega}$.

(When **V** and **W** are respectively the quasivarieties of all K-structures and all
L-structures, we say that Γ is a word-construction **from** K **to** L.) For each
structure A in **V** there is a structure $\Gamma(A)$ in **W** defined as follows. Let X be

(3.7)   the set of all expressions $t(\bar{a})$, where $t(\bar{x})$ is a term of $J, \bar{a}$ is
         a tuple from A and $A \models \Gamma_t^{\mathrm{gen}}(\bar{a})$.

Let Φ be the set of all atomic sentences of the form

(3.8) $\phi(t_0(\bar{a}_0), \ldots, t_{k-1}(\bar{a}_{k-1}))$ where $\phi(t_0(\bar{x}_0), \ldots, t_{k-1}(\bar{x}_{k-1}))$ is a
         formula ψ as in (3.6) and $A \vDash \bigwedge_{i<k} \Gamma_{t_i}^{\mathrm{gen}}(\bar{a}_i) \wedge$
         $\Gamma_\psi^{\mathrm{rel}}(\bar{a}_0, \ldots, \bar{a}_{k-1})$. (Note: each $t_i(\bar{a}_i)$ appears in $\phi(t_0(\bar{a}_0),$
         $\ldots, t_{k-1}(\bar{a}_{k-1}))$ as a parameter from X, not as a compound
         term.)

Then $\langle X; \Phi \rangle$ is a **W**-presentation. We define $\Gamma(A)$ to be the structure in **W**
presented by $\langle X; \Phi \rangle$. Thus each element of $\Gamma(A)$ is of the form $t^\sim$ where t
is a word $\in X$ and $\sim$ is the equivalence relation defined by the set of
equations in Φ.

Note that if one of the formulas $\Gamma_t^{\mathrm{gen}}, \Gamma_\psi^{\mathrm{rel}}$ is $\bot$, then it plays no role in the
definition of $\Gamma(A)$. The set of all those formulas $\Gamma_t^{\mathrm{gen}}, \Gamma_\psi^{\mathrm{rel}}$ which are not $\bot$ is
called the **schedule** of Γ.

Example 4: *Replica functors*. Suppose Δ is a replica functor from a class **V** of
K-structures to a quasivariety **W** of K^+-structures; let T be the defining
axioms for **W**. Then we can describe the action of Γ on structures in **V** as a
word-construction in the following way. The signature J_Γ is empty. The
formula Γ_x^{gen} is $x = x$; all the remaining formulas Γ_t^{gen} are $\bot$. The formulas
$\Gamma_{\psi(\bar{x})}^{\mathrm{rel}}$ are given thus. For each atomic formula ψ of L, $\Gamma_{\psi(\bar{x})}^{\mathrm{rel}}$ is $\psi(\bar{x})$. The
formula $\Gamma_{P(x)}^{\mathrm{rel}}$ is $x = x$. The remaining formulas $\Gamma_\psi^{\mathrm{rel}}$ are $\bot$. It's straightforward to check that if A is a structure in **V**, then $\Gamma(A)$ is the structure in **W**
presented by $\langle \mathrm{dom}(A); \mathrm{diag}^+ A \cup \{P(a): a \in \mathrm{dom}(A)\} \rangle$, so that $\Gamma(A) = \Delta(A)$.

Example 5: *Interpretations*. Every interpretation Δ of L in K can be seen as
a word-construction Γ from the class **V** of all K-structures to the class **W** of
all L-structures, as follows. If Δ is n-dimensional, let J_Γ consist of K
together with a single new n-ary function symbol F. The formula $\Gamma_{F(\bar{x})}^{\mathrm{gen}}$ is
$\partial_\Delta(\bar{x})$ and the remaining formulas Γ_t^{gen} are $\bot$. For each unnested atomic
formula $\phi(y_0, \ldots, y_{k-1})$ of L, $\Gamma_{\phi(F(\bar{x}_0), \ldots, F(\bar{x}_{k-1}))}^{\mathrm{rel}}$ is $\phi_\Delta(\bar{x}_0, \ldots, \bar{x}_{k-1})$; the
remaining formulas $\Gamma_\psi^{\mathrm{rel}}$ are $\bot$. If A is an L-structure and $\Delta(A)$ is defined,
then $\Gamma(A)$ is isomorphic to $\Delta(A)$ in an obvious way.

Immediately from the definition, we can prove a reduction theorem which
is the exact analogue of the reduction theorem for interpretations (Theorem
5.3.2).

Theorem 9.3.2. *Let* **V** *and* **W** *be quasivarieties in signatures K and L
respectively, and Γ a word-construction from* **V** *to* **W**. *Then for every formula
$\phi(y_0, \ldots, y_{k-1})$ of $L_{\infty\omega}$ and every K-tuple $(t_0, \ldots, t_{k-1})$ of terms in J_Γ, there
is a formula $\phi_\Gamma^{t_0 \cdots t_{k-1}}(\bar{x}_0, \ldots, \bar{x}_{k-1})$ of $K_{\infty\omega}$ such that for every structure A in*
V *and all tuples $\bar{a}_0, \ldots, \bar{a}_{k-1}$ from A such that $A \vDash \Gamma_{t_i}^{\mathrm{gen}}(\bar{a}_i)$ for all $i < k$,*

(3.9) $\Gamma A \vDash \phi(t_0)(\bar{a}_0)^\sim, \ldots, t_{k-1}(\bar{a}_{k-1})^\sim) \Leftrightarrow A \vDash \phi_\Gamma^{t_0 \cdots t_{k-1}}(\bar{a}_0, \ldots, \bar{a}_{k-1})$.

Proof. The construction of $\phi_\Gamma^{t_0 \cdots t_{k-1}}$, or ϕ_Γ^t for short, is very much like the construction of ϕ_Γ in the proof of Theorem 5.3.2. Just two cases need special treatment. Suppose ΓA is defined by a presentation $\langle X; \Phi \rangle$ as in (3.7) and (3.8).

ϕ *atomic*. Then $\Gamma A \models \phi(t_0(\bar{a}_0)^\sim, \ldots, t_{k-1}(\bar{a}_{k-1})^\sim)$ if and only if some finite subset of $\mathrm{Th}(\mathbf{W}) \cup \Phi$ implies $\phi(t_0(\bar{a}_0), \ldots, t_{k-1}(\bar{a}_{k-1}))$. So $k-1$) will be a disjunction (in general infinite) of formulas of the form

$$(3.10) \qquad \exists \bar{z} \left(\bigwedge_{i<p} \Gamma_{s_i}^{\mathrm{gen}}(\bar{x}_0, \ldots, \bar{x}_{k-1}, \bar{z}) \wedge \bigwedge_{j<q} \Gamma_{\psi_j}^{\mathrm{rel}}(\bar{x}_0, \ldots, \bar{x}_{k-1}, \bar{z}) \right)$$

such that each formula (3.10) puts into Φ a finite set of formulas which together with $\mathrm{Th}(\mathbf{W})$ imply $\phi(t_0(\bar{x}_0), \ldots, t_{k-1}(\bar{x}_{k-1}))$.

ϕ *of the form* $\exists z\ \psi(y_0, \ldots, y_{k-1}, z)$. Then

$$\Gamma A \models \phi(t_0(\bar{a}_0)^\sim, \ldots, t_{k-1}(\bar{a}_{k-1})^\sim)$$

if and only if there are a term $r(\bar{w})$ of L, terms $s_0, \ldots, s_{j-1}$ in J_Γ and tuples $\bar{b}_0, \ldots, \bar{b}_{j-1}$ in A, such that $A \models \bigwedge_{i<j} \Gamma_{s_i}^{\mathrm{gen}}(\bar{b}_i)$ and $\Gamma A \models \psi(t_0(\bar{a}_0)^\sim, \ldots, t_{k-1}(\bar{a}_{k-1})^\sim, r(s_0(\bar{b}_0)^\sim, \ldots, s_{j-1}(\bar{b}_{j-1})^\sim))$. So $\phi_\Gamma^t(\bar{x}_0, \ldots, \bar{x}_{k-1})$ will be a disjunction (in general infinite) of formulas of the form

$$(3.11) \qquad \exists \bar{z}_0 \ldots \bar{z}_{j-1} \left(\bigwedge_{i<j} \Gamma_{s_i}^{\mathrm{gen}}(\bar{z}_i) \wedge \psi_\Gamma'^{\,ts}(\bar{x}_0, \ldots, \bar{x}_{k-1}, \bar{z}_0, \ldots, \bar{z}_{j-1}) \right)$$

where ψ' is $\psi(y_0, \ldots, y_{k-1}, r(\bar{w}))$.

Likewise if ϕ begins with a universal quantifier. $\qquad \square$

We want an estimate of how much more complicated ϕ_Γ^t can be than ϕ. Let K' be the smallest first-order-closed sublanguage of $K_{\infty\omega}$ which contains the schedule of Γ. Then we define $|\Gamma|$ to be $|K'| + |L| + |J_\Gamma|$. One can check that if $\lambda \geq |\Gamma|$ and ϕ is a formula of $L_{\lambda^+\omega}$ then ϕ_Γ^t is a formula of $K_{\lambda^+\omega}$. So we have a preservation theorem as follows. We say that Γ **preserves** $\equiv_{\kappa\omega}$ if $A \equiv_{\kappa\omega} B$ implies $\Gamma(A) \equiv_{\kappa\omega} \Gamma(B)$ whenever A and B are in the domain of Γ.

Corollary 9.3.3. *Let* Γ *be a word-construction from* $\mathbf{V}$ *to* $\mathbf{W}$*, and* λ *a cardinal* $\geq |\Gamma|$*. Then* Γ *preserves* $\equiv_{\lambda^+\omega}$*.*

Proof. Apply the theorem to sentences of $K_{\lambda^+\omega}$. $\qquad \square$

In particular if $A \equiv_{\infty\omega} B$ then $\Gamma A \equiv_{\infty\omega} \Gamma B$. But note that λ^+ in the corollary is always uncountable; word-constructions generally don't preserve elementary equivalence. (See Exercises 2, 3.)

The construction of $\Gamma(A)$ from A is very explicit and uniform. With a little generalisation, word-constructions are probably the best way to make precise the intuitive notion of a 'concrete' construction. The little generalisation is to allow terms that use infinite sequences of elements of A – as in formal power series rings $k[[X]]$ over a field k, where each element has infinitely many coefficients in the ground field k. In any case we have the following; its proof is uninteresting.

Theorem 9.3.4. *If* Γ *is a word-construction from* **V** *to* **W***, then the map* $A \mapsto \Gamma(A)$ *on* **V** *is defined by a* Σ_1 *formula of set theory, with parameters the signature of* **W***, the set of axioms of* **W** *and the set* $\Gamma = \langle J_\Gamma, \Gamma^{gen}, \Gamma^{rel} \rangle$. $\qquad\square$

Presenting word-constructions

A word-construction is defined only on structures, not on maps, But in favourable cases we can extend its definition to all homomorphisms as well. **Positive primitive** (p.p.) formulas were defined in section 2.5.

Let us say that a word-construction Γ from **V** to **W** is **presenting** if every formula Γ_t^{gen} is either $\perp$ or a (finite) conjunction of p.p. formulas, and every formula Γ_ψ^{rel} is a disjunction (possibly infinite, possibly empty) of p.p. formulas.

Suppose that Γ is presenting and $f: A \to B$ is a homomorphism between structures A and B in **V**. Let $\langle X(A), \Phi(A) \rangle$ and $\langle X(B), \Phi(B) \rangle$ be the **W**-presentations which define $\Gamma(A)$ and $\Gamma(B)$ respectively. Then for each term t of J_Γ and tuple $\bar{a}$ of elements of A, $A \vDash \Gamma_t^{gen}(\bar{a})$ implies $B \vDash \Gamma_t^{gen}(f\bar{a})$, and so f induces a map $f': X(A) \to X(B)$. By the same argument with Γ^{rel}, this map f' induces a map $f'': \Phi(A) \to \Phi(B)$, and so there is an induced homomorphism $\Gamma(f): \Gamma(A) \to \Gamma(B)$. Thus *a presenting word-construction* Γ *induces a functor defined on homomorphisms*. We call this functor the **associated functor** of Γ, and there is no danger of confusion if we call it Γ too.

The associated functor of the word-construction Γ in Example 4 turns out to be the replica functor Δ. Indeed its definition is just the definition of Δ, thinly disguised.

Not all word-constructions are presenting, and not all presenting word-constructions define replica functors. But there is a neat criterion for a functor between quasivarieties to be the associated functor of a presenting word-construction (up to isomorphism). To state it, we need to study filtered colimits. This complication seems to be necessary for good preservation results.

Filtered colimits

Let **V** be a quasivariety. Suppose **S** is a set of structures in **V** which contains at least one example of each isomorphism type of finitely presented structure in **V**. Then each structure in **V** can be described in terms of **S**. There are several ways of achieving this. The method described below does it uniformly for every structure in **V**, without having to make any arbitrary choices.

The key notion is that of a *filtered colimit*. Since this is not a text of category theory, I give an informal definition. A **filtered diagram in V** consists of a picture D which shows some structures in **V** (labelled as A_i where i ranges over some index set $I = I(D)$) and some homomorphisms between them (so that each homomorphism is labelled as being from some A_i to some A_j), and specifies that certain homomorphisms are equal, subject to the following conditions:

(3.12) given indices i and j in I, there are an index k in I and homomorphisms $f: A_i \to A_k$ and $g: A_j \to A_k$;

(3.13) given i, j in I and any two homomorphisms $f: A_i \to A_j, g: A_i \to A_j$, there is a homomorphism $h: A_j \to A_k$ such that $hf = hg$.

A **right cone** of the filtered diagram D is a pair (η, B) where B is a structure in **V** and η is a map taking each $i \in I$ to a homomorphism $\eta_i: A_i \to B$, in such a way that for every homomorphism $f: A_i \to A_j$ in D we have $\eta_i = \eta_j \cdot f$. A **colimit** of the filtered diagram D is a right cone (η, B) of D which is 'as close to D as possible', in the following sense: if (ζ, C) is any right cone of D, then there is a unique homomorphism $g: B \to C$ such that for every $i \in I$, $\zeta_i = g \cdot \eta_i$.

A filtered diagram need not have a colimit. But if it does have a colimit (η, B), then category arguments quickly show that the colimit structure B is determined up to isomorphism by D. A **filtered colimit** is a colimit of a filtered diagram; it is a **filtered colimit of** structures in a class **S** if all the substructures A_i in the filtered diagram are in **S**.

Filtered colimits behave well in quasivarieties, as the next two lemmas show.

Lemma 9.3.5. *Let* **V** *be a quasivariety and* D *a filtered diagram. Then* D *has a colimit in* **V**.

Proof. We can assume that the domains of all the structures A_i in D are pairwise disjoint. Let X be the union $\bigcup\{\mathrm{dom}(A_i): i \in I(D)\}$ and let Φ be $\bigcup\{\mathrm{diag}^+ A_i: i \in I(D)\}$. Let Ψ be the set of all equations $a = c$ where $f: A_i \to A_j$ is a map in D, a is an element of A_i and c is $f(a)$. Let B be the structure in **V** presented by $\langle X; \Phi \cup \Psi \rangle$. Since $\mathrm{diag}^+ A_i \subseteq \Phi$, there is a unique induced homomorphism $\eta_i: A_i \to B$. The equations in Ψ ensure that these maps commute so as to make (η, B) a right cone for D. Finally if (ζ, C) is another right cone for D, we interpret each constant $c \in \mathrm{dom}(A_i)$ as a name of the element $\zeta_i(c)$ in D, and this makes C into a model of $\Phi \cup \Psi$.

The induced homomorphism $f: B \to C$ satisfies the commutativity requirements, and it is uniquely determined since the elements named by constants
in X generate B. So (η, B) is a colimit of D. $\square$

For category-minded readers I remark that the argument of Lemma 9.3.5
works for any diagram in $\mathbf{V}$; the filter properties (3.12), (3.13) are irrelevant.
Hence $\mathbf{V}$ is cocomplete. The adjoint functor theorem gives a totally different
proof of this fact.

Direct limits are the special case of filtered colimits where everything
commutes. Thus the next lemma is an analogue of Exercise 2.4.4.

Lemma 9.3.6. *Let $\mathbf{V}$ be a quasivariety of K-structures, and let $\mathbf{S}$ be a class of
structures in $\mathbf{V}$, which contains at least one isomorphic copy of each finitely
presented structure in $\mathbf{V}$. Then in a uniform way, each structure in $\mathbf{V}$ can be
expressed as a filtered colimit of structures in $\mathbf{S}$.*

Proof. Let B be a structure in $\mathbf{V}$. We build a filtered diagram D. As index
set I we take the set of all homomorphisms $i: A \to B$ where A is a finitely
presented structure in $\mathbf{S}$, and we put $A_i = A$. We include a homomorphism
$f: A_i \to A_j$ in the diagram D if and only if $i = j \cdot f$. The diagram states that
two maps are equal exactly when they are equal. I leave it as an exercise to
check that this works. $\square$

We say that a functor Γ from a class $\mathbf{V}$ to a class $\mathbf{W}$ **preserves filtered
colimits** if the following holds: If (η, B) is a colimit of a filtered diagram D in
$\mathbf{V}$, then $(\Gamma\eta, \Gamma(B))$ is a colimit of $\Gamma(D)$ in $\mathbf{W}$. (Here $(\Gamma\eta, \Gamma(B))$ is the result
of applying Γ to all structures and maps in (η, B).)

The next theorem gives an algebraic sufficient condition for a construction
to be a word-construction. It includes the fact that replica functors preserve
filtered colimits; but category theorists will note that since replica functors are
left adjoints, they preserve all colimits whatever.

Let L and L^+ be signatures with $L \subseteq L^+$, and suppose one of the symbols
in L^+ but not in L is a 1-ary relation symbol R. The **relativised reduction**
from a class $\mathbf{V}$ of L^+-structures to a class $\mathbf{W}$ of L-structures is the function
which takes each L^+-structure A to the substructure A_P of $A|L$ with domain
P^A. When we speak of the relativised reduction from $\mathbf{V}$ to $\mathbf{W}$, this is
understood to imply that A^P is well-defined for every A in $\mathbf{V}$. The relativised
reduction can be regarded as a functor which is defined on homomorphisms,
since every homomorphism $f: A \to B$ in $\mathbf{V}$ induces a homomorphism
$f_P: A_P \to B_P$. This functor preserves filtered colimits (Exercise 12).

Theorem 9.3.7. *Let Γ be a functor from a quasivariety $\mathbf{V}$ of signature K to a quasivariety $\mathbf{W}$ of signature L. Then the following are equivalent (up to isomorphism of structures – or for category theorists, up to natural isomorphism of functors).*

(a) *Γ preserves filtered colimits.*

(b) *There is a quasivariety $\mathbf{U}$ such that Γ can be written as $\Xi\Delta$ where $\Delta: \mathbf{V} \to \mathbf{U}$ is a replica functor and $\Xi: \mathbf{U} \to \mathbf{W}$ is a relativised reduction.*

(c) *Γ is the associated functor of a presenting word-construction.*

Proof. (a) $\Rightarrow$ (b). Suppose that Γ preserves filtered colimits. Let Π be the class of finite $\mathbf{W}$-presentations; we can make Π into a set by identifying two presentations when they differ only in the choice of parameters. Index the presentations in Π as $\langle \bar{c}_k; \Phi_k \rangle$, and let C_k be the structure presented by $\langle \bar{c}_k; \Phi_k \rangle$. Let $\mathbf{S}$ be the set of all the structures C_k.

We introduce a new signature J. Its symbols are those of L, together with a new 1-ary relation symbol E (for 'exists'), and a family of new function symbols $F_{k,c}$, one for each presentation $\langle \bar{c}_k; \Phi_k \rangle$ and each element c of $\Delta(C_k)$; the arity of $F_{k,c}$ is the length of $\bar{c}_k$. We shall describe a quasivariety $\mathbf{U}$ in the signature J; the replica functor Δ of (b) will be from $\mathbf{V}$ to $\mathbf{U}$. Intuitively, if A is a structure in $\mathbf{V}$, then the element $F_{k,c}(\bar{a})$ will be the image of c under the homomorphism $\Delta(f): \Delta(C_k) \to \Delta(A)$ where $f: C_k \to A$ is got by taking the elements named by $\bar{c}_k$ to the elements $\bar{a}$. The formula $E(x)$ will express that x is an element of $\Gamma(A)$; the relativised reduction will consist of passing from a structure B in $\mathbf{U}$ to the L-structure B_E.

The quasivariety $\mathbf{U}$ will have one set of laws which express that $F_{k,c}(\bar{a})$ exists whenever there is a homomorphism $f: C_k \to A$ taking $\bar{c}_k$ to $\bar{a}$. These laws are easy to write down from the definition of C_k:

(3.14) $\forall \bar{x} (\bigwedge \Phi_k(\bar{x}) \to E(F_{k,c}(\bar{x})))$ for all k and c.

A second set of laws say that whenever there is a homomorphism $f: C_k \to A$ taking $\bar{c}_k$ to $\bar{a}$, then any atomic formula satisfied by $\bar{c}_k$ in C_k is also satisifed by $f(\bar{c}_k)$ in A:

(3.15) $\forall \bar{x}\ \phi(F_{k,c_0}(\bar{x}_0), \ldots, F_{k,c_{n-1}}(\bar{x}_{n-1}))$
             whenever ϕ is an atomic formula and $C_k \vDash \phi(\bar{c}_k)$.

Finally there is a third set of laws which express when $F_{k,c}(\bar{a}) = F_{k',c'}(\bar{a}')$ holds. By the properties of the diagram in Lemma 9.3.5, it holds if there is some $C_{k''}$ with presentation $\langle \bar{c}_k \,\hat{}\, \bar{c}_{k'} \,\hat{}\, \bar{d}; \sum(\bar{c}_k, \bar{c}_{k'}, \bar{d}) \rangle$, with $\sum(\bar{x}, \bar{y}, \bar{z})$ containing $\Phi_k(\bar{x})$ and $\Phi_{k'}(\bar{y})$, such that some homomorphism $h: C_{k''} \to A$ takes the elements named $\bar{c}_k, \bar{c}_{k'}$ to the elements $\bar{a}, \bar{a}'$ respectively, and if $f: C_k \to C_{k''}$ and $g: C_{k'} \to C_{k''}$ are the induced maps, then $\Delta(f)(c) = \Delta(g)(c')$. There may be many candidates for $C_{k''}$; for each one there is a corresponding law of $\mathbf{W}$:

(3.16) $\forall \bar{x}\,\bar{y}\,\bar{z}(\bigwedge\sum(\bar{x},\,\bar{y},\,\bar{z}) \rightarrow F_{k,c}(\bar{x}) = F_{k',c'}(\bar{y})).$

This defines the quasivariety **W**.

(b) $\Rightarrow$ (c) can be checked directly.

(c) $\Rightarrow$ (a). Let D be a filtered diagram in **V** with colimit (η, B). By the uniqueness of colimits, we can suppose that (η, B) is the colimit given by the proof of Lemma 9.3.5 above. Since Γ is a functor, it preserves commutativity of diagrams, and so $(\Gamma_\eta, \Gamma(B))$ is a right cone of $\Gamma(D)$. To prove that $(\Gamma\eta, \Gamma(B))$ is a colimit cone, it will suffice if we show that for every atomic formula $\psi(x_0, \ldots, x_{n-1})$ of K and every n-tuple $\bar{b}$ of elements of $\Gamma(B)$ such that $\Gamma(B) \vDash \psi(\bar{b})$, there is some A_i in D such that $\Gamma(\eta_i)(\bar{c}) = \bar{b}$ for some elements of $\bar{c}$ of $\Gamma(A_i)$, and $\Gamma(A_i) \vDash \psi(\bar{c})$.

Let $\langle X, \Phi \rangle$ be the presentation used to define $\Gamma(B)$, and T the axioms defining **W**. Express $\bar{b}$ as $\bar{r}(\bar{t}(\bar{d}))$ where $\bar{r}$ are terms of L, $\bar{t}$ are terms of J_Γ and $\bar{d}$ is a tuple of elements of B. We can dismiss the terms $\bar{r}$ by absorbing them into ψ. Since $\Gamma(B) \vDash \psi(\bar{b})$, we can suppose (after adjusting ψ to add more items to $\bar{d}$ if necessary) that there is a finite subset $\Phi'(\bar{d})$ of Φ such that $T \vdash \forall\bar{x}(\bigwedge\Phi'(\bar{x}) \rightarrow \psi(\bar{t}(\bar{x})))$. Since Γ is a presenting word-construction, each formula in Φ' is accounted for by some finite set of atomic sentences true in B. But D is filtered, and so there is some A_i in D with elements $\bar{a}$ such that $\eta_i(\bar{a}) = \bar{d}$ and all the sentences $\Phi'(\bar{a})$ are in the presentation defining $\Gamma(A_i)$. Then $\Gamma(A_i) \vDash \psi(\bar{t}(\bar{a}))$, and $\Gamma(\eta_i)$ takes the elements $\bar{t}(\bar{a})$ to the elements $\bar{t}(\bar{d}) = \bar{b}$. This is what we had to show. $\square$

The proof of Theorem 9.3.7 tells us more.

Corollary 9.3.8. *Suppose* **V** *and* **W** *are quasivarieties of countable signature, and* Γ *is a functor from* **V** *to* **W** *which preserves filtered colimits. Then* Γ *preserves* $\equiv_{\omega_1\omega}$.

Proof. Since the signatures are countable, the set **S** and the schedule of the word-construction Γ as defined in the proof of Theorem 9.3.7 are countable, and the schedule lies in $L_{\omega_1\omega}$. The proof of Theorem 9.3.2 does the rest. $\square$

In case any enthusiasts for algebraic specification have wandered this far into the text, let me remark that functors as in (b) of the theorem are very nearly the functors that one uses in algebraic specification. To get exactly there, one should allow many-sorted structures and partial functions, and one should add a condition which amounts to saying that the set **S** in the proof is recursively given. The replica functor is a free functor, and the items which are thrown away by the relativised reduction are what are known as the 'hidden' features of the specification (Thatcher, Wagner & Wright [1978]; see also Bergstra & Tucker [1987]). Clause (a) expresses a continuity requirement which any believable notion of datatype must meet: information depends on only finitely many facts at a time. From this point of view, the theorem proves that initial algebra specification, in spite of its *ad hoc* appearance, is in fact the most general possible form of specification. (It also proves that the Russians got there first.)

In the Appendix, section A.5, we shall discuss a heuristic principle in algebraic geometry which is credited to Lefschetz. Corollaries 9.3.3 and 9.3.8 give two formalisations of Lefschetz' principle. See Exercise 10 below for yet another version, perhaps the most useful.

Another direction

If Γ is a word-construction but not presenting, then we can define a functor associated with Γ; but it is only defined on those homomorphisms which preserve all the formulas in Γ^{gen} and Γ^{rel}. The most interesting case is where the formulas in Γ^{gen} and Γ^{rel} are all quantifier-free; let us call these **Feferman word-constructions**. I state the main properties of Feferman word-constructions without proof.

Theorem 9.3.9. (a) *A Feferman word-construction has an asociated functor which is defined on all embeddings, takes embeddings to embeddings, and preserves direct limits of embeddings.*

(b) *Let Γ be any word-construction with domain $\mathbf{V}$ in signature L. By expanding L so as to atomise all the formulas in the schedule of Γ, we can form a class $\mathbf{V}' = \{A^{\#} : A \in \mathbf{V}\}$ and a Feferman word-construction Γ' with domain $\mathbf{V}'$, such that for each structure A in $\mathbf{V}$, $\Gamma(A) = \Gamma'(A^{\#})|L$.* ☐

Exercises for section 9.3

1. Show that the composition of two replica functors is a replica functor.

2. Give an example of a replica functor which doesn't preserve elementary equivalence. [$\mathbf{V}$ has one constant symbol 0, one 1-ary function symbol S and no axioms; $\mathbf{W}$ has in addition 1-ary relation symbols P, Q and axioms $Q(0)$, $\forall x(Q(x) \to Q(Sx))$.]

The next exercise gives a more interesting answer to the same question.
*3. Show that there are elementarily equivalent algebraically closed fields A, B such that $A[X] \not\equiv B[X]$.

4. Show that if Γ is a word-construction from $\mathbf{U}$ to $\mathbf{V}$, and Δ is a word-construction from $\mathbf{V}$ to $\mathbf{W}$, then we can explicitly define a word-construction $\Delta \circ \Gamma$ which acts on structures in $\mathbf{U}$ as the composition of Γ and Δ (up to isomorphism).

5. Let Γ be a word-construction from signature K to signature L, and let A be a K-structure. Show that $|\Gamma A| \leq |A| + |\Gamma|$.

6. Complete the proof of Lemma 9.3.6. *(This is a typical comma category construction; see Mac Lane [1971] pp. 46ff.)*

7. Let **V** be the variety of abelian groups and B an abelian group. Show directly that the map $A \mapsto A \otimes B$ can be written as a presenting word-construction from **V** to **V**.

8. Let **V** and **W** be quasivarieties, and Γ a functor from **V** to **W** which is defined on embeddings (not on all homomorphisms). Show that if Γ preserves direct limits of embeddings then Γ is the associated functor of a word-construction from **V** to **W** (not in general either presenting or Feferman).

9. Let **V** and **W** be quasivarieties and Γ a functor from **V** to **W** which is defined on embeddings and takes embeddings to embeddings, and preserves direct limits of embeddings. Show by a back-and-forth argument that Γ preserves $\equiv_{\infty\omega}$.

10. Let **W** be any quasivariety and let **U** be the class of algebraically closed fields of a fixed characteristic and infinite transcendence degree. Show that if Γ is a functor from **U** to **W** which preserves direct limits, then Γ preserves $\equiv_{\infty\omega}$. *This is not a special case of Corollary 9.3.3: Γ is in general not expressible as a word-construction.*

11. Show directly that every replica functor preserves coproducts.

12. Show that every relativised reduction functor preserves filtered colimits.

13. Give an example of a functor from a quasivariety **V** to a quasivariety **W** which preserves filtered colimits but not coproducts (and hence is not a replica functor). [The boolean power functor $M[-]^*$ of section 9.7 below is an example: if 2 is the two-element boolean algebra and M is, say, the infinite cyclic group in the variety of abelian groups, then $M[2*2]^* = M[2]^* = M \neq M*M$.]

9.4 Reduced products

Reduced products are homomorphic images of direct products, got by factoring out a filter over the index set (see section 6.2 for filters over sets). If the filter is an ultrafilter, the reduced product is known as an ultraproduct. Ultraproducts are intimately connected with the compactness theorem, and so I treat them separately in the next section. Reduced products in general are less interesting. But they appear in a representation theorem for quasi-varieties (Theorem 9.4.7 below). The challenge of describing the first-order theories of reduced products led to some powerful work in the 1960s; see Chapter 6 of Chang & Keisler [1973].

We say that a filter $\mathcal{F}$ over a set I is λ-**complete** if for every family G of fewer than λ sets in $\mathcal{F}$, the intersection $\bigcap G$ is also in $\mathcal{F}$. Countably complete (i.e. ω_1-complete) ultrafilters are rather rare. By contrast it's very easy to find λ-complete filters for any cardinal λ. One consequence is that the theory of quasivarieties is not greatly affected if we allow infinitary axioms, or even functions and relations of infinite arity. I explore this briefly in the exercises.

Reduced products defined

Let L be a first-order language, I a non-empty set and $(A_i: i \in I)$ a family of non-empty L-structures. Let $\phi(\bar{x})$ be a formula of L and $\bar{a}$ a tuple of elements of the product $\prod_I A_i$. We define the **boolean value** of $\phi(\bar{a})$, in symbols $\|\phi(\bar{a})\|$, to be the set $\{i \in I: A_i \vDash \phi(\bar{a}(i))\}$.

This definition is lifted almost directly from George Boole's first logical monograph, published in 1847. In essence he pointed out the laws

$$(4.1) \|\phi \wedge \psi\| = \|\phi\| \cap \|\psi\|, \qquad \|\phi \vee \psi\| = \|\phi\| \cup \|\psi\|, \qquad \|\neg \phi\| = I \backslash \|\phi\|,$$

and used them to show that both logic and set theory provide interpretations of his boolean calculus.

The analogue of (4.1) for the existential quantifier should say that $\|\exists x \, \phi(x)\|$ is the union of the sets $\|\phi(a)\|$ with a in $\prod_I A_i$. But in fact something stronger is true, both for $\prod_I A_i$ and for some of its substructures C. We say that C **respects** $\exists$ if for every formula $\phi(x)$ of L with parameters from C.

$$(4.2) \qquad \|\exists x \, \phi(x)\| = \|\phi(a)\| \qquad \text{for some element } a \text{ of } C.$$

It is clear that $\prod_I A_i$ respects $\exists$: for each $i \in \|\exists x \, \phi(x)\|$, choose an element a_i such that $A_i \vDash \phi(a_i)$, and take the element a of $\prod_I A_i$ such that $a(i) = a_i$ for each $i \in \|\exists x \, \phi(x)\|$. (Here we invoke the axiom of choice.)

As in section 6.2, a **filter** over a non-empty set I is a non-empty set $\mathcal{F}$ of subsets of I such that

$$(4.3) \quad X, Y \in \mathcal{F} \Rightarrow X \cap Y \in \mathcal{F}; \ X \in \mathcal{F}, X \subseteq Y \subseteq I \Rightarrow Y \in \mathcal{F};$$
$$\text{and } \varnothing \notin \mathcal{F}.$$

In particular $I \in \mathcal{F}$ by the second part of (4.3) and the fact that $\mathcal{F}$ is not empty.

Let I be a non-empty set, $(A_i: i \in I)$ a family of non-empty L-structures and $\mathcal{F}$ a filter over I. We form the product $\prod_I A_i$, and using $\mathcal{F}$ we define an equivalence relation $\sim$ on dom $\prod_I A_i$ by

$$(4.4) \qquad\qquad a \sim b \qquad \text{iff} \qquad \|a = b\| \in \mathcal{F}.$$

We verify that $\sim$ is an equivalence relation. Reflexive: for each element a of $\prod_I A_i$, $\|a = a\| = I \in \mathcal{F}$. Symmetry is clear. Transitive: $\|a = b\| \cap \|b = c\| \subseteq \|a = c\|$, so that if $\|a = b\|, \|b = c\| \in \mathcal{F}$, then $\|a = c\| \in \mathcal{F}$ by (4.3). Thus $\sim$

is an equivalence relation. We write $a/\mathcal{F}$ for the equivalence class of the element a.

We define an L-structure D as follows. The domain $\text{dom}(D)$ is the set of equivalence classes $a/\mathcal{F}$ with $a \in \text{dom}\prod_I A_i$. For each constant symbol c of L we put

(4.5) $\qquad c^D = a/\mathcal{F} \qquad$ where $a(i) = c^{A_i}$ for each $i \in I$.

Next let F be an n-ary function symbol of L, and $a_0, \ldots, a_{n-1}$ elements of $\prod_I A_i$. We define

(4.6) $\quad F^D(a_0/\mathcal{F}, \ldots, a_{n-1}/\mathcal{F}) = b/\mathcal{F}$

$\qquad$ where $b(i) = F^{A_i}(a_0(i), \ldots, a_{n-1}(i))$ for each $i \in I$.

It has to be checked that (4.6) is a sound definition. Suppose $a_j \sim a'_j$ for each $j < n$. Then by (4.3) there is a set $X \in \mathcal{F}$ such that $X \subseteq \|a_j = a'_j\|$ for each $j < n$. It follows that $X \subseteq \|F(a_0, \ldots, a_{n-1}) = F(a'_0, \ldots, a'_{n-1})\|$, which justifies the definition. Finally if R is an n-ary relation symbol of L and $a_0, \ldots, a_{n-1}$ are elements of $\prod_I A_i$, then we put

(4.7) $\quad (a_0/\mathcal{F}, \ldots, a_{n-1}/\mathcal{F}) \in R^D \qquad$ iff $\qquad \|R(a_0, \ldots, a_{n-1})\| \in \mathcal{F}$.

Again (4.3) shows that this definition is sound.

We have defined an L-structure D. This structure is called the **reduced product** of $(A_i : i \in I)$ over $\mathcal{F}$, in symbols $\prod_I A_i/\mathcal{F}$. The effect of definitions (4.5)–(4.7) is that for every unnested atomic formula $\phi(\bar{x})$ of L and every tuple $\bar{a}$ of elements of $\prod_I A_i$.

(4.8) $\qquad \prod_I A_i/\mathcal{F} \vDash \phi(\bar{a}/\mathcal{F}) \qquad$ iff $\qquad \|\phi(\bar{a})\| \in \mathcal{F}$.

Note that $\prod_I A_i$ itself is just the reduced product $\prod_I A_i/\{I\}$, so that every direct product is a reduced product.

Our first result says that taking reduced products commutes with taking relativised reducts. This generalises Theorem 9.1.3 and is important for the same reasons.

Theorem 9.4.1. *Let L and L^+ be signatures and P a 1-ary relation symbol of L^+. Let $(A_i : i \in I)$ be a non-empty family of non-empty L^+-structures such that $(A_i)_P$ is defined (see section 5.1) and $\mathcal{F}$ a filter over I. Then $(\prod_I A_i/\mathcal{F})_P = \prod_I((A_i)_P)/\mathcal{F}$.*

Proof. Define $f : \prod_I((A_i)_P)/\mathcal{F} \to \prod_I A_i/\mathcal{F}$ by taking any element $a/\mathcal{F}$ of $\prod_I((A_i)_P)/\mathcal{F}$ to the corresponding element $a/\mathcal{F}$ of $\prod_I A_i/\mathcal{F}$. One can check from the definition of reduced products that this definition is sound, and that f is an embedding with image $(\prod_I A_i/\mathcal{F})_P$. $\qquad\square$

When all the structures A_i are equal to a fixed structure A, we call $\prod_I A/\mathcal{F}$ the **reduced power** $A^I/\mathcal{F}$. There is an embedding $e : A \to A^I/\mathcal{F}$ defined by

$e(b) = a/\mathcal{F}$ where $a(i) = b$ for all $i \in I$. The fact that e is an embedding follows from the next lemma, but it's easy to check directly. We call e the **diagonal embedding**.

Recall that a **positive primitive (p.p.)** formula is a first-order formula of the form $\exists \bar{y} \bigwedge \Phi$ where Φ is a set of atomic formulas.

Lemma 9.4.2. *Let L be a signature and $\phi(\bar{x})$ a p.p. formula of L. Let $(A_i : i \in I)$ be a non-empty family of non-empty L-structures and $\bar{a}$ a tuple of elements of $\prod_I A_i$. Let $\mathcal{F}$ be a filter over I. Then*

$$(4.9) \qquad \prod_I A_i/\mathcal{F} \vDash \phi(\bar{a}/\mathcal{F}) \qquad \text{if and only if} \qquad \|\phi(\bar{a})\| \in \mathcal{F}.$$

Proof. We go by induction on the complexity of ϕ. As in the proof of Lemma 9.1.4, we can assume that ϕ is unnested. Then (4.9) for atomic formulas is just (4.8).

If (4.9) holds for formulas $\phi(\bar{x})$, $\psi(\bar{x})$ then it holds for their conjunction. From left to right, suppose $\prod_I A_i/\mathcal{F} \vDash (\phi \wedge \psi)(\bar{a}/\mathcal{F})$. Then by assumption $\|\phi(\bar{a})\|$ and $\|\psi(\bar{a})\|$ are both in $\mathcal{F}$. It follows that $\|(\phi \wedge \psi)(\bar{a})\| \in \mathcal{F}$ by the first parts of (4.1) and (4.3). From right to left, if $\|(\phi \wedge \psi)(\bar{a})\| \in \mathcal{F}$ then $\|\phi(\bar{a})\| \in \mathcal{F}$ by the second part of (4.3), since $\|(\phi \wedge \psi)(\bar{a})\| \subseteq \|\phi(\bar{a})\|$. The rest is clear.

If (4.9) holds for $\psi(\bar{x}, \bar{y})$ then it holds for $\exists \bar{y} \, \psi(\bar{x}, \bar{y})$. From left to right, suppose $\prod_I A_i/\mathcal{F} \vDash \exists \bar{y} \, \psi(\bar{a}/\mathcal{F}, \bar{y})$. Then there are elements $\bar{b}$ of $\prod_I A_i$ such that $\prod_I A_i/\mathcal{F} \vDash \psi(\bar{a}/\mathcal{F}, \bar{b}/\mathcal{F})$, so that $\|\psi(\bar{a}, \bar{b})\| \in \mathcal{F}$ by assumption. Since $\|\psi(\bar{a}, \bar{b})\| \subseteq \|\exists \bar{y} \, \psi(\bar{a}, \bar{y})\|$, it follows that $\|\exists \bar{y} \, \psi(\bar{a}, \bar{y})\| \in \mathcal{F}$ by (4.3). Conversely suppose $\|\exists \bar{y} \, \psi(\bar{a}, \bar{y})\| \in \mathcal{F}$. Since $\prod_I A_i$ respects $\exists$, there are elements $\bar{b}$ of $\prod_I A_i$ such that $\|\psi(\bar{a}, \bar{b})\| = \|\exists \bar{y} \, \psi(\bar{a}, \bar{y})\|$; whence $\prod_I A_i/\mathcal{F} \vDash \psi(\bar{a}/\mathcal{F}, \bar{b}/\mathcal{F})$ by assumption. Hence $\prod_I A_i \vDash \exists \bar{y} \, \psi(\bar{a}/\mathcal{F}, \bar{y})$. $\qquad\square$

We say that a sentence ϕ is **preserved in reduced products** if every reduced product of models of ϕ is also a model of ϕ. According to our next result, Horn sentences are preserved in reduced products.

Theorem 9.4.3. *Let L be a first-order language and $\phi(\bar{x})$ a Horn formula of L. Let $(A_i : i \in I)$ be a non-empty family of non-empty L-structures, $\mathcal{F}$ a filter over I and $\bar{a}$ a tuple of elements of $\prod_I A_i$. Then*

$$(4.10) \qquad \text{If} \quad \|\phi(\bar{a})\| \in \mathcal{F} \quad \text{then} \quad \prod_I A_i/\mathcal{F} \vDash \phi(\bar{a}/\mathcal{F}).$$

Proof. By induction on the complexity of ϕ, just as Theorem 9.1.5 followed from Lemma 9.1.4. $\qquad\square$

A deep result of Keisler and Galvin states as follows.

Theorem 9.4.4. *A first-order formula is preserved in reduced products if and only if it is logically equivalent to a Horn formula.*

Proof. This is Theorem 6.2.5′ on p. 366 of Chang & Keisler [1973].   □

The corresponding result for universal Horn sentences is not at all deep. I devote the rest of this section to it.

Universal Horn theories

We saw in Theorem 9.2.5 that a class **K** of structures is axiomatised by universal Horn sentences if and only if **K** is closed under substructures, products and isomorphic copies. What if one wants the Horn sentences to be first-order? We shall show that this is equivalent to replacing 'product' by 'reduced product'.

But first a combinatorial lemma on filters.

Lemma 9.4.5. *Let I be an infinite set. Then there is a filter $\mathscr{F}$ over I containing sets $X(j)$ ($j \in I$) such that for each $i \in I$ the set $\{j \in I : i \in X(j)\}$ is finite.*

Proof. Clearly it is enough if we prove the lemma for some set J which has the same cardinality as I. Let J be the set of all finite subsets of I. For each $i \in I$, let $X(i)$ be $\{j \in J : i \in j\}$, so that $j \in X(i) \Leftrightarrow i \in j$. If $i_0, \ldots, i_{n-1}$ are in I then $X(i_0) \cap \ldots \cap X(i_{n-1})$ is not empty, since it is the set of all $j \in J$ with $i_0, \ldots, i_{n-1} \in j$. Let $\mathscr{F}$ be the set of all subsets Y of J such that for some finite set $\{i_0, \ldots, i_{n-1}\} \subseteq I$, $X(i_0) \cap \ldots \cap X(i_{n-1}) \subseteq Y$. Use a bijection between I and J to relabel the sets $X(i)$ as $X(j)$ ($j \in J$). Then $\mathscr{F}$ meets our requirements.   □

A filter $\mathscr{F}$ as in Lemma 9.4.5 is said to be **regular**.

Lemma 9.4.6. *Let L be a first-order language, $\mathbf{K}$ a non-empty class of non-empty L-structures and A an L-structure. Then the following are equivalent.*
(a) *A is embeddable in some reduced product of structures in $\mathbf{K}$.*
(b) *If χ is any universal Horn sentence of L which is true in all structures in $\mathbf{K}$, then $A \vDash \chi$.*

Proof. (a) ⇒ (b) is by Theorem 9.4.3 and Corollary 2.4.2(b). For the converse, suppose A satisfies (b). Let $\bar{a}$ list the elements of A without repetition, and form $\operatorname{diag}^+ A$ in the language $L(\bar{a})$ got by adding $\bar{a}$ to L as parameters. Let Θ be the set of all negated atomic sentences of $L(\bar{a})$ which are true in A.

We claim first that if X is a finite subset of diag$^+ A$, then there is a model $(B_X, \bar{b}_X)$ of $X \cup \Theta$ where B_X is a product of structures in $\mathbf{K}$.

To prove this, let X be $\{\phi_0(\bar{a}), \ldots, \phi_{k-1}(\bar{a})\}$. Let $\neg \theta(\bar{a})$ be any sentence in Θ. Then $A \vDash \phi_0(\bar{a}) \wedge \ldots \wedge \phi_{k-1}(\bar{a}) \wedge \neg \theta(\bar{a})$. By assumption on A, this implies that the universal Horn sentence $\forall \bar{x}(\phi_0(\bar{x}) \wedge \ldots \wedge \phi_{k-1}(\bar{x}) \rightarrow \theta(\bar{x}))$ is false for some structure C_θ in $\mathbf{K}$, and so there are elements $\bar{c}_\theta$ of C_θ such that, forming the $L(\bar{a})$-structure $(C_\theta, \bar{c}_\theta)$,

$$(4.11) \qquad (C_\theta, \bar{c}_\theta) \vDash \phi_0(\bar{a}) \wedge \ldots \wedge \phi_{k-1}(\bar{a}) \wedge \neg \theta(\bar{a}).$$

By the same argument, the universal Horn sentence $\forall \bar{x}(\phi_0(\bar{x}) \wedge \ldots \wedge \phi_{k-1}(\bar{x}) \rightarrow \bot)$ is false from some structure $C_\bot$ in $\mathbf{K}$, and so there are elements $\bar{c}_\bot$ of $C_\bot$ such that

$$(4.12) \qquad (C_\bot, \bar{c}_\bot) \vDash \phi_0(\bar{a}) \wedge \ldots \wedge \phi_{k-1}(\bar{a}).$$

Form the product $(B_X, \bar{b}_X) = (\prod_\Theta (C_\theta, \bar{c}_\theta)) \times (C_\bot, \bar{c}_\bot)$. By Theorem 9.1.3, $B_X = (\prod_\Theta C_\theta) \times C_\bot$, and so B_X is a product of structures in $\mathbf{K}$. By Lemma 9.4.2, $(B_X, \bar{b}_X) \vDash X \cup \Theta$. The claim is proved.

Next we claim that there is a model $(D, \bar{d})$ of diag(A) such that D is a reduced product of structures in $\mathbf{K}$.

If diag$^+ A$ has the good luck to be finite, the first claim already gives what we want. Suppose then that diag$^+ A$ is infinite. By Lemma 9.4.5, putting $I = $ diag$^+ A$, there is a filter $\mathcal{F}$ over I which contains sets X_ϕ ($\phi \in$ diag$^+ A$), such that if $i \in I$ then the set $Z_i = \{\phi : i \in X_\phi\}$ is finite. Define $(D_i, \bar{d}_i)$ to be $(B_{Z_i}, \bar{b}_{Z_i})$, and put $(D, \bar{d}) = \prod_I (D_i, \bar{d}_i)/\mathcal{F}$. Now if $\phi \in$ diag$^+ A$ and $i \in X_\phi$ then $i \in \|\phi\|$ in the product $\prod_I (D_i, \bar{d}_i)$; so $\|\phi\| \in \mathcal{F}$ and thus $(D, \bar{d}) \vDash \phi$ by Lemma 9.4.2. Likewise $(D, \bar{d}) \vDash \Theta$ by Theorem 9.4.3, since every structure $(D_i, \bar{d}_i)$ is a model of Θ. Thus $(D, \bar{d}) \vDash$ diag(A). Now $(D, \bar{d})$ is a reduced product of products of structures in $\mathbf{K}$. I leave it to the reader to verify (Exercise 1) that $(D, \bar{d})$ can be recast as a reduced product of structures in $\mathbf{K}$. This proves the second claim.

Let $(D, \bar{d})$ be as in the second claim. By the diagram lemma (Lemma 1.4.2) there is an embedding of A into D. This proves (a) as required. $\qquad \square$

Thanks to Lemma 9.4.6 we have a model-theoretic characterisation of first-order universal Horn theories.

Theorem 9.4.7. *Let L be a first-order language and $\mathbf{K}$ a class of L-structures. Then the following are equivalent.*
(a) $\mathbf{K}$ *is closed under substructures, reduced products and isomorphic copies.*
(b) $\mathbf{K}$ *is the class of all models of a universal Horn theory in L.*

Proof. (b) $\Rightarrow$ (a) is by Corollary 2.4.2(b) and Theorem 9.4.3. Conversely suppose (a) holds, and let T be the set of all universal Horn sentences of L

which are true in every structure in **K**. It suffices to show that if A is a model of T then $A \in \mathbf{K}$. If there is an empty L-structure, every $\forall_1$ sentence of L is trivially true in it; so T is the set of all universal Horn sentences which are true in every non-empty structure in **K**. Thus by the lemma, every model of T is embeddable in a reduced product of members of **K** and hence is a member of **K**.   □

Corollary 9.4.8. *Let L be a first-order language and T a universal Horn theory in L. Let ϕ be a sentence of L. Then the following are equivalent.*
(a) *All substructures of reduced products of models of $T \cup \{\phi\}$ are models of ϕ.*
(b) *ϕ is equivalent modulo T to a sentence of L which is a conjunction of universal Horn sentences.*

Proof. (a) ⇒ (b). Let U be the set of all universal Horn consequences of $T \cup \{\phi\}$ in L. If (a) holds, then by Lemma 9.4.6, every model of U is a model of ϕ. Now use compactness. (b) ⇒ (a) is clear.   □

Theorem 9.4.9. *Let L be a first-order language and $\mathbf{K}$ a class of L-structures containing the trivial product. Then $\mathbf{ISP}(\mathbf{K})$ (the smallest quasivariety containing $\mathbf{K}$) is the class of all structures which are embeddable into reduced products of structures in $\mathbf{K}$.*

Proof. Let $\mathbf{K}'$ be the class of all structures which are embeddable into reduced products of structures in **K**. Any quasivariety containing **K** must contain $\mathbf{K}'$. By Lemma 9.4.6, $\mathbf{K}'$ is the class of all models of T, where T is the set of all universal Horn sentences true in every structure in **K**. These sentences are all strict since **K** contains 1. So $\mathbf{K}'$ is a quasivariety.   □

Exercises for section 9.4

1. Let **K** be a class of L-structures. Show that a reduced product of products of structures in **K** is isomorphic to a reduced product of structures in **K**.

2. Let L be a first-order language and **K** a class of L-structures. Show that **K** is a quasivariety if and only if **K** is closed under substructures, reduced products and isomorphic copies, and contains the trivial product.

3. Let L be a first-order language. Let Φ be the smallest class of formulas of L such that (1) every atomic formula of L is in Φ, (2) the conjunction of any two formulas in Φ is in Φ, (3) if ϕ is in Φ then so are $\forall x\, \phi$ and $\exists x\, \phi$, and (4) if $\psi(\bar{x}, \bar{y})$ and $\chi(\bar{x}, \bar{y})$ are in Φ then so is the formula $\exists \bar{y}\, \psi \wedge \forall \bar{y}(\psi \rightarrow \chi)$. The formulas in Φ are known as the *h-formulas* of L. Show that if $\phi(\bar{x})$ is an h-formula and $(A_i : i \in I)$ is a non-empty

family of non-empty L-structures, $\mathcal{F}$ a filter over I and $\bar{a}$ a tuple of elements of $\prod_I A_i$, then $\|\phi(\bar{a})\| \in \mathcal{F} \Leftrightarrow \prod_I A_i/\mathcal{F} \vDash \phi(\bar{a}/\mathcal{F})$.

4. Let ϕ be a sentence which axiomatises the class of boolean algebras that are not atomless. Show that ϕ is preserved in products but not in reduced products.

*Let A and B be L-structures. We say that B is a **strong homomorphic image** of A if there is a surjective homomorphism $f: A \to B$ such that for every relation symbol R of L and every tuple $\bar{b}$ in B such that $B \vDash R(\bar{b})$ there is $\bar{a}$ in A such that $A \vDash R(\bar{a})$ and $f\bar{a} = \bar{b}$.*

5. Let L be a first-order language and $\mathbf{K}$ a class of L-structures. Show that the following are equivalent. (a) $\mathbf{K}$ is closed under substructures, reduced products, the trivial product 1 and strong homomorphic images. (b) $\mathbf{K}$ is the class of models of a set of universal Horn sentences of L of the form $\forall \bar{x}_0 \; \ldots \; \bar{x}_{n-1}(R_0(\bar{x}_0) \wedge \; \ldots \; \wedge R_{n-1}(\bar{x}_{n-1}) \to \psi)$ where ψ is atomic, $R_0, \ldots, R_{n-1}$ are relation symbols and the variables $\bar{x}_0, \ldots, \bar{x}_{n-1}$ are all distinct; n may be 0.

6. Let $(A_i: i \in I)$ be a non-empty family of non-empty L-structures, and $\mathcal{F}$ a filter over I. Show that the reduced product $\prod_I A_i/\mathcal{F}$ is the direct limit of a diagram of structures $\prod_J A_i$ with $J \in \mathcal{F}$, where the index set of the diagram is $\mathcal{F}$ ordered by reverse inclusion, and if $J' \supseteq J$ then the map $\prod_{J'} A_i \to \prod_J A_i$ is the projection.

7. Show that the reduced product $A = \prod_I A_i/\mathcal{F}$ can be embedded in a product as follows. Let S be the set of ultrafilters over I which extend $\mathcal{F}$ (so that S is the Stone space of the boolean algebra $\mathcal{P}(I)/\mathcal{F}$). For each $U \in S$, let $f_U: A \to \prod_I A_i/U$ be the natural homomorphism. The maps f_U induce an embedding of A into $\prod_X(\prod_I A_i/U)$. *Using Łoś's theorem, Theorem 9.5.1 below, one can show that for each tuple $\bar{a}$ in A and each first-order formula $\phi(\bar{x})$, the value $\|\phi(\bar{a})\|$ is a clopen subset of S. This allows one to treat reduced products as a special case of the boolean products mentioned in section 9.7 below.*

8. Let $\mathcal{F}$ be a filter over the set I, and for each $i \in I$ let $\mathcal{D}_i$ be a filter over the set J_i; we suppose that the sets J_i are disjoint. Let J be $\bigcup_{i \in I} J_i$, and for each $j \in J$ let A_j be a non-empty L-structure. Show that $\prod_I(\prod_{j \in J_i} A_j/\mathcal{D}_i)/\mathcal{F}$ can be written as a reduced product $\prod_J A_j/\mathcal{F}'$. [Put $\mathcal{F}' = \{X \subseteq J: \{i \in I: X \cap J_i \in \mathcal{D}_i\} \in \mathcal{F}\}$.]

For the following exercises, λ is a regular cardinal. We allow function and relation symbols of any arity $< \lambda$. Universal quantifiers $\forall \bar{x}$ can have any number of variables in $\bar{x}$. A λ-reduced product is a reduced product $\prod_I A_i/\mathcal{F}$ where $\mathcal{F}$ is λ-complete.

*9. Show that if I is an infinite set such that $|I|^{<\lambda} = |I|$, then there is a λ-complete filter $\mathcal{F}$ over I containing a family of sets $X(i)$ $(i \in I)$ such that for each $i \in I$, $|\{j \in I: i \in X(j)\}| < \lambda$.

*10. By a **basic λ-Horn formula** we mean a formula of the form $\bigwedge \Phi \to \psi$, where Φ is a set of fewer than λ atomic formulas and ψ is atomic or $\bot$. By a **universal λ-Horn formula** we mean a formula of the form $\forall \bar{x} \, \phi$ where ϕ is basic λ-Horn. Let $\mathbf{K}$ be a class of L-structures. Show that the following are equivalent. (a) **K** is closed under substructures, isomorphic copies and λ-reduced products. (b) **K** is the class of all models of some set of universal λ-Horn sentences.

*11. A λ-**Horn formula** is a formula of the form $Q_0x_0Q_1x_1 \ldots \bigwedge\Phi$ where Φ is a set of fewer than λ basic Horn formulas, each Q_i is either $\forall$ or $\exists$, and the quantifier string is well-ordered of length $< \lambda$. (See Example 2 in section 3.4 above on how to read such quantifiers.) Show that λ-Horn sentences are preserved in λ-reduced products.

*12. Let L and L^+ be signatures with $L \subseteq L^+$. Let T be a λ-Horn theory in L and ϕ an atomic formula of L^+. Suppose that for every pair of models A, B of T, each homomorphism $f \colon A|L \to B|L$ preserves ϕ. Show that ϕ is equivalent modulo T to a formula of the form $\exists \bar{x} \bigwedge\Phi$ where Φ is a set of fewer than λ atomic formulas of L.

9.5 Ultraproducts

We recall from section 6.2 that an **ultrafilter** over a non-empty set I is a filter $\mathcal{U}$ over I such that

(5.1)   for every set $X \subseteq I$, exactly one of $X, I \backslash X$ is in $\mathcal{U}$.

By Theorem 6.2.1, every filter over I (and more generally every subset of $\mathcal{P}(I)$ which has the finite intersection property) can be extended to an ultrafilter over I.

An **ultraproduct** is a reduced product $\prod_I A_i/\mathcal{U}$ in which $\mathcal{U}$ is an ultrafilter over the set I. Likewise an **ultrapower** is a reduced power $A^I/\mathcal{U}$ where $\mathcal{U}$ is an ultrafilter over I.

It's slightly paradoxical that three leading textbooks of model theory (Bell & Slomson [1969], Chang & Keisler [1973], Kopperman [1972]) concentrate on the ultraproduct construction, and yet there is almost no theorem of model theory that needs ultraproducts for its proof. (Apart from theorems about ultraproducts, that is.)

The paradox is not hard to resolve. First a historical point: these three textbooks all came in the wake of some dramatic applications of ultraproducts in set theory. Ultraproducts are still an indispensable tool of large cardinal theory. But leaving aside history, there are several distinctive features of ultraproducts that give them an honoured place in model theory. Thus,

(5.2)   taking ultraproducts commutes with taking relativised reducts
          (this is a special case of Theorem 9.4.1, except that now by
          Łoś's theorem, Theorem 9.5.1 below, we can take the relativi-
          sing formula to be any first-order formula),

(5.3)   ultrapowers give arbitrarily large elementary extensions of
          infinite structures (see Corollary 9.5.5 below; in section 10.2
          we shall be able to improve 'arbitrarily large' to 'arbitrarily
          highly saturated'),

(5.4)   two structures are elementarily equivalent if and only if they
         have isomorphic ultrapowers (Theorem 9.5.7 below),

(5.5)   ultraproducts compute limits in the space $S_0(\varnothing)$ of complete
         theories (see Theorem 9.5.9 below).

Some people have hoped, when they learned of (5.3) and (5.4), that
ultraproducts might make it possible to do model theory without all the
complexities of logic. Disillusionment set in when they found themselves
faced with the complexities of boolean algebra instead.

Let me take points (5.3)–(5.5) in turn, and see what they are good for. All
of them rest on the following fundamental theorem.

Theorem 9.5.1. (*Łoś's theorem*). *Let L be a first-order language, $(A_i: i \in I)$ a
non-empty family of non-empty L-structures and $\mathcal{U}$ an ultrafilter over I. Then
for any formula $\phi(\bar{x})$ of L and tuple $\bar{a}$ of elements of $\prod_I A_i$,*

(5.6) $$\prod_I A_i/\mathcal{U} \vDash \phi(\bar{a}/\mathcal{U}) \quad \text{if and only if} \quad \|\phi(\bar{a})\| \in \mathcal{U}.$$

Proof. Comparing with the proof of Theorem 9.4.3, we see that only one
more thing is needed: assuming that (5.6) holds for ϕ, we have to deduce it
for $\neg \phi$ too. But this is easy by (5.1):

(5.7) $$\prod_I A_i/\mathcal{U} \vDash \neg \phi(\bar{a}/\mathcal{U}) \Leftrightarrow \|\phi(\bar{a})\| \notin \mathcal{U} \Leftrightarrow \|\neg \phi(\bar{a})\| \in \mathcal{U}. \qquad \square$$

Corollary 9.5.2. *If $A^I/\mathcal{U}$ is an ultrapower of A, then the diagonal map
$e: A \to A^I/\mathcal{U}$ is an elementary embedding.*

Proof. Immediate. $\square$

By the usual manipulation (see Exercise 1.2.4(b)), Corollary 9.5.2 allows
us to regard A as an elementary substructure of $A^I/\mathcal{U}$. So ultrapowers give
elementary extensions. To show that they can be arbitrarily large when A is
infinite, we first adapt a lemma from section 9.4.

Lemma 9.5.3. *Let I be an infinite set. Then there is an ultrafilter $\mathcal{U}$ over I
containing sets X_j $(j \in I)$ such that for each $i \in I$ the set $\{j: i \in X_j\}$ is finite.*

Proof. Take the filter $\mathcal{F}$ of Lemma 9.4.5, and extend it to an ultrafilter by
Theorem 6.2.1. $\square$

An ultrafilter $\mathcal{U}$ with the property of Lemma 9.5.3 is said to be **regular**.

Theorem 9.5.4. *Let L be a first-order language, A an L-structure, I an infinite set and $\mathcal{U}$ a regular ultrafilter over I.*

(a) *If $\phi(x)$ is a formula of L such that $|\phi(A)|$ is infinite, then $|\phi(A^I/\mathcal{U})| = |\phi(A)|^{|I|}$.*

(b) *If $\Phi(\bar{x})$ is a type over $\text{dom}(A)$ with respect to A, and $|\Phi| \leqslant |I|$, then some tuple $\bar{a}$ in $A^I/\mathcal{U}$ realises Φ.*

Proof. (a) We first prove $\leqslant$. By Łoś's theorem (Theorem 9.5.1) each element of $\phi(A^I/\mathcal{U})$ is of the form $b/\mathcal{U}$ for some b such that $\|\phi(b)\| \in \mathcal{U}$. Since we can change b anywhere outside a set in $\mathcal{U}$ without affecting $b/\mathcal{U}$, we can choose b so that $\|\phi(b)\| = I$. This sets up an injection from $\phi(A^I/\mathcal{U})$ to the set $\phi(A)^I$ of all maps from I to $\phi(A)$.

Next we prove $\geqslant$. Since $\mathcal{U}$ is regular, there are sets X_i ($i \in I$) in $\mathcal{U}$ such that for each $j \in I$ the set $Z_j = \{i \in I: j \in X_i\}$ is finite. For each $j \in I$, let μ_j be a bijection taking the set $\phi(A)^{Z_j}$ (of all maps from Z_j to $\phi(A)$) to $\phi(A)$. Such a μ_j exists since $\phi(A)$ is infinite. For each function $f: I \to \phi(A)$, define f^μ to be the map from I to $\phi(A)$ such that

(5.8) for each $j \in I$, $f^\mu(j) = \mu_j(f|Z_j)$.

Each function $f^\mu: I \to \phi(A)$ is an element of A^I, and by Łoś's theorem $f^\mu/\mathcal{U} \in \phi(A^I/\mathcal{U})$. So it remains only to show that if f, g are distinct maps from I to $\phi(A)$ then $f^\mu/\mathcal{U} \neq g^\mu/\mathcal{U}$. Suppose then that $f(i) \neq g(i)$ for some $i \in I$. It follows that $f|Z_j \neq g|Z_j$ whenever $i \in Z_j$, i.e. whenever $j \in X_i$. Hence $X_i \subseteq \|f^\mu \neq g^\mu\|$, and so $f^\mu/\mathcal{U} \neq g^\mu/\mathcal{U}$ since $X_i \in \mathcal{U}$.

(b) Since $\mathcal{U}$ is regular, there is a family $\{X_\phi: \phi \in \Phi\}$ of sets in $\mathcal{U}$, such that for each $i \in I$ the set $Z_i = \{\phi \in \Phi: i \in X_\phi\}$ is finite. Since Φ is a type over $\text{dom}(A)$, for each $i \in I$ there is a tuple $\bar{a}_i$ in A which satisfies Z_i. Let $\bar{a}$ be the tuple in A^I such that $\bar{a}(i) = \bar{a}_i$ for each i. Then for each formula ϕ in Φ, if $i \in X_\phi$ then $\phi \in Z_i$ and so $A \vDash \phi(\bar{a}_i)$. Thus $X_\phi \subseteq \|\phi(\bar{a})\|$, and by Łoś's theorem (Theorem 9.5.1) we deduce that $A^I/\mathcal{U} \vDash \phi(\bar{a})$. $\square$

Corollary 9.5.2 and Theorem 9.5.4(a) give us (5.3) at once. In fact they give us the following stronger statement, for which no other proof is known.

Corollary 9.5.5. *Let L be a first-order language, A an L-structure and κ an infinite cardinal. Then A has an elementary extension B such that for every formula $\phi(\bar{x})$ of L, $|\phi(B)|$ is either finite or equal to $|\phi(A)|^\kappa$.* $\square$

The next application also has no other known proof. It is more complicated than Corollary 9.5.5, but also more useful. The finite cover property was defined in section 4.3 (and used in section 4.7).

Corollary 9.5.6. *Let L be a first-order language and T a complete theory in L which has infinite models. Suppose T is λ-categorical for some $\lambda \geqslant \max((2^\omega)^+, |L|)$. Then T doesn't have the finite cover property.*

Proof. I use Shelah's definition of the finite cover property, as in section 4.3. This is harmless since the hypothesis on T implies that T is stable (see Theorems 12.2.2 and 12.2.3). But in any case the proof adapts at once to Keisler's definition (given at Exercise 4.3.10).

Let A be a model of T of cardinality at least λ. If T has the finite cover property, then there is a formula $\phi(\bar{x}, \bar{y}, \bar{z})$ of L such that for each tuple $\bar{c}$ in A, $\phi(\bar{x}, \bar{y}, \bar{c})$ defines an equivalence relation $E_{\bar{c}}$, and for each $n < \omega$ there is a tuple $\bar{c}_n$ such that the number of equivalence classes of $E_{\bar{c}_n}$ is finite and at least n. In some suitable finite slice B of A^{eq}, the equivalence classes of each $E_{\bar{c}}$ form a set of elements $X_{\bar{c}}$; $X_{\bar{c}}$ is definable in terms of $\bar{c}$.

Let $\mathcal{U}$ be a regular ultrafilter over ω and $(X_m : m < \omega)$ a descending chain of sets in $\mathcal{U}$ with empty intersection. Choose a tuple $\bar{c}/\mathcal{U}$ in B^ω so that for each $i < \omega$, if $i \in X_n \setminus X_{n+1}$ then $\bar{c}(i) = \bar{c}_n$. For every $n < \omega$, the set

(5.9) $\| \phi(\bar{x}, \bar{y}, \bar{c})$ defines an equivalence relation whose set $X_{\bar{c}}$ of

equivalence classes has more than n elements$\|$

contains all but finitely many $i \in \omega$. So by Łoś's theorem, $X_{\bar{c}}$ is an infinite set consisting of the equivalence classes of $\phi(\bar{x}, \bar{y}, \bar{c})$ in $B^\omega/\mathcal{U}$. But $X_{\bar{c}}$ is also a relativised reduct of $B^\omega/\mathcal{U}$ (with parameters $\bar{c}$). Hence we can apply (5.2) to deduce that $|X_{\bar{c}}| = |\prod_\omega X_{\bar{c}(i)}/\mathcal{U}| \leq \prod_\omega |X_{\bar{c}(i)}| \leq 2^\omega$. So in $A^\omega/\mathcal{U}$, the number of equivalence classes of $\phi(\bar{x}, \bar{y}, \bar{c})$ is infinite but $\leq 2^\omega$.

Now by the downward Löwenheim–Skolem theorem, $A^\omega/\mathcal{U}$ has an elementary substructure of cardinality λ in which the number of equivalence classes of $\phi(\bar{x}, \bar{y}, \bar{c})$ is infinite but $\leq 2^\omega$. But one easily applies the compactness theorem to construct a model of T of cardinality λ in which for every tuple $\bar{d}$, the number of equivalence classes of $\phi(\bar{x}, \bar{y}, \bar{d})$ is either finite or at least λ (see Exercise 6.1.10). This contradicts the assumption that T is λ-categorical. $\square$

We turn to (5.4). A slightly stronger statement is the following.

Theorem 9.5.7 (Keisler–Shelah theorem). *Let L be a signature and let A, B be L-structures. The following are equivalent.*
(a) $A \equiv B$.
(b) *There are a set I and an ultrafilter $\mathcal{U}$ over I such that $A^I/\mathcal{U} \cong B^I/\mathcal{U}$.*

Proof. The proof uses some quite difficult combinatorics. It can be found in the proof (but not the statement) of Theorem 6.1.15 in Chang & Keisler [1973]. $\square$

The Keisler–Shelah theorem was an impressive solution of a natural problem. But it hasn't led to much new information. The following application is typical in two ways: it uses (5.2), and there is also a straightforward proof by elementary means (see Exercise 6.6.1).

Corollary 9.5.8 *(Robinson's joint consistency lemma).* *Let* L_1 *and* L_2 *be first-order languages and* $L = L_1 \cap L_2$. *Let* T_1 *and* T_2 *be consistent theories in* L_1 *and* L_2 *respectively, such that* $T_1 \cap T_2$ *is a complete theory in* L. *Then* $T_1 \cup T_2$ *is consistent.*

Proof. Let A_1, A_2 be models of T_1, T_2 respectively. Then since $T_1 \cap T_2$ is complete, $A_1 | L \equiv A_2 | L$. By the Keisler–Shelah theorem there is an ultra-filter $\mathcal{U}$ over a set I such that $(A_1 | L)^I / \mathcal{U} \cong (A_2 | L)^I / \mathcal{U}$. Corollary 9.5.2 tells us that $A_1^I / \mathcal{U} \vDash T_1$ and $A_2^I / \mathcal{U} \vDash T_2$. By (5.2), $A_1^I / \mathcal{U}$ is an expansion of $(A_1 | L)^I / \mathcal{U}$. But also (5.2) tells us that $A_2^I / \mathcal{U}$ is an expansion of an isomorphic copy of $(A_1 | L)^I / \mathcal{U}$. So we can use $A_2^I / \mathcal{U}$ as a template to expand $A_1^I / \mathcal{U}$ to a model of T_2. ☐

Finally we consider (5.5). The wording of (5.5) calls for some explanation.

Let L be a first-order language, and let S be the set of all theories in L which are of the form $\text{Th}(A)$ for some L-structure A. In other words, S is the set $S_0(\varnothing)$ of Example 1 in section 6.3. Let X be a subset of S and T a set of sentences of L. Let us call T a **limit point** of X if

(5.10)   for every sentence ϕ of L, exactly one of ϕ, $\neg \phi$ is in T, and

(5.11)   for every finite $T_0 \subseteq T$ there is $T' \in X$ with $T_0 \subseteq T'$.

The next theorem is one way of showing that such a set T is in fact an element of S. Then (5.11) says that T is a limit point of X in the topology that we gave to S in section 6.3.

Theorem 9.5.9. *Let* L *be a first-order language,* **K** *a class of* L-*structures and* T *a limit point of* $\{\text{Th}(A): A \in \mathbf{K}\}$. *Then* T *is* $\text{Th}(B)$ *for some ultraproduct* B *of structures in* **K**.

Proof. The proof is a variant of that of Theorem 9.5.4(b). Let $\mathcal{U}$ be a regular ultrafilter over the set T. Then there is a family $\{X_\phi : \phi \in T\}$ of sets in $\mathcal{U}$, such that for each $i \in T$ the set $Z_i = \{\phi \in T: i \in X_\phi\}$ is finite. Since T is a limit point of $\{\text{Th}(A): A \in \mathbf{K}\}$, for each $i \in T$ there is a structure $A_i \in \mathbf{K}$ such that $A_i \vDash Z_i$. Put $B = \prod_T A_i / \mathcal{U}$. If $i \in X_\phi$ then $\phi \in Z_i$ and so $A_i \vDash \phi$; hence $X_\phi \subseteq \|\phi\|$ for each sentence ϕ in T. It follows by Łoś's theorem (Theorem 9.5.1) that $B \vDash T$, and so $T = \text{Th}(B)$ by (5.10). ☐

Readers who enjoy going round in circles should spare a minute to deduce the compactness theorem from Theorem 9.5.9. The rest of us will move on to deduce a criterion for first-order axiomatisability.

First-order axiomatisability

Corollary 9.5.10. *Let L be a first-order language and $\mathbf{K}$ a class of L-structures. Then the following are equivalent.*

(a) $\mathbf{K}$ *is axiomatisable by a set of sentences of L.*

(b) $\mathbf{K}$ *is closed under ultraproducts and isomorphic copies, and if A is an L-structure such that some ultrapower of A lies in $\mathbf{K}$, then A is in $\mathbf{K}$.*

Proof. (a) $\Rightarrow$ (b) follows at once from Theorem 9.5.1 and Corollary 9.5.2.

Conversely suppose (b) holds, and let T be the set of all sentences of L which are true in every structure in $\mathbf{K}$. To prove (a) it suffices to show that any model A of T lies in $\mathbf{K}$.

We begin by showing that $\mathrm{Th}(A)$ is a limit point of $\{\mathrm{Th}(C): C \in \mathbf{K}\}$. For this, let U be a finite set of sentences of L which are true in A. Then $\bigwedge U$ is a sentence ϕ which is true in A, and so $\neg\phi \notin T$ since A is a model of T. It follows by the definition of T that some structure in $\mathbf{K}$ is a model of ϕ. Thus $\mathrm{Th}(A)$ is a limit point of $\{\mathrm{Th}(C): C \in \mathbf{K}\}$. By Theorem 9.5.9 we deduce that A is elementarily equivalent to some ultraproduct of structures in $\mathbf{K}$, and hence (by (b)) to some structure B in $\mathbf{K}$. By the Keisler–Shelah theorem (Theorem 9.5.7) it follows that some ultrapower of A is isomorphic to an ultrapower of B, and so by (b) again, A is in $\mathbf{K}$. ☐

Corollary 9.5.10 is handy for proving that certain classes are not first-order axiomatisable. It is not the best method for all cases – one can sometimes reach home faster by the route of Corollary 3.3.3 above. But let me illustrate it by redeeming a promise made in section 2.8 above. In that section we showed that under certain conditions on a ring R, some classes of left R-modules are first-order axiomatisable. I promised converses.

Theorem 9.5.11. *Let R be a ring.*

(a) *Let I be a finitely generated left ideal of R. Then the class of I-injective left R-modules is first-order definable (and likewise first-order axiomatisable) if and only if I is finitely presented.*

(b) *Let $\mathbf{K}$ be the class of left R-modules which are I-injective for all finitely generated left ideals I of R. Then $\mathbf{K}$ is first-order axiomatisable if and only if R is left coherent.*

(c) *Let $\mathbf{K}'$ be the class of injective left R-modules. Then $\mathbf{K}'$ is first-order axiomatisable if and only if R is left noetherian.*

Proof. Right to left in all parts were proved in Theorem 2.8.3. We prove the converses.

First we show that if I is a finitely generated but not finitely presented left ideal of R, then there is a family of injective left R-modules with an ultraproduct which is not I-injective. By Corollary 9.5.10 this proves left to right in (a) and (b).

In the notation of section 2.8, let $\bar{r} = (r_0, \ldots, r_{n-1})$ be generators of I. Write Δ for $\Delta_I(\bar{x})$. Since I is not finitely presented, Δ is infinite and for each finite subset $\Phi(\bar{x})$ of Δ there is a left R-module M_Φ with an n-tuple of elements $\bar{a}_\Phi$ such that $M \vDash \bigwedge\Phi(\bar{a}_\Phi) \wedge \neg\bigwedge\Delta(\bar{a}_\Phi)$. For each such Φ, let N_Φ be an injective hull of M_Φ. Let $\mathcal{U}$ be a regular ultrafilter over Δ. There are sets X_ϕ in $\mathcal{U}$ ($\phi \in \Delta$) such that for each $i \in \Delta$ the set $Z_i = \{\phi : i \in X_\phi\}$ is finite. For each $i \in \Delta$, let $(B_i, \bar{b}_i)$ be $(N_{Z_i}, \bar{a}_{Z_i})$, and put $(C, \bar{c}) = \prod_\Delta (B_i, \bar{b}_i)/\mathcal{U}$. Then C is an ultraproduct of injective left R-modules. But C is not I-injective. To show this, consider any $\phi \in \Delta$. We have by construction $X_\phi \subseteq \|\phi(\bar{b})\|$ where $\bar{b}(i) = \bar{b}_i$ for each $i \in \Delta$. Hence $C \vDash \phi(\bar{c})$ by Łoś's theorem, since $\bar{c} = \bar{b}/\mathcal{U}$. Therefore $C \vDash \bigwedge\Delta(\bar{c})$. If C was I-injective, it would follow by Lemma 2.8.2 that for some element d of $\prod_\Delta B_i$, writing $\bar{c} = (c_0, \ldots, c_{n-1})$,

(5.12) $\qquad C \vDash r_0 \cdot (d/\mathcal{U}) = c_0 \wedge \ldots \wedge r_{n-1} \cdot (d/\mathcal{U}) = c_{n-1}$.

So by Łoś's theorem again there is some $i \in \Delta$ such that in B_i we have $-1 \cdot d(i) = b_{n-1}(i)$, where $(b_0, \ldots, b_{n-1})$ is $\bar{b}_i$. By definition of Δ this implies that $B_i \vDash \bigwedge\Delta(\bar{b}_i)$, and hence $M_{Z_i} \vDash \bigwedge \Delta(\bar{a}_{Z_i})$ since Δ is quantifier-free. This contradicts the choice of M_{Z_i} and $\bar{a}_{Z_i}$. Hence C is not I-injective.

Finally we prove left to right in (c). Suppose R is not left noetherian. We shall find a left R-module which is not injective but has an injective elementary extension.

By an argument of H. Bass and Z. Papp (cf. Theorem 4.1 in Sharpe & Vámos [1972]) there is a family $(M_j : j \in J)$ of injective left R-modules such that the direct sum $A = \bigoplus_{j \in J} M_j$ is not injective. We show that

(5.13) A is I'-injective for every finitely generated left ideal I' of R.

A model-theoretic proof goes by noting that the sentence in (8.13) of section 2.8 is Horn and hence preserved in the product $\prod_J M_j$ by Theorem 9.1.5; but by Lemma A.1.6(a), $\prod_J M_j \equiv \bigoplus_J M_j = A$.

Noting that R is infinite, let $\mathcal{U}$ be a regular ultrafilter over $\mathrm{dom}(R)$, and put $B = A^{\mathrm{dom}(R)}/\mathcal{U}$. We finish the proof by showing that B is injective. Let I be a left ideal of R and $f : I \to B$ a homomorphism. Let $\Phi(x)$ be the set of all formulas $rx = f(r)$ with r in R. To show that B is I-injective, we need to extend f to a homomorphism $g : R \to B$, and this is equivalent to finding an element of B which realises Φ. By (5.13), every finite subset of Φ is realised already in A, and so we need only quote Theorem 9.5.4(b). $\qquad\square$

Exercises for section 9.5

Throughout these exercises we assume that all structures are non-empty.

1. Show that if $\mathcal{U}$ is a principal ultrafilter then the ultraproduct $\prod_I A_i/\mathcal{U}$ is isomorphic to one of the A_i.

2. Show that if $|A_i| \leqslant |B_i|$ for all $i \in I$ then $|\prod_I A_i/\mathcal{U}| \leqslant |\prod_I B_i/\mathcal{U}|$.

An ultrafilter $\mathcal{U}$ over a set I is λ-complete if for every set X of fewer than λ sets in $\mathcal{U}$, the intersection $\cap X$ is also in $\mathcal{U}$. Every ultrafilter is ω-complete. If there are any non-principal ω_1-complete ultrafilters, then a measurable cardinal exists (see Jech [1978] Chapter 5).

3. Show that no regular ultrafilter over an infinite set is ω_1-complete.

4. Show that if the ultrafilter $\mathcal{U}$ is not ω_1-complete, then every ultraproduct $\prod_I A_i/\mathcal{U}$ has cardinality $< \omega$ or $\geqslant 2^\omega$.

5. Show that if B is an ultraproduct of finite structures then $|B|$ is either finite or $\geqslant 2^\omega$.

6. Show that the following conditions on an ultrafilter $\mathcal{U}$ over I are equivalent. (a) $\mathcal{U}$ is not ω_1-complete. (b) There are disjoint non-empty sets $X_i \subseteq I$ ($i < \omega$) such that for each $n < \omega$, $\bigcup_{i \geqslant n} X_i \in \mathcal{U}$. (c) The ultrapower $(\omega, <)^I/\mathcal{U}$ is not well-ordered.

7. An ultrafilter over a set I is said to be **uniform** if every set in the ultrafilter has cardinality $|I|$. Show that every regular ultrafilter over an infinite set is uniform.

8. Show that if I is an infinite set then there are $2^{2^{|I|}}$ regular ultrafilters over I. [Take a regular ultrafilter $\mathcal{U}$ over I. For any ultrafilter $\mathcal{F}$ over I, let $\mathcal{F}^+$ be an ultrafilter over $I \times I$ which contains all the sets $X \times Y$ with $X \in \mathcal{U}$ and $Y \in \mathcal{F}$. Show that $\mathcal{F}^+$ is a regular ultrafilter over $I \times I$, and that if $\mathcal{F}_1$, $\mathcal{F}_2$ are distinct ultrafilters over I then $\mathcal{F}_1^+ \neq \mathcal{F}_2^+$. Now quote Exercise 6.2.10.]

9. (Frayne's theorem) Show that two L-structures A and B are elementarily equivalent if and only if A is elementarily embeddable in some ultrapower of B. (Give a direct proof without quoting the Keisler–Shelah theorem.)

10. Let **K** be a class of L-structures. Show (a) **K** is first-order axiomatisable if and only if **K** is closed under ultraproducts and under elementary equivalence, (b) **K** is first-order definable if and only if both **K** and its complement in the class of L-structures are closed under ultraproducts and elementary equivalence.

11. Use the compactness theorem to deduce each of Robinson's joint consistency lemma (Corollary 9.5.8) and Craig's interpolation theorem (Theorem 6.6.3) from the other.

12. Let L be a signature, **K** a class of L-structure and A an L-structure. Show that the following are equivalent. (a) Every $\forall_1$ sentence in Th(**K**) is true in A. (b) A is embeddable in an ultraproduct of structures in **K**.

13. Show that if $\mathcal{U}$ is a regular ultrafilter over a set I of cardinality κ then the structure $(\kappa^+, <)$ is embeddable in $(\omega, <)^I/\mathcal{U}$.

*An L-structure A is said to be λ-**universal** if every L-structure of cardinality $< \lambda$ which is elementarily equivalent to A is elementarily embeddable in A; see section 10.1 below.*

14. Let $\mathcal{U}$ be an ultrafilter over a set I of cardinality κ. Show that $\mathcal{U}$ is regular if and only if for every signature L with $|L| \leq \kappa$ and every L-structure A, $A^I/\mathcal{U}$ is κ^+-universal. [For right to left, consider a structure A and a type $\Phi(x)$ over $\varnothing$ with respect to A, such that Φ has cardinality κ and is not realised in A. If b realises Φ in $A^I/\mathcal{U}$, for each formula ϕ in Φ consider $\{i \in I: A \models \phi(b(i))\}$.]

15. Let G be a group and $\mathbf{K}$ a family of normal subgroups of G; suppose that for any finite set X of elements of G there is $N \in \mathbf{K}$ such that $xN \neq yN$ for all distinct x and y in X. Show that G is embeddable in an ultraproduct of groups of form G/N with N in $\mathbf{K}$.

16. (a) Show that every ultraproduct of a family of modules is a direct summand of an ultrapower of the product of the family. [Put $B = \prod_I A_i$. Using canonical projections and embeddings, define linear maps $f: B^I \to \prod_I A_i$ and $g: \prod_I A_i \to B^I$ so that fg is the identity. Show that the same equation holds for the induced maps between $B^I/\mathcal{U}$ and $\prod_I A_i/\mathcal{U}$.] (b) Show the same with 'direct sum' for 'product'.

*17. Show that the following are equivalent for any commutative ring R. (a) The class of injective left R-modules is first-order definable. (b) R is noetherian and has only finitely many prime ideals. (c) R is noetherian and the class of flat left R-modules is first-order definable.

18. (a) Let L be a signature and A an L-structure. Let $\mathbf{K}$ be a set of substructures of A such that if X is any finite set of elements of A then there is a structure B in $\mathbf{K}$ with $X \subseteq \text{dom}(B)$. Show that A is embeddable in an ultraproduct of structures in $\mathbf{K}$. (b) Show that every divisible torsion-free abelian group is elementarily embeddable in an ultraproduct of finite groups. [Arrange that for each prime p, almost all the groups A_i are p-divisible and have no non-zero elements of order $< p$.]

19. Let ϕ be an $\forall_2$ sentence in the first-order language of abelian groups. Show that if ϕ is true in some finitely generated abelian group, then ϕ is true in some finite abelian group. [Suppose ϕ is true in some group $A \oplus B$ where A is finite and B is a direct sum of copies of $\mathbb{Z}$. If $B = \mathbb{Z}^n$, embed $A \oplus B \to A \oplus \mathbb{Q}^n \to (A \oplus \mathbb{Q}^n) \oplus B$ in the obvious way; repeat ω times and take the union, which has the form $A \oplus \mathbb{Q}^{(\omega)}$. Now assuming B is not trivial, using section A.2 below, each $A \oplus \mathbb{Q}^k \oplus B$ is elementarily equivalent to $A \oplus B$, hence also a model of ϕ. But ϕ is $\forall_2$ and hence preserved in unions of chains; so $A \oplus \mathbb{Q}^{(\omega)} \models \phi$. Then use Exercise 18 above. For a counterexample with $\exists_2$, consider 'No element has order 2 but some element is not divisible by 2.']

To find examples of classes which are closed under ultraproducts but not under elementary equivalence, one should remember (5.2) and look at pseudo-elementary classes. On (a) see Theorem A.2.9, on (b) see Exercise 3.5.7 and on (c) see Example 2 in section 5.2.

20. Show that each of the following classes is closed under ultraproducts. (a) The class of groups which are isomorphic to the multiplicative groups of fields. (b) The class of left primitive rings. [A ring R is left primitive iff it has a maximal regular right ideal ρ such that $(\rho:R) = (0)$; see Herstein [1968] p. 40.] (c) The class of linear orderings which are isomorphic to the orderings of ordered abelian groups.

9.6 The Feferman–Vaught theorem

The Feferman–Vaught technique is a way of discovering the first-order theories of complex structures by analysing their components. This description is a little vague, and in fact the Feferman–Vaught technique itself has something of a floating identity. It works for direct products, as we shall see. Clever people can make it work in other situations too.

Let me sketch the technique. Until further notice, L is a first-order language with finite signature. According to the Fraïssé-Hintikka theorem (Theorem 3.3.2), for every $r < \omega$ and every tuple $\bar{x}$ of variables we can effectively find a finite family of formulas of L, $\theta_0(\bar{x}), \ldots, \theta_{k-1}(\bar{x})$, such that each $\theta_j(\bar{x})$ is unnested and of quantifier rank $\leqslant r$, each unnested formula $\phi(\bar{x})$ of L of quantifier rank $\leqslant r$ is effectively reducible to a logically equivalent formula which is a disjunction of some of the θ_j, and

(6.1) $\vdash \forall \bar{x}$ (exactly one out of $\theta_0(\bar{x}), \ldots, \theta_{k-1}(\bar{x})$ holds).

We call these formulas the **game-normal formulas** for r and $\bar{x}$. We assume that for any fixed r and $\bar{x}$ there is an agreed order in which these formulas should be listed.

By Theorem 2.6.1, every formula of L is effectively reducible to an unnested formula. So if we want to known whether a given formula ϕ holds of some elements of a structure, we can suppose that ϕ is unnested, say of quantifier rank $\mathrm{qr}(\phi) = r$, and compute the corresponding game-normal formulas for r. The Feferman–Vaught method determines whether ϕ holds in the complex structure, in terms of which of the game-normal formulas holds at each component.

Let us make this more precise in the case of direct products. Suppose I is the index set and $X_0, \ldots, X_{k-1}$ are subsets of I. Then there is a natural first-order language for talking about the structure $(I, X_0, \ldots, X_{k-1})$: its signature has just k relation symbols, $R_0, \ldots, R_{k-1}$ to name $X_0, \ldots, X_{k-1}$ respectively. We write K_k for this language. As in section 9.4 above, if $\phi(\bar{x})$ is a formula of L and $\bar{a}$ a tuple of elements of the product $\prod_I A_i$ of L-structures, we write $\|\phi(\bar{a})\|$ for the set $\{i \in I: A_i \vDash \phi(\bar{a}(i))\}$.

Theorem 9.6.1 (*Feferman–Vaught theorem for direct products*). *Let L be a first-order language of finite signature. Then there is an algorithm which computes, for each unnested formula $\phi(\bar{x})$ of L, a sentence ϕ^* of K_k (for*

appropriate k) such that if $(A_i: i \in I)$ is any non-empty family of L-structures, $\bar{a}$ is any tuple of elements of $\prod_I A_i$, and $\theta_0(\bar{x}), \ldots, \theta_{k-1}(\bar{x})$ are the game-normal formulas of L for $\mathrm{qr}(\phi)$ and $\bar{x}$, then

$$(6.2) \qquad \prod_I A_i \vDash \phi(\bar{a}) \Leftrightarrow (I, \|\theta_0(\bar{a})\|, \ldots, \|\theta_{k-1}(\bar{a})\|) \vDash \phi^*.$$

We shall prove Theorem 9.6.1 as a corollary of a more general result about substructures of products.

Let I be a non-empty set. Then a **boolean algebra of subsets of** I is a set B of subsets of I which contains I and is closed under $\cap$, $\cup$ and complement $(I \backslash X)$; B forms a boolean algebra under the partial ordering $\subseteq$.

Let $(A_i: i \in I)$ be a family of L-structures and C a substructure of the product $\prod_I A_i$. Let B be a boolean algebra of subsets of I. We say that C is **B-valued** if

(6.3)   for every formula $\phi(\bar{x})$ of L and tuple $\bar{a}$ of elements of C,
         $\|\phi(\bar{a})\| \in B$.

We say that C **admits gluing over** B if for every set $X \in B$ and any pair of elements a, b of C there is an element c of C such that

(6.4)   for all $i \in X$, $c(i) = a(i)$; and for all $i \in I \backslash X$, $c(i) = b(i)$.

It is clear that if C admits gluing over B, and $X_0, \ldots, X_{n-1}$ are disjoint elements of B and $a_0, \ldots, a_{n-1}$ are elements of C, then C has an element c such that $c(i) = a_j(i)$ whenever $i \in X_j$.

One can easily see that the product $\prod_I A_i$ itself is $\mathcal{P}(I)$-valued and admits gluing over $\mathcal{P}(I)$. But there are plenty of other examples. For example, in a power A^I let C be the substructure consisting of all those elements a of A^I such that $\{a(i): i \in I\}$ is finite. Then C is $\mathcal{P}(I)$-valued and admits gluing over $\mathcal{P}(I)$. In section 9.7 we shall see how to generalise this example to any boolean algebra of subsets of I.

Suppose I is a non-empty set, B is a boolean algebra of subsets of I and $X_0, \ldots, X_{k-1} \in B$. Then we can extend the language K_k in order to talk about B, as follows. We allow 1-ary relation variables $P_0, P_1, \ldots$ as well as the relation symbols $R_0, \ldots, R_{k-1}$, and we introduce quantifiers $\forall P, \exists P$ which bind relation variables P. The resulting language is the monadic second-order language K_k^{mon}; see section 2.8 above. If ϕ is a sentence of K_k^{mon}, then we write

$$(6.5) \qquad\qquad (I, X_0, \ldots, X_{k-1}; B) \vDash \phi$$

to mean that ϕ is true in $(I, X_0, \ldots, X_{k-1})$ when we interpret $\forall P$ (resp. $\exists P$) as meaning 'for every set P in B' (resp. 'for some set P in B'). For example we can express that B is atomless by writing

$$(6.6) \qquad \forall P_0(\exists x\, P_0 x \rightarrow \exists P_1(\exists y(P_0 y \wedge P_1 y) \wedge \exists z(P_0 z \wedge \neg P_1 z))).$$

Paraphrased in boolean terms, this sentence says $\forall P(P \neq 0 \rightarrow \exists Q(P \cap Q \neq 0 \wedge P \cap (I\backslash Q) \neq 0))$.

It will be helpful to pick out some formulas of K_k^{mon} which speak only about the boolean algebra properties of B. Using Q to stand for any relation symbol or relation variable, we define the **boolean formulas** of K_k^{mon} to be the formulas which can be got from formulas of the form

(6.7) $Q_1 \subseteq Q_2$ (i.e. $\forall x(Q_1 x \rightarrow Q_2 x)$)

by using at most $\wedge$, $\vee$, $\neg$ and quantification of relation variables. Thus for example '$Q \neq 0$' can be expressed by the boolean formula $\neg \forall P(Q \subseteq P)$. Likewise '$Q_1 \cap Q_2 = Q_3$' can be written $\forall P(P \subseteq Q_3 \leftrightarrow P \subseteq Q_1 \wedge P \subseteq Q_2)$, which is clearly equivalent to a boolean formula.

Now we can state our most general form of the Feferman–Vaught theorem. Recall from section 9.4 that a substructure C of $\prod_I A_i$ **respects** $\exists$ if for every first-order formula $\phi(x)$ with parameters from C there is an element a of C such that $\|\exists x\, \phi\| = \|\phi(a)\|$.

Theorem 9.6.2 *(General Feferman–Vaught theorem). Let L be a first-order language of finite signature. There is an algorithm which computes, for each unnested formula $\phi(\bar{x})$ of L, a boolean sentence ϕ^b of K_k^{mon} (for the appropriate k) such that if $\theta_0(\bar{x}), \ldots, \theta_{k-1}(\bar{x})$ are the game-normal formulas for $\mathrm{qr}(\phi)$ and $\bar{x}$, then*

(6.8)   *if I is a non-empty set, B is a boolean algebra of subsets of I, $(A_i : i \in I)$ is a family of L-structures, C is a B-valued substructure of $\prod_I A_i$, C respects $\exists$ and admits gluing over B, and $\bar{a}$ is a tuple of elements of C, then*

$$C \vDash \phi(\bar{a}) \Leftrightarrow (I, \|\theta_0(\bar{a})\|, \ldots, \|\theta_{k-1}(\bar{a})\|; B) \vDash \phi^b.$$

Proof. By induction on the complexity of ϕ. Keep an eye on the construction of game-normal formulas in the proof of Theorem 3.3.2. If you find the argument hard going, try it out in the special case where I is the set $\{0, 1\}$.

Suppose first that $\phi(\bar{x})$ is atomic. Then the game-normal formulas are conjunctions of literals; ϕ^b should express that at every index i, $\bar{a}(i)$ satisfies a conjunction containing ϕ. So for ϕ^b we write a boolean sentence expressing '$R_{i_0} \cup \ldots \cup R_{i_{m-1}} = I$' where $\theta_{i_0}, \ldots, \theta_{i_{m-1}}$ are the game-normal formulas which have ϕ as a conjunct.

Next suppose $\phi(\bar{x})$ is a conjunction $\psi(\bar{x}) \wedge \chi(\bar{x})$. If both ψ and χ have quantifier rank r, then we simply take $(\psi \wedge \chi)^b$ to be $\psi^b \wedge \chi^b$. If one of them, say χ, has quantifier rank $s < r$, then we first have to reduce the game-normal formulas of rank s to disjunctions of game-normal formulas of rank r and make corresponding adjustments to χ^b. The reduction to rank r is possible by Theorem 3.3.2(c), and I leave the adjustment of χ^b as an exercise.

The argument is the same for disjunctions. Negations are easy: we take $(\neg \phi)^b$ to be $\neg (\phi^b)$. We eliminate $\forall x$ by writing it as $\neg \exists x \neg$. This leaves just one case, namely a formula of the form $\exists y \, \phi(\bar{x}, y)$.

By induction hypothesis we suppose that we have found $\phi(\bar{x}, y)^b$. Let $\theta_0(\bar{x}, y), \ldots, \theta_{k-1}(\bar{x}, y)$ be the game-normal formulas for $r - 1$ and $\bar{x}y$. Putting $k' = 2^k$, we list the subsets of k as $s_0, \ldots, s_{k'-1}$. The jth game-normal formula for r and $\bar{x}$, call it $\psi_j(\bar{x})$, is

$$(6.9) \qquad \bigwedge_{h \in s_j} \exists y \, \theta_h(\bar{x}, y) \wedge \forall y \bigvee_{h \in s_j} \theta_h(\bar{x}, y).$$

Thus for any product $\prod_I A_i$ and tuple $\bar{a}$ from $\prod_I A_i$,

$$(6.10) \qquad i \in \|\psi_j(\bar{a})\| \Leftrightarrow s_j = \{h < k : i \in \|\exists y \, \theta_h(\bar{a}, y)\|\}.$$

We take $(\exists y \, \phi(\bar{x}, y))^b$ to be a boolean sentence of K_k^{mon} which is true of a structure $(I, X_0, \ldots, X_{k'-1}; B)$ if and only if

(6.11) there are disjoint sets $I_0, \ldots, I_{k-1}$ in B with union I, such that $(I, I_0, \ldots, I_{k-1}; B) \vDash \phi^b$, and for each $h < k$, $I_h \subseteq \bigcup \{X_j : h \in s_j\}$.

We must check that this choice of $(\exists y \, \phi(\bar{x}, y))^b$ works. In other words, we must show that if C and $\bar{a}$ are as assumed in (6.8), then $C \vDash \exists y \, \phi(\bar{a}, y)$ if and only if (6.11) holds where each X_j is $\|\psi_j(\bar{a})\|$.

For these values of X_j, the last clause in (6.11) can be simplified. For each $h < k$ and each $i \in I$ we have

$$(6.12) \qquad i \in \bigcup \{\|\psi_j(\bar{a})\| : h \in s_j\} \Leftrightarrow \text{for some } j < k', \, i \in \|\psi_j(\bar{a})\| \text{ and } h \in s_j$$
$$\Leftrightarrow i \in \|\exists y \, \theta_h(\bar{a}, y)\|.$$

To go from right to left in the second equivalence, choose j so that the right-hand side of (6.10) holds.

Suppose first that $C \vDash \exists y \, \phi(\bar{a}, y)$. Then there is an element c of C such that $C \vDash \phi(\bar{a}, c)$. For each $h < k$ put $I_h = \|\theta_h(\bar{a}, c)\|$. Then each I_h is in B since C is B-valued. We have $(I, I_0, \ldots, I_{k'-1}; B) \vDash \phi^b$ by induction hypothesis, and it is trivial that $I_h \subseteq \|\exists y \, \theta_h(\bar{a}, c)\|$ for each $h < k$.

Conversely suppose that $(I, \|\psi_0(\bar{a})\|, \ldots, B)$ is a structure $(I, X_0, \ldots; B)$ which satisfies (6.11), and let $I_0, \ldots, I_{k-1}$ be as in (6.11). By (6.12) and the last clause of (6.11), $I_h \subseteq \|\exists y \, \theta_h(\bar{a}, y)\|$. Since C respects $\exists$, there is an element c_h of C such that $\|\exists y \, \theta_h(\bar{a}, y)\| = \|\theta_h(\bar{a}, c_h)\|$. Since C admits gluing over B, there is an element c of C such that for each $h < k$ and each $i \in I_h, c(i) = c_h(i)$. By (6.1) it follows that $I_h = \|\theta_h(\bar{a}, c)\|$ for each $h < k$, and hence $(I, \|\theta_0(\bar{a}, c)\|, \ldots; B) \vDash \phi^b$ by (6.11). So $C \vDash \phi(\bar{a}, c)$ by induction hypothesis, whence $C \vDash \exists y \, \phi(\bar{a}, c)$ as required. $\qquad \square$

The sequence of formulas $(\phi^b; \theta_0, \ldots, \theta_{k-1})$ of the theorem is known as the **determining sequence** for ϕ.

Theorem 9.6.1 follows at once from Theorem 9.6.2 and the fact below. By a **fundamental sentence** of K_k we mean a sentence of the form $\exists_{=n} x (R'_0 x \wedge \ldots \wedge R'_{k-1} x)$, where $n < \omega$ and each $R'_i x$ is either $R_i x$ or $\neg R_i x$.

Fact 9.6.3. *There is an algorithm that finds, for each boolean sentence ϕ of K_k^{mon}, a sentence ϕ^* of K_k which is a boolean combination of fundamental sentences and agrees with ϕ on all structures of the form $(I, X_0, . . ., X_{k-1}; \mathcal{P}(I))$.*

Proof. This is an old result of Skolem [1919]. See Exercise 1 below.   □

Corollaries

What does the Feferman–Vaught theorem tell us about a product of two structures?

Corollary 9.6.4. *Let L be a first-order language, let A and B be L-structures and let $\phi(\bar{x})$ be a formula of L. Then there is a finite set $\{(\theta_i(\bar{x}), \chi_i(\bar{x})): i < n\}$ of pairs of formulas of L, such that for all tuples $\bar{a} = (a_0, a_1, \ldots), \bar{b} = (b_0, b_1, \ldots)$ from A, B respectively,*

(6.13) $A \times B \vDash \phi((a_0, b_0), (a_1, b_1), \ldots) \Leftrightarrow$ *for some $i < n$,*
          $A \vDash \theta_i(\bar{a})$ *and* $B \vDash \chi_i(\bar{b})$.

Proof. Unpick what the theorem says.   □

The next corollary could have been proved directly by the technique of Theorem 3.3.10.

Corollary 9.6.5. (a) *If I is a non-empty set and for each $i \in I$, A_i and B_i are elementarily equivalent L-structures, then $\prod_I A_i \equiv \prod_I B_i$.*
     (b) *If I is a non-empty set and for each $i \in I$, A_i and B_i are L-structures with $A_i \preccurlyeq B_i$, then $\prod_I A_i \preccurlyeq \prod_I B_i$.*

Proof. Immediate from Theorem 9.6.1.   □

Corollary 9.6.6. *Let L be a first-order language with finite signature, and suppose $r < \omega$. Let $\theta_0, \ldots, \theta_{k-1}$ be the game-normal sentences of L for quantifier rank r. Then there is a positive integer n, effectively computable from L and r, such that, for any unnested sentence ϕ of L of quantifier rank $\leqslant r$ and any product $\prod_I A_i$ of L-structures, the question whether or not $\prod_I A_i \vDash \phi$ holds can be reduced, effectively and independent of $\prod_I A_i$, to the questions*

(6.14) *for each $h < k$, is the cardinality of $\|\theta_h\|$ equal to $0, \ldots, n$ or is it greater than n?*

Proof. Find ϕ^b as in Theorem 9.6.2, hence ϕ^* as in Fact 9.6.3, and choose n so that if ϕ^* contains a fundamental sentence beginning with $\exists_{=m}$ then $m \leqslant n$. □

Corollary 9.6.7. *For every first-order sentence ϕ there is a positive integer n, which we can compute effectively from ϕ, such that for every structure A, if $A^n \vDash \phi$ then $A^I \vDash \phi$ for every set I of cardinality $\geqslant n$.*

Proof. This follows from the previous corollary, since we can effectively reduce ϕ to unnested form. □

We say that a sentence ϕ is **preserved in products** if a product $\prod_I A_i$ is a model of ϕ whenever the factors A_i are models of ϕ. Likewise we say that ϕ is **preserved in products of models of** T if the same holds provided that all the A_i are models of T. Likewise we define when ϕ is **preserved in powers**, etc.

Corollary 9.6.8. *Let T be a decidable theory in a first-order language L. Then we can effectively determine, for any sentence ϕ of L, whether or not ϕ is preserved in products of models of T.*

Proof. Consider a sublanguage L' of L with finite signature, a number $r < \omega$, and the set Θ of game-normal sentences of L for r. Since T is decidable, we can determine which are the sentences θ in Θ such that some model of T satisfies θ. But then by Corollary 9.6.6 we can effectively discover, for each subset Ψ of Θ, whether or not $\bigvee \Psi$ is preserved in products of models of T. Now use the Fraïssé–Hintikka theorem (Theorem 3.3.2). □

Readers of Corollary 9.6.8 should bear in mind that there is no algorithm for determining whether a given first-order sentence is preserved in products (Machover [1960], Almagambetov [1965]). Anybody who tries to extract such an algorithm from Corollary 9.6.8 will run into the following difficulty. Suppose θ_0 and θ_1 are distinct game-normal sentences, and the proof of Corollary 9.6.6 establishes that if $A \vDash \theta_0$ then $A^2 \vDash \theta_1$. Does this show that θ_0 is not preserved in products? Unfortunately not, because θ_0 may have no models. This is why T in Corollary 9.6.8 has to be decidable.

Now it follows from Theorem 9.6.1 that for any fixed $r < \omega$, as far as sentences of quantifier rank $\leqslant r$ are concerned, the theory of $A \times B$ is completely determined by the theories of A and B. Hence if ϕ and ψ are game-normal sentences of quantifier rank r, then there is a unique game-normal sentence θ of quantifier rank r such that if $A \vDash \phi$ and $B \vDash \psi$ then $A \times B \vDash \theta$. We write this game-normal sentence as $\phi \times \psi$.

Corollary 9.6.9. *Let L be a first-order language and ϕ a sentence of L. Then ϕ is preserved in powers if and only if ϕ is logically equivalent to a disjunction of a finite number of sentences of L which are preserved in products.*

Proof. Right to left is easy. In the other direction, we can suppose without loss that L has finite signature and ϕ is unnested and of quantifier rank r. Let Θ be the set of game-normal sentences of L for rank r, and let Ψ be the set of all sentences θ in Θ such that $\theta \vdash \phi$ and θ has a model. For each $\theta \in \Psi$ let Φ_θ be the set of all the sentences $\theta, \theta \times \theta, \theta \times \theta \times \theta$ etc. Then each sentence θ' in Φ_θ has a model B which is a power of a model of ϕ and hence itself a model of ϕ; since θ' is game-normal, it follows that θ' must imply ϕ, and so we have $\Phi_\theta \subseteq \Psi$. Also by its definition, $\bigvee\Phi_\theta$ is preserved in products of finitely many structures, and so by another application of Fact 9.6.3, $\bigvee\Phi_\theta$ is preserved in arbitrary products. Then ϕ is logically equivalent to the disjunction $\bigvee\{\bigvee\Phi_\theta : \theta \in \Psi\}$. $\square$

Exercises for section 9.6

1. Show that for every formula $\phi(Y_0, \ldots, Y_{m-1})$ of K_k^{mon} without first-order free variables, we can effectively find a formula which is equivalent to $\phi(Y_0, \ldots, Y_{m-1})$ in $\mathscr{P}(I)$ and is a boolean combination of formulas of the form $\exists_{=n} x\ (R_0' x \wedge \cdots \wedge R_{k-1}' x \wedge Y_0' x \wedge \cdots \wedge Y_{m-1}' x)$, where $n < \omega$ and $R_i' x$ (resp. $Y_i' x$) is either $R_i x$ or $\neg R_i x$ (resp. either $Y_i x$ or $\neg Y_i x$).

2. Show that for each first-order sentence ϕ we can compute a natural number n such that for all families $(A_i : i \in I)$, if $\prod_I A_i \vDash \phi$ then there is a subset J of I, with $|J| \leq n$, such that for all sets J' with $J \subseteq J' \subseteq I$, we have $\prod_{J'} A_i \vDash \phi$.

3. Show if $A \equiv A \times B \times C$ then $A = A \times B$.

4. Let L be a first-order language of finite signature, $r < \omega$ and Φ the set of those game-normal sentences of L for quantifier rank r which have models. For $\phi, \theta \in \Phi$ define $\phi \leqslant \theta$ if there are L-structures A, B such that $A \vDash \phi$ and $A \times B \vDash \theta$. (a) Show that $\leqslant$ is a partial ordering of Φ. (b) Show that for each $\theta \in \Phi$ the sentence θ^0, viz. $\bigvee\{\phi \in \Phi : \theta \leqslant \phi\}$, is preserved in products. (c) Show that every sentence in Φ is logically equivalent to a boolean combination of sentences of the form θ^0 with $\theta \in \Phi$. (d) Show that every sentence of L is logically equivalent to a boolean combination of sentences which are preserved in products.

5. Show that every Horn sentence is logically equivalent to a boolean combination of strict Horn sentences. [See Exercise 9.1.9.]

6. Let L be a first-order language and T a theory in L which is preserved in products. Let S be a complete theory which extends T in L; we say that S is **indecomposable** if when S can be written as $S = \text{Th}(\prod_I B_i)$ then $S = \text{Th}(B_i)$ for some $i \in I$. Show that every model of T is elementarily equivalent to a product of models of indecomposable complete extensions of T.

7. Let L be a first-order language of finite signature. Show that the sets of sentences of L which are (a) preserved in products, (b) preserved in powers, are both effectively enumerable.

8. Let $(A_i : i \in I)$ be a non-empty family of L-structures and C a substructure of $\prod_I A_i$ which respects $\exists$. Let ϕ be a Horn sentence of L such that $A_i \vDash \phi$ for every $i \in I$. Show that $\prod_I A_i \vDash \phi$.

9. Show that if ϕ in Theorem 9.6.2 is an unnested formula of quantifier rank r beginning with an existential quantifier, then there are formulas $\theta_0, \ldots, \theta_{k-1}$ of quantifier rank r and beginning with existential quantifiers, such that the conclusion of the theorem holds. [Use formulas $\exists y\, \theta'$ where θ' are the game-normal formulas of the previous step in the inductive proof of the theorem; (6.11) can be simplified.]

10. Show that if ϕ in Theorem 9.6.2 is an $\exists_n$ (resp. $\forall_n$) formula, then there are $\exists_n$ (resp. $\forall_n$) formulas $\theta_0, \ldots, \theta_{k-1}$ such that the conclusion of the theorem holds. [Combine the previous exercise with Exercise 3.3.2.]

11. Show that if $\mathscr{F}$ is a filter over the set I, $(A_i : i \in I)$ is a family of L-structures and $\bar{a}$ is a tuple of elements of $\prod_I A_i/\mathscr{F}$, then for every unnested first-order formula $\phi(\bar{x})$ of L,

$$\prod_I A_i/\mathscr{F} \vDash \phi(\bar{a}) \Leftrightarrow (\mathscr{P}(I), \|\theta_0(\bar{a})\|, \ldots, \|\theta_{k-1}(\bar{a})\|)/\mathscr{F} \vDash \phi^{\text{b}},$$

where ϕ^{b} is read in the natural way as a statement about the boolean algebra $\mathscr{P}(I)/\mathscr{F}$ with distinguished elements $\|\theta_0(\bar{a})\|/\mathscr{F}$. [One route is Exercise 9.4.7.]

12. Suppose the boolean algebra $\mathscr{P}(I)/\mathscr{F}$ is atomless. (a) Show that if A is either finite or a countable ω-categorical structure, then the reduced power $A^I/\mathscr{F}$ is ω-categorical. (b) Give an example of finite structures A_i $(i \in I)$ such that $\prod_I A_i/\mathscr{F}$ is infinite and not ω-categorical.

13. Let T be a first-order theory and let $\mathbf{K}$ be the class of all direct products of models of T. Show that if U is a theory such that every finite subset of U has a model in $\mathbf{K}$, then U has a model in $\mathbf{K}$.

14. Let A be the multiplicative semigroup of positive integers. Show that $\text{Th}(A)$ is decidable. [A is the ωth power of the structure $(\omega, +)$. Use Theorem 3.3.8.]

Disjoint sums $A + B$ were defined at Exercise 3.2.8.

*15. Suppose L and L' are disjoint signatures with no function symbols. Let A, B be respectively an L-structure and an L'-structure. (a) Show that every first-order formula ϕ of signature $L + L'$ is equivalent in $A + B$ to a boolean combination ψ of first-order formulas relativised to A and first-order formulas relativised to B; ψ is independent of the choice of A and B. (b) Show that this need not be true if we drop 'first-order'. [Let L be the signature of sets, with $\in$ for membership, and L' a copy of L. Writing V_α for the universe of sets hereditarily of rank $< \alpha$ (as in Exercise 4.3.1), find a formula $\phi_\alpha(x, y)$ of $(L + L')_{\omega_1\omega}$ which expresses in $V_\alpha + V_\alpha$ an isomorphism between the two copies of V_α, by induction on $\alpha < \omega_1$. Then putting $\alpha = \omega + 3$, let B be a countable structure $L_{\omega_1\omega}$-equivalent to V_α, and compare $V_\alpha + V_\alpha$ with $V_\alpha + B$.]

9.7 Boolean powers

If M is an L-structure and B is a boolean algebra, we can combine them to form an L-structure $M[B]^*$ known as the *boolean power* of M by B. Boolean powers pass the first test of a good construction: they are used to prove interesting theorems which don't mention them. These include theorems about decidability, ω-categoricity and the existence of many non-isomorphic models.

There are several ways of defining boolean powers, and each has its uses. I give two. The first construction builds $M[B]^*$ as a substructure of a direct power of M; this is handy for Feferman–Vaught results. The second describes $M[B]^*$ as a presenting word-construction on B; this helps for finding many non-isomorphic structures. We use the letters $M, N, \ldots$ for structures and $A, B, \ldots$ for boolean algebras.

First construction

We assume for the moment that B is a boolean algebra of sets. More precisely, let I be a non-empty set and B a subset of $\mathcal{P}(I)$ containing I and closed under $\cap$, $\cup$ and complementation $I \backslash X$. By a **partition of unity in B** we mean a finite set of non-empty elements $X_0, \ldots, X_{n-1}$ of B such that $X_0 \cup \ldots \cup X_{n-1} = I$ and the X_i are pairwise disjoint.

Now consider a signature L and a non-empty L-structure M. We say that an element a of the direct power M^I is **constant on** an element X if there is c in M such that $a(i) = c$ for all $i \in X$. We say that a is **based on** a partition of unity π in B if a is constant on each set in π. Note that if W is a finite set of elements which are all based on partitions of unity in B, then there is a unique smallest partition of unity π such that every element of W is based on π.

One easily checks that the elements of M^I which are based on partitions of unity in B form a substructure of M^I. We write $M[B]^*$ for this substructure.

It is known as the **boolean power of** M **by** B. (Some authors say **bounded boolean power**, to distinguish it from the unbounded boolean powers which appear in Exercise 5 below.) The diagonal embedding $e: M \to M^I$ embeds M into $M[B]^*$.

If a is an element of $M[B]^*$ and c an element of M, let us write $a^{-1}(c)$ for the set $\{i \in I: a(i) = c\}$. Then for each element a of $M[B]^*$ there is a unique finite set of elements $c_0, \ldots, c_{n-1}$ of M such that $a^{-1}(c_0), \ldots, a^{-1}(c_{n-1})$ forms a partition of unity in B.

Lemma 9.7.1. *Let* A, B *be boolean algebras of subsets of the non-empty sets* I, J *respectively. Let* $f: A \to B$ *be a homomorphism of boolean algebras. Then for every non-empty* L-*structure* M, f *induces a homomorphism* $M[f]^*: M[A]^* \to M[B]^*$. *If* f *is an embedding (resp. an isomorphism) then so is* $M[f]^*$.

Proof. For any element a of $M[A]^*$, choose elements $c_0, \ldots, c_{n-1}$ of M so that $a^{-1}(c_0), \ldots, a^{-1}(c_{n-1})$ form a partition of unity in A. Then (disregarding empty sets) $f(a^{-1}(c_0)), \ldots, f(a^{-1}(c_{n-1}))$ form a partition of unity in B. Define $M[f]^*(a)$ to be the element b of $M[B]^*$ such that $b^{-1}(c_j) = f(a^{-1}(c_j))$ for each $j < n$. Then clearly $M[f]^*$ maps $\mathrm{dom}\, M[A]^*$ into $\mathrm{dom}\, M[B]^*$.

Let $\bar{a} = (a_0, \ldots, a_{k-1})$ be a tuple of elements of $M[A]^*$. When $i \in I$, we write $\bar{a}(i)$ for the tuple of values $(a_0(i), \ldots, a_{k-1}(i))$. There is a partition of unity in A, say $\pi = (X_0, \ldots, X_{n-1})$, such that all the elements of $\bar{a}$ are based on π. For each $j < n$ let $\bar{c}_j$ be the k-tuple giving the values of $\bar{a}$ on the set X_j. Suppose now that $\psi(\bar{x})$ is an atomic formula of L such that $M[A]^* \vDash \psi(\bar{a})$. Then $M^I \vDash \psi(\bar{a})$ and so $M \vDash \psi(\bar{a}(i))$ for each $i \in I$. It follows that $M \vDash \psi(\bar{c}_j)$ for each $j < n$. Hence $M^J \vDash \psi(M[f]^*\bar{a})$ and so $M[B]^* \vDash \psi(M[f]^*\bar{a})$. This shows that $M[f]^*$ is a homomorphism. The last sentence of the lemma follows by the same argument in the other direction, using the fact that when f is an embedding, each set $f(X_j)$ is also non-empty. $\square$

It follows from Lemma 9.7.1 that $M[B]^*$ really depends only on the isomorphism type of the boolean algebra B and not on the way it is expressed as a boolean algebra of sets. By Stone duality (see Bell & Machover [1977] Chapter 4 or Burris & Sankappanavar [1981] IV.4), every boolean algebra is isomorphic to a boolean algebra of sets. So our first construction has defined $M[B]^*$ for every boolean algebra B. Our second construction will work direct from the boolean algebra.

Remark. Speaking of Stone duality, let $S(A)$ be the Stone space of ultrafilters on the boolean algebra A. Then every element a of $M[A]^*$ determines a map $a^*: S(B) \to \mathrm{dom}\,(M)$ as follows. Choose elements $c_0, \ldots, c_{n-1}$ of M so that $a^{-1}(c_0), \ldots a^{-1}(c_{n-1})$ form a partition of unity in A. Then for every element p of $S(A)$, put $a^*(p) = c_i$ iff $c_i \in p$. Giving $\mathrm{dom}\,(M)$ the discrete topology, this makes a^* a continuous map from $S(A)$ to $\mathrm{dom}\,(M)$. Conversely every continuous map from $S(A)$ to $\mathrm{dom}\,(M)$ has only finitely many elements in its image (by the compactness of $S(A)$), and the pre-image of each element is a clopen subset of $S(A)$, so that between them the pre-images describe a partition of A. Thus there is a natural correspondence between the elements of $M[A]^*$ and the continuous maps from $S(A)$ to $\mathrm{dom}\,(M)$. Sheaf-theoretic writers often prefer to describe $M[A]^*$ in terms of the continuous maps.

Meanwhile we record some simple consequences of the definition of boolean powers.

Theorem 9.7.2. *Let M be a non-empty L-structure.*

(a) *Suppose B is a boolean algebra. If $|M| \geqslant 2$ and at least one of M or B is infinite, then $|M[B]^*| = \max(|M|, |B|)$.*

(b) *If B is a boolean algebra and $L^- \subseteq L$ then $M[B]^*|L^- = (M|L)[B]^*$.*

(c) *$M[-]^*$ is a functor from the variety of boolean algebras to the class of all L-structures and homomorphisms; this functor preserves filtered colimits.*

Proof. All easy. $\qquad\qquad\qquad\qquad\qquad\qquad\qquad\qquad\qquad\qquad\qquad$ ☐

Part (c) of the theorem implies that $M[-]^*$ can be written as a presenting word-construction, by Theorem 9.3.7. Our second construction will show how to do this. (The functor is not a replica functor, by Exercise 9.3.13.)

Second construction

The elements $M[B]^*$ shall be the sequences $(a_0, \ldots, a_{n-1}; X_0, \ldots, X_{n-1})$ such that n is a positive integer, $a_0, \ldots, a_{n-1}$ are distinct elements of M and $X_0, \ldots, X_{n-1}$ are elements $\neq 0$ in B satisfying $X_0 \vee \ldots \vee X_{n-1} = 1$ and $X_i \wedge X_j = 0$ when $i < j$. If R is a 2-ary relation symbol of L and $(\bar{a}; \bar{X}) = (a_0, \ldots, a_{n-1}; X_0, \ldots, X_{n-1})$, $(\bar{a}', \bar{X}') = (a_0', \ldots, a_{m-1}'; X_0', \ldots, X_{m-1}')$ are elements of $M[B]^*$, then we define $R^{M[B]^*}$ so that

(7.1) $M[B]^* \vDash R((\bar{a}, \bar{X}), (\bar{a}', \bar{X}'))$ if and only if

  for all $i < n$ and all $j < m$, if $X_i \wedge X_j' \neq 0$ then $M \vDash R(a_i, a_j')$.

Similar definitions apply to the other symbols of L. I leave it to the reader to check that there is a natural isomorphism between this $M[B]^*$ and the $M[B]^*$ of the previous construction.

This second construction can be expressed as a word-construction Γ from the variety $\mathbf{V}$ of boolean algebras to the quasivariety $\mathbf{W}$ of all L-structures, as follows. The signature J_Γ is got by adding to L a set of new function symbols $F_{\bar{a}}$, one for each non-empty non-repeating tuple $\bar{a}$ of elements of M; the arity of $F_{\bar{a}}$ is equal to the length of $\bar{a}$. For each such n-tuple $\bar{a}$,

(7.2) $\Gamma^{\text{gen}}_{F[\bar{a}]}(\bar{x})$ is '$x_0 \vee \ldots \vee x_{n-1} = 1$ and for all $i < j < n$, $x_i \wedge x_j = 0$'.

Then if $\bar{X}$ is a partition of unity in the boolean algebra B, $F_{\bar{a}}(\bar{X})$ will represent the element $(\bar{a}, \bar{X})$. Note that (7.2) allows $x_i = 0$ for some i; but there is no harm in this.

Now suppose R is a 2-ary relation symbol of L and $F_{\bar{a}}(\bar{X})$, $F_{\bar{a}'}(\bar{X}')$ are two elements of $M[B^*]$. Then by (7.1),

(7.3) $M[B]^* \vDash R(F_{\bar{a}}(\bar{X}), F_{\bar{a}'}(\bar{X}'))$ $\Leftrightarrow$ for all i and j, if
$M \vDash \neg R(a_i, a'_j)$ then $X_i \wedge X'_j = 0$.

From (7.3) we can read off what $\Gamma^{\text{rel}}_{R(F[\bar{a}],F[\bar{a}'])}$ must be: it is the conjunction of those equations $x_i \wedge x'_j = 0$ such that $M \vDash \neg R(a_i, a'_j)$. Similarly all the other formulas Γ^{rel}_ψ will be conjunctions of equations. With these definitions, $\Gamma(B) = M[B]^*$ for any boolean algebra B. In short, after a little checking, we get the following.

Theorem 9.7.3. *Let L be a signature, M and L-structure, $\mathbf{W}$ the quasivariety of all L-structures and $\mathbf{V}$ the variety of boolean algebras. Then the construction $M[-]^*$ is the associated functor of a presenting word-construction Γ with $|\Gamma| \leqslant |M| + |L|$. All the formulas in Γ^{gen} and Γ^{rel} are finite conjunctions of equations, so that Γ is a Feferman word-construction too.* $\qquad\square$

Equivalence

Corollary 9.7.4. *If $\kappa \geqslant |M| + \omega$ and A, B are $L_{\kappa^+\omega}$-equivalent boolean algebras, then $M[A]^* \equiv_{\kappa^+\omega} M[B]^*$.*

Proof. This follows from Theorem 9.7.3 by Corollary 9.3.3, noting that by Theorem 9.7.2(b) we can assume L has cardinality at most κ. $\qquad\square$

This argument won't cover elementary equivalence. But another argument will, by way of the following lemma and the ideas of section 9.6.

Lemma 9.7.5. *Let M be a non-empty L-structure and B a boolean algebra of subsets of the non-empty set I.*

(a) As a substructure of M^I, $M[B]^*$ is B-valued, respects $\exists$ and admits gluing over B.

(b) Every Horn sentence of L which is true in M is also true in $M[B]^*$.

Proof. (a) is straightforward from the definitions. For (b) we copy the proofs of Lemma 9.1.4 and Theorem 9.1.5. First we show that if $\psi(\bar{x})$ is a positive primitive formula of L and $\bar{a}$ is a tuple of elements of $M[B]^*$ (on the first construction), then

(7.4) $M[B]^* \vDash \psi(\bar{a})$ if and only if   for every $i \in I$, $M \vDash \psi(\bar{a}(i))$.

The fact that C respects $\exists$ is precisely what we need in order to deal with the existential quantifiers in ψ. Now (b) follows as for Theorem 9.1.5.   □

In particular every boolean power $M[B]^*$ lies in the quasivariety generated by M.

Theorem 9.7.6. (a) *If* $M \equiv N$ *and* $A \equiv B$ *then* $M[A]^* \equiv N[B]^*$.
    (b) *If* $M \preccurlyeq N$ *and* $A \preccurlyeq B$ *then* $M[A]^*$ *is elementarily embedded in* $N[B]^*$.

Proof. (a) is by Lemma 9.7.5(a) and the Feferman–Vaught theorem (Theorem 9.6.2) applied to a first-order sentence ϕ. Note that each $\|\psi_i\|$ is either 1 or 0, the same for both M and N since $M \equiv N$. Then since the sentence ϕ^b is boolean, it makes a first-order statement about the boolean algebras A and B, which must hold in A if and only if it holds in B.
    (b) is similar.   □

Isomorphism

Of course if $A \cong B$ then $M[A]^* \cong M[B]^*$. But some stronger statements are true. First we consider ω-categoricity.

Theorem 9.7.7. *Let* L *be a first-order language and* M *an* L-*structure. Suppose either that* M *is finite with more than one element, or that* M *is countable and* $\mathrm{Th}(M)$ *is* ω-*categorical. Let* B *be a countable atomless boolean algebra. Then* $\mathrm{Th}(M[B]^*)$ *is* ω-*categorical.*

Proof. By Theorem 9.7.2(a), $M[B]^*$ is a countable structure. Using Theorem 7.3.1 (the theorem of Engeler, Ryll-Nardzewski and Svenonius) when M is infinite, we know that for every $n < \omega$, $\mathrm{Aut}(M)$ has only finitely many orbits in its action on n-tuples of elements of M. By Theorem 7.3.1 again, it suffices to show that the same holds for $M[B]^*$ as well.
    We use the first construction of $M[B]^*$. Fixing n, list the orbits of n-tuples of elements of M under $\mathrm{Aut}(M)$ as $\varnothing_0, \ldots, \varnothing_{k-1}$. To each n-tuple $\bar{a}$ of elements of $M[B]^*$ we assign a **colour** $\subseteq k$, viz. the set $\{j < k: \text{for some } i \in I, \bar{a}(i) \text{ lies in } \varnothing_j\}$. There are finitely many colours. We shall show that if $\bar{a}$ and $\bar{b}$ have the same colour then some automorphism of $M[B]^*$ takes $\bar{a}$ to $\bar{b}$.
    If $\bar{a}$ and $\bar{b}$ have the same colour, then there are partitions of unity in B,

say $X_0, \ldots, X_{m-1}$ and $Y_0, \ldots, Y_{m-1}$, such that for each $j < m$, if $i \in X_j$ and $i' \in Y_j$ then $\bar{a}(i)$ and $\bar{b}(i')$ lie in the same orbit $\varnothing_{j'}$. Since B is countable and atomless, there is an automorphism α of B which takes each X_j to Y_j. (See Example 4 in section 3.2 above.) By Lemma 9.7.1 this automorphism induces an automorphism of $M[B]^*$, so we may assume that $X_j = Y_j$ for each $j < m$. Also we may refine the partition $X_0, \ldots, X_{m-1}$ so that $\bar{a}$ and $\bar{b}$ are both constant on each partition set. For each $j < m$ there is an automorphism σ_j of M which takes $\bar{a}(i)$ to $\bar{b}(i)$ whenever $i \in X_j$. Now one easily checks that

(7.5)   there is an automorphism σ of $M[B]^*$ such that for each element a of $M[B]^*$, each $j < n$ and each $i \in X_j$, $(\sigma a)(i) = \sigma_j(a(i))$.

By (7.5), $\sigma \bar{a} = \bar{b}$ as required. $\square$

Corollary 9.7.8. *Let T be a Horn theory in a first-order language L. Suppose*
(a) *all infinite models of T are elementarily equivalent, and*
(b) *T has a finite model with more than one element.*
Then T is ω-categorical.

Proof. Let M be a finite model of T with more than one element. By the theorem, $\mathrm{Th}(M[B]^*)$ is ω-categorical. By Lemma 9.7.2(a) and Lemma 9.7.5, $M[B]^*$ is an infinite model of T. $\square$

Our next and last result is at the other extreme from Theorem 9.7.7. There are a number of results which state that under certain conditions, $M[A]^* \cong M[B]^*$ implies $A \cong B$. Since we have large families of non-isomorphic boolean algebras (for example by Corollary 11.3.7 below), these results give us large families of non-isomorphic structures in the quasivarieties generated by M.

I quote one such theorem. Its proof uses results from Chapter 11 below. By Exercise 9.1.15(b), the hypothesis on M holds whenever M is a non-abelian group.

Theorem 9.7.9. *Let M be an L-structure and κ a regular cardinal $> \max(|M|, |L|)$. Suppose that there are a conjunction $\psi(\bar{x}, \bar{y})$ of atomic formulas of L, and tuples $\bar{a}_0, \bar{a}_1, \bar{b}_0, \bar{b}_1$ in M such that*

(7.6) $M \vDash \psi(\bar{a}_m, \bar{b}_n)$ *iff* $m \leqslant n$.

Then there is a family $(B_i : i < 2^\kappa)$ of boolean algebras such that for all $i < j < 2^\kappa$,
(a) *$M[B_i]^* \equiv_{\infty\omega} M[B_j]^*$,*
(b) *neither of $M[B_i]^*$, $M[B_j]^*$ is embeddable in the other,*
(c) *$|M[B_i]^*| = \kappa$.*

Proof. Take the family $(\eta_i: i < 2^\kappa)$ of linear orderings of cardinality κ as in the proof of Theorem 11.3.6. Let Γ be the interval-algebra construction from section 11.2 which takes each linear ordering to a boolean algebra; write B_i for the algebra $\Gamma(\eta_i)$. By Theorem 11.2.5, Γ is a word-construction, and by Theorem 9.7.3 the boolean power construction $M[-]^*$ is a word-construction too. So their composition is a word-construction Δ; one can check that $|\Delta| \leqslant \max(|M|, |L|)$. Hence by Theorem 11.3.6, the structures $M[B_i]^*$ all have cardinality $\leqslant \kappa$, they are back-and-forth equivalent, and if $i \neq j$ then no set of tuples from $M[B_j]$ is ordered in order-type η_i by any formula of $L_{\infty\omega}$.

So to complete the argument, it's enough to show that if η is a linear ordering, then $M[\Gamma(\eta)]^*$ contains a family of tuples ordered in order-type η by some quantifier-free formula of L (so that an embedding will take these tuples to tuples ordered in the same order-type by the same formula). We show this as follows, using the Remark after Lemma 9.7.1 to describe elements of $M[\Gamma(\eta)]^*$ as continuous maps from $S(\Gamma(\eta))$ to dom M.

Put $B = \Gamma(\eta)$ and regard η as a linearly ordered subset of B. Let $\bar{a}_0, \bar{a}_1, \bar{b}_0, \bar{b}_1$ be as in the theorem. For each element d of η in B, let $\bar{c}_d, \bar{e}_d$ be the tuples in $M[B]^*$ such that for each ultrafilter p on B,

$$(7.7) \quad \bar{c}_d(p) = \begin{cases} \bar{a}_1 \\ \bar{a}_0 \end{cases} \quad \text{and} \quad \bar{e}_d(p) = \begin{cases} \bar{b}_1 & \text{if } d \in p, \\ \bar{b}_0 & \text{otherwise.} \end{cases}$$

Suppose now that $d < d'$ in η. Then $\bar{c}_d, \bar{e}_d$ agree with $\bar{c}_{d'}, \bar{e}_{d'}$ on all ultrafilters p except those containing d' but not d, and for these p we have

$$(7.8) \qquad M \vDash \psi(\bar{c}_d(p), \bar{e}_{d'}(p)) \wedge \neg \psi(\bar{c}_{d'}(p), \bar{e}_d(p)).$$

It follows that $M[B]^* \vDash \psi(\bar{c}_d, \bar{e}_{d'}) \wedge \neg\psi(\bar{c}_{d'}, \bar{e}_d)$. So $(\bar{c}_d, \bar{e}_d: d \in \eta)$ is ordered by the quantifier-free formula $\theta(\bar{x}_0\bar{x}_1, \bar{y}_0\bar{y}_1)$, viz. $\neg\psi(\bar{y}_0, \bar{x}_1)$. $\qquad\square$

Exercises for section 9.7

1. Let M be an L-structure. (a) If B is the two-element boolean algebra, show that $M[B]^* \cong M$. (b) If B and C are boolean algebras, show that $M[B \times C]^* \cong M[B]^* \times M[C]^*$.

2. Let M be an L-structure and B, C boolean algebras. Show that $(M[B]^*)[C]^* \cong M[B[C]^*]^*$.

3. If M is a finite L-structure and B is a boolean algebra, show that $M[B]^*$ is residually finite. (This means that if $\phi(\bar{x})$ is an atomic formula and $\bar{a}$ is a tuple of elements of $M[B]^*$ such that $M[B]^* \vDash \neg\phi(\bar{a})$, then there is a homomorphism $f: M[B]^* \to N$ such that N is finite and $N \vDash \neg\phi(f\bar{a})$.)

4. Show that Theorem 9.7.7 remains true if B is a countable boolean algebra with at most finitely many atoms.

5. We say that a boolean algebra B is **complete** if every subset of B has a supremum in the lattice ordering of B. Let B be a boolean algebra of subsets of a set I, and suppose B is complete. (This doesn't imply that the supremum of every family of elements of B is its union.) A **partition** of B is a set π of disjoint elements of B such that $\bigcup \pi = I$. Let M be an L-structure. We define the **complete boolean power** $M[B]$ to be the substructure of M^I whose domain is the set X of all elements a such that $\{a^{-1}(m): m \in \text{dom}(M)\}$ is a partition of B. (a) Show that X is the domain of a substructure of M^I. (b) Show that $M[B]$ is a B-valued substructure of M^I which respects $\exists$ and admits gluing over B. (c) Show that $M[B]^* \preccurlyeq M[B]$.

6. Let M be an L-structure such that $\text{Th}(M)$ is decidable, and let $\mathbf{K}$ be a class of boolean algebras such that $\text{Th}(\mathbf{K})$ is decidable. Show that $\text{Th}\{M[B]^*: B \in \mathbf{K}\}$ is decidable. [Use Feferman–Vaught.]

7. Show that when M is finite, there is a first-order interpretation which, for any boolean algebra B, interprets $M[B]^*$ in B.

*8. Show that if F is a finite field then every F-algebra is (as a ring) of form $F[B]^*$ for some boolean algebra B.

*9. Show that if G is a countable characteristically simple non-abelian group which has a proper subgroup of finite index, then G is isomorphic to a boolean power of some finite simple non-abelian group.

*10. Show that if M is a module and B, C are any two boolean algebras of the same cardinality, then $M[B]^* \cong M[C]^*$.

*11. Show that if B and C are any two boolean algebras, then $B[C]^*$ is isomorphic to the coproduct $B*C$ in the variety of boolean algebras.

History and bibliography

Section 9.1.   Direct products over an arbitrary index set seem to have come from group theory, where they are also known as **Cartesian products** and as **complete direct sums**. McKinsey [1943] studied basic Horn formulas and universal first-order Horn formulas; to this paper we owe Theorem 9.1.5 for universal first-order Horn formulas and a simple version of Corollary 9.1.7. Horn [1951] gave Theorem 9.1.5 in full.

Exercise 6 is from group theory; see Corollary 15.72 in H. Neumann [1967]. Exercise 8: Henschen & Wos [1974]. Exercise 11: Appel [1959]. Other people have found other solutions, e.g. Marongiu & Tulipani [1986]. Exercise 13: Lyndon [1959c]. Exercise 14: Mal'tsev [1959], Vaught [1962]. Exercise 15: Baldwin & Lachlan [1973]; Baldwin & McKenzie [1982] gives more examples in terms of the universal-algebraic commutator calculus. Exercise 16: Baldwin & Lachlan [1973]. Exercise 20: (a) Hule [1978], (b) Baldwin [1973a].

Section 9.2. We are in the heartlands of universal algebra. Cohn [1981] Chapters 3, 4 and Henkin, Monk & Tarski [1971] contain surveys of this area; Taylor [1979] examines varieties in greater depth. The first attempt at a systematic account of definition by generators and relations was probably Grassmann [1844]; but at that date the distinction between structures and languages was too primitive to allow meaningful theorems. W. Dyck [1882] proved some highly meaningful theorems in the special case of groups, and the definition of presentations is essentially his. G. Birkhoff [1935b] introduced the notion of varieties and proved Corollary 9.2.8 (for structures in algebraic signatures). Theorem 9.2.2 and the notion of a quasivariety are from A. I. Mal'tsev [1956a]; Theorem 9.2.2(c) was made explicit by Chudnovskiĭ [1968]. See Mahr & Makowsky [1983] for a discussion of Theorems 9.2.2 and 9.2.5 in connection with specification languages.

A fascinating question not touched on here is to determine what language is needed to axiomatise the class of free structures in a given quasivariety. See Eklof & Mekler [1990] for more information on this.

Exercise 1: Vaught [1962]. Exercise 8 is proved (but not stated) by Vaught in Tarski & Vaught [1957]. Exercise 11: for a classification of $|L|^+$-categorical quasivarieties, see Givant [1979], Palyutin [1975]. Exercise 12: Mal'tsev [1970] p. 295. Exercise 15: McKinsey [1943]. Exercise 16: Goetz & Ryll-Nardzewski [1960]. Exercise 17: (a) is from Jónsson & Tarski [1961] and (b) from Baldwin & MacKenzie [1982]. Exercise 19: Fisher [1977b]; cf. Makowsky [1985a]. Exercise 21(b): Sikorski [1950]. Exercise 22: these are a small part of Shelah's two proofs (Shelah [1975a], Hodges [1981a]) that if λ is a singular cardinal and A is an abelian group of cardinality λ in which every subgroup of smaller cardinality is free, then A is free. Exercise 23: See MacHenry [1960].

Section 9.3. A special but typical case of replica functors was described by Mal'tsev ([1958a], section 11.3 of [1970]). (I warmly thank Aleksandr Ryaskin, Kairtaĭ Shegirov, Sergeĭ Starchenko, Helene Kremer and Alëna Stepanova for the gift of a copy of Mal'tsev [1970].) Word-constructions are from Hodges [1976a], which gives Corollary 9.3.3, Theorem 9.3.4, Theorem 9.3.7 and Exercise 7. Theorem 9.3.2 generalises an argument of Chang [1968a] on the Ehrenfeucht–Mostowski construction, which is a word-construction; See Corollary 11.2.6 below. Feferman [1972] studies functors which preserve embeddings and directed colimits. Theorem 9.3.9(b) is from Eklof [1975]. For the categorical background to this section, a good reference is Mac Lane [1971] Chapters IV (for left adjoints of forgetful functors), IX (for filtered colimits), X (for the proof of Theorem 9.3.7).

Exercise 1: Essentially Theorem 3 on p. 291 of Mal'tsev [1970]. Exercise 3: Scott [1958]. Exercise 9: Feferman [1972]. Exercise 10: Eklof [1973]; cf. the application in Murthy & Swan [1976].

Section 9.4. Reduced products are a natural generalisation of ultraproducts, which came earlier – see the references on section 9.5. Tarski's group at Berkeley discussed reduced products in the late 1950s, and the main outcome was Frayne, Morel & Scott [1962]. This paper includes Theorem 9.4.3, which is due to Chang. Theorem 9.4.4 was proved by Keisler [1965c] assuming the continuum hypothesis, and Galvin [1970] found an indirect way of removing this assumption. Shelah [1971a] indicated a direct proof not assuming the continuum hypothesis. Theorem 9.4.7 and Exercise 2 are from Mal'tsev [1958b], [1966].

There are properties of varieties which don't lift to infinitary varieties. Nelson [1974] finds an infinitary variety which doesn't contain a simple algebra; for (finitary) varieties this is impossible by Magari [1969].

Exercise 3: Palyutin [1980]. Exercise 4: Chang & Morel [1958]. Exercise 5: Dixon [1977]. Exercise 7: Burris & Werner [1979]. Exercise 8: Frayne, Morel & Scott [1962]. Exercise 10: Banaschewski & Herrlich [1976]. Exercise 11: Hodges & Shelah [1981] (which also contains infinitary Horn analogues of Lyndon's theorem on surjective homomorphisms and the Beth definability theorem). Exercise 12: Isbell [1973] and Hodges [1980b].

Section 9.5. Special cases or slight variants of ultraproducts appear in Skolem [1934], Hewitt [1948] and Arrow [1950]. Łoś [1955b] described the general construction and proved Theorem 9.5.1. Kochen [1961] and Frayne, Morel & Scott [1962] proved a number of basic results on ultraproducts, including Corollary 9.5.2 and Corollary 9.5.10; the latter paper gives a version of Theorem 9.5.4 too. Corollary 9.5.5 is from Chang & Keisler [1962], Corollary 9.5.6 from Keisler [1967] and Theorem 9.5.9 from Chang & Keisler [1966]. Keisler [1964c] studied regular filters; his paper [1964b] discusses good ultrafilters, which are an important strengthening of the notion of regular filters and have close links with saturation (see Chapter 10 below). The Keisler–Shelah theorem (Theorem 9.5.7) was proved by Keisler [1964b] using good ultrafilters and the GCH; Shelah [1971a] gave a different proof not using the GCH. Corollary 9.5.8 is from Abraham Robinson [1956a]. Theorem 9.5.11 is from Eklof & Sabbagh [1971].

Chang & Keisler [1973] contains a definitive account of ultraproducts; it would surely have been the last word on the subject if Shelah ([1978a] Chapter 6) hadn't become interested in saturation of ultrapowers. Eklof [1977] is a highly readable survey on applications of ultraproducts, though most of his examples don't really need ultraproducts.

Exercise 4: Frayne, Morel & Scott [1962]. Exercise 6: Keisler [1962]; cf. Scott [1961]. Exercise 8: Keisler [1964b], [1964c]. Exercises 9, 10: Frayne, Morel & Scott [1962] and Kochen [1961]. Exercise 14: Keisler [1967]. Exercise 15: Kegel & Wehrfritz [1973] Theorem 1.L.8. Exercise 16: Eklof & Sabbagh [1971] proof of Lemma 3.11. Exercise 17: Jensen & Vámos [1979] Theorem 2.8. Exercise 19: Kueker [1973].

Section 9.6. Mostowski [1952a] proved Corollaries 9.6.5 and 9.6.7 for direct powers and remarked that there was an extension of these results to direct products; Vaught carried out the extension in [1954b]. Feferman proved a version of Theorem 9.6.2 in 1953, for the case where all the structures A_i are identical; Vaught removed this restriction in 1956. Feferman & Vaught [1959] published the results jointly, using a notion of 'generalised product'. The version in the text above, which uses the sheaf-theoretic notions of respecting ∃ and gluing in place of Feferman and Vaught's explicit description of the products, follows Comer [1974]. For other variants of a sheaf-theoretic kind, see Weispfenning [1975] and Volger [1976].

Exercise 1: Skolem [1919]. Exercise 2: Vaught [1954b]. Exercise 3: Galvin [1970]. Exercises 4, 5: Galvin [1970] proved the stronger statement that every first-order sentence is logically equivalent to a boolean combination of strict Horn sentences. Exercise 6: Garavaglia [1979]. Exercise 7: Oberschelp [1958], Almagambetov [1965]. Exercise 10: Weinstein [1965]. Exercise 11: Weinstein in Galvin [1970]. Exercise 12: Waszkiewicz & Węglorz [1969]. Exercise 13: Makkai [1965]. Exercise 14: Skolem [1930], but this proof was suggested by Mostowski [1952a]. Exercise 15(b): Malitz [1971].

Section 9.7. Burris [1983] is a history of boolean constructions, with full references. Boolean powers were first defined by Arens & Kaplansky [1948] when M is a ring, and Philip Hall (unpublished but by 1959; see Apps [1982]) when M is a group. Foster [1961] gave the general definition. Theorem 9.7.6 is from Wojciechowska [1969]. Theorem 9.7.7 and its corollary appeared in

Baldwin & Lachlan [1973]. Theorems on non-isomorphism of boolean powers go back to Tarski [1957] and Jónsson [1957]; see for example Burris [1975] and Lawrence [1981a]. Theorem 9.7.9 is tidied up from Hodges [1984b]; Grossberg & Shelah [199?] announce the existence of many pairwise non-embeddable boolean powers in singular cardinalities too.

Much of this section can be generalised to boolean products where the factors may be distinct. The generalisation is worthwhile because there are interesting classes where every structure can be represented as a boolean product of structures from some fixed set. See for example Burris & Werner [1979].

Exercises 1, 2, 6: Burris [1975]. Exercise 3: Apps [1983b] in a comparison of different ways of constructing ω-categorical groups. Exercise 5: (a) Foster [1953], (c) Banaschewski & Nelson [1980]. Exercise 7: Ershov [1967b]. Exercise 8: Arens & Kaplansky [1948]. Exercise 9: Philip Hall; see Apps [1983c]. Exercise 10: Bergman [1972]. Exercise 11: Quackenbush [1972].

10

Saturation

> I have made numerous composites of various groups of convicts ... The first set of portraits are those of criminals convicted of murder, manslaughter, or robbery accompanied with violence. It will be observed that the features of the composites are much better looking than those of the components. The special villainous irregularities have disappeared, and the common humanity that underlies them has prevailed.
>
> *Francis Galton, Inquiries into human faculty and its development (1907)*

A saturated model of a complete theory T is a 'most typical' model of T. It has no avoidable asymmetries; unlike the man of Devizes, it's not short at one side and long at the other. (Or like the man of the Nore, it's the same shape behind as before.) We create saturated models by amalgamating together all possible models, rather in the spirit of Fraïssé's construction from section 7.1.

Although in a sense these are typical models, in another sense every saturated model A has some quite remarkable properties. Every small enough model of T is elementarily embeddable in A – this is called *universality*. Every type-preserving map between small subsets of A extends to an automorphism of A – this is *strong homogeneity*. We can expand A to a model of any theory consistent with $\mathrm{Th}(A)$ – this is *resplendence*.

These properties, particularly the resplendence and the strong homogeneity, make saturated models a valuable tool for proving facts about the theory T. For example saturated models can be used for proving preservation and interpolation theorems. The universality makes saturated models suitable as work-spaces: we choose a large saturated model M, and all the other models that we consider are regarded as elementary substructures of M. A model M used as a work-space is said to be a *big* or *monster* model of T.

Sadly there is no guarantee that T has a saturated model, unless T happens to be stable. One response to this is to declare yourself a formalist and announce that infinitary mathematics is meaningless anyway. For those of us unwilling to take this easy escape route, it's more sensible to weaken the notion of saturation, and the first natural way to do this is to think of λ-*saturated* models. As section 10.1 explains, these models are saturated with respect to any set of fewer than λ parameters. The good news is that every structure has λ-saturated elementary extensions, for any cardinal λ. The bad news is that λ-saturation is too weak for many applications; it doesn't guarantee resplendence or any degree of strong homogeneity.

So we have to try again, aiming somewhere between saturation and λ-saturation. In this chapter I describe two notions that between them give us everything we need. The first is λ-*bigness*, which is resplendence after adding fewer than λ parameters. The second is *special models*, which are limits of chains of highly saturated models.

The last two sections of the chapter describe two other weakenings of the notion of saturation, got by limiting the classes of formulas that we look at. Recursively saturated models turn up in the study of models of arithmetic. Atomic compact models were invented by the algebraists and are important in the structure theory of modules, where they are better known as pure-injective modules.

10.1 The great and the good

In arguments which involve several structures and maps between them, things usually go smoother when the maps are inclusions. There are at least two good mathematical reasons for this. First, *if the maps are inclusions, then diagrams automatically commute*. And second, *if A is a substructure of B, then we can specify A by giving B and* dom(A); *there is no need to describe the relations of A as well as those of B.*

Thoughts of this kind have led to the use of *big models*, sometimes known as *monster models*. Informally, a big model is a structure M such that every commutative diagram of structures and maps that we want to consider is isomorphic to a diagram of inclusions between substructures of M. Of course a structure M with this property can't exist. It would have to contain isomorphic copies of all structures, and so its domain would be a proper class and not a set.

So we draw in our horns and demand something less. For the moment let us say that M is **splendid** if the following holds.

(1.1) Suppose L^+ is a first-order language got by adding a new
        relation symbol R to L. If N is an L^+-structure such that
        $M \equiv N|L$, then we can interpret R by a relation S on the
        domain of M so that $(M, S) \equiv N$.

Informally this says that M is compatible with any extra structural features which are consistent with Th(M).

Example 1: *Equivalence relations.* Let M be a structure consisting of an equivalence relation with two equivalence classes, whose cardinalities are ω

and ω_1. Then M is not splendid, because we can take an elementary extension N where the two equivalence classes have the same size, and add a bijection between these classes.

For any cardinal λ, we shall say that M is λ-**big** if $(M, \bar{a})$ is splendid whenever $\bar{a}$ is a sequence of fewer than λ elements of M. Thus splendid $= 0$-big.

One can define a **big model** (or **monster model**) to be a model which is λ-big for some cardinal λ which is 'large enough to cover everything interesting'. This is vague, but in practice there is no need to make it more precise. In stability theory one is interested in the models of some complete first-order theory T; the usual habit is to choose a big model of T without specifying how large λ is.

It will emerge that every structure has λ-big elementary extensions for any λ. The proof is technical and doesn't illuminate the use of these structures. So I postpone it to the next section, and turn to some important properties which are closely related to bigness.

λ-saturation and λ-homogeneity

I recall some definitions from section 6.3. Let A be an L-structure and X a set of elements of A; write $L(X)$ for the first-order language formed from L by adding constants for the elements of X. If $n < \omega$, then a **complete n-type** over X with respect to A is a set of the form $\{\varphi(x_0, \ldots, x_{n-1}): \varphi$ is in $L(X)$ and $B \vDash \varphi(\bar{b})\}$ where B is an elementary extension of A and $\bar{b}$ is an n-tuple of elements of B. We write this n-type as $\mathrm{tp}_B(\bar{b}/X)$, and we say that $\bar{b}$ **realises** this n-type **in** B. We write $S_n(X; A)$ for the set of all complete n-types over X with respect to A. A **type** is an n-type for some $n < \omega$.

Now let λ be a cardinal. We say that A is λ-**saturated** if

(1.2)   for every set X of elements of A, if $|X| < \lambda$ then all complete 1-types over X with respect to A are realised by elements in A.

We say that A is **saturated** if A is $|A|$-saturated. We say that A is λ-**homogeneous** if

(1.3)   for every pair of sequences $\bar{a}$, $\bar{b}$ of length less than λ, if $(A, \bar{a}) \equiv (A, \bar{b})$ and d is any element of A, then there is an element c such that $(A, \bar{a}, c) \equiv (A, \bar{b}, d)$.

We say that A is **homogeneous** if A is $|A|$-homogeneous.

The literature contains far too many different concepts called homogeneity, and muddles have occurred. One should probably refer to λ-homogeneous structures as **elementarily λ-homogeneous**. But this is a mouthful, and throughout this chapter there is no danger of confusion.

Finally we say that A is λ-**universal** if

(1.4)   if B is any L-structure of cardinality $< \lambda$ and $B \equiv A$, then B
         is elementarily embeddable in A.

Immediately from the definitions, we have the following.

Lemma 10.1.1. *Suppose A is λ-big and $\kappa < \lambda$. Then A is κ-big. Similarly with '-saturated', '-homogeneous' or '-universal' in place of '-big'.*   ☐

The simplest links between these concepts run as follows:

$$\lambda\text{-homogeneous}$$
$$\nearrow$$
(1.5) $\lambda\text{-big} \Rightarrow \lambda\text{-saturated}$
$$\searrow$$
$$\lambda\text{-universal}$$

Let us prove these implications. (One can say more. For example in Corollary 10.4.12 it will emerge that if A is saturated then A is $|A|$-big.)

Theorem 10.1.2. *Suppose A is λ-big. Then A is λ-saturated.*

Proof. Suppose A is a λ-big L-structure. Let $\bar{a}$ be a sequence of fewer than λ elements of A; let B be an elementary extension of A and b an element of A. We must show that $\mathrm{tp}_B(b/\bar{a})$ is realised in A. Let L^+ be the first-order language formed by adding a 1-ary relation symbol R to L, and make B into an L^+-structure B^+ by interpreting R as the singleton $\{b\}$. By λ-bigness there is a relation S on the domain of A, such that $(A, S, \bar{a}) \equiv (B^+, \bar{a})$. Now $B^+ \vDash$ 'Exactly one element satisfies Rx', and so S is a singleton $\{c\}$. Clearly c realises $\mathrm{tp}_B(b/\bar{a})$.   ☐

Our next lemma shows that λ-saturation is a strengthening of λ-homogeneity.

Lemma 10.1.3. *Let A be an L-structure and λ a cardinal. The following are equivalent.*
(a) *A is λ-saturated.*
(b) *For every L-structure B and every pair of sequences $\bar{a}$, $\bar{b}$ of elements of A, B respectively, if $\bar{a}$ and $\bar{b}$ have the same length $< \lambda$ and $(A, \bar{a}) \equiv (B, \bar{b})$, and d is any element of B, then there is an element c of A such that $(A, \bar{a}, c) \equiv (B, \bar{b}, d)$.*

Proof. (a) $\Rightarrow$ (b). Assume (a), and suppose $\bar{a}$, $\bar{b}$ are as in the hypothesis of (b). By elementary amalgamation (Theorem 6.4.1) there are an elementary extension D of A and an elementary embedding $f: B \to D$ such that $f\bar{b} = \bar{a}$. Now if d is any element of B, then $(D, \bar{a}, fd) \equiv (B, \bar{b}, d)$ since f is elementary. But $\bar{a}$ contains fewer than λ elements and A is λ-saturated, so that A contains an element c such that $\mathrm{tp}_A(c/\bar{a}) = \mathrm{tp}_D(fd/\bar{a})$. Then $(A, \bar{a}, c) \equiv (D, \bar{a}, fd) \equiv (B, \bar{b}, d)$ as required.

The implication (b) $\Rightarrow$ (a) is immediate from the definitions.   □

Theorem 10.1.4. *If A is λ-saturated then A is λ-homogeneous.*

Proof. The definition of λ-homogeneity is the special case of Lemma 10.1.3(b) where $A = B$.   □

Lemma 10.1.3 can be applied over and over again, to build up maps between structures, as follows.

Lemma 10.1.5. *Let L be a first-order language and A an L-structure.*

(a) *Suppose A is λ-saturated, B is an L-structure and $\bar{a}$, $\bar{b}$ are sequences of elements of A, B respectively such that $(A, \bar{a}) \equiv (B, \bar{b})$. Suppose $\bar{a}$, $\bar{b}$ have length $< \lambda$, and let $\bar{d}$ be a sequence of elements of B, of length $\leqslant \lambda$. Then there is a sequence $\bar{c}$ of elements of A such that $(A, \bar{a}, \bar{c}) \equiv (B, \bar{b}, \bar{d})$.*

(b) *The same holds if we replace 'λ-saturated' by 'λ-homogeneous' and add the assumption that $A = B$.*

Proof. We prove (a); the proof of (b) is similar. By induction we shall define a sequence $\bar{c} = (c_i : i < \lambda)$ of elements of A so that

(1.6)   for each $i \leqslant \lambda$, $(A, \bar{a}, \bar{c}|i) \equiv (B, \bar{b}, \bar{d}|i)$.

For $i = 0$, (1.6) says that $(B, \bar{b}) \equiv (A, \bar{a})$, which is given by assumption. There is nothing to do at limit ordinals, since any formula of L has only finitely many free variables. Suppose then that $\bar{c}|i$ has just been chosen and $i < \lambda$. Since A is λ-saturated and $\bar{c}|i$ has length $< \lambda$, Lemma 10.1.3 gives us an element c_i of A such that $(A, \bar{a}, \bar{c}|i, c_i) \equiv (B, \bar{b}, \bar{d}|i, d_i)$.   □

Theorem 10.1.6. *Let L be a first-order language and A a λ-saturated L-structure. Then A is λ^+-universal.*

Proof. We have to show that if B is an L-structure of cardinality $\leqslant \lambda$ and $B \equiv A$, then there is an elementary embedding $e: B \to A$. List the elements of B as $\bar{d} = (d_i : i < \lambda)$; repetitions are allowed. By the lemma there is a sequence $\bar{c}$ of elements of A such that $(B, \bar{d}) \equiv (A, \bar{c})$. It follows by the elementary diagram lemma (Lemma 2.5.3) that there is an elementary embedding of B into A taking $\bar{d}$ to $\bar{c}$.   □

In fact λ-saturation is exactly λ-homogeneity plus λ-universality. I leave this as an exercise (Exercise 7 below).

Lemma 10.1.5 implies that when λ is infinite, the definition of λ-saturation need not be limited to types in one variable.

Theorem 10.1.7. *Let L be a first-order language, A an L-structure, λ an infinite cardinal and $\bar{y}$ any tuple of variables. Suppose A is λ-saturated. Let $\bar{a}$ be a sequence of fewer than λ elements of A, and $\Phi(\bar{x}, \bar{y})$ a set of formulas of L such that for each finite subset Ψ of Φ, $A \vDash \exists \bar{y} \bigwedge \Psi(\bar{a}, \bar{y})$. Then there is a tuple $\bar{b}$ of elements of A such that $A \vDash \bigwedge \Phi(\bar{a}, \bar{b})$.*

Proof. By the compactness theorem (see Theorem 6.3.1(a)) there is certainly an elementary extension B of A containing a tuple $\bar{d} = (d_0, \ldots, d_{m-1})$ such that $B \vDash \bigwedge \Phi(\bar{a}, \bar{d})$. Now $(A, \bar{a}) \equiv (B, \bar{a})$, and since λ is infinite, the sequence $\bar{d}$ has fewer than λ elements. So by Lemma 10.1.5 there is a tuple $\bar{c}$ of elements of A such that $(A, \bar{a}, \bar{c}) \equiv (B, \bar{a}, \bar{d})$. Hence $A \vDash \bigwedge \Phi(\bar{a}, \bar{c})$ as required. $\qquad\square$

The idea of Lemma 10.1.5 can be run back and forth between two structures.

Theorem 10.1.8. *Let A and B be elementarily equivalent L-structures of the same cardinality λ.*

  (a) *If A and B are both saturated then $A \cong B$.*

  (b) *If A and B are both homogeneous and realise the same n-types over $\varnothing$ for each $n < \omega$, then $A \cong B$.*

Proof. (a) First assume that λ is infinite. List the elements of A as $(a_i: i < \lambda)$ and those of B as $(b_i: i < \lambda)$. We claim that there are sequences $\bar{c}$, $\bar{d}$ of elements of A and B respectively, both of length λ, such that for each $i \leqslant \lambda$,

$$(1.7) \qquad\qquad (A, \bar{a}|i, \bar{c}|i) \equiv (B, \bar{d}|i, \bar{b}|i).$$

The proof is by induction on i. Again the case $i = 0$ is given in the theorem hypothesis, and there is nothing to do at limit ordinals. When (1.7) has been established for some $i < \lambda$, fewer than λ parameters have been chosen (since λ is infinite). We use the saturation of B to find d_i such that $(A, \bar{a}|i, a_i, \bar{c}|i) \equiv (B, \bar{d}|i, d_i, \bar{b}|i)$; then we use the saturation of A to find c_i such that $(A, \bar{a}|i, a_i, \bar{c}|i, c_i) \equiv (B, \bar{d}|i, d_i, \bar{b}|i, b_i)$. At the end of the construction, the diagram lemma gives us an embedding $f: A \to B$ such that $f\bar{a} = \bar{d}$ and $f\bar{c} = \bar{b}$. The embedding is onto B since $\bar{b}$ includes all the elements of B.

When λ is finite, Theorem 10.1.6 tells us that there is an elementary embedding $f: A \to B$. Then f must be an isomorphism since A and B both have cardinality λ.

(b) The argument here is similar but with a couple of extra twists. We can assume that λ is infinite. We begin by establishing

(1.8)   if $i < \lambda$ and $\bar{b}$ is a sequence in B of length i, then there is a
         sequence $\bar{a}$ of elements of A such that $(A, \bar{a}) \equiv (B, \bar{b})$. (And
         the same with A and B transposed.)

The proof is by induction on i. Since the hypotheses are symmetrical in A and B, we only need prove (1.8) one way round.

When i is finite, (1.8) is given by the theorem hypothesis. When i is infinite we distinguish two cases. First suppose that i is a cardinal. Then we build up $\bar{a}$ so that for each $j < i$,

(1.9) $(A, \bar{a}|j) \equiv (B, \bar{b}|j)$.

The theorem hypothesis gives the case $j = 0$. When j is a limit ordinal there is nothing to do. Given (1.9) for j, we find a_j as follows. By induction hypothesis, since $|j + 1| < i$, there is a sequence $\bar{c} = (c_k : k \leqslant j)$ in A such that $(A, \bar{c}) \equiv (B, \bar{b}|(j + 1))$. Then $(A, \bar{a}|j) \equiv (A, \bar{c}|j)$, and so by the homogeneity of A there is a_j such that $(A, \bar{a}|j, a_j) \equiv (A, \bar{c}) \equiv (B, \bar{b}|(j + 1))$, giving (1.9) for $j + 1$. This establishes (1.8) when i is a cardinal. Finally when i is not a cardinal, we reduce to the case where it is a cardinal by rearranging the elements of $\bar{b}$ into a sequence of order-type $|i|$.

To prove the theorem, we go back and forth as in (a). For example, to find d_i we first use (1.8) to choose a sequence $\bar{e}$ in D such that $(A, \bar{a}|i, a_i, \bar{c}|i) \equiv (B, \bar{e})$, and then we use the homogeneity of B to find d_i so that $(B, \bar{e}) \equiv (B, \bar{d}|i, d_i, \bar{b}|i)$. $\square$

Examples

Example 2: *Finite structures.* If the structure A is finite, then any structure elementarily equivalent to A is isomorphic to A. It follows that A is λ-big for all cardinals λ. In particular A is saturated and homogeneous. This is a freak case, but it explains why the word 'infinite' will keep appearing in this chapter.

Example 3: *Vector spaces.* Let K be a field, λ an infinite cardinal $\geqslant |K|$ and A a vector space of dimension λ over K. Then A is λ-big. For this, let L be the language of A, and L^+ a language got by adding one relation symbol R. Let $\bar{a}$ be a sequence of fewer than λ elements of A, and B an L^+-structure with elements $\bar{b}$ such that $(B|L, \bar{b}) \equiv (A, \bar{a})$. The language of $(B, \bar{b})$ has cardinality at most λ, and so by using compactness and the downward Löwenheim–Skolem theorem we can suppose without loss that $B|L$ is also a vector space of cardinality $|A|$ and dimension λ over K. (If $|A| > |K|$ it suffices to get $|B| = |A|$, since then B must also have dimension λ over K. If $|A| = |K|$, we need to realise λ types of linearly independent elements in B, which we can do by compactness.) Then linear algebra shows that $(B|L, \bar{b}) \cong (A, \bar{a})$, and the isomorphism carries R^B across to A as required.

If A is infinite but has dimension less than λ, then A is no longer λ-saturated: consider a basis $\bar{a}$ of A and a set of formulas which express 'x is not linearly dependent on $\bar{a}$'. However, every vector space is ω-homogeneous, since an isomorphism between finite-dimensional subspaces always extends to an automorphism of the whole space.

Example 4: *Algebraically closed fields*. An argument like that of Example 3 shows that every algebraically closed field A of infinite transcendence degree over the prime field is $|A|$-big and hence saturated. (When the transcendence degree is finite, the field is homogeneous but not saturated.) André Weil [1946] proposed using algebraically closed fields of infinite transcendence degree as big models for field theory; he called them **universal domains**. In fact his proposal only codified what was already common practice – when we add roots of a polynomial, where are they supposed to come from?

Eample 5: *Countable ω-categorical structures*. Exercise 7.2.11 showed that every countable ω-categorical structure is saturated.

Example 6: *Dense linear orderings without endpoints*. The previous example shows that every countable dense linear ordering without endpoints is ω-saturated. Suppose A is an uncountable dense linear ordering without endpoints. Then A is certainly ω-saturated (see Exercise 8 below). Suppose that in fact A is ω_1-saturated. Then A has an element x sitting in each gap of the form

$$(1.10) \qquad a_0 \leqslant a_1 \leqslant a_2 \leqslant \ldots < x < \ldots \leqslant b_2 \leqslant b_1 \leqslant b_0$$

where a_i and b_i ($i < \omega$) are elements of A. Now it follows by Cantor's famous argument for the reals that A must have at least 2^ω elements. Closer inspection shows that A is not much like the ordering of the reals. No strictly increasing sequence (a_i: $i < \omega$) of elements of A has a supremum in A. For if b was its supremum, we could get a contradiction by taking $b_i = b$ for each $i < \omega$ in (1.10). Saturated dense linear orderings without endpoints, of cardinality ω_α, are known as η_α-**sets**; they were investigated by Hausdorff [1908].

Preservation of bigness

When one takes a 'large' structure and builds another structure out of it, one often wants to know whether the new structure is still 'large'. Table 10.1 shows which of the main types of construction preserve the largeness

Table 10.1

	λ-big	λ-saturated	λ-homo-geneous
1. Union of elementary chains of length κ, cf$(\kappa) \geqslant \lambda$	No	Yes	Yes
2. Adding $< \lambda$ parameters	Yes	Yes	Yes
3. Definitional expansion	Yes	Yes	Yes
4. Substructures with $\varnothing$-definable domain	Yes	Yes	Yes
5. Interpretations (λ large) and reducts	Yes	Yes	No
6. Direct products $A \times B$	No	Yes	No
7. Direct products with any number of factors	No	No	No

properties that were defined in this section. Except for Theorem 10.1.9 below, I leave the proofs to the exercises. It makes sense to ask whether other more specialised constructions (such as tensor products) preserve saturation. These questions can involve quite delicate algebra.

Theorem 10.1.9. *Let K and L be first-order languages, Γ an interpretation of L in K and A a K-structure such that ΓA is defined. For every infinite cardinal λ, if A is λ-saturated then so is ΓA. In particular every reduct or relativised reduct of A is λ-saturated.*

Proof. Suppose Γ is n-dimensional and let $\sim : \partial_\Gamma(A^n) \to \text{dom}(\Gamma A)$ be the coordinate map. Let X be a set of fewer than λ elements of ΓA and $\Phi(\bar{y})$ a type over X with respect to ΓA. For each element b of X, choose a tuple $\bar{a}_b$ in A such that $\bar{a}_b^{\sim} = b$. Let Y be a set of fewer than λ elements of A such that all the tuples $\bar{a}_b$ are tuples of elements of Y; this is possible since λ is infinite. Each formula in $\Phi(\bar{y})$ is of the form $\phi(\bar{y}, \bar{b})$ for some formula $\phi(\bar{y}, \bar{z})$ of L and some tuple $\bar{b}$ in X. Write $\Phi_\Gamma(\bar{x})$ for the set of formulas $\phi_\Gamma(\bar{x}, \bar{a}_{\bar{b}})$ with $\phi(\bar{y}, \bar{b})$ in $\Phi(\bar{y})$.

We claim that Φ_Γ is a type over Y in A. By the reduction theorem (Theorem 5.3.2), if $\phi(\bar{y}, \bar{b})$ is in Φ then

(1.11)   for all $\bar{c}$ from $\partial_\Gamma(A^n)$, $\Gamma A \vDash \phi(\bar{c}^{\sim}, \bar{b}) \Leftrightarrow A \vDash \phi_\Gamma(\bar{c}, \bar{a}_{\bar{b}})$.

In particular if $\Phi_0 = \Phi_0(\bar{y}, \bar{b})$ is a finite subset of Φ, then by assumption some tuple $\bar{c}^{\sim}$ satisfies $\bigwedge \Phi_0$ in ΓA, so $\bar{c}$ satisfies $\bigwedge \{\phi_\Gamma(\bar{x}, \bar{a}_{\bar{b}}) : \phi(\bar{y}, \bar{b}) \in \Phi_0\}$ in A. This proves the claim.

Now since A is λ-saturated, some tuple $\bar{c}$ satisfies Φ_Γ, and the reduction theorem shows at once that $\bar{c}^{\sim}$ satisfies Φ. $\square$

Corollary 10.1.10. *Suppose B is interpretable in A, and B is infinite. Suppose also that A is saturated (for example A is $|A|$-categorical; see Corollary 12.2.13.) Then $|A| = |B|$.*

Proof. By Exercise 5.3.1, $|B| \leqslant |A|$. By the theorem, B is $|A|$-saturated and hence (since B is infinite) $|B| \geqslant |A|$. $\qquad\square$

When $\lambda > |K| + |L|$, the result of Theorem 10.1.9 holds also for λ-bigness in place of λ-saturation, but this will be easier to show when we have special structures (Exercise 10.4.11).

Exercises for section 10.1

1. Suppose λ is an infinite cardinal and A is a λ-saturated L-structure. Show that if E is an equivalence relation on n-tuples of elements of A which is first-order definable with parameters, then the number of equivalence classes of E is either finite or $\geqslant \lambda$. (In particular if X is a subset of $\mathrm{dom}(A)$ which is first-order definable with parameters, then $|X|$ is either finite or $\geqslant \lambda$.)

2. Let A be an L-structure and λ a cardinal $> |A|$. Show that the following are equivalent. (a) A is λ-big. (b) A is λ-saturated. (c) A is finite.

3. We define λ^- to be μ if λ is a successor cardinal μ^+, and λ otherwise. Show that if an L-structure A is not λ-saturated, then for all κ with $\max(|L|, \lambda^-) \leqslant \kappa < |A|$ there are elementary substructures of A of cardinality κ which are not λ-saturated.

A structure A is said to be **strongly λ-homogeneous** *if for every pair of sequences $\bar{a}$, $\bar{b}$ of fewer than λ elements of A, if $(A, \bar{a}) \equiv (A, \bar{b})$ then there is an automorphism of A which takes $\bar{a}$ to $\bar{b}$.*
4. (a) Show that if A is λ-big then A is strongly λ-homogeneous. [It suffices to find an elementary extension A' of A with an automorphism taking $\bar{a}$ to $\bar{b}$. This can be done with repeated applications of elementary amalgamation (Theorem 6.4.1) back and forth. Alternatively use special models; see section 10.4 below.] (b) Show that if A is strongly λ-homogeneous then A is λ-homogeneous. (c) Show that if A is $|A|$-homogeneous then A is strongly $|A|$-homogeneous.

5. Let λ be an infinite cardinal, A a λ-saturated structure and X a set of fewer than λ elements of A. Show (a) if an element a of A is not algebraic over X, then infinitely many elements of A realise $\mathrm{tp}_A(a/X)$, (b) if an element a of A is not definable over X, then at least two elements of A realise $\mathrm{tp}_A(a/X)$, (c) $\mathrm{ACL}(X) = \mathrm{acl}(X)$ and $\mathrm{DCL}(X) = \mathrm{dcl}(X)$.

6. Show that if A is a λ-big L-structure and $\bar{a}$ is a sequence of fewer than λ elements of A, then $(A, \bar{a})$ is λ-big. Show that the same holds for λ-saturation, λ-homogeneity and strong λ-homogeneity.

7. Show that the following are equivalent, where λ, μ are any cardinals with $\min(\lambda, \omega) \leqslant \mu \leqslant \lambda^+$. (a) A is λ-saturated. (b) A is λ-homogeneous and μ-universal. [For (b) $\Rightarrow$ (a), adapt the proof of Theorem 10.1.8(b).]

8. Show that if A and B are elementarily equivalent structures and A is ω-saturated, then B is ω-saturated if and only if it is back-and-forth equivalent to A.

9. Show that if $(A_i: i < \kappa)$ is an elementary chain of λ-saturated structures and $\mathrm{cf}(\kappa) \geq \lambda$ then $\bigcup_{i<\kappa} A_i$ is λ-saturated. Show the same for λ-homogeneity.

10. Show that the result of Exercise 9 fails for λ-bigness and strong λ-homogeneity. [Take a countable structure consisting of an equivalence relation with just two equivalence classes, both infinite. Form a chain of extensions of length ω_1, by adding elements to just one of the equivalence classes.]

11. Show that if L and L^+ are languages with $L \subseteq L^+$, and A is a λ-big L^+-structure, then $A|L$ is a λ-big L-structure. Show the same for λ-saturation. On the other hand, show that if A is strongly λ-homogeneous, it need not follow that $A|L$ is λ-homogeneous. [For λ-big, use Theorem 6.6.1.]

12. Show that if L, L^+ are first-order languages with $L \subseteq L^+$, A is an L^+-structure, $A|L$ is λ-big and A is a definitional expansion of $A|L$ (see section 2.6), then A is λ-big. Show that the same holds for λ-saturation, λ-homogeneity and strong λ-homogeneity.

13. Let L be a first-order language. Show that if A is a λ-big L-structure and $\phi(x)$ is a formula of L such that $\phi(A)$ is the domain of a substructure B of A, then B is λ-big. Show that the same holds for λ-saturation, λ-homogeneity and strong λ-homogeneity.

14. Let L be a first-order language and A a non-empty L-structure. (a) Show that A^{eq} is never 0-saturated. (b) Show that if λ is infinite and A is λ-saturated, then every finite slice of A^{eq} is λ-saturated, and moreover A^{eq} can be made λ-saturated by adding λ elements which are not in any sort.

15. Show that if A and B are λ-big structures, then the disjoint sum of A and B (see Exercise 3.2.8) is λ-big. Show the same for λ-saturation, λ-homogeneity and strong λ-homogeneity.

*16. (a) Show that if A and B are λ-saturated L-structures, then $A \times B$ is a λ-saturated L-structure. [Use the previous exercise: $A \times B$ is interpretable in the disjoint sum of A and B.] (b) Let A be the field of complex numbers and B the field of algebraic numbers in characteristic 0. Show that for every cardinal $\lambda \leq \omega$, A and B are strongly λ-homogeneous but $A \times B$ is not 1-homogeneous. [Consider idempotent elements e and the type '$e \cdot x = x$ and x is not algebraic'.] (c) Let A and B be uncountable algebraically closed fields of the same characteristic but different cardinalities. Show that A and B are splendid (i.e. 0-big) but $A \times B$ is not. (d) Let A be the direct product of cyclic groups C_n $(1 \leq n < \omega)$ where C_n has order n. Show that A is not 0-saturated. [Only the identity is divisible by every prime.]

*A structure A is said to be λ-**compact** if every type (not necessarily complete) of cardinality < λ over a set of elements of A is realised in A.*

17. (a) Show that if L is a first-order language and $\lambda > |L|$, then an L-structure λ is λ-saturated if and only if it is λ-compact. (b) Show that every structure is ω-compact. (c) Show that for every infinite cardinal λ, a λ-saturated structure is λ-compact. (d) Give an example of an $|L|$-compact L-structure which is not $|L|$-saturated.

18. Suppose the L-structure A is λ-compact, $\bar{a}$ is a tuple from A, $\psi(\bar{x}, \bar{y})$ is a formula of L and $\Phi(\bar{x}, \bar{y})$ is a set of fewer than λ formulas of L. Show that if $A \vDash \forall \bar{x}(\bigwedge \Phi(\bar{x}, \bar{a}) \to \psi(\bar{x}, \bar{a}))$, then there is a finite subset Φ_0 of Φ such that $A \vDash \forall \bar{x}(\bigwedge \Phi_0(\bar{x}, \bar{a}) \to \psi(\bar{x}, \bar{a}))$.

19. Let L be a countable first-order language and A a countable atomic L-structure. Show that A is homogeneous.

20. Show that if A is ω-saturated and of cardinality λ, then A is definitionally equivalent to an L-structure for some first-order language L of cardinality $\leq \lambda$. [Use Exercise 6.2.9 to show that there are at most λ $\varnothing$-definable relations on A.]

*21. Let T be a complete theory in a countable first-order language. Show that up to isomorphism, the number of homogeneous models of T of cardinality ω is either 2^ω or $\leq \omega$.

The next exercise picks up the notion of Cantor–Bendixson rank from sections 3.4 and 5.6; it was used in Lemma 5.6.2(e).

22. Let L be a first-order language and A an ω-saturated L-structure. Show that if B is an elementary extension of A and $\phi(\bar{x})$ is a formula of L with parameters in A, then the Cantor–Bendixson rank of ϕ is the same whether we compute it in A or in B. [Use induction on the rank.]

The final exercise shows that a natural extension of the notion of λ-big – allowing us to expand to a model of a theory with many new symbols – is not really an extension at all.

23. Let L and L^+ be first-order languages with $L \subseteq L^+$. Let λ be an infinite cardinal such that the number of symbols in the signature of L^+ but not in L is less than λ. Show that if A is an L^+-structure and B is a λ-big L-structure such that $A|L \equiv B$, then B can be expanded to a structure $B' \equiv A$. [Adding a pairing function to L^+, we can suppose that all the other new symbols of L^+ are 1-ary relation symbols. They can then be coded up by a single 2-ary relation symbol and a set of fewer than λ constants.]

10.2 Big models exist

We shall show that for every structure A and every cardinal λ, there is an elementary extension of A which is λ-big. The proof will illustrate a common model-theoretic trick: we name the elements of a structure before we build the structure. This helps us to plan the construction in advance.

The cardinal $\mu^{<\lambda}$ is the sum of all cardinals μ^κ with $\kappa < \lambda$; so for example if $\lambda = \kappa^+$ then $\mu^{<\lambda}$ is just μ^κ. (Fact 10.4.1 below lists some facts of cardinal arithmetic that involve $\mu^{<\lambda}$.)

Existence of λ-big models

Theorem 10.2.1. *Let L be a first-order language, A an L-structure and λ a regular cardinal $> |L|$. Then A has a λ-big elementary extension B such that $|B| \leq |A|^{<\lambda}$.*

Proof. If A is finite then A is already λ-big for any cardinal λ. So we can assume henceforth that A is infinite.

Let C and D be structures. We shall say that D is an **expanded elementary extension** of C if D is an expansion of some elementary extension of C. An **expanded elementary chain** is a chain $(C_i : i < \kappa)$ of structures such that whenever $i < j < \kappa$, C_j is an expanded elementary extension of C_i. Using the Tarski–Vaught theorem on elementary chains (Theorem 2.5.2) it's not hard to see that each expanded elementary chain has a union D which is an expanded elementary extension of every structure in the chain. We write $\bigcup_{i<\kappa} C_i$ for the union of the expanded elementary chain $(C_i : i < \kappa)$.

Put $\mu = (|A| + |L|^+)^{<\lambda}$. Then $\mu = \mu^{<\lambda} \geq \lambda$. The ordinal $\mu^2 \cdot \lambda$ consists of $\mu \cdot \lambda$ copies of μ laid end to end. The object will be to construct B (or rather, an expansion of B) as the union of an expanded elementary chain $(A_i : 0 < i < \mu \cdot \lambda)$ where for each $i < \mu \cdot \lambda$ the domain of A_i is the ordinal $\mu \cdot i$. Then B will have cardinality $|\mu^2 \cdot \lambda| = \mu$ as required. The ordinals $< \mu^2 \cdot \lambda$ will be called **witnesses**. We can regard them either as elements of the structure to be built, or as new constants which will be used as names of themselves.

Since A is infinite, we can suppose without loss that A has cardinality μ (by the upward Löwenheim–Skolem theorem, Corollary 6.1.4). Then we can identify $\text{dom}(A)$ with the ordinal $\mu \cdot 1 = \mu$ and put $A_1 = A$. At limit ordinals $\delta < \mu$ we put $A_\delta = \bigcup_{0 < i < \delta} A_i$. It remains to define A_{i+1} when A_i has been defined; this is where the work is done.

Suppose L_0 is a first-order language and L', L'' are first-order languages got by adding new relation symbols R', R'' respectively to L_0. We say that theories T', T'' in L', L'' respectively are **conjugate** if T'' comes from T' by replacing R' by R'' throughout. We can list as $((X_i, T_i) : 0 < i < \mu \cdot \lambda)$ the set of 'all' pairs (X_i, T_i) where X_i is a set of fewer than λ witnesses, and T_i is a complete theory in the first-order language L_i formed by adding to L the witnesses in X_i and one new relation symbol R_i. Here 'all' means that for each such pair (X, T) there is a pair (X_i, T_i) with $X = X_i$ and T conjugate to

T_i. To check the arithmetic, note first that for each cardinal $v < \lambda$, the number of sets X consisting of v witnesses is $\mu^v = \mu$, and the number of complete theories T (up to conjugacy) in the language got by adding X and a relation symbol R to L is at most $2^{|L|+v} \leqslant \mu^{<\lambda} = \mu$. So the total number of pairs that we need is at most $\mu \cdot \lambda = \mu$. The listing can be done so that

(2.1)   the relation symbols R_i are all distinct, and

(2.2)   up to conjugacy, each possible pair (X, T) appears as (X_i, T_i) cofinally often in the listing.

In fact $\mu \cdot \lambda$ consists of λ blocks of length μ; we can make sure that each (X, T) appears at least once – up to conjugacy – in each of these blocks.

Now we define A_{i+1} inductively, assuming A_i has been defined with domain $\mu \cdot i$. Consider the pair (X_i, T_i). If some witness $\geqslant \mu \cdot i$ appears in X_i, then we take A_{i+1} to be an arbitrary elementary extension of A_i with domain $\mu \cdot (i + 1)$ (which is possible by Exercise 6.1.11).

Suppose then that every witness in X_i is an element of A_i. If the theory T_i is inconsistent with the elementary diagram of A_i, then again we take A_{i+1} to be an arbitrary elementary extension of A_i with domain $\mu \cdot (i + 1)$.

Finally suppose that every witness in X_i is an element of A_i, and T_i is consistent with the elementary diagram of A_i. Then some expanded elementary extension D of A_i is a model of T_i. By the downward Löwenheim–Skolem theorem we can suppose that D has cardinality μ, and so again (after adding at most μ elements if necessary) we can identify the elements of D with the ordinals $< \mu \cdot (i + 1)$. This done, we take A_{i+1} to be D.

Thus the chain $(A_i: 0 < i < \mu \cdot \lambda)$ is defined, and we put $B^+ = \bigcup_{0 < i < \mu \cdot \lambda} A_i$ and $B = B^+ | L$. The structure B^+ is an expanded elementary extension of A, so B is an elementary extension of A. Since B is the union of a chain of length μ in which every structure has cardinality μ, B has cardinality μ.

It remains to show that B is λ-big. Suppose $\bar{a}$ is a sequence of fewer than λ elements of B, and C is a structure with a new relation symbol R, such that $(C | L, \bar{c}) \equiv (B, \bar{a})$ for some sequence $\bar{c}$ in C. Adjusting C, we can suppose without loss that $\bar{c}$ is $\bar{a}$. Since $\mu \cdot \lambda$ is an ordinal of cofinality λ and λ is regular, there is some $j < \mu \cdot \lambda$ such that all the witnesses in $\bar{a}$ are less than j, and thus $(C | L, \bar{a}) \equiv (A_j, \bar{a})$. By (2.2) there is some $i \geqslant j$ such that T_i is conjugate to $\mathrm{Th}(C, \bar{a})$. Then $\mathrm{Th}(A_i | L, \bar{a}) \cup T_i$ is consistent, and so by (2.1) and Theorem 6.6.1, T_i is consistent with the elementary diagram of A_i. So by construction A_{i+1} is a model of T_i, and hence so is B^+. Thus B expands to a model of T_i as required. $\square$

We deduce the following.

Corollary 10.2.2. Let A be an L-structure and λ a cardinal $\geqslant |L|$. Then A has a λ^+-big (and hence λ^+-saturated) elementary extension of cardinality $\leqslant |A|^\lambda$.

Proof. Direct from the theorem.   □

In section 6.4 there was advance notice of a result saying that one can realise many complete types together. Corollary 10.2.2 establishes this. It's worth remarking that if we only want λ^+-saturation, not λ^+-bigness, then the elementary amalgamation theorem can be used in place of Theorem 6.6.1.

Corollary 10.2.3. *Let λ be any cardinal. Then every structure is elementarily equivalent to a λ-big structure.*

Proof. This follows from Corollary 10.2.2 and Lemma 10.1.1.   □

Corollary 10.2.3 is important. It tells us that if we want to classify the models of a first-order theory T up to elementary equivalence, it's enough to choose a cardinal λ and classify the λ-big models up to elementary equivalence. Since the λ-big models of T may be a much better behaved collection than the models of T in general, this is real progress. For example if R is an infinite ring, $|R|^+$-saturated left R-modules are algebraically compact (this is a special case of Theorem 10.7.3 below). A good structure theory is known for such modules.

If the generalised continuum hypothesis (GCH) holds, then $\lambda^{<\lambda} = \lambda$ for every regular cardinal λ. So by Theorem 10.2.1, if λ is a regular cardinal $> |L|$ and A is an L-structure of cardinality $\leq \lambda$, then A has a λ-big elementary extension B of cardinality at most λ. In particular, we have the following.

Corollary 10.2.4. *If the GCH holds, every structure has a saturated elementary extension.*   □

When the GCH fails, we may have to fall back on something less than saturation. See section 10.4 below.

Our proof of Theorem 10.2.1 was rather like the proof of the existence of existentially closed models in section 8.1: we kept adding new pieces until the process closed off. There is another argument which gives λ^+-saturated elementary extensions in one blow. It uses ultraproducts. See Chang & Keisler [1973] Theorem 6.1.8.

Existence of λ-homogeneous models

Since every λ-big structure is λ-homogeneous, Theorem 10.2.1 creates λ-homogeneous elementary extensions too. But if all we want is λ-homogeneity, we can generally get it with a smaller structure than Theorem 10.2.1 offers. This is useful.

Theorem 10.2.5. *Let L be a first-order language, A an L-structure and λ a regular cardinal. Then A has a λ-homogeneous elementary extension C such that $|C| \leqslant (|A| + |L|)^{<\lambda}$.*

Proof. By Theorem 10.2.1 we have a λ-big elementary extension B of A; never mind its cardinality. Write ν for $(|A| + |L|)^{<\lambda}$, noting that $\nu \geqslant \lambda$. (Otherwise $\nu = \nu^{<\lambda} = (\nu^{<\lambda})^{\nu} \geqslant 2^{\nu} > \nu$.) If D is any elementary substructure of B with cardinality at most ν, we can find a structure D^* with $D \preccurlyeq D^* \preccurlyeq B$ so that

(2.3)   if $\bar{a}$ and $\bar{b}$ are two sequences of elements of D, both of length $< \lambda$, and $(D, \bar{a}) \equiv (D, \bar{b})$, then for every element c of D there is an element d of D^* such that $(D, \bar{a}, c) \equiv (D^*, \bar{b}, d)$.

We can find D^* as the union of a chain of elementary substructures of B, taking one such substructure for each triple $(\bar{a}, \bar{b}, c)$ such that $(D, \bar{a}) \equiv (D, \bar{b})$ and c is in D. Such a chain is automatically elementary. As we move one step up the chain, we choose the next structure so that it contains some d with $(D, \bar{a}, c) \equiv (B, \bar{b}, d)$. This is possible since B is λ-homogeneous. The number of triples $(\bar{a}, \bar{b}, c)$ is at most $\nu^{<\lambda} = \nu$, and each structure in the chain can be chosen of cardinality at most ν; so the union D^* can be found with cardinality at most ν.

Now we build a chain $(A_i : i < \lambda)$ of elementary substructures of B, so that for each $i < \lambda$, A_{i+1} is A_i^*. At limit ordinals we take unions. Let C be $\bigcup_{i<\lambda} A_i$. Then C has cardinality at most $\nu \cdot \lambda = \nu$. If $(C, \bar{a}) \equiv (C, \bar{b})$ where $\bar{a}$ and $\bar{b}$ are sequences of length $< \lambda$ in C, and c is an element of C, then since λ is regular, all of $\bar{a}$, $\bar{b}$ and c must lie within some A_i, so that A_{i+1} contains d with $(C, \bar{a}, c) \equiv (A_{i+1}, \bar{a}, c) \equiv (A_{i+1}, \bar{b}, d) \equiv (C, \bar{b}, d)$. Thus C is λ-homogeneous as required. $\square$

This theorem most often appears as (a) of the corollary below. Strictly (b) is not a corollary but an analogue. We start with Exercise 10.1.4(a) and form a chain as in the proof of Theorem 10.2.5, adding enough elements to allow the required automorphisms. I leave the details as Exercise 5.

Corollary 10.2.6. *Let A be an infinite L-structure and μ a cardinal $\geqslant |A| + |L|$.*

(a) A has an ω-homogeneous elementary extension of cardinality μ. In particular every complete and countable first-order theory with infinite models has a countable homogeneous model.

(b) A has an elementary extension B of cardinality μ which is strongly ω-homogeneous (i.e. if $\bar{a}$, $\bar{b}$ are tuples in B such that $(B, \bar{a}) \equiv (B, \bar{b})$, then there is an automorphism of B taking $\bar{a}$ to $\bar{b}$). $\square$

Saturated models of stable theories

Stable theories always have saturated models in many cardinalities. To apply the next result, note that if λ and μ are any infinite cardinals, then $(\lambda^\mu)^+$ is a regular cardinal $\lambda' \geqslant \lambda$ such that $(\lambda')^\mu = \lambda'$ (see Exercise 8 below).

Theorem 10.2.7. *Let L be a first-order language and T a complete theory in L with infinite models. Let λ be a regular cardinal $\geqslant |L|$ such that T is λ-stable; then T has a saturated model of cardinality λ. In particular if T is stable then T has a saturated model of each regular cardinality λ with $\lambda^{|L|} = \lambda$, and if T is totally transcendental then T has a saturated model of each regular cardinality $\lambda \geqslant |L|$.*

Proof. We build an elementary chain $(A_i : i \leqslant \lambda)$ of models of T, all of cardinality λ. The model A_0 is arbitrary, and at limit ordinals we take unions. When A_i has been chosen, by λ-stability there are only λ complete types over $\mathrm{dom}(A_i)$ with respect to A_i; let A_{i+1} be an elementary extension of A_i of cardinality λ which realises all these types.

Put $B = A_\lambda$. We claim that B is saturated. For suppose X is a set of elements of B of cardinality $< \lambda$ and $p(x)$ is a complete type over X. Then since λ is regular, there is some A_i which contains all of X. We can extend p to a complete type $q(x)$ over $\mathrm{dom}(A_i)$, and by construction some element b in A_{i+1} realises q. So b realises p as required. $\square$

Stability theory has several improvements of Theorem 10.2.7, taking into account singular cardinals as well as regular. The one which follows is perhaps the simplest, but miles of sophistication separate it from the results above. I omit the proof, except to say that it involves making a large number of attempts at an element which will realise the right type, and using stability to show that they can't all be wrong.

Theorem 10.2.8. *Let L be a first-order language and T a complete totally transcendental theory in L. Then for every cardinal $\lambda > |L|$, T has a saturated model of cardinality λ.*

Proof. Harnik [1975]. $\square$

Exercises for section 10.2

The first three exercises give consequences of the fact that countable complete theories have countable homogeneous models.

1. (a) Let T be a countable first-order theory with infinite models. Show that T has a countable strongly ω-homogeneous model. (b) Show that if the continuum hypothesis

fails, then there is a countable first-order theory with infinite models but with no strongly ω_1-homogeneous model of cardinality ω_1. [Consider a very symmetrical tree with 2^ω branches.]

2. Let T be a complete theory in a countable first-order language. Suppose T has infinite models, and there is a finite set of types of T, such that all countable models of T realising these types are isomorphic. Show that T is ω-categorical. [There is a countable strongly ω-homogeneous model in which all these types are realised; so we have an ω-categorical theory got by adding finitely many parameters to T.]

3. Show that if T is a countable complete first-order theory, then the number of countable models of T, counted up to isomorphism, is not 2. [If it isn't 1, then by the theorem of Engeler, Ryll-Nardzewski and Svenonius (Theorem 7.3.1), T has a non-principal type $p(\bar{x})$ over the empty set. If it is 2, then apply the previous exercise.]

4. Show that if $2 < n < \omega$ then there is a countable complete first-order theory T such that up to isomorphism, T has exactly n countable models. [Start with the theory of a dense linear ordering without endpoints. To get three models, add constants for some sequence of elements of order-type ω. All known solutions to this exercise are variants of this. It is an open question whether a countable first-order theory with a finite number > 1 of countable models must have the strict order property.]

5. Show that Theorem 10.2.5 holds with 'strongly λ-homogeneous' in place of 'λ-homogeneous'. (This is perceptibly harder than Theorem 10.2.5, though the extra difficulty is mostly book-keeping.)

*6. In elementary analysis classes one proves that if an ordered field has the property (*) 'Every non-empty bounded set has a least upper bound' then the field is archimedean ordered (i.e. for every element a there is a natural number $> a$). Show that in this theorem the condition (*) can't be replaced by its consequence (**) 'For every countable family of closed intervals, if each finite subfamily has non-empty intersection then so does the whole family'.

*7. Show that if L is a countable signature, $(A_i: i < \omega)$ is a non-empty family of non-empty L-structures and $\mathscr{F}$ is the filter of cofinite subsets of ω, then the reduced product $\prod_{i<\omega} A_i / \mathscr{F}$ is ω_1-saturated. [Adding parameters, we can reduce it to proving 0-saturation, i.e. satisfying a type $\Phi(\bar{x})$ over the empty set. Consider Exercise 9.6.11; the relevant formulas on the reduced power set algebra can be assumed quantifier-free, hence disjunctions of conjunctions of literals. Use König's tree lemma to choose one disjunct from each of these formulas.]

8. Show that if λ and μ are infinite cardinals such that $\lambda^\mu = \lambda$, then $(\lambda^+)^\mu = \lambda^+$. [$\lambda^\mu = \lambda$ implies $\mu < \mathrm{cf}(\lambda)$ by König's theorem.]

9. Let L be a first-order language, T an $\forall_2$ theory in L, A an L-structure which is a model of T and λ a regular cardinal $> |L|$. Show that there is an e.c. model B of T such that $A \subseteq B$, $|B| \leqslant |A|^{<\lambda}$ and

(2.4)   for every sequence $\bar{b}$ of $< \lambda$ elements of B, every e.c. model C of T extending B, and every element c of C, there is an element d of B such that $(B, \bar{b}, d) \equiv_1 (C, \bar{b}, c)$.

[Mingle the proofs of Theorem 10.2.1 and Theorem 8.2.1.] *In Abraham Robinson's terminology, an e.c. model of T which satisfies (2.4) when $\lambda = \omega$ is said to be* **existentially universal**. *An* **infinite-generic** *model of T is a model which is an elementary substructure of an existentially universal model.*

10. Let T be an $\forall_2$ theory in a first-order language L. (a) Show that every model of T can be embedded in some infinite-generic model of T. (b) Show that if A and B are infinite-generic models of T with $A \subseteq B$ then $A \preccurlyeq B$. [In fact if $\bar{a}$ is any tuple of elements of A, then $(A, \bar{a})$ and $(B, \bar{a})$ are back-and-forth equivalent.] (c) Show that if T is companionable, then the infinite-generic models of T are exactly the e.c. models of T.

11. Let L be a first-order language and T an $\forall_2$ theory in L. Suppose A is a model of T and λ is an infinite cardinal. Show that there is an e.c. model B of T with $A \subseteq B$, such that (a) any two tuples in B which realise the same maximal $\exists$-type are in the same orbit of $\mathrm{Aut}(B)$, and (b) if $\bar{a}$, $\bar{b}$ are sequences of fewer than λ elements of B such that $(B, \bar{a}) \equiv_1 (B, \bar{b})$, and c is an element of some e.c. model C of T with $B \subseteq C$, then there is d in B such that $(B, \bar{a}, c) \equiv_1 (C, \bar{a}, d)$. [The proof of (a) is the analogue of Exercise 5, using Exercise 9 in place of Theorem 10.2.1. The result holds also if instead of tuples in (a) we have sequences of length $< \lambda$.]

10.3 Syntactic characterisations

Lemma 10.1.5 is handy for setting up maps between λ-saturated structures. Sometimes these maps give intuitive proofs of preservation theorems. Two examples will serve (with more in the Exercises). After these examples we shall extract some general principles for getting syntactic characterisations out of saturated models. The principles take the form of a game of infinite length.

Our first example shows how to embed a structure in a λ-saturated structure. Recall that if A and B are L-structures, then '$A \Rrightarrow_1 B$' means that for every $\exists_1$ first-order sentence ϕ of L, if $A \vDash \phi$ then $B \vDash \phi$.

Theorem 10.3.1. *Let L be a first-order language. Let A and B be L-structures, and suppose B is $|A|$-saturated and $A \Rrightarrow_1 B$. Then A is embeddable in B.*

Proof. List the elements of A as $\bar{a} = (a_i: i < \lambda)$ where $\lambda = |A|$. We claim that there is a sequence $\bar{b} = (b_i: i < \lambda)$ of elements of B such that

(3.1)   for each $i \leqslant \lambda$, $(A, \bar{a}|i) \Rrightarrow_1 (B, \bar{b}|i)$.

The proof is by induction on i. When $i = 0$, $A \Rrightarrow_1 B$ by assumption. When i is a limit ordinal, (3.1) holds at i provided it holds at all smaller ordinals.

This leaves the case where i is a successor ordinal $j + 1$. Let $\bar{x}$ be the sequence of variables $(x_\alpha : \alpha < i)$, and let $\Phi(\bar{x}, y)$ be the set of all $\exists_1$ formulas $\phi(\bar{x}, y)$ of L such that $A \vDash \phi(\bar{a}|i, a_i)$. For each finite set of formulas $\phi_0, \ldots,$ ϕ_{n-1} from Φ, we have $A \vDash \exists y \bigwedge_{k<n} \phi_k(\bar{a}|j, y)$. But $\exists y \bigwedge_{k<n} \phi_k$ is equivalent to an $\exists_1$ formula, and so $B \vDash \exists y \, \phi(\bar{b}|j, y)$ by inductive assumption. It follows by Theorem 6.3.1(a) that $\Phi(\bar{b}|j, y)$ is a type with respect to B. Since $j < \lambda$ and B is λ-saturated, this type is realised in B, say by an element b_j. Then $(A, \bar{a}|i) \Rrightarrow_1 (B, \bar{b}|i)$ as required. This proves the claim.

Hence $(A, \bar{a}) \Rrightarrow_1 (B, \bar{b})$. It follows by the diagram lemma that there is an embedding $f : A \to B$ such that $f\bar{a} = \bar{b}$. $\qquad \square$

This theorem leads at once to a new proof of the dual of the Łoś–Tarski theorem (cf. Exercise 6.5.1).

Corollary 10.3.2. *Let L be a first-order language, T a theory in L and $\Phi(\bar{x})$ a set of formulas of L (where the sequence $\bar{x}$ may be infinite). Suppose that whenever A and B are models of T with $A \subseteq B$, and $\bar{a}$ is a sequence of elements of A such that $A \vDash \bigwedge \Phi(\bar{a})$, we have $B \vDash \bigwedge \Phi(\bar{a})$. Then Φ is equivalent modulo T to a set $\Psi(\bar{x})$ of $\exists_1$ formulas of L.*

Proof. Putting new constants for the variables $\bar{x}$, we can suppose that the formulas in Φ are sentences. Let Ψ be the set of all $\exists_1$ sentences ψ of L such that $T \cup \Phi \vdash \psi$. It suffices to show that $T \cup \Psi \vdash \bigwedge \Phi$. If $T \cup \Psi$ has no models then this holds trivially. If $T \cup \Psi$ has models, let B' be one. By Corollary 10.2.2, B' is elementarily equivalent to a λ-saturated structure B where $\lambda \geqslant |L|$. Write U for the set of all $\forall_1$ sentences of L which are true in B. Then $T \cup \Phi \cup U$ has a model. (For otherwise by the compactness theorem there is a finite subset $\{\theta_0, \ldots, \theta_{m-1}\}$ of U such that $T \cup \Phi \vdash \neg\theta_0 \vee \ldots \vee \neg\theta_{m-1}$. Then $\neg\theta_0 \vee \ldots \vee \neg\theta_{m-1}$ is equivalent to a sentence in Ψ, and hence it is true in B' and B; contradiction.) Let A be a model of $T \cup \Phi \cup U$ of cardinality $\leqslant |L|$. By the choice of U, $A \Rrightarrow_1 B$. So A is embeddable in B by Theorem 10.3.1, and it follows that $B \vDash \bigwedge \Phi$ since Φ is a set of $\exists_1$ sentences. Thus $B' \vDash \bigwedge \Phi$ as required. $\qquad \square$

Lyndon's theorem

Our second application is a useful preservation result about positive occurrences of relation symbols in formulas.

Let $f : A \to B$ be a homomorphism of L-structures and R a relation symbol of L. We shall say that f **fixes** R if for every tuple $\bar{a}$ of elements of

A, $A \vDash R\bar{a}$ if and only if $B \vDash Rf\bar{a}$. In this definition we allow R to be the equality symbol $=$. Thus f fixes $=$ if and only if f is injective; f fixes all relation symbols if and only if f is an embedding. What formulas are preserved by a homomorphism which fixes certain relations but not others?

Let ϕ be a formula and R a relation symbol. We say that R is **positive in** ϕ if ϕ can be brought to negation normal form (see Exercise 2.6.2) in such a way that there are no subformulas of the form $\neg R\bar{t}$.

Note that up to logical equivalence, a formula ϕ is positive (in the sense we defined in section 2.4) if and only if every relation symbol, including $=$, is positive in ϕ. (See Exercises 2.4.15 and 2.6.3.)

We aim to prove the following.

Theorem 10.3.3. *Let L be a first-order language, Σ a set of relation symbols of L (possibly including $=$) and $\phi(\bar{x})$ a formula of L in which every relation symbol in Σ is positive.*

(a) *If $f: A \to B$ is a surjective homomorphism of L-structures, and f fixes all relation symbols (including possibly $=$) which are not in Σ, then f preserves ϕ.*

(b) *Suppose that every surjective homomorphism between models of T which fixes all relation symbols not in Σ preserves ϕ. Then ϕ is equivalent modulo T to a formula $\psi(\bar{x})$ of L in which every relation symbol in Σ is positive.*

Proof. (a) is a straightforward variant of Theorem 2.4.3. To prove (b), we start along the same track as the proof of Corollary 10.3.2. Replacing the variables $\bar{x}$ by distinct new constants, we can assume that ϕ is a sentence. Let Θ be the set of all formulas of L in which every relation symbol in Σ is positive.

We shall use Θ in the same way as we used $\exists_1$ in Theorem 10.3.1. Thus if C and D are any L-structures, we write $(C, \bar{c}) \Rrightarrow_\Theta (D, \bar{d})$ to mean that if $\theta(\bar{x})$ is any formula in Θ such that $C \vDash \theta(\bar{c})$, then $D \vDash \theta(\bar{d})$. In particular $C \Rrightarrow_\Theta D$ means that every sentence in Θ which is true in C is also true in D. In place of Theorem 10.3.1, we shall show the following.

Lemma 10.3.4. *Let L, Σ and Θ be as in the theorem. Let λ be a cardinal $\geq |L|$, and suppose A and B are λ-saturated structures such that $A \Rrightarrow_\Theta B$. Then there are elementary substructures A', B' of A, B respectively, and a surjective homomorphism $f: A' \to B'$ which fixes all relation symbols not in Σ.*

Proof of lemma. We shall build up sequences $\bar{a}$, $\bar{b}$ of elements of A, B respectively, both of length λ, in such a way that

(3.2)   for every $i \leq \lambda$, $(A, \bar{a}|i) \Rrightarrow_\Theta (B, \bar{b}|i)$, and

(3.3) $\bar{a}$ (resp. $\bar{b}$) is the domain of an elementary substructure of A
           (resp. B).

The construction will be by induction on i, as in the proof of Theorem 10.3.1.

There is one main difference from the proof of Theorem 10.3.1. In that proof, each a_j was given and we had to find an element b_j to match. Here we shall sometimes choose the b_j first and then look for an answering a_j. One can think of the process as a back-and-forth game of length λ between A and B: player $\forall$ chooses an element a_j (or b_j), and player $\exists$ has to find a corresponding element b_j (or a_j). Player $\exists$ wins iff (3.2) holds after λ steps.

We shall show that player $\exists$ can always win this game. At the beginning of the game, $A \Rightarrow_\Theta B$ by assumption. If i is a limit ordinal and $(A, \bar{a}|j) \Rightarrow_\Theta (B, \bar{b}|j)$ for all $j < i$, then $(A, \bar{a}|i) \Rightarrow_\Theta (B, \bar{b}|i)$ since all formulas are finite. So again we are led to the case where i is a successor ordinal $j + 1$. There are two situations, according as player $\forall$ chooses from A or from B.

Suppose first that player $\forall$ has just chosen a_j from A. Let $\Phi(\bar{x}, y)$ be the set of all formulas $\phi(\bar{x}, y)$ in Θ such that $A \models \phi(\bar{a}|j, a_j)$. Since Φ is closed under conjunctions and existential quantification, exactly the same argument as in the proof of Theorem 10.3.1 shows that $\Phi(\bar{b}|j, y)$ is a type over $\bar{b}|j$ with respect to B, and so there is an element b in B such that $(A, \bar{a}|j, a_j) \Rightarrow_\Theta (B, \bar{b}|i, b)$ as required. Let player $\exists$ choose b_j to be this element b.

Second, suppose player $\forall$ chose b_j from B, so that player $\exists$ must find a suitable a_j. The argument is just the same but from right to left, using the set $\{\neg\theta : \theta \in \Theta\}$ in place of Θ.

So player $\exists$ can be sure of winning. This takes care of (3.2). To make (3.3) true too, we issue some instructions to player $\forall$. As the play proceeds, he must keep a note of all the formulas of the form $\phi(\bar{a}|i, y)$, with ϕ in L, such that $A \models \exists y\, \phi(\bar{a}|i, y)$. For each such formula he must make sure that at some stage j later than i, he chooses a_j so that $A \models \phi(\bar{a}|i, a_j)$. He must do the same with B. At the end of the play, (3.3) will hold by the Tarski–Vaught criterion, Theorem 2.5.1.

Finally suppose the game is played, and sequences $\bar{a}$, $\bar{b}$ satisfying (3.2) and (3.3) have been found. Let A', B' be the substructures of A, B with domains listed by $\bar{a}$, $\bar{b}$ respectively. Since all atomic formulas of L are in Θ, the diagram lemma gives us a homomorphism $f: A' \to B'$ such that $f\bar{a} = \bar{b}$. Clearly f is surjective. If R is a relation symbol not in Σ, then the formula $\neg R\bar{z}$ is in Θ, and so (3.2) implies that f fixes R.   □ Lemma.

The rest of the argument is much as in the proof of Corollary 10.3.2, and I leave it to the reader.   □ Theorem.

Probably the best known corollary of Theorem 10.3.3 is the following converse of Theorem 2.4.3(b).

Corollary 10.3.5 *(Lyndon's preservation theorem). Let T be a theory in a first-order language L and $\phi(\bar{x})$ a formula of L which is preserved by all surjective homomorphism between models of T. Then ϕ is equivalent modulo T to a positive formula $\psi(\bar{x})$ of L.*

Proof. Let Σ in the theorem be the set of all relation symbols of L, including the symbol $=$. $\square$

Using the same argument, we can replace ϕ and ψ in this corollary by sets Φ, Ψ of formulas; see Exercise 1 below.

Keisler games

In several of the arguments of this section and section 10.1, we inductively constructed a sequence $\bar{a}$ of elements of a structure A, using some saturation property of A. The sequence $\bar{a}$ had to satisfy certain conditions. Games are a handy way of organising arguments of this type.

In a Keisler game on a structure A, the two players $\forall$ and $\exists$ take turns to pick elements of A. Player $\exists$'s aim is to define some new relations on A, and these new relations must satisfy some conditions which depend on the choices of player $\forall$.

Let L be a first-order language and λ an infinite cardinal. A **Keisler sentence** of length λ in L is an infinitary expression of the form

$$(3.4) \qquad Q_0 x_0 Q_1 x_1 \ldots Q_i x_i \, (i < \lambda) \ldots \bigwedge \Phi$$

where each Q_i is either $\forall$ or $\exists$, and Φ is a set of formulas $\phi(x_0, x_1, \ldots)$ of L. If χ is the Keisler sentence (3.4) and A is an L-structure, then the **Keisler game** $G(\chi, A)$ is played as follows. There are λ steps. At the ith step, one of the players chooses an element a_i of A; player $\forall$ makes the choice if Q_i is $\forall$, and player $\exists$ makes it otherwise. At the end of the play, player $\exists$ wins if $A \vDash \bigwedge \Phi(a_0, a_1, \ldots)$. We define

$$(3.5) \qquad\qquad\qquad A \vDash \chi$$

to mean that player $\exists$ has a winning strategy for the game $G(\chi, A)$.

A **finite approximation** to the Keisler sentence (3.4) is a sentence $\bar{Q} \bigwedge \Psi$, where Ψ is a finite subset of Φ and $\bar{Q}$ is a finite subsequence of the quantifier prefix in (3.4), containing quantifiers to bind all the free variables of Ψ. We write app(χ) for the set of all finite approximations to the Keisler sentence χ.

These definitions adapt in an obvious way to give **Keisler formulas** $\chi(\bar{w})$ and games $G(\chi(\bar{w}), A, \bar{c})$. Thus $A \vDash \chi(\bar{c})$ holds if player $\exists$ has a winning strategy for $G(\chi(\bar{w}), A, \bar{c})$. In particular, let χ be the Keisler sentence (3.4) and let α be an ordinal $< \lambda$. Then we write $\chi^\alpha(x_i : i < \alpha)$ for the Keisler formula got from χ by removing the quantifiers $Q_i x_i \, (i < \alpha)$.

The following lemma tells us that we can detach the leftmost quantifier $Q_0 x_0$ of a Keisler sentence and treat it exactly like an ordinary quantifier. Of course the lemma generalises to Keisler formulas $\chi(\bar{w})$ too.

Lemma 10.3.6. *With the notation above, we have*
$$(3.6) \qquad A \vDash \chi \qquad iff \qquad A \vDash Q_0 x_0 \chi^1(x_0).$$

Proof. Suppose first that Q_0 is $\forall$. If $A \vDash \chi$, then the initial position in $G(\chi, A)$ is winning for player $\exists$, so that every choice a of player $\forall$ puts player $\exists$ into winning position in $G(\chi^1, A, a)$, whence $A \vDash \chi^1(a)$; so $A \vDash \forall x_0 \chi^1(x_0)$. The converse, and the corresponding arguments for the case $Q_0 = \exists$, are similar. $\square$

The next theorem makes the crucial connection between Keisler games and saturation.

Theorem 10.3.7. *Let A be a non-empty L-structure, λ an infinite cardinal and χ a Keisler sentence of L of length λ.*
  (a) *If $A \vDash \chi$ then $A \vDash \bigwedge \mathrm{app}(\chi)$.*
  (b) *If $A \vDash \bigwedge \mathrm{app}(\chi)$ and A is λ-saturated then $A \vDash \chi$.*

Proof. (a) We show that if $\alpha < \lambda$ and $\theta(x_i: i < \alpha)$ is a finite approximation to χ^α, and $\bar{a}$ is a sequence of elements of A such that $A \vDash \chi^\alpha(\bar{a})$, then $A \vDash \theta(\bar{a})$. The proof is by induction on the number n of quantifiers in the quantifier prefix of θ. We write χ as in (3.4).

If $n = 0$ then θ is a conjunction of formulas $\phi(x_i: i < \alpha)$ from Φ. If $A \vDash \chi^\alpha(\bar{a})$ then player $\exists$ has a winning strategy for $G(\chi^\alpha, A, \bar{a})$, and it follows that $A \vDash \theta(\bar{a})$.

Suppose $n > 0$ and the quantifier prefix of θ begins with a universal quantifier $\forall x_\beta$. Then $\beta \geqslant \alpha$ and we can write θ as $\forall x_\beta \theta'(x_i: i \leqslant \beta)$. (Note that none of the variables x_i with $i > \alpha$ are free in θ.) If $A \vDash \chi^\alpha(\bar{a})$, then player $\exists$ has a winning strategy for $G(\chi^\alpha, A, \bar{a})$; let the players play this game through the steps $Q_i x_i$ ($\alpha \leqslant i < \beta$), with player $\exists$ using her winning strategy, and let $\bar{b}$ be the sequence of elements chosen. (This is possible because A is not empty.) Then $A \vDash \chi^\beta(\bar{a}, \bar{b})$, and hence $A \vDash \forall x_\beta \chi^{\beta+1}(\bar{a}, \bar{b}, x_\beta)$ by Lemma 10.3.6. So for every element c of A, $A \vDash \chi^{\beta+1}(\bar{a}, \bar{b}, c)$, which by induction hypothesis implies $A \vDash \theta'(\bar{a}, \bar{b}, c)$. Thus $A \vDash \theta(\bar{a})$. The argument when θ begins with an existential quantifier is similar.

Finally putting $\alpha = 0$, we have (a) of the theorem.

(b) Assume A is λ-saturated and $A \vDash \bigwedge \mathrm{app}(\chi)$. Then player $\exists$ should adopt the following rule for playing $G(\chi, A)$: always choose so that for each $\alpha < \lambda$, if $\bar{b}$ is the sequence of elements chosen before the αth step, then

$A \vDash \bigwedge \mathrm{app}(\chi^{\alpha})(\bar{b})$. If she succeeds in following this rule until the end of the game, when a sequence $\bar{a}$ of length λ has been chosen, then $A \vDash \bigwedge \Phi(\bar{a})$, so that she wins. We only have to show that she can follow the rule.

Suppose then that she has followed this rule up to the choice of $\bar{b} = (b_i : i < \alpha)$, so that $A \vDash \bigwedge \mathrm{app}(\chi^{\alpha})(\bar{b})$. First suppose that Q_{α} is $\exists$. Without loss write any finite approximation θ to χ^{α} as $\exists x_{\alpha} \theta'(x_i : i \leqslant \alpha)$. To maintain the rule, player $\exists$ has to choose an element b_{α} so that $A \vDash \theta'(\bar{b}, b_{\alpha})$ for each $\theta \in \mathrm{app}(\chi^{\alpha})$. Now A is λ-saturated and $\bar{b}$ has length less than λ. Hence we only need show that if $\{\theta_0, \ldots \theta_{n-1}\}$ is a finite set of formulas in $\mathrm{app}(\chi^{\alpha})$, then $A \vDash \exists x_{\alpha}(\theta_0' \wedge \ldots \wedge \theta_{n-1}')(\bar{b}, x_{\alpha})$. But clearly there is some finite approximation θ to χ^{α} which begins with $\exists x_{\alpha}$ and is such that θ' implies $\theta_0', \ldots, \theta_{n-1}'$. By assumption $A \vDash \theta(\bar{b})$, in other words $A \vDash \exists x_{\alpha} \theta'(\bar{b}, x_{\alpha})$. This completes the argument when Q_{α} is $\exists$.

Next suppose that Q_{α} is $\forall$, and let $\theta'(x_i : i \leqslant \alpha)$ be a finite approximation to $\chi^{\alpha+1}$. Then $\forall x_{\alpha} \theta'$ is a finite approximation to χ^{α}, and so by assumption $A \vDash \forall x_{\alpha} \theta'(\bar{b}, x_{\alpha})$. Hence $A \vDash \theta'(\bar{b}, b_{\alpha})$ regardless of the choice of b_{α}. So player $\forall$ can never break player $\exists$'s rule.

The reader can check that limit ordinals are no threat to player $\exists$'s rule. So she can follow the rule and win. Therefore $A \vDash \chi$. $\square$

Theorem 10.3.7 generalises straightforwardly to Keisler formulas $\chi(\bar{w})$ with fewer than λ free variables $\bar{w}$. In this setting, part (a) of the theorem reads

(3.7)   If $A \vDash \chi(\bar{b})$ then $A \vDash (\bigwedge \mathrm{app}(\chi))(\bar{b})$,

and likewise with part (b).

Theories admitting group actions

Keisler games will be used in sections 10.6 and 12.1 below. For the moment, here is a typical application. We use the games to find an explicit syntactic characterisation of a model-theoretic property.

Let A be an L-structure and G a group. An **action** of G on A is a map $f : G \times \mathrm{dom}(A) \to \mathrm{dom}(A)$ (where we write ga for $f(g, a)$) such that for each g in G the map $a \mapsto ga$ is an automorphism of A, and moreover the following laws hold: $1a = a$, $(gh)(a) = g(h(a))$ for all elements a of A and all g, h in G. The action is **faithful** if for every $g \neq 1$ in G there is a in A such that $ga \neq a$. We say that G **acts faithfully on** A if there is a faithful action of G on A.

Theorem 10.3.8. *Let L be a first-order language and G a finite group. Then there is a theory $U = U_G$ in L such that the following are equivalent for every theory T in L.*

(a) $T \cup U$ has a model.

(b) T has a model on which G acts faithfully.

Moreover we can describe U explicitly from G.

Proof. For simplicity I take the case where L is countable and G is the Klein four-group; it illustrates everything except the messiness of the general case. We aim to write down a Keisler sentence χ such that if two players play $G(\chi, A)$ and player $\exists$ wins, then the elements chosen during the play will form an elementary substructure of A on which G acts faithfully.

The sentence χ will be of length ω. A play of $G(\chi, A)$ should produce an array of elements of A as follows:

(3.8)

a_0	a_1	a_2	a_3	$\cdots$
b_0	b_1	b_2	b_3	$\cdots$
c_0	c_1	c_2	c_3	$\cdots$

with the following properties:

(3.9) $\{a_i : i < \omega\}$ is the domain of an elementary substructure B of A;

(3.10) for all $i < j < \omega$, $a_i = a_j \Leftrightarrow b_i = b_j \Leftrightarrow c_i = c_j$.

Then the maps $a_i \mapsto b_i$ and $a_i \mapsto c_i$ are maps f and g which are well-defined on $\mathrm{dom}(B)$.

(3.11) $f^2 = g^2 = (fg)^2 = $ identity (and in particular $\{a_i : i < \omega\} = \{b_i : i < \omega\} = \{c_i : i < \omega\}$).

(3.12) $a_0 \neq b_0$, $a_1 \neq c_1$ and $b_2 \neq c_2$ (so that none of f, g, fg is the identity).

(3.13) f and g are automorphisms of B.

It remains to write down a Keisler sentence χ which guarantees all this.

The sentence will be of the form

(3.14) $$\exists x_0 \exists y_0 \exists z_0 \exists x_1 \exists y_1 \exists z_1 \ldots \bigwedge_{i < \omega} \phi_i(x_0, \ldots, z_i).$$

The variables x_i, y_i, z_i will supply the elements a_i, b_i, c_i respectively. So the set of formulas $\Phi = \{\phi_i : i < \omega\}$ must contain all the following formulas. For (3.9) and the last part of (3.11), if $\psi(x_0, y_0, z_0, x_1, \ldots, z_{k-1}, x)$ is any formula of L, then Φ must contain $\exists x\, \psi(x_0, \ldots, z_{k-1}, x) \to \psi(x_0, \ldots, z_{k-1}, x_i)$ for some $i \geq k$. For the first equation of (3.11), Φ must contain all the formulas $(y_i = x_j \to x_i = y_j)$ with $i, j < \omega$; similar formulas deal with the rest of (3.11). Clause (3.12) needs the formulas $x_0 \neq y_0$, $x_1 \neq z_1$, $y_2 \neq z_2$. Finally for (3.10) and (3.13) we put into Φ all formulas of the form $\psi(x_{i_0}, \ldots, x_{i_{m-1}}) \leftrightarrow \psi(y_{i_0}, \ldots, y_{i_{m-1}})$ (and likewise with z for y) where ψ are unnested atomic formulas of L.

We take U to be the theory app(χ). Using Theorem 10.3.7 it is easy to check that an ω-saturated structure A is a model of app(χ) if and only if it has a finite or countable elementary substructure on which G acts faithfully.

Essentially the same recipe works for any finitely generated group, though in general one has to replace (3.11) and (3.12) by infinitely many equations and inequations. I leave it to the reader to extend the result to arbitrary groups, for example using the idea of Corollary 6.6.8.   □

Up to equivalence of theories, the theory U described in the proof above consists of all sentences of the form

$$(3.15) \quad \exists x_0 y_0 z_0((\exists x_0 \eta_0 \to \eta_0) \wedge$$
$$\exists x_1 y_1 z_1((\exists x_1 \eta_1 \to \eta_1) \wedge$$

$$\cdots$$

$$\exists x_n y_n z_n((\exists x_n \eta_n \to \eta_n) \wedge$$
$$\bigwedge_{j<m} ((\psi_j(x_0, \ldots, x_n) \leftrightarrow \psi_j(y_0, \ldots, y_n)) \wedge$$
$$(\psi_j(x_0, \ldots, x_n) \leftrightarrow \psi_j(z_0, \ldots, z_n))) \wedge$$
$$\bigwedge_{i,j \leqslant n} ((x_i = y_j \leftrightarrow x_j = y_i) \wedge (x_i = z_j \leftrightarrow x_j = z_i)) \wedge$$
$$\bigwedge_{i,j,k,l \leqslant n} (z_i = x_j \wedge y_j = x_k \wedge z_k = x_l \to y_l = x_i) \wedge$$
$$(x_0 \neq y_0 \wedge x_1 \neq z_1 \wedge y_2 \neq z_2) \ldots),$$

where $\eta_i(x_0, y_0, z_0, \ldots, x_{i-1}, y_{i-1}, z_{i-1}, x_i)$ $(i \leqslant n)$ are any formulas of L and $\psi_j(x_0, \ldots, x_n)$ $(j < m)$ are unnested atomic formulas of L.

In section 11.2 we shall see that every first-order theory with infinite models has a model on which the automorphism group $\mathrm{Aut}(\mathbb{Q}, <)$ of the rational numbers acts faithfully.

Exercises for section 10.3

1. Let T be a theory in a first-order language L and $\Phi(\bar{x})$ a set of formulas of L such that if $f: A \to B$ is any surjective homomorphism between models A, B of T and $A \vDash \bigwedge \Phi(\bar{a})$ then $B \vDash \bigwedge \Phi(\bar{b})$. Show that Φ is equivalent modulo T to a set $\Psi(\bar{x})$ of positive formulas of L.

The next exercise extracts the crucial facts about saturation used in the constructions in Theorem 10.3.1 and Lemma 10.3.4.
2. Let L be a first-order language and Θ a set of formulas of L. If A, B are L-structures, we write $(A, \bar{a}) \Rrightarrow_\Theta (B, \bar{b})$ to mean that for every formula θ in Θ, if $A \vDash \theta(\bar{a})$ then $B \vDash \theta(\bar{b})$. Show (a) if Θ is closed under conjunction, disjunction and both existential and universal quantification, and A and B are λ-saturated L-structures

with $\lambda \geqslant |L|$ such that $A \Rightarrow_\Theta B$, then there are sequences $\bar{a}$, $\bar{b}$ in A, B respectively, both of length λ, such that $\bar{a}$, $\bar{b}$ list the domains of elementary substructures A', B' of A, B respectively, and $(A, \bar{a}) \Rightarrow_\Theta (B, \bar{b})$, (b) if Θ is closed under conjunction and existential quantification, and A and B are L-structures such that $A \Rightarrow_\Theta B$ and B is $|A|$-saturated, then there are sequences $\bar{a}$, $\bar{b}$ in A, B respectively such that $\bar{a}$ lists the domain of A and $(A, \bar{a}) \Rightarrow_\Theta (B, \bar{b})$.

3. Let L be a first-order language containing a 1-ary relation symbol P. Let A and B be elementarily equivalent L-structures of cardinality λ. Suppose that B is saturated, but make the following weaker assumption on A: if X is any set of fewer than λ elements of A, and $\Phi(x)$ is a type over X with respect to A, which contains the formula $P(x)$, then Φ is realised in A. Show that there is an elementary embedding $f: A \to B$ which is a bijection from P^A to P^B.

4. Let L be a first-order language, let A and B be $|L|$-saturated L-structures, and suppose that every sentence of $\text{Th}(A)$ which is either positive or $\forall_1$ is in $\text{Th}(B)$. Show that there are elementary substructures A', B' of A, B respectively, a surjective homomorphism $f: A' \to B'$ and an embedding $e: B' \to A$. [Build up sequences $\bar{a}$, $\bar{b}$, $\bar{c}$ so that the right relationships hold between $(A, \bar{a})$, $(B, \bar{b})$ and $(A, \bar{c})$.]

5. Let L be a first-order language and T a theory in L. Show that the following are equivalent, for every sentence ϕ of L. (a) For every model A of T and endomorphism $e: A \to A$, if ϕ is true in A then ϕ is true in the image of A. (b) ϕ is equivalent modulo T to a positive boolean combination of positive sentences of L and $\forall_1$ sentences of L. [Use the previous exercise.]

6. Let L be a first-order language whose symbols include a 2-ary relation symbol $<$, and let T be a theory in L and $\phi(\bar{x})$ a formula of L. In the terminology of Exercise 2.4.8, show that the following are equivalent. (a) If A and B are models of T and A is an end-extension of B, then for every tuple $\bar{b}$ of elements of B, $B \vDash \phi(\bar{b})$ implies $A \vDash \phi(\bar{a})$. (b) ϕ is equivalent modulo T to a Σ_1^0 formula $\psi(\bar{x})$ of L.

7. Let L be a first-order language whose symbols include a 2-ary relation symbol $<$, and let T be a theory in L which implies '$<$ linearly orders the set of all elements, with no last element'. Recall from Exercise 2.4.12 the notion of a **cofinal substructure**; let Φ be defined as in that exercise, except that the formulas in Φ are required to be first-order. Show (a) if A and B are models of T, A and B are $|L|$-saturated and $A \Rightarrow_\Phi B$, then there is an embedding of an elementary substructure A' of A onto a cofinal substructure of an elementary substructure B' of B, (b) if ϕ is a formula of L, and f preserves ϕ whenever f is an embedding of a model of T onto a cofinal substructure of a model of T, then ϕ is equivalent modulo T to a formula in Φ.

8. Let L be a first-order language whose symbols include a 2-ary relation symbol R, and let T be a theory in L. Suppose that T implies that R expresses a reflexive symmetric relation. If A is a model of T, we define a relation $\sim$ on $\text{dom}(A)$ by '$\sim$ is the smallest equivalence relation containing R^A'. A **closed** substructure of A is a

substructure whose domain is a union of equivalence classes of $\sim$. We define Θ to be the least set of formulas of L such that (i) Θ contains all quantifier-free formulas, (ii) Θ is closed under disjunction, conjunction and existential quantification, and (iii) if $\phi(\bar{x}yz)$ is in Θ, and y occurs free in ϕ, then the formula $\forall z(Ryz \to \phi)$ is in Θ. Show (a) if A and B are models of T, A and B are $|L|$-saturated and $A \Rightarrow_\Theta B$, then there is an embedding of an elementary substructure A' of A onto a closed substructure of an elementary substructure B' of B, (b) if ϕ is a formula of L, and f preserves ϕ whenever f is an embedding of a model of T onto a closed substructure of a model of T, then ϕ is equivalent modulo T to a formula in Θ.

Contrast the next exercise with Corollary 6.5.3.
9. Let ϕ be the sentence $\exists xy(Rxy \land \forall z(Rxz \to \exists t(Rxt \land Rzt)))$. Show (a) if a structure A is a homomorphic image of a model of ϕ, then A contains elements a_i ($i < \omega$), not necessarily distinct, such that $A \vDash R(a_0, a_i) \land R(a_i, a_{i+1})$ whenever $0 < i < \omega$, (b) if a structure B contains arbitrarily long finite sequences like the sequence of length ω in (a), then some elementary extension of B is a homomorphic image of a model of ϕ, (c) by (a) and (b), the class of homomorphic images of models of a first-order sentence need not be closed under elementary equivalence.

10. Let L be the first-order language of groups and T the theory of groups (a finite theory). It is known that the set of consequences of T in L is not recursive (e.g. by the Novikov–Boone theorem, Theorem 12.12 in Rotman [1973]). Let S be the set of all sentences of L which are true in the one-element group. (a) Show that S is a recursive set. (b) Show that if ϕ is in S, then $\neg \phi$ is equivalent modulo T to a positive sentence of L if and only if $T \vdash \phi$. (c) Show that there is no algorithm to determine whether a given sentence in S is a consequence of T. (d) Deduce that there is no algorithm to determine, for any first-order sentence ψ, whether or not ψ is equivalent modulo T to a positive first-order sentence.

*11. Let L be a first-order language with at least one relation or function symbol of arity ≥ 2, and let $\mathbf{K}$ be the class of L-structures which admit an automorphism of order 2. Show that if $n < \omega$, then there is no set T of sentences of L of quantifier rank $\leq n$ such that for all L-structures A, $A \vDash T \Leftrightarrow T$ is elementarily equivalent to some B in $\mathbf{K}$.

Subdirect products and special Horn sentences were defined at Exercise 9.1.13.
12. Show that if T is a first-order theory then T is preserved in subdirect products if and only if T is equivalent to a set of special Horn sentences.

10.4 Special models

In ZFC there is no guarantee that every consistent first-order theory has a saturated model. (See Exercises 6–8 below.) This is a nuisance, but one can find ways around it. Here are three.

The first is the *absoluteness* method. To prove a theorem by this method, we prove it *assuming the generalised continuum hypothesis* (GCH). Then we show that the truth or falsity of the theorem can't depend on whether the GCH holds. (By Shoenfield's absoluteness lemma this is the case when the theorem can be written as an arithmetical statement, or even as a Σ_2^1 statement, about the natural numbers. See 3.17 in Mansfield & Weitkamp [1985].) Section 6.2 of Chang & Keisler [1973] contains an example worked out in detail.

The second method is to use *resplendent models* in place of saturated ones. These models always exist provided that the theory is consistent, but their uses are more limited. Resplendence behaves rather like a recursive analogue of bigness. See section 10.6 below.

In this section we study the third approach, namely *special models*. Every saturated model is special, but not vice versa. Special models always exist, and for many purposes they are just as good as saturated ones, though their definition is a little less neat.

For any cardinality λ and ordinal i, we define the cardinal $\beth_i(\lambda)$ by induction on i: $\beth_0(\lambda) = \lambda$, $\beth_{i+1}(\lambda) = 2^{\beth_i(\lambda)}$, $\beth_\delta(\lambda) = \bigcup_{i<\delta} \beth_i(\lambda)$ for limit δ. We write $\beth_i$ for $\beth_i(\omega)$. Thus $\beth_0, \beth_1, \beth_2, \ldots$ are ω, 2^ω, $2^{2^\omega}, \ldots$. The symbol $\beth$ (the second letter of the Hebrew alphabet) is pronounced 'bet' or 'beth' according as you prefer modern or classical Hebrew, and the cardinals $\beth_i$ are known as the **beths**. The cardinals of the form $\beth_\delta$ for a limit ordinal δ are known as **limit beths**. A **strong limit number** is a number which is ω or a limit beth.

Fact 10.4.1. *Let λ be an infinite cardinal.*

(a) $\lambda^{<\lambda} = \lambda$ *if and only if λ is regular and $2^{<\lambda} = \lambda$.*

(b) *If λ is strongly inaccessible (i.e. either ω or a regular limit beth) then* $\lambda^{<\lambda} = \lambda$.

(c) *If λ is regular and the GCH holds then $\lambda^{<\lambda} = \lambda$.*

(d) *λ is a limit beth if and only if λ is uncountable and $2^\mu < \lambda$ whenever* $\mu < \lambda$.

(e) *If $\beth_\delta$ is a limit beth then there is a strictly increasing sequence $(\mu_i : i < \mathrm{cf}\delta)$ of regular cardinals such that $\beth_\delta = \sum_{i<\mathrm{cf}\delta} \mu_i = \sum_{i<\mathrm{cf}\delta} 2^{\mu_i}$.*

Proof. See section 6 of Jech [1978].   □

Existence of special models

Let A be a structure of cardinality λ. By a **specialising chain** for A we mean an elementary chain $(A_\kappa : \kappa$ a cardinal $< \lambda)$ such that $A = \bigcup_\kappa A_\kappa$, and for each κ, A_κ is κ^+-saturated. (This is not a slip of the pen. We index the chain by cardinals, not ordinals.) We say that A is **special** if there is a specialising chain for A.

Theorem 10.4.2. (a) *Every saturated structure is special.*

(b) *Let A be an infinite L-structure and λ a strong limit number $> |A| + |L|$. Then A has a special elementary extension of cardinality λ.*

(c) *Let T be a theory with infinite models, in a first-order language L, and let λ be a strong limit number $> |L|$. Then T has a special model of cardinality λ.*

(d) *Every consistent first-order theory has a special model.*

Proof. (a) If A is saturated, then A is κ^+-saturated for every cardinal $\kappa < |A|$. Hence we can take the specialising chain to be the constant chain with each A_κ equal to A.

(b) Let λ be the strong limit number $\beth_\delta$. By Fact 10.4.1(e) there is a strictly increasing sequence $(\mu_i : i < \mathrm{cf}\delta)$ of cardinals such that $\beth_\delta = \sum_{i<\mathrm{cf}\delta} \mu_i = \sum_{i<\mathrm{cf}\delta} 2^{\mu_i}$. By throwing away an initial part of this sequence if necessary, we can suppose that $\mu_0 \geqslant |A| + |L|$. We claim that there is an elementary chain $(B_i : i < \mathrm{cf}\delta)$ such that $B_0 = A$, each B_{i+1} is μ_i^+-saturated, and for each i $|B_{i+1}| = 2^{\mu_i}$. This is proved by induction on i. At successor ordinals we use Corollary 10.2.2. At limit ordinals j we take unions, noting that if $|B_{i+1}| = 2^{\mu_i}$ for all $i < j$, then $|\bigcup_{i<j} B_i| = \sup(2^{\mu_i} : i < j) \leqslant 2^{\mu_j}$.

When the chain has been constructed, put $B = \bigcup_{i<\mathrm{cf}\delta} B_i$. Then B is an elementary extension of A of cardinality $\sum_{i<\mathrm{cf}\delta} 2^{\mu_i} = \lambda$. To find a specialising chain for B, let f be a map from the set K of cardinals $< \lambda$ to the ordinal $\mathrm{cf}\delta$, such that for each $\kappa \in K$, $\mu_{f(\kappa)} \geqslant \kappa$, and $\kappa \leqslant \nu$ implies $f(\kappa) \leqslant f(\nu)$. Then the chain $(B_{f(\kappa)+1} : \kappa \in K)$ will serve.

(c) follows at once from (b). For (d) we need only add that finite structures are saturated, by Example 2 in section 10.1. $\qquad\square$

There are plenty of strong limit numbers. Let λ be any cardinal, and let $\beth_i$ be the first beth which is $\geqslant \lambda$. Then $\beth_{i+\omega}$ is a strong limit number $\geqslant \lambda$.

Uniqueness of special models

By the back-and-forth arguments of section 10.1 we can show that if A and B are elementarily equivalent special structures of the same cardinality, then $A \cong B$. This generalises Theorem 10.1.8.

Let A be a special L-structure of cardinality λ. Write K for the set of all cardinals $< \lambda$, and let $(A_\kappa : \kappa \in K)$ be a specialising chain for A. By a **cautious enumeration** of A we mean a sequence $\bar{a}$ of length λ which lists the elements of A in such a way that for every cardinal $\kappa \in K$, $\bar{a}|\kappa^+$ lies inside A_κ.

Lemma 10.4.3. *Let A be a special structure (with a given specialising chain). Then there exists a cautious enumeration of A.*

Proof. Let $\lambda = |A|$ and let $(A_\kappa : \kappa \in K)$ be the given specialising chain. If A is the empty L-structure, the empty sequence is a cautious enumeration. Assume then that A_0 has at least one element a_0, and let $(b_j : i < \lambda)$ be an enumeration of dom(A) without repetitions. Define $\bar{a} = (a_i : i < \lambda)$ by induction on i, as follows: for each $i > 0$, let a_i be the first b_k $(k < \lambda)$ which is in $A_{|i|}$ and is not equal to any a_j with $j < i$, if there is such a b_k; let a_i be a_0 otherwise.

To show that $\bar{a}$ is a cautious enumeration of A, it suffices to prove that every element of A is in $\bar{a}$. Let b_k be a counterexample, and let κ be the least cardinal in K such that b_k is in A_κ. Then by construction the elements a_i with $\kappa \leqslant i < \lambda$ are all distinct and of form b_j with $j < k$. Since there are fewer than λ elements b_j with $j < k$, this is clearly impossible. □

Theorem 10.4.4. *Let A and B be elementarily equivalent structures of the same cardinality λ, and suppose A and B are both special. Then $A \cong B$.*

Proof. Choose specialising chains $(A_\kappa : \kappa \in K)$, $(B_\kappa : \kappa \in K)$ for A, B respectively. By the lemma, there are cautious enumerations $\bar{a}$, $\bar{b}$ of A, B respectively. We claim that there are a sequence $\bar{c} = (c_i : i < \lambda)$ of elements of A, and a sequence $\bar{d} = (d_i : i < \lambda)$ of elements of B, such that for each $i \leqslant \lambda$,

$$(4.1) \qquad (A \; \bar{a}|i, \bar{c}|i) \equiv (B, \bar{d}|i, \bar{b}|i),$$

and moreover each c_i (resp. d_i) lies in $A_{|i|}$ (resp. $B_{|i|}$).

The claim is proved by induction on i. When $i = 0$ it states that $A \equiv B$, which we have assumed. There is nothing to prove at limit ordinals. When i is a successor ordinal $j + 1$, we have by induction hypothesis that $(A, \bar{a}|j, \bar{c}|j) \equiv (B, \bar{d}|j, \bar{b}|j)$, $\bar{a}|j$ and $\bar{c}|j$ lie in $A_{|j|}$, and $\bar{d}|j$ and $\bar{b}|j$ lie in $B_{|j|}$. Hence

$$(4.2) \qquad (A_{|j|}, \bar{a}|j, \bar{c}|j) \equiv (B_{|j|}, \bar{d}|j, \bar{b}|j).$$

Since $A_{|j|}$ is $|j|^+$-saturated and b_j is in $B_{|j|}$, there is an element c_j of $A_{|j|}$ such that $(A_{|j|}, \bar{a}|j, \bar{c}|i) \equiv (B_{|j|}, \bar{d}|j, \bar{b}|i)$. Similarly there is an element d_j of $B_{|j|}$ such that $(A_{|j|}, \bar{a}|i, \bar{c}|i) \equiv (B_{|j|}, \bar{d}|i, \bar{b}|i)$, whence (4.1) holds for i. This proves the claim.

Now apply the claim with $i = \lambda$. By the diagram lemma there is an embedding $f : A \to B$ such that $f\bar{a} = \bar{d}$ and $f\bar{c} = \bar{b}$. Since $\bar{b}$ lists all the elements of B, f is an isomorphism. □

Further properties

Theorem 10.4.5. (a) *If A is special of cardinality λ and $\bar{a}$ is a sequence of length $< \mathrm{cf}(\lambda)$ then $(A, \bar{a})$ is special.*

(b) *Every reduct of a special structure is special.*

Proof. For (a), take a specialising chain $(A_\kappa : \kappa \in K)$ for A. There is some $\mu < \lambda$ such that all of $\bar{a}$ lies within A_μ. Then $(A_{\max(\kappa,\mu)} : \kappa \in K)$ is a specialising chain for $(A, \bar{a})$ (see Exercise 10.1.6). (b) is immediate from the definition. $\qquad\square$

Corollary 10.4.6. *If A is special of cardinality λ then A is strongly λ-homogeneous.*

Proof. Suppose $\bar{a}$ and $\bar{b}$ are tuples of length $< \mathrm{cf}(\lambda)$ such that $(A, \bar{a}) \equiv (A, \bar{b})$. By Theorem 10.4.5, $(A, \bar{a})$ and $(A, \bar{b})$ are special. Hence they are isomorphic by Theorem 10.4.4. $\qquad\square$

Keisler games work for special models too. Suppose A is a special structure with specialising chain $(A_\kappa : \kappa \in K)$. Then we alter the instructions for the game $G(\chi, A)$ a little: for each $i < \lambda$ we add the requirement that the player who chooses a_i must choose it from $A_{|i|}$. Let us write $A \vDash_{\mathrm{sp}} \chi$ to mean that player $\exists$ has a winning strategy in this revised game. We have the following.

Theorem 10.4.7. *Let A be a special L-structure of infinite cardinality λ, and χ a Keisler sentence of L of length λ. Then $A \vDash_{\mathrm{sp}} \chi$ if and only if $A \vDash \bigwedge \mathrm{app}(\chi)$.* $\qquad\square$

Theorem 10.4.7 is a powerful machine. Let us use it straight away to derive one of the most interesting properties of special models.

Suppose L and L^+ are first-order languages with $L \subseteq L^+$. If A is an L-structure and T is a theory in L^+ such that $\mathrm{Th}(A) \cup T$ is consistent, then there is some model B of T with $A \equiv B|L$. When can we do better than this and have $A = B|L$? Theorem 10.4.9 says that we can do it provided that A is special. This rests on a lemma.

Lemma 10.4.8. *Let L, L^+ be first-order languages with $L \subseteq L^+$, T a theory in L^+ and λ a cardinal $\geq |L^+|$. Then there is a Keisler sentence χ of L of length λ such that*

(4.3)   *if an L^+-structure B is a model of T of cardinality ≥ 2, then $B \vDash \chi$,*

(4.4)   *if C is an L-structure of cardinality $\leq \lambda$, and $C \vDash \chi$, then some expansion of C is a model of T.*

In (4.4) we can replace $\vDash$ by $\vDash_{\mathrm{sp}}$ when C is special.

Proof. The translations of sections 2.6 and 3.1 allow us to adjust T into a new theory T^* in a first-order language L^* extending L^+. First, we can add two new constant symbols *true* and *false*, and use them to replace all relation symbols of $L^+\backslash L$ by function symbols; see Exercise 5. (The theory T^* will then contain the sentence *true* $\neq$ *false*. This is why the structure B in (4.3) has to have at least two elements.) Second, by skolemising we can suppose that T^* consists of $\forall_1$ sentences. Third, by Theorem 2.6.1 we can assume that T^* is unnested. Every model of T^* is a model of T, and every model of T with at least two elements can be expanded to a model of T^*.

Our Keisler sentence χ will be of the form

$$(4.5) \qquad\qquad \forall x_0 \exists y_0 \forall x_1 \exists y_1 \forall x_2 \ldots \bigwedge \Phi.$$

In a play of the resulting game, we shall write $a_0, a_1, \ldots$ for player $\forall$'s successive choices, and $b_0, b_1, \ldots$ for those of player $\exists$. We shall write $\bar{a}$ for $(a_0, a_1, \ldots)$, and likewise $\bar{b}$ for $(b_0, b_1, \ldots)$.

The intention will be that when we play the game $G(\chi, A)$ on an L-structure A, the following things happen. Each quantifier $\exists y_i$ will be linked to some unnested term $F(\bar{u})$ of L^*, where $\bar{u}$ is a tuple of variables x_j that appear before y_i in (4.5). When player $\exists$ chooses the element b_i for y_i, she will be choosing the value of the term $F(\bar{u})$ for the assignments already made to the variables in $\bar{u}$; so player $\exists$ will decide what the function F^A shall be.

Let Θ be the set of all unnested terms of the forms $F(\bar{u})$ or c, where F is a function symbol of $L^*\backslash L$, $\bar{u}$ is a tuple of variables x_i and c is a constant of $L^+\backslash L$. Choose an injective map $f: \Theta \to \lambda$ so that

$$(4.6) \quad \text{if } f(t) = i \text{ then for every variable } x_j \text{ in } t \text{ we have } j \leqslant i.$$

Since $\lambda \geqslant |L^+|$, it is clear that such a map f can be found.

In (4.5), the formulas in Φ are of two kinds. The first kind will make sure that the interpretations of the function symbols of $L^*\backslash L$ are well-defined. Suppose F is one of these function symbols, and consider two terms $F(\bar{u})$ and $F(\bar{v})$ in Θ. We require that if the same assignments are made to $\bar{u}$ and to $\bar{v}$, then $F(\bar{u})$ and $F(\bar{v})$ are given the same value too; in other words,

$$(4.7) \qquad\qquad \bar{u} = \bar{v} \to y_{f(F(\bar{u}))} = y_{f(F(\bar{v}))}.$$

(Here $\bar{u} = \bar{v}$ is shorthand for a conjunction of equations $u_0 = v_0$ etc.) There are corresponding formulas for the constants c of $L^+\backslash L$ as well. All these formulas will appear in Φ.

Second, consider any sentence $\forall \bar{z}\, \psi(\bar{z})$ in T^*, where ψ is quantifier-free and unnested. Let $\bar{x}$ be any tuple of variables x_j, of the same length as $\bar{z}$ and possibly with repetitions. Let $\psi_{\bar{x}}$ be the formula which is got from $\psi(\bar{x})$ as follows: replace every term $F(\bar{w}) \in \Theta$ by the variable $y_{f(F(\bar{w}))}$, and every term $c \in \Theta$ by the variable $y_{f(c)}$. These formulas $\psi_{\bar{x}}$, for all choices of sentence $\forall \bar{z}\, \psi$ and tuple $\bar{x}$, make up the rest of Φ.

This defines the Keisler sentence χ.

We prove (4.3). Since B has at least two elements, we can assume that B is also a model of T^*. The following is a winning strategy for player $\exists$. When

i is not in the image of f, she can choose any element b_i for y_i. When i is $f(F(\bar{u}))$, let her replace each variable x_j in $\bar{u}$ by the corresponding element a_j (which has already been chosen, by (4.6)). If $\bar{d}$ is the resulting tuple of elements of B, she must choose b_i to be $F^B(\bar{d})$. This ensures that all the formulas (4.7) will be satisfied at the end of the game. Since B is a model of T^* and T^* is an $\forall_1$ theory, all the formulas $\psi_{\bar{x}}$ will be satisfied too.

We prove (4.4). Suppose players $\forall$ and $\exists$ play $G(\chi, C)$, where C has cardinality at most λ, and let player $\exists$ use her winning strategy. Player $\forall$ can play so that every element of C appears among his moves. Let the resulting play be $\bar{a}$, $\bar{b}$. Then $\bar{a}$ lists all the elements of C. We can define the new symbols of $L^+\backslash L$ as follows. For each function symbol F of arity n and each n-tuple $\bar{u}$ of variables x_i, we want to determine the element $t(\bar{a})$ where t is $F(\bar{u})$. We choose the value to be $b_{f(t)}$. Since player $\exists$ wins this play, the formulas (4.7) guarantee that F is well-defined. Similarly we choose the value of a constant c to be $b_{f(c)}$. With these definitions, C becomes an L^*-structure B. All the sentences of T^* are true in B, because of the formulas $\psi_{\bar{x}}$ in Φ. So $B \vDash T$ as required.

Finally suppose C is special in (4.4). We only need to check that player $\forall$ can choose a cautious enumeration of C, and Lemma 10.4.3 guarantees this.

$\qquad\qquad\qquad\qquad\qquad\qquad\qquad\qquad\qquad\qquad\qquad\qquad\qquad\qquad$ $\square$

When $\lambda = \omega$, the Keisler sentence χ in this proof is in fact a Svenonius sentence; see Example 2 in section 3.4.

Theorem 10.4.9. *Let L and L^+ be first-order languages with $L \subseteq L^+$, and T a theory in L^+. Let A be an L-structure which is either finite, or special and of cardinality $\geq |L^+|$. If $\mathrm{Th}(A) \cup T$ is consistent, then A can be expanded to form a model of T.*

Proof. Suppose first that A is finite. Add constants for the elements of A. Then using compactness, one easily checks that $\mathrm{eldiag}(A) \cup T$ is consistent; let B be a model of it. Since $\mathrm{Th}(A)$ contains a sentence giving the cardinality of A, $B|L$ is isomorphic to A. So we can choose B to be an expansion of A.

Next suppose that A is infinite. Let χ be the Keisler sentence of length $|A|$ as in the lemma. By Theorem 10.4.7 it suffices to show that $A \vDash \bigwedge \mathrm{app}(\chi)$. Since $\mathrm{Th}(A) \cup T$ is consistent, it has a model B, and $B \vDash \chi$ by the lemma. So $B \vDash \bigwedge \mathrm{app}(\chi)$ by Theorem 10.3.7(a), and thus $A \vDash \bigwedge \mathrm{app}(\chi)$ since $A \equiv B|L$.

$\qquad\qquad\qquad\qquad\qquad\qquad\qquad\qquad\qquad\qquad\qquad\qquad\qquad\qquad$ $\square$

Two easy corollaries, generalising Theorems 10.1.6 and 10.3.1, follow.

Corollary 10.4.10. *Let A be a special structure of cardinality λ. Then A is λ^+-universal.*

Proof. Let C be elementarily equivalent to A and of cardinality $\leqslant \lambda$. Adding constants for all elements of C, let T be eldiag(C). Then by Theorem 10.4.9, some expansion B of A is a model of T. So by the elementary diagram lemma B is elementarily embeddable in A. $\qquad\square$

Corollary 10.4.11. *Let A be a special L-structure of cardinality λ, and C an L-structure of cardinality $\leqslant \lambda$ such that $C \Rightarrow_1 A$. Then C can be embedded in A.*

Proof. The argument is the same, using diag(C) for eldiag(C). $\qquad\square$

Our third corollary gives new information about saturated structures too.

Corollary 10.4.12. (a) *Every special structure of cardinality λ is cf(λ)-big.*
   (b) *Every saturated structure of cardinality λ is λ-big (and hence strongly λ-homogeneous, etc.).*

Proof. By the theorem, every special structure of cardinality λ is 0-big. If we add fewer than cf(λ) parameters to a special structure of cardinality λ, it remains special. If we add fewer than λ parameters to a saturated structure of cardinality λ, it remains saturated (even when λ is singular). $\qquad\square$

Exercises for section 10.4

1. Show (a) for all ordinals α, β and cardinals λ, $\beth_{\alpha+\beta}(\lambda) = \beth_{\beta}(\beth_{\alpha}(\lambda))$, (b) for every infinite cardinal λ, $\beth_{\lambda} \geqslant \lambda$.

2. Let A be a structure of cardinality λ which is special. Show that every relation in A which is first-order definable with parameters has cardinality either $< \omega$ or $= \lambda$.

3. Show that if λ is a regular cardinal and A is a special structure of cardinality λ then A is saturated.

4. Suppose A and B are special L-structures, and every positive sentence true in A is also true in B. Suppose either that A and B have the same cardinality, or that B is finite. Show that there is a surjective homomorphism from A to B. Use this to give another proof of Lyndon's preservation theorem, Corollary 10.3.5.

5. Let L be a signature. For each n-ary relation symbol R of L, introduce a new n-ary function symbol F_R. Let L^f be the signature got from L by replacing each relation symbol R by F_R, and adding two new constant symbols *true* and *false*. If A is an L-structure with more than one element, we form an L^f-structure A^f by choosing any two distinct elements c, d of A to be named by *true* and *false* respectively, and

putting $F_R^{A^f}(\bar{a}) = c$ if $A \vDash R(\bar{a})$, $= d$ otherwise. If ϕ is any formula of signature L, we form ϕ^f by replacing each subformula of the form $R(\bar{t})$ by the equation $F_R(\bar{t}) = true$. Show that for any formula $\phi(\bar{x})$ of signature L and any tuple $\bar{a}$ of elements of A, $A \vDash \phi(\bar{a}) \Leftrightarrow A^f \vDash \phi^f(\bar{a})$.

6. Let T be the complete theory of some linear ordering with no endpoints. (Or more generally let T be any complete theory with the strict order property.) Show that if λ is a singular cardinal then T has no saturated model of cardinality λ.

7. Let T be a complete theory with the independence property, in a countable first-order language. Show that for every infinite cardinal λ, a λ-saturated model of T must have cardinality at least $2^{<\lambda}$.

8. By a theorem of Hugh Woodin, assuming the existence of some large cardinal (a supercompact is more than enough), there is a model M of ZFC in which $2^\mu = \mu^{++}$ for every infinite cardinal μ. We can suppose that M has no uncountable strong inaccessibles. Write down a countable first-order theory T which has no saturated models in M. [Use the two preceding exercises.]

9. (a) Under the hypotheses of Theorem 10.4.9, show that if T has a special model of cardinality $|A|$ then A can be expanded to form a special model of T. (b) Give an example of a structure A and a theory T as in Theorem 10.4.9, where A is special but can't be expanded to a special model of T.

10. Show that if A is a special structure and B is interpretable in A, then B is special. Deduce that if A is special and B is infinite and interpretable in A, then $|A| = |B|$.

11. Let K, L be first-order languages and λ a cardinal $> |K| + |L|$. Let A be a K-structure and B an L-structure which is interpretable in A. Show that if A is λ-big then B is λ-big. [Suppose $C|L \equiv B$. Absorbing any parameters into A, let Γ be the interpretation. Choose a strong limit number $\kappa > |K| + |L|$. Let A' be special of cardinality κ and $\equiv A$; likewise let C' be special of cardinality κ and $\equiv C$. Then $C'|L$ can be identified with $\Gamma A'$. Write a theory T^* in a language expanding K which describes C' in terms of A'. Then A' can be expanded to a model of T^*, and hence so can A by Exercise 10.1.23, inducing an expansion of B.]

12. Show that every special structure is strongly ω-homogeneous.

13. Let L and L^+ be first-order languages, where $L \subseteq L^+$ and L^+ has a 1-ary relation symbol P. Let A be an L^+-structure of cardinality $\geq |L|$ such that P^A is the domain of an elementary substructure A_P of $A|L$. Show that A has an elementary extension B of cardinality $|A|$ such that (a) B and B_P (the substructure of $B|L$ with domain P^B) are both strongly ω-homogeneous, and (b) $B|L$ and B_P realise the same types over $\varnothing$. [If B is special then B and B_P are strongly ω-homogeneous. To bring the cardinality down to λ, cast an eye over the proof of Theorem 10.2.5.]

10.5 Definability

There are three main definability theorems for complete first-order theories. They all have the following setting:

(5.1) L and L^+ are first-order languages with $L \subseteq L^+$, and R is a relation symbol in L^+; T is a complete theory in L^+. (As usual, 'complete' is shorthand for 'complete and consistent'.)

All three theorems say 'If in every model A of T, the relation R^A is in some way determined by $A|L$, then T contains some kind of definition of R in terms of the symbols of L, and vice versa'. The difference between the theorems is that in the first, R^A is completely determined by $A|L$, in the second it is determined up to some uniformly bounded number of possibilities (and the bound must be finite), while in the third it is determined up to $|A|$ possibilities.

All three theorems also have another equivalent condition in terms of automorphisms of a single structure. If A is a structure and G is a group of permutations of $\mathrm{dom}(A)$, we say that two relations X and Y are **conjugates under** G if there is a permutation $\sigma \in G$ such that $\sigma(X) = Y$. The 'automorphism' condition says that there is a suitably big model A in which the relation R^A has few conjugates under $\mathrm{Aut}(A|L)$.

In this section we prove the first and second definability theorems. The third is tougher; I let it rest until section 12.4 below.

First definability theorem

Theorem 10.5.1. *Let* L, L^+ *be first-order languages with* $L \subseteq L^+$, λ *an infinite cardinal greater than the number of symbols in* $L^+\backslash L$, *and* R *a relation symbol of* L^+. *Let* T *be a complete theory in* L^+. *Then the following are equivalent.*

(a)   *If A, B are any two models of T with $A|L = B|L$, then $R^A = R^B$.*

(b1) *There is a model A of T such that $A|L$ is either λ-big or saturated of cardinality $\geqslant |L^+|$, and if A' is any expansion of $A|L$ to a model of T then $R^{A'} = R^A$.*

(b2) *There is a model A of T such that $A|L$ is either λ-big or saturated of cardinality $\geqslant |L^+|$, and if A' is any expansion of $A|L$ to a model of T and e is any automorphism of $A|L$, then $e(R^{A'}) = R^{A'}$.*

(c)   *R is explicitly definable in T in terms of L.*

Proof. (c) $\Rightarrow$ (a) and (b1) $\Rightarrow$ (b2) are clear, and (a) $\Rightarrow$ (b1) holds by Corollary 10.2.2. To prove (b2) $\Rightarrow$ (c), assume (b2) holds with a model A, but (c) fails. Let $\bar{x}$ be a tuple of distinct variables whose length is the arity of R. Taking tuples $\bar{c}$ and $\bar{d}$ of new constants, consider the theory

(5.2) $T \cup \{\phi(\bar{c}) \leftrightarrow \phi(\bar{d}): \phi(\bar{x})$ a formula of $L\} \cup \{R\bar{c}, \neg R\bar{d}\}$.

We claim that (5.2) has a model. For if not, then compactness gives us a finite set of formulas $\phi_0, \ldots, \phi_{n-1}$ of L such that

(5.3) $T \vdash \forall \bar{x}\bar{y}\left(\bigwedge_{i<n} (\phi_i(\bar{x}) \leftrightarrow \phi_i(\bar{y})) \rightarrow (R\bar{x} \rightarrow R\bar{y})\right).$

It follows that according to T, we know whether a tuple satisfies R as soon as we know which of the formulas ϕ_i it satisfies. So R does after all have an explicit definition in T in terms of L; contradiction. Hence (5.2) has a model as claimed.

Since T is complete, $A|L \equiv C|L$. So we can expand $A|L$ into a model A' of (5.2), quoting Exercise 10.1.23 if $A|L$ is λ-big and Theorem 10.4.9 if $A|L$ is saturated of cardinality $\geq |L^+|$. Let $\bar{a}$, $\bar{b}$ be respectively the tuples $\bar{c}^{A'}$, $\bar{d}^{A'}$. Then $(A|L, \bar{a}) \equiv (A|L, \bar{b})$. But $A|L$ is ω-big (quoting Corollary 10.4.12 in the saturated case) and hence strongly ω-homogeneous by Exercise 10.1.4. So there is an automorphism e of $A|L$ which takes $\bar{a}$ to $\bar{b}$. But $A' \vDash R\bar{a} \wedge \neg R\bar{b}$ contradicting (b2). $\square$

We already knew the equivalence of (a) and (c) from Beth's definability theorem (Theorem 6.6.4), which doesn't even need T to be complete. But the added clause (b2) allows us to draw two useful corollaries about automorphisms. Each implies the other, and together they are known as **Svenonius' theorem**.

Corollary 10.5.2. *Let L, L^+ be first-order languages with $L \subseteq L^+$, and R a relation symbol of L^+. Let T be a complete theory in L^+. Suppose that for every model A of T and every automorphism e of $A|L$, $e(R^A) = R^A$. Then R is explicitly definable in T in terms of L.*

Proof. The set of consequences of T in L has an ω-big model; so the hypothesis implies (b2) of the theorem. $\square$

Corollary 10.5.3. *Let L, L^+ be first-order languages with $L \subseteq L^+$, and R a relation symbol of L^+. Let T be a theory in L^+ which is consistent but not necessarily complete. Suppose that for every model A of T and every automorphism e of $A|L$, $e(R^A) = R^A$. Then there is a finite set of formulas $\phi_0(x), \ldots, \phi_{k-1}(\bar{x})$ of L such that T implies the sentence*

(5.4) $\bigvee_{i<k} \forall \bar{x}(R\bar{x} \leftrightarrow \phi_i(\bar{x})).$

Proof. By the previous corollary, every completion S of T implies some explicit definition of R of the form $\forall \bar{x}(R\bar{x} \leftrightarrow \phi_S(\bar{x}))$. So T implies the disjunction of all these explicit definitions, and hence by compactness it implies a disjunction of finitely many of them. □

In stability theory one is constantly looking for definitions of relations inside big models, and the following version of Theorem 10.5.1 is particularly helpful.

Corollary 10.5.4. *Let M be a big model in a first-order language L. Let X be a set of elements of M, and R a relation on the domain of M. Suppose that both the following hold.*
(a) *For every automorphism e of M, if e fixes X pointwise then e fixes R setwise. (In symbols,* $\mathrm{Aut}(M)_{(X)} \subseteq \mathrm{Aut}(M)_{\{R\}}$.*)*
(b) *R is definable in M with parameters.*
Then R is X-definable.

Proof. This is a typical 'big model' argument. Since we are not told the λ for which M is λ-big, we are allowed to assume that it's infinite and greater than $|X|$. Then if $\bar{a}$ is a sequence listing the elements of X, $(M, \bar{a})$ is still λ-big and hence ω-big.

Suppose now that R is defined by a formula $\phi(\bar{x}, \bar{b})$ where $\bar{b}$ is a tuple of parameters from M. Let T be $\mathrm{Th}(M, \bar{a}, \bar{b}, R)$. Let $(M, \bar{a}, \bar{b}', R')$ be any expansion of $(M, \bar{a})$ to a model of T (using Exercise 10.1.23 again). We claim that $R = R'$. Now $\mathrm{tp}(\bar{b}/X) = \mathrm{tp}(\bar{b}'/X)$, and so (since $(M, \bar{a})$ is ω-big) there is an automorphism f of $(M, \bar{a})$ which takes $\bar{b}$ to $\bar{b}'$ and hence takes R to R'. But by (a), this implies that $R = R'$. This proves the claim, and it follows at once from Theorem 10.5.1(b2) that R is X-definable. □

Example 1: *Centraliser-connected components of stable groups.* Suppose the big model M is stable (section 6.7) and group-like (section 5.7). The **centraliser-connected component** M^c of M is the intersection of all subgroups of finite index which are centralisers of elements of M. Then M^c is a characteristic subgroup; all automorphisms of M fix it setwise. Also the centraliser of an element b is a definable subgroup, using the element b as parameter. So by the Baldwin–Saxl lemma (Lemma 6.7.7), M^c is the intersection of finitely many definable subgroups, and hence it is definable with parameters. Both the conditions of Corollary 10.5.4 are met, and so we deduce that M^c is *definable without parameters*.

We used the next corollary for eliminating imaginaries in section 4.4.

Corollary 10.5.5. *Let T be a theory in a first-order language L. Suppose that for every model A of T, every equivalence formula θ of A and every tuple $\bar{a}$ in A, there is a tuple $\bar{b}$ in A such that*

(5.5)   *for every automorphism g of A, g fixes the set $\theta(A^n, \bar{a})$
          (setwise) if and only if $g\bar{b} = \bar{b}$.*

Then T has semi-uniform elimination of imaginaries.

Proof. By the argument of Theorem 4.4.1 we need only prove semi-uniform
elimination of imaginaries for equivalence formulas. Let M be a big model of
T, $\theta(\bar{x}, \bar{y})$ an equivalence formula of M and $\bar{a}$ a tuple in M. Take a tuple $\bar{b}$
as in (5.5). By Corollary 10.5.4, $\theta(M^n, \bar{a})$ is definable in M with $\bar{b}$ as
parameters, and so for some formula $\phi(\bar{x}, \bar{z})$ of L we have

(5.6) $M \vDash \forall \bar{x}(\theta(\bar{x}, \bar{a}) \leftrightarrow \phi(\bar{x}, \bar{b}))$.

Let $\Phi(\bar{z})$ be the type of $\bar{b}$ over the empty set with respect to M. We claim
that if $\bar{b}'$ is any tuple such that

(5.7) $M \vDash \bigwedge \Phi(\bar{b}') \wedge \forall \bar{x}(\theta(\bar{x}, \bar{a}) \leftrightarrow \phi(\bar{x}, \bar{b}'))$,

then $\bar{b} = \bar{b}'$. For since $(M, \bar{b}) \equiv (M, \bar{b}')$ and M is big, there is some
automorphism g of M which takes $\bar{b}$ to $\bar{b}'$. By (5.6) and (5.7), g fixes
$\theta(M^n, \bar{a})$ setwise, and so by (5.4), $\bar{b} = \bar{b}'$. Thus

(5.8) $M \vDash \forall \bar{z}\bar{z}'(\bigwedge \Phi(\bar{z}) \wedge \bigwedge \Phi(\bar{z}') \wedge \forall \bar{x}(\theta(\bar{x}, \bar{a}) \leftrightarrow \phi(\bar{x}, \bar{z})) \wedge$
          $\forall \bar{x}(\theta(\bar{x}, \bar{a}) \leftrightarrow \phi(\bar{x}, \bar{z}')) \to \bar{a} = \bar{z}')$.

Since M is big, it follows (see Exercise 10.1.18) that for some formula
$\chi(\bar{z}) \in \Phi$, we can replace $\phi(\bar{x}, \bar{z})$ in (5.6) by $\phi(\bar{x}, \bar{z}) \wedge \chi(\bar{z})$, with the effect
that $\bar{b}$ is the unique tuple in M which makes (5.6) true.

It follows that

(5.9) $T \vdash$ 'θ defines an equivalence relation' $\to$

$$\forall \bar{y} \left(\bigvee_{\phi \text{ in } L} \exists \bar{z} \; (\text{'}\bar{z} \text{ is the unique tuple such that} \right.$$

$$\left. \forall \bar{x}(\theta(\bar{x}, \bar{y}) \leftrightarrow \phi(\bar{x}, \bar{z}))\text{'}) \right).$$

By compactness the disjunction can be shrunk to a finite set $\{\phi_i : i < m\}$. As
in (4.3) of section 4.4, we want just one of the ϕ_i to work in any instance.
This can be arranged by replacing each formula ϕ_i by a formula $(\phi_i \wedge$ 'ϕ_i
works for defining $\theta(\bar{x}, \bar{y})$, and no formula ϕ_j ($j < i$) does'). $\qquad \square$

Second definability theorem

Suppose L, L^+ are first-order languages with $L \subseteq L^+$, R is a relation symbol
of L^+ and T is a complete theory in L^+. We say that R is **almost definable**
in T in terms of L if there is a formula $E(\bar{x}, \bar{y})$ of L such that $T \vdash$ 'E defines
an equivalence relation with n equivalence classes' for some $n < \omega$, and in
every model A of T, R^A is a union of equivalence classes of E^A. An
equivalence relation with finitely many classes is known as a **finite equivalence
relation**.

Theorem 10.5.6. *Let* L, L^+ *be first-order languages with* $L \subseteq L^+$, *and* R *a relation symbol of* L^+. *Let* T *be a complete theory in* L^+, *and* λ *an uncountable cardinal greater than the number of symbols in* $L^+ \backslash L$. *Then the following are equivalent.*

(a) *If* A_i $(i \in I)$ *are any family of models of* T *with* $A_i | L = A_j | L$ *for all* i, $j \in I$, *then there are only finitely many distinct relations* R^{A_i} $(i \in I)$.

(b) *There is a model* A *of* T *such that* $A | L$ *is either* λ-*big or saturated of cardinality* $\geq |L^+|$, *and if* A' *is any expansion of* $A | L$ *to a model of* T *then* $R^{A'}$ *has only finitely many conjugates under* $\mathrm{Aut}(A | L)$.

(c) R *is almost definable in* T *in terms of* L.

Proof. (c) $\Rightarrow$ (a) is clear, and (a) $\Rightarrow$ (b) holds by Corollary 10.2.2. Before we close the circle, it will be helpful to show that (b) implies (a).

Assume (b) holds with a model A. For any number $n < \omega$, let T_n be a copy of T got by replacing R by a new symbol R_n, and leaving the other symbols unchanged. We claim that there is a largest n such that the theory

(5.10) $T_0 \cup \ldots \cup T_{n-1} \cup \{\exists \bar{x}(R_i(\bar{x}) \leftrightarrow \neg R_j(\bar{x})): i < j < n\}$

is consistent. For if not, $\bigcup_{n<\omega} T_n$ is consistent, so by Theorem 10.4.2 it has a special model B of cardinality $> |T|$. The reducts $(B|L, R_n^B)$ are elementarily equivalent since T is complete, and by Theorem 10.4.5(b) they are special. So by Theorem 10.4.4 there are isomorphisms $i_n: (B|L, R_0^B) \rightarrow (B|L, R_n^B)$. Write $T^@$ for the complete theory of the structure $(B, i_1, i_2, \ldots)$. Since T is complete, $T^@$ has the same consequences in L as T, and so A can be expanded to a model $A^@$ of $T^@$, either by bigness or by saturation. Then $R^{A^@}$ has infinitely many conjugates under $\mathrm{Aut}(A|L)$, contradicting the choice of A. This proves the claim.

Henceforth let n be the largest n such that (5.10) is consistent. Then (a) follows, with n as a uniform upper bound on the number of distinct relations R^{A_i}.

Still assuming (b), we prove (c). Let $\bar{x}$, $\bar{y}$ be tuples of distinct variables whose length is the arity of R. Let T^* be (5.10) together with the definition

(5.11) $\forall \bar{x} \bar{y} \left(E(\bar{x}, \bar{y}) \leftrightarrow \bigwedge_{i<n} (R_i(\bar{x}) \leftrightarrow R_i(\bar{y})) \right).$

In any model C of T^*, E^C is an equivalence relation with at most 2^n equivalence classes.

We show that E is definable in T^* in terms of L. For this we use Theorem 10.5.1 above. Let C and C' be two models of T^* with $C|L = C|L'$. Then by the definition of T^*, the two sets $\{R_i^C: i < n\}$ and $\{R_i^{C'}: i < n\}$ are both the set of all possible interpretations of R in expansions of $C|L$ to models of T, and hence $E^C = E^{C'}$ as required.

If we now replace E by the formula of L which defines it in T^*, it follows at once that R is almost definable in T in terms of L. $\qquad \square$

Now that we have Theorem 10.5.6, we can quickly find several other equivalent conditions.

Corollary 10.5.7. *As in the theorem, let L, L^+ be first-order languages with $L \subseteq L^+$, and R a relation symbol of L^+. Let T be a complete theory in L^+. Then the following are each equivalent to (a)–(c) of the theorem.*
(d) *There are $k < \omega$ and formulas $\sigma(\bar{z})$ and $\phi_i(\bar{x}, \bar{z})$ of L $(i < k)$ such that $T \vdash \exists \bar{z} \sigma(\bar{z})$ and $T \vdash \forall \bar{z}(\sigma(\bar{z}) \to \bigvee_{i<k} \forall \bar{x}(R\bar{x} \leftrightarrow \phi_i(\bar{x}, \bar{z})))$.*
(e) *If A is any model of T and B an elementary substructure of A, then R^A is definable by a formula $\phi(\bar{x}, \bar{b})$ where ϕ is in L and $\bar{b}$ are in B.*
(f) *There is some fixed upper bound (independent of the choice of A) on the number of conjugates of R^A under $\mathrm{Aut}(A|L)$ in any model A of T.*
(g) *In every model A of T, R^A has finitely many conjugates under $\mathrm{Aut}(A|L)$.*

Proof. First we prove $(d) \Rightarrow (e) \Rightarrow (f) \Rightarrow (g)$. A compactness argument shows that (f) implies (g). Next, (e) implies (f). For let B be some fixed model of T and A any model of T. Some special elementary extension C of A contains a copy of B as an elementary substructure, and so by (e) the number of conjugates of R^C under $\mathrm{Aut}(C|L)$ is at most the number of formulas of L with parameters from B, i.e. $|B| + |L|$. Since C is special, it follows that R^A has at most $|B| + |L|$ conjugates under $\mathrm{Aut}(A|L)$ too. Finally (d) implies (e) at once.

Now we can join up the ends. Since $|L^+|^+$-big models exist, (g) implies (b). At the other end, suppose (c) holds, so that R is almost definable in T in terms of L; let $E(\bar{x}, \bar{y})$ be the formula of L which defines the required equivalence relation. Since E has finitely many equivalence classes, each of these classes is represented in every model of T. Let $\sigma(\bar{z})$ say '$\bar{z}$ is a list $\bar{z}_0 \ldots \bar{z}_{n-1}$ of representatives of all the equivalence classes of E', and for each subset S of n let $\phi_S(\bar{x}, \bar{z})$ say '$R\bar{x}$ holds if and only if $E(\bar{x}, \bar{z}_i)$ holds for some $i \in S$'. Thus we have (d).   □

Apart from condition (b) in Theorem 10.5.6, nothing changes when we drop the assumption that T is complete.

Corollary 10.5.8. *Let L, L^+ be first-order languages with $L \subseteq L^+$, and R a relation symbol of L^+. Let T be a consistent theory in L^+. Then all the conditions (a), (c), (d), (e), (f), (g) of Theorem 10.5.6 and Corollary 10.5.7 remain equivalent.*

Proof. Each of (a), (e), (f), (g) implies that every completion of T satisfies the conditions of the theorem. So by Corollary 10.5.7(d) and compactness, T

implies a finite disjunction $\bigvee_{j<m}\{\exists\bar{z}_j\sigma_j(\bar{z}_j) \land \forall\bar{z}_j(\sigma_j(\bar{z}_j) \rightarrow \bigvee_{i<k}\forall\bar{x}(R\bar{x} \leftrightarrow \phi_{ij}(\bar{x}, \bar{z}_j)))\}$. Write $\sigma(\bar{z}_0 \ldots \bar{z}_{m-1})$ for the conjunction of all the formulas $\exists\bar{z}_j\sigma_j(\bar{z}_j) \rightarrow \sigma_j(\bar{z}_j)$. Then T implies

$$\forall\bar{z}_0 \ldots \bar{z}_{m-1}\left(\sigma(\bar{z}_0 \ldots \bar{z}_{m-1}) \rightarrow \bigvee_{j,i} \forall\bar{x}(R\bar{x} \leftrightarrow \phi_{ij}(\bar{x}, \bar{z}_j))\right),$$

giving (d). One easily checks that (d) implies (a), which completes the circle. A similar argument deals with (c). $\qquad\square$

The following corollary is important in the foundations of stability theory. Let A be an L-structure, X a set of elements of A and R a relation on $\text{dom}(A)$. We say that R is **almost definable over** X **in** A if there is a formula $E(\bar{x}, \bar{y})$ of L with parameters in X, which defines an equivalence relation E^A in A with finitely many classes, and R is a union of equivalence classes of E^A.

Corollary 10.5.9. *Let M be a big model in a first-order language L. Let X be a set of elements of M, and R a relation on the domain of M. Suppose that both the following hold.*
(a) *The number of conjugates of R under automorphisms of M which fix X pointwise is finite. (In group-theoretic terms,* $\text{Aut}(M)_{(X)} \cap \text{Aut}(M)_{\{R\}}$ *has finite index in* $\text{Aut}(M)_{(X)}$.)
(b) *R is definable in M with parameters.*
Then R is almost definable over X in M.

Proof. This is the analogue of Corollary 10.5.4, and its proof is similar. $\qquad\square$

Exercises for section 10.5

1. With reference to Corollary 10.5.7, show (i) in condition (d) we can always choose $\sigma(\bar{z})$ so that $\exists\bar{z}\,\sigma(\bar{z})$ is a logical truth, (ii) nevertheless we can't replace (d) by the simpler condition: there are $k < \omega$ and formulas $\phi_i(\bar{x}, \bar{z})$ of L $(i < k)$ such that $T \vdash \forall\bar{z} \bigvee_{i<k} \forall\bar{x}(R\bar{x} \leftrightarrow \phi_i(\bar{x}, \bar{z}))$.

Working in a big model M of a theory in the language L, we say that a formula $\phi(\bar{x}, \bar{b})$ with parameters $\bar{b}$ is **almost over** *a set X if there is a formula $E(\bar{x}, \bar{y})$ of L with parameters from X, which defines in M an equivalence relation with a finite number of equivalence classes, and for every two tuples $\bar{a}$, $\bar{a}'$ of elements of M, $M \vDash E(\bar{a}, \bar{a}') \rightarrow (\phi(\bar{a}, \bar{b}) \leftrightarrow \phi(\bar{a}', \bar{b}))$.*
2. Show that $\phi(\bar{x}, \bar{b})$ is almost over X if and only if the number of conjugates of $\phi(\bar{x}, \bar{b})$ under $\text{Aut}(M)_{(X)}$ is finite.

3. (Still in the big model M.) Show that if a formula $\phi(\bar{x}, \bar{y}, \bar{b})$ is almost over X and defines an equivalence relation E in M, then there is a formula with parameters from X which defines an equivalence relation refining E.

4. Let R be a relation on the domain of a big L-structure A, and X a set of elements of A. Show that R has finitely many conjugates under $\mathrm{Aut}(A)_{(X)}$ if and only if in A^{eq}, R is definable over $\mathrm{acl}(X)$.

5. Let L and L^+ be first-order languages with $L \subseteq L^+$, and suppose L^+ contains relation symbols $R_0, \ldots, R_{n-1}$. Let T be a complete theory in L^+. Show that the following are equivalent. (a) None of $R_0, \ldots, R_{n-1}$ are almost definable in T in terms of L. (b) There is a family of infinitely many models A_i $(i < \omega)$ of T such that if $i < j < \omega$ then $A_i | L = A_j | L$ but $R_m^{A_i} \neq R_m^{A_j}$ for each $m < n$.

An automorphism σ of a structure A is **strong over** *the set of elements X if σ setwise fixes each equivalence class of every finite equivalence relation which is definable over X. We write $\mathrm{Autf}(A)_{(X)}$ for the set of strong automorphisms of A over X. (The f is for 'fort' – this notation came from France.)*

*6. (a) Show that $\mathrm{Autf}(A)_{(X)}$ is always a normal subgroup of $\mathrm{Aut}(A)_{(X)}$. (b) We write $G(A/\bar{a})$ for $\mathrm{Aut}(A)_{(\bar{a})}/\mathrm{Autf}(A)_{(\bar{a})}$. Show that if $\bar{a}$ is a sequence of elements of A, then $G(A/\bar{a})$ is determined up to isomorphism by $\mathrm{Th}(A, \bar{a})$; in fact $G(A/\bar{a})$ is the automorphism group of the algebraic closure of $\bar{a}$ in A^{eq}. (c) Show that if A is a countable ω-categorical structure then $G(A/\varnothing)$ has cardinality either a positive integer or 2^ω; show that all these possibilities occur.

10.6 Resplendence

Let L be a first-order language. We say that an L-structure A is **resplendent** if

(6.1)   for every first-order language $L^+ \supseteq L$, every formula $\phi(\bar{x})$ in L^+ and every tuple $\bar{b}$ of elements of A, if some elementary extension B of A can be expanded to an L^+-structure B^+ so that $B^+ \vDash \phi(\bar{b})$, then A can be expanded to an L^+-structure A^+ so that $A^+ \vDash \phi(\bar{a})$.

Resplendence is the same thing as 0-bigness for a single sentence at a time.

Resplendent structures have two main uses. First, Barwise and Schlipf [1976] proposed them as a teaching aid. They do give tidy proofs of elementary theorems, and I shall illustrate that in a moment. Second, they are prominent in the model theory of arithmetic. At first sight this is a specialised field. But if we are studying expansions of a structure, we may as well expand it right away to a model of arithmetic, and so in a way the specialised methods are available for general use. The proof of the Kleene–Makkai theorem (Theorem 10.6.3) is a typical example.

There has also been some recent work to see how resplendence fits in among the other notions of this chapter. The deepest result is that of Buechler [1984]: every resplendent model of an ω-stable theory is homogeneous. Early claims that resplendent models would be 'useful in model theory' led to a reaction from Poizat [1986], who maintained that the notion 'n'aura jamais de signification pour un mathématicien normal'.

Theorem 10.6.1. *Let L be a countable language and A an L-structure. Then A has an elementary extension of cardinality $|A|$ which is resplendent.*

Proof. If A is finite then trivially A is resplendent, since any structure elementarily equivalent to A is isomorphic to A. Suppose then that A is infinite. Repeat the construction of Theorem 10.2.1. Only $|L|$ stages are needed, one for each sentence of a possible L^+ extending L. So the resulting structure has cardinality $|A|$ as required. $\square$

Suppose that L is a recursive first-order language. We say that an L-structure A is **recursively saturated** if

(6.2)   for every recursive set $\Phi(x, \bar{y})$ of formulas of L and every tuple $\bar{b}$ in A, if $\Phi(x, \bar{b})$ is a type with respect to A then it is realised in A.

Just as λ-bigness implies λ-saturation, so resplendence implies recursive saturation. (But where did the recursion theory spring from? The notion of resplendence seems purely model-theoretic. I can shed no light on this.)

Theorem 10.6.2. *Let L be a recursive first-order language and A an L-structure.*

   (a) *If A is resplendent then A is recursively saturated.*

   (b) *If A is recursively saturated and at most countable then A is resplendent.*

Proof. (a) Finite structures are saturated, and so we can assume that A is infinite. Suppose $\bar{b}$ is a tuple of elements of A and $\Phi(x, \bar{b})$ is a recursive type with respect to A. Introduce a new constant c; we have to show that $(A, \bar{b})$ can be expanded to a model of the theory $\Phi(c, \bar{b})$. Here we quote a theorem of Kleene and Makkai (Theorem 10.6.3 below), which says that since $\Phi(c, \bar{b})$ is recursive, there is a first-order sentence σ, maybe with extra symbols, such that an infinite structure B is a model of $\Phi(c, \bar{b})$ if and only if B can be expanded to a model of σ. Now since $\Phi(x, \bar{b})$ is a type, it is realised by an element of some elementary extension $(B, \bar{b})$ of $(A, \bar{b})$. So $(B, \bar{b})$ expands to a model of $\Phi(c, \bar{b})$, and hence also to a model of σ. Since A is resplendent, this implies that $(A, \bar{b})$ expands to a model of σ, say with c naming an element a. By choice of σ, the element a realises $\Phi(x, \bar{b})$.

   (b) Suppose A is recursively saturated and at most countable, $\bar{a}$ is a tuple of elements of A, and some elementary extension $(A', \bar{a})$ of $(A, \bar{a})$ can be expanded to a model of a sentence ϕ of a first-order language extending L.

We must show that $(A, \bar{a})$ can be expanded to a model of ϕ too. If A is finite, $A' = A$ and we are done. Henceforth we suppose that A has cardinality ω. Recursive saturation is not affected by adding a finite number of parameters; so we can replace A by $(A, \bar{a})$ and assume that ϕ has no parameters.

To show that A can be expanded to a model of ϕ, we use the proof of Lemma 10.4.8 above. By that lemma there is a Keisler sentence χ of length ω, such that every infinite model of ϕ is a model of χ, and every countable L-structure which is a model of χ can be expanded to a model of ϕ. Let χ be $\forall x_0 \exists y_1 \forall x_2 \ldots \bigwedge \Phi$. Inspection of the proof of Lemma 10.4.8 shows that we can arrange that Φ is a recursive set of formulas. By assumption $A' \vDash \chi$, and so $A' \vDash \bigwedge \mathrm{app}(\chi)$ by Theorem 10.3.7(a). Since $A \preccurlyeq A'$, we infer that $A \vDash \bigwedge \mathrm{app}(\chi)$. The proof of Theorem 10.3.7(b) tells us how to deduce that $A \vDash \chi$. Namely, we need only show that for each odd $i < \omega$ and i-tuple $\bar{b}$ in A, if $A \vDash \bigwedge \mathrm{app}(\chi^i)(\bar{b})$ then there is an element realising the type $\{\theta(\bar{b}, y_i) : \exists y_i \, \theta(x_0, \ldots, x_{i-1}, y_i) \in \mathrm{app}(\chi^{i+1})\}$. Each of these types is recursive.   □

Easy proofs of easy results

We want to prove the Łoś–Tarski theorem. As in section 6.5 above, the main task is to show that

(6.3)   if B and C are L-structures with $C \Rrightarrow_1 B$, then some structure $C' \equiv C$ is embeddable in some structure $B' \equiv B$.

For simplicity we suppose that L is a recursive first-order language. (If we only want to show that sentences preserved in substructures are elementarily equivalent to $\forall_1$ sentences, we can make do with a finite signature, and then L will certainly be recursive.) Also we can assume without loss that B and C are finite or countable.

We build a structure A with two disjoint parts: a copy of B and a copy of C. By Theorem 10.6.1, A has an elementary extension $A' = (B', C')$ which is resplendent and at most countable. By Theorem 10.6.2(a), A' is recursively saturated. We have $B' \equiv B$ and $C' \equiv C$ by the relativisation theorem (Theorem 5.1.1). Let $\bar{c} = c_0 c_1 \ldots$ be a sequence listing the elements of C'; we need to find a sequence $\bar{b} = b_0 b_1 \ldots$ of elements of B' such that for every $\exists_1$ formula $\phi(x_0, \ldots, x_{n-1})$ of L,

(6.4) $C' \vDash \phi(c_0, \ldots, c_{n-1}) \Rightarrow B' \vDash \phi(b_0, \ldots, b_{n-1})$.

Relativising, we can write (6.4) as

(6.5) $A' \vDash \phi^{C'}(c_0, \ldots, c_{n-1}) \rightarrow \phi^{B'}(b_0 \ldots, b_{n-1})$.

We go by induction. Suppose $b_0, \ldots, b_{n-1}$ have been chosen. Let $\Phi(x_0, \ldots, x_n, y_0, \ldots, y_n)$ be the recursive set of formulas

(6.6) $\{\phi^{C'}(x_0, \ldots, x_n) \rightarrow \phi^{B'}(y_0, \ldots, y_n): \phi(x_0, \ldots, x_n)$ is an $\exists_1$
formula of $L\}$.

It suffices to check that $\Phi(c_0, \ldots, c_n, b_0, \ldots, b_{n-1}, x)$ is a type, since then we can invoke recursive saturation and choose b_n to be an element of A' which realises this type. But if $\Phi(c_0, \ldots, c_n, b_0, \ldots, b_{n-1}, x)$ is not a type, then there is a finite set of $\exists_1$ formulas $\phi_i(x_0, \ldots, x_n)$ $(i < k)$ such that

(6.7) $C' \vDash \bigwedge_{i<k} \phi_i(c_0, \ldots, c_n)$ but $B' \vDash \neg \exists x \bigwedge_{i<k} \phi_i(b_0, \ldots, b_{n-1}, x)$.

Since $\exists x \bigwedge_{i<k} \phi_i(x_0, \ldots, x_{n-1}, x)$ is an $\exists_1$ formula, (6.7) contradicts the inductive assumption (6.4). This concludes the proof.

The reference to Theorem 10.6.2(a) contained a hidden appeal to the Kleene–Makkai theorem. To avoid this, one can show directly that every countable structure has a countable recursively saturated elementary extension. But then of course any appeal to resplendence would have to be justified by climbing back again through the other half of Theorem 10.6.2. It could be hard to use these methods in an elementary class without taking quite a lot for granted. Still, the approach is neat. Schlipf [1978] gives several further examples of this style of reasoning.

The Kleene–Makkai theorem

Theorem 10.6.3 (Theorem of Kleene and Makkai). *Let L be a first-order language with finite signature and T a recursive theory in L. Then there are a first-order language $L^+ \supseteq L$ and a sentence σ of L^+ such that for every infinite L-structure A, $A \vDash T$ if and only if some expansion of A is a model of σ.*

Proof. I give the broad outline; there are several details to be checked. We can take some of the weight off our feet by quoting Theorem 5.2.1(b), which says that relativised expansions will serve in place of expansions. Thus it suffices to find a suitable sentence σ which describes a two-sorted structure; the first sort will be the L-structure A and the second will be a model of arithmetic. The expanded language L^+ will have relation symbols to describe the two sorts, and symbols 0, S (successor), $+$ and $\cdot$ for talking about numbers. It will also have a pairing function π on the domain of A, and a 2-ary relation symbol Sat. As a shorthand, if $\bar{a}$ is an n-tuple $(a_0, \ldots, a_{n-1})$ of elements of A, then I write $\pi\bar{a}$ for the element $\pi(\ldots \pi(\pi(a_0, a_1), a_2), \ldots), a_{n-1})$.

The language L is recursive, so that every expression e of L has a Gödel number $GN(e)$. If B is any L^+-structure, we say that Sat^B is a **satisfaction predicate** if for every formula $\phi(x_0, \ldots, x_{n-1})$ of L and every n-tuple $\bar{a}$ from the first sort of B,

(6.8) $B \vDash \phi(\bar{a}) \Leftrightarrow B \vDash Sat(\mathrm{GN}(\phi), \pi(\bar{a}))$.

Since L has finite signature, we can write down a sentence θ_0 of L^+ which expresses the usual inductive definition of satisfaction (clauses (1.5)–(1.10) of section 2.1), using the symbol Sat. Since T is recursive, we can write down a formula $\psi(x)$ of arithmetic such that $\psi(\mathbb{N})$ is exactly the set of Gödel numbers of sentences in T. Let θ_1 be the sentence

(6.9)   ($\forall x$ of first sort)($\forall n$ of second sort)($\psi(n) \to Sat(n, x)$).

Let σ be the conjunction of θ_0, θ_1 and Robinson's arithmetic (Q in section 5.5).

   If A is a model of T, then we can make A into an L^+-structure A^+ by adding the natural numbers as second sort, and interpreting Sat as $\vDash$. Clearly then $A^+ \vDash \sigma$. The other direction needs more care. Suppose B is an expansion of A to an L^+-structure, and $B \vDash \sigma$. Then in general the second sort of B will contain some nonstandard numbers as well as the standard ones. However, Σ_1^0 formulas are preserved in end-extensions (Exercise 2.4.8), and every true Σ_1^0 sentence of arithmetic is provable from Robinson's arithmetic. So we should take care that all the universal quantifiers of Sat which range over natural numbers are bounded. Then σ will imply that *at least* all the sentences in T are true in A; and this is enough.   □

Exercises for section 10.6

1. Show that if A is countable, recursively saturated and contains (as relativised reducts) two structures B and C with $B \equiv C$, then $B \cong C$. [To avoid special models etc., use recursive saturation to show that B and C are back-and-forth equivalent.]

2. Use the previous exercise to give a simple proof of Craig's interpolation theorem.

3. Improving Theorem 10.6.2(b), show that if A is a recursively saturated and countable L-structure, then for every sentence σ of a language L^+ extending L (maybe with parameters from A), if some elementary extension of A can be expanded to a model of σ, then A can be expanded to a resplendent model of σ. [Build up a chain of languages (L_i: $i < \omega$) so that each L_i has new constants $c_0, \ldots, c_{i-1}$ and also new relation symbols. Inductively build a corresponding chain (T_i: $i < \omega$) of theories so that $\bigcup_{i<\omega} T_i$ will be the complete diagram of a resplendent structure whose elements are $c_0, c_1, \ldots$. The trick is to use recursive saturation to choose elements $a_0, a_1, \ldots$ of A as you go, so that each T_i contains eldiag($A, a_0, \ldots, a_{i-1}$) and each element of A appears as an a_i. Thus $\bigcup_{i<\omega} T_i$ will describe an expansion of A.]

4. Peano arithmetic has definable Skolem functions (see section 3.1 above). If A is a model of Peano arithmetic and X is a set of elements of A, write $\langle X \rangle$ for the closure of X under the definable Skolem functions. Show that if A is nonstandard and X is finite then $\langle X \rangle$ is a model of Peano arithmetic which is not recursively saturated.

5. Let us say that a formula is n-**prenex** if it has the form $Q_1 \ldots Q_n\phi$ where each Q_i is a quantifier $\forall x_i$ or $\exists x_i$, and ϕ is quantifier-free. Let A be a nonstandard model of Peano arithmetic and $\Phi(x)$ a type over $\varnothing$ which is recursive and consists of n-prenex formulas for some fixed n. Show that Φ is realised in A. [A satisfaction predicate for one-variable n-prenex formulas is definable in the language of arithmetic. For every standard n there is an element satisfying the first n formulas in Φ. Now use overspill; see section 11.4.]

6. Show that if $(A, +, *)$ is a model of Peano arithmetic then both $(A, +)$ and $(A, *)$ are recursively saturated. [For $(A, +)$ use quantifier elimination (see Theorem 3.3.8) to express any recursive type for $(A, +)$ as a recursive quantifier-free type, and use a universal predicate for quantifier-free formulas. The argument for $(A, *)$ is similar.]

7. Let L be a recursive language. Show that there is an r.e. theory T in a recursive language $L^+ \supseteq L$, such that for every L-structure A, A is recursively saturated if and only if A can be expanded to a model of T.

*8. Show that if L is a first-order language of finite signature and T is an unstable complete theory in L, then every uncountable homogeneous resplendent model of T is saturated. [This needs the Ehrenfeucht–Mostowski theorem, Theorem 11.2.8 below, to show that T has models with indiscernible sequences ordered by a formula in the order-type of the rationals. Using the previous exercise, find an r.e. theory T_1 in a language extending L, so that if A is any model of T_1 then A is an expansion of a recursively saturated structure with an indiscernible sequence S of tuples ordered in the order-type of the rationals, and named by constants $\bar{a}_i$ $(i < \omega)$. Every set X of natural numbers can be coded up as a cut in S. If $A|L$ is ω_1-homogeneous, some element c lies in the cut, and X is decodable from c. Thus recursive saturation implies every n-type over $\varnothing$ is realised, so that Theorem 10.1.8 applies.]

9. Let L be a recursive first-order language and T a recursive theory in L. Show that there are a first-order language $L^+ \supseteq L$ and a sentence σ of L^+ such that for every sentence ϕ of L, $T \vdash \phi$ if and only if $\sigma \vdash \phi$.

10. Let L be a recursive first-order language and **K** a class of infinite L-structures. Show that the following are equivalent. (a) **K** is the class of all models of some recursive theory in L. (a) **K** is both a PC class (see section 5.2) and a first-order axiomatisable class of L-structures.

10.7 Atomic compactness

Atomic compactness is saturation for positive primitive formulas, but with an extra twist: there is no bound on the number of parameters.

For a model theorist the most interesting examples of atomic compactness are atomic compact modules, or (to use their more familiar name) pure-injective modules. There are several reasons for this interest. One is that every module has an atomic compact elementary extension (by Theorem 10.7.3 below); this provides a bridge between algebraic and first-order properties. Another is that atomic compact modules have nice decompositions, generalising the well-known Matlis–Papp decomposition theorem for injective modules over a noetherian ring. See Appendix, section A.1 and its references.

Here I confine myself to characterisations and existence proofs. The next step in the general theory would be to look at pure-injective hulls and other kinds of minimal atomic compact extension. Unfortunately there are several ways of defining these extensions; they are subtly inequivalent, and the right choice depends on the particular theory you have in mind. During the 1970s there was vigorous work in this area by universal algebraists; it is well surveyed in Wenzel [1979].

Let L be a first-order language and A an L-structure. Let us say that a set $\Phi(\bar{x})$ of formulas of L with parameters in A is **finitely realised** in A if for every finite subset Ψ of Φ, $A \vDash \exists \bar{x} \bigwedge \Psi$. Here $\bar{x}$ need not be finite. We say that A is **atomic compact** if for every set Φ of p.p. formulas of L with parameters in A, with any number of free variables, if Φ is finitely realised in A then Φ is realised in A. (When L has no relation symbols, atomic compact structures are also known as **equationally compact** structures.)

Our first theorem characterises atomic compact structures in various ways. '$A \Rrightarrow_1^+ B$' means that every p.p. sentence which is true in A is also true in B. An embedding $e: A \to B$ is **pure** if for every p.p. formula $\phi(\bar{x})$ and every tuple $\bar{a}$ in A, $B \vDash \phi(ea) \Rightarrow A \vDash \phi(\bar{a})$ (see section 2.5).

Theorem 10.7.1. *Let L be a first-order language and A an L-structure. Write $L(A)$ for the language got by adding the elements of A to L as parameters. The following are equivalent.*

(a) *A is atomic compact.*

(b) *Every set $\Phi(x)$ of p.p. formulas of $L(A)$, with just the variable x free, which is finitely realised in A is realised in A.*

(c) *Every set of atomic formulas of $L(A)$ in any number of variables which is finitely realised in A is realised in A.*

(d) *If C is any L-structure and $\bar{c}$ a sequence of elements of C, and $f: \langle \bar{c} \rangle_C \to A$ is a homomorphism such that $(C, \bar{c}) \Rrightarrow_1^+ (A, f\bar{c})$, then there is a homomorphism $g: C \to A$ extending f.*

(e) *If $f: A \to C$ is a pure embedding then there is a homomorphism $g: C \to A$ such that $gf = 1_A$.*

(f) *If $f: A \to C$ is an elementary embedding then there is a homomorphism $g: C \to A$ such that $gf = 1_A$.*

Proof. Trivially (a) implies both (b) and (c).

(b) $\Rightarrow$ (d). Let $f: \langle \bar{c} \rangle_C \to A$ be given as in (d). List as $\bar{b} = (b_i: i < \gamma)$ the elements of C. By induction on i, choose a sequence $\bar{a} = (a_i: i < \gamma)$ of elements of A so that

(7.1)   for every p.p formula $\phi(\bar{x}, \bar{y})$ of L, if $C \vDash \phi(\bar{c}, \bar{b})$ then
         $A \vDash \phi(f\bar{c}, \bar{a})$.

The argument is just as for Theorem 10.3.1, but with p.p. formulas. By the diagram lemma (Lemma 1.4.2) there is a homomorphism $g: C \to A$ such that $g\bar{c} = f\bar{c}$.

(d) implies (e) by taking $\langle \bar{c} \rangle_C$ to be the image of f. (e) clearly implies (f). To deduce (a) from (f), consider a set $\Phi(\bar{x}, \bar{y})$ of p.p. formulas of L, and a sequence $\bar{a}$ listing the elements of A, and suppose $\Phi(\bar{x}, \bar{a})$ is finitely realised in A. Then A has an elementary extension B with elements $\bar{b}$ such that $B \vDash \bigwedge \Phi(\bar{b}, \bar{a})$, by the elementary diagram lemma. (We proved this in Theorem 6.3.1(a), with the irrelevant assumption that $\bar{x}$ is a tuple.) By (f) there is a homomorphism $g: B \to A$ such that $g\bar{a} = \bar{a}$, and thus $A \vDash \bigwedge \Phi(g\bar{b}, \bar{a})$ since homomorphisms preserve p.p. formulas. So $g\bar{b}$ realises $\Phi(\bar{x}, \bar{a})$ in A.

Finally suppose (c) holds; we deduce (e). Given a pure embedding $f: A \to C$, let $\bar{a}$ list the elements of A and let $\bar{c}$ list the elements of C. Find a set $\Phi(\bar{x}, \bar{y})$ of atomic formulas of L such that $\Phi(\bar{c}, \bar{a}) = \mathrm{diag}^+ C$, using each element a of A as a name for the element fa. Since f is pure, $\Phi(\bar{x}, \bar{a})$ is finitely realised in A. So by (c) there are $\bar{b}$ in A such that $A \vDash \bigwedge \Phi(\bar{b}, \bar{a})$. The diagram lemma yields a homomorphism $g: C \to A$ such that $g\bar{c} = \bar{b}$ and $gf\bar{a} = \bar{a}$, as required. $\square$

A module with property (e) of the theorem is said to be **pure-injective**. Equivalently, a module A is pure-injective iff A is a direct summand of every module that contains it as a pure submodule. Pure-injective modules are sometimes known as **algebraically compact** modules.

Where do we find atomic compact structures? The next theorem gives us a few. (B is a **retract** of A if $B \subseteq A$ and there is a homomorphism $f: A \to B$ which is the identity on B.)

Theorem 10.7.2. (a) *Every finite structure is atomic compact.*

(b) *Every direct product of atomic compact structures is atomic compact.*

(c) *Every retract of an atomic compact structure is atomic compact.*

Proof. (a) If A is finite and $f: A \to C$ is an elementary embedding then f is an isomorphism, so that A is atomic compact by Theorem 10.7.1(f).

(b) By (a) we need only consider non-trivial products $\prod_I A_i$ of atomic compact structures. Write p_i for the ith canonical projection. Let $f: \prod_I A_i \to C$ be a pure embedding. Listing as $\bar{a}$ the elements of $\prod_I A_i$, for each $j \in I$ we have $(C, f\bar{a}) \Rightarrow_1^+ (\prod_I A_i, \bar{a}) \Rightarrow_1^+ (A_j, p_j\bar{a})$. So by Theorem 10.7.1(d) there is for each j a homomorphism $g_j \colon C \to A_j$ such that $g_j f = p_j$. By Theorem 9.1.1(b) the maps g_j induce a homomorphism $g \colon C \to \prod_I A_i$ so that for each j, $g_j = p_j g$. Then $p_j g f = g_j f = p_j$ for each j, whence $gf = 1_A$, proving by Theorem 10.7.1(e) that $\prod_I A_i$ is atomic compact.

(c) is left to the reader.   □

For some types of structure, Theorem 10.7.2 exhausts the possibilities. Thus no infinite field is atomic compact: if A is an infinite field then A has a proper elementary extension C, and no homomorphism $g \colon C \to A$ can be the identity on A. For a similar reason, no existentially closed group is atomic compact, since any such group has proper pure extensions which are simple groups (see Exercise 9 below).

However, there is a model-theoretic hypothesis that gives plenty of atomic compact structures. Let us say that an L-structure A is **p.p. normal** if it obeys the following condition:

(7.2)   for all p.p. formulas $\phi(\bar{x}, \bar{y})$ of L and all tuples $\bar{a}$, $\bar{b}$ in A, if
         there is $\bar{c}$ in A such that $A \vDash \phi(\bar{a}, \bar{c}) \wedge \phi(\bar{b}, \bar{c})$, then for all $\bar{d}$
         in A, $A \vDash \phi(\bar{a}, \bar{d}) \leftrightarrow \phi(\bar{b}, \bar{d})$.

Every module is p.p. normal. For if A is a module and $\phi(\bar{x}, \bar{y})$ is a p.p. formula without parameters, then the relation defined by ϕ in A is a subgroup G of the product group $(A, +)^n$ for some n (see the proof of Corollary A.1.2). So the hypothesis $A \vDash \phi(\bar{a}, \bar{c}) \wedge \phi(\bar{b}, \bar{c}) \wedge \phi(\bar{a}, \bar{d})$ says that $\bar{a}^\frown \bar{c}$, $\bar{b}^\frown \bar{c}$ and $\bar{a}^\frown \bar{d}$ are all in the group G, and it follows that $\bar{b}^\frown \bar{d} = -\bar{a}^\frown \bar{c} + \bar{b}^\frown \bar{c} + \bar{a}^\frown \bar{d} \in G$, whence $A \vDash \phi(\bar{b}, \bar{d})$.

By a similar argument, if A is a structure such that all powers A^I of A are finite or stable, then A is p.p. normal; see Exercise 9.1.15.

Theorem 10.7.3. *Suppose the L-structure A is p.p. normal and $|L|^+$-saturated. Then A is atomic compact.*

Proof. We prove (f) of Theorem 10.7.1. Let $f \colon A \to C$ be an elementary extension. List as $\bar{a}$, $\bar{c}$ the elements of A, C respectively. We claim that a sequence $\bar{b}$ can be chosen in A so that for all $j \leqslant \text{length}(\bar{b})$,

(7.3)   if $\phi(\bar{x}, \bar{y})$ is a p.p. formula of L and $C \vDash \phi(\bar{a}, \bar{c}|j)$ then $A \vDash$
         $\phi(\bar{a}, \bar{b}|j)$.

The claim is proved by induction on j, along the lines of Lemma 10.1.5.

We only need to think about the successor ordinals. Suppose (7.3) holds for j; we want to find b_j so that it holds also for $j + 1$. Consider the set $\Psi(z)$ of all p.p. formulas $\psi(\bar{a}, \bar{b}|j, z)$ such that $C \vDash \psi(\bar{a}, \bar{c}|j, c_j)$. The set Ψ is determined by a certain subset containing at most $|L|$ formulas. To see this, consider any p.p. formula $\theta(\bar{x}, z)$ of L and two tuples $\bar{d}$, $\bar{e}$ from A. Assumption (7.2) implies that if $A \vDash \theta(\bar{d}, b)$, then $A \vDash \theta(\bar{e}, b)$ if and only if

$A \vDash \exists z(\theta(\bar{d}, z) \wedge \theta(\bar{e}, z))$; and likewise in C since $A \preccurlyeq C$. Thus in Ψ we only need consider one formula corresponding to each p.p. formula of L. Since A is $|L|^+$-saturated, it follows that Ψ is realised in A. This proves the claim. The theorem follows by mapping $\bar{c}$ to $\bar{b}$. $\qquad \square$

Exercises for section 10.7

1. Add to Theorem 10.7.1 two more equivalent conditions: (g) every set of positive formulas of $L(A)$, in any number of variables, which is finitely realised in A is realised in A; (h) if $f: A \to C$ is the diagonal embedding of A in an ultrapower C of A then there is a homomorphism $g: C \to A$ such that $gf = 1_A$.

2. Let L be a signature without relation symbols. A **compact** L-structure is an L-structure A whose domain is a Hausdorff topological space, in such a way that for each function symbol F of L, F^A is a continuous function. Show that every compact L-structure is atomic compact.

*Let A and B be L-structures. We say that B is a **pure-essential** extension of A if B is a pure extension of A and for every homomorphism $f: B \to C$, if the restriction of f to A is a pure embedding of A into C, then f is an embedding. (This notion comes from abelian group theory, where it seems less arbitrary than it does here!)*

3. Let L be a first-order language, T an equational theory in L and A a model of T. Show that the following are equivalent. (a) A is atomic compact. (b) A has no proper pure-essential extensions which are models of T.

4. Show that if $A \subseteq B$ and $A \times B$ is atomic compact then A is atomic compact.

5. Suppose T is a complete theory in a first-order language L, the class of models of T is closed under non-empty products (e.g. T is a Horn theory), and T is λ-categorical for every $\lambda > |L|$. Show that every model of T is atomic compact. [Every model of T is p.p. normal, by Exercise 9.1.15 and Theorem 12.2.3. Then use Theorem 10.7.3 and Exercise 4 above.]

6. Let X be a set and G a set of subsets of X. We form a structure $A = \langle X, R_0, R_1, \ldots \rangle$ by listing the sets in G as 1-ary relations R_0, R_1 etc. (a) Show that A is atomic compact if and only if every family in G with the finite intersection property has non-empty intersection. (b) Use Exercise 1(g) above to deduce the Alexander subbase theorem: if G is a subbase of closed sets for a topology on X and every family in G with the finite intersection property has non-empty intersection, then X forms a compact space.

7. Let B be a boolean algebra (regarded as a complemented lattice). Show that the following are equivalent. (a) B is atomic compact. (b) B is complete. (c) For every set Φ of equations of the form $a \wedge x = b$ with a, b in B, if Φ is finitely realised in B then Φ is realised in B.

8. Show that if A is a finite structure and B is a complete boolean algebra, then the boolean power $A[B]^*$ is atomic compact. [Use Exercise 9.7.7.]

9. Show that no existentially closed group is atomic compact. [Extend the e.c. group G to a larger e.c. group H. Then H is simple (by Exercise 8.2.13) and G is pure in H.]

*In the next exercise we assume that L is a first-order language and T an $\forall_1$ theory in L. Let A be a model of T, $\phi(\bar{x})$ a p.p. formula of L and $\bar{a}$ a tuple of elements of A. We say that A is $\phi(\bar{a})$-**irreducible** if*

> $A \vDash \neg\phi(\bar{a})$, and for every model B of T and homomorphism $f: A \to B$, either f is an embedding or $B \vDash \phi(f\bar{a})$.

*We say that A is **pure-irreducible** if A is $\phi(\bar{a})$-irreducible for some p.p. ϕ and tuple $\bar{a}$.*

10. Let A be a model of T. (a) Suppose $\phi(\bar{x})$ is a p.p. formula of L and $\bar{a}$ is a tuple of elements of A such that $A \vDash \neg\phi(\bar{a})$. Show that there are a model B of T and a surjective homomorphism $h: A \to B$ such that B is $\phi(h\bar{a})$-irreducible. (b) Show that there is a family $(B_i: i \in I)$ of pure-irreducible models of T, together with surjective homomorphisms $f_i: A \to B_i$, such that the induced product map $f: A \to \prod_I B_i$ is a pure embedding.

11. Let $\mathbb{Z}$ be the ring of integers. Find a family Φ of ω_1 equations over $\mathbb{Z}$ in uncountably many variables, such that Φ has no solution in $\mathbb{Z}$ but every countable subset of Φ has one. [Let P be the set of primes. By a theorem of Sierpiński (Exercise 4.2.4) there is a family $(P_i: i < \omega_1)$ of infinite subsets of P such that the intersection of any two sets in the family is finite. For all $i < j < \omega_1$ choose primes $p(i, j) \in P_j \backslash P_i$ so that $p(i, j) \neq p(i', j)$ whenever $i \neq i'$. The set Φ consists of the equations $x_j - x_i = 1 + p(i, j)y_{ij}$ $(i < j < \omega_1)$.]

History and bibliography

Section 10.1. Shelah [1978a] worked with a big model; he required it to be saturated (so it may not exist). The definition of λ-big given here is new. Morley & Vaught [1962] introduced saturated models as a special case of Jónsson's [1956], [1960] generalisation of Fraïssé limits to higher cardinalities. Soon afterwards Keisler [1961a] came on λ-compact structures (he called them 'λ-replete') by way of ultraproducts. The notions of λ-universality and λ-homogeneity are from the paper of Morley & Vaught, adapting similar but different notions in Jónsson's papers. Shelah (e.g. [1978a]) introduced strong λ-homogeneity; see Kueker & Steitz [199?a] for some situations (involving stable structures) where we can extend elementary maps on sets of $|A|$ elements to automorphisms of A. Theorem 10.1.8(b) is from Keisler & Morley [1967].

Exercise 20: Keisler [1971b]. Exercise 21: Lo [1983].

Section 10.2. One should look at the papers of Jónsson and Morley & Vaught in the references for section 10.1 to see the history of these constructions. Theorem 10.2.8 is a weak form of Theorem III.3.12 in Shelah [1978a]; cf. Harnik [1975].

Exercise 3: Vaught [1961]. Exercise 4: Ehrenfeucht, as reported in Vaught [1961]. For further work on characterising those theories which have just finitely many countable models, see Benda [1974], Pillay [1978], [1980], [1989b], Tsuboi [1985], Woodrow [1976, 1978]. Exercise 7: Jónsson & Olin [1968]. Exercises 9, 10: Robinson [1971b]. Abraham Robinson originally defined infinite-generic models in a quite different way from this, using a version of forcing. Fisher (unpublished, about 1980) showed that Robinson's notion was equivalent to the definition given above. For more on infinite-generic models, see Hirschfeld & Wheeler [1975].

Section 10.3. The style of argument in Theorems 10.3.1 and 10.3.3, going back and/or forth one element at a time, was developed by Chang & Keisler [1966] for proving preservation theorems. Keisler games and their finite approximations appear in Keisler [1965a], and Keisler [1965b] exploits these games to prove a wide range of syntactic characterisations. Theorem 10.3.3 and Exercise 1 are from Lyndon [1959a]; Lyndon's proof was by cut elimination – the approach by saturated models is essentially due to Keisler [1961b]. Corollary 10.3.5 was known to Łoś [1955a] and Mal'tsev [1956b]. Theorem 10.3.8 is based on Anapolitanos [1979].

Exercise 3: Chang [1965]. Exercise 5: D. Scott (unpublished), Węglorz [1967]. Exercise 6: Feferman [1968b]. Exercise 7: Motohashi [1986]. Exercise 8: Compton [1983]. Exercises 9, 10: Lyndon [1959c]. Exercise 11: Anapolitanos & Väänänen [1980]. Exercise 12: Lyndon [1959b].

Section 10.4. Special structures are from Morley & Vaught [1962], who proved Theorem 10.4.4. Chang & Keisler [1973] simplified the definition. Theorem 10.4.9 for saturated structures is from Shelah [1978a] (p. 8), who gives a remarkably short proof using stability theory. For special structures it was proved by Kueker [1970] under the assumption that L^+ comes from L by adding a single relation symbol; but by Exercise 10.1.23 this case already implies a large part of the theorem.

Exercise 4: Morley & Vaught [1962]. Exercise 8: Woodin's result will appear in Cummings & Woodin [199?].

Section 10.5. The equivalence of (a) and (c) in Theorem 10.5.1 is Beth's theorem (Beth [1953]). Corollaries 10.5.2 and 10.5.3 are from Svenonius [1959b]; the use of a big model is hidden in Svenonius' proof, and it only came to public view when Shelah proposed to do stability theory inside a big model. Theorem 10.5.6 and its Corollary 10.5.7 have a similar history: the equivalence of (a), (d) and (g) was proved by Kueker [1970] (with T not necessarily complete), but it was only the use of big models which led to the full theorem being stated. See Harnik & Harrington [1984] for a version of it. The notion of almost definability is from Shelah [1978a] p. 93.

Exercise 6: Lascar [1982]; in his appendix Lascar gives an example for (c) with $|G(A, \varnothing)| = 2^\omega$, due to Cherlin. It seems that for ω-categorical theories T at least, smallness of the groups $G(A/X)$ (A a model of T and X finite) makes it easier to recover T from the category of models of T and elementary embeddings. For more on this theme, see Lascar [1989].

Section 10.6. During the early 1970s, recursively saturated structures were discovered in slightly different guises by Schlipf (see Barwise & Schlipf [1976]), Ressayre [1977], both in connection with admissible set theory, and Wilmers (see Kaye [1991] Chapter 13) in connection with Scott sets in models of arithmetic (Scott [1962]). Theorem 10.6.2 is due to Barwise (in Barwise & Schlipf [1976]), and part (b) independently to Ressayre [1977]. The prototype for Theorem 10.6.3 was the Bernays–Gödel axiomatisation of set theory, which reduces an axiom schema to a finite set of axioms. Kleene's [1952a] form of the result was Exercise 9, and Makkai [1964b] proved the full theorem. Kaye [1991] Theorem 15.11 gives more details of the proof in the context of models of arithmetic.

Exercise 1: Barwise & Schlipf [1976]. Exercise 3: Ressayre [1977]. Exercise 4: Krajewski; see Kotlarski [1980]. Exercise 5 is essentially in A. Robinson [1963b]. Exercise 6: Lipshitz & Nadel [1978]. Exercises 7, 8: Knight [1986b]. Exercise 10: This slightly extends a result in Craig & Vaught [1958].

Section 10.7. Irving Kaplansky [1954] proved that all compact abelian groups allow a certain decomposition; he called a group 'algebraically compact' if it has this decomposition. His algebraically compact abelian groups are exactly our atomic compact groups, though it was Mycielski [1964] who saw that there is a purely model-theoretic definition. Theorem 10.7.1 (except for (b)) and Exercise 1 are from Węglorz [1966]; (b) is from Mycielski & Ryll-Nardzewski [1968]. Theorem 10.7.2 is from Mycielski [1964]. Important references for the 1970s work are Walter Taylor [1971], [1972], McKenzie & Shelah [1974], Fisher [1977a].

Exercise 3: Maranda [1960] for the abelian group case. Exercises 4, 5: Węglorz [1976] extending Baldwin [1974]. Exercise 6: Haley [1979]. Exercise 7: condition (b) is from Węglorz [1966], (c) from Abian [1970]. Exercise 8: Burris [1975]. Exercise 10: Taylor [1972]. Exercise 11: McKenzie [1971b].

11

Combinatorics

William Byrd, Non vos relinquam.

In that curious way in which branches of mathematics come together for a time and drift apart again, set theory and model theory cohabited intensely during the 1960s and early 1970s, then separated. Both sides gained something permanent from the relationship. On one side the set theorists got game quantifiers, ultraproducts of models of set theory and the Silver indiscernibles, and certainly Jensen's work on two-cardinal theorems in the constructible universe paid off handsomely in combinatorial principles. On the other side the model theorists got Hanf numbers, omitting types theorems, two-cardinal theorems and the use of indiscernibles for proving non-structure theorems. A parting gift from model theory to set theory was proper forcing, which first appeared in a paper of Shelah [1980b] about counting numbers of models.

Over the same period, Abraham Robinson's nonstandard analysis fed some powerful methods from model theory into calculus and measure theory. This work has a very different feel to it. In the first place it paid almost nothing back into model theory, and today nonstandard analysis is hardly reckoned a branch of model theory at all. (In section 11.4 I have confined myself to some examples of the same approach in algebra and field theory; they are a little closer to home.) In the second place it hardly mentioned properties of cardinal numbers, so that it had more appeal for mathematicians whose set theory stopped short of forcing and large cardinals.

But there was a common theme to both nonstandard analysis and the combinatorial work of the first paragraph. In both cases one used the fact that *the languages of model theory are bad at characterising infinite cardinals*. Thus one can build ordered structures which are elementarily equivalent (in the appropriate language) to some cardinal but are not even well-ordered. This idea lies behind most sections of this chapter.

Today there is relatively little traffic between the set theorists and the model theorists, and I think it is virtually all from set theory to model theory. Shelah continues to prove non-structure theories, as always using the sharpest tools available. There have been some bids to apply measure-theoretic or topological ideas in model theory (see part E of Barwise & Feferman [1985]); these things have not come into centre stage, but they might at any time.

The passage of Byrd's motet quoted at the head of this chapter is – with allowances for artistry – an indiscernible sequence of length four. Stretched to ω, it represents the legend that Jesus ascended into heaven.

11.1 Indiscernibles

There's a party game where each of the players has to talk for one minute without repeating themselves. It's hard; change the minute to an hour and it's impossible. The theorems of this section will make the point mathematically. If you have a large structure built out of a small range of materials, the structure has to contain many repetitions.

Let X be a set, $<$ a linear ordering of X and k a positive integer. We write $[X]^k$ for the set of all $<$-increasing k-tuples of elements of K. Here '$<$-increasing' will always mean 'strictly increasing in the ordering $(X, <)$'. The notation $[X]^k$ leaves out $<$; we can write $[(X, <)]^k$ when we need to specify it.

Let f be a map whose domain is $[X]^k$. We say that a subset Y of X (or more explicitly a subordering $(Y, <)$ of $(X, <)$) is f-**indiscernible** if for any two $<$-increasing k-tuples $\bar{a}$, $\bar{b}$ from Y, $f(\bar{a}) = f(\bar{b})$; in other words, if f is constant on $[Y]^k$.

Here is an important special case. Let A be an L-structure. Let k be a positive integer and Φ a set of formulas of L, all of the form $\phi(x_0, \ldots, x_{k-1})$. Suppose X is a set of elements of A and $<$ is a linear ordering of X. Then we say that $(X, <)$ is a Φ-**indiscernible sequence** in A if for every formula $\phi(x_0, \ldots, x_{k-1})$ in Φ and every pair $\bar{a}, \bar{b} \in [X]^k$,

$$(1.1) \qquad\qquad A \vDash \phi(\bar{a}) \leftrightarrow \phi(\bar{b}).$$

We can choose a function $f: [X]^k \to 2^{|L|}$ so that

(1.2)   for all $\bar{a}$ and $\bar{b}$ in $[X]^k$,
$$f(\bar{a}) = f(\bar{b}) \Leftrightarrow (1.1) \text{ holds for all } \phi \text{ in } \Phi.$$

Then $(X, <)$ is f-indiscernible if and only if it is Φ-indiscernible.

Example 1: *Bases of free structures*. Let A be a free algebra in a variety (see section 9.2 above) and X a basis of A – for example, let A be a vector space and X a basis of A. Let $<$ be any linear ordering of X. Then X is $\phi(x_0, \ldots, x_{k-1})$-indiscernible for every formula ϕ in the first-order language of L. For let $\bar{a}$, $\bar{b}$ be any two strictly increasing k-tuples from $(X, <)$. Since X is a basis of A, there is an automorphism of A which takes $\bar{a}$ to $\bar{b}$. Hence $A \vDash \phi(\bar{a})$ implies $A \vDash \phi(\bar{b})$.

The set X in this example is indiscernible in a very strong sense. Suppose A is any structure, X a set of elements of A linearly ordered by $<$, and $(X, <)$ is a $\{\phi\}$-indiscernible sequence simultaneously for every first-order formula $\phi(\bar{x})$ of L (with any number of variables); then we say that $(X, <)$ is an **indiscernible sequence** in A. We say that X is an **indiscernible set** in A if $(X, <)$ is an indiscernible sequence for every linear ordering $<$ of X. (**Warning**: set theorists call an indiscernible sequence a **set of indiscernibles**.)

Example 1 shows that a basis of a free algebra in a variety is always an indiscernible set. The next example shows the same thing for bases of minimal sets.

Example 2: *Independent sets in minimal sets*. It follows at once from Lemma 4.5.6 above that any independent set in a $\varnothing$-definable minimal set is an indiscernible set. For example a transcendence basis in an algebraically closed field is an indiscernible set.

In this section we shall prove a number of results about the existence of indiscernibles. Results of this kind are known as **partition theorems**, because a map f with domain $[X]^k$ acts as a partition of $[X]^k$. This branch of mathematics is called **Ramsey theory**, because it was first discovered by Frank Ramsey in the 1920s.

Partition theorems in model theory fall broadly into two groups, the stability type and the Ramsey type. Example 2 illustrates the stability type: indiscernibles exist because of a peculiarity of the structure involved. I give another example of the stability type in Exercises 12 and 13 below. Results of this kind are vitally important in stability theory. They tell us that models of stable theories contain large sets which are very like bases – and then other results from stability tell us how the rest of the structure sits over such a set.

By contrast the Ramsey type theorems hold for reasons of cardinality alone, and they are independent of the theory of the structure. Our first example shall be Ramsey's original theorem.

Ramsey's theorem

To begin with some notation, let λ, μ and v be cardinals and k a positive integer. We write

$$(1.3) \qquad\qquad \lambda \to (\mu)_v^k$$

to mean that if X is any linearly ordered set of cardinality λ, and $f: [X]^k \to v$, then there is a subset Y of X of cardinality μ such that f is constant on $[Y]^k$. Facts of the form (1.3) are known as **Erdős–Rado partition relations**. The notation was chosen by Paul Erdős and Richard Rado so that if (1.3) holds, then it still holds when λ on the left is raised and μ, v and k on the right are lowered.

Lemma 11.1.1. *Let λ, μ, v, λ', μ', v' be cardinals and k, k' positive integers such that $\lambda \le \lambda'$, $\mu \ge \mu'$, $v \ge v'$ and $k \ge k'$. If $\lambda \to (\mu)_v^k$ then $\lambda' \to (\mu')_{v'}^{k'}$.*

Proof. Immediate from the definitions. ☐

Lemma 11.1.2. *To check that $\lambda \to (\mu)_v^k$, it suffices to show that if $g: [\lambda]^k \to v$ then there is a subset Y of λ of order-type μ such that g is constant on $[Y]^k$. (Here as always, we use the natural well-ordering of λ.)*

Proof. Again immediate; if X has cardinality λ and $f: [(X, <)]^k \to v$, let $<$ well-order X in order-type λ and define $g: [(X, <)]^k \to v$ so that $g(\bar a) = f(\bar a')$ when $\bar a'$ is $\bar a$ re-ordered into the $<$ ordering. (Really f is a mapping on k-element sets; the ordering doesn't matter.) ☐

The most important partition theorem, and the first to be discovered, runs as follows.

Theorem 11.1.3 *(Ramsey's theorem, infinite form).* *For all positive integers k, n we have $\omega \to (\omega)_n^k$.*

Proof. By induction on k. When $k = 1$, the theorem says that if ω is partitioned into at most m parts (with m finite), then at least one of the parts is infinite. This is true and it has a name: the pigeonhole principle.

Suppose then that $k > 1$, and let a map $f: [\omega]^k \to n$ be given. Let A be the structure built as follows.

(1.4)   dom$(A) = \omega$. There are names $0, 1, \ldots$ for the elements of
         ω. There are a relation symbol $<$ and a function symbol F,
         to represent the usual ordering of ω and the function f
         respectively. (Put $F^A(a_0, \ldots, a_{k-1}) = 0$ when $(a_0, \ldots, a_{k-1})$
         is not an increasing k-tuple.)

By the compactness theorem there is a proper elementary extension B of A.
Now $A \vDash \forall x(F(x) < n)$, and so the same holds in B. Also $A \vDash$ '$<$ linearly
orders all the elements', and for each natural number m, $A \vDash$ 'the element m
has exactly m $<$-predecessors'. So any element of B which is not in A must
come $<$-after all the elements of A. Take such an element and call it ∞.

We shall choose natural numbers $m(0), m(1), \ldots$ inductively, so that for
each $i < \omega$,

(1.5) $m(j) < m(i)$ for each $j < i$,

(1.6)   for all $j_0 < \ldots < j_{k-2} < i$, $B \vDash$
         $F(m(j_0), \ldots, m(j_{k-2}), m(i)) = F(m(j_0), \ldots, m(j_{k-2}), \infty)$.

Suppose $m(0), \ldots, m(i - 1)$ have been chosen. Then we can write a first-
order formula $\phi(x)$, using the constants and the symbols $<$ and F, which
expresses that $x > m(i - 1)$, and for all $j_0 < \ldots < j_{k-2} < i$, the value of
$F^B(m(j_0), \ldots, m(j_{k-2}), x)$ is $F^B(m(j_0), \ldots, m(j_{k-2}), \infty)$. (There are only
finitely many conditions here, and for each choice of $j_0, \ldots, j_{k-2}$ the number
$F^B(m(j_0), \ldots, m(j_{k-2}), \infty)$ is $< n$, so that it is named by a constant.) Now
clearly $B \vDash \phi(\infty)$, and so $B \vDash \exists x\, \phi(x)$. Since $A \preccurlyeq B$, it follows that $A \vDash \phi(m)$
for some natural number m. Put $m(i) = m$. This completes the choice of the
numbers $m(i)$. Put $W = \{m(i) : i < \omega\}$.

Define a map $g : [W]^{k-1} \to n$ by

(1.7) $$g(\bar{b}) = f(\bar{b}{}^{\frown}c) \text{ for every tuple } \bar{b}{}^{\frown}c \in [W]^k.$$

By (1.6), g is well-defined. By induction hypothesis there is an infinite subset
Y of W such that g is constant on $[Y]^{k-1}$. Now let $\bar{a} = (y_0, \ldots, y_{k-1})$ and
$\bar{b} = (z_0, \ldots, z_{k-1})$ be sets in $[Y]^{k-1}$ with $y_0 < \ldots < y_{k-1}$ and $z_0 < \ldots$
$< z_{k-1}$. Taking any c which is $> \max(y_{k-1}, z_{k-1})$,

(1.8) $$f(\bar{a}) = f(y_0, \ldots, y_{k-2}, c) = g(y_0, \ldots, y_{k-2})$$
         $$= g(z_0, \ldots, z_{k-2}) = f(\bar{b}).$$

Hence f is constant on $[Y]^k$ as required.   □

An infinite cardinal λ is said to be **weakly compact** if $\lambda \to (\lambda)_2^2$. Rather
surprisingly this implies $\lambda \to (\lambda)_n^k$ for all positive integers k and n; see
Exercises 8 and 9 below. A weakly compact cardinal must be strongly
inaccessible, and hence it's consistent with Zermelo–Fraenkel set theory that
the only weakly compact cardinal is ω. (The name comes from the fact that
an analogue of the compactness theorem holds for $L_{\lambda\lambda}$ whenever λ is weakly

compact. However, the implication only runs in one direction; compactness for $L_{\lambda\lambda}$ doesn't imply that λ is weakly compact. See Dickmann [1975] Chapter 3.)

Corollary 11.1.4 (*Ramsey's theorem, finite form*). *For all positive integers* k, m, n *there is a positive integer* l *such that* $l \to (m)_n^k$.

Proof. Fix k and n. Suppose for contradiction that there is no positive integer l such that $l \to (m)_n^k$. Let $\mathbb{N}$ be the structure of the natural numbers, $(\omega, 0, S, +, \cdot)$ (where S is the successor function). Let T be the following first-order theory, in the language of $\mathbb{N}$ with extra symbols c and F:

(1.9) $\mathrm{Th}(\mathbb{N}) \cup \{F$ is a function from $[\{0, \ldots, c-1\}]^k$ to $n\}$
        $\cup \{$There is no increasing sequence $\bar{a}$ of m distinct elements,
            all in $\{0, \ldots, c-1\}$, such that F is constant on
            $[\bar{a}]^k\} \cup \{\text{'}j < c\text{'}: j$ a natural number$\}$.

Every finite subset U of T has a model: choose c to be a natural number h greater than every number mentioned in U, and let F be a function which shows that $h \not\to (m)_n^k$. So by the compactness theorem, (1.9) has a model A. The infinite Ramsey theorem applied to $F^A: [\{a: A \vDash a < c^A\}]^k \to \{0^A, \ldots, n^A - 1\}$ gives us an infinite set Y such that F^A is constant on $[Y]^k$, which contradicts (1.9). $\qquad \square$

A typical application of Ramsey's theorem is the following.

Corollary 11.1.5. *Let* $(P, \leqslant)$ *be an infinite partially ordered set. Then either* P *contains an infinite linearly ordered set, or* P *contains an infinite set of pairwise incomparable elements.*

Proof. Linearly order the elements of P in any way at all, say by an ordering $<$. Define the map $F: [P]^2 \to 3$ as follows, wherever $a < b$ in P: $F(a, b) = 0$ if $a < b$, $F(a, b) = 1$ if $b < a$, $F(a, b) = 2$ if a and b are incomparable. Then apply the infinite Ramsey theorem. $\qquad \square$

An uncountable partition theorem

The proof of Theorem 11.1.3 adapts to prove several other results in the same vein. I give one example which uses ideas from the previous chapter. We need a preliminary lemma.

Lemma 11.1.6. *Let* A *be an* L-*structure and* λ *a cardinal such that* $|L| \leqslant \lambda$ *and* $|A| > 2^\lambda$. *Then there is a sequence* $\bar{b} = (b_i: i < \lambda^+)$ *of distinct elements of* A *such that for each* $i < j < \lambda^+$,

(1.10) $$\mathrm{tp}_A(b_i/\bar{b}|i) = \mathrm{tp}_A(b_j/\bar{b}|i).$$

($\mathrm{tp}_A(b/\bar{c})$ *is the complete type of the element b over the elements $\bar{c}$ with respect to A; see section 6.3.*)

Proof. By an easy variation on the proof of Theorem 10.2.5, we can find an elementary substructure B of A of cardinality $\leq 2^\lambda$, such that if Z is any set of $\leq \lambda$ elements of B and a is any element of A, then the type $\mathrm{tp}_A(a/Z)$ is realised in B. Since $|B| \leq 2^\lambda < |A|$, there is an element ∞ of A which is not in B. Choose inductively a sequence $\bar{b} = (b_i: i < \lambda^+)$ of elements of B so that for each $i < \lambda^+$,

(1.11) $$\mathrm{tp}_B(b_i/\bar{b}|i) = \mathrm{tp}_A(\infty/\bar{b}|i).$$

Clearly this sequence gives the lemma. To check that the b_i are distinct, suppose $k < i$ but $b_k = b_i$. Then by (1.11), $b_k = \infty$, which is impossible since b_k is in B. $\qquad\square$

Recall the cardinal notation $\beth_{k-1}(\lambda)$ from section 10.4 above.

Theorem 11.1.7 (*Erdős–Rado theorem*). *Let λ be an infinite cardinal and k a positive integer. Then $\beth_{k-1}(\lambda)^+ \to (\lambda^+)_\lambda^k$.*

Proof. We go by induction on k. When $k = 1$, the theorem states that if λ^+ is partitioned into at most λ sets, then some set in the partition has cardinality at least λ^+. This is true because λ^+ is regular.

Assume then that $k > 1$. Let X be a linearly ordered set of cardinality $> \beth_{k-1}(\lambda)$ and f a map from $[X]^k$ to λ. By Lemma 11.1.2 we can assume X is well-ordered; then throwing out some elements of X if necessary, we can identify X with the cardinal $\beth_{k-1}(\lambda)^+$. Construct a language L in which we can name f and the ordinals $< \lambda$, so that there is an L-structure $A = (X, f, 0, 1, 2, \ldots)$. (Put $f(\bar{b}) = 0$ when the items in $\bar{b}$ are not all distinct.) Then $|L| = \lambda$. By Lemma 11.1.6 there is a sequence $(b_i: i < \beth_{k-2}(\lambda)^+)$ of distinct elements of A such that (1.10) holds whenever $i < j < \beth_{k-2}(\lambda)^+$. Put $W = \{b_i: i < \beth_{k-2}(\lambda)^+\}$.

We finish exactly as in the proof of Ramsey's theorem, defining a map $g: [W]^{k-1} \to \lambda$ by

(1.12) $$g(\bar{b}) = f(\bar{b}^\frown c) \text{ where } \bar{b}^\frown c \in [W]^k.$$

This definition of $g(\bar{b})$ is sound by (1.10). $\qquad\square$

I record the simplest non-trivial case of Theorem 11.1.7.

Corollary 11.1.8. *For every infinite cardinal λ, $(2^\lambda)^+ \to (\lambda^+)_\lambda^2$.* $\qquad\square$

Spectra

To close this section, here is a short application of Ramsey's theorem to a counting problem in finite model theory.

Let T be a theory, L its signature and κ a cardinal. We write $I(\kappa, T)$ for the number of non-isomorphic L-structures of cardinality κ which are models of T. (The notation should mention L. But one can take L to be the smallest signature containing T.) The function $I(-, T)$ is called the **spectrum function** of T.

There are a number of natural questions that one can ask about the spectrum function. For example if ϕ is a first-order sentence, let us write Sch(ϕ) (after Heinrich Scholz [1952]) for the set of natural numbers n such that ϕ has a model with just n elements. Thus Sch(ϕ) $= \{n < \omega : I(n, \phi) \neq 0\}$. Let us say that a subset of ω is a **spectrum set** if it has the form Sch(ϕ) for some first-order sentence ϕ. Scholz asked what sets are spectrum sets. This problem is still unsolved. For instance we still don't know whether $\omega \backslash X$ is a spectrum set whenever X is a spectrum set. (The problem is connected with computation theory. Jones & Selman [1974] show that spectrum sets are exactly those subsets of ω which are acceptable by a nondeterministic Turing machine in time 2^{cx} where c is constant and x is the length of the input.)

Theorem 11.1.9. *Let L be a first-order language whose signature consists of finitely many relation symbols (a 'finite relational language'). Let T be an $\forall_1$ theory in L, and suppose that T has a model C which contains an isomorphic copy of every finite model of T. Then for all $m < n < \omega$ we have $I(m, T) \leqslant I(n, T)$.*

Proof. We prove that $I(n, T) \leqslant I(n + 1, T)$, by induction on n. Our restriction on the language L implies that each of these numbers $I(n, T)$ is finite.

Suppose $I(n, T) = k$. Let $A_0, \ldots, A_{k-1}$ be non-isomorphic models of T of cardinality n. We can suppose that each A_i is a substructure of C, and that C is countable. Let W be the union of the domains of the A_i; then W is a finite subset of dom(C). Put $X = \text{dom}(C) \backslash W$ and let $<$ be a linear ordering of X in order type ω. Define a function f on $[X]^k$ in such a way that the following holds.

(1.13)   Suppose $\bar{a}$ and $\bar{b}$ are in $[X]^k$, $f(\bar{a}) = f(\bar{b})$, $j \leqslant k$ and Z is a set of j elements of W. Then the two substructures whose domains are $Z \cup \{\bar{a}|(n - j)\}$ and $Z \cup \{\bar{b}|(n - j)\}$ respectively are isomorphic.

By Ramsey's theorem there is an infinite subset Y of X such that f is constant on $[Y]^k$.

Now for each $i < k$ we choose a substructure A_i' of C which is isomorphic to A_i and has domain of the form $Z \cup Q$, where Z is a subset of W and Q is a subset of Y; choose A_i' so that the cardinality m_i of Z is as small as possible. Let B_i be a substructure of C which is got by adding one more element of Y to A_i', so that B_i has cardinality $n + 1$. Then B_i is a model of T since T is an $\forall_1$ theory (see Corollary 2.4.2).

We claim that if $i \neq j$ then B_i is not isomorphic to B_j. For suppose $B_i \cong B_j$. Without loss assume $m_i \leq m_j$. The structure A_j is embeddable in B_j, and hence it must also be isomorphic to a substructure A'' of B_i. Write the domain of B_i as $Z_i \cup Q_i$ where $Z_i \subseteq W$ and $Q_i \subseteq Y$. Since m_j was chosen as small as possible and $|Z_i| = m_i \leq m_j$, A'' must contain the whole of Z_i. But then the choice of Y shows that A'' must be isomorphic to A_i, contradicting that $A_i \not\cong A_j$. This proves the claim. It follows at once that $I(n + 1, T) \geq k$. $\square$

Exercises for section 11.1

1. Show that $6 \to (3)_2^2$. [Suppose each pair in $[6]^2$ is coloured either red or green. Renumbering if necessary, we can suppose that $(0, 1)$, $(0, 2)$ and $(0, 3)$ all have the same colour. Consider the colours of $(1, 2)$, $(1, 3)$ and $(2, 3)$.] *This result suggests a game: the players RED and GREEN take turns to colour pairs in $[6]^2$, and the first player to complete a triangle in his or her colour loses. See Gardner [1973].*

2. Show that if the structure A is stable and $(X, <)$ is an infinite indiscernible sequence in A, then X is an indiscernible set in A.

3. Let A be an infinite structure with finite relational signature. Suppose that for every $n < \omega$, all n-element substructures of A are isomorphic. Show that there is a linear ordering $<$ such that $(\text{dom}(A), <)$ is an indiscernible sequence. [Use the compactness theorem.]

4. Show (a) every finite or cofinite subset of ω is a spectrum set, (b) every infinite arithmetic progression is a spectrum set, (c) the set of prime powers is a spectrum set, (d) the intersection of two spectrum sets is a spectrum set, (e) the union of two spectrum sets is a spectrum set.

5. Suppose n is a positive integer, X is an infinite linearly ordered set, C is a finite set and $f: [X]^n \to C$ is a surjective function. Show that there are infinite subsets $Y_0, \ldots,$ Y_{k-1} of X, and a listing of the elements of C as $c_0, \ldots, c_{k-1}$, in such a way that if $i < j < k$ then the image of $[Y_i]^n$ under f includes c_i but not c_j. [This is hidden in the proof of Theorem 11.1.9, if one notes that the structures A_j embeddable in B_i are at most A_i and those A_j for which $m_j < m_i$.]

The next result is intermediate between the finite and the infinite forms of Ramsey's theorem. By Paris & Harrington [1977] it can be stated in the language of first-order arithmetic but not proved from the first-order Peano axioms.

6. Prove that for all positive integers k, m, n there is a positive integer l such that if $[l]^k = P_0 \cup \ldots \cup P_{n-1}$ then there are $i < n$ and a set $X \subseteq l$ of cardinality at least m, such that $[X]^k \subseteq P_i$ and $|X| \geq \min(X)$. [Assuming otherwise for a fixed k, m and n, use König's tree lemma (Exercise 5.6.5) to find a partition of $[\omega]^k$ into at most n pieces, such that every l is a counterexample to the theorem. Now apply Ramsey's theorem to this partition.]

The next result is an uncountable analogue of B. H. Neumann's lemma, Lemma 4.2.1 above. In fact Neumann's lemma has a similar proof by Ramsey's theorem.

7. Let λ be an infinite cardinal, and G a group which can be written as a union of λ cosets $H_i a_i$ ($i < \lambda$) of subgroups H_i. Show that if $H_0 x_0$ is not contained in the union of the remaining cosets, then $(G: H_0) \leq 2^\lambda$. [Otherwise choose $(2^\lambda)^+$ coset representatives y_i of H_0. Let g be an element which is only in $H_0 x_0$, and map $f: (i, j) \mapsto$ (least k such that $y_i y_j^{-1} \in H_k x_k g^{-1}$). Apply Corollary 11.1.8 to find $Y = \{y_0', y_1', y_2'\}$ such that f is constant on Y, to get $x_k g^{-1} \in H_k$ for some $k > 0$, contradiction.]

*A cardinal λ is said to have the **tree property** if the following holds: for every tree T of height λ (in the sense of Exercise 2.8.2), if T has fewer than λ nodes at height α for each $\alpha < \lambda$, then T has a branch of length λ. In particular König's tree lemma (Exercise 5.6.5) says that ω has the tree property.*

*8. Show that if λ is a strongly inaccessible cardinal and λ and has the tree property, then $\lambda \to (\lambda)_v^k$ for every positive integer k and every cardinal $v < \lambda$. [Given a partition of $[\lambda]^k$ into v parts, build a tree of increasing sequences $s = (\alpha_i: i < \gamma)$ with $\gamma < \lambda$ and each $\alpha_i < \lambda$, so that if $x_0 < \ldots < x_{k-1} < x_k$ in s then $(x_0, \ldots, x_{k-1})$ and $(x_0, \ldots, x_{k-2}, x_k)$ lie in the same partition set. The tree property then finds a sequence of length λ with this property, and we can finish as in the proof of Theorem 11.1.3. This was Ramsey's original argument.]

*A cardinal λ is said to have the **order property** if the following holds: if $(X, <)$ is a linearly ordered set with $|X| = \lambda$, then X has a subset of cardinality λ which is either well-ordered or inversely well-ordered by $<$. Thus for example ω has the order property, because every infinite ordering contains a subset of order type either ω or ω^* (ω reversed).*

*9. (a) Show that if $\lambda \to (\lambda)_2^2$ then λ has the order property. (b) Show that if λ is infinite and has the order property then λ is strongly inaccessible. (c) Show that if λ is regular and has the order property then λ has the tree property. [Order the branches lexicographically.]

*10. Let δ be a limit ordinal. A **colouring** of $[\delta]^2$ is a map f from $[\delta]^2$ to some finite set C of 'colours'. The colouring is **additive** if there is a map $\pi: C \times C \to C$ such that if $a < b < c$ in δ then $f(a, c) = \pi(f(a, b), f(b, c))$. Show that if f is an additive colouring of $[\delta]^2$, then there is a cofinal subset of δ which is f-indiscernible.

*11. Let λ be an infinite cardinal. Show that for every map $f: [\lambda]^2 \to 2$, there is $Y \subseteq \lambda$ such that *either* $|Y| = \lambda$ and $f(\bar{a}) = 1$ for all $\bar{a}$ in $[Y]^2$, *or* $|Y| = \omega$ and $f(\bar{a}) = 0$ for all $\bar{a}$ in $[Y]^2$.

12. Let L be a countable first-order language, λ an infinite cardinal and $(A_i : i < \lambda^+)$ a continuous elementary chain of L-structures with union M. For each $i < \lambda^+$, let b_i be an element of A_{i+1} which is not in A_i. Suppose that there is a map d taking each formula $\phi(\bar{x}, \bar{y})$ of L to a formula $d\phi(\bar{y})$ of L with parameters from A_0, and for each $i < \lambda^+$, d is a definition schema of $\mathrm{tp}_M(b_i/\mathrm{dom}\, A_i)$ (see section 6.7). Show that $\{b_i : i < \lambda^+\}$ is an indiscernible set. [By induction on k show that any two increasing k-tuples have the same type over $\mathrm{dom}(A_i)$. At the induction step, first consider $(k + 1)$-tuples which differ only in the last item. Then take $(k + 1)$-tuples which have the same last item, using d and induction hypothesis.]

13. Let L be a countable first-order language, λ an infinite cardinal of uncountable confinality, M a λ-stable structure and X a set of λ^+ elements of M. Show that some subset of X of cardinality λ^+ is indiscernible. [Write M as the union of an elementary chain of substructures A_i of cardinality λ, as in the previous exercise, in such a way that each set $\mathrm{dom}(A_{i+1})\backslash\mathrm{dom}(A_i)$ contains an element b_i of X. Let S be the set of ordinals $< \lambda^+$ which have cofinality ω_1. By Corollary 6.7.11 each type $\mathrm{tp}_M(b_i/\mathrm{dom}\, A_i)$ $(i \in S)$ has a definition schema d_i whose formulas take their parameters from $\mathrm{dom}(A_j(i))$ for some $j(i) < i$. By Födor's lemma (Fact 11.3.3(b)) there is a subset R of S, of cardinality λ^+, such that $j(i)$ is constant for all $i \in R$. Use λ-stability (carefully) to cut down to a subset of cardinality λ^+ on which d_i is constant too.]

*14. Let A be an L-structure and X an infinite set of elements of A. Let $<_0$ and $<_1$ be two linear orderings of X, and suppose that both $(X, <_0)$ and $(X, <_1)$ are indiscernible sequences in A. Suppose also that some formula $\phi(x_0, \ldots, x_{k-1})$ is satisfied by increasing sequences in $(X, <_0)$ but not by all k-tuples of distinct elements of X. Show that $<_1$ comes from $<_0$ by at most (a) reversing the order, (b) taking an initial segment to the end (so that $\eta + \zeta$ becomes $\zeta + \eta$), and (c) removing at most $k - 1$ elements from each end. [The same argument gave Example 3 in section 5.1.]

11.2 Ehrenfeucht–Mostowski models

Andrzej Ehrenfeucht and Andrzej Mostowski had the ingenious idea of building structures around linearly ordered sets, so that properties of the linearly ordered sets would control the properties of the resulting structures. For readers familiar with categories, the natural way to describe the Ehrenfeucht–Mostowski construction is as a functor from the category of linear orderings to the category of L-structures for some signature L. But the discussion below will not assume anything from category theory.

Definition of EM functors

A linearly ordered set, or a **linear ordering** as we shall say for brevity, is an ordered pair $(\eta, <^\eta)$ where η is a set and $<^\eta$ is an irreflexive linear ordering of η. Often we refer to the pair simply as η. We use variables η, ζ, ξ to stand for linear orderings. As in the previous section, we write $[\eta]^k$ for the set of all increasing k-tuples from η (i.e. all finite sequences $(a_0, \ldots, a_{k-1})$ of elements of η such that $a_0 <^\eta \ldots <^\eta a_{k-1}$). An **embedding** of the linear ordering η in the linear ordering ξ is a map f from η to ξ such that $x <^\eta y$ implies $fx <^\xi fy$.

We shall say that the structure A **contains** the linear ordering η if every element of η is an element of A; there need not be any other connection between the ordering relation $<^\eta$ and the structure A.

Let L be a language (for example a first-order language, or one of the form $L_{\infty\omega}$). An **Ehrenfeucht–Mostowski functor** in L, or **EM functor** for short, is defined to be a function F which takes each linear ordering η to an L-structure $F(\eta)$ so that the following three axioms are satisfied.

(2.1)   For each linear ordering η, the structure $F(\eta)$ contains η as a set of generators.

(2.2)   For each embedding $f: \eta \to \xi$ there is an embedding $F(f): F(\eta) \to F(\xi)$ which extends f.

(2.3) F is functorial; i.e. for all embeddings $f: \eta \to \xi$ and $g: \xi \to \zeta$, $F(gf) = F(g) \cdot F(f)$, and for every linear ordering η, $F(1_\eta) = 1_{F(\eta)}$.

For example, if f is an automorphism of the linear ordering η then $F(f): F(\eta) \to F(\eta)$ is an automorphism extending f. The automorphism group of η is embedded in the automorphism group of $F(\eta)$.

By (2.1), $F(\eta)$ contains η. We call η the **spine** of $F(\eta)$. Since the spine generates $F(\eta)$, every element of $F(\eta)$ is of the form $t^{F(\eta)}(\bar{a})$ for some term $t(x_0, \ldots, x_{k-1})$ of L and some increasing tuple $\bar{a} \in [\eta]^k$.

The two central properties of EM functors are known in the trade as **sliding** (i.e. we can slide elements up and down the spine without noticing) and **stretching** (i.e. we can pull out the spine into as long an ordering as we like). Theorems 11.2.1 and 11.2.4 will make this precise.

Theorem 11.2.1 (Sliding). *Let F be an EM functor in L, and let $\bar{a}$, $\bar{b}$ be increasing k-tuples from linear orderings η, ξ respectively. Then for every quantifier-free formula $\phi(x_0, \ldots, x_{k-1})$ of L we have $F(\eta) \vDash \phi(\bar{a})$ $\Leftrightarrow F(\xi) \vDash \phi(\bar{b})$.*

Proof. Find a linear ordering ζ and embeddings $f: \eta \to \zeta$ and $g: \xi \to \zeta$ such that $f\bar{a} = g\bar{b}$. Consider the diagram

(2.4) $$F(\eta) \xrightarrow{F(f)} F(\zeta) \xleftarrow{F(g)} F(\xi).$$

Assuming that $F(\eta) \vDash \phi(\bar{a})$ and recalling that embeddings preserve quantifier-free formulas (Theorem 2.4.1), we have $F(\zeta) \vDash \phi(f\bar{a})$ by $F(f)$. So $F(\zeta) \vDash \phi(g\bar{b})$, and hence $F(\xi) \vDash \phi(\bar{b})$ by $F(g)$. $\qquad\qquad\square$

In the previous section we defined 'ϕ-indiscernible sequence'.

Corollary 11.2.2. *If F is an EM functor and ζ an ordering, then ζ is a ϕ-indiscernible sequence in $F(\zeta)$ for every quantifier-free formula ϕ.* ☐

Suppose A is an L-structure and η is a linear ordering contained in A. We define the **theory** of η in A, $\mathrm{Th}(A, \eta)$, to be the set of all first-order formulas $\phi(x_0, \ldots, x_{n-1})$ of L such that $A \vDash \phi(\bar{a})$ for every increasing n-tuple $\bar{a}$ from η. (If formulas of $L_{\omega_1\omega}$ were allowed, we would write $\mathrm{Th}_{\omega_1\omega}(A, \eta)$, and so on.) The **theory** of the EM functor F in L, $\mathrm{Th}(F)$, is defined to be the set of all first-order formulas $\phi(x_0, \ldots, x_{n-1})$ of L such that for *every* linear ordering η and every increasing n-tuple $\bar{a}$ from η, $F(\eta) \vDash \phi(\bar{a})$.

Since every first-order formula has just finitely many free variables, Theorem 11.2.1 tells us that $\mathrm{Th}(F)$ contains exactly the same quantifier-free formulas as $\mathrm{Th}(F(\eta), \eta)$ for any infinite linear ordering η. We can say more.

Lemma 11.2.3. *Let F be an EM functor in the first-order language L; suppose η is an infinite linear ordering and ϕ is an $\forall_1$ sentence of L which is true in $F(\eta)$. Then $\phi \in \mathrm{Th}(F)$.*

Proof. Suppose ϕ is $\forall \bar{x}\, \psi(\bar{x})$ with ψ quantifier-free. Let ζ be any linear ordering and $\bar{a}$ a tuple of elements of $F(\zeta)$; we must show $F(\zeta) \vDash \psi(\bar{a})$. Since ζ generates $F(\zeta)$, there is some finite subordering ζ_0 of ζ with $\bar{a}$ in $F(\zeta_0)$. Since η is infinite, there is an embedding $f: \zeta_0 \to \eta$. By assumption $F(f)\bar{a}$ satisfies ψ in $F(\eta)$, so $F(\zeta_0) \vDash \psi(\bar{a})$ since ψ is quantifier-free; then $F(\zeta) \vDash \psi(\bar{a})$ likewise. ☐

Lemma 11.2.3 implies that $\mathrm{Th}(F)$ is recoverable from $\mathrm{Th}(F(\omega), \omega)$. But in fact the whole of F is recoverable up to isomorphism from $F(\omega)$.

Theorem 11.2.4 (Stretching). *Let L be a signature; let A be any L-structure containing the linear ordering ω as a set of generators. If ω is a ϕ-indiscernible sequence in A for all atomic formulas ϕ of L, then there is an EM functor F in L such that $A = F(\omega)$. This functor F is unique up to natural isomorphism of functors (i.e. if G is any other EM functor with this property, then for each linear ordering η there is an isomorphism $i_\eta: F(\eta) \to G(\eta)$ which is the identity on η).*

Proof. To construct F, take any ordering η and write $L(\eta)$ for L with the elements of η added as new constants. We shall define a set $S(\eta)$ of atomic sentences of $L(\eta)$. Let ϕ be any atomic sentence of $L(\eta)$. Then ϕ can be written as $\psi(\bar{c})$ for some atomic formula $\psi(\bar{x})$ of L and some increasing tuple $\bar{c}$ from η. We put ϕ into $S(\eta)$ if $\psi(\bar{x}) \in \mathrm{Th}(A, \omega)$. The choice of ψ here is not unique (there could be redundant variables in $\bar{x}$), but an easy sliding argument shows that the definition is sound.

We claim that $S(\eta)$ is =-closed in $L(\eta)$ (see section 1.5). Clearly $S(\eta)$ contains $t = t$ for each closed term t, since $x_0 = x_0 \in \mathrm{Th}(A, \omega)$. Suppose $S(\eta)$ contains both $\psi(s(\bar{c}), \bar{c})$ and $s(\bar{c}) = t(\bar{c})$, where $\psi(s(\bar{x}), \bar{x})$ is an atomic formula of L and $\bar{c}$ is increasing in η. Then for every increasing tuple $\bar{a}$ from ω, $A \vDash \psi(s(\bar{a}), \bar{a}) \wedge s(\bar{a}) = t(\bar{a})$, so that $\psi(t(\bar{x}), \bar{x}) \in \mathrm{Th}(A, \omega)$ and hence $\psi(t(\bar{c}), \bar{c}) \in S(\eta)$. This proves the claim.

Now define $F(\eta)$ to be the L-reduct of the canonical model of $S(\eta)$. Since $x_0 = x_1 \notin \mathrm{Th}(A, \omega)$, the elements $a^{F(\eta)}$ with a in η are pairwise distinct, and hence we can identify each $a^{F(\eta)}$ with a. Then $F(\eta)$ contains η as a set of generators. Let $f \colon \eta \to \xi$ be an embedding of linear orderings. Then for each atomic formula $\psi(\bar{x})$ of L and each increasing tuple $\bar{a}$ from η,

$$(2.5) \qquad F(\eta) \vDash \psi(\bar{a}) \Leftrightarrow \psi(\bar{x}) \in \mathrm{Th}(A, \omega) \Leftrightarrow F(\xi) \vDash \psi(f\bar{a}).$$

It follows by the diagram lemma (Lemma 1.4.2) that we can define an embedding $F(f) \colon F(\eta) \to F(\xi)$ by putting $F(f)(t^{F(\eta)}\bar{a}) = t^{F(\xi)}f\bar{a}$, for each term $t(\bar{x})$ of L and each increasing tuple $\bar{a}$ from η. This definition satisfies (2.3), so that F is an EM functor.

We constructed F so that $\mathrm{Th}(F)$ agrees with $\mathrm{Th}(A, \omega)$ in all atomic formulas of L. Let G be any other EM functor with this property. Then for every linear ordering η, every atomic formula $\psi(\bar{x})$ of L and every increasing tuple $\bar{a}$ from η, $F(\eta) \vDash \psi(\bar{a})$ if and only if $G(\eta) \vDash \psi(\bar{a})$. Since η generates both $F(\eta)$ and $G(\eta)$, it follows that we can define an isomorphism $i_\eta \colon F(\eta) \to G(\eta)$ by putting $i_\eta(t^{F(\eta)}\bar{a}) = t^{G(\eta)}\bar{a}$. Taking t to be x_0, i_η is the identity on η.

By the same argument, $F(\omega)$ can be identified with A. $\qquad\square$

This lemma recalls a well-known fact of universal algebra: the free algebras in a variety are completely determined by the free algebra on countably many generators (see Exercise 5 in section 9.2 above). Actually this fact is a special case of our lemma. For any variety **V**, define $F(\eta)$ to be the free algebra in **V** with basis the elements of η. Then F is an EM functor. This is an unilluminating example, because in this case the ordering relation $<^\eta$ plays no role whatever. In many applications, the structures $F(\eta)$ will be models of an unstable theory and the spine η will be linearly ordered by a formula of L.

EM functors can be written as word-constructions. This will be useful in the next section.

Theorem 11.2.5. *Let F be an EM functor in L, and let K be the signature of linear orderings. Then there is a word-construction Γ from K to L, such that*

(a) $|\Gamma| = |L|$ and the schedule of Γ consists of quantifier-free first-order formulas of K,

(b) the associated functor of Γ agrees with F (up to natural isomorphism) on linear orderings and embeddings of linear orderings.

Proof. Take J_Γ to be the empty signature and $\Gamma_x^{\text{gen}}(x)$ to be the formula $x = x$. This ensures that for every linear ordering η, $\Gamma(\eta)$ will be generated by the elements of η. For each atomic formula $\phi(y_0, \ldots, y_{k-1})$ of L, if $\phi \in \text{Th}(F)$ then let $\Gamma_\phi^{\text{rel}}(x_0, \ldots, x_{k-1})$ be the formula $x_0 < \ldots < x_{k-1}$; if $\phi \notin \text{Th}(F)$ let Γ_ϕ^{rel} be $\bot$. This ensures that $\text{Th}(\Gamma(\eta), \eta)$ will be the closure of $\text{Th}(F)$ under logical consequences, in other words it will be $\text{Th}(F)$. $\qquad\square$

Corollary 11.2.6. If F is an EM functor in the signature L, then F preserves $\equiv_{\lambda^+\omega}$ for every infinite cardinal λ.

Proof. This follows by Corollary 9.3.3. $\qquad\square$

Finding Ehrenfeucht–Mostowski models

If T is a theory, an **Ehrenfeucht–Mostowski model** of T is a structure of the form $F(\eta)$ which is a model of T, where F is a EM functor. (In practice, reducts of $F(\eta)$ are also known as Ehrenfeucht–Mostowski models.) How do we find Ehrenfeucht–Mostowski models of a given theory?

The answer depends on the kind of theory. For first-order theories there are two steps: first skolemise, then use Ramsey's theorem. The next two lemmas give the details. The procedure for infinitary theories is much the same, but using more sophisticated partition theorems in place of Ramsey's theorem; see section 11.6 below.

Lemma 11.2.7. Let F be an EM functor and suppose $\text{Th}(F(\omega))$ is a Skolem theory. Then for every first-order formula $\phi(\bar{x})$, either ϕ or $\neg\phi$ is in $\text{Th}(F)$. In particular all the structures $F(\eta)$ are elementarily equivalent, and in each structure $F(\eta)$, η is an indiscernible sequence.

Proof. A Skolem theory is axiomatised by a set of $\forall_1$ sentences, and modulo the theory, every formula is equivalent to a quantifier-free formula. Now quote Lemma 11.2.3 and Theorem 11.2.1. $\qquad\square$

Theorem 11.2.8 (Ehrenfeucht–Mostowski theorem). Let L be a first-order language and A an L-structure such that $\text{Th}(A)$ is a Skolem theory. Suppose A contains an infinite linear ordering η. (The ordering relation $<^\eta$ need not have anything to do with A.) Then there is an EM functor F in L whose theory contains $\text{Th}(A, \eta)$.

Proof. Let $\bar{c}$ be a sequence $(c_i : i < \omega)$ of pairwise distinct constants not in L, and write $L(\bar{c})$ for the language got by adding the constants c_i to L. Let T be the following set of sentences of $L(\bar{c})$:

(2.6) $\phi(\bar{a}) \leftrightarrow \phi(\bar{b})$ for each first-order formula $\phi(x_0,$
                                                 $\ldots, x_{k-1})$ of L and all $\bar{a}, \bar{b} \in [\bar{c}]^k$;

(2.7) $\phi(c_0, \ldots, c_{k-1})$ for each formula $\phi(x_0, \ldots, x_{k-1})$
                                                 $\in \mathrm{Th}(A, \eta)$.

We claim that T has a model.

The claim follows by the compactness theorem if we show that every finite subset of T has a model. Let U be a finite subset of T. The formulas $\phi(\bar{x})$ in (2.6), (2.7) which occur in U can be listed as $\phi_0, \ldots, \phi_{m-1}$ for some finite m, and for some finite k the new constants which occur in U are all among $c_0, \ldots, c_{k-1}$. By adding redundant variables at the right-hand end, we can write each of the formulas ϕ_i as $\phi(x_0, \ldots, x_{k-1})$. Now if $\bar{a}, \bar{b} \in [\eta]^k$, write $\bar{a} \sim \bar{b}$ if

(2.8) $A \vDash \phi_j(\bar{a}) \Leftrightarrow A \vDash \phi_j(\bar{b})$ for every $j < m$.

Then $\sim$ is an equivalence relation on $[\eta]^k$ with a finite number of equivalence classes. Hence by Ramsey's theorem (Theorem 11.1.3) there is an increasing sequence $\bar{e} = (e_j : j < 2k)$ in η such that any two increasing k-tuples from $\bar{e}$ lie in the same equivalence class of $\sim$. Interpreting each c_j as e_j $(j < k)$, we make A into a model of U. (We chose $\bar{e}$ of length $2k$ to allow space for any redundant variables in the formulas ϕ_i.) The claim is proved.

Now let B be any model of T. Since the formula $x_0 \neq x_1$ is in $\mathrm{Th}(A, \eta)$, the elements c_i^B are pairwise distinct. Hence we can identify each c_i^B with the number i, so that B contains ω. Let $B|L$ be the L-reduct of B and let C be the substructure of $B|L$ generated by ω. By (2.7), $\mathrm{Th}(A, \eta) \subseteq \mathrm{Th}(B|L, \omega)$. In particular $\mathrm{Th}(B|L)$ is a Skolem theory, so that $C \preccurlyeq B|L$. It follows that $\mathrm{Th}(A, \eta) \subseteq \mathrm{Th}(C, \omega)$. By (2.6), ω is an indiscernible sequence in C. The theorem follows by Theorem 11.2.4 and Lemma 11.2.7. $\square$

Theorem 11.2.8 tells us that every theory with infinite models has Ehrenfeucht–Mostowski models with spines of any order types we care to choose – though we may have to skolemise the theory before we construct the models.

Features of Ehrenfeucht–Mostowski models

Ehrenfeucht–Mostowski models of a theory T form the 'thinnest possible' models of T – a kind of opposite to saturated models. The next result is one way of making this precise.

Theorem 11.2.9. *Let L be a first-order language, T a Skolem theory in L and A an Ehrenfeucht–Mostowski model of T.*

(a) *For every $n < \omega$, the number of complete types $\in S_n(T)$ which are realised in A is at most $|L|$.*

(b) *If the spine of A is well-ordered and X is a set of elements of A, then the number of complete 1-types over X which are realised in A is at most $|L| + |X|$.*

Proof. Let F be an EM functor, and suppose $A = F(\eta)$.

(a) Taking a fixed $n < \omega$, let $\bar{a} = (a_0, \ldots, a_{n-1})$ be an n-tuple of elements of A. Since η generates A, for each $i < n$ we can choose a term $t_i(\bar{y}_i)$ of L and an increasing tuple $\bar{b}_i$ of elements of η, such that a_i is $t_i^A(\bar{b}_i)$. By adding redundant variables to the terms t_i, we can suppose without loss that the tuples $\bar{b}_i$ are all equal; write them $\bar{b}$.

Suppose now that $\bar{b}'$ is an increasing tuple of elements of η, of the same length as $\bar{b}$. Write $\bar{a}' = (a_0', \ldots, a_{n-1}')$ for the n-tuple where each a_i' is $t_i^A(\bar{b}')$. Let $\phi(x_0, \ldots, x_{n-1})$ be any formula of L. By the indiscernibility of η we have

$$(2.9) \qquad A \vDash \phi(t_0(\bar{b}), \ldots, t_{n-1}(\bar{b})) \Leftrightarrow A \vDash \phi(t_0(\bar{b}'), \ldots, t_{n-1}(\bar{b}')).$$

So $A \vDash \phi(\bar{a}) \Leftrightarrow A \vDash \phi(\bar{a}')$. It follows that the complete type of $\bar{a}$ is determined once we know the terms $t_0, \ldots, t_{n-1}$. But there are only $|L|$ ways of choosing these terms.

(b) Assume η is a cardinal κ. Let X be any set of elements of $F(\kappa)$. For each element a of X we can choose a representation of the form $t_a^{F(\kappa)}(\bar{b}_a)$ where $t_a(\bar{x})$ is a term of L and $\bar{b}_a$ is an increasing tuple from κ. Let W be the smallest subset of κ such that all $\bar{b}_a$ $(a \in X)$ lie in W. Then $|W| \leq |X| + \omega$. Let $s(\bar{y})$ be any term of L. By indiscernibility, for each increasing tuple $\bar{c}$ from κ the type of the element $s^{F(\kappa)}(\bar{c})$ over X is completely determined by the positions of the elements of $\bar{c}$ relative to the elements of W in κ. Since κ is well-ordered, there are at most $|W| + \omega$ ways that $\bar{c}$ can lie relative to W. So the elements $s^{F(\kappa)}(\bar{c})$ with $\bar{c}$ increasing in κ account for at most $|W| + \omega$ complete types over X. There are at most $|L|$ terms $s(\bar{y})$, yielding a total of $|W| + |L| = |X| + |L|$ types of elements over X. $\qquad \square$

Corollary 11.2.10. *Suppose T is λ-categorical for some $\lambda > |L|$. Then T is κ-stable for every κ, $|L| \leq \kappa < \lambda$. In particular if L is countable and T is λ-categorical for some uncountable λ, then T is totally transcendental.*

Proof. By the theorem, T has a model A of cardinality λ such that for every set X of at least $|L|$ elements of A, the number of complete 1-types over X realised in A is $|X|$. So by λ-categoricity, every model of T of cardinality λ

has this property. It follows at once that T is κ-stable for each κ with $|L| \leq \kappa < \lambda$. The final sentence then follows by Theorem 6.7.5.   □

We can use Theorem 11.2.9 to prove an easy result about the number of models of a complete theory in a given cardinality.

Corollary 11.2.11. *Let L be a countable first-order language and T a theory in L with infinite models. Suppose that for some $n < \omega$ the Stone space $S_n(T)$ is uncountable. Then for every infinite cardinal λ, T has at least 2^ω models of cardinality λ which are pairwise non-isomorphic.*

Proof. Let λ be any infinite cardinal. Choose $n < \omega$ so that $S_n(T)$ is uncountable; by Theorem 6.3.4, $S_n(T)$ has cardinality 2^ω. Form a first-order language L^* from L by adding n new constants $c_0, \ldots, c_{n-1}$, and write $\bar{c}$ for the tuple $(c_0, \ldots, c_{n-1})$.

For each type $p(\bar{x})$ in $S_n(T)$, let T_p be the theory $T \cup p(\bar{c})$. By Theorem 3.1.2 there is a skolemisation T_p^+ of T_p in a countable first-order language L_p^+ extending L^*. By Theorem 11.2.8 there is an Ehrenfeucht–Mostowski model B_p of T_p^+ whose spine has cardinality λ; then B_p has cardinality $\lambda + |L_p^+| = \lambda$ by Theorem 1.2.3. Let A_p be the reduct $B_p|L$. By Theorem 11.2.9, the n-tuples of elements of B_p realise only countably many complete types, and so the same is true of A_p. But the n-tuple $\bar{c}^{B_p}$ realises the type p (in B_p and hence also in A_p).

Now we count on our fingers. Between them the n-tuples in all the models A_p realise every type in $S_n(T)$, and this set of types has cardinality 2^ω. But only countably many types in $S_n(T)$ are realised in any one model A_p. So among the models A_p there must be at least 2^ω models realising different types, and clearly these models are not isomorphic.   □

In many applications of Ehrenfeucht–Mostowski models one wants to know how the order-type of the spine is reflected in order-types of other parts of the model. This is a complicated matter; the next section will prove a deep theorem about it. But the following fact is useful and quite easy to extract.

Theorem 11.2.12. *Let L be a first-order language, F an EM functor in L and $t(x_0, \ldots, x_{n-1})$ a term of L. Let '$x < y$' be a quantifier-free formula of L which, according to $\mathrm{Th}(F)$, linearly orders all the elements of the form $t(\bar{a})$ with $\bar{a}$ increasing in the spine. Then there are a number $k \leq n$, an injective map $f: k \to n$ and a map $s: k \to \{1, -1\}$ such that if η is any linear ordering, writing $<^1$ for $<^\eta$ and $<^{-1}$ for $>^\eta$,*

(2.10)   *for any two tuples $\bar{a} = (a_0, \ldots, a_{n-1})$ and $\bar{b} = (b_0, \ldots, b_{n-1})$
          in $[\eta]^n$, $F(\eta) \vDash t(\bar{a}) < t(\bar{b})$ iff for some $i < k$, $a_{f(0)} = b_{f(0)}, \ldots,$
          $a_{f(i-1)} = b_{f(i-1)}$ and $a_{f(i)} <^{s(i)} b_{f(i)}$.*

Proof. Everything here is expressed by quantifier-free formulas, so we can choose η to taste. For convenience we take η to be the ordering of the rationals. The theorem claims among other things that the $<$-ordering of $t^{F(\eta)}(\bar{a})$ and $t^{F(\eta)}(\bar{b})$ depends only on the relative order of a_i and b_i in η for each $i < n$; the relative order of a_i and b_j when $i \neq j$ is irrelevant. To see that this must be so, use the density of the rationals to find a sequence of tuples $\bar{a} = \bar{a}(0), \ldots, \bar{a}(m) = \bar{b}$ so that

(2.11) for all $l < m$ and all $i < n$, $\bar{a}(l)_i \gtreqless \bar{a}(l+1)_i \Leftrightarrow a_i \gtreqless b_i$, and

(2.12) if $l < m$ and $i < j < n$, then $\bar{a}(l)_i$ and $\bar{a}(l)_j$ are both $<^\eta \bar{a}(l+1)_i$ and $\bar{a}(l+1)_j$.

It's not hard to see that $t^{F(\eta)}(\bar{a}) \gtreqless t^{F(\eta)}(\bar{b})$ if and only if $t^{F(\eta)}(\bar{a}(l)) \gtreqless t^{F(\eta)}(\bar{a}(l+1))$ for all $l < m$, and hence the same holds for each $l < m$ separately. The claim follows by (2.12).

We assign to each $i < n$ a parity $p(i)$, which is either -1, 0 or 1 as follows. Let $\bar{a}$ and $\bar{b}$ be any two increasing n-tuples from η which differ only at the ith place, and suppose $a_i < b_i$. We put $p(i) = -1$, 0 or 1 according as $t^{F(\eta)}(\bar{a}) > t^{F(\eta)}(\bar{b})$, $t^{F(\eta)}(\bar{a}) = t^{F(\eta)}(\bar{b})$ or $t^{F(\eta)}(\bar{a}) < t^{F(\eta)}(\bar{b})$. We say that a set $X \subseteq n$ is a **majority** if the following holds for any increasing tuples $\bar{a}$, $\bar{b}$ from η:

(2.13) if for every $i \in X$, $a_i <^\eta b_i$, $a_i = b_i$ or $a_i >^\eta b_i$ according as $p(i) = 1, 0$ or -1, then $t^{F(\eta)}(\bar{a}) < t^{F(\eta)}(\bar{b})$.

Certainly the set $\{i: p(i) \neq 0\}$ is a majority; this follows from the definition of p and some sliding. Also if X is a majority and $X \subseteq W \subseteq n$ then W is a majority.

We claim that if $X = Y \cup Z \subseteq \{i: p(i) \neq 0\}$ and $Y \cap Z = \varnothing$, then either Y or Z is a majority. For this, choose increasing n-tuples $\bar{a}$, $\bar{b}$, $\bar{c}$ from η so that

(2.14)
$$i \in Y \Rightarrow b_i <^{p(i)} a_i <^{p(i)} c_i,$$
$$i \in Z \Rightarrow a_i <^{p(i)} c_i <^{p(i)} b_i,$$
$$i \notin X \Rightarrow a_i = b_i = c_i.$$

Since $a_i <^{p(i)} c_i$ for all $i \in X$, we have $t^{F(\eta)}(\bar{a}) < t^{F(\eta)}(\bar{c})$. So it follows that either (1) $t^{F(\eta)}(\bar{b}) < t^{F(\eta)}(\bar{c})$ or (2) $t^{F(\eta)}(\bar{a}) < t^{F(\eta)}(\bar{b})$. In case (1), the fact that $b_i <^{p(i)} c_i$ when $i \in Y$ but $c_i <^{p(i)} b_i$ when $i \in Z$ shows that Y is a majority. Likewise Z is a majority in case (2). The claim is proved.

It follows that some singleton $\{i_0\}$ is a majority. Take $f(0)$ to be i_0. Repeat the argument, restricting to n-tuples which agree at i and considering sets $X \subseteq n \setminus \{i_0\}$, to find $\{i_1\}$. Put $f(1) = i_1$, and repeat until the set $\{i: p(i) \neq 0\}$ is exhausted. Finally put $s(i) = p(f(i))$ for each i in the domain k of f. $\qquad \square$

Interval algebras

To close this section, here is an example of an EM functor (suggested by Robert Bonnet). It is simple and elegant, and has a rather different flavour from the skolemised examples that one usually meets in model theory.

Let η be a linear ordering. A **left-closed right-open interval** of η is a subset of η of one of the forms

$$(2.15) \qquad \eta,$$

$$(2.16) \qquad (-\infty, a) = \{b \in \eta: b <^\eta a\},$$

$$(2.17) \qquad [a, \infty) = \{b \in \eta: a \leqslant^\eta b\},$$

$$(2.18) \qquad [a, c) = \{b \in \eta: a \leqslant^\eta b <^\eta c\}.$$

The **interval algebra** of η, $B\langle \eta \rangle$, is the boolean algebra whose elements are the finite unions of left-closed right-open intervals of η, ordered by inclusion.

Interval algebras make an EM functor F in the language L of boolean algebras (with symbols $\wedge$, $\vee$ and complementation $'$) as follows. Given a linear ordering η, identify each element a of η with the interval $(-\infty, a)$ in $B\langle \eta \rangle$, so that $<^\eta$ agrees with the boolean algebra ordering $<$. Since $[a, \infty) = (-\infty, a)'$ and $[a, c) = (-\infty, c) \wedge [a, \infty)$, η generates the whole of $B\langle \eta \rangle$. Put $F(\eta) = B\langle \eta \rangle$.

If $f: \eta \to \zeta$ is an embedding, then by our identification of a with $(-\infty, a)$, f takes each interval $(-\infty, a)^\eta$ to $(-\infty, fa)^\zeta$, and this induces a boolean algebra embedding $F(f): F(\eta) \to F(\zeta)$ with the functorial property.

Exercises for section 11.2

1. Give an example to show that if F is an EM functor and $f: \omega \to \omega$ an order-preserving map, $F(f): F(\omega) \to F(\omega)$ need not preserve $\forall_1$ first-order formulas.

2. Show that if T is any first-order theory with infinite models and G is a group of automorphisms of a linear ordering η, then there is a model A of T which contains η, such that G is the restriction to η of a subgroup of $\mathrm{Aut}(A)$. In particular show that T has a model on which the automorphism group of the ordering $(\mathbb{Q}, <)$ of the rationals acts faithfully.

3. Show that if T is a skolemised stable theory with infinite models, and λ is a cardinal $\geqslant |L|$, then T has a model A of cardinality λ such that the symmetric group on λ is embeddable in the automorphism group of A. [Use Exercise 11.1.2.]

4. Show that if T is the theory of algebraically closed fields of some fixed characteristic, then no skolemisation of T is stable. [Use the previous exercise and the theorem of Artin and Schreier, that no algebraically closed field has an automorphism of finite order > 2; see the last theorem in Jacobson [1964].] *Zil'ber (unpublished) asks whether it is possible to interpret a real-closed field in every skolemisation of an algebraically closed field.*

5. Let F be an EM functor in the first-order language L, with skolemised theory. Show that if η is any infinite linear ordering and X is any set which is first-order definable in $F(\eta)$ with parameters, then X has cardinality either $|\eta|$ or $\leq |L|$.

6. Let L and L^+ be first-order languages with $L \subseteq L^+$, and suppose every symbol in L^+ but not in L is a relation symbol. Let F be an EM functor in L and T an $\forall_1$ theory in L^+ which is consistent with $\mathrm{Th}(F(\omega))$. Show that there is an EM functor F^+ in L^+ such that for each linear ordering η, $F^+(\eta)$ is a model of T and $F(\eta) = F^+(\eta)|L$.

7. Let L be a first-order language containing a 2-ary relation symbol $<$, and A an L-structure such that $<^A$ linearly orders the elements of A in order-type κ for some infinite cardinal κ. Writing η for the ordering $(\mathrm{dom}\, A, <^A)$, show that $\mathrm{Th}(A, \eta)$ contains the following formulas: (i) $x_0 < x_1$; (ii) formulas indicating that for each term $t(x_0, \ldots, x_{n-1})$ of L, in the notation of Theorem 11.2.12, $f(0) > f(i)$ for all $i\, (0 < i < k)$, and $s(i) = 1$ for all $i < k$ (what are these formulas?); (iii) in the same notation, the formula $t(x_0, \ldots, x_{f(0)}, \ldots, x_{n-1}) < x_{f(0)+1}$ for each term $t(x_0, \ldots, x_{n-1})$.

8. Let L be a first-order language containing a 2-ary relation symbol $<$, and F an EM functor for L which contains all the formulas (i), (ii) of the preceding exercise. Show that if λ is any infinite cardinal, then in $F(\lambda)$ there are no $<$-descending sequences of length $|L|^+$, the spine is cofinal and no element α of the spine has more than $|\alpha| + |L|$ predecessors in the $<$-ordering.

A linear ordering η is said to be λ-like if η has cardinality λ and no element of η has λ predecessors in η.

9. Let L be a first-order language containing a 2-ary relation symbol $<$, and T a theory in L which implies that every model A is linearly ordered by $<^A$. We say that A is λ-like if the ordering $<^A$ is λ-like. Show that if T has a model whose order-type is λ-like for some λ such that $\lambda \to (\lambda)^2_2$ (for instance a model of order-type ω), then T has models with κ-like order-type for each cardinal $\kappa > |L|$. [Adjust the two preceding exercises.]

*10. Show that every countable boolean algebra is isomorphic to the interval algebra of some countable linear ordering. [Build the algebra as the union of a chain of finite subalgebras. For a finite interval algebra, the atoms are the minimal intervals.]

*11. Show that the interval algebra construction is a presenting word-construction (see section 9.3).

11.3 EM models of unstable theories

Often the best applications of mathematical ideas are not so much applications as adaptations. In this section we use Ehrenfeucht–Mostowski functors to prove some facts about models of an unstable theory. There are two groups of results to be proved. The first group are about the bad behaviour of heir–coheir amalgams of unstable structures. The second group, which fall under the general head of *non-structure* theorems, tell us that the models of an unstable theory are hard to classify. For both groups of results, the first step is to redefine Ehrenfeucht–Mostowski functors.

Let n be a positive integer and A an L-structure. Then an **ordering of n-tuples** in A is a sequence $(\bar{a}_i : i \in \eta)$ indexed by a linear ordering η, where each $\bar{a}_i$ is an n-tuple of elements of A. We always assume that $i \neq j$ implies $\bar{a}_i \neq \bar{a}_j$, unless anything is said to the contrary. We shall use the letters I, J to stand for orderings of n-tuples. If $I = (\bar{a}_i : i \in \eta)$, we write $\bar{a}_i <^I \bar{a}_j$, or simply $\bar{a}_i < \bar{a}_j$, to mean $i <^\eta j$.

Let $\phi(\bar{x}_0, \ldots, \bar{x}_{k-1})$ be a formula of L in which each $\bar{x}_i$ is an n-tuple of variables. We say that the ordering I of n-tuples is ϕ-**indiscernible in** A if whenever $\bar{a}_0 <^I \ldots <^I \bar{a}_{k-1}$ and $\bar{b}_0 <^I \ldots <^I \bar{b}_{k-1}$, we have $A \vDash \phi(\bar{a}_0, \ldots, \bar{a}_{k-1}) \Leftrightarrow A \vDash \phi(\bar{b}_0, \ldots, \bar{b}_{k-1})$. An ordering of n-tuples which is ϕ-indiscernible for all first-order formulas ϕ of L (with their variables partitioned into n-tuples) is said to be an **indiscernible sequence of n-tuples** in A. $\mathrm{Th}(A, I)$ is the set of all first-order formulas $\phi(\bar{x}_0, \ldots, \bar{x}_{k-1})$ such that for all $\bar{a}_0 <^I \ldots <^I \bar{a}_{k-1}$, $A \vDash \phi(\bar{a}_0, \ldots, \bar{a}_{k-1})$.

It should be clear how to define an **EM functor on n-tuples**. The defining axioms are as before, except that the spine of $F(\eta)$ is now an ordering $(\bar{a}_i : i \in \eta)$ of n-tuples, and the elements occurring in the various $\bar{a}_i$s generate $F(\eta)$; if $(\bar{a}_i : i \in \eta)$ and $(\bar{b}_j : j \in \xi)$ are the spines of $F(\eta)$ and $F(\xi)$ respectively, then an embedding $f : \eta \to \xi$ induces an embedding $F(f) : F(\eta) \to F(\xi)$ by $F(f)\bar{a}_i = \bar{b}_{f(i)}$.

Suppose T is an unstable theory in a first-order language L. By the definition of instability in section 6.7, there are $n < \omega$, a formula $\phi(\bar{x}, \bar{y})$ of L and a model A of T containing n-tuples $\bar{a}_i$ $(i < \omega)$ such that for all $i, j < \omega$,

(3.1) $A \vDash \phi(\bar{a}_i, \bar{a}_j) \Leftrightarrow i < j.$

Let us say that such a formula ϕ **witnesses the instability of** T. The proof of the Ehrenfeucht–Mostowski theorem (Theorem 11.2.8) gives us an EM functor F on n-tuples, with the property that for every linear ordering η, if the spine of $F(\eta)$ is $(\bar{b}_i : i \in \eta)$, then for all i, j in η,

(3.2) $F(\eta) \vDash \phi(\bar{b}_i, \bar{b}_j) \Leftrightarrow i <^\eta j.$

For short I shall call F an **EM functor for ϕ and T**.

Amalgams of unstable structures

It follows at once from Corollary 6.7.12 that if a complete theory T is stable, then heir–coheir amalgams are unique in the following sense.

(3.3)   Let A, B and C be models of T with $A \leqslant B$ and $A \leqslant C$. If
         $B \leqslant D$ and $g: C \to D$ form an heir–coheir amalgam of B and
         C over A, then for all formulas $\phi(\bar{x}, \bar{y})$ of L and all tuples $\bar{b}$
         in B and $\bar{c}$ in C, the question whether $D \vDash \phi(\bar{b}, g\bar{c})$ is
         independent of the choice of D and g.

Our first result is the converse of this. If T is unstable then heir–coheir
amalgams aren't unique.

Theorem 11.3.1. *Let T be an unstable complete theory in a first-order language
L. Then (3.3) fails.*

Proof. Let $\phi(\bar{x}, \bar{y})$ be a formula of L which witnesses the instability of T.
Let T^+ be a skolemisation of T, and let F be an EM functor for ϕ and T^+.
Choose suborderings η, ζ, θ, ξ of the rationals so that

$$\eta = (-\infty, 0) + (1, \infty),$$
$$\zeta = (-\infty, 0] + (1, \infty),$$
$$\theta = (-\infty, 0) + [1, \infty),$$
$$\xi = (-\infty, 0] + [1, \infty).$$

Then we have a commutative diagram (3.4) of models of T^+:

(3.4)

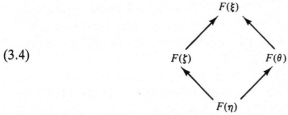

where all the structures are models of T^+ and the maps are elementary
inclusions. The spine of $F(\xi)$ is an ordering of n-tuples $(\bar{a}_i : i \in \xi)$, and the
spines of $F(\eta)$ etc. are the corresponding suborderings.

First, (3.4) is an heir–coheir amalgam. Any statement of the form
$F(\xi) \vDash \phi(\bar{b}, \bar{c})$, where $\bar{b}$ is from $F(\zeta)$ and $\bar{c}$ from $F(\theta)$, can be written in the
form

(3.5) $$F(\xi) \vDash \psi(\bar{a}_{i_0}, \ldots, \bar{a}_{i_{m-1}}, \bar{a}_{j_0}, \ldots, \bar{a}_{j_{k-1}})$$

where $i_0 < \ldots < i_{m-1}$ in ζ and $j_0 < \ldots < j_{k-1}$ in θ. We have to prove that
$_{-1}$) in (3.5) can be replaced by a tuple of tuples which lies in $F(\eta)$. For this
it's enough to show that if (3.5) holds and $1 = j_l$ for some $l < k$, then we can
replace j_l by some rational $j'_l > 1$ without altering the relative order of i_0,
$\ldots, i_{m-1}, j_0, \ldots, j_{k-1}$. This is clearly true, since 1 doesn't occur in ζ.
    Contrast (3.4) with another amalgam:

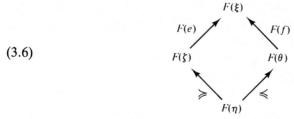

(3.6)

where $e(0) = 1$, $f(1) = 0$ and e, f are the identity elsewhere. By the same argument as before, (3.6) is heir–coheir (except that this time we have to shift j_l downwards below 0). But the L-reducts of these amalgams contradict (3.3), since $F(\xi)|L \vDash \phi(\bar{a}_0, \bar{a}_1) \wedge \neg \phi(\bar{a}_{e(0)}, \bar{a}_{f(1)})$. $\qquad\square$

A **coheir–heir amalgam** is the same thing as an heir–coheir amalgam, but reversed from left to right. In stable theories every heir–coheir amalgam is also a coheir–heir amalgam; this is one form of the fundamental result known as **forking symmetry**. The proof of Theorem 11.3.1 adapts to show that if T is unstable, then some heir–coheir amalgams of models of T fail to be coheir–heir. I leave this as Exercise 2.

Non-structure theorems

Suppose **K** is a class of structures. A *structure theorem* for **K** is a theorem which says that the structures in **K** can be classified in some reasonable way. For example the structure theorem for finitely generated abelian groups says that each such group is a direct sum of finitely many cyclic groups of infinite or prime-power order, and the number of direct summands of each order is uniquely determined by the group.

Saharon Shelah has proposed an opposite kind of theorem, which he calls *non-structure theorems*. A non-structure theorem for **K** says that the structures in **K** can't be classified reasonably. How would one set about proving such a theorem? Probably the most telling approach is to find *two structures in **K** which are not isomorphic but are extremely hard to tell apart*. For example the two structures might be of the same cardinality, satisfying exactly the same sentences of some powerful infinitary language, but fail to be isomorphic.

Another possible approach is to find a large number of similar but non-isomorphic structures in **K**, for example 2^κ pairwise non-isomorphic structures of the same cardinality κ. This is perhaps less convincing as a non-structure theorem, because it shows only that there are a lot of structures to be classified, not that there is no reasonable classification. On the other hand the combinatorial methods used for this kind of result can often be bullied into proving more. For example they sometimes yield a large family of structures in **K** such that there are no non-trivial homomorphisms between

structures in the family, or a large structure in **K** which has few endomorphisms.

Shelah has devoted much of his career to proving theorems of all these
types, sometimes for specific classes in algebra and sometimes in a general
model theoretic setting. There is room here for just one example, but it's a
typical one. We start by considering a family of linear orderings which are
very hard to tell apart. Then we apply a word-construction Γ, and we check
that the resulting structures $\Gamma(\eta)$ are still hard to tell apart but not
isomorphic. Taking the case where Γ is an EM functor for an unstable theory,
we get a general non-structure theorem for all unstable theories.

Non-structure theorems nearly always need a heavy dose of infinitary
combinatorics. So I take this opportunity to assemble some useful set theory.
Throughout these definitions, κ is a fixed uncountable regular cardinal.

Let X be a subset of κ. The set ∂X of **limit points** of X is the set of all
limit ordinals $\delta < \kappa$ such that $\sup(X \cap \delta) = \delta$. The set X is said to be **closed**
if $\partial X \subseteq X$. The set X is said to be **unbounded** if $\sup(X) = \kappa$, and a **club** if it
is both closed and unbounded. The set X is said to be **fat** if it contains a
club, and **thin** if its complement is fat. The set X is said to be **stationary** if it
is not thin, i.e. if it meets every club.

Fact 11.3.2. (a) *An intersection of fewer than κ clubs is again a club.*

(b) *If for each $i < \kappa$ a club C_i is given, then the diagonal intersection
$\{i < \kappa$: for each $j < i$, $i \in C_j\}$ is again a club.*

(c) *If $h: \kappa \to \kappa$ is any function, the set of closure points of h (i.e. the set
$\{i < \kappa$: for all $j < i$, $h(j) < i\})$ is a club.*

(d) *If X is any unbounded subset of κ then ∂X is a club.*

Proof. Exercise. Theorem 3.6.2 should get you started. $\qquad\qquad\square$

Fact 11.3.3. (a) *Let μ be a regular cardinal $< \kappa$. Then the set of limit ordinals
$< \kappa$ which have cofinality μ is stationary.*

(b) (Födor's lemma) *If S is a stationary set and $f: S \to \kappa$ is a map such that
$f(i) < i$ for all i, then there is a stationary set $R \subseteq S$ such that f is constant on
R.*

(c) (Solovay's lemma) *Every stationary set can be partitioned into κ
stationary sets.*

(d) *If S is stationary, then there are 2^κ subsets S_i $(i < 2^\kappa)$ of S such that if
$i \neq j$ then $S_i \backslash S_j$ is stationary.*

Proof. (a) is an exercise and (b) is a paraphrase of Fact 11.3.2(b). (c) is more
substantial; see pp. 433f of Jech [1978]. Finally to prove (d), use Solovay's
lemma to partition S into stationary sets P_α, Q_α $(\alpha < \kappa)$ of S, and for each
subset X of κ let S_X be $\bigcup_{\alpha \in X} P_\alpha \cup \bigcup_{\alpha \notin X} Q_\alpha$. $\qquad\qquad\square$

Baumgartner's orderings

We begin with a non-structure theorem for linear orderings. Its proof already contains most of the ingredients that we shall need for the general case. Notation is as on p. xiii. The linear orderings constructed in the proof of the next theorem were studied by James Baumgartner.

Theorem 11.3.4. *Let κ be a regular uncountable cardinal. Then there is a family $(\eta_i : i < 2^\kappa)$ of linear orderings such that*
(a) each η_i has cardinality κ and is dense without first or last element,
(b) if $i \neq j$ then η_i is not embeddable in η_j.

Recall that $^{\leq\omega}\kappa$ is the set of all sequences of length $\leq \omega$ consisting of ordinals $< \kappa$. We write σ, τ etc. for elements of $^{\leq\omega}\kappa$. We define an ordering $<$ on $^{\leq\omega}\kappa$ as follows:

(3.7) $\sigma < \tau$ iff either τ is a proper initial segment of σ, or there is some $i < \mathrm{lh}(\sigma) \cap \mathrm{lh}(\tau)$ such that $\sigma | i = \tau | i$ but $\sigma(i) < \tau(i)$:

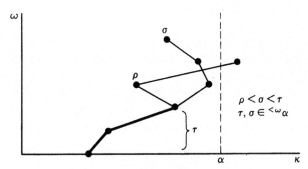

If $X \subseteq {}^{\leq\omega}\kappa$ and $\sigma, \tau \in {}^{\leq\omega}\kappa$, we write

(3.8) $\sigma \sim_X \tau$ iff   for all $\rho \in X$, $\sigma \geq \rho \Leftrightarrow \tau \geq \rho$.

Clearly $\sim_X$ is an equivalence relation on $^{\leq\omega}\kappa$.

If $X \subseteq {}^{\leq\omega}\kappa$ and $\alpha < \kappa$, then we write $X|\alpha$ for $X \cap ({}^{\leq\omega}\alpha)$.

Lemma 11.3.5. *Suppose that κ is an uncountable regular cardinal, $X \subseteq {}^{\leq\omega}\kappa$ and $\alpha < \kappa$.*

(a) The relation $\sim_{X|\alpha}$ restricted to $^{<\omega}\kappa$ has fewer than κ equivalence classes.

(b) Suppose also that X is closed under initial segments, δ is an ordinal such that $\alpha < \delta < \kappa$, no element of X contains any items in the interval $[\alpha, \delta)$, and σ and τ both lie in $^{\leq\omega}\delta$. Then

(3.9) $\sigma \sim_{X|\alpha} \tau \Leftrightarrow \sigma \sim_X \tau$

Proof of lemma. (a) Suppose to the contrary that $\sim_{X|\alpha}$ has at least κ equivalence classes represented in $^{<\omega}\kappa$. Then since κ has uncountable cofinality, there is a least ordinal $n < \omega$ such that if Y_n is the set of all elements of $^{<\omega}\kappa$ of length n, then κ equivalence classes are represented in Y_n. But induction on n shows that there are fewer than κ equivalence classes in Y_n.

(b) Right to left in (3.9) is trivial, and the other direction is straightforward from the definition of $<$. □ Lemma

Proof of Theorem 11.3.4. Let S be the set of all ordinals $< \kappa$ which have cofinality ω. Then S is stationary by Fact 11.3.3(a), and by Fact 11.3.3(d) there are 2^κ subsets of S, $S_i\,(i < 2^\kappa)$, such that whenever $i \neq j$, $S_i \backslash S_j$ is stationary in κ. For each $\delta \in S$, choose an increasing sequence σ_δ of length ω which converges to δ. The set $^{\leqslant\omega}\kappa$ is linearly ordered as in (3.7); define η_i to be the subordering consisting of $^{<\omega}\kappa$ and the elements $\sigma_\delta\,(\delta \in S_i)$. Then clearly η_i has cardinality κ, and it is not hard to check that η_i is dense without endpoints. This gives (a).

We turn to the proof of (b). Consider any two distinct ordinals $i, j < 2^\kappa$, and suppose $e: \eta_i \rightarrow \eta_j$ is an embedding. We claim that there is a club C of ordinals $\gamma < \kappa$ such that

(3.10) if $\alpha < \gamma$ and $\sigma \in {}^{<\omega}\alpha$, then for every ordinal $\beta > \alpha$ there is $\beta' < \gamma$ such that $e(\sigma^\frown\beta) \sim_{\eta_j|\alpha} e(\sigma^\frown\beta')$.

To see this, consider any $\alpha < \kappa$. The set $^{<\omega}\alpha$ has cardinality $< \kappa$. By Lemma 11.3.5(a) the number of $\sim_{\eta_j|\alpha}$-equivalence classes on $^{<\omega}\kappa$ is less than κ; since η_j contains fewer than κ infinite sequences lying inside $^{<\omega}\alpha$, it follows that the number of $\sim_{\eta_j|\alpha}$-equivalence classes on η_j is also less than κ. In particular if $\sigma \in {}^{<\omega}\alpha$, then there are fewer than $\kappa \sim_{\eta_j|\alpha}$-equivalence classes of elements of the form $e(\sigma^\frown\beta)$. Hence there is for each $\gamma < \kappa$ an ordinal $h(\gamma) < \kappa$ such that for all $\alpha < \gamma$, all $\sigma \in {}^{<\omega}\alpha$ and all $\beta < \kappa$ there is $\beta' < h(\gamma)$ such that $e(\sigma^\frown\beta) \sim_{\eta_j|\alpha} e(\sigma^\frown\beta')$. Take C to be the set of closure points of h; then C is a club by Fact 11.3.2(c).

Also there is a club D of ordinals γ such that

(3.11)   if $\alpha < \gamma$ then for every $\sigma \in \eta_i|\alpha$, $e(\sigma) \in \eta_j|\gamma$.

The proof of this is left to the reader.

Thus $\partial C \cap D$ is a club by Fact 11.3.2(a, d). Also $S_i \backslash S_j$ is stationary, and so we can choose an element δ of $S_i \backslash S_j$ which is in ∂C and in D. Since $\delta \notin S_j$, there is $\alpha < \delta$ such that no item of $e(\sigma_\delta)$ lies in the interval $[\alpha, \delta)$. Since δ is in ∂C, there is γ in C such that $\alpha < \gamma < \delta$. Since σ_δ is cofinal in δ, there is a least $n < \omega$ such that $\sigma_\delta(n) \geqslant \gamma$. Let τ be the sequence $\sigma_\delta|n$ and let β be $\sigma_\delta(n)$, so that $\tau \in {}^{<\omega}\alpha$ and $\sigma_\delta|(n+1)$ is $\tau^\frown\beta$. Let α' be $\max(\alpha, \sigma_\delta(n-1)+1)$ $(= \alpha$ if $n = 0)$; then $\alpha' < \gamma$, so by (3.10) there is $\beta' < \gamma \leqslant \beta$ such that

(3.12) $e(\tau\hat{\ }\beta) \sim_{\eta_j|\alpha'} e(\tau\hat{\ }\beta')$.

Also we have

(3.13) $\tau\hat{\ }\beta' < \sigma_\delta < \tau\hat{\ }\beta$,

since $\beta' < \beta$ and $\tau\hat{\ }\beta$ is an initial segment of σ_δ.

But since $\delta \in D$, both $e(\tau\hat{\ }\beta)$ and $e(\tau\hat{\ }\beta')$ lie in $\eta_j|\delta$. Since $e(\sigma_\delta)$ has no items in the interval $[\alpha', \delta)$, it follows from (3.12) and (3.9) (taking X as the set of initial segments of $e(\sigma_\delta)$) that

(3.14) $e(\tau\hat{\ }\beta) < e(\sigma_\delta) \Leftrightarrow e(\tau\hat{\ }\beta') < e(\sigma_\delta)$.

Comparing (3.13) with (3.14) we deduce that e is not an embedding. This contradiction completes the proof.   □

Non-isomorphic structures

With the proof of Theorem 11.3.4 preparing the way, the next theorem should be quite easy.

Theorem 11.3.6. *Let L be a signature, K the signature of linear orderings and Γ a word-construction from K to the class of all L-structures (or in fact to a quasivariety of signature L – the proof is the same). Then for every regular cardinal $\kappa > |\Gamma|$ there is a family $(\eta_i : i < 2^\kappa)$ of dense linear orderings η_i without endpoints, such that*

(a) *each $\Gamma(\eta_i)$ has cardinality $\leqslant \kappa$,*

(b) *for all $i < j < 2^\kappa$, $\Gamma(\eta_i) \equiv_{\infty\omega} \Gamma(\eta_j)$,*

(c) *if $i \neq j$ then no set of tuples in $\Gamma(\eta_j)$ is ordered in order-type η_i by any formula $\psi(\bar{x}_0, \bar{x}_1)$ of $L_{\infty\omega}$.*

Proof. We take the η_i $(i < 2^\kappa)$ to be exactly as in the proof of Theorem 11.3.4. Since each linear ordering η_i has κ elements, each $\Gamma(\eta_i)$ has cardinality at most κ (see Exercise 9.3.5). Since each η_i is dense without endpoints, we have $\eta_i \equiv_{\infty\omega} \eta_j$ for all $i < j < 2^\kappa$ (by Corollary 3.5.3 together with Example 3 in section 3.2), and hence $\Gamma(\eta_i) \equiv_{\infty\omega} \Gamma(\eta_j)$ by Corollary 9.3.3. It remains to prove (c).

To contradict (c), suppose that $i \neq j$, ψ is a formula of $L_{\infty\omega}$ and f is a map which takes each $\sigma \in \eta_i$ to a tuple $f(\sigma)$ in $\Gamma(\eta_j)$, such that for all $\sigma, \tau \in \eta_i$,

(3.15) $\sigma < \tau \Leftrightarrow \Gamma(\eta_j) \vDash \psi(f(\sigma), f(\tau))$.

Each element c of $\Gamma(\eta_j)$ can be written in the form $c = r(t_0(\bar{a}_0), \ldots, t_{n-1}(\bar{a}_{n-1}))\hat{\ }$ where r is a term of L, $t_0, \ldots, t_{n-1}$ are terms of J_Γ and $\bar{a}_0, \ldots, \bar{a}_{n-1}$ are tuples of elements of η_j; rearranging the schedule of Γ a little if necessary, we can assume that $\bar{a}_0, \ldots, \bar{a}_{n-1}$ are all equal to some increasing tuple $\bar{a}$ in η_j. Likewise with a touch of shorthand we can write the tuple $f(\sigma)$

as $\bar{t}_\sigma(e(\sigma))^\sim$ where $\bar{t}_\sigma$ is a tuple of terms of $L \cup J_\Gamma$ and $e(\sigma)$ is an increasing tuple from η_j.

The situation is more complicated than in Theorem 11.3.4 in three ways. First we now have the terms $\bar{t}_\sigma$ to contend with, and second, each $e(\sigma)$ is not a single element of η_j but a tuple. The third difference – that each $f(\sigma)$ is a tuple of elements instead of a single one – actually causes us no more trouble than a few bars on the tops of letters.

We deal with the second point first. Let $\bar{\sigma} = (\sigma_0, \ldots, \sigma_{k-1})$ and $\bar{\tau} = (\tau_0, \ldots, \tau_{k-1})$ be increasing k-tuples in $^{\leq\omega}\kappa$ and let X be a subset of $^{\leq\omega}\kappa$. We define

(3.16) $\bar{\sigma} \sim_X \bar{\tau} \Leftrightarrow$ for all $h < k$ and all $\rho \in X$, $\sigma_h \geq \rho$ if and only if $\tau_h \geq \rho$.

As before and for the same reason, if $\alpha < \delta < \kappa$, X is a set of sequences in $^{\leq\omega}\kappa$ which contain no items in the interval $[\alpha, \delta)$, X is closed under initial segments and σ, τ are increasing k-tuples of sequences in $^{\leq\omega}\delta$, then

(3.17) $\bar{\sigma} \sim_{X|\alpha} \bar{\tau} \Leftrightarrow \bar{\sigma} \sim_X \bar{\tau}$.

(Compare Lemma 11.3.5.)

Now in place of (3.10) we claim that there is a closed unbounded set C of ordinals $\gamma < \kappa$ such that

(3.18)   if $\alpha < \gamma$ and $\sigma \in {^{<\omega}\alpha}$, then for every ordinal $\beta > \alpha$ there is

$\beta' < \gamma$ such that $e(\sigma\hat{\ }\beta) \sim_{\eta_j|\alpha} e(\sigma\hat{\ }\beta')$ and $\bar{t}_{\sigma\hat{\ }\beta} = \bar{t}_{\sigma\hat{\ }\beta'}$.

The proof is as before, taking into account the fact that there are fewer than κ tuples of terms of $L \cup J_\Gamma$. We choose D as at (3.11) (except that e now means something different).

We choose δ, α, γ, β, τ, β' as before, except that now (3.18) also gives us $\bar{t}_{\tau\hat{\ }\beta} = \bar{t}_{\tau\hat{\ }\beta'}$. As before, (3.12) and (3.13) hold. By (3.12) and (3.17) it follows as before that the elements $e(\tau\hat{\ }\beta)$, $e(\sigma_\delta)$ are in the same relative order in η_j as the elements $e(\tau\hat{\ }\beta')$, $e(\sigma_\delta)$, and so (using Example 3 in section 3.2 again)

(3.19) $(\eta_j, e(\tau\hat{\ }\beta), e(\sigma_\delta)) \equiv_{\infty\omega} (\eta_j, e(\tau\hat{\ }\beta'), e(\sigma_\delta))$.

By Theorem 9.3.1 for word-constructions we infer that

(3.20) $\Gamma(\eta_j) \vDash \psi(\bar{t}_{\tau\hat{\ }\beta}(e(\tau\hat{\ }\beta))^\sim, \bar{t}_{\sigma_\delta}(e(\sigma_\delta))^\sim) \Leftrightarrow$

$\Gamma(\eta_j) \vDash \psi(\bar{t}_{\tau\hat{\ }\beta}(e(\tau\hat{\ }\beta'))^\sim, \bar{t}_{\sigma_\delta}(e(\sigma_\delta))^\sim)$.

Since $\bar{t}_{\tau\hat{\ }\beta} = \bar{t}_{\tau\hat{\ }\beta'}$, (3.20) reduces to

(3.21) $\Gamma(\eta_j) \vDash \psi(f(\tau\hat{\ }\beta), f(\sigma_\delta)) \Leftrightarrow \Gamma(\eta_j) \vDash \psi(f(\tau\hat{\ }\beta'), f(\sigma_\delta))$.

Taken with (3.13), this contradicts (3.15); so (c) is proved.   □

An immediate corollary is the following.

Corollary 11.3.7. *In every uncountable cardinality κ there is a family of 2^κ atomless boolean algebras, none of which can be embedded in any other algebra of the family.*

Proof. Let Γ be the interval algebra construction at the end of section 11.2; by Theorem 11.2.5 it's a word-construction. Let $(\eta_i: i < 2^\kappa)$ be the linear orderings of the theorem. Since each η_i is embeddable in the boolean algebra $\Gamma(\eta_i)$, $\Gamma(\eta_i)$ must also have cardinality κ. The fact that the η_i are dense and without endpoints quickly implies that the $\Gamma(\eta_i)$ are atomless. If there is an embedding of $\Gamma(\eta_i)$ in $\Gamma(\eta_j)$, it restricts to an embedding of η_i into the partial ordering of $\Gamma(\eta_j)$, so that $i = j$ by the theorem. $\square$

See section 9.7 for another corollary in the same vein, constructing large families of pairwise non-embeddable boolean powers.

Theorem 11.3.6 is really a theorem of set-theoretic algebra. It never mentions first-order logic. Nevertheless its best-known corollary is the following result of Shelah in first-order model theory.

Theorem 11.3.8. *Let T be an unstable theory in a first-order language L. Then for every regular cardinal $\kappa > |L|$, there is a family $(A_i: i < 2^\kappa)$ of L-structures which are models of T of cardinality κ, such that for all $i \neq j$, A_i is not elementarily embeddable in A_j.*

Proof. Let T^+ be a skolemisation of T in a first-order language L^+ of the same cardinality as L. Then by Theorem 11.2.8 and the instability of T, there is (to use the terminology of the beginning of this section) an EM functor F on n-tuples for '$\bar{x}_0 < \bar{x}_1$' and T^+, where '$\bar{x}_0 < \bar{x}_1$' is a formula of L which orders the spine of each $F(\eta)$.

We claim that F can be written as a word-construction Γ. The proof is almost the same as that of Theorem 11.2.5, except that now the spine consists of n-tuples. For this, take J_Γ to consist of n 1-ary function symbols $\iota_0, \ldots, \iota_{n-1}$, and write Γ so that if $(\bar{a}_i: i \in \eta)$ is the spine of $F(\eta)$, then each $\bar{a}_i$ is the n-tuple $(\iota_0(i)^\sim, \ldots, \iota_{n-1}(i)^\sim)$. The rest is routine.

Now take the linear orderings η_i as in Theorem 11.3.6, and for each $i < 2^\kappa$ let A_i be $F(\eta_i)|L$. If $i \neq j$ and there is an elementary embedding of A_i into A_j, then in $F(\eta_j)$ the image of η_i is ordered by $<$, which is impossible by Theorem 11.3.6(c). $\square$

It's tiresome to plug the gap in Theorem 11.3.8 left by the singular cardinals, not least because it needs different arguments in different models of set theory. Exercise 5 takes care of some cases, and for the rest we must fall back on the following result (whose proof I omit).

Theorem 11.3.9. *Let L be a signature and F an EM functor in L on n-tuples, such that for some quantifier-free formula $\phi(\bar{x}, \bar{y})$, $\vdash \forall \bar{x}\,\bar{y}(\phi(\bar{x}, \bar{y}) \rightarrow \neg\phi(\bar{y}, \bar{x}))$ and ϕ linearly orders the spine of each structure $F(\eta)$. Then for* *every cardinal* $\kappa > |L|$ *there is a family of* 2^κ *linear orderings* η *of cardinality* κ *such that the structures* $F(\eta)$ *are pairwise not isomorphic.* $\square$

Recently Shelah announced the definitive result in this direction, as follows. The proof is not yet published.

Theorem 11.3.10. *Let* T *be an unsuperstable theory in a first-order language* L. *Then for every cardinal* $\kappa > |L|$, *there is a family* $(A_i : i < 2^\kappa)$ *of* L-*structures which are models of* T *of cardinality* κ, *such that for all* $i \neq j$, A_i *is not elementarily embeddable in* A_j. $\square$

It may be worth remarking that Theorem 11.3.10 is not just a firework display. Some form of it was needed for showing that a first-order theory T which is categorical in some cardinal greater than $|T|$ must be superstable. See Theorem 12.2.3 below.

Taking a last glance back at Shelah's non-structure programme, we naturally ask whether these theorems can be improved to make the structures equivalent in some stronger language than $L_{\infty\omega}$. The answer is yes, and the proof is a more vigorous application of the same general ideas. To be specific, Shelah [1987a] shows that if L is a first-order language and T a complete theory in L which is not superstable, then for every regular cardinal $\kappa > |L|$, T has two models of cardinality κ which are $L_{\infty\kappa}$-equivalent but not isomorphic. For unstable theories, Hyttinen & Tuuri [1991] extend this result to even stronger languages defined by games of uncountable length, assuming the generalised continuum hypothesis.

Exercises for section 11.3

1. Show that in (3.1) we can suppose without loss that $\phi(\bar{x}, \bar{y})$ defines an asymmetric relation on the set of n-tuples of elements of A.

2. Let L be a first-order language and T an unstable complete theory in L. Show that there are models A, B, C, D of T with $A \leqslant B \leqslant D$ and $A \leqslant C \leqslant D$, forming an heir–coheir amalgam over A, such that the mirror image (with B and C transposed) is not an heir–coheir amalgam. [In the notation of Theorem 11.3.1, take η and ξ as before, but put $\zeta = (-\infty, 0) \cup [1/2] \cup (1, \infty)$ and $\theta = (-\infty, 0] \cup [1, \infty)$. Put $A = F(\eta)$, $B = F(\zeta)$, $C = F(\theta)$, $D = F(\xi)$. Consider a formula which says that $\bar{a}_{1/2}$ lies strictly between $\bar{a}_0$ and $\bar{a}_1$, noting that $\bar{a}_0$ and $\bar{a}_1$ lie in the same position relative to all the tuples in the spine of A. Caution: we are trying to move $\bar{a}_{1/2}$ into $F(\eta)$, but not necessarily into the spine of $F(\eta)$.]

3. Prove Facts 11.3.2 and 11.3.3 (excluding Fact 11.3.3(c)).

*4. Let $(\eta_i : i < 2^\kappa)$ be the orderings constructed in the proof of Theorem 11.3.4. Show (a) for each i, ω_1^* is not embeddable in η_i. (b) no η_i is the union of fewer than κ well-orderings.

*5. Prove the following variant of Theorem 11.3.6. Let L be a signature and Γ a word-construction from the class of linear orderings to the class of all L-structures. Let μ be a regular cardinal $> |\Gamma|$, and let κ be a cardinal such that $\mu \leqslant \kappa$. Then there is a family $(\zeta_i : i < 2^\mu)$ of dense linear orderings ζ_i without endpoints, such that (a) each $\Gamma(\zeta_i)$ has cardinality at most κ. (b) for all $i < j < 2^\mu$, $\Gamma(\zeta_i) \equiv_{\infty\omega} \Gamma(\zeta_j)$. (c) if $i \neq j$ then no set of tuples in $\Gamma(\zeta_j)$ is ordered in order-type ζ_i by any formula $\psi(\bar{x}_0, \bar{x}_1)$ of $L_{\infty\omega}$. [Take $(\eta_i : i < 2^\mu)$ as in the proof of Theorem 11.3.6 with μ in place of κ. For each $i < 2^\mu$, let ζ_i be $\eta_i + \kappa$.]

*6. Let T be a first-order theory with infinite models, and λ a regular cardinal $> |T|$; suppose that λ is regular but not weakly compact. Show that T has a model A of cardinality λ such that if ψ is a formula of L, X is a set of elements of A and $(X, <)$ is a ψ-indiscernible sequence of order type λ, then ψ doesn't distinguish the order of elements in $(X, <)$. [Use the fact (Exercises 11.1.8 and 11.1.9) that there is a linear ordering of cardinality λ in which neither λ nor λ^* is embeddable.]

*7. Suppose T is a theory in a first-order language L, and $\phi(x)$, $\psi(x, y)$ are formulas of L such that in every model A of T, $\phi(A)$ is infinite, has the same cardinality as A and is linearly ordered by $\psi(A^2)$. Show that for every cardinal $\lambda > |L|$ which is regular but not weakly compact, T has two models A, B of cardinality λ such that every model elementarily embeddable in both A and B must have cardinality $< \lambda$. [Use the previous exercise.]

*8. Let L be a countable first-order language and T a theory in L with the strict order property. Show that for every uncountable linear ordering η there is a model $E(\eta)$ of T of the same cardinality as η, such that if η and ζ are two uncountable linear orderings, then $E(\eta)$ is elementarily embeddable in $E(\zeta)$ if and only if η is embeddable in ζ. [Apply an EM functor F to right–left symmetric linear orderings, using Theorem 11.2.12 and information about how elements given by different terms are related. Take a subordering ξ of the reals which has cardinality ω_1, contains 0 and is invariant under multiplication by -1. If η is any linear ordering, let $\xi^{[\eta]}$ be the ordering consisting of all maps from η to ξ which differ from 0 at a finite number of places, ordered lexicographically. Map each η to $F(\xi^{[\eta]})$.]

*9. Let L be a countable first-order language and T a theory in L whose models are partially ordered by a formula $\psi(x, y)$ of L, such that some model of T contains an infinite set linearly ordered by ψ. Assuming the generalised continuum hypothesis, prove that for every infinite successor cardinal κ, T has a family of 2^κ models of cardinality κ such that any structure elementarily embeddable in two structures of the family has cardinality $< \kappa$. [Use the result of Bonnet [1980]: assuming GCH, in each infinite successor cardinality κ there is a family of 2^κ linear orderings such that every linear ordering embeddable or reverse-embeddable in two or more orderings in the family has cardinality $< \kappa$. Combine this with Theorem 11.2.12.]

11.4 Nonstandard methods

In section 11.1 we proved Ramsey's theorem by building a structure that wasn't ω but thought it was. More prosaically, we wrote down a set of axioms which has an intended model A – in our case the natural numbers with some relations on them – and we took a proper elementary extension B of A. Then we used properties of B to prove a fact about A. The elements of B which are not in the intended model A are said to be **nonstandard**, and proofs which use nonstandard elements are said to be **nonstandard**.

Abraham Robinson's **nonstandard analysis** uses nonstandard arguments to prove theorems of calculus, measure theory and the like. This is now a specialist field with its own methods and conferences; readers who want to pursue it might consult Keisler [1976], Hurd & Loeb [1985], Cutland [1988] or the third edition of Chang & Keisler [1990]. Here we shall take a brief look at another area of nonstandard methods: nonstandard algebra. But before we do that, let me make some general remarks about nonstandard methods.

A nonstandard universe

The usual setting for nonstandard arguments is as follows. We have an infinite structure M which is a standard mathematical object, such as the natural number structure $\mathbb{N} = (\omega, 0, 1, +, \cdot, <)$ or the field $\mathbb{R}$ of real numbers. We also have some proper elementary extension $*M$ (read as **pseudo-M**) of M; $*M$ is a reduct or a relativised reduct of some structure B:

(4.1)
$$
\begin{array}{c}
B \\
\uparrow \\
\mid \text{ relativised} \\
\mid \text{ reduct} \\
\mid \\
\mid \\
M \quad < \quad *M
\end{array}
$$

The structure B is called the **nonstandard universe**. We prove facts about B by using two kinds of information: first, $*M$ contains some nonstandard elements, and second, $*M$ is an elementary extension of M. In many nonstandard arguments one assumes that B (and hence also $*M$ by Theorem 10.1.9) is λ-saturated for some suitable cardinal λ.

Nonstandard universes are often constructed in one of the two following styles – we shall see examples of both.

First style. For each element i of some index set I, an L-structure A_i is given which has M as a relativised reduct. We take B to be the ultraproduct $\Pi_I A_i/\mathcal{U}$ for some non-principal ultrafilter $\mathcal{U}$ over I. Then $*M$ is $M^I/\mathcal{U}$ (since taking ultraproducts commutes with taking relativised reducts, by Theorem 9.4.1), so that $*M$ is an elementary extension of M by Corollary 9.5.2.

Algebraists tend to like this construction because it gives a concrete feel to the universe B. We can ensure that $*M$ is a proper extension of M by taking I to be of cardinality $\geq |A|$ and $\mathcal{U}$ regular (see Theorem 9.5.4). With a suitable choice of ultrafilter we can make B and $*M$ λ-saturated for some given cardinal λ too.

Second style. A structure A is given, which has M as a relativised reduct. Using the upward Löwenheim–Skolem theorem (Corollary 6.1.4), or Corollary 10.2.2 if we want λ-saturation, we form an elementary extension B of A. Then the formula which gave M in A gives a structure $*M$ in B, and so $M \preccurlyeq *M$ by Corollary 5.1.2:

(4.2)
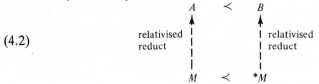

A special case is where A is M and B is just an elementary extension of M. We can ensure that $*M$ is a proper extension of M either by a compactness argument or by taking B to be $|M|^+$-saturated.

What's the idea behind nonstandard methods? I don't want to say anything that limits the imagination of the user; but in practice there are some family features in common between many nonstandard arguments. The central point is that *by passing to an elementary extension we can change a statement of the form $\forall x \exists y\, \phi$ to a statement of the form $\exists y \forall x\, \psi$.*

Example 1: *Trees*. Let A be the following partially ordered set. The elements of A are the finite sequences of 0's and 1's; if σ, τ are elements of A then $\sigma < \tau$ if and only if τ is a proper end-extension of σ. We have

(4.3) $A \vDash \forall x \exists y\, x < y.$

Note the placing of the quantifiers. Now suppose B is a proper elementary extension of A, and let ∞ be a nonstandard (i.e. new) element of B. Counting levels from the bottom of the tree, at each finite level there are just finitely many elements in A; since we can say this with a first-order sentence, the same is true in B. So ∞ must be infinitely far above the bottom level. At each finite level there must be just one element of A which is $< \infty$ in B. So ∞ picks out a branch in A, and we can say

(4.4)   *there is* an element ∞ of B such that *at every finite* level there
          is a unique element of A below ∞.

Thus a single nonstandard element encodes an infinite set of standard elements.

Our proof of Ramsey's theorem in section 11.1 was really Example 1 in disguise. Ramsey himself built up a tree of numbers and then chose a branch through the tree. We chose a branch at the outset, by picking an infinite element ∞. It took a little work to decode the branch from the element.

Purely for exposition, I group the applications below under the three heads.

Applications 1: choice principles

In section 3.1 we saw how to pass from $\forall x \exists y\, \phi(x, y)$ to $\exists F \forall x\, \phi(x, F(x))$ by introducing Skolem functions. Skolemisation is different from nonstandard analysis, because it adds new functions F^A to the structure A, not new elements ∞. But the switch of quantifiers normally needs the axiom of choice, whether it's done with new functions or new elements. Our first application of nonstandard methods is really a way of introducing the axiom of choice.

Consider a structure A which consists of a cardinal λ and the set $\mathcal{P}(\lambda)$ of all subsets of λ, with the membership relation $\in$ between elements and subsets of λ, and a 1-ary relation symbol P to pick out the set λ. Thus $A_P = \lambda$. Let B be an elementary extension of A such that B_P contains elements not in A_P. Then B is a nonstandard universe, and B_P is $*\lambda$. Choose a nonstandard element ∞ of $*\lambda$, and define

$$(4.5) \qquad \mathcal{U} = \{a \subseteq \lambda : B \vDash \text{`}\infty \in a\text{'}\}.$$

This makes sense since each subset a of λ is an element of A and hence of B too.

We claim that $\mathcal{U}$ is a non-principal ultrafilter over λ. Suppose a, b and c are subsets of λ with $a \cap b \subseteq c$. Then

$$(4.6) \qquad A \vDash \forall x (x \in a \wedge x \in b \rightarrow x \in c).$$

Since $A \preccurlyeq B$, the same is true of B. So if a and b are in $\mathcal{U}$, then computing in B we have $\infty \in a$ and $\infty \in b$, so $\infty \in c$ and hence $c \in \mathcal{U}$. A similar argument shows that for every subset a of λ, exactly one of a and $\lambda \backslash a$ lies in $\mathcal{U}$. Finally $\mathcal{U}$ is not principal; for suppose some element a of $\mathcal{U}$ is a singleton $\{\beta\}$. Then $B \vDash \text{`}\infty \in a\text{'}$, which is clearly impossible since $A \vDash \forall x (x \in a \rightarrow x = \beta)$.

This style of argument has other applications in set theory. If the elementary extension B is proper but of the same cardinality as A, this fact can sometimes be used to show that the cardinal λ is a 'large cardinal', either measurable or greater. See for example Solovay, Reinhardt & Kanamori [1978].

Applications 2: overspill

Suppose M is the structure $\mathbb{N}$ of natural numbers, and $\phi(y)$ is a first-order formula such that

(4.7) $\mathbb{N} \vDash \forall x \exists y \, (x < y \wedge \phi(y))$.

Then an easy compactness argument gives us an elementary extension $*M$ of M with a nonstandard element b satisfying ϕ; in other words

(4.8)   there is y in $*M$ such that for all natural numbers x,
         $*M \vDash x < y \wedge \phi(y)$.

This is the same reversal of quantifiers that we saw in Example 1. But now we can get extra mileage from the fact that the induction axiom holds in $\mathbb{N}$.

Lemma 11.4.1. *In the setting of (4.2) above, suppose M is $\mathbb{N}$. Then the set ω of all elements of M is not first-order definable in B with parameters.*

Proof. Suppose to the contrary that $\phi(x, \bar{y})$ is a first-order formula and $\bar{b}$ are parameters in B such that $\omega = \phi(B, \bar{b})$. Let $P(x)$ be the formula which picks out M in A. We have

(4.9) $B \vDash \exists \bar{y}(\phi(0, \bar{y}) \wedge \forall x(\phi(x, \bar{y}) \rightarrow \phi(x + 1, \bar{y})) \wedge \exists z(P(z)$
         $\wedge \neg \phi(z, \bar{y})))$.

Since $A \preccurlyeq B$, it follows that (4.9) holds with A in place of B. But this contradicts the induction axiom on $\mathbb{N}$. $\square$

Theorem 11.4.2. *In the setting of (4.2) above with $M = \mathbb{N}$, suppose X is a set which is first-order definable in B with parameters. Then the following are equivalent.*
(a) *X contains arbitrarily large numbers from M.*
(b) *For every nonstandard number y in $*M$, X contains some nonstandard number $< y$.*

Proof. If (a) holds but not (b), let y be a counterexample to (b). Then we can define ω in B as follows:

(4.10) $z \in \omega \Leftrightarrow z$ is in $*M$ and there is $x \in X$ with $z < x < y$.

A similar argument shows that (b) implies (a). $\square$

The implication (a) $\Rightarrow$ (b) in Theorem 11.4.2 is known as **overspill**, and its converse (b) $\Rightarrow$ (a) is **underspill**. Overspill constantly appears in the model theory of arithmetic, and there are analogues of it in nonstandard analysis.

The following application is interesting in its own right, though it uses less than the full strength of overspill. In many nonstandard arguments one begins by taking a highly saturated nonstandard universe. The next result shows that in nonstandard models of arithmetic, a certain amount of saturation comes for free. We assume that the language L is a first-order language of finite signature, so that each of its expressions e has a Gödel number $GN(e)$. We also assume that L contains a 1-ary relation symbol P. By an n-**formula** we shall mean a formula of the form $(\forall x_0 \in P)(\exists y_0 \in P)(\forall x_1 \in P) \ldots$ $(\forall x_{n-1} \in P)(\exists y_{n-1} \in P)\psi$ where ψ is unnested and quantifier-free.

Corollary 11.4.3. *Suppose the L-structure A has the natural numbers $\mathbb{N}$ as a relativised reduct, $A_P = \mathbb{N}$, and let B be an elementary extension of A in which $*\mathbb{N}$ is a proper extension of $\mathbb{N}$. For some $n < \omega$, let $\Phi(x, \bar{y})$ be an arithmetically definable set of n-formulas $\phi(x, \bar{y})$ of L. If $\bar{b}$ is a tuple of elements of B such that each finite subset of $\Phi(x, \bar{b})$ is realised by an element of B, then $\Phi(x, \bar{b})$ is realised in B.*

Proof. For each n and each tuple $\bar{x}$ of variables, there is a formula $Sat_n(v, \bar{x})$ of L which defines a satisfaction predicate for n-formulas $\phi(\bar{x})$, as in section 10.6. Thus if $\phi(\bar{x})$ is an n-formula, we have

(4.11) $$A \vDash \forall \bar{x}(Sat_n(GN(\phi), \bar{x}) \leftrightarrow \phi(\bar{x})),$$

and so the same sentence holds in B too. The definition of Sat_n is by induction on n, and I leave it to the reader. Now let $\psi(x)$ define in A the set of Gödel numbers of formulas in $\Phi(x, \bar{y})$. Consider the following formula $\chi(w, x, \bar{y})$:

(4.12) $$(\forall z < w)(\psi(z) \rightarrow Sat_n(z, x, \bar{y})).$$

By assumption there are arbitrarily large natural numbers n such that $B \vDash \exists x \chi(n, x, \bar{b})$. Hence by overspill there is some nonstandard number ∞ in B_P such that $B \vDash \exists x \chi(\infty, x, \bar{b})$. Choose an element a such that $B \vDash \chi(\infty, a, \bar{b})$. If ϕ is any formula in Φ, then $GN(\phi)$ satisfies ψ and is less than ∞, so that we can deduce $B \vDash Sat_n(GN(\phi), a, \bar{b})$ and hence $B \vDash \phi(a, \bar{b})$. Thus a realises $\Phi(x, \bar{b})$ in B. $\qquad\qquad\square$

Applications 3: uniform bounds

Our next example is adapted from effective algebra. Let K be a field and $\bar{X} = (X_0, \ldots, X_{k-1})$ a k-tuple of indeterminates. Suppose I is an ideal of the ring $K[\bar{X}]$, generated by polynomials $q_0, \ldots, q_{m-1}$ each of which has total degree $\leq d$. For the nonce I write $I \downarrow \mu$ to mean the set of all polynomials p in $K[\bar{X}]$ which can be written as $\sum_{i<m} q_i r_i$ where each polynomial r_i has total degree $\leq \mu$. An old result of Grete Hermann [1926] states as follows.

Fact 11.4.4. *For every choice of positive integers k and d we can compute a* *number* $\mu = \mu(k, d)$ *such that if* K, $\bar{X}$ *and* I *are as above, then every polynomial in* I *of total degree* $\leqslant d$ *lies in* $I \downarrow \mu$. $\qquad\Box$

As model theorists we may or may not be impressed that the bound $\mu(k, d)$ can be computed. But we should certainly be interested in the fact that it exists, because it tells us a fact about definability in fields. Namely, let L be the language of rings. Every polynomial in $K[\bar{X}]$ with total degree $\leqslant d$ can be written as $p(\bar{a}, \bar{X})$ where $\bar{a}$ is a tuple listing the coefficients of the monomials. Then equations between polynomials $p(\bar{a}, \bar{X})$ can be expressed as quantifier-free formulas of L relating their coefficients $\bar{a}$.

Corollary 11.4.5. *For any positive integers* d, k *and* m, *there is an* $\exists_1$ *formula of the language of rings which expresses, given any field* K *and polynomials* p, $q_0, \ldots, q_{m-1}$ *in* $K[X_0, \ldots, X_{k-1}]$, *all of total degree* $\leqslant d$, *the statement 'p is in the ideal generated by* $q_0, \ldots, q_{m-1}$'.

Proof. By the Fact, p is in this ideal if and only if there are $r_0, \ldots, r_{m-1}$ all of total degree $\leqslant \mu(k, d)$, such that $p = \sum_{i < m} q_i r_i$. We can write this out as an $\exists_1$ formula relating the coefficients. $\qquad\Box$

Hermann's proof is rather laborious, and it was natural to look for improvements. Abraham Robinson [1973b], using nonstandard methods, found a shorter proof that the bound $\mu(k, d)$ exists, but without any indication of how it can be computed. Not long afterwards, Seidenberg [1974] published an even shorter proof which is completely effective and gives an easy bound. So far this is not much of an advertisement for nonstandard methods. However, Robinson [1973a] had already described his proof as 'only a beginning'. Tragically he died within two years of making that remark, and it was left to van den Dries to take up the challenge. I shall report Seidenberg's argument, and then van den Dries' nonstandard version of it.

We prove Fact 11.4.4 by induction on $k \geqslant 0$, the case $k = 0$ being trivial. We suppose we are given positive integers k, d, a field K and polynomials p, $q_0, \ldots, q_{m-1}$ of total degree $\leqslant d$ in $K[\bar{X}]$, where $\bar{X} = (X_0, \ldots, X_{k-1})$. There are only a finite number of monomials in $\bar{X}$ with degree $\leqslant d$, and so we can bound m in terms of k and d. By assumption $p = \sum_{i < m} q_i r_i$ for some r_i in $K[\bar{X}]$, and we want to show that the r_i can be chosen to be of total degree less than some $\mu(k, d)$.

The first step is to notice that we can assume K is infinite. For otherwise replace K by a transcendental extension $K(t)$. Assuming the result is proved in the infinite case, we have $r_i(t)$ $(i < m)$ of total degree $\leqslant \mu(k, d)$ in $\bar{X}$ which satisfy $p = \sum_{i < m} q_i r_i(t)$, and we get the required polynomials $\in K[\bar{X}]$ by looking at the terms of the $r_i(t)$ in which t doesn't appear.

Next we take the equation $p = \sum_{i<m} q_i r_i$, and we apply a linear transformation to $K[\bar{X}]$ as follows. Replace each X_j $(j < k - 1)$ by $X_j + \lambda_j X_{k-1}$ for some λ_j in K. Since K is infinite, we can choose the λ_j so that after this transformation, the coefficient of the leading term of q_0 as a polynomial in X_{k-1} lies in K. Note now that we have an equation $0 = \sum_{i<m} q_i r_i'$ where $r_0' = q_1$, $r_1' = -q_0$ and $r_2' = \ldots = r_{m-1}' = 0$. Adding an appropriate multiple of $(r_0', \ldots, r_{m-1}')$ to $(r_0, \ldots, r_{m-1})$, we can bring r_1 to degree $\leq d$ in X_{k-1}; by the same argument we can do the same for $r_2, \ldots, r_{k-1}$. Having done this, we take the equation $p = \sum_{i<m} q_i r_i$ and separate out the coefficients of the powers of X_{k-1}. This gives us a set of at most $d + 1$ equations in which X_{k-1} doesn't occur, and we finish by quoting the induction hypothesis.

We turn to van den Dries' reconstruction. Taking a fixed k and d, suppose Fact 11.4.4 fails. Then

(4.13)   for all positive integers μ there are a field K_μ and polynomials $p_\mu, q_{0\mu}, \ldots, q_{m-1,\mu}$ of total degree $\leq d$, such that if I_μ is the ideal generated by $q_{0\mu}, \ldots, q_{m-1,\mu}$ in $K[\bar{X}]$ then $p_\mu \in I_\mu \backslash I_\mu \downarrow \mu$.

(We saw that m can be chosen from k and d.) Let A be a universe of sets which contains all the fields K_μ and the polynomial rings $K_\mu[\bar{X}]$. This means for example that A contains the set ω of natural numbers and a map which takes each polynomial to its total degree $< \omega$. For each μ let A_μ be A with symbols picking out $K_\mu, p_\mu, q_{0\mu}, \ldots, q_{m-1,\mu}$ and ω (together with the operations for addition etc.). Let B be a non-principal ultraproduct $\prod_\omega A_n / \mathcal{U}$. Then B is a nonstandard universe containing the nonstandard number system $*\omega = \omega^\omega / \mathcal{U}$.

The symbol which picked out K_μ in each A_μ picks out a field K in B; this follows from Łoś's theorem (Theorem 9.5.1), since we can write down a first-order formula which expresses 'x is a field', along the lines of the usual definition of fields. Then in B the first-order definition of the polynomial ring over K in the indeterminates $\bar{X}$ yields an object $*K[\bar{X}]$. (This is $*(K[\bar{X}])$, not $(*K)[\bar{X}]$.) What can we say about $*K[\bar{X}]$?

Certainly $*K[\bar{X}]$ was a commutative ring, since each $K_\mu[\bar{X}]$ was a commutative ring – this follows by Łoś's theorem again, using the fact that the axioms for commutative rings are first-order. Also each element of $*K[\bar{X}]$ is a 'polynomial' with a total 'degree' in $*\omega$. The snag is that $*\omega$ contains nonstandard numbers; in each A_μ every element of ω was the degree of some polynomial in $K_\mu[\bar{X}]$, and so (by Łoś again) each element of $*\omega$ is the total 'degree' of some 'polynomial' in $*K[\bar{X}]$.

Terminological aside. In nonstandard arguments one often distinguishes between **internal** objects which are first-order definable in the nonstandard universe, and **external** objects which aren't. Thus $*K[\bar{X}]$ is internal and $K[\bar{X}]$ is external. (But note that some authors use the words 'internal' and 'external' in subtler ways than this.) Likewise one distinguishes what is true 'inside B' from what is true 'in the real world': thus inside B, $*K[\bar{X}]$ is a polynomial ring.

The symbol which picked out p_μ in each A_μ will now pick out a polynomial p in B. But here we are in better shape; each p_μ had total degree $\leq d$ and d is finite, so by Łoś's theorem again, p has total degree $\leq d$. The same holds for polynomials $q_0, \ldots, q_{m-1}$. For each μ there are $r_{i\mu}$ $(i < m)$ such that $p_\mu = \sum_{i<m} q_{i\mu} r_{i\mu}$, and so by Łoś again there are r_i in $*K[\bar{X}]$ such that $p = \sum_{i<m} q_i r_i$. However, for each μ at least one of the $r_{i\mu}$ must have total degree $> \mu$ by (4.13), and so at least one of the r_i must have infinite total degree (again by Łoś, need I say?).

Notice that we have carried out exactly the usual nonstandard reversal of quantifiers. Assumption (4.13) has become '*There exists* a nonstandard $*K[\bar{X}]$ with elements $p, q_0, \ldots, q_{m-1}$ such that *for all standard μ . . .*'.

Now we get a contradiction by proving

(4.14)   if $f, g_0, \ldots, g_{n-1}$ are *any* polynomials $\in K[\bar{X}]$ such that the equation $f = \sum_{i<n} g_i h_i$ can be solved for some $h_i \in *K[\bar{X}]$, then these h_i differ from some solution in $K[\bar{X}]$ by a linear combination of some solutions in $K[\bar{X}]$ of the corresponding homogeneous equation $0 = \sum_{i<n} g_i h_i'$.

The proof of (4.14) is exactly Seidenberg's argument.

At first sight van den Dries' argument is just Seidenberg's with some set-theoretic nonsense attached. The major advantage of van den Dries' version is that (4.14) is exactly the statement that the ring $*K[\bar{X}]$ is faithfully flat over the ring $K[\bar{X}]$ (Bourbaki [1972] section I.3.7). This is an algebraic statement with a clear intuitive meaning. In a more complicated example, that could make all the difference between finding a proof and not finding it.

The major disadvantage of van den Dries' version is that it carries the burden of forming an ultraproduct. This disadvantage is not too serious; that part of the argument is highly predictable, and after a little practice one can dismiss it with sentence or two.

It would be on my conscience if I finished this section without mentioning two beautiful applications of nonstandard methods. The first is a reworking by van den Dries & Wilkie [1984] of Gromov's proof that every finitely generated group of polynomial growth has a nilpotent subgroup of finite index. They replace Gromov's limits of systems of standard metric spaces by a single nonstandard metric space. The effect is to turn a difficult proof into something quite intuitive.

The second application is in field theory. A field A is said to be **hilbertian** if

(4.15)   for every finite set $\{p_i(t, x): i < n\}$ of irreducible polynomials
            over A, there is an element a of A such that each of the
            polynomials $p_i(a, x)$ is irreducible over A.

This definition goes back to work of Hilbert. Gilmore and Robinson showed
that a field A is hilbertian if and only if there is an elementary extension B of
A containing an element b which is transcendental over A, such that the field
$A(b)$ is algebraically closed in B. This characterisation is so neat and simple
that one is half tempted to say it justifies the definition (4.15). Using it,
Roquette was able to give an elegant proof of an old result of Franz and
Inaba, that every function field over an infinite field is hilbertian.

    These two applications prompt one last comment. Why is it that non-
standard methods are used so often for giving new proofs of old results?
Can't they give us anything new? The answer is that mathematicians using
nonstandard methods are finding new things all the time, of course. But it
would make very little sense to claim that 'there are theorems which can be
proved with nonstandard methods but can't be proved without them'. In the
first place there is no sharp boundary between nonstandard and standard
methods. In the second place, Henson & Keisler [1986] made the question
precise for nonstandard analysis, and concluded that every statement of
analysis provable by nonstandard methods can also be proved using higher-
order comprehension axioms instead.

Exercises for section 11.4

*Here is a nonstandard proof of the Robinson joint consistency lemma (for countable
theories).*
1. Let L_1 and L_2 be two countable first-order languages and $L = L_1 \cap L_2$. Let T_1
and T_2 be consistent theories in L_1 and L_2 respectively, such that $T_1 \cap T_2$ is a
complete theory T in L. In some countable model U of set theory, let A and B be
respectively models of T_1 and T_2, and let σ be (in U) a map which takes each $r < \omega$
to a winning strategy σ_r for player $\exists$ in the game $EF_r(A|L, B|L)$ (see Corollary 3.3.3).
Let V be a countable proper elementary extension of U in which the set of natural
numbers, $*\omega$, is a proper extension of ω. Consider a nonstandard element ∞ of $*\omega$,
and use the strategy σ_∞ to prove that $*A|L \cong *B|L$. *This argument is interesting
because of a variant. Suppose $\mathcal{L}$ is a logic which contains first-order logic, and such that
every countable model of a theory in $\mathcal{L}$ has a proper countable extension which is also a
model of this theory. Then elementary equivalence implies $\mathcal{L}$-equivalence; for otherwise
consider a sentence ϕ of $\mathcal{L}$ and replace T_1, T_2 in the argument above by $T \cup \{\phi\}$ and
$T \cup \{\neg\phi\}$ to get a contradiction.*

*2. Let λ be a strongly inaccessible cardinal. Show that the following are equivalent.
(a) $\lambda \to (\lambda)_2^2$. (b) If ϕ is any sentence of $L_{\lambda^+\omega}$ and A is a model of ϕ, then some proper
extension of A is also a model of ϕ. (c) If B is a boolean algebra of cardinality λ, and
for each $i < \lambda$, b_i is an element of B and X_i is a subset with $b_i = \sup(X_i)$, then B has

a non-principal ultrafilter $\mathcal{U}$ with the following property: for all $i < \lambda$, if $b_i \in \mathcal{U}$ then X_i meets $\mathcal{U}$.

3. Let V be a transitive model of ZFC and $j: V \to M$ a proper elementary embedding of V into a transitive submodel M of V, where both j and M are definable in V. (a) Show that j must move at least one ordinal upwards. [Use induction on rank.] (b) Show that if α is the least ordinal which is moved by f, and $\mathcal{U}$ is the set of all subsets X of α in V such that $\alpha \in j(X)$, then $\alpha > \omega$ and $\mathcal{U}$ is (in V) a non-principal $|\alpha|$-complete ultrafilter over α. ($|\alpha|$-complete means that if $\kappa < |\alpha|$ and $(X_i : i < \kappa)$ is a family of members of $\mathcal{U}$, then $\bigcap_{i<\kappa} X_i \in \mathcal{U}$.) [If $Y = \bigcap_{i<\kappa} X_i$ then $j(Y) = \bigcap_{i<\kappa} j(X_i)$.]

4. In a nonstandard universe B, let $*\mathbb{R}$ be a proper elementary extension of the field $\mathbb{R}$ of real numbers. An **infinitesimal** is an element of $*\mathbb{R}$ which is > 0 but $< 1/n$ for every standard n. (a) Using the embedding of $\mathbb{N}$ in $\mathbb{R}$, show that $*\mathbb{R}$ contains elements which are $>$ every standard number, and deduce that $*\mathbb{R}$ contains infinitesimals. (b) Two elements a, b of $*\mathbb{R}$ are said to be **near** each other if $|b - a| < \alpha$ for some infinitesimal α. Show that nearness is an equivalence relation.

5. In the setting of Exercise 4, suppose that the structure $\mathbb{R}$ also carries a symbol for a function $f: I \to \mathbb{R}$ where I is an interval of $\mathbb{R}$, and let $*f: *I \to *\mathbb{R}$ be the function named in $*\mathbb{R}$ by the same symbol. (a) Show that $*f$ extends f. (b) Show that f is continuous on I if and only if for every element α of $*I$ which is near an element b of I, $*f(\alpha)$ is near $f(b)$. [Use overspill.] (c) Show that f is uniformly continuous on I if and only if for all elements α, β in $*I$, if α is near β then $*f(\alpha)$ is near $*f(\beta)$. [Overspill again.] (d) Show that if I is a closed and bounded interval then every element in $*I$ is near an element of I. (e) Deduce that every continuous real function on a closed and bounded interval is uniformly continuous.

*6. In a nonstandard universe B, let K be an internal field and $X_0, \ldots, X_{k-1}$ indeterminates. Show that the finitely generated field extension $*K(X_0, \ldots, X_{k-1})$ in the sense of B is a regular extension of the (external) field $K(X_0, \ldots, X_{k-1})$.

11.5 Defining well-orderings

Through most of this section we make the following assumptions.

(5.1) L is a first-order language which has among its symbols a 1-ary relation symbol P and a 2-ary relation symbol $<$. T is a theory in L which contains a sentence expressing '$<$ is a linear ordering of the elements x such that Px'.

If A is a model of T, then $<^A$ linearly orders P^A. We shall speak of the order type of $<^A$ as the **order type** of the structure A. Likewise we shall say A is **well-ordered** if $<^A$ is a well-ordering. Sometimes we shall have an infinitary sentence in place of T, but the same terminology applies.

If T has a model whose order type is infinite, then T has a model whose

order type is not well-ordered. This is an easy consequence of the compactness theorem (see Exercise 6.1.12). But if we restrict ourselves to models of
T which omit certain types, then the position alters dramatically. For
example, suppose α is an ordinal and L contains names for all ordinals $< \alpha$.
Let T be the complete theory of the structure $(\alpha, <)$, and for each ordinal
$\beta \leq \alpha$ let Φ_β be the type

$$(5.2) \qquad \{'x < \beta'\} \cup \{'x \neq \gamma': \gamma < \beta\}.$$

(When $\beta = \alpha$ we leave out the first formula.) Then the only models of T
which omit all the types Φ_β $(\beta \leq \alpha)$ are $(\alpha, <)$ and the structures isomorphic
to it.

So we can drastically alter the descriptive power of a first-order theory by
restricting to those models which omit certain types. This is hardly surprising:
Chang's reduction (Theorem 2.6.7) showed that every sentence of $L_{\infty\omega}$ can
be paraphrased as a first-order theory together with statements that certain
types are omitted.

Nevertheless we shall show that if $\mathbf{K}$ is the class of order types of those
models of some fixed first-order theory T which omit all the types in some
fixed set S, then $\mathbf{K}$ can't be the class of all well-orderings. If $\mathbf{K}$ contains a
large enough ordinal (and 'large enough' can be measured in terms of T and
S), then $\mathbf{K}$ must contain some order type which is not well-ordered. This is a
powerful combinatorial theorem – it can be used to set a bound on the
expressive power of several different kinds of logic.

Finding non-well-ordered models

Until further notice (5.1) holds and S is a set of types of T. We write
$\mathbf{K}(T, S)$ for the class of those models of T which omit every type in S.

Theorem 11.5.1. Let λ be $\max(|L|, |S|)$. If for every ordinal $\alpha < (2^\lambda)^+$,
$\mathbf{K}(T, S)$ contains a well-ordered structure of order type $\geq \alpha$, then $\mathbf{K}(T, S)$
contains a structure which is not well-ordered.

Proof. We can suppose without loss that T is a Skolem theory (by the
skolemisation theorem, Theorem 3.1.2). Introduce new constants $(c_n: n < \omega)$;
let L_n be L with the constants $c_0, \ldots, c_{n-1}$ added, and put $L_\omega = \bigcup_{n<\omega} L_n$.
Thus L_0 is L. The aim will be to build an L_ω-structure B which is a model of
T omitting all the types in S, and such that $c_0^B >^B c_1^B >^B \ldots$.

Assume that for every ordinal $\alpha < (2^\lambda)^+$ there is a well-ordered model A_α
in $\mathbf{K}(T, S)$ with order type $\geq \alpha$. For each $n < \omega$ we shall find a sequence of
L_n-structures $(A_\alpha^n: \alpha < (2^\lambda)^+)$ so that

(5.3) each A_α^n is an expansion of some A_β^{n-1}, and $A_\alpha^n \models P(c_{n-1})$ (at
         $n = 0$, A_α^0 is one of the A_β's),

(5.4)   for each α, $A_\alpha^n \vDash c_0 > \ldots > c_{n-1}$, and the set of elements which are $< c_{n-1}$ in A_α^n has order type $\geq \alpha$ (at $n = 0$, A_α^0 has order type $\geq \alpha$),

(5.5)   for each $m < \omega$, each type $\Phi(x_0, \ldots, x_{m-1}) \in S$ and each m-tuple $\bar{t}$ of terms $t_i(y_0, \ldots, y_{n-1})$ of L, there is a formula $\phi_{\Phi, \bar{t}} \in \Phi$ such that $A_\alpha^n \vDash \neg \phi_{\Phi, \bar{t}}(t_0(\bar{c}), \ldots, t_{m-1}(\bar{c}))$ for each $\alpha < (2^\lambda)^+$ (where $\bar{c}$ is $(c_0, \ldots, c_{n-1})$).

When n is 0, (5.5) holds if the structures A_α^0 are all elementarily equivalent. But as α ranges through $(2^\lambda)^+$, there are at most $2^{|L|} \leq 2^\lambda$ theories $\mathrm{Th}(A_\alpha)$. We choose a theory U which appears cofinally in this list, and we take each A_α^0 to be some model A_β of U with $\beta \geq \alpha$. Then (5.3)–(5.5) are satisfied.

When the A_α^n have been defined, we choose A_α^{n+1} as follows. Let C_α be $A_{\alpha+1}^n$ with the symbol c_n added in such a way that $c_{n-1} > c_n$ (if $n > 0$) and the set of elements $< c_n$ has order type $\geq \alpha$; this is possible by (5.4). Writing $\bar{c}$ for $(c_0, \ldots, c_n)$, let $\Phi(x_0, \ldots, x_{m-1})$ be a type $\in S$ and $\bar{t} = (t_0, \ldots, t_{m-1})$ an m-tuple of terms $t_i(y_0, \ldots, y_n)$ of L. Since C_α omits Φ by (5.4), there is a formula ϕ in Φ such that $C_\alpha \vDash \neg \phi(t_0(\bar{c}), \ldots, t_{m-1}(\bar{c}))$; choose one such formula ϕ and call it $\phi_{\Phi, \bar{t}}^\alpha$. By the definition of λ, the number of pairs $(\Phi, \bar{t})$ is at most λ and for each pair the number of possible choices for ϕ is at most λ. So the number of possible maps $(\Phi, \bar{t}) \mapsto \phi_{\Phi, \bar{t}}^\alpha$ is at most 2^λ. Since the number of structures C_α is $(2^\lambda)^+$, there is a cofinal subset Z of $(2^\lambda)^+$ so that the choice of formulas $\phi_{\Phi, \bar{t}}^\alpha$ is the same for all $\alpha \in Z$. Renumbering, we can suppose that Z is the whole of $(2^\lambda)^+$. Then we define A_α^{n+1} to be C_α for each $\alpha < (2^\lambda)^+$, to get (5.3)–(5.5) everywhere.

Let T^* be the set of all sentences ψ of L_ω such that for some $n < \omega$, all the structures $A_\alpha^n (\alpha < (2^\lambda)^+)$ are models of ψ. Then $T \subseteq T^*$, and hence T^* is also a Skolem theory since L_ω comes from L by adding constants (see Exercise 3.1.1). Clearly every finite subset of T^* has a model, and so by the compactness theorem T^* has a model D. By (5.4), D is not well-ordered. The substructure of D generated by the empty set is an elementary substructure of D, since T^* is a Skolem theory. So we can assume that every element of D has the form $t(c_0, \ldots, c_{n-1})^D$ for some term t of L. By (5.5) it follows that D omits every type in S. Thus the reduct $D|L$ is in $\mathbf{K}(T, S)$ and is not well-ordered. $\square$

The theorem immediately allows us to define the **well-ordering numbers** $\delta(\kappa, \lambda)$ as follows. (In a moment we shall calculate some more precise values for $\delta(\kappa, \lambda)$ in particular cases.)

Corollary 11.5.2. *For every pair of cardinals κ, λ there is a least ordinal $\delta = \delta(\kappa, \lambda)$ with the following property.*

(5.6)   *Let L and T be as in (5.1) with $|L| \leq \kappa$, and let S be a set of
         at most λ types of T. Let $\mathbf{K}(T, S)$ be the class of models of T
         which omit every type in S, and suppose that for every ordinal
         $\alpha < \delta$ there is a structure in $\mathbf{K}(T, S)$ which is well-ordered
         and has order type $\geq \alpha$. Then $\mathbf{K}(T, S)$ contains a structure
         which is not well-ordered.*

When κ is infinite, the ordinal $\delta(\kappa, \lambda)$ is a limit ordinal $< (2^{\kappa+\lambda})^+$.

Proof. When κ is finite, (5.6) holds vacuously; so we assume henceforth that
κ is infinite. The theorem shows that the ordinal $(2^{\kappa+\lambda})^+$ has the property
(5.6), so that $\delta(\kappa, \lambda)$ exists and is $\leq (2^{\kappa+\lambda})^+$. We show that $\delta(\kappa, \lambda) < (2^{\kappa+\lambda})^+$.
Let us call two signatures **essentially the same** if they have the same number
of constants, the same number of n-ary relation symbols and the same
number of n-ary function symbols for each $n < \omega$. There is a set Ω of 2^κ
signatures of cardinality κ such that every signature of cardinality κ is
essentially the same as one in Ω. In each signature in Ω there are at most 2^κ
distinct first-order theories and 2^κ distinct types; so the number of ways of
choosing a theory T and a set S of at most λ types in this signature is at most
$2^\kappa \times (2^\kappa)^\lambda = 2^{\kappa+\lambda}$. Now for each signature in Ω and each pair (T, S) of this
signature, there is an ordinal $\beta < (2^{\kappa+\lambda})^+$ such that either $\mathbf{K}(T, S)$ contains
no well-ordered structures of order type $\geq \beta$, or $\mathbf{K}(T, S)$ contains non-well-
ordered structures. Then $\delta(\kappa, \lambda)$ is at most the sup of these ordinals β, which
is $< (2^{\kappa+\lambda})^+$ since $(2^{\kappa+\lambda})^+$ is regular.

To show that $\delta(\kappa, \lambda)$ is a limit ordinal, suppose $\alpha < \delta(\kappa, \lambda)$. We see as
follows that $\alpha + 1$ must be $< \delta(\kappa, \lambda)$ too. By the definition of $\delta(\kappa, \lambda)$ there is
some $\mathbf{K}(T, S)$ in a signature of cardinality $\leq \kappa$, with $|S| \leq \lambda$, such that all
structures in $\mathbf{K}(T, S)$ are well-ordered, and at least one, say A, has order
type $\geq \alpha$. Now write down a theory T^+ in a first-order language L^+ which
contains L and a new 1-ary relation symbol Q, so that T^+ expresses 'The set
of elements satisfying Q forms a model of T, and the set of elements
satisfying P consists of those in Q together with one extra element, which
comes at the end in the $<$-ordering'. With the obvious choice of S^+, this
gives us a family $\mathbf{K}(T^+, S^+)$ of well-ordered structures including one of order
type $\geq \alpha + 1$. $\square$

We put $\delta(\kappa) = \delta(\kappa, \kappa)$, and we call $\delta(\kappa)$ the **well-ordering number** of κ.

Theorem 11.5.3. (a) $\delta(\kappa) = \delta(\kappa, \mu)$ *for every* μ, $1 \leq \mu \leq \kappa$.
  (b) *If κ is an infinite cardinal then* $\kappa^+ \leq \delta(\kappa) < (2^\kappa)^+$
  (c) $\delta(\omega) = \omega_1$.

Proof. (a) It suffices to show that $\delta(\kappa, 1) \geq \delta(\kappa, \kappa)$. Suppose $\alpha < \delta(\kappa, \kappa)$, and let $\mathbf{K}(T, S)$ witness this. Without loss we can add a pairing function to T and assume that every tuple in S is 1-ary. Expand the language by adding (i) a constant $c_{\psi, \Phi}$ for each type $\Phi \in S$ and formula $\psi(x) \in \Phi$, (ii) a 1-ary relation symbol P_Φ for each type $\Phi \in S$, (iii) a 1-ary relation symbol Q and (iv) a 2-ary relation symbol Sat. Write sentences to express that every constant $c_{\psi, \Phi}$ satisfies $P_\Phi(x)$ but not $P_\Theta(x)$ when $\Phi \neq \Theta$, that every element satisfying P_Φ satisfies Q, that $Sat(c_{\psi, \Phi}, x)$ is equivalent to $\psi(x)$, and that for each $\Phi \in S$ there is no element satisfying $\forall y (P_\Phi(y) \to Sat(y, x))$. Let T' be the resulting theory and S' the singleton set containing the type $\{Q(x)\} \cup \{x \neq c_{\psi, \Phi} : \Phi \in S$ and $\psi \in \Phi\}$. Then every structure in $\mathbf{K}(T, S)$ corresponds to a structure in $\mathbf{K}(T', S')$ with the same order type, and vice versa.

(b) Corollary 11.5.2 gave the second inequality. For the first, suppose $\alpha < \kappa^+$. By Exercise 2.2.10 there is a sentence ϕ_α of $L_{\kappa^+\omega}$ which characterises the ordinal α; Chang's reduction turns ϕ_α into a theory T and a set of types S of cardinality $\leqslant \kappa$ such that all the structures in $\mathbf{K}(T, S)$ have order type α. This shows that $\delta(\kappa) > \alpha$; so $\delta(\kappa) \geqslant \kappa^+$.

(c) We alter the proof of Theorem 11.5.1 as follows. We begin with a battery of structures $(A_\alpha : \alpha < \omega_1)$, and we replace $(2^\lambda)^+$ by ω_1 everywhere in the proof. There are countably many pairs $(\Phi, \bar{\imath})$ to be dealt with; we can alter (5.5) so that for each $n < \omega$ it mentions just one pair $(\Phi_n, \bar{\imath}_n)$. Then each of the C_α's chooses a formula $\phi_{\Phi_n, \bar{\imath}_n}$. Since there are only countably many formulas to choose from, there must be uncountably many C_α which choose the same formula.   ☐

Definability in infinitary languages

Corollary 11.5.2 applies at once to $L_{\infty\omega}$.

Theorem 11.5.4. *Let κ be an infinite cardinal. Then the class of well-orderings is not definable in any language $L_{\kappa^+\omega}$. More precisely, if L is a signature containing a symbol $<$, and $\phi(x)$, ψ are respectively a formula and a sentence of $L_{\kappa^+\omega}$ such that ψ has a model A in which $<^A$ is a linear ordering of $\phi(A)$ in order type $\geqslant \delta(\kappa)$, then ψ has a model B in which $<^B$ is a linear ordering of $\phi(B)$ and is not well-ordered.*

Proof. Let ψ be a sentence of $L_{\kappa^+\omega}$. The number of symbols occurring in ϕ and ψ is at most κ, so that we lose nothing if we assume $|L| \leqslant \kappa$. Chang's reduction (Theorem 2.6.7) gives us a signature $L^+ \supseteq L$, a first-order theory T in L and a set S of first-order types in L^+ such that the models of $\psi \wedge \forall x (Px \leftrightarrow \phi(x))$ are precisely the L-reducts of the models of T which omit every type in S. The cardinality of L^+ is at most $|L| + ($the number of subformulas of $\psi \wedge \phi) \leqslant \kappa$, using Exercise 2.1.9. The number of types in S is at most the number of infinitary conjunctions and disjunctions in $\psi \wedge \phi$, so again it is $\leqslant \kappa$. Corollary 11.5.2 does the rest. ☐

Theorem 11.5.4 allows us to show that various classes are not definable by sentences of $L_{\infty\omega}$; in this it's the counterpart of the compactness theorem for first-order logic. For example Exercise 3.2.10 reported that any ring which is back-and-forth equivalent to a left noetherian ring is again left noetherian, from which it follows that the class of left noetherian rings is axiomatised by a proper class of sentences of $L_{\infty\omega}$ where L is the signature of rings. The next corollary shows that this proper class can't be cut down to a set.

Corollary 11.5.5. *There is no sentence of $L_{\infty\omega}$ which defines the class of noetherian commutative rings. (In fact this class is not even a PC_Δ class in any $L_{\kappa^+\omega}$.)*

Proof. Aiming for a contradiction, we assume that there is a sentence ψ of $L_{\infty\omega}$ such that the class of noetherian commutative rings is the class of relativised reducts B_P of models B of ψ. There is some infinite cardinal κ such that ψ is a sentence of $L_{\kappa^+\omega}$.

Let A be any noetherian commutative ring. Then A has no infinite ascending chains of ideals. It follows that we can rank the proper ideals I of A by ordinals, so that for each ideal I the rank of I is $\sup\{\text{rank}(J) + 1 \colon J$ is an ideal properly extending $I\}$. Then we can incorporate A into a structure M which also contains two defined sets *Ideals* and *Ranking* as follows:

(5.7)   there is a relation R of M such that for each $i \in Ideals$ the
set $\{a \in A \colon R(a, i)\}$ is an ideal (call it R_i) of A;
there is a relation $<$ which linearly orders *Ranking*;
there is a map F which maps *Ideals* onto *Ranking*;
if $i \in Ideals$ and $w < F(i)$ then there is $j \in Ideals$ such that
$F(j) = w$ and $R_i \subset R_j$.

(Take *Ideals* to be the set of ideals of A and *Ranking* to be the set of ranks of ideals; F is the ranking function.) The facts in (5.7) can all be written as a first-order sentence χ_0 which is true in M. Also by exploiting ψ we can find a sentence χ_1 of $L'_{\kappa^+\omega}$ (for the appropriate signature L') which is true in M and expresses that A is a noetherian commutative ring. Write χ_2 for $\chi_0 \wedge \chi_1$.

Now by a theorem of Gordon & Robson ([1973] p. 66) and Gulliksen [1974], for every ordinal α there is a noetherian commutative ring A with ideals of rank $\geqslant \alpha$. (In fact for 'ideal' we could read 'prime ideal' throughout – but for our purposes this stronger result is unnecessary.) Hence χ_2 has models M in which $<^M$ well-orders $Ranking^M$ in arbitrarily large order type. By Theorem 11.5.4 it follows that χ_2 has a model N in which $<^N$ doesn't well-order W^N.

Consider this model N. Certainly $<^N$ linearly orders $Ranking^N$. By χ_1, N incorporates a noetherian commutative ring B, and there is a map F^N mapping a subset of the set of ideals of B onto $Ranking^N$ in such a way that if $w <^N F^N(I)$ in $Ranking^N$ then there is some ideal J properly extending I, such that $F^N(J) = w$. But since B is noetherian, it follows that $Ranking^N$ must be well-ordered. This contradiction proves the theorem. $\qquad\square$

Other applications

There is plenty more to be said about the ideas of this section. I mention two themes, rather too briefly.

The first is *inductive definitions*. In section 2.8 we defined the expression $\mathrm{LFP}_{P,\bar{x}}\phi$ where $\phi(\bar{x})$ is a formula in which the n-ary relation symbol P is positive. In a structure A, $\mathrm{LFP}_{P,\bar{x}}\phi$ defines the least fixed point $\mathrm{lfp}(\pi)$ of the operation π defined by

(5.8) $\pi(X) = \{\bar{a}: (A, X, \ldots) \vDash \phi(\bar{a})\}$ when P is interpreted as a
         name for the set $X \subseteq (\mathrm{dom}\ A)^n$.

We write this operation π as $\pi_{P,\phi}$. The least ordinal α such that $\pi^{\alpha+1}(\varnothing) = \pi^\alpha(\varnothing)$ is called the **closure ordinal** of π. We saw in Exercise 2.4.17 that if ϕ is an $\exists_1$ first-order formula then the least fixed point of π is $\pi^\omega(\varnothing)$, which implies that the closure ordinal of π is at most ω. But if ϕ has other forms, the closure ordinal of π may be much larger than ω. Indeed by Exercise 2.8.11 there need not be an upper bound to the closure ordinal of π independently of the structure A.

Corollary 11.5.6. *Let κ be an infinite cardinal and $\phi(\bar{x})$ a formula of $L_{\kappa^+\omega}$ in which the n-ary relation symbol P is positive. Let ψ and $\theta(\bar{x})$ be respectively a sentence and a formula of $L_{\kappa^+\omega}$ with the following property:*

(5.9)   *if A is any model of ψ, then θ defines the least fixed point of
         $\pi_{P,\phi}$ on A.*

Then there is an ordinal $\gamma < \delta(\kappa)$ such that for every model A of ψ, the closure ordinal of $\pi_{P,\phi}$ is $\leq \gamma$.

Proof sketch. The proof is very much the same as that of Corollary 11.5.5, which could be regarded as a special case. Assuming (5.9), we can write a sentence χ of $L'_{\kappa^+\omega}$ (where L' is a signature extending L) so that each model of χ consists of a model A of ψ together with a set $Ranking$ linearly ordered by $<$, and a map which correlates each element α of $Ranking$ to the set of those elements of $\theta(A)$ which lie in $\pi^{\alpha+1}(\varnothing)\backslash\pi^\alpha(\varnothing)$. In fact (5.9) will ensure that $Ranking$ is well-ordered, so that this condition makes sense. If the conclusion fails, then for every $\gamma < \delta(\kappa)$ we have a model of χ whose order type is an ordinal $\geq \gamma$, and so Corollary 11.5.2 finds a model of χ which is not well-ordered; contradiction. □

Arguments like the one above are sometimes explained by saying that there must be a *formal proof* that θ defines $\mathrm{lfp}(\pi)$, and then an ordinal bound can be extracted from the proof. Certainly there are cases where an argument along those lines gives useful information. But it can hardly be the truth behind Corollary 11.5.6, since our treatment has never come within ten miles of a formal proof in $L_{\kappa^+\omega}$.

Example 1: *Euclid's algorithm.* Let R be a ring. For any set X of elements of R, we define X' to be the set $X \cup \{x \in R : (\forall y \in R)\ (\exists w \in R) y + xw \in X \cup \{0\}\}$. Thus $X' = \pi(X)$ where $\phi(x)$ is $Px \vee \forall y \exists w\ (P(y + xw) \vee y + xw = 0)$ and π is $\pi_{P,\phi}$. The set $\pi^1(\varnothing)$ is the set of units of R, and an easy inductive argument shows that $0 \notin \mathrm{lfp}(\pi)$. To say that $R\setminus\{0\} = \mathrm{lpf}(\pi)$ is the same thing as saying that Euclid's algorithm works for R. More precisely, suppose $R\setminus\{0\} = \mathrm{lfp}(\pi)$ and define $|a|$ to be the least ordinal α such that $a \in \pi^{\alpha+1}(\varnothing)$ if $a \neq 0$, and -1 if $a = 0$. Suppose a and b are non-zero elements of R, say with $|b| = \beta$. Then by definition of π there is an element q in R such that $|a + bq| < \beta$. (Conversely one can show that if R satisfies Euclid's algorithm for some valuation function v, then $R\setminus\{0\} = \mathrm{lfp}(\pi)$ and $v(a) \geq |a|$ for all elements a of R. I leave this as an exercise.) Accordingly we say that R **satisfies the transfinite division algorithm** when $R\setminus\{0\} = \mathrm{lfp}(\pi)$.

Samuel [1968] asked whether in every ring which satisfies the transfinite division algorithm, the closure ordinal is at most ω. Notice that the formula ϕ contains a universal quantifier, so that the argument of Exercise 2.4.17 doesn't apply; and indeed Jategaonkar [1969] gave examples of noncommutative rings satisfying the transfinite division algorithm with arbitrarily high closure ordinals. For commutative rings Samuel's question remains open. What Corollary 11.5.6 tells us is that if **K** is a class of rings defined by a sentence of $L_{\omega_1\omega}$, and every ring in **K** satisfies the transfinite division algorithm, then some ordinal below ω_1 uniformly bounds the closure ordinals for these rings.

To get sharper general results on closure ordinals, one should work in admissible set theory. By that route we can bring the ordinal in Example 1 down below Church–Kleene constructive ω_1 when **K** is defined by a sentence in the smallest admissible set above ω – this is still much too high for an answer to Samuel's question!

The second theme is *approximations to Vaught sentences*. Vaught sentences are a generalisation of the Svenonius sentences of Example 2 in section 3.4. A **Vaught sentence** of signature L is a sentence ϕ of the form

(5.10) $\forall y_0 \bigwedge\limits_{j_1 \in J_1} \exists z_2 \bigvee\limits_{i_3 \in I_3} \forall y_4 \bigwedge\limits_{j_5 \in J_5} \exists z_6 \bigvee\limits_{i_7 \in I_7} \forall y_8 \dots$

$\bigwedge\limits_{n<\omega} \psi^{(n)}_{j_1 i_3 j_5 i_7 \dots} (y_0, z_2, y_4, \dots, \bar{x})$

where J_1, I_3, J_5, $\dots$ are sets, and for every $n < \omega$ and every choice of indices $j_k \in J_k$, $i_k \in I_k$ (for all possible $k < n$), $\psi^{(n)}_{j_1 i_3 j_5 \dots}$ is a formula of $L_{\infty\omega}$ whose free variables are the y_k and the z_k with $k < n$. The first line of (5.10) is called the **Vaught prefix** of the Vaught sentence ϕ. The second line is the **matrix** of the sentence, and the formulas $\psi^{(n)}_{j_1 i_3 j_5 \dots}$ are its **matrix formulas**.

We interpret this Vaught sentence ϕ in an L-structure A by means of a **Vaught game** $\mathrm{Va}(\phi, A)$ of length ω, as follows. First player $\forall$ chooses an element a_0 of A, then he chooses an element j_1 of J_1, then player $\exists$ chooses an element b_2 of A, then she chooses an element i_3 of I_3, then player $\forall$ chooses an element a_4 of A, and so on. At the end of the play, player $\exists$ wins the game if

(5.11) $A \models \psi^{(n)}_{j_1 i_3 j_5 \dots} (a_0, b_2, a_4, \dots)$ for every $n < \omega$;

otherwise player $\forall$ wins. We write $A \models \phi$ to mean that player $\exists$ has a winning strategy for this game $\mathrm{Va}(\phi, A)$.

The obvious analogue of Theorem 3.4.5 holds, with the same proof. A position in the game can be thought of as a pair $(\bar{i}\bar{j}, \bar{a}\bar{b})$ of tuples representing the choices made so far. There is a formula $\theta^{n,\alpha}_{\bar{i}\bar{j}}(\bar{x})$ of $L_{\infty\omega}$ such that for every L-structure A and tuple $\bar{a}\bar{b}$ from A,

(5.12) $A \models \theta^{n,\alpha}_{\bar{i}\bar{j}}(\bar{a}\bar{b}) \Leftrightarrow \mathrm{rank}(\bar{i}\bar{j}, \bar{a}\bar{b}) \geqslant \alpha$ in $\mathrm{Va}(\phi, A)$.

The sentence $\theta^{0,\alpha}$ is written ϕ^α and called the α-**th approximation** to ϕ. The analogue of Corollary 3.4.6 holds (and again for exactly the same reason).

Corollary 11.5.7 (Vaught's covering theorem). Let ϕ and ψ be Vaught sentences of L, and κ an infinite cardinal such that all the conjunctions and disjunctions occurring in ϕ or ψ have cardinality $\leqslant \kappa$. If $\{\phi, \psi\}$ has no model then for some $\alpha < \delta(\kappa)$, $\{\phi^\alpha, \psi\}$ already has no model.

Proof sketch. This is really the same theorem over again. By interweaving ϕ and ψ into one Vaught sentence, we reduce to showing that if ϕ has no model, then some approximation ϕ^α with $\alpha < \delta(\kappa)$ already has no model. Suppose to the contrary that for every $\alpha < \delta(\kappa)$ there is a model A_α of ϕ^α. Using (5.12) we can write a sentence χ of $L_{\kappa^+\omega}$ which says 'I incorporate an L-structure A and a linearly ordered set $Ranking$, together with functions which rank positions in a play of $\mathrm{Va}(\phi, A)$.' Then as before, χ has well-ordered models whose order types are not bounded below $\delta(\kappa)$, so that χ must have a non-well-ordered model incorporating an L-structure B. By Corollary 3.4.4 it follows that player $\exists$ has a winning strategy for $\mathrm{Va}(\phi, B)$, so that ϕ does after all have a model. $\square$

In the hands of experts, Vaught's covering theorem provides a uniform way of proving definability, interpolation and preservation results for infinitary logics, and for $L_{\omega_1\omega}$ in particular. I am bound to say that in my experience there are cleaner and less error-prone ways of getting the same results – notably the consistency property method used by Makkai [1969] and Keisler [1971a].

Exercises for section 11.5

1. Let L be a first-order language, T a theory in L and S a set of types of T. Let δ be a limit ordinal of cofinality $\leqslant \kappa$, and suppose that for every ordinal $\alpha < \delta$, $\mathbf{K}(T, S)$ contains a structure of order type α. Show that there are a first-order theory L', a theory T' in L' and a set S' of types of T' such that $|L'| = |L|$, $|T'| = |T|$, $|S'| = |S|$ and $\mathbf{K}(T', S')$ contains a structure of order type δ. [A model of T' should consist of an indexed family of models of T, together with an ordering whose proper initial segments are mapped into the order types of the models of T.]

2. Show that if κ is a strong limit number of cofinality ω then $\delta(\kappa) = \kappa^+$.

3. Show that if κ is an infinite cardinal then $\delta(\kappa, 2^\kappa) = (2^\kappa)^+$. [If $\alpha < (2^\kappa)^+$, associate to each $\beta < \alpha$ a distinct subset S_β of κ. Introduce 1-ary relation symbols P_i for all $i < \kappa$, and use omitting types to say that if $\beta < \gamma < \alpha$, then for all x and y, if $P_i(x) \Leftrightarrow i \in S_\beta$ and $P_j(y) \Leftrightarrow j \in S_\gamma$ then $x < y$.]

*4. Show that the class of left perfect rings is not the class of relativised reducts A_P of models A of some sentence of $L_{\infty\omega}$. [Let A be a ring and $J(A)$ its Jacobson radical. Inductively define $J_0(A) = \{0\}$; $J_{\alpha+1}(A) = $ the right ideal such that $J_{\alpha+1}(A)/J_\alpha(A)$ is the socle of the right A-module $J(A)/J_\alpha(A)$; $J_\delta(A) = \bigcup_{\alpha<\delta} J_\alpha(A)$ when δ is a limit ordinal. The least ordinal α such that $J_\alpha(A) = J(A)$ is the **Loewy length** of $J(A)$. By Osofsky [1971] there are left perfect rings whose Jacobson radicals have arbitrarily high Loewy length.]

5. Let L be a first-order language of cardinality λ and T a totally transcendental complete theory in L. Show that T has Morley rank $< \lambda^+$. [Working in an ω-saturated model A of T, assume that for every ordinal $\alpha < \lambda^+$ there is a formula $\phi_\alpha(x)$ with parameters in A which has Morley rank $\geqslant \alpha$, and aim to construct a tree of formulas as in Lemma 5.6.1(b). The procedure apes the proof of Theorem 11.5.1. First prune the family of formulas ϕ_α ($\alpha < \lambda^+$) so that they are all of the form $\psi(x, \bar{a})$ with the same formula ψ of L. Then splitting each $\phi_{\alpha+1}(A)$ into two disjoint sets of Morley rank $\geqslant \alpha$, prune the sequence again so that the same formulas of L are involved in each splitting. Etc. ω times. At the end a tree of formulas of L has been constructed, and parameters can be found in A by ω-saturation.]

*6. In the context of Example 1, show the following. Suppose v is a map from the non-zero elements of the ring R to the ordinals, and suppose that whenever a, $b \in R$

and $b \neq 0$, there are $q, r \in R$ such that $a = bq + r$ and $v(r) < v(b)$. Then $R \setminus \{0\} = \mathrm{lfp}(\pi)$ and $v(a) \geq |a|$ for all elements a of R.

7. Show that every sentence of $L_{\kappa^+\omega}$ is equivalent to a Vaught sentence of L in which all the conjunctions and disjunctions in the prefix are over sets of cardinality at most κ, and all the formulas in the matrix are literals.

8. Consider Vaught sentences of L such that (i) all the matrix formulas are first-order, and (ii) in the prefix, no universal quantifiers $\forall y$ occur, all the disjunctions $\bigvee_{i \in I}$ are over index sets I of cardinality at most λ and all the conjunctions $\bigwedge_{j \in J}$ are over index sets J of cardinality $< \kappa$. (a) Show that if ϕ is such a sentence in a signature with a symbol $<$, and ϕ has well-ordered models of cardinality $\geq \alpha$ for every $\alpha < (\lambda^{<\kappa})^+$, then ϕ has models which are not well-ordered. (b) Show that if ϕ has no models, then for some $\alpha < (\lambda^{<\kappa})^+$, the approximation ϕ^α already has no models. *This is why we get the bound λ^+ in Exercise 5.*

11.6 Infinitary indiscernibles

In sections 11.2 and 11.3 we described how to construct and use EM functors for models of first-order theories. The process has two parts, an Input step and an Output step. At the Input step one takes a structure A which contains a linear ordering η, and one uses Ramsey's theorem and stretching to build an EM functor F whose theory contains all first-order formulas satisfied by increasing tuples from η in A. At the Output step, one chooses orderings ζ so that the structures $F(\zeta)$ are interesting. It's time we tried our hand at doing the same thing with infinitary theories.

The generalisation is surprisingly easy. The Output step works exactly as before. For the Input step, Ramsey's theorem is no longer any help when the language is uncountable and compactness fails. Instead we shall use the Erdős–Rado theorem from section 11.1 and the well-ordering number from section 11.5. The main theorem, Theorem 11.6.1 below, is very much like the Ehrenfeucht–Mostowski theorem (Theorem 11.2.8), except that the ordering η has to be rather long. Applications range from pure theory (an upward Löwenheim–Skolem theorem for $L_{\kappa^+\omega}$, Corollary 11.6.3 below) to algebra (construction of many pairwise non-embeddable existentially closed skew fields, among other examples).

Finding the indiscernibles

Let L be a first-order-closed fragment of $L_{\infty\omega}$ (see section 2.1 for the definition). We say that a linear ordering η contained in an L-structure A is k-**indiscernible**^{L} in A if for all increasing k-tuples $\bar{a}$, $\bar{b}$ from η and all formulas $\phi(x_0, \ldots, x_{k-1})$ of L, $A \vDash \phi(\bar{a}) \Leftrightarrow A \vDash \phi(\bar{b})$. We say that η is

indiscernible^L in A if it is k-indiscernible^L in A for all $k < \omega$. Analogous definitions apply to orderings of n-tuples in A. We drop the superscript ^{L} when the context allows. By analogy with the first-order case, we write $\mathrm{Th}_L(A)$ for the set of all sentences of L which are true in A, $\mathrm{Th}_L(A, \eta)$ for the set of all formulas of L which are satisfied by increasing tuples from η in A, and $\mathrm{Th}_L(F)$ for the intersection of all sets $\mathrm{Th}_L(F(\eta), \eta)$.

Recall from section 11.5 the well-ordering number $\delta(\kappa)$, and from Exercise 3.1.10 the notion of a Skolem theory in a first-order-closed fragment of $L_{\infty\omega}$.

Theorem 11.6.1. *Let L be a first-order-closed fragment of $L_{\infty\omega}$ of cardinality κ. Let A be an L-structure which contains a linear ordering η of order type $\beth_{\delta(\kappa)}$, and suppose that $\mathrm{Th}_L(A)$ is a Skolem theory in L. Then there is an EM functor F in L such that $\mathrm{Th}_L(F)$ contains all variable-finite formulas of L which are in $\mathrm{Th}_L(A, \eta)$.*

Proof. We first isolate the combinatorial core of the proof. We assign a rank to each pair (B, ζ), where B is an L-structure and ζ an infinite linear ordering contained in B, as follows.

(6.1)   Each pair (B, ζ) has rank ≥ 0.
          (B, ζ) has rank $\geq \alpha + 1$ iff for every $k < \omega$ there is an infinite k-indiscernible^L subordering θ of ζ such that (B, θ) has rank $\geq \alpha$.
          For limit δ, (B, ζ) has rank $\geq \delta$ iff (B, ζ) has rank $\geq \alpha$ for every $\alpha < \delta$.

By this definition, some pairs (B, ζ) may have a rank ∞, i.e. greater than any ordinal. We allow this.

Lemma 11.6.2. *If B contains a well-ordering ζ of length $\beth_{\omega\alpha}(\kappa)$ then (B, ζ) has rank $\geq \alpha$.*

Proof of lemma. By induction on α. The cases $\alpha = 0$ or a limit ordinal are trivial. Assume $\alpha = \gamma + 1$ and suppose that ζ is well-ordered with length $\geq \beth_{\omega\alpha}(\kappa)$. Consider any $k < \omega$, and let $f : [\zeta]^k \to 2^\kappa$ be a map such that if $\bar{a}, \bar{b} \in [\zeta]^k$ and $\phi(x_0, \ldots, x_{k-1})$ is a formula of L then $f(\bar{a}) = f(\bar{b})$ implies that $B \vDash \phi(\bar{a}) \to \phi(\bar{b})$. By the Erdős–Rado theorem (Theorem 11.1.7), $\beth_{\omega\alpha}(\kappa) \to (\beth_{\omega\gamma}(\kappa))^k_{2^\kappa}$. Hence there is a subordering θ of ζ, of length $\beth_{\omega\gamma}(\kappa)$, such that f is constant on $[\theta]^k$. So θ is k-indiscernible in B, and by induction hypothesis (B, θ) has rank $\geq \gamma$. Therefore (B, ζ) has rank $\geq \alpha$. $\square$ Lemma.

The lemma implies that (A, η) has rank at least the well-ordering number $\delta(\kappa)$, as follows. By Corollary 11.5.2 and Theorem 11.5.3(b), $\delta(\kappa)$ is a limit ordinal $> \kappa$. Let α be any ordinal $< \delta(\kappa)$. The length of η is $\beth_{\delta(\kappa)} > \beth_{\kappa + \omega\alpha} \geqslant \beth_{\omega\alpha}(\kappa)$, so (A, η) has rank $\geqslant \alpha$ by the lemma.

Now let T be a theory which says the following:

(6.2) J is a set of elements linearly ordered by a relation $<$;

$f: I \to J$ is a surjective map;

P is an L-structure;

for each $i \in I$, ζ_i is an infinite ordering contained in P, and $\mathrm{Th}_L(P, \zeta_i)$ contains all variable-finite formulas of $\mathrm{Th}_L(A, \eta)$;

for all $j_1 < j_2$ in J, if $f(i_2) = j_2$ then for every $k < \omega$ there is i_1 such that $f(i_1) = j_1$, ζ_{i_1} is a subordering of ζ_{i_2}, and ζ_{i_1} is k-indiscernible^L in P.

Since L is a fragment of cardinality κ, it lies in $L_{\kappa^+\omega}$. It's not hard to check that T can be written as a single sentence of some language $L'_{\kappa^+\omega}$, where L' comes from L by adding finitely many new symbols such as J, $<$, I, f, P and a 3-ary relation symbol ζ.

We claim that T has a model C in which J^C is well-ordered with order type $\delta(\kappa)$. For this, put $J^C = \delta(\kappa)$, $P^C = A$, and for each $\alpha < \delta(\kappa)$ let the linear orderings ζ_i^C with $f^C(i) = \alpha$ be all the infinite suborderings ζ of η such that (A, ζ) has rank $\geqslant \alpha$. The only part of T which is not immediately verified is that $f^C: I^C \to J^C$ is surjective. This holds because (A, η) has rank $\geqslant \delta(\kappa)$.

By the definition of $\delta(\kappa)$ it follows that T has a model D in which J^D is not well-ordered. Pick an infinite descending sequence $j_0 >^D j_1 >^D \ldots$ in J^D. By induction on n, choose for each $n < \omega$ an element i_n of I^D so that $f^D(i_n) = j_n$, ζ_j^D is n-indiscernible^L in P^D, and if $n > 0$ then ζ_j^D is a subordering of $\zeta_{j_{n-1}}^D$. For each n, let $\Psi_n(x_0, \ldots, x_{n-1})$ be the set of formulas $\psi(x_0, \ldots, x_{n-1})$ such that $D \vDash \text{`}\psi \in \mathrm{Th}_L(P, \zeta_j)\text{'}$. Introduce a sequence $\bar{c} = (c_n : n < \omega)$ of new constants, and write $L(\bar{c})$ for L with these new constants added. Adding just the first n constants gives $L(c_0, \ldots, c_{n-1})$, or L_n for short.

We claim that for each n, $\Psi_n(\bar{c})$ is a Hintikka set in L_n. Since ζ_j^D is n-indiscernible^L in P^D, $\Psi_n(\bar{c})$ is a consistent theory in L_n which contains either ϕ or $\neg\phi$ for every sentence ϕ of L_n. Also $\Psi_n(\bar{c})$ contains all the Skolem sentences of $\mathrm{Th}_L(A)$. Finally if $\Psi_n(\bar{c})$ contains a sentence $\bigvee \Phi(\bar{c})$ with Φ in L, then since all increasing n-tuples from ζ_j satisfy the same formulas from Φ, there must be some $\phi(\bar{x}) \in \Phi$ such that $\phi(\bar{c}) \in \Psi_n(\bar{c})$. So the claim follows by Theorem 2.3.4.

Clearly $\Psi_n(\bar{c}) \subseteq \Psi_{n+1}(\bar{c})$ for all $n < \omega$. So the claim and Exercise 2.3.6 imply that $\bigcup_{n<\omega} \Psi_n(\bar{c})$ is a Hintikka set in $L(\bar{c})$, so that it has a model B. Write ζ for the ordering $(c_n^B : n < \omega)$ in B, and let E be the L-elementary substructure of $B|L$ generated by ζ. Then ζ is an indiscernible sequence in E of order type ω, and ζ generates E. It follows by stretching (Theorem 11.2.4) that there is an EM functor F in L with $F(\omega) = E$. Also $\text{Th}_L(E, \zeta)$ contains all the variable-finite formulas in $\text{Th}_L(A, \eta)$, including all the Skolem sentences. It follows (just as in Lemma 11.2.7 for the first-order case) that all the variable-finite formulas of $\text{Th}_L(A, \eta)$ are in $\text{Th}_L(F)$.   ☐ Theorem

There are variations on Theorem 11.6.1. The most important is that we can replace η in the theorem by an ordering of n-tuples, so that F becomes an EM functor on n-tuples.

Hanf numbers

Our first application of the theorem is an upward Löwenheim–Skolem theorem for single sentences of $L_{\kappa^+\omega}$.

The family of all languages $L_{\kappa^+\omega}$, as L ranges over all possible signatures, is known as the **logic** $\mathcal{L}_{\kappa^+\omega}$. The **Hanf number** of this logic $\mathcal{L}_{\kappa^+\omega}$ is defined to be the least cardinal λ such that if L is any signature and ϕ any sentence of $L_{\kappa^+\omega}$ which has a model of cardinality $\geq \lambda$, then ϕ has arbitrarily large models. How do we know there is such a cardinal? After all, there is a proper class of possible signatures L.

A little thought shows first that any sentence of a language $L_{\kappa^+\omega}$ uses at most κ symbols from L. Moreover the logical properties of the sentence aren't altered if we replace each n-ary relation symbol by a new n-ary relation symbol, and likewise with the other symbols of L. So we really only need to consider a *set* of signatures. For each of these signatures we only need concern ourselves with a *set* of sentences, for similar reasons. Now for each sentence ϕ there is a least cardinal $\lambda(\phi)$ such that either ϕ has arbitrarily large models or ϕ has no models of cardinality $\geq \lambda(\phi)$. The sup of $\lambda(\phi)$ as ϕ ranges over a set of sentences is a well-defined cardinal. So the Hanf number exists. This argument is due to William Hanf [1960].

Corollary 11.6.3. *The Hanf number of* $L_{\kappa^+\omega}$ *is* $\beth_{\delta(\kappa)}$.

Proof. First we show that the Hanf number is $\leq \beth_{\delta(\kappa)}$. Let ϕ be a sentence of $L_{\kappa^+\omega}$ with a model A of cardinality $\geq \beth_{\delta(\kappa)}$. Let η be a well-ordering of the elements of A. We can find a first-order-closed fragment L' of cardinality κ, such that ϕ is in L', and such that A can be expanded to an L'-structure A' with $\text{Th}_{L'}(A')$ a Skolem theory in L' (see Exercise 3.1.10). By the theorem there is an EM functor F such that for any ordering ζ, $F(\zeta)$ is a model of $\text{Th}_{L'}(A')$. Then for any cardinal μ, $F(\mu)$ is a model of ϕ of cardinality at least μ.

Next we show that the Hanf number is $\geq \beth_{\delta(\kappa)}$. Since $\delta(\kappa)$ is a limit ordinal (see Corollary 11.5.2), it suffices to take an arbitrary ordinal $\beta < \delta(\kappa)$ and show that there is a sentence ϕ of some $L_{\kappa^+\omega}$ with models of cardinality $\geqslant \beth_\beta$ but none of cardinality $\geqslant \beth_{\delta(\kappa)}$. Since $\delta(\kappa) > \kappa$ (Theorem 11.5.3(b)), we can assume without loss that $\beta \geqslant \kappa$. By the definition of $\delta(\kappa)$, there is a sentence ψ of $L'_{\kappa^+\omega}$, for some language L' containing a symbol $<$, such that

(6.3)   in every model A of ψ, $<^A$ is well-ordered, and ψ has
         models A in which $<^A$ has length $> \beta$.

With the aid of ψ we can write a sentence ϕ of some $L_{\kappa^+\omega}$ which says the following:

(6.4) J is a set of elements linearly ordered by a relation $<$ which
         satisfies ψ;
         $f: I \to J$ is a surjective map;
         for all $j \in J$ and i_1, $i_2 \in I$, if $f(i_1) = f(i_2) = j$ but $i_1 \neq i_2$ then
         there is $i_0 \in I$ such that $f(i_0) < j$ and $R(i_0, i_1) \leftrightarrow \neg R(i_0, i_2)$;
         g is a bijection between I and the universe.

Then ϕ has a model B of cardinality $\geqslant \beth_\beta$ as follows. Take a model B' of ψ in which $<^{B'}$ has order-type $\gamma > \beta$. By the downward Löwenheim–Skolem theorem for $L_{\kappa^+\omega}$ (Corollary 3.1.6), we can assume that B' has cardinality $|\gamma|$. Build B around B' so that $<^B = <^{B'}$, each $(f^B)^{-1}(n)$ has cardinality 2^n ($n < \omega$) and each $(f^B)^{-1}(\omega + j)$ has cardinality $\beth_j$. Then I^B has cardinality $\sum_{j<\gamma}\beth_j \geqslant \beth_\beta$, and we can find a bijection $g^B: I^B \to \mathrm{dom}(B)$.

  However, any model C of ϕ has cardinality $< \beth_{\delta(\kappa)}$. For by the definition of $\delta(\kappa)$, (6.3) implies that $<^C$ must have length $\alpha < \delta(\kappa)$. By induction on γ, if j is the γth element of J^C under $<^C$, then $(f^C)^{-1}(j)$ has cardinality at most $\beth_\gamma$, and so I^C and hence C itself have cardinality $\leqslant \beth_\alpha < \beth_{\delta(\kappa)}$.   □

Algebraic applications

A locally finite group G is said to be **universal** if (a) every finite group is embeddable in G and (b) if $f: H \to K$ is an isomorphism between finite subgroups of G then some inner automorphism of G extends f. It follows at once that G is an ultrahomogeneous group whose age is the class of finite groups. Such groups are known to exist in all infinite cardinalities; in fact every existentially closed locally finite group is universal (see Exercise 7). So we saw in Example 1 of section 7.1 that the countable universal locally finite group must be unique up to isomorphism. It was natural to ask whether the same holds in higher cardinalities. The next theorem says no.

Theorem 11.6.4. *Let* **K** *be any one of the following classes:*
(a) *the existentially closed members of some non-abelian quasivariety of groups;*
(b) *universal locally finite groups;*
(c) *existentially closed skew fields.*

Then for every uncountable regular cardinal κ there is a family $(A_i: i < 2^\kappa)$ of algebras in **K**, *all of cardinality κ, such that if $i \neq j$ then A_i is not embeddable in A_j.*

Proof. This is a paradigm example of the format described at the beginning of this section. The Input step is to show that each class **K** contains an algebra A in which some very long well-ordered family of n-tuples has its ordering expressed by a formula ψ. The Output step is Theorem 11.3.6.

Let **K** be one of the classes (a)–(c). Then **K** is defined by a sentence ϕ of $L_{\omega_1\omega}$. For universal locally finite groups this is immediate from the definition, and for classes (a) and (c) it follows from the characterisation of existentially closed models in Theorem 8.2.4. By Theorem 11.5.3(c) the well-ordering number $\delta(\omega)$ is ω_1. Put $\lambda = \beth_{\omega_1}$.

We claim that in all three cases (a)–(c), **K** contains a structure A in which a sequence of tuples is ordered in order-type λ by ψ. To begin with case (a), let G be any non-abelian group. Then by Exercise 9.1.15 the power G^λ contains a sequence of tuples which is ordered in order-type λ by a quantifier-free first-order formula $\psi(\bar{x}, \bar{y})$. Every variety is closed under unions of chains, and so by Theorem 8.2.1 we can extend G^λ to a group A in **K**. Quantifier-free formulas are preserved in extensions, so that A contains a sequence of tuples ordered in order type λ by a quantifier-free formula. This proves the claim for case (a).

If moreover the non-abelian group G is finite then G^λ is locally finite (see Exercise 9.1.6). The class of locally finite groups is also closed under unions of chains, and one can show (Exercise 7) that existentially closed locally finite groups are universal. So we have the claim for class (b) too.

For case (c) we first construct an ordered group G in which some long sequence of tuples is ordered by a quantifier-free formula. That can be done as follows. Let G be the nilpotent group of class 2 generated by elements $a(i)$ $(i < \lambda)$ and c, subject to the conditions

$$(6.5) \qquad\qquad [a(i), a(j)] = c \qquad (i < j < \lambda).$$

Then each element of G can be written uniquely as a finite product

$$(6.6) \qquad\qquad a(i_1)^{n_1} \ldots a(i_k)^{n_k} \cdot c^m \qquad (i_1 < \ldots < i_k).$$

If this element is g, and h is $a(i_1)^{n_1'} \ldots a(i_k)^{n_k'} \cdot c^{m'}$, we put $g < h$ iff $(n_1, \ldots, n_k, m)$ precedes $(n_1', \ldots, n_k', m')$ in the lexicographic ordering by first differences. This makes G an ordered group. The sequence $((a(i), c): i < \lambda)$ is ordered by a formula $\psi(x_0, x_1; y_0, y_1)$, viz. $[x_0, y_0] = x_1$. By the theorem of Mal'tsev and Neumann (see Cohn [1981] p. 276), G can be embedded into the multiplicative group of a skew field A'. Then by Theorem 8.2.1 again, A' can be extended to an existentially closed skew field. Since ψ is quantifier-free, it is preserved in extensions. Thus the claim is proved in all cases.

In any of cases (a)–(c), let A be a structure in $\mathbf{K}$ containing a sequence of tuples ordered in order type λ by ψ. We can expand A to an L^+-structure A^+, where L^+ is a countable first-order-closed fragment containing both ϕ and ψ, and the L^+-theory of A^+ is skolemised (see Exercise 3.1.10). By Theorem 11.6.1 there is an EM functor F in L^+ such that the L^+-theory of F contains both ϕ and ψ. The theorem now follows at once from Theorem 11.3.6, by constructing structures $F(\eta_i)$ and considering their reducts to the language of $\mathbf{K}$. $\qquad\qquad\qquad\qquad\qquad\qquad\qquad\qquad\qquad\qquad\qquad\square$

Shelah & Ziegler [1979] show that if G is any existentially closed group, then in every uncountable cardinality κ there are 2^κ non-isomorphic groups which are back-and-forth equivalent to G. It is not known whether this result generalises to other varieties of groups.

Exercises for section 11.6

1. Sharpen Corollary 11.6.3 as follows. Suppose L is a first-order-closed fragment of $L_{\kappa^+\omega}$, and ϕ is a sentence of L. Show that if for every cardinal $\lambda < \beth_{\delta(\kappa)}$ there is a model of ϕ which has cardinality $\geq \lambda$, then ϕ has arbitrarily large models.

2. Let L be a countable first-order language, T a theory in L and $\Phi(\bar{x})$ a type of T. Show that if for every cardinal $\lambda < \beth_{\omega_1}$, T has a model of cardinality $\geq \lambda$ which omits Φ, then T has arbitrarily large models which omit Φ.

3. Show that $\beth_{\omega_1}$ is the Hanf number for countable theories (not sentences) of weak second-order logic $\mathcal{L}_w^{II}$. [The Hanf number is at most $\beth_{\omega_1}$ by Corollary 11.6.3. To show it's at least this great, prove that each countable ordinal α is a PC_Δ class in the sense of $\mathcal{L}_w^{II}$. Countable conjunctions can be expressed by coding each conjunct as an element of ω.]

Given cardinals κ, λ, define $\mu(\kappa, \lambda)$ to be the least cardinal μ such that the following holds: for every first-order language L of cardinality $\leq \kappa$, every theory T in L and every set P of at most λ types of L, if T has models of arbitrarily large cardinality $< \mu(\kappa, \lambda)$ which omit all the types in P, then T has models of arbitrarily large cardinality which omit all the types in P. The cardinals $\mu(\kappa, \lambda)$ are known as **Morley numbers**.

4. Show that for every infinite cardinal κ, $\mu(\kappa, 1)$ is equal to the Hanf number of $L_{\kappa^+\omega}$. [Use Chang's reduction, Theorem 2.6.7.]

5. Show that for all $\lambda > 0$, $\mu(\kappa, \lambda) = \beth_{\delta(\kappa,\lambda)}$ where $\delta(\kappa, \lambda)$ is as in Corollary 11.5.2.

*6. Assuming that ZFC has a model, show that for every cardinal λ there is a model M of ZFC of cardinality at least λ such that the ordinals of M have no subset of order type ω_1.

*7. Let G be a locally finite group. Show that the following are equivalent. (a) G is universal locally finite. (b) G is existentially closed in the class of locally finite groups. (c) Every finite group is embeddable in G; and if H, K are isomorphic finite subgroups of G, then there is an inner automorphism of G which takes H to K. [(a) $\Rightarrow$ (b) is by Example 2 in section 8.1.1. For (b) $\Rightarrow$ (c), show that if G_0 is a finite group and $f: H \to K$ an isomorphism between subgroups of G_0, then there is a finite group $G_1 \supseteq G_0$ in which f extends to an inner automorphism. This can be proved using permutation representations of H and K. For (c) $\Rightarrow$ (a) the central point is that every automorphism of a finite subgroup H of G extends to an inner automorphism of G. If G contains the holomorph J of H, this is trivial; but in fact it's enough that G contains an isomorphic copy of J.]

8. (a) Let L be a signature, T a first-order theory in L and A an L-structure of cardinality $\kappa \geqslant |L|$. Let $\mathbf{K}(A, T)$ be the class of all L-structures C which are models of T such that there is a surjective homomorphism $f: C \to A$. Show that there are a signature L^* with $|L^*| = \lambda$, and a sentence ϕ of $L^*_{\kappa^+\omega}$ such that $\mathbf{K}(A, T)$ is the class $\{B|L: B \models \phi\}$. Deduce that if $\mathbf{K}(A, T)$ contains structures of arbitrarily large cardinalities $< \beth_{\delta(\kappa)}$, then $\mathbf{K}(A, T)$ contains arbitrarily large structures. (b) For a suitable countable L and T in (a), show that for every countable ordinal α there is a countable L-structure A such that $\mathbf{K}(A, T)$ contains structures of cardinality $\geqslant \beth_\alpha$ but not arbitrarily large structures.

*9. Let L be a first-order language without relation symbols and $\mathbf{V}$ a variety in L. An L-structure A is said to be **subdirectly irreducible** if there are distinct elements a, b of A such that every homomorphism $f: A \to B$ in which $f(a) \neq f(b)$ is an embedding. Show that the following are equivalent. (a) $\mathbf{V}$ contains arbitrarily large subdirectly irreducible structures. (b) $\mathbf{V}$ contains subdirectly irreducible structures of cardinality $> 2^{|L|}$. (c) There is a positive formula ϕ of L such that

$$\vdash \forall yz(\exists x \, \phi(x, x, y, z) \to y = z),$$

and for every finite n there is a structure B in $\mathbf{V}$ containing elements $a \neq b$ such that $B \models \exists x_0 \ldots x_{n-1} \bigwedge_{i<j<n} \phi(x_i, x_j, a, b)$. [For (b) $\Rightarrow$ (c), let A be subdirectly irreducible of cardinality $> 2^{|L|}$. By Corollary 11.1.8 there are a sequence $(c_i: i < \lambda^+)$ of elements of A and a positive formula ϕ such that $A \models \phi(c_i, c_j, a, b) \, (i < j < \lambda^+)$ but $\vdash \forall yz(\exists x \, \phi(x, x, y, z) \to y = z)$. For (c) $\Rightarrow$ (a), if μ is any cardinal, introduce constants $c_i \, (i < \mu)$ and take a model A of the sentences $\phi(c_i, c_j, a, b) \, (i < j < \mu)$ and $a \neq b$ in $\mathbf{V}$. Factor out a maximal congruence θ allowing $a \neq b$, so that A/θ is subdirectly irreducible.]

10. Let L, L^+ and P be as in section 5.1, with L^+ countable. For a fixed L-structure A we consider the class $\mathbf{K}$ of all L^+-structures B such that $B_P = A$. Show that if $\mathbf{K}$ contains structures of cardinality $\geqslant \beth_\omega(|A|)$ but not arbitrarily large structures, then there is a countable theory T in a first-order language extending L^+, such that in every model C of T with $C_P \cong A$, the structure $(\omega, <)$ is also a relativised reduct of C. [Adapt the proof of Lemma 11.6.2; arrange that there are models in which the set J has arbitrarily large finite cardinality (since $\mathbf{K}$ contains a structure of cardinality $\geqslant \beth_\omega(|A|)$) but none in which it is infinite (otherwise $\mathbf{K}$ would contain arbitrarily large structures).]

History and bibliography

Section 11.1. Ramsey proved Theorem 11.1.3 and Corollary 11.1.4 in his [1930]. With this paper he invented the partition calculus. Being in logic, the paper attracted little attention until Szekeres rediscovered and applied the finite Ramsey theorem (see Erdős & Szekeres [1935]). See Graham, Rothschild & Spencer [1980] on other analogues of the finite Ramsey theorem; the affine and vector space Ramsey theorems in their Chapter 2 and the graph-theoretic Ramsey theorem of Nesetril & Rödl in their Chapter 5 have found model-theoretic applications. Dushnik & Miller [1941] extended the infinite Ramsey theorem by proving Exercise 12. Erdős [1942] followed with Corollary 11.1.8. By [1956] Erdős and Rado were able to present a substantial body of infinitary partition theorems (including Theorem 11.1.7, which is part of their Theorem 39), and the subject has been lively ever since. See Erdős, Hajnal, Máté & Rado [1984] on the infinitary partition calculus. Our proof of Theorem 11.1.7 is due to Simpson [1970]; in this and in his work [1985] on Nash–Williams theory he shows how methods from logic can be used to tidy up complicated arguments in combinatorics. Theorem 11.1.9 is from Pouzet [1976]; see Macpherson, Mekler & Shelah [1991] for a partial analogue counting infinite substructures of a given structure.

Following work of Gaifman and Rowbottom, Silver [1971] showed that if there is a measurable cardinal, then the set-theoretic universe contains a sequence of ordinals which includes all uncountable cardinals and is indiscernible in Gödel's constructible universe; from this it follows that the cardinal ω_1 in the universe is strongly inaccessible in the constructible universe. These indiscernibles are known as the **Silver indiscernibles**, and the set of formulas which are satisfied in the constructible universe by increasing tuples of Silver indiscernibles is known as $O^\#$ (O-sharp). Coded up as a set of Gödel numbers, $O^\#$ is a nonconstructible Δ_3^1 subset of ω (Solovay [1967]). This work is deep and beautiful, and it undoubtedly gave a boost to the study of indiscernibles in the 1970s. But I have not pursued it here, since it is now generally reckoned to belong to set theory rather than model theory. See Jech [1978] Chapter 5.

Exercise 5: Pouzet [1976]. Exercise 6: Paris & Harrington [1977]. Exercise 7: Tomkinson [1986]; cf. Chatzidakis, Pappas & Tomkinson [1990]. Exercise 9: (b) Erdős & Tarski [1962]; (c) Hanf [1964]; our proof is from Hajnal [1964]. Exercise 10: Büchi & Siefkes [1973], Shelah [1975d]. Exercise 11: Dushnik & Miller [1941]. Exercises 12, 13: Shelah [1978a] Theorem III.4.23; cf. Grossberg [199?] for a strengthening. Exercise 14: Hodges, Lachlan & Shelah [1977].

Section 11.2.   Ehrenfeucht and Mostowski introduced their functors in [1956]; my proof of Theorem 2.2.8 follows theirs. Gaifman [1967] hit on the same idea later from another direction which avoids Ramsey's theorem; see Keisler [1970] for yet another approach. Corollary 11.2.6 is from Chang [1968a]. Theorem 11.2.9 and Corollary 11.2.10 are from Morley [1965a]; every treatment of uncountably categorical theories starts with essentially this argument. Theorem 11.2.9(b) also lies behind Rowbottom's [1971] proof (from the early 1960s) that if there is a measurable cardinal, then for every infinite cardinal κ the constructible universe contains only κ subsets of κ. Corollary 11.2.11 is from Ehrenfeucht [1958]. Theorem 11.2.12 appeared in Hodges [1969] together with a corresponding result describing how the elements given by different terms are related. These results were not published until Hodges & Shelah [1991]; meanwhile they were rediscovered by Charretton & Pouzet [1983a], [1983b]. Interval algebras and Exercise 10 appear in Mostowski & Tarski [1939].

Exercise 2: Ehrenfeucht & Mostowski [1956]. Exercise 4: Ash & Nerode [1975]. Exercise 6: This develops an idea of Charretton [1979]. Exercises 7–9: for $\lambda = \omega$, the result of Exercise 9 can be derived from MacDowell & Specker [1961]; I think the general case must be due to Helling [1966] by essentially the argument given in Exercises 7, 8. The chief interest of this argument lies in Silver's [1971] brilliant adaptation of it to models of set theory; see the references to section 11.1.

Section 11.3.   Theorem 11.3.1 and Exercise 2 are from Hodges [1984a]. Poizat [1981] in effect proves both results, though less directly, when he shows that an unstable theory has a complete type (over a model) with more than one coheir. The idea of using Ehrenfeucht–Mostowski models to construct non-isomorphic models of an unstable theory belongs to Ehrenfeucht [1957]. I recall that in 1967 I was looking for ways of constructing many models of non-categorical theories, and Keisler pointed me to Ehrenfeucht's method. However, the breakthrough came in 1970 when Shelah [1971e] introduced Ehrenfeucht–Mostowski models on n-tuples, and used them to prove that an unstable theory in a first-order language L has 2^κ pairwise non-isomorphic models of every cardinality $\kappa > |L|$. In [1974b] he proved the same for unsuperstable theories. The argument for Theorem 11.3.8 using Baumgartner's orderings is from Shelah [1978a] Theorem VIII.2.2(1); the observation (Theorem 11.3.6) that it works for word-constructions generally is new here. Meanwhile Baumgartner [1976] independently discovered this family of orderings. Theorem 11.3.9 is from Shelah [1971e].

Shelah formulated the idea of non-structure theorems in about 1971, and published it in [1976b]. In this paper and in his [1985] he puts the view that the main question is the number of non-isomorphic models. Two writers who remained unconvinced were Hodges [1987] and Lascar [1987b]. Shelah has a large number of papers applying non-structure techniques to construct, for example, abelian p-groups and boolean algebras with few automorphisms. See his [1975c], [1979], [1986].

Exercise 4: Baumgartner [1976]. Exercise 5: Shelah [1978a] Theorem VIII.2.2(1). Exercises 6, 7: Ehrenfeucht [1957] at accessible cardinals, Hodges [1973] in general. Exercise 8: Charretton & Pouzet [1983b]. Exercise 9: Charretton & Pouzet [1983a].

Section 11.4. We get nonstandard methods by operating on the tools of model theory (the natural numbers, the set-theoretic universe) just like the other objects that these tools are applied to. This is a very twentieth-century idea: in 1915 Ivan Puni exhibited at a show in St Petersburg a picture consisting of a hammer and some pieces of cardboard. One might have expected that ideas of this kind would have filtered gradually into model theory from various sources. But in fact nonstandard methods appeared quite suddenly in 1960, and they are entirely due to the genius of Abraham Robinson [1961]. Possibly his anti-platonist views [1965] helped him to break down the distinction between tools and objects; the same is surely true of Skolem [1934], who introduced nonstandard models of first-order arithmetic (but without any suggestion that they could be used to prove useful theorems of standard mathematics).

The idea of getting ultrafilters out of proper extensions can be traced to Rabin [1959], and its use in large cardinal theory stems from Keisler & Tarski [1964]. The name 'overspill' is due to Machover, but the idea appears constantly in Robinson's work on nonstandard analysis. Friedman [1973] used a version of Corollary 11.4.3.

Robinson [1973b] urged the use of model theory as a way to get bounds in the theory of polynomials without going through the calculations of Hermann [1926]. Our nonstandard proof of Fact 11.4.4 is from van den Dries [1978] and van den Dries & Schmidt [1984]. The Gilmore–Robinson characterisation of hilbertian fields was extracted from Gilmore & Robinson [1955] in his proof of the quoted result of Franz [1931] (in the separable case) and Inaba [1944] (in general). In fairness to Henson & Keisler [1986] I should mention their own conclusion that 'there are theorems which can be proved with nonstandard analysis but cannot be proved without it'. On my reading, their theorem proves the opposite of this.

Exercise 1: The argument sketched after the exercise is from Lindström [1969]. Exercises 2, 3: see Jech [1978] sections 28, 29, 32 and his references. Exercises 4, 5: this is a typical if simple-minded example of Robinson's [1966] nonstandard analysis. Exercise 6: Roquette [1975].

Section 11.5. The theory of analytic sets in classical descriptive set theory had some influence on several things in this chapter, and particularly in this section, but I am not sure how precise this influence was. The main theorems of this section and the next were originally proved in the wrong order; first Morley [1965b] gave a version of Theorem 11.6.1, and then López-Escobar [1966] used Morley's work to prove the first statement of Theorem 11.5.4. I think it was Morley & Morley [1967] who first got things the right way round. Then Barwise & Kunen [1971] gave a definitive treatment, studying what $\delta(\kappa)$ can be in different models of set theory. The separation of κ and λ in $\delta(\kappa, \lambda)$ appears in section VIII.5 of Shelah [1978a], together with some calculations of bounds.

Kuratowski [1937] proved (in topological language) that the class of all well-orderings is not definable by a sentence of $L_{\omega_1\omega}$. Karp [1965] showed the same for sentences of $L_{\infty\omega}$. Theorem 11.5.4 proves more: it shows that the class is not even a PC'_Δ class in any $L_{\kappa^+\omega}$ (where PC'_Δ is defined as in section 5.2), answering a question of Mostowski.

Corollary 11.5.5 was conjectured by Sabbagh & Eklof [1971]; in fact they sketched the model-theoretic part of the proof, but then we had to wait for the algebraists to do their bit. Corollary 11.5.6 is from Barwise & Moscho-vakis [1978] (for countable admissible languages, but the proof is the same). Corollary 11.5.7 was proved by Vaught [1973] (again for the countable case). It was applied by Makkai [1973], Harnik & Makkai [1976] and Harnik [1979].

Exercise 3: Shelah [1978a] Theorem VII.5.5. Exercise 4: Sabbagh & Eklof [1971]. Exercise 5: Lachlan [1971]. Exercise 6: Cohn [1971] §2.9. Exercise 8: see Green [1978].

Section 11.6. Morley [1965b] proved Exercise 2 and a corresponding result for uncountable first-order languages. Then Chang [1968b] and López-Escobar [1966] independently found Exercise 4. The connection between Hanf numbers and well-ordering numbers is from Barwise & Kunen [1971]. Shelah [1972a] introduces an analogue of stability for infinitary theories and uses it to find indiscernibles.

Theorem 11.6.4(b) is from Macintyre & Shelah [1976], who have it also for singular cardinals (the model-theoretic part of their proof is mostly irrelevant – there is no need to count types). Theorem 11.6.4(a, c) are routine examples of Shelah's non-structure machinery, but I thank Benjamin Baumslag and Paul Cohn for asking these questions. Eklof [1985] discusses applications of infinitary model theory to algebra.

Exercise 3: Helling in López-Escobar [1966]. Exercise 5: Shelah [1978a] Theorem VII.5.4. Exercise 6: Hodges [1972]. Exercise 7: (a) ⇔ (c) P. Hall [1959]; (a) ⇔ (b) Macintyre & Shelah [1976]. Exercise 9: Taylor [1972]; cf. McKenzie & Shelah [1974] for a similar result on simple models. Exercise 10: Morley [1970b].

12

Expansions and categoricity

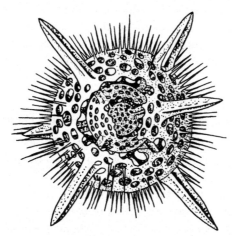

Skeleton of radiolarian Actinomma, from P. Tasch, Paleobiology of the invertebrates, Copyright © J. Wiley & Sons, Inc., 1973.

This is not a book about model-theoretic classification. Nevertheless it would be odd to close without some discussion of uncountably categorical theories, since they have stood at the centre of model-theoretic research over the last thirty or more years. Various questions about them have been asked and answered; the outcome is that we understand fairly well what their models look like.

The family of models of an uncountably categorical theory contains, counting up to isomorphism, just one structure in each uncountable cardinality. One can show that each model is built in layers. The inner layer – let us call it the *nucleus* – is a geometry; it can have any infinite dimension (possibly any finite dimension too), and its dimension determines its size. The other layers are built cumulatively on top of the nucleus in such a way that they are completely determined by the nucleus. The reason why there is only one model in each uncountable cardinality λ is that there is only one nucleus of dimension λ, and the nucleus determines everything else.

So the classification of models of uncountably categorical theories falls into two parts. First we classify the possibly nuclei, and second we classify the ways of building the rest of a model over a nucleus. The nuclei are in fact strongly minimal sets, and so our study of these sets in Chapter 4 has already made some headway on the first part. (Models of an uncountable theory which is categorical in higher cardinalities also have nuclei; they need not be strongly minimal sets, but they always carry some kind of geometry.)

Not so much is known about the second part – the way in which the layers are built up around the nucleus. But we know much more than we did fifteen years ago. We can extract some information from the fact that the outer layers never have higher cardinality than the inner ones; as the jargon goes, models of an uncountably categorical theory are *one-cardinal*. Another approach is to study the automorphisms of the outer layers when the inner layers are fixed; this leads to questions of cohomology.

12.1 One-cardinal and two-cardinal theorems

A **two-cardinal theorem** is a theorem which tells us that under certain conditions, a theory has a model A in which the $\varnothing$-definable subsets $\phi(A)$, $\psi(A)$ defined by two given formulas $\phi(x)$, $\psi(x)$ have different cardinalities. A **one-cardinal theorem** is a theorem with the opposite conclusion. For example if κ and λ are any two infinite cardinals with $\kappa \leqslant \lambda$, then there is a group G of cardinality λ whose centre has cardinality κ; this is a two-cardinal theorem. But if G is a finite-dimensional general linear group over an infinite field, then G has the same cardinality as its centre; this is a one-cardinal theorem.

The following theorem, a memorable result of Vaught, was the first model-theoretic two-cardinal theorem to be proved. Its proof uses the notion of homogeneous structures as in section 10.1 above.

Theorem 12.1.1 (*Vaught's two-cardinal theorem*). *Let L be a countable first-order language, $\phi(x)$ and $\psi(x)$ two formulas of L and T a theory in L. Then the following are equivalent.*
(a) *T has a model A in which $|\phi(A)| \leqslant \omega$ but $|\psi(A)| = \omega_1$.*
(b) *T has a model A in which $|\phi(A)| < |\psi(A)| \geqslant \omega$.*
(c) *T has models A, B such that $B \leqslant A$ and $\phi(A) = \phi(B)$ but $\psi(A) \neq \psi(B)$.*

Proof. (a) $\Rightarrow$ (b) is trivial. (b) $\Rightarrow$ (c): assuming (b), when $|\phi(A)|$ is infinite, let B be any elementary substructure of A which contains $\phi(A)$ and has cardinality $|\phi(A)|$.

(c) $\Rightarrow$ (a). Assume (c). Form a first-order language L^+ by adding to L a 1-ary relation symbol P, and let A^+ be the L^+-structure got from A by interpreting P as dom(B). This puts us into the setting of section 5.1; in the notation of that section $(A^+)_P = B$. Now A is infinite since $\psi(A) \neq \psi(B)$ and $B \leqslant A$. Hence A^+ has an elementary substructure C of cardinality ω. By the relativisation theorem (Theorem 5.1.1), C_P is defined, $C_P \leqslant C|L$, $\phi(C_P) = \phi(C|L)$ and $\psi(C_P) \neq \psi(C|L)$, since all these facts are recorded in Th (C).

We claim that C has a countable elementary extension D such that $D|L$ and D_P are homogeneous, and for every $n < \omega$, each n-type over $\varnothing$ which is realised in $D|L$ is also realised in D_P. (This makes sense since $D_P \leqslant D|L$).

To construct D we build a countable chain $C = C_0 \leqslant C_1 \leqslant \ldots$ as follows. When C_i has been constructed, we find a countable elementary extension C_{i+1} of C_i such that

(1.1)   if $\bar{a}$, $\bar{b}$ are tuples of elements of C_i with $(C_i|L, \bar{a}) \equiv (C_i|L, \bar{b})$, and d is any element of C_i, then there is an element c of C_{i+1} such that $(C_{i+1}|L, \bar{a}, c) \equiv (C_{i+1}|L, \bar{b}, d)$,

(1.2)   if $\bar{a}$, $\bar{b}$ are tuples of elements of $(C_i)_P$ with $(C_i|L, \bar{a}) \equiv (C_i|L, \bar{b})$, and d is any element of C_i, then there is an element c of $(C_{i+1})_P$ such that $(C_{i+1}|L, \bar{a}, c) \equiv (C_{i+1}|L, \bar{b}, d)$.

Thus, let E be an ω-saturated elementary extension of C_i; then $E|L$ and E_P are ω-saturated elementary extensions of $C_i|L$ and $(C_i)_P$ respectively, by Theorem 10.1.9. So we can choose C_{i+1} to be a suitable elementary substructure of E. Finally we put $D = \bigcup_{i<\omega} C_i$, and it readily follows that $D_P = \bigcup_{i<\omega} (C_i)_P$. Both D and D_P are countable. By (1.1) and (1.2) respectively, $D|L$ and D_P are both homogeneous. By (1.2) and induction on n, every n-tuple over $\varnothing$ which is realised in $D|L$ is also realised in D_P. The claim is proved.

If D is as in the claim, then Theorem 10.1.8(b) tells us that $D|L$ and D_P are isomorphic. Changing notation, write A for $D|L$ and B for D_P. Since D was an elementary extension of C, clause (c) of the theorem now holds with A and B countable, ω-homogeneous and isomorphic.

We build an elementary chain $(A_i : i < \omega_1)$ as follows, so that each structure A_i is isomorphic to A. We put $A_0 = B$. When A_i has been defined, we choose A_{i+1} and an isomorphism $f : A \to A_{i+1}$ so that the image of B under f is A_i. At a limit ordinal δ we put $A_\delta = \bigcup_{i<\delta} A_i$. The definition of homogeneity implies at once that the union of a countable elementary chain of countable homogeneous structures is homogeneous; so A_δ is homogeneous, and clearly it realises the same types over $\varnothing$ as do the structures A_i $(i < \delta)$. So by Theorem 10.1.8(b) again, $A_\delta \cong A$.

Put $A_{\omega_1} = \bigcup_{i<\omega_1} A_i$. By the construction, $\phi(A_{\omega_1}) = \phi(A_i)$ for each $i < \omega_1$, so that $\phi(A_{\omega_1})$ is at most countable. But for each $i < \omega_1$, $\psi(A_i) \subset \psi(A_{i+1})$, and so $\psi(A_{\omega_1})$ is uncountable. This proves (a) of the theorem. $\qquad\square$

Suppose T is a theory in a language L. A formula $\phi(x)$ of L is said to be **two-cardinal** for T if there is a model A of T such that $|A|$ and $|\phi(A)|$ are distinct; otherwise it is **one-cardinal** for T. We say the theory T is **two-cardinal** if there is a two-cardinal formula for T, and **one-cardinal** otherwise. A **Vaught pair** for the formula ϕ is a pair of structures A, B such that $B \leqslant A$, $B \neq A$ and $\phi(B) = \phi(A)$. For a countable first-order theory T, the implication (c) $\Rightarrow$ (b) in Vaught's two-cardinal theorem says that if some formula $\phi(x)$ has a Vaught pair of models of T, then ϕ is two-cardinal for T and so T is a two-cardinal theory.

In the late 1960s and early 1970s there was something of a gold-rush for generalisations of Vaught's two-cardinal theorem. One can separate out three strands in the work of that time.

First there were attempts to prove Vaught's two-cardinal theorem with other pairs of cardinals in place of ω and ω_1. It seems that Vaught had already taken the best pickings; nearly all the theorems one might hope to prove along these lines turn out to be independent of Zermelo–Fraenkel set theory. But two theorems worth knowing are Chang's two-cardinal theorem (Theorem 12.1.3 below) and Vaught's theorem on cardinals far apart (Theorem 12.1.4). Already by 1970 this area had been almost completely abandoned to the set theorists, and specifically to Ronald Jensen, who proved some startling and difficult two-cardinal theorems in the constructible universe. (For example he showed [1972] that in the constructible universe, if κ is any infinite cardinal then Theorem 12.1.1 holds with κ, κ^+ in place of ω, ω_1 in (a).)

In the second place, rather general methods were devised for constructing proper elementary extensions of countable structures, so that certain definable sets would grow in the process and others would stay constant. Keisler [1970] pioneered this approach in his work on the logics $L(Q_\alpha)$ – see Exercise 2.8.8 above. Shelah devised some powerful machinery for handling questions of this kind. A typical example is his proof [1977b], assuming the continuum hypothesis, that for every countable existentially closed group G there is a group M of cardinality ω_1 which is back-and-forth equivalent to G and is complete (i.e. every automorphism is inner), but M has no uncountable abelian subgroup.

And third, it was found that one could get dramatic improvements of Vaught's result by restricting attention to stable theories. For example Shelah [1969] proved the following.

Theorem 12.1.2. *Let L be a first-order language, T a stable complete theory in L and $\phi(x)$, $\psi(x)$ formulas of L. Suppose that T has a model A in which*

$\omega \leqslant |\phi(A)| < |\psi(A)|$. *Then for any two cardinals κ, λ with $|L| \leqslant \lambda < \kappa$ there is a model B of T with $|\phi(A)| = \lambda$ and $|\psi(A)| = \kappa$.* $\qquad\square$

So for stable theories, if there is any restriction at all on the gap between $|\phi(A)|$ and $|\psi(A)|$, these two cardinals must be equal. This conclusion ties ϕ and ψ very close together; for example if ϕ has finite Morley rank, then so has ψ. We shall prove this as Corollary 12.1.6 below.

Further two-cardinal theorems

The most direct generalisation of Vaught's two-cardinal theorem is the following result of Chang. Its proof is based on that of Vaught's theorem, but using saturated structures in place of homogeneous ones. Hence it only works when saturated structures exist, and this puts a heavy restriction on the cardinals involved. However, if the generalised continuum hypothesis holds, then Chang's result does generalise Vaught's from ω to any regular cardinal. (see Fact 10.4.1 for the strength of the assumption on κ.)

Theorem 12.1.3 (*Chang's two-cardinal theorem*). *Let κ be a cardinal such that $\kappa = \kappa^{<\kappa}$, and L be a first-order language with $|L| \leqslant \kappa$, and let $\phi(x)$ be a formula of L and T a theory in L. If ϕ is two-cardinal for T then there is a model A of T of cardinality κ^+ with $|\phi(A)| = \kappa$ or finite.*

Proof. Theorem 7.2.7 and Exercise 7.2.16 in Chang & Keisler [1973]. Cf. Exercises 1, 2 below. $\qquad\square$

Vaught's theorem and Chang's theorem both speak about small gaps between cardinalities. There are cases where we can pull apart the cardinalities of $\phi(A)$ and $\psi(A)$ by a certain amount, but only within limits. (For example, suppose A is a linear ordering and P^A is a dense subordering.) The next theorem discusses this situation. There is no loss in replacing the formulas ϕ and ψ by 1-ary relation symbols P and Q.

Theorem 12.1.4 (*Vaught's theorem on cardinals far apart*). *Let L be a first-order language whose symbols include 1-ary relation symbols P and Q, and let T be a theory in L. Suppose that for every $n < \omega$, T has a model B such that $|P^B| \geqslant \beth_n(|Q^B| + |L|)$ and Q^B is infinite. Then for all $\kappa \geqslant \lambda \geqslant |L|$, T has a model A such that $|P^A| = \kappa$ and $|Q^A| = \lambda$.*

A natural way to try to prove this would be to find a suitable EM functor F. We could get $|P^A| = \kappa$ by taking $A = F(\kappa)$, and by bolstering up the language with parameters we might also be able to ensure that $|Q^A| = \lambda$. In fact this method works reasonably well, and the second proof (due to Morley) of Theorem 12.1.4 ran along these lines.

But a more versatile approach is to replace the parameters by a second indiscernible sequence, independent of the first. This idea leads us to define **EM functors on pairs of orderings**, generalising the definition in section 11.2 in the obvious way. If F is such a functor in a language L, then for each pair of linear orderings (η, ζ), $F(\eta, \zeta)$ is an L-structure containing η and ζ, $\eta \cup \zeta$ generates $F(\eta, \zeta)$, and any pair of embeddings $f: \eta \to \eta'$ and $g: \zeta \to \zeta'$ extends to an embedding $F(f, g): F(\eta, \zeta) \to F(\eta', \zeta')$ in a functorial way. The argument of Theorem 11.2.1 shows that η, ζ will be independently indiscernible in $F(\eta, \zeta)$ for quantifier-free formulas, in the following sense.

Let $\phi(\bar{x}; \bar{y})$ be a formula of L, with its variables grouped into left $\bar{x}$ and right $\bar{y}$. We say that the pair of orderings η, ζ contained in the L-structure A is **independently ϕ-indiscernible** if for all increasing tuples $\bar{a}, \bar{a}'$ in η and all increasing tuples $\bar{b}, \bar{b}'$ in ζ, of appropriate lengths,

$$(1.3) \qquad A \vDash \phi(\bar{a}; \bar{b}) \Leftrightarrow A \vDash \phi(\bar{a}'; \bar{b}) \Leftrightarrow A \vDash \phi(\bar{a}'; \bar{b}').$$

We say that η, ζ are **independently indiscernible** in A if they are independently ϕ-indiscernible for all first-order formulas $\phi(\bar{x}; \bar{y})$ of L. We define $\mathrm{Th}(F)$ to be the set of first-order formulas $\phi(\bar{x}; \bar{y})$ of L such that for any orderings η and ζ, and any increasing tuples $\bar{a}, \bar{b}$ from η, ζ respectively, $F(\eta, \zeta) \vDash \phi(\bar{a}; \bar{b})$.

I leave it to the reader to work out exact analogues of the results in section 11.2 for the functors $F(-, -)$. In any case we shall save time by always assuming that $\mathrm{Th}(F)$ is skolemised, so that η, ζ are always independently indiscernible in $F(\eta, \zeta)$.

Proof of Theorem 12.1.4. Without loss assume that T is skolemised. Let Φ be the set of all formulas of L of the following forms:

$$(1.4) \qquad \phi(x_0, \ldots, x_{n-1}; \bar{y}) \leftrightarrow \phi(x_{i_0}, \ldots, x_{i_{n-1}}; \bar{y})$$
$$\text{whenever } i_0 < \ldots < i_{n-1} < \omega,$$

$$(1.5) \qquad \phi(\bar{x}; y_0, \ldots, y_{m-1}) \leftrightarrow \phi(\bar{x}; y_{j_0}, \ldots, y_{j_{m-1}})$$
$$\text{whenever } j_0 < \ldots < j_{m-1} < \omega,$$

$$(1.6) \qquad Px_0 \wedge Qy_0,$$

$$(1.7) \qquad Q(t(x_0, \ldots, x_{n-1}; \bar{y})) \to t(x_0, \ldots, x_{n-1}; \bar{y}) = t(x_{i_0}, \ldots, x_{i_{n-1}}; \bar{y})$$
$$\text{whenever } i_0 < \ldots < i_{n-1} < \omega.$$

Let $(c_n : n < \omega)$ and $(d_n : n < \omega)$ be new constants, and let U be T together with the sentences got by putting c_i, d_i for x_i, y_i respectively throughout Φ.

We claim that U has a model. By compactness it suffices to show this for any finite subset U_0 of U.

First we tackle (1.4). For any $k < \omega$, the number of complete k-types over Q^B which are realised in B is at most $2^{|Q^B|+|L|}$. Putting $\mu = 2^{|Q^B|+|L|}$, we know from the theorem assumption that P^B has cardinality

$\geqslant \beth_k(\mu) \geqslant \beth_{k-1}(\mu)^+$. By the Erdős–Rado theorem (Theorem 11.1.7), $\beth_{k-1}(\mu)^+ \to (\mu^+)^k_\mu$. We infer that B contains a sequence $(a_\alpha : \alpha < \mu^+)$ of elements of P^B such that any two increasing k-tuples from $(a_\alpha : \alpha < \mu^+)$ realise the same k-type over Q^B. (In the jargon of section 11.6, $(a_\alpha : \alpha < \mu^+)$ is k-indiscernible over Q^B.)

Next, by Ramsey's theorem we find, for any finite set of formulas $\phi_0(\bar{x}; \bar{y}), \ldots, \phi_{l-1}(\bar{x}; \bar{y})$ with $\bar{x} = (x_0, \ldots, x_{k-1})$ and $\bar{y} = (y_0, \ldots, y_{k-1})$, a sequence $(b_n : n < \omega)$ of elements of Q^B such that for any $j < l$ and any two increasing k-tuples $\bar{b}', \bar{b}''$ from $(b_n : n < \omega)$,

$$(1.8) \qquad A \vDash \phi_j(a_{i_0}, \ldots, a_{i_{k-1}}; \bar{b}') \leftrightarrow \phi_j(a_{i_0}, \ldots, a_{i_{k-1}}; \bar{b}'')$$

$$\text{whenever } i_0 < \ldots < i_{k-1} < \omega.$$

For this, note that by our choice of $(a_\alpha : \alpha < \mu^+)$ it suffices to ensure (1.8) when $i_0, \ldots, i_{k-1}$ are $0, \ldots, k-1$.

With suitable choices of k and the formulas $\phi_0, \ldots, \phi_{l-1}$, we satisfy all the sentences in U_0 which come from $(1.4), (1.5), (1.6)$ or T if we interpret each c_n as a_n and each d_n as b_n.

What about the sentences which come from (1.7)? Suppose the term t is $t(x_0, \ldots, x_{k'-1}; \bar{y})$ where $2k' \geqslant k$, and $\bar{b}'$ is any increasing tuple from $(b_n : n < \omega)$. Then we assert that by the choice of $(a_\alpha : \alpha < \mu)$, just one of the following holds:

(1.9) For all increasing tuples $\bar{a}'$ from $(a_\alpha : \alpha < \mu^+)$, $t^B(\bar{a}', \bar{b}')$ $\notin Q^B$.

(1.10) For all increasing tuples $\bar{a}', \bar{a}''$ from $(a_\alpha : \alpha < \mu^+)$, $t^B(\bar{a}'; \bar{b}') = t^B(\bar{a}''; \bar{b}') \in Q$.

(1.11) For all pairwise disjoint increasing tuples $\bar{a}', \bar{a}''$ from $(a_\alpha : \alpha < \mu^+)$, $t^B(\bar{a}', \bar{b}')$ and $t^B(\bar{a}'', \bar{b}')$ are distinct elements of Q^B.

This follows from the k-indiscernibility of $(a_\alpha : \alpha < \mu^+)$ over Q^B. But (1.11) is impossible since $|Q^B| < \mu$. Hence by taking k large enough, we ensure that all sentences of U_0 which come from (1.7) are satisfied too.

Thus U has a model A. It follows, just as in the proof of the Ehrenfeucht–Mostowski theorem (Theorem 11.2.8), that there is an EM functor F on pairs of orderings, such that $T \cup \Phi \subseteq \text{Th}(F)$. (We find F by taking $F(\omega, \omega)$ to be the L-reduct of a substructure of A, and then stretching.)

For any cardinals $\kappa \geqslant \lambda \geqslant |L|$, consider $F(\kappa, \lambda)$ (where κ and λ are considered as disjoint sets inside the structure!). By (1.6), all the elements of κ satisfy Px and all the elements of λ satisfy Qy. Every element satisfying Qy is of the form $t^{F(\kappa,\lambda)}(\bar{a}, \bar{b})$ for some increasing tuples $\bar{a}, \bar{b}$ in κ, λ respectively, and by (1.7) the element is independent of the choice of $\bar{a}$. So the total number of elements satisfying Qy is at most $|L| \cdot \lambda = \lambda$. $\qquad\square$

The same game can be played in cascades. If L has 1-ary relation symbols P, Q and R, and for every $n < \omega$, T has models B with $|P^B| \geq \beth_n(|Q^B| + |L|)$, $|Q^B| \geq \beth_n(|R^B| + |L|)$ and $|R^B| \geq \omega$, then for all $\kappa \geq \lambda \geq \mu \geq |L|$ there are models A with $|P^A| = \kappa$, $|Q^A| = \lambda$, $|R^A| = \mu$. For this we use EM functors on triples, $F(\eta, \zeta, \theta)$, in which each of η, ζ, θ is indiscernible over the union of the other two. And so on.

Unless we restrict to stable theories, the bounds $|P^B| \geq \beth_n(|Q^B| + |L|)$ in the hypothesis of Theorem 12.1.4 are in general best possible (see Exercise 3).

A more unexpected two-cardinal theorem appears in Shelah [1975b]. Let L be a first-order language whose symbols include two 1-ary relation symbols P and Q; suppose that T is a theory in L, and for every $n < \omega$ there is a model A of T such that $n^n \leq |Q^A|^n \leq |P^A| < \omega$. Then Shelah shows that if κ and λ are any cardinals with $|L| \leq \lambda \leq \kappa$, such that there is a tree with $\leq \lambda$ nodes and $\geq \kappa$ branches, then T has a model B with $|Q^B| = \lambda$ and $|P^B| = \kappa$.

Layerings

To the end of the section, L is a first-order language and T is a theory in L. If $\phi(x)$ and $\psi(x)$ are formulas of L, we write $\psi \leq \phi$ to mean that for every pair of models A, B of T with $B \leq A$, if $\phi(A) = \phi(B)$ then $\psi(A) = \psi(B)$. We shall give a syntactic characterisation of the relation $\leq$. (Recall that we met the relation $\leq$ in Theorem 12.1.1 above. But what follows is independent of that theorem, and unlike that theorem it is not limited to countable languages.)

If $\phi(x)$ is a formula of L, we write $(\forall x \in \phi)\chi$ and $(\exists x \in \phi)\chi$ for $\forall x(\phi \to \chi)$ and $\exists x(\phi \wedge \chi)$ respectively. A **layering by** $\phi(x)$ is a formula $\theta(x)$ of the following form, for some positive integer n:

(1.12) $(\exists y_0 \in \phi)\forall z_0((\exists z_0\eta_0 \to \eta_0) \to$

$\qquad (\exists y_1 \in \phi)\forall z_1((\exists z_1\eta_1 \to \eta_1) \to$

$\qquad \cdots$

$\qquad (\exists y_{n-1} \in \phi)\forall z_{n-1}((\exists z_{n-1}\eta_{n-1} \to \eta_{n-1}) \to$

$\qquad x = z_0 \vee \ldots \vee x = z_{n-1} \ldots),$

where each η_i is a formula $\eta_i(y_0, z_0, y_1, \ldots, y_i, z_i)$ of L.

Theorem 12.1.5. *Let L and T be as above, and let $\phi(x)$, $\psi(x)$ be formulas of L such that $T \vdash \exists x\, \phi$. Then $\psi \leq \phi$ if and only if there exists a layering $\theta(x)$ by ϕ such that $T \vdash \forall x(\psi \to \theta)$.*

Proof. The easy direction is from right to left. Suppose A, B are models of T

with $B \leqslant A$, and θ is a layering by ϕ such that $T \vdash \forall x(\psi \to \theta)$. Suppose also that $\phi(A) = \phi(B)$. Let a be any element of $\psi(A)$; we must show that a is in $\psi(B)$. Now by induction on $i < n$, where n is as in (1.12), we can choose elements b_i in $\phi(A) = \phi(B)$ and c_i in B so that

$$(1.13) \qquad B \vdash (\exists z_i\, \eta_i \to \eta_i)(b_0, c_0, \ldots, b_i, c_i).$$

(This is trivial. Since $T \vdash \exists x\, \phi$, we can find elements b_j in $\phi(A)$. When b_0, $\ldots$, b_i have been chosen, if $B \vdash \exists z_i\, \eta_i(b_0, \ldots, b_i)$ we have the required c_i; otherwise any element c_i in B will do.) Since $A \vDash \theta(a)$, it follows that a is one of $c_0, \ldots, c_{n-1}$, so that a is in B. Since $\psi(B) = \mathrm{dom}\,(B) \cap \psi(A)$, we have $\psi(A) = \psi(B)$.

For the converse we use Keisler games, slightly adapted. Assume $\psi \leqslant \phi$, and let λ be a strong limit number $> |L|$. In (3.4) of section 10.3, let each of the quantifiers $Q_i x_i$ have one of the forms $\forall x_i$, $\exists x_i$, $(\forall x_i \in \phi)$, $(\exists x_i \in \psi)$. We already know what the first two kinds of quantifier mean in the corresponding game on a structure A. When $Q_i x_i$ is $(\forall x_i \in \phi)$ (resp. $(\exists x_i \in \psi)$), player $\forall$ (resp. $\exists$) chooses the element a_i at step i, but now a_i must be an element of $\phi(A)$ (resp. $\psi(A)$). The rest is as before.

We shall need distinct variables y_i, z_i $(i < \lambda)$. We shall also need a list $(\eta_i : i < \lambda)$ of formulas of L such that

(1.14) each η_i is of the form $\chi_i(y_0, z_0, y_1, z_1, \ldots, y_i, z_i)$,

(1.15) if $\bar{y}$ (resp. $\bar{z}$) is a tuple of variables y_j (resp. z_j) and $\eta(x, \bar{y}, \bar{z})$ is any formula of L, then there is $i < \lambda$ such that $\eta(z_i, \bar{y}, \bar{z})$ is η_i and z_i is not in $\bar{z}$.

It is clear that such a list can be made.

Let χ be the Keisler sentence

$$(1.16) \quad (\exists x \in \psi)(\forall y_0 \in \phi)\exists z_0 \ldots (\forall y_i \in \phi)\exists z_i \ldots$$
$$\left(\bigwedge_{i < \lambda} (x \neq z_i) \wedge \bigwedge_{i < \lambda} (\exists z_i\, \eta_i \to \eta_i) \right).$$

We claim that $T \cup \mathrm{app}\,(\chi)$ has no model.

Assume for contradiction that $T \cup \mathrm{app}\,(\chi)$ has a model. Then by Theorem 10.4.2(b), $T \cup \mathrm{app}\,(\chi)$ has a model A which is special and of cardinality $\leqslant \lambda$. For convenience I assume that A is saturated, so that by Theorem 10.3.7(b), player $\exists$ has a winning strategy in the game $G(\chi, A)$. (Strictly we should use Theorem 10.4.7 instead, and a slightly more complicated game, but the difference is uninteresting.) Let the two players play this game; let player $\exists$ play to win, and let player $\forall$ choose so that his moves exhaust $\phi(A)$. Suppose the resulting play is

$$(1.17) \qquad a, b_0, c_0, \ldots, b_i, c_i, \ldots$$

and let X be the set $\{c_i : i < \lambda\}$. By (1.15) and the second conjunction in χ, if $\eta(z)$ is any formula with parameters from among the b_i and the c_i, and

$A \vDash \exists z\, \eta$, then $A \vDash \eta(c)$ for some $c \in X$. Hence X is the domain of an elementary substructure B of A (using the Tarski–Vaught criterion, Exercise 2.5.1). Hence also every b_i is in X, and so $\phi(B) = \phi(A)$ in view of player $\forall$'s moves. But by the first conjunction in χ, $a \notin X$ and hence $\psi(B) \neq \psi(A)$. Thus $\psi \not\leqslant \phi$; contradiction. The claim is proved.

So by compactness there are sentences $\xi_0, \ldots, \xi_{m-1}$ in app (χ) such that $T \vdash \neg(\xi_0 \wedge \ldots \wedge \xi_{m-1})$. But by the definition of app (χ), there is some single sentence ξ in app (χ) which implies all of $\xi_0, \ldots, \xi_{m-1}$, so that $T \vdash \neg \xi$. An easy rearrangement of $\neg \xi$ brings it to the form $\forall x(\psi \to \theta)$ where θ is a layering by ϕ. □

As promised, we use Theorem 12.1.5 to prove a result on Morley ranks.

Corollary 12.1.6. *Let T be a complete theory in a first-order language L, and suppose the formula $\phi(x)$ of L is one-cardinal for T.*
  (a) *If ϕ has finite Morley rank then so does T.*
  (b) *If ϕ has Morley rank $< \infty$ then so does T.*

Proof. We prove (a); the proof of (b) is much the same.

Let A be a model of T. Since ϕ is a one-cardinal formula, $(x = x) \leqslant \phi$. So by the theorem there is a layering $\theta(x)$ by ϕ such that $T \vdash \forall x\, \theta$. Referring to (1.12), write $\theta_i(x, y_0, z_0, \ldots, y_{i-1}, z_{i-1})$ for θ with the first i lines removed. Thus θ_0 is θ. We claim that for every $i \leqslant n$, there is $m_i < \omega$ such that for all $b_0, \ldots, b_{i-1}$ in $\phi(A)$ and all $c_0, \ldots, c_{i-1}$ in A, the Morley rank of $\theta_i(x, b_0, c_0, \ldots, b_{i-1}, c_{i-1})$ is at most m_i.

The claim is proved by downwards induction from n to 0. The formula $\theta_n(x, b_0, \ldots, c_{n-1})$ is simply $x = c_0 \vee \ldots \vee x = c_{n-1}$, which has Morley rank $\leqslant 0$. Assuming the claim holds for $i+1$, we consider the formula $\psi(x, y_i, z_i)$, viz.

(1.18) $(\exists z_i\, \eta_i(b_0, \ldots, c_{i-1}, y_i, z_i) \to \eta_i(b_0, \ldots, c_{i-1}, y_i, z_i))$

$\to \theta_{i+1}(x, b_0, \ldots, c_{i-1}, y_i, z_i),$

noting that $\theta_i(x, b_0, \ldots, c_{i-1})$ is $(\exists y_i \in \phi)\forall z_i\, \psi(x, y_i, z_i)$. Let b be any element of $\phi(A)$. We shall put a bound on the Morley rank of $\forall z_i\, \psi(x, b, z_i)$. Take any element c of A, such that $A \vDash \eta_i(b_0, \ldots, c_{i-1}, b, c)$ if possible. Then $\psi(A, b, c)$ is $\theta_{i+1}(A, b_0, \ldots, c_{i-1}, b, c)$, which has Morley rank $\leqslant m_{i+1}$ by induction hypothesis. Hence

(1.19) $\forall z_i\, \psi(x, b, z_i)$ has Morley rank $\leqslant m_{i+1}$.

Now $\exists x(\forall z_i\, \psi(x, y, z_i) \wedge \phi(y))$ has finite Morley rank since ϕ has. From this and (1.19) it follows at once by Erimbetov's inequality (Lemma 5.6.8) that there is a uniform finite upper bound m_i on the rank of $\exists y(\forall z_i\, \psi(x, y, z_i) \wedge \phi(y))$, which is equivalent to $\theta_i(x, b_0, \ldots, c_{i-1})$. This proves the claim.

By the claim, $\theta_0(x)$ has finite Morley rank. But since $A \vDash \forall x\, \theta$, θ_0 is equivalent to $x = x$. □

Exercises for section 12.1

1. Prove Theorem 12.1.3 for the case where $\kappa > |L|$. [Follow the proof of (c) $\Rightarrow$ (a) in Theorem 12.1.1, but using saturated models of cardinality κ in place of countable homogeneous models. The only thing that goes wrong is that unions C_δ of elementary chains $(C_i: i < \delta)$ of saturated structures are in general not saturated. To keep the chain going, it suffices to find an elementary embedding $f: C_\delta \to C_0$ which is the identity on $\phi(C_\delta) = \phi(C_0)$. For this use Exercise 10.3.3. We must show that if $\Phi(x)$ is a type in C_δ over fewer than κ parameters, and $\phi(x) \in \Phi$, then C_δ realises Φ. We can suppose there is a relation symbol R which indexes by elements of $\phi(C_0)$ all the finite subsets of $\phi(C_0)$ (and no doubt other sets as well) as $X(a)$ ($a \in \phi(C_0)$). For each finite subset Ψ of Φ choose an element c_Ψ of $\phi(C_0)$ realising Ψ. By saturation of some C_i ($i < \delta$), choose an element a_Ψ of $\phi(C_0)$ such that $X(a_\Psi)$ is a set which contains $c_{\Psi'}$ for all finite subsets Ψ' of Φ containing Ψ, and such that every element in $X(a_\Psi)$ satisfies Ψ. Then by saturation of C_0, find an element in the intersection of the sets $X(a_\Psi)$.]

2. Show that Theorem 12.1.3 can be reduced to the case where $\kappa > |L|$. [If $\omega < \kappa = |L|$, use a countable language to code up the formulas of L as elements of $\phi(A)$.]

3. Given $n < \omega$, write down a theory T in a first-order language L containing a 1-ary relation symbol P, such that T has models A of cardinality $\geqslant \beth_n$ in which $|P^A| = \omega$, but not arbitrarily large models B with $|P^B| = \omega$. [For $\beth_1$, arrange that the number of elements of A is at most the number of subsets of P^A.]

4. Let L be a first-order-closed fragment of $L_{\infty\omega}$ of cardinality κ, where L contains 1-ary relation symbols P and Q, and let σ be a sentence of L. Suppose that for every $\alpha < \delta(\kappa)$, σ has a model B with $|P^B| \geqslant \beth_\alpha(|Q^B| + \kappa)$. Show that for every cardinal $\lambda \geqslant \kappa$, T has a model A with $|P^A| = \lambda$ and $|Q^A| = \kappa$. [This needs a revision of the proof of Theorem 11.6.1. Redefine 'η is k-indiscernible^L in A' by 'the elements of η satisfy P, and for all increasing k-tuples $\bar{a}$, $\bar{b}$ from η and all formulas $\phi(x_0, \ldots, x_{k-1})$ of L with parameters from Q^A, $A \vDash \phi(\bar{a}) \Leftrightarrow A \vDash \phi(\bar{b})$'. The revised version of Theorem 11.6.1 then yields an EM functor F with the property that $P(x)$ and the formulas (1.7) of the present section are in $\mathrm{Th}(F)$.]

5. Show that Theorem 12.1.4 remains true if we replace P and Q by sets of formulas $\Phi(x)$ and $\Psi(x)$, and the condition 'for every $n < \omega$' by 'for every ordinal $n < \delta(|L|)$'. [Use the previous exercise.]

*6. Let L and L^+ be countable first-order languages with $L \subseteq L^+$, and let T be a theory in L^+. Let κ and λ be uncountable cardinals, and suppose $\lambda = \beth_{\omega_1 \cdot \alpha}$ for some positive ordinal α. Show that if for all models A, B of T of cardinality κ we have $A|L \cong B|L$, then the same holds with λ in place of κ. [It suffices to show that if T has any models A of cardinality λ with $A|L$ unsaturated, then T has a model C of cardinality κ such that $C|L$ is not ω_1-saturated. A type $\Phi(x)$ over fewer than λ parameters in $A|L$ contains fewer than λ sentences; adding symbols to A if necessary, code up these sentences as elements of a set Q^A of cardinality $< \lambda$, and note that $\lambda \geqslant \beth_{\omega_1}(|P^A|)$.]

7. Let L be a countable first-order language, T a theory in L and $\phi(x)$, $\psi(x)$ formulas of L. Suppose that for every model A of T, $|\psi(A)| \leqslant |\phi(A)| + \omega$. Show that there is a polynomial $p(x)$ with integer coefficients, such that for every model A of T, if $|\phi(A)| = m < \omega$ then $|\psi(A)| \leqslant p(m)$. [Use Theorem 12.1.5.]

8. Let L be a first-order language, T a complete theory in L with infinite models, and $\phi(x)$, $\psi(x)$ formulas of L. By a **stratification** of ψ over ϕ in a model A of T we mean a formula $\sigma(x, y)$ of L with parameters from A, such that $A \vDash \forall x(\psi(x) \leftrightarrow \exists y(\sigma(x, y) \wedge \phi(y)))$; we call the stratification σ **algebraic** if for every element b of $\phi(A)$, the set $\{a : A \vDash \sigma(a, b)\}$ is finite. Show (a) even when $\psi \leqslant \phi$, there need not be an algebraic stratification of ψ over ϕ in any model of T, (b) if A is a model of T, $\psi \leqslant \phi$ and $\psi(A)$ is infinite, then there are a formula $\rho(x)$ of L with parameters from A such that $\rho(A)$ is an infinite subset of $\psi(A)$, and an algebraic stratification of ρ over ϕ in A. [Referring to the formula $\theta(x)$ of (1.12), write $\theta_k(x, y_0, z_0, y_1, \ldots, z_{k-1}, y_k)$ for the formula which results if we delete everything up to ($\exists y_k \in \phi$) inclusive. *Case 1*. For each $b \in \phi(A)$ there are only finitely many $a \in \psi(A)$ such that $A \vDash \theta_0(a, b)$. Then put $\rho = \psi$ and $\sigma(x, y) = \theta_0(x, y)$. *Case 2*. There is $b_0 \in \phi(A)$ such that the set $X_0 = \{a : A \vDash \theta_0(a, b_0)\}$ is infinite, but for each $b \in \phi(A)$ there are only finitely many $a \in X_0$ such that $A \vDash \theta_1(a, b_0, c_0, b)$ (where c_0 is some fixed element of A such that $A \vDash \eta_0(b_0, c_0)$ if there is such an element). Then put $\rho(x) = \theta_0(x, b_0)$ and $\sigma(x, y) = \theta_1(x, b_0, c_0, y)$. Etc. up to case n.]

9. Let T be a stable complete first-order theory which has the finite cover property. Show that T^{eq} (the theory of A^{eq} where A is any model of T) is a two-cardinal theory. [This was the proof of Corollary 9.5.6.]

*10. Give an example of a stable complete countable first-order theory which has the finite cover property but no two-cardinal formula. [Hrushovski (unpublished) has shown that no example can be superstable. Ziegler's example is as follows. For some prime p let $\mathbb{Z}_{(p)}$ be the localisation of the ring $\mathbb{Z}$ at p, and for each $i < \omega$ let N_i be a copy of $\mathbb{Z}_{(p)}$. Put $N = \bigoplus_{i<\omega} N_i$. Partition $\mathrm{dom}(N)$ into sets I_n ($0 < n < \omega$) so that each I_n has cardinality n, and put $P = \bigcup_{0<n<\omega}(\bigoplus_{i \in I_n} N_i + pN)$. Let T be $\mathrm{Th}(N, P)$.]

11. For the present exercise, a (κ, κ^+)-**tree** is a partially ordered set $(P, \leqslant)$ together with a linear ordering ξ and a map $f : P \to \xi$ such that (i) on each branch (i.e. maximal linearly ordered subset) of $(P, \leqslant)$, f is an order-embedding onto an initial segment of ξ, (ii) each element of ξ has at most κ predecessors and at most κ pre-images under f, (iii) P has cardinality κ^+. A (κ, κ^+)-tree is called a **special κ^+-Aronszajn tree** if ξ is

well-ordered. (a) Show that there is a first-order theory T in a first-order language L one of whose symbols is a 1-ary relation symbol P, such that for every infinite cardinal κ, T has a model A with $|A| = \kappa^+$ and $|P^A| = \kappa$ if and only if there is a (κ, κ^+)-tree. (b) Show that if a (κ, κ^+)-tree exists, then so does a special κ^+-Aronszajn tree. (c) It is known that special ω_1-Aronszajn trees exist, but if it is consistent that a Mahlo cardinal exists, then it is consistent that there is no special ω_2-Aronszajn tree. Draw the appropriate conclusion about the assumptions in Chang's two-cardinal theorem.

12.2 Categoricity

Throughout this section, L is a first-order language and T is a complete theory in L. Recall that a theory T is λ-**categorical** if T has, up to isomorphism, exactly one model of cardinality λ. A structure A is λ-**categorical** if $\mathrm{Th}(A)$ is λ-categorical.

Model theorists have felt for a long time that λ-categoricity, at least for large enough λ, is a very strong condition on a theory T, so that one should be able to prove a good many theorems about such theories and their models. And indeed this has been one of the most fertile branches of model theory since the early 1960s. Many of the results use methods from stability which go far beyond this book. Nevertheless the facts collected in this section, with or without their proofs, should be part of the general education of every model theorist.

Since I am mostly giving results and not proofs, I have not felt bound by the logical order in which the results are proved. In practice there is a good deal of bootstrapping: a theorem is proved in a weak form in special cases, the result is used to prove a second theorem, and the second theorem leads back to a full proof of the first theorem.

The spectrum of categoricity

The one major result of this section which is straightforwardly provable within the limits of this book is the upward direction of Morley's theorem, and I sketch a proof in the exercises. Morley's theorem is the case of Theorem 12.2.1 where L is countable.

Theorem 12.2.1 (*Morley–Shelah theorem*). *Let λ and μ be any two cardinals, both $> |L|$. If T is λ-categorical then T is μ-categorical.* $\square$

Hence if T is λ-categorical for some cardinal $\lambda > |L|$, then by Corollary 9.5.6, T doesn't have the finite cover property.

We say that T if **uncountably categorical** if T is countable and μ-categorical for every uncountable μ. We say that T is **totally categorical** if T is countable and μ-categorical for every infinite μ.

The theory of (non-empty) dense linearly ordered sets without first or last element is ω-categorical but not uncountably categorical. Example 3 in section 3.2 proves the ω-categoricity. To see that the theory is not ω_1-categorical, consider the orderings $\eta \cdot \omega_1$ and $\eta \cdot (\omega_1^*)$ where η is the ordering of the rationals and ω_1^* is ω_1 backwards.

The theory of algebraically closed fields of a fixed characteristic is uncountably categorical by Corollary 4.5.7. It is not ω-categorical, because there are countable models with any transcendence degree $\leq \omega$.

The theory of infinite vector spaces over a field F is λ-categorical for every infinite cardinal $\lambda > |F|$, by Theorem 4.7.3 or a direct algebraic argument. If F is finite, the theory is countable and totally categorical.

These three examples leave one question open, which the next theorem answers.

Theorem 12.2.2. *If L is uncountable and T is λ-categorical for some $\lambda \leq |L|$, then T is definitionally equivalent to some theory of cardinality $< \lambda$, and hence T is μ-categorical for every $\mu \geq |L|$.* $\qquad\square$

In section 7.3 we said some things about ω-categorical theories. For the rest of this section we need categoricity in at least one uncountable cardinal, in order to use the following fundamental theorem.

Theorem 12.2.3. *Suppose T is λ-categorical for some $\lambda > |L|$.*
   (a) *T is superstable.*
   (b) *If L is countable then T is totally transcendental.*

Proof. (a) uses Theorem 11.3.10, which I stated without proof. (b) is easier; we proved it in Corollary 11.2.10. $\qquad\square$

Prime models

We shall need the notion of a prime model over a set. Let C be a model of T and X a set of elements of C; let the sequence $\bar{a}$ list the elements of X. A **prime model** over X (or over $\bar{a}$) in C is an elementary substructure A of C containing X, such that

(2.1)   if B is any model of T and $\bar{b}$ any sequence of elements of B such that $(C, \bar{a}) \equiv (B, \bar{b})$, then there is an elementary embedding of $(A, \bar{a})$ into $(B, \bar{b})$.

If X is empty then C is redundant in this definition; a **prime model over the empty set** is a model of T which is elementarily embeddable in every model of T.

Theorem 12.2.4. *Let T be λ-categorical for some $\lambda > |L|$; let C be a model of T and X a set of elements of C. Then there is a prime model A over X in C. It is unique in the following sense: if $\bar{a}$ is a sequence listing the elements of X, D is a model of T and $\bar{b}$ is a sequence of elements of D such that $(C, \bar{a}) \equiv (D, \bar{b})$, and B is a prime model over $\bar{b}$ in D, then $(A, \bar{a}) \cong (B, \bar{b})$.* $\square$

This theorem has completely different proofs for the two sides of the Buechler dichotomy, Theorem 12.2.9 below. In the totally transcendental case one can prove it directly; see Exercise 1 below. The following example shows that prime models are not necessarily unique in a stronger sense. We say that a model A of T is **minimal over** a set X of elements of A if there is no proper elementary substructure of A which contains X.

Example 1: *Models which are prime but not minimal.* Let L be the language of empty signature and T the theory which says there are infinitely many elements. Then T is totally categorical. If C is any model of T, then any countable subset of dom(C) forms a prime model over the empty set in T. Thus we have prime models A_0, A_1 with A_0 a proper elementary substructure of A_1; in other words, A_1 is not a minimal model of T.

There are not many examples of this phenomenon; one can show (Laskowski [1988]) that if a theory T is λ-categorical for some $\lambda > |L|$, and T has some model which is prime over a set but not minimal over that set, then T must be totally transcendental and ω-categorical. For example let T be the theory of infinite vector spaces over a field F. If F is infinite, then the prime model of T over the empty set is minimal, and it consists of a 1-dimensional space. If F is finite we have a situation like Example 1.

Theorem 12.2.5. *Suppose T is λ-categorical for some $\lambda > |L|$, and let A be the prime model of T over the empty set.*

(a) *If $\bar{a}$ is a sequence listing the elements of A, then $\mathrm{Th}(A, \bar{a})$ is also λ-categorical for some $\lambda > |L|$, and its prime model over the empty set is the unique model of $\mathrm{Th}(A, \bar{a})$ where every element is named by a constant.*

(b) *If T is totally transcendental, then A is atomic (i.e. every tuple in A realises a principal complete type over the empty set; see section 7.2).* $\square$

Theorem 12.2.5 is used to tidy up the proofs in the totally transcendental case. Suppose A is a model of T. We can think of the prime model C over the empty set as an elementary substructure of A. If $\phi(x)$ is a formula of L with parameters $\bar{c}$ from C, we can add the parameters to the language. Assuming T is totally transcendental, (b) tells us that these parameters $\bar{c}$ realise a principal complete type over the empty set, and accordingly

Th $(A, \bar{c})$ is called a **principal extension** of T. Principal extensions of T are completely innocent: if $\psi(\bar{x})$ isolates the type of $\bar{c}$ over $\varnothing$, then $T \vdash \exists \bar{x} \, \psi$, so every model B of T contains a tuple $\bar{d}$ realising this same type. Then $(A, \bar{c}) \equiv (B, \bar{d})$, so that B expands to a model of Th $(A, \bar{c})$.

Example 2: *A principal extension.* Let T be the theory which says that E is an equivalence relation with two classes, both infinite, and R is a symmetric relation describing a bijection between the two classes. Then T is totally categorical. Without parameters we can't name either equivalence class. But T has only one complete type $p(x)$ over the empty set. If A is any model of T and c an element of A, then Th (A, c) is a principal extension of T, and thanks to c we can name the two equivalence classes. This gives us a $\varnothing$-definable strongly minimal set.

The minimal set

In the most familiar examples which are λ-categorical for some $\lambda > |L|$, every model is prime over a strongly minimal set. For almost strongly minimal theories this is clear. Here is a further example.

Example 3: *A totally categorical theory which is not almost strongly minimal.* As in Example 1 of section 4.7, let p be a prime and A the direct sum of infinitely many cyclic groups of order p^2. We saw that T is not almost strongly minimal. Every model of T is a direct sum of cyclic groups of order p^2, and the number of summands is the dimension of the socle as a vector space over the p-element field. So T is totally categorical. However, the socle is strongly minimal and the whole model is prime over the socle.

In fact we can show, assuming λ-categoricity for some $\lambda > |L|$, that every model with a strongly minimal set is prime over that set. The proof needs a couple of lemmas.

Lemma 12.2.6. *Let T be a stable theory and $\phi(x)$ a formula of L. If T has models A, B such that A is a proper elementary substructure of B and $\phi(A) = \phi(B)$ and $\phi(A)$ is infinite, then for every pair of cardinals $\lambda > \mu \geqslant |L|$, T has a model C with $|\phi(C)| = \mu$ and $|C| = \lambda$.* $\qquad \square$

Lemma 12.2.7. *If T is λ-categorical for some $\lambda > |L|$, then no model of T has a two-cardinal formula defining an infinite set (even allowing parameters).*

Proof. Suppose for contradiction that T is λ-categorical for some $\lambda > |L|$, but there is a model C of T' such that $\omega \leqslant |\phi(C, \bar{c})| < |C|$, where ϕ is some formula of L and $\bar{c}$ is a tuple of elements of C. By Theorems 12.2.3 and 6.7.2, T' is stable. So by the previous lemma, T' has a model A of cardinality λ with a tuple $\bar{a}$ of elements such that $|\phi(A, \bar{a})|$ has cardinality $|L|$. But by compactness (Exercise 6.1.10), T' has a model B of cardinality λ in which $\phi(B, \bar{b})$ is either finite or of cardinality λ for each tuple $\bar{b}$ in B. Clearly A and B are not isomorphic. $\qquad\square$

Theorem 12.2.8. *Let T be λ-categorical for some $\lambda > |L|$. Suppose that for some model A of T, there is a strongly minimal formula $\phi(x)$ with parameters $\bar{a}$ in A.*

(a) *T has finite Morley rank.*

(b) *$(A, \bar{a})$ is prime over $\phi(A)$.*

Proof. Let $\bar{a}$ be the parameters in ϕ, and write T' for $\text{Th}(A, \bar{a})$. By Lemma 12.2.7 and Corollary 12.1.6, $\text{Th}(A, \bar{a})$ has finite Morley rank; so does T, by Exercise 5.6.1. By Theorem 12.2.4 (or Exercise 1 below) there is a prime model B of T' over $\phi(A)$ in $(A, \bar{a})$. If B is not the whole of $(A, \bar{a})$, then by Lemma 12.2.6, T' has models in which ϕ is a two-cardinal formula, contradicting Lemma 12.2.7. So B is $(A, \bar{a})$ and thus $(A, \bar{a})$ is prime over $\phi(A)$. $\qquad\square$

Unfortunately not all theories which are λ-categorical for some $\lambda > |L|$ have a strongly minimal set.

Example 4: *A theory T of cardinality 2^ω which is $(2^\omega)^+$-categorical, such that no finite extension of T by constants has a strongly minimal formula.* Let p be a prime and for each $i < \omega$ let C_i be a copy of the cyclic group $Z(p)$ of order p. Let X be the direct product $\prod_{i<\omega} C_i$. Let L be the first-order language whose signature consists of a 2-ary function symbol $+$, countably many 2-ary relation symbols E_i ($i < \omega$) and a constant c_a for each element a of X. Thus $|L| = 2^\omega$. Let A be the following L-structure. The domain of A is X; every element of A is a sequence $a = (a_0, a_1, \ldots)$. Each constant c_a stands for the element a. Each symbol E_i stands for the equivalence relation which holds between a and b if and only if $a_j = b_j$ for all $j < i$. Regarding X as an additive abelian group, $+$ stands for the addition operation. Put $T = \text{Th}(A)$.

Suppose B is any model of T of cardinality $(2^\omega)^+$. If a, b are any elements of B, we write $a \sim b$ iff $B \vDash E_i(a, b)$ for all $i < \omega$. Then $\sim$ is an equivalence relation, and it has 2^ω equivalence classes, each of them represented by some element c^B with c a constant of L. Let c_0 be the constant naming the constant element $(0, 0, \ldots)$ in A. Then the equivalence class $c_0^{\sim}$ of c_0^B under $\sim$ is an elementary abelian $Z(p)$-group. If c is a constant of L, then the map $a \mapsto a +^B c^B$ is a bijection between the equivalence classes $c_0^{\sim}$ and $c^{\sim}$. Hence each of these classes has cardinality $(2^{\omega})^{+}$. Thus the abelian group $c_0^{\sim}$ is completely determined, and so in turn is the whole of B. This proves that T is $(2^{\omega})^{+}$-categorical.

Finally T is not totally transcendental; the equivalence relations E_i see to that, noting Lemma 5.6.5. We show that if A is any model of T and $\bar{a}$ a tuple of elements of A, then $\mathrm{Th}(A, \bar{a})$ has no strongly minimal formula. Since T is stable by Theorem 12.2.3, $\mathrm{Th}(A, \bar{a})$ is stable too. If T' had a strongly minimal formula $\phi(x)$, then by Theorem 12.2.8, $\mathrm{Th}(A, \bar{a})$ would be totally transcendental; but this is impossible since T is not totally transcendental.

Note that in Example 4, every model of T does indeed contain a vector space, namely the equivalence class $c_0^{\sim}$. But $c_0^{\sim}$ is defined by an infinite conjunction of formulas, not by a single formula.

A complete type $p(x)$ over a set of elements in some model of a theory T is said to be **regular** if for every model A of T, the set $p(A)$ of elements realising p carries a geometry defined by a dependence relation (in analogy to the geometry on $\phi(A)$ when ϕ is a strongly minimal formula). This definition is loose; see any textbook of stability theory for the precise definition. In Example 4 there is a regular type which expresses '$x \sim c_0$ and $x \neq c_0$'.

Theorem 12.2.9 (Buechler's dichotomy). *Let T be λ-categorical for some $\lambda > |L|$. Then one of the following holds.*

(a) *T is totally transcendental and has a strongly minimal formula whose parameters realise a principal complete type over $\varnothing$.*

(b) *T is not totally transcendental, but T has a regular type over the prime model over $\varnothing$, and the geometry of this type is locally modular (see section 4.6).*   □

In case (a) we can pass to a principal extension of T and suppose that the strongly minimal set has no parameters; recall the comments after Theorem 12.2.5.

Corollary 12.2.10. *Let T be an uncountably categorical theory, or more generally a totally transcendental theory which is λ-categorical for some $\lambda > |L|$. Then T has finite Morley rank.*

Proof. By the theorem and Theorems 12.2.3 and 12.2.8.   □

Now we can describe all the models of T.

Theorem 12.2.11. *Let T be λ-categorical for some $\lambda > |L|$. Then there is an elementary chain $(A_i: i$ an ordinal$)$ of models of T with the following properties.*

(a) *Every model of T is isomorphic to A_i for some unique i.*

(b) *A_0 is prime over the empty set.*

Let $\Phi(x)$ be either a strongly minimal formula with parameters in A_0, or a regular type over A_0 whose geometry is locally modular, according to the two cases in Theorem 12.2.9. Write $\Phi(A)$ for the set of elements of A which satisfy Φ.

(c) *Each model A_i is prime over $\Phi(A_i)$.*

(d) *For each ordinal i we can choose a basis X_i for the geometry of $\Phi(A_i)$, so that $X_i \subseteq X_j$ when $i < j$.* $\qquad\square$

Each model A_i in Theorem 12.2.11 has a dimension, namely the cardinality of X_i. When A_i has dimension $\mu > |L|$, A_i is up to isomorphism the unique model of T of cardinality μ.

By looking at the construction of the models A_i in Theorem 12.2.11, one can read off some more information.

Theorem 12.2.12. *Let T be a λ-categorical theory for some $\lambda > |L|$.*

(a) *If A is a model of dimension $|A|$, and in particular if A is a model of cardinality $> |L|$, then A is saturated.*

(b) *The number of models of T of cardinality $|L|$, counting up to isomorphism, is either 1 or $|L|$.*

(c) *If L is countable then every model of T is homogeneous (in the sense of section 10.1).* $\qquad\square$

Corollary 12.2.13. *If A is an $|A|$-categorical L-structure then A is saturated.*

Proof. If A is finite, this is by Exercise 10.1.2. If A is countable, it's by Exercise 7.2.11. Finally if A is uncountable, then by Theorem 12.2.2 we can suppose that $|L| < |A|$, so that Theorem 12.2.12(a) applies. $\qquad\square$

Structure theory

By Theorem 12.2.11, the problem of describing the models A of a theory T which is λ-categorical for some $\lambda > |L|$ reduces to two parts. First we need a description of $\Phi(A)$. Second, we need to know explicitly how one constructs the prime model over $\Phi(A)$.

When the theory T is totally categorical, $\Phi(A)$ is strongly minimal and ω-categorical. So by Lemmas 4.5.5, 4.5.6 and Corollary 7.3.2, its geometry is a homogeneous locally finite combinatorial geometry of infinite dimension. The Cherlin–Mills–Zil'ber theorem (Theorem 4.6.1) describes all such geometries: they are either disintegrated, or projective over a finite field, or affine over a finite field. If we drop the assumption of ω-categoricity, other geometries appear. They include that of algebraically closed fields of a given characteristic (Evans & Hrushovski [1991]) and the puzzling 'free' examples of Hrushovski [1988], but no overall classification is known.

In the non-totally-transcendental case, $\Phi(A)$ has a locally modular geometry. So by Theorem 4.6.5, either this geometry is disintegrated or a refinement of Theorem 4.8.1 (Hrushovski [1987]) finds an infinite abelian group implicit in $\Phi(A)$. This suggests that Example 4 is typical of this case; but again no overall classification is known.

Even when we know the geometry of $\Phi(A)$, the structure on $\Phi(A)$ is by no means completely determined. Of course there is not the slightest hope of pinning down the language, but one may still be able to say things about the definable relations on $\Phi(A)$. The problem points towards group theory and geometrical combinatorics. (See for example the end of Zil'ber [1984b], or in the disintegrated locally finite case, Hrushovski [1989b] and Evans [199?].)

We turn to the prime model over $\Phi(A)$. One constructs it by successively realising appropriate types in a well-ordered sequence. This construction by itself throws very little light on the shape of the model A. Happily in the totally transcendental case there is another approach, which is implicit already in the proof of Theorem 12.2.11.

Theorem 12.2.14. *Suppose T is a totally transcendental theory which is λ-categorical for some $\lambda > |L|$, and let $\phi(x)$ be a strongly minimal formula without parameters. Passing to a definitional extension, we can suppose that $\phi(x)$ is $P(x)$ for some 1-ary relation symbol P, and that for each model A of T there is a relativised reduct A_P with domain P^A, such that the $\varnothing$-definable relations of A_P are exactly the restrictions to P^A of the $\varnothing$-definable relations of A. Then every automorphism of A_P extends to an automorphism of A.*

We shall pursue this further in section 12.5 below. Meanwhile it raises an obvious question: how many different automorphisms of A are there which extend a given automorphism of A_P? Equivalently, what is the automorphism group $\mathrm{Aut}\,(A)_{(P^A)}$? The following result of Zil'ber [1984c] threw a sudden shaft of light on this question.

Theorem 12.2.15. *Let T be a totally transcendental theory which is λ-categorical for some $\lambda > |L|$.*

(a) *There are a finite set of formulas $\phi_0, \ldots, \phi_{m-1}$ of L with parameters, such that $\phi_0(A) \subseteq \ldots \subseteq \phi_{m-1}(A)$ for every model A of T, ϕ_0 is strongly minimal with parameters in the prime model over $\varnothing$, and $\phi_{m-1}(A)$ is the whole of A.*

(b) *Let A be any model of T, and let $\psi(A)$ be a subset of $\phi_{i+1}(A)$ which is definable with parameters in $\phi_i(A)$, and is minimal with this property. Then A^{eq} has a relativised reduct with parameters in $\phi_i(A)$, which consists of (i) a* *group G and (ii) a faithful left action of G on $\psi(A)$, such that the maps $a \mapsto g(a)$ ($g \in G$) are exactly the permutations of $\psi(A)$ induced by automorphisms of A which fix $\phi_i(A)$ pointwise.* ☐

The group G is called the **binding group** of $\psi(A)$ over $\phi_i(A)$. If T in Theorem 12.2.15 is ω-categorical, then the sets $\psi(A)$ in (b) will be precisely the sets of the form $p(A)$ where $p(x)$ is a complete type over $\phi_i(A)$ containing the formula ϕ_{i+1}; the automorphisms of A which fix $\phi_i(A)$ pointwise act transitively on such sets $p(A)$. So in this case the binding group acts regularly on $\psi(A)$.

Now one can show that if $\phi(A)$ is a strongly minimal set with disintegrated geometry, then no infinite group is interpretable in $\phi(A)$ (see for example Theorem A.6.9 for the case where $\phi(A)$ is just an infinite set). We deduce the following.

Corollary 12.2.16. *If T is totally categorical and has a strongly minimal set with disintegrated geometry, then T is almost strongly minimal.* ☐

Very recently Lascar announced a result which will certainly be useful for structure theory.

Theorem 12.2.17. *If A is a countable model of a totally categorical theory (or more generally, an ω-stable ω-categorical theory) then A has the small index property.* ☐

Axiomatisations

A theory T is said to be **quasi-finitely axiomatisable** if T is definitionally equivalent to a theory consisting of a single sentence together with all the sentences $\exists_{\geqslant n} x\, x = x$ ($n < \omega$). The following theorem answers an old unpublished conjecture of Vaught.

Theorem 12.2.18. *Suppose T is totally categorical.*

(a) T is not finitely axiomatisable. (In fact this holds whenever T is ω-stable and ω-categorical.)

(b) T is quasi-finitely axiomatisable. ☐

For uncountably categorical theories, Theorem 12.2.18(a) fails. The example below is not the easiest one to prove, but of those I have seen, it is the most economical and the easiest to picture.

Example 5: *An uncountably categorical theory which is finitely axiomatisable.* We describe a model A of T. The domain of A is $\mathbb{Z} \times \mathbb{Z} \times \{0, 1\}$, and we think of this as a subset of Euclidean space $\mathbb{R}^3$. We write each element a of A as (a^0, a^1, a^2) (the superscripts are indices, not powers!). There are a 1-ary relation symbol *Diag* and 1-ary function symbols π_0, π_1 (projections), *Up*, *Right*, *Swap* and μ (for Möbius). The structure A is defined by (2.2)–(2.5).

(2.2) $A \vDash Diag\,(a)$ $\Leftrightarrow$ $a^0 = a^1$ and $a^2 = 0$.

(2.3) $A \vDash \pi_0(a) = b$ $\Leftrightarrow$ $b = (a^0, a^0, 0)$.

        $A \vDash \pi_1(a) = b$ $\Leftrightarrow$ $b = (a^1, a^1, 0)$.

(2.4) $A \vDash Right\,(a) = b$ $\Leftrightarrow$ $b = (a^0 + 1, a^1, a^2)$.

        $A \vDash Up\,(a) = b$ $\Leftrightarrow$ $b = (a^0, a^1 + 1, a^2)$

        $A \vDash Swap\,(a) = b$ $\Leftrightarrow$ $b = (a^0, a^1, 1 - a^2)$.

(2.5) $A \vDash \mu(a) = b$ $\Leftrightarrow$ either $a^0 = a^1$ and $b = a$,

                                    or $a^0 > a^1$ and $b = (a^1, a^0, a^2)$,

                                    or $a^0 < a^1$ and $b = (a^1, a^0, 1 - a^2)$.

We take T to be the statements (2.6)–(2.17), all of which can be written as first-order sentences true in A.

(2.6)   *Right, Up* and *Swap* each permute the universe (i.e. they are invertible).

(2.7)   *Right, Up* and *Swap* commute with each other.

(2.8) $\pi_0^2 = \pi_0$ and $\pi_1^2 = \pi_1$.

(2.9) π_0 and π_1 commute with each other and with *Swap*.

(2.10)   For all x, $Diag\,(\pi_0(x))$ and $Diag\,(\pi_1(x))$.

(2.11)   For all x_0, x_1 in *Diag* there are exactly two elements y such that $\pi_0(y) = x_0$ and $\pi_1(y) = x_1$; if y_0 is one of them, $Swap\,(y_0)$ is the other.

(2.12)   For all x, $Swap^2\,(x) = x \neq Swap\,(x)$.

(2.13)   The function $Up\,(Right\,(x))$ restricted to *Diag* is a permutation of *Diag* without fixed points.

(2.14)   For all $x, \pi_0(Right\,(x)) = Up\,(Right\,(\pi_0(x)))$ and $\pi_0(Up\,(x)) = \pi_0(x)$;   likewise $\pi_1(Up\,(x)) = Up\,(Right\,(\pi_1(x)))$ and $\pi_1(Right\,(x)) = \pi_1(x)$.

(2.15)   For all x, at most one of $Diag\,(x)$ and $Diag\,(Swap\,(x))$ holds.

(2.16)   If $Diag\,(x)$ then $\pi_0(x) = \pi_1(x) = x$.

(2.17)   If $Diag\,(x)$ or $Diag\,(Swap\,(x))$ then $\mu(x) = x$ and $\mu(Up\,(x)) = Swap\,(Right\,(x))$ and $\mu(Right(x)) = Up(x)$; if neither $Diag\,(Right\,(x))$ nor $Diag\,(Swap\,(Right\,(x)))$ then $\mu(Right\,(x)) = Up\,(\mu(x))$;

if neither $Diag(x)$ nor $Diag(Swap(x))$ nor $Diag(Up(x))$ nor $Diag(Swap(Up(x))$ then $\mu(Up(x)) = Right(\mu(x))$.

We show by a series of claims that T is uncountably categorical.

Suppose B is a model of T. We say that two elements of B are **in the same component** if one can be reached from the other by applying the maps $Right$, Up, $Swap$ and their inverses (here we use (2.6)). This is an equivalence relation on $\text{dom}(B)$; the **components** of B are the equivalence classes. Each component consists of two copies of $\mathbb{Z} \times \mathbb{Z}$, one above the other.

(2.18) If a is an element of B such that $F(a) = a$ where F is some word in $Right$, Up, $Swap$ and their inverses, then $F(b) = b$ for all b in the component of a.

This follows at once from (2.7). We write $NE(x)$ for $Up(Right(x))$.

(2.19) Suppose n is a positive integer. Let X be the component of some element a of $Diag$ such that $NE^n(a) = a$. Then for each element b of X, $Up^n(b)$ and $Right^n(b)$ are equal to each other, and also to one or other of b and $Swap(b)$.

First note that by (2.18), $NE^n(b) = b$ for all $b \in X$. Now by (2.14), $\pi_0(Right^n(b)) = NE^n(\pi_0(b))$, which $= \pi_0(b)$ by (2.10) and the assumption. Also $\pi_1(Right^n(b)) = \pi_1(b)$ by (2.14). Hence $Right^n(b)$ is either b or $Swap(b)$, by (2.11). Similarly with $Up^n(b)$. Finally if $Right^n(b) = b$, then $Right^n(b) = Right^n(NE^n(b))$ (by (2.18)) $= Up^n Right^n Right^n(b)$ (by (2.7)) $= Up^n(b)$; while if $Right^n(b) = Swap(b)$, then $Right^n(b) = Up^n Right^n Right^n(b) = Up^n Right^n Swap^n(b) = Up^n Swap(Right^n(b))$ (by (2.7)) $= Up^n Swap^2(b) = Up^n(b)$ (by (2.12)).

(2.20) Suppose n is a positive integer and a an element of $Diag$ such that $Diag(Right^n(a))$ or $Diag(Up^n(a))$ or $Diag(Swap(Right^n(a)))$ or $Diag(Swap(Up^n(a)))$. Then $NE^n(a) = a$.

First suppose $Diag(Right^n(a))$. By (2.16) we have $Right^n(a) = \pi_0(Right^n(a)) = NE^n(\pi_0(a)) = NE^n(a)$. So by (2.6) and (2.7), $a = Up^n(a)$, whence $Diag(Up^n(a))$ and thus $a = Right^n(a)$ by the same argument, so that $NE^n(a) = Right^n(a) = a$. The other cases are similar, noting that by (2.14) and (2.16), $Diag(x)$ if and only if $Diag(Swap(x))$.

(2.21) There are no element a of $Diag$ and positive integer n such that $NE^n(a) = a$.

For otherwise let X be the component of a, and apply (2.18) and (2.19). Let n be the least possible. By (2.13), $n \geqslant 2$; by (2.20), if $0 < k < n$ then $Up^k(a)$ and $Swap(Up^k(a))$ are not in $Diag$. Write σ for the identity map or $Swap$, according as $Up^n(a)$ is a or $Swap(a)$. We shall trace $\mu(Up^{-1}(a))$ around the Möbius loop and get a contradiction. Thus: $\mu(Up^{-1}(a)) = \mu(Up^{n-1}\sigma(a)) =$

$Right^{n-2} \mu(Up\, \sigma(a))$ by   (2.17) $= Swap\,(Right^{n-1} \mu\sigma(a))$ by   (2.17)
$= Swap\,(Right^{n-1}\, \sigma(a)) = Swap\,(Right^{-1}\,(a))$.   But   also $\mu(Up^{-1}\,(a)) =$
$\mu(Right\,(NE^{-1}\,(a))) = Up\,(NE^{-1}(a))$ by   (2.17) $= Right^{-1}\,(a)$.   So
$Swap\,(Right^{-1}\,(a)) = Right^{-1}\,(a)$, contradicting (2.12).

(2.22) If $x = Right^m\, Up^n\,(x)$ for some integers $m,$ n then
$\quad\quad m = n = 0$; $x \neq Swap\,(Right^m\, Up^n\,(x))$ for all integers $m,\, n$.

If $x = Right^m\, Up^n\,(x)$ with $n \neq 0$ then $\pi_1(x) = \pi_1\, Up^n\,(x) = NE^n\, \pi_1(x)$ by
(2.14), which by (2.10) contradicts (2.21). Similarly $m = 0$. If $x = Swap$
$(Right^m\, Up^n\,(x))$ then the same argument gives $\pi_1(x) = Swap\,(NE^n\, \pi_1(x))$,
which contradicts (2.10), (2.13) and (2.15).

(2.23) If C and D are two models of T of the same uncountable
$\quad\quad$ cardinality, then there is a bijection $f: Diag^C \to Diag^D$ which
$\quad\quad$ commutes with NE.

By (2.13), $Diag^C$ is a union of orbits of NE. By (2.21) each of these orbits is
an infinite cycle. By (2.10) and (2.11), $Diag^C$ has the same cardinality as C,
and hence it splits into $|C|$ orbits. The same is true in D too; so f is easily
found.

(2.24) If $x,$ y are elements such that $\pi_0(x)$ and $\pi_0(y)$ are in the
$\quad\quad$ same orbit of NE, and $\pi_1(x)$ and $\pi_1(y)$ are in the same orbit
$\quad\quad$ of NE, then x and y lie in the same component.

Suppose for example that $\pi_0(x) = \pi_0(y)$ and $\pi_1(x) = NE^k\,(\pi_1(y))$. Then
$\pi_0(x) = \pi_0(Up^k\,(y))$ and $\pi_1(x) = \pi_1(Up^k\,(y))$ by (2.14), so $x = Up^k\,(y)$ or
$Swap\,(Up^k\,(y))$ by (2.11).

(2.25) If $C,$ $D,$ f are as in (2.23), then f extends to an isomorphism
$\quad\quad$ g from C to D.

In each orbit X of NE on $Diag^C$, pick an element $c(X)$. If X_0 and X_1 are
orbits, pick an element c_{01} such that $\pi_0(c_{01}) = c(X_0)$ and $\pi_1(c_{01}) = c(X_1)$;
when $X_0 = X_1$, c_{01} should be $c(X_0)$, as is possible by (2.16). Do the same
for the orbits Y of NE on $Diag^D$, picking elements $d(X)$, d_{01}. Now for any
pair of orbits $X_0,$ X_1 and any word F in $Right,$ Up and $Swap$, put
$g(F^C(c_{01})) = F^D(d_{01})$. This defines a bijection from dom(C) to dom(D) by
(2.10), (2.22) and (2.24). Clearly g extends f and is an isomorphism with
respect to $Right,$ Up and $Swap$. It remains only to show that g is an
isomorphism with respect to $\pi_0,$ π_1 and μ. For π_0 and π_1 this is easily proved
using (2.14), and for μ we use (2.17).

One other result in this area is worth noting.

Theorem 12.2.19. *For every n there is a countable almost strongly minimal
structure A such that* $\mathrm{Th}\,(A)$ *is not axiomatised by an $\forall_n$ theory.* $\square$

Exercises for section 12.2

The first six exercises form a proof of Morley's theorem for countable theories. Throughout these exercises, T is a complete theory in a countable first-order language L, and T is λ-categorical for some uncountable cardinal λ. Every known proof starts with Theorem 12.2.3(b), which gives that T is ω-stable; you may assume this result.

1. Show that if A is a model of T and X a set of elements of A, then there is a prime model over X in A. [Use Exercise 5.6.13 to construct a sequence $(a_i: i < \alpha)$ of elements of A such that for each i, a_i realises a principal type over $X \cup \{a_j: j < i\}$, and for every formula $\phi(x)$ of L with parameters in $X \cup \{a_i: i < \alpha\}$, there is $k < \alpha$ such that $A \vDash \phi(a_k)$.]

2. Show that there is a unique prime model over the empty set, and this model is atomic. [Theorems 7.2.2 and 7.2.3.]

3. Show that if T is ω_1-categorical, then there are no models $A \prec B$ of T and formula $\phi(x)$ with parameters in A, such that $\phi(A) = \phi(B)$ and $\phi(A)$ is infinite. [Theorem 12.1.1.]

The next exercise yields Theorem 12.2.9(a) when the language is countable. A different argument using the finite cover property is needed in the general case.

4. Suppose T is ω_1-categorical. Show that if A_0 is the prime model of T over $\varnothing$, then there is a strongly minimal set $\phi(A_0)$ where ϕ has parameters from A_0. [Take the definable set $\mathrm{dom}(A_0)$ and chop as often as possible into two disjoint definable subsets. The resulting set $\phi(A_0)$ is minimal. To show it's strongly minimal, take any elementary extension B of A_0 and a formula $\psi(x, \bar{b})$ which chops $\phi(B)$ into two infinite pieces. Replacing ψ by $\neg \psi$ if necessary, it follows that for every $n < \omega$ there is a tuple $\bar{a}$ in A_0 such that $\phi(A_0) \cap \psi(A_0, \bar{a})$ has finite cardinality $> n$. Use this with compactness to get a model C with a proper elementary substructure C_0 and elements $\bar{c}$ of C_0 such that $\phi(C) \cap \psi(C, \bar{c}) = \phi(C_0) \cap \psi(C_0, \bar{c})$, contradicting the previous exercise.]

Henceforth let ϕ be the formula of Exercise 4; we assume that the parameters of ϕ are absorbed into the language L, so that the new theory T is a principal extension of the old T.

5. Assuming T is ω_1-categorical, show (a) if A is any model of T and $B \preccurlyeq A$ is prime over $\phi(A)$, then $A = B$, and B is the prime model over $\phi(A)$ constructed as in Exercise 1 [If $A \neq B$ we contradict Exercise 3. If B' is constructed as in Exercise 1, we have $B' = A = B$.], (b) if A and B are models of T and $\bar{c}, \bar{d}$ are sequences listing the elements of $\phi(A)$, $\phi(B)$ respectively, then $(A, \bar{c}) \equiv (B, \bar{d})$ implies that there is an isomorphism i taking the prime model over $\phi(A)$ in A to the prime model over $\phi(B)$ in B, and $i\bar{c} = \bar{d}$ [Use the construction of the prime models.].

6. Assuming T is ω_1-categorical, show that T is uncountably categorical. [Suppose A, B are two models of the same uncountable cardinality λ. Fix a bijection f from a basis X of $\phi(A)$ to a basis Y of $\phi(B)$, and extend f to a bijection g from $\phi(A)$ to $\phi(B)$ by

Lemmas 4.5.5 and 4.5.6. Since A is prime over $\phi(A)$, g extends to an elementary embedding of A into B; $g(A)$ must be the whole of B by Exercise 3.]

7. Let λ be an uncountable cardinal. Give an example of a theory T in a language of cardinality λ, such that T is λ^+-categorical but T has a model of cardinality λ which is not homogeneous. [A vector space of dimension ω over a field of cardinality λ.]

8. Show that the theory in Example 5 is definitionally equivalent to a finite set of first-order sentences, each of which uses at most three variables (though possibly with several occurrences), in a signature consisting of 1-ary and 2-ary relation symbols. [For *Right*, *Up*, *Swap* and μ use 2-ary relation symbols expressing the graphs of the functions. For π_0 and π_1 introduce two equivalence relations E_0, E_1 so that $E_i(x, y)$ means $\pi_i(x) = \pi_i(y)$; this will deal with (2.11). Likewise one can handle (2.17) by introducing new 2-ary relation symbols for the graph of the function $\mu(Right\,(x))$, etc.] *By Exercise 3.3.12 there is no such theory using just two variables.*

12.3 Cohomology of expansions

Throughout this section the following assumptions are in force. We shall find several situations where they apply.

(3.1) L and L^+ are first-order languages with $L \subseteq L^+$; one of the symbols of L^+ is a 1-ary relation symbol P; B is an L^+-structure, and A is the substructure of $B|L$ with domain P^B.

(3.2)   Every automorphism of A extends to an automorphism of B.

In this setting, every automorphism α of B induces an automorphism β of A by restriction; we write $\mu(\alpha)$ for β. Then $\mu: \mathrm{Aut}(B) \to \mathrm{Aut}(A)$ is a group homomorphism, and by (3.2) it is surjective. The kernel of μ is the group $\mathrm{Aut}(B)_{(A)}$ of all automorphisms of B which fix A pointwise. We write $\iota: \mathrm{Aut}(B)_{(A)} \to \mathrm{Aut}(B)$ for the inclusion map. Thus we have a short exact sequence of groups:

(3.3) $$\mathrm{Aut}(B)_{(A)} \overset{\iota}{\rightarrowtail} \mathrm{Aut}(B) \overset{\mu}{\twoheadrightarrow} \mathrm{Aut}(A)$$

We call (3.3) the **short exact automorphism sequence** of B.

The map μ is surjective. So there exist maps $\sigma: \mathrm{Aut}(A) \to \mathrm{Aut}(B)$ such that $\mu\sigma$ is the identity on $\mathrm{Aut}(A)$. Such a map σ is called a **section** of μ. If σ is also a group embedding, we call it a **splitting** of μ (or of the short exact sequence). We call it a **continuous splitting** if it is a continuous group embedding, in the sense of the topology on automorphism groups which we defined in section 4.1. We say that B is **natural over** A (or **over** P) if there is a splitting of μ, and **continuously natural over** A if there is a continuous splitting. These notions will be important in what follows, so I give two examples at once.

Example 1: *A continuously natural extension.* Let B be the direct product of two cyclic groups of order 4, and A the direct product of two cyclic groups of order 2, lying inside B in the obvious way. Writing $\mathbb{Z}_n$ for the ring of integers mod n, we can express B as a $\mathbb{Z}_4$-module $\mathbb{Z}_4 a \oplus \mathbb{Z}_4 b$, so that $2a$ and $2b$ generate A. We can represent automorphisms of B by 2×2 matrices over $\mathbb{Z}_4$, and automorphisms of A by 2×2 matrices over $\mathbb{Z}_2$. Automorphisms of A lift to B by the homomorphism generated by

$$(3.4) \qquad \begin{pmatrix} 0 & 1 \\ 1 & 0 \end{pmatrix} \mapsto \begin{pmatrix} 0 & 1 \\ 1 & 0 \end{pmatrix}, \qquad \begin{pmatrix} 0 & 1 \\ 1 & 1 \end{pmatrix} \mapsto \begin{pmatrix} 0 & 1 \\ 3 & 3 \end{pmatrix}$$

We have continuity for free since B is finite. Note that the choice of (3.4) is not at all obvious. When automorphisms lift, they may do so for strange reasons.

Example 2: *An extension which is not natural.* The structure B has six elements, a, b, c, d, u, v. The structure A consists of the two elements u and v. In B there are a relation Q such that $Q(a, u)$, $Q(b, v)$, $Q(c, u)$ and $Q(d, v)$ hold, and a relation R such that $R(a, b)$, $R(b, c)$, $R(c, d)$ and $R(d, a)$ hold:

(3.5)

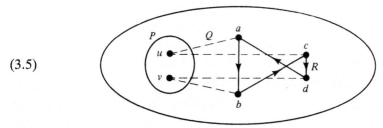

$\mathrm{Aut}(A)$ is cyclic of order 2, generated by a transposition α. Clearly α extends to an automorphism of B. In fact it extends to two different automorphisms, namely β which takes (a, b, c, d) to (b, c, d, a), and γ which takes (a, b, c, d) to (d, a, b, c). However, both β and γ have order 4, so there is no group embedding σ of $\mathrm{Aut}(A)$ into $\mathrm{Aut}(B)$ which takes α to either β or γ.

Exercise 1 shows that we can have B natural but not continuously natural over A. One has to work harder to find an example where B is countable; see Evans & Hewitt [1990]. Our first lemma tells us where not to look for such an example.

Lemma 12.3.1. *Suppose both A and B are countable, B is natural over A and A has the small index property. Then B is continuously natural over A.*

Proof. This is immediate from Lemma 4.2.6. □

In Theorem 4.1.4 we saw that the topology on the automorphism group of a structure with domain Ω makes it a closed subgroup of the symmetric group on Ω. This applies to $\text{Aut}(A)$ and $\text{Aut}(B)$. It also applies to $\text{Aut}(B)_{(A)}$, since that group is $\text{Aut}(B, \bar{a})$ where $\bar{a}$ is any sequence listing all the elements of A.

Lemma 12.3.2. *In the short exact automorphism sequence, μ is a continuous homomorphism, and $\text{Aut}(B)_{(A)}$ has the topology inherited from $\text{Aut}(B)$ (so that ι is a continuous embedding).*

Proof. A typical basic open subgroup of $\text{Aut}(A)$ is $\text{Aut}(A)_{(X)}$ where X is a finite set of elements of A. Clearly μ maps the open subgroup $\text{Aut}(B)_{(X)}$ of $\text{Aut}(B)$ into $\text{Aut}(A)_{(X)}$. The basic open subgroups of $\text{Aut}(B)_{(A)}$ are of the form $\text{Aut}(B)_{(A)(Y)} = \text{Aut}(B)_{(A)} \cap \text{Aut}(B)_{(Y)}$ where Y is a finite set of elements of B. $\qquad\square$

By the relativisation theorem (Theorem 5.1.1), if $\theta(\bar{x})$ is any formula of L then there is a formula $\theta^P(\bar{x})$ of L^+ such that

(3.6)   for all tuples $\bar{a}$ in A, $A \vDash \theta(\bar{a}) \Leftrightarrow B \vDash \theta^P(\bar{a})$.

It would be useful to have a map in the other direction too, from L^+ to L. We shall say that B has the **reduction property** (over A) if for every formula $\phi(\bar{x})$ of L^+ there is a formula $\phi^\circ(\bar{x})$ of L such that

(3.7)   for all tuples $\bar{a}$ in A, $B \vDash \phi(\bar{a}) \Leftrightarrow A \vDash \phi^\circ(\bar{a})$.

(In other words, if R is any relation which is $\varnothing$-definable in B, then the restriction of R to A is $\varnothing$-definable in A.) The map $\phi \mapsto \phi^\circ$ is called the **reduction map**. In section 12.5 we shall find a categoricity assumption which implies that B has the reduction property. Meanwhile here is a simpler sufficient condition; in fact it gives rather more than the reduction property.

Theorem 12.3.3. *In the setting of (3.1) and (3.2), suppose that B is at most countable and A is either finite, or countable and ω-categorical.*
   (a) *B has the reduction property.*
   (b) *For every tuple $\bar{b}$ in B there is a tuple $\bar{a}$ in A such that for every formula $\phi(\bar{x}, \bar{y})$ of L^+, the restriction of $\phi(B^n, \bar{b})$ to A is of the form $\psi(A^n, \bar{a})$ for some formula ψ of L.*

Proof. (a) Let $\phi(\bar{x})$ be a formula of L^+ with $\bar{x}$ an n-tuple. The set $Y = \phi(B^n) \cap (\text{dom } A)^n$ is a union of orbits under $\text{Aut}(B)$. But by (3.2) the orbits of $\text{Aut}(B)$ on tuples in A are exactly the same as the orbits of $\text{Aut}(A)$. So by Theorem 7.3.1 or the finiteness of A, Y is a union of some of the finitely many orbits of $\mathrm{Aut}(A)$ on n-tuples, and hence Y is $\varnothing$-definable in A.

(b) Put $H = \mathrm{Aut}(B)_{(\bar{b})}$. Then H has at most countable index in $\mathrm{Aut}(B)$, and so its image $\mu(H)$ has at most countable image in $\mathrm{Aut}(A)$. Accordingly the closure of its image, $\mathrm{Cl}(\mu(H))$, is a closed subgroup of at most countable index in $\mathrm{Aut}(A)$ and hence is open by Theorem 4.2.3. Thus there is a tuple $\bar{a}$ in A such that $\mathrm{Cl}(\mu(H)) \supseteq \mathrm{Aut}(A)_{(\bar{a})}$, and so $\mu(H)$ is dense in $\mathrm{Aut}(A)_{(\bar{a})}$. (This was Exercise 4.2.3.) Now for any formula $\phi(\bar{x}, \bar{y})$ of L^{+}, $\phi(B^{n}, \bar{b}) \cap (\mathrm{dom}\,A)^{n}$ is a union of orbits of $\mathrm{Aut}(B, \bar{b})$ on A, and we have just shown (cf. Lemma 4.1.5(d)) that these orbits are unions of those of $\mathrm{Aut}(A, \bar{a})$. Since $(A, \bar{a})$ is finite or ω-categorical, we finish as in (a). $\qquad\square$

The questions that we shall ask in the situation of (3.1) and (3.2) are all to do with reconstructing B from A.

Coordinatisation

The simplest kind of reconstruction is when the whole of B is explicitly definable from A. But what does it mean to say that B is explicitly definable from A? The reader should try a few definitions; the chances are they will be equivalent to the following.

Assuming (3.1) and (3.2) as always, we say that the structure B is **coordinatised** (over A) if

(3.8)   every element of B is in the definable closure of $\mathrm{dom}(A)$ in B, and moreover B has the reduction property.

If B is coordinatised, then every element b of B is pinned down as the unique element b such that $B \vDash \phi_{b}(b, \bar{a}_{b})$, where $\phi_{b}(x, \bar{y})$ is some formula of L^{+} and $\bar{a}_{b}$ is some tuple in A. We say that $\bar{a}_{b}$ are **coordinates** for b. (There may be several possible choices of coordinates for $\bar{b}$, even with the same formula ϕ_{b}.)

Note that a formula $\phi(x, \bar{y})$ of L^{+} and a tuple $\bar{a}$ in A together pin down some element of B in this way if and only if $B \vDash \exists_{=1} x\, \phi(x, \bar{a})$, i.e. if and only if $A \vDash \psi^{\circ}(\bar{a})$ where $\psi(\bar{y})$ is the formula $\exists_{=1} x\, \phi(x, \bar{y})$ and $^{\circ}$ is the reduction map. In a sense, A knows what the elements of B are. We can take this idea further and translate properties of elements of B into properties of their coordinates.

Suppose for example that R is a 2-ary relation symbol of L^{+}, and b, b' are elements of B. Then $B \vDash R(b, b')$ if and only if $A \vDash \chi^{\circ}(\bar{a}_{b}, \bar{a}_{b'})$, where $\chi(\bar{y}, \bar{y}')$ is the formula

(3.9) $\forall x x'(\phi_{b}(x, \bar{y}) \wedge \phi_{b'}(x', \bar{y}') \to R(x, x'))$.

In short, if B is coordinatised over A, then A determines B completely, up to isomorphism over A.

By a **coordinatisation of B over A** we mean an expansion $B^\#$ of B which is coordinatised over A. We say that B is **coordinatisable over A** if there is a coordinatisation of B over A. (Gauri Gupta suggests the name should be 'coordinable over'. Maybe he is right.)

Suppose $B^\#$ is a coordinatisation of B over A. Then $B^\#$ defines a map taking each automorphism α of A to an automorphism $\alpha^\#$ of B, as follows. Let b be any element of B, and working in $B^\#$ suppose that the formula $\phi(x, \bar{y})$ and the tuple $\bar{a}$ from A pin down b. Write $\psi(\bar{y})$ for $\exists_{=1} x \, \phi(x, \bar{y})$ and $^\circ$ for the reduction map. We have $A \vDash \psi^\circ(\bar{a})$, and so $A \vDash \psi^\circ(\alpha\bar{a})$ since $\alpha \in \text{Aut}(A)$. Let $\alpha^\#(b)$ be the unique element b' of B such that $B^\# \vDash \phi(b', \alpha\bar{a})$. I leave it to the reader to check, using the reduction map, that this definition of $\alpha^\#$ doesn't depend on the choice of ϕ and $\bar{a}$, that $\alpha^\#$ is an automorphism of $B^\#$ and hence of B, and that $\alpha^\#$ extends α.

Theorem 12.3.4. *Assuming (3.1) and (3.2) as always, suppose B is coordinatisable over A. Then B is continuously natural over A; in fact the map $^\#$ defined above is a continuous splitting of the short exact automorphism sequence.*

Proof. Continuity is easy: if X is a finite set of elements of B, choose a finite set Y of elements of A such that each element in X has a tuple of coordinates lying in Y. Then if α fixes Y pointwise, $\alpha^\#$ will fix X pointwise. It remains to prove that $^\#$ is a group homomorphism. Let α and β be automorphisms of A. Then taking b, ϕ and $\bar{a}$ as above, ϕ and $\alpha\bar{a}$ pin down $\alpha^\#(b)$, so by definition $\beta^\#(\alpha^\#(b))$ is the unique element b'' such that $B^\# \vDash \phi(b'', \beta(\alpha\bar{a}))$. But this element is also $(\beta\alpha)^\#(b)$ by definition. $\qquad \square$

Thus for example the structure of Example 2 is not coordinatisable, because it is not natural.

In favourable cases, Theorem 12.3.4 has a converse.

Theorem 12.3.5. *Let B be either finite, or countable and ω-categorical. Suppose B is continuously natural over A. Then B is coordinatisable over A.*

Proof. I prove it for the ω-categorical case. Let σ be a continuous splitting of μ. Let b be any element of B. Then by continuity, $\sigma^{-1}(\text{Aut}(B)_{(b)})$ contains some open subgroup $\text{Aut}(A)_{(\bar{a})}$ of A, where $\bar{a}$ is a tuple in A. Let R_b be the orbit of the tuple $\bar{a}^\frown b$ under $\sigma(\text{Aut}(A))$. We can list the relations R_b as $\{P_i : i < \omega\}$. Let $B^\#$ be B with the relations P_i $(i < \omega)$ added, and let $L^\#$ be the first-order language appropriate for $B^\#$. Since the sets P_i were orbits under $\sigma(\mathrm{Aut}\,(A))$, every automorphism in $\sigma(\mathrm{Aut}\,(A))$ is an automorphism of $B^{\#}$, and in particular $B^{\#}$ still satisfies (3.1) and (3.2) over A.

By the choice of the relations R_b, every element of $B^{\#}$ lies in the definable closure of $\mathrm{dom}\,(A)$ in $B^{\#}$. It remains to show that $B^{\#}$ has the reduction property. Let $\phi(\bar{x})$ be any formula of $L^{\#}$. If $\bar{a}$ is any tuple of elements of A and α any automorphism of A, then α lifts to an automorphism $\sigma(\alpha)$ of $B^{\#}$; so $B^{\#} \vDash \phi(\bar{a})$ if and only if $B^{\#} \vDash \phi(\alpha\bar{a})$. Thus the set $\{\bar{a}$ in $A: B^{\#} \vDash \phi(\bar{a})\}$ is a union of orbits of $\mathrm{Aut}\,(A)$, and hence by ω-categoricity it is defined by some formula ϕ° of L. This defines a reduction map for $B^{\#}$. $\quad\square$

It would be pleasant if we could cut down the set of relations P_i in the proof above to a finite set. Unfortunately this is impossible in general; consider the case where B is a direct sum of infinitely many cyclic groups, $\mathbb{Z}_2^{(\omega)} \oplus \mathbb{Z}_3^{(\omega)}$, and A is $\mathbb{Z}_2^{(\omega)}$.

Cohomology and definability

The connection between cohomology and definability runs deeper than the results above. For example the following result can be used to classify totally categorical theories.

Theorem 12.3.6. *In the setting of (3.1) and (3.2), let B be a countable totally categorical structure, and suppose the short exact automorphism sequence of B splits. Suppose $\mathrm{Aut}\,(B)_{(A)}$ is abelian, and let K be a subgroup of $\mathrm{Aut}\,(B)_{(A)}$ which is closed and normal in $\mathrm{Aut}\,(B)$.*

(a) There is a natural bijection between the closed subgroups G of $\mathrm{Aut}\,(B)$ such that (3.3) induces a short exact sequence

(3.10)
$$K \rightarrowtail G \xrightarrow{\;\mu|G\;} \mathrm{Aut}\,(A)$$

(counted up to conjugacy in $\mathrm{Aut}\,(B)$), and the elements of the first cohomology group $H^1(\mathrm{Aut}\,(A), \mathrm{Aut}\,(B)_{(A)}/K)$. (For this to make sense, an action of $\mathrm{Aut}\,(A)$ on $\mathrm{Aut}\,(B)_{(A)}$ has to be defined from (3.3).)

(b) If B^* is a totally categorical expansion of B such that every automorphism of A extends to B^*, then $\mathrm{Aut}\,(B^*)$ is a subgroup G of $\mathrm{Aut}\,(B)$ as in (3.10), with $K = \mathrm{Aut}\,(B^*)_{(A)}$. Moreover each short exact sequence (3.10) arises from such a B^*. $\quad\square$

I omit the proof, but here is a brief account of what the theorem is good for. It was intended for use in the setting of Theorem 12.2.14; the structures B, B^* are totally categorical and all contain the same strongly minimal set A. Under suitable hypotheses, the kernel group K of (3.10) is a closed subgroup of a product of groups interpretable in A, by Theorem 12.2.15. The action of $\mathrm{Aut}\,(A)$ on these interpreted groups induces an action of $\mathrm{Aut}\,(A)$ on K; it's

exactly the action that cohomologists extract from the short exact sequence
(3.10). So from the strongly minimal set A alone we can already read off
$\text{Aut}(A)$, the possibilities for K, and the actions of $\text{Aut}(A)$ on each possible
K. The theorem then tells us how to find at least the automorphism groups
of the possible structures B^*; since these structures are countable and
ω-categorical, their automorphism groups determine them up to interpretability.

Set-theoretic definability

Theorem 12.3.4 should be compared with Lemmas 2.1.1 and 4.3.1. Both
these lemmas gave a purely structural condition which implies that some
expansion A^+ of a structure A is not definable in terms of A. The structural
condition was that for any tuple $\bar{a}$ of parameters in A, some automorphism of
$(A, \bar{a})$ was not an automorphism of $(A^+, \bar{a})$. (The parameters were needed
because we were discussing definability with parameters.)

The situation of Theorem 12.3.4 is the same after a fashion: B is an
expansion of A, but with extra elements added. The difference is that we
have assumed right from the outset that every automorphism of A *does*
extend to an automorphism of B. The non-definability in Theorem 12.3.4 has
a different cause from that in Lemma 4.3.1.

It happens that there is a connection with the axiom of choice in both
cases. Many applications of the axiom of choice have to do with adding some
extra relation to a structure – for example an ultrafilter to a boolean algebra.
The axiom of choice is needed because the new relation is not definable from
the given structure, in a broad set-theoretic sense of 'definable'. But not all
applications of the axiom of choice have this form.

Example 3: *Algebraic closures of fields yet again.* One of the first applications
of the axiom of choice in algebra was Steinitz' proof [1910] that a field whose
elements can be well-ordered has an algebraic closure. Some form of the
axiom of choice really is needed here: ZF has models in which there are fields
with no algebraic closures. But this has nothing to do with adding extra
relations to the given field. If A is a field and B is its algebraic closure, then
every automorphism of A extends to an automorphism of B.

We write ZFU (Zermelo–Fraenkel set theory with urelements) for the set
theory which is like ZF except that it allows a set of elements which are not
sets. These elements are known as **urelements**; they have no members
themselves, but they can be members of sets. We have to weaken the
extensionality axiom to say that any two *sets* with the same members are
equal. (See Jech [1973b].)

Theorem 12.3.7. *Assuming (3.1) and (3.2), let B be a structure such that for every tuple $\bar{a}$ in A, $(B, \bar{a})$ is not natural over $(A, \bar{a})$. Then there is a model V of ZFU containing a copy A' of A, such that no structure in V is isomorphic to B over A'.*   □

Example 3: *continued.* Artin & Schreier [1927] show that an algebraically closed field can never have an automorphism of finite order > 2. Assuming this, choose a field A such that for every tuple $\bar{a}$ in A, $(A, \bar{a})$ has an automorphism α of order 3; for example A can be a pure transcendental extension of infinite dimension over a prime field. If B is any algebraic closure of A, then α can't be extended to an automorphism of B of order 3. It follows that B is not natural over A. So by Theorem 12.3.7, there is a model of ZFU containing a field with no algebraic closure.

There are difficulties about transferring Theorem 12.3.7 to ZF. The problem is that if we take a generic copy A' of A inside a model of ZF, the elements of A' all have distinct structures as sets, and it's not clear that this fails to provide enough leverage to build a copy of B over A'. Pincus [1972] gave a sufficient condition for instances of Theorem 12.3.7 to be transferable to ZF. Pincus' condition involves bounding the cardinality of B in terms of that of A, and it is satisfied in the case of algebraic closures of fields.

There is a related set-theoretic question that arises even when we assume the axiom of choice. We can write down a set-theoretic formula $\phi(x, y)$ which says 'x is a field and y is an algebraic closure of x'. The question is whether we can uniformise ϕ – is there a formula which picks out, for each field x, *exactly one* algebraic closure of x? If there is such a formula for a model M of ZFC, then it expresses a weak form of global choice in M. One can show that there is a model of set theory in which there is no such formula; again the salient fact is that there are fields whose algebraic closures are not natural over them (even allowing parameters).

Exercises for section 12.3

1. Let the structure A consist of a countable set with no further structure. Let B consist of A $(= B_P)$ together with the set of all subsets of A, with the membership relation $\in$ between elements and subsets of A. Show that B is natural over A, but not continuously.

2. Show that (3.1) and the reduction property together don't imply (3.2). [Let B be the Fraïssé limit of the class of all finite structures C where P^C is a subset of $\mathrm{dom}(C)$ and R^C is a subset of $P^C \times (\mathrm{dom}(C) \backslash P^C)$. There are no non-trivial $\varnothing$-definable relations on B_P; only countably many subsets of P^B are of the form $\{a : B \vDash R(a, b)\}$ for a fixed b.]

3. Assuming (3.1) and (3.2), show that if B is countable, $\bar{x}$ is an n-tuple of variables and $\phi(\bar{x})$ is a formula of $L^+_{\omega_1\omega}$, then there is a formula $\psi(\bar{x})$ of $L_{\omega_1\omega}$ which defines on A exactly the restriction of $\phi(B^n)$ to dom (A).

4. In the setting of (3.1) and (3.2), suppose that A is saturated, and for every $\varnothing$-definable relation R on B, the restriction of R to A is definable in A with parameters. Show that B has the reduction property. [Use Corollary 10.5.4.]

5. Let $^\#$ be the map defined before Theorem 12.3.4. Show that $\alpha^\#$ is well-defined, and that for every automorphism α of A, $\alpha^\#$ is an automorphism of $B^\#$ which extends α.

6. Show that if A and B satisfy (3.1) and (3.8), then they satisfy (3.2) too.

*7. Writing $\mathbb{Z}_n$ for the cyclic group of order n, show that for every cardinal $\kappa > 1$, the direct sum $\mathbb{Z}_{25}^{(\kappa)}$ of κ copies of $\mathbb{Z}_{25}$ is not natural over $\mathbb{Z}_5^{(\kappa)}$.

8. Show that the axiom of choice is needed for constructing the divisible hulls of abelian groups. [You may prove it for ZFU and trust that there is a suitable transfer principle. Use the previous exercise, noting that the divisible hull of $\mathbb{Z}_5^{(\kappa)}$ contains $\mathbb{Z}_{25}^{(\kappa)}$ as a definable subset.]

12.4 Counting expansions

We return for one last time to a familiar topic: definability. The setting will be as follows:

(4.1) J, K and L are first-order languages with $J \subseteq K \subseteq L$; the signature of K has just finitely many symbols not in J; T is a theory in L which has infinite models.

We shall be constructing families of L-structures which are models of T, are identical in their J-reducts and differ from each other in their K-reducts. Let us say that two L-structures A, B are **K-equivalent** if $A|K = B|K$. Likewise if R is a symbol of K, we shall say that A and B are **R-equivalent** if $R^A = R^B$. In this terminology, we shall be constructing many models of T which are isomorphic and J-equivalent but not K-equivalent; any two of them are R-inequivalent for some symbol R of K.

According to (4.1), the signature of K has just finitely many symbols which are not in J. Then up to choice of variables there are just finitely many unnested atomic formulas of K which are not in J. List them as $\phi_0(\bar{x}_0), \ldots,$ $\phi_{m-1}(\bar{x}_{m-1})$. For present purposes a **definition of ϕ_i over J** is a formula of the form $\forall \bar{x}_i(\phi_i(\bar{x}_i) \leftrightarrow \theta(\bar{y}, \bar{x}_i))$ where θ is some first-order formula of J. A **definition of K over J** is a conjunction $\sigma_0(\bar{y}) \wedge \ldots \wedge \sigma_{m-1}(\bar{y})$ where each σ_i is a definition of ϕ_i over J.

Theorem 12.4.1. *Let λ be an infinite cardinal. Suppose we are in the setting of (4.1) where J, K and L are first-order languages of cardinality $\leq \lambda$. Then the following are equivalent.*

(a) *There is a family of more than λ isomorphic and J-equivalent models of T of cardinality λ which are pairwise not K-equivalent.*

(b) *There is a family of more than λ J-equivalent models of T of cardinality λ which are pairwise not K-equivalent.*

(c) *It is false that there are $k < \omega$ and definitions $\delta_0(\bar{y})$, $\ldots$, $\delta_{k-1}(\bar{y})$ of K over J such that $T \vdash \bigvee_{j<k} \exists \bar{y}\, \delta_i(\bar{y})$.*

If T is complete, then (a)–(c) are equivalent to the following.

(d) *If A is any model of T, then there are elementary extensions B, C of A which are J-equivalent but not K-equivalent.*

Proof. By Exercise 2.6.1 there is no loss in assuming that the signature of L contains no function or constant symbols. The implications (a) $\Rightarrow$ (b) $\Rightarrow$ (c) in the theorem are clear; so also is the implication (d) $\Rightarrow$ (c).

We begin by proving (d) under the assumption that (c) holds and T is complete. Under this assumption, if $\delta(\bar{y})$ is any definition of K over J then $T \vdash \neg \exists \bar{y}\, \delta$. If R is any relation symbol which is in L but not in J, write R' for a new copy of R; if R is a relation symbol of J, let R' be R itself. Write L' for the language got from L by replacing each symbol R by R'. If U is a theory in L, write U' for the corresponding theory in L'; likewise with formulas of L. List the relation symbols which are in K but not in J as R_0, $\ldots$, R_{m-1}.

Let A be a model of T. Introduce constants for the elements of A, together with tuples $\bar{c}_0$, $\ldots$, $\bar{c}_{m-1}$ of new constants, and consider the theory

$$(4.2) \quad \mathrm{eldiag}\,(A) \cup \mathrm{eldiag}\,(A)' \cup \left\{ \bigvee_{i<m} (R_i(\bar{c}_i) \leftrightarrow \neg R_i'(\bar{c}_i)) \right\}$$

where the constants for elements of A are not changed in $\mathrm{eldiag}\,(A)'$. If (4.2) has a model, then by the elementary diagram lemma (Lemma 2.5.3) it has a model D such that $A \preccurlyeq D|L$. Write B for $D|L$, and let C be $D|L'$ with each relation symbol R' changed back to R. Let $\bar{b}_i$ be the elements of D named by the constants $\bar{c}_i$. Then B and C are elementary extensions of A, $B|J = C|J$, and there is $i < m$ such that $B \vDash R_i(\bar{b}) \Leftrightarrow C \nvDash R_i(\bar{b}_i)$; in short (d) holds. Thus we need only show that (4.2) has a model.

Suppose it has no model. Then by compactness there is $\phi(\bar{a})$ in $\mathrm{eldiag}\,(A)$ (with $\phi(\bar{y})$ in L) such that for each $i < m$,

$$(4.3) \qquad\qquad \phi(\bar{a}),\, R_i(\bar{c}_i) \vdash \phi'(\bar{a}) \rightarrow R_i'(\bar{c}_i).$$

Using Craig's interpolation theorem (Theorem 6.6.3) just as in the proof of Beth's definability theorem (Theorem 6.6.4), it follows that for some formula $\theta_i(\bar{x}_i, \bar{y})$ of J,

(4.4) $\phi(\bar{a}) \vdash \forall \bar{x}_i (R_i \bar{x}_i \leftrightarrow \theta_i(\bar{x}_i, \bar{a}))$,

so that $A \vDash \exists \bar{y} \forall \bar{x}_i (R_i \bar{x}_i \leftrightarrow \theta_i(\bar{x}_i, \bar{y}))$. Since this holds for all $i < m$, A satisfies some definition of K over J, contradicting our assumptions. This completes the proof of (d).

Finally we prove (c) $\Rightarrow$ (a). Assuming (c), we can extend T to a complete theory for which (c) still holds, so that we have (d) too. The rest of the proof falls into two parts. The first part is to construct a tree which will serve as a climbing frame; the second is to put suitable models at the nodes of the tree.

For present purposes a **tree** is a partially ordered set $(P, \leqslant)$ where for each element p of P the set $p^\wedge = \{q \in P : q < p\}$ is well-ordered by $\leqslant$. The elements in P are called **nodes**. The **height** of a node p is the order-type of $p^\wedge$. The **height** of the tree is the least ordinal greater than the heights of all nodes. The αth **level** of the tree is the set of all nodes of height α. A **branch** of the tree is a maximal linearly ordered subset of P. We say that a node in a tree **splits** if it has at least two immediate successors.

Lemma 12.4.2. *Let λ be an infinite cardinal. Then there is a tree $(P, \leqslant)$ such that*

(a) *$(P, \leqslant)$ has more than λ branches and at most λ nodes,*

(b) *the height of $(P, \leqslant)$ is a cardinal $\leqslant \lambda$,*

(c) *$(P, \leqslant)$ has just one node of height 0,*

(d) *for each limit ordinal $\delta < \lambda$, if p and q are nodes of height δ and $p^\wedge = q^\wedge$, then $p = q$,*

(e) *each node of P has at most two immediate successors.*

Proof of lemma. If α is an ordinal, write $^\alpha 2$ for the set of all sequences of 0's and 1's, of length α. Write $^{<\alpha}2$ for $\bigcup_{\beta < \alpha} {}^\beta 2$. Let μ be the least cardinal such that $2^\mu > \lambda$. Then $\mu \leqslant \lambda$. Let the set Q be $^{<\mu}2$, and if p and q are sequences $\in P$, put $p \leqslant q$ iff p is an initial segment of q. The number of nodes of $(Q, \leqslant)$ is $|\bigcup_{\alpha < \mu} {}^\alpha 2| \leqslant \mu\lambda \leqslant \lambda$. Each branch of $(Q, \leqslant)$ is the set of proper initial segments of some sequence $\in {}^\mu 2$; conversely if s is a sequence $\in {}^\mu 2$, then the set of proper initial segments of s forms a branch of $(Q, \leqslant)$. Putting these two facts together, the tree has $|{}^\mu 2| = 2^\mu > \lambda$ branches, each of length μ. $\square$ Lemma 12.4.2.

We turn to the construction of models. The argument is largely book-keeping.

Let $(P, \leqslant)$ be the tree promised by the lemma above; write μ for its height. For each ordinal $\gamma < \mu$, we write $P^{(\gamma)}$ for $\{p \in P : p$ has height $\leqslant \gamma\}$. For each node p in P and each relation symbol R which is in L but not in J, we introduce a new copy $R_{\rightarrow p}$; when R is a relation symbol in J, we write $R_{\rightarrow p}$ for R itself. We write $L_{\rightarrow p}$ for the language got from L by replacing each symbol R by $R_{\to p}$. Thus every formula ϕ or theory in U in L translates into a formula $\phi_{\to p}$ or theory $U_{\to p}$ in $L_{\to p}$.

Again when $\gamma < \mu$, let $L^{(\gamma)}$ be the first-order language whose signature consists of the following symbols:

(4.5) all the symbols $R_{\to p}$ with p of height $\leqslant \gamma$;
       for each ordinal $\alpha \leqslant \gamma$, a 1-ary relation symbol ∂_α;
       for each pair of distinct nodes p, q of the same height $\leqslant \gamma$ in
       P, a 2-ary relation symbol I_{pq}.

Using $L^{(\gamma)}$ we can write down a theory $T^{(\gamma)}$ which expresses that any model A of $T^{(\gamma)}$ has the following properties (4.6)–(4.9).

(4.6) For each node p of $P^{(\gamma)}$, if α is the height of p, then the set
       of elements $\{x : \partial_\alpha(x)\}$ is the domain of a model $A_{\to p}$ of
       $T_{\to p}$.

We write $A_{\leftarrow p}$ for the L-structure got from $A_{\to p}$ by replacing each symbol $R_{\to p}$ by R. (So the subscripts $\leftarrow$, $\to$ indicate that the language is L or a copy of L respectively.)

(4.7) If $p \leqslant q$ in $P^{(\gamma)}$, then $A_{\leftarrow p} \leqslant A_{\leftarrow q}$.

(4.8) If p, q are any two distinct nodes of the same height $\alpha \leqslant \gamma$,
       then there are a symbol R of K and a tuple $\bar{a}$ in $\{x : \partial_\alpha(x)\}$
       such that $A_{\to p} \vDash R_{\to p}(\bar{a}) \Leftrightarrow A_{\to q} \nvDash R_{\to q}(\bar{a})$.

(4.9) If p, q are two disinct nodes of the same height $\alpha \leqslant \gamma$, then
       I_{pq} defines a permutation of $\mathrm{dom}(A)$ which is an automorph-
       ism between $A_{\leftarrow p}$ and $A_{\leftarrow q}$; and moreover if $\alpha < \alpha' \leqslant \gamma$ and
       p', q' are nodes of height α' which are above p, q
       respectively, then $I_{pq} \subseteq I_{p'q'}$.

We put $L^! = \bigcup_{\gamma < \mu} L^{(\gamma)}$ and $T^! = \bigcup_{\gamma < \mu} L^{(\gamma)}$. To prove (a), it suffices to find a model A of $T^!$ of cardinality λ in which $\{a : A \vDash \partial_0(a)\}$ also has cardinality λ. For suppose A is such a model. By (4.6), $A_{\leftarrow p} \vDash T$ for each node p of P. By (4.7), if β is any branch of $(P, \leqslant)$, the structures $(A_{\leftarrow p} : p \in \beta)$ form an elementary chain whose union we write A_β. Each structure A_β has cardinality λ, since $A_{\leftarrow 0} \leqslant A_{\leftarrow \beta}$. By (4.8), if β, γ are distinct branches of $(P, \leqslant)$ then A_β and A_γ are not K-equivalent. Finally by (4.9), if β and γ are distinct branches of $(P, \leqslant)$ then the union of the relations I_{pq} (where $\alpha < \lambda$ and p, q are respectively the nodes of height α in β, γ) forms an automorphism from $A_{\leftarrow \beta}$ to $A_{\leftarrow \gamma}$.

The language $L^!$ has cardinality λ. So in order to find a model A of $T^!$ of the form required, recalling that T has infinite models by (4.1), we need only show that $T^!$ has a model. By compactness this reduces to the case where $\lambda = \omega$. Thus we can assume that $(P, \leqslant)$ has height ω. By stretching $(P, \leqslant)$ out as in (4.10)

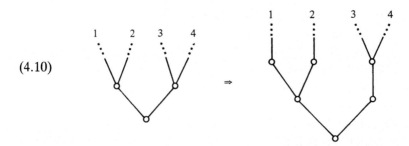

(4.10)

we can assume too that at each level $< \omega$, exactly one node of $(P, \leqslant)$ has two immediate successors. (Necessarily the others each have just one immediate successor, by Lemma 12.4.2(e).)

Now $T^! = \bigcup_{n<\omega} T^{(n)}$. So to prove that $T^!$ has a model, we need only show by induction on n that each theory $T^{(n)}$ has a model. The theory $T^{(0)}$ has a model, since it is T in disguise.

Suppose then that $T^{(n)}$ has a model A; we shall show that $T^{(n+1)}$ has a model. Let the nodes of height n be $p_0, \ldots, p_n$, and let those of height $n+1$ be $q_0, \ldots, q_{n+1}$, numbered so that for each $i < n$, q_i is the unique successor of p_i, and q_n, q_{n+1} are the successors of p_n. As above, the structure A yields structures $A_{\to p}$ for each node p of height $\leqslant n$.

The task is to find an $L^{(n+1)}$-structure D such that (writing $f_{i,j}$ for $I^A_{p_i p_j}$ and $g_{i,j}$ for $I^D_{q_i q_j}$)

(4.11)   for every node p of height $\leqslant n+1$, the set of elements
         $\{x: \partial_n(x)\}$ is the domain of a model $D_{\to p}$ of $T_{\to p}$ (as above,
         we write $D_{\leftarrow p}$ for its translation into an L-structure),

(4.12)   for every node p of height $\leqslant n$, $D_{\to p} = A_{\to p}$,

(4.13)   if $i \leqslant n$ then $D_{\leftarrow p_i} \leqslant D_{\leftarrow q_i}$, and $D_{\leftarrow p_n} \leqslant D_{\leftarrow q_{n+1}}$,

(4.14)   there are a symbol R of $K \backslash J$ and a tuple $\bar{d}$ in D such that
         $D_{\leftarrow q_n} \vDash R(\bar{d}) \Leftrightarrow D_{\leftarrow q_{n+1}} \nvDash R(\bar{d})$,

(4.15)   for all $i < j \leqslant n$, $g_{i,j}$ is an isomorphism from $D_{\leftarrow q_i}$ to $D_{\leftarrow q_j}$
         extending $f_{i,j}$,
         for each $i < n$, $g_{i,n+1}$ is an isomorphism from $D_{\leftarrow q_i}$ to $D_{\leftarrow q_{n+1}}$
         extending $f_{i,n}$,
         $g_{n,n+1}$ is an isomorphism from $D_{\leftarrow q_n}$ to $D_{\leftarrow q_{n+1}}$ extending the
         identity on $p_{\to p_n}$.

The first step is to find D so that the structures $D_{\to q}$ are defined and satisfy (4.11)–(4.14). When $n = 0$, we do this readily by (d) of the theorem. When $n > 0$, we first build $D_{\leftarrow q_n}$ and $D_{\leftarrow q_{n+1}}$ together by (d) of the theorem, we translate them into $D_{\to q_n}$ and $D_{\to q_{n+1}}$, and finally we quote Theorem 6.6.1 to combine these with $D_{\to p_i}$ when $i < n$. This gives D satisfying (4.11)–(4.14).

Now let κ be some strong limit number of cofinality $> |A|$, and by

Theorem 10.4.2 let D' be a special elementary extension of D with cardinality κ. List the elements of A as $\bar{a}$. We have for all $i < j \leqslant n$

(4.16) $\qquad (D'_{\leftarrow q_i}, \bar{a}) \equiv (A'_{\leftarrow p_i}, \bar{a}) \equiv (A_{\leftarrow p_j}, f_{i,j}\bar{a}) \equiv (D'_{\leftarrow q_j}, f_{i,j}\bar{a})$.

Since the structures at either end of (4.16) are special of the same cardinality, they are isomorphic by Theorem 10.4.4, so that each $f_{i,j}$ extends to an isomorphism $g_{i,j}$ from $D'_{\leftarrow q_i}$ to $D'_{\leftarrow q_j}$. We find $g_{i,j}$ similarly when $i < j = n + 1$ (taking the two cases $i < n, i = n$). Rename D' as D. Let ∂^D_{n+1} be dom (D), and let each $I^D_{q_i q_j}$ be $g_{i,j}$. Then (4.11)–(4.15) all hold, and we are through. $\qquad\qquad\qquad\qquad\qquad\qquad\qquad\qquad\qquad\qquad$ ☐ Theorem 12.4.1

What if we want relativised reducts in place of the reducts $A|J$ in Theorem 12.4.1? Consider the following situation.

(4.17) L and L^+ are two first-order languages with $L \subseteq L^+$. P is a 1-ary relation symbol in $L^+ \backslash L$. T^+ is a theory in L^+, such that if B is any model of T^+ then P^B is the domain of a substructure of $B|L$; write B_P for this substructure (as in section 5.1).

If B and B' are two models of T^+ with $B_P = B'_P$, and Q is a relation symbol of T^+, then we say that B and B' are Q-**equivalent on** P if Q^B and $Q^{B'}$ have the same restriction to the set P^B. For tidiness suppose that $T^+ \vdash \forall\bar{x}(Q\bar{x} \to$ 'all the elements in $\bar{x}$ satisfy Px').

Corollary 12.4.3. *Let L, L^+, T^+, P and Q be as above. Let λ be a cardinal $\geqslant |L^+|$. Then the following are equivalent.*
(a) *For every L-structure A of cardinality λ, if $\mathbf{K}$ is the class of models B of T^+ with $B_P = A$, the number of Q-equivalence classes on P among structures B in $\mathbf{K}$ is at most λ.*
(b) *There are $m < \omega$ and formulas $\psi_0(\bar{x}, \bar{y}), \ldots, \psi_{m-1}(\bar{x}, \bar{y})$ of L such that $T^+ \vdash \bigvee_{j<m}(\exists\bar{y} \in P)(\forall\bar{x} \in P)(Q\bar{x} \leftrightarrow \psi^P_j(\bar{x}, \bar{y}))$.*

Proof. By Theorem 5.2.1 there are a first-order language L^* (with $|L^*| \leqslant |L^+|$) containing L and Q, and a theory T^* in L^* such that an L^*-structure C is a model of T^* if and only if there is some model B of T^+ such that C is (B_P, Q^B). Then (a) is equivalent to the following:

(4.18)   for every L-structure A of cardinality λ, the number of Q-equivalence classes of models of T^* which are expansions of A is $\leqslant \lambda$.

By Theorem 12.4.1 (with $L, L \cup \{Q\}, L^+$ for J, K, L respectively) it follows that (a) is also equivalent to the following:

(4.19)   there are $m < \omega$ and formulas $\psi_0(\bar{x}, \bar{y}), \ldots, \psi_{m-1}(\bar{x}, \bar{y})$ of L such that $T^* \vdash \bigvee_{i<m} \exists\bar{y}\forall\bar{x}(Q\bar{x} \leftrightarrow \psi_i(\bar{x}, \bar{y}))$.

I leave it to the reader to show that (4.19) is equivalent to (b), using the relativisation theorem (Theorem 5.1.1). □

Exercises for section 12.4

1. In Theorem 12.4.1 and Corollary 12.4.3, show that if λ is a strong limit number (for example ω), then the family of models of T in the theorem can be chosen to have cardinality 2^{λ}.

2. In the setting of Theorem 12.4.1, suppose every model of T has at least two elements. Show that we can replace (c) in the theorem by the clause (c') There is no definition $\delta(\bar{y})$ of K over J such that $T \vdash \exists \bar{y}\, \delta$. [Use the time-sharing trick of Exercise 2.1.6.]

12.5. Relative categoricity

Throughout this section the following assumptions hold:

(5.1) L and L^{+} are first-order languages with $L \subseteq L^{+}$, and one of the symbols in L^{+} is a 1-ary relation symbol P;

T is a complete theory in L^{+} such that (using the terminology of section 5.1) the P-part B_P is defined for every model B of T.

Note that since T is complete, B_P is defined either for all models B of T or for none, depending on whether T contains the admissibility conditions for relativisation; see section 5.1. The relativisation theorem (Theorem 5.1.1) tells us that in this setting there is a map $\theta(\bar{x}) \mapsto \theta^{P}(\bar{x})$ such that for every model B of T and every tuple $\bar{b}$ in B_P, $B \vDash \theta^{P}(\bar{b}) \Leftrightarrow B_P \vDash \theta(\bar{b})$; θ^{P} is known as the **relativisation** of θ. We write T^{P} for $\mathrm{Th}(B_P)$ where B is any model of T; by the relativisation theorem, T^{P} depends only on T.

We say that T is **relatively categorical** if

(5.2)   for all models B, B' of T, if $B_P = B'_P$ then there is an isomorphism from B to B' which extends the identity on B_P.

Note that if T is relatively categorical, B and B' are models of T and $i: B_P \to B'_P$ is an isomorphism, then i extends to an isomorphism $i^{+}: B \to B'$. (This is clear if you have the right mental picture, but the formal proof can be elusive – see Exercise 1 below.)

Sources of relative categoricity

There are three reasons to be interested in relatively categorical theories. The first has to do with Beth's theorem on definability. Suppose $T \vdash \forall x\, Px$. Then the condition (5.2) says that if A is an L-structure, there is at most one way of expanding A to form an L^+-structure A^+ which is a model of T. In the language of section 6.6, the condition (5.2) says that the symbols of $L^+\backslash L$ are implicitly defined by T in terms of L. According to Beth's theorem (Theorem 6.6.4), this holds if and only if the symbols of $L^+\backslash L$ are explicitly defined by T in terms of L. If we drop the condition that $T \vdash \forall x\, Px$, then the passage from A to A^+ involves adding new elements as well as new symbols, and it's no longer obvious what an 'explicit definition' should be. Is there a good analogue of Beth's theorem when new elements are allowed?

The second source of relative categoricity is the following example.

Example 1: *Strongly minimal sets in uncountably categorical structures.* Suppose T is an uncountably categorical theory in a countable first-order language L^+. We saw in Theorem 12.2.9 that there is no serious loss in supposing that T has a strongly minimal formula of the form Px. Take L to be L^+; then after passing to a suitable definitional extension, T is relatively categorical by Theorem 12.2.14. This viewpoint is natural when we ask how models of T are built up from their strongly minimal sets; we made a first approach to this question in section 12.3.

And thirdly, one natural way to generalise categoricity theory is to study structures which contain a certain *standard part*, for example a copy of the natural numbers. Instead of demanding that all the models of T of cardinality λ are isomorphic, we could require that all the models B of T of cardinality λ *such that B_P is some given structure* are isomorphic. Thoughts in this direction lead naturally to Shelah's theory of *stability over a predicate*. I hope I offend nobody if I say that this theory is not yet well understood by anybody except Shelah. But we shall use some results from it below. In Shelah's setting it is natural to loosen the definition of relative categoricity as follows.

Let λ and μ be cardinals with $\lambda \leqslant \mu$, and suppose (5.1) holds. We say that the theory T is (λ, μ)-**categorical** if whenever B and B' are models of T of cardinality μ, B_P has cardinality λ and $i : B_P \to B'_P$ is an isomorphism, then i extends to an isomorphism $i^+ : B \to B'$. We say that T is **relatively λ-categorical** when the same holds but without the requirement that B and B' are of cardinality μ. Thus T is relatively categorical if and only if T is relatively λ-categorical for every cardinal λ.

Example 2: *μ-categorical theories.* Ordinary μ-categoricity reduces to $(0, \mu)$-categoricity as follows. Suppose L_0 is a first-order language and T_0 a theory in L_0 with infinite models. Take a new 1-ary relation symbol P, form L^+ by adding P to L_0, form T by adding to T_0 the sentence $\forall x \neg Px$, and let L be the language with empty signature. Then for every infinite cardinal μ, T is $(0, \mu)$-categorical if and only if T_0 is μ-categorical (i.e. if and only if T_0 has up to isomorphism exactly one model of cardinality μ).

Example 3: *An (ω, ω)-categorical but not relatively ω-categorical theory.* Let A_1 and A_2 be direct sums of countably many copies of the two-element cyclic group $\mathbb{Z}_2$, and let B be a direct sum of countably many copies of the three-element group $\mathbb{Z}_3$. Let L be the language of abelian groups and let L^+ be L with the symbol P added. Let G be the group $A_1 \oplus A_2 \oplus B$, made into an L^+-structure by taking G_P to be the subgroup A_1. Then the theory $T = \mathrm{Th}(G)$ is (ω, ω)-categorical but not (λ, μ)-categorical for any other pair λ, μ of infinite cardinals.

All the examples so far have been stable. This was an accident; I set it to rights at once.

Example 4: *Affine transformations of a field.* This example will incorporate Shelah's all-purpose counterexample, Example 2 in section 4.1. Instead of taking an abelian group and hiding the 0, we take a field F and hide both the 0 and the 1. The elements of the structure B will be those of F together with a copy of each element of F; B_P is F itself, including the field operations. The elements of B not in P^B form a copy F' of F, but not with the field operations. Instead there is a relation to express the function σ as follows, where I write a' for the copy in F' of the element a in F. If c_0', c_1' are distinct elements of F', then for every element a of F, $\sigma(c_0', c_1', a) = (a(c_1 - c_0) + c_0)'$. Put $T = \mathrm{Th}(B)$. If C is any model of T, then C_P is a field. We get a copy of C_P by choosing any two distinct elements c_0, c_1 not in P^C, and mapping each element a of C_P to $\sigma^C(c_0, c_1, a)$; note that 0 goes to c_0 and 1 goes to c_1. Hence T is relatively categorical. If C is infinite, the number of automorphisms of C which fix C_P pointwise is $|C|$ – one for each ordered pair of distinct elements outside the P-part. The field is arbitrary, so that T will generally be unstable. Note that no element outside the P-part is algebraic over the P-part.

There are examples of countable theories which are relatively ω-categorical but not relatively ω_1-categorical. But they are distinctly difficult to find. Shelah wrote one down (in Shelah & Hart [1990]) in order to persuade Leo Harrington that the proofs in stability theory over a predicate really need to be as complicated as they are.

Reduction and definability

We say that the theory T has the **uniform reduction property** (over P and L) if for every formula $\phi(\bar{x})$ of L^+ there is a formula $\phi^\circ(\bar{x})$ of L such that if B is any model of T and $\bar{b}$ any tuple from B_P, then

(5.3) $B \vDash \phi(\bar{b}) \;\Leftrightarrow\; B_P \vDash \phi^\circ(\bar{b}).$

The map $\phi \mapsto \phi°$ is called the **reduction map**. (Thus T has the uniform reduction property if and only if one or all of its models have the reduction property in the sense of section 12.3. I say 'uniform' pedantically, to emphasise that the same reduction map works for all models of T. Of course this is automatic, since T is complete.)

Lemma 12.5.1. *Under assumption (5.1), suppose that the theory T has models B with B_P infinite. Suppose there is a cardinal $\lambda \geqslant |L^+|$ such that*

(5.4)   *for all models B of T of cardinality λ with $|B_P| = \lambda$, every automorphism of B_P extends to an automorphism of B.*

(This holds for example if T is (λ, λ)-categorical.) Then T has the uniform reduction property.

Proof. Suppose the conclusion is false; let $\phi(\bar{x})$ be a counterexample. Then we claim that the following theory is consistent, where $\bar{c}$, $\bar{d}$ are tuples of new constants:

(5.5) $T \cup \{\theta^P(\bar{c}) \leftrightarrow \theta^P(\bar{d}): \theta(\bar{x})$ is a formula of $L\} \cup \{\phi(\bar{c}) \wedge \neg \phi(\bar{d})\}$.

For suppose (5.5) is inconsistent. Then by compactness and the lemma on constants (Lemma 2.3.2) there are formulas $\theta_0(\bar{x}), \ldots, \theta_{k-1}(\bar{x})$ of L such that

(5.6) $$T \vdash \forall \bar{x} \bar{y} \left(\bigwedge_{i < k} (\theta_i^P(\bar{x}) \leftrightarrow \theta_i^P(\bar{y})) \to (\phi(\bar{x}) \leftrightarrow \phi(\bar{y})) \right).$$

Since T is complete, it follows that in any model B of T we know whether or not $B \vDash \phi(\bar{b})$ as soon as we know which of the formulas $\theta_0, \ldots, \theta_{k-1}$ are satisfied by $\bar{b}$ in B_P. But then it's easy to put together a formula $\phi°$ to prove the conclusion of the lemma. The claim holds.

By the claim, there is a model of B of T such that B_P contains two tuples $\bar{a}$ and $\bar{b}$ for which $(B_P, \bar{a}) \equiv (B_P, \bar{b})$ but $B \vDash \phi(\bar{a}) \wedge \neg \phi(\bar{b})$. By the lemma assumptions on T, we can arrange that $|B| = |B_P| = \lambda$. By Exercise 10.4.13 there is an elementary extension C of B of cardinality λ such that C_P is strongly ω-homogeneous. Since C_P is strongly ω-homogeneous, it has an automorphism s which takes $\bar{a}$ to $\bar{b}$. By (5.4), s extends to an automorphism s^+ of C. Hence $(C, \bar{a}) \equiv (C, \bar{b})$, since $s^+ \bar{a} = \bar{b}$. But $B \preccurlyeq C$ and so $C \vDash \phi(\bar{a}) \wedge \neg \phi(\bar{b})$; contradiction.   □

The next result tells us that if T is (λ, λ)-categorical, then types over the P-part are definable. More precisely let us say that **types over the P-part are definable** for T if for every formula $\phi(\bar{x}, \bar{y})$ of L^+ there is a formula $d\phi(\bar{y}, \bar{z})$ of L such that for each model B of T and each tuple $\bar{b}$ in B, there is a tuple $\bar{c}$ in B_P such that for every tuple $\bar{a}$ in B_P,

(5.7) $B \vDash \phi(\bar{b}, \bar{a}) \Leftrightarrow B_P \vDash d\phi(\bar{a}, \bar{c})$.

This is a notion in the spirit of stability theory (cf. (7.9) in section 6.7 above), but note that the theory T^P may be unstable as in Example 4 above. Actually we shall assume a little less than (λ, λ)-categoricity. If B, B' are two structures with $B_P = B'_P$, let us say that B and B' are **equivalent over** P if there is an isomorphism $e: B \to B'$ which is the identity on B_P. Equivalence over P is an equivalence relation.

Lemma 12.5.2. *In the setting (5.1), let λ be a cardinal $\geqslant |L^+|$ such that if $(B_i: i \in I)$ are any models of T of cardinality λ, and the structures $(B_i)_P$ are all equal to some fixed structure A of cardinality λ, then there are at most λ equivalence classes over P among the B_i $(i \in I)$. Then types over the P-part are definable for T.*

Proof. Consider a formula $\phi(\bar{x}, \bar{y})$ of L^+. Let $\bar{c}$ be a tuple of new constants, of the same length as $\bar{x}$, and Q a new relation symbol whose arity is the length of $\bar{y}$. Add $\bar{c}$ and Q to L^+, forming a first-order language $L^\#$. Let $T^\#$ be T together with the sentence $\forall \bar{y}(Q\bar{y} \leftrightarrow \phi(\bar{c}, \bar{y}))$. Condition (a) of Corollary 12.4.3 holds with $T^\#$ for T^+. So condition (b) holds too, and the lemma follows. Note that since the structures B_P are infinite, we can use Exercise 12.4.2 to get $\mathrm{d}\phi$ of the form $\psi(\bar{y}, \bar{a})$ where $\psi(\bar{y}, \bar{z})$ depends only on T and ϕ. $\qquad \square$

Under stronger assumptions we can sharpen the preceding lemma to say that *types are isolated over* P. This means that if B is any model of T and $\bar{b}$ is any tuple in B, then there is a formula $\psi(\bar{x})$ of L^+ with parameters in B_P, such that

(5.8)   for every formula $\phi(\bar{x})$ of L^+ with parameters from B_P
$\qquad B \vDash \phi(\bar{b})$ if and only if $B \vDash \forall \bar{x}(\psi \to \phi)$.

In the terminology of section 7.2, this is the same as saying that the type of $\bar{b}$ over the empty set is principal over the theory of B with parameters for the elements of B_P.

Lemma 12.5.3. *Under the assumptions (5.1), suppose L^+ is countable, B_P is always infinite and T is (ω, ω)-categorical. Then types are isolated over P.*

Proof. Suppose not; find a counterexample consisting of a model B of T and a tuple $\bar{b}$ in B. Then for every formula $\phi(\bar{x}, \bar{y})$ of L^+ and every tuple $\bar{a}$ in B_P such that $B \vDash \phi(\bar{a}, \bar{b})$, there are a formula $\psi(\bar{z}, \bar{y})$ of L^+ and a tuple $\bar{d}$ in B_P such that $B \vDash \exists \bar{y}(\phi(\bar{a}, \bar{y}) \wedge \psi(\bar{d}, \bar{y})) \wedge \exists \bar{y}(\phi(\bar{a}, \bar{y}) \wedge \neg \psi(\bar{d}, \bar{y}))$. So the tuple $\bar{b}$ is already a counterexample to the claim in some countable elementary substructure of B. We can assume then that B is countable.

Let $\bar{c}$ be a sequence listing the elements of B_P, and let $L^+(\bar{c})$ be L^+ with these elements added as parameters. Let Φ be the set of formulas of $L^+(\bar{c})$ satisfied by $\bar{b}$ in $(B, \bar{c})$. By assumption Φ is a non-principal complete type. Let Ψ be the set of formulas

$$(5.9) \qquad \{c \neq y : c \in P^B\} \cup \{Py\}.$$

Since $\text{Th}(B, \bar{c})$ is complete and has a model omitting Ψ, Ψ is non-principal over $\text{Th}(B, \bar{c})$. So by the omitting types theorem (Theorem 7.2.1) there is a model of $\text{Th}(B, \bar{c})$ which omits both Φ and Ψ. Let $(D, \bar{d})$ be such a model. Then D is a model of T, and $\bar{d}$ lists the elements of D_p since Ψ is omitted. So there is an isomorphism $f : B_P \to D_P$ taking $\bar{c}$ to $\bar{d}$. Since T is (ω, ω)-categorical, f extends to an isomorphism $g : B \to D$. But then $g\bar{a}$ realises Φ in $(D, \bar{d})$; contradiction. $\qquad \square$

With what we know so far, we can write down characterisations of (ω, ω)-categorical and relatively ω-categorical theories when the language is countable.

Theorem 12.5.4. *In the setting of (5.1), suppose L^+ is a countable language and B_P is always infinite. Then the theory T is (ω, ω)-categorical if and only if*
(a) *T has the uniform reduction property over P and L, and*
(b) *types are isolated over P.*
The theory T is relatively ω-categorical if and only if it satisfies (a), (b) and
(c) *$P(x)$ is a one-cardinal formula for T.*

Proof. First suppose T is (ω, ω)-categorical. Then (a) holds by Lemma 12.5.1, and (b) by Lemma 12.5.3. Conversely suppose (a) and (b) hold, and let B, C be countable models of T with $B_P = C_P$ a countable structure. Let $\bar{a}$ be a sequence listing the elements of B_P. By the uniform reduction property, $(B, \bar{a}) \equiv (C, \bar{a})$. The type of any tuple of elements of $(B, \bar{a})$ or $(C, \bar{a})$ over the empty set is principal, by (b). It follows that $(B, \bar{a})$ and $(C, \bar{a})$ are elementarily equivalent countable atomic structures, and so by Theorem 7.2.3 there is an isomorphism $f : (B, \bar{a}) \to (C, \bar{a})$. Then f is an isomorphism from B to C which extends the identity on B_P, proving that T is (ω, ω)-categorical.

Suppose next that (c) fails. Then by Vaught's two-cardinal theorem (Theorem 12.1.1) there are models $B < C$ of T with $B_P = C_P$, $|B| = \omega$ and $|C| = \omega_1$, and this contradicts relative ω-categoricity. Finally suppose (a), (b) and (c) all hold. Then T is (ω, ω)-categorical, and if B is any model of T with B_P countable then B is countable too. It follows at once that T is relatively ω-categorical. $\qquad \square$

I quote one further result from Shelah's theory of stability over a predicate. The hypothesis can be weakened, but the reader will have to refer to Shelah's papers for details.

Lemma 12.5.5. *Suppose T is relatively categorical. Then for any two models, B, C of T, if e is an elementary embedding of B_P into C_P then e extends to an elementary embedding of B into C.* ☐

Coordinatisation

Let us come back to the question of generalising Beth's theorem. If T is relatively categorical, is it true that every model B of T is explicitly describable in terms of B_P? Before we can answer this question, we need a sensible notion of 'explicitly describable'. The following definitions suggest themselves.

We still assume (5.1). The theory T is said to be **coordinatised** if every model B of T is coordinatised over B_P in the sense of section 12.3; in other words, if

(5.10) T has the uniform reduction property, and for every model B
          of T, all the elements of B lie in the definable closure of
          dom (A).

A **coordinatisation** of T is a coordinatised theory $T^{\#}$ in a first-order language $L^{\#}$ extending T, such that

(5.11)   every model B of T can be expanded to a model $B^{\#}$ of $T^{\#}$.

(Thus (5.1) holds for $T^{\#}$ with $L^{\#}$ in place of L^{+}. In the terminology of section 12.3, the structure $B^{\#}$ is a **coordinatisation** of B.) We shall say that T is **coordinatisable** if it has some coordinatisation.

Lemma 12.5.6. *Suppose T is coordinatised. If A and B are models of T, then any elementary embedding $e: A_P \to B_P$ extends to an elementary embedding $e^{+}: A \to B$; if e is an isomorphism then so is e^{+}.*

Proof. The definition of e^{+} is very much like the definition of $\alpha^{\#}$ for Theorem 12.3.4, and I leave the details to the reader. ☐

Theorem 12.5.7. *If T is coordinatisable, then T is relatively categorical and all its models are continuously natural over P.*

Proof. Let $T^{\#}$ be a coordinatisation of T. If B and C are models of T, then by (5.11) they can be expanded to models $B^{\#}$, $C^{\#}$ of $T^{\#}$ which are respectively coordinatisations of B, C. By the lemma, every isomorphism $i: B_P \to C_P$ extends to an isomorphism $i^{\#}: B^{\#} \to C^{\#}$. This proves relative categoricity. The rest is immediate from Theorem 12.3.4. ☐

From the point of view of Beth's theorem, the proof of Theorem 12.5.7 proves too much. In some sense it tells us that 'explicitly definable' (i.e.

coordinatisable) implies 'implicitly definable' (i.e. relatively categorical). But there is no hope of proving the converse, because we have already seen that there are relatively categorical theories whose models are not natural over P. Take for example the theory of the structure in Example 2 of section 12.3 above. So relative categoricity certainly doesn't imply coordinatisability, and thus our attempt at a generalisation of Beth's theorem breaks down.

Nevertheless coordinatisability is worth having, and there are some sufficient conditions for getting it. I prove two results of this form. The first assumes that every model of T is rigid over its P-part, and the second assumes naturality plus a little extra.

Let us say that T is **rigidly relatively categorical** if T is relatively categorical and for every model B of T, if i is an automorphism of B which is the identity on B_P, then i is the identity. (In short, B is rigid over B_P.)

Theorem 12.5.8 (*Gaifman's coordinatisation theorem*). *Under assumption (5.1), suppose also that L is countable and B_P is always infinite. Then the following are equivalent.*
(a) *T is rigidly relatively categorical.*
(b) *T is coordinatised.*

Proof. The argument from (b) to (a) is straightforward and I omit it. In the other direction, we know already from Lemma 12.5.1 that T has the uniform reduction property. It remains to show that if B is a model of T, then every element of B is in the definable closure of P_B.

Let B be a model of T and b an element of B. Write $\bar{a}$ for a sequence listing all the elements of P^B. By Lemma 12.5.3, the type of b over B_P is isolated, say by a formula $\psi(x, \bar{a})$ where $\psi(x, \bar{y})$ is in L. (Of course only finitely many items in $\bar{a}$ occur in $\psi(x, \bar{a})$.) We claim that $B \vDash \exists_{=1} x\, \psi(x, \bar{a})$. For suppose not. Then there is an element $c \neq b$ in B such that $B \vDash \psi(c, \bar{a})$. By choice of $\psi(x, \bar{a})$, we have $(B, \bar{a}, b) \equiv (B, \bar{a}, c)$. Let B' be a countable elementary substructure of B containing b, c and those elements of B_P which occur in ψ; let $\bar{a}'$ be a list of the elements of B'_P. Then $(B', \bar{a}', b) \equiv (B', \bar{a}', c)$. By Lemma 12.5.3 again, the structure $(B', \bar{a}')$ is atomic. Since B' is countable, it follows as in Theorem 7.2.3 that there is an automorphism f of B' taking b to c and fixing $\bar{a}'$ pointwise. This contradicts the rigid relative categoricity of T. The claim is proved, and with it the theorem.

$\square$

We turn to our second positive result. The hypotheses are less special than they might seem. In the setting of Example 1 above, where T is a totally categorical theory and B_P is a strongly minimal set in the countable model B of T, Theorem 12.2.17 told us that B_P has the small index property.

Naturality sometimes holds in this setting, and sometimes it fails; we saw some examples in section 12.3. The theory T^P was defined at the beginning of this section.

Theorem 12.5.9. *Assuming (5.1), suppose that T is a countable relatively categorical theory, T^P is ω-categorical, and if B is any countable model of T then*
(a) *B is natural over P and*
(b) *B_P has the small index property.*
Then T is coordinatisable.

Proof. Since T is relatively categorical and T^P is ω-categorical, there is just one countable model B of T up to isomorphism. By (a), (b), Lemma 12.3.1 and Theorem 12.3.5, B is coordinatisable over B_P. Thus there are a first-order language $L^{\#}$ extending L^+, an expansion $B^{\#}$ of B to an $L^{\#}$-structure, and a set $\{\phi_i(x, \bar{y}): i < \omega\}$ of formulas of $L^{\#}$ such that

(5.12)   for every element b there are $i < \omega$ and a unique $\bar{a}$ in B_P
            such that b is the unique element in $\phi_i(B^{\#}, \bar{a})$.

Moreover $B^{\#}$ has the reduction property, so that writing $T^{\#}$ for $\text{Th}(B^{\#})$, $T^{\#}$ is a complete theory which extends T and has the uniform reduction property. We can suppose that $L^{\#}$ is countable, since $L^{\#}$ only needs to contain L^+ and the formulas ϕ_i.

We claim that for some finite m, $T^{\#}$ contains the sentence

(5.13) $\forall x \bigvee_{i < m} (\exists \bar{y} \in P)(\text{'}x \text{ is the unique element such that } \phi_i(x, \bar{y})\text{')}.$

For suppose not. Then by compactness there is some model C of $T^{\#}$ containing an element c such that for all $i < \omega$,

(5.14) $C \vDash \neg(\exists \bar{y} \in P)(\text{'}c \text{ is the unique element such that } \phi_i(x, \bar{y})\text{')}.$

By the downward Löwenheim–Skolem theorem we can suppose that C is countable, and so by ω-categoricity we can assume $C_P = B_P$. Using the reduction property, we can map each element b of $B^{\#}$ to an element $e(b)$ of C which satisfies the same definition over B_P, and this map e elementarily embeds $B^{\#}$ in C. Since $P(x)$ is a one-cardinal formula for T (by Theorem 12.5.4), it follows that $eB^{\#}$ is the whole of C, contradicting (5.14).

This proves the claim, and it follows at once that $T^{\#}$ is coordinatised. We have one more thing to show, namely that every model of T can be expanded to a model of $T^{\#}$.

Let C be a model of T, and let D be an elementary extension of C which is special. By Theorem 10.4.9, D can be expanded to a model $D^{\#}$ of $T^{\#}$. Let X be the definable closure of the set P^C in $D^{\#}$. We claim that X is the domain of an elementary substructure E of $D^{\#}$ with $P^E = P^C$.

For the first part of this claim we use the Tarski–Vaught criterion in the form of Exercise 2.5.1. Suppose $\bar{e}$ is a tuple from X and $\psi(\bar{x}, y)$ a formula of $L^{\#}$ such that $D^{\#} \vDash \exists y \, \psi(\bar{e}, y)$. Then there are $i < m$ and a tuple $\bar{d}$ in $D_P^{\#}$ such that the unique element of $D^{\#}$ satisfying $\phi_i(x, \bar{d})$ also satisfies $\psi(\bar{e}, y)$. But $C_P \leqslant D_P$ since $C \leqslant D$, and so by the reduction property there is $\bar{d}'$ already in C_P such that the unique element b' of $D^{\#}$ satisfying $\phi_i(x, \bar{d}')$ satisfies $\psi(\bar{e}, y)$. By definition of X, b' lies in X. Thus $X = \mathrm{dom}(E)$ for some $E \leqslant D^{\#}$. To finish the proof of the claim, we certainly have $P^C \subseteq P^E$. If e is any element of P^E, then it is definable, say by $\phi_j(x, \bar{a})$ with $\bar{a}$ in C_P. Using the reduction property, we can express in E_P that b is defined by $\phi_j(x, \bar{a})$. Since $C_P \leqslant D_P^{\#}$ and $C_P \subseteq E_P \leqslant D_P^{\#}$, it follows that some element b' in C_P is defined by $\phi_j(x, \bar{a})$, and then $b = b' \in P^C$.

This proves the claim. By relative categoricity $E|L^+$ is isomorphic to C, and hence C can be expanded to a model of $T^{\#}$. $\qquad\square$

We have already seen in Theorem 12.3.4 that condition (a) is needed in Theorem 12.5.9. Evans & Hewitt [1990] give an example which shows that (b) can't be dropped completely; perhaps some weakening of it will suffice.

One last comment is in order. Let us say that B is **finitely coordinatised** over P if it is coordinatised, and moreover there is a finite set of formulas $\phi_i(x, \bar{y})$ $(i < m)$ of L^+ such that each element of B is the unique element of $\phi_i(B, \bar{a})$ for some $i < m$ and some tuple $\bar{a}$ in B_P. We say that B is **finitely coordinatisable** over P if some expansion of B is finitely coordinatised over P. The proof of Theorem 12.5.9 showed that every model of T is finitely coordinatisable.

If B is finitely coordinatisable over P (and B_P has more than one element), then B is interpretable in B_P. This is straightforward, but see Exercises 13, 14 for the details.

Example 5: _Direct sums of cyclic groups._ Consider the cyclic groups $\mathbb{Z}_2$ and $\mathbb{Z}_4$ of orders 2 and 4 respectively, and let B be the group $\bigoplus_{i<\omega}\mathbb{Z}_4$ with a relation symbol P picking out $\bigoplus_{i<\omega}\mathbb{Z}_2$. Let T be $\mathrm{Th}(B)$. Then T is relatively categorical, T^P is ω-categorical and (either by Fact 4.2.11 or by Theorem 12.2.17) B_P has the small index property. But by Fact 5.7.8, B is not interpretable in A. It follows by Theorem 12.5.9 that B is not natural over B_P. This example is interesting because it shows that there is two-way traffic over the bridge between naturality and coordinatisability. One can give a direct proof that B is not natural over B_P, using a rather heavy amount of linear algebra.

Thus we finish as a model theorist should, proving a purely algebraic fact by considering first-order definability.

Exercises for section 12.5

1. Show that if T is relatively categorical, B and B' are models of T and $i: B_P \to B'_P$ is an isomorphism, then i extends to an isomorphism $i^+: B \to B'$.

2. Show that if T has the uniform reduction property over P and L, and for each countable model B of T, every element of B is algebraic over B_P, then T is (ω, ω)-categorical.

3. Show that if T is (ω, ω)-categorical and A, B are two models of T with $|A| = \omega$ and $A_P \preceq B_P$, then the inclusion of A_P in B_P can be extended to an elementary embedding of A into B.

4. Suppose (5.1) holds and L^+ is countable. Suppose also that for any models A, B of T, every elementary embedding $e: A_P \to B_P$ extends to an elementary embedding of A into B. Show that T is relatively categorical. [Show that Px can't be two-cardinal; deduce that if $e: A_P \to B_P$ is the identity, its extension to A must be onto B.]

5. Give an example of a countable theory T as in (5.1) such that Px is one-cardinal and T has uniform reduction over P, but T is not relatively ω-categorical. [We regard $^\omega 2$ as an abelian group with pointwise addition (mod 2). Let η be any sequence $\in {}^\omega 2$ which is not eventually 0. The model B_η of T consists of those sequences $\in {}^\omega 2$ which are either eventually 0 or eventually equal to η; P picks out the sequences which are eventually 0. There is a symbol $+$ for the abelian group operation, and for each $\sigma \in {}^{<\omega}2$ there is a symbol Q_σ picking out those sequences which have σ as an initial segment. We put $L = L^+$. For uniform reduction use Lemma 12.5.1. For the failure of relative ω-categoricity, compare B_η, B_ζ where $\eta \neq \zeta$.]

6. In the situation of (5.1), suppose L^+ is countable and relatively categorical. Show that if B is a model of T and $A \equiv B_P$, then $A = C_P$ for some model C of T. [When A is countable, show that eldiag$(A)^P \cup T$ has a model which omits the type $\{Px\} \cup \{x \neq a: a \in \text{dom}(A)\}$. In the general case, go by induction on $|A|$, writing A as the union of an elementary chain of smaller structures, and using Lemma 12.5.5 to fit together the extensions of structures in the chain.]

*7. Let L be the language of abelian groups and L^+ the language got by adding a 1-ary relation symbol P to L. Suppose B is an abelian group, made into an L^+-structure so that B_P is a subgroup A of B. Show that Th(B) is (ω, ω)-categorical if and only if (a) B/A has bounded exponent and (b) for every prime p and natural number n, the set $p^n B \cap A$ is $\varnothing$-definable in A.

*8. Let L be the language of linear orderings and let L^+ be L with a 1-ary relation symbol P added. Let the L^+-structure B be a linear ordering, and suppose that Th(B) is (ω, ω)-categorical. Show that every model of Th(B_P) is of the form C_P for some model C of Th(B). [The proof that I have uses the classification of ω-categorical linear orderings; see Theorem A.6.15.]

9. Give an example of a countable theory T as in (5.1) such that (a) T is relatively categorical, (b) every model of T is continuously natural, and (c) T is not coordinatisable. [Example 5 in section 12.2 incorporates Baĭzhanov's example: the Möbius function is not needed.]

10. Show that Gaifman's coordinatisation theorem implies the following form of Beth's theorem. Suppose L and L^+ are countable first-order theories with $L \subseteq L^+$, and T is a complete theory in L^+, such that for any two models A, B of T, if $A|L = B|L$ then $A = B$. Then every symbol in $L^+ \setminus L$ is explicitly definable by T in terms of L.

*Adapting (5.10), let us say that T is **finitely coordinatised** if there is a finite set of formulas $\phi_i(x, \bar{y})$ $(i < m)$ of L such that for every model B of T, each element of b is the unique element of $\phi_i(B, \bar{a})$ for some $i < m$ and some tuple $\bar{a}$ in B_P.* **Finite coordinatisation** *is defined similarly.*
11. Show that in Theorem 12.5.8 we can add a third condition: (c) T is finitely coordinatised.

12. We assume (5.1) with L countable, and we suppose that T has infinite models. Show that the following are equivalent. (a) T is relatively categorical, and for every model B of T, the number of automorphisms of B which are the identity on B_P is at most $|B|$. (b) T has a finite coordinatisation $T^\#$ where the formulas making any model $B^\#$ definable over B_P are in L with finitely many new constants added.

13. Suppose we are in the situation of (3.1) and (3.2) of section 12.3, B is finitely coordinatisable over A, and A has at least two elements. We write Δ for the relativised reduction from B to A. (a) Show that there is an interpretation Γ of B in A such that $\Delta \circ \Gamma$ is a self-interpretation of A in A. [Let $B^\#$ be the finitely coordinatised expansion of B. By time-sharing and the assumption on A we can suppose that there is a single formula $\phi(x, \bar{y})$ of $L^\#$ such that every element b of B is pinned down by $\phi(x, \bar{a}_b)$ for some $\bar{a}_b$ in A. There is a coordinate map which takes each $\bar{a}_b$ to b, and the reduction map $^\circ$ then provides the interpretation Γ. The formula ϕ° gives the homotopy.] (b) Show that if moreover B is finitely coordinatised over A, then Γ and Δ form a bi-interpretation.

14. Suppose (3.1) of section 12.3 holds, and Δ is the relativised reduction from B to A. Suppose there is an interpretation Γ of B in A such that $\Delta \circ \Gamma$ is a self-interpretation. (a) Show that B is finitely coordinatisable over A. [Expand to $B^\#$ by adding the coordinate map. The reduction theorem (Theorem 5.3.2) then lives up to its name by providing the reduction map.] (b) Show that if moreover Γ and Δ form a bi-interpretation, then B is finitely coordinatised over A.

History and bibliography

Section 12.1. Vaught's two-cardinal theorem (Theorem 12.1.1) is from Morley & Vaught [1962] (not Vaught [1961]). Chang's two-cardinal theorem

(Theorem 12.1.3) followed in [1965]. Vaught [1965] proved Theorem 12.1.4 for countable languages; Morley [1965b] then found a slicker proof of the full theorem by Ehrenfeucht–Mostowski methods. For further two-cardinal theorems that depend on the properties of particular cardinals, see Keisler [1966b], [1968], [1971a], Schmerl [1972], Shelah [1972b], [1975b], [1977a], Schmerl & Shelah [1972] and the survey by Schmerl [1985]. The second half of Hodges [1985] is a survey of ways of making some things big and other things small. Devlin [1984] describes Jensen's combinatorial principles and how they can be used to prove two-cardinal theorems.

Theorem 12.1.5 (for countable language) appeared in unpublished notes of Haim Gaifman in about 1974; a version of it is implicit in §3 of Baldwin & Lachlan [1971]. Shelah [1978a] section V.6 develops the same theme with stable theories in mind (though unfortunately this is not the clearest section of his book). Keisler [1966a] earlier gave a different syntactic characterisation of two-cardinal formulas. Corollary 12.1.6 is from Erimbetov [1975] when the language is countable; without this restriction Shelah [1978a] Lemma V.7.11 gives part (b), and presumably Shelah knew (a) too. Erimbetov [1985] proves an analogous theorem with stability in place of total transcendence or finite Morley rank.

Exercise 1: Chang [1965]. Exercise 2: This trick is derived from Vaught [1965]. Exercise 3: Raphael Robinson in Morley & Vaught [1962]. Exercises 4, 5: Morley [1965b]. Exercise 6: Keisler [1971a] Theorem 24; cf. also Chudnovskiĭ [1970], Shelah [1969]. Exercise 7: Shelah [1972b]; this proof is by Gaifman. Exercise 8: Tuschik [1985] proved it for countable L. Exercise 10: Ziegler [1988]. Exercise 11: Mitchell [1972].

Section 12.2. Veblen [1904] introduced the notion of a categorical set of axioms, i.e. a set whose models are all isomorphic. The notion was studied by Tarski [1934] too, but inconclusively. Not that this should surprise us; apart from the complete theories of finite structures, virtually the only examples of categoricity come from infinitary logics and second-order logic, where the known model theory is still not rich. Vaught [1954a] and Łoś [1954a] opened the subject up by defining λ-categoricity and asking some natural questions.

For a countable language, Theorem 12.2.1 was proved by Morley [1965a]; the question had been posed by Łoś [1954a]. The general case is due to Shelah [1974a], and there is a useful exposition by Saffe [1984]. Theorem 12.2.2 is from Keisler [1971b] together with Shelah [1978a] Theorem IX.1.19. The arguments of Morley and Shelah included Theorem 12.2.3. Theorem 12.2.4 is due to Morley [1965a] in the totally transcendental case and Laskowski [1988] in the locally modular case of the Buechler dichotomy.

Baldwin & Lachlan [1971] contained a more complicated example than Example 3 of an uncountably categorical non-almost-minimal theory. This paper was also the first to use the absence of two-cardinal formulas in λ-categorical theories. Lemma 12.2.6 is from Shelah's PhD thesis; see Shelah [1978a] Conclusion V.6.14. Example 4 reconstructs an example which Shelah showed me in 1981. Buechler's dichotomy (Theorem 12.2.9) is from Buechler [1985]. In the totally transcendental case, the existence of strongly minimal formulas whose parameters realise a principal type is from Baldwin & Lachlan [1971], and the finiteness of Morley rank (Corollary 12.2.10) from Baldwin [1973b] and independently Zil'ber [1974]. Theorems 12.2.11 and 12.2.12 are from Baldwin & Lachlan [1971] in the totally trancendental case and Laskowski [1988] in the locally modular case; the saturation of all models of cardinality $> |L|$ was a guiding principal already in Morley [1965a] and Shelah [1974a].

References for contributions of Zil'ber and Hrushovski are in the text. (See also Hrushovski [1989c] for a survey in depth.) Corollary 12.2.16 is from Ahlbrandt [1987]. Theorem 12.2.17 is from Hodges, Hodkinson, Lascar & Shelah [199?]. Theorem 12.2.18(a) is due to Zil'ber [1980b], [1981] for totally transcendental theories, and to Cherlin, Harrington & Lachlan [1985] for ω-stable ω-categorical theories. Theorem 12.2.18(b) was proved by Hrushovski [1989b] following special cases in Ahlbrandt & Ziegler [1986]. Example 5 is due to Baĭsalov [1991]; it is a variant of the first example of a finitely axiomatisable uncountably categorical structure, which was found by Peretyat'kin [1980]. Theorem 12.2.19 is from Marker [1989].

Exercise 1: Morley [1965a]. Exercises 2–8: Baldwin & Lachlan [1971]. Exercise 7: Shelah [1978a] Lemma IX.1.17.

Section 12.3. Within model theory this line of enquiry began with Theorem 12.3.4 (without the continuity) and Example 2, which are from Hodges [1976a]. The importance of the topology in this context was pointed out by several people, including Ahlbrandt & Ziegler [199?], Evans and Hrushovski; Theorem 12.3.4 in full appears in Hodges, Hodkinson & Macpherson [1990] together with Lemma 12.3.1 and Theorem 12.3.5. Example 1 is one of several examples of naturality handled by an argument of Steve Donkin (on p-adic representations) in Evans, Hodges & Hodkinson [1991], which also contains Exercise 7. Theorem 12.3.3(b) is related to a result of Macpherson & Steinhorn [199?]; cf. Kueker & Steitz [199?b]. Theorem 12.3.6 is from Ahlbrandt & Ziegler [199?], except for the second sentence of (b), which is from Hodges & Pillay [199?].

The set-theoretic questions about the axiom of choice are older: Läuchli [1962] essentially used Theorem 12.3.7 to construct a model of ZFU in which there was a field with no algebraic closure. Models of ZF containing fields with no algebraic closure were given by Pincus [1972] and Hodges [1976b]; Pincus used his transfer principle to carry Läuchli's result over to ZF. The use of the Artin–Schreier lemma was suggested by Pincus (in a letter to me in 1973). Hodges & Shelah [1986], [199?] discuss the set-theoretic definability of algebraic constructions more generally.

Exercise 2: The example is from a remark of Hrushovski to Ziegler. Exercise 8: Hodges [1980a]. Pope [1982] showed that the weak choice axiom 'Every set can be linearly ordered' is not strong enough to prove that every abelian group has a divisible hull; it is not known whether the boolean prime ideal theorem suffices.

Section 12.4. The earliest form of Theorem 12.4.1 was due to Chang [1964] and Makkai [1964a]. Putting $K = L$, they showed it for some cardinal λ, not for every cardinal $\lambda \geqslant |L|$. Both originally assumed the generalised continuum hypothesis; this was removed by Vaught (see Chang's paper). Reyes [1970] proved the theorem in more cardinalities (for example with $\lambda = \omega$ for countable languages). Shelah [1971b] showed it for all $\lambda \geqslant |L|$. He also let L loose from K, which is trivial in the argument but important for applications, such as Corollary 12.4.3 which is from Pillay & Shelah [1985].

Exercise 1: Chang [1964] for some λ, Shelah [1971b] in general. Exercise 2: Shelah [1971b].

Section 12.5. The notion of relative categoricity appears first in Gaifman [1974a]. The notion of uniform reducibility is from Feferman [1974], which proves a weak version of Lemma 12.5.1 (but for some infinitary languages too; cf. also Barwise [1973]). The full lemma appears in Pillay & Shelah [1985], together with Lemma 12.5.2. Lemma 12.5.3 is from unpublished notes of Gaifman; Shelah [1986] contains generalisations. Theorem 12.5.4 is from Hodges, Hodkinson & Macpherson [1990], and independently Villemaire [1988]; see also the analysis in Pillay [1983b]. Lemma 12.5.5 is still unpublished, but see Shelah [1986] Chapter 2. Theorem 12.5.7 is from Hodges [1976a]. Theorem 12.5.8 is due independently to Gaifman [1974b] and Dale Myers (both unpublished). Theorem 12.5.9 is from Hodges, Hodkinson & Macpherson [1990].

Exercises 3–6: Pillay [1977]. Exercise 7: Villemaire [1990]. Exercise 9: Baĭzhanov [1990]. Exercises 11, 12: Gaifman [1974b].

APPENDIX: EXAMPLES

And thirteenthly . . ..

From a sermon of Meister Eckhardt.

This appendix is something of a ragbag. It assembles results on various interesting theories, from modules to linear orderings. The section on nilpotent groups proves a deep theorem of Alan Mekler, who died while this book was in proof; I dedicate the section to his memory. The section on groups mentions some themes which dominated model-theoretic research in the 1980s.

A.1 Modules

I discuss only left modules over a fixed ring with 1, so that the language and axioms of (2.21) in section 2.2 are appropriate. Also the results will all be concerned with the logical classification of modules. For example I say virtually nothing about the structure theory of pure injective modules, the Ziegler toplogy or representation types, three topics which have generated a quantity of recent research. Prest [1988] is full of up-to-date information.

Quantifier elimination

Most work on the model theory of modules begins with the Baur–Monk quantifier elimination theorem, Corollary A.1.2 below. The heart of the proof lies in Theorem A.1.1 (which is discussed by Gute & Reuter [1990]). I follow the argument of Monk [1975], though he stated it only for abelian groups.

Let L be a first-order language whose signature includes symbols $\cdot$, $^{-1}$, 1 as in the language of groups. We shall say that an L-structure A is **group-like** if $\cdot^A$, $^{-1A}$ and 1^A form a group on $\mathrm{dom}(A)$. (This is a slightly narrower definition than in section 5.7 above.) By a **basic formula** we shall mean a p.p. formula $\phi(x_0, \ldots, x_{n-1})$ of L such that $\phi(A^n)$ forms a subgroup of the product group A^n. By an **invariant sentence** we mean a sentence which expresses a fact of the form '$(G:H \cap G) \le m$' where G, H are subgroups of A defined by basic formulas, and $m < \omega$.

Theorem A.1.1. *Let $\phi(x_0, \ldots, x_{n-1}, y)$ be a boolean combination of basic formulas for the group-like L-structure A. Then there is a formula $\phi^*(x_0, \ldots, x_{n-1})$, depending only on L and not on A, which is a boolean combination of basic formulas and invariant sentences, and is equivalent in A to $\exists y \, \phi(x_0, \ldots, x_{n-1})$.*

Proof. We begin with some remarks on basic formulas $\psi(x_0, \ldots, x_n)$ for a fixed n – for brevity let us call them n-**basic**. First, a conjunction of n-basic formulas is again n-basic (after bringing the quantifiers to the front). Second, if $\psi(x_0, \ldots, x_n)$ is n-basic then $\exists x_n \, \psi$ is $(n-1)$-basic. Third, the formula $x_0 = 1 \land \ldots \land x_{n-1} = 1 \land x_n = x_n$ (write it σ_n) is n-basic; it defines the group $1 \times \ldots \times 1 \times A$. It follows that if ψ is n-basic, then so is $\psi \land \sigma_n$; so $\psi(1, \ldots, 1, x_0)$ is 0-basic. And fourth, if $a_0, \ldots, a_{n-1}$ are elements of A then $\{a_0\} \times \ldots \times \{a_{n-1}\} \times A$ is a right (and left) coset of $1 \times \ldots \times 1 \times A$ in A^{n+1}; so the formula $\psi(a_0, \ldots, a_{n-1}, x_n)$ defines either the empty set or a right coset of the subgroup defined by $\psi(1, \ldots, 1, x_n)$.

For the proof we shall take an arbitrary tuple $\bar{d}$ from A, and we shall show that $\exists y \, \phi(\bar{d}, y)$ is equivalent in A to a boolean combination of invariant sentences and formulas $\theta(\bar{d})$ with θ basic. This boolean combination will be written so as to cover all cases, so that it will be independent of the choice of A and $\bar{d}$.

Write the formula $\exists y \, \phi(\bar{x}, y)$ as

(1.1) $\exists y(\psi_0(\bar{x}, y) \land \ldots \land \psi_{m-1}(\bar{x}, y) \land \neg \chi_0(\bar{x}, y) \land \ldots$

           $\land \neg \chi_{n-1}(\bar{x}, y))$,

where $\psi_0, \ldots, \psi_{m-1}, \chi_0, \ldots, \chi_{n-1}$ are basic formulas. Since a conjunction of basic formulas can be reduced to a single one, there is no loss in replacing (1.1) by

(1.2) $\exists y(\psi(\bar{x}, y) \land \neg \chi_0(\bar{x}, y) \land \ldots \land \neg \chi_{n-1}(\bar{x}, y))$

where $\psi, \chi_0, \ldots, \chi_{n-1}$ are basic formulas. Replacing each χ_i by $\chi_i \land \psi$, we can assume that in (1.2), each formula χ_i implies ψ.

Let G be the subgroup $\psi(1, \ldots, 1, A)$ of A and for each $i < n$ let H_i be the subgroup $\chi_i(1, \ldots, 1, A)$ of A. By our assumption, each H_i is a subgroup of G. We have for any n-tuple $\bar{d}$ from A

(1.3) $A \vDash \neg \phi(\bar{d})$ iff $\psi(\bar{d}, A) \subseteq \bigcup_{i<n} \chi_i(\bar{d}, A)$.

Now by the remarks at the beginning of the proof. $\psi(\bar{d}, A)$ is either empty or a right coset of G, and each $\chi_i(\bar{d}, A)$ is either empty or a coset of H_i. We can distinguish the various possibilities by the conditions $\exists y\, \psi(\bar{d}, y)$, $\exists y\, \chi_i(\bar{d}, y)$. If $\psi(\bar{d}, A)$ is empty then the right-hand side of (1.3) is true trivially. If $\chi_i(\bar{d}, A)$ is empty then we can leave it out of the union of the right-hand side of (1.3). So with no loss we can assume henceforth that $\psi(\bar{d}, A)$ and all the sets $\chi_i(\bar{d}, A)$ are non-empty.

Choose elements $c, b'_0, \ldots, b'_{n-1}$ of A so that $\psi(\bar{d}, A) = Gc$ and for each $i < n, \chi_i(\bar{d}, A) = H_i b'_i$. Then the right-hand side of (1.3) is equivalent to the group-theoretic statement $G \subseteq \bigcup_{i<n}(H_i b'_i c^{-1})$. Writing b_i for $b'_i c^{-1}$, we finally reduce the right-hand side of (1.3) to

$$(1.4) \qquad\qquad G \subseteq \bigcup_{i<n} H_i b_i.$$

Here we invoke B. H. Neumann's lemma, Lemma 4.2.1. By that lemma, if (1.4) holds then it still holds if we delete $H_i b_i$ whenever $(G : H_i) > n!$. The statement '$(G : H_i) \leqslant n!$' can be written as an invariant sentence. So again without loss we can suppose that in (1.4), $(G : H_i) \leqslant n!$ for each subgroup H_i $(i < n)$. Put $H = \bigcap_{i<n} H_i$. If X is a union of right cosets of H, write $N(X)$ for the number of distinct cosets of H which lie in X.

By elementary combinatorics we have

$$(1.5) \quad G \subseteq \bigcup_{i<n} H_i b_i \Leftrightarrow N(G) \leqslant N\left(\bigcup_{i<n} H_i b_i\right)$$
$$\Leftrightarrow N(G) \leqslant \sum_{1 \leqslant k \leqslant n} (-1)^{k-1}\left\{\sum_{\substack{J \subseteq n, \\ |J|=k}} N\left(\bigcap_{i \in J} H_i b_i\right)\right\}.$$

Now $\bigcap_{i \in J} H_i b_i$ is either empty or a union of cosets of H, and in the latter case $N(\bigcap_{i \in J} H_i b_i)$ is equal to $(\bigcap_{i \in J} H_i : H)$. So by (1.5) we can reduce (1.4) to a boolean combination of statements of the forms

$$(1.6) \qquad\qquad \left(\bigcap_{i \in J} H_i : H\right) \leqslant j, \qquad \bigcap_{i \in J} H_i b_i \neq \varnothing.$$

The first of these statements translates at once into an invariant sentence. The second is equivalent to $\bigcap_{i \in J} H_i b'_i \neq \varnothing$, i.e. to $\exists y\, \bigwedge_{i \in J} \chi_i(\bar{d}, y)$, which is a basic condition on $\bar{d}$. This completes the proof. $\qquad\square$

Let R be a ring, L the language of left R-modules and A a left R-module. We apply Theorem A.1.1 to the additive group of A. The **Baur–Monk invariants** of A are the numbers $|\phi(A)/\psi(A) \cap \phi(A)|$ with $\phi(x)$ and $\psi(x)$ p.p. formulas of L; we write them $\mathrm{Inv}(\phi, \psi, A)$. These invariants are taken to be either finite numbers or ∞ (where ∞ includes all infinite cardinalities). An **invariant sentence** is a sentence which expresses that some Baur–Monk invariant has value $\leqslant m$ for a particular $m < \omega$.

Corollary A.1.2 (Baur–Monk quantifier elimination theorem, Baur [1976], L. Monk [1975]). *Let R be a ring and L the language of left R-modules. Then* *for every formula $\phi(\bar{x})$ of L there is a formula $\psi(\bar{x})$ which is a boolean combination of p.p. formulas and invariant sentences, and which is equivalent to ψ in all left R-modules.*

Proof. Let Φ be the set of formulas of L which are boolean combinations of p.p. formulas and invariant sentences. By Lemma 2.3.1 it suffices to show that for every formula $\psi(\bar{x}, y)$ in Φ, the formula $\exists y\, \psi$ is equivalent in A to some formula $\chi(\bar{x})$ in Φ. This follows at once from Theorem A.1.1, provided we show that for every p.p. formula $\phi(x_0, \ldots, x_{n-1})$ of L, $\phi(A^n)$ is a subgroup of the additive group of A.

We show it. Every atomic formula $\psi(x_0, \ldots, x_{n-1})$ of L is equivalent, modulo the theory of left R-modules, to an equation $\sum_{i<n} r_i x_i = 0$, where r_i are ring elements. So the formula ϕ can be read as saying

(1.7) there are elements $\bar{y}$ such that $M\bar{x} + N\bar{y} = 0$,

where M and N are matrices over R and $\bar{x}, \bar{y}$ are read as column matrices of module elements. If $A \vDash \phi(\bar{a})$ and $A \vDash \phi(\bar{b})$, then there are column matrices $\bar{c}, \bar{d}$ of module elements such that

(1.8) $$M\bar{a} + N\bar{c} = 0 = M\bar{b} + N\bar{d}.$$

Adding the two equations in (1.8), $M(\bar{a} + \bar{b}) + N(\bar{c} + \bar{d}) = 0$, and so $A \vDash \phi(\bar{a} + \bar{b})$. This shows that $\phi(A^k)$ is closed under $+$. By this and similar arguments, $\phi(A^n)$ is a subgroup of A^n. (It need not be a submodule unless R is commutative.) $\square$

In fact the result of Corollary A.1.2 works for any group-like structure A with the property that p.p. formulas $\phi(x_0, \ldots, x_{n-1})$ without parameters define subgroups of A^n. Examples are an abelian group with a family of subgroups distinguished by 1-ary relation symbols, and an abelian group with a function symbol for an endomorphism of the group.

Let A and B be left R-modules. We say that A is a **pure submodule** of B if $A \subseteq B$ and the inclusion map is a pure embedding (see section 2.5). If A is a pure submodule of B, then for every p.p. formula $\phi(\bar{x})$, $\phi(A^k) = \phi(B^k) \cap (\operatorname{dom} A)^k$.

Corollary A.1.3 *(Sabbagh [1971]. Let A and B be left R-modules; suppose that A is a pure submodule of B and $A \equiv B$. Then $A \preccurlyeq B$.*

Proof. A pure embedding preserves p.p. formulas (by Theorem 2.4.1) and their negations (since it is pure). $\square$

For vector spaces V we have already proved more than Corollary A.1.2: Th(V) has quantifier elimination (see Example 1 in section 8.4). By the

Baur–Monk theorem, this is equivalent to saying that in a vector space, every p.p. formula is equivalent to a quantifier-free formula. For what rings R do all left R-modules have this property? The next theorem gives the answer. (The equivalence (a) $\Leftrightarrow$ (b) is from Eklof & Sabbagh [1971]; the rest is by Hodges and is reported in Weispfenning [1985], who also provides a more effective proof.)

Theorem A.1.4. *The following conditions on a ring R are equivalent*
(a) *R is von Neumann regular.*
(b) *Every embedding between left R-modules is pure.*
(c) *Every p.p. formula $\phi(x_0, \ldots, x_{n-1})$ is equivalent, modulo the theory of left R-modules, to a finite conjunction of equations of the form $\sum_{i<n} r_i x_i = 0$.*
(d) *Every left R-module has quantifier elimination.*
(e) *Every two-generator left R-module has quantifier elimination.*

Proof. (a) $\Rightarrow$ (b): see Eklof & Sabbagh [1971] Prop. 3.25.

(b) $\Rightarrow$ (c). Write T for the theory of left R-modules. We show first that every p.p. formula $\phi(\bar{x})$ is equivalent modulo T to a positive quantifier-free formula. By Theorem 8.4.9 it suffices to show that if A and B are left R-modules, $\bar{a}$ is a sequence of elements of A and e: $\langle \bar{a} \rangle_A \to B$ is a homomorphism, then e extends to a homomorphism $f: A \to D$ where D is some pure extension of B. To show this we take D to be the pushout of A and B over $\langle \bar{a} \rangle_A$. By (b), B is pure in D.

Now let $\psi(\bar{x})$ be a positive quantifier-free formula which is equivalent to ϕ modulo T. We can write ψ as a disjunction of conjunctions of atomic formulas, $\psi_0 \vee \ldots \vee \psi_{k-1}$. Then $T \vdash \forall \bar{x}(\phi \to \psi_0 \vee \ldots \vee \psi_{k-1})$. But T and ϕ are both Horn, so by McKinsey's lemma (Corollary 9.1.7), $T \vdash \forall \bar{x}(\phi \to \psi_i)$ already for some ψ_i, proving (d).

(d) $\Rightarrow$ (e) is clear.

(e) $\Rightarrow$ (a). Consider $r \in R$ and the elements $a = (r, 0)$, $b = (0, 1r)$ of the left R-module $M = R \oplus Rr$. There is an isomorphism $f: \langle a \rangle_M \to \langle b \rangle_M$ taking a to b; so a and b satisfy the same quantifier-free formulas, and hence by (e) the same first-order formulas. Since $M \vDash \exists x\, rx = a$, it follows that $M \vDash \exists x\, rx = b$; in other words there is an element (s, tr) of M such that $(rs, rtr) = (0, r)$. Thus $r = rtr$ as required. $\square$

Theories of modules

The next few lemmas are useful for calculating Baur–Monk invariants $\text{Inv}(\phi, \psi, A)$ in specific cases (as for example abelian groups in section A.2 below).

Lemma A.1.5. *Let* $(A_i: i \in I)$ *be a non-empty family of left R-modules. Then for every Baur–Monk invariant* F,

(1.9) $$F\left(\prod_I A_i\right) = \sup \left\{ \prod_{i \in J} F(A_i) : J \text{ a finite subset of } I \right\}$$

$$= F\left(\bigoplus_I A_i\right).$$

(Here we write $\prod_{i<n} k_i$ for $k_0 \times \ldots \times k_{n-1}$.)

Proof. Let $\phi(x)$ and $\psi(x)$ be p.p. formulas of L. Then Lemma 9.1.4 shows that $\phi(\prod_I A_i) = \prod_I \phi(A_i)$, and the same applies to ψ. Hence by an abelian group calculation

(1.10) $$\phi\left(\prod_I A_i\right) / (\phi \wedge \psi)\left(\prod_I A_i\right) = \prod_I (\phi(A_i)/(\phi \wedge \psi)(A_i)).$$

This implies the first equation of the lemma, for the invariant $F = \text{Inv}(\phi, \psi, -)$. The calculation for $\oplus$ is similar. $\square$

Lemma A.1.6. *Let* $(A_i: i \in I)$ *and* $(B_i: i \in I)$ *be two non-empty families of left R-modules.*
   (a) *If* $A_i \equiv B_i$ *for each* $i \in I$, *then* $\bigoplus_I A_i \equiv \prod_I B_i$.
   (b) *If* $A_i \preccurlyeq B_i$ *for each* $i \in I$, *then* $\bigoplus_I A_i \preccurlyeq \prod_I B_i$.

Proof. (a) The previous lemma shows that $\bigoplus_I A_i$ and $\prod_I B_i$ have the same Baur–Monk invariants. So $\bigoplus_I A_i \equiv \prod_I B_i$ by Corollary A.1.2.
   (b) One checks easily that $\bigoplus_I A_i$ is a pure submodule of $\prod_I A_i$, and by Feferman–Vaught (Corollary 9.6.5), $\prod_I A_i \preccurlyeq \prod_I B_i$. Now apply Corollary A.1.3. $\square$

Lemma A.1.7. *Let* A *be a left R-module. Then the following are equivalent*:
(a) $A \equiv A \oplus A$;
(b) $A \equiv A^{(\omega)}$ *(the direct sum of countably many copies of* A*)*;
(c) *every Baur–Monk invariant of* A *is either* 1 *or* ∞.

Proof. Use Corollary 9.6.7. $\square$

Recently Villemaire [1992] added a comment to this lemma (cf. Felgner [1980a] for the case of abelian groups):

Theorem A.1.8. *Let* R *be a ring and* T *the complete theory of some left R-module. Then* T *is a Horn theory (up to equivalence) if and only if the class of models of* T *is closed under direct products.* $\square$

Lemma A.1.9. *Let A be a left R-module and B a pure submodule of A. Then for every Baur–Monk invariant F, $F(A) = F(B \oplus (A/B)) = F(B) \cdot F(A/B)$.*

Proof. Form the structure $D = (A, B, A/B)$ and let D' be a $|R|^+$-saturated elementary extension of D. Then D is of the form $(A', B', A'/B')$ where A' is a left R-module and B' is pure in A'. Since B' is a relativised reduct of D', it is $|R|^+$-saturated (Theorem 10.1.9) and hence atomic compact (Theorem 10.7.3), so that by Theorem 10.7.1 the short exact sequence $B' \rightarrowtail A' \twoheadrightarrow A'/B'$ splits. Thus we have a commuting diagram

(1.11)

$$
\begin{array}{ccccc}
B' & \longrightarrow & B' \oplus C' & \longrightarrow & C' \\
\uparrow & & \uparrow & & \uparrow \\
& & & & \\
B & \longrightarrow & A & \longrightarrow & A/B
\end{array}
$$

where the vertical maps are elementary embeddings (by Corollary 5.1.2) and the horizontal maps are the obvious ones. Now apply Lemma A.1.5. (I thank Gabriel Sabbagh for this proof.) ☐

I note here a few more results on definability of classes of modules. Theorem 9.5.11 showed that the class of injective left R-modules is first-order axiomatisable if and only if R is left noetherian. There are further results of this kind in Chapter 6 of Jensen & Lenzing [1989].

The next result is essentially in Prest's Ph.D. thesis; see Prest [1979]. The proof of (c) uses torsion theory.

Theorem A.1.10. *Let **K** be a class of left R-modules and L the signature of left R-modules.*

*(a) **K** is a variety if and only if there is a two-sided ideal I of R such that **K** is axiomatised by the theory of left R-modules together with the set of sentences $\forall x\, rx = 0 \ (r \in I)$.*

*(b) We say that **K** is **closed under extensions** if **K** is closed under isomorphic copies and for any pair $A \subseteq B$ of left R-modules, if A and B/A are in **K** then B is in **K**. Then **K** is a variety which is closed under extensions if and only if: there is a two-sided ideal I of R such that **K** is axiomatised by the theory of left R-modules together with the sentences $\forall x\, rx = 0 \ (r \in I)$, and $I^2 = I$.*

(c) If I is an ideal of R then we can read '$Ix = 0$' as the conjunction $\bigwedge \{rx = 0;\, r \in I\}$. The following are equivalent.

    *(i) **K** is closed under products, submodules and extensions.*

    *(ii) For some set S of left ideals of I of R, **K** is axiomatised by the following sentences of $L_{\infty\omega}$: $\forall x(Ix = 0 \to x = 0) \ (I \in S)$.* ☐

Tyukavkin [1982] proved the following

Theorem A.1.11. *Let R be a ring. Then R is an infinite simple von Neumann regular ring if and only if the theory of all nonzero left R-modules is model-complete.* $\qquad\square$

Eklof & Sabbagh [1971] showed the following

Theorem A.1.12. *Let R be a ring. Then the theory of left R-modules is companionable if and only if R is coherent.* $\qquad\square$

Stability classification of modules

In recent years stability theory has made the running in the model theory of modules. One good reason for this is the following result of Fisher [1972].

Theorem A.1.13. *Every module is stable.*

Proof. Let A be a left R-module and L the language of left R-modules. By Lemma 6.7.6, Exercise 6.7.9 and the Baur–Monk quantifier elimination theorem, it suffices to show that if $\phi(\bar{x}, \bar{y})$ is a p.p. formula of L then ϕ is stable. In fact we shall prove something stronger: ϕ has ladder index $\leqslant 2$. As in (1.7) above, we can read $\phi(\bar{x}, \bar{y})$ as

$$(1.12) \qquad \text{there is } \bar{z} \text{ such that } K\bar{x} + M\bar{y} + N\bar{z} = 0,$$

where K, M, N are matrices over R, 0 is a column vector of zeros, and $\bar{x}$ $\bar{y}$, $\bar{z}$ are read as column vectors.

Suppose $A \vDash \phi(\bar{a}_i, \bar{b}_j)$ whenever $i \leqslant j < 2$. Then by (1.12) there are tuples $\bar{c}_{00}$, $\bar{c}_{01}$, $\bar{c}_{11}$ in A such that

$$(1.13) \qquad\qquad K\bar{a}_0 + M\bar{b}_0 + N\bar{c}_{00} = 0,$$

$$(1.14) \qquad\qquad K\bar{a}_0 + M\bar{b}_1 + N\bar{c}_{01} = 0,$$

$$(1.15) \qquad\qquad K\bar{a}_1 + M\bar{b}_1 + N\bar{c}_{11} = 0.$$

Adding (1.13) to (1.15) and subtracting (1.14), we get

$$(1.16) \qquad K\bar{a}_1 + M\bar{b}_0 + N(\bar{c}_{00} - \bar{c}_{01} + \bar{c}_{11}) = 0,$$

proving that $A \vDash \phi(\bar{a}_1, \bar{a}_0)$. $\qquad\square$

We turn to totally transcendental modules. The equivalence between (b), (c) and (d) in Theorem A.1.14 below was proved by Gruson & Jensen [1976] and Zimmermann [1977]; Garavaglia [1979] supplied (a). These modules have a nice decomposition theory, directly generalising the classical decomposition theory for injective modules over a noetherian ring (e.g. Sharpe & Vámos [1972] Chapter 4). Garavaglia [1980] gives a largely model-theoretic proof of this decomposition, and his argument is a good illustration of the general point that stable theories tend to have nice structure theories. The structure theory for totally transcendental modules is now a special case of the elegant structure theory for pure injective modules. This theory has a long history, culminating in the model-theoretic treatment by Ziegler [1984]; cf. also Facchini [1985].

Theorem A.1.14. *The following conditions on an infinite left R-module A are equivalent, where L is the language of left R-modules.*
(a) *A is totally transcendental.*
(b) *There are no infinite descending chains of subgroups of A defined by p.p. formulas of L.*
(c) *Every direct sum of copies of A is pure injective.*
(d) *There is a cardinal λ such that every power A^I is a direct sum of left R-modules with $\leq \lambda$ generators.*

Proof. The model-theoretic part is (a) $\Leftrightarrow$ (b). The implication (a) $\Rightarrow$ (b) is by Theorem 5.7.2. In the other direction, suppose for contradiction that (a) fails but (b) holds. Thus A is not totally transcendental; we can suppose without loss that A is ω-saturated. Note first that (b) implies there are no infinite descending chains of sets $\phi(A, \bar{a})$ with ϕ a p.p. formula. This is because $\phi(A, \bar{a})$ is either empty or a coset of $\phi(A, 0, \ldots, 0)$; so if $\phi(A, \bar{a}) \supset \psi(A, \bar{b}) \neq \varnothing$ where both ϕ and ψ are p.p. formulas, then $\phi(A, 0, \ldots, 0) \supset \psi(A, 0, \ldots, 0)$.

It follows that there is some p.p. formula $\phi(x, \bar{a})$ of L with Morley rank ∞, such that for all p.p. formulas $\psi(x, \bar{b})$ of L with Morley rank ∞, $\psi(A, \bar{b}) \subseteq \phi(A, \bar{a})$ implies $\psi(A, \bar{b}) = \phi(A, \bar{a})$. Playing the Cantor–Bendixson game on $V = \phi(A, \bar{a})$ and using the Baur–Monk quantifier elimination, we find p.p. definable subsets $X, Y_0, \ldots, Y_{k-1}, X', Y_0', \ldots, Y_{l-1}'$ of V such that $X \cap (V \backslash Y_0) \cap \ldots \cap (V \backslash Y_{k-1})$ and $X' \cap (V \backslash Y_0') \cap \ldots \cap (V \backslash Y_{l-1}')$ are disjoint subsets of V of infinite Morley rank. Choose k and then l as small as possible. Since $X \subseteq V$ and X has infinite Morley rank, $X = V$ by choice of ϕ; similarly $X' = V$. Since the sets W, W' are disjoint, $V \subseteq \bigcup_i Y_i \cup \bigcup_j Y_j'$, so that at least one set Y_i or Y_j' has infinite Morley rank; say it is Y_0. By the same argument as before, $Y_0 = V$ and we have a contradiction to the choice of k or l. $\qquad \square$

The next theorem was assembled by Rothmaler [1983]; the equivalence of (c) and (d) is from Shelah [1976b].

Theorem A.1.15. *For any ring R the following are equivalent.*
(a) *The class of pure injective left R-modules is first-order axiomatisable.*
(b) *Every left R-module is totally transcendental.*

(c) *Every left R-module is superstable.*

(d) *There is a cardinal λ such that every left R-module is a direct sum of left R-modules with at most λ generators.*   □

In the same vein, we have the following.

Theorem A.1.16 *(Prest [1984]). For any ring R the following are equivalent.*

(a) *Every infinite left R-module has finite Morley rank.*

(b) *R has finite representation type* (as defined e.g. in Curtis & Reiner [1962]).   □

Unfortunately no good descriptions are known for $|R|^+$-categorical modules. But Baur [1975] gave a good characterisation of ω-categorical modules over a finite or countable ring.

Theorem A.1.17. *An R-module A is ω-categorical if and only if A is a direct sum $B \oplus C_0 \oplus \ldots \oplus C_{n-1}$ where B is a finite R-module and each C_i is a direct sum of infinitely many copies of a finite indecomposable pure injective R-module.*   □

To this Pillay [1984] added the following.

Theorem A.1.18. *Let R be a countable ring and T the complete theory of some infinite R-module. Then the number of countable models of T, counting up to isomorphism, is either 1 or infinite.*   □

In this last theorem, we know that 'infinite' can at least be ω or 2^ω. But in spite of the best efforts of Ziegler [1984], the possibility of ω_1 is still open. In other words, Vaught's conjecture for modules is unsolved. Very recently Buechler [1990] proved Vaught's conjecture for unidimensional theories by reducing to the case of modules with some additional relation symbols for subgroups.

If S is a monoid (a set with an associative multiplication and two-sided identity 1), we define an S-**system** to be a set X together with a left action of S on X (thus if $s, t \in S$ and $x \in X$, we have $s(t(x)) = (st)(x)$ and $1(x) = x$). So S-systems are the analogue of left R-modules for a ring R, when one replaces R by S. Some of the results of this section carry over to S-systems – for example the criterion for companionability, Theorem A.1.12; see Gould [1987]. Some don't – for example S-systems can be unstable.

A.2 Abelian groups

For the whole of this section, L is the first-order language of abelian groups and T_{ab} is the theory of abelian groups, (2.19) in section 2.2. Since abelian groups are $\mathbb{Z}$-modules, several of the results of the previous section carry over to them at once. With the Baur–Monk quantifier elimination in mind, we start by asking what a p.p. formula can express in an abelian group.

Lemma A.2.1 (Prüfer). *Every p.p. formula $\phi(\bar{x})$ of L is equivalent modulo T_{ab} to a finite conjunction of formulas of the forms '$t(\bar{x}) = 0$' where t is a term, and '$p^n | t(\bar{x})$' where t is a term, p a prime and n a positive integer. Furthermore '$p^n | t(x)$' (with a single variable x) is equivalent modulo T_{ab} either to $0 = 0$ or to a formula '$p^n | p^m x$' with $0 \leqslant m < n$.*

Proof. As in (1.7) above, $\phi(\bar{x})$ can be written as
$$(2.1) \qquad \text{There is } \bar{y} \text{ such that } M\bar{x} = N\bar{y}$$
where M and N are matrices of integers and $\bar{x}$, $\bar{y}$ are read as column matrices. Now there are invertible matrices P and Q of integers such that PNQ is a diagonal matrix (see Cohn [1974] p. 279, Jacobson [1974] p. 176). So (2.1) is equivalent to
$$(2.2) \qquad \text{There is } \bar{z} \text{ such that } PM\bar{x} = PNQ\bar{z}.$$
But (2.2) is equivalent to a conjunction of formulas of the form $n | t(\bar{x})$, where n are the numbers on the main diagonal of PNQ. Clearly the formula $0 | t(\bar{x})$ is equivalent to $t(\bar{x}) = 0$, and $1 | t(\bar{x})$ is equivalent to $0 = 0$. The rest is an exercise in elementary number theory. $\qquad\square$

Theorem A.2.2 (Szmielew's definability theorem, Szmielew [1955]). *Let A be an abelian group and R an n-ary relation on $\mathrm{dom}(A)$ which is first-order definable without parameters. Then R is a boolean combination of relations of the forms $\phi(A^n)$ where $\phi(\bar{x})$ is either $t(\bar{x}) = 0$ or $p^m | t(\bar{x})$ for some term t, prime p and positive integer m.*

Proof. By the lemma and the Baur–Monk theorem, Corollary A.1.2. $\qquad\square$

The meaning of the invariant sentences

If A is an abelian group, we write $t(A)$ for the torsion subgroup of A. Then $t(A)$ is a direct sum of p-groups; we write A_p for the p-component of A, so that $t(A) = \bigoplus_{p \text{ prime}} A_p$. We write $\mathbb{Z}(p^k)$ for the cyclic group of order p^k, and $\mathbb{Z}(p^\infty)$ for the Prüfer p-group. We write $\mathbb{Z}_{(p)}$ for the localisation of the integers $\mathbb{Z}$ at the prime p, i.e. the additive group of all rational numbers with denominator not divisible by p. We write $\mathbb{Q}$ for the additive group of all rationals. We write $A^{(\kappa)}$ for the direct sum of κ copies of A. (See Fuchs [1970] for background on abelian groups.)

By a **Szmielew group** we mean a finite or countable abelian group of the form

$$(2.3) \qquad \bigoplus_{p \text{ prime}} \left[\bigoplus_{n>0} \mathbb{Z}(p^n)^{(\kappa_{p,n-1})} \oplus \mathbb{Z}(p^\infty)^{(\lambda_p)} \oplus \mathbb{Z}_{(p)}^{(\mu_p)} \right] \oplus \mathbb{Q}^{(\nu)},$$

where $\kappa_{p,n-1}, \lambda_p, \mu_p, \nu$ are cardinals $\leq \omega$.

Lemma A.2.3. *Every abelian group is elementarily equivalent to a Szmielew group.*

Proof. By the downward Löwenheim–Skolem theorem (Corollary 3.1.5) it suffices to show this for an abelian group A which is at most countable. We shall find a Szmielew group B which has the same Baur–Monk invariants as A; then the Baur–Monk theorem (Corollary A.1.2) implies that $A \equiv B$.

By Lemma A.1.9, if A has a pure subgroup C then A has the same Baur–Monk invariants as $C \oplus A/C$. By Lemma A.1.6, the Baur–Monk invariants of a direct sum are determined by those of the summands.

Thus $t(A)$ is a pure subgroup of A, and so A has the same invariants as $t(A) \oplus A/t(A)$. Each primary summand A_p of $t(A)$ has a pure subgroup A'_p which is a direct sum of cyclic subgroups of p-power orders, such that A_p/A'_p is divisible and hence a direct sum of copies of $\mathbb{Z}(p^\infty)$ (see Fuchs [1970] Theorem 32.3). This takes care of the torsion part of (2.3).

As for the torsion-free part $A/t(A)$, we can first decompose it into $R \oplus \mathbb{Q}^{(\nu)}$ where R is reduced (i.e. has no non-trivial divisible subgroup; see Fuchs [1970] Theorem 21.3). It remains only to show that a torsion-free reduced abelian group R has the same Baur–Monk invariants as some direct sum of groups of the form $\mathbb{Z}_{(p)}$.

In a torsion-free group, if $n \neq 0$ and $na = nb$ then $a = b$; similarly $nm|na$ implies $m|a$ when $n \neq 0$. Put against Lemma A.2.1, this implies that in R, every p.p. formula $\phi(x)$ of L is equivalent to a conjunction of formulas of the form $p^n|x$, and hence to a single formula $k|x$. The set of elements of an abelian group A which are divisible by k is written kA. In this notation every Baur–Monk invariant of R is of the form $|jA/kA|$ for some integers j, k with $j|k$. We can write a sequence of integers $j = j_0, j_1, \ldots, j_{i-1}, j_i = k$ so that each j_{h+1} is of the form $p_h j_h$ for some prime p_h. Then $|jR/kR|$ is the product of the cardinals $|j_h R/j_{h+1}R| = |j_h R/p_h j_h R| = |R/p_h R|$.

It follows that the Baur–Monk invariants of R are determined by the cardinals $|R/pR|$ with p prime. Each such cardinal is either 1, ∞ or a positive power of p.

Now each $\mathbb{Z}_{(p)}$ is a reduced torsion-free abelian group, $|\mathbb{Z}_{(p)}/p\mathbb{Z}_{(p)}| = p$ and $|\mathbb{Z}_{(p)}/q\mathbb{Z}_{(p)}| = 1$ for every prime $q \neq p$. Hence, invoking Lemma A.1.5, we can assemble a reduced torsion-free abelian group with the same Baur–Monk invariants as R by taking a direct sum of groups of the form $\mathbb{Z}_{(p)}$, as in (2.3). $\square$

One can have two Szmielew groups which are elementarily equivalent but not isomorphic, as the next lemma shows. We say that an abelian group A has **bounded exponent** if there is some $n < \omega$ such that $nA = \{0\}$; otherwise A has **unbounded exponent**.

Lemma A.2.4. *For any abelian group A the following are equivalent.*
(a) *A has unbounded exponent.*
(b) *$A \equiv A \oplus \mathbb{Q}$.*
(c) *$A \equiv A \oplus \mathbb{Q}^{(\omega)}$.*

Proof. It is clear that (b) and (c) each imply (a). Assuming (a), we shall prove that for some abelian group B, $A \equiv B \oplus \mathbb{Q}^{(\omega)}$. Then by Theorem 3.3.10 (or the Feferman–Vaught theorem), $A \oplus \mathbb{Q} \equiv B \oplus \mathbb{Q}^{(\omega)} \oplus \mathbb{Q} \cong B \oplus \mathbb{Q}^{(\omega)} \equiv A$, which proves (b), and a similar argument gives (c).

By diagram arguments (Lemmas 1.4.2 and 2.5.3), if eldiag $(A) \cup \text{diag } \mathbb{Q}^{(\omega)}$ has a model, then some elementary extension of A has a subgroup isomorphic to $\mathbb{Q}^{(\omega)}$ and hence is of the form $B \oplus \mathbb{Q}^{(\omega)}$ (see Fuchs [1970] Theorem 21.2). So it suffices to show that eldiag $(A) \cup \text{diag } \mathbb{Q}^{(\omega)}$ has a model.

Suppose for contradiction that this theory has no model. Then by the compactness theorem and the structure theorem for finitely generated abelian groups (Fuchs [1970] Theorem 15.5), there are a finite direct sum $\bigoplus_{i<n} \Theta a_i$ of copies of the integers Θ and a finite set Δ of sentences from diag $\bigoplus_{i<n} \Theta a_i$ such that eldiag $(A) \cup \Delta$ has no model. We can suppose that the only parameters which appear in Δ are $a_0, \ldots, a_{n-1}$; then any equation in Δ is a consequence of T_{ab}. Hence Δ is equivalent to a finite set of sentences of the form $k_0 a_0 + \ldots + k_{n-1} a_{n-1} \neq 0$, where $k_0, \ldots, k_{n-1}$ are integers which are not all zero. But since A has unbounded exponent, we can find an element b of A of high enough order, so that we can make all these finitely many inequations true by taking $a_0, \ldots, a_{n-1}$ to be suitable multiples of b. Thus eldiag $(A) \cup \Delta$ does have a model; contradiction. $\square$

When the unboundedness is concentrated in the reduced part of some A_p, then we can alter the group in other ways without changing its theory. If A is an abelian group and p a prime, we say that A has **bounded p-length** if there is a finite upper bound on the orders of those elements of A_p which are not divisible by p. Otherwise we say that A has **unbounded p-length**.

Lemma A.2.5. *If p is a prime and A an abelian group of unbounded p-length, then $A \equiv A \oplus \mathbb{Z}(p^\infty)^{(\omega)} \equiv A \oplus \mathbb{Z}_{(p)}^{(\omega)}$.*

Proof. The argument to show that $A \equiv A \oplus \mathbb{Z}(p^\infty)^{(\omega)}$ is very much the same as the proof of Lemma A.2.4, and I leave the details to the reader.

To show that $A \equiv A \oplus \mathbb{Z}_{(p)}^{(\omega)}$, first consider the theory

(2.4) $\text{Th}(A) \cup \{p \nmid c_i : i < \omega\} \cup \left\{ \sum_{i < k} n_i c_i \neq 0: k < \omega \text{ and} \right.$

$$n_0, \ldots, n_{k-1} \text{ are integers not all zero} \Big\}.$$

Since A has unbounded p-length, a compactness argument shows that (2.4) has a model C'; put $C = C' \lfloor L$. Then as in the proof of Lemma A.2.3, $A \equiv C \equiv t(C) \oplus C/t(C)$. Put $B = C/t(C)$. By (2.4), B/pB is infinite. It follows that if we replace B by an elementarily equivalent Szmielew group B_1 as in the proof of Lemma A.2.3, the direct summand $\mathbb{Z}_{(p)}$ will occur infinitely often in B_1, so that $B_1 \cong B_1 \oplus \mathbb{Z}_{(p)}^{(\omega)}$. Then by Theorem 3.3.10 we infer that $A \equiv A \oplus \mathbb{Z}_{(p)}^{(\omega)}$. $\square$

By these two lemmas, the Szmielew group elementarily equivalent to a given abelian group is not necessarily uniquely determined. Let us say that the group (2.3) is a **strict Szmielew group** if (1) the cardinal v is either 0 or ω, (2) if the part preceding '$\oplus \mathbb{Q}^{(v)}$' has unbounded exponent, then $v = 0$, and (3) for each prime p, if the group has unbounded p-length then $\lambda_p = \mu_p = 0$.

Lemmas A.2.3–A.2.5 imply that every abelian group is elementarily equivalent to some strict Szmielew group. We shall describe a particular set of Baur–Monk invariants which are enough to determine a strict Szmielew group up to isomorphism. It will follow that these invariants determine any abelian group up to elementary equivalence.

Szmielew invariants

For any abelian group A and prime p, we write $A[p]$ for the subgroup $\{a: pa = 0\}$. Note that $(p^k A)[p]$, or $p^k A[p]$ as we shall write it, is $\phi(A)$ where ϕ is the p.p. formula $px = 0 \wedge p^k | x$.

We define the **Szmielew invariants** of an abelian group A as follows:

(2.5) $U(p, n; A) = |p^n A[p]/p^{n+1} A[p]|$ (p prime, $n \geq 0$),

(2.6) $D(p, n; A) = |p^n A[p]|$ (p prime, $n \geq 0$),

(2.7) $\text{Tf}(p, n; A) = |p^n A/p^{n+1} A|$ (p prime, $n \geq 0$),

(2.8) $\text{Exp}(p, n; A) = |p^n A|$ (p prime, $n \geq 0$).

Here U is for Ulm (since the invariants (2.5) are just the Ulm invariants; see Fuchs [1970] p. 154), D is for Divisible, Tf for Torsion-free and Exp for Exponent. It is clear that all of these invariants are in fact Baur–Monk invariants. Hence each one of them has corresponding invariant sentences; we call these the **Szmielew invariant sentences**. By judicious choice of short exact sequences one can prove some relationships between the Szmielew invariants:

for example $D(p, n; A) = U(p, n; A) \cdot D(p, n + 1; A)$ and $\mathrm{Tf}(p, n; A) = U(p, n; A) \cdot \mathrm{Tf}(p, n + 1; A)$.

Lemma A.2.6. *A strict Szmielew group is determined up to isomorphism by its Szmielew invariants.*

Proof. Let A be a strict Szmielew group. We write A as in (2.3). Table A1 shows the Szmielew invariants of all the direct summands of A. I leave it to the reader to check the table. In the calculations below we reckon that $n^\omega = \infty$ for every positive integer n.

By the table and Lemmas A.1.5 and A.1.6 we have

(2.9) $U(p, n; A) = p^{\kappa_{p,n}}$ for each prime p, and $n \geqslant 0$.

So we can read off the reduced torsion part of A from its U invariants. Note that A has unbounded p-length if and only if $\kappa_{p,n} \neq 0$ for infinitely many n, or equivalently if and only if $U(p, n; A) \neq 1$ for infinitely many n.

From the table again we can read off

(2.10) $D(p, n; A) = p^{\Sigma_{k \geqslant n} \kappa_{p,k}} \cdot p^{\lambda_p}$ for each prime p, and $n \geqslant 0$.

If A has unbounded p-length then $\lambda_p = 0$ since A is strict Szmielew. But if A has bounded p-length then for all large enough n the first factor of (2.10) reduces to 1, so that we can read off λ_p. Similarly the table tells us

(2.11) $\mathrm{Tf}(p, n; A) = p^{\Sigma_{k \geqslant n} \kappa_{p,k}} \cdot p^{\mu_p}$ for each prime p, and $n \geqslant 0$.

By the same argument as for the D invariants, these invariants tell us what μ_p must be.

There only remains the invariant v. Let B be the part of (2.3) to the left of '$\oplus \mathbb{Q}^{(v)}$', so that $A = B \oplus \mathbb{Q}^{(v)}$. From the invariants above, we already know whether or not B has bounded exponent. If B has unbounded exponent then $v = 0$ since A is strict Szmielew. On the other hand if B has bounded exponent, then from their definition (2.8) it is clear that the invariants $\mathrm{Exp}(p, n; A)$ tell us whether or not A has bounded exponent. If it has then $v = 0$; otherwise $v = \infty$. $\qquad\square$

The main results of Szmielew [1955] follow almost at once.

Theorem A.2.7 (Szmielew's elementary classification theorem). *For every abelian group A, $\mathrm{Th}(A)$ is determined by the Szmielew invariants of A. Equivalently, every sentence of the first-order language of abelian groups is equivalent, modulo the theory of abelian groups, to a boolean combination of Szmielew invariant sentences.*

Proof. By Lemmas A.2.3–A.2.5 there is a strict Szmielew group B which is elementarily equivalent to A. By Lemma A.2.6, $\mathrm{Th}(B)$ is determined by the

Table A1

	$\mathbb{Z}(q^k)$	$\mathbb{Z}(q^\infty)$	$\mathbb{Z}(q)$	$\mathbb{Q}$
$U(p, n; -)$	$\begin{cases} p & (p = q,\ n = k-1); \\ 1 & (\text{otherwise}) \end{cases}$	1	1	1
$D(p, n; -)$	$\begin{cases} p & (p = q,\ n < k); \\ 1 & (\text{otherwise}) \end{cases}$	$\begin{cases} p & (p = q); \\ 1 & (\text{otherwise}) \end{cases}$	1	1
$\text{Tf}(p, n; -)$	$\begin{cases} p & (p = q,\ n < k); \\ 1 & (\text{otherwise}) \end{cases}$	1	$\begin{cases} p & (p = q); \\ 1 & (\text{otherwise}) \end{cases}$	1
$\text{Exp}(p, n; -)$	$\begin{cases} p^{k-n} & (p = q,\ n < k); \\ q^k & (\text{otherwise}) \end{cases}$	∞	∞	∞

Szmielew invariants of B, which are the same as those of A since they are defined by the Szmielew invariant sentences. The second half of the theorem follows from the first by the compactness theorem (Exercise 6.1.2). □

Eklof & Fisher [1972] give a quite different proof of Theorem A.2.7. They show that if A is a pure injective abelian group, then the Szmielew invariants determine enough of the structure of A to fix $\mathrm{Th}(A)$; in fact one can read this off from the known structure theory of pure injective abelian groups, as in Fuchs [1970] §40. Using Theorem 10.7.3, every abelian group has a pure injective elementary extension.

Theorem A.2.8 *(Szmielew's decidability theorem)*. *The first-order theory of abelian groups is decidable.*

Proof. Let ϕ be a sentence of λ; we must determine whether ϕ is true in all abelian groups. There are two steps. First, by Theorem A.2.7 there is a boolean combination ψ of Szmielew invariant sentences such that $T_{\mathrm{ab}} \vdash \phi \leftrightarrow \psi$. Since T_{ab} is recursively axiomatised, the set of its consequences is recursively enumerable by Corollary 6.1.3, so that we can effectively find a suitable ψ. (This is a lazy argument. By using the proof of Theorem A.1.1 and being more careful about reducing Baur–Monk invariants to Szmielew invariants we can find ψ as a primitive recursive function of ϕ.)

The second step is to test whether ψ is true in every abelian group. It is enough if we can tell, given a finite set of Szmielew invariant sentences and negations of such sentences, whether they can all be true together in some Szmielew group. But we can read off that information from Table A1. □

Some classes of abelian groups

First a result on non-axiomatisability.

Theorem A.2.9 *(Kogalovskiĭ [1961], Sabbagh [1969], Adler [1978])*. *The class of all groups which are multiplicative groups of fields is not first-order axiomatisable.*

Proof. Consider an algebraically closed field K of prime characteristic and infinite transcendence degree. Since K is algebraically closed, its multiplicative group $K^\times$ is divisible. The torsion part of $K^\times$ is the multiplicative group A of the subfield of algebraic elements, and so $K^\times = A \oplus \mathbb{Q}^{(\omega)}$. Now $\mathbb{Q} \preccurlyeq \mathbb{Q}^{(\omega)}$, for example by Lindström's test (Theorem 8.3.4), and so $A \oplus \mathbb{Q} \preccurlyeq A \oplus \mathbb{Q}^{(\omega)}$ by Lemma A.1.6. So it suffices to show that there is no field F with multiplicative group $F^\times$ isomorphic to $A \oplus \mathbb{Q}$.

Let F be such a field. Let t be an element of F corresponding to 1 in $\mathbb{Q}$. Certainly F has prime characteristic; for if it has characteristic 0, then there must be some integers m, n such that $2^m = 3^n$ in F, which is impossible by the fundamental theorem of arithmetic. So t is transcendental, and thus $t + 1$ is expressible as $a \cdot t^q$ for some algebraic a and some rational q. But then t is algebraic, which is a contradiction. $\square$

We turn to the stability classification. A technical lemma will help us.

Lemma A.2.10. *Let A be any abelian group.*
(a) *If A is infinite, the number of complete 1-types over* $\mathrm{dom}(A)$ *with respect to A is at least* $|A| \cdot \prod_{p \text{ prime}} \prod_{n < \omega} |p^n A / p^{n+1} A|$.
(b) *For all primes p and all $n < \omega$,* $\mathrm{Tf}(p, n; A) = U(p, n; A) \cdot \mathrm{Tf}(p, n + 1; A)$.

Proof. (a) There are $|A|$ types of elements of A; so it suffices to show that the number of types is at least $\prod_{p \text{ prime}} \prod_{n < \omega} |p^n A / p^{n+1} A|$. Let R be the set of all maps ρ which take each prime p and each $n < \omega$ to an element $\rho(p, n)$ of $A/p^n A$ in such a way that $\rho(p, n + 1)$ maps to $\rho(p, n)$ under the natural homomorphism $A/p^{n+1} A \to A/p^n A$. Then $|R|$ is clearly $\prod_{p \text{ prime}} \prod_{n < \omega} |p^n A / p^{n+1} A|$. We shall find a distinct type corresponding to each map ρ in R, as follows.

For each prime p and each n, let $\phi_{\rho,p,n}$ be the formula $p^n | (x - a)$ where a is some element of A such that $\rho(p, n) = a + p^n A$; let $\Phi_\rho(x)$ be the set of all these formulas $\phi_{\rho,p,n}$. To show that $\Phi_\rho(x)$ is a type, we have to find, for a finite set $p_0, \ldots, p_{i-1}$ of distinct primes together with numbers $n_0, \ldots, n_{i-1} < \omega$ and elements $a_0, \ldots, a_{i-1}$ of A, an element b of A such that $p_i^{n_i} | (b - a_i)$ for each i. Euclid's algorithm obliges. Thus each Φ_ρ is a type, and we can extend it to a complete type Ψ_ρ over $\mathrm{dom}(A)$. If $\rho \neq \rho'$ it is not hard to see that $\Psi_\rho \neq \Psi_{\rho'}$.

(b) is by the short exact sequence

(2.12) $p^n A[p]/p^{n+1} A[p] \rightarrowtail p^n A/p^{n+1} A \twoheadrightarrow p^{n+1} A/p^{n+2} A,$

where the second map is multiplication by p. $\square$

Theorem A.2.11 *(Rogers [1977] for (b), Macintyre [1971a] for (c)).* *Let A be an infinite abelian group. Then the following are equivalent.*
(a) *A is totally transcendental.*
(b) *There are just finitely many pairs p, n such that $\mathrm{Tf}(p, n; A) \neq 1$.*
(c) *A is the direct sum of a divisible group and a group of bounded exponent.*

Proof. (a) $\Rightarrow$ (b). Suppose there are infinitely many pairs p, n with $\mathrm{Tf}(p, n; A) \neq 1$, and B is a countable group elementarily equivalent to A.

Then $|p^n B/p^{n+1} B| = \mathrm{Tf}(p, n; B)$ for each such pair p, n, and so Lemma A.2.10 tells us that there are 2^ω complete types over $\mathrm{dom}(B)$. Hence B is not ω-stable, and Theorem 6.7.5 implies that A is not totally transcendental.

(b) $\Rightarrow$ (c). If A is a Szmielew group, then we can read this off from (b) and Table A1; say $A = C \oplus D$ where C has exponent k and D is divisible. Then for any group A' elementarily equivalent to A, kA' is divisible and hence $A' = C' \oplus kA'$ for some group C'. Since $kC' \subseteq kA'$, we have $kC' = \{0\}$ and hence A' is as required.

(c) $\Rightarrow$ (a). Assume (c). By the proof of (b) $\Rightarrow$ (c), any elementary extension of A is also a direct sum of a divisible group and a group of bounded exponent. Hence it suffices if we prove that if A' is a countable elementary subgroup of A, then the elements of A realise at most countably many 1-types over $\mathrm{dom}(A')$. We have $A = C \oplus \bigoplus_p D_p \oplus E$ where C has bounded exponent, each D_p is a direct sum of copies of $\mathbb{Z}(p^\infty)$ and E is a direct sum of copies of $\mathbb{Q}$. There is no harm in extending A' within A; so by an elementary chain argument we can suppose that $A' = C' \oplus \bigoplus_p D'_p \oplus E'$ where $C' \subseteq C$, each $D'_p \subseteq D_p$ and $E' \subseteq E$. These subgroups must then be direct summands. We reach (a) by counting orbits of $\mathrm{Aut}(A)_{(\mathrm{dom}\, A')}$. $\qquad\square$

Theorem A.2.12 (Macintyre [1971a]). *Let A be an infinite abelian group. Then the following are equivalent.*

(a) *A is uncountably categorical.*

(b) *$A = B \oplus C$ where B is finite and C is of one of the two following forms:*
   (i) *C is a direct sum of copies of some group $\mathbb{Z}(p^k)$;*
   (ii) *C is divisible and for each prime p, C has only finitely many elements of order p.*

Proof. The proof of (b) $\Rightarrow$ (a) is straightforward. In the other direction, (a) implies that A is totally transcendental by Theorem 12.2.3, so we can use the previous theorem. We can suppose that A is countable. There are two cases.

Case 1: A has bounded exponent. Then the divisible part of A is empty, so A is simply a direct sum of copies of some finite number of cyclic groups of prime-power order. If there are two distinct such groups which appear infinitely often in the direct sum, we have two ways of bumping the cardinality up to ω_1. This gives (i).

Case 2: A has unbounded exponent. Then by Lemma A.2.4 we can raise the cardinality of A by adding copies of $\mathbb{Q}$. Hence by categoricity we can't raise the cardinality in any other way. Thus by Lemma A.1.5, no infinite direct sums $B^{(\kappa)}$ appear except where B is torsion-free divisible. This gives (ii). $\qquad\square$

Theorem A.2.13 (Rogers [1977]). *Let A be an infinite abelian group. Then the following are equivalent.*

(a) *A is superstable*.

(b) *There are only finitely many pairs p, n (p prime, $n < \omega$) such that $\text{Tf}(p, n; A)$ is infinite*.

Proof. (a) $\Rightarrow$ (b). Assume (b) fails. Let κ be a cardinal such that $\kappa^\omega > \kappa$ (for example a cardinal of cofinality ω). Let A' be a Szmielew group of cardinality κ which is elementarily equivalent to A, with the property that all the invariants $\kappa_{p,n}$, λ_p, μ_p, ν are either finite or equal to κ. Again by comparing Lemma A.2.10 with Table A1, we see that there are κ^ω complete 1-types over $\text{dom}(A')$. This implies that A is not superstable.

I omit the proof of (b) $\Rightarrow$ (a). $\qquad\Box$

Next come two oddments that should have a use somewhere.

Theorem A.2.14 *(Ershov [1963], Basarab [1975])* *Let ϕ be a sentence in the first-order language of abelian groups*.

(a) *ϕ is true in some finite abelian group if and only if ϕ is true in some abelian group A such that $D(p, n; A) = \text{Tf}(p, n; A)$ for all p and n*.

(b) *ϕ is true in some abelian torsion group if and only if the same holds with $=$ replaced by $\geqslant$*.

Hence in both cases it is decidable whether or not ϕ has the stated property. $\Box$

We say an abelian group A has **bounded p-length** if there is a finite m such that $U(p, n; A) = 1$ for all $n \geqslant m$.

Theorem A.2.15 *(Baudisch [1984], slightly sharpening Jackson [1973])*. *For any abelian group A the following are equivalent*:

(a) *for every abelian group B, $A \equiv B$ implies $A/t(A) \equiv B/t(B)$*;

(b) *A has bounded p-length for all primes p; and either $t(A)$ has bounded exponent or there is a prime p such that $\text{Tf}(p, n; A) > 1$ for all n*;

(c) *for all abelian groups B and C, if $A \equiv B$ then $A \otimes C \equiv B \otimes C$*.

If $A_1, \ldots, A_n$ each satisfy (a)–(c) above and $A_1 \equiv B_1, \ldots, A_n \equiv B_n$, then $A_1 \otimes \ldots \otimes A_n \equiv B_1 \otimes \ldots \otimes B_n$; similarly with $\leqslant$ for $\equiv$. $\Box$

Two further topics

It would be crass to leave abelian groups without mentioning two other areas of model-theoretic work on them, though there is only space here for the highlights.

A good deal of work has been done on descriptions of abelian groups in stronger languages than first-order. For example, how accurately can one define the class of free abelian groups by means of sentences of a logical language? Eklof [1974b] contains the following result of Kueker.

Theorem A.2.16. *An abelian group* A *is back-and-forth equivalent to a free abelian group if and only if every countable subgroup of* A *is free.* $\square$

Baudisch [1981] and Eklof & Mekler [1981] studied the theory of abelian groups in stationary logic. A startling recent result of Mekler and Oikkonen, based on Hyttinen & Tuuri [1991], puts logical limits to possible extensions of Ulm's theorem. It states that if the generalised continuum hypothesis holds, then for any uncountable regular cardinal λ, isomorphism of abelian p-groups of cardinality λ is not expressible in a certain very strong infinitary language based on games of infinite length.

The other topic that must be mentioned is the first-order theory of ordered abelian groups (see Example 2 in section 5.2). Gurevich [1964] proved that this theory is decidable. After seeing Paul Cohen's quantifier elimination for p-adic number fields, he looked for an analogous result, and found the following.

Theorem A.2.17 (Gurevich [1977]). *The theory of ordered abelian groups has quantifier elimination in a monadic second-order language with second-order variables ranging over convex subgroups.* $\square$

The difficulties of this area have scared people away, regrettably. One brave writer who followed Gurevich's lead was Schmitt [1984]. Ordered abelian groups are all unstable, so that their model theory is very different from that of abelian groups. But Gurevich & Schmitt [1984] showed that they don't have the independence property.

A.3 Nilpotent groups of class 2

The aim of this section is to show that every structure of finite signature is interpretable in a nilpotent group of class ≤ 2, by an interpretation which preserves many interesting model-theoretic properties. This interpretation is due to Alan Mekler [1981].

Throughout the section, p is an odd prime. We write N_2^p for the variety of nilpotent groups of class 2 and exponent p. This is in some sense the simplest non-abelian variety of groups. Recall that in a nilpotent group of class 2 we have $[[x, y], z] = 1$ for all elements x, y and z. It follows that $[x \cdot y, z] = [x, z] \cdot [y, z]$ and $[x, y \cdot z] = [x, y] \cdot [x, z]$, and for every positive integer α, $[x^\alpha, y] = [x, y^\alpha] = [x, y]^\alpha$.

Let A be a graph with at least two vertexes, regarded as a structure in the style of Example 1 of section 1.1. We say that two vertexes a, b of A are **joined** if there is an edge from a to b in A. We define a group $G(A)$ in the variety N_2^p by the following presentation. The generators of $G(A)$ are the vertexes of A; the relations are the equations $[a, b] = 1$ where a, b are vertexes of A which are joined. This group is not hard to describe explicitly: take the description in Exercise 9.2.23 of the free group in N_2^p with basis the set of vertexes of A, and cancel the direct summands generated by the commutators $[a, b]$ where a and b are joined in A.

The map $A \mapsto G(A)$ is a word-construction which will not greatly concern us, except to note that each vertex a of A gives rise to an element a^* of $G(A)$. Our first aim is to find a construction that recovers A from $G(A)$. In fact Corollary A.3.11 will say that if the graph A has the following properties (3.1) and (3.2), then A is interpretable in $G(A)$.

(3.1)   There are at least two vertexes; and for any two distinct vertexes a, b there is some vertex c which is joined to a but not to b.

(3.2)   There are no triangles or squares in A; more precisely, if a_0, ..., a_{m-1} are distinct vertexes such that each a_i $(i < m - 1)$ is joined to a_{i+1} and a_0 is joined to a_{m-1}, then $m \neq 3, 4$.

We say that the graph A is **nice** if it has these two properties.

Lemma A.3.1. *There is a first-order sentence ϕ such that a graph A is nice if and only if $A \vDash \phi$.*

Proof. Clear.   □

Write Z for the centre $Z(G(A))$ of the group $G(A)$.

Lemma A.3.2. *Suppose A is a nice graph.*

(a) *Every element g of $G(A)$ can be written as*

$$g = (a_0^*)^{\alpha_0} \cdot \ldots \cdot (a_{m-1}^*)^{\alpha_{m-1}} \cdot c$$

where a_0, ..., a_{m-1} are distinct vertexes of A, c is in the commutator subgroup $G(A)'$, and $0 \leq \alpha_i < p$ for each $i < m$.

(b) *$G(A)'$ is an elementary abelian p-group with basis the set of commutators $[a^*, b^*]$ such that a and b are vertexes of A which are not joined.*

(c) *$G(A)/G(A)'$ is an elementary abelian p-group with basis the set of elements $a^*G(A)'$ such that a is a vertex of A. In particular the indices α_0, ..., α_{m-1} in (a) are unique.*

(d) *$Z = G(A)'$.*

Proof. (a) Every element g can be written as a product of elements of the form a^* with a a vertex of A. We have $x \cdot y = y \cdot x \cdot [x, y]$, and since $G(A)$ is nilpotent of class ≤ 2, $[x, y] \in Z$. Hence we can group together the occurrences of any one vertex, and pull the commutators through to the end of the expression for g.

(b) and (c) follow from the structure theorem for free groups in N_2^p (see Exercise 9.2.23), noting that $G(A)'$ is the image of the commutator subgroup of the N_2^p-free group on the set of vertexes.

(d) $G(A)' \subseteq Z$ since $G(A)$ is nilpotent of class $\leqslant 2$. For the converse, suppose $g \in Z$. Write $g = (a_0^*)^{\alpha_0} \cdot \ldots \cdot (a_{m-1}^*)^{\alpha_{m-1}} \cdot c$ as in (a), with m as small as possible. It suffices to show that $m = 0$. Assume to the contrary that $m > 0$. Then by (3.1) there is a vertex b which is not joined to a_0. The commutator $[g, b^*]$ is equal to $\prod_{i<m} [a_i^*, b^*]^{\alpha_i}$ by the laws of N_2^p, and by (b) this product is not 1 since $[a_0^*, b^*] \neq 1$. Hence g is not central; contradiction. $\square$

By (c) of the lemma, we can identify each vertex a of A with the element a^* of $G(A)$, so that in future we can drop the asterisk $*$. Note that two vertexes commute if and only if they are joined. By (c) and (d), the elements aZ form a basis of $G(A)/Z$. It follows that if we represent g as in (a) of the lemma, with all the α_i non-zero, then the set $\{a_0, \ldots, a_{m-1}\}$ is uniquely determined by g; we call it the **support** of g, and we call m the **length** of g.

At first sight the length of an element of $G(A)$ doesn't look first-order definable. But we shall see that it is; in fact the vertexes of A are for practical purposes identifiable as the elements of length 1, and this will lead to an interpretation of A in $G(A)$. For the proof we shall need to examine three equivalence relations on dom $(G(A))$. We define, for any elements g, h of $G(A)$,

(3.3) $g \sim h \Leftrightarrow C(g) = C(h)$, where $C(g)$ is the centraliser of g in $G(A)$,

$g \approx h \Leftrightarrow$ modulo Z, each of g and h is a power of the other, i.e. for some $c \in Z$ and some α $(0 \leqslant \alpha < p)$, $h = g^\alpha \cdot c$,

$g \equiv_Z h \Leftrightarrow gZ = hZ$.

These three equivalence relations are clearly $\varnothing$-definable.

Lemma A.3.3. *Assume A is nice. For any elements g, h of $G(A)$, $g \equiv_Z h \Rightarrow g \approx h \Rightarrow g \sim h$.*

Proof. The first implication is trivial. For the second, assume $h = g^\alpha \cdot c$ with $c \in Z$ and $0 \leqslant \alpha < p$, and $f \in C(g)$. Then $[h, f] = [g, f]^\alpha = 1$. $\square$

In order to get a purchase on the relation $\sim$, we need to study when two elements of $G(A)$ commute. The following lemma completely answers this question.

Lemma A.3.4. *Assume A is nice. Let $g = a_0^{\alpha_0} \cdot \ldots \cdot a_{m-1}^{\alpha_{m-1}} \cdot c$ and $h = a_0^{\beta_0} \cdot \ldots \cdot a_{m-1}^{\beta_{m-1}} \cdot d$ be elements of $G(A)$, where $a_0, \ldots, a_{m-1}$ are distinct vertexes*

of A, each index α_i, β_i is ≥ 0 and $< p - 1$, and c, d are in Z. Then g and h commute if and only if the following holds:

(3.4) *if $i < j < m$ and a_i, a_j don't commute in A, then $\alpha_i\beta_j \equiv \alpha_j\beta_i$* (mod p).

Proof. The commutator $[g, h]$ is

$$\prod_{i,j<m} [a_i^{\alpha_i}, a_j^{\beta_j}] = \prod_{i,j<m} [a_i, a_j]^{\alpha_i\beta_j} = \prod_{i<j<m} [a_i, a_j]^{\alpha_i\beta_j - \alpha_j\beta_i}.$$

By Lemma A.3.2(b), this vanishes if and only if for all $i < j < m$ such that a_i doesn't commute with a_j, we have $\alpha_i\beta_j - \alpha_j\beta_i \equiv 0$ (mod p). □

We analyse the situation of Lemma A.3.4 more closely.

Lemma A.3.5. *Suppose A is nice, and g, h are non-central elements of $G(A)$ which commute. Write $g = a_0^{\alpha_0} \cdot \ldots \cdot a_{m-1}^{\alpha_{m-1}} \cdot c$ and $h = a_0^{\beta_0} \cdot \ldots \cdot a_{m-1}^{\beta_{m-1}} \cdot d$ as in the previous lemma, with m as small as possible. Partition m as $W \cup X \cup Y$ where $W \cup X$ (resp. $W \cup Y$) is the set of $i < m$ such that $\alpha_i \neq 0$ (resp. $\beta_i \neq 0$).*

(a) For all i, $j < \mu$, if i and j lie in different partition sets then a_i commutes with a_j.

(b) At least one and at most two of W, X, Y are non-empty. If only one partition set is non-empty, it is W.

(c) Suppose at least two partition sets are non-empty. Then at least one of these sets is a singleton. If i, j are distinct elements of the other partition set, then a_i doesn't commute with a_j.

(d) Suppose $X \cup Y$ is not empty; or more generally suppose that for all elements j of W except the first, there is some $i \in W$ such that $i < j$ and a_i doesn't commute with a_j. Then there is γ, $0 < \gamma < p$, such that for all $i \in W$, $\beta_i = \gamma\alpha_i$.

Proof. (a) Suppose $i < j < m$ and a_i doesn't commute with a_j. Then $\alpha_i\beta_j \equiv \alpha_j\beta_i$ (mod p) by Lemma A.3.4. In particular suppose $i \in Y$, so that $\alpha_i = 0$ and $\beta_i \neq 0$. Then $\alpha_j = 0$, so $j \notin W$, whence $j \in Y$. Likewise with X in place of Y.

(b) Since g and h are non-central, $m > 0$ so that at least one of the partition sets is non-empty. For the same reason, if W is empty then X and Y are non-empty. If $i \in W$, $j \in X$ and $k \in Y$, then by (a), a_i, a_j and a_k are all pairwise joined, which puts a triangle into A, contradicting (3.2).

(c) Suppose for example that $i \neq j$ in W and $k \neq l \in X$. By (a), each of a_i and a_j must be joined to each of a_k and a_l; but this puts a square into A, contradicting (3.2). Suppose then that $W = \{i, j\}$ and $X = \{k\}$. By (3.2) again, there are no triangles in A, and so a_i doesn't commute with a_j.

(d) Suppose l is the first element of W. Choose γ so that $\beta_l = \gamma\alpha_l$

(mod p). We go by induction on $j \in W$, $j > l$. By assumption there is $i \in W$, $i < j$, such that a_i doesn't commute with a_j. By induction hypothesis $\beta_i \equiv \gamma\alpha_i$ (mod p), and by Lemma A.3.4, $\alpha_i\beta_j \equiv \alpha_j\beta_i$ (mod p). Hence $\alpha_i\beta_j \equiv \alpha_j\gamma\alpha_i$ (mod p). Since $i \in W$, we can cancel α_i to get $\beta_j \equiv \gamma\alpha_j$. $\qquad\square$

Lemma A.3.5$^+$. *In the previous lemma, suppose moreover that $g \sim h$. Then besides (a)–(d) we have the following.*
  (e) *W is not empty.*
  (f) *If W is a singleton then $X = Y = \varnothing$.*

Proof. (e) Suppose X contains at least two elements i, j. Then for any choice of α and β, $a_i^\alpha \cdot a_j^\beta$ commutes with h by Lemma A.3.4; but by the same lemma we can choose α and β so that $a_i^\alpha \cdot a_j^\beta$ fails to commute with g, contradicting that $g \sim h$. Likewise with Y for X. Next suppose $X = \{i\}$ and $Y = \{j\}$. Then W is empty by (b). By (3.1) there is a vertex which commutes with a_i but not with a_j; this contradicts that $g \sim h$. Thus we have shown that $|X \cup Y| \leqslant 1$. It follows that if W is empty, then at least one of g, h has length 0 and hence is central; contradiction.

(f) Suppose $W = \{i\}$ and $Y = \{j\}$. By (3.1) there is a vertex b which is joined to a_i but not to a_j; so b commutes with g but not with h, contradicting that $g \sim h$. $\qquad\square$

From the two lemmas above, we shall read off a description of all the equivalence classes of the relation $\sim$. Let us say that an element g is **isolated** if every non-central element of A which commutes with g is $\approx$-equivalent to g. By Lemma A.3.3, every class of $\sim$ is a union of classes of $\approx$; we shall say that an element g is of **type** q if q is the number of $\approx$-classes in the $\sim$-class of g. We shall say that g is of **type** q$^\iota$ (resp. q^v) if g if of type q and isolated (resp. of type q and not isolated). Observe that for every positive integer q, the classes of elements of type q, type q^ι and type q^v are each $\varnothing$-definable.

Lemma A.3.6. *Every vertex of A is of type 1^v.*

Proof. Let a_0 be a vertex of A By (3.1), a_0 commutes with some other vertex of A, and so by Lemma A.3.2(c), a_0 is not isolated. Let g, h as in Lemma A.3.5$^+$ be respectively a_0 and an element which is $\sim$-equivalent to a_0. Then in the notation of that lemma, $W = \{0\}$ by (e), hence $X = Y = \varnothing$ by (f) of the lemma. So $g \approx h$, proving that g is of type 1. $\qquad\square$

Lemma A.3.7. *If g is an element of length 2, say with support $\{a_0, a_1\}$, and a_0 commutes with a_1, then g is of type $p - 1$.*

Proof. Put $g = a_0^{\alpha_0} \cdot a_1^{\alpha_1} \cdot c$ where c is in Z. First suppose that h is a non-central element which commutes with g. In the notation of Lemma A.3.5, put $h = a_0^{\beta_0} \cdot \ \ldots \ \cdot a_{m-1}^{\beta_{m-1}} \cdot d$. We show that $Y = \varnothing$. If not, then by Lemma A.3.5(c), 0 and 1 lie in different partition sets, so that W and X are not empty, contradicting (b). Thus $Y = \varnothing$, and so the support of h is a subset of $\{a_0, a_1\}$.

If $h \sim g$ then by Lemma A.3.5$^+$(f), h also has support $\{a_0, a_1\}$. Conversely we show that if g' is any element with support $\{a_0, a_1\}$, then $g' \sim g$. For suppose h commutes with g; then we have just shown that h has support $\subseteq \{a_0, a_1\}$, so that h commutes with g' too. Thus $C(g) \subseteq C(g')$ and so $g \sim g'$ by symmetry. The $\approx$-classes of elements $\sim$-equivalent to g are represented by the $p - 1$ elements $a_0 \cdot a_1^{\alpha}$ $(0 < \alpha < p)$. $\qquad\square$

Lemma A.3.8. *Suppose g is an element of length $\geqslant 2$ with support $\{a_0, \ldots, a_{m-1}\}$, and a_m is a vertex which commutes with all of $a_0, \ldots, a_{m-1}$ and is distinct from all of them. Then a_m is the unique vertex which commutes with all a_i $(i < m)$. For every element h of $G(A)$ the following are equivalent.*
(a) *$h \sim g$.*
(b) *h is an element of length $\geqslant 2$ which commutes with g.*
(c) *h is of the form $g' \cdot a_m^{\beta} \cdot d$ where $g' \approx g$, $0 \leqslant \beta < p$ and $d \in Z$.*
In particular both g and $g \cdot a_m^{\beta}$ (for any β) are of type p.

Proof. The uniqueness of a_m follows from the fact that A has no triangles and no squares, according to (3.2). Put $g = a_0^{\alpha_0} \cdot \ \ldots \ \cdot a_{m-1}^{\alpha_{m-1}} \cdot c$ with $c \in Z$.

(a) $\Rightarrow$ (b). If $h \sim g$ then h certainly commutes with g but is not central. By Lemma A.3.5$^+$(a,f) the length of h is at least 2.

(b) $\Rightarrow$ (c). Assuming (b), write $h = a_0^{\beta_0} \cdot \ \ldots \ \cdot a_{n-1}^{\beta_{n-1}} \cdot d'$ with $d' \in Z$ and $n \geqslant m$. In the notation of Lemma A.3.5, we have $W \cup X = m$. Suppose $j \in Y$. Then by Lemma A.3.5(a), a_j commutes with all a_i $(i < m)$ and hence $j = m$. Thus the support of h is a subset of $\{a_0, \ldots, a_m\}$.

We claim that $W = m$. If $Y = \varnothing$ but $W \neq m$, then by Lemma A.3.5(a,c) there is $j < m$ such that a_j commutes with all a_i $(i < m)$, which we have seen is impossible. If $Y \neq \varnothing$ and $W \neq m$, then by Lemma A.3.5(b), W must be empty, contradicting that h has length $\geqslant 2$. This proves the claim. Then by Lemma A.3.5(d), h is $g^{\gamma} \cdot a_m^{\beta} \cdot d$ for some $\gamma \neq 0$, some β and some d in Z.

(c) $\Rightarrow$ (a). Assuming (c), if $\beta = 0$ then $h \approx g$ and so $h \sim g$. Suppose then that $\beta \neq 0$. We show first that $C(g) \subseteq C(h)$. Let x commute with g. If x is a vertex b which commutes with g, then by Lemma A.3.5(a), b commutes with a_i for each $i < m$; so b must be a_m, and thus it commutes with h. This takes care of the case where x has length $\leqslant 1$. If x has length $\geqslant 2$ then the implication (b) $\Rightarrow$ (c) shows that x is of the form $g'' \cdot a_m^{\delta} \cdot d''$ where $g'' \approx g$ and $d'' \in Z$, whence x commutes with h too.

It remains to show that $C(h) \subseteq C(g)$. Suppose x is non-central and commutes with h; write $x = a_0^{\delta_0} \cdot \ldots \cdot a_{n-1}^{\delta_{n-1}} \cdot d'$ with $m < n$. Put $W = \{i < n : a_i$ is in the supports of h and $x\}$, $X = \{i < n : a_i$ is only in the support of $h\}$ and $Y = \{i < n : a_i$ is only in the support of $x\}$. Then $W \cup X = m + 1$. By Lemma A.3.5(a), if $j \in Y$ then a_j commutes with all a_i $(i < m)$, and so $j = m$, contradiction. Therefore Y is empty. So W is not empty, since x is not central. If X is not empty, then by Lemma A.3.5(c), $W = \{0, \ldots, m-1\}$ and $X = \{m\}$, and it follows by Lemma A.3.5(d) that $x \approx g$, so that x commutes with g. If X is empty, then for some σ, $x \cdot a_m^\sigma$ has support $\{a_0, \ldots, a_{m-1}\}$ and commutes with h, so that $x \cdot a_m^\sigma$ commutes with g by the argument just given, and hence also x commutes with g. $\qquad\square$

Lemma A.3.9. *Suppose g is an element of length ≥ 2 with support $\{a_0, \ldots, a_{m-1}\}$ and there is no vertex which commutes with all of $a_0, \ldots, a_{m-1}$. Then for every non-central element h the following are equivalent.*
(a) $h \sim g$.
(b) h *commutes with* g.
(c) $h \approx g$.
Thus g is of type 1^t.

Proof. Write $g = a_0^{\alpha_0} \cdot \ldots \cdot a_{m-1}^{\alpha_{m-1}} \cdot c$. Since a_0 doesn't commute with all elements of the support, we can suppose without loss that a_0 and a_1 fail to commute. Also by the fact (3.2) that there are no squares in A, there is at most one i with $2 \leq i < m$ such that a_i commutes with both a_0 and a_1; we can number the vertexes so that if there is such an i, it is $m - 1$. Then

(3.5)   for every $j < m$ and $j > 0$, there is $j' < j$ such that $a_{j'}$ and a_j do not commute.

Let $h = a_0^{\beta_0} \ldots a_{n-1}^{\beta_{n-1}} \cdot d$ (with $m \leq n$ and d in Z) be a non-central element; we can assume $\beta_m, \ldots, \beta_{n-1}$ are all strictly between 0 and p. (a) $\Rightarrow$ (b) is immediate and (c) $\Rightarrow$ (a) follows from Lemma A.3.3.

(b) $\Rightarrow$ (c). Assume h commutes with g. Consider any $i < n$, $i \geq m$. By the assumption on g, there is some $j < m$ such that a_i and a_j fail to commute. Hence $\alpha_i \beta_j \equiv \beta_i \alpha_j \pmod{p}$ by Lemma A.3.4. Since $i \geq m$, $\beta_i \neq 0$ and $\alpha_i = 0$, so $\alpha_j = 0$, contradiction. This shows that $m = n$. In the notation of Lemma A.3.5, $W \cup X = m$ and $Y = \varnothing$. Now if W and X are both non-empty, then by Lemma A.3.5(c), one of them is a singleton, so by Lemma A.3.5(a) there is a vertex which commutes with all of $a_0, \ldots, a_{m-1}$, contradiction. So one of W and X is empty; it must be X since h is not central. Now Lemma A.3.5(d) and (3.5) imply that $h \approx g$. $\qquad\square$

Summarising, we have the following.

Theorem A.3.10. *Let A be a nice graph.*

(a) *Every non-central element of $G(A)$ is of type 1, $p-1$ or p.*

(b) *An element of $G(A)$ is $\approx$-equivalent to a vertex if and only if it is of type 1^v.*

(c) *For every element g of type p, there is an element b of type 1^v which commutes with g; b is unique up to $\sim$-equivalence.* □

This leads to our first main result: A is recoverable from $G(A)$.

Corollary A.3.11. *There is an interpretation Γ without parameters, such that $\Gamma G(A) \cong A$ for each nice graph A. Thus Γ is right total from the class N_2^p to the class of nice graphs.*

Proof. The map taking each vertex a of A to its $\sim$-class $a^\sim$ is injective, and by Theorem A.3.10(b) there is a formula without parameters which picks out the elements $\sim$-equivalent to a vertex. If a and b are vertexes, and $g \sim a$ and $h \sim b$, then a and b are joined in A if and only if g and h commute. □

A little bonus.

Corollary A.3.12 (Ershov [1974]). *The theory of N_2^p is hereditarily undecidable.*

Proof. Exercise 5.5.9 verified that the graphs in Theorem 5.5.1 are nice. Use Theorem 5.5.1 and Corollary A.3.11 to interpret the natural numbers in a model of N_2^p, and quote Fact 5.5.6 and Theorem 5.5.7. □

From now on, we know that the map $G(A) \mapsto A$ preserves those model-theoretic properties which are preserved by interpretations. The startling fact is that the converse map $A \mapsto G(A)$ preserves many of these properties too. To see this, we shall study the first-order theory T_{ng} of the class $\mathbf{K} = \{G(A): A \text{ is a nice graph}\}$. Not every model of T_{ng} is of the form $G(A)$ with A a nice graph; in fact the infinite saturated models of T_{ng} are never of this form. But we shall see that each saturated model of T_{ng} consists of a structure in $\mathbf{K}$ with some extra pieces added on, and these extra pieces can be kept under control.

The first step is to write down ten axiom schemas which are contained in T_{ng}. As we go, we verify that each of these axiom schemas is true of structures in $\mathbf{K}$.

Axiom 1. *The axioms for nilpotent groups of class 2 and exponent p.*

We write *vertex* (x) for a formula which expresses 'x is an element of type 1^v', and *edge* (x, y) for a formula which expresses 'x and y are elements satisfying *vertex* which commute and are not $\sim$-equivalent'. By Lemma A.3.6, $\sim$ coincides with $\approx$ on vertexes.

Axiom 2. *Modulo the equivalence relation* $\sim$, *the formulas 'vertex' and 'edge' define a nice graph.*

Axiom 3. *Every non-central element is of exactly one of the four types* 1^v, 1^t, $p - 1$ *and* p; *the elements of type* $p - 1$ *and* p *are not isolated.*

Axiom 4. *Every element of type* $p - 1$ *is the product of two* $\sim$-*inequivalent elements of type* 1^v.

Axiom 5. *For every element* g *of type* p, *the elements which commute with* p *are precisely* (i) *the elements* $\sim$-*equivalent to* g *and* (ii) *an element* b *of type* 1^v *together with the elements* $\sim$-*equivalent to* b.

If b and g are related as in Axiom 5, we say that b is a **handle** of g. If g and g' are elements of type g with handles b and b' respectively, we say that g and g' **have the same handle** if $b \sim b'$. The handle of g is definable from g, up to $\sim$-equivalence.

Axiom 6. *Suppose* $g_0, \ldots, g_{k-1}$ *are elements of type* p *with the same handle* b. *Then the product* $g_0 \cdot \ldots \cdot g_{k-1}$ *is either an element of type* p *with handle* b, *or an element of type* 1^v, *or a central element.*

Proof in $G(A)$ for a nice graph A. The supports of $g_0, \ldots, g_{k-1}$ all consist of elements which commute with b, and hence so does the support of $g_0 \cdot \ldots \cdot g_{k-1}$. $\qquad\square$

Axiom 7. *Suppose that* $a_0, \ldots, a_{k-1}$ *are pairwise* $\sim$-*inequivalent elements of type* 1^v, *and* $g_0, \ldots, g_{m-1}$ *are elements of type* p *with distinct handles. Suppose that modulo the centre, there is a non-trivial equation* $E = 1$ *relating the* a_i's *and the* g_i's. *Then modulo the centre, one of the elements* g_i *is equal to a product of at most* $k + m - 1$ *elements of type* 1^v.

Proof in $G(A)$ for a nice graph A. The elements a_i are independent modulo the centre, by Lemma A.3.2(c,d). Hence at least one g_i is not 1; let it be g_0. For each $i < m$, write S_i for the support of g_i. Since by (3.2) there are no squares in A, and the elements g_i have distinct handles, if $i < i' < m$ then S_i and $S_{i'}$ have at most one element in common.

Now we consider each vertex a in S_0. The occurrences of a in E multiply out to 1. Hence either a occurs in some S_i $(i \neq 0)$, or one of the a_i is a non-zero power of a. But each S_i can contain at most one element of S_0. Hence S_0 has cardinality at most $k + m - 1$. The required equation expresses g_0 in terms of its support. $\qquad\square$

In the remaining three axioms and henceforth, 'independent' means 'linearly independent in the relevant elementary abelian p-group regarded as a vector space over the p-element field'. The next two axioms need a small calculation.

Lemma A.3.13. *Let A be a nice graph, and suppose $g_0, \ldots, g_{m-1}$ are elements of $G(A)$ such that $\prod_{i<j<m} [g_i, g_j]^{\alpha_{ij}} = 1$. Let b be a vertex of A and β_i the index of b in g_i (see Lemma A.3.2(c)). Then $[b, g] = 1$ where $g = \prod_{i<m} g_i^{\Sigma_{j<i}\alpha_{ji}\beta_j - \Sigma_{i<j}\alpha_{ij}\beta_j}$.*

Proof. Write $g_i = b^{\beta_i} \cdot x_i$. Then

$$(3.6) \qquad 1 = \prod_{i<j<m} [b, x_j]^{\alpha_{ij}\beta_i} \cdot \prod_{i<j<m} [x_i, b]^{\alpha_{ij}\beta_j} \cdot \prod_{i<j<m} [x_i, x_j]^{\alpha_{ij}}.$$

Noting that $[b, g_i] = [b, x_i]$ and using Lemma A.3.2(b) to restrict to the terms involving b, this implies

$$(3.7) \qquad 1 = \prod_{j<m} [b, g_j]^{\Sigma_{i<j}\alpha_{ij}\beta_i} \cdot \prod_{i<m} [b, g_i]^{-\Sigma_{i<j}\alpha_{ij}\beta_j}$$
$$= \prod_{i<m} [b, g_i]^{\Sigma_{j<i}\alpha_{ji}\beta_j - \Sigma_{i<j}\alpha_{ij}\beta_j}$$
$$= [b, g]. \qquad\square$$

Axiom 8. *Let $g_0, \ldots, g_{k-1}$ be a set of non-isolated elements which are independent modulo the centre, and let $Y = \{g_k, \ldots, g_{m-1}\}$ be a set of elements of type 1^t. Suppose that modulo the centre, no non-trivial combination of elements of Y is equal to a product of at most $k + 1$ non-isolated elements. Then in any true equation $\prod_{i<j<m} [g_i, g_j]^{\alpha_{ij}} = 1$, we have $\alpha_{ij} \equiv 0$ (mod p) whenever $i \neq j$ and $i \geq k$.*

Proof in $G(A)$ when A is a nice graph. Note first of all that the assumptions imply that $g_0, \ldots, g_{m-1}$ are independent modulo the centre. Now consider any vertex b and apply Lemma A.3.13. By that lemma, the element $g = \prod_{i<m} g_i^{\Sigma_{j<i}\alpha_{ji}\beta_j - \Sigma_{i<j}\alpha_{ij}\beta_j}$ commutes with b and hence is non-isolated (either because it is not $\sim$-equivalent to b, or because it is). Moreover we have $g \cdot \prod_{i<k} g_i^{\Sigma_{i<j}\alpha_{ij}\beta_j - \Sigma_{j<i}\alpha_{ji}\beta_j} = \prod_{i \geq k} g_i^{\Sigma_{j<i}\alpha_{ji}\beta_j - \Sigma_{i<j}\alpha_{ij}\beta_j}$. By the assumption on Y it follows that for each $i \geq k$, $\Sigma_{i<j}\alpha_{ij}\beta_j - \Sigma_{j<i} \alpha_{ji}\beta_j = 0$. So, fixing $i \geq k$, we have

$$(3.8) \qquad \prod_{i<j}(b^{\beta_j})^{\alpha_{ij}} = \prod_{j<i}(b^{\beta_j})^{\alpha_{ij}}$$

Multiplying together the equations (3.8) for all the relevant vertexes b, we get

$$(3.9) \qquad \prod_{i<j}g_j^{\alpha_{ij}} = \prod_{j<i}g_j^{\alpha_{ij}}\cdot c$$

for some central element c. By the independence of the elements g_j modulo the centre, it follows that $\alpha_{ij}\equiv 0 \pmod p$ for each $j\neq i$. $\qquad\square$

Axiom 9. *Let $\{g_0,\ldots,g_{k-1}\}$ be a set of elements of type $1^\vee$ and $\{g_k,\ldots, g_{m-1}\}$ a set of elements of type p. Suppose that $g_0,\ldots,g_{m-1}$ are independent modulo the centre. Suppose that no non-trivial combination of the elements $g_k, \ldots, g_{m-1}$ is equal to an element whose support has at most $k+m+1$ elements. Suppose $\prod_{i<j<m}[g_i,g_j]^{\alpha_{ij}} = 1$. Then for all $i < j < m$, $[g_i,g_j]^{\alpha_{ij}} = 1$.*

Proof in $G(A)$ when A is a nice graph. Let b be any vertex. As in Lemma A.3.13 again, we have $[b,g]=1$ where $g=\prod_i g_i^{\sum_{j<i}\alpha_{ji}\beta_j-\sum_{i<j}\alpha_{ij}\beta_j}$. Let R be the set of $i<m$ such that $[b,g_i]\neq 1$, and put

$$(3.10) \qquad g' = \prod_{i\in R}g_i^{\sum_{j<i}\alpha_{ji}\beta_j-\sum_{i<j}\alpha_{ij}\beta_j}$$

Then $[b,g']=1$. In particular the support of g' contains no vertexes which fail to commute with b.

We claim that for each $i\in R$, there is at most one element a_i of the support of g_i such that $[b,a_i]=1$. If $i<k$ this is immediate. If $k\leqslant i<m$, by Theorem A.3.10(c) there is a unique vertex d_i which commutes with all the vertexes in the support of g_i. Since $[b,g_i]\neq 1$, b is not d_i. Hence if b commutes with two or more elements of the support of g_i, we have a square in A, contradicting (3.2). This proves the claim.

It follows that the support of g' has cardinality at most m. By the assumption on $g_k,\ldots,g_{m-1}$ and the equation (3.10) it follows that for all $i\geqslant k$ with $i\in R$, $\sum_{j<i}\alpha_{ji}\beta_j-\sum_{i<j}\alpha_{ij}\beta_j=0$. Write $\alpha_i(b) = \sum_{j<i}\alpha_{ji}\beta_j-\sum_{i<j}\alpha_{ij}\beta_j$. We have shown that if b fails to commute with g_i then $\alpha_i(b)=0$.

Now consider any i such that $k\leqslant i<m$, and let h_i be $\prod_{j<i}g_j^{\alpha_{ji}}\cdot\prod_{i<j}g_j^{-\alpha_{ij}}$. One readily checks that for any vertex b, the index of b in h_i is $\alpha_i(b)$. Since g_i is of type p, there is a unique vertex b_i which commutes with g_i (by Theorem A.3.10(c)). Thus $h_i = b_i^{\alpha_i(b_i)}\cdot\prod_{b'}b'^{\alpha_i(b')}\cdot c = b_i^{\alpha_i(b_i)}\cdot c$ for some central element c, where b' are the vertexes in the support of h_i which don't commute with g_i. Consider the equation

$$(3.11) \qquad \prod_{j<i}g_j^{\alpha_{ji}}\cdot\prod_{i<j}g_j^{-\alpha_{ij}} = b_i^{\alpha_i(b_i)}\cdot c \quad\text{(where i is fixed).}$$

If b_i is not $\sim$-equivalent to any of the elements $g_0, \ldots, g_{k-1}$, then the assumption on $g_k, \ldots, g_{m-1}$ immediately gives that all the indices α_{ij}, α_{ji} are 0, so $[g_i, g_j]^{\alpha_{ij}} = 1$ for each $j \neq i$. If b_i is $\sim$-equivalent, say, to g_l, then the same argument gives that $[g_i, g_j]^{\alpha_{ij}} = 1$ for all $j \neq i, l$; but then $[g_i, g_l] = 1$. $\square$

Axiom 10. *Let g be an element of type 1^v and h an element of type p which commutes with g. For each integer $m \geq 3$, if h is not a product of $\leq m$ elements of type 1^v, then there are at least $m - 1$ other elements of type p, pairwise not $\sim$-equivalent, which commute with g.*

Proof in $G(A)$ when A is a nice graph. The element h must be a product of a set W of n pairwise non-$\sim$-equivalent elements of type 1^v, for some $n > m$, and g must be $\sim$-equivalent to the unique vertex which commutes with everything in W. Then if b is any element of W which is not $\sim$-equivalent to g, the element $\prod_{w \in W, w \neq b} w$ is another element of type p which commutes with g. $\square$

Let A be a nice graph. We shall say that a graph A^+ is a **cover of** A if $A \subseteq A^+$ and for every vertex b which is in A^+ but not in A, *either*

(3.12)   there is a vertex a in A such that (1) a is the unique vertex which is joined to b in A^+, and (2) a is joined to infinitely many vertexes in A,

or

(3.13) b is not joined to any other vertex.

We shall say that the graph A^+ is a λ-**cover of** A if (i) A^+ is a cover of A, (ii) for every vertex a of A, the number of new vertexes of A^+ which are joined to a is λ if a is joined to infinitely many vertexes in A, and 0 otherwise, and (iii) the number of new vertexes of A^+ which are not joined to any other vertex is λ if A is infinite and 0 otherwise.

The elements of a cover will correspond to certain elements of a model G of T_{ng}. By a 1^v-**transversal** we mean a set consisting of one representative of each $\sim$-class of elements of type 1^v. An element of G is **proper** if it is not a product of elements of type 1^v. By a p-**transversal** we mean a set X of representatives of $\sim$-classes of proper elements of type p, which is maximal with the property that if Y is a finite subset of X and all elements of Y have the same handle, then Y is independent modulo the subgroup of G generated by elements of type 1^v. By a 1^l-**transversal** we mean a set of representatives of $\sim$-classes of proper elements of type 1^l, which is maximal independent modulo the subgroup generated by $Z(G)$ and the elements of type 1^v or p. A **transversal** is a set which is a union of a 1^v-transversal, a p-transversal and a 1^l-transversal.

Theorem A.3.14. *Let G be a model of Axioms 1–10 and X a transversal of G. Write the 1^v, p and 1^ι parts of X as X_{1^v}, X_p and X_{1^ι} respectively.*

    (a) *The graph $\Gamma(G)$ is well-defined and nice; call it A. The elements of A correspond naturally to the elements of X_{1^v}.*

    (b) *The elements of X are independent in G modulo $Z(G)$.*

    (c) *$\langle X \rangle \cong G(A^+)$ for some cover A^+ of A.*

    (d) *$G = \langle X \rangle \times H$ for some elementary abelian p-group H.*

Proof. (a) The graph $\Gamma(G)$ exists and is nice by Axiom 2. The map from X_{1^v} to $\mathrm{dom}(A)$ is got by identifying each element of the transversal with its $\sim$-equivalence class.

    (b) By definition of a 1^ι-transversal, we need only concern ourselves with X_{1^v} and X_p. Suppose we have a non-trivial equation relating the elements of X_{1^v} to those of X_p modulo $Z(G)$. Group the elements of X_p according to their handles, getting an equation $h = \prod_{i<m} g_i \pmod{Z(G)}$ where h is a product of elements of X_{1^v}, each g_i is a product of elements of X_p with the same handle, and $g_i \notin Z(G)$. Since the equation is not trivial and X_{1^v} is a 1^v-transversal, m is positive. By Axioms 6 and 7, g_0 is (modulo $G(A)$) a product of finitely many elements of type 1^v, contradicting the definition of p-transversals.

    (c) First we define a graph structure on the transversal X. We join the vertexes a and b if and only if $a \neq b$ and a commutes with b. Let A^+ be this graph. Then by (a), the graph A is naturally identified with the part of A^+ which comes from X_{1^v}. Suppose b is an element of X_p. Then by Axiom 5, b is joined to a single element a of A and not to any other element of X. By Axiom 10, a is joined to infinitely many elements in A. Since the elements of X_{1^ι} are isolated, they are not joined to anything. Hence A^+ is a cover of A.

    It remains to check that the only equations between commutators are those implied by the true equations $[a, b] = 1$, where a is of type 1^v and b is of type either 1^v or p. It suffices to show the following, where some fixed linear ordering of X has been chosen.

(3.14)   Suppose $\prod_i [a_i, b_i]^{\alpha_i} = 1$, where the commutators $[a_i, b_i]$ are distinct and for each i, a_i precedes b_i in X. Then for each i, $[a_i, b_i]^{\alpha_i} = 1$.

This follows from Axioms 8 and 9.

    (d) We show first that X generates G modulo $Z(G)$. By Axiom 3, every non-central element is of one of the types 1^v, 1^ι, $p - 1$, p. By choice of X and the definition of $\sim$, every element of type 1^v is in the subgroup $\langle X \rangle$ of G generated by X. It follows by Axiom 4 that every element of type $p - 1$ is in $\langle X \rangle$ too. The choice of X now implies first that every element of type p is in $\langle X \rangle$, and second that every element of type 1^ι is in $\langle X \rangle$.

Since G is nilpotent of class 2, the commutator subgroup $\langle X \rangle'$ of $\langle X \rangle$ lies in $Z(G)$, and so we can write $Z(G)$ as $\langle X \rangle' \times H$ for some elementary abelian group H. If g is any element of G, then we have shown that g can be written as $h \cdot c$ with $h \in \langle X \rangle$ and $c \in Z(G)$; then c in turn can be written as $h' \cdot d$ with $h' \in \langle X \rangle'$ and $d \in Z(G)$. So $G = \langle X \rangle H$. Finally suppose $x \in \langle X \rangle \cap H$. Since x is in H, it lies in the centre of G and hence in the centre of $\langle X \rangle$. But $\langle X \rangle \cong G(A^+)$, and so the proof of Lemma A.3.2(d) shows that x lies in $\langle X \rangle'$, whence $x = 1$. Thus $G = \langle X \rangle \times H$ as required. $\qquad\square$

Corollary A.3.15. *Let λ be an infinite cardinal and G a special model of T_{ng} (or of Axioms 1–10) of cardinality λ. Then $G = G(A^+) \times H$ where A^+ is a λ-cover of the nice graph $\Gamma(A)$ (which also has cardinality λ), and H is an elementary abelian p-group of cardinality λ.*

Proof. The facts that A^+ is a λ-cover of A and that H has cardinality λ both follow directly from Exercise 10.4.2. $\qquad\square$

Corollary A.3.16. *If G_1 and G_2 are special models of T_{ng} of the same infinite cardinality λ, then every isomorphism from $\Gamma(G_1)$ to $\Gamma(G_2)$ lifts to an isomorphism from G_1 to G_2. In particular every automorphism of $\Gamma(G_1)$ extends to an automorphism of G_1.*

Proof. Immediate from the preceding corollary. $\qquad\square$

It follows that G preserves elementary equivalence, rather surprisingly.

Corollary A.3.17. *Suppose A and B are elementarily equivalent nice graphs. Then $G(A) \equiv G(B)$.*

Proof. We can suppose that both A and B are infinite. Choose a strong limit cardinal $\lambda > |G(A)| + |G(B)|$. Invoking Theorem 10.4.2(b), let G and H be special structures of cardinality λ which are elementary extensions of $G(A)$, $G(B)$ respectively. By Theorem 5.3.2, $\Gamma(G) \equiv \Gamma(G(A)) \cong A \equiv B \cong \Gamma(G(B)) \equiv \Gamma(H)$. By Exercise 10.4.10, both $\Gamma(G)$ and $\Gamma(H)$ are special structures of cardinality λ. So by Theorem 10.4.4, $\Gamma(G) \cong \Gamma(H)$, and hence $G \cong H$ by the previous corollary. Thus $G(A) \equiv G \cong H \equiv G(B)$. $\qquad\square$

Corollary A.3.18. *Axioms 1–10 axiomatise the theory T_{ng}.*

Proof. Let G be any model of Axioms 1–10, and let A be the nice graph $\Gamma(G)$. Choose a large enough strong limit cardinal λ, and let G_1 be a special elementary extension of G in cardinality λ, and G_2 a special elementary extension of $G(A)$ in cardinality λ. We have $\Gamma(G_1) \equiv \Gamma(G)$ and $\Gamma(G_2) \equiv \Gamma(G(A))$ since Γ is an interpretation. But $\Gamma(G) = A \cong \Gamma(G(A))$, so $\Gamma(G_1)$ and $\Gamma(G_2)$ are elementarily equivalent special structures of the same cardinality. Thus they are isomorphic, and so are G_1 and G_2 by Corollary A.3.16.   □

Finally we show that Γ preserves stability properties in both directions.

Corollary A.3.19. *Let C be an infinite nice graph and λ an uncountable cardinal. Then $\mathrm{Th}\,(C)$ is λ-stable if and only if $\mathrm{Th}\,(G(C))$ is λ-stable.*

Proof. Since C is interpretable in $G(C)$, the implication from right to left is by Exercise 6.7.15. In the other direction, assume $\mathrm{Th}\,(C)$ is λ-stable and let G be a model of $\mathrm{Th}\,(G(C))$ of cardinality λ. We shall show that there are only λ complete types over G. Let μ be a strong limit cardinal $> \lambda$ and let M be a special elementary extension of G of cardinality μ. Choose 1^v-transversals X_{1^v} for G and Y_{1^v} for M with $X_{1^v} \subseteq Y_{1^v}$. Likewise choose p-transversals X_p, Y_p and 1^{ι}-transversals $X_{1^{\iota}}$, Y_{1^v} for G, M respectively with $X_p \subseteq Y_p$ and $X_{1^{\iota}} \subseteq Y_{1^{\iota}}$.

As in Theorem A.3.14, these transversals give rise to nice graphs $A = \Gamma(G)$ and $B = \Gamma(M)$ with covers A^+, B^+ respectively. Note that $A \preccurlyeq B \equiv C$ since Γ preserves elementary equivalence and elementary embeddings. Since M is special, the cover B^+ is a λ-cover. Thus we can write G as $G(A) \times H_1$ and M as $G(B) \times H_2$ where H_1 and H_2 are elementary abelian p-groups. Non-trivial elements of H_1 are not products of commutators in G, so they are not products of commutators in M either; it follows that we can arrange that $H_1 \subseteq H_2$.

We define a set W in M as follows. Let U be a subset of M which contains a representative of every type of tuple in B over A; by the λ-stability of $\mathrm{Th}\,(C)$ we can choose U to have cardinality λ. Let V_{1^v}, $V_{1^{\iota}}$, V_{ab} be subsets of $Y_{1^v} \backslash X_{1^v}$, $Y_{1^{\iota}} \backslash H_{1^{\iota}}$, $H_2 \backslash H_1$ respectively, each of cardinality ω. Let V_p be a subset of $Y_p \backslash X_p$ of cardinality ω containing, for each element a of V_{1^v}, ω elements which commute with a. Let W be the subgroup of M generated by G, U, V_{1^v}, V_p, $V_{1^{\iota}}$ and V_{ab}. Thus $|W| = \lambda$. To prove the theorem it suffices to show that every orbit of $\mathrm{Aut}\,(M)_{(\mathrm{dom}\,G)}$ meets W.

Let g be an element of M. Then g can be written as a product of elements of the transversal $Y_{1^v} \cup Y_p \cup Y_{1^{\iota}}$ and elements of H_2. Let P be this set of elements, and let Q be the set of all elements a of V_{1^v} such that some element of $P \cap V_p$ commutes with a. List $(P \cap V_{1^v}) \cup Q$ as a tuple $\bar{b}$ of elements of B, and let $\bar{b}'$ be a tuple in U such that $\mathrm{tp}_B(\bar{b}'/\mathrm{dom}\,(A)) = \mathrm{tp}_B(\bar{b}/\mathrm{dom}\,(A))$. Since B is special by Exercise 10.4.10, there is an automorphism s_1 of B which fixes A pointwise and takes $\bar{b}$ to $\bar{b}'$. We can extend

s to an automorphism s_2 of B^+ which takes all elements of $P \backslash H_2$ into W. By Corollary A.3.15 above, s_2 extends to an automorphism s_3 of $G(B)$, and we easily extend s_3 to an automorphism s_4 of M which takes $P \cap H_2$ into W. $\square$

Thus by Theorem 6.7.2 and the definition of superstability in section 6.7, if A is an infinite nice graph, then A is stable if and only if $G(A)$ is stable, and A is superstable if and only if $G(A)$ is superstable. By Theorem 6.7.5, A is totally transcendental if and only if $G(A)$ is totally transcendental.

Theorem 5.5.1 and Exercise 5.5.9 together showed that for every first-order language L of finite signature, the class of infinite L-structures is bi-interpretable with a class of infinite nice graphs. This means in particular that any property which L-structures can have, and which is preserved by bi-interpretations, is also a property of some infinite nice graph. Then Mekler's interpretation will often find us a nilpotent group of class 2 and exponent p with this same property. For example it gives us nilpotent groups which are superstable but not totally transcendental. (See Baudisch [1989] for further examples.) In a word, Murphy's law holds for the model theory of nilpotent groups.

A.4 Groups

This appendix is an incomplete guide to the many and varied links between model theory and groups. Broadly I start with constructions and move on to classifications.

Speaking of classification, one can find three different kinds of interaction between model theory and groups. In the first place there are questions of the form 'Take an idea from group theory and an idea from model theory, mix them and see what happens'. This sounds a cynical description, but in fact some of the best work in the area was originally done in this spirit. For example it includes, I would say, most of the work on existentially closed structures in various classes of group.

In the second place there are situations where a question arose in group theory and methods of model theory were used to answer it. The earliest work in applied model theory, Mal'tsev's paper [1940] on linear groups, falls under this head. Shelah has offered several examples, including his positive solution [1980a] of Kurosh's problem (whether there is an uncountable group whose proper subgroups are all countable), and his proof [1987c], answering a question of Rips, that every uncountable group G has at least $|G|$ pairwise non-conjugate subgroups. But I think it is fair to say that most of Shelah's contributions to group theory, including these two, owe more to combinatorial set theory than to model theory.

Thirdly there are solutions of model-theoretic questions by means of results from group theory, usually permutation groups and geometry. The Cherlin–Mills–Zil'ber theorem (Theorem 4.6.1) is an obvious instance; several items in section 12.3 come into the same category. There is a puzzling feature of these interactions between groups and models: very often the theorems have a purely model-theoretic proof by the side of their group-theoretic proof. An earlier instance was B. H. Neumann's lemma, Lemma 4.2.1; it lies behind several model-theoretic results such as Theorem 6.4.5, though in fact Theorem 6.4.5 was originally proved by a model-theoretic argument making no mention of groups. Some of us believe that this area lies, in the words of Borovik quoted below, 'at the point where logic, algebra and geometry meet', and that future mathematicians may well see the field as a branch of mathematics in its own right.

A model-theoretic construction: existentially closed groups

W. R. Scott [1951] defined existentially closed groups in analogy to algebraically closed fields. At that date there was no reason to link this construction with logic. The link came through Abraham Robinson's work on existentially closed structures at Yale in 1970, and Macintyre's work [1972b] on questions raised by Robinson's research group.

First, here are some typical properties of all e.c. groups.

Theorem A.4.1. *Let G be an e.c. group.*

(a) *(B. H. Neumann [1952]) G is simple.*

(b) *(B. H. Neumann [1973]) G is not finitely generated.*

(c) *(B. H. Neumann [1973]) Every finitely generated group with solvable word problem is embeddable in G.*

(d) *(Macintyre [1972b]) G has a finitely generated subgroup with unsolvable word problem.*

(e) *(Macintyre [1972b]) G is ultrahomogeneous.*

(f) *(Baumslag, Dyer & Heller [1980]) G is acyclic (i.e. has trivial singular homology over $\mathbb{Z}$).* □

A large part of the model theory of e.c. groups can be read off from the following decisive result of Ziegler [1980].

Theorem A.4.2. *In the first-order language L of groups, let $\bar{x}$ be a tuple of variables and $\Phi(\bar{x})$ a recursively enumerable set of quantifier-free Horn formulas. Then there is a p.p. formula $\psi(\bar{x})$ whose resultant is equivalent to $\bigwedge \Phi$, so that ψ is equivalent to $\bigwedge \Phi$ in all e.c. groups.* □

Hodges [1985] gives a proof of Theorem A.4.2, together with a number of applications. One of them is the next result, which distinguishes some e.c. groups from others.

Theorem A.4.3 *(Belegradek [1978b], Ziegler [1980]). There are continuum many pairwise elementarily non-equivalent e.c. groups; in fact these groups differ in what $\forall_3$ sentences are true in them.* ☐

Theorem A.4.4 *(Hodges [1985] Theorem 4.1.6; cf. Hickin [1990] for a quicker proof without the elementary equivalence). There is a family $(G_i : i < 2^\omega)$ of countable elementarily equivalent e.c. groups, such that if $i < j < 2^\omega$ then any finitely generated group embeddable in both G_i and G_j has solvable word problem.* ☐

Both these two theorems rest on a method of Macintyre [1972a] for omitting types in e.c. groups.

There is further information about the group-theoretic properties of e.c. groups in Hickin & Macintyre [1980], Higman & Scott [1988] and Ziegler [1980].

E.c. groups in other classes of groups have been studied. Thus for nilpotent groups of a given class, see Hodges [1985], Maier [1983], [1989], Saracino [1976], Saracino & Wood [1979]. On e.c. torsion-free nilpotent groups, see Baumslag & Levin [1976], and Maier [1984]. On e.c. soluble groups see Saracino [1974]. On e.c. lattice-ordered groups, possibly abelian, see Glass & Pierce [1980a], [1980b]. Maier [1987] proposes an 'algebraic analogue' of existential closure and applies it to some classes of groups which are not first-order axiomatisable.

Interesting constructions of uncountable e.c. groups appear in Shelah [1977b] and Shelah & Ziegler [1979]. Leinen [1986] and others have studied uncountable e.c. groups in locally finite group classes (see also Macintyre & Shelah [1976], quoted at Theorem 11.6.4(b) above). Recent papers in the same vein, with references to the earlier literature, are Hickin [1988] and Hickin & Phillips [1990].

Classification, elementary equivalence

One of the oldest topics in model theory is classification up to elementary equivalence. There are any number of group-theoretic questions that one can ask under this head. For example, in about 1950 Tarski asked whether all free groups on more than one generator are elementarily equivalent. This question has remained open for far too long. Virtually all we know is the result of Merzlyakov [1966] that any two free groups on more than one generator agree on positive first-order sentences; Sacerdote [1973] gives a proof by small cancellation theory.

More recent results on first-order classification include the two following.

Theorem A.4.5. (a) *(Oger [1983])* *If* G *and* H *are groups such that* $G \times \mathbb{Z} \equiv H \times \mathbb{Z}$ *then* $G \equiv H$.
    (b) *(Oger [1991])* *If* G *and* H *are finitely generated nilpotent groups, then* $G \times \mathbb{Z} \cong H \times \mathbb{Z}$ *if and only if* $G \equiv H$. $\square$

Theorem A.4.6 (Sabbagh [1987]). *A polycyclic group elementarily equivalent to its profinite completion is abelian-by-finite.* $\square$

Felgner [1980b] studied the first-order theories of FC-groups, i.e. groups in which every conjugacy class is finite. This class contains all abelian groups and all finite groups. Cherlin & Schmitt [1983] discussed the first-order theories of topological abelian groups.

Glass [1989] reports results of Gurevich, Khisamiev and Kokorin on elementary theories of ordered and lattice-ordered groups.

The model theory of profinite groups has been studied mostly for its applications in field theory – see Fried & Jarden [1986]. Cherlin, van den Dries & Macintyre [1982] devised a language for profinite groups which allows one to talk about their finite images; this language is in some sense dual to a first-order language. In this context they proved analogues (or duals, maybe) of several results in first-order model theory. Chatzidakis [1984] developed this 'profinite model theory'.

Stable groups

Cherlin [1979] conjectured: *Every simple ω-stable group is an algebraic group over an algebraically closed field.* This has become known as **Cherlin's conjecture**. Zil'ber [1977a] had asked a similar question. Both authors had in mind the analogies between algebraic groups and groups of finite Morley rank (though their emphases were different – Zil'ber thought of Morley rank as a notion of dimension, while Cherlin looked for analogues of the Bruhat decomposition).

Borovik [1984], who already had a deep understanding of finite linear groups, put another gloss on the matter. I quote:

> The aim of this work, and of its proposed continuation, is to show that the structure of the known part of the theory of finite groups is not an accident – it pretty well exactly reflects the structure of the objects being studied, namely linear algebraic groups. We hope we can induce specialists in finite group theory to take an interest in the theory of uncountably categorical groups, which lie at the point where logic, algebra and geometry meet. . . .

The contents of section I [of this paper] run parallel to the first chapters of any textbook of algebraic groups . . . .

These remarks set in motion **Borovik's programme**, which is to rework the classification of finite simple groups, but for groups of finite Morley rank instead.

Nobody expected instant success for either Cherlin's conjecture or Borovik's programme. Nevertheless one case of Cherlin's conjecture did fall into place at once, though it threw little light on the general problem.

Theorem A.4.7 *(Thomas [1983a]). A stable locally finite simple group is a Chevalley group over the algebraic closure of a finite field.*

Proof. The group must be linear; I omit the proof of this, except to note that it uses known group-theoretic results together with Lemma 6.7.7 (to derive the descending chain condition on centralisers) and Exercise 3.1.14. An infinite locally finite simple linear group G is either a Chevalley group or a twisted Chevalley group over a locally finite field K. (This was proved by Thomas [1983b], and independently by Belyaev and Shute.) In the untwisted case, the field K is interpretable in the group G. Since the group is stable, so is the field, and hence by Theorem A.5.16 the field is algebraically closed. In the twisted case we are given a non-trivial automorphism α of K, and we can interpret the fixed field of α in G; this must again be algebraically closed, which is impossible. The interpretations need detailed examination of the root systems of the Chevalley groups. $\square$

Another partial positive answer was found by Berline [1986], extending ideas from Cherlin [1979], as follows.

Theorem A.4.8. *Let G be a superstable simple group of Lascar rank $\omega^{\alpha} \cdot 3$ for some ordinal α, and suppose G has a definable subgroup of rank $\omega^{\alpha} \cdot 2$. Then G is isomorphic to $\mathrm{PSL}_2(F)$ for some algebraically closed field F.* $\square$

A good deal is known about stable groups; many of the results hold equally well for group-like structures. Two good references are Poizat [1987a] and Nesin & Pillay [1989]. Here I do no more than pick out some highlights, starting with strong stability assumptions and gradually weakening them. Strong minimality is the obvious place to begin; people have suggested an analogy between the following result and the theorem (see Humphreys [1975] p. 131) that a connected one-dimensional algebraic group is either the additive or the multiplicative group of the underlying field.

Theorem A.4.9 (*Reineke [1975]*). *Let G be an infinite group. The following are equivalent.*

(a) *G is a minimal structure (in the sense of section 4.5).*

(b) *G is strongly minimal.*

(c) *G is abelian, and is either a divisible group in which for each prime p the p-group direct summand has finite rank, or a direct sum of cyclic groups of order p for some prime p.* □

One step weaker than strongly minimal is finite Morley rank. Zil'ber's indecomposability theorem (Theorem 5.7.12, an analogue of Chevalley's theorem on p. 55 of Humphreys [1975]) is about this property, and has been very influential. From work of Borovik and colleagues we also know the following.

Theorem A.4.10. *Let G be an infinite group of finite Morley rank.*

(a) *(**Borovik & Poizat [1990]**) All maximal 2-subgroups are locally finite and conjugate.*

(b) *(**Borovik [1989]**) Let K be a conjugacy class of elements of G, such that any two elements of K generate a 2-group. Then the normal subgroup generated by K is nilpotent-by-finite.*

(c) *(**Aguzarov, Farey & Goode [1991]**) G has an infinite number of conjugacy classes.* □

Borovik calls attention to the analogy between the Baer–Suzuki theorem and (b) of this theorem. Poizat & Wagner [199?] develop the idea further. There are also penetrating results of Nesin on groups of finite Morley rank, including a model-theoretic generalisation of the Lie–Kolchin theorem (Nesin [1989] as analysed by Poizat [1987a] p. 93).

For ω-categorical theories, totally transcendental coincides with superstable. However, it's still an open question whether every stable ω-categorical group is superstable.

Theorem A.4.11 (*Baur, Cherlin & Macintyre [1979]*). *If G is ω-categorical and superstable then G is abelian-by-finite.* □

Theorem A.4.12 (*Macpherson [1988]*). *If G is a group which is ω-categorical and doesn't have the strict order property, then G is nilpotent-by-finite.* □

Baur, Cherlin and Macintyre (and independently Felgner [1978]) showed that a stable ω-categorical group must be nilpotent-by-finite. However, this now falls out immediately from a result of pure group theory.

Theorem A.4.13 (Kegel [1989]). *If G is a group which has bounded exponent and descending chain condition on centralisers, then G is nilpotent-by-finite.* □

Moving on to superstable groups in general, we have the following analysis by Baudisch [1990].

Theorem A.4.14. *If G is a superstable group, then there are subgroups $\{1\} = H_0 < \ldots < H_r < G$, each first-order definable with parameters, such that each H_{i+1}/H_i is abelian or simple, and H_r is of finite index in G.* □

This entails the result of Gibone [1976] that free groups on more than one generator are unsuperstable. See also Poizat [1983a].

Groups interpretable in other structures

The following theorem was proved in 1986 by Ehud Hrushovski, answering a question of Poizat. In 1982 van den Dries (unpublished) proved it in characteristic 0, by reducing to work of Weil [1955] on group chunks.

Theorem A.4.15. *Let K be an algebraically closed field and G a group interpretable in K with parameters. Then G is isomorphic to an algebraic group over K; the isomorphism is expressible by a first-order formula.*

Proof. See Poizat [1987a] p. 141ff or Bouscaren [1989b]. The proof uses the machinery of stable groups. □

The search for analogues of this theorem has been quite successful, as the next two results bear witness.

Theorem A.4.16 (Pillay [1989a]). *Let Q_p be the p-adic number field, and let G be a group which is injectively interpretable (with parameters) in Q_p. Then there is a definable covering of G by a finite number of definable open subsets of Q_p^n for some fixed n, which induces on G the structure of a p-adic analytic group. Moreover any two such structures on G are analytically isomorphic.* □

Theorem A.4.17 (Pillay [1990]). *Every group injectively interpretable in a differentially closed field is differentially algebraic.* □

There are things to be said about groups interpretable in other kinds of structure besides fields. Poizat showed (Theorem A.6.9) that no infinite group is interpretable in a linear ordering. The following result of Macpherson [1991] has a beautiful proof using the vector space Ramsey theorem.

Theorem A.4.18. *If L is a relational language of finite signature and A is a countable ultrahomogeneous L-structure, then no infinite group is interpretable in A.* □

Groups with large automorphism groups

The class of all finite ultrahomogeneous groups (i.e. groups G such that Th(G) has elimination of quantifiers) was described by Cherlin, Felgner and Πeter Neumann, in part independently. The full list appears as Theorem D of Felgner [1986] (where the word 'possibly' against item (11) should be deleted). Saracino & Wood [1982] construct continuum many countable nilpotent groups of class 2 and exponent 4 whose theories have quantifier elimination.

Several results above mention ω-categoricity. See also section 7.3 for references to the classification and construction of countable ω-categorical groups.

A.5 Fields

In the 1970s John Bell had a student who wouldn't touch model theory, because 'It's just talking about algebraically closed fields and pretending not to'. (He later became a stockbroker.) The complaint is well taken. Algebraically closed fields are a paradigm for two of the main concerns of model theory: first the theory of existential closure and quantifier elimination, and second the theory of uncountable categoricity.

The first of these two shows has been on the road longer and has led to more new developments within field theory itself. Much of this work is both attractive as mathematics and useful as a source of examples. Earlier versions of this book contained several sections on this material, but I had to cut for lack of space. Happily I can refer readers to Fried & Jarden [1986] for a definitive account.

It remains to be seen whether the ideas coming from categoricity and stability theory will have a similar impact within field theory. Watch this space.

Fields in the language of rings

The class of fields is defined by a finite set of $\forall_2$ first-order axioms in the language of rings ((2.24) in section 2.2). In this language a substructure of a field is the same thing as an integral domain. If we want the substructures to be subfields, one way is to add a symbol $^{-1}$ with the axiom

$$(1.5) \qquad 0^{-1} = 0 \wedge \forall x(x \neq 0 \rightarrow x^{-1} \cdot x = 1)$$

replacing the last axiom in (2.24) of section 2.2. But an integral domain determines its field of fractions in a very easy way, and so it makes little difference which approach we take. Henceforth I assume the language is that of rings unless we say otherwise.

The class of algebraically closed fields is axiomatised by an infinite set of $\forall_2$ first-order sentences, (2.26) in section 2.2, call it T_{acl}. By Galois theory it's not hard to show that every finite subset of T_{acl} has a model which is not algebraically closed. (For a suitable finite set X of primes, take a field which is closed under finite extensions whose degree is divisible only by primes in X).

Theorem A.5.1. *The theory of algebraically closed fields is model-complete.*

Proof. Abraham Robinson [1956b] gave four proofs, noting that this theorem is essentially the same as Hilbert's Nullstellensatz. Probably the simplest proof (not one of his four) starts with Steinitz' theorem that any two algebraically closed fields of the same characteristic and transcendence degree are isomorphic. (This is straightforward algebra – the two fields are algebraic closures of pure transcendental extensions of the same prime field, with transcendence bases of the same cardinality. We can't simply quote Corollary 4.5.7 here because our proof of that corollary assumed quantifier elimination). Thus let T_χ be the theory of algebraically closed fields of characteristic χ. By Steinitz, T_χ is an uncountably categorical theory with no finite models. So T_χ is model-complete by Lindström's test (Theorem 8.3.4). Finally T_{acl} is model-complete because fields embed only into other fields of the same characteristic. $\qquad\square$

Better than Theorem A.5.1, the theory of algebraically closed fields has quantifier elimination. Geometers know this from Chevalley's theorem (see Humphreys [1975] section 4.4) that the image of a constructible set under a morphism of varieties is again constructible. (A **constructible set** is a relation defined by a quantifier-free formula.) But it also has a model-theoretic proof, as follows.

Corollary A.5.2. *The theory of algebraically closed fields is the model companion of the theory of fields. It has quantifier elimination and is decidable. The theory of algebraically closed fields of a fixed characteristic is complete.*

Proof. Every integral domain can be extended to an existentially closed field (Theorem 8.2.1), and e.c. fields are algebraically closed (see Example 1 in section 8.1). So $(T_{acf})_\forall$ is the theory of integral domains (see Corollary 6.5.3), and by Theorem A.5.1 it follows that T_{acf} is the model companion of the theory of fields. By elementary algebra the class of fields has the amalgamation property, and so T_{acf} has quantifier elimination by Theorem 8.4.1(d).

If T_χ is the theory of algebraically closed fields of characteristic χ, then T_χ is model-complete and has a model which is embeddable in every model. So T_χ is complete by Theorem 8.3.5, and hence decidable since its language is recursive. Thus T_{acl} is a decidable theory by Exercise 6.1.18. (Namely, for every ϕ not deducible from T_{acl}, $\neg\phi$ is deducible from some T_χ.) $\qquad\square$

Ziegler [1982] shows that no finite subset of T_{acl} is decidable. Fischer & Rabin [1974] give a lower bound on the time taken by any decision procedure for the theory of algebraically closed fields of a fixed characteristic; it's exponential in the length of the sentence. Nevertheless Heintz [1983] finds a reasonably fast quantifier elimination procedure based on geometrical intersection theory; see also Chistov & Grigor'ev [1984]. Rose [1980] shows that if A is an algebraically closed field and $\bar{a}$ is a tuple of elements of A, then $\mathrm{Th}(A, \bar{a})$ is decidable.

Every field with quantifier elimination is either algebraically closed or finite. This was proved by Macintyre [1971b] using ideas from stability. Macintyre, McKenna & van den Dries [1983] gave a more purely algebraic proof; I follow Wheeler's [1979a] exposition.

Theorem A.5.3. *A field A has quantifier elimination if and only if either A is finite or A is algebraically closed.*

Proof. We showed in Example 2 of section 8.4 that every finite or algebraically closed field has quantifier elimination. For the converse, suppose A has quantifier elimination and is infinite. Since A is infinite, by taking an elementary extension we can suppose that A has infinite transcendence degree.

(5.2)   For every positive integer n and every element a of A, A contains a root of $X^n = a$.

For let θ be an element which is algebraically independent of a. Then so is θ^n, and $X^n - \theta^n$ has the root θ. By quantifier elimination it follows that every element algebraically independent of a has an nth root, since all such elements satisfy the same quantifier-free formulas with parameter a. In particular $a\theta^n$ has an nth root $b\theta$ with $b^n = a$.

(5.3)   Let n be a positive integer and $a_1, \ldots, a_n$ algebraically independent elements of A. Then the polynomial $y^n + a_1 y^{n-1} + \ldots + a_n$ has a root in A.

For suppose $b_0, \ldots, b_{n-1}$ are algebraically independent elements of A, and let $p(X)$ be the polynomial $\prod_{i<n}(X - b_i)$, whose roots are all in A. We can write $p(X)$ as $y^n + s_1 y^{n-1} + \ldots + s_n$ where $s_1, \ldots, s_n$ are algebraically independent (they are $\pm$ the symmetric functions on $b_0, \ldots, b_{n-1}$; see Jacobson [1974] p. 135). Again quote quantifier elimination, using the fact that $a_1, \ldots, a_n$ satisfy the same quantifier-free formulas as $s_1, \ldots, s_n$.

Now suppose A is not algebraically closed. Then, noting that A is perfect by (5.2), some irreducible polynomial p of degree n over A has n roots $c_0, \ldots, c_{n-1}$. Choose algebraically independent elements $t_0, \ldots, t_{n-1}$ of A, and form the polynomial $F(y, t_0, \ldots, t_{n-1}) = \prod_{i<n}(y - \sum_{j<n}(t_j c_i^j))$. We can write this polynomial as $q(y) = y^n + g_1 y^{n-1} + \ldots + g_n$ where each g_i is in $A[t_0, \ldots, t_{n-1}]$.

We claim that the g_i are algebraically independent. For this it suffices to show that they generate a field of transcendence degree n over A, and in turn this follows if each t_j is algebraic over $A(g_1, \ldots, g_n)$. Let the roots of $q(y)$ be $r_0, \ldots, r_{n-1}$. Then

$$
(5.4) \qquad
\begin{pmatrix}
1 & c_0 & \cdots & c_0^{n-1} \\
\vdots & & & \vdots \\
1 & c_{n-1} & \cdots & c_{n-1}^{n-1}
\end{pmatrix}
\begin{pmatrix}
t_1 \\
\vdots \\
t_n
\end{pmatrix}
=
\begin{pmatrix}
r_0 \\
\vdots \\
r_{n-1}
\end{pmatrix}
$$

The matrix on the left is invertible, since it is the Vandermonde matrix of p. So by inverting the equation (5.4) we get the t_j as linear functions of the roots r_i, which are certainly algebraic over $A(g_1, \ldots, g_n)$. This proves the claim.

By the claim and (5.3), $q(y)$ has a root d in A. Then $d = \sum_{j<n}(t_j c_i^j)$ for some $i < n$, and hence c_i has degree at most $n - 1$ over A; contradiction. $\qquad\square$

We turn to categoricity.

Theorem A.5.4. *The theory of algebraically closed fields of a given characteristic is ω_1-categorical and hence totally transcendental. In fact algebraically closed fields are strongly minimal, and every relation which is first-order definable with parameters in an algebraically closed field has finite Morley rank.*

Proof. We have already used the ω_1-categoricity, noting that it is a theorem of Steinitz. Strong minimality follows from the quantifier elimination by Example 1 in section 4.5, and the finite Morley rank follows by Theorem 5.6.10 and Lemma 5.6.8. Incidentally this theorem is the origin of the name 'totally transcendental'. $\qquad\square$

In fact we can give a purely algebraic definition of the Morley rank of a definable relation, as follows. Every such relation can be written as $\phi(A^n, \bar{a})$ where ϕ is a quantifier-free formula in disjunctive normal form, say $(\phi_0 \vee \ldots \vee \phi_{m-1})$. The Morley rank of $\phi(A^n, \bar{a})$ is the maximum of the Morley ranks of the $\phi_i(A^n, \bar{a})$. So we can suppose that ϕ is a conjunction of equations and inequations; since fields are integral domains, we can suppose there is at most one inequation, bringing ϕ to the form $\bigwedge_{i<k} p_i(\bar{x}, \bar{a}) = 0 \wedge q(\bar{x}, \bar{a}) \neq 0$. If $\phi(A^n, \bar{a})$ is not empty, its Morley rank is unaltered if we forget the inequation. This reduces to the case where $\phi(A^n, \bar{a})$ is a variety; taking disjunctions again, we can assume it is irreducible. Its Morley rank is then the Krull dimension of its defining ideal. See Berline [1983]. Note that this gives a purely algebraic description of the non-forking relation ψ (see section 4.7) in an algebraically closed field.

Macintyre [1971b] showed that every totally transcendental infinite field is algebraically closed. Cherlin & Shelah [1980] extended this to superstable fields, while Zil'ber [1977a] showed that every ω_1-categorical division ring is an algebraically closed field. The full theorem below is due to Cherlin; see Berline & Lascar [1986]; it is one of the major results of stability theory. Pillay & Sokolović [1992] give a slick proof using the ideas of Theorem A.5.3 above.

Theorem A.5.5. *Every infinite superstable division ring is an algebraically closed field.* □

It follows at once that every strongly minimal field is algebraically closed. An open question of Reineke asks whether 'strongly minimal' can be replaced here by 'minimal'.

Since algebraically closed fields are strongly minimal, they have a geometry in the sense of section 4.6. We have already noted in Theorem 4.6.3 that this geometry can never be made modular by adding a finite set of parameters. In fact more is true: if A is a totally transcendental structure which is an expansion of an algebraically closed field, then the geometry of strongly minimal sets in A can also never be made modular by adding finitely many parameters. This fact is an important source of information about what can be interpreted in such structures.

Turning to matters of interpretability, we have seen in Theorem 4.4.6 that the theory of algebraically closed fields has uniform elimination of imaginaries. The next result appears in unpublished notes of van den Dries from 1982; Poizat [1987a] p. 150 (also [1988]) gives a proof.

Theorem A.5.6. *Let K be an algebraically closed field and A an infinite field* *which is interpretable in K with parameters. Then A is isomorphic to K; the isomorphism is expressible by a first-order formula.*   □

Zil'ber has conjectured that if A is a structure interpretable in an algebraically closed field B, then either A has locally modular geometry or some algebraically closed field (which must be B by Theorem A.5.6) is interpretable in A. Several positive results are known in this direction, for example the following.

Theorem A.5.7 *(Marker & Pillay [1990]; cf. also Martin [1988], Rabinovich [1992]). Let K be the field of complex numbers and A a reduct of a definitional expansion of K; assume that $+$ from K is definable in A. Then the following are equivalent.*
(a) *The geometry of A is not locally modular.*
(b) *The multiplication of K is definable in A with parameters.*   □

One can ask many other questions about what sets and relations are definable in an algebraically closed field – in other words, what relations are constructible. For example van den Dries, Marker & Martin [1989] analyse the equivalence relations definable on an algebraically closed field. One of their results is that there is no ∅-definable equivalence relation on the elements of an algebraically closed field, such that each equivalence class has just two elements. In other words, the number of elements of an algebraically closed field is odd!

In section 11.4 we saw that results on effective bounds in the theory of polynomials can sometimes be used to prove that certain relations on a field are first-order definable. Generally these bounds are uniform for all fields, so that algebraic closure is irrelevant.

Finally there is an old heuristic principle which Weil [1946] called **Lefschetz' principle**. According to Weil (p. 242f),

> ... for a given value of the characteristic p, every result, involving only a finite number of points and varieties, which can be proved for some choice of the universal domain ... is true without restriction; *there is but one algebraic geometry of characteristic p*, for each value of p, and not one algebraic geometry for each choice of the universal domain. In particular, as S. Lefschetz has observed on various occasions, whenever a result, involving only a finite number of points and varieties, can be proved in the 'classical case' where the universal domain is the field of all complex numbers, it remains true whenever the characteristic is 0 . . ..

This looks as if it should be a model-theoretic principle, and several writers have suggested what model-theoretic principle it might be. The most convincing proposals are those of Barwise & Eklof [1969] using $L_{\omega_1\omega}$, and of Eklof [1973] using $L_{\infty\omega}$; see Exercise 9.3.10 for the latter.

Other classes in the language of fields

The first generalisation is the class of separably closed fields.

Theorem A.5.8 *(Ershov [1967a]). Let A be a separably closed field of characteristic p. The theory of A is determined by the degree $[A:A^p]$ (which can be ∞ or any power of p).* □

The degree $[A:A^p]$ is known as the **degree of imperfection** of A, or as the **Ershov invariant** of A. Ershov's proof also showed the following.

Theorem A.5.9. *The theory of separably closed fields of a fixed characteristic and Ershov invariant is model-complete if one adds predicates to express p-independence.* □

In fact with a little care one can get quantifier elimination, by adding functions $F_i(x, y_0, \ldots, y_{n-1})$ to express 'the ith term of the expression of x in terms of monomials in $y_0, \ldots, y_{n-1}$, assuming $y_0, \ldots, y_{n-1}$ are p-independent'. See Delon [1988].

Theorem A.5.10 *(Delon [1988]). The theory of separably closed fields has uniform elimination of imaginaries.* □

Theorem A.5.11 *(Macintyre and Shelah in Wood [1979]). Separably closed fields are stable (but by Theorem A.5.5 above, they are superstable if and only if they are algebraically closed).* □

Can one usefully generalise beyond the separably closed fields? Ax [1968] answered this question by introducing the class of pseudo-algebraically closed fields. To define these we need some algebra. Let A be a field. An ideal I of the polynomial ring $A[\bar{X}]$ is an **absolutely prime ideal over** A if BI is prime in $B[\bar{X}]$ for any field B extending A (or equivalently, if CI is prime in $C[\bar{X}]$ where C is the algebraic closure of A). A **point** of I is a tuple $\bar{a}$ such that $p(\bar{a}) = 0$ for every $p \in I$. The next lemma says that the relation of purity between fields is really a geometric notion.

Lemma A.5.12. *Let B be a field and A a subfield of B. The following are equivalent.*
(a) *A is pure in B.*
(b) *B is a regular extension of A; and every absolutely prime ideal over A which has a point in B already has a point in A.*
(c) *Some extension of B is an elementary extension of A.*

Proof. Wheeler [1979b] proves (a) $\Leftrightarrow$ (b). (c) $\Rightarrow$ (a) follows from existential amalgamation (Theorem 6.5.1) and Rabinowitsch's trick (Example 1 in section 8.1). $\qquad\qquad\square$

Theorem A.5.13. *Let A be a field. Then the following are equivalent.*
(a) *If B is any regular field extension of A and the degree of B over A is finite, then A is pure in B.*
(b) *Every absolutely prime ideal over A has a point in A. (In the geometers' terminology, every non-empty absolutely irreducible variety over A has a rational point.)*
(c) *For every absolutely irreducible polynomial $p(x, y)$ over A which is monic in Y, and every non-zero polynomial $q(x)$ over A, $A \vDash \exists xy(p(x, y) = 0 \wedge q(x) \neq 0)$.*
(d) *For any field B, if B is a regular extension of A then some extension of B is an elementary extension of A.* $\qquad\qquad\square$

We say that a field is **pseudo-algebraically-closed** (PAC) if it satisfies any of the equivalent conditions (a)–(d). Clause (c) can be turned into a first-order axiomatisation of PAC fields by an infinite set of $\forall_2$ sentences. Clause (b) shows at once that algebraically closed fields are PAC. In fact there are many ways to find PAC fields; I quote three in the next theorem. Parts (b) and (c) both rest on Weil's Riemann hypothesis for curves over a finite field.

Theorem A.5.14. (a) *Separably closed fields are PAC.*
(b) *Locally finite fields are PAC.*
(c) *(Ax [1968]) Ultraproducts of finite fields are PAC.*
(d) *(Ershov [1980b]) If A is any field, then there is a PAC field which is a regular extension of A.* $\qquad\qquad\square$

There is an attractive general theory of PAC fields, leading to a classification up to elementary equivalence in terms of their absolute Galois groups. See Fried & Jarden [1986] (Chapter 10 onwards) for details. Let me mention two results which use PAC fields (at least in their proofs) but are useful for people who don't know what PAC means. The first is a fact about definability in finite fields.

Theorem A.5.15 (Chatzidakis, van den Dries & Macintyre [1992]). *Let $\phi(\bar{x}, \bar{y})$ be a formula of the language of rings, where $\bar{x} = (x_0, \ldots, x_{m-1})$ and $\bar{y} = (y_0, \ldots, y_{n-1})$. Then there are a positive constant C and a finite set D of pairs (d, μ) with $d \in \{0, \ldots, n\}$ and μ a positive rational, such that for each finite field $A = \mathbb{F}_q$ and each m-tuple $\bar{a}$ from A, putting $l = |\phi(\bar{a}, A^n)|$, either $l = 0$ or $|l - \mu q^d| \leqslant Cq^{(d-1/2)}$ for some $(d, \mu) \in D$.* $\qquad\qquad\square$

The second is a result on stability.

Theorem A.5.16. *(Duret [1980])*. *A field which is PAC but not separably closed has the independence property, and hence is unstable. In particular a stable infinite locally finite field is algebraically closed.* ☐

Speaking of locally finite fields, the following easy fact is worth noting.

Theorem A.5.17. *There are no ω-categorical fields.*

Proof. Let F be an ω-categorical field. Then every finitely generated subfield of F is finite and hence has cyclic multiplicative group. Since F is infinite, its multiplicative group must have elements of arbitrarily high finite order, contradicting uniform local finiteness. ☐

Changing the language: (1) ordered fields

In retrospect it seems curious that the first significant logical results in field theory were for the reals, not the complex numbers. Adding to the language of rings a symbol $\leq$ for a linear ordering relation, Tarski [1931], [1951] proved the following.

Theorem A.5.18. *The theory of real-closed fields is decidable and has quantifier elimination.* ☐

Tarski's proof was by the method of quantifier elimination; see section 2.7. A faster modern proof is sketched in section 8.4. Theorem A.5.18 is important for robotics – not just the decidability, but the fact that quite fast algorithms exist for useful families of sentences. Recently the *Journal of Symbolic Computation* devoted two issues to a survey of the topic: Arnon & Buchberger [1988].

Tarski's proof of Theorem A.5.18 also gave some purchase on the class of definable relations in the real numbers, and this was the beginning of real algebraic geometry. Van den Dries [1988a] surveys this work of Tarski and some of its consequences. In fact van den Dries has been instrumental in getting model theorists to think about this area, and O-minimal structures (Pillay & Steinhorn [1986], etc.) were one of the outcomes.

In his monograph [1951], Tarski asked whether his quantifier elimination procedure could be extended to deal with the reals in a more expressive language. One case that he mentioned is the reals with a 1-ary relation symbol for the algebraic reals.

Theorem A.5.19. (a) *(A. Robinson [1959c])* *If A is a real-closed field with a 1-ary relation symbol picking out a dense real-closed subfield, then* Th(A) *is decidable.*

(b) *(Baur [1982a])* The theory of all structures (A, X), *where A is a real-closed field and X is a real-closed subfield, is undecidable.* □

Robinson's positive result above was extended in several different directions. Macintyre [1975] generalised the notion of 'dense'; see also Baur [1982b]. Keisler [1964a], Jarden [1976] and Mekler [1979] considered algebraically closed fields with a distinguished algebraically closed subfield.

Tarski [1951] also asked what happens if we add a symbol for exponentiation 2^x. Osgood (see van den Dries [1982a]) proved that the resulting theory doesn't have quantifier elimination. But as this book goes to press, Wilkie [199?] announces that this theory is model-complete.

It makes good sense to ask similar questions about the theory of the real numbers or the complex numbers with other symbols added. Some results are known. For example van den Dries [1988b] showed that we still have model-completeness if we add to the reals a symbol for e^x on the interval $[0, 1]$ and a symbol for sin on the interval $[0, \pi]$. Zil'ber (unpublished) showed that if we add to the complex numbers a relation symbol for the roots of unity, the resulting theory is totally transcendental.

If one doesn't extend the language, Tarski's theorem is best possible.

Theorem A.5.20 *(Macintyre, McKenna & van den Dries [1983])* If A is an *ordered field in the language of ordered fields, and* $\mathrm{Th}(A)$ *has quantifier elimination, then A is real-closed.* □

In [1958] Abraham Robinson considered how his diagram methods could be used to get a model-theoretic proof of the main results of Tarski [1951]. But Robinson's main interest in this direction was to offer a more conceptual proof [1955] of Artin's solution of Hilbert's Seventeenth Problem on definite forms. He was able to use model-theoretic methods to get new results on the existence of uniform bounds.

Changing the language: (2) valued fields

The model theory of p-adic number fields Q_p sprang into life rather dramatically in the mid 1960s with the solution by Ax, Kochen and Ershov of a problem of Artin about numbers of solutions of equations over Q_p. This was work of some depth. Macintyre [1986] has performed a public service by analysing the themes behind it.

Ax and Kochen, and independently Ershov, answered Artin's question by developing a model theory of valued fields. The language of valued fields is the language of rings with one extra 1-ary relation symbol V to stand for the valuation ring. A heuristic principle is that if A is a henselian field, then the theory of A in this language is determined by the theories of the residue class field and the value group; this principle is sometimes known as the **Ax–Kochen–Ershov principle**. Two instances are the following.

Theorem A.5.21 *(Ax & Kochen [1965b], [1966], Ershov [1965b], [1966]). In the language of valued fields, the theory of each p-adic number field Q_p is decidable and model-complete.* □

Theorem A.5.22 *(Ax & Kochen [1965a], [1966], Ershov [1965b], [1966]). In the language of valued fields, let A and B be two valued fields with elementarily equivalent value groups and elementarily equivalent residue class fields of characteristic 0. Then $A \equiv B$.* □

In particular suppose we take an ultraproduct A of the fields Q_p by a non-principal ultrafilter. Then A will be elementarily equivalent to a formal power series field. Artin had asked whether the fields Q_p are C_2, i.e. whether for every positive integer d, every homogeneous polynomial of degree d with more than d^2 variables has a non-trivial zero in Q_p. The answer is known to be yes for the corresponding ultraproduct B of formal power series fields $\mathbb{F}_p[[X]]$, since it is true for each $\mathbb{F}_p[[X]]$ separately by a result of Lang. But by Theorem A.5.22. $A \equiv B$. So for each choice of d the statement must be true for A too, and hence by Łoś's theorem it must be true for all but finitely many Q_p. (It was already known that there are some sporadic counter-examples.)

Soon afterwards Cohen [1969] proved the same conclusion by an entirely un-model-theoretic quantifier elimination. Cohen's proof allows one in principle to calculated what are the finitely many rogue primes for Artin's problem.

Prestel & Roquette [1984] offer a proof of Theorem A.5.21. According to taste, one can prove Theorem A.5.22 in either a very model-theoretic way (Chang & Keisler [1973] Theorem 5.4.12) or a very algebraic way (Ax [1971]).

Both Ershov and Ax & Kochen made heavy use of a cross section from the residue field. Macintyre took a different direction. Starting with the language of rings, he introduced for each positive n a 1-ary relation symbol P_n to stand for the set of all non-zero field elements of the form a^n.

Theorem A.5.23 *(Macintyre [1976]). In the language just described, $\mathrm{Th}(Q_p)$ has quantifier elimination.* □

Using the same language, van den Dries [1984] then proved the following.

Theorem A.5.24. $\mathrm{Th}(Q_p)$ *has definable Skolem functions.* □

These last two results bore fruit at once in Denef's [1984] proof of the rationality of certain Poincaré series arising from algebraic sets over Q_p, answering a question of Serre and Oesterlé. In fact Serre and Oesterlé had asked the same question for analytic sets too, and this inspired Denef & van den Dries [199?] to lift the whole argument to an analytic setting. The first step was to prove a quantifier elimination theorem for a language appropriate for talking about analytic sets.

Changing the language: (3) differential fields

Abraham Robinson [1959b] studied fields in which there is a derivation operation d satisfying the laws $d(x + y) = d(x) + d(y), d(xy) = xd(y) + yd(x)$. Such fields are known as **differential fields**. Existentially closed differential fields are said to be **differentially closed**.

Theorem A.5.25 (*A. Robinson [1959b]*). *The theory of differentially closed fields is the model companion of the theory of differential fields. Every differential field A has a differentially closed extension which is algebraic over A in the sense of differential polynomials.*   □

One early application of stability theory was the following result.

Theorem A.5.26. *Over every differential field A of characteristic 0 there is a unique differentially closed field B which is prime over A (which means here that every embedding of A into a differentially closed field C can be extended to an embedding of B into C). In general B is not minimal over A.*

Proof. Blum [1977] showed that the theory of differentially closed fields of characteristic 0 is totally transcendental (it has Morley rank ω). Shelah [1972c] proved the uniqueness of prime models over any set in models of a totally transcendental theory.   □

The situation in characteristic p is more complicated; see Shelah [1973a] and Wood [1973], [1974].

Pillay & Sokolović [1992] make substantial progress towards a proof of the conjecture that every superstable differential field is differentially closed, adding 'we believe the full conjecture to be false'.

A.6 Linear orderings

The material below is grouped under four heads. The *Feferman–Vaught analysis* studies how the first-order theory of a linear ordering is related to the theories of intervals that make up the ordering. *Interpretability* is about structures interpretable in linear orderings, and structures in which linear orderings are interpretable. *Decidability* and *classification* are what the names imply.

It's convenient to prove several of the results in a broader setting. Let λ be a cardinal. A λ-**ordering** is a linear ordering in which 1-ary relation symbols P_i $(i < \lambda)$ pick out subsets. We write L for the first-order language of linear orderings and L_λ for the first-order language of λ-orderings. A 0-ordering is the same thing as a linear ordering, and $L_0 = L$.

If A and B are λ-orderings, we form the λ-ordering $A + B$ by laying a copy of B after a copy of A. More generally, suppose ζ is a linear ordering, and for each $i \in \zeta$ a λ-ordering B_i is given. Then we form the λ-ordering $A = \sum_{i \in \zeta} B_i$ as follows. We can suppose that $\mathrm{dom}(B_i)$ is disjoint from $\mathrm{dom}(B_j)$ whenever $i \neq j$ (after taking isomorphic copies of the structures B_i if necessary). Then $\mathrm{dom}(A)$ is $\bigcup\{\mathrm{dom}(B_i): i \in \zeta\}$. If $i < j$ in ζ, $a \in \mathrm{dom}(B_i)$ and $b \in \mathrm{dom}(B_j)$, then we put $a <^A b$; for elements within any one structure B_i, the relations of A agree exactly with those of B_i.

We write $B \cdot \zeta$ for $\sum_{i \in \zeta} B_i$ when each λ-ordering B_i is equal to B. We write B^* for the λ-ordering B run backwards, i.e. with $>^B$ in place of $<^B$. We say that C is a **subinterval** of A if A can be written as $B + C + D$; B and D can be empty.

The Feferman–Vaught analysis

First a triviality.

Lemma A.6.1. *Let A be a λ-ordering, a and b elements of A with $a < b$, and B the substructure of A whose domain is the interval (a, b). Then B is interpretable in A with parameters a, b.* □

It's more interesting to go in the other direction. Suppose we are considering a λ-ordering of the form $\sum_{i \in \zeta} B_i$, and ψ is a sentence of L_λ. Then we write $\|\psi\|$ for the set $\{i \in \zeta: B_i \vDash \psi\}$.

Theorem A.6.2 (Feferman–Vaught theorem for λ-orderings, where $\lambda < \omega$). *Let $k < \omega$, and let $\psi_0, \ldots, \psi_{\mu-1}$ be the game-normal sentences of L_λ for quantifier rank k (see section 3.3). Then there is an algorithm which computes, for each sentence ϕ of L_λ with quantifier rank k, a sentence ϕ^{fv} of L_μ such that*

(6.1)   *if ζ is a linear ordering and for each $i \in \zeta$, B_i is a λ-ordering, then*

$$\sum_{i \in \zeta} B_i \vDash \phi \Leftrightarrow (\zeta, <^\zeta, \|\psi_0\|, \ldots, \|\psi_{\mu-1}\|) \vDash \phi^{\mathrm{fv}}.$$

Moreover the algorithm is uniform in k. □

In lieu of a proof I make four remarks on this result. The first is that there is really no restriction to finite λ. If λ is infinite, we consider only the symbols P_i which occur in ϕ, and this straight away gives us a language of finite signature.

Second, we can state the theorem in a more elaborate form where there are parameters in ϕ, and in place of each B_i we have $(B_i, \bar{b})$ where $\bar{b}$ are those parameters of ϕ which lie in B_i. In fact one needs to tickle up the theorem in this way in order to prove it. However, the stronger version already follows at once from the version I gave: instead of a parameter b, we can equally well introduce a 1-ary relation symbol P which is true only of b. This trick will be used in our first corollary below.

Third, the proof runs exactly like the back-and-forth arguments of section 3.3, or indeed like the proof of the Feferman–Vaught theorem, Theorem 9.6.1, which is the same idea pursued more energetically. Versions of the Feferman–Vaught theorem for sums of orderings appeared before the version for products. (They can be found in Beth [1954] and Fraïssé [1955]. Feferman & Vaught [1959] discuss how to handle sums of orderings within their framework.)

Fourth, without descending far into details, the interesting case of the proof is where we have found $\chi(\bar{x}, y)^{\mathrm{fv}}$ and we want to construct $(\exists y \, \chi(\bar{x}, y))^{\mathrm{fv}}$. This sentence $(\exists y \, \phi)^{\mathrm{fv}}$ should say 'there is an element i in ζ such that for some element b of B_i, $(\zeta, <^{\zeta}, \|\psi_0(\bar{a}_0^*)\|, \ldots, \|\psi_{\mu-1}(\bar{a}_{s-1}^*)\|) \vDash \phi^{\mathrm{fv}}$ where $\psi_0, \ldots, \psi_{m-1}$ are the appropriate game-normal formulas'. We only need consider a single element b at time, and not a choice of elements b_i from all B_i. This is why ϕ^{fv} in Theorem A.6.2 is first-order, whereas ϕ^{b} in the Feferman–Vaught theorem (Theorem 9.6.2) was second-order monadic.

We draw some corollaries.

Corollary A.6.3. *Let B_i, B_i' be λ-orderings and $\bar{b}_i$, $\bar{b}_i'$ tuples such that $(B_i, \bar{b}_i) \equiv (B_i', \bar{b}_i')$. Write $\bar{c}$ for a well-ordered sequence consisting of all the items in the tuples $\bar{b}_i$ (i in ζ), and $\bar{c}'$ for a sequence which is like $\bar{c}$ but with the items of $\bar{b}_i'$ in place of the corresponding items of $\bar{b}_i$ for each i in ζ. Then $(\sum_{i\in\zeta} B_i, \bar{c}) \equiv (\sum_{i\in\zeta} B_i', \bar{c}')$.*

Proof. See the first and second remarks above. $\qquad\square$

Corollary A.6.4. *For each element i of ζ, let B_i and B_i' be λ-orderings.*
  (a) *If $B_i \equiv B_i'$ for each i, then $\sum_{i\in\zeta} B_i \equiv \sum_{i\in\zeta} B_i'$.*
  (b) *If $B_i \preccurlyeq B_i'$ for each i, then $\sum_{i\in\zeta} B_i \preccurlyeq \sum_{i\in\zeta} B_i'$.*

Proof. Immediate from the previous corollary. $\qquad\square$

For example it follows that if A is a non-empty λ-ordering with no last element, then A has a proper elementary extension B in which all the new elements come later than all elements of A. For this, take a sequence $(a_i: i < \alpha)$ of elements which is cofinal in A, and by compactness let C be an elementary extension of A in which there are elements later than all elements in A. By Lemma A.6.1, each interval $[a_i, a_{i+1})^A$ is an elementary substructure of the corresponding interval $[a_i, a_{i+1})^C$. Using Corollary A.6.4, replace each of the intervals $[a_i, a_{i+1})^C$ in C by its elementary substructure (and likewise for the initial interval $(-\infty, a_0)$).

Corollary A.6.5 (Rubin [1974]). *Let A and C be λ-orderings. Then for every formula $\phi(\bar{x})$ of the language L_λ with parameters from A and C, and every λ-ordering B, there is a formula $\phi^*(\bar{x})$ of L_λ such that if $\bar{b}$ is any tuple in B, we have*

(6.2) $$A + B + C \vDash \phi(\bar{b}) \Leftrightarrow B \vDash \phi^*(\bar{b}).$$

Proof. We can disguise the parameters and the tuple $\bar{b}$ by using the second remark above; so without loss ϕ is a sentence of L_λ. By (6.1) the left-hand side of (6.2) is equivalent to a boolean combination of conditions of the form '$B \vDash \psi_i$', where $\psi_0, \ldots, \psi_{\mu-1}$ are the game-normal sentences of the theorem. So ϕ^* can be written as a boolean combination of the ψ_i's. $\qquad\square$

The next corollary was probably discovered by several people. It appears in Kamp's analysis [1968] of linear orderings in connection with the logical properties of the temporal operators *since* and *until*. One can prove it much like Corollary A.6.5.

Corollary A.6.6. *In L_λ, let T be the theory of linear orderings. Every formula $\phi(x_0, \ldots, x_{m-1})$ of L_λ, with $m \geqslant 2$, is equivalent modulo T to a boolean combination of formulas $\psi(x_i, x_j)$ of L_λ with at most two free variables.* $\qquad\square$

Interpretability

If an infinite linear ordering is interpretable (with or without parameters) in the structure A, then A is unstable; in fact A has the strict order property. This is immediate from the definitions (section 6.7). The converse is clearly false: the Fraïssé limit of the class of finite partial orderings has the strict order property, but no infinite linear ordering is interpretable in it.

By Poizat [1981], no λ-ordering has the independence property.

What can be interpreted in a linear ordering?

We have the following quite extraordinary result of Lachlan [1987b] (cf. Hrushovski [1989b]).

Theorem A.6.7. *Let A be a countable model of a totally categorical theory, and suppose that the geometry of the strongly minimal set of A is disintegrated. Then A is interpretable with parameters in the ordering of the rationals.* ☐

The interpretation must in general be many-dimensional, as one sees from Example 3 in section 5.1.

On the other hand no infinite group can be interpreted in a λ-ordering. For this we need a lemma. If $\bar{a}$ is an m-tuple and X is a subset of m, we write $\bar{a}|X$ for the sequence $(a_{i_0}, \ldots, a_{i_{k-1}})$ where $i_0, \ldots, i_{k-1}$ are the elements of X in increasing order.

Lemma A.6.8. *Let m and n be positive integers, A a linear ordering and $\bar{a}$, $\bar{b}$ respectively an m-tuple and an n-tuple from A. Then there is a subset X of m, of cardinality at most $2n$, such that if $(A, \bar{a}) \equiv (A, \bar{a}')$ and $(A, \bar{a}|X, \bar{b}) \equiv (A, \bar{a}'|X, \bar{b})$ then $(A, \bar{a}, \bar{b}) \equiv (A, \bar{a}', \bar{b}')$.*

Proof. We can suppose without loss that the tuple $\bar{a}$ is $(a_0, \ldots, a_{m-1})$ where $a_0 < \ldots, < a_{m-1}$, and likewise $\bar{a}'$ is $(a'_0, \ldots, a'_{m-1})$. For simplicity we first prove the case where $n = 1$ so that $\bar{b}$, $\bar{b}'$ are single elements b, b'. If b is some a_i then the result is clear. Suppose then that $a_i < b < a_{i+1}$. Let B, C, D be the suborderings of A consisting of $(-\infty, a_i]$, (a_i, a_{i+1}), $[a_{i+1}, \infty)$ respectively; let B', C', D' be the corresponding suborderings using $\bar{a}'$ in place of $\bar{a}$. Assume

(6.3) $(A, \bar{a}) \equiv (A, \bar{a}')$ and $(A, a_i, a_{i+1}, b) \equiv (A, a'_i, a'_{i+1}, b')$.

Then by Lemma A.6.1,

(6.4) $(B, a_0, \ldots, a_i) \equiv (B', a'_0, \ldots, a'_i), (C, b) \equiv (C', b')$ and
          $+1, \ldots, a'_{m-1})$.

So by Corollary A.6.3 above, $(A, \bar{a}, b) \equiv (A, \bar{a}', b')$. This proves the lemma with $X = \{i, i + 1\}$. The cases where $b < a_0$ and $b > a_{m-1}$ are similar.

The argument when $n > 1$ runs the same way, but A may have to be broken up into more intervals according to where the items in $\bar{b}$ lie. ☐

Theorem A.6.9 (Poizat [1987b]). *Let A be a λ-ordering. Then no infinite group is interpretable in A with parameters. (Hence no infinite group is interpretable in an infinite set, since an infinite set is a reduct of an infinite linear ordering.)*

Proof. Suppose to the contrary that A is a model of T and G is an infinite group interpretable in A with parameters, by an interpretation Γ. As above, we can absorb the parameters into the 1-ary relations of L_λ. Replacing A by an elementary extension of A if necessary (and noting Theorem 5.3.4), we can use the Ehrenfeucht–Mostowski method as in section 11.3 to find

n-tuples $(\bar{a}_q : q \in \mathbb{Q})$ in A which form an indiscernible sequence in A with the order-type of the rationals, such that $\bar{a}_q/=_\Gamma$ are distinct elements of G. Write b_q for $\bar{a}_q/=_\Gamma$, and for each $i < \omega$ choose an n-tuple $\bar{c}_i$ in A so that

(6.5) $$b_0 \cdot \ldots \cdot b_i = \bar{c}_i/=_\Gamma.$$

By Lemma A.6.8 there are some i and $j \leqslant i$ such that the type of $\bar{c}_i$ over $\bar{a}_0$, $\ldots$, $\bar{a}_i$ is determined by its type over $\bar{a}_0, \ldots, \bar{a}_{j-1}, \bar{a}_{j+1}, \ldots, \bar{a}_i$ and the theory of the indiscernible sequence. Let $\phi(x_0, \ldots, x_{i+1})$ be the group-theoretic formula $x_0 \cdot \ldots \cdot x_i = x_{i+1}$. Then we have

(6.6) $$A \vDash \phi_\Gamma(\bar{a}_0, \ldots, \bar{a}_i, \bar{c}_i)$$

and so by choice of i and j,

(6.7) $$A \vDash \phi_\Gamma(\bar{a}_0, \ldots, \bar{a}_{j-1}, \bar{a}_{j+1/2}, \bar{a}_{j+1}, \ldots, \bar{a}_i, \bar{c}_i).$$

Then $b_0 \cdot \ldots \cdot b_i = \bar{c}_i/=_\Gamma = b_0 \cdot \ldots \cdot b_{j-1} \cdot b_{j+1/2} \cdot b_{j+1} \cdot \ldots \cdot b_i$, so $b_j = b_{j+1/2}$, contradiction. $\qquad\square$

Decidability

Ehrenfeucht [1959] proved the following.

Theorem A.6.10. *For every finite λ, the theory of λ-orderings is decidable.*

I sketch a proof, mainly following Läuchli & Leonard [1966]. It introduces some ideas of wider interest.

Let S be a finite set of λ-orderings. We define a λ-ordering σS, known as the **shuffle** of S. Write $\mathbb{Q}$ for the rationals in their usual ordering, and suppose that for each rational number q a λ-ordering B_q in S is chosen, so that

(6.8)   for each B in S, every interval of the rationals contains some q with $B = B_q$.

We define σS to be the λ-ordering $\sum_{q \in \mathbb{Q}} B_q$. It's an exercise (much like Exercise 3.2.5) to show that (6.8) determines σS up to isomorphism by S.

Lemma A.6.11. *For a fixed k, let Φ be the set of game-normal sentences of L_λ for quantifier rank k.*

(a) We can compute, for each pair (ψ, ψ') of sentences in Φ, a sentence $\psi + \psi'$ in Φ such that if ζ, ζ' are any λ-orderings and $A \vDash \psi$ and $B \vDash \psi'$, then $A + B \vDash \psi + \psi'$.

(b) We can compute, for each sentence ψ in Φ, a sentence $\psi \cdot \omega$ in Φ such that if B is any λ-ordering and $B \vDash \psi$, then $B \cdot \omega \vDash \psi \cdot \omega$. (Likewise a sentence $\psi \cdot \omega^*$ for $B \cdot \omega^*$.)

(c) We can compute, for any finite set of sentences $\psi_0, \ldots, \psi_{n-1}$ in Φ, a

sentence $\sigma(\{\psi_0, \ldots, \psi_{n-1}\})$ *in* Φ *such that if* $B_0, \ldots, B_{n-1}$ *are* λ-*orderings which are models of* $\psi_0, \ldots, \psi_{n-1}$ *respectively, then* $\sigma(\{B_0, \ldots, B_{n-1}\})$ *is a model of* $\sigma(\{\psi_0, \ldots, \psi_{n-1}\})$.

Moreover the algorithms for finding $\psi + \psi'$, $\psi \cdot \omega$, $\sigma(\{\psi_0, \ldots, \psi_{n-1}\})$ *are uniform in* k.

Proof. (a) Theorem A.6.2 with $\zeta = 2$ tells us that for each $\chi \in \Phi$ we can compute a subset $\Psi(\chi)$ of $\Phi \times \Phi$ such that for any λ-orderings A and B, $A + B \vDash \chi$ if and only if $A \vDash \psi$ and $B \vDash \psi'$ for some pair $(\psi, \psi') \in \Psi(\chi)$. If ψ and ψ' are sentences in Φ which are true in some λ-orderings, then there is a unique χ such that (ψ, ψ') lies in $\Psi(\chi)$.

(b) In the notation of (6.1), ψ is one of the formulas $\psi_0, \ldots, \psi_{\mu-1}$; let it be ψ_i. The problem is to find the unique ϕ in Φ such that $(\omega, <, X_0, \ldots, X_{\mu-1}) \vDash \phi^{\mathrm{fv}}$, where $X_i = \omega$ and $X_j = \varnothing$ whenever $i \neq j < \mu$. We can do this provided that $\mathrm{Th}(\omega, <)$ is decidable. But that was Corollary 3.3.9.

(c) is similar. $\qquad\qquad\qquad\qquad\qquad\qquad\qquad\qquad\qquad\qquad\qquad\qquad\square$

We shall use the notation $\psi + \psi'$, $\psi \cdot \omega$, $\sigma(X)$ of this lemma. The following corollary is a detour, but we shall use some ideas from its proof.

Corollary A.6.12. (a) *For a fixed finite* λ, *the theory of the class of finite* λ-*orderings is decidable.*

(b) *For a fixed finite* λ, *the theory of* λ-*orderings of the form* $(\omega, <, X_0, \ldots, X_{\lambda-1})$ *(where* $<$ *is the usual ordering on* ω*) is decidable.*

Proof. (a) Consider $k < \omega$, and let Φ be the set of game-normal sentences of L_λ for quantifier rank k. By direct computation we can find the set Φ_0 of those sentences in Φ which are true in some λ-ordering with at most one element. Let Φ_1 be the least subset of Φ which contains Φ_0 and is closed under $+$. By Lemma A.6.11(a), Φ_1 is computable from k. Also if Ψ is a subset of Φ, then $\bigvee \Psi$ is true in some finite λ-ordering if and only if $\Psi \cap \Phi_1 \neq \varnothing$. This tells us which game-normal formulas are true in some λ-ordering. Now use the Fraïssé–Hintikka theorem (Theorem 3.3.2).

(b) As in part (a), we can compute the set Φ_2 of those sentences in Φ which are true in some non-empty finite λ-ordering. Let Φ_3 be the set of all sentences in Φ of the form $\psi + \psi' \cdot \omega$ with ψ in Φ_1 and ψ' in Φ_2. Then Φ_3 is the set of all sentences in Φ which are true in some λ-ordering of the form $A + B \cdot \omega$, where A is finite and B is finite and non-empty. It follows that every sentence in Φ_3 is true in some λ-ordering of the form $(\omega, <, X_0, \ldots, X_{\lambda-1})$. To complete the argument as in (a), we only need to show that every λ-ordering A of the form $(\omega, <, X_0, \ldots, X_{\lambda-1})$ satisfies some sentence in Φ_3.

We prove it as follows. Let A be as above. In the notation of Ramsey's theorem (section 11.1 above), define a map $f: [\omega]^2 \to \Phi_2$ as follows: if $m < n$ then $f(m, n)$ is the unique sentence in Φ_2 which is true of the subinterval $[m, n)$ of A. By Ramsey's theorem, Theorem 11.1.3, there is an infinite subset Y of ω such that f takes a constant value of ψ' on $[Y]^2$. Let m be the least element of Y and ψ the sentence in Φ_1 which is true of the interval $[0, m)$ of A. Then A can be written as $C + \sum_{i<\omega} B_i$ where $C \vDash \psi$ and for all $i < \omega$, $B_i \vDash \psi'$. So A satisfies $\psi + \psi' \cdot \omega$ as required. $\qquad\square$

The **Läuchli–Leonard set** of λ-orderings, M_λ, is defined to be the smallest set of λ-orderings such that (1) every λ-ordering with at most one element is in M_λ, (2) if A and B are in M_λ then $A + B$ is in M_λ, (3) if A is in M_λ then both $A \cdot \omega$ and $A \cdot \omega^*$ are in M_λ, and (4) if S is a finite subset of M_λ then σS is in M_λ.

Lemma A.6.13. *For each $\lambda < \omega$, the theory of the class M_λ is decidable.*

Proof. This is like the proof of Corollary A.6.12, but using the shuffle as well. $\qquad\square$

Now we prove Theorem A.6.10. Take any $k < \omega$, and let Ψ be the set of all game-normal sentences of L_λ for quantifier rank k which are true in at least one λ-ordering in M_λ. By Lemma A.6.11 we can compute Ψ from k. By the Fraïssé–Hintikka theorem again, the theorem follows if we show that every λ-ordering satisfies some sentence in Ψ.

Call a λ-ordering **good** if it satisfies some sentence in Ψ, and **very good** if every subinterval of it is good. We shall show that every λ-ordering A is good. By the downward Löwenheim–Skolem theorem (Corollary 3.1.5) we can assume without loss that A is at most countable.

Define a relation $\sim$ on $\mathrm{dom}(A)$ by

(6.9) $\quad a \sim b$ iff every subinterval of the interval (a, b) in A (or (b, a) if $b < a$) is very good.

Since Ψ is closed under $+$, $\sim$ is an equivalence relation on $\mathrm{dom}(A)$. Since A is at most countable, any equivalence class of $\sim$ can be written as an interval in A of the form $\sum_{i \in \omega^* + \omega} B_i$ where each B_i is very good. By (1)–(3) in the definition of M_λ, together with an application of Ramsey's theorem as in the proof of Corollary A.6.12(b), it follows that the equivalence class is very good too. We can write A as $\sum_{j \in \zeta} B'_j$ where each B'_j is an equivalence class of $\sim$. Since Ψ is closed under $+$, no element of the linear ordering ζ has an immediate successor in ζ; in other words, ζ is either 0 or 1 or a countable dense ordering. We shall show that ζ is not countable and dense. It will follow at once that ζ is one equivalence class, so that A is very good.

Suppose to the contrary that ζ is countable and dense. There are a finite subset Ψ_1 of Ψ and an open interval ξ in ζ such that each sentence in Ψ_1 is true 'densely often' among the B'_j with j in ξ, but no other sentence of Ψ is true in any B'_j with j in ξ. Then by (4) in the definition of M_λ, $\sum_{j \in \xi} B'_j$ is very good, so that it must lie inside a single equivalence class of $\sim$. This contradiction completes the proof.   ☐ Theorem A.6.10

In the same vein one can use the Hausdorff rank (Exercise 3.4.13) to prove the decidability of the theory of scattered orderings. (Rosenstein [1982] Theorem 13.50 points out that the theory of scattered orderings is the complement of r.e.)

Läuchli [1968] proved that the weak monadic second-order theory of λ-orderings is decidable, by a version of the same argument. Yet another variant gives that the monadic second-order theory of *countable* linear orderings is decidable – but since we lack the downward Löwenheim–Skolem theorem for monadic second-order theories, this tells us nothing about the monadic second-order theory of linear orderings in general. (In fact Shelah [1975d] shows that the continuum hypothesis implies the monadic second-order theory of linear orderings is undecidable.) We also get a decidability result by taking the second-order variables to range over intervals of the orderings; bearing in mind that every countable boolean algebra is the interval algebra of some linear ordering (Mostowski & Tarski [1939]), this yields the decidability of the theory of boolean algebras (which Tarski [1949c] proved by quantifier elimination – see Chang & Keisler [1973] section 5.5 for a proof.) Yet another variation takes on board third-order quantifiers, and deduces that the theory of countable boolean algebras with second-order quantifiers ranging over ideals is decidable. Gurevich [1985] surveys the monadic second-order theories of linear orderings.

Elementary classification and numbers of models

In section 7.2 the following special case of Vaught's conjecture was mentioned. The proof is difficult and throws up a good deal of extra information about the model theory of linear orderings.

Theorem A.6.14 (Rubin [1974]). *If T is the complete theory of some infinite linear ordering, then the number of countable models of T up to isomorphism is either 1 or 2^ω.*   ☐

The next result was the first non-trivial classification of the ω-categorical models of a first-order theory.

Theorem A.6.15 *(Rosenstein [1969]).* *The countable ω-categorical linear order-ings are exactly those infinite orderings which are got from* 1 *by a finite number of applications of* + *and shuffle.*   ☐

All infinite λ-orderings are unstable. So if A is an infinite λ-ordering, then by Theorem 11.3.10, for every uncountable cardinal κ there is a family of 2^{κ} λ-orderings $(A_i: i < 2^{\kappa})$ elementarily equivalent to A, such that if $i \neq j$ then A_i is not elementarily embeddable in A_j. (By Theorem 2.7.1 we can replace 'elementarily embeddable' by 'embeddable'.)

Finally there are a number of papers which study the classification of linear orderings by equivalence in various languages. Rosenstein [1982] surveys these results.

For example Doner, Mostowski & Tarski [1978] (reporting results of Mostowski and Tarski from 1938–41) classify ordinals by their first-order theories. They show that every ordinal is elementarily equivalent to an ordinal $< \omega^{\omega} \cdot 2$, and that every sentence of the first-order language of linear orderings which is true in some ordinal is true in some ordinal $< \omega^{\omega}$. Kino [1966] shows that if κ is any regular cardinal, then every sentence of $L_{\kappa\kappa}$ which is true in some ordinal is true in some ordinal $< \kappa^{\kappa}$. This is useful as a test of the expressive strength of $L_{\kappa\kappa}$, in rather the same spirit as Theorem 11.5.4 for $L_{\kappa^{+}\omega}$.

REFERENCES

The numbers in square brackets at the end of each entry refer to pages in the text. To that extent, this is an author index as well as a bibliography. I have indicated – to the best of my knowledge – how the authors refer to themselves. (But some authors have chosen to use pseudonyms: Ivan Aguzarov, R. E. Farey, John B. Goode, Hans B. Gute, K. K. Reiter.)

The bibliography of Addison, Henkin & Tarski [1965] lists virtually all publications in model theory up to 1964, including itself.

Abian, A. 1970. On the solvability of infinite systems of Boolean polynomial equations. *Colloq. Math.*, **21**, 27–30. [534]

Ackermann, W. (Wilhelm Ackermann) 1954. *Solvable cases of the decision problem*. Amsterdam: North-Holland. [84]

Addison, J. W., Henkin, L. A. & Tarski, A. (eds) 1965. *The theory of models, Proc. 1963 Internat. Symposium at Berkeley*. Amsterdam: North-Holland. [83]

Adler, A. (Allan Adler) 1978. On the multiplicative semigroups of rings. *Comm. Algebra*, **6**, 1751–3. [669]

Aguzarov, I., Farey, R. E. & Goode, J. B. 1991. An infinite superstable group has infinitely many conjugacy classes. *J. Symbolic Logic*, **56**, 618–23. [693]

Ahlbrandt, G. (Gisela Ahlbrandt) 1987. Almost strongly minimal totally categorical theories. In *Logic Colloquium '85*, ed. The Paris Logic Group, pp. 17–31. Amsterdam: North-Holland. [651]

Ahlbrandt, G. & Ziegler, M. 1986. Quasi finitely axiomatizable totally categorical theories. *Ann. Pure Appl. Logic*, **30**, 63–82. [197, 222, 261, 358, 359, 651]

—199?. What's so special about $(\mathbb{Z}/4\mathbb{Z})^{\omega}$? *Arch. Math. Logic* (to appear). [651]

Albert, M. H. & Burris, S. N. 1988. Bounded obstructions, model companions, and amalgamation bases, *Z. Math. Logik Grundlag. Math.*, **34**, 109–15. [411]

Almagambetov, Zh. A. (Ж. А. Алмагамбетов) 1965. О классах акхиом, замкнутых относительно заданных приведенных произведений и степеней (On classes of axioms closed under given reduced products and powers). *Algebra i Logika*, **4** (3), 71–7. [463, 476]

Anapolitanos, D. A. 1979. Automorphisms of finite order. *Z. Math. Logik Grundlag. Math.*, **25**, 565–75. [533]

Anapolitanos, D. A. & Väänänen, J. A. (Jouko A. Väänänen) 1980. On the axiomatizability of the notion of an automorphism of a finite order. *Z. Math. Logik Grundlag. Math.*, **26**, 433–7. [533]

Appel, K. I. (Kenneth I. Appel) 1959. Horn sentences in identity theory. *J. Symbolic Logic*, **24**, 306–10. [473]

Apps, A. B. 1982. Boolean powers of groups. *Math. Proc. Camb. Phil. Soc.*, **91**, 375–95. [358, 476]

— 1983a. $\aleph_0$-categorical finite extensions of Boolean powers. *Proc. London Math. Soc.*, **47**, 385–410. [358]

— 1983b. On $\aleph_0$-categorical class two groups. *J. Algebra*, **82**, 516–38. [344, 477]

— 1983c. On the structure of $\aleph_0$-categorical groups. *J. Algebra*, **81**, 320–39. [358, 477]

Aquinas, T. (Thomas Aquinas) 1921. *Summa theologiae*. English translation, London: Burns Oates & Washbourne. [21]

Arens, R. F. & Kaplansky, I. 1948. Topological representations of algebras. *Trans. Amer. Math. Soc.*, **63**, 457–81. [476, 477]

Arnon, D. S. & Buchberger, B. 1988. Algorithms in real algebraic geometry. Reprinted from *J. Symbolic Computation*, **5** (1, 2), (1988). London: Academic Press. [703]

Arrow, K. J. (Kenneth J. Arrow) 1950. A difficulty in the concept of social welfare. *J. Political Economy*, **58**, 328–46. [475]

Artin, E. & Schreier, O. 1927. Algebraische Konstruktion reeller Körper. *Abh. Math. Sem. Univ. Hamburg*, **5**, 83–115. [386, 409, 631]

Ash, C. J. (Christopher J. Ash) 1971. Undecidable $\aleph_0$-categorical theories. Abstract. *Not. Amer. Math. Soc.*, **18**, 423. [359]

Ash, C. J. & Nerode, A. (Anil Nerode) 1975. Functorial properties of algebraic closure and Skolemization. *Logic Paper* **20**, Dept. Math., Monash University. [595]

Ash, C. J. & Rosenthal, J. W. (John Rosenthal) 1980. Some theories associated with algebraically closed fields. *J. Symbolic Logic*, **45**, 359–62. [199]

Ax, J. (James Ax) 1968. The elementary theory of finite fields. *Ann. Math.*, **88**, 239–71. [701, 702]

— 1971. A metamathematical approach to some problems in number theory. In *1969 Number Theory Institute*, Proc. Symp. in Pure Math. XX, ed. D. J. Lewis, pp. 161–90. Providence, RI: Amer. Math. Soc. [319, 705]

Ax, J. & Kochen, S. 1965a. Diophantine problems over local fields I. *Amer. J. Math.*, **87**, 605–30. [705]

— 1965b. Diophantine problems over local fields II: A complete set of axioms for p-adic number theory. *Amer. J. Math.*, **87**, 631–48. [705]

— 1966. Diophantine problems over local fields III: Decidable fields. *Ann. Math.*, **83**, 437–56. [705]

Bacsich, P. D. (Paul D. Bacsich) 1972. Cofinal simplicity and algebraic closedness. *Algebra Universalis*, **2**, 354–60. [409]

— 1973. Defining algebraic elements. *J. Symbolic Logic*, **38**, 93–101. [197]

— 1975. The strong amalgamation property. *Colloq. Math.*, **33**, 13–23. [320, 411]

Bacsich, P. D. & Rowlands Hughes, D. (Dafydd Rowlands Hughes) 1974. Syntactic characterisations of amalgamation, convexity and related properties. *J. Symbolic Logic*, **39**, 433–51. [411]

Baïsalov, E. R. (Е. Р. Байсалов) 1991. Пример теории, отвечающий на вопрос Б. С. Байжанова (Example of a theory which answers a question of B. S. Baïzhanov). *Algebra i Logika*, **30**, 15–16. [651]

Baïzhanov, B. S. (Б. С. Байжанов) 1990. Группы автоморфизмов и координатизируемость (Automorphism groups and coordinatisability). In *Soviet–French Colloquium in Theory of Models* (Russian), p. 4. Karaganda: Karaganda State Univ. [652]

Baldwin, J. T. (John T. Baldwin) 1970. A note on definability in totally transcendental theories. Abstract. *Not. Amer. Math. Soc.*, **17**, 1087. [321]

— 1972. Almost strongly minimal theories, I, II. *J. Symbolic Logic*, **37**, 487–93 and 657–60. [200]

— 1973a. A sufficient condition for a variety to have the amalgamation property. *Colloq. Math.*, **28**, 181–3. [474]

— 1973b. α_T is finite for $\aleph_1$-categorical T. *Trans. Amer. Math. Soc.*, **181**, 37–51. [651]

— 1974. Atomic compactness in $\aleph_1$-categorical Horn theories. *Fundamenta Math.*, **83**, 263–8. [534]

— 1978. Some EC$_\Sigma$ classes of rings. *Z. Math. Logik Grundlag. Math.*, **24**, 489–92. [83]

— 1985. Definable second-order quantifiers. In Barwise & Feferman [1985], pp. 445–77. [86]

— 1988. *Fundamentals of stability theory*. Berlin: Springer. [306, 322]

— 1989. Some notes on stable groups. In Nesin & Pillay [1989], pp. 100–16. [132]

Baldwin, J. T., Blass, A. R., Glass, A. M. W. & Kueker, D. W. 1973. A 'natural' theory without a prime model. *Algebra Universalis*, **3**, 152–5. [358]

Baldwin, J. T. & Kueker, D. W. 1980. Ramsey quantifiers and the finite cover property. *Pacific J. Math.*, **90**, 11–19. [198]

— 1981. Algebraically prime models. *Ann. Math. Logic*, **20**, 289–330. [378, 410]

Baldwin, J. T. & Lachlan, A. H. 1971. On strongly minimal sets. *J. Symbolic Logic*, **36**, 79–96. [82, 199, 650, 651]

— 1973. On universal Horn classes categorical in some infinite power. *Algebra Universalis*, **3**, 98–111. [474, 477]

Baldwin, J. T. & McKenzie, R. N. 1982. Counting models in universal Horn classes. *Algebra Universalis*, **15**, 359–84. [474]

Baldwin, J. T. & Rose, B. I. 1977. $\aleph_0$-categoricity and stability of rings. *J. Algebra*, **45**, 1–16. [358]

Baldwin, J. T. & Saxl, J. (Jan Saxl) 1976. Logical stability in group theory. *J. Austral. Math. Soc. Ser. A*, **21**, 267–76. [321]

Baldwin. J. T. & Shelah, S. 1985. Second order quantifiers and the complexity of theories. *Notre Dame J. Formal Logic*, **26**, 229–303. [86]

Ball, R. N. 1984. Distributive Cauchy lattices. *Algebra Universalis*, **18**, 134–74. [86]

Banaschewski, B. & Herrlich, H. 1976. Subcategories defined by implications. *Houston J. Math.*, **2**, 105–13. [475]

Banaschewski, B. & Nelson, E. 1980. Boolean powers as algebras of continuous functions. *Dissertationes Math. (Warsaw)*, **179**. [477]

Bankston, P. J. & Ruitenburg, W. B. G. (Paul Bankston, Wim Ruitenburg) 1990. Notions of relative ubiquity for invariant sets of relational structures. *J. Symbolic Logic*, **55**, 948–86. [352]

Baranskiĭ, V. A. (В. А. Баранский) 1983. Об алгебраических системах, элементарная теория которых совместима с произвольной группой. *Algebra i Logika*, **22** (6), 599–607. *Trans. as* Algebraic systems whose elementary theory is compatible with an arbitrary group. *Algebra and Logic*, **22** (1983), 425–31. [197]

Barwise, K. J. (Jon Barwise) 1972. Absolute logic and $L_{\infty\omega}$. *Ann. Math. Logic.*, **4**, 309–40. [129]

— 1973. A preservation theorem for interpretations. In *Cambridge Summer School in Mathematical Logic*, ed. A. R. D. Mathias & H. Rogers, Lecture Notes in Math. 337, pp. 618–21. Berlin: Springer. [320, 652]

— 1975. *Admissible sets and structures*. Berlin: Springer. [83, 129]

Barwise, K. J. & Eklof, P. C. 1969. Lefschetz's principle. *J. Algebra*, **13**, 554–70. [700]

Barwise, K. J. & Feferman, S. (eds) 1985. *Model-theoretic logics*. New York: Springer. [28, 86, 319, 536]

Barwise, K. J. & Kunen, K. 1971. Hanf numbers for fragments of $L_{\infty\omega}$. *Israel J. Math.*, **10**, 306–20. [597]

Barwise, K. J. & Moschovakis, Y. N. 1978. Global inductive definability. *J. Symbolic Logic*, **43**, 521–34. [597]

Barwise, K. J. & Robinson, A. 1970. Completing theories by forcing, *Ann. Math. Logic*, **2**, 119–42. Reprinted in A. Robinson [1979a], pp. 219–42. [409]

Barwise, K. J. & Schlipf, J. S. 1976. An introduction to recursively saturated and resplendent models. *J. Symbolic Logic*, **41**, 531–6. [522, 534]

Basarab, Ş. A. (Şerban A. Basarab) 1975. The models of the elementary theory of abelian finite groups (Romanian, English summary). *Stud. Cerc. Mat.*, **27**, 381–6. [672]

Baudisch, A. (Andreas Baudisch) 1981. The elementary theory of Abelian groups with μ-chains of pure subgroups. *Fundamenta Math.*, **112**, 147–57. [673]

— 1984. Tensor products of modules and elementary equivalence. *Algebra Universalis*, **19**, 120–7. [672]

— 1989. Classification and interpretation. *J. Symbolic Logic*, **54**, 138–59. [688]

— 1990. On superstable groups. *J. London Math. Soc.*, **42**, 452–64. [694]

Baumgartner, J. E. (James E. Baumgartner) 1976. A new class of order types. *Ann. Math. Logic*, **9**, 187–222. [595, 596]

Baumslag, B. & Levin, F. (Benjamin Baumslag, Frank Levin) 1976. Algebraically closed torsion-free nilpotent groups of class 2. *Comm. Algebra*, **4**, 533–60. [690]

Baumslag, G., Dyer E. & Heller, A. 1980. The topology of discrete groups. *J. Pure Appl. Algebra*, **16**, 1–47. [689]

Baur, W. (Walter Baur) 1975. $\aleph_0$-categorical modules. *J. Symbolic Logic*, **40**, 213–20. [358, 662]

— 1976. Elimination of quantifiers for modules. *Israel J. Math.*, **25**, 64–70. [655]

— 1982a. Die Theorie der Paare reell abgeschlossener Körper. *L'Enseignement Math.*, **30**, 25–34. [704]

— 1982b. On the elementary theory of pairs of real closed fields. II. *J. Symbolic Logic*, **47**, 669–79, [704]

Baur, W., Cherlin, G. L. & Macintyre, A. J. 1979. Totally categorical groups and rings. *J. Algebra*, **57**, 407–40. [261, 263, 358, 693]

Bauval, A. (Anne Bauval) 1985. Polynomial rings and weak second-order logic. *J. Symbolic Logic*, **50**, 953–72. [86]

Belegradek, O. V. (О. В. Белеградек) 1978a. О нестабильных теориях групп (On unstable theories of groups). *Izv. Vyssh. Uchebn. Zaved. Matematika*, **22** (8), 41–4. [321]

— 1978b. элементарные свойства алгебраически замкнутых групп (Elementary properties of algebraically closed groups). *Fundamenta Math.*, **98**, 83–101. [690]

— 1988. Теория моделей локально свободных алгебр (The theory of models of locally free algebras). In Теория моделей и её применения (*Theory of models and its applications*), pp. 3–25. Novosibirsk: Nauka. *English summary*: Some model theory of locally free algebras. In *6th Easter Conference on Model Theory*, ed. B. I. Dahn & H. Wolter, pp. 28–32. Berlin: Humboldt-Univ. [85, 321]

— 199?. The Mal'cev correspondence revisited. In *Conf. on Algebra honoring A. Mal'cev (Novosibirsk, August 21–26, 1989)*, ed. L. A. Bokut *et al.*, Contemporary Mathematics. Providence, RI: Amer. Math. Soc. [262]

Bell, J. L. & Machover, M. 1977. *A course in mathematical logic*. Amsterdam: North-Holland. [467]

Bell, J. L. & Slomson, A. B. 1969. *Models and ultraproducts*. Amsterdam: North-Holland. [449]

Beltrami, E. (Eugenio Beltrami) 1868. Saggio di interpretazione della geometria non-euclidea. *Giornale di Matematiche*, **6**, 284–312. [260]

Benda, M. (Miroslav Benda) 1974. Remarks on countable models. *Fundamenta Math.*, **81**, 107–19. [533]

Bergman, G. M. (George Bergman) 1972. Boolean rings of projection maps. *J. London Math. Soc.*, **4**, 593–8. [477]

Bergstra, J. A. & Meyer, J. 1983. On specifying sets of integers. *J. Informat. Proc. Cybernetics – EIK*, **20**, 531–41. [428]

Bergstra, J. A. & Tucker, J. V. 1987. Algebraic specifications of computable and semicomputable data types. *Theoretical Comp. Sci.*, **50**, 137–81. [439]

Berline, C. (Chantal Berline) 1983. Déviation des types dans les corps algébriquement clos. In *Théories stables 3*, ed. B. Poizat, pp. 3.01–3.10. Paris: Université Pierre et Marie Curie. [699]

— 1986. Superstable groups; a partial answer to conjectures of Cherlin and Zil'ber. *Ann. Pure Appl. Logic*, **30**, 45–61. [692]

Berline, C. & Lascar, D. 1986. Superstable groups. *Ann. Pure Appl. Logic*, **30**, 1–43. [699]

Bernays, P. (Paul Bernays) 1942. Review of Max Steck, Ein unbekannter Brief von Gottlob Frege. *J. Symbolic Logic*, **7**, 92–3. [83]

Beth, E. W. (Evert W. Beth) 1953. On Padoa's method in the theory of definition. *Nederl. Akad. Wetensch. Proc. Ser. A*, **56** (= *Indag. Math.*, **15**), 330–9. [321, 533]

— 1954. Observations métamathématiques sur les structures simplement ordonnés. In *Applications scientifiques de la logique mathématique, Collection de Logique Math.*, Série A. Fasc. 5, pp. 29–35. Paris and Louvain. [708]

— 1955. Semantic entailment and formal derivability. *Mededelingen der Koninklijke Nederlandse Akademie van Wetenschappen, Afd. Letterkunde*, **18**, 309–42. [84]

Birch, B. J., Burns, R. G., Macdonald, S. O. & Neumann, P. M. 1976. On the orbit-sizes of permutation groups containing elements separating finite subsets. *Bull. Austral. Math. Soc.*, **14**, 7–10. [198]

Birkhoff, G. (Garrett Birkhoff) 1935a. Abstract linear dependence and lattices. *Amer. J. Math.*, **57**, 800–4. [199]

— 1935b. On the structure of abstract algebras. *Proc. Camb. Phil. Soc.*, **31**, 433–54. [474]

Blass, A. R. (Andreas Blass) 1979. Injectivity, projectivity, and the axiom of choice. *Trans. Amer. Math. Soc.*, **255**, 31–59. [86]

Blass, A. R., Gurevich, Y. S. & Kozen, D. 1985. A zero-one law for logic with a fixed-point operator. *Information and Control*, **67**, 70–90. [86, 359]

Blum, L. (Lenore Blum) 1977. Differentially closed fields: a model-theoretic tour. In *Contributions to algebra (Collection of papers dedicated to Ellis Kolchin)*, pp. 37–61. New York: Academic Press. [706]

Boffa, M. (Maurice Boffa) 1986. L'élimination des inverses dans les groupes. *C.R. Acad. Sci. Paris*, **303**, 587–9. [410]

Bolzano, B. 1837. *Wissenschaftslehre*. Trans. as *Theory of science*, ed. R. George (1972). Berkeley: Univ. California Press. [83]

Bonnet, R. (Robert Bonnet) 1980. Very strongly rigid Boolean Algebra, continuum discrete set condition, countable antichain condition (I). *Algebra Universalis*, **11**, 341–64. [566]

Boole, G. (George Boole) 1847. *The mathematical analysis of logic*. Cambridge: Macmillan, Barclay & Macmillan. [319]

Börger, E. (Egon Börger) 1989, *Computability, complexity, logic*. Amsterdam: North-Holland. [234]

Borovik, A. V. (А. В. Боровик) 1984. Теория конечных групп и несчётно категоричные группы (The theory of finite groups and uncountably categorical groups). Preprint of Acad. Sci. USSR, Siberian Division. [691]

— 1989. Sylow theory for groups of finite Morley rank (Russian). *Sibirsk. Mat. Zh.*, **30** (6), 52–7. *Trans. as* Sylow theory for groups of finite Morley rank. *Siberian Math. J.*, **30** (1990), 873–7. [693]

Borovik, A. V. & Poizat, B. 1990. Tores et *p*-groupes. *J. Symbolic Logic*, **55**, 478–91. [693]

Botto Mura, R. & Rhemtulla, A. 1977. *Orderable groups*. Lecture notes in pure and applied math. 27. New York: Marcel Dekker. [260]

Bourbaki, N. (Nicolas Bourbaki) 1972. *Commutative algebra*. Paris: Hermann, and Reading, Mass.: Addison-Wesley. [79, 574]

Bouscaren, E. (Elisabeth Bouscaren) 1989a. Dimensional Order Property and pairs of models. *Ann. Pure Appl. Logic*, **41**, 205–31. [359]

— 1989b. Model theoretic versions of Weil's theorem on pregroups. In Nesin & Pillay [1989], pp. 177–85. [694]

— 1989c. The group configuration – after E. Hrushovski. In Nesin & Pillay [1989], pp. 199–209. [200]

Bouscaren, E. & Poizat, B. 1988. Des belles paires aux beaux uples. *J. Symbolic Logic*, **53**, 434–42. [321]

Brumfiel, G. W. 1979. *Partially ordered rings and semi-algebraic geometry*. London Math. Soc. Lecture Note Series 37. Cambridge: Cambridge Univ. Press. [388]

Bryars, D. A. (David A. Bryars) 1973. On the syntactic characterization of some model theoretic relations. Ph.D. thesis. Bedford College, London University. [358, 411]

Büchi, J. R. (J. Richard Büchi) 1960. Weak second order arithmetic and finite automata. *Z. Math. Logik Grundlag. Math.*, **6**, 66–92. [86]

— 1962. On a decision method in restricted second order arithmetic. In *Logic, Methodology and Philosophy of Science*, ed. E. Nagel *et al.*, pp. 1–11. Stanford: Stanford Univ. Press. [86]

Büchi, J. R. & Siefkes, D. (Dirk Siefkes) 1973. Axiomatization of the monadic second order theory of ω_1. In *Decidable theories, II. Monadic second order theory of all countable ordinals*, ed. G. H. Müller & D. Siefkes, Lecture Notes in Math. 328, pp. 129–217. Berlin: Springer. [594]

Buechler, S. (Steven Buechler) 1984. Recursive definability and resplendency in ω-stable theories. *Israel J. Math.*, **49**, 26–34. [522]

— 1985. The geometry of weakly minimal types. *J. Symbolic Logic*, **50**, 1044–53. [651]

— 1990. Vaught's conjecture for unidimensional theories. In *Soviet-French Colloquium in Theory of Models* (Russian), p. 54. Karaganda: Karaganda State University. [339, 662]

— 199?. Vaught's conjecture for unidimensional theories. Preprint. [339]

Burris, S. N. (Stanley Burris) 1975. Boolean powers. *Algebra Universalis*, **5**, 341–60. [222, 477, 534]

— 1983. Boolean constructions. In *Universal algebra and lattice theory, Proceedings, Puebla 1982*, Lecture Notes in Math. 1004, pp. 67–90. Berlin: Springer. [476]

Burris, S. N. & Sankappanavar, H. P. 1981. *A course in universal algebra*. New York: Springer. [467]

Burris, S. N. & Werner, H. (Heinrich Werner) 1979. Sheaf constructions and their elementary properties. *Trans. Amer. Math. Soc.*, **248**, 269–309. [475, 477]

Cameron, P. J. (Peter J. Cameron) 1976. Transitivity of permutation groups on unordered sets. *Math. Z.*, **148**, 127–39. [260]

— 1984. Aspects of the random graph. In *Graph theory and combinatorics*, ed. B. Bollobás, pp. 65–79. London: Academic Press. [359]

— 1990. *Oligomorphic permutation groups*. London Math. Soc. Lecture Note Series 152. Cambridge: Cambridge Univ. Press. [134, 197, 352, 358, 359, 409]

Cantor, G. (Georg Cantor) 1895. Beiträge zur Begründung der transfiniten Mengenlehre, I. *Math. Annalen*, **46**, 481–512. [128]

Carson, A. B. (Andrew B. Carson) 1973. The model completion of the theory of commutative regular rings. *J. Algebra*, **27**, 136–46. [394]

Chandra, A. & Harel, D. (Ashok Chandra, David Harel) 1982. Structure and complexity of relational queries. *J. Comp. System Sci.*, **25**, 99–128. [86]

Chang, C. C. (Chen Chung Chang) 1959. On unions of chains of models. *Proc. Amer. Math. Soc.*, **10**, 120–7. [320]

— 1964. Some new results in definability. *Bull. Amer. Math. Soc.*, **70**, 808–13. [652]

— 1965. A note on the two cardinal problem. *Proc. Amer. Math. Soc.*, **16**, 1148–55. [533, 650]

— 1968a. Infinitary properties of models generated from indiscernibles. In *Logic, methodology and philosophy of science III*, ed. B. van Rootselaar & J. F. Staal, pp. 9–21. Amsterdam: North-Holland. [474, 595]

— 1968b. Some remarks on the model theory of infinitary languages. In *The syntax and semantics of infinitary languages*, ed. K. J. Barwise, Lecture Notes in Math. 72. pp. 36–63. Berlin: Springer. [84, 129, 597]

Chang, C. C. & Keisler, H. J. 1962. Applications of ultraproducts of pairs of cardinals to the theory of models. *Pacific J. Math.*, **12**, 835–45. [475]

— 1966. *Continuous model theory*. Princeton, NJ: Princeton Univ. Press. [475, 533]

— 1973. *Model theory*. Amsterdam: North-Holland. [358, 412, 441, 445, 449, 452, 476, 492, 507, 533, 603, 705, 714]

— 1990. Third edition of Chang & Keisler [1973]. [567]

Chang, C. C. & Morel, A. C. 1958. On closure under direct product. *J. Symbolic Logic*, **23**, 149–54. [475]

Charretton, C. (Christine Charretton) 1979. *Type d'ordre des modèles non-standards*. Thèse de Troisième Cycle. Lyon I: Université Claude-Bernard. [595]

Charretton, C. & Pouzet, M. 1983a. Chains in Ehrenfeucht–Mostowski models. *Fundamenta Math.*, **118**, 109–22. [595, 596]

— 1983b. Comparaison des structures engendrées par des chaînes. In *Proc. Dietrichshagen Easter Conference on Model Theory 1983*, pp. 17–27. Berlin: Humboldt-Univ. [595, 596]

Chatzidakis, Z. (Zoé Chatzidakis) 1984. *Model theory of profinite groups*. Doctoral dissertation. Yale. [691]

Chatzidakis, Z., van den Dries, L. P. D. & Macintyre, A. J. 1992. Definable sets over finite fields. *J. Reine Angew. Math.*, **427**, 107–35. [702]

Chatzidakis, Z., Pappas, P. & Tomkinson, M. J. (Peter Pappas) 1990. Separation theorems for infinite permutation groups. *Bull. London Math. Soc.*, **22**, 344–8. [594]

Cherlin, G. L. (Gregory Cherlin) 1973. Algebraically closed commutative rings. *J. Symbolic Logic*, **38**, 493–9. [411]

— 1979. Groups of small Morley rank. *Ann. Math. Logic*, **17**, 1–28. [262, 691, 692]

— 1980a. On $\aleph_0$-categorical nilrings. *Algebra Universalis*, **10**, 27–30. [358]

— 1980b. On $\aleph_0$-categorical nilrings. II. *J. Symbolic Logic*, **45**, 291–301. [358]

Cherlin, G. L., van den Dries, L. P. D. & Macintyre, A. J. 1982. The elementary theory of regularly closed fields. Preprint. [691]

Cherlin, G. L., Harrington, L. A. & Lachlan, A. H. (Leo Harrington) 1985. $\aleph_0$-categorical, $\aleph_0$-stable structures. *Ann. Pure Appl. Logic*, **28**, 103–35. [199, 200, 651]

Cherlin, G. L. & Lachlan, A. H. 1986. Stable finitely homogeneous structures. *Trans. Amer. Math. Soc.*, **296**, 815–50. [330]

Cherlin, G. L. & Reineke, J. 1976. Categoricity and stability of commutative rings. *Ann. Math. Logic*, **10**, 367–99. [321]

Cherlin, G. L. & Rosenstein, J. G. 1978. On $\aleph_0$-categorical abelian by finite groups. *J. Algebra*, **53**, 188–226. [358]

Cherlin, G. L. & Schmitt, P. H. 1983. Locally pure topological abelian groups: Elementary invariants. *Ann. Pure Appl. Logic*, **24**, 49–85. [691]

Cherlin, G. L. & Shelah, S. 1980. Superstable fields and groups. *Ann. Math. Logic*, **18**, 227–70. [262, 699]

Cherlin, G. L. & Volger, H. 1984. Convexity properties and algebraic closure operators. In *Models and Sets, Proc. Logic Colloq. Aachen 1983*, Part I, ed. G. H. Müller & M. M. Richter, Lecture Notes in Math. 1103, pp. 113–46. Berlin: Springer. [320]

Chevalley, C. (Claude Chevalley) 1951. *Théorie des groupes de Lie*. Tome II, *Groupes algébriques*. Paris: Hermann. [262]

Chistov, A. L. & Grigor'ev, D. Yu. 1984. Complexity of quantifier elimination in the theory of algebraically closed fields. In *Mathematical foundations of computer science*, ed. M. P. Chytil & V. Koubek, Lecture Notes in Comp. Sci. 176, pp. 17–31. Berlin: Springer. [697]

Chudnovskiĭ, G. V. (Г. В. Чудновский) 1968. Некоторые результаты в теории бесконечно длинных выражений. *Dokl. Akad. Nauk SSSR*, **179** (6), 1286–8. *Trans. as* Some results in the theory of infinitely long expressions. *Soviet Mathematics Doklady*, **9** (1968), 556–9. [474]

— 1970. Вопросы теории моделей, связанные категоричностью. *Algebra i Logika*, **9**, 80–120. *Trans. as* Problems of the theory of models, related to categoricity. *Algebra and Logic*, **9** (1970), 50–74. [650]

Church, A. (Alonzo Church) 1936. A note on the Entscheidungsproblem. *J. Symbolic Logic*, **1**, 40–1. Correction, *ibid*, 101–2. [44, 93]

Clark, K. L. 1978. Negation as failure. In *Logic and databases*, ed. H. Gallaire & J. Minker, pp. 293–322. New York: Plenum Press. [83]

Cleave, J. P. (John P. Cleave) 1969. Local properties of systems. *J. London Math. Soc.*, **44**, 121–30 (= **1** (1969), 121–30). Addendum, *J. London Math. Soc.*, **1** (1969), 384. [321]

Cohen, P. J. (Paul Cohen) 1969. Decision procedures for real and p-adic fields. *Comm. Pure Appl. Math.*, **22**, 131–51. [705]

Cohn, P. M. (Paul M. Cohn) 1959. On the free product of associative rings. *Math. Z.*, **71**, 380–98. [394]

— 1971. *Free rings and their relations*. London: Academic Press. [597]

— 1974. *Algebra*, volume I. London: John Wiley. [166, 663]

— 1976. *Morita equivalence and duality*. London: Queen Mary College Mathematics Notes. [219]

— 1977. *Algebra*, volume II. London: John Wiley. [79, 253, 365, 386]

— 1981. *Universal algebra*. Dordrecht: Reidel. [166, 474, 591]

Cole, J. C. & Dickmann, M. A. 1972. Non-axiomatizability results in infinitary languages for higher-order structures. In *Conference in Mathematical Logic – London '70*, ed. W. A.

Hodges, Lecture Notes in Math. 255, pp. 29–41. Berlin: Springer. [128]

Comer, S. D. (Stephen D. Comer) 1974. Elementary properties of structures of sections. *Bol. Soc. Mat. Mexicana*, **19**, 78–85. [476]

Compton, K. J. (Kevin J. Compton) 1983. Some useful preservation theorems. *J. Symbolic Logic*, **48**, 427–40. [533]

Covington, J. (Jacinta Covington) 1989. A universal structure for *N*-free graphs. *Proc. London Math. Soc.*, **58**, 1–16. [411]

Coxeter, H. S. M. 1961. *Introduction to geometry*. New York: John Wiley. [197]

Coxeter, H. S. M. & Moser, W. O. J. (William O. J. Moser) 1957. *Generators and relations for discrete groups*. Berlin: Springer. [132]

Craig, W. (William Craig) 1953. On axiomatizability within a system. *J. Symbolic Logic*, **18**, 30–2. [319]

— 1957. Three uses of the Herbrand–Gentzen theorem in relating model theory and proof theory. *J. Symbolic Logic*, **22**, 269–85. [320]

Craig, W. & Vaught, R. L. 1958. Finite axiomatizability using additional predicates. *J. Symbolic Logic*, **23**, 289–308. [534]

Crapo, H. H. & Rota, G.-C. (Henry H. Crapo, Gian-Carlo Rota) 1970. *On the foundations of combinatorial theory: Combinatorial geometries*. Cambridge, Mass.: MIT Press. [171]

Cummings, J. W. & Woodin, W. H. (James W. Cummings, W. Hugh Woodin) 199?. *The singular cardinal problem*. Book in preparation. [533]

Curtis, C. W. & Reiner, I. 1962. *Representation theory of finite groups and associative algebras*. New York: J. Wiley and Sons. [662]

Cutland, N. J. (Nigel J. Cutland) (ed.) 1988. *Nonstandard analysis and its applications*. Cambridge: Cambridge Univ. Press. [567]

Davis, M. D. (Martin D. Davis) 1983. The prehistory and early history of automated deduction. In *Automation of reasoning* 1, ed. J. Siekmann & G. Wrightson, pp. 1–28. Berlin: Springer. [20]

Day, A. (Alan Day) 1983. Geometrical applications in modular lattices. In *Universal algebra and lattice theory*, ed. R. S. Freese & O. C. Garcia, Lecture Notes in Math. 1004, pp. 111–41. Berlin: Springer. [260]

Delon, F. (Françoise Delon) 1981. Indécidabilité de la théorie des anneaux de séries formelles à plusieurs indéterminées. *Fundamenta Math.*, **112**, 215–29. [262]

— 1988. Idéaux et types sur les corps séparablement clos. *Mém. Soc. Math. France*, **33**, *Suppl. au Bulletin de la S.M.F.* **116**, fasc. 3. [701]

Denef, J. 1984. The rationality of the Poincaré series associated to the *p*-adic points on a variety. *Invent. Math.*, **77**, 1–23. [706]

Denef, J. & van den Dries, L. P. D. 199?. *P*-adic and real subanalytic sets. Preprint. [706]

Denes, J. 1973. Definable automorphisms in model theory I. Abstract. *J. Symbolic Logic*, **38**, 354. [197]

Devlin, K. J. (Keith Devlin) 1984. *Constructibility*. Berlin: Springer. [650]

Dickmann, M. A. (Max Dickmann) 1975. *Large infinitary languages: model theory*. Amsterdam: North-Holland. [540]

Dixon, J. D., Neumann, P. M. & Thomas, S. R. (John D. Dixon) 1986. Subgroups of small index in infinite symmetric groups. *Bull. London Math. Soc.*, **18**, 580–6. [198]

Dixon, P. G. (Peter G. Dixon) 1977. Classes of algebraic systems defined by universal Horn sentences. *Algebra Universalis*, **7**, 315–39. [475]

Doner, J. E., Mostowski, A. & Tarski, A. (John E. Doner) 1978. The elementary theory of well-ordering – a metamathematical study. In *Logic Colloquium 77*, ed. A. J. Macintyre *et al.*, pp. 1–54. Amsterdam: North-Holland. [86, 715]

van den Dries, L. P. D. (Lou van den Dries) 1978. *Model theory of fields: Decidability, and bounds for polynomial ideals*. Doctoral thesis. Univ. Utrecht. [596]

— 1982a. Remarks on Tarski's problem concerning ($\mathbb{R}$, +, ·, exp). In *Logic Colloquium, '82*, ed. G. Lolli *et al.*, pp. 97–121. Amsterdam: North-Holland. [82, 704]

— 1982b. Some applications of a model-theoretic fact to (semi-)algebraic geometry. *Indag. Math.*,

44, 397–401. [388, 410]

— 1984. Algebraic theories with definable Skolem functions. *J. Symbolic Logic*, **49**, 625–9. [128, 199, 705]

— 1988a. Alfred Tarski's elimination theory for real closed fields. *J. Symbolic Logic*, **53**, 7–19. [410, 703]

— 1988b. On the elementary theory of restricted elementary functions. *J. Symbolic Logic*, **53**, 796–808. [704]

van den Dries, L. P. D., Marker, D. E. & Martin, G. A. 1989. Definable equivalence relations on algebraically closed fields. *J. Symbolic Logic*, **54**, 928–35. [700]

van den Dries, L. P. D. & Schmidt, K. (Karsten Schmidt) 1984. Bounds in the theory of polynomial rings over fields. A nonstandard approach. *Invent. Math.*, **76**, 77–91. [596]

van den Dries, L. P. D. & Wilkie, A. J. 1984. Gromov's theorem on groups of polynomial growth and elementary logic. *J. Algebra*, **89**, 349–74. [574]

Droste, M. & Göbel, R. (Manfred Droste, Rüdiger Göbel) 199?. A categorical theorem on universal objects and its applications in abelian group theory and computer science. In *Conf. on Algebra honoring A. Mal'cev (Novosibirsk, August 21–26, 1989)*, ed. L. A. Bokut *et al.*, Contemporary Mathematics. Providence, RI: Amer. Math. Soc. [357]

Droste, M., Holland, W. C., & Macpherson, H. D. 1989. Automorphism groups of infinite semilinear orders (II). *Proc. London Math. Soc.*, **58**, 479–94. [147]

Duret, J.-L. (Jean-Louis Duret) 1977. Instabilité des corps formellement réels. *Canad. Math. Bull.*, **20**, 385–7. [322]

— 1980. Les corps faiblement algébriquement clos non séparablement clos ont la propriété d'indépendance. In *Model theory of algebra and arithmetic*, ed. L. Pacholski *et al.*, Lecture Notes in Math. 834, pp. 136–62. Berlin: Springer. [703]

Dushnik, B. & Miller, E. W. 1941. Partially ordered sets. *Amer. J. Math.*, **63**, 605. [594]

Dyck, D. (Walther Dyck) 1882. Gruppentheoretische Studien. *Math. Ann.*, **20** 1–44. [474]

Ebbinghaus, H. D., Flum, J. & Thomas, W. (Heinz-Dieter Ebbinghaus, Jörg Flum) 1984. *Mathematical logic*. New York: Springer. [40]

Ehrenfeucht, A. (Andrzej Ehrenfeucht) 1957. On theories categorical in power. *Fundamenta Math.*, **44**, 241–8. [595, 596]

— 1958. Theories having at least continuum many non-isomorphic models in each infinite power. Abstract. *Not. Amer. Math. Soc.*, **5**, 680. [595]

— 1959. Decidability of the theory of one linear ordering relation. Abstract. *Not. Amer. Math. Soc.*, **6**, 268–9. [86, 711]

— 1961. An application of games to the completeness problem for formalized theories. *Fundamenta Math.*, **49**, 129–41. [128]

— 1972. There are continuum ω_0-categorical theories. *Bull. Acad. Polon. Sci. Sér. Math. Astron. Phys.*, **20**, 425–7. [359]

Ehrenfeucht, A. & Mostowski, A. 1956. Models of axiomatic theories admitting automorphisms. *Fundamenta Math.*, **43**, 50–68. [595]

Eklof, P. C. (Paul C. Eklof) 1973. Lefschetz's principle and local functors. *Proc. Amer. Math. Soc.*, **37**, 333–9. [475, 700]

— 1974a. Algebraic closure operators and strong amalgamation bases. *Algebra Universalis*, **4** 89–98. [411]

— 1974b. Infinitary equivalence of abelian groups. *Fundamenta Math.*, **81**, 305–14. [672]

— 1975. Categories of local functors. In *Model theory and algebra: A memorial tribute to Abraham Robinson*, ed. D. H. Saracino & V. B. Weispfenning, Lecture Notes in Math. 498, pp. 91–116. Berlin: Springer. [475]

— 1977. Ultraproducts for algebraists. In *Handbook of mathematical logic*, ed. K. J. Barwise, pp. 105–37. Amsterdam: North-Holland. [476]

— 1980. *Set theoretic methods in homological algebra and abelian groups*. Séminaire de Mathématiques Supérieures. Université de Montréal. [129]

— 1985. Applications to algebra. In Barwise & Feferman [1985], pp. 423–41. [598]

Eklof, P. C. & Fisher, E. R. 1972. The elementary theory of abelian groups. *Ann. Math. Logic*, **4**, 115–71. [669]

Eklof, P. C. & Mekler, A. H. 1981. Infinitary stationary logic and abelian groups. *Fundamenta Math.*, **112**, 1–15. [673]

— 1990. *Almost free modules: set-theoretic methods*. Amsterdam: North-Holland. [129, 474]

Eklof, P. C. & Sabbagh, G. 1971. Model-completions and modules. *Ann. Math. Logic*, **2**, 251–95. [86, 410, 411, 476, 657, 660]

van Emden, M. H & Kowalski, R. A. 1976. The semantics of predicate logic as a programming language. *J. Associ. Comp. Mach.*, 733–42. [417]

Engeler, E. (Erwin Engeler) 1959. A characterization of theories with isomorphic denumerable models. Abstract. *Not. Amer. Math. Soc.*, **6**, 161. [358]

— 1961. Unendliche Formeln in der Modelltheorie. *Z. Math. Logik Grundlag. Math.*, **7**, 154–60. [84]

Engelking, R. & Karłowicz, M. 1965. Some theorems of set theory and their topological consequences. *Fundamenta Math.*, **57**, 275–85. [319]

Erdős, P. (Paul Erdős) 1942. Some set-theoretical properties of graphs. *Revista Universidad Nacional de Tucumán, Ser. A*, **3**, 363–7. [594]

Erdős, P., Hajnal, A., Máté, A. & Rado, R. (Andras Hajnal, Attila Máté, Richard Rado) 1984. *Combinatorial set theory: partition relations for cardinals*. Amsterdam: North-Holland. [594]

Erdős, P. & Rado, R. 1956. A partition calculus in set theory. *Bull. Amer. Math. Soc.*, **62**, 427–89. [594]

Erdős, P. & Rényi, A. 1963. Asymmetric graphs. *Acta Math. Acad. Sci. Hungar.*, **14**, 295–315. [359]

Erdős, P. & Szekeres, G. (George Szekeres) 1935. A combinatorial problem in geometry. *Compositio Math.*, **2**, 464–70. [594]

Erdős, P. & Tarski, A. 1962. On some problems involving inaccessible cardinals. In *Essays on the foundations of mathematics*, ed. Y. Bar-Hillel *et al.*, pp. 50–82. Jerusalem: Magnus Press. [319, 594]

Erimbetov, M. M. (М. М. Еримбетов) 1975. О полных теориях с 1-кардинальными формулами (Complete theories with 1-cardinal formulas). *Algebra i Logika*, **14**, 245–57. [262, 650]

— 1985. Связь мощностей формульных подмножеств со стабильностью формул (A connection between the cardinalities of definable sets and the stability of formulas). *Algebra i Logika*, **24**, 627–30. [650]

Ershov, Yu. L. (Ю. Л. Ершов) 1963. Разрешимость элементарных теорий некоторых классов абелевых групп (Decidability of the elementary theories of some classes of abelian groups). *Algebra i Logika*, **1** (6), 37–41. [319, 672]

— 1965a. Неразрешимость некоторых полей. *Dokl. Akad. Nauk SSSR*, **161** (1), 27–9. *Trans. as* Undecidability of certain fields. *Soviet Math. Dokl.*, **6** (1965), 349–52. [262]

— 1965b. Об элементарной теории максимальных нормированных полей (On the elementary theory of maximal normed fields). *Algebra i Logika*, **4** (3), 31–70. [705]

— 1966. Об элементарной теории максимальных нормированных полей II (On the elementary theory of maximal normed fields II). *Algebra i Logika*, **5** (1), 5–40. [705]

— 1967a. Fields with a solvable theory. *Soviet Math. Dokl.*, **8** (1967) 575–6. [701]

— 1967b. Об элементарных теориях многообразий Поста (On the elementary theories of Post varieties). *Algebra i Logika*, **6** (5), 7–15. [477]

— 1974. Theories of nonabelian varieties of groups. In *Proc. Tarski Symposium*, ed. L. A. Henkin *et al.*, pp. 255–64. Providence, RI: Amer. Math. Soc. [221, 261, 680]

— 1980a. Проблемы разрешимости и конструктивные модели (*Decision problems and constructive models*). Moscow: Nauka. [261]

— 1980b. Регулярно замкнутые поля. *Dokl. Akad. Nauk SSSR*, **251**, 783–5. *Trans. as* Regularly closed fields. *Soviet Math. Dokl.*, **21** (1980), 510–2. [702]

Ershov, Yu, L., Lavrov, I. A., Taĭmanov, A. D. & Taĭtslin, M. A. 1965. Элементарные теории. *Uspekhi Mat. Nauk*, **20**, 4 (124), 37–108. *Trans. as* Elementary theories. *Russian Math. Surveys*, **20** (1965), 4, 35–105. [262]

Ershov, Yu, L. & Taĭtslin, M. A. 1963. Неразрешимость некоторых теорий (The undecidability of certain theories). *Algebra i Logika*, **2** (5), 37–42. [261]

Etchemendy, J. (John Etchemendy) 1988. Tarski on truth and logical consequence. *J. Symbolic Logic*, **53**, 51–79. [83]

Evans, D. M. (David M. Evans) 1986a. Homogeneous geometries. *Proc. London Math. Soc.*, **52**, 305–27. [199]

— 1986b. Subgroups of small index in infinite general linear groups. *Bull. London Math. Soc.*, **18**, 587–90. [198]

— 1987. A note on automorphism groups of countably infinite structures. *Arch. Math. (Basel)*, **49**, 479–83. [198]

— 1991. The small index property for infinite dimensional classical groups. *J. Algebra*, **136**, 248–64. [198]

— 199?. Some subdirect products of finite nilpotent groups. Preprint. [618]

Evans, D. M. & Hewitt, P. R. (Paul R. Hewitt) 1990. Counterexamples to a conjecture on relative categoricity. *Ann. Pure Appl. Logic*, **46**, 201–9. [197, 359, 625, 647]

Evans, D. M., Hodges, W. A. & Hodkinson, I. M. 1991. Automorphisms of bounded abelian groups. *Forum Mathematicum*, **3**, 523–41. [198, 651]

Evans, D. M. & Hrushovski, E. 1991. Projective planes in algebraically closed fields. *Proc. London Math. Soc.*, **62**, 1–24. [200, 618]

Evans, D. M., Pillay, A. & Poizat, B. 1990. Le groupe dans le groupe. *Algebra i Logika*, **29** (3), 368–78. *Trans. as* A group in a group. *Algebra and Logic*, **29**, 244–52. [198, 254, 263]

Facchini, A. 1985. Decompositions of algebraically compact modules. *Pacific J. Math.*, **116**, 25–37. [661] .

Fagin, R. (Ronald Fagin) 1976. Probabilities on finite models. *J. Symbolic Logic*, **41**, 50–8. [359]

Feferman, S. (Solomon Feferman) 1968a. Lectures on proof theory. In *Proc. Summer School in Logic, Leeds, 1967*, ed. M. H. Löb, Lecture Notes in Math. 70, pp. 1–107. Berlin: Springer. [410]

— 1968b. Persistent and invariant formulas for outer extensions. *Compositio Math.*, **20**, 29–52. [533]

— 1972. Infinitary properties, local functors, and systems of ordinal functions. In *Conference in Mathematical Logic – London '70*, ed. W.A. Hodges, Lecture Notes in Math. 255, pp. 63–97. Berlin: Springer. [475]

— 1974. Two notes on abstract model theory, I: Properties invariant on the range of definable relations between structures. *Fundamenta Math.*, **82**, 153–65. [652]

Feferman, S. & Vaught, R. L. 1959. The first order properties of products of algebraic systems. *Fundamenta Math.*, **47**, 57–103. [476, 708]

Felgner, U. (Ulrich Felgner) 1976. On $\aleph_0$-categorical extra-special p-groups. *Logique et Analyse*, **71–72**, 407–28. Reprinted in *Six days of model theory*, ed. P. Henrard (1977), pp. 175–96. Albeuve: Editions Castella. [344]

— 1977. Stability and $\aleph_0$-categoricity of nonabelian groups. In *Logic Colloquium '76*, ed. R. O. Gandy & J. M. E. Hyland, pp. 301–24. Amsterdam: North-Holland. [358]

— 1978. $\aleph_0$-categorical stable groups. *Math. Z.*, **160**, 27–49. [358, 693]

— 1980a. Horn-theories of abelian groups. In *Model theory of algebra and arithmetic*, ed. L. Pacholski *et al.*, Lecture Notes in Math. 834, pp. 163–73. Berlin: Springer. [658]

— 1980b. The model theory of FC-groups. In *Mathematical logic in Latin America*, ed. A. I. Arruda *et al.*, pp. 163–90. Amsterdam: North-Holland. [691]

— 1986. The classification of all quantifier-eliminable FC-groups. In *Atti del Congresso 'Logica e Filosofia della Scienza, oggi', San Gimignano 1983*, Vol. I. pp. 27–33. Bologna: CLUEB. [695]

Ferrante, J. & Rackoff, C. W. 1979. *The computational complexity of logical theories*. Lecture Notes in Math. 718. Berlin: Springer. [128]

Fischer, M. J. & Rabin, M. O. (Michael J. Fischer) 1974. Super-exponential complexity of Presburger arithmetic. In *Complexity of computation*, ed. R. M. Karp, pp. 27–41. Providence, RI: Amer. Math. Soc. [108, 697]

Fisher, E. R. (Edward R. Fisher) 1972. Powers of saturated modules. Abstract. *J. Symbolic Logic*, **37**, 777. [660]

— 1977a. Abelian structures. I. In *Abelian Group Theory, 2nd New Mexico State Univ. Conference 1976*, ed. D. Arnold *et al.*, Lecture Notes in Math. 616, pp. 270–322. Berlin: Springer. [534]

— 1977b. Vopěnka's principle, category theory and universal algebra. Abstract. *Not. Amer. Math. Soc.*, **24**, A-44. [474]

Foster, A. L. 1953. Generalized 'Boolean' theory of universal algebras, Part I: Subdirect sums and normal representation theorem. *Math. Z.*, **58**, 306–36. [477]

— 1961. Functional completeness in the small. Algebraic structure theorems and identities. *Math. Ann.*, **143**, 29–53. [476]

Fraenkel, A. A. (Abraham A. Fraenkel) 1922. Der Begriff 'definit' and die Unabhängigkeit des Auswahlsaxioms. *Sitzungsberichte Preuss. Akad. Wiss. (Phys.-Math. Klasse)*, **21**, 253–7. *Trans. as* The notion 'definite' and the independence of the axiom of choice. In *From Frege to Gödel*, ed. J. van Heijenoort (1967), pp. 285–9. Cambridge, Mass: Harvard Univ. Press. [198]

Fraïssé, R. (Roland Fraïssé) 1954. Sur l'extension aux relations de quelques propriétés des ordres. *Ann. Sci. École Norm. Sup.*, **71**, 363–88. [357]

— 1955. Sur quelques classifications des relations, basées sur des isomorphismes restreints. II. Applications aux relations d'ordre, et construction d'exemples montrant que ces classifications sont distinctes. *Alger-Math.*, **2**, 273–95. [128, 708]

— 1956a. Etude de certains opérateurs dans les classes de relations, définis à partir d'isomorphismes restreints. *Z. Math. Logik Grundlag. Math.*, **2**, 59–75. [129]

— 1965b. Application des γ-opérateurs au calcul logique du premier échelon. *Z. Math. Logik Grundlag. Math.*, **2**, 76–92. [128]

Franz, W. 1931. Untersuchungen sum Hilbertschen Irreduzibilitätssatz. *Math. Z.*, **33**, 275–93. [596]

Frasnay, C. (Claude Frasnay) 1965. Quelques problèmes combinatoires concernant les ordres totaux et les relations monomorphes. *Ann. Inst. Fourier (Grenoble)*, **15** (ii), 415–524. [260]

— 1984. Relations enchaînables, rangements et pseudo-rangements. In *Orders: description and roles (L'Arbresle 1982)*, ed. M. Pouzet & D. Richard, pp. 237–68. Amsterdam: North-Holland. [260]

Frayne, T. E., Morel, A. C. & Scott, D. S. (Thomas E. Frayne) 1962. Reduced direct products. *Fundamenta Math.*, **51**, 195–228. [475, 476]

Frege, G. (Gottlob Frege) 1879. *Begriffsschrift, eine der arithmetischen nachgebildete Formelsprache des reinen Denkens*. Halle. *Trans. as* Begriffsschrift, a formula language, modeled upon that of arithmetic, for pure thought. In *From Frege to Gödel*, ed. J. van Heijenoort (1967), pp. 5–82. Cambridge, Mass.: Harvard Univ. Press. [83]

— 1971. *On the foundations of geometry, and formal theories of arithmetic*, Trans. E. W. Kluge. New Haven, Conn.: Yale Univ. Press. [21]

Fried, M. D. & Jarden, M. (Michael D. Fried, Moshe Jarden) 1986. Field arithmetic. Berlin: Springer. [691, 695, 702]

Friedman, H. M. (Harvey M. Friedman) 1973. Countable models of set theories. In *Cambridge Summer School in Math. Logic*, ed. A. R. D. Mathias & H. Rogers, Lecture Notes in Math. 337, pp. 539–73. Berlin: Springer. [596]

Frink, O. 1946. Complemented modular lattice and projective spaces of infinite dimension. *Trans. Amer. Math. Soc.*, **60**, 452–67. [199]

Fuchs, L. (László Fuchs) 1960. *Abelian groups*. Oxford: Pergamon Press. [260]

— 1970. *Infinite abelian groups I*. New York: Academic Press. [663, 664, 665, 666, 669]

Fuks-Rabinovich, D. I. (Д. И. Фукс-Рабинович) 1940. О непростоте локально свободных групп (On the non-simplicity of locally free groups). *Mat. Sbornik*, **7**, 327–8. [321]

Gaifman, H. (Haim Gaifman) 1967. Uniform extension operators for models and their applications. In *Sets, models and recursion theory*, ed. J. N. Crossley, pp. 122–55. Amsterdam: North-Holland. [595]

— 1974a. Operations on relational structures, functors and classes, I. In *Proc. Tarski Symposium*, ed. L. A. Henkin *et al.*, pp. 21–39. Providence, RI: Amer. Math. Soc. [652]

— 1974b. Some results and conjectures concerning definability questions. Preprint. [652]

— 1976. Models and types of Peano's arithmetic. *Ann. Math. Logic*, **9**, 223–306. [198]

Gale, D: & Stewart, F. M. (David Gale, Frank M. Stewart) 1953. Infinite games with perfect information. In *Contributions to the theory of games II, Ann. Math. Studies*, **28**, 245–66. [129]

Galvin, F. (Fred Galvin) 1970. Horn sentences. *Ann. Math. Logic*, **1**, 389–422. [475, 476]

Garavaglia, S. (Steven Garavaglia) 1979. Direct product decomposition of theories of modules. *J. Symbolic Logic*, **44**, 77–88. [476, 660]

— 1980. Decomposition of totally transcendental modules. *J. Symbolic Logic*, **45**, 155–64. [660]

Gardner, M. (Martin Gardner) 1973. Sim, Chomp and Race Track: new games for the intellect (and not for Lady Luck). *Scientific American*, **228** (1), 108–12. [543]

Gergonne, J. D. 1818. Essai sur la théorie des définitions. *Annales de Math. Pures et Appl.*, **9**, 1–35. [321]

Gibone, P. 1976. On the stability of free groups. Abstract. *Not. Amer. Math. Soc.*, **23**, No. 4, A-449, [694]

Gilmore, P. C. & Robinson, A. 1955. Metamathematical considerations on the relative irreducibility of polynomials. *Canad. J. Math.*, **7**, 483–9. Reprinted in A. Robinson [1979a] pp. 348–54. [596]

Givant, S. R. (Steven R. Givant) 1979. Universal Horn classes categorical or free in power. *Ann. Math. Logic*, **15**, 1–53. [199, 474]

Glass, A. M. W. 1989. The universal theory of lattice-ordered abelian groups. In *Actes de la Journée Algèbre Ordonnée, Le Mans 1987*, ed. M. Giraudet. Assoc. Française d'Algèbre Ordonnée. [691]

Glass, A. M. W. & Pierce, K. R. (Keith R. Pierce) 1980a. Existentially complete abelian lattice-ordered groups. *Trans. Amer. Math. Soc.*, **261**, 255–70. [690]

— 1980b. Existentially complete lattice-ordered groups. *Israel J. Math.*, **36**, 257–72. [690]

Glassmire, W. (W. Glassmire Jr.) 1971. There are $2^{\aleph_0}$ countably categorical theories. *Bull. Acad. Polon. Sci. Sér. Sci. Math. Astronom. Phys.*, **19**, 185–90. [359]

Glebskiĭ, Yu. V., Kogan, D. I., Liogon'kiĭ, M. I. & Talanov, V. A. (Ю. В. Глебский, Д. И., Коган, М. И. Лиогонький, В. А. Таланов) 1969. Объем и доля выполнимости формул исчисления предикатов (The extent and degree of satisfiability of formulas of the restricted predicate calculus). *Kibernetika (Kiev)*, **2**, 17–27. [359]

Gödel, K. (Kurt Gödel) 1929. *Über die Vollständigkeit des Logikkalküls*. Doctoral dissertation. Univ. Vienna. Reprinted with translation in Kurt Gödel, *Collected works*, vol. 1, ed. S. Feferman *et al.*, pp. 60–101. New York: Oxford Univ. Press. [318]

— 1930. Die Vollständigkeit der Axiome des logischen Funktionenkalküls. *Monatshefte für Math. u. Phys.*, **37**, 349–60. [261, 318]

— 1931a. Eine Eigenschaft der Realisierungen des Aussagenkalküls. *Ergebnisse Math. Kolloq.*, **3**, 20–1. [21, 318]

— 1931b. Über formal unentscheidbare Sätze der Principia Mathematica und verwandter Systeme I. *Monatshefte für Math. u. Phys.*, **38**, 173–98. [36, 84]

— 1944. Russell's mathematical logic. In *The philosophy of Bertrand Russell*, ed. Paul A. Schilpp, pp. 125–53. New York: Tudor Publishing Co. [86]

Goetz, A. & Ryll-Nardzewski, C. 1960. On bases of abstract algebras. *Bull. Acad. Polon. Sci. Sér. Sci. Math.*, **8**, 157–61. [474]

Gordon, R. & Robson, J. C. (Robert Gordon, J. Christopher Robson) 1973. *Krull dimension*. Mem. Amer. Math. Soc. 133. Providence, RI: Amer. Math. Soc. [581]

Gould, V. A. R. (Victoria A. R. Gould) 1987. Model companions of *S*-systems. *Quart. J. Math. Oxford*, **38**, 189–211. [662]

Graham, R. L., Rothschild, B. L. & Spencer, J. H. (Ronald L. Graham, Bruce L. Rothschild, Joel H. Spencer) 1980. *Ramsey theory*. New York: Wiley. [594]

Grandjean, E. (Etienne Grandjean) 1983. Complexity of the first-order theory of almost all finite structures. *Information and Control*, **57**, 180–204. [359]

Grassmann, H. G. (Hermann Günther Grassmann) 1844. *Die lineale Ausdehnungslehre, ein neuer Zweig der Mathematik dargestellt und durch Anwendungen auf die übrigen Zweige der Mathe-*

matik, wie auch auf Statik, Mechanik, die Lehre von Magnetismus und die Krystallonomie erläutert. Stettin. [474]

Green, J. (Judy Green) 1978. κ-Suslin logic. *J. Symbolic Logic*, **43**, 659–66. [597]

Grilliot, T. J. (Thomas J. Grilliot) 1972. Omitting types: application to recursion theory. *J. Symbolic Logic*, **37**, 81–9. [358]

Grossberg, R. P. (Rami P. Grossberg) 199?. Indiscernible sequences in stable models. Preprint. [594]

Grossberg, R. P. & Shelah, S. 1983. On universal locally finite groups. *Israel J. Math.*, **44**, 289–302. [411]

— 199?. On the number of nonisomorphic models of an infinitary theory which has the infinitary order property, Part B. Preprint. [477]

Gruenberg, K. W. & Weir, A. J. (Karl W. Gruenberg) 1977. *Linear geometry*. New York: Springer. [172]

Gruson, L. & Jensen, C. U. (Laurent Gruson, Christian U. Jensen) 1976. Deux applications de la notion de *L*-dimension. *C. R. Acad. Sci. Paris Sér. A-B*, **282**, A23–4. [660]

Grzegorczyk, A. 1968. Logical uniformity by decomposition and categoricity in $\aleph_0$. *Bull. Acad. Polon. Sci. Sér. Sci. Math. Astron. Phys.*, **16**, 687–92. [358]

Gulliksen, T. H. (Tor H. Gulliksen) 1974. The Krull ordinal, coprof, and Noetherian localizations of large polynomial rings. *Amer. J. Math.*, **96**, 324–39. [581]

Gurevich, Y. S. (Ю. С. Гуревич, Yuri Gurevich) 1964. Элементарные свойства упорядоченных абелевых групп. *Algebra i Logika*, **3** (1), 5–39. *Trans. as* Elementary properties of ordered Abelian groups, *A.M.S. Translations*, **46** (1965), 165–92. [261, 410, 673]

— 1977. Expanded theory of ordered Abelian groups. *Ann. Math. Logic*, **12**, 193–228. [673]

— 1985. Monadic second-order theories. In Barwise & Feferman [1985], pp. 479–506. [76, 714]

Gurevich, Y. S. & Harrington, L. A. 1982. Trees, automata, and games. In *Proc. ACM Symp. on Theory of Computing*, ed. H. R. Lewis, pp. 60–5. San Francisco: Assoc. Computing Machinery. [86]

Gurevich, Y. S. & Schmitt, P. H. 1984. The theory of ordered abelian groups does not have the independence property. *Trans. Amer. Math. Soc.*, **284**, 171–82. [673]

Gute, H. B. & Reuter, K. K. 1990. The last word on elimination of quantifiers in modules. *J. Symbolic Logic*, **55**, 670–3. [653]

Hajnal, A. (Andras Hajnal) 1964. Remarks on the theory of W. P. Hanf. *Fundamenta Math.*, **54**, 109–13. [594]

Hajnal, A. & Kertész, A. 1972. Some new algebraic equivalents of the axiom of choice. *Publ. Math. Debrecen*, **19**, 339–40. [319]

Haley, D. K. (David K. Haley) 1979. *Equational compactness in rings*. Lecture Notes in Math. 745. Berlin: Springer. [534]

Hall, M. (Marshall Hall) 1943. Projective planes. *Trans. Amer. Math. Soc.*, **54**, 229–77. [409]

Hall, P. (Philip Hall) 1959. Some constructions for locally finite groups. *J. London Math. Soc.*, **34**, 305–19. [330, 598]

Halmos, P. R. (Paul R. Halmos) 1963, *Lectures on Boolean algebras*. Princeton: Van Nostrand. [274]

Halpern, J. D. & Lévy, A. (James D. Halpern, Azriel Levy) 1971. The Boolean prime ideal theorem does not imply the axiom of choice. In *Axiomatic set theory*, ed. D. S. Scott, pp. 83–134. Providence, RI: Amer. Math. Soc. [273]

Hanf, W. P. (William, P. Hanf) 1960. Models of languages with infinitely long expressions. In *Abstracts of contributed papers from the First Logic, Methodology and Philosophy of Science Congress*, vol. 1, p. 24. Stanford University. [589]

— 1964. On a problem of Erdős and Tarski. *Fundamenta Math.*, **53**, 325–34. [594]

Harnik, V. (Victor Harnik) 1975. On the existence of saturated models of stable theories. *Proc. Amer. Math. Soc.*, **52**, 361–7. [494, 532]

— 1979. Refinements of Vaught's normal form theorem. *J. Symbolic Logic*, **44**, 289–306. [597]

Harnik, V. & Harrington, L. A. 1984. Fundamentals of forking, *Ann. Pure Appl. Logic*, **26**, 245–86. [198, 534]

Harnik, V. & Makkai, M. 1976. Applications of Vaught sentences and the covering theorem. *J. Symbolic Logic*, **41**, 171–87. [597]

Hasenjaeger, G. F. R. (Gisbert F. R. Hasenjaeger) 1953. Eine Bemerkung zu Henkin's Beweis für die Vollständigkeit des Prädikatenkalküls. *J. Symbolic Logic*, **18**, 42–8. [261]

Hausdorff, F. (Felix Hausdorff) 1908. Grundzüge einer Theorie der geordneten Mengen. *Math. Annalen*, **65**, 435–505. [129, 485]

— 1914. *Grundzüge der Mengenlehre*. Leipzig: Veit. [128]

— 1936. Über zwei Sätze von G. Fichtenholz und L. Kantorovitch. *Studia Math.*, **6**, 18–19. [319]

Heintz, J. (Joos Heintz) 1983. Definability and fast quantifier elimination in algebraically closed fields. *Theoretical Comp. Sci.*, **24**, 239–77. [697]

Helling, M. 1966. *Model-theoretic problems for some extensions of first-order languages*. Doctoral dissertation. Univ. California, Berkeley. [595]

Henkin, L. A. (Leon Henkin) 1949. The completeness of the first-order functional calculus. *J. Symbolic Logic*, **14**, 159–66. [318]

— 1950. Completeness in the theory of types. *J. Symbolic Logic*, **15**, 81–91. [319]

— 1954. Metamathematical theorems equivalent to the prime ideal theorems for Boolean algebras. Preliminary report. *Bull. Amer. Math. Soc.*, **60**, 387–8. [319]

Henkin, L. A., Monk, J. D. & Tarski, A. 1971. *Cylindrical algebras*, Part 1. Amsterdam: North-Holland. [474]

Henkin, L. A. & Mostowski, A. 1959. Review of Malt'sev [1941]. *J. Symbolic Logic*, **24**, 55–7. [318]

Henrard, P. (Paul Henrard) 1973. Le 'forcing-compagnon' sans 'forcing'. *C.R. Acad. Sci. Paris Sér. A-B*, **276**, A821–2. [409]

Henschen, L. & Wos, L. 1974. Unit refutations and Horn sets. *J. Assoc. Comp. Mach.*, **21**, 590–605. [473]

Hensel, K. 1907. Über die arithmetischen Eigenschaften der Zahlen. *Jahresber. der D.M.V.*, **16**, 299–319, 388–93, 474–96. [409]

Henson, C. W. (C. Ward Henson) 1971. A family of countable homogeneous graphs. *Pacific J. Math.*, **38**, 69–83. [359]

— 1972. Countable homogeneous relational structures and $\aleph_0$-categorical theories. *J. Symbolic Logic*, **37**, 494–500. [359]

Henson, C. W. & Keisler, H. J. 1986. On the strength of nonstandard analysis. *J. Symbolic Logic*, **51**, 377–86. [575, 596]

Herbrand, J. (Jacques Herbrand) 1930. Recherches sur la théorie de la démonstration. *Trav. Soc. Sci. Lett. Varsovie Cl. III*, **33**, 1–128. Trans. with commentary in *Logical writings, Jacques Herbrand*, ed. W. D. Goldfarb (1971). Dordrecht: Reidel. [22, 390]

Hermann, G. (Grete Hermann) 1926, Die Frage der endlich vielen Schritte in der Theorie der Polynomideale. *Math. Ann.*, **95**, 736–88. [571, 596]

Herstein, I. N. 1968. *Noncommutative rings*. Math. Assoc. America, John Wiley. [211, 458]

Hewitt, E. 1948, Rings of real-valued continuous functions. *Trans. Amer. Math. Soc.*, **64**, 45–99. [475]

Hickin, K. K. (Kenneth K. Hickin) 1973. Countable type local theorems in algebra. *J. Algebra*, **27**, 523–37. [130]

— 1988. Some applications of tree-limits to groups. Part I. *Trans. Amer. Math. Soc.*, **305**, 797–839. [690]

— 1990. Review of Higman & Scott [1988]. *Bull. Amer. Math. Soc.*, **23**, 242–9. [690]

Hickin, K. K. & Macintyre, A. J. 1980. Algebraically closed groups: embeddings and centralizers. In *Word Problems II*, ed. S. I. Adian *et al.*, pp. 141–55. Amsterdam: North-Holland. [690]

Hickin, K. K. & Phillips, R. E. 1973. Local theorems and group extensions. *Proc. Camb. Philos. Soc.*, **73**, 7–20. [129, 130]

— 1990. Universal locally finite extensions of groups. In *Proc. Second Internat. Group Theory Conference, Bressanone 1989*, pp. 143–71. Supplemento ai *Rendiconti del Circolo Matem. di Palermo*. [690]

Higman, G. (Graham Higman) 1961. Subgroups of finitely presented groups. *Proc. Roy. Soc.*

London Ser. A, **262**, 455–75. [331]

Higman, G., Neumann, B. H. & Neumann, H. 1949. Embedding theorems for groups. *J. London Math. Soc.*, **24**, 247–54. [411]

Higman, G. & Scott, E. (Elizabeth Scott) 1988. *Existentially closed groups*. Oxford: Clarendon Press. [690]

Hilbert, D. (David Hilbert) 1893. Über die vollen Invariantensystem. *Math. Ann.*, **42**, 313–73. [409]

— 1899. *Grundlagen der Geometrie*. Leipzig: Teubner. Trans. as *Foundations of geometry* (1971). La Salle, Ill.: Open Court. [83, 260]

— 1926. Über das Unendliche. *Math. Annalen*, **95**, 161–190. *Trans. as* On the infinite. In *From Frege to Gödel*, ed. J. van Heijenoort (1967), pp. 369–92. Cambridge, Mass.: Harvard Univ. Press. [30]

Hilbert, D. & Ackermann, W. 1928. *Grundzüge der Theoretischen Logik*. Berlin: Springer. [86]

Hilbert, D. & Bernays, P. 1934. *Grundlagen der Mathematik*, I. Berlin: Springer. [84]

Hintikka, J. (Jaakko Hintikka) 1953. Distributive normal forms in the calculus of predicates. *Acta Philosophica Fennica*, **6**. [128]

— 1955. Form and content in quantification theory. *Acta Philosophica Fennica*, **8**, 11–55. [84]

Hirschelmann, A. (Arnulf Hirschelmann) 1972. An application of ultra-products to prime rings with polynomial identities. In *Conference in Mathematical Logic – London '70*, ed. W. A. Hodges, Lecture Notes in Math. 255, pp. 145–8. Berlin: Springer. [260]

Hirschfield, J. & Wheeler, W. H. (Joram Hirschfeld) 1975. *Forcing, arithmetic, division rings*. Lecture Notes in Math. 454. Berlin: Springer. [533]

Hodges, W. A. (Wilfrid Hodges) 1969. *Some questions on the structure of models: the Ehrenfeucht–Mostowski method of constructing models*. D.Phil. thesis. Oxford University. [595]

— 1972. On order-types of models. *J. Symbolic Logic*, **37**, 69–70. [598]

— 1973. Models in which all long indiscernible sequences are indiscernible sets. *Fundamenta Math.*, **78**, 1–6. [596]

— 1974. Six impossible rings. *J. Algebra*, **31**, 218–44. [198]

— 1976a. A normal form for algebraic constructions II. *Logique et Analyse* **71–72**, 429–87. Reprinted in *Six days of model theory*, ed. P. Henrard (1977), pp. 197–255. Albeuve: Editions Castella. [260, 474, 651, 652]

— 1976b. On the effectivity of some field constructions. *Proc. London Math. Soc.*, **32**, 133–62. [198, 652]

— 1976c. Läuchli's algebraic closure of Q. *Math. Proc. Camb. Phil. Soc.*, **79**, 289–97. [198]

— 1980a. Constructing pure injective hulls. *J. Symbolic Logic*, **45**, 544–8. [652]

— 1980b. Functorial uniform reducibility. *Fundamenta Math.*, **108**, 77–81. [475]

— 1981a. In singular cardinality, locally free algebras are free. *Algebra Universalis*, **12**, 205–20. Errata, *Algebra Universalis*, **19** (1984), 135. [129, 474]

— 1981b. Encoding orders and trees in binary relations. *Mathematika*, **28**, 67–81. [321]

— 1984a. Models built on linear orderings. In *Orders: description and roles*, ed. M. Pouzet & D. Richard, pp. 207–34. Amsterdam: North-Holland. [595]

— 1984b. On constructing many non-isomorphic algebras. In *Universal Algebra and its links with logic, algebra, combinatorics and computer science*, ed. P. Burmeister *et al.*, pp. 67–77. Berlin: Heldermann Verlag. [477]

— 1985. *Building models by games*. London Math. Soc. Student Texts no. 2. Cambridge: Cambridge Univ. Press. [319, 323, 358, 409, 411, 650, 689, 690]

— 1986. Truth in a structure. *Proc. Aristotelian Soc.*, **86**, 135–51. [82]

— 1987. What is a structure theory? *Bull. London Math. Soc.*, **19**, 209–37. [306, 596]

— 1989. Categoricity and permutation groups. In *Logic Colloquium '87, Granada 1987*, ed. H.-D. Ebbinghaus *et al.*, pp. 53–72. New York: Elsevier. [359]

— 199?. Existentially closed groups and determinacy. *J. Algebra* (to appear). [411]

Hodges, W. A., Hodkinson, I. M., Lascar, D. & Shelah, S. 199?. The small index property for ω-stable ω-categorical structures and for the random graph. *J. London Math. Soc.* (to appear). [352, 651]

Hodges, W. A., Hodkinson, I. M. & Macpherson, H. D. 1990. Omega-categoricity, relative categoricity and coordinatisation. *Ann. Pure Appl. Logic*, **46**, 169–99. [197, 198, 359, 651, 652]

Hodges, W. A., Lachlan, A. H. and Shelah, S. 1977. Possible orderings of an indiscernible sequence. *Bull. London Math. Soc.*, **9**, 212–15. [260, 594]

Hodges, W. A. & Pillay, A. 199?. Cohomology of structures and some problems of Ahlbrandt and Ziegler. *J. London Math. Soc.*, (to appear). [651]

Hodges, W. A. and Shelah, S. 1981. Infinite games and reduced products. *Ann. Math. Logic*, **20**, 77–108. [475]

— 1986. Naturality ànd definability I. *J. London Math. Soc.*, **33**, 1–12. [652]

— 1991. There are reasonably nice logics. *J. Symbolic Logic*, **56**, 300–22. [595]

— 199?. Naturality and definability II. Preprint. [652]

Hodkinson, I. M. & Macpherson, H. D. 1988. Relation structures determined by their finite induced substructures. *J. Symbolic Logic*, **53**, 222–30. [358]

Hodkinson, I. M. & Shelah, S. 199?. A construction of many uncountable rings using SFP domains and Aronszajn trees. *Proc. London Math. Soc.* (to appear). [409]

Horn, A. (Alfred Horn) 1951. On sentences which are true of direct unions of algebras. *J. Symbolic Logic*, **16**, 14–21. [473]

Hrushovski, E. (Ehud Hrushovski) 1987. Locally modular regular types. In *Classification theory*, ed. J. T. Baldwin, Lecture Notes in Mathematics 1292, pp. 132–64. Berlin: Springer. [200, 618]

— 1988. Construction of a strongly minimal set. Preprint. [199, 618]

— 1989a. Almost orthogonal regular types. *Ann. Pure Appl. Logic*, **45**, 139–55. [262]

— 1989b. Totally categorical structures. *Trans. Amer. Math. Soc.*, **313**, 131–59. [618, 651, 709]

— 1989c. Unidimensional theories: An introduction to geometric stability theory. In *Logic Colloquium '87, Granada 1987*, ed. H.-D. Ebbinghaus *et al.*, pp. 73–103. New York: Elsevier. [651]

— 1990. Unidimensional theories are superstable. *Ann. Pure Appl. Logic*, **50**, 117–38. [131]

— 199?a. Strongly minimal expansions of algebraically closed fields. Preprint. [199]

— 199?b. Unimodular minimal structures. *J. London Math. Soc.* (to appear). [199]

Hrushovski, E. & Pillay, A. 1987. Weakly normal groups. In *Logic Colloquium '85*, ed. The Paris Logic Group. pp. 233–44. Amsterdam: North-Holland. [256]

Huber-Dyson, V. (Verena Huber-Dyson) 1964. On the decision problem for theories of finite models. *Israel J. Math.*, **2**, 55–70. [319]

Hughes, D. R. & Piper, F. C. (Daniel R. Hughes, Fred C. Piper) 1973. *Projective planes*. New York: Springer. [223, 226]

Hule, H. (Harald Hule) 1978. Relations between the amalgamation property and algebraic equations. *J. Austral. Math. Soc. Ser. A*, **25**, 257–63. [474]

Humphreys, J. E. (James E. Humphreys) 1975. *Linear algebraic groups*. New York: Springer. [262, 692, 693, 696]

Huntington, E. V. (Edward V. Huntington) 1904a. Sets of independent postulates for the algebra of logic. *Trans. Amer. Math. Soc.*, **5**, 288–309. [83, 319]

— 1904b. *The continuum and other types of serial order, with an introduction to Cantor's transfinite numbers*. Cambridge, Mass: Harvard Univ. Press. [128]

— 1924. A new set of postulates for betweenness, with proof of complete independence. *Trans. Amer. Math. Soc.*, **26**, 257–82. [83]

Hurd, A. E. & Loeb, P. A. (Albert E. Hurd, Peter A. Loeb) 1985. *An introduction to nonstandard real analysis*. Orlando, Fla: Academic Press. [567]

Hyttinen, T. & Tuuri, H. (Tapani Hyttinen, Heikki Tuuri) 1991. Constructing strongly equivalent nonisomorphic models for unstable theories. *Ann. Pure Appl. Logic*, **52**, 203–48. [123, 565, 673]

Immerman, N. (Neil Immerman) 1982. Upper and lower bounds for first order expressibility. *J. Comp. System Sci.*, **25**, 76–98. [129]

— 1986. Relational queries computable in polynomial time. *Information and Control*, **68**, 86–104. [86]

Inaba, E. 1944. Über den Hilbertschen Irreduzibilitätssatz. *Japan. J. Math.*, **19**, 1–25. [596]

Isbell, J. R. (John R. Isbell) 1973. Functorial implicit operations. *Israel J. Math.*, **15**, 185–8. [475]

Ivanov, A. A. (A. A. Иванов) 1986. Модельное пополнение теории унарных алгебр (The model completion of the theory of unary algebras). *Sibirsk. Mat. Zh.*, **27** (2), 205–7. [410]

Jackson, S. C. (Stephen Colin Jackson) 1973. *Model theory of abelian groups*. Ph.D. thesis. Bedford College, London University. [672]

Jacobson, N. (Nathan Jacobson) 1964. *Lectures in abstract algebra III, Theory of fields and Galois theory*. Princeton, NJ: Van Nostrand. [253, 554]

— 1974. *Basic algebra I*. San Francisco: W. H. Freeman. [386, 663, 698]

— 1980. *Basic algebra II*. San Francisco: W. H. Freeman. [79, 123, 362, 365]

Jarden, M. (Moshe Jarden) 1976. Algebraically closed fields with distinguished subfields. *Arch. Math. (Basel)*, **27**, 502–5. [704]

Jategaonkar, A. V. (Arun V. Jategaonkar) 1969. A counter-example in homological algebra and ring theory. *J. Algebra*, **12**, 418–40. [583]

Jech, T. J. (Thomas Jech) 1973a. Some combinatorial problems concerning uncountable cardinals. *Ann. Math. Logic*, **5**, 165–98. [130]

— 1973b. *The axiom of choice*. Amsterdam: North-Holland. [150, 630]

— 1978. *Set theory*. New York: Academic Press. [429, 456, 507, 559, 594, 597]

Jensen, C. U. & Lenzing, H. (Christian U. Jensen, Helmut Lenzing) 1989. *Model theoretic algebra, with particular emphasis on fields, rings, modules*. New York: Gordon & Breach. [86, 659]

Jensen, C. U. & Vámos, P. (Peter Vámos) 1979. On the axiomatizability of certain classes of modules. *Math. Z.*, **167**, 227–37. [476]

Jensen, R. B. (Ronald B. Jensen) 1972. The fine structure of the constructible hierarchy. *Ann. Math. Logic*, **4**, 229–308. Erratum p. 443. [602]

Jevons, W. S. 1864. *Pure logic, or The logic of quality apart from quantity*, London. [319]

Ježek, J. 1985. Elementarily non-equivalent infinite partition lattices. *Algebra Universalis*, **20**, 132–3. [261]

Jones, N. D. & Selman, A. L. (Neil D. Jones, Alan L. Selman) 1974. Turing machines and the spectra of first-order formulas. *J. Symbolic Logic*, **39**, 139–50. [542]

Jónsson, B. (Bjarni Jónsson) 1956. Universal relational systems. *Math. Scand.*, **4**, 193–208. [532]

— 1957. On isomorphism types of groups and other algebraic systems. *Math. Scand.*, **5**, 224–9. [477]

— 1960. Homogeneous universal relational systems. *Math. Scand.*, **8**, 137–42. [532]

Jónsson, B. & Olin, P. (Philip Olin) 1968. Almost direct products and saturation. *Compositio Math.*, **20**, 125–32. [533]

Jónsson, B. & Tarski, A. 1961. On two properties of free algebras. *Math. Scand.*, **9**, 95–101. [474]

Kaiser, K. (Klaus Kaiser) 1969. Über eine Verallgemeinerung der Robinsonschen Modellvervollständigung I. *Z. Math. Logik Grundlag. Math.*, **15**, 37–48. [409]

Kamp, J. A. W. 1968. *Tense logic and the theory of linear order*. Doctoral dissertation. Univ. California at Los Angeles. [709]

Kantor, W. M., Liebeck, M. W. & Macpherson, H. D. (William M. Kantor, Martin W. Liebeck) 1989. $\aleph_0$-categorical structures smoothly approximated by finite substructures. *Proc. London Math. Soc.*, **59**, 439–63. [200]

Kaplansky, I. (Irving Kaplansky) 1954. *Infinite abelian groups*. Ann Arbor, Mich.: Univ. Michigan Press. [534]

Karp, C. R. (Carol R. Karp) 1964. *Languages with expressions of infinite length*, Amsterdam: North-Holland. [82]

— 1965. Finite-quantifier equivalence. In Addison, Henkin & Tarski [1965], pp. 407–12. [129, 597]

Kaufmann, M. J. & Shelah, S. (Matthew J. Kaufmann) 1985. On random models of finite power and monadic logic. *J. Symbolic Logic*, **54**, 285–93. [359]

Kaye, R. (Richard Kaye) 1991. *Models of Peano arithmetic*. Oxford: Oxford Univ. Press. [534]

Kegel, O. H. (Otto H. Kegel) 1989. Four lectures on Sylow theory in locally finite groups. In

Group Theory, Proc. Singapore Group Theory Conf., ed. Kai Nah Cheng & Yu Kaiang Leong. Berlin: de Gruyter. [694]

Kegel, O. H. & Wehrfritz, B. A. F. (Bertram A. F. Wehrfritz) 1973. *Locally finite groups*. Amsterdam: North-Holland. [94, 476]

Keisler, H. J. (H. Jerome Keisler) 1960. Theory of models with generalized atomic formulas. *J. Symbolic Logic*, **25**, 1–26. [320]

— 1961a. Replete relational systems. Abstract. *Not. Amer. Math. Soc.*, **8**, 63. [532]

— 1961b. Properties preserved under reduced products II. Abstract. *Not. Amer. Math. Soc.*, **8**, 64. [533]

— 1962. Some applications of the theory of models to set theory. In *Logic, methodology and philosophy of science*, ed. E. Nagel *et al.*, pp. 80–6. Stanford: Stanford Univ. Press. [476]

— 1964a. Complete theories of algebraically closed fields with distinguished subfields. *Michigan Math. J.*, **11**, 71–81. [704]

— 1964b. Good ideals in fields of sets. *Ann. Math.*, **79**, 338–59. [475, 476]

— 1964c. On cardinalities of ultraproducts. *Bull. Amer. Math. Soc.*, **70**, 644–7. [475, 476]

— 1965a. Finite approximations of infinitely long formulas. In Addison, Henkin & Tarski [1965], pp. 158–69. Amsterdam: North-Holland. [533]

— 1965b. Some applications of infinitely long formulas. *J. Symbolic Logic*, **30**, 339–49. [533]

— 1965c. Reduced products and Horn classes. *Trans. Amer. Math. Soc.*, **117**, 307–28. [475]

— 1966a. First order properties of pairs of cardinals. *Bull. Amer. Math. Soc.*, **72**, 141–4. [650]

— 1966b. Some model-theoretic results for ω-logic. *Israel J. Math.*, **4**, 249–61. [650]

— 1967. Ultraproducts which are not saturated. *J. Symbolic Logic*, **32**, 23–46. [155, 198, 475, 476]

— 1968. Models with orderings. In *Logic, methodology and philosophy of science III*, ed. B. van Rootselaar & J. F. Staal, pp. 35–62. Amsterdam: North-Holland. [650]

— 1970. Logic with the quantifier 'there exist uncountably many'. *Ann. Math. Logic*, **1**, 1–93. [129, 594, 602]

— 1971a. *Model theory for infinitary logic*. Amsterdam: North-Holland. [585, 650]

— 1971b. On theories categorical in their own power. *J. Symbolic Logic*, **36**, 240–4. [532, 650]

— 1973. Forcing and the omitting types theorem. In *Studies in model theory*, ed. M. D. Morley, MAA Studies in Math. Vol. 8, pp. 96–133. Buffalo, NY: Math. Assoc. Amer. [409]

— 1976. *Foundations of infinitesimal analysis*. Boston: Prindle, Weber & Schmidt. [567]

Keisler, H. J. & Morley, M. D. 1967. On the number of homogeneous models of a given power. *Israel J. Math.*, **5**, 73–8. [532]

Keisler, H. J. & Tarski, A. 1964. From accessible to inaccessible cardinals. *Fundamenta Math.*, **53**, 225–308. [596]

Kemeny, J. G. (John G. Kemeny) 1956. A new approach to semantics, parts I, II. *J. Symbolic Logic*, **21**, 1–27 and 149–161. [22]

Kino, A. (Akiko Kino) 1966. On definability of ordinals in logic with infinitely long expressions. *J. Symbolic Logic*, **31**, 365–75. [715]

Kirchner, C. (Claude Kirchner) (ed.) 1990. *Unification*. London: Academic Press. [22]

Kleene, S. C. (Stephen Cole Kleene) 1934. Proof by cases in formal logic. *Ann. Math.*, **35**, 529–44. [83]

— 1943. Recursive predicates and quantifiers. *Trans. Amer. Math. Soc.*, **53**, 41–73. [31]

— 1952a. Finite axiomatizability of theories in the predicate calculus using additional predicate symbols. *Two papers on the predicate calculus*. *Mem. Amer. Math. Soc.*, **10**, 27–68. [534]

— 1952b. *Introduction to metamathematics*. Amsterdam: North-Holland. [22, 261]

Klein, F. (Felix Klein) 1872. Vergleichende Betrachtungen über neuere geometrische Forschungen. Republished in *Collected mathematical works*, F. Klein (1973). Heidelberg: Springer. [131]

Knaster, B. 1928. Un théorème sur les fonctions d'ensembles. *Ann. Soc. Polon. Math.*, **6**, 133–4. [84]

Knebusch, M. (Manfred Knebusch) 1972. On the uniqueness of real closures and the existence of real places. *Comm. Math. Helv.*, **47**, 260–9. [386]

Knight, J. F. (Julia F. Knight) 1986a. Effective construction of models. In *Logic Colloquium '84*, ed. J. B. Paris *et al.*, pp. 105-19. Amsterdam: North-Holland. [261]

— 1986b. Saturation of homogeneous resplendent models. *J. Symbolic Logic*, **51**, 222-4. [534]

Knight, J. F., Pillay, A. & Steinhorn, C. I. (Charles I. Steinhorn) 1986. Definable sets in ordered structures II. *Trans. Amer. Math. Soc.*, **295**, 593-605. [158]

Kochen, S. B. (Simon B. Kochen) 1961. Ultraproducts in the theory of models. *Ann. Math.*, **74**, 221-61. [475, 476]

Kogalovskiĭ, S. R. (С. Р. Когаловский) 1961. О мультипликативных полугруппах колец. *Dokl. Akad. Nauk SSSR*, **140**, 1005-7. *Trans. as* On multiplicative semigroups of rings. *Soviet Math. Dokl.*, **2** (1961), 1299-1301. [669]

Kolaitis, Ph. G. (Phokion G. Kolaitis) 1985. Game quantification. In Barwise & Feferman [1985], pp. 365-421. [113]

König, D. (Denes König) 1926. Sur les correspondances multivoques des ensembles. *Fundamenta Math.*, **8**, 114-34. [262]

Kopperman, R. D. 1972. *Model theory and its applications*. Boston: Allyn & Bacon. [449]

Kopperman, R. D. & Mathias, A. R. D. (Ralph D. Kopperman, Adrian Mathias) 1968. Some problems in group theory. In *The syntax and semantics of infinitary languages*, ed. K. J. Barwise, Lecture Notes in Math. 72, pp. 131-8. Berlin: Springer. [83, 130]

Kotlarski, H. 1980. On Skolem ultrapowers and their nonstandard variants. *Z. Math. Logik Grundlag. Math.*, **26**, 227-36. [534]

Krasner, M. (Marc Krasner) 1938. Une généralisation de la notion de corps. *J. Math. Pures Appl.*, **17**, 367-85. [197]

Kronecker, L. (Leopold Kronecker) 1882. Grundzüge einer arithmetischen Theorie der algebraischen Grössen. *Crelle's J.*, **92**, 1-122. [22, 410]

Kueker, D. W. (David W. Kueker) 1968. Definability, automorphisms, and infinitary languages. In *The syntax and semantics of infinitary languages*, ed. K. J. Barwise, Lecture Notes in Math. 72, pp. 152-65. Berlin: Springer. [198]

— 1970 Generalized interpolation and definability. *Ann. Math. Logic*, **1**, 423-68. [533]

— 1973. A note on the elementary theory of finite abelian groups. *Algebra Universalis*, **3**, 156-9. [476]

— 1977. Countable approximations and Löwenheim-Skolem theorems. *Ann. Math. Logic*, **11**, 57-103. [129]

— 1981. $L_{\infty\omega_1}$-elementarily equivalent models of power ω_1. In *Logic year 1979-80*, ed. M. Lerman *et al.*, Lecture Notes in Math. 859, pp. 120-31. Berlin: Springer. [129]

Kueker, D. W. & Laskowski, M. C. 1992. On generic structures. *Notre Dame J. Formal Logic*, **33**, 175-83. [357]

Kueker, D. W. & Steitz, P. (Philip Steitz) 199?a. Saturated models of stable theories. Preprint. [532]

— 199?b. Stabilizers of definable sets in homogeneous models. Preprint. [651]

Kunen, K. (Kenneth Kunen) 1987. Negation in logic programming. *J. Logic Programming*, **4**, 289-308. [85]

Kuratowski, C. 1937. Les types d'ordre définissables et les ensembles Boreliens. *Fundamenta Math.*, **28**, 97-100. [597]

— 1958. *Topologie* I. Warsaw: Państwowe Wydawnictwo Naukowe. [129, 197]

Lachlan, A. H. (Alistair H. Lachlan) 1971. The transcendental rank of a theory. *Pacific J. Math.*, **37**, 119-22. [597]

— 1972. A property of stable theories. *Fundamenta Math.*, **77**, 9-20. [321]

— 1978. Skolem functions and elementary extensions. *J. London Math. Soc.*, **18**, 1-6. [91]

— 1980. Singular properties of Morley rank. *Fundamenta Math.*, **108**, 145-57. [243]

— 1987a. Homogeneous structures. In *Proc. Internat. Congress of Mathematicians, Berkeley 1986*, pp. 314-21. [307]

— 1987b. Structures coordinatized by indiscernible sets. *Ann. Pure Appl. Logic*, **34**, 245-73. [709]

— 1990. Complete coinductive theories I. *Trans. Amer. Math. Soc.*, **319**, 209-41. [321]

Lachlan, A. H. & Woodrow, R. E. 1980. Countable homogeneous undirected graphs. *Trans. Amer. Math. Soc.*, **262**, 51-94. [330]

Lang, S. (Serge Lang) 1958. *Introduction to algebraic geometry*. New York: Interscience. [153, 160]

Langford, C. H. 1926. Some theorems on deducibility. *Ann. Math.*, **28**, 16–40. [85, 128, 260, 261]

Lascar, D. (Daniel Lascar) 1982. On the category of models of a complete theory. *J. Symbolic Logic*, **47**, 249–66. [359, 534]

— 1985. Why some people are excited by Vaught's conjecture. *J. Symbolic Logic*, **50**, 973–82. [339]

— 1987a. *Stability in model theory*. Pitman Monographs and Surveys in Pure and Applied Mathematics 36. Harlow: Longman Scientific & Technical. [82, 306, 322]

— 1987b. Théorie de la classification. *Séminaire BOURBAKI*, 39ème année (1986–7) no. 682, June 1987; *Astérisque*, **152–3**, **5** (1988), 253–261. [596]

— 1989. Le demi-groupe des endomorphismes d'une structure $\aleph_0$-catégorique. In *Actes de la Journée Algèbre Ordonnée (Le Mans, 1987)*, ed. M. Giraudet, pp. 33–43. [534]

— 1991. Autour de la propriété du petit indice. *Proc. London Math. Soc.*, **62**, 25–53. [197]

Lascar, D. & Poizat, B. 1979. An introduction to forking. *J. Symbolic Logic*, **44**, 330–50. [320]

Laskowski, M. C. (Michael Chris Laskowski) 1988. Uncountable theories that are categorical in a higher power. *J. Symbolic Logic*, **53**, 512–30. [131, 613, 650, 651]

Läuchli, H. (Hans Läuchli) 1962. Auswahlaxiom in der Algebra. *Comment. Math. Helv.*, **37**, 1–18. [198, 652]

— 1964. The independence of the ordering principle from a restricted axiom of choice. *Fundamenta Math.*, **54**, 31–43. [359]

— 1968. A decision procedure for the weak second order theory of linear order. In *Contributions to mathematical logic, Hannover 1966*, ed. H. A. Schmidt *et al.*, pp. 189–97. Amsterdam: North-Holland. [714]

— 1971. Coloring infinite graphs and the Boolean prime ideal theorem. *Israel J. Math.*, **9**, 422–9. [319]

Läuchli, H. & Leonard, J. 1966. On the elementary theory of linear order, *Fundamenta Math.*, **59**, 109–16. [711]

Lavrov, I. A. (И. А. Лавров) 1963. Эффективная неотделимость множества тождественно истинных и множества конечно опровержимых формул некоторых элементарных теорий (Effective inseparability of the set of identically true formulas and the set of formulas with finite counterexamples for certain elementary theories). *Algebra i Logika*, **2** (1), 5–18. [261, 262]

Lawrence, J. (John Lawrence) 1981a. Boolean powers of groups. *Proc. Amer. Math. Soc.*, **82**, 512–15. [477]

— 1981b. Primitive rings do not form an elementary class. *Comm. Algebra*, **9**, 397–400. [129]

Leinen, F. (Felix Leinen) 1986. Existentially closed locally finite p-groups. *J. Algebra*, **103**, 160–83. [690]

Lindström, P. (Per Lindström) 1964. On model-completeness. *Theoria*, **30**, 183–96. [409, 410]

— 1969. On extensions of elementary logic. *Theoria*, **35**, 1–11. [597]

Lipshitz, L. M. & Nadel, M. (Leonard M. Lipshitz) 1978. The additive structure of models of arithmetic. *Proc. Amer. Math. Soc.*, **68**, 331–6. [534]

Lipshitz, L. M. & Saracino, D. H. 1973. The model companion of the theory of commutative rings without nilpotent elements. *Proc. Amer. Math. Soc.*, **38**, 381–7. [394]

Lo, L. (Lo Libo) 1983. On the number of countable homogeneous models. *J. Symbolic Logic*, **48**, 539–41. [532]

López-Escobar, E. G. K. (Edgar G. K. López-Escobar) 1966. On defining well-orderings. *Fundamenta Math.*, **59**, 13–21. [597, 598]

Łoś, J. (Jerzy Łoś) 1954a. On the categoricity in power of elementary deductive systems and some related problems. *Colloq. Math.*, **3**, 58–62. [319, 650]

— 1954b. On the existence of linear order in a group. *Bull. Acad. Polon. Sci.*, (3) **2**, 21–3. [319, 321]

— 1955a. On the extending of models (I). *Fundamenta Math.*, **42**, 38–54. [320, 533]

— 1955b. Quelques remarques, théorèmes et problèmes sur les classes définissables d'algèbres. In

Mathematical interpretation of formal systems, ed. L. E. J. Brouwer *et al.*, pp. 98–113. Amsterdam: North-Holland. [475]

Łoś, J. & Suszko, R. 1955. On the extending of models (II) Common extensions. *Fundamenta Math.*, **42**, 343–7. [320]

— 1957. On the extending of models (IV) Infinite sums of models. *Fundamenta Math.*, **44**, 52–60. [320]

Löwenheim, L. (Leopold Löwenheim) 1915. Über Möglichkeiten im Relativkalkül. *Math. Ann.*, **76**, 447–70. *Trans. as* On possibilities in the calculus of relatives. In *From Frege to Gödel*, ed. J. van Heijenoort (1967), pp. 232–51. Cambridge, Mass.: Harvard Univ. Press. [84, 128, 261]

Lyndon, R. C. (Roger C. Lyndon) 1959a. An interpolation theorem in the predicate calculus. *Pacific J. Math.*, **9**, 129–42. [533]

— 1959b. Properties preserved in subdirect products. *Pacific J. Math.*, **9**, 155–64. [533]

— 1959c. Properties preserved under homomorphism. *Pacific J. Math.*, **9**, 143–54. [474, 533]

Lyndon, R. C. & Schupp, P. E. (Paul E. Schupp) 1977. *Combinatorial group theory*. Berlin: Springer. [366, 402]

MacDowell, R. & Specker, E. 1961. Modelle der Arithmetik. In *Infinitistic Methods, Warsaw 1959*, pp. 257–63. Oxford: Pergamon Press and Warsaw: Państwowe Wydawnictwo Naukowe. [595]

MacHenry, T. S. 1960. The tensor product and the 2nd nilpotent product of groups. *Math. Z.*, **73**, 134–45. [474]

Machover, M. (Moshé Machover) 1960. A note on sentences preserved under direct products and powers. *Bull. Acad. Polon. Sci. Math.*, **8**, 519–23. [463]

Macintyre, A. J. (Angus Macintyre) 1971a. On ω_1-categorical theories of abelian groups. *Fundamenta Math.*, **70**, 253–70. [263, 670, 671]

— 1971b. On ω_1-categorical theories of fields. *Fundamenta Math.*, **71**, 1–25. [263, 697, 699]

— 1972a. Omitting quantifier-free types in generic structures. *J. Symbolic Logic*, **37**, 512–20. [409, 690]

— 1972b. On algebraically closed groups. *Ann. Math.*, **96**, 53–97. [409, 411, 689]

— 1975. Dense embeddings I: A theorem of Robinson in a general setting. In *Model theory and algebra: A memorial tribute to Abraham Robinson*, ed. D. H. Saracino & V.B. Weispfenning, Lecture Notes in Math. 498, pp. 200–19. Berlin: Springer. [704]

— 1976. On definable subsets of *p*-adic fields. *J. Symbolic Logic*, **41**, 605–10. [705]

— 1977. Model completeness. In *Handbook of mathematical logic*, ed. K. J. Barwise, pp. 139–80. Amsterdam: North-Holland [379]

— 1984. *Notes on real exponentiation, from a graduate course at Urbana*. Oxford: Mathematical Institute. [85]

— 1986. Twenty years of *p*-adic model theory. In *Logic Colloquium '84*, ed. J. B. Paris *et al.*, pp. 121–53. Amsterdam: North-Holland. [704]

Macintyre, A. J., McKenna, K. & van den Dries, L. (Kenneth McKenna) 1983. Elimination of quantifiers in algebraic structures. *Advances in Math.*, **47**, 74–87. [697, 704]

Macintyre, A. J. & Rosenstein, J. G. 1976. $\aleph_0$-categoricity for rings without nilpotent elements and for Boolean structures. *J. Algebra*, **43**, 129–54. [358]

Macintyre, A. J. & Shelah, S. 1976. Uncountable universal locally finite groups. *J. Algebra*, **43**, 168–75. [598, 690]

McKenzie, R. N. (Ralph McKenzie) 1971a. On elementary types of symmetric groups. *Algebra Universalis*, **1**, 13–20. [261]

— 1971b. $\aleph_1$-incompactness of Z. *Colloq. Math.*, **23**, 199–202. [534]

McKenzie, R. N. & Shelah, S. 1974. The cardinals of simple models for universal theories. In *Proc. Tarski Symposium*, ed. L. A. Henkin *et al.*, pp. 53–74. Providence, RI: Amer. Math Soc. [534, 598]

McKinsey, J. C. C. 1943. The decision problem for some classes of sentences without quantifiers. *J. Symbolic Logic*, **8**, 61–76. [473,474]

McLain, D. H. 1959. Local theorems in universal algebras. *J. London Math Soc.*, **34**, 177–84. [321]

Mac Lane, S. (Saunders Mac Lane) 1938. A lattice formulation for transcendence degrees and

p-bases. *Duke Math. J.*, **4**, 455–68. [199]

— 1971. *Categories for the working mathematician*. New York: Springer. [441, 475]

Macpherson, H. D. (Dugald Macpherson) 1983. *Enumeration of orbits of infinite permutation groups*. D.Phil thesis. Oxford University. [359]

— 1986a. Groups of automorphisms of $\aleph_0$-categorical structures. *Quart. J. Math. Oxford*, **37**, 449–65.

— 1986b. Homogeneity in infinite permutation groups. *Periodica Math. Hungar.*, **17**, 211–33. [260]

— 1988. Absolutely ubiquitous structures and $\aleph_0$-categorical groups. *Quart. J. Math. Oxford*, **39**, 483–500. [358, 693]

— 1991. Interpreting groups in ω-categorical structures. *J. Symbolic Logic*, **56**, 1317–24. [694]

Macpherson, H. D., Mekler, A. H. & Shelah, S. 1991. The number of infinite substructures. *Math. Proc. Camb. Phil. Soc.*, **109**, 193–209. [594]

Macpherson, H. D. & Steinhorn, C. I. 199?. Paper on C-minimality. In preparation. [651]

Magari, R. 1969. Una dimostrazione del fatto che ogni varietà ammette algebre semplici. *Ann. Univ. Ferrara Sez. VII*, **14**, 1–4. [475]

Magidor, M. (Menachem Magidor) 1971. There are many normal ultrafilters corresponding to a supercompact cardinal. *Israel J. Math.*, **9**, 186–92. [129]

Mahlo, P. 1911. Über lineare transfinite Mengen. *Berich. Verh. Sächsischen Akad. Wissensch. Leipzig, Math. Phys. Kl.*, **63**, 187–225. [129]

Mahr, B. & Makowsky, J. A. 1983. Characterizing specification languages which admit initial semantics. In *Proc. CAAP '83*, ed. G. Ausiello & M. Protasi, Lecture Notes in Comp. Sci. 159, pp. 300–16. Berlin: Springer. [474]

Maier, B. J. (Berthold J. Maier) 1983. On existentially closed and generic nilpotent groups. *Israel J. Math.*, **46**, 170–88. [690]

— 1984. Existentially closed torsion-free nilpotent groups of class three. *J. Symbolic Logic*, **49**, 220–30. [690]

— 1987. On countable locally described structures. *Ann. Pure Appl. Logic*, **35**, 205–46. [690]

— 1989. On nilpotent groups of exponent p. *J. Algebra*, **127**, 279–89. [690]

Makinson, D. C. 1969. On the number of ultrafilters of an infinite Boolean algebra. *Z. Math. Logik Grundlag. Math.*, **15**, 121–2. [319]

Makkai, M. (Mihaly Makkai, Michael Makkai) 1964a. On a generalization of a theorem of E.W. Beth. *Acta Math. Acad. Sci. Hungar.*, **15**, 227–35. [652]

— 1964b. On PC$_\Delta$-classes in the theory of models. *Matematikai Kutató Intézetének Közleményei*, **9**, 159–94. [260, 534]

— 1965. A compactness result concerning direct products of models. *Fundamenta Math.*, **57**, 313–25. [476]

— 1969. On the model theory of denumerably long formulas with finite strings of quantifiers. *J. Symbolic Logic*, **34**, 437–59. [585]

— 1973. Vaught sentences and Lindström's regular relations. In *Cambridge Summer School in Mathematical Logic*, ed. A. R. D. Mathias & H. Rogers, Lecture Notes in Math. 337, pp. 622–60. Berlin: Springer. [597]

— 1977. Admissible sets and infinitary logic. In *Handbook of mathematical logic*, ed. K. J. Barwise, pp. 233–81. Amsterdam: North-Holland. [129]

— 1981. An example concerning Scott heights. *J. Symbolic Logic*, **46**, 301–18. [129]

Makowsky, J. A. (Johann A. Makowsky) 1974. On some conjectures connected with complete sentences. *Fundamenta Math.*, **81**, 193–202. [200]

— 1985a. Vopěnka's principle and compact logics. *J. Symbolic Logic*, **50**, 42–8. [474]

— 1985b. Compactness, embeddings and definability. In Barwise & Feferman [1985], pp. 645–716. [286]

Makowsky, J. A. & Shelah, S. 1979. The theorems of Beth and Craig in abstract model theory, I: The abstract setting. *Trans. Amer. Math. Soc.*, **256**, 215–39. [302]

Malitz, J. I. (Jerome I. Malitz) 1971. Infinitary analogs of theorems from first order model theory. *J. Symbolic Logic*, **36**, 216–28. [476]

Malitz, J. I. & Reinhardt, W. N. (William N. Reinhardt) 1972. Maximal models in the language with quantifier 'There exist uncountably many'. *Pacific J. Math.*, **40**, 139–55. [288]

Mal'tsev, A. I. (А. И. Mal'cev, А. И. Мальцев) 1936. Untersuchungen aus dem Gebiete der mathematischen Logik. *Mat. Sbornik*, **1** (43), 323–36. Trans. in Mal'tsev [1971], pp. 1–14. [2, 21, 318]

— 1940. Об изоморфном представлении бесконечных групп матрицами. *Mat. Sbornik*, **8** (50), 405–22. *Trans.* as On the faithful representation of infinite groups by matrices. *A.M.S. Translations*, (2) **45** (1965), pp. 1–18. [321, 688]

— 1941. Об одном общем методе получения локальных теорем теории групп. Uchenye Zapiski Ivanov. Ped. Inst. (Fiz-mat. Fakul'tet), **1**, no. 1, 3–9. *Trans.* as A general method for obtaining local theorems in group theory. In Mal'tsev [1971], pp. 15–21. [260, 318, 321]

— 1956a. Квазипримитивные классы абстрактных алгебр. *Dokl. Akad. Nauk SSSR*, **108**, 187–9. *Trans.* as Quasiprimitive classes of abstract algebras. In Mal'tsev [1971], pp. 27–31. [474]

— 1956b. Подпрямые произведения моделей. *Dokl. Akad. Nauk SSSR*, **109**, 264–6. *Trans.* as Subdirect products of models. In Mal'tsev [1971], pp. 32–6. [533]

— 1958a. Структурная характеристика некоторых классах алгебр. *Dokl. Akad. Nuak SSSR*, **120**, 29–32. *Trans.* as The structural characterization of certain classes of algebras. In Mal'tsev [1971], pp. 56–60. [431, 474]

— 1958b. О некоторых классах моделей. *Dokl. Akad. Nauk. SSSR*, **120**, 245–8. *Trans.* as Certain classes of models. In Mal'tsev [1971], pp. 61–5. [475]

— 1959. Модельные соответствия. *Izv. Akad Nauk SSSR (ser. mat.)*, **23**, 313–36. *Trans.* as Model correspondences. In Mal'tsev [1971], pp. 66–94. [321, 474]

— 1960. Об одном соответствии между кольцами и группами. *Mat. Sbornik*, **50** (92), 257–66. *Trans.* as A correspondence between rings and groups. In Mal'tsev [1971], pp. 124–37. [220, 261, 262]

— 1961a. Об элементарных свойствах линейных групп. In *Some problems of mathematics and mechanics* (Russian), pp. 110–32. Novosibirsk: Akad. Nauk SSSR (Sibirsk. otdel.). *Trans.* as Elementary properties of linear groups. In Mal'tsev [1971], pp. 221–47. [83]

— 1961b. Об элементарных теориях локально свободных универсальных алгебр. *Dokl. Akad. Nauk SSSR*, **138**, 1009–12. *Trans.* as On the elementary theories of locally free universal algebras. *Soviet Math Dokl.*, **2** (3) (1961), 768–71. [83, 85]

— 1962. Аксиоматизируемые классы локально свободных алгебр некоторых типов. *Sibirsk. Mat. Zh.*, **3**, 729–43. *Trans.* as Axiomatizable classes of locally free algebras of various types. In Mal'tsev [1971], pp. 262–81. [83, 85]

— 1966. Несколько замечаний о квазимногообразиях алгебраических систем. *Algebra i Logika*, **5** (3), 3–9. *Trans.* as A few remarks on quasivarieties of algebraic systems. In Mal'tsev [1971], pp. 416–21. [475]

— 1970. Алгебраические системы. Moscow: Nauka. Trans. as *Algebraic systems* (1973). Berlin: Springer. [474, 475]

— 1971. *The metamathematics of algebraic systems. Collected papers: 1936–1967*, trans. Benjamin Franklin Wells III. Amsterdam: North-Holland.

Mansfield, R. & Weitkamp, G.L. (Richard Mansfield, Galen L. Weitkamp) 1985. *Recursive aspects of descriptive set theory*. New York: Oxford Univ. Press. [80, 507]

Maranda, J. M. 1960. On pure subgroups of abelian groups. *Arch. Math. (Basel)*, **11**, 1–13. [534]

Marcus, L. G. (Leo G. Marcus) 1972. A minimal prime model with an infinite set of indiscernibles. *Israel J. Math.*, **11**, 180–3. [358]

— 1975. A type-open minimal model. *Arch. Math. Logik Grundlagenforsch.*, **17**, 17–24. [320]

Marczewski, E. 1951. Sur les congruences et les propriétés positives d'algèbres abstraites. *Colloq. Math.*, **2**, 220–8. [84, 320]

Marker, D. E. (David E. Marker) 1987. A strongly minimal expansion of (ω, s). *J. Symbolic Logic*, **52**, 205–7. [199]

— 1989. Non Σ_n axiomatizable almost strongly minimal theories. *J. Symbolic Logic*, **54**, 921–7. [651]

Marker, D. E. & Pillay, A. 1990. Reducts of $(C, +, \cdot)$ which contain $+$. *J. Symbolic Logic*, **55**, 1243–51. [700]

Marongiu, G. & Tulipani, S. (Sauro Tulipani) 1986. Horn sentences of small size in identity theory. *Z. Math. Logik Grundlag. Math.*, **32**, 439–44. [474]

Marsh, W. E. (William E. Marsh) 1966. *On ω_1-categorical and not ω-categorical theories.* Dissertation. Dartmouth College. [82, 199]

Martin, D. A. (Donald A. Martin) 1975. Borel determinacy. *Ann. Math.*, **102**, 363–71. [129]

Martin, G. A. (Gary A. Martin) 1988. Definability in reducts of algebraically closed fields. *J. Symbolic Logic*, **53**, 188–99. [700]

Mayer, L. L. (Laura L. Mayer) 1988. Vaught's conjecture for O-minimal theories. *J. Symbolic Logic*, **53**, 146–59. [339]

Mekler, A. H. (Alan H. Mekler) 1979. Model complete theories with a distinguished substructure. *Proc. Amer. Math. Soc.*, **75**, 294–9. [704]

— 1980. On residual properties. *Proc. Amer. Math. Soc.*, **78**, 187–8. [130]

— 1981. Stability of nilpotent groups of class 2 and prime exponent. *J. Symbolic Logic*, **46**, 781–8. [232, 673]

— 1982. Primitive rings are not definable in $L_{\infty\infty}$. *Comm. Alg.*, **10**, 1689–90. [129]

— 1984. Stationary logic of ordinals. *Ann. Pure Appl. Logic*, **26**, 47–68. [128]

Mekler, A. H., Pelletier, D. H. & Taylor, A. D. (Donald H. Pelletier, Alan D. Taylor) 1981. A note on a lemma of Shelah concerning stationary sets. *Proc. Amer. Math. Soc.*, **83**, 764–8. [130]

Merzlyakov, Yu. I. (Ю. И. Мерзляков) 1966. Позитивные формулы на свободных группах (Positive formulas of free groups). *Algebra i Logika,* **5** (4), 25–42. [690]

Mitchell, W. J. (William J. Mitchell) 1972. Aronszajn trees and the independence of the transfer property. *Ann. Math. Logic*, **5**, 21–46. [650]

Monk, J. D. (J. Donald Monk) 1989. *Handbook of boolean algebras,* vol. 1. Amsterdam: Elsevier. [319]

Monk, L. 1975. *Elementary-recursive decision procedures.* Ph.D. dissertation. Univ. California, Berkeley. [653, 655]

Montague, R. (Richard Montague) 1955. Well-founded relations: generalizations of principles of induction and recursion. *Bull. Amer. Math. Soc.*, **61**, 443. [84]

— 1965. Interpretability in terms of models. *Indag. Math.*, **27**, 467–76. [262]

Morel, A. C. (Anne C. Morel) 1968. Structure and order structure in Abelian groups. *Colloq. Math.*, **19**, 199–209. [207]

Morley, M. D. (Michael Morley) 1965a. Categoricity in power. *Trans. Amer. Math. Soc.*, **114**, 514–38. [82, 129, 262, 320, 321, 595, 650, 651]

— 1965b. Omitting classes of elements. In Addison, Henkin & Tarski [1965], pp. 265–73. [597, 650]

— 1970a. The number of countable models. *J. Symbolic Logic*, **35**, 14–18. [207, 339, 340]

— 1970b. The Löwenheim-Skolem theorem for models with standard part. In *Symposia Math. V (INDAM, Rome 1969/70)*, pp. 43–52. London: Academic Press. [598]

Morley, M. D. & Morley, V. 1967. The Hanf number for κ-logic. Abstract. *Not. Amer. Math. Soc.*, **14**, 556. [597]

Morley, M. D. & Vaught, R. L. Homogeneous universal models. *Math. Scand.*, **11**, 37–57. [532, 533, 649, 650]

Mortimer, M. E. 1974. *Some topics in model theory.* Ph.D. thesis. Bedford College, London University. [410]

— 1975. On languages with two variables. *Z. Math. Logik Grundlag. Math.*, **21**, 135–40. [129]

Moschovakis, Y. N. (Yiannis N. Moschovakis) 1974. *Elementary induction on abstract structures.* Amsterdam: North-Holland. [129]

Mostowski, A. (Andrzej Mostowski) 1937. Abzählbare Boolesche Körper und ihre Anwendung auf die allgemeine Metamathematik. *Fundamenta Math.*, **29**, 34–53. [320]

— 1939. Über die Unabhängigkeit des Wohlordnungssatzes vom Ordnungsprinzip. *Fundamenta Math.*, **32**, 201–52. [198].

— 1952a. On direct products of theories. *J. Symbolic Logic*, **17**, 1–31. [128, 476]

— 1952b. On models of axiomatic systems. *Fundamenta Math.*, **39**, 133–58. [22, 261]

— 1957. On a generalization of quantifiers. *Fundamenta Math.*, **44**, 12–36. [86]

Mostowski, A. & Tarski, A. 1939. Boolesche Ringe mit geordneter Basis. *Fundamenta Math.*, **32**, 69–86. [595, 714]

Motohashi, N. (Nobuyoshi Motohashi) 1986. Preservation theorem and relativization theorem for cofinal extensions. *J. Symbolic Logic*, **51**, 1022–8. [533]

Muchnik, A. A. (А. А. Мучник) 1985. Games on infinite trees and automata with dead-ends. A new proof for the decidability of the monadic second order theory of two successors (Russian). *Semiotika i Informatika*, **24**, 16–40. [86]

Murthy, M. P. & Swan, R. G. (M. Pavaman Murthy, Richard G. Swan) 1976. Vector bundles over affine surfaces. *Invent. Math.*, **36**, 125–65. [475]

Mycielski, J. (Jan Mycielski) 1964. Some compactifications of general algebras. *Colloq. Math.*, **13**, 1–9. [534]

— 1965. On unions of denumerable models. *Algebra i Logika*, **4** (2), 57–8. [129]

Mycielski, J. & Ryll-Nardzewski, C. 1968. Equationally compact algebras II. *Fundamenta Math.*, **61**, 271–81. Errata *Fundamenta Math.*, **62** (1968), 309. [534]

Nadel, M. (Mark Nadel) 1972. An application of set theory to model theory. *Israel J. Math.*, **11**, 386–93. [129]

— 1985. $L_{\omega_1\omega}$ and admissible fragments. In Barwise & Feferman [1985], pp. 271–316. [129]

Nagel, E. (Ernest Nagel) 1939. The formation of modern conceptions of formal logic in the development of geometry. *Osiris*, **7**, 142–224. [321]

Nelson, E. (Evelyn Nelson) 1974. Not every equational class of infinitary algebras contains a simple algebra. *Colloq. Math.*, **30**, 27–30. [475]

Nesin, A. H. (Ali H. Nesin) 1989. Solvable groups of finite Morley rank. *J. Algebra*, **121**, 26–39. [263, 693]

Nesin, A. H. & Pillay, A. (eds) 1989. *The model theory of groups*. Notre Dame Mathematical Lectures 11. Notre Dame, Ind.: Univ. Notre Dame Press. [692]

Neumann, B. H. (Bernhard H. Neumann) 1943. Adjunction of elements to groups. *J. London Math. Soc.*, **18**, 4–11. [410]

— 1952. A note on algebraically closed groups. *J. London Math. Soc.*, **27**, 247–9. [409, 689]

— 1954. Groups covered by permutable subsets. *J. London Math. Soc.*, **29**, 236–48. [197]

— 1959. Permutational products of groups. *J. Austral. Math. Soc. Ser. A*, **1**, 299–310. [330]

— 1973. The isomorphism problem for algebraically closed groups. In *Word problems*, ed. W. W. Boone *et al.*, pp. 553–62. Amsterdam: North-Holland. [689]

Neumann, H. (Hanna Neumann) 1967. *Varieties of groups*. New York: Springer. [473]

Neumann, P. M. (Peter M. Neumann, П. М. Neumann) 1976. The structure of finitary permutation groups. *Archiv der Math. (Basel)*, **27**, 3–17. [198]

— 1985. Some primitive permutation groups. *Proc. London Math. Soc.*, **50**, 265–81. [199]

Newelski, L. (Ludomir Newelski) 1987. Omitting types and the real line. *J. Symbolic Logic*, **52**, 1020–6. [358]

Nurtazin, A. T. (А. Т. Нуртазин) 1990. Об автоморфизмах счётных моделей несчётно категоричных теорий (On automorphisms of countable models of uncountably categorical theories). In *Theory of models* (Russian), ed. B. S. Baĭzhanov *et al.*, pp. 62–71. Alma-Ata: Kazakh State University. [199]

Oberschelp, A. (Arnold Oberschelp) 1958. Über die Axiome produkt-abgeschlossener arithmetischer Klassen. *Arch. Math. Logik Grundlagenforsch.*, **4**, 95–123. [476]

Oger, F. (Francis Oger) 1983. Cancellation and elementary equivalence of groups. *J. Pure Appl. Algebra*, **30**, 293–9. [691]

— 1991. Cancellation and elementary equivalence of finitely generated finite-by-nilpotent groups. *J. London Math. Soc.*, **44**, 173–83. [691]

Olson, L. 1973. An elementary proof that elliptic curves are abelian varieties. *Enseignement Math.*, **19**, 173–81. [219]

Osofsky, B. L. (Barbara L. Osofsky) 1971. Loewy length of perfect rings. *Proc. Amer. Math. Soc.*, **28**, 352–4. [585]

Oxtoby, J. C. (John C. Oxtoby) 1971. *Measure and category*. New York: Springer. [358]

Padoa, A. (Alessandro Padoa) 1901. Essai d'une théorie algébrique des nombres entiers, précédé d'une introduction logique à une théorie déductive quelconque. *Bibliothèque du Congrès internat. de philosophie, Paris, 1900*, vol. 3, pp. 309–65. Paris: Armand Colin. Partial trans. in *From Frege to Gödel*, ed. J. van Heijenoort (1967), pp. 119–23. Cambridge, Mass.: Harvard Univ. Press. [84]

Palyutin, E. A. (E. A. Палютин) 1975. Описание категоричных квазимногообразий. *Algebra i Logika*, **14**, 145–85. *Trans. as* The description of categorical quasivarieties. *Algebra and Logic*, **14** (1975), 86–111. [199, 474]

— 1980. Категоруные хорновы классы, 1. *Algebra i Logika*, **19**, 582–614. *Trans. as* Categorical Horn classes I. *Algebra and Logic*, **19**, 377–400. [475]

Paris, J. B. & Harrington, L. A. (Jeffrey B. Paris) 1977. A mathematical incompleteness in Peano Arithmetic. In *Handbook of mathematical logic,* ed. K. J. Barwise, pp. 1133–42. Amsterdam: North-Holland. [543,594]

Park, D. M. R. 1964a. *Set theoretic constructions in model theory*. Doctoral dissertation. MIT, Cambridge, Mass. [198, 320]

— 1964b. Intersection properties of first order theories. Abstract. *J. Symbolic Logic*, **29**, 219–20. [320, 411]

Peirce, C. S. (Charles Sanders Peirce) 1870. Description of a notation for the logic of relatives, resulting from an amplification of the conceptions of Boole's calculus of logic. *Mem. Amer. Academy of Arts and Sciences*, **9**, 317–78. [319]

— 1885. On the algebra of logic: a contribution to the philosophy of notation. *Amer. J. Math.*, **7**, 180–202. [84, 85, 86]

Peretyat'kin, M. G. (M. Г. Перетятькин) 1973. О полных теориях с конечным числом счётных моделей (On complete theories with a finite number of countable models). *Algebra i Logika*, **12**, 550–76. [359]

— 1980. Пример ω_1-категоричной полной конечно аксиоматизируемой теории. *Algebra i Logika*, **19** (3), 314–47. *Trans. as* An example of an ω_1-categorical complete finitely axiomatizable theory. *Algebra and Logic*, **19** (1980), 202–29. [651]

Phillips, R. E. (Richard E. Phillips) 1971. Countably recognizable classes of groups. *Rocky Mountain J. Math.*, **1**, 489–97. [130]

Pillay, A. (Anand Pillay) 1977. *Gaifman operations, minimal models and the number of countable models*. Ph.D. thesis. Bedford College, London University. [652]

— 1978. Number of countable models. *J. Symbolic Logic*, **43**, 492–6 [358, 533]

— 1980. Instability and theories with few models. *Proc. Amer. Math. Soc.*, **80**, 461–8. [533]

— 1983a. *An introduction to stability theory*. Oxford: Clarendon Press. [322]

— 1983b. $\aleph_0$-categoricity over a predicate. *Notre Dame J. Formal Logic*, **24**, 527–36. [652]

— 1984. Countable modules, *Fundamenta Math.*, **121**, 125–32. [662]

— 1989a. On fields definable in Q_p. *Arch. Math. Logic*, **29**, 1–7. [694]

— 1989b. Stable theories, pseudoplanes and the number of countable models. *Ann. Pure Appl. Logic*, **43**, 147–60. [533]

— 1990. Differentially algebraic group chunks. *J. Symbolic Logic*, **55**, 1138–42. [694]

Pillay, A. & Shelah, S. 1985. Classification over a predicate I. *Notre Dame J. Formal Logic*, **26**, 361–76. [652]

Pillay, A. & Sokolović, Ž. 1992. Superstable differential fields. *J. Symbolic Logic*, **57**, 97–108. [699, 706]

Pillay, A. & Steinhorn, C. I. 1984. Definable sets in ordered structures, *Bull. Amer. Math. Soc.*, **11**, 159–62. [82]

— 1986. Definable sets in ordered structures I. *Trans. Amer. Math. Soc.*, **295**, 565–92. [82, 703]

Pincus, D. (David Pincus) 1972. Zermelo–Fraenkel consistency results by Fraenkel–Mostowski methods. *J. Symbolic Logic*, **37**, 721–43. [631, 652]

— Two model theoretic ideas in independence proofs. *Fundamenta Math.*, **92**, 113–30. [359]

Pinus, A. G. (A. Г. Пинус) 1988. Elementary equivalence of partition lattices (Russian). *Sibirsk. Mat. Zh.*, **29**, 211–12. *Trans., Siberian Math. J.*, **29**, 507–8. [261]

Plotkin, J. M. (Jacob Manuel Plotkin) 1990. Who put the back in back-and-forth? *J. Symbolic*

Logic, **55**, 444–5. [128]

Point, F. & Prest, M. Y. (Françoise Point) 1988. Decidability for theories of modules. *J. London Math. Soc.*, **38**, 193–206. [261]

Poizat, B. (Bruno Poizat) 1981. Théories instables. *J. Symbolic Logic*, **46**, 513–22. [595, 709]

— 1983a. Groupes stables, avec types génériques réguliers. *J. Symbolic Logic*, **48**, 339–55. [262, 694]

— 1983b. Paires de structures stables. *J. Symbolic Logic*, **48**, 239–49. [198]

— 1983c. Une théorie de Galois imaginaire. *J. Symbolic Logic*, **48**, 1151–70. [198]

— 1984. Deux remarques à propos de la propriété de recouvrement fini. *J. Symbolic Logic*, **49**, 803–7. [198]

— 1985. *Cours de théorie des modèles*. Villeurbanne: Nur al-Mantiq wal-Ma'rifah (obtainable from author at Université Paris 6). [198, 320, 322]

— 1986. Malaise et guérison. In *Logic Colloquium '84*, ed. J. B. Paris *et al.*, pp. 155–63. Amsterdam: North-Holland. [523]

— 1987a. *Groupes stables*. Villeurbanne: Nur al-Mantiq wal-Ma'rifah (obtainable from author at Université Paris 6). [263, 692, 693, 694, 699]

— 1987b. A propos de groupes stables. In *Logic Colloquium '85*, ed. The Paris Logic Group, pp. 245–65. Amsterdam: North-Holland. [710]

— 1988. MM. Borel, Tits, Zil'ber et le général nonsense. *J. Symbolic Logic*, **53**, 124–31. [699]

Poizat, B. & Wagner, F. O. (Frank Olaf Wagner) 199?. Sous-groupes périodiques d'un groupe stable. Preprint. [693]

Pope, A. L. (Alun L. Pope) 1982. *Some applications of set theory to algebra*. Ph.D. thesis. Bedford College, London University. [86, 652]

Pospíšil, B. 1937. Remark on bicompact spaces. *Ann. Math.*, **38**, 845–6. [319]

Post, E. L. (Emil Post) 1921. Introduction to a general theory of elementary propositions. *Amer. J. Math.*, **43**, 163–85. Reprinted in *From Frege to Gödel*, ed. J. van Heijenoort (1967), pp. 264–83. Cambridge, Mass.: Harvard Univ. Press. [82]

Pouzet, M. (Maurice Pouzet) 1972a. Modèle universel d'une théorie *n*-complète. *C.R. Acad. Sci. Paris Sér. A-B*, **274**, A433–6. [409]

— 1972b. Modèle universel d'une théorie *n*-complète: Modèle uniformément préhomogène. *C.R. Acad. Sci. Paris Sér. A-B*, **274**, A695–8. [409, 411]

— 1972c. Modèle universel d'une théorie *n*-complète: Modèle préhomogène. *C.R. Acad. Sci. Paris Sér A-B*, **274**, A813–16. [409]

— 1976. Application d'une propriété combinatoire des parties d'un ensemble aux groupes et aux relations. *Math. Z.*, **150**, 117–34. [594]

Prenowitz, W. 1943. Projective geometries as multigroups. *Amer. J. Math.*, **65**, 235–56. [199]

Presburger, M. 1930. Über die Vollständigkeit eines gewissen Systems der Arithmetik ganzer Zahlen, in welchem die Addition als einzige Operation hervortritt. In *Sprawozdanie z I Kongresu Mat. Krajów Słowiańskich*, Warsaw, pp. 92–101. [85, 128]

Prest, M. Y. (Michael Y. Prest) 1979. Torsion and universal Horn classes of modules. *J. London Math. Soc.*, **19**, 411–16. [659]

— 1984. Rings of finite representation type and modules of finite Morley rank. *J. Algebra*, **88**, 502–33. [662]

— 1988. *Model theory and modules*. London Math. Soc. Lecture Note Series 130. Cambridge: Cambridge Univ. Press. [653]

Prestel, A. (Alexander Prestel) 1984. *Lectures on formally real fields*. Lecture Notes in Math. 1093. Berlin: Springer. [410]

Prestel, A. & Roquette, P. 1984. *Formally p-adic fields*. Lecture Notes in Math. 1050. Berlin: Springer. [705]

Quackenbush, R. W. 1972. Free products of bounded distributive lattices. *Algebra Universalis*, **2**, 393–4. [477]

Quine, W. V. O. (Willard Van Orman Quine) 1960. Variables explained away. *Proc. Amer. Philos. Soc.*, **104**, 343–7. [82]

Rabin, M. O. (Michael O. Rabin) 1959. Arithmetical extensions with prescribed cardinality.

Nederl. Akad. Wetensch. Proc. Ser. A, **62** (= *Indag. Math.*, **21**), 439–46. [596]

— 1962. Non-standard models and independence of the induction axiom. In *Essays on the foundations of mathematics*, ed. Y. Bar-Hillel *et al.*, pp. 287–99. Amsterdam: North-Holland. [409]

— 1965a. A simple method for undecidability proofs and some applications. In *Logic, Methodology and Philosophy of Science, Proc. 1964 Internat. Congress*, ed. Y. Bar-Hillel, pp. 58–68. Amsterdam: North-Holland. [262]

— 1965b. Universal groups of automorphisms of models. In Addison, Henkin & Tarski [1965], pp. 274–84. [321]

— 1969. Decidability of second-order theories and automata on infinite trees. *Trans. Amer. Math. Soc.*, **141**, 1–35. [76, 86]

— 1977. Decidable theories. In *Handbook of mathematical logic*, ed. K. J. Barwise, pp. 595–629. Amsterdam: North-Holland. [86]

Rabinovich, E. B. (E. Б. Рабинович) 1977. Embedding theorems and de Bruijn's problem for bounded symmetric groups (Russian). *Dokl. Akad. Nauk. Belor. S.S.R.*, **21** (9), 784–7. [198]

Rabinovich, E. D. (Evgenia D. Rabinovich) 1992. *Definability of a field in sufficiently rich incidence systems*. QMW Maths Notes. London: School of Math. Sciences, Queen Mary and Westfield College. [700]

Rabinowitsch, J. L. 1930. Zum Hilbertschen Nullstellensatz. *Math. Ann.*, **102**, 520 [409]

Rado, R. (Richard Rado) 1964. Universal graphs and universal functions. *Acta Arith.*, **9**, 331–40. [359]

Ramsey, F. P. (Frank P. Ramsey) 1930. On a problem of formal logic. *Proc. London Math. Soc.*, **30**, 264–86. [594]

— 1978. *Foundations: Essays in philosophy, logic, mathematics and economics*. Ed. D. H. Mellor. London: Routledge & Kegan Paul. [82]

Rasiowa, H. & Sikorski, S. (Helena Rasiowa) 1950. A proof of the completeness theorem of Gödel. *Fundamenta Math.*, **37**, 193–200. [358]

— 1963. *The mathematics of metamathematics*. Warsaw: Państwowe Wydawnictwo Naukowe. [319]

Rav, Y. (Yehuda Rav) 1977. Variants of Rado's selection lemma and their applications. *Math. Nachr.*, **79**, 145–65. [319]

Reineke, J. (Joachim Reineke) 1975. Minimale Gruppen. *Z. Math. Logik Grundlag. Math.*, **21**, 357–9. [693]

Ressayre, J. P. 1977. Models with compactness properties relative to an admissible language. *Ann. Math. Logic*, **11**, 31–55. [534]

Reyes, G. E. 1970. Local definability theory. *Ann. Math. Logic*, **1**, 95–137. [197, 198, 652]

Robinson, A. (Abraham Robinson) 1949. *On the metamathematics of algebraic systems*. Ph.D. thesis. Birkbeck College, London University. [84]

— 1950. On the application of symbolic logic to algebra. In *Proc. Internat. Congress of Math., Cambridge Mass. 1950*, vol. 1, pp. 686–94. Reprinted in A. Robinson [1979a], pp. 3–11. [22, 319]

— 1951. On axiomatic systems which possess finite models. In *Methodos*, **3**, 140–9. Reprinted in A. Robinson [1979a], pp. 322–31. [85]

— 1955. On ordered fields and definite functions, *Math. Ann.*, **130**, 257–271. Reprinted in A. Robinson [1979a], pp. 355–69. [704]

— 1956a. A result on consistency and its application to the theory of definition. *Indag. Math.*, **18**, 47–58. Reprinted in A. Robinson [1979a], pp. 87–98. [22, 320, 476]

— 1956b. *Complete theories*. Amsterdam: North-Holland. [22, 84, 85, 320, 409, 696]

— 1956c. Completeness and persistence in the theory of models. *Z. Math. Logik Grundlag. Math.*, **2**, 15–26. Reprinted in A. Robinson [1979a], pp. 108–119. [22]

— 1957. Some problems of definability in the lower predicate calculus. *Fundamenta Math.*, **44**, 309–29. Reprinted in A. Robinson [1979a], pp. 375–95. [409]

— 1958. Relative model-completeness and the elimination of quantifiers. *Dialectica*, **12**, 394–407. Reprinted in A. Robinson [1979a], pp. 146–59. [410, 704]

— 1959a. Obstructions to arithmetical extension and the theorem of Łoś and Suszko. *Indag. Math.*, **21**, 489–95. Reprinted in A. Robinson [1979a], pp. 160–6. [320]

— 1959b. On the concept of a differentially closed field. *Bull. Res. Council Israel, section F*, **8**, 113–28. Reprinted in A. Robinson [1979a], pp. 440–55. [706]

— 1959c. Solution of a problem of Tarski. *Fundamenta Math.*, **47**, 179–204. Reprinted in A. Robinson [1979a], pp. 414–39. [703]

— 1961. Non-standard analysis. *Indag. Math.*, **23**, 432–40. Reprinted in A. Robinson [1979b], pp. 3–11. [596]

— 1963a. *Introduction to model theory and to the metamathematics of algebra*. Amsterdam: North-Holland. [320]

— 1963b. On languages which are based on nonstandard arithmetic. *Nagoya Math. J.*, **22**, 83–117. Reprinted in A. Robinson [1979b]. [534]

— 1965. Formalism 64. In *Proc. Internat. Congress for Logic, Methodology and Philos. Sci., Jerusalem 1964*, pp. 228–46. Amsterdam: North-Holland. Reprinted in A. Robinson [1979b], pp. 505–23. [596]

— 1966. *Non-standard analysis*. Amsterdam: North-Holland. [597]

— 1971a. Forcing in model theory. In *Symposia Math. Vol. V (INDAM, Rome 1969/70)*, pp. 69–82. London: Academic Press. Reprinted in A. Robinson [1979a], pp. 205–18. [358, 409]

— 1971b. Infinite forcing in model theory. In *Proc. Second Scand. Logic Sympos., Oslo 1970*, pp. 317–40. Amsterdam: North-Holland. Reprinted in A. Robinson [1979a], pp. 243–66. [409, 533]

— 1973a. Metamathematical problems. *J. Symbolic Logic*, **38**, 500–16. Reprinted in A. Robinson [1979a], pp. 43–59. [572]

— 1973b. On bounds in the theory of polynomial ideals. In *Selected questions of algebra and logic (A collection dedicated to the memory of A. I. Mal'tsev)* (Russian), pp. 245–52. Novosibirsk: Izdat. 'Nauka' Sibirsk. Otdel. Reprinted in A. Robinson [1979a], pp. 482–9. [572, 596]

— 1979a. *Selected papers of Abraham Robinson*, Vol. 1, *Model theory and algebra*. Ed. H. J. Keisler *et al.* Amsterdam: North-Holland.

— 1979b. *Selected papers of Abraham Robinson*, Vol. 2, *Nonstandard analysis and philosophy*. Ed. H. J. Keisler *et al.* Amsterdam: North-Holland.

Robinson, A. & Zakon, E. (Elias Zakon) 1960. Elementary properties of ordered abelian groups. *Trans. Amer. Math. Soc.*, **96**, 222–36. Reprinted in A. Robinson [1979a], pp. 456–70. [410]

Robinson, J. A. 1965. A machine oriented logic based on the resolution principle. *J. Assoc. Comp. Mach.*, **12**, 23–41. [22]

Robinson, J. B. (Julia Robinson) 1949. Definability and decision problems in arithmetic. *J. Symbolic Logic*, **14**, 98–114. [261]

— 1959. The undecidability of algebraic rings and fields. *Proc. Amer. Math. Soc.*, **10**, 950–7. [261]

— 1965. The decision problem for fields. In Addison, Henkin & Tarski [1965], pp. 299–311. [319]

Robinson, R. M. (Raphael M. Robinson) 1951. Undecidable rings. *Trans. Amer. Math. Soc.*, **70**, 137–59. [83, 220, 261]

Rogers, H. (Hartley Rogers, Jr) 1956. Certain logical reduction and decision problems. *Ann. Math.*, **64**, 264–84. [262]

— 1967. *Theory of recursive functions and effective computability*. New York: McGraw-Hill. [215]

Rogers, P. K. (Patricia K. Rogers) 1977. *Topics in the model theory of abelian and nilpotent groups*. Ph.D. thesis. Bedford College, London University. [670, 671]

Roquette, P. (Peter Roquette) 1967. On class field towers. In *Algebraic number theory*, ed. J. W. S. Cassels & A. Fröhlich, pp. 231–49. London: Academic Press. [150]

— 1975. Nonstandard aspects of Hilbert's irreducibility theorem. In *Model theory and algebra: A memorial tribute to Abraham Robinson*, ed. D. H. Saracino & V. B. Weispfenning, Lecture Notes in Math. 498, pp. 231–75. Berlin: Springer. [597]

Rose, B. I. (Bruce Rose) 1980. On the model theory of finite-dimensional algebras. *Proc. London Math. Soc.*, **40**, 21–39. [697]

Rosenstein, J. G. (Joseph G. Rosenstein) 1969. $\aleph_0$-categoricity of linear orderings. *Fundamenta*

Math., **64**, 1–5. [358, 715]

— 1973. $\aleph_0$-categoricity of groups. *J. Algebra*, **25**, 435–67. Correction *J. Algebra*, **48** (1977), 236–40. [358]

— 1976. $\aleph_0$-categoricity is not inherited by factor groups. *Algebra Universalis*, **6**, 93–5. [358]

— 1982. *Linear orderings*. New York: Academic Press. [714, 715]

Rosser, J. B. (J. Barclay Rosser) 1935. A mathematical logic without variables. *Ann. Math.*, **36**, 127–50. [83]

Rothmaler, P. (Philipp Rothmaler) 1983. Some model theory of modules I: On total transcendence of modules. *J. Symbolic Logic*, **48**, 570–4. [661]

— 1991. A trivial remark on purity. In *Proc. Ninth Easter Conference on Model Theory, Gosen, April 1–6, 1991*, ed. H. Wolter, p. 127. Berlin: Fachbereich Math. der Humboldt-Univ. [84]

Rotman, J. J. (Joseph J. Rotman) 1973. *The theory of groups*. Boston: Allyn & Bacon. [506]

Rowbottom, F. (Frederick Rowbottom) 1964. The Łoś conjecture for uncountable theories. Abstract. *Not. Amer. Math. Soc.*, **11**, 248. [321]

— 1971. Some strong axioms of infinity incompatible with the axiom of constructibility. *Ann. Math. Logic*, **3**, 1–44. [595]

Rubin, H. & Rubin, J. E. (Herman Rubin, Jean E. Rubin) 1985. *Equivalents of the axiom of choice*. 2nd ed. Amsterdam: North-Holland. [273]

Rubin, M. (Mattatyahu Rubin) 1974. Theories of linear order. *Israel J. Math.*, **17**, 392–443. [339, 709, 714]

— 1977. Vaught's conjecture for linear orderings. Abstract. *Not. Amer. Math. Soc.*, **24**, A-390. [339]

— 1979. On the automorphism groups of homogeneous and saturated Boolean algebras. *Algebra Universalis*, **9**, 54–86. [138]

— 1989a. On the reconstruction of Boolean algebras from their automorphism groups. In *Handbook of Boolean Algebras*, ed. J. D. Monk, pp. 549–606. Amsterdam: Elsevier. [138, 223]

— 1989b. On the reconstruction of topological spaces from their groups of homeomorphisms. *Trans. Amer. Math. Soc.*, **312**, 487–538. [138]

— 199?a. On the reconstruction of $\aleph_0$-categorical structures from their automorphism groups. Preprint. [138]

— 199?b. On the reconstruction of trees from their automorphism groups. Preprint. [138]

Rudin, W. (Walter Rudin) 1964. *Principles of mathematical analysis*. New York: McGraw-Hill. [353]

Rumely, R. S. (Robert S. Rumely) 1980. Undecidability and definability for the theory of global fields. *Trans. Amer. Math. Soc.*, **262**, 195–217. [261]

Russell, B. (Bertrand Russell) 1903. *The principles of mathematics*, Vol. I. Cambridge: Cambridge Univ. Press. [83]

Ryll-Nardzewski, C. (Czesław Ryll-Nardzewski) 1959. On the categoricity in power $\le \aleph_0$. *Bull. Acad. Polon. Sci. Sér. Sci. Math. Astronom. Phys.*, **7**, 545–8. [358]

Sabbagh, G. (Gabriel Sabbagh) 1969. How not to characterize the multiplicative groups of fields. *J. London Math. Soc.*, **1**, 369–70. [669]

— 1971. Sous-modules purs, existentiellement clos et élémentaires. *C. R. Acad. Sci. Paris Sér. A-B*, **272**, A1289–92. [656]

— 1976. Caractérisation algébrique des groupes de type fini ayant un problème de mots résoluble (théorème de Boone-Higman, travaux de B. H. Neumann et Macintyre). In *Séminaire Bourbaki 1974/75*, Exposés 453–470, Exp. 457, Lecture Notes in Math. 514, pp. 61–80. Berlin: Springer. [409]

— 1987. Propriétés élémentaires des groupes polycycliques et de certains de leurs invariants. *C. R. Acad. Sci. Paris Sér. I*, **305**, 101–3. [691]

Sabbagh, G. & Eklof, P. C. 1971. Definability problems for modules and rings. *J. Symbolic Logic*, **36**, 623–49. [128, 597]

Sacerdote, G. S. (George S. Sacerdote) 1973. Elementary properties of free groups. *Trans. Amer. Math. Soc.*, **178**, 127–38. [690]

Saffe, J. (Jürgen Saffe) 1984. Categoricity and ranks. *J. Symbolic Logic*, **49**, 1379–82. [650]

Samuel, P. (Pierre Samuel) 1968. Unique factorization. *Amer. Math. Monthly*, **75**, 945–52. [583]

Saracino, D. H. (Daniel H. Saracino) 1973. Model companions for $\aleph_0$-categorical theories. *Proc. Amer. Math. Soc.*, **39**, 591–8. [411]

— 1974. Wreath products and existentially complete solvable groups. *Trans. Amer. Math. Soc.*, **197**, 327–39. [690]

— 1975. A counterexample in the theory of model companions. *J. Symbolic Logic*, **40**, 31–4. [410]

— 1976. Existentially complete nilpotent groups. *Israel J. Math.*, **25**, 241–8. [690]

Saracino, D. H. & Wood, C. 1979. Periodic existentially closed nilpotent groups. *J. Algebra*, **58**, 189–207. [344, 690]

— 1982. QE nil-2 groups of exponent 4. *J. Algebra*, **76**, 337–52. [695]

Schlipf, J. S. (John S. Schlipf) 1978. Toward model theory through recursive saturation. *J. Symbolic Logic*, **43**, 183–206. [525]

Schmerl, J. H. (James H. Schmerl) 1972. An elementary sentence which has ordered models. *J. Symbolic Logic*, **37**, 521–30. [650]

— 1977. On $\aleph_0$-categoricity and the theory of trees. *Fundamenta Math.*, **94**, 121–8. [358]

— 1980. Decidability and $\aleph_0$-categoricity of theories of partially ordered sets. *J. Symbolic Logic*, **45**, 585–611. [358, 359]

— 1984. $\aleph_0$-categorical partially ordered sets. In *Orders: description and roles (L'Arbresle 1982)*, ed. M. Pouzet & D. Richard, pp. 269–85. Amsterdam: North-Holland. [358]

— 1985. Transfer theorems and their applications to logics. In Barwise & Feferman [1985], pp. 177–209. [650]

Schmerl, J. H. & Shelah, S. 1972. On power-like models for hyperinaccessible cardinals. *J. Symbolic Logic*, **37**, 531–7. [650]

Schmitt, P. H. (Peter H. Schmitt) 1984. Model- and substructure-complete theories of ordered abelian groups. In *Models and Sets, Proc. Logic Colloq. Aachen 1983 Part I*, ed. G. H. Müller & M. M. Richter, Lecture Notes in Math. 1103, pp. 389–418. Berlin: Springer. [673]

Scholz, H. (Heinrich Scholz) 1952. Ein ungelöstes Problem in der symbolischen Logik. *J. Symbolic Logic*, **17**, 160. [542]

Schönfinkel, M. (Moses Schönfinkel) 1924. Über die Bausteine der mathematischen Logik. *Math. Annalen*, **92**, 305–16. *Trans. as* On the building blocks of mathematical logic. In *From Frege to Gödel*, ed. J. van Heijenoort (1967), pp. 357–66. Cambridge, Mass.: Harvard Univ. Press. [82]

Schröder, E. (Ernst Schröder) 1895. *Vorlesungen über die Algebra der Logik*, vol. 3. Leipzig. [21, 82]

Schuppar, B. (Berthold Schuppar) 1980. Elementare Aussagen zur Arithmetik und Galoistheorie von Funktionenkörpern. *J. Reine Angew. Math.*, **313**, 59–71. [261]

Schütte, K. (Kurt Schütte) 1956. Ein System des verknüpfenden Schliessens. *Arch. Math. Logik Grundlagenforsch.*, **2**, 55–67. [84]

Scott, D. S. (Dana S. Scott) 1954. Prime ideal theorems for rings, lattices and Boolean algebras. *Bull. Amer. Math. Soc.*, **60**, 390. [319]

— 1958. Definability in polynomial rings. Abstract. *Not. Amer. Math. Soc.*, **5**, 221–2. [475]

— 1961. Measurable cardinals and constructible sets. *Bull. Acad. Polon. Sci. Sér. Math. Astronom. Phys.*, **9**, 521–4. [476]

— 1962. Algebras of sets binumerable in complete extensions of arithmetic. In *Recursive Function Theory*, Proc. Sympos. Pure Math., pp. 117–21. Providence, RI: Amer. Math. Soc. [534]

— 1965. Logic with denumerably long formulas and finite strings of quantifiers. In Addison, Henkin & Tarski [1965], pp. 329–41. [129]

Scott, D. S. & Tarski, A. 1958. The sentential calculus with infinitely long expressions. *Colloq. Math.*, **6**, 165–70. [82]

Scott, W. R. 1951. Algebraically closed groups. *Proc. Amer. Math. Soc.*, **2**, 118–21. [373, 409, 689]

— 1964. *Group theory*. Englewood Cliffs, NJ: Prentice-Hall. [145, 373]

Seidenberg, A. 1974. Constructions in algebra. *Trans. Amer. Math. Soc.*, **197**, 273–313. [572]

Semmes, S. W. (Stephen William Semmes) 1981. Endomorphisms of infinite symmetric groups. *Abstracts Amer. Math. Soc.*, **2**, 426. [198]

Shafarevich, I. R. 1977. *Basic algebraic geometry*. Berlin: Springer. [206]

Sharpe, D. W. & Vámos, P. (David W. Sharpe) 1972. *Injective modules*. Cambridge: Cambridge Univ. Press. [455, 660]

Shelah, S. (Saharon Shelah) 1969. Stable theories. *Israel J. Math.*, **7**, 187–202. [306, 321, 602, 650]

—1970. On languages with non-homogeneous strings of quantifiers. *Israel J. Math.*, **8**, 75–9. [113]

—1971a. Every two elementarily equivalent models have isomorphic ultrapowers. *Israel J. Math.*, **10**, 224–33. [475, 476]

—1971b. Remark to 'Local definability theory' of Reyes. *Ann. Math. Logic*, **2**, 441–7. [652]

—1971c. Stability, the f.c.p., and superstability; model theoretic properties of formulas in first order theory. *Ann. Math. Logic*, **3**, 271–362. [321]

—1971d. The number of non-almost isomorphic models of *T* in a power, *Pacific J. Math.*, **36**, 811–18. [129]

—1971e. The number of non-isomorphic models of an unstable first-order theory. *Israel J. Math.*, **9**, 473–87. [595]

—1972a. A combinatorial problem; stability and order for models and theories in infinitary languages. *Pacific J. Math.*, **41**, 247–61. [597]

—1972b. On models with power-like orderings. *J. Symbolic Logic*, **37**, 247–67. [650]

—1972c. Uniqueness and characterization of prime models over sets for totally transcendental first-order theories. *J. Symbolic Logic*, **37**, 107–13. [706]

—1973a. Differentially closed fields. *Israel J. Math.*, **16**, 314–28. [706]

—1973b. First order theory of permutation groups. *Israel J. Math.*, **14**, 149–62. Errata *Israel J. Math.*, **15** (1973), 437–41. [261]

—1973c. There are just four second-order quantifiers. *Israel J. Math.*, **15**, 282–300. [86]

—1974a. Categoricity of uncountable theories. In *Proc. Tarski Symposium*, ed. L. A. Henkin *et al.*, pp. 187–203. Providence, RI: Amer. Math. Soc. [650, 651]

—1974b. Why there are many nonisomorphic models for unsuperstable theories. In *Proc. Internat. Congress of Mathematicians, Vancouver 1974*, Vol. 1, ed. R. D. James, pp. 259–63. Canadian Mathematical Congress. [595]

—1975a. A compactness theorem for singular cardinals, free algebras, Whitehead problem and transversals. *Israel J. Math.*, **21**, 319–49. [474]

—1975b. A two-cardinal theorem. *Proc. Amer. Math. Soc.*, **48**, 207–13. [606, 650]

—1975c. Existence of rigid like families of abelian *p*-groups. In *Model theory and algebra: A memorial tribute to Abraham Robinson*, ed. D. H. Saracino & V. B. Weispfenning, Lecture Notes in Math. 498, pp. 384–402. Berlin: Springer. [596]

—1975d. The monadic theory of order. *Ann. Math.*, **102**, 379–419. [86, 594, 714]

—1976a. Interpreting set theory in the endomorphism semi-group of a free algebra or in a category. *Ann. Sci. Univ. Clermont Math.*, **13**, 1–29. [261]

—1976b. The lazy model-theoretician's guide to stability. *Logique et Analyse*, **71–72**, 241–308. Reprinted in *Six days of model theory*, ed. P. Henrard (1977), pp. 9–76. Albeuve: Editions Castella. [596, 661]

—1977a. A two cardinal theorem and a combinatorial theorem. *Proc. Amer. Math. Soc.*, **62**, 134–6. [650]

—1977b. Existentially-closed groups in $\aleph_1$ with special properties. *Bull. Soc. Math. Grèce*, **18**, 17–27. [602, 690]

—1978a. *Classification theory and the number of non-isomorphic models*. Amsterdam: North-Holland. [155, 198, 200, 262, 317, 321, 322, 358, 476, 532, 533, 534, 594, 595, 596, 597, 598, 650, 651]

—1978b. On the number of minimal models. *J. Symbolic Logic*, **43**, 475–80. [340]

—1979. Boolean algebras with few endomorphisms. *Proc. Amer. Math. Soc.*, **14**, 135–42. [596]

—1980a. A problem of Kurosh, Jónsson groups and applications. In *Word Problems* II, ed. S. I.

Adian *et al.*, pp. 373–94. Amsterdam: North-Holland. [688]

— 1980b. Independence results. *J. Symbolic Logic*, **45**, 563–73. [535]

— 1985. Classification of first order theories which have a structure theorem. *Bull. Amer. Math. Soc.*, **12**, 227–32. [306, 596]

— 1986. *Around classification theory of models*. Lecture Notes in Math. 1182. Berlin: Springer. [596, 652]

— 1987a. Existence of many $L_{\infty,\lambda}$-equivalent, non-isomorphic models of T of power λ. *Ann. Pure Appl. Logic*, **34**, 291–310. [123, 565]

— 1987b. Taxonomy of universal and other classes. In *Proc. Internat. Congress of Mathematicians (Berkeley, Ca. 1986)*, vol. 1, pp. 154–62. Providence, RI: Amer. Math. Soc. [306]

— 1987c. Uncountable groups have many nonconjugate subgroups. *Ann. Pure Appl. Logic*, **36**, 153–206. [688]

— 1990. Second edition of Shelah [1978a], with four new chapters. [306, 339]

Shelah, S., Harrington, L. A. & Makkai, M. 1984. A proof of Vaught's conjecture for ω-stable theories. *Israel J. Math.*, **49**, 259–80. [339]

Shelah, S. & Hart, B. (Bradd Hart) 1990. Categoricity over P for first order T or categoricity for $\phi \in L_{\omega_1\omega}$ can stop at $\aleph_k$ while holding for $\aleph_0, \ldots, \aleph_{k-1}$. *Israel J. Math.*, **70**, 219–35. [197, 640]

Shelah, S. & Spencer, J. H. 1988. Zero–one laws for space random graphs. *J. Assoc. Math. Stats.*, **1**, 97–115. [359]

Shelah, S., Tuuri, H. & Väänänen, J. A. 199?. On the number of automorphisms of uncountable models. Preprint. [197]

Shelah, S. & Ziegler, M. 1979. Algebraically closed groups of large cardinality. *J. Symbolic Logic*, **44**, 522–32. [592, 690]

Shoenfield, J. R. (Joseph R. Shoenfield) 1967. *Mathematical logic*. Reading, Mass.: Addison-Wesley. [41]

— 1971. A theorem on quantifier elimination. In *Symposia Math. (vol. V, INDAM 1969/70 Rome)*, pp. 173–6. London: Academic Press. [410]

— 1977. Quantifier elimination in fields. In *Non-classical logics, model theory and computability*, ed. A. I. Arruda *et al.*, pp. 243–52. Amsterdam: North-Holland. [410]

Sierpiński, W. 1928. Sur une décomposition d'ensembles. *Monatshefte Math. u. Phys.*, **35**, 239–42. [198]

Sikorski, R. (Roman Sikorski) 1950. Independent fields and cartesian products. *Studia Math.*, **11**, 171–84. [474]

— 1964. *Boolean algebras*. Berlin: Springer. [274]

Silver, J. H. (Jack H. Silver) 1971. Some applications of model theory in set theory. *Ann. Math. Logic*, **3**, 45–110. [594, 595]

Simmons, H. (Harold Simmons) 1973. An omitting types theorem with an application to the construction of generic structures. *Math. Scand.*, **33**, 46–54. [409]

— 1976. Counting countable e.c. structures. *Logique et Analyse*, **71–72**, 309–57. Reprinted in *Six days of model theory*, ed. P. Henrard (1977), pp. 77–125. Albeuve: Editions Castella. [409]

Simpson, S. G. (Stephen G. Simpson) 1970. Model-theoretic proof of a partition theorem. Abstract. *Not. Amer. Math. Soc.*, **17**, 964. [594]

— 1985. Bqo theory and Fraïssé's conjecture. In Mansfield & Weitkamp [1985], pp. 124–38. [594]

Skolem, Th. (Thoralf Skolem) 1919. Untersuchungen über die Axiome des Klassenkalküls und über Produktations- und Summationsprobleme, welche gewisse Klassen von Aussagen betreffen. *Videnskapsselskapets Skrifter, I. Matem.-naturv. klasse*, no. 3. Reprinted in Skolem [1970], pp. 67–101. [84, 462, 476]

— 1920. Logisch-kombinatorische Untersuchungen über die Erfüllbarkeit oder Beweisbarkeit mathematischer Sätze nebst einem Theoreme über dichte Mengen. *Videnskapsselskapets Skrifter, I. Matem.-naturv. klasse* I no. 4, 1–36. Reprinted in Skolem [1970], pp. 103–36. [84, 128]

— 1922. Einige Bemerkungen zur axiomatischen Begründung der Mengenlehre. In *Proc. 5th Scand. Math. Congress, Helsinki*, pp. 217–32. Reprinted in Skolem [1970], pp. 137–52. [128]

— 1928. Über die mathematische Logik. *Norsk Matematisk Tidsskrift*, **10**, 125–42. Reprinted in Skolem [1970], pp. 189–206. [85]

— 1930. Über einige Satzfunktionen in der Arithmetik. *Skrifter Vitenskapsakademiet i Oslo* 1, 1–28. Reprinted in Skolem [1970], pp. 281–306. [84, 476]

— 1934. Über die Nichtcharakterisierbarkeit der Zahlenreihe mittels endlich oder abzählbar unendlich vieler Aussagen mit ausschliesslich Zahlenvariablen. *Fundamenta Math.*, **23**, 150–61. Reprinted in Skolem [1970], pp. 355–66. [37, 318, 475, 596]

— 1955. A critical remark on foundational research. *Kongelige Norske Videnskabsselskabs Forhandlinger, Trondheim*, **28**, no. 20, 100–5. Reprinted in Skolem [1970], pp. 581–6. [318]

— 1958. Une relativisation des notions mathématiques fondamentales. In *Colloques internationaux du Centre National de la Recherche Scientifique, Paris*, pp. 13–18. Reprinted in Skolem [1970], pp. 633–8. [318]

— 1970. *Selected works in logic*, ed. J. E. Fenstad. Oslo: Universitetsforlaget.

Solovay, R. M. (Robert M. Solovay) 1967. A nonconstructible Δ_3^1 set of integers. *Trans. Amer. Math. Soc.* **127**, 50–75. [594]

Solovay, R. M., Reinhardt, W. N. & Kanamori, A. (Akihiro Kanamori) 1978. Strong axioms of infinity and elementary embeddings. *Ann. Math. Logic*, **13**, 73–116. [569]

von Staudt, C. (Christian von Staudt) 1847. *Geometrie der Lage*. Nuremberg. [260]

Steel, J. (John Steel) 1978. On Vaught's conjecture. In *Cabal Seminar 76–77*, ed. A. S. Kechris & Y. N. Moschovakis, Lecture Notes in Math. 689, pp. 193–208. Berlin: Springer. [339]

Steinitz, E. (Ernst Steinitz) 1910. Algebraische Theorie der Körper. *J. Reine Angew. Math.*, **137**, 167–309. [168, 383, 409, 630]

Stone, M. H. 1935. Subsumption of Boolean algebras under the theory of rings. *Proc. Nat. Acad. Sci. USA*, **21**, 103–5. [84]

— 1936. The theory of representations for Boolean algebras. *Trans. Amer. Math. Soc.*, **40**, 37–111. [84, 319]

Surma, S. J. (Stanisław J. Surma) 1982. On the origin and subsequent applications of the concept of the Lindenbaum algebra. In *Logic, methodology and philosophy of science*, ed. L. J. Cohen *et al.*, pp. 719–34. Amsterdam: North-Holland. [320]

Svenonius, L. 1959a. ℵ₀-categoricity in first-order predicate calculus. *Theoria*, **25**, 82–94. [358]

— 1959b. A theorem on permutations in models. *Theoria*, **25**, 173–8. [533]

— 1965. On the denumerable models of theories with extra predicates. In Addison, Henkin & Tarski [1965], pp. 376–89. [129]

Szmielew, W. 1955. Elementary properties of Abelian groups. *Fundamenta Math.*, **41**, 203–71. [663, 667]

Taĭmanov, A. D. (А. Д. Тайманов) 1962. Характеристики аксиоматизируемых классов моделей (Characteristics of axiomatisable classes of models). *Algebra i Logika*, **1** (4), 5–31. [128]

Taĭtslin, M. A. (М. А. Тайцлин) 1962. Эффективная неотделимость множества тождественно истинных и множества конечно опровержимых формул элементарной теории структур (Effective inseparability of the sets of identically true and finitely refutable formulae of elementary lattice theory). *Algebra i Logika*, **1** (3), 24–38. [261]

— 1963. Неразрешимость элементарных теорий некоторых классов конечных коммутативных ассоциативных колец (Undecidability of the elementary theories of certain classes of finite commutative associative rings). *Algebra i Logika*, **2** (3), 29–51. [262]

— 1974. Экзистенциально замкнутые коммутативные кольца (Existentially closed commutative rings). In *Third All-Union Conference on Mathematical Logic*, pp. 213–15. Novosibirsk. [411]

Tarski, A. (Alfred Tarski) 1929. Sur les groupes de Abel ordonnés. *Ann. Soc. Polon. Math.*, **7**, 267–8. [319]

— 1930a. Über einige fundamentalen Begriffe der Metamathematik. *C. R. Séances Soc. Sci. Let. Varsovie, Classe III*, **23**, 22–9. Trans. as On some fundamental concepts of metamathematics. In Tarski [1956], pp. 30–7. [83, 85, 319]

— 1930b. Une contribution à la théorie de la mesure. *Fundamenta Math.*, **15**, 42–50. [319]

— 1931. Sur les ensembles définissables de nombres réels. I. *Fundamenta Math.*, **17**, 210–39.

Trans. as On definable sets of real numbers. In Tarski [1956], pp. 110–42. [85, 703]

— 1934. Z badań metodologicznych nad definjowalnością terminów. *Rev. Philos.*, **37**, 438–60. *Trans. as* Some methodological investigations on the definability of concepts. In Tarski [1956], pp. 296–319. [650]

— 1935. Der Wahrheitsbegriff in den formalisierten Sprachen. *Studia Philos.*, **1**, 261–405. *Trans. as* The concept of truth in formalized languages. In Tarski [1956], pp. 152–278. [82]

— 1935 + 6. Grundzüge des Systemenkalküls. *Fundamenta Math.*, **25**, 503–26 and **26**, 283–301. *Trans. as* Foundations of the calculus of systems. In Tarski [1956], pp. 342–83. [85, 319]

— 1936. Über den Begriff der logischen Folgerung. In *Actes Congrès Internat. de Philosophie Scientifique*, **7**, pp. 1–11. Paris: Hermann. *Trans. as* On the concept of logical consequence. In Tarski [1956], pp. 409–20. [83]

— 1949a. Arithmetical classes and types of mathematical systems. Abstract. *Bull. Amer. Math. Soc.*, **55**, 63, 1192. [85]

— 1949b. Metamathematical aspects of arithmetical classes and types. Abstract. *Bull. Amer. Math. Soc.*, **55**, 63–4, 1192. [85]

— 1949c. Arithmetical classes and types of Boolean algebras. Abstract. *Bull. Amer. Math. Soc.*, **55**, 64, 1192. [85, 714]

— 1949d. Arithmetical classes and types of algebraically closed and real-closed fields. Abstract. *Bull. Amer. Math. Soc.*, **55**, 64, 1192. [85]

— 1951. *A decision method for elementary algebra and geometry* (prepared for publication with the assistance of J. C. C. McKinsey). 2nd edition. Berkeley: Univ. California Press. [85, 703, 704]

— 1954. Contributions to the theory of models. I. *Konink. Nederl. Akad. Wetensch. Proc. Ser. A*, **57** (= *Indag. Math.*, **16**), 572–81; II, ibid. 582–8. [84, 260, 320, 321]

— 1955. A lattice-theoretical fixpoint theorem and its applications. *Pacific J. Math.*, **5**, 285–309. [84]

— 1956. *Logic, semantics, metamathematics, Papers from 1923 to 1938.* Trans. J. H. Woodger. Oxford: Clarendon Press. Revised edition with analytic index, ed. J. Corcoran (1983). Indianapolis: Hackett Publishing Co.

— 1957. Remarks on direct products of commutative semigroups. *Math. Scand.*, **5**, 218–23. [477]

— 1958. Some model-theoretical results concerning weak second-order logic. Abstract. *Not. Amer. Math. Soc.*, **5**, 673. [86]

Tarski, A. & Lindenbaum, A. 1927. Sur l'indépendance des notions primitives dans les systèmes mathématiques. *Ann. Soc. Polon. Math.*, **5**, 111–13. [84].

Tarski, A., Mostowski, A. & Robinson, R. M. 1953. *Undecidable theories.* Amsterdam: North-Holland. [83, 219, 220, 234, 260, 261, 319]

Tarski, A. & Vaught, R. L. 1957. Arithmetical extensions of relational systems. *Comp. Math.*, **13**, 81–102. [82, 84, 128, 318, 474]

Taylor, W. F. (Walter F. Taylor) 1971. Some constructions of compact algebras. *Ann. Math. Logic*, **3**, 395–435. [534]

— 1972. Residually small varieties. *Algebra Universalis*, **2**, 35–53. [534, 598]

— 1979. Equational logic. *Houston J. Math. Survey.* [474]

Thatcher, J. W., Wagner, E. G. & Wright, J. B. 1978. Data type specification: parameterization and the power of specification techniques. In *10th Symposium on Theory of Computing*, pp. 119–32. [439]

Thomas, S. R. (Simon Thomas) 1983a. *Classification theory of simple locally finite groups.* Ph.D. dissertation. Bedford College, London University. [692]

— 1983b. The classification of the simple periodic linear groups. *Arch. Math. (Basel)*, **41**, 103–16. [128, 692]

— 1986. Groups acting on infinite dimensional projective spaces. *J. London Math. Soc.*, **34**, 265–73. [359]

— 1991. Reducts of the random graph. *J. Symbolic Logic*, **56**, 176–81. [260]

Tolstykh, V. A. (В. А. Толстых) 1990. On elementary types of general linear groups. In *Soviet–French Colloquium in Theory of Models* (Russian), pp. 48–9. Karaganda: Karaganda State Univ. [261]

Tomkinson, M. J. (Michael J. Tomkinson) 1986. Groups covered by abelian subgroups. In *Proc. of Groups–St. Andrews 1985*, ed. C. M. Campbell & E. F. Robertson, London Math. Soc. Lecture Note Ser. 121, pp. 332–4. Cambridge: Cambridge Univ. Press. [594]

— 1987. Groups covered by finitely many cosets or subgroups. *Comm. Algebra*, **15**, 845–59. [198]

Trakhtenbrot, B. A. (Б. А. Трахтенброт) 1953. О рекурсивной неотделимости (On recursive inseparability). *Dokl. Akad. Sci. SSSR*, **88**, 953–6. [261]

Truss, J. K. 1985. The group of the countable universal graph. *Math. Proc. Camb. Philos. Soc.*, **98**, 213–45. [359]

— 1989. Infinite permutation groups II; subgroups of small index. *J. Algebra*, **120**, 494–515. [198]

Tsuboi, A. (Akito Tsuboi) 1985. On theories having a finite number of nonisomorphic countable models. *J. Symbolic Logic*, **50**, 806–8. [533]

— 1988. On a property of ω-stable solvable groups. *Arch. Math. Logic*, **27**, 193–7. [263]

Tsuzuku, T. 1982. *Finite groups and finite geometries*. Cambridge: Cambridge Univ. Press. [172]

Tuschik, H. P. (H. Peter Tuschik) 1985. Algebraic connections between definable predicates. In *Proc. Third Easter Conference on Model Theory*, pp. 215–26. Berlin: Sektion Math. der Humboldt-Univ. zu Berlin. [650]

Tyukavkin, L. V. (Л. В. Тюкавкин) 1982. О модельной полноте некоторых теорий модулей. *Algebra i Logika*, **21** (1), 73–83. *Trans. as* Model completeness of certain theories of modules. *Algebra and Logic*, **21** (1982), 50–7. [659]

Ulam, S. (Stanisław Ulam) 1930. Zur Masstheorie in der allgemeinen Mengenlehre. *Fundamenta Math.*, **16**, 140–50. [130]

Ulm, H. 1933. Zur Theorie der abzählbar-unendlichen abelschen Gruppen. *Math. Ann.*, **107**, 774–803. [128]

Urbanik, K. 1963. A representation theorem for v^*-algebras. *Fundamenta Math.*, **52**, 291–317. [261]

Vaught, R. L. (Robert L. Vaught) 1954. Applications of the Löwenheim–Skolem–Tarski theorem to problems of completeness and decidability. *Nederl. Akad. Wetensch. Proc. Ser. A*, **57** (= *Indag. Math.*, **16**), 467–72. [319, 650]

— 1954b. On sentences holding in direct products of relational systems. In *Proc. Int. Congr. Math., Amsterdam 1954*, vol. 2, p. 409. Groningen. [476]

— 1956. On the axiom of choice and some metamathematical theorems. *Bull. Amer. Math. Soc.*, **62**, 262–3. [319]

— 1961. Denumerable models of complete theories. In *Infinitistic methods, Proc. Symp. Foundations of Math., Warsaw 1959*, pp. 303–21. Warsaw: Państwowe Wydawnictwo Naukowe. [320, 358, 533, 649]

— 1962. The elementary character of two notions from general algebra. In *Essays on the foundations of mathematics*, ed. Y. Bar-Hillel *et al.*, pp. 226–33. Amsterdam: North-Holland. [321, 474]

— 1965. A Löwenheim–Skolem theorem for cardinals far apart. In Addison, Henkin & Tarski [1965], pp. 390–401. [650]

— 1966. Elementary classes closed under descending intersection. *Proc. Amer. Math. Soc.*, **17**, 430–3. [411]

— 1973. Descriptive set theory in $L_{\omega_1 \omega}$. In *Cambridge Summer School in Mathematical Logic*, ed. A. R. D. Mathias & H. Rogers, Lecture Notes in Math. 337, pp. 574–98. Berlin: Springer. [597]

— 1987. Alfred Tarski's work in model theory. *J. Symbolic Logic*, **51**, 869–82. [85]

Veblen, O. 1904. A system of axioms for geometry. *Trans. Amer. Math. Soc.*, **5**, 343–81. [650]

Videla, C. (Carlos Videla) 1987. Elementarily equivalent fields with inequivalent perfect closures. *Proc. Amer. Math. Soc.*, **99**, 171–5. [260]

Villemaire, R. (Roger Villemaire) 1988. *Aleph-zero-categoricity over a predicate*. Doctoral thesis, Univ. Tübingen. [652]

— 1990. Abelian groups $\aleph_0$-categorical over a subgroup. *J. Pure Appl. Algebra*, **69**, 193–204. [263, 652]

— 1992. Theories of modules closed under direct products. *J. Symbolic Logic*, **57**, 515–21. [658]

Vinner, S. 1972. A generalization of Ehrenfeucht's game and some applications. *Israel J. Math.*, **12**, 279–98. [129]

Volger, H. (Hugo Volger) 1976. The Feferman–Vaught theorem revisited. *Colloq. Math.*, **36**, 1–11. [476]

van der Waerden, B. L. 1949. *Modern algebra*, Vol. 1. New York: Frederick Ungar Publ. Co. [166]

Waszkiewicz, J. & Węglorz, B. (Jan Waszkiewicz) 1969. On ω_0-categoricity of powers. *Bull. Acad. Polon. Sci. Sér. Sci. Math.*, **17**, 195–9. [358, 476]

Węglorz, B. (Bogdan Węglorz) 1966. Equationally compact algebras (I). *Fundamenta Math.*, **59**, 289–98. [534]

— 1967. Some preservation theorems. *Colloq. Math.*, **17**, 269–76. [533]

— 1976. A note on 'Atomic compactness in $\aleph_1$-categorical Horn theories' by John T. Baldwin. *Fundamenta Math.*, **93**, 181–3. [534]

Weil, A. (André Weil) 1946. *Foundations of algebraic geometry*. Amer. Math. Soc. Colloquium Publics. vol. 29. New York: Amer. Math. Soc. [153, 160, 485, 700]

— 1955. On algebraic groups of transformations. *Amer. J. Math.*, **77**, 203–71. [694]

Weinstein, J. M. 1965. *First order properties preserved by direct product*. Ph.D. thesis. Univ. Wisconsin, Madison, Wis. [129, 476]

Weispfenning, V. B. (Volker B. Weispfenning) 1975. Model-completeness and elimination of quantifiers for subdirect products of structures. *J. Algebra*, **36**, 252–77. [476]

— 1985. Quantifier elimination for modules. *Arch. Math. Logik Grundlagenforsch.*, **25**, 1–11. [657]

Wenzel, G. H. (Günter H. Wenzel) 1979. Equational compactness. In *Universal algebra*, G. Grätzer, pp. 417–47. New York: Springer. [528]

Wheeler, W. H. (William H. Wheeler) 1979a. Amalgamation and elimination of quantifiers for theories of fields. *Proc. Amer. Math. Soc.*, **77**, 243–50. [411, 697]

— 1979b. Model-complete theories of pseudo-algebraically closed fields. *Ann. Math. Logic*, **17**, 205–26. [702]

Whitehead, A. N. & Russell, B. (Alfred North Whitehead) 1910. *Principia mathematica*, vol. I. Cambridge: Cambridge Univ. Press. [82, 85]

Whitney, H. 1935. On the abstract properties of linear dependence. *Amer. J. Math.*, **57**, 509–33. [199]

Wiegand, R. (Roger Wiegand) 1978. Homeomorphisms of affine surfaces over a finite field. *J. London Math. Soc.*, **18**, 28–32. [358]

Wilkie, A. J. 199?. Model completeness results for expansions of the real field I: restricted Pfaffian functions; II: the exponential function. Preprint.

Wilson, J. S. (John S. Wilson) 1982. The algebraic structure of $\aleph_0$-categorical groups. In *Proc. of Groups–St Andrews 1981*, ed. C. M. Campbell & E. F. Robertson, London Math. Soc. Lecture Note Ser. 121, pp. 345–58. Cambridge: Cambridge Univ. Press. [358]

Winter, D. J. 1968. Representations of locally finite groups. *Bull. Amer. Math. Soc.*, **74**, 145–8. [94]

Wittgenstein, L. (Ludwig Wittgenstein) 1961. *Tractatus logico-philosophicus*. Trans. D. F. Pears & B. F. McGuinness. London: Routledge & Kegan Paul. [22, 82]

Wojciechowska, A. (Anna Wojciechowska) 1969. Generalized limit powers. *Bull. Acad. Polon. Sci. Sér. Sci. Math. Astronom. Phys.*, **17**, 121–2. [476].

Wood, C. (Carol Wood) 1973. The model theory of differential fields of characteristic $p \neq 0$. *Proc. Amer. Math. Soc.*, **40**, 577–84. [706]

— 1974. Prime model extensions for differential fields of characteristic $p \neq 0$. *J. Symbolic Logic*, **39**, 469–77. [706]

— 1979. Notes on the stability of separably closed fields. *J. Symbolic Logic*, **44**, 412–16. [701]

Woodrow, R. E. (Robert E. Woodrow) 1976. A note on countable complete theories having three isomorphism types of countable models. *J. Symbolic Logic*, **41**, 672–80. [533]

— 1978. Theories with a finite number of countable models. *J. Symbolic Logic*, **43**, 442–55. [533]

— 1979. There are four countable ultrahomogeneous graphs without triangles. *J. Combinatorial*

Theory Ser. B, **27**, 168–79. [357]

Ziegler, M. (Martin Ziegler) 1975. A counterexample in the theory of definable automorphisms. *Pacific J. Math.*, **58**, 665–8. [197]

— 1980. Algebraisch abgeschlossene Gruppen. In *Word Problems* II, ed. S. I. Adian *et al.*, pp. 449–576. Amsterdam: North-Holland. [394, 409, 689, 690]

— 1982. Einige unentscheidbare Körpertheorien. *L'Enseign. Math.*, **28**, 269–80. [697]

— 1984. Model theory of modules. *Ann. Pure Appl. Logic*, **26**, 149–213. [661, 662]

— 1988. Ein stabiles Modell mit der finite cover property aber ohne Vaughtsche Paare. In *Proc. Sixth Easter Conference in Model Theory*, ed. B. I. Dahn *et al.*, pp. 179–85. Berlin: Humboldt-Univ. [650]

Zil'ber, B. I. (Б. И. Зильбер) 1974. О ранге трансцендентности формул $\aleph_1$-категоричной теории (On the transcendence rank of formulas of an $\aleph_1$-categorical theory). *Mat. Zametki*, **15** (2), 321–9. [651]

— 1977a. Группы и кольца, теория которых категорична. *Fundamenta Math.*, **95**, 173–88. *Trans. as* Groups and rings whose theory is categorical. *A.M.S. Translations*, (2) **149** (1991), 1–16. [83, 180, 262, 263, 691, 699]

— 1977b. Строение моделей категоричных теорий и проблема конечной аксиоматизируемости (The construction of models of categorical theories and the problem of finite axiomatisability). Kemerovo: Kemerovo State University. [198, 262, 263]

— 1980a. Сильно минимальные счетно категоричные теории (Strongly minimal countably categorical theories). *Sibirsk. Mat. Zh.*, **21**, 98–112. [199]

— 1980b. Totally categorical theories: structural properties and the non-finite axiomatizability. In *Model theory of algebra and arithmetic*, ed. L. Pacholski *et al.*, Lecture Notes in Math. 834, pp. 381–410. Berlin: Springer. [131, 200, 651]

— 1981. О решении проблемы конечной аксиоматизируемости для теорий, категоричных во всех бесконечных мощностых. In Исследования по теоретическому программированию, Alma-Ata, pp. 69–74. *Trans. as* On a solution of the problem of finite axiomatizability for theories categorical in all infinite powers. *Amer. Math. Soc. Transl.*, (2) **135** (1987), 13–17. [131, 200, 651]

— 1984a. Сильно минимальные счетно категоричные теории. II (Strongly minimal countably categorical theories. II). *Sibirsk. Mat. Zh.*, **25** No. 3, 71–88. [199, 200]

— 1984b. Сильно минимальные счетно категоричные теории. III (Strongly minimal countably categorical theories. III). *Sibirsk. Mat. Zh.*, **25** No. 4, 63–77. [199, 200, 618]

— 1984c. The structure of models of uncountably categorical theories. In *Proc. Internat. Congress of Math. 1983*, pp. 359–68. Warsaw: Państwowe Wydawnictwo Naukowe. [199, 618]

— 1988. Finite homogeneous geometries. In *Proc. Sixth Easter Conference in Model Theory*, ed. B. I. Dahn *et al.*, pp. 186–208. Berlin: Humboldt-Univ. [199]

Zil'ber, B. I. & Smurov, V. P. (В. П. Смуров) 1988. О минимальных структурах (On minimal structures). In *Ninth All-Union Conference on Mathematical Logic* (Russian), p. 65. Leningrad: Nauka. [199]

Zimmermann, W. (Wolfgang Zimmermann) 1977. Rein injektive direkte Summen von Moduln. *Comm. Algebra*, **5**, 1083–117. [660]

INDEX TO SYMBOLS

INDEX

Some brackets have been added to determine the alphabetic listing. Thus 'ω-categorical' (without brackets) is listed as if it was 'Omega-categorical', and '(λ-)categorical' (with brackets) is listed as if it was 'Categorical'.